Basic Pro/ENGINEER® in 20 Lessons

Louis Gary Lamit
De Anza College

with technical assistance provided by
James Gee, William Jorgenson, John I. Shull
De Anza College

PWS PUBLISHING COMPANY
I(T)P *An International Thomson Publishing Company*
Boston • Albany • Bonn • Cincinnati • Detroit • London • Madrid
Melbourne • Mexico City • New York • Pacific Grove • Paris
San Francisco • Singapore • Tokyo • Toronto • Washington

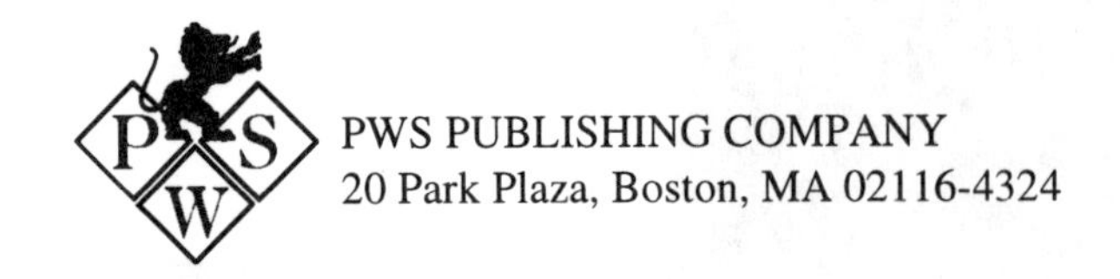

PWS PUBLISHING COMPANY
20 Park Plaza, Boston, MA 02116-4324

Parametric Technology Corporation
128 Technology Drive
Waltham, MA 02154
617-398-5000

Some material in this book was provided by CADTRAIN, Inc. and is based on their CBT product for Pro/ENGINEER®. We wish to thank CADTRAIN for the use of their product, COAch for Pro/ENGINEER®.
CADTRAIN, Inc.
2429 West Coast Highway
Newport Beach, CA 92663
800-631-5757

Sponsoring Editor: *Bill Barter*
Assistant Editor: *Suzanne Jeans*
Technology Editor: *Susan Garland*
Marketing Manager: *Nathan Wilbur*
Production Manager: *Elise Kaiser*
Cover Designer: *Julia Gecha*
Manufacturing Buyer: *Andrew Christensen*
Text Printer: *Courier-Westford*
Cover Printer: *Phoenix Color Corp.*
Cover Photo: *Courtesy of PTC*

Printed and bound in the United States of America.
98 99 00 01 — 10 9 8 7 6 5 4 3

Library of Congress Cataloging-in-Publication Data
Lamit, Louis Gary.
Basic Pro/ENGINEER in 20 lessons, version 18 / Louis Gary Lamit and technical assistance provided by James Gee, William Jorgenson, and John I. Shull.
p. cm.
Includes index.
ISBN 0-534-95068-X
1. Pro/Engineer. 2. Computer-aided design. 3. Mechanical drawing. 4. Computer graphics. I. Title
TA174.L34 1997
620'.0042'028553042--dc21 97-35748
CIP

For more information, contact:
PWS Publishing Company
20 Park Plaza
Boston, MA 02116

International Thomson Publishing Europe
Berkshire House 168-173
High Holborn
London WC1V 7AA
England

Thomas Nelson Australia
102 Dodds Street
South Melbourne, 3205
Victoria, Australia

Nelson Canada
1120 Birchmont Road
Scarborough, Ontario
Canada M1K 5G4

International Thomson Editores
Campos Eliseos 385, Piso 7
Col. Polanco
11560 Mexico D.F., Mexico

International Thomson Publishing GmbH
Königswinterer Strasse 418
53227 Bonn, Germany

International Thomson Publishing Asia
221 Henderson Road
#05-10 Henderson Building
Singapore 0315

International Thomson Publishing Japan
Hirakawacho Kyowa Building, 31
2-2-1 Hirakawacho
Chiyoda-ku, Tokyo 102
Japan

Dedication

To Dennis for the great friend he has been all these years

About the Author

Louis Gary Lamit is the former department head of drafting and CAD facility manager and is currently an instructor at De Anza College in Cupertino, California, where he teaches computer-aided drafting and design using AutoCAD and Pro/ENGINEER, as well as basic drafting.

Mr. Lamit has worked as a drafter, designer, numerical control (NC) programmer, and engineer in the automotive, aircraft, and piping industries. A majority of his work experience is in the area of mechanical and piping design. Mr. Lamit started as a drafter in Detroit (as a job shopper) working for the automobile industry doing tooling, dies, jigs and fixture layout, and detailing at Koltanbar Engineering, Tool Engineering, Time Engineering, and Premier Engineering for Chrysler, Ford, AMC, and Fisher Body. Mr. Lamit has worked at Remington Arms and Pratt & Whitney Aircraft as a designer, and at Boeing Aircraft and Kollmorgan Optics as an NC programmer and aircraft engineer.

Since leaving industry, Mr. Lamit has taught at all levels (Melby Junior High School, Warren, Michigan; Carroll County Vocational Technical School, Carrollton, Georgia; Heald Engineering College, San Francisco, California; Cogswell Polytechnical College, San Francisco, and Cupertino, California; Mission College, Santa Clara, California; Santa Rosa Junior College, Santa Rosa, California; Northern Kentucky University, Highland Heights, Kentucky; and De Anza College, Cupertino, California).

Mr. Lamit has written a number of textbooks, including ***Industrial Model Building***, with Engineering Model Associates, Inc. (1981), ***Piping Drafting and Design*** (1981), ***Descriptive Geometry*** (1983), and ***Pipe Fitting and Piping Handbook*** (1984) for Prentice-Hall; ***Drafting for Electronics***, with Sandra Lloyd (3rd edition, 1998) and ***CADD***, with Vernon Paige (1987) were published by Charles Merrill (Macmillan-Prentice Hall Publishing). ***Technical Drawing and Design*** (1994), ***Principals of Engineering Drawing***, with Kathy Kitto (1994), ***Engineering Graphics and Design*** with Kathy Kitto (1997), and ***Engineering Graphics and Design with Graphical Analysis***, with Kathy Kitto, (1997) were published by West Publishing (ITP/Delmar)

Mr. Lamit received a BS degree from Western Michigan University in 1970 and did masters work at Wayne State University and Michigan State University. He has also done graduate work at the University of California at Berkeley and holds an NC programming certificate from Boeing Aircraft.

Contents

Part Two Assemblies

Part Three Generating Drawings

Appendixes

Index

Preface

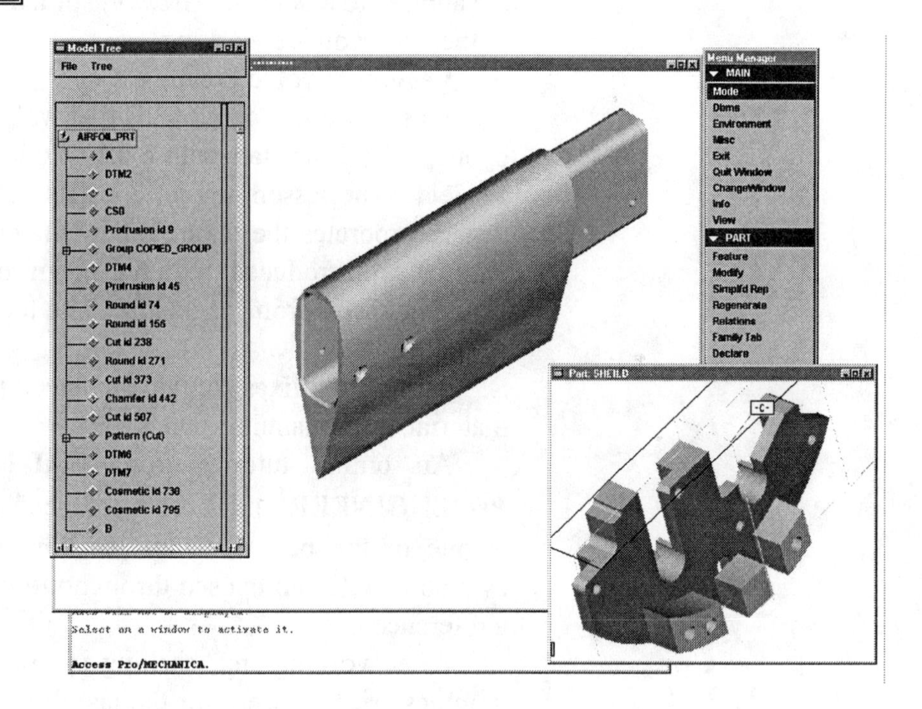

Pro/ENGINEER® is one of the most widely used CAD/CAM software programs in the world today. This book introduces you to the basics of the program and enables you to build on these basic commands to expand your knowledge beyond the scope of the book.

Pro/ENGINEER (Pro/E) changes and improves with a new version every six months. The book does not attempt to cover all of Pro/E's features, only to provide an introduction to the software, make you reasonably proficient in its use and establish a firm basis for exploring and growing with the program as you use it in your career or classroom. For information on new releases of Pro/E, as well as data files and other useful tools pertaining to this book, point your web browser to **www.pws.com/ge/lamit.html**.

The basic premise of this book is that the more parts, assemblies, and drawings you create using Pro/E, the better you learn the software. With this in mind, each lesson introduces a new set of commands (with an incrementally harder set of parts), building on previous lessons. The parts created in the first 13 lessons are used later in the text to create assemblies and generate drawings. This procedure allows you to work with actual completed parts, assemblies, and drawings in a short lesson format, instead of building large complex projects where new commands may get lost in the process.

The book is divided into three parts: **Part One--Creating Parts**, **Part Two--Assemblies**, and **Part Three--Generating Drawings**. Each **Part** has a variety of individual **lessons**.

Every lesson introduces a new set of commands and concepts that are applied to a *part*, an *assembly*, or a *drawing*, depending on where in the book you are working.

Lessons involve creating a new part, an assembly, or a drawing using a set of Pro/E commands that walk you through the process step by step. Lessons start with a set of objectives and end with a **lesson project**. The lesson project consists of a part, assembly, or drawing that incorporates the lesson's new material and expands on and uses previously introduced material from other lessons. Projects use planning sheets from Appendix D as tools for establishing the design intent.

The **Appendixes** contain advanced projects, a glossary, reference materials, and planning sheets.

An online tutorial from **CADTRAIN** called **COAch for Pro/ENGINEER®** has been referenced throughout the book. I feel it is one of the best ways of teaching CAD/CAM software that is available. **COAch** is used throughout this book within lessons and as a reference.

After working and teaching in drafting, design, and engineering graphics oriented areas for the last thirty years I feel that Pro/E is one of the most comprehensive, intuitive, productive, and stimulating programs I have experienced. This book was written so that students and professionals will have a timely Pro/E software manual. The book can be used as a home study guide for those wishing to expand their knowledge of Pro/E, a training guide, or a reference for the Pro/E user.

Pro/E runs on either a Windows or Unix platform.

Engineering students and professionals with technical training in Pro/E are finding that knowledge of this powerful tool can open up a whole new career path with unlimited possibilities.

If you wish to contact me concerning questions, changes, additions, suggestions, comments, etc., please send them via email to the following address:

llamit@ix.netcom.com (Lamit and Associates)
lamit@columbia.atc.fhda.edu (De Anza College)

Acknowledgments

I would like to thank the following for assistance in preparing this manuscript:

- James Gee
- John Shull
- Ken Louie
- Will Jorgenson
- Sanjiv Sheth
- Sunmoon Suhaimi
- Francis Nicholson
- Roenna Del Rosario
- Steven Koko Washington

and Jason Perry, Design Engineer, PPC for valuable comments during the review of the manuscript.

I would also like to thank the following for the support and materials granted the author:

Parametric Technology Corporation for their support and timely software updates.

- Dave Pettine *Parametric Technology Corporation*
- Larry Fire *Parametric Technology Corporation*

CADTRAIN makers of COAch for Pro/ENGINEER® for their Pro/ENGINEER tutorial software.

- Dennis Ftajic CADTRAIN
- Kathy Bennett CADTRAIN
- Ron Gates CADTRAIN

Sections

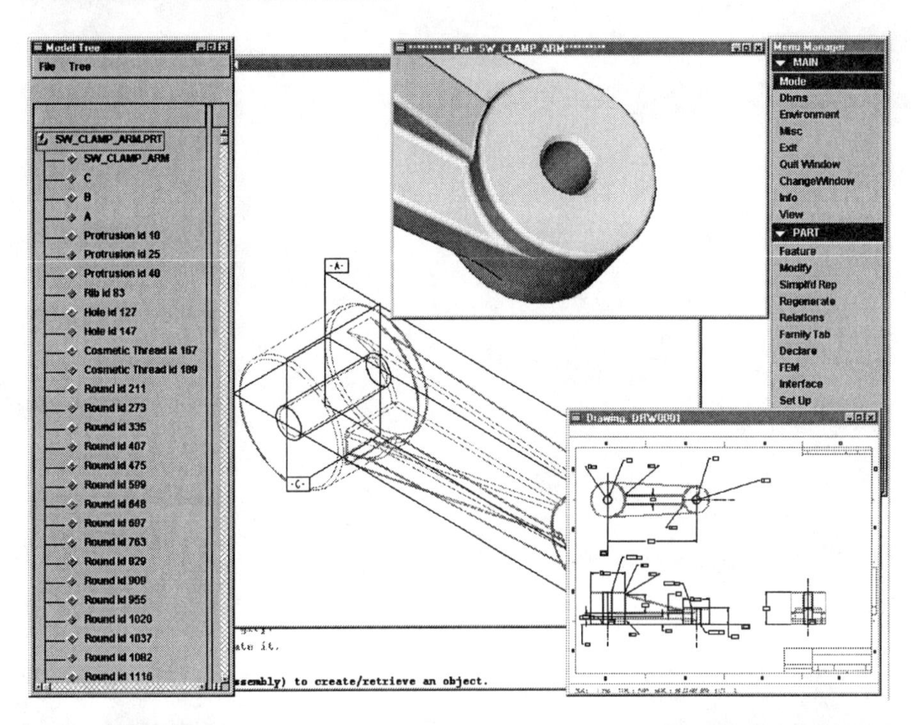

Part Design

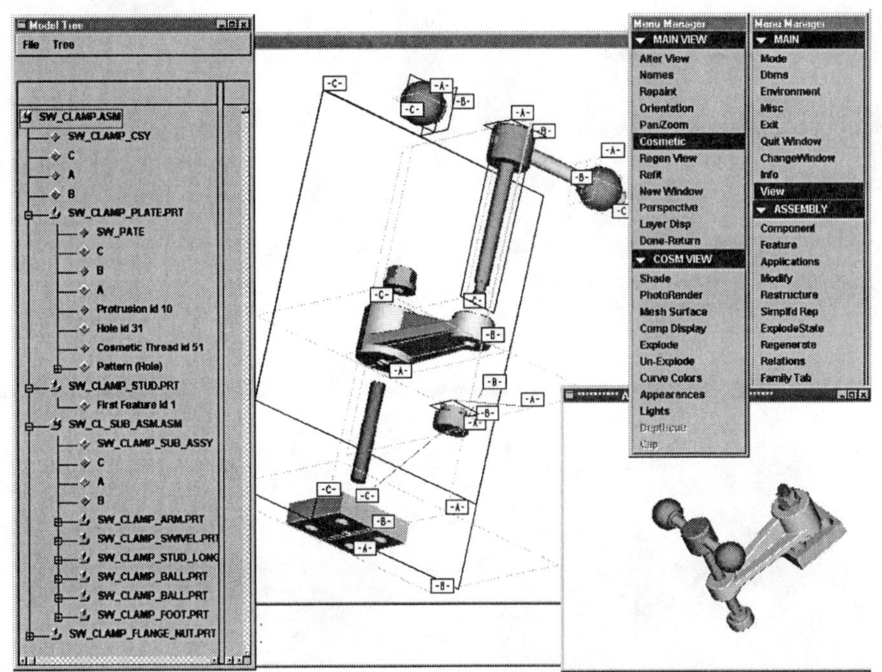

Assembly Design

Section 1

Introduction

This work-text introduces the basic concepts of parametric design using Pro/ENGINEER to create and document individual parts, assemblies, and drawings. **Parametric** can be defined as *"any set of physical properties whose values determine the characteristics or behavior of something."* **Parametric design** enables you to generate a variety of information about your design -- its mass properties, a drawing, or a base model. To get this information, you must first model your part design (Fig. 1.1).

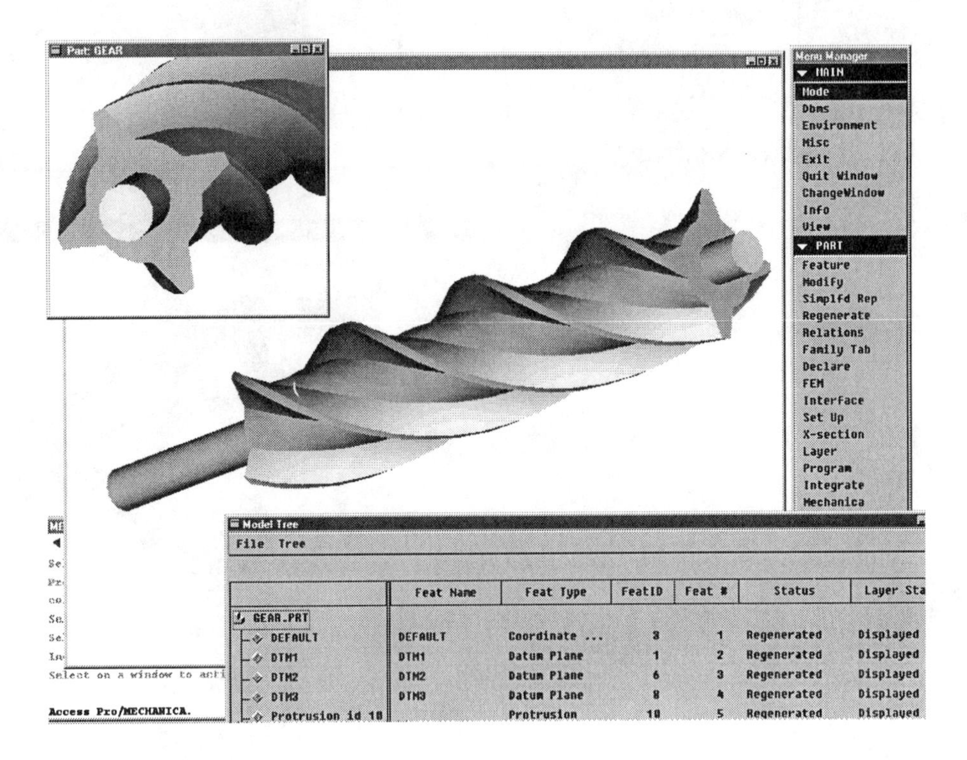

Figure 1.1
Part Design

This section is intended to introduce you to parametric modeling philosophies used in Pro/ENGINEER.

Feature-Based Modeling Parametric design represents solid models as combinations of engineering features (Fig. 1.2).

Creation of Assemblies Just as features are combined into parts, parts may be combined into assemblies, as shown in Figure 1.3.

Capturing Design Intent The ability to incorporate engineering knowledge successfully into the solid model is an essential aspect of parametric modeling (Fig. 1.4).

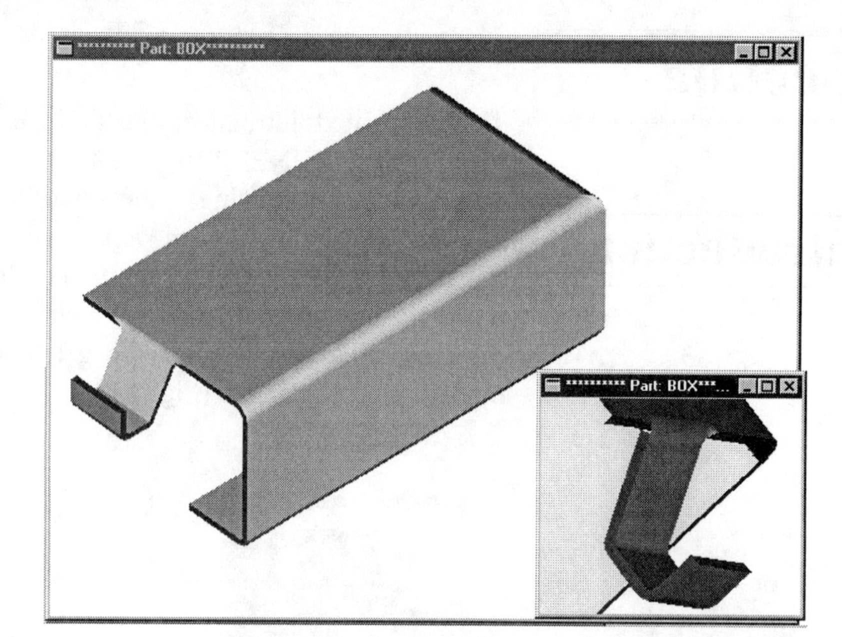

Figure 1.2
Feature-based Modeling

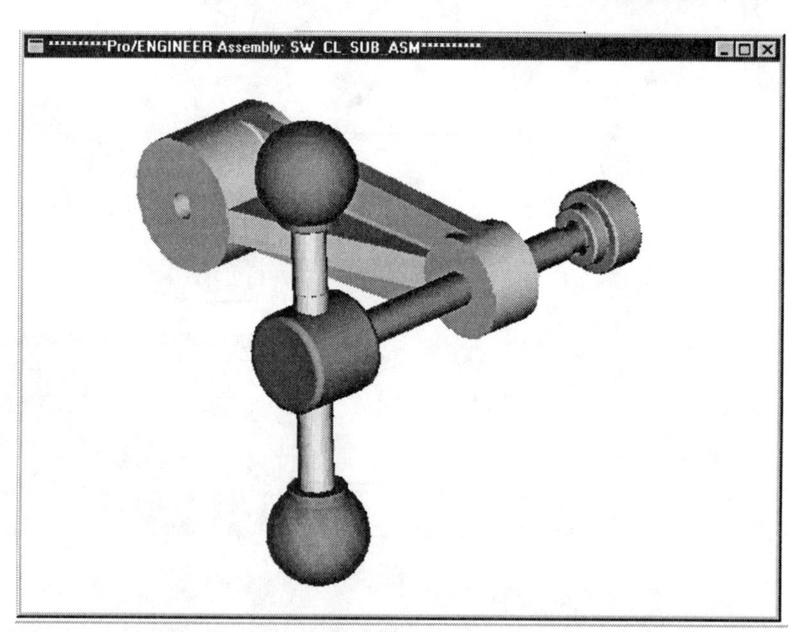

Figure 1.3
Creation of Assemblies

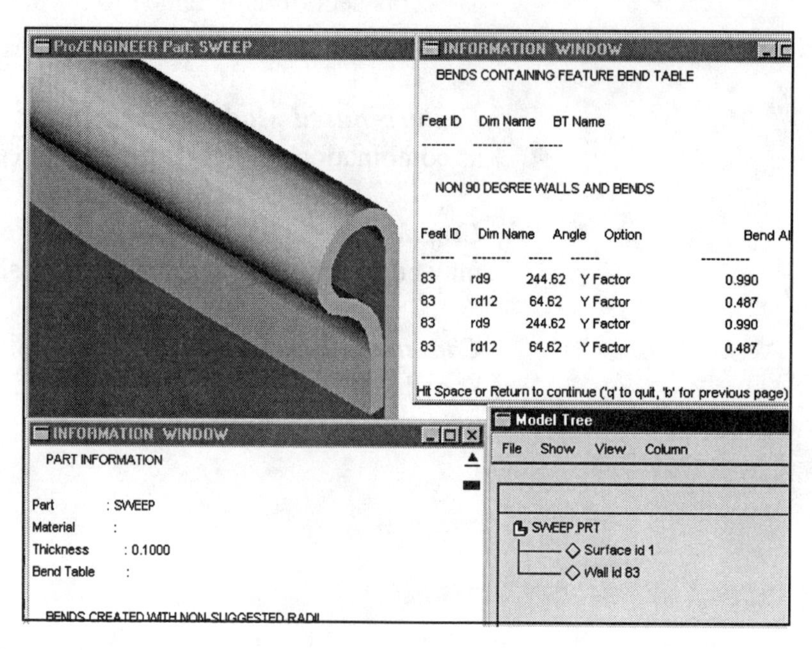

Figure 1.4
Capturing Design Intent

These methodologies are the principal aspects of successful parametric solid modeling. In Figure 1.5 illustrates the role of each in the modeling process.

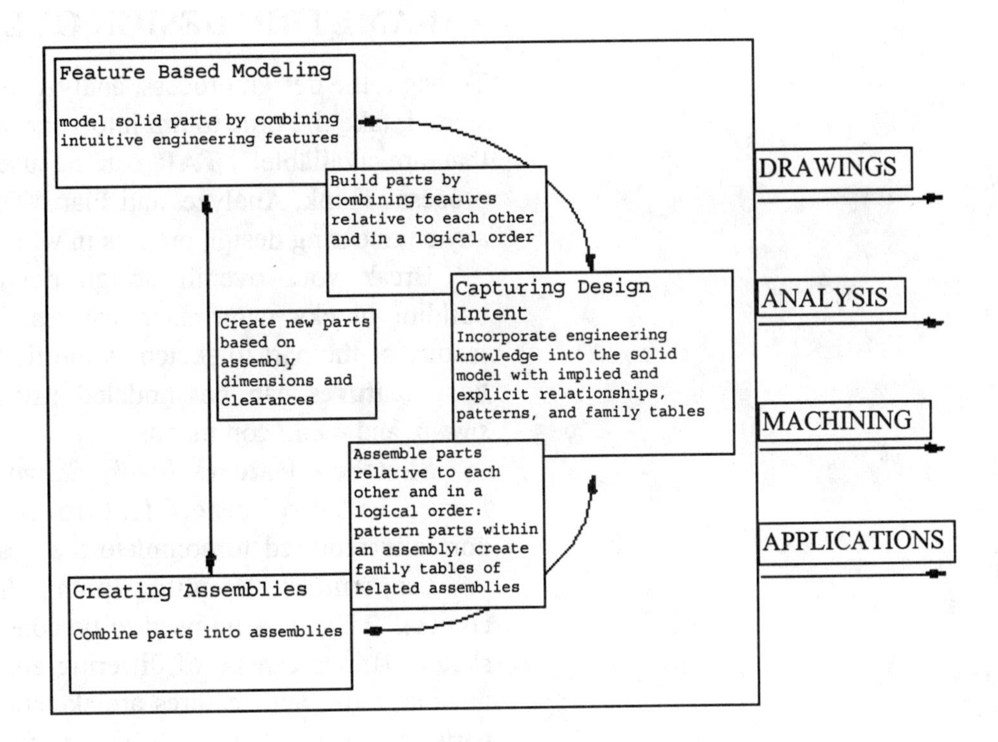

Figure 1.5
Parametric
Design Methodology

Modeling vs. Drafting

A primary and essential difference between parametric design and traditional computer-aided drafting packages is that parametric design models are three dimensional. Designs increasingly are represented in the form of solid models that capture design intent as well as design geometry. Engineering designs today are constructed as mathematical solid models instead of as 2D drawings. A solid model is one that represents a shape as a 3D object having volume and mass properties.

There are two main reasons for the move to solid models. First of all, solid modeling software packages can serve as an easy means of portraying parts for study by cross-functional concurrent-engineering teams. The solid model can be understood by even non-technical members of the team, such as those from the marketing and sales departments.

Second, the capabilities of solid modelers have been upgraded so the model can represent not only the geometry of the part being designed, but also the intent of the designer. This facility is of most significance when the designer needs to make changes to the part geometry.

The designer must make far fewer changes in later-generation parametric solid models that capture design intent than in previous CAD/CAM modeling software. In parametric design, drawings are produced as views of the 3D model, rather than the other way around.

Parametric design models are not drawn so much as they are *sculpted* from solid volumes of materials.

PARAMETRIC DESIGN OVERVIEW

To begin the design process, analyze your design. Before any work is started, take the time to ***tap*** into your own knowledge bank and others that are available. **TAP** can be used to remind yourself of this process: **T**hink, **A**nalyze, and **P**lan. These three steps are essential to any engineering design process in which you may be involved.

Break your overall design down into its basic components, building blocks, or primary features. Identify the most fundamental feature of the part to sketch as the first or base feature. A variety of **base features** can be modeled using *protrusion-extrude, revolve, sweep,* and *blend* commands.

Sketched features (*neck, flange,* and *cut*) and pick-and-place features called **referenced features** (*holes, rounds,* and *chamfers*) are normally required to complete the design. With the **SKETCHER**, you use familiar 2D entities (points, lines, circles, arcs, splines, and conics). There is no need to be concerned with the accuracy of the sketch. Lines can be of differing angles, arcs and circles can have unequal radii, and features are sketched with no regard to the actual parts' dimensional sizes. In fact, exaggerating the difference between entities that are similar but not exactly the same is actually a far better practice when using the SKETCHER.

The software enables you to apply logical geometric constraints to the sketch. **Constraints** clean up the sketch geometry according to the software assumptions. **Geometry assumptions** and constraints close ends of connected lines, align parallel lines, and snap sketched lines to horizontal and vertical orthogonal orientations. Additional constraints are added by means of **parametric dimensions** to control the size and shape of the sketch.

Feature-Based Modeling

Features are the basic building blocks you use to create a part. Features "understand" their fit and function as though they have "smarts" built into the features themselves. For example, a hole, neck, or cut feature knows its shape and location and that it has a negative volume. As you modify a feature, the entire part automatically updates after regeneration. The idea behind feature-based modeling is that the designer constructs a part so that it is composed of individual features that describe how the geometry is supposed to behave in the event its dimensions change. This happens quite often in industry, as in the case of a design change.

Parametric modeling is the term used to describe the capturing of design operations as they take place, as well as future modifications and editing that takes place on the design (Fig. 1.6). The order of the design operations is significant. Suppose that a designer specifies that two surfaces are parallel such that surface 2 is parallel to surface 1. Therefore, if surface 1 moves, then surface 2 moves along with it to maintain the specified design relationship. Surface 2 is a **child** of surface 1 in this example. Parametric modelers allow the designer to **reorder** the steps in the part's creation.

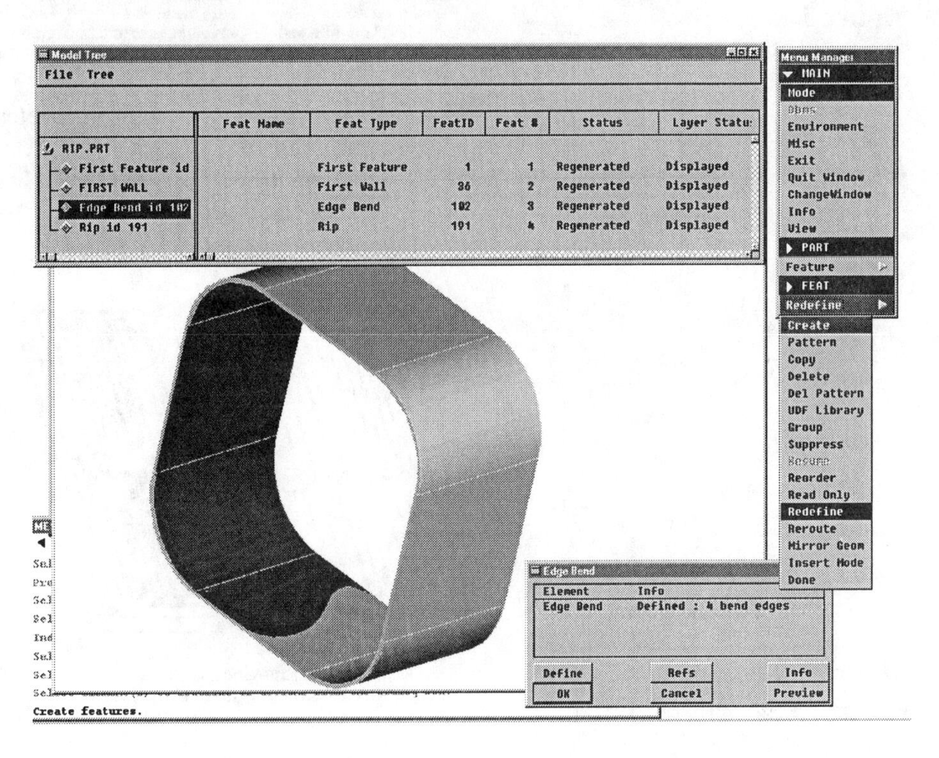

Figure 1.6
Editing a Design

The "chunks" of solid material from which parametric design models are constructed are called **features**. Features generally fall into one of the following categories:

Base Feature The base feature may be either a sketched feature or datum plane(s) referencing the default coordinate system. The base feature is important because all future model geometry will reference this feature directly or indirectly; it becomes the root feature. Changes to the base feature will affect the geometry of the entire model (Fig. 1.7).

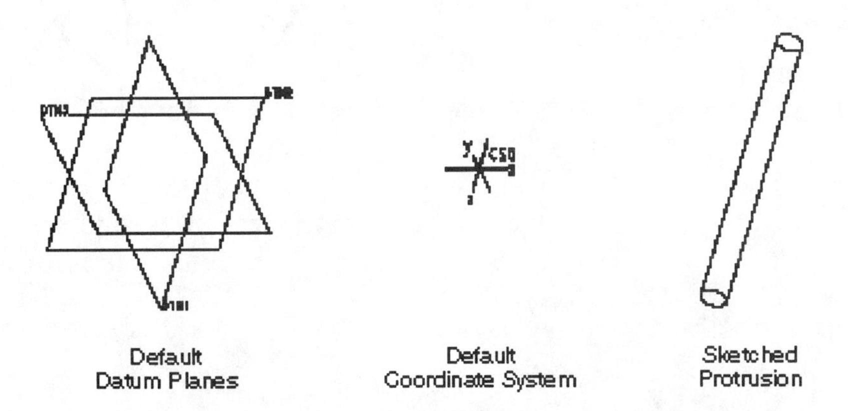

Figure 1.7
Base Features

Sketched Features Sketched features are created by extruding, revolving, blending, or sweeping a sketched cross section. Material may be added or removed by protruding or cutting the feature from the existing model (Fig. 1.8).

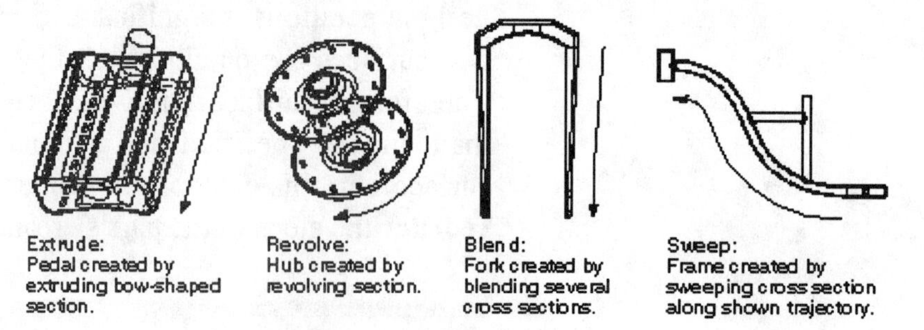

Figure 1.8
Sketched Features

Referenced Features Referenced features reference existing geometry and employ an inherent form; they do not need to be sketched. Some examples of referenced features are rounds, drilled holes, and shells (Fig. 1.9).

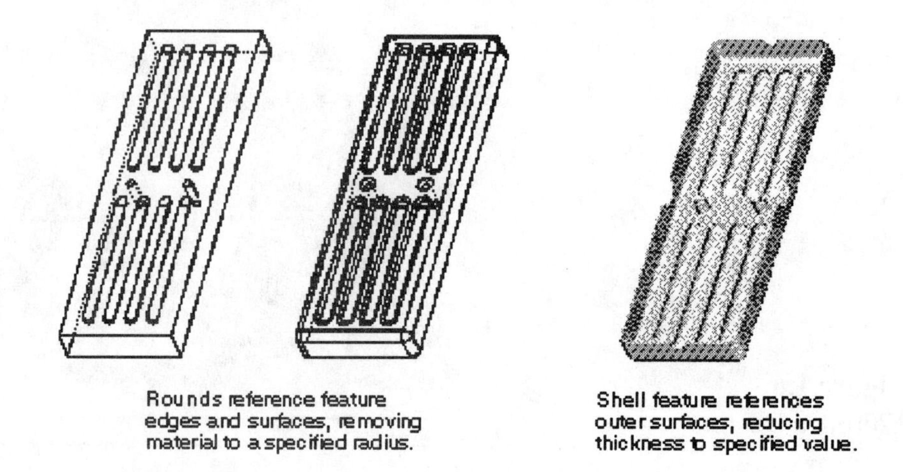

Figure 1.9
Referenced Features

Datum Features Datum features, such as planes, axes, curves, and points, are generally used to provide sketching planes and contour references for sketched and referenced features. Datum features do not have physical volume or mass, and may be visually hidden without affecting solid geometry (Fig. 1.10).

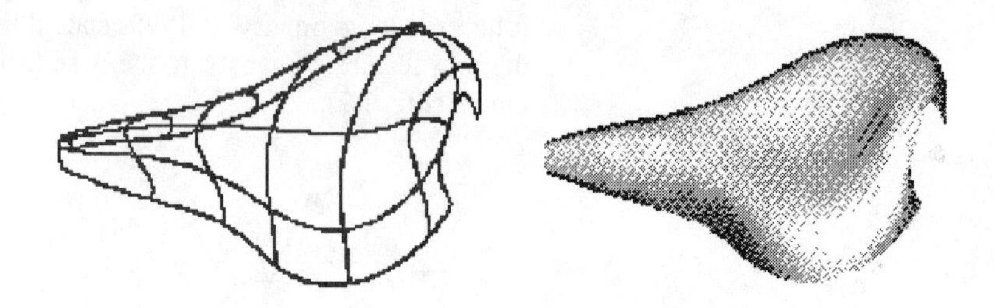

Figure 1.10
Datum Curves

The various types of features are used as building blocks in the progressive creation of solid parts. Figure 1.11 shows base features, datum features, sketched features, and referenced features.

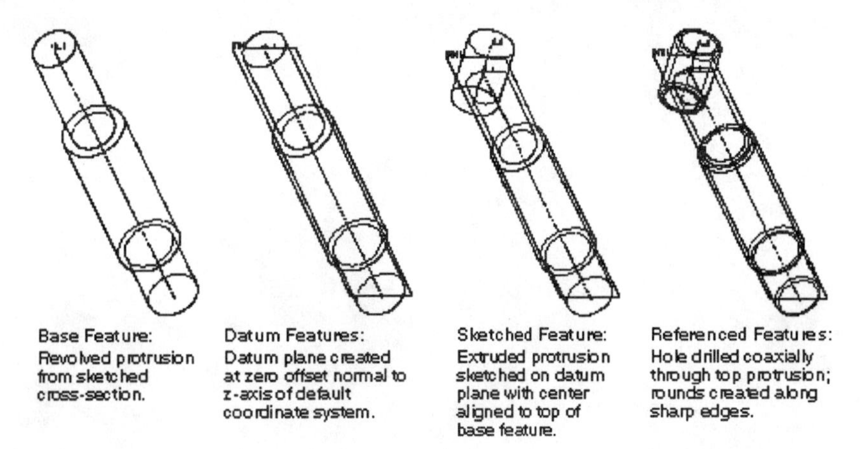

Figure 1.11
Feature Types

Conclusion

As you progress through the text's lessons, you will be prompted to return to a referenced section to review. At this point in the process of learning Pro/E, you cannot expect to understand and apply every concept involved in this very complex software. The sections in the text are really references that are utilized throughout each lesson's step-by-step construction and each lesson's project. You will become more familiar with the concepts and capabilities of Pro/E as you complete each project. Reading the reference sections will help in this understanding, but nothing takes the place of actual practice. In reality, the lessons are presented in a building block format that increases your knowledge incrementally and without unnecessary, complicated discussions. Feel free to return to the various sections to reaffirm material that you encounter throughout the lessons.

Besides referring to the sections, you should get comfortable with the online documentation available in Pro/E. **Pro/HELP** can be accessed at any time during a working Pro/E session (Fig. 1.12).

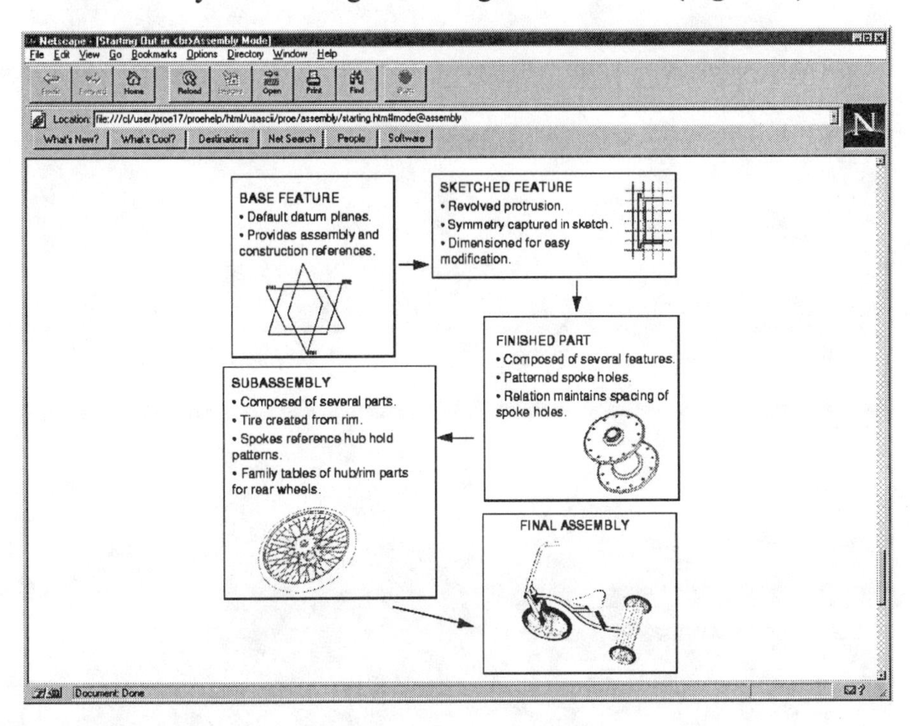

Figure 1.12
Pro/HELP

Section 2

Using the Text

The text is divided into three parts: **Part One--Creating Parts, Part Two--Assemblies**, and **Part Three--Generating Drawings**. Each **part** has a variety of individual **lessons**. Every lesson introduces a new set of commands and concepts that are applied to a *part*, an *assembly*, or a *drawing*, depending on where in the text you are working.

Lessons involve creating a new part, an assembly, or a drawing using a set of commands that walk you through the process step by step. Lessons start with a set of objectives and end with a new project that requires you to apply the material you have just learned in that lesson (and building on previous lessons).

Each lesson ends with a **Lesson Project**. The lesson project consists of a part, assembly, or drawing that incorporates the lesson's new material and expands on and uses previously introduced material from other lessons. Student projects use planning sheets from Appendix D as tools for establishing the design intent.

The following page shows the typical lesson page layout. The smaller, left-hand column holds symbols that prompt you, at appropriate places, with a variety of messages, including when to save, ask for help, set up, configure, implement engineering change orders, incorporate hints, and apply notes. Sample menus will also be shown in this column.

Design Intent Planning Sheets

Appendix D provides a variety of sketching formats for planning your design. These sheets are referred to in the text as **DIPS** (**D**esign **I**ntent **P**lanning **S**heets). The design intent of a feature, a part, or an assembly should be established before any work is done with Pro/E.

A few minutes of planning and sketching on paper will save countless hours of redoing your design on the computer system. Skipping this step in the design of a feature, part, or assembly is a recipe for disaster. In industry, there are thousands of stories of how a designer or engineer created a graphically correct part or assembly that looked visually correct. Upon closer examination, the model or assembly had too many or too few datum planes and a variety of parent-child relationships that were nothing but an example of the designer's incorrect use of Pro/E.

A variety of other problems will occur downstream in the design and manufacturing process, including possible feature failures that result when minor engineering change orders are introduced. Without proper process planning, organization, and a well-defined design intent, the model is useless. In most cases it would take more time to reorder, modify, redefine, and reroute the model to correct poor design habits than it would be to remodel the part.

In order to get you used to the fact that changes are part of the design process, engineering change orders (ECOs) are introduced in the lesson step-by-step assignments and in the lesson projects.

Symbols	Meanings
 **Dbms ⇒ Save ⇒ enter** **Purge ⇒ enter ⇒** **Done-Return**	***DBMS*** The **DBMS** symbol will remind you to save your file or design work as you complete a set of tasks or commands. **Save** and **purge** your file every few minutes.
? Pro/HELP	***Help*** Move the cursor and **highlight** the command that you need online help with, then press the right button on your mouse.
Set Up ⇒ Units ⇒ **Millimeter ⇒ Done**	***Commands*** Commands are boldface and sometimes enclosed in a box. The ⇒ symbol means to initiate the next command pick. When the command line has a ⇒ **enter**, it means to press the keyboard's Enter key.
CONFIG.PRO **def_layer_datum**	***Config.pro*** Configuration file settings are given to change default settings and help you become comfortable with customizing Pro/ENGINEER. From the **Main Menu**, pick **Misc ⇒ Edit Config**.
E C O	***ECO*** **Engineering change orders** are introduced at various times throughout the lessons. Changes may entail adding *parametric relations*, using *insert mode*, *redefining* the feature, *modifying* dimension values, *rerouting*, or *reordering* features.
HINT **DATUM PLANES** will be the first feature on all parts and assemblies.	***Hints*** Hints to commands or procedures are provided to assist you in completing each lesson.
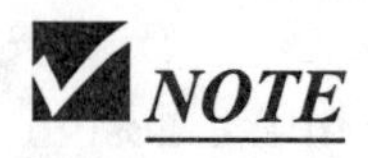 ***NOTE***	***Notes*** Notes are given to explain an aspect of design intent.
EDG REFERENCE **Engineering Graphics and Design** by L. Lamit and K. Kitto Read Chapters: See Pages:	***EDG REFERENCES*** EGD references are to the chapter and page numbers in *Engineering Graphics And Design* by Lamit and Kitto, West/Delmar Publishing Co., 1997, that can be consulted for more information on a lesson part, project, or concept.

Section 3

Fundamentals

The design of parts and assemblies and the creation of related drawings form the foundation of engineering graphics. With Pro/ENGINEER many of the previous steps in the design process have been eliminated, streamlined, altered, refined, or expanded. The model you create as a part that must be manufactured forms the basis for all engineering and design functions. The part model contains the geometric data describing the part's features, but also includes nongraphical information embedded in the design itself. The part, its associated assembly, and the graphic documentation (drawings) are parametric. The physical properties described in the part drive (determine the characteristics and behavior of) the assembly and the drawing. Any data established in the assembly mode in turn determines that aspect of the part and subsequently the drawings of the part and the assembly. In other words, all of the information contained in the part, the assembly, and the drawing is interrelated, interconnected, and parametric (Fig. 3.1).

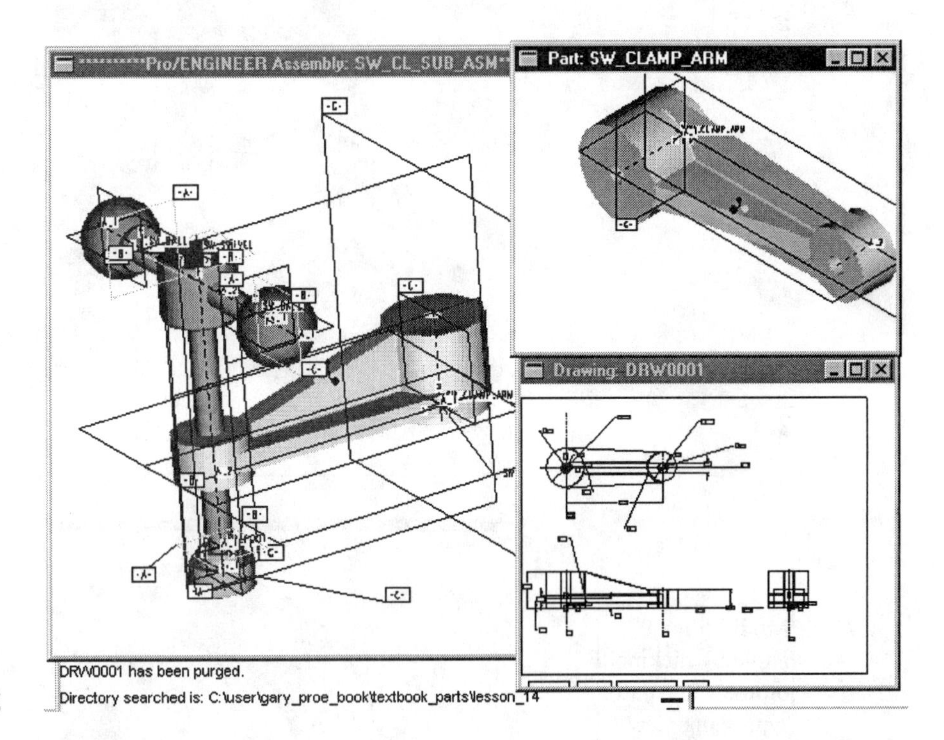

Figure 3.1
Part, Assembly, and Drawing

In most cases, the part will be the first component of this interconnected process. Therefore, in this text the first set of lessons (1-13) covers part design.

PART DESIGN

The part function in Pro/E is used to design components. Parts are started by sketching the basic form to produce the part's "base feature."

During part design (Fig. 3.2), you can accomplish the following:

* Define the base feature
* Define and redefine construction features to the base feature
* Modify the dimensional values of part features (Fig. 3.3)
* Embed design intent into the model using tolerance specifications and dimensioning schemes
* Create detail drawings of the part
* Create pictorial and shaded views of the component
* Create part families (family tables)
* Perform mass properties analysis and clearance checks
* List part, features, layer, and other model information
* Measure and calculate model features

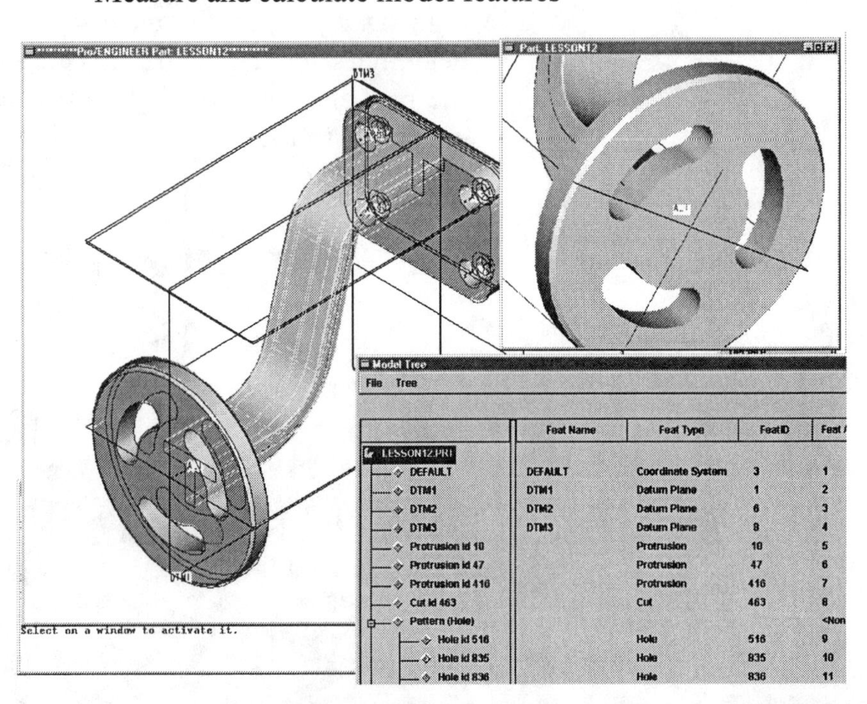

Figure 3.2
Part Design

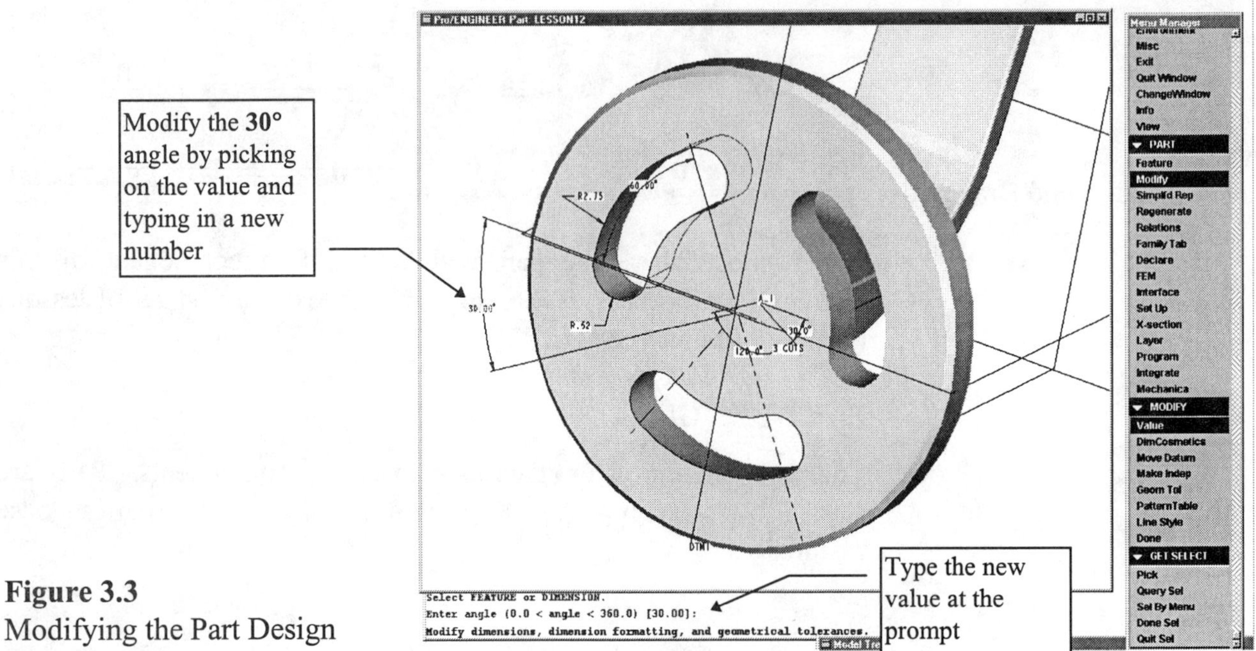

Figure 3.3
Modifying the Part Design

Establishing Part Features

The design of any part requires that the part be *confined*, *restricted*, *constrained*, and *referenced*. In parametric design the easiest method to establish and control the geometry of your part design is to use three datum planes. Pro/E allows you to use the **primary datum** to start your base feature. By creating the default datum planes **(DTM1, DTM2,** and **DTM3** in Pro/ENGINEER) you can constrain your design in all three directions. In Figure 3.4, three **default datum planes** and a **default coordinate system** have been created. Note that in the **Model Tree window** they have become the first features of the part, meaning they will be the **parents** of the construction features that follow.

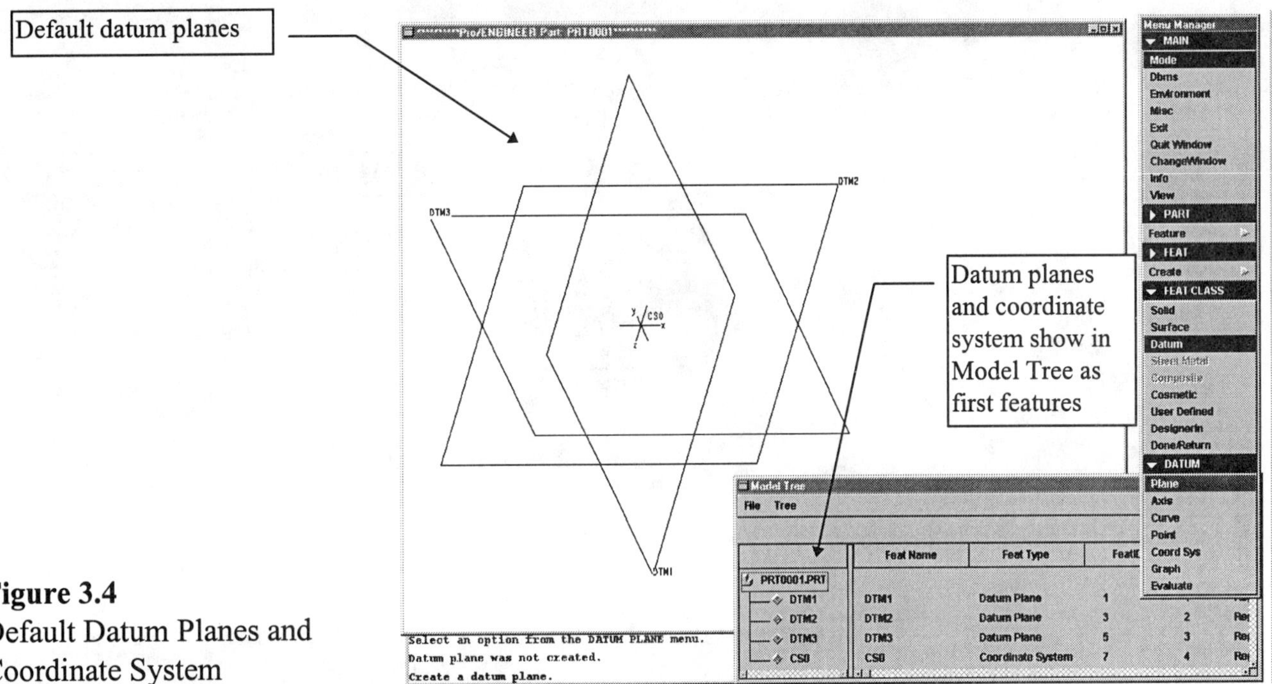

Figure 3.4
Default Datum Planes and Coordinate System

In order to see how datum planes work in the design of a part, try a simple exercise. Take a book or a box and put it on the floor of a room in your house or school. Choose the most important plane. You have now established **datum A**, or the **primary datum** plane (**DTM3** in many of the lessons and projects in the text). Take the book or box and slide it up to a wall near the corner of the room. Choose the longest or second most important plane. You have now established **datum B**, the **secondary datum** plane (**DTM2** for many of the lessons and projects in the text). Lastly, shove the book or box up against the other wall. You have now established **datum C**, or the **tertiary datum** plane (**DTM1** for many of the lessons and projects in the text). The book is now constrained by three planar surfaces. With a couple of clamps you can secure the part to these walls and machine it as if it were on a milling table.

Although this exercise and description is simplified and will not work for many parts, it does demonstrate the process of establishing your part in space using datums. In Pro/E you may use any of the datums as sketching planes or for that matter any of the part planes for sketching geometry. Any number of other datums can be introduced into the part as required for feature creation, assembly operations, or manufacturing applications. The bracket in Figure 3.5 has four datum planes. Datum 4 is the fourteenth feature of this part and is the parent of the three "slot" cuts (Fig. 3.6).

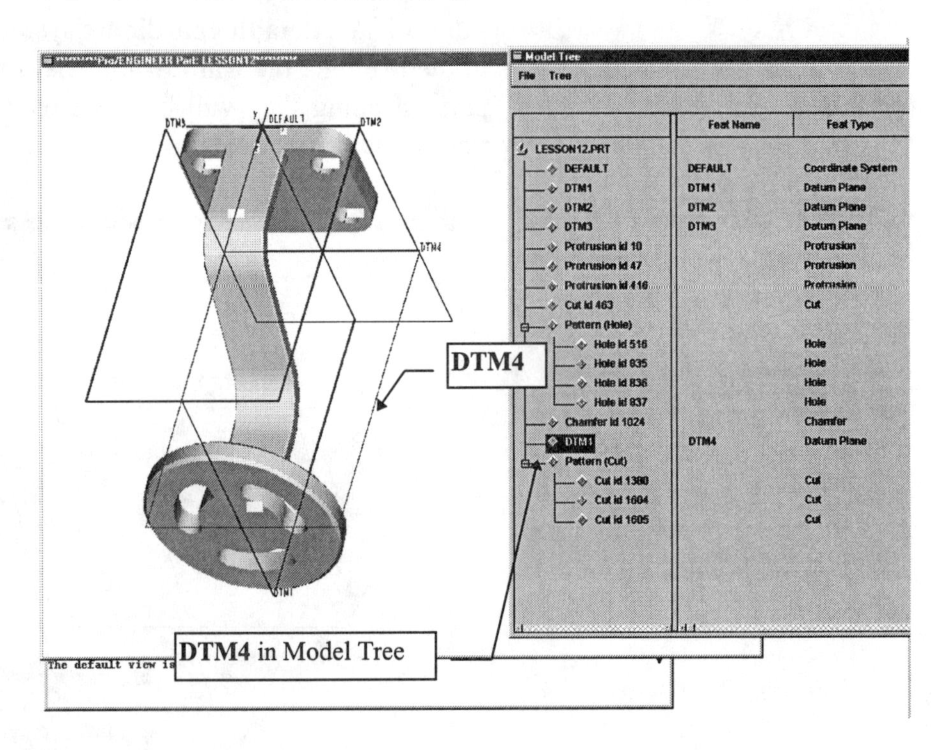

Figure 3.5
DTM4

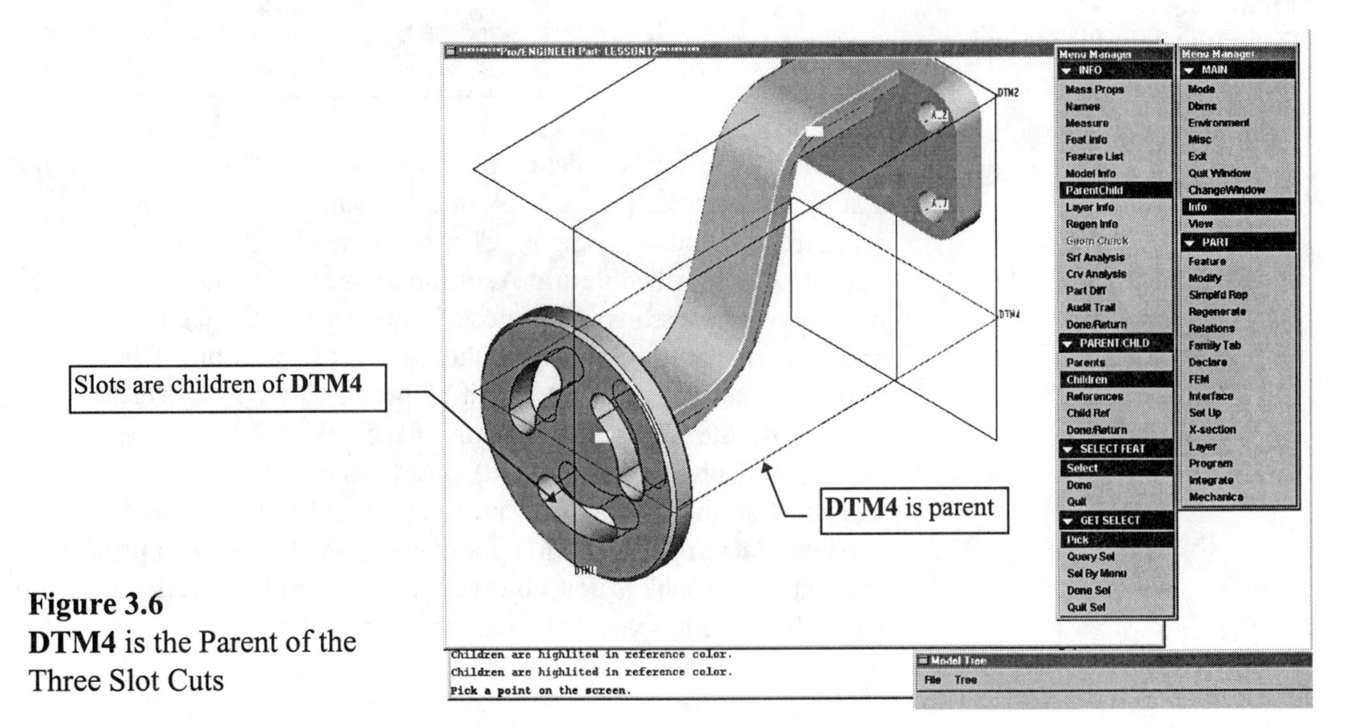

Figure 3.6
DTM4 is the Parent of the Three Slot Cuts

Datum Features

Datum features are planes, axes, and points you use to place geometric features on the active part. Previously we have discussed *default* datums. Datums other than defaults can be created at any time during the design process, as was done in the construction of the bracket shown in Figure 3.6, where datum 4 was introduced and then used to create the slots. There are three (primary) types of datum features: **datum planes**, **datum axes**, and **datum points**. (There are also ***datum curves*** and ***default datum coordinate systems***). You can display all types of datum features, but they do not define the surfaces or edges of the part or add to its mass properties. In Figure 3.7 the cell phone has a variety of datum planes used in its creation.

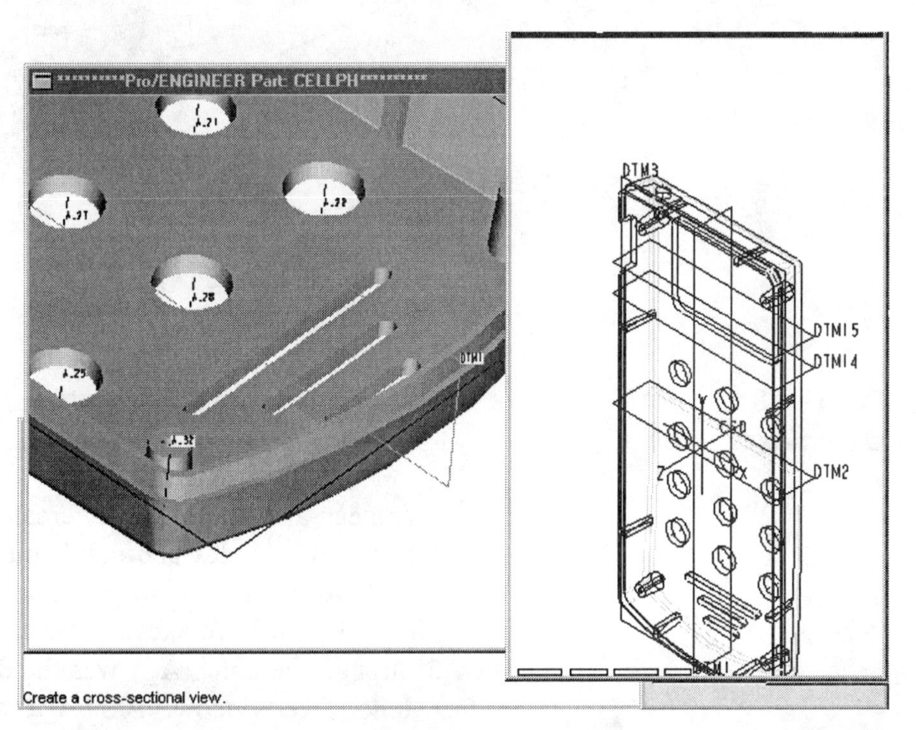

Figure 3.7
Datums in Part Design

Datum planes are infinite planes located in 3D model mode and associated with the part that was active at the time of their creation. To select a datum plane, you can pick on its name or anywhere on the perimeter edge. Datum Planes are *parametric* -- geometrically associated with the part. Parametric datum planes are associated with and dependent upon the edges, surfaces, vertices, and axes of a part. For example, a datum plane placed parallel to a planar face and on the edge of a part moves whenever the edge moves and rotates about the edge if the face moves. As you create parametric datum planes, relationships to the active part are determined by defining combinations of a placement option that link the datum plane to the part.

Datum planes are used to create a reference on a part where one does not already exist. For example, you can sketch or place features on a datum plane when there is no appropriate planar surface. You can also dimension to a datum plane as if it were an edge.

A datum is created by specifying constraints that locate it with respect to existing geometry. For example, a datum plane might be made to pass through the axis of a hole and parallel to a planar surface. Chosen constraints must locate the datum plane relative to the model without ambiguity. You can also use and create datums in assembly mode.

Besides datum planes, datum axes and datum points can be created to assist in the design process. In Figure 3.8, a datum axis has been created through the cylindrical feature.

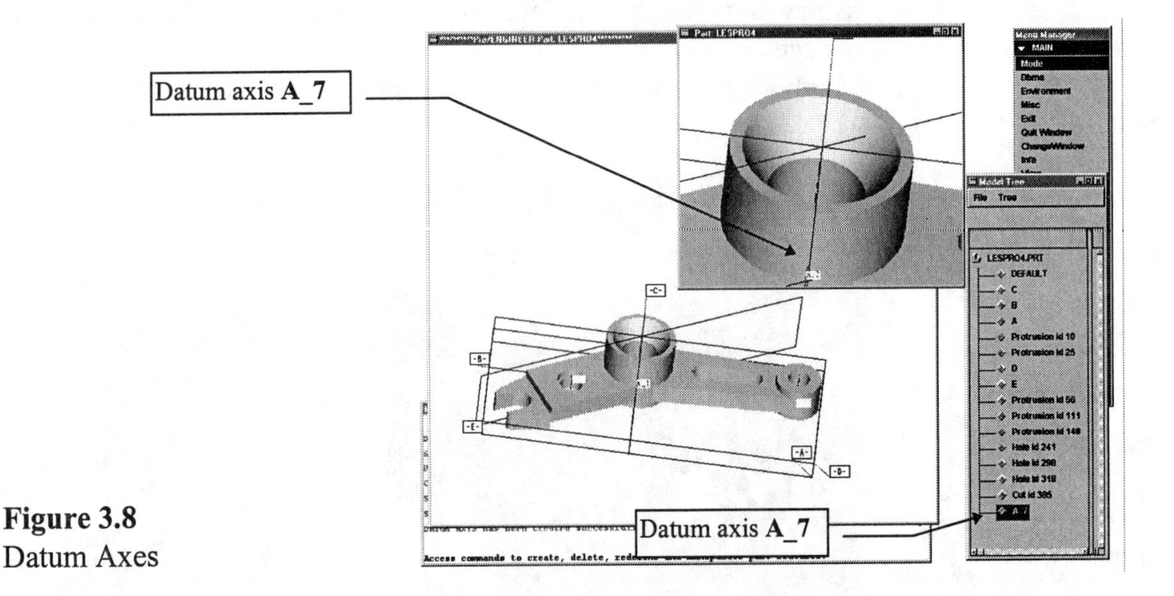

Figure 3.8
Datum Axes

You can also automatically create datum axes through cylindrical features such as holes and solid round features by setting this as a default in your system configuration file (the holes in the cell phone in Figure 3.7 all have axes). The part in Figure 3.9 shows **A_1** and **A_2** through the hole. **A_1** was the default axis and doesn't show in the Model Tree window. Axis **A_2** was created as a datum axis and therefore is a feature shown in the Model Tree window.

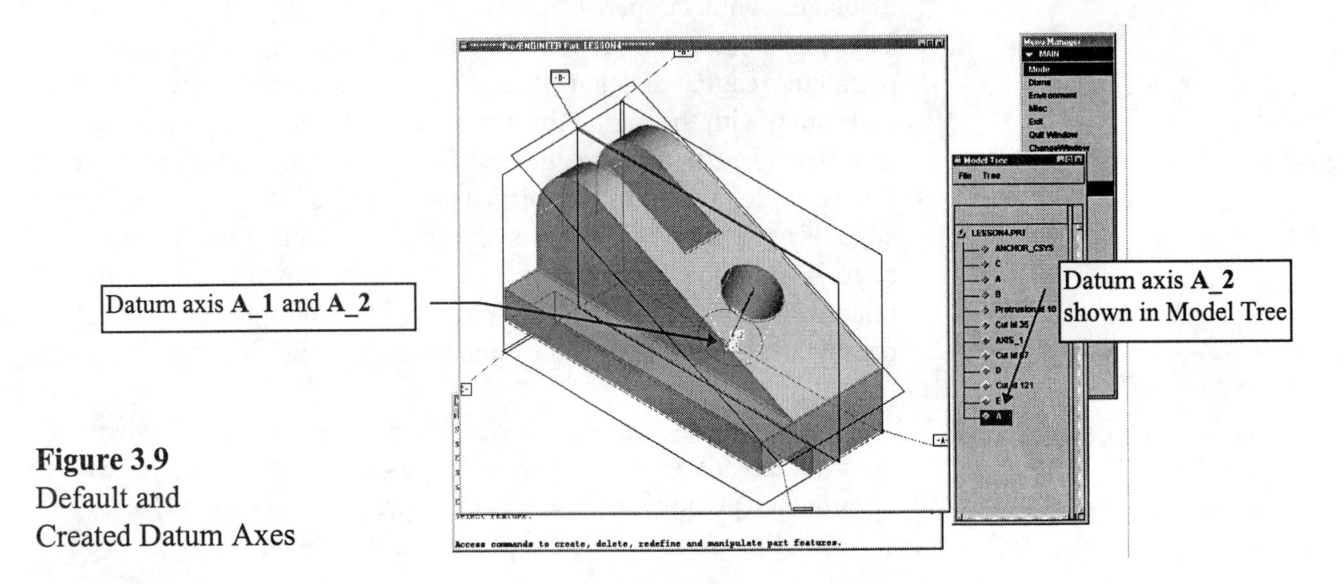

Figure 3.9
Default and
Created Datum Axes

Parent-Child Relationships

Because solid modeling is a cumulative process, certain features must, by necessity, precede others. Those that follow must rely on previously defined features for dimensional and geometric references. The relationships between features and those that reference them are termed **parent-child relationships**. Because children reference parents, features can exist without children, but children cannot exist without their parents. Using **Model Info**, Pro/E lists out the models information, including the parent-child relationships (Fig. 3.10)

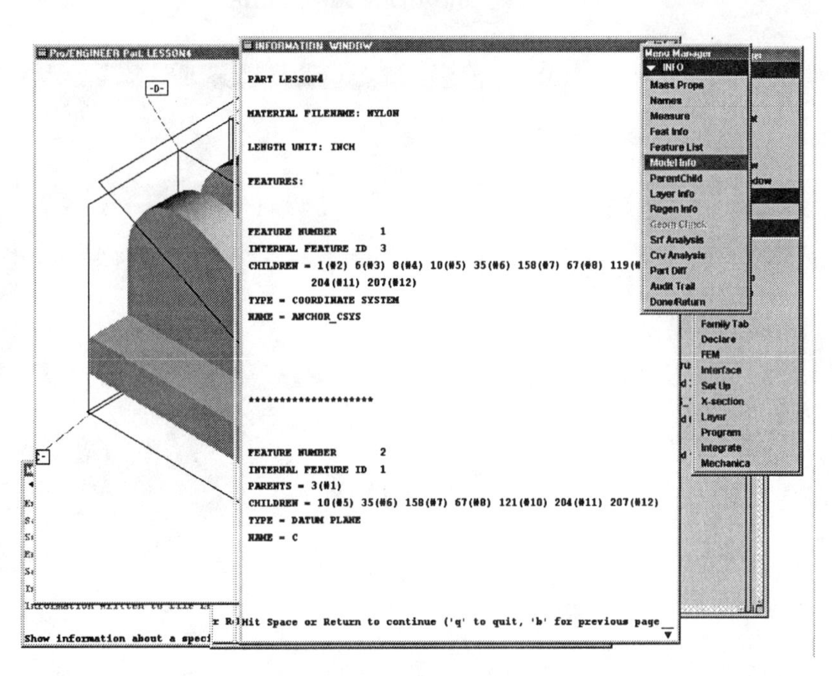

Figure 3.10
Parent-Child Relationships

To get the parent-child information for a feature, give the following command:

Info ⇒ **ParentChild** ⇒ **Children** ⇒ (select and pick the feature, as shown in Figure 3.11)

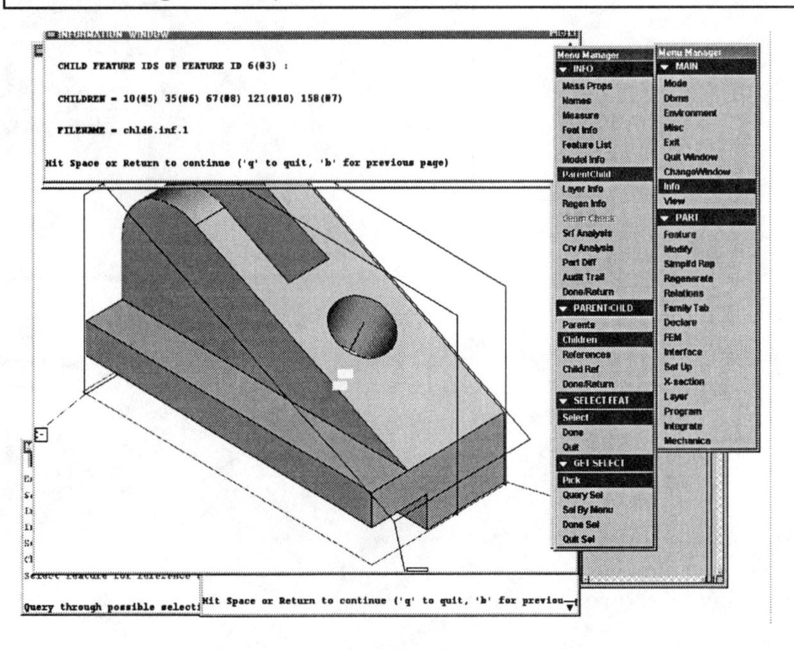

Figure 3.11
Parent-Child Information

The parent-child relationship is one of the most powerful aspects of parametric design. When a parent feature is modified, its children are automatically recreated to reflect the changes in the parent feature's geometry. It is essential to reference feature dimensions so that design modifications are correctly propagated through the model/part. As an example, the modification to the length of a part is automatically propagated through the part and will affect all children of the modified feature. This is shown in Figures 3.12 and 3.13, where the base feature of the part has been modified in width, length, and front cut height.

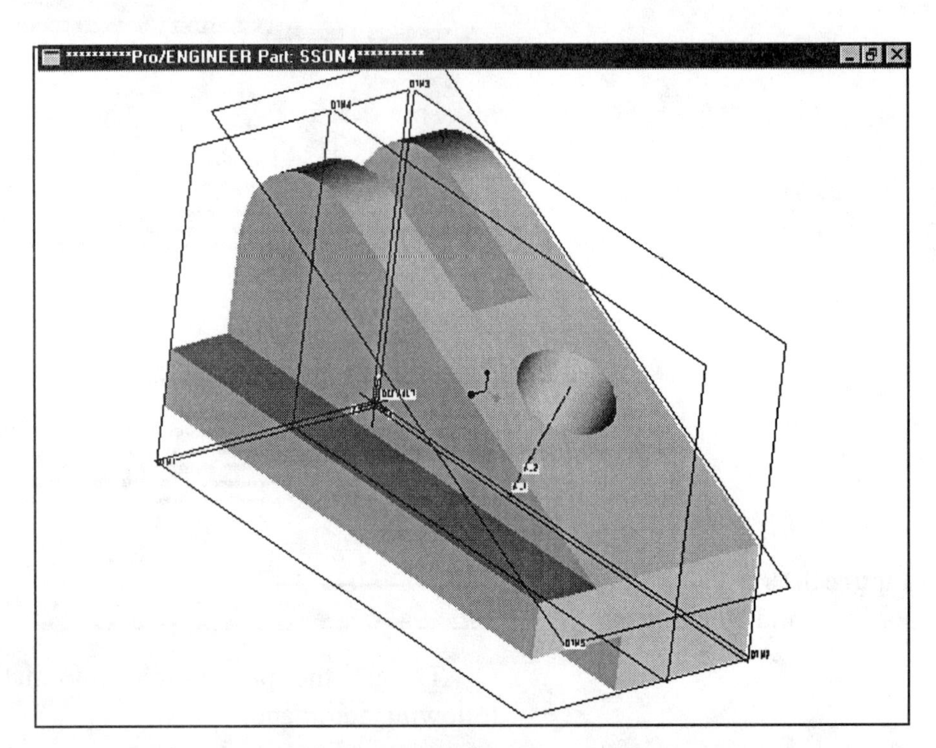

Figure 3.12
Original Design

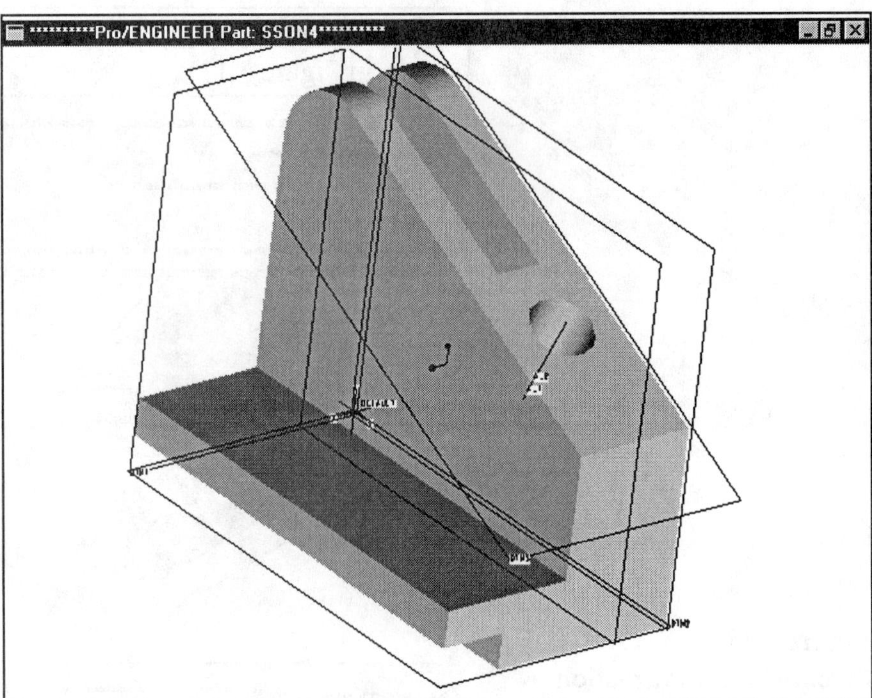

Figure 3.13
Modified Base Feature

Capturing Design Intent

A valuable characteristic of any design tool is its ability to **render** the design and at the same time capture its **intent**. Parametric methods depend on the sequence of operations used to construct the design. The software maintains a *history of changes* the designer makes in specific parameters. The point of capturing this history is to keep track of operations that depend on each other. Whenever Pro/E is told to change a specific dimension, it can update all operations that are referenced to that dimension.

For example, a circle representing a bolt hole may be constructed so that it is always concentric to a circular slot: If the slot moves, so does the bolt circle. Parameters are usually displayed in terms of dimensions or labels and serve as the mechanism by which geometry is changed. The designer can change parameters manually by changing a dimension or can reference them to a variable in an equation (**relation**) that is solved either by the modeling program itself or by external programs such as spreadsheets.

Parametric modeling is particularly useful in modeling whole **families** of similar parts and in rapidly modifying complex 3D designs. It is most effective in working with designs where changes are likely to consist of dimensional changes rather than radically different geometry.

Feature-based modeling refers to the construction of geometry as a combination of **form features**. The designer specifies features in engineering terms, such as holes, slots, or bosses, rather than geometric terms, such as circles or boxes.

Features can also store nongraphic information. This information can be used in activities such as drafting, numerical control (**NC**), finite-element analysis (**FEA**), and kinematics analysis.

The concept of capturing design intent is based on incorporating engineering knowledge into a model by establishing and preserving certain geometrical relationships. The wall thickness of a pressure vessel, for example, should be proportional to its surface area and should remain so, even as its size changes. Parametric design captures these relationships in several ways:

Implicit Relationships Implicit relationships occur when new model geometry is sketched and dimensioned relative to existing features and parts. An implicit relationship is established, for instance, when the section sketch of a tire (Fig. 3.14) uses rim edges as a reference.

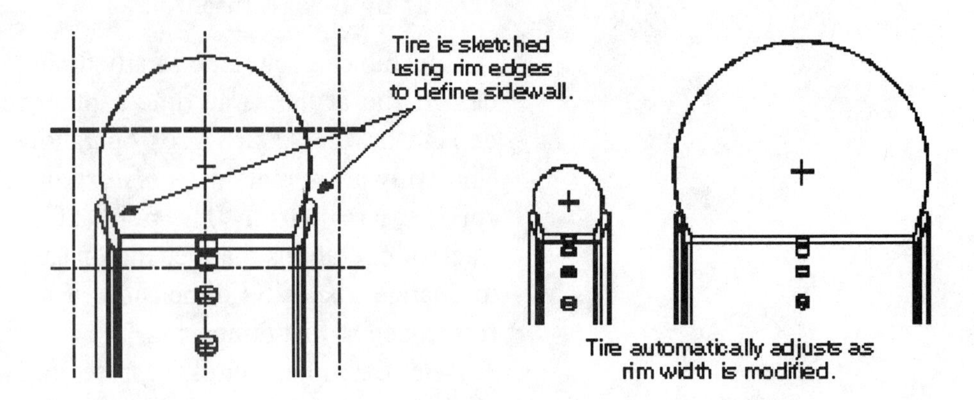

Figure 3.14
Tire and Rim

Patterns Design features often follow a geometrically predictable pattern. Features and parts are patterned in parametric design by referencing either construction dimensions or existing patterns. One example of patterning is a wheel hub with spokes (Fig. 3.15). First, the spoke holes are radially patterned. The spokes can then be strung by referencing this pattern.

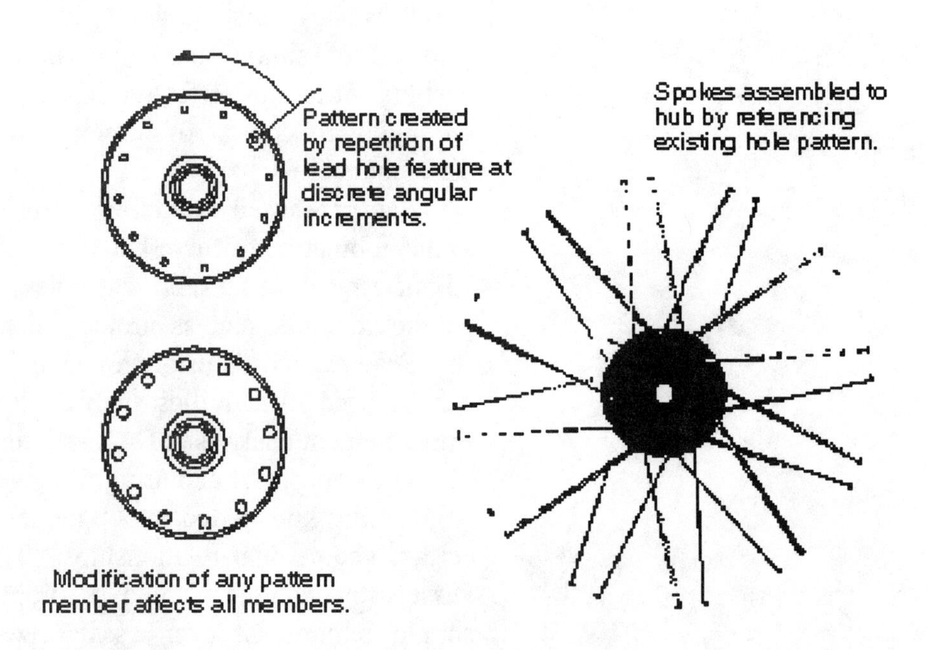

Figure 3.15
Patterns

Modifications to a pattern member affects all members of that pattern. This helps capture design intent by preserving the duplicate geometry of pattern members.

Explicit Relations Where as implicit relationships are implied by the feature creation method, an explicit relation is mathematically entered by the users. This equation is used to relate part an feature dimensions in the desired manner. An explicit relation might be used, for example, to ensure that any number of spoke holes will be evenly spaced around a wheel hub (Fig. 3.16).

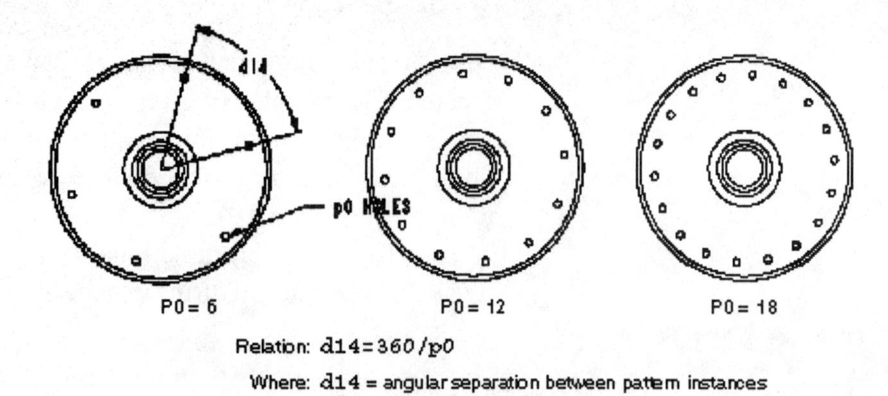

Figure 3.16
Adding Relations

Family Tables Family tables are used to create part families from generic models by tabulating dimensions or the presence of certain feature or parts. A family table might be used, for example, to catalog a series of wheel rims with varying width and diameter (Fig. 3.17).

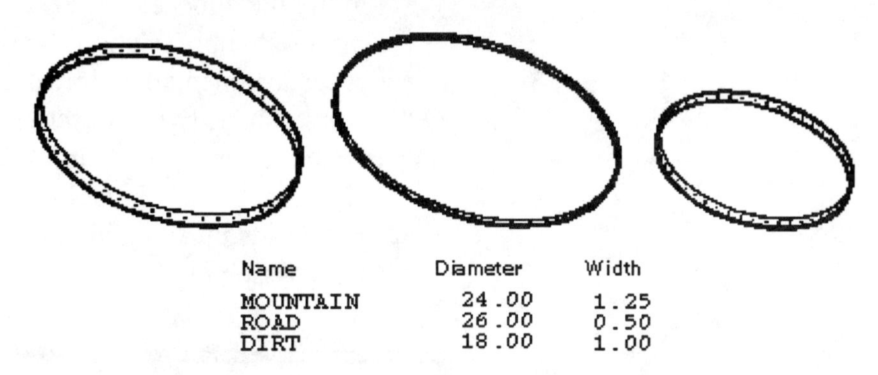

Name	Diameter	Width
MOUNTAIN	24.00	1.25
ROAD	26.00	0.50
DIRT	18.00	1.00

Figure 3.17
Family Table

The modeling task is to incorporate the features and parts of a complex design while properly capturing design intent to provide flexibility in modification. Parametric design modeling is a synthesis of physical and intellectual design (Fig. 3.18).

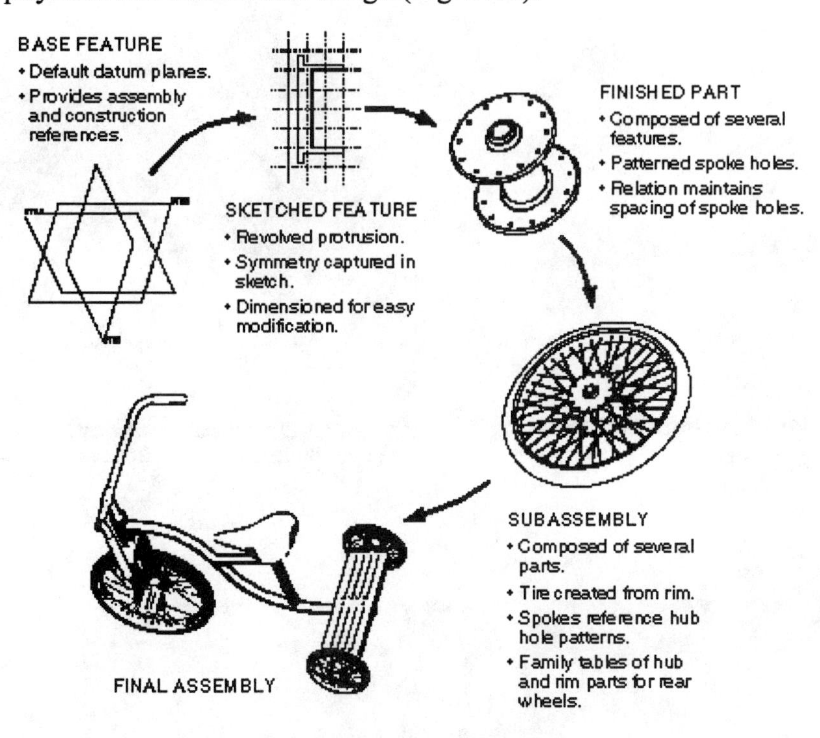

Figure 3.18
Relations

ASSEMBLIES

Just as parts are created from related features, so **assemblies** are created from related parts. The progressive combination of subassemblies, parts, and features into an assembly creates parent-child relationships based on the references used to assemble each component (Fig. 3.19).

The *Assembly* functionality is used to assemble existing parts and subassemblies. During assembly creation, you have the option to:

* Simplify a view of a large assembly by creating a simplified representation
* Perform automatic or manual placement of component parts
* Create an exploded view of the component parts
* Perform analysis, such as mass properties and clearance checks
* Modify the dimensional value of component parts
* Define assembly relations between component parts
* Create documentation drawings of the assembly
* Create a shaded view of the assembly
* Use the assembly as a sub-assembly
* Perform automatic interchange of component parts
* Create parts in Assembly mode
* Create Assembly features

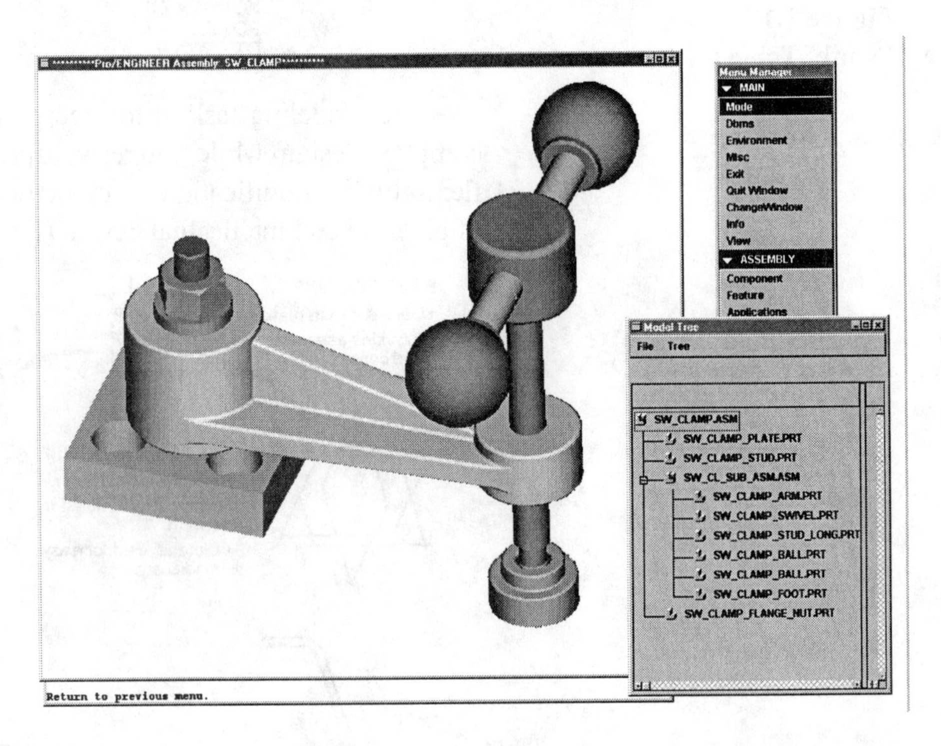

Figure 3.19
Clamp Assembly and Model Tree

As features can reference part geometry, parametric design also allows creation of parts referencing assembly geometry. Assembly mode allows the designer to both fit parts together, and design parts based on how they should fit together.

In Figure 3.19, the assembly of the clamp was shown. Figure 3.20 shows the clamp subassembly. A *Bill of Materials* report is generated for the assembly in Figure 3.21. **Info ⇒ BOM** is chosen from the INFORMATION menu.

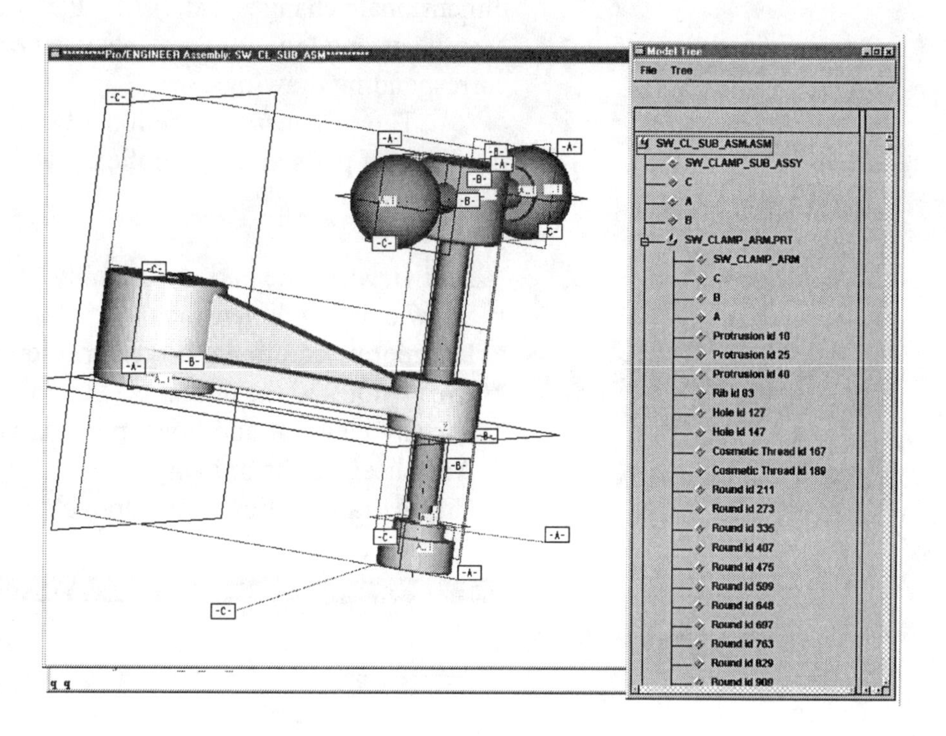

Figure 3.20
Clamp Subassembly

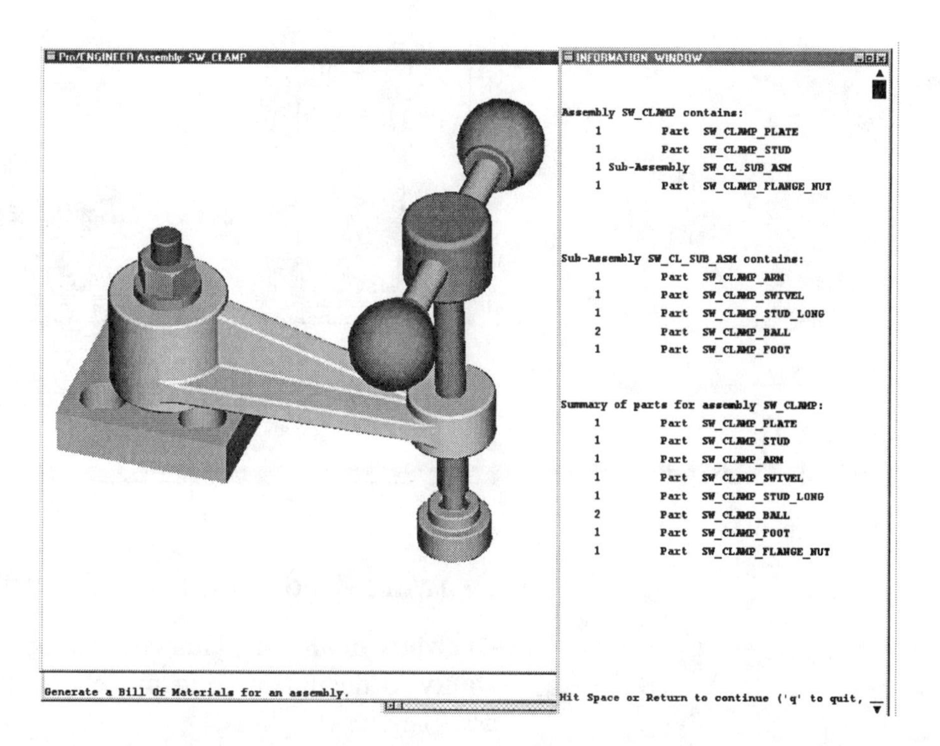

Figure 3.21
Clamp Assembly and BOM

DRAWINGS

You can create **drawings** of all parametric design models or by importing files from other systems. All model views in the drawing are **associative**; if you change a dimensional value in one view, other drawing views update accordingly. Moreover, drawings are associated with their parent models. Any dimensional changes made to a drawing are automatically reflected in the model. Any changes made to the model (i.e., addition of features, deletion of features, dimensional changes, etc.) in Part, Sheet Metal, Assembly, or Manufacturing modes are also automatically reflected in their corresponding drawings.

The *Drawing* functionality is used to create annotated drawings of parts and assemblies. During drawing creation, you have the options to:

* Add views of the part or assembly
* Show existing dimensions
* Incorporate additional driven or reference dimensions
* Create notes to the drawing
* Display views of additional parts or assemblies
* Add sheets to the drawing
* Create draft entities on the drawing

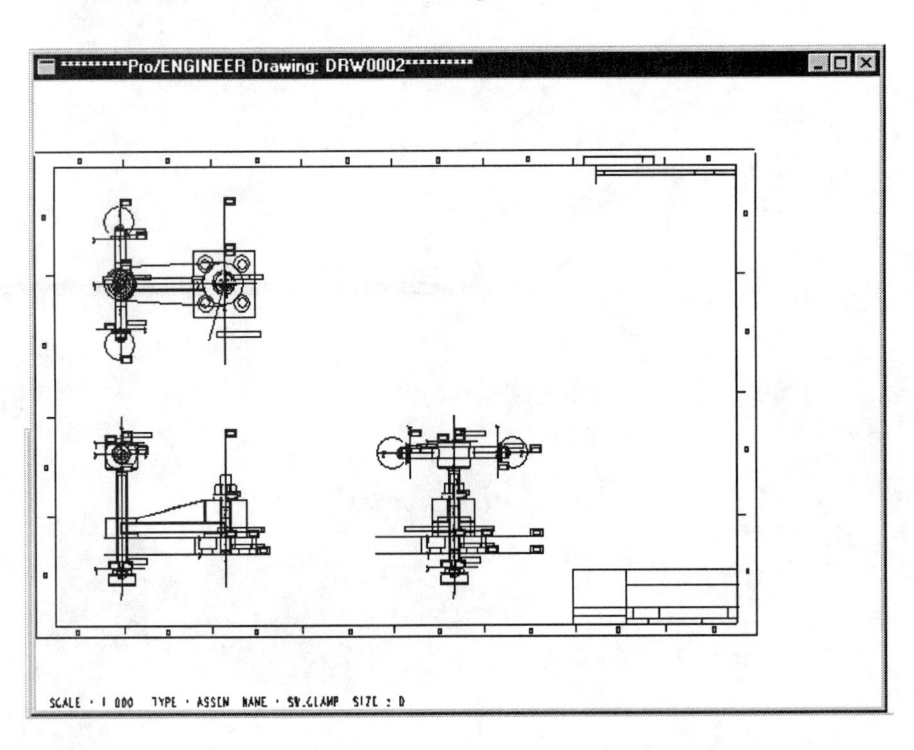

Figure 3.22
Assembly Drawing

Drawing Mode and Basic Parametric Design

Drawing mode in parametric design provides you with the basic ability to document solid models in drawings that share a two-way associativity (Fig. 3.22).

Changes that are made to the model in Part or Assembly modes will cause the drawing to automatically update and reflects the changes. Any changes made to the model in Drawing mode will be immediately visible on the model in Part and Assembly modes. The part shown in Figure 3.23 has been detailed in Figure 3.24

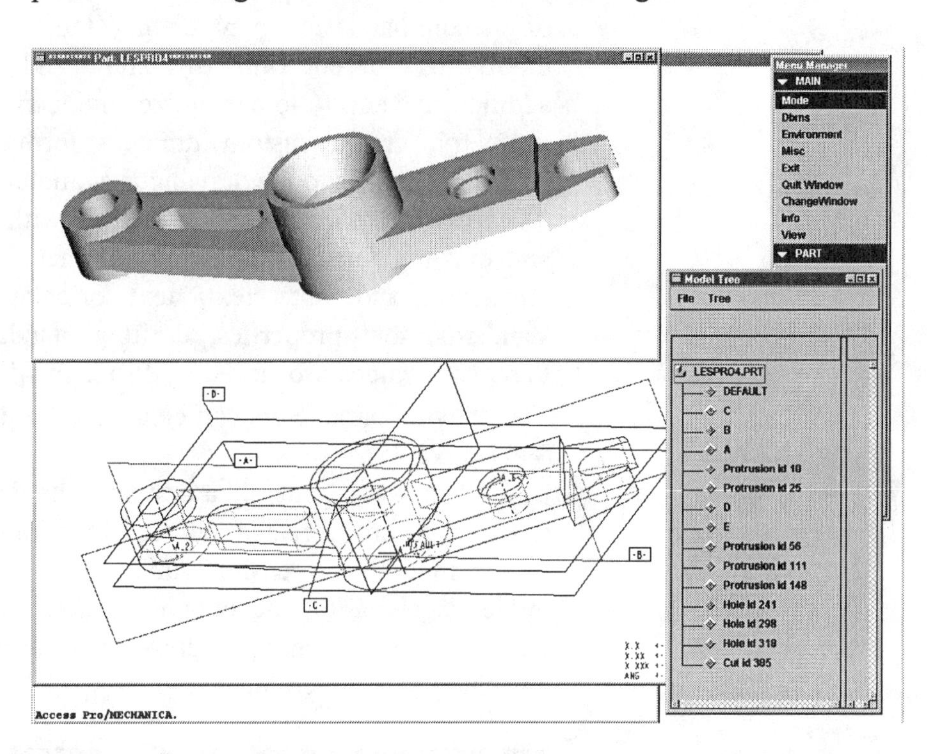

Figure 3.23
Angle Frame Model

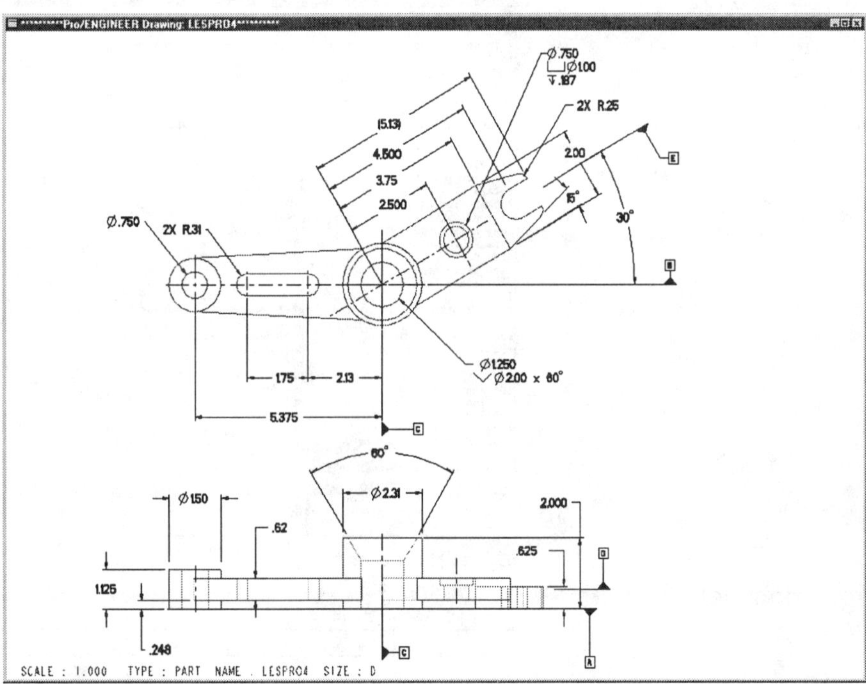

Figure 3.24
Angle Frame Drawing

Basic Pro/E (without the optional module Pro/DETAIL) allows you to create drawing views of one or more models in a number of standard view with dimensions.

You may annotate the drawing with notes, manipulate the dimensions, and use layers to manage the display of different items on the drawing. The optional module **Pro/DETAIL** may be used to extend the drawing capability or as a stand-alone module allowing you to create, view, and annotate models and drawings.

Pro/DETAIL supports additional view types, multi-sheets, and offers commands for manipulating items in the drawing, adding and modifying different kinds of textural and symbolic information. In addition, the ability to customize engineering drawings with sketched geometry, create custom drawing formats, and make numerous cosmetic changes to the drawing is available.

Drawing parameters are saved with each individual drawing and drawing format. Drawing parameters determine the height of dimension and note text, text orientation, geometric tolerance standards, font properties, drafting standards, and arrow lengths. Parameter values are stored by Pro/E in files with extensions of **.dtl**. The drawing parameters can be altered by picking **Set Up ⇒ Modify Val** (Fig. 3.25).

When you regenerate a drawing, the drawing and the model that it represents are recreated, not simply redrawn. This means that if any of the model's dimension values were changed while in drawing mode, regenerating the drawing causes the model to update these changes. The regenerated drawing displays the updated model and any changes that were made to it.

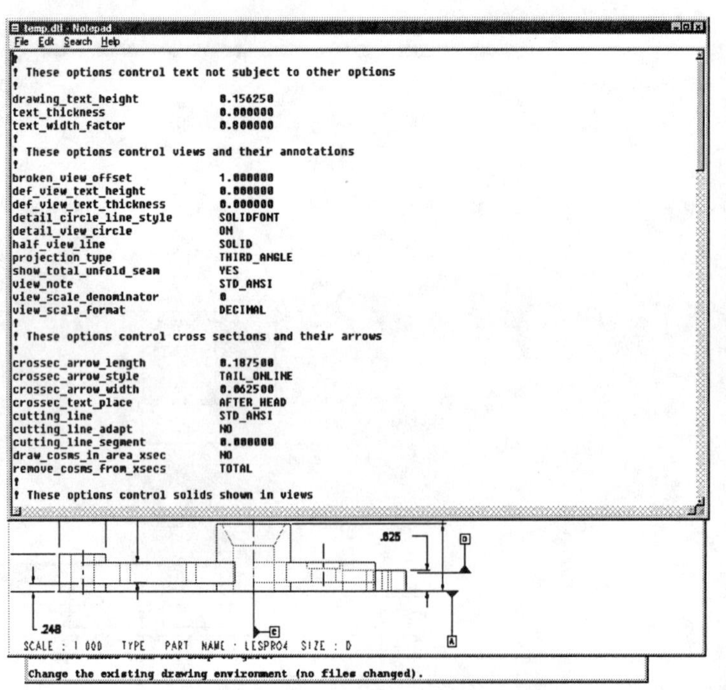

Figure 3.25
Setting **DTL** File Configuration Defaults with **Modify Val**

MANUFACTURING AND PARAMETRIC DESIGN

Parametric design systems also provide the tools to program and simulate numerical control manufacturing processes (Fig. 3.26). The information created can be quickly updated should the engineering design model change. NC programs in the form of ASCII cutter location (**CL**) data files, tool lists, operation reports, and in-process geometry can be generated.

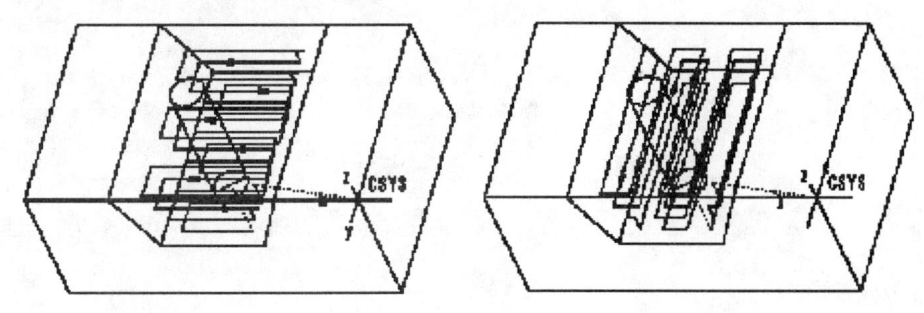

Figure 3.26
Numerical Control Machining Simulation

Pro/MANUFACTURING

Pro/MANUFACTURING software creates the data necessary to drive an NC machine tool to machine a part. It does this by providing the tools to let the manufacturing engineer follow a logical sequence of steps to progress from a design model to an ASCII CL data file that can be post-processed into NC machine data (Fig. 3.27).

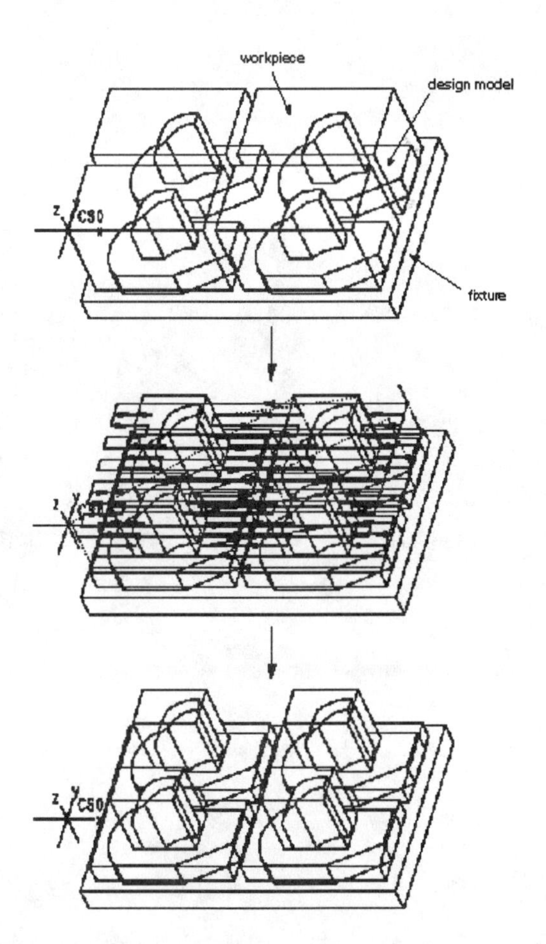

Figure 3.27
Multiple Part Machining using Pro/MANUFACTURING

Section 4

Utilities

There are several utilities used during the design of parts, assemblies, and drawings. The abilities to access and change directories, edit a trail file, give system-level commands, and edit and load new configuration settings are essential to the efficient and accurate creation of project databases.

Operating Modes

The **Mode** command in the **MAIN** menu allows you to enter the major modes of Pro/E operation. When this command is chosen, a submenu appears with the various operational modes. The modes available on your system are dependent on the modules purchased, licensed, and installed on your workstation. In Figure 4.1 the mode command was given and the module selection menu appears. Note that the system has a variety of modules available, including, **Sheet Metal**, **Cast**, **Composite**, **Diagram**, **Dieface**, **Mold**, and **Markup**.

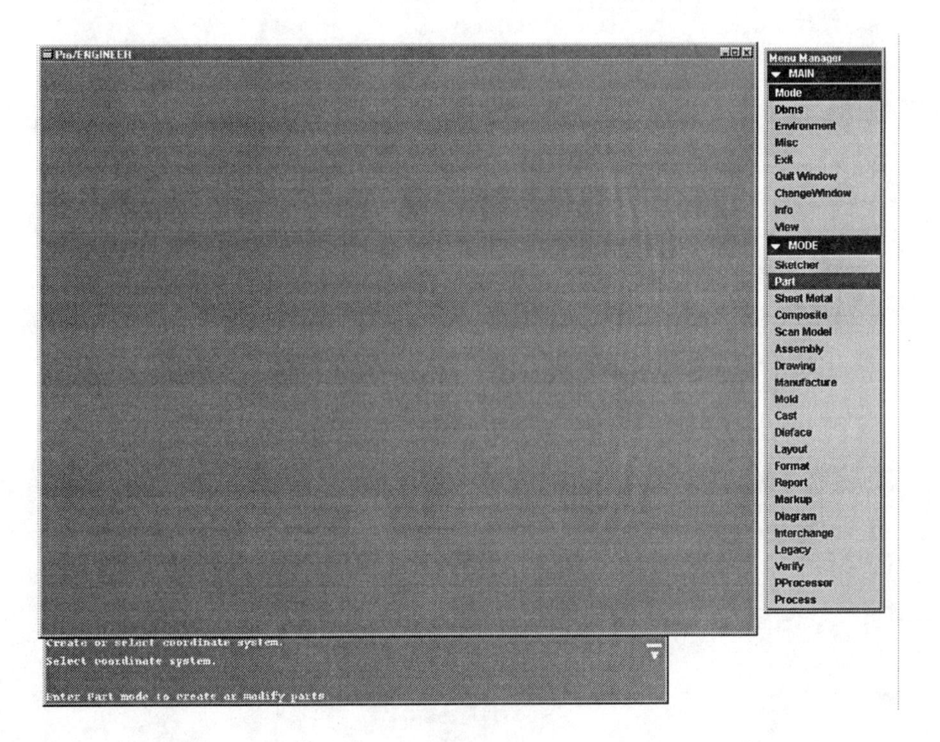

Figure 4.1
Mode Command

More than one module can be active in a Pro/E session at one time. You can have a part, the assembly where it is used, and its drawing on the screen simultaneously. In Figure 4.2, part, assembly, and drawing modes have been activated. You can move between the windows and work on a drawing, a part, or the assembly. Since Pro/E is *parametric*, any editing or modifications made to the drawing will reflect in the assembly and part modes after regeneration. The same is true of part mode and assembly mode changes.

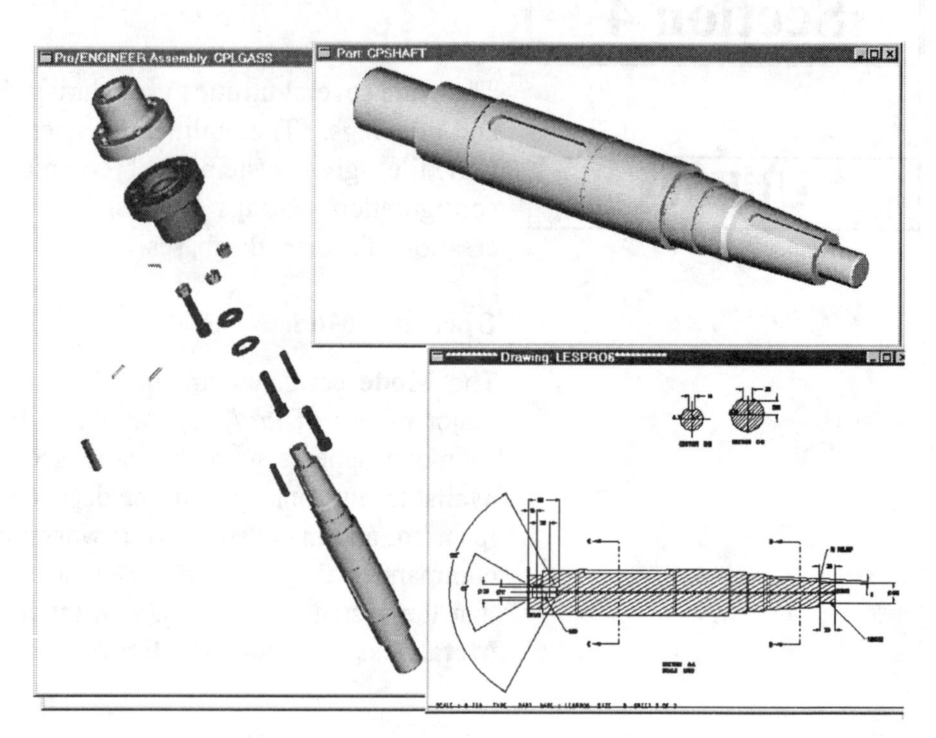

Figure 4.2
Part, Assembly, and Drawing Modes Open in the Same Session

The primary modules covered in this text are:

Part This mode is used to create parts from *section sketches*, modify parts, and add features to parts (Fig. 4.3) . It is the module that you will be working in most of the time. This mode is used in Lessons 1 through 13.

Assembly This mode is used to assemble components and is covered in Lessons 14 and 15.

Drawing This mode is used to create fully dimensioned drawings of parts and assemblies. The drawing mode is used for Lessons 16 through 20.

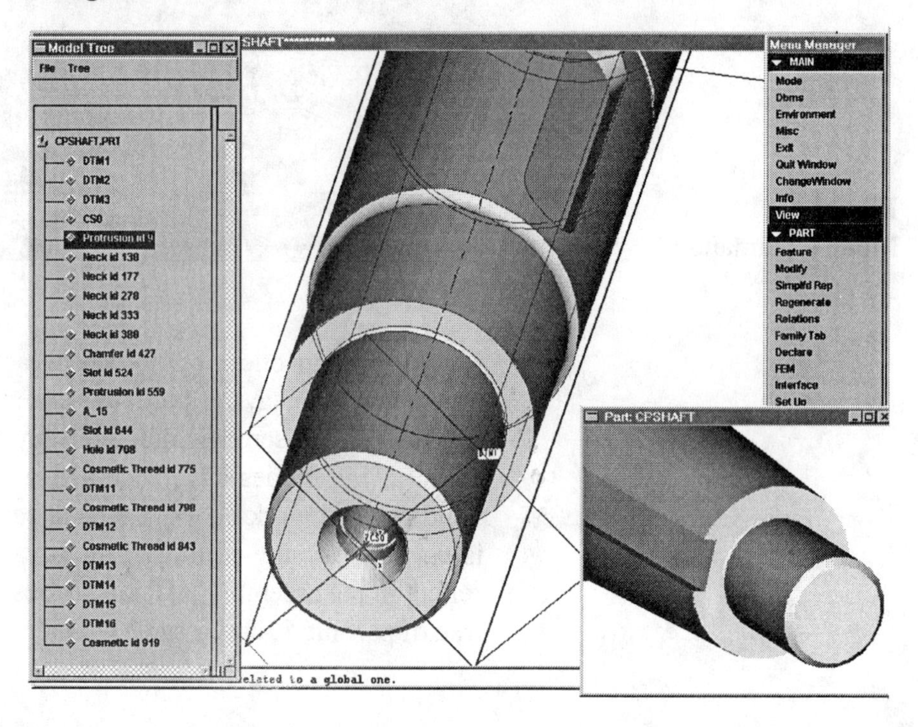

Figure 4.3 Part Mode

File Naming

The rules for file naming, such as file name length and the effects of uppercase/lowercase names, are dependent on the operating system. Be aware of these rules before you file "objects" in Pro/E. Generally, all UNIX file names and directory names must be lowercase. DOS names can be only **8** characters long and have only one extension with three characters.

The following is an example of default-naming conventions that are used by Pro/E to aid file management. Refer to Appendix A in *Fundamentals of Pro/E* for other naming conventions.

xxx.sec	Cross section
s2d###.sec	Default cross section
xxx.prt	Part
prt###.prt	Default part
xxx.drw	Drawing
drw###.drw	Default drawing
xxx.frm	Drawing format
xxx.asm	Assembly
asm###.asm	Default assembly
color.map	Shaded view color settings
names.inf	File used for Names INFO listing
rels.inf	Temporary file for relations listing
layer_all.inf	File used for Layer All INFO listing
layer_#.inf	File used for Layer ID INFO listing
partname.inf	Temporary file used for Part INFO listing
partname.m_p	Temporary file used for Part Mass Properties listing
assembyname.inf	Temporary file used for Assembly INFO listing
feature.inf	Temporary file used for Feature INFO listing
config.pro	Default configuration options file
trail.txt	Default name for Trail file
xxxx.cfg	Model Tree Settings file

Because Pro/E adds a period "." between the file name entered and the appropriate extension, do not use a period "." in any file names created.

File Management

Within the Pro/E environment, files are managed through the **DBMS** menu. This menu item is found in the **MAIN** menu (Fig. 4.4). The submenu contained under DBMS has the a variety of options (where "object" may be a part, assembly, drawing, layout, sketch, or manufacturing model). *You can abort the function by pressing the <ESC> key.* The following options are available:

Save Files the object to the disk.
Save As Duplicates and files current objects with a new file name.
Backup Saves a backup copy of an object to a specified directory.
Rename Changes the file name of the current object to a new name. Have the part and its assembly active (in memory) before changing names!
Erase Removes an object from workstation working memory.
EraseNotDisplay Removes all objects from workstation memory *except* those that are currently displayed and any objects referenced by the displayed objects.
Purge Deletes all but the last object version from disk. Use this every time you save, and it will eliminate all versions except the one you just saved. If you do not do this, then every time you save the object, you will have created a file reflecting the condition of the object at that time. If you save 27 times, then you have 27 different versions!

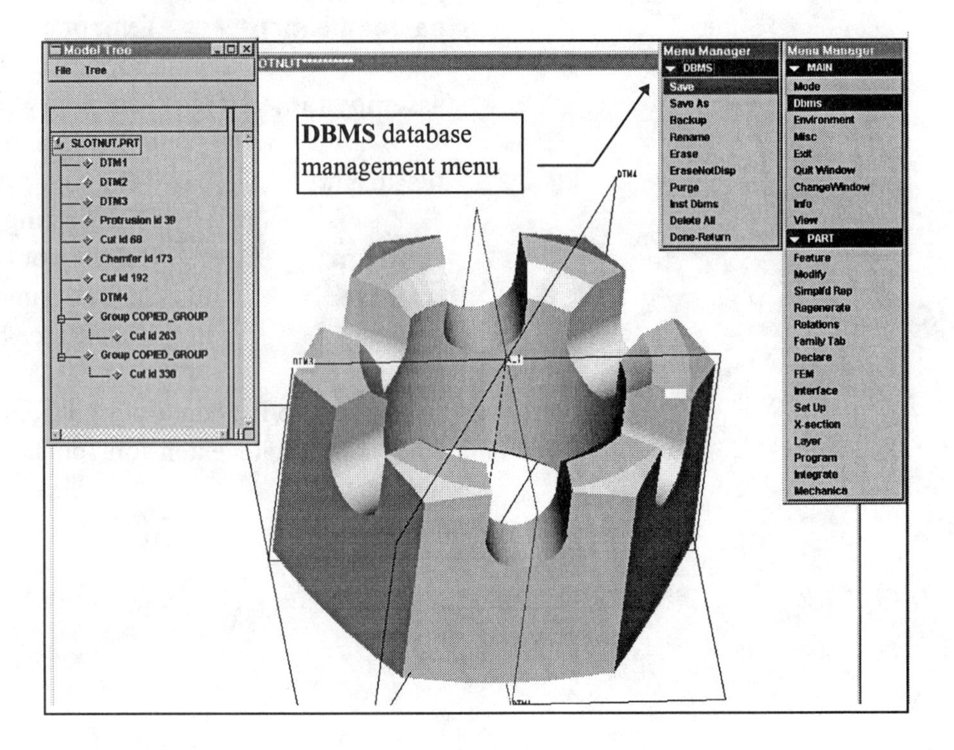

Figure 4.4
DBMS Menu

Inst Dbms Accesses the family table DBMS function.

Delete All Deletes all current and saved object versions from the disk and workstation memory. *Be very careful using this command since it deletes all versions of the object.*

In all Pro/E modes (Sketcher, Part, Assembly, Drawing, etc.), files can be created, retrieved, imported, and listed through an ENTERSECTION, ENTERPART, ENTERASSY, or ENTERDRAWING menu, depending on what mode you are in. In order to work in these modes, a new file must be created or an existing file retrieved. If different versions of a file exist, entering its name, extension, and version number will retrieve it. As a matter of practice, get used to using **Search/Retr**. This will allow you to navigate the current directory and change to another directory. If you use **Retrieve** and give a wrong object name or you are in the wrong directory, Pro/E may take a considerable amount of time searching for the file.

Choosing **List** from the menu will display the files that have been created in that mode and note any files that not yet been stored using **Dbms**. There is a configuration option that will prompt you to file only unsaved files upon exiting Pro/E.

Choosing **Search/Retr** from the menu (Fig. 4.5) displays a list of all of the files of the desired type and directories the user can pick from instead of typing in the name of the file to be retrieved.

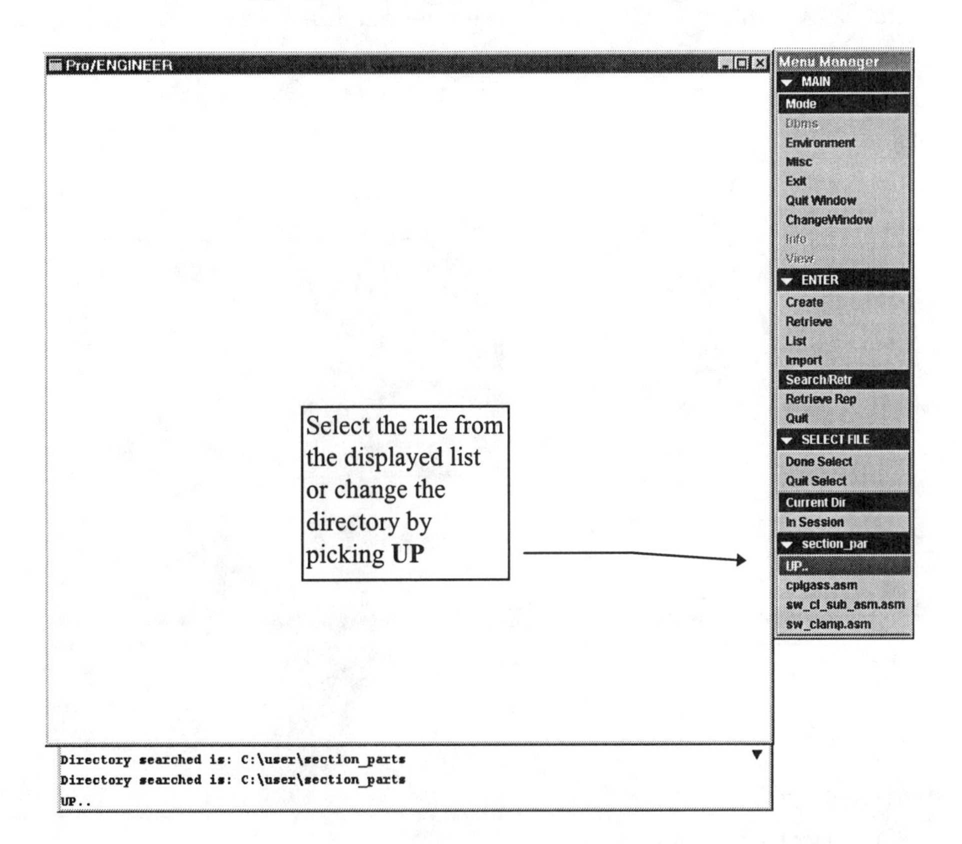

Figure 4.5
Using Search/Retr

OPERATING SYSTEM ACCESS

To access the operating system or perform a system command, choose **Misc** (Fig. 4.6) from the MAIN menu. The MISC menu enables you to access various applications, including the following:

List Dir Lists the contents of the current working directory (Fig. 4.6). This is one of the most used system commands.

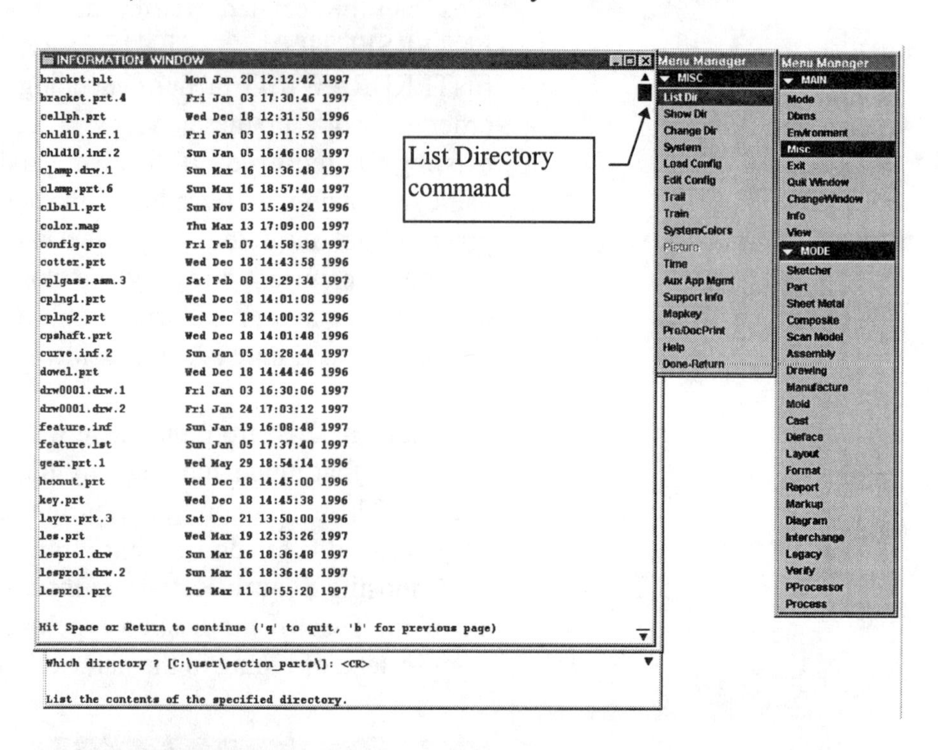

Figure 4.6
List Dir Command

Show Dir Shows the current working directory at the command line at the bottom left of the screen (Fig. 4.7).

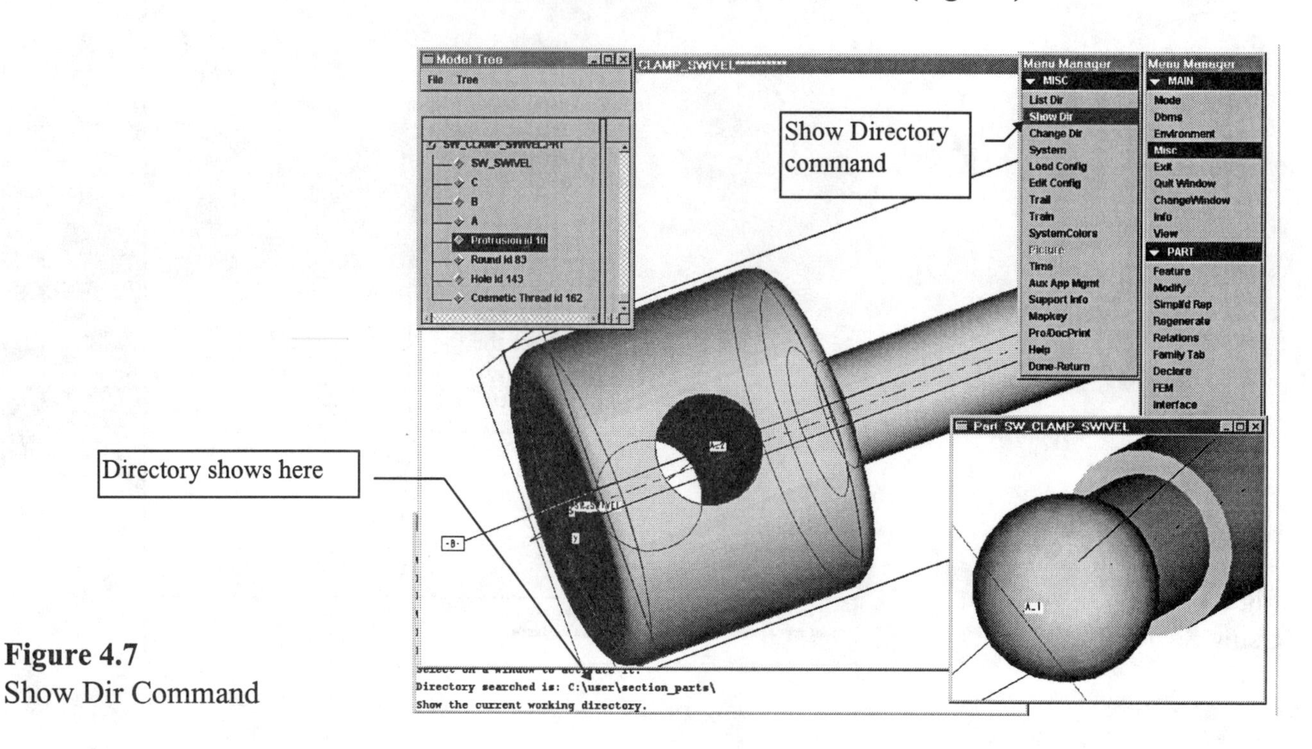

Figure 4.7
Show Dir Command

Change Dir Changes to another working directory (Fig. 4.8). Enter a **?** at the prompt to see current options. By picking **UP** you go to the next-higher-level directory. In this example the directories have been named **Lessons**. To make another directory the working directory, pick it from the choices and then choose **HERE**.

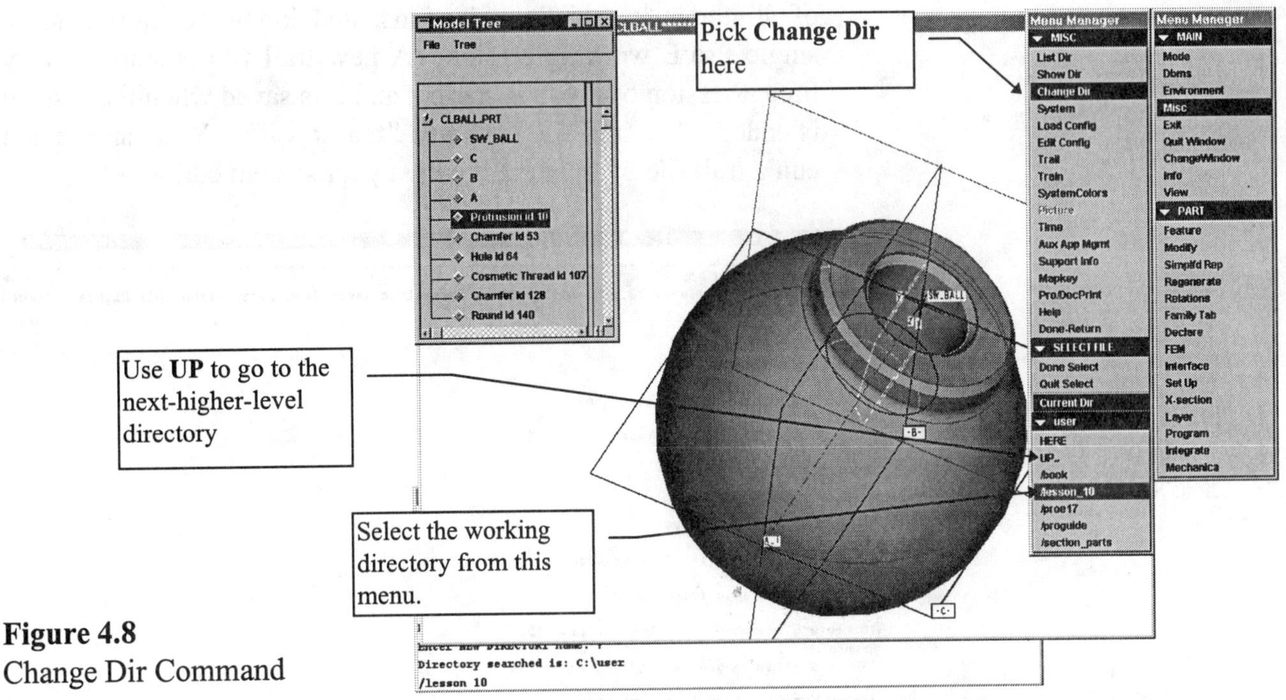

Figure 4.8
Change Dir Command

System Enters the system window without exiting Pro/E (Fig. 4.9). This window allows you access to the system-level commands. UNIX or Windows system commands can be entered here. The type and level of commands here are determined by your system privileges.

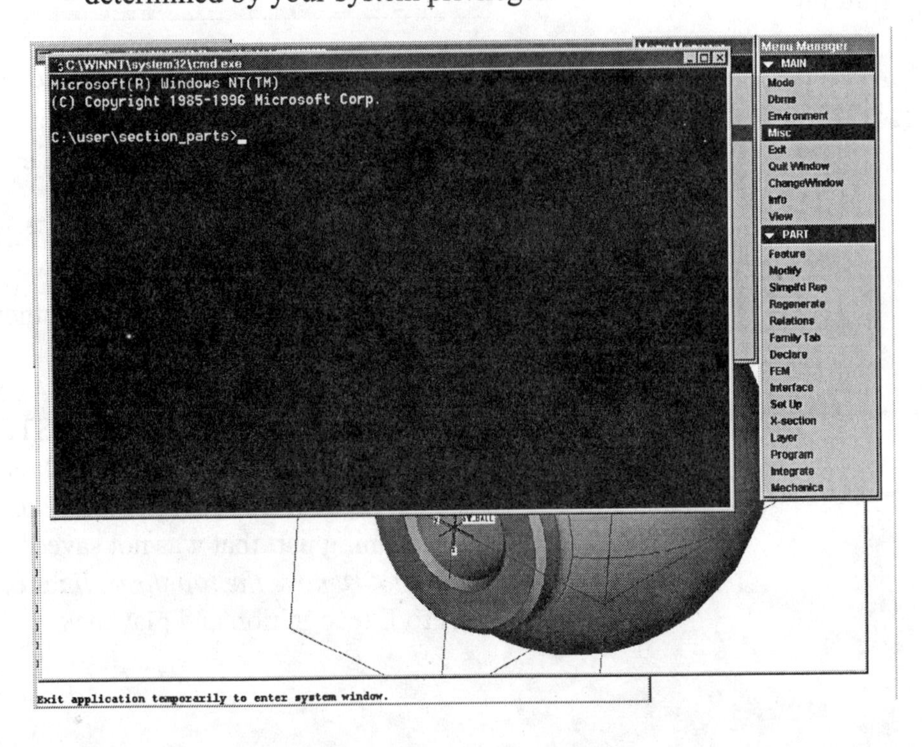

Figure 4.9
System Window

Load Config Loads a configuration file (this is explained in detail in Section V).

Edit Config Edits a configuration file. If you use Pro/TABLE, this allows you to see all possible configuration options as well as applicable values (see Section V).

Trail Runs a Pro/E trail file (Fig. 4.10). A **trail file** is a record of all the menu picks, selections, and keyboard entries for a single Pro/E working session. A new trail file is started every time a session of Pro/E is started, and it is saved when the session is ended. The trail file is named **"trail.txt.#"**. You can see and edit a trail file by opening it with a your system editor.

```
trail.txt - Notepad
File Edit Search Help
!trail file version No. 993
!Pro/ENGINEER  TM  Release 18.0  (c) 1988-96 by Parametric Technology Corporation  All Rights Reserve
!Select a menu item.
#MISC
#CHANGE DIR
!Enter NEW DIRECTORY name:
?
!Directory searched is: C:\user\book
#UP..
!Directory searched is: C:\user
#QUIT SELECT
#DONE-RETURN
#MISC
#CHANGE DIR
!Enter NEW DIRECTORY name:
?
!Directory searched is: C:\user\book
#UP..
!Directory searched is: C:\user
#/lesson_10
!Directory searched is: C:\user\lesson_10
#HERE
!Successfully changed to C:\user\lesson_10\ directory.
#LOAD CONFIG
!Enter Configuration file name [config.pro]:

!Configuration file config.pro has been loaded.
#LOAD CONFIG
!Enter Configuration file name [config.pro]:
text_config.pro
!Configuration file TEXT_CONFIG.PRO has been loaded.
#SYSTEMCOLORS
#RESTORE
!Directory searched is: C:\user\lesson_10
#syscol.scl
```

Figure 4.10
Trail File

The file can be replayed to retrieve information that was lost in a Pro/E session (an object that was not saved to disk). Choose **Misc** ⇒ **Trail** and enter the trail file name to execute the file.

When using trail files remember the following rules:

1. Copy the trail file to something other than "trail.txt". Do not keep *trail* as the file name.
2. Edit the copied file as required. As an example, remove commands such as #EXIT and #SYSTEM.
3. Replay only what is needed. If three parts were created and two were saved, then it is only necessary to replay the trail file to recreate the remaining part that was not saved.
4. *Do not remove the top three lines of the trail file*; they are required for Pro/E recognition and playback.

Train Runs a Pro/E training trail file. You can edit and save it as a training file for education or training programs.

System Colors Modifies the colors used by Pro/E for background (Fig. 4.11), geometry, letters, etc.

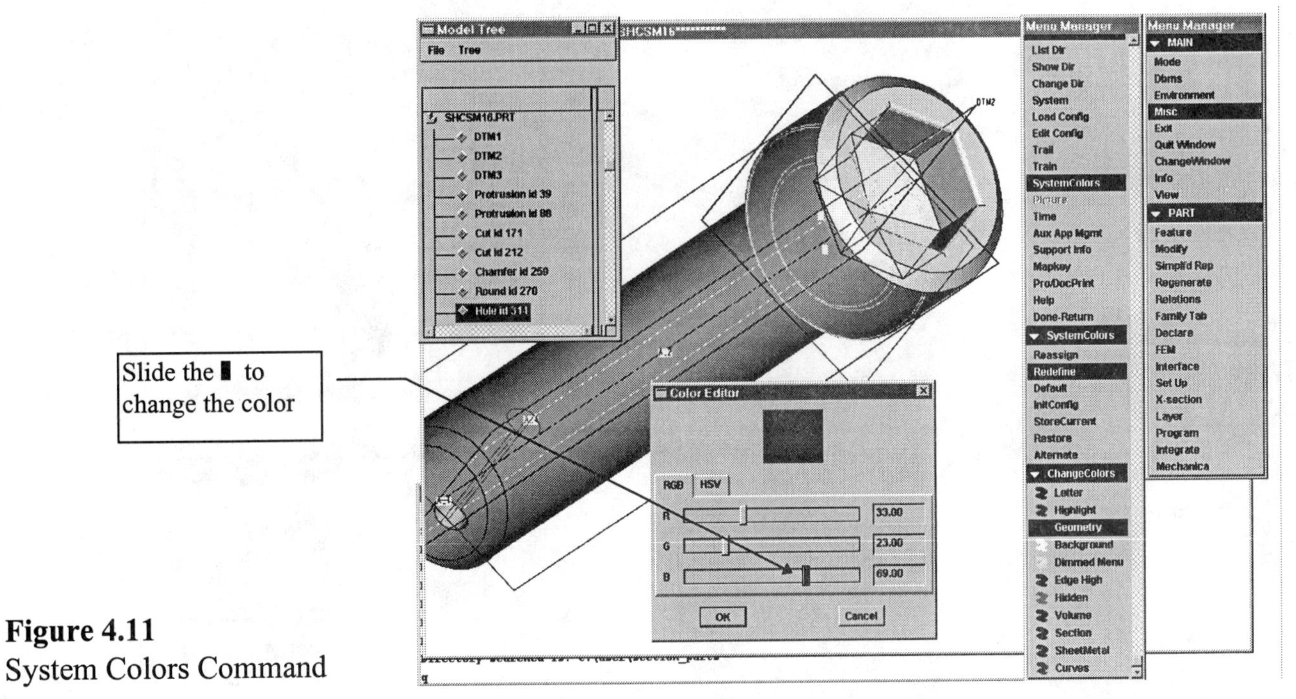

Figure 4.11
System Colors Command

Support Info Provides information about your system and current Pro/E products installed on your system, including addresses and phone and FAX numbers for support (Fig. 4.12).

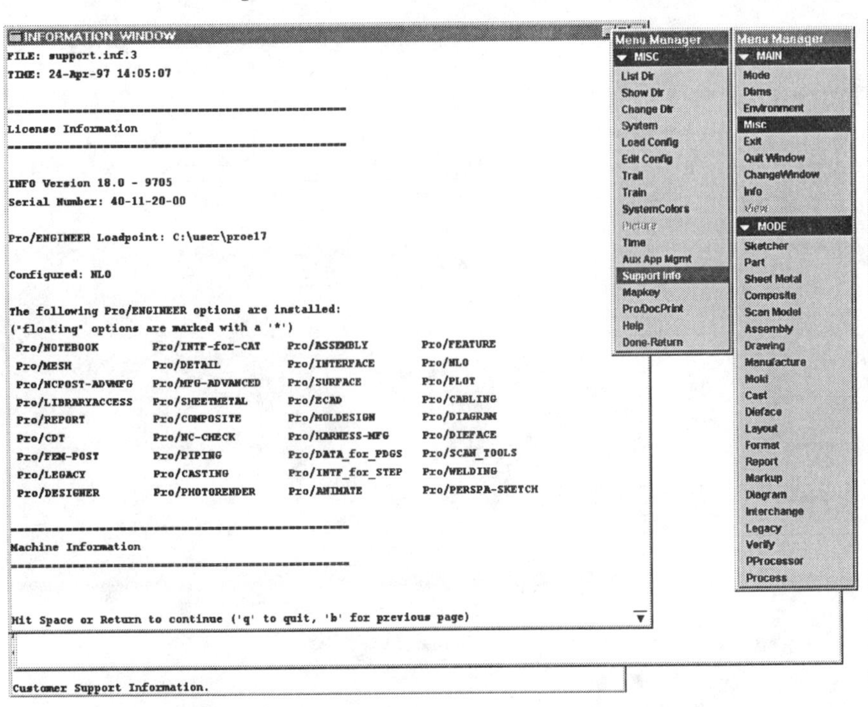

Figure 4.12
Support Info

Time Displays the current time at the command line.

Mapkey Creates a macro interactively (Fig. 4.13).

Aux Appl Mgmt Manages auxiliary applications (Fig. 4.14).

Pro/Doc Print Prints a Pro/Help document (Fig. 4.15).

Figure 4.13
Mapkey

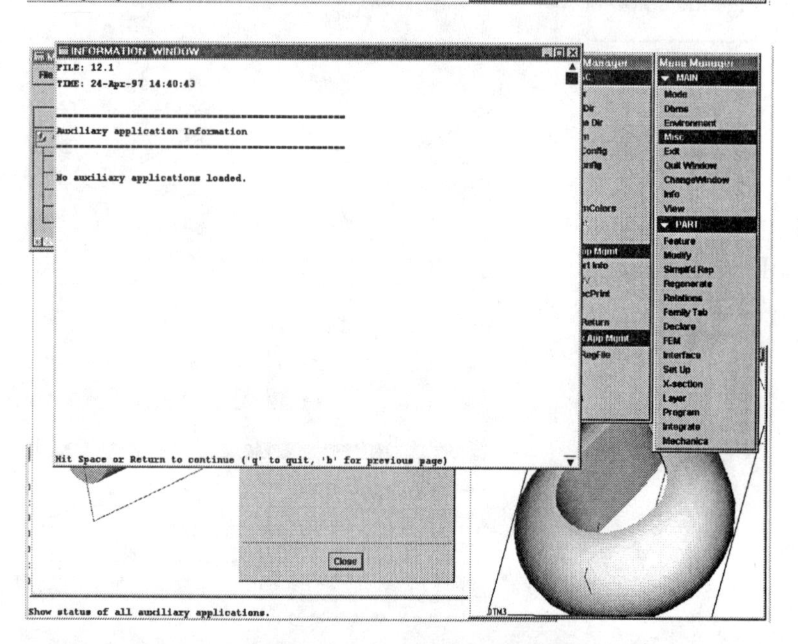

Figure 4.14
Auxiliary Application Manager

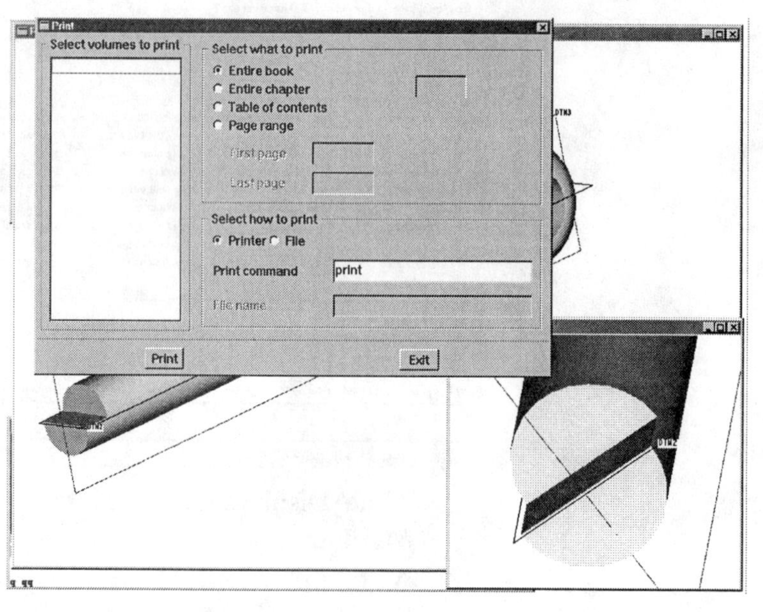

Figure 4.15
Pro/E Document Printer

Section 5

Configuration Files and Mapkeys

A **configuration file** allows you to configure (set up) a Pro/E working session to your project or company requirements. A configuration file can set school or company standards for storage, formatting, establishing default units for new parts (such as millimeters instead of inches), etc. Configuration files are important in establishing the location of directories that contain your library items.

Each user can have an individual configuration file, but establishing a system wide configuration file is a way to establish project standards.

You can control the environment in which Pro/E runs, that is, specify the way your models are oriented, plotting configurations, environment settings, table editor, system colors, and so on. There are two ways to set the working environment: *editing the configuration file* and *using the environment menu.*

Using the Configuration File *config.pro*

You can enter the settings into this configuration file, before starting Pro/E. These settings will then be activated every time you start a new session. If you do not enter a particular setting in the configuration file and there is no master configuration file, then Pro/E will use the system default. Figure 5.1 shows a typical personalized configuration file.

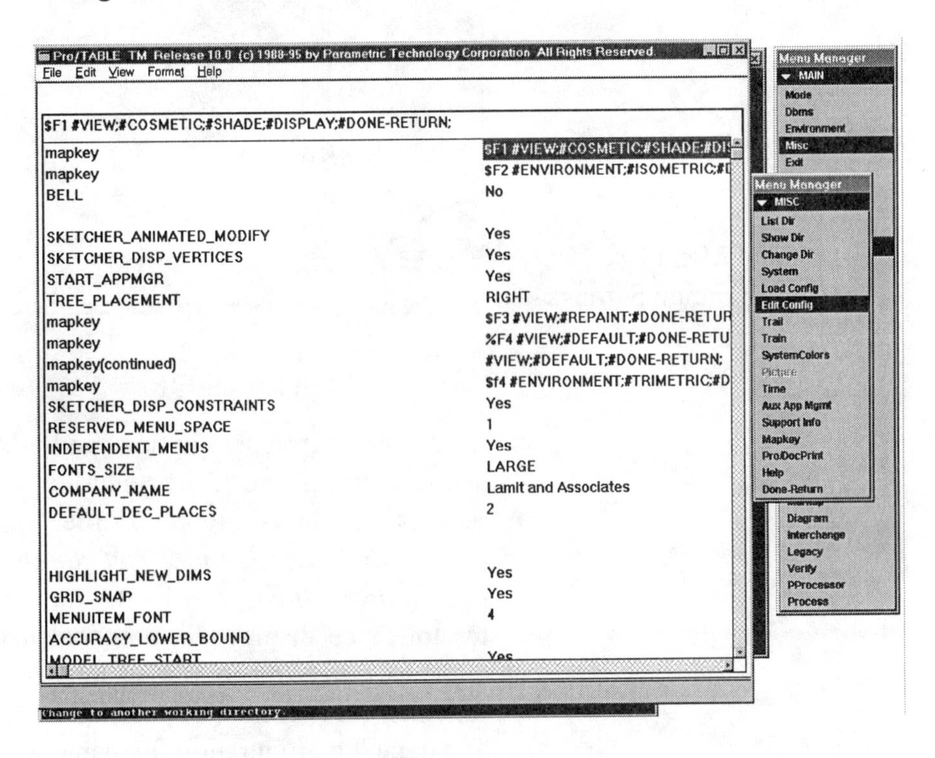

Figure 5.1
Configuration File

Using the Environment Menu

Frequently used configuration settings are included in the ENVIRONMENT menu, which is accessible from every Pro/E mode. This enables you to change the settings easily during the current session. When you enter Pro/E, the values of the environmental variables are set to those in your *config.pro* file; otherwise, they are set to system defaults. In Figure 5.2 the environment menu is shown (note that the large font size needed to capture this image has truncated the menu name and removed a few of the possible settings). To change a setting, just pick the item, a check ✓ will appear to activate this capability. To remove a setting, pick the item and the check will disappear.

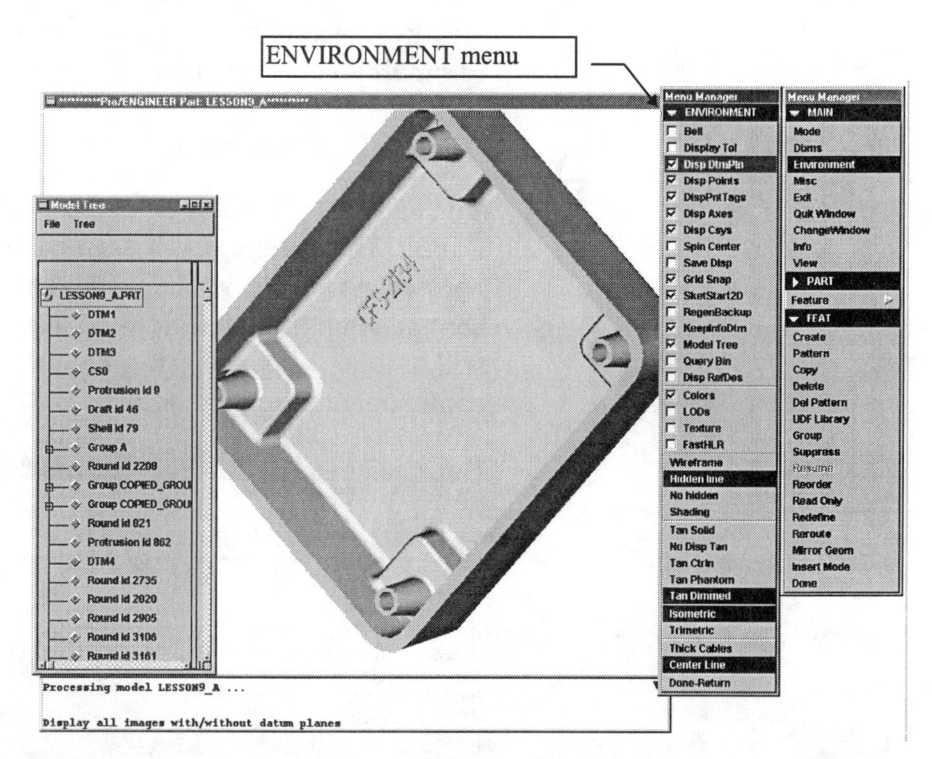

Figure 5.2
Environment Menu Used to Alter Configuration Settings

Setting Configuration Files

All of the environment options and other global settings can be preset before starting Pro/E by entering the relevant options and their settings into the configuration file. *If Pro/TABLE is not available, a system window comes up. You can then edit your config.pro file with your system editor.* The configuration files can also include settings for tolerance display, formats, calculation accuracy, the number of digits used in Sketcher dimensions etc. Pro/E assumes default values for variables that can be specified in the configuration file but are not. The default configuration file name is *config.pro*.

You may add comments to the configuration file by entering an [!] at the beginning of a line (Fig. 5.3).

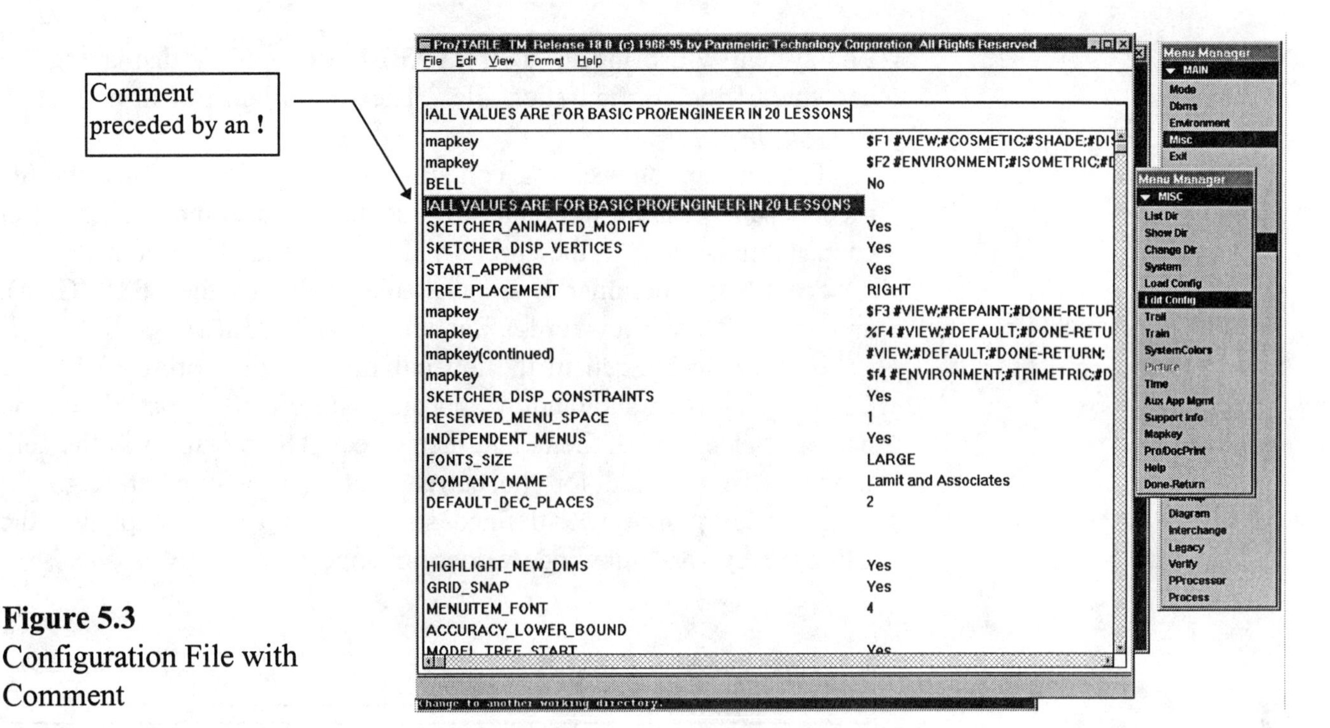

Figure 5.3
Configuration File with Comment

Pro/E can read configuration files from several areas. If a particular option is present in more than one configuration file, the latest value is the one used by the system.

At startup, Pro/E first reads in a protected system configuration file called *config.sup*. Next, it reads configuration files from the following directories, in the order in which they are listed:

1. *Load/point*/text - (*loadpoint* is the directory in which Pro/E is installed.) Your system administrator may have put a configuration file in this location to support your organization's standards for formats and libraries. Anyone accessing Pro/E from this loadpoint will use the values in this file.
2. Login directories - This is the home directory for your login ID. Placing your configuration file here lets you start Pro/E from any directory without having a copy of the file in directory.
3. Startup directory - This is your current or working directory, when you start Pro/E. The local *config.pro* (in your startup directory) is the last to be read; therefore, it will override any conflicting *config.pro* option entries.
4. Lastly, Pro/E uses the default values. These values are built into the software and automatically take effect, unless a configuration file option specifies otherwise.

Editing a Configuration File

To edit a configuration file during a Pro/E session, choose **Edit Config** from the MISC menu, enter the file name (*config.pro* is the default), and press <Enter> on the keyboard.

The system will bring up a **Pro/TABLE** subwindow displaying the contents of the configuration file. The file is laid out in cells, two cells per line.

The first cell on each line contains the name of an options, and the second cell contains its value. If the line is a comment (the first character must be "!"), then Pro/TABLE still treats it as two cells.

Pro/TABLE commands are available, including the <F3> (Goto), and <F4> (Choose keywords) function keys (Fig. 5.4).

If you select a cell in the left column and then press <F4>, the **Choose Keyword** subwindow appears, with a list of possible option names. Select one of them, and then select **OK** to enter it in the cell. If the value required for the current option is a number (e.g., for "angular_tol") or a user-defined string, such as a "Mapkey", the **Choose Keyword** subwindow does not appear.

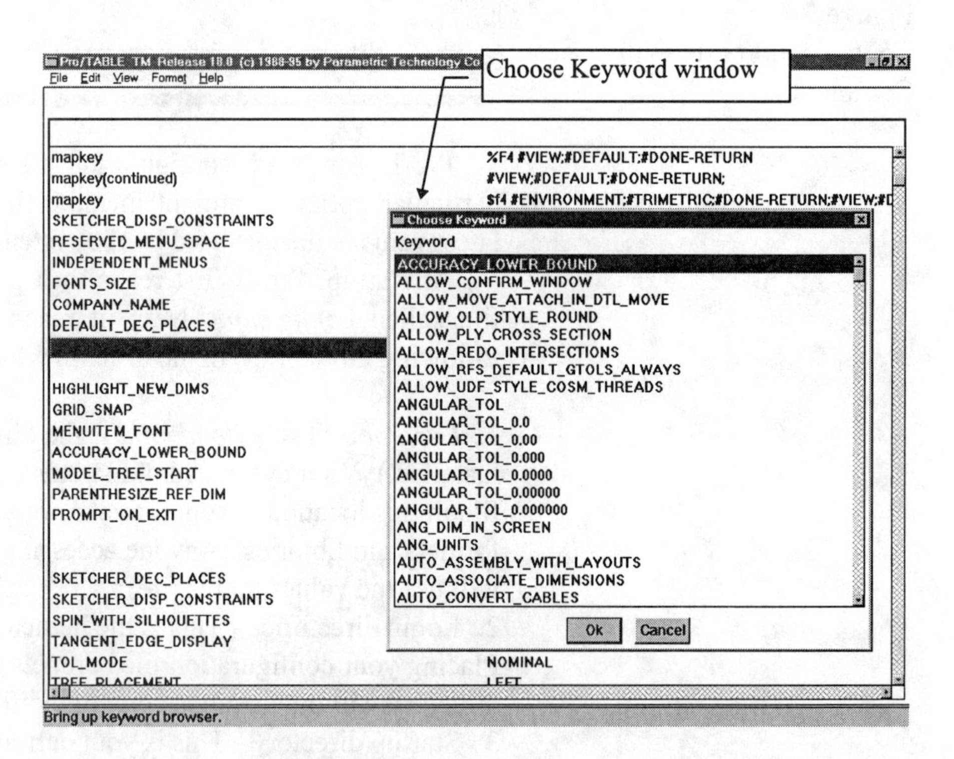

Figure 5.4
Choose Keyword Window

After highlighting a cell in the left-hand column (Fig. 5.4), press the **F4** key; the Choose Keyword menu appears. Pick **OK**, and the *config.pro* setting will appear in the left-hand column (Fig. 5.5). Now select a cell in the right column, and press **F4**. The **Choose Keyword** subwindow appears, with a list of possible (non-numerical) values for the option in corresponding left cell. You can select one of them by picking **OK** to enter it in the cell.

In Figure 5.5 the 3D_MENUS_BUTTONS configuration file option was chosen, and the **Choose Keyword** submenu gives the choice of **Yes** or **No**.

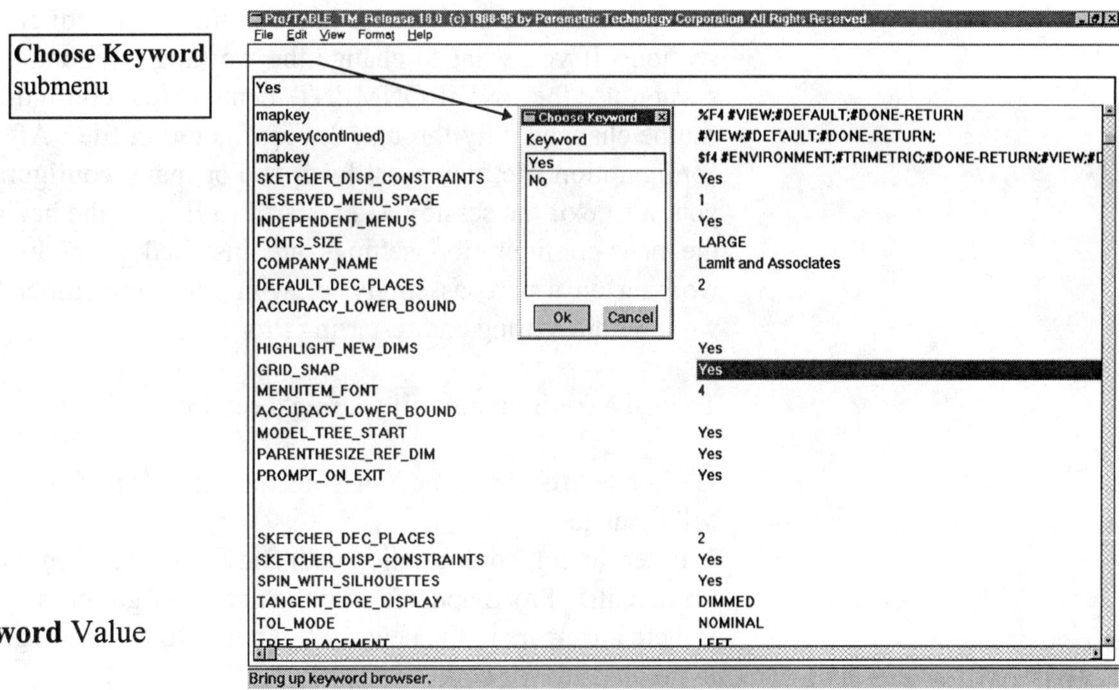

Figure 5.5
Choose Keyword Value Submenu

A configuration file line cannot be more than 80 characters long. After creating the required configuration options and setting their values, pick **File** from the menu and **Exit** and **Save** the new settings (Fig. 5.6).

Again, if Pro/TABLE is not available, a system window comes up. You can then edit your *config.pro* file with your system editor.

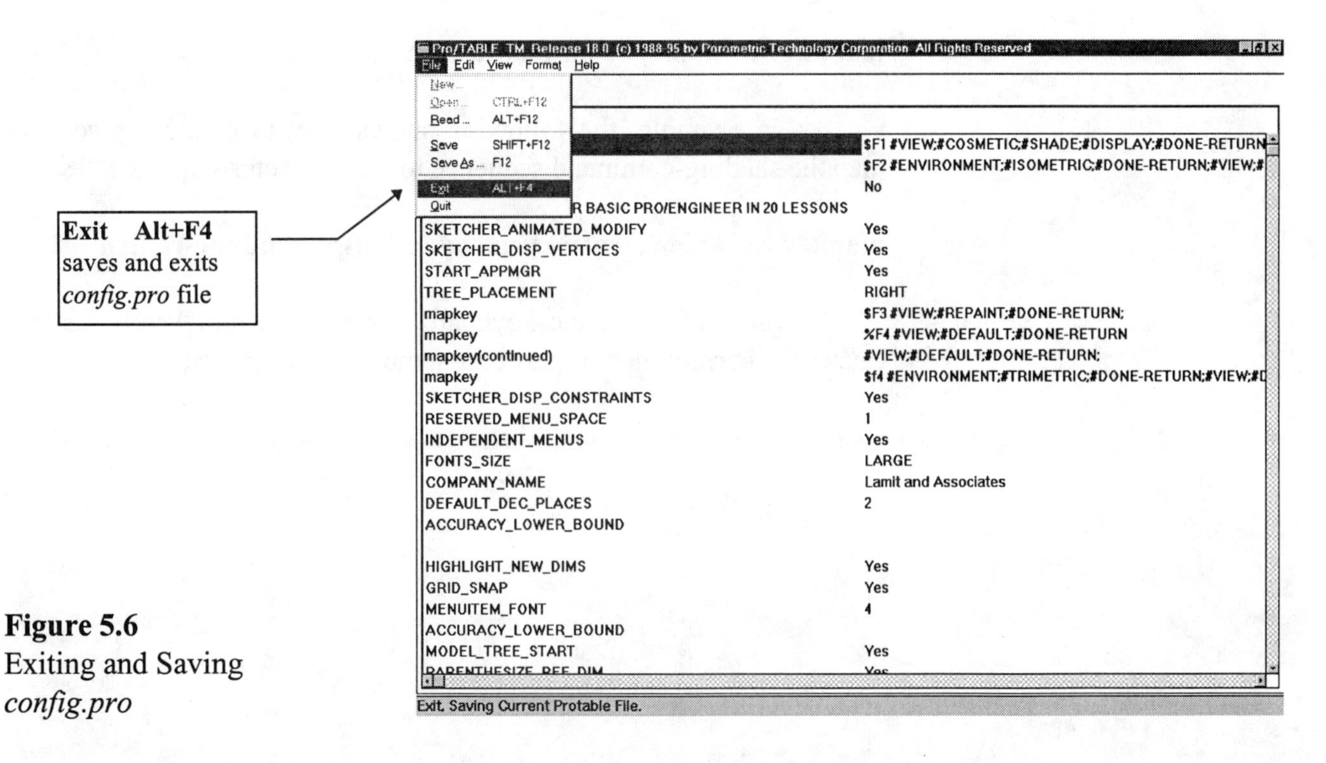

Figure 5.6
Exiting and Saving
config.pro

Loading a Configuration File

After you have finished editing, load the file using the **Load Config** option in the MISC menu to enable the new settings. Default configuration file settings are made in the process of starting a Pro/E session. If you want to change the working environment during the session use the ENVIRONMENT menu. Most configuration options can be changed only through the configuration file. After editing the configuration file, you must load it. For many configuration options you must exit the session and restart Pro/E from the beginning so that the new configuration settings are installed properly. If you are working on a part, assembly, drawing, etc., remember to save your work before exiting and restarting Pro/E.

To load a configuration file during a session:

1. Choose **Misc** from the MAIN menu, then **Load Config** from the MISC menu.
2. Enter the full configuration path and file name (*config.pro* being the default). Pro/E updates with the new configuration file settings. If there are errors in the new configuration file, a message is displayed in the start-up window.

MAPKEYS

You can create keyboard macros, which map frequently used command sequences to certain keyboard keys. The macros are saved in the **configuration file**. Each macro begins on a new line. Keyboard macros require the following format:

mapkey keyname #command;#command; . . .

As an example, the following shows how to create a macro to map the shading-command sequence to the character sequence "**vs**":

mapkey vs #view;#cosmetic;#shade;#display;#done-return

Entering [**vs**] from the keyboard during a current Pro/E session causes the current model (part or assembly) to be shaded.

The last line in Figure 5.7 creates a macro to save and purge the current model with the function key **F5**:

mapkey $F5 #dbms; #save; #purge; #done-return

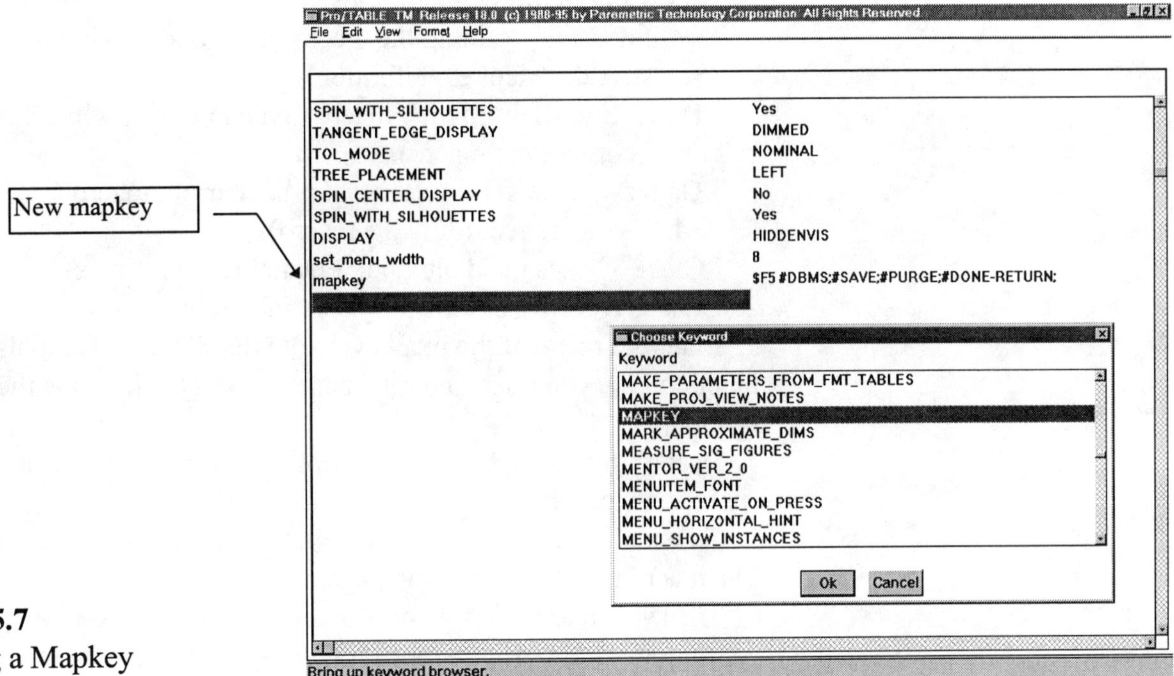

Figure 5.7
Creating a Mapkey

Creating Macros

You can create keyboard macros in two ways:

1. Create the macro interactively during the current session. Pro/E records the keystrokes. Upon completion, Pro/E appends the macro to the configuration file in your current directory.
2. Edit the configuration file.

When you are creating keyboard macros, you must maintain the following syntax requirements:

1. The mapped keys can be either **function keys** or regular **keyboard characters**. You cannot use mixed sets.
2. The "$" character is reserved for specifying function keys. If you wish to use a function key, you must enter its name as [**$Fn**], where **n** is the key number. You are limited by the number of function keys on your keyboard, usually twelve. (Pro/E supports up to twenty function keys).

Creating Macros Interactively

Starting with Pro/E 17, a macro can be created interactively by following these steps (Figs. 5.8 and 5.9):

1. Choose **Mapkey** from the MISC menu. The Mapkey dialog box appears with the following options:
   - **Define** Starts defining the macro (available only before you start the Mapkey definition).
   - **Done** Ends defining the macro (available only while you are recording the macro).
   - **Cancel** Cancels the definition of the current macro (available while you are recording the macro).
   - **Close** Closes the dialog box (available at all times).
2. Choose **Define** from the Mapkey dialog box.
3. Enter the name of the mapkey. Note that if you wish to use a function key, you must enter its name as [**$Fn**], when **n** is the function key number.
4. Record the macro by choosing options from the menus in the appropriate order.
5. To end the definition, choose **Done** from the Mapkey dialog box. The macro is now defined in session.
6. The system asks you if you want to save the macro in a configuration file. If you reply [y], the system asks you to enter the name of a configuration file (*config.pro* is the default). Pro/E then appends the macro to the configuration file. If you reply [**n**], the macro is defined only in the current session. Pro/E does *not* append it to a configuration file. When you end the current Pro/E session, the macro disappears.
7. To define another macro, repeat steps 2 through 6.
8. Choose **Close** to end this process.

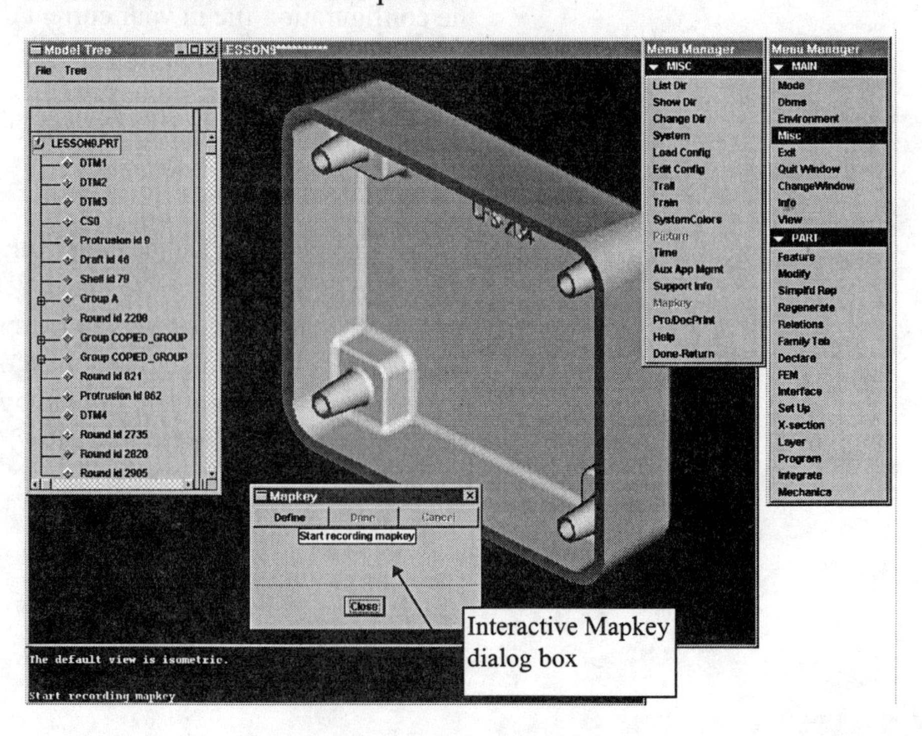

Figure 5.8 Creating a Mapkey Interactively

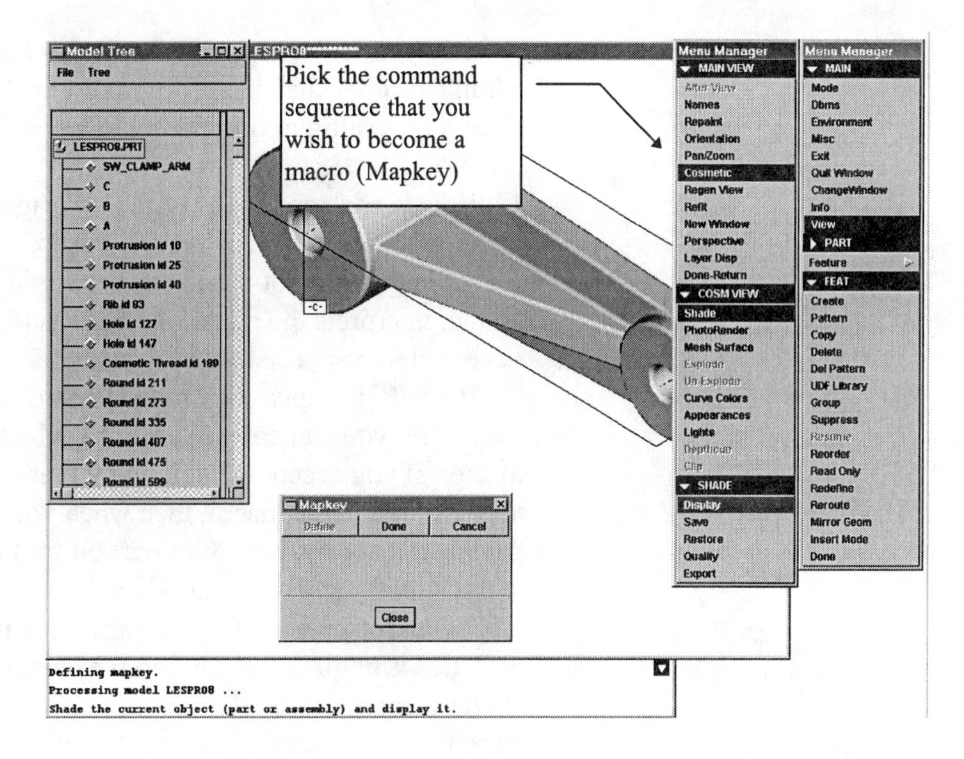

Figure 5.9
Mapkey Creation

The Mapkey dialog box can be left open and will not affect the operation of Pro/E. You can see and edit your interactively created mapkey by picking **Misc** ⇒ **Edit Config** ⇒ **enter**. Figure 5.10 shows the open configuration file and the new **vp** mapkey.

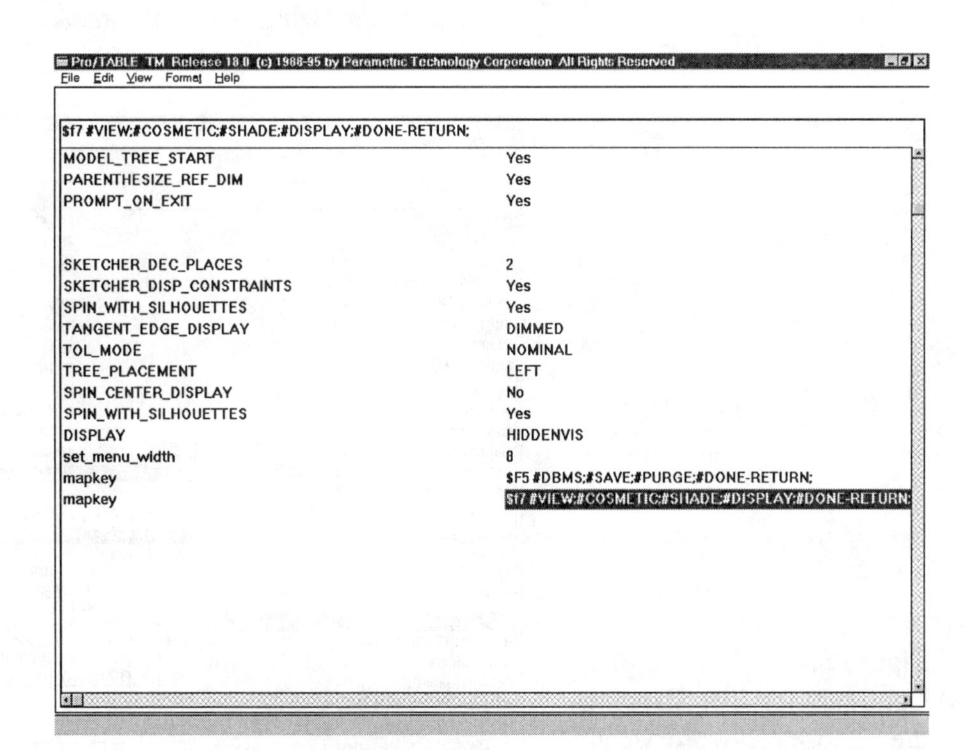

MODEL_TREE_START	Yes
PARENTHESIZE_REF_DIM	Yes
PROMPT_ON_EXIT	Yes
SKETCHER_DEC_PLACES	2
SKETCHER_DISP_CONSTRAINTS	Yes
SPIN_WITH_SILHOUETTES	Yes
TANGENT_EDGE_DISPLAY	DIMMED
TOL_MODE	NOMINAL
TREE_PLACEMENT	LEFT
SPIN_CENTER_DISPLAY	No
SPIN_WITH_SILHOUETTES	Yes
DISPLAY	HIDDENVIS
set_menu_width	8
mapkey	$F5 #DBMS;#SAVE;#PURGE;#DONE-RETURN;
mapkey	$f7 #VIEW;#COSMETIC;#SHADE;#DISPLAY;#DONE-RETURN;

Figure 5.10
Configuration File Showing the Interactively Created Mapkey

Mapkey Syntax

A variety of rules must be applied when creating macros or when editing macros that were originally created *interactively* or through editing the configuration file (*config.pro*):

1. Precede each command with a "#" sign.
2. Semicolons ";" must be used to separate commands.
3. When the first not-space character in a field is not a "#" sign, the system interprets the rest of the field as if you had entered it from the keyboard in response to a *prompt*.

   If Pro/E prompts you for a keyboard input while you are creating the macro, you can enter a response, which is then recorded in the macro. If you accept the default by pressing **Enter**, the response is *not* recorded in the macro, then when you execute the macro, Pro/E pauses at that point and prompts you for keyboard input.
4. Leading spaces are ignored by the system.
5. Except for keyboard input, entries are not case sensitive
6. There is no practical limit to the length of a macro (the maximum length of an interactively created macro is 2048 characters). Use the backslash character (\ ) as a continuation character. Do *not* have any characters or spaces after the backslash.

   For example, the following macro "**vs**" was created as shown:

mapkey vs #view; #default; #done-return #view; \ #cosmetic; #shade; #display; #done-return

Appendix B has a sample mapkey file that you can create and save in your current directory.

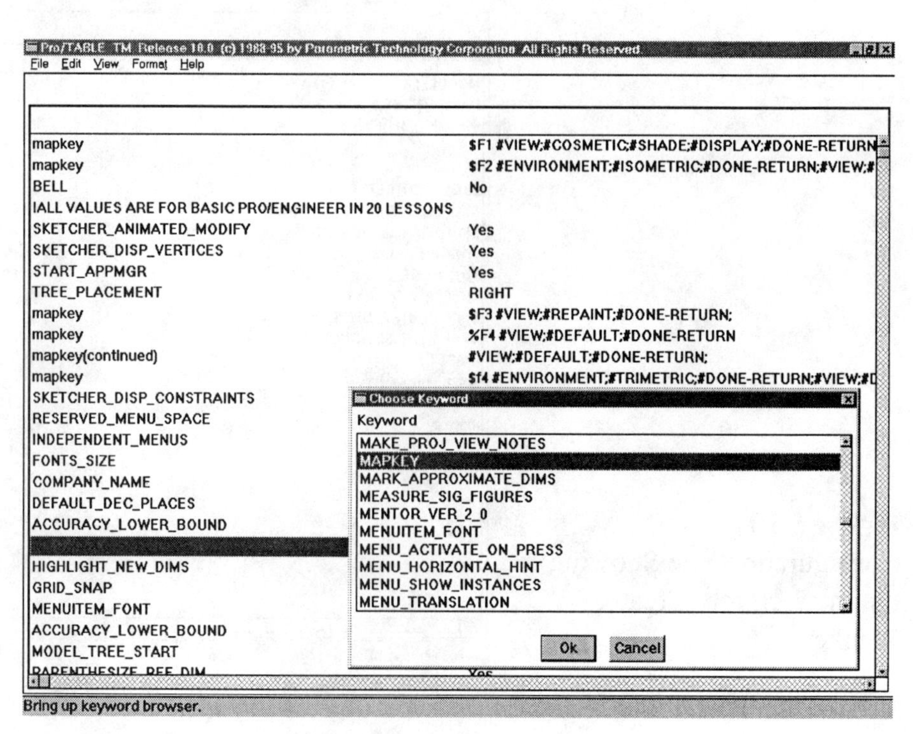

Figure 5.11
Configuration File Showing the Choose Keyword Menu Used to Create a Mapkey

Section 6

Environment

Pro/E lets you work in a multiple-window environment where you may have anywhere from 2 to 10 or more windows open at a time, depending on your computer system's capabilities. You start working in the **Main working** window by default. You can easily change your working window from one to another window.

If the Main window is being used and you start a **Mode** that requires a new window, a new, smaller window is created for you to work in. As an example, when you are working on a part, if you then create or retrieve a different part without first quitting the Main window, the new part will be created or retrieved in a new window. In Figure 6.1, the swivel was retrieved first (main window) and then the arm was retrieved (smaller window).

The **MAIN menu** contains two window options: **ChangeWindow** and **Quit Window**. You can work in any window by picking **ChangeWindow** from the MAIN menu and picking inside the window in which you wish to work.

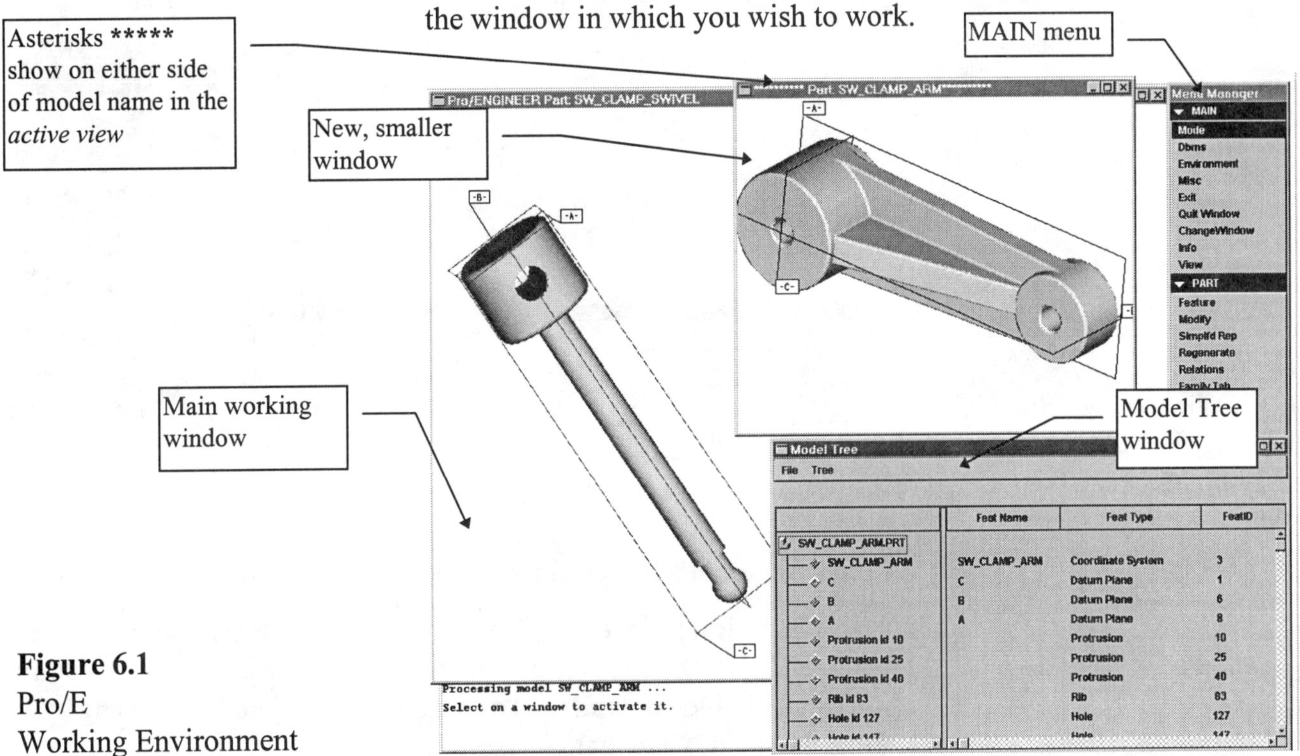

Figure 6.1
Pro/E
Working Environment

Quit Window blanks out the currently active window. When multiple windows are open, **Quit Window** is used to clear the active window. The system *does not* automatically assign an active window from the remaining windows. **ChangeWindow** must be used to activate one of the remaining windows.

The active window will have asterisks on either side of the objects name: **********SW_CLAMP_ARM**********

The **Model Tree window** for the part in the selected window reflects the active part or assembly (Fig. 6.1, lower right corner).

Screen Layout

Upon starting a Pro/E session, the workstation screen is divided into three main areas:

Main working window A majority of graphical work is accomplished in this window (**Pro/ENGINEER** in Fig. 6.2).
Pop-up menus Pro/E modes and commands are selected using the mouse in these menus (**MAIN** and **MODE** in Fig. 6.2).
MESSAGE WINDOW Pro/E prompts and operation information are displayed here (**MESSAGE WINDOW** in Fig. 6.2).

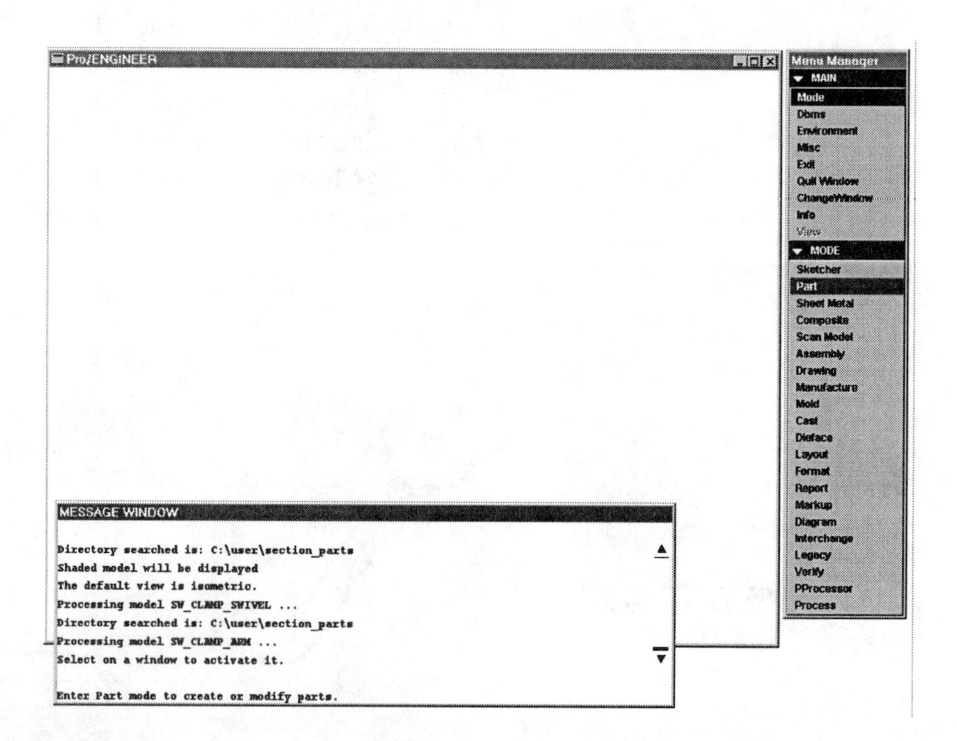

Figure 6.2
Main Window, MESSAGE WINDOW, MAIN Menu, and MODE Menu

Menu Boxes and Dialog Boxes

Menu and **dialog boxes** are used for *creating* and *redefining* features in Pro/E. You can create or redefine a feature by selecting information blocks called "options" and "elements" using the following menus and dialog boxes:

Feature Option Menus During the **Feature ⇒ Create** sequence, you can select solid feature "options," such as **Extrude**, **Revolve**, **Sweep**, and **Blend**, and whether the feature will be **Solid** or **Thin** (Fig. 6.3). *Options cannot be redefined.*
Feature Elements Menus Control the process of defining the elements required to create the feature, such as **Direction** and **Depth** (for example, **Depth** from the **SPEC TO** menu, as shown in Fig. 6.4). *Elements can be redefined.*

Dialog Boxes Provide a visual method of displaying and changing the feature **Elements** and their current status. Information regarding the feature and its references can be accessed. The **Preview** option provides a view of the feature or operation prior to creation, allowing you either to accept the design or redefine the feature's elements. In Figure 6.4, the **CUT: Extrude** dialog box is displayed.

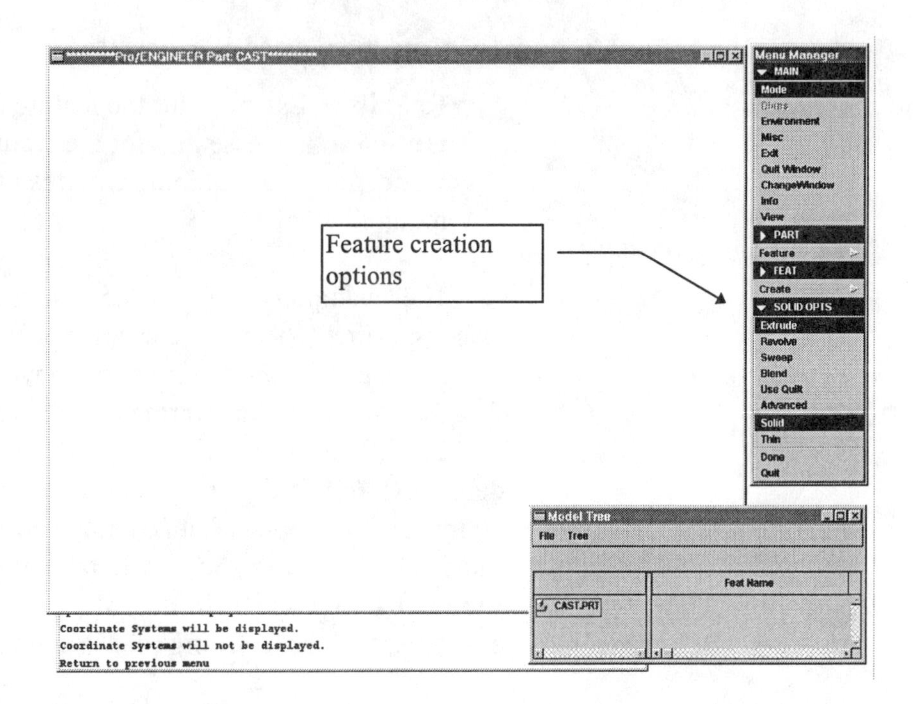

Figure 6.3
Feature Options Menu

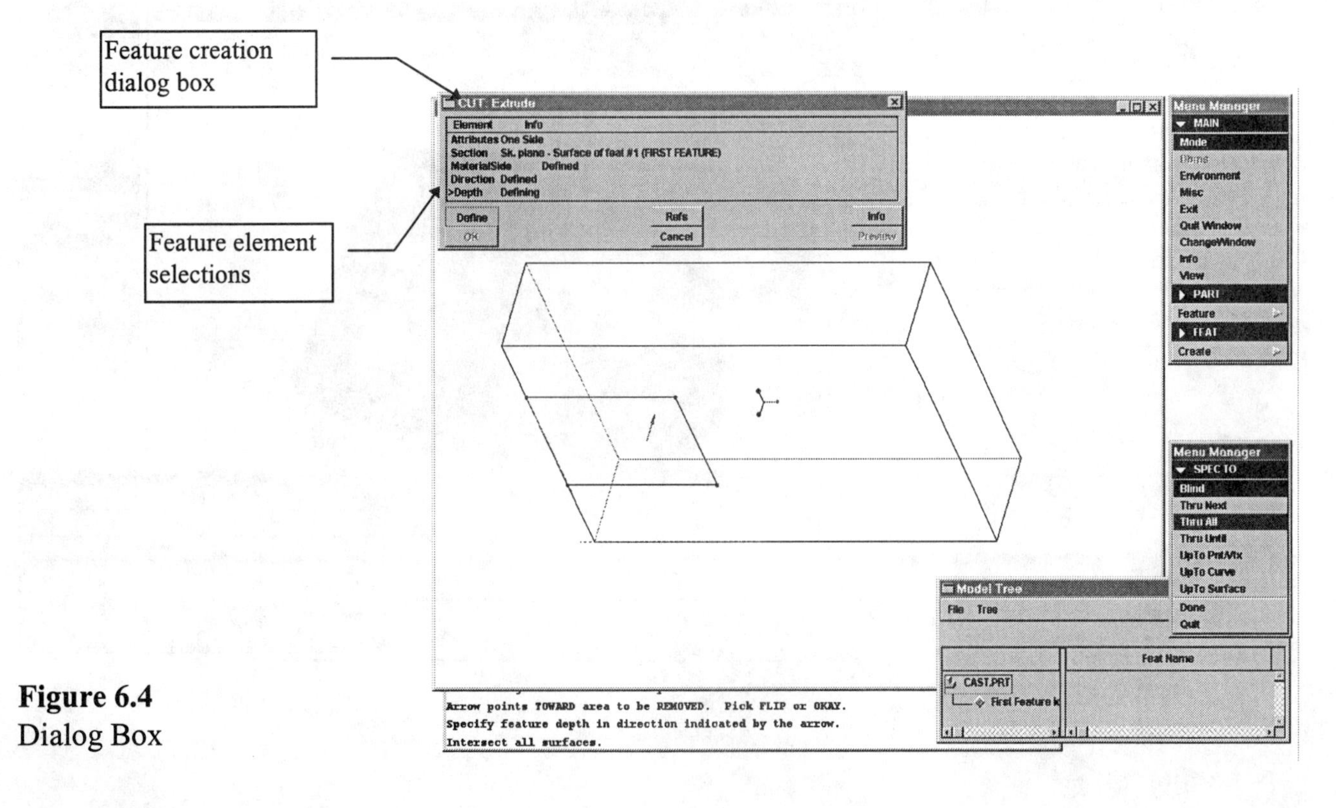

Figure 6.4
Dialog Box

FEATURE CREATION

When creating a feature, values are required for the feature options and elements. The options specified will determine the elements required to complete the feature. As an example, the cut feature in Figure 6.4 was extruded; its elements are **Attributes**, **Section**, **MaterialSide**, **Direction**, and **Depth** and a sweep cut will have the elements of **Trajectory**, **Section**, and **MaterialSide**.

There are three types of elements:

Required Essential for the feature definition
Optional Not essential for the feature definition
Conditional Required as a result of the selection of another required element

The example in Figure 6.5 shows a feature being added to an existing part. The feature class has been defined as **Solid** and the feature type as a **Shell**. Once you have specified the feature options, Pro/E will display the information dialog box.

The dialog box title is based on the feature type and option defined (**Shell** in Fig. 6.5) and contains an Element list and action buttons. The Element list (**Remove Surfs**, **Thickness to Define**, and **Spec Thick** in Fig. 6.5) is a tabulated list box that contains the **Element name** and **Info** describing the current state of the features elements, such as **Defining** (or **Undefined**), **Required**, and **Optional**.

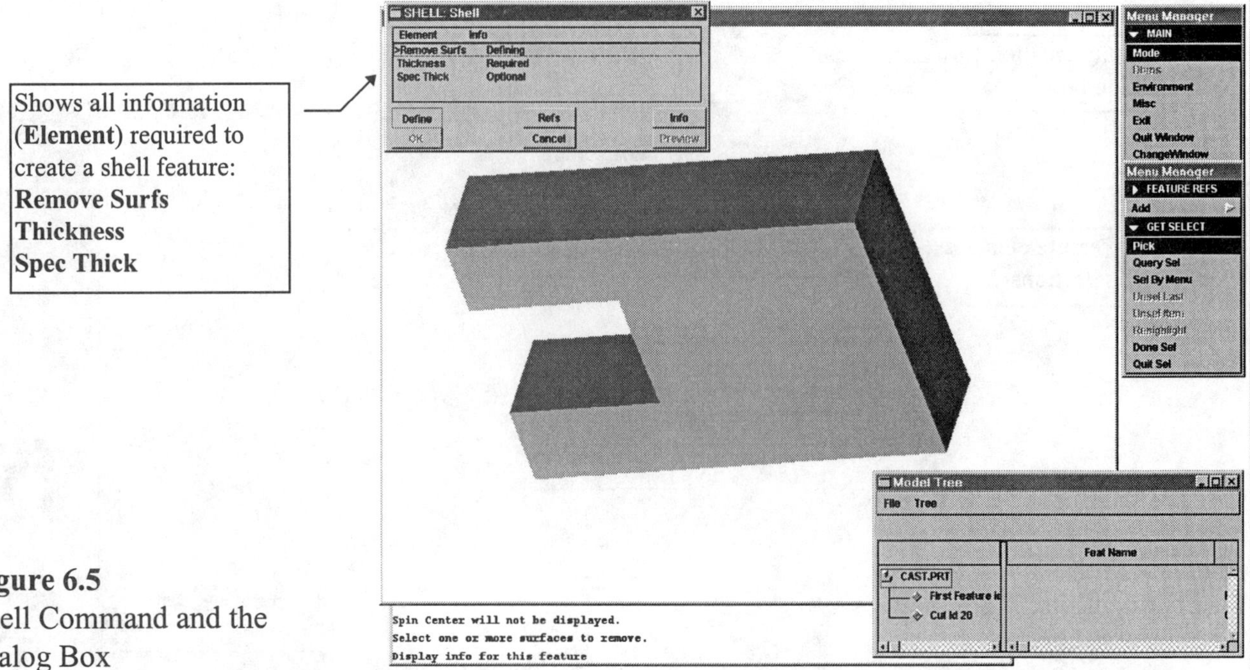

Figure 6.5
Shell Command and the Dialog Box

The dialog box action buttons (Fig. 6.6) enable you to perform a variety of functions:

Define Defines or *redefines* all marked elements for the feature, working from the top to the bottom of the element list. Pro/E displays the menus for the elements currently being processed and displays messages and required prompts in the message window.

Refs Displays the references for the selected elements and allows the user to step through each reference by navigating through the SHOW REFS menu.

Info Displays information on the feature being created.

OK Completes feature creation and closes the dialog box.

Cancel Aborts the current feature definition process and closes the dialog box.

Preview Displays the feature geometry as it will be built with the current definitions. Preview has been used in Figure 6.6, where the shell command was given, and the Preview button allows you to see the feature as it will be created before accepting the design input. At this time you could change (define/redefine) the shell thickness or change-add reference surfaces to be removed.

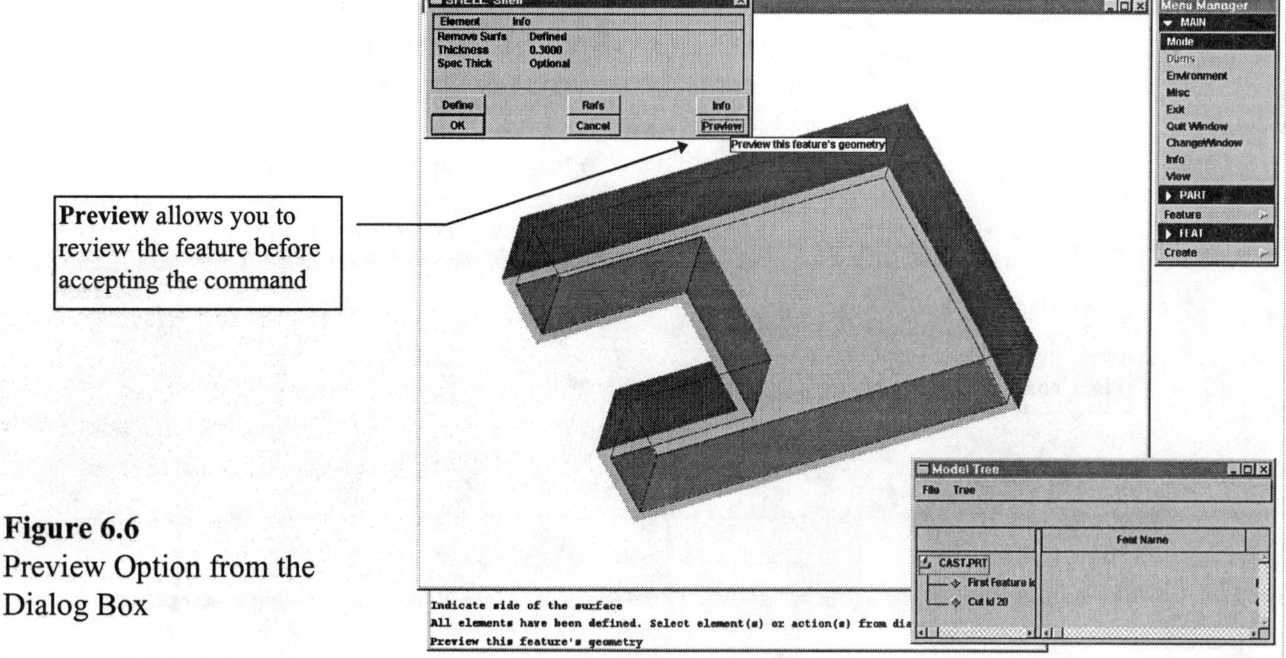

Figure 6.6
Preview Option from the Dialog Box

Choosing Dialog Box Items

Elements listed in the dialog box are selected by moving the cursor over one or more items and clicking with the left mouse button. To select one element at a time, move the cursor over one element and click once with the left mouse button.

For a range selection, move the cursor to the top or bottom of the range of elements. Click and hold the left mouse button down, and then drag the cursor to the opposite end of the range to select. Another method is to move the cursor to the top or bottom of the range, and then click once. To select more than one element, move the cursor to the element on the opposite end of the range and, while pressing the **Shift** key, click once with the left mouse button.

Action buttons in the dialog boxes are selected using the left mouse button. For example, a user would select the **Define** button, after highlighting elements in the tabulated list to redefine those elements and then select **OK** to complete the feature. Double-clicking on an element in the tabulated list will automatically initiate the **Define** action button.

System Messages

System messages appear in the **MESSAGE WINDOW**, at the bottom of the main working window. The messages have the following functions:

* Request additional information to complete a command.
* Provide a *one-line help message* about a menu being chosen. Help messages appear in yellow at the bottom of the Message window (**"Place coaxially to an axis"** in Fig. 6.7)
* Provide information about the status of an operation in progress. These appear in white letters as they scroll the message window.

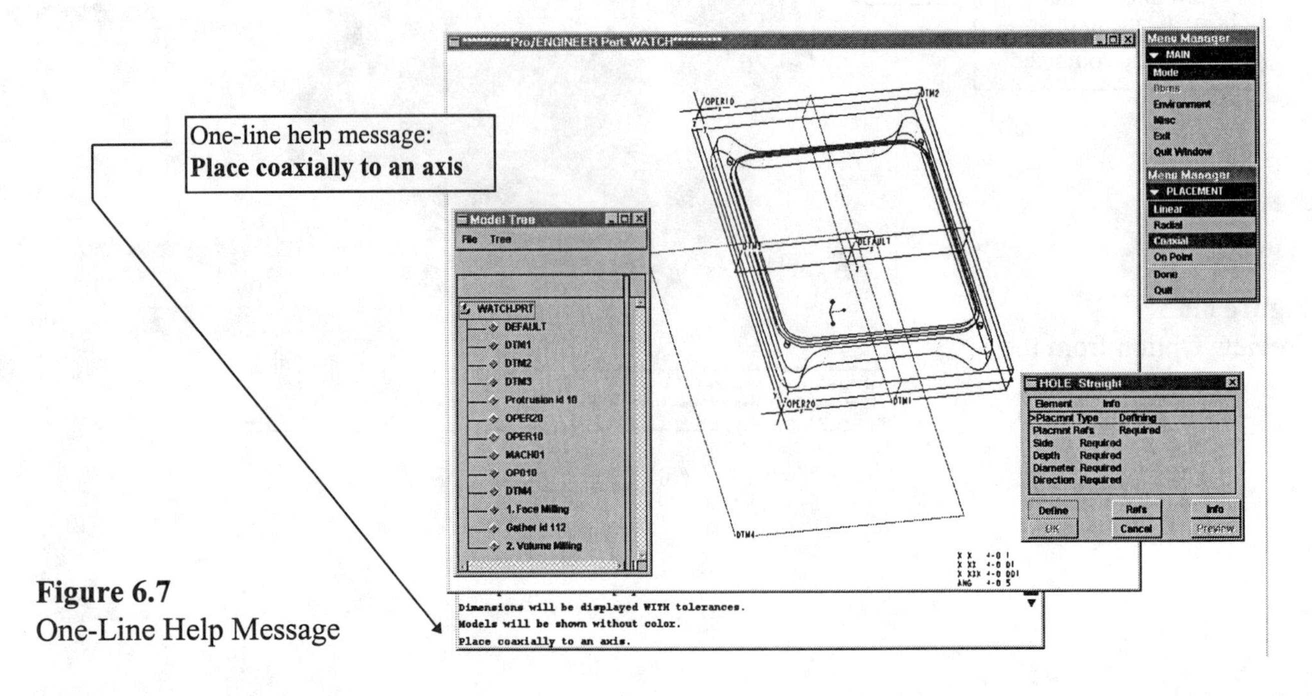

Figure 6.7
One-Line Help Message

When Pro/E is providing information, you can proceed with another function and the message will disappear. If Pro/E is asking for additional information, provide the information by entering it from the keyboard or indicating it with a mouse pick.

A bell will sound to notify you when input is needed (unless the bell has been turned off in the ENVIRONMENT menu manually or as a *config.pro* setting; most users turn the bell off, since it is extremely annoying, especially if there are 30 users in the same room!). When data is required to be entered, a prompt ending with a colon will appear in the Message window. When data input is required, all other functions are temporarily disabled until data entry is complete.

The **MESSAGE WINDOW** can be expanded using the workstation's window management functions to "pop" the window in front (Fig. 6.8). The main window can overlap the MESSAGE WINDOW, showing up to five (5) message lines at start up, though the default is one line.

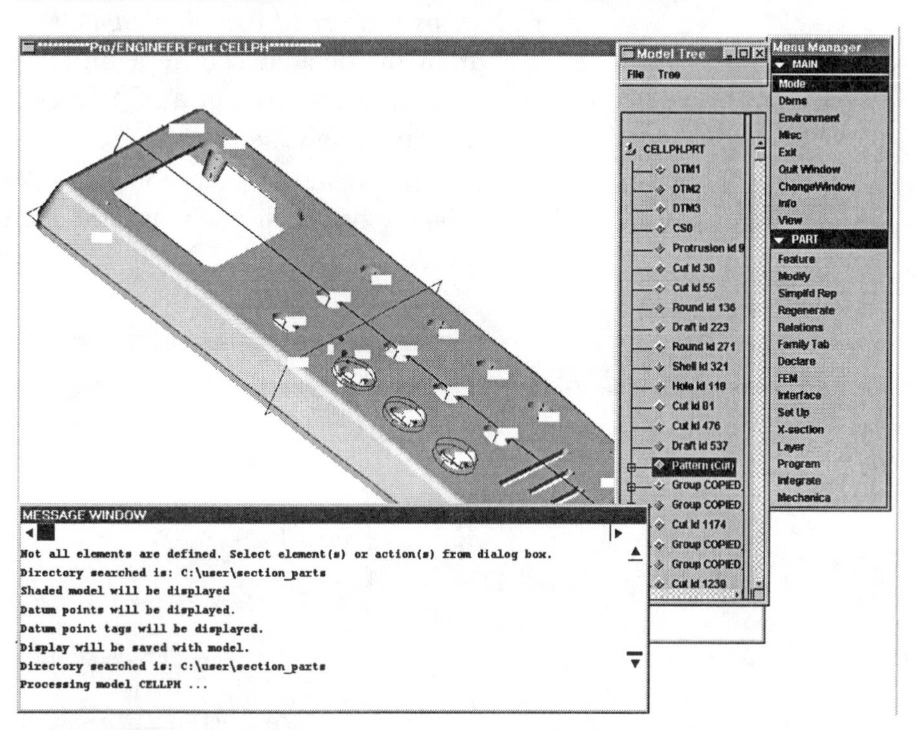

Figure 6.8
Expanded
MESSAGE WINDOW

THE MAIN MENU

The **MAIN** menu contains access to the major modes of Pro/E operation and other system-related functions:

Mode Allows you to access major Pro/E modes. It brings up the MODE menu that allows you to choose the desired mode.
Dbms Used to manage Pro/E files.
Environment Affects the operating modes. It is used interactively to change the working environment (e.g., display mode settings, prompt bell status, file storage settings).

Info This menu item lets you access information functions without having to quit your current Pro/E action.

Misc Gives you access to the workstation's operating system functions without exiting Pro/E, and allows the rerun of journal and training files in the Pro/E environment.

Exit Ends the Pro/E working session.

Quit Window Quits from the current window being used.

ChangeWindow Changes the window being worked in, from one to another.

View Changes the view of an object.

Choosing Menu Items

Menu items are selected via a three-button mouse. To choose a menu item, move the cursor over the item until it is highlighted. Click the *left* mouse button to activate the selection. The menu item is then highlighted in a different color and the function is executed. *Submenu items that are already highlighted are the default options.* To change from the default, choose a different item in that submenu. Menu selections that are not available during an operation appear dimmed and are disabled.

In Figure 6.9, the command **View** ⇒ **Cosmetic** ⇒ **Shade** ⇒ **Display** has been given and the model is shaded in the small active window.

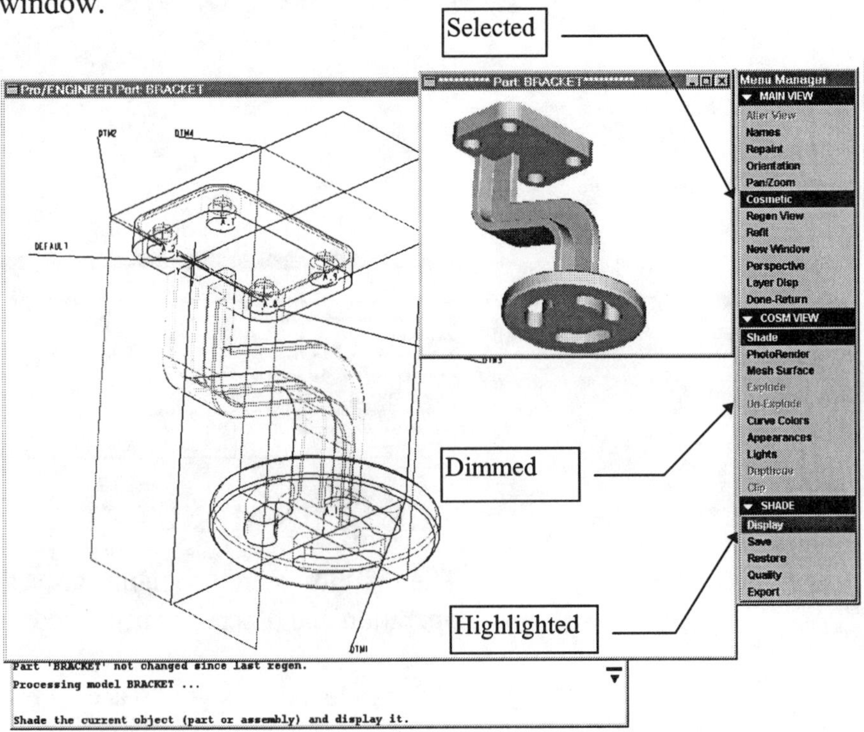

Figure 6.9
Selected, Highlighted, and Dimmed Menu Items

ONLINE HELP (OnLine Documentation with Pro/HELP)

? Pro/HELP

?GetHelp

When you scroll the cursor up and down the menu items, a one-line description of the menu item that is highlighted appears in the system MESSAGE WINDOW at the bottom left of the screen.

If more detailed help is needed, then, with the menu item highlighted, click the right mouse button to access **online documentation**. In Assembly mode (Fig. 6.10), the item **Modify** was highlighted. Clicking the right mouse button brought scrolled pages of the **Assembly Modeling User's Guide** to the screen. You can scroll to other pages, or go to a table of contents, index, or do a search by scrolling to the menu items at the very top or bottom of the online documentation.

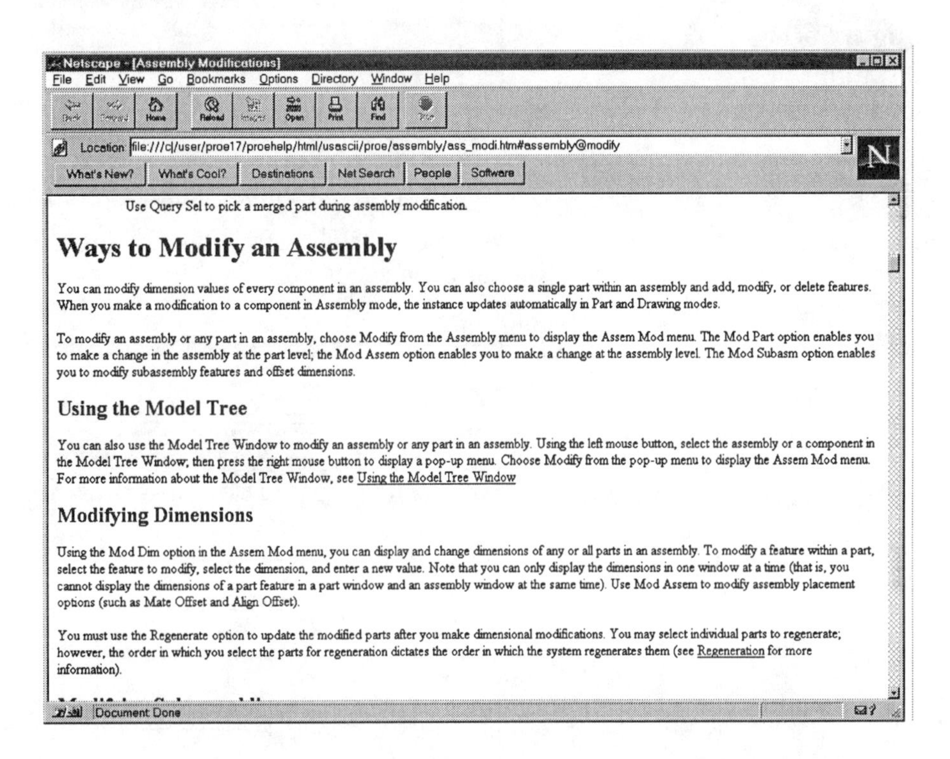

Figure 6.10
Pro/HELP

Pro/HELP is a tool that allows you to view Pro/E user guides and support documentation on your computer screen. When installed and configured on your system, Pro/HELP opens to the appropriate document and page to provide information and help whenever the right mouse button is clicked on a menu item.

This gives you access to various user guides and libraries installed on your system. You have the ability to:

Browse User Guide documentation
Establish bookmarks
Search for keywords
Print pages (Fig. 6.11)

The Print dialog box is activated when choosing **Print**, found along the top of the Netscape Menu (Fig. 6.11).

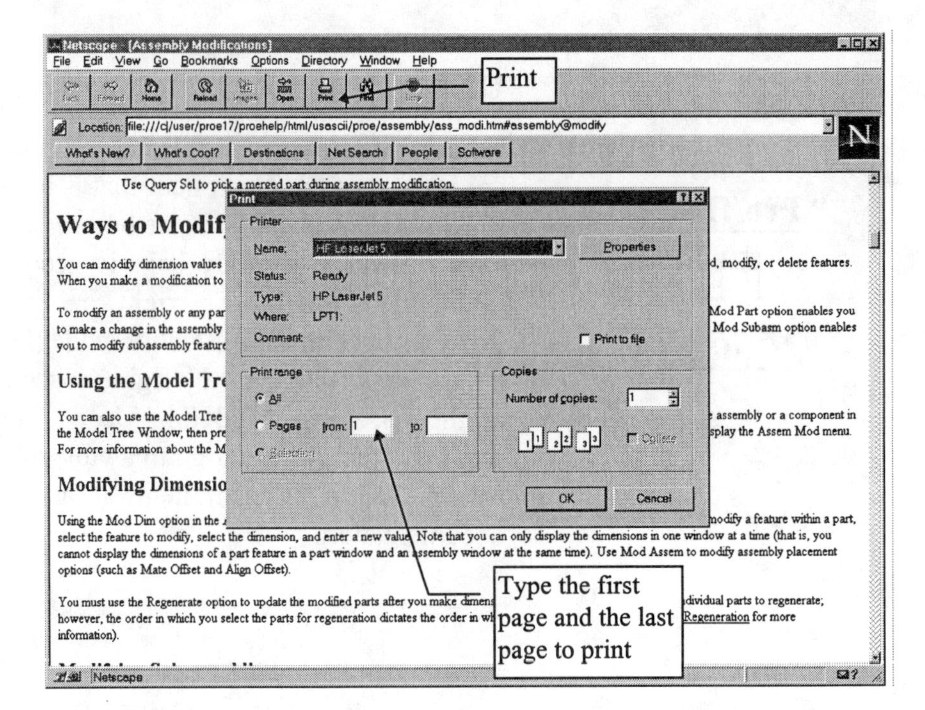

Figure 6.11
Printing
Pro/HELP Pages

Besides providing help, Pro/HELP can access any library that you have installed on your system. In Figure 6.12, the Tooling library has been accessed.

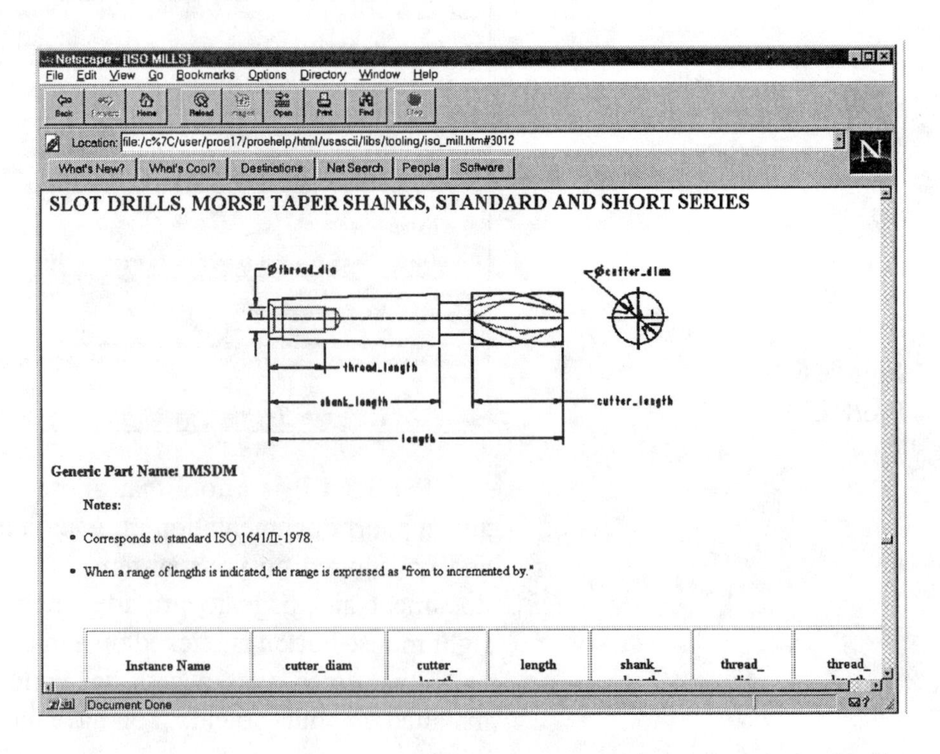

Figure 6.12
Pro/HELP
Tooling Library

Throughout the text you will be prompted to access help for the items that you will be learning in that lesson. There is no way that this text can cover all the commands and capabilities of Pro/E. You must use part of every session looking up information on commands and discovering the wide range of options available in Pro/E. It is the only way to expand your understanding of Pro/E and become a more complete and competent user.

Model Tree

The **Model Tree** displays your part or assembly structure. Each feature (Fig. 6.13) or part (if it's a component of an assembly) is displayed along with information about the object or feature.

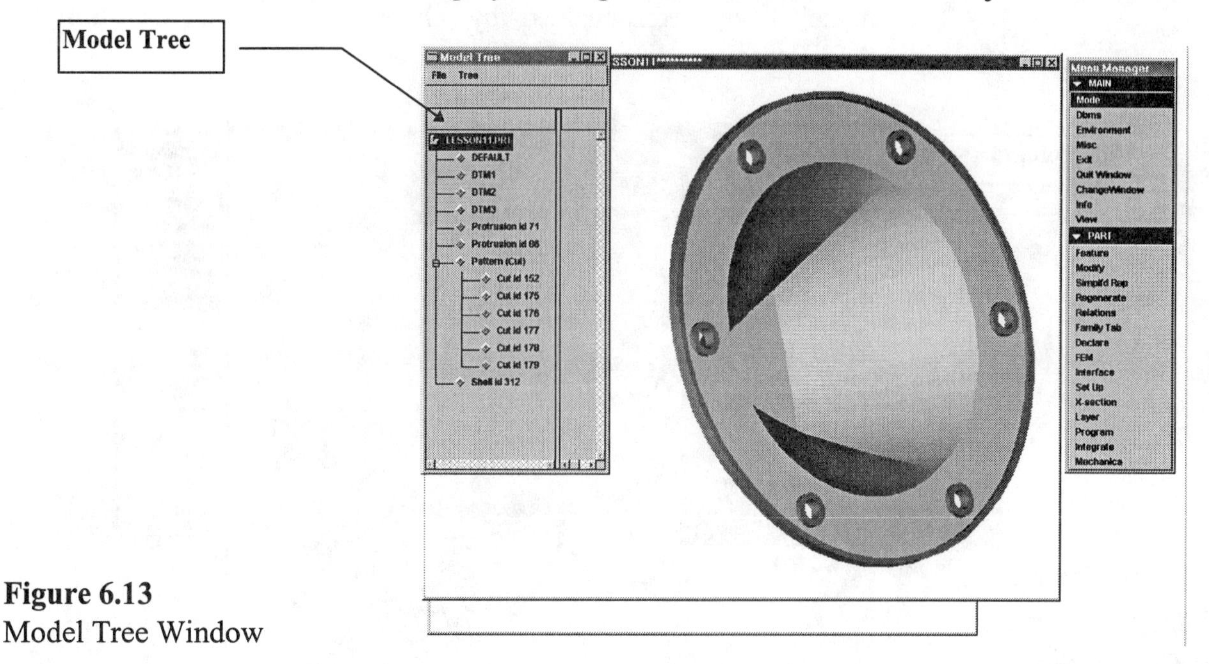

Figure 6.13
Model Tree Window

In Figure 6.14, an assembly is shown with its model tree displayed. The assembly and its components are listed in a tree structure.

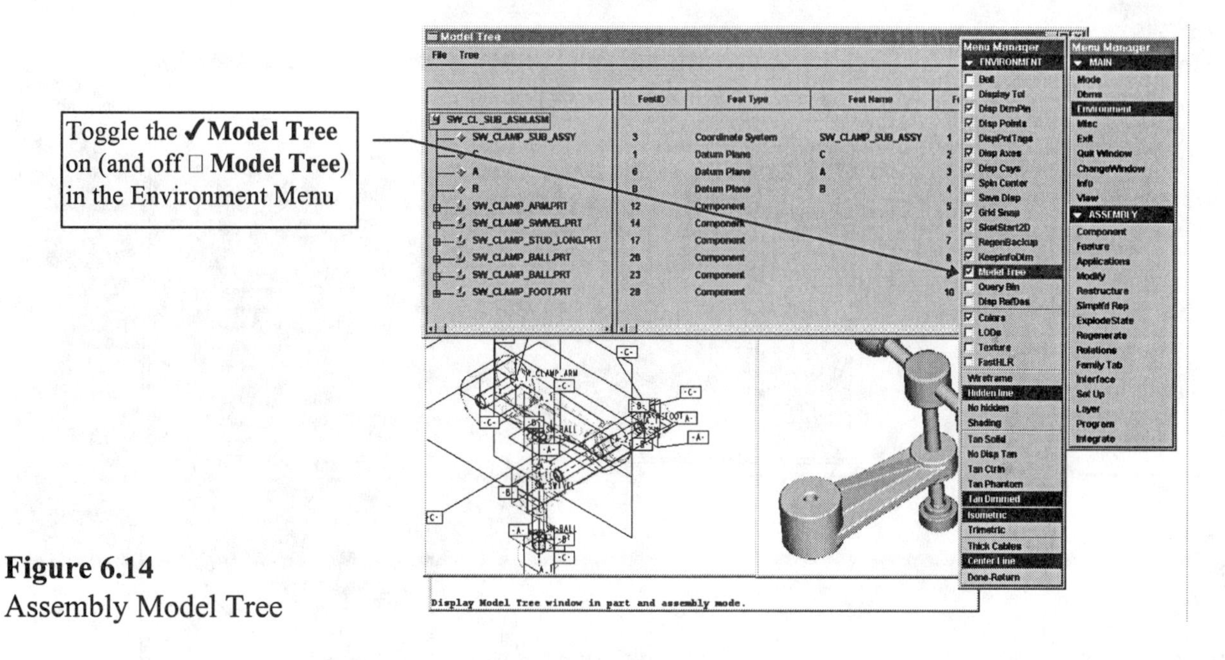

Figure 6.14
Assembly Model Tree

The model tree can be disabled by toggling off (□) the check mark (✓) in the ENVIRONMENT menu.

Besides the feature or part listing in the model tree, a variety of information columns can be displayed. Information columns can be added or removed and their format altered. The format can also be saved for later use. The model tree in the Part mode was shown in Figure 6.13. In Figure 6.15, columns for information about the part model have been enabled by choosing **Tree** ⇒ **Columns** ⇒ **Add/Remove.**

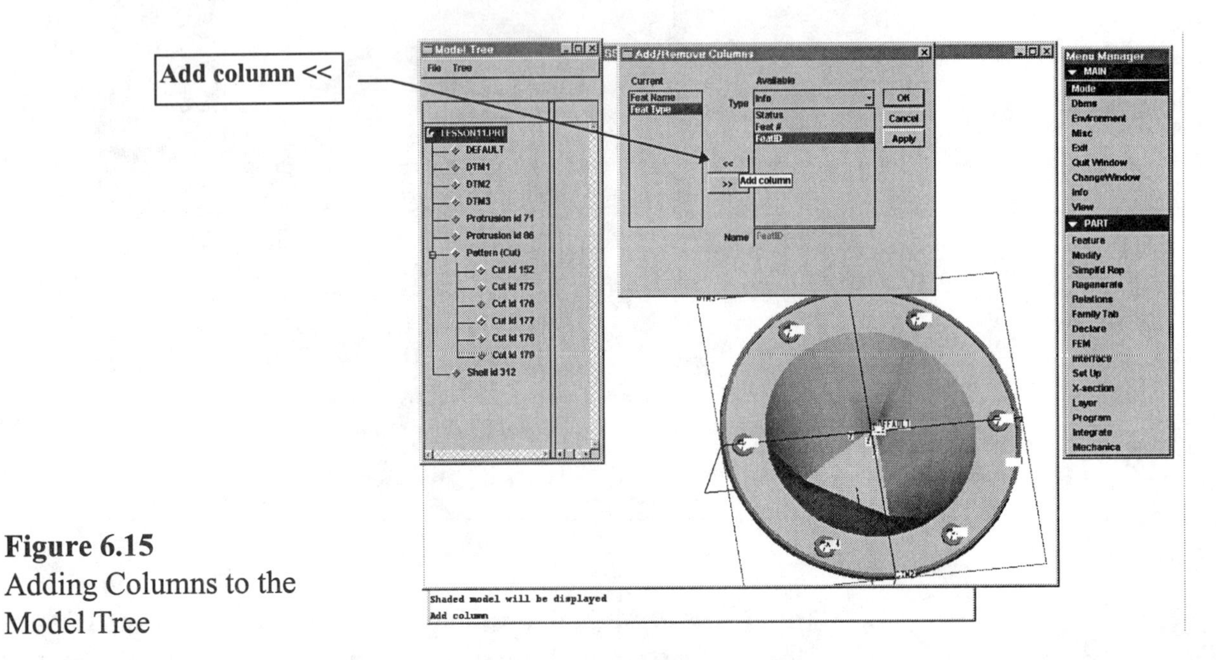

Figure 6.15
Adding Columns to the Model Tree

After the columns have been added, the model tree looks like the screen capture in Figure 6.16.

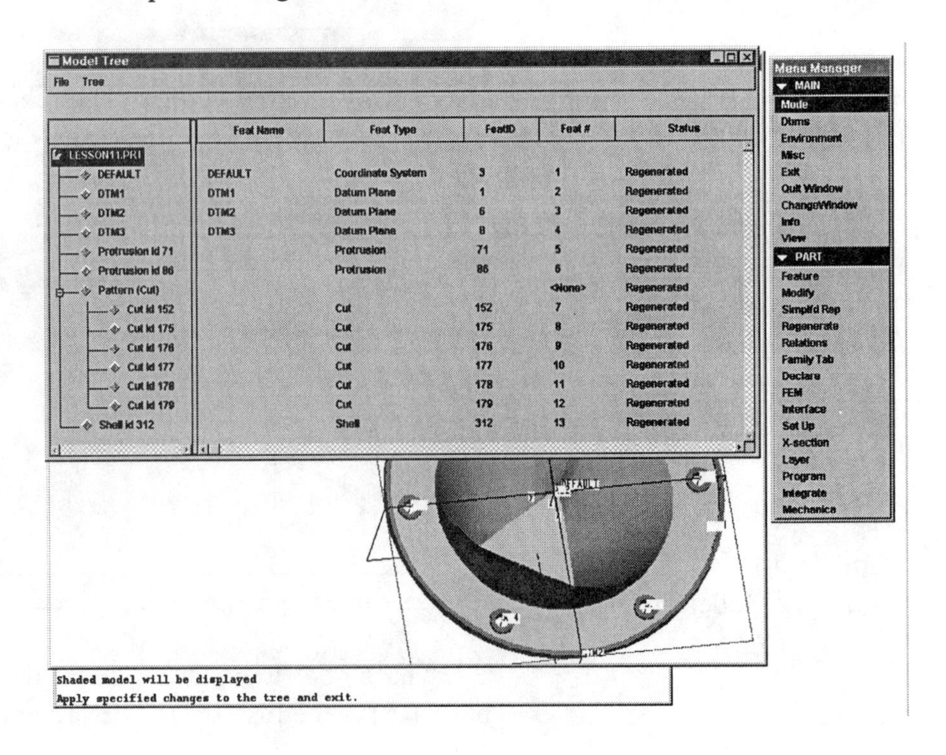

Figure 6.16
Expanded Model Tree

The columns can be altered by picking **Tree ⇒ Columns ⇒ Format** and changing the width of each category (Fig. 6.17).

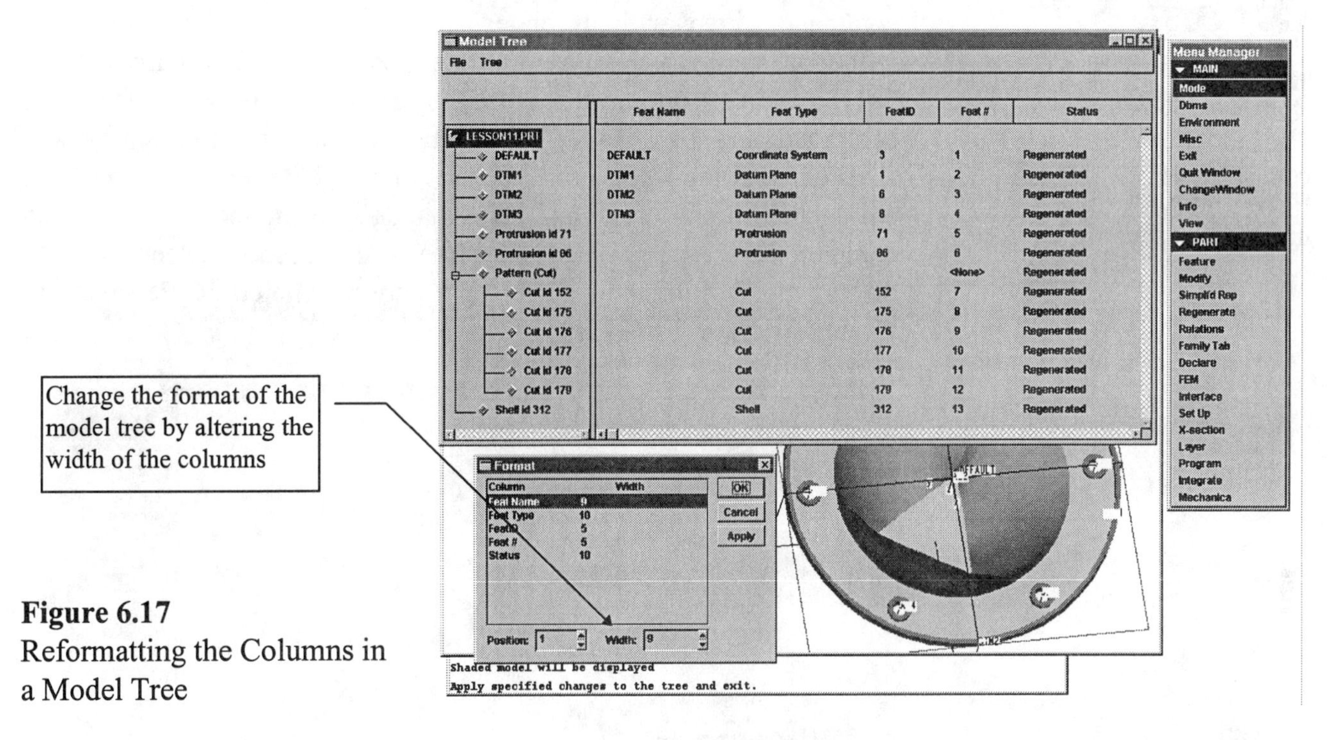

Figure 6.17
Reformatting the Columns in a Model Tree

Since the window is interactive, you can select objects directly from it, or by selecting **Sel By Menu** from the GET SELECT menu and then picking a component or feature.

Figure 6.18 is an example of selecting parts of an assembly from the model tree. The **SW_CLAMP_ARM.PRT** has been selected and is shown highlighted.

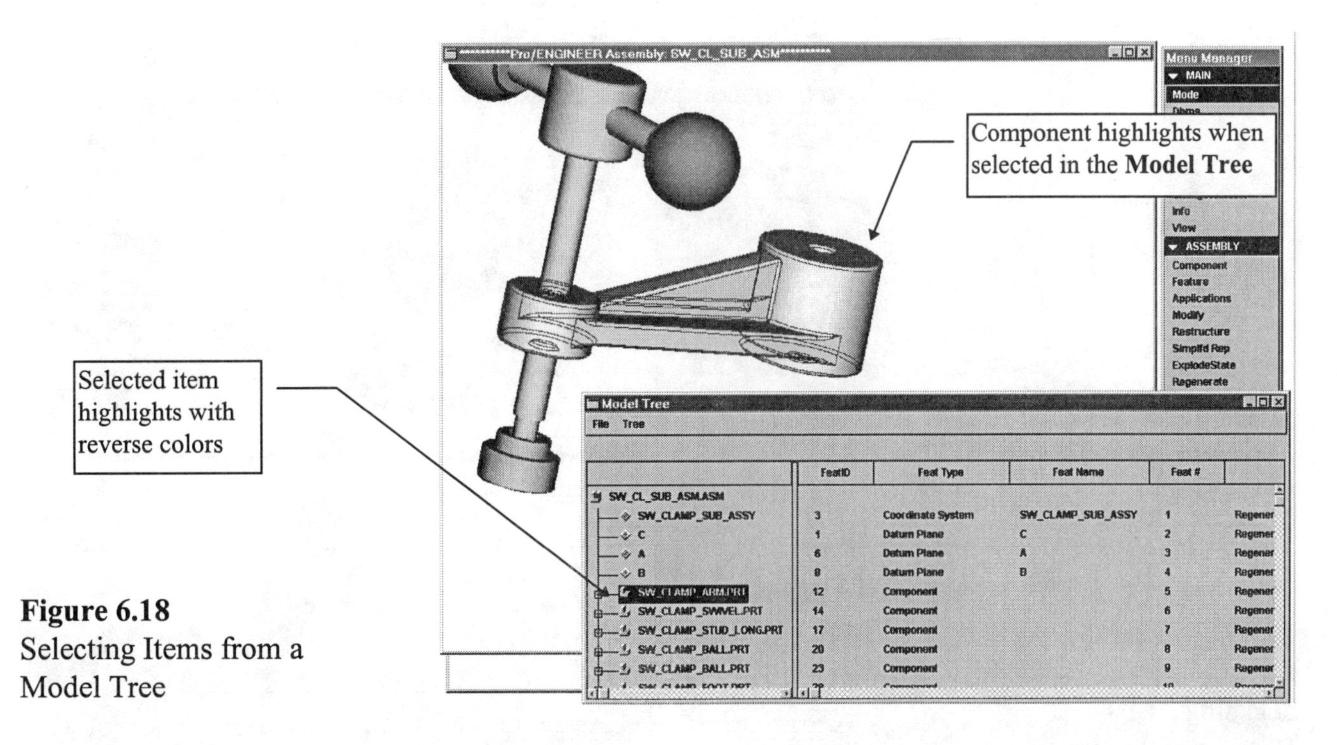

Figure 6.18
Selecting Items from a Model Tree

The model tree is also used to complete a variety of commands formerly available only through the menu structure. Commands such as **Modify**, **Redefine**, **Suppress**, **Reroute**, **Info**, and **Delete** can be chosen directly from the model tree. The feature is selected and the command given without choosing from the menu structure and without selecting the feature from the model itself. In Figure 6.19 the **Shell** feature was selected with the *left mouse button*. The **Info** command and the **Model Info** option were then highlighted and selected with the *right mouse button*. For the Assembly mode (Fig. 6.20), the model tree can be expanded to include the selected components features. The model tree is also used to select a variety of commands similar to those for a part but including **Pattern** and **Create Feature** (Fig. 6.20).

NOTE

Choose the feature from the model tree

Highlight and select from the cascading choices

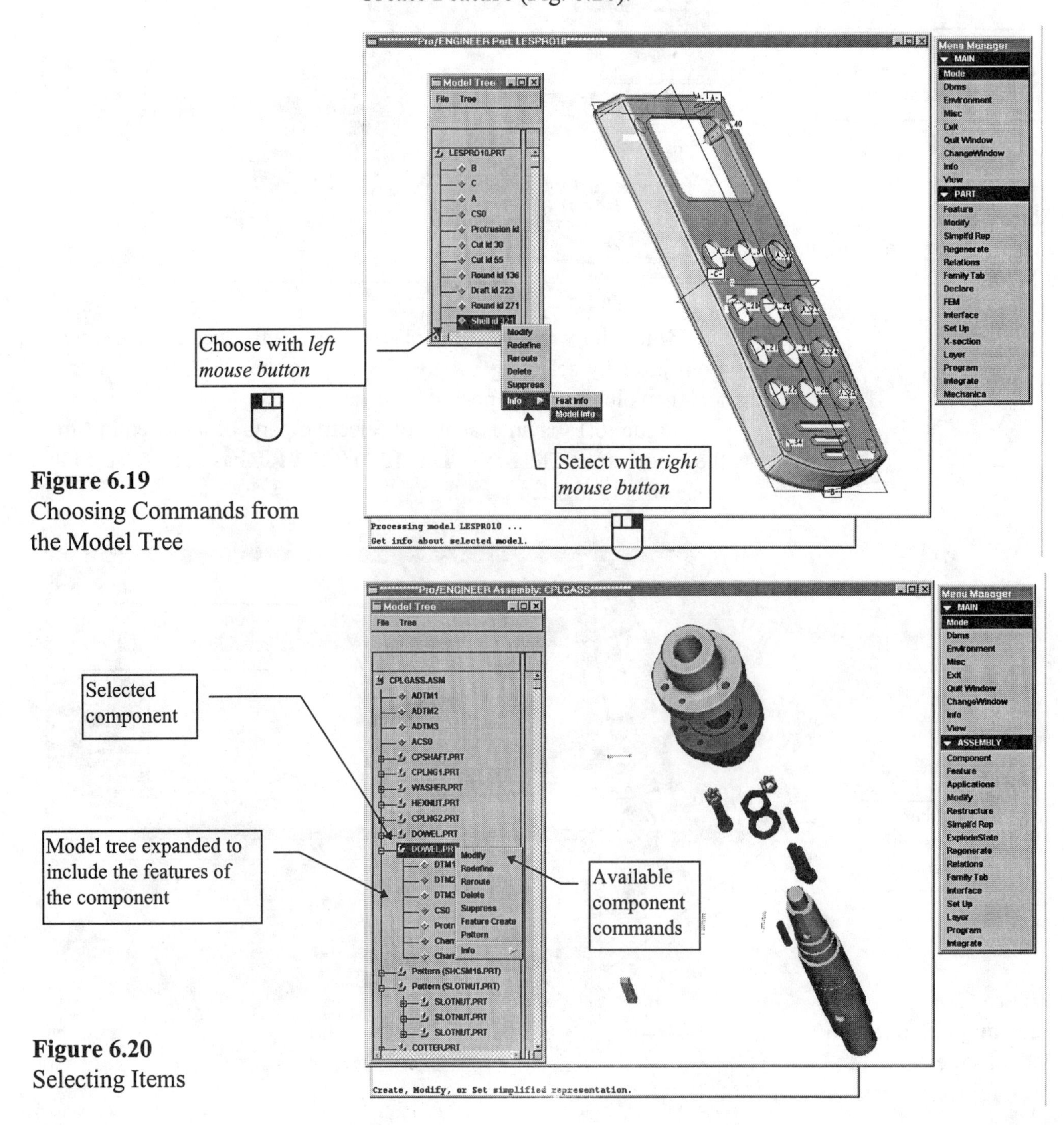

Figure 6.19
Choosing Commands from the Model Tree

Figure 6.20
Selecting Items

Section 7

Setup and Information

There are several capabilities that are used during the design of parts, assemblies, and drawings. In this section, **Info**, and **Set Up** functionality are covered. Before a project is started, whether it be a part or an assembly, you must set up the working standards for that project. Units must be selected. Standards are set using ANSI (ASME) or ISO/DIN. Materials should be selected for the part. Parameters are input. Accuracy is set. Geometric tolerances are established, and other specific design settings are organized and input at this point.

Information about the assembly or part can be requested at any time in the design process. The units, accuracy, and other settings established with the **Set Up** command will of course now be reflected when information is requested about the model. Mass properties (Fig. 7.1), surface analysis (Fig. 7.2), geometric tolerances, measurements, and a variety of information listings are available with this command.

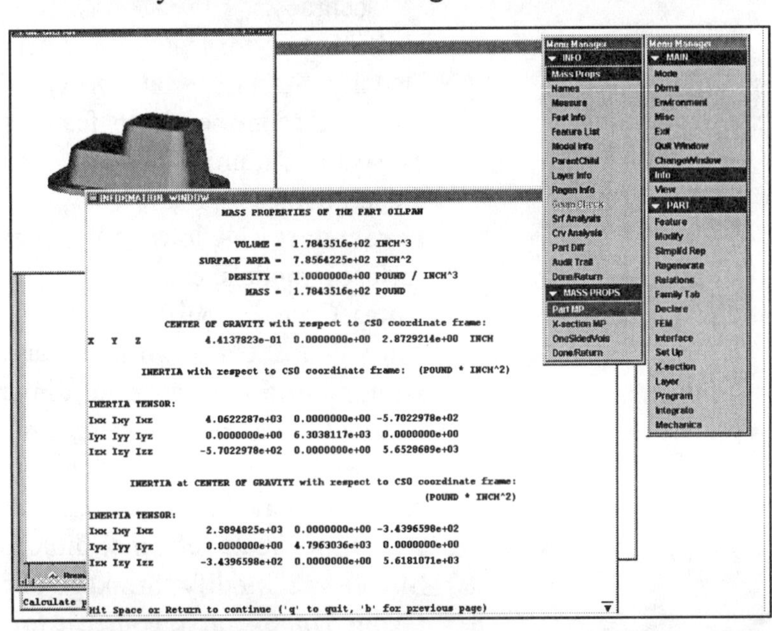

Figure 7.1
Mass Properties

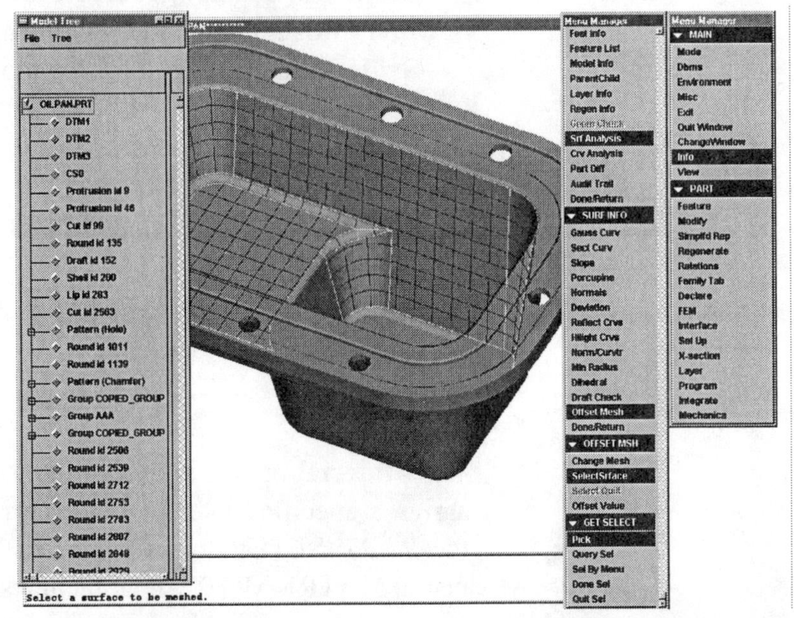

Figure 7.2
Surface Analysis

PART AND ASSEMBLY SETUP

In *part mode and assembly mode*, the SETUP menu allows you to define various attributes of the model. The most important of these capabilities are discussed in detail in this section. The following is a listing of options for part and assembly setup:

Material Create and modify material data files.
Accuracy
- **Relative** Modify relative part accuracy (Part mode). This is a computational accuracy of Pro/E geometry calculations and has a default value of **0.0012**. Part accuracy is relative to the size of the part. Increasing the part's accuracy increases regeneration time.
- **Absolute** Enabled by the configuration option "enable_absolute_accuracy". In most cases use relative accuracy.

Units Set up for features and offsets.
Density Set a material density value to be assigned to the part.
Name Assign names to features, assembly members, etc. You can name datum planes and coordinate systems instead of using default names.
Parameters Allows parameters to be assigned to features, surfaces, models, etc.
Notes Create, modify, or remove notes associated with the model.
Mass Props Create a file of mass properties.
Dim Bound Change a dimension regeneration value from nominal to its upper, lower, or midpoint tolerance boundary. This is used in assemblies to compute clearance/interference checks and on part models to perform tolerance stack-up studies.
Ref Dim Create reference dimensions for the model.
Shrinkage Modify shrinkage of part dimensions.
Geom Tol Specify geometric tolerances for surfaces and features.
Surf Finish Define surface finish symbols for the part model.
Grid Define a 3D grid for the model.
X-Section Create, modify and display assembly cross sections.
Declare Establish declarations to a layout.
Tol Setup Specify tolerance standards (ANSI, ISO/DIN).
Interchange Shows information or removes references to interchange groups.
Zone Define zones for assembly simplified representations.
Envelope Create or modify envelope components for assemblies.

Material

The **Material** option is used to create and modify material data. To assign a specific material to the current part, choose **Material** from the PART SETUP menu. This calls up the material management menu, MATRL MGT, which includes the following options:

Define Define the properties of a new material. The system editor displays a default specification file that must be edited in order to add the desired values for the material parameters (Fig. 7.3).

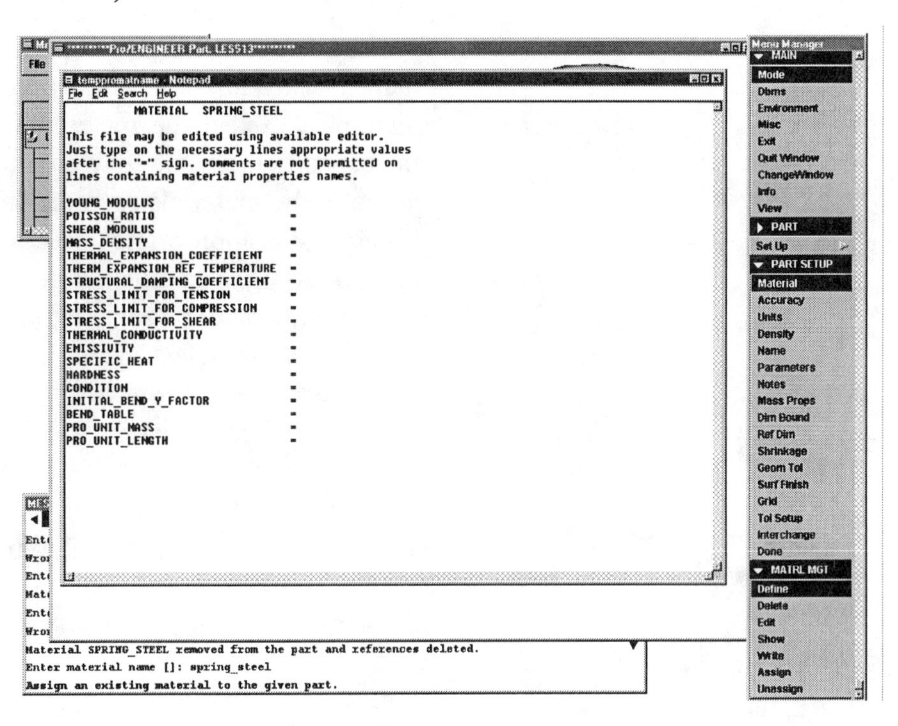

Figure 7.3
Material Defining

Different materials can be created by building new material files with the system editor. The file name for material properties is *materialname.mat*, where *materialname* is the name of the material. When you define a new material, edit the default specification file to include the desired values for the material (Fig. 7.4).

Material Steel_1040		
Young_Modulus	=	29000000
Poisson_Ratio	=	0.27
Shear_Modulus	=	11000000
Mass_Density	=	0.00879
Thermal_Expansion_Coefficient	=	6.78
Thermal_Expansion_Ref_Temperature	=	32.0
Structural_Damping_Coefficient	=	0.01
Stress_Limit_For_Tension	=	36000
Stress_Limit_For_ Compression	=	36000
Stress_Limit_For_ Shear	=	36000
Thermal_Conductivity	=	
Emissivity	=	
Specific_Heat	=	
Hardness	=	
Condition	=	
Initial_Bend_Y_Factor	=	
Bend_Table	=	steel_1040

Figure 7.4
Steel Material File

Delete Remove a material from the part's internal database. Pick the material from the MAT_LIST menu.

Show Displays the material specification file in an information window. The material must be in the part's internal database. Select the material from the MAT_LIST menu.

Write Write material properties from the part to a disk file named *materialname.mat* (*materialname* is the name of the material). Select the material from the MAT_LIST, menu and then specify the name by which the material will be stored. After defining the materials required in your design, create a materials library in the appropriate directory.

Assign Assigns an existing material to the part. This material is used in all analysis calculations of the part. From the USE MATER menu, select the material list to choose from:

> **From Part** From the part's internal database, use the material as listed in the MAT_LIST namelist menu (Fig. 7.5).
>
> **From File** Use material that is stored in a disk file. Provide the name of the material to be retrieved

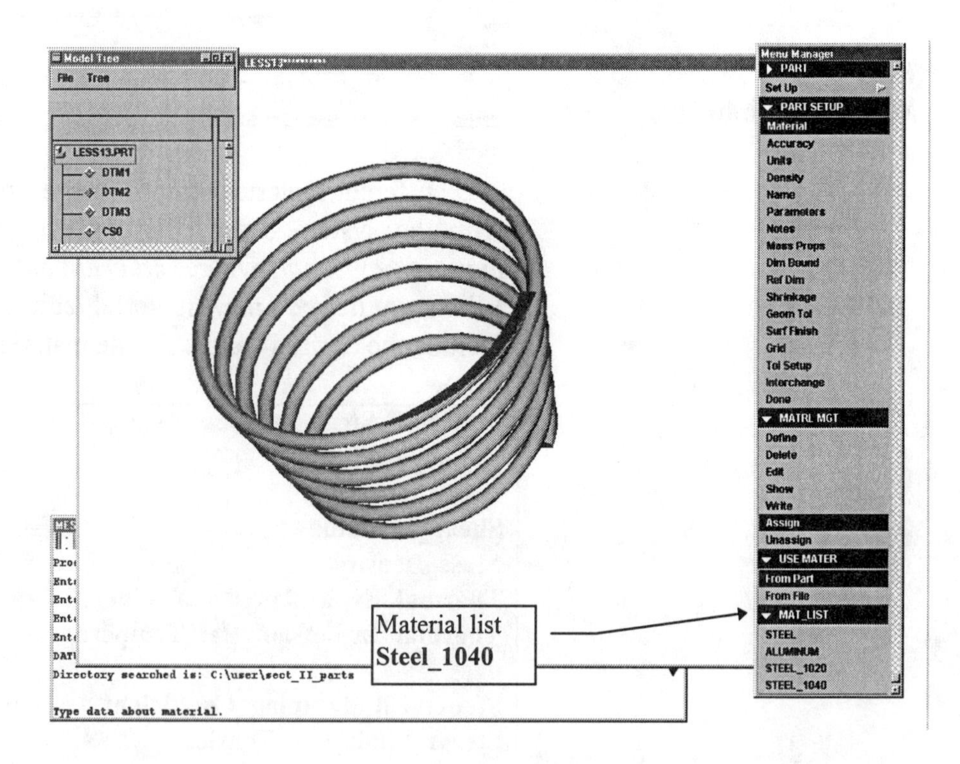

Figure 7.5
Assigning Material to a Part

Unassign Unassign the currently assigned part material.

Changing Material Parameters In A Part

You can specify or change material parameters by choosing **Material** ⇒ **Edit** to edit a material stored in the part's internal database or by choosing **Material** ⇒ **Assign** to change assigned parameters in the material file. After regeneration the materials values are updated.

UNITS

The **Units** option under the PART SETUP menu enables you to specify the dimension units to be associated with the part (Fig. 7.6). The **Units** option under the ASSEM SETUP menu governs the units of assembly features, placement offsets, and explode distances.

Units for a part are established at startup of Pro/E by means of the configuration file option "pro_unit_length". The default can be set to:

unit_inch	**unit_foot**	**unit_mm**	**unit_cm**	**unit_m**

Rules about the use of units:

1. Since relations are not scaled along with the model, modifying units may invalidate relations. Non parametric features, such as cosmetic, and IGES models are also not scaled with the change of units.
2. A part can have only one set of units.
3. Unless otherwise set by the pro_unit_length configuration file, parts default to inches and pounds.
4. The internal units for an assembly are those of its base component. If, however, the units of the base component have been changed, the assembly units will not automatically change. You must specify the units of the assembly again.
5. You cannot change the units of an assembly that contains assembly features that intersect a part.
6. User-defined parameters have no units. The appropriate conversion factors in relations must be included.
7. Cross sections do not have units.

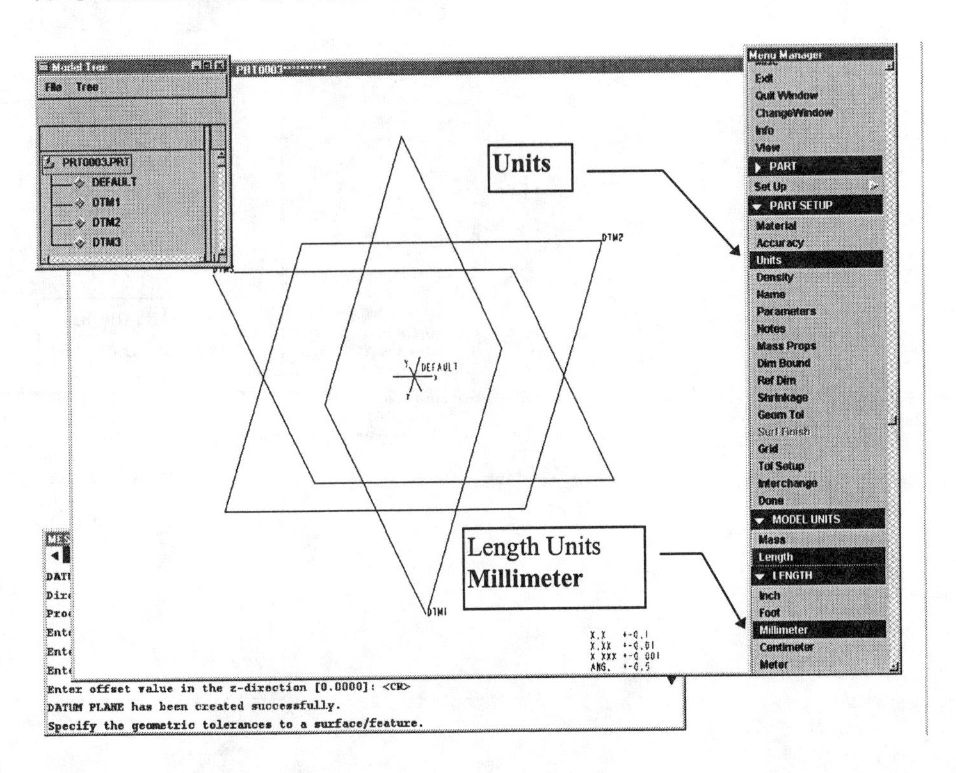

Figure 7.6
Setting Units at Beginning of Part Creation

Specifying units

To specify the units of a part:

1. Choose **Set Up ⇒ Units.**
2. For length units, choose **Length** and one of (**Inch, Foot, Millimeter, Centimeter, Meter,** or **Other unit**). Once selected, these become the active, displayed units. For mass units, choose **Mass** and one of (**Ounce, Pound, Ton, Gram, Kilogram, Other Unit**).
3. If a model exists when length units are changed, the SCALE menu appears (Fig. 7.7):
   Same Size Tells Pro/E to keep the model the *same physical size*, therefore the values of the dimensions will change.
   Same Dims This option tells Pro/E to keep the dimension values the same, therefore the *size of the model changes*.
   User Scale Scales the part by a specified amount. Enter the scale factor in terms of a size relative to the current part (e.g., **1.0** will not change the part's current size, **.25** makes the part ¼ (**25%**) of its current size).
4. Choose one of the preceding options, and then choose **Done** from the SCALE menu.

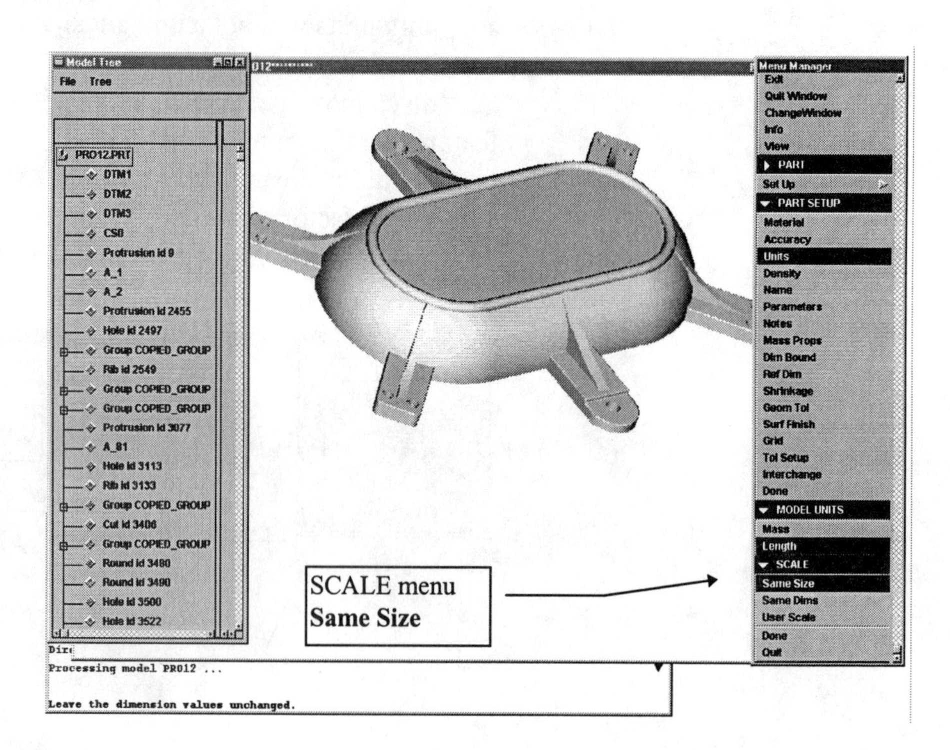

Figure 7.7
Changing Part Units

Dimension Tolerance Setup for Parts or Assemblies

The application of dimension tolerances is governed by either ANSI or ISO/DIN standards. By default, the configuration file option “tolerance_standard” is set to ANSI. ANSI dimension tolerances are assigned according to the number of digits specified. ISO/DIN dimension tolerances are driven by a set of ISO/DIN tolerance tables. To change to ISO/DIN tolerances, change the “tolerance_standard” option to ISO/DIN.

Setting Up Dimension Tolerances

To use the **Tol Setup** command in Part or Assembly mode (Fig. 7.8) do the following:

1. Choose **Set Up** in the PART menu or ASSEMBLY menu.
2. Choose **Tol Setup** in the PART SETUP menu or ASSEM SETUP menu.
3. The TOL SETUP menu displays with the following options:

    Standard Change the tolerance standard of the current model from the TOL STANDARD menu.
    ANSI Switch the standard from ISO/DIN to ANSI. New tolerances are determined for all dimensions based upon the number of digits in the dimensions. Tolerance tables, if any, will be deleted.
    ISO/DIN Switches the standard from ANSI to ISO/DIN. The system tolerance tables and any available user-defined tolerance tables are loaded.

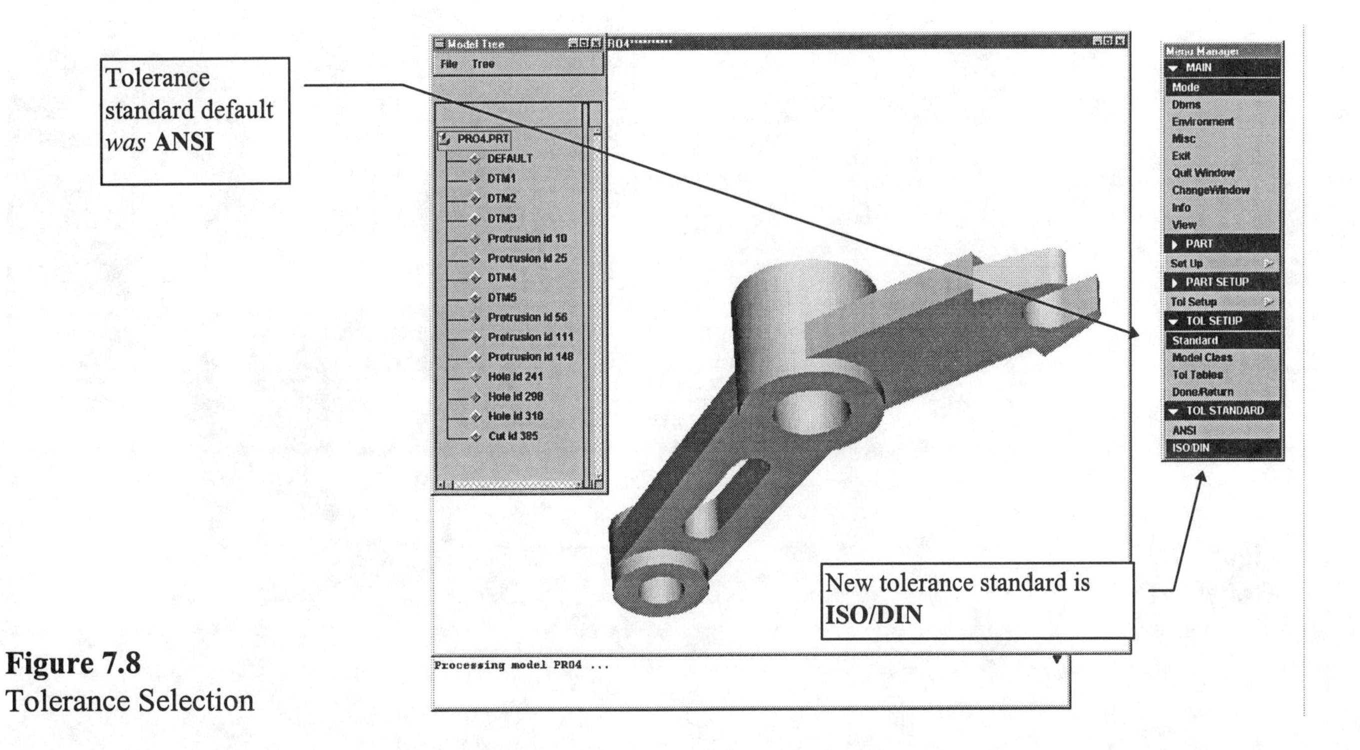

Figure 7.8
Tolerance Selection

INFORMATION

Choosing **Info** brings up the INFO menu that provides options to obtain information about the geometric properties of models, names of models and sections, clearance between parts or surfaces, etc. Figure 7.9 shows the command option **Model Info** in the INFO menu.

A variety of different types of analysis are available, including:

Mass property computation
File name listing
Measure clearance & interference
Measurement
Feature information
Feature list
Model (part, assembly) information
Layer information
Curve analysis
Regeneration information
Bill of materials (Assembly mode)
Sketcher section information
Part difference information
Audit trail
Parent/Child information
Geometry checking
Surface analysis

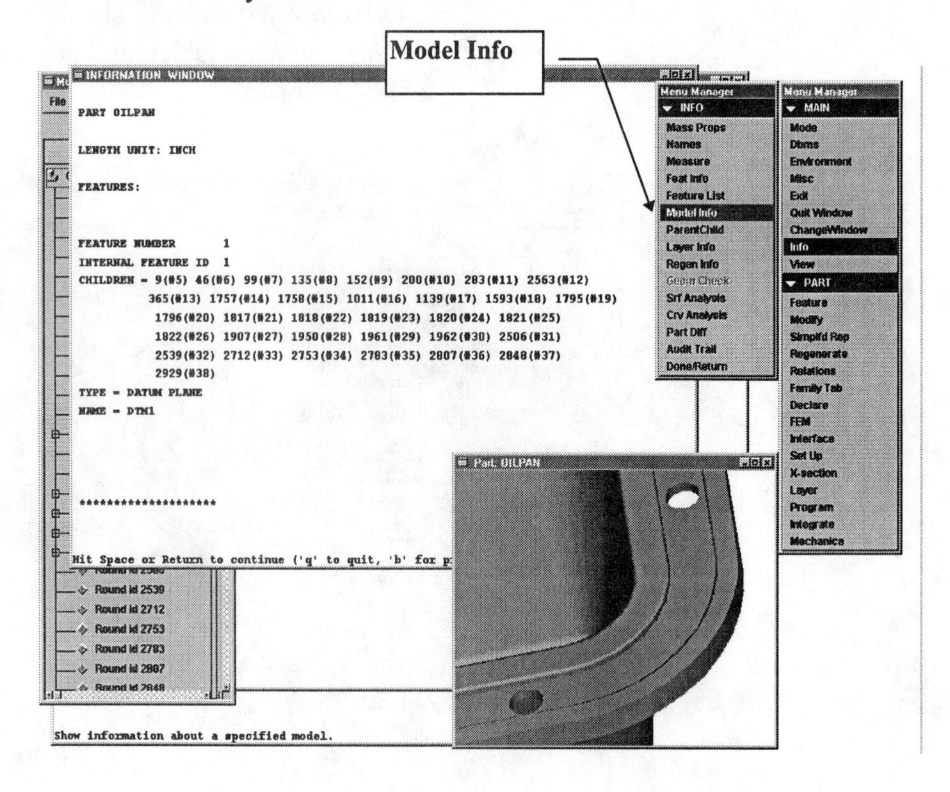

Figure 7.9
Model Info

The Info Menu Options

The **Info** option under the MAIN menu can be used to compute mass properties, measure distances, clearances, and interferences, and obtain model information. The **BOM** (bill of materials) option is available only in Assembly mode. The **Regen Info** option is available in both Part mode and Assembly mode. **Mass Property Computations** for parts, assemblies, and cross sections can be computed by choosing **Mass Props** from the INFO menu.

Measure

The **Measure** command in the INFO menu is used to analyze model and draft geometry. It is available in all modes. Figure 7.10 shows the **Measure ⇒ Area ⇒ Query Sel ⇒ Accept** command given and the circular surface of the boss selected and accepted.

Measuring highlighted surface area

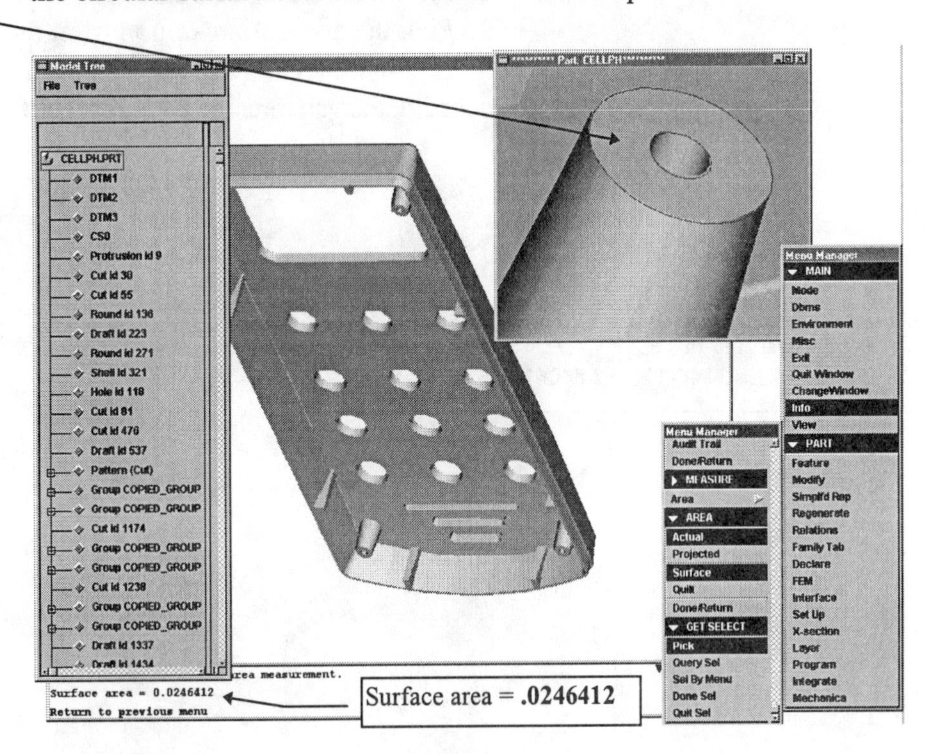

Figure 7.10
Measure

Curve/Edge

Part edges and datum curves are measured using the **Curve/Edge** option from the MEASURE menu. Available **Info** options for edge or curve information include:

Length Measures and displays the length of an edge or a curve segment.
Type Displays the type of the edge or curve.
Normal You have to select or create a coordinate system to measure normals. Displays in red the normal vector to the edge or curve at the selected point.

Tangent Displays in red the tangent vector (first derivative) to the edge or curve at the selected point. The coordinates of the tangent vector are displayed in the message window. You have to select or create a coordinate system to measure tangents.

Curvature Calculates and displays the curvature of the edge or curve at the selected point.

Radius Calculates the radius at the selected point on the curve or edge.

All Displays the combined information provided by Length, Type, Normal, Tangent, and Curvature. In Figure 7.10, the cell phone's front edge was selected and information is displayed in the Information window.

To measure an edge or datum curve:

1. Pick **Curve/Edge** from the MEASURE menu.
2. Pick the appropriate option from the INFO CURVE menu.
3. If **Normal, Tangent,** or **All** (Fig. 7.11) is chosen, select or create a coordinate system using the GET COORDS menu options.

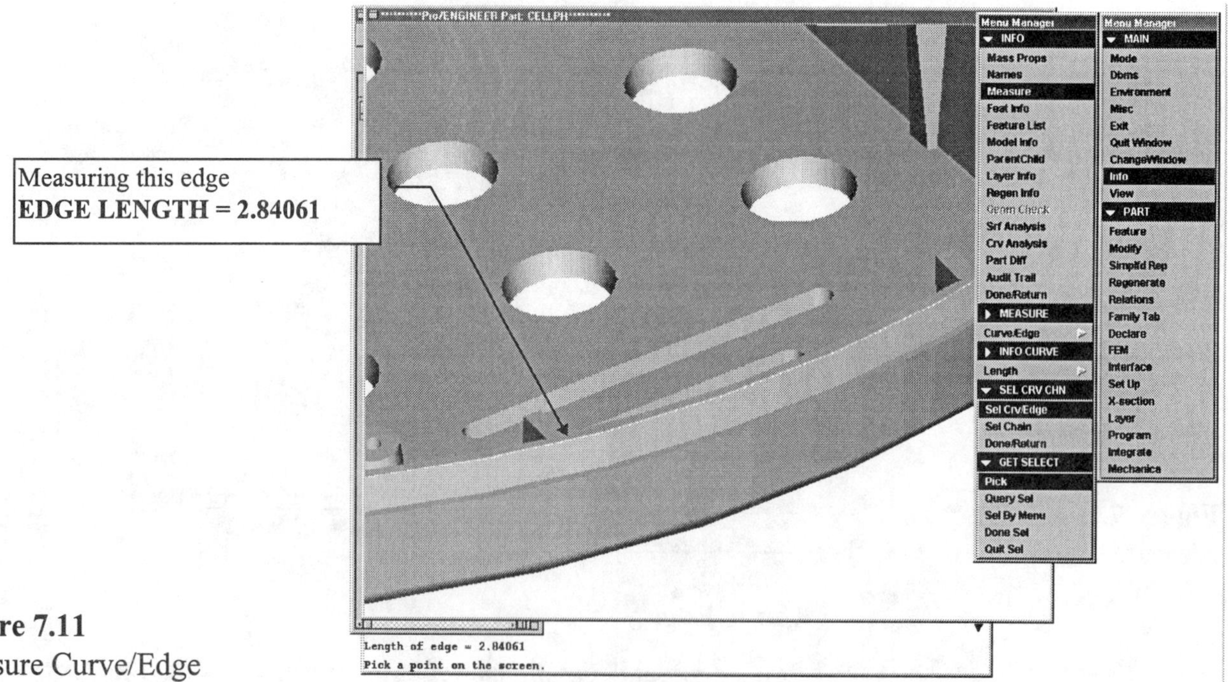

Figure 7.11
Measure Curve/Edge

4. If **Curvature** or **Radius** is chosen, select an option from the POINT OPT menu:

    Select The measurements will be made for the pick point.

    End point The measurements will be made for the nearest endpoint of the edge or curve segment.

5. Select the edge or curve to be measured. Use the **Sel Chain** command when measuring the length of several curves/edges lying on the same surface.

Angle

The **Angle** command is used to measure the angle between axes, planar curves, and planar nonlinear edges. The angle measured depends on where you select the edges. To measure angles between planes, do the following:

1. Pick **Angle** from the MEASURE menu.
2. Select **Plane**, select the plane surface to measure from, and then select the second plane surface. The selected geometry highlights in blue, and the angle will be displayed in the MESSAGE WINDOW.

When you measure the angle to a nonlinear edge or planar curve, the system calculates the angle to the plane in which the indicated edge or curve lies. When computing the value of an angle between two planes (Fig. 7.12) or between a plane and a line, the system selects the smaller of the two possible angle values. In this example the angle measurement between the angled surface and the front ledge surface is **25°**.

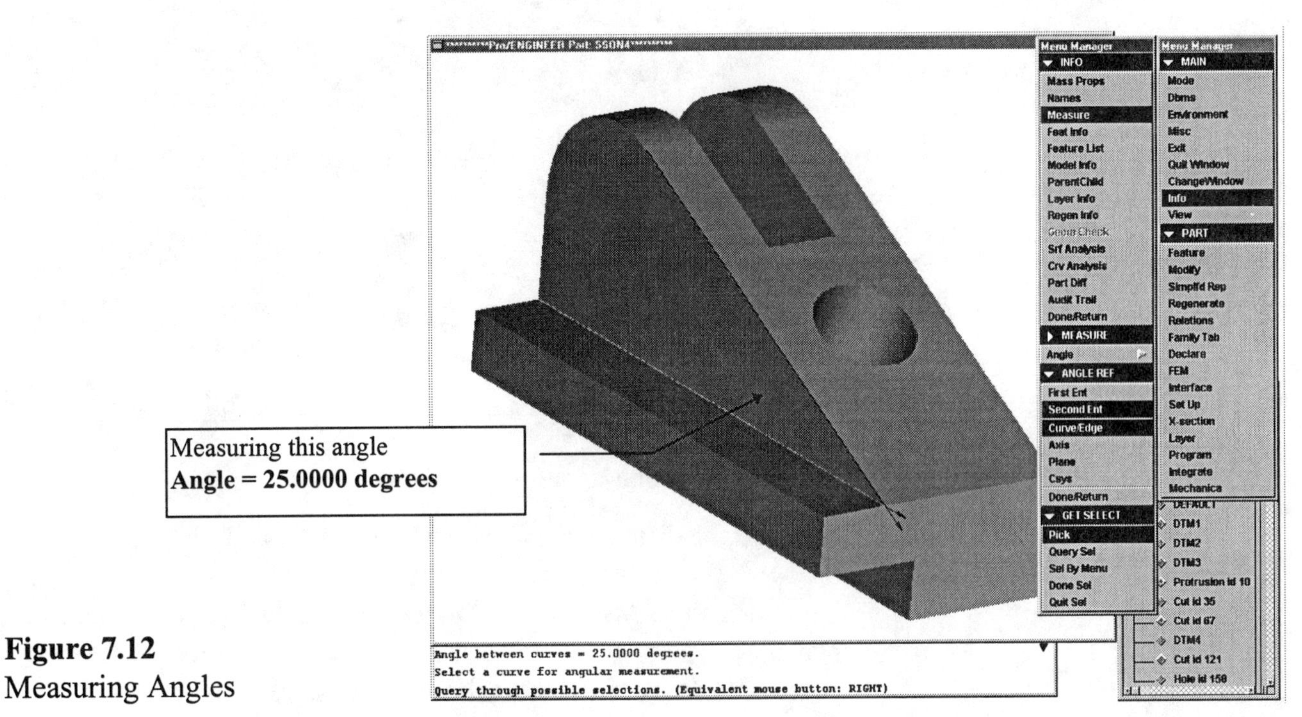

Figure 7.12
Measuring Angles

Distance

Distances are measured with respect to a basis entity. The *basis entity* is the one from which you measure, that is, the first entity selected when you start measuring **Distance**. After the basis entity is selected, you can make as many measurements as you like by selecting various entities to which to measure. Each distance will be calculated with respect to the first entity, until you restart the measuring process by selecting a new basis entity.

When measuring distances, you have to specify the entity type before selecting an entity from or to which to measure. The entity types are:

Point A point on the part surface, or a datum point.
Vertex A vertex of the part (Fig. 7.13)
Plane A planar part surface or datum plane.
Axis An axis of a feature or datum axis.
Coord Sys A coordinate system. When you select **Coord Sys** as the entity type, the GET COORDS menu is displayed. You can select or create a coordinate system to measure "**From**" or "**To**."

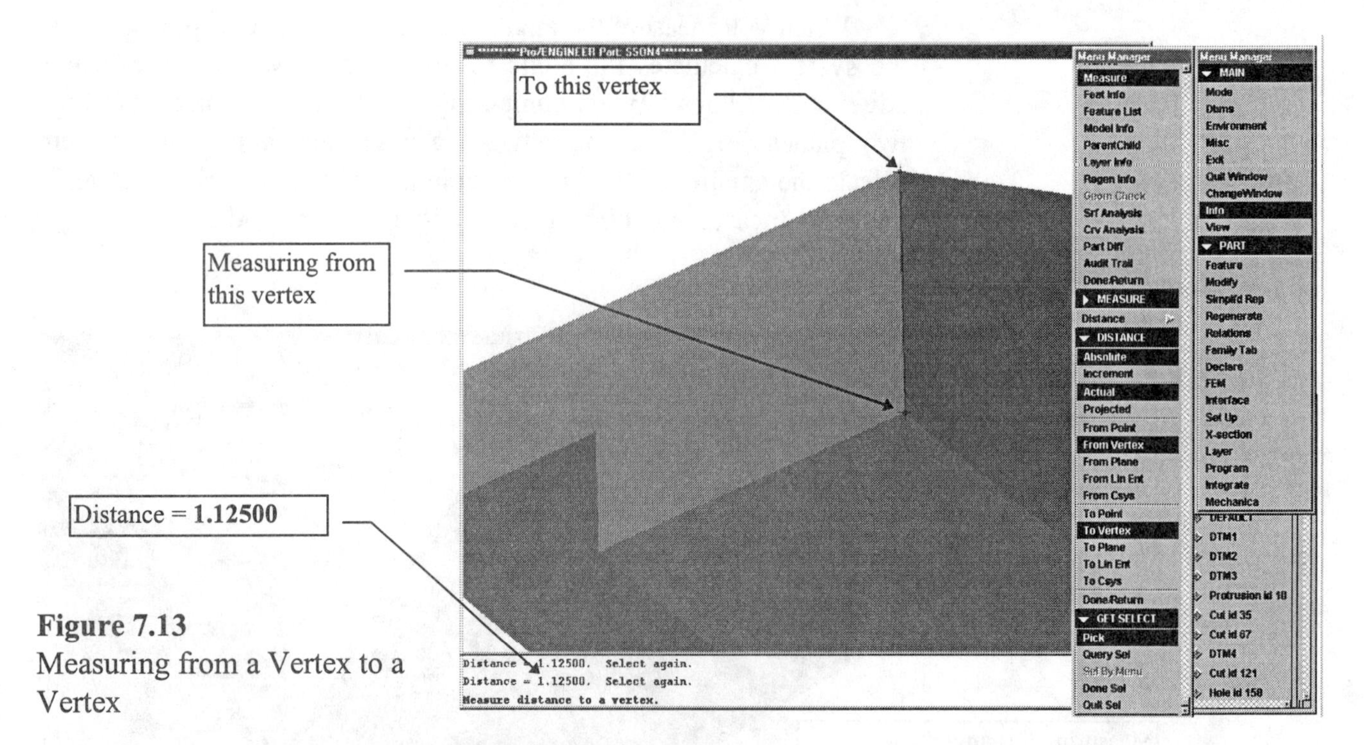

Figure 7.13
Measuring from a Vertex to a Vertex

Measuring Distance

To measure the distance between any two entities:

1. Choose **Info** ⇒ **Measure** ⇒ **Distance**. The DISTANCE menu displays.
2. Select absolute or incremental coordinates:

   Absolute Display the measured distance with respect to the basis coordinate system.

   Increment Display both the measured distance and the coordinate incremental (e.g., **dx**, **dy**, **dz**) with respect to the basis coordinate system. Both the **From** and **To** options are inaccessible until you specify a coordinate system from which to determine the measurements; you will have to select both **From** and **To** options for each measurement. In doing so, you can change the entity type before each selection.

3. Choose the appropriate **From** option to establish the basis entity.
   From Point - Measure from an arbitrary point.
   From Vertex - Measure from a vertex.
   From Plane - Measure from a plane.
   From Axis - Measure from an axis
   From Csys - Measure from a coordinate system. The GET COORD S menu will display.
4. Choose the appropriate **To** option.
   To Point - Measure to an arbitrary point.
   To Vertex - Measure to a vertex.
   To Plane - Measure to a plane.
   To Axis - Measure to an axis
   To Csys - measure to a coordinate system. The GET COORD S menu will display.
5. Make as many measurements as you want from the basis entity. You can change the type of entity you are measuring to at any time.
6. To restart the measuring process for a new basis entity, repeat the process from step 2, choosing a new basis entity option (i.e., **From**).
7. To end the measuring process, choose **Done/Return.**

In Figure 7.14, **Info ⇒ Measure ⇒ Distance ⇒ From Csys ⇒ Sel By Menu ⇒ Select** (pick the coordinate system) ⇒ **Done ⇒ To Plane** (select the bottom circular plane of the part) was used to derive the **.750000** distance.

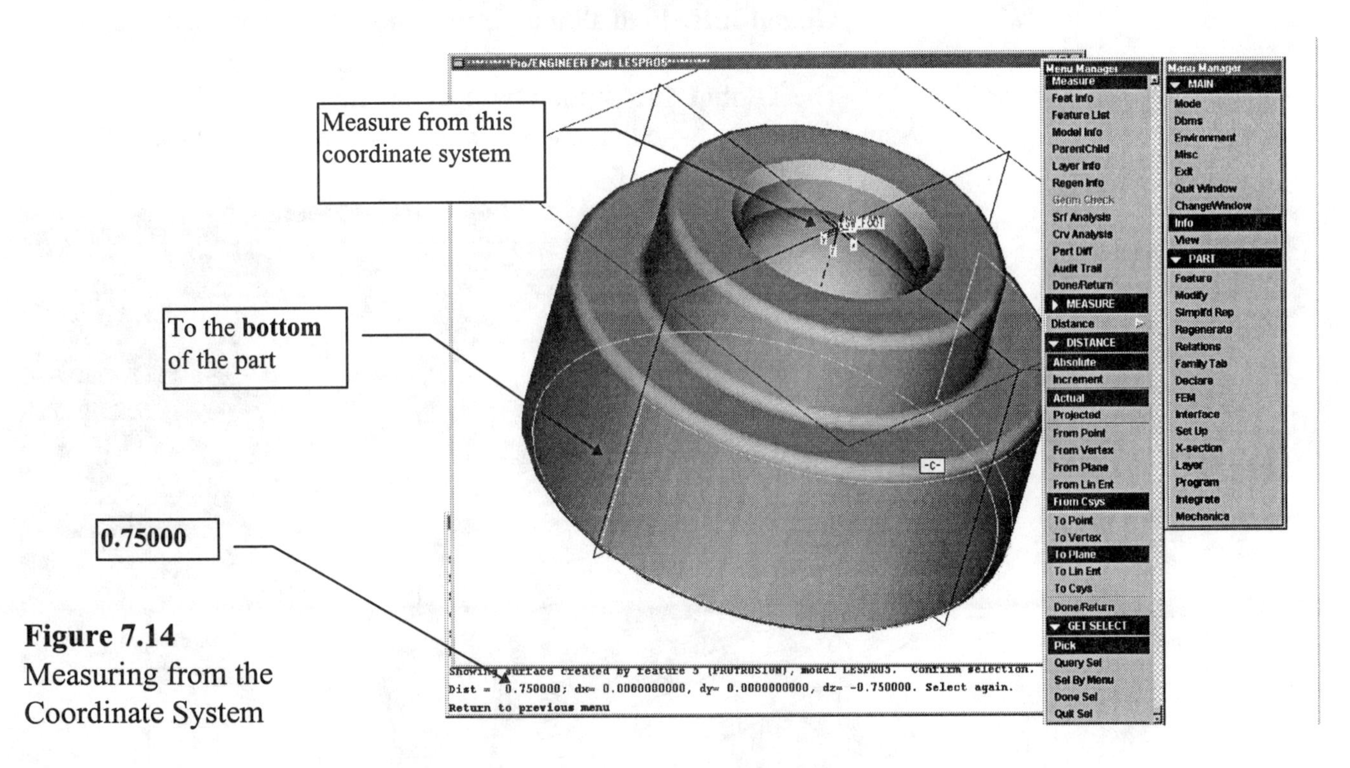

Figure 7.14
Measuring from the Coordinate System

Clearance and Interference Calculations

The **Clear/Intf** option calculates and displays either the clearance distance or interference between any combination of subassemblies, parts, surfaces, cables, or entities. Figure 7.15 shows the menu structure for measuring clearance and interference.

If the objects selected do not interfere, the minimum clearance is displayed graphically as a red line. A small red circle with crosshairs will display at each end of the line to identify the location at which the clearance is being measured. The clearance value is displayed in the MESSAGE WINDOW. If there is interference between the two surfaces, the system highlights the volume of interference and provides the value or highlights the curve or point of intersection, as appropriate for the items selected. The following is the procedure to determine a clearance or interference:

1. Pick **Measure** from the INFO menu.
2. Select **Clear/Intf** from the MEASURE menu.
3. Choose the desired option from the CLEAR/INTF menu:

   Pairs Get clearance or volume of interference between pairs of any combination of subassemblies, parts, surfaces, cables, and/or entities.
   Global Clr Find all pairs of parts or subassemblies that have clearances less than a specified clearance distance.
   Global Intf Find all interfering pairs of parts or subassemblies.

The **Global Intf** option was given in Figure 7.15 (the clamp subassembly). Note that one set of pairs has an interference.

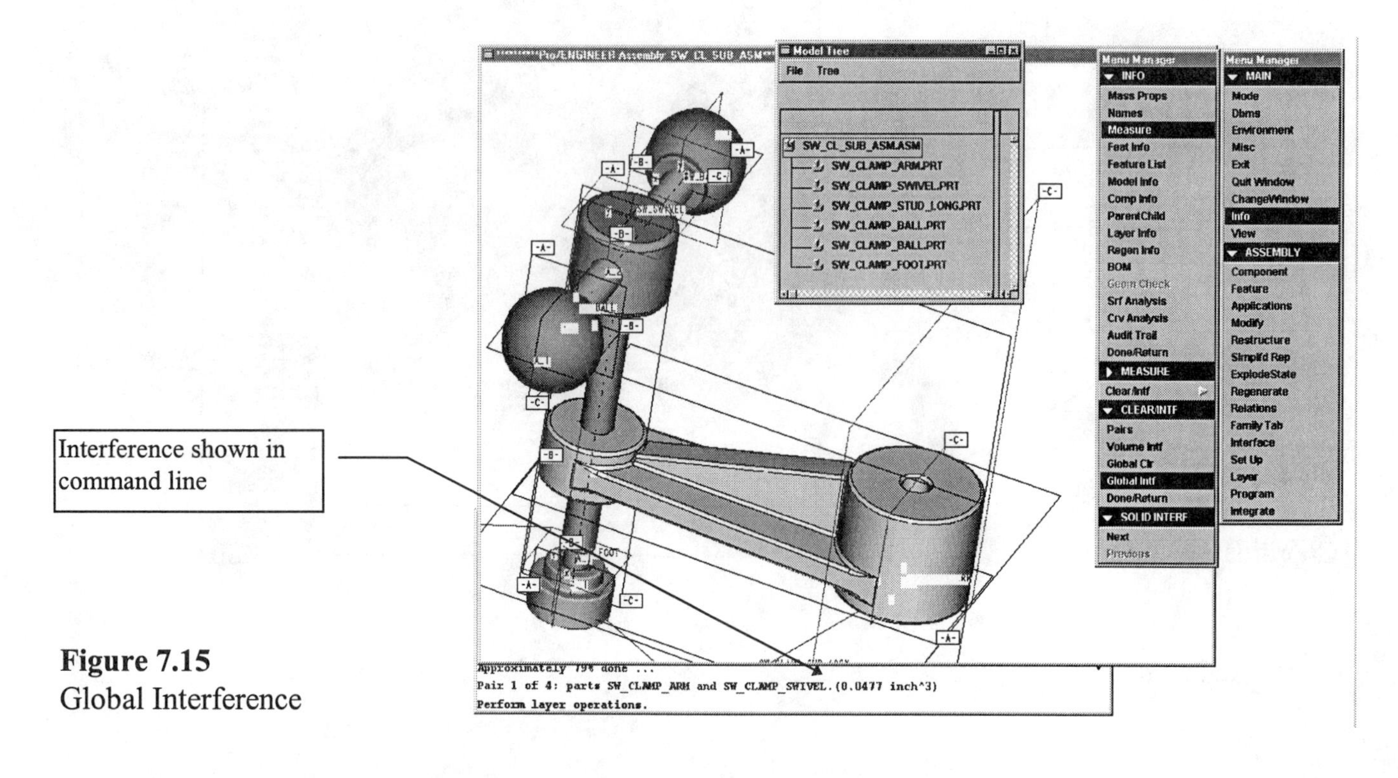

Figure 7.15
Global Interference

Global Clearance within an Assembly

This measurement command is used to identify, in an entire assembly, all components parts or sub-assemblies for which clearances are less than or equal to a clearance distance that is specified in the design (Fig. 7.16):

1. Select **Measure** from the INFO menu.
2. Pick **Clear/Intf** from the MEASURE menu.
3. Choose **Global Clr**. The GLOBAL SETUP menu displays. The options will appear in mutually exclusive pairs. Select the desired pair options and then choose **Done/Return.**

   Pair 1:
   Subasms Only Does a global check for clearance between all sub-assemblies, but not within individual sub-assemblies.
   Parts Only Performs global checking for clearance between parts in the assembly.
   Pair 2:
   ExcludQuilts Computes the clearance between solids.
   IncludeQuilts Not available for global clearance processing.

4. When prompted in the MESSAGE WINDOW, enter the clearance distance. The Pro/E will determine if any components of the assembly are within the specified clearance distance. Interference's are included.
5. Use **Next** or **Previous** in the GLOBAL CLR menu to step through the display of identified pairs (Fig. 7.16). To exit the process, choose **Done/Return** in the CLEAR/INTF menu.

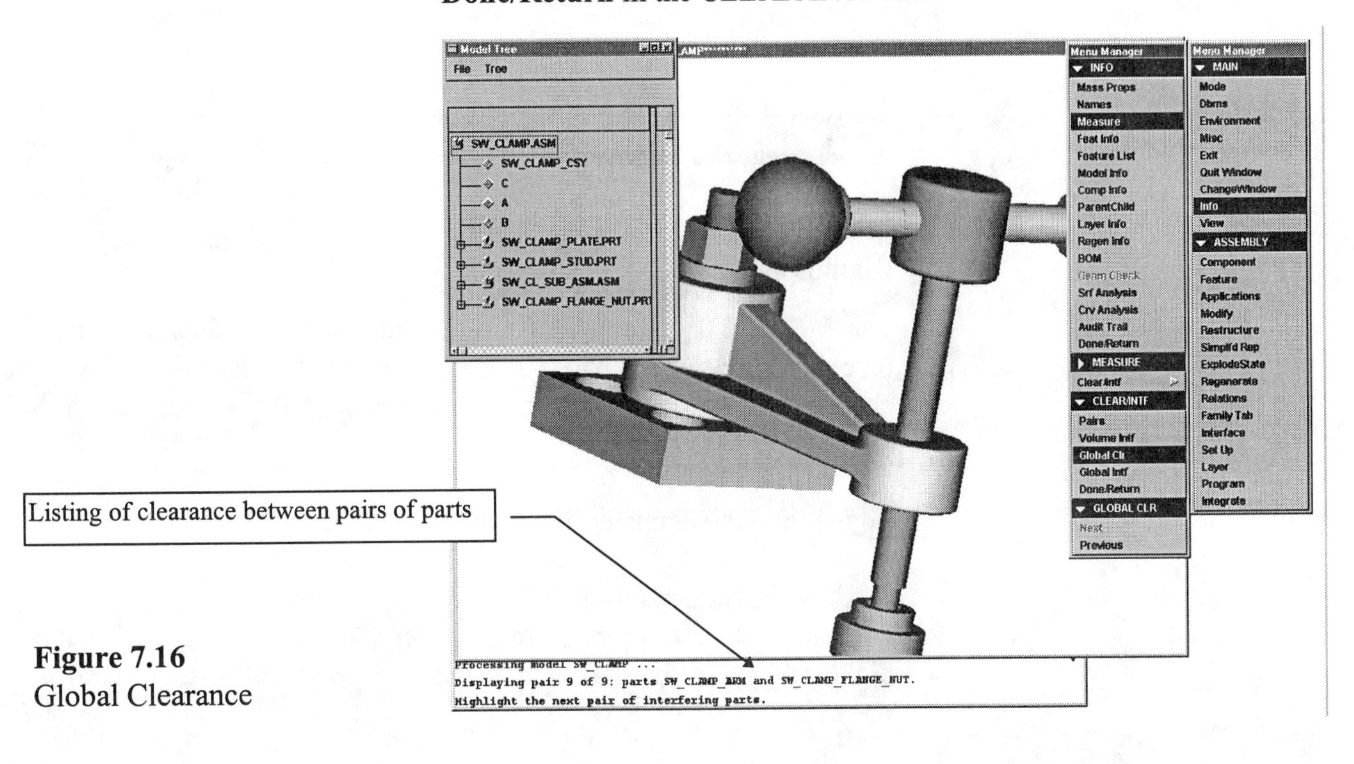

Figure 7.16
Global Clearance

Surface Area

The MEASURE command **Area** measures the surface area of any surface on the part or a datum surface.

1. Pick **Area** from the MEASURE menu. Choose an option:
   Actual Calculates the area of a surface or quilt.
   Projected Specify a projection direction and calculate the area of a surface or quilt projected in that direction.
   Surface Calculates the surface area of a face.
   Quilt Calculates the surface area of a quilt.
2. Select a surface for area measurement. The surface selected will be highlighted in red (Fig. 7.17).

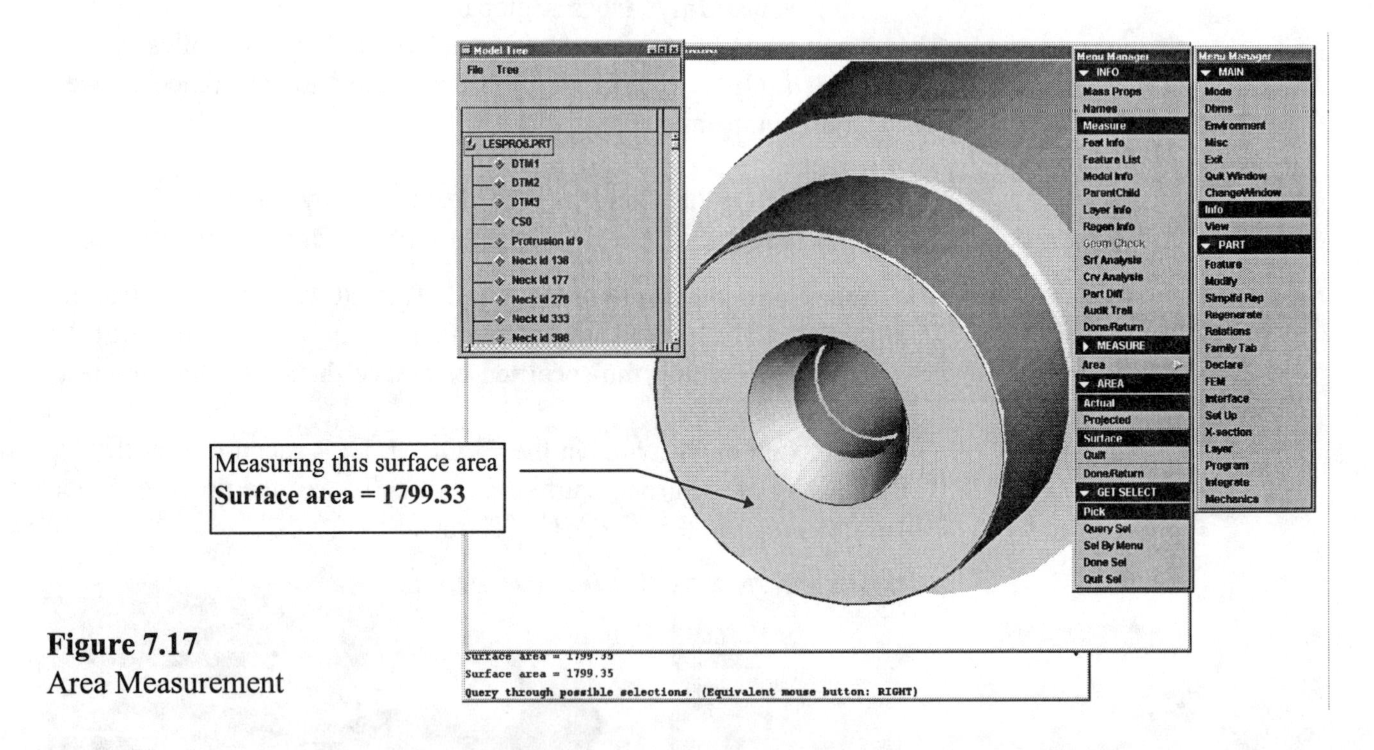

Figure 7.17
Area Measurement

Diameter

The MEASURE command **Diameter** measures the diameter of any revolved surface of a part. The edge of the counterbored hole was chosen in Figure 7.18 and the diameter of **1.00000** is displayed in the MESSAGE WINDOW.

To measure the diameter of a revolved feature:

1. Choose **Diameter** from the MEASURE menu.
2. Select a feature to measure diameter. The surface selected will be highlighted, and the diameter *at the pick point* will displayed in the message window.

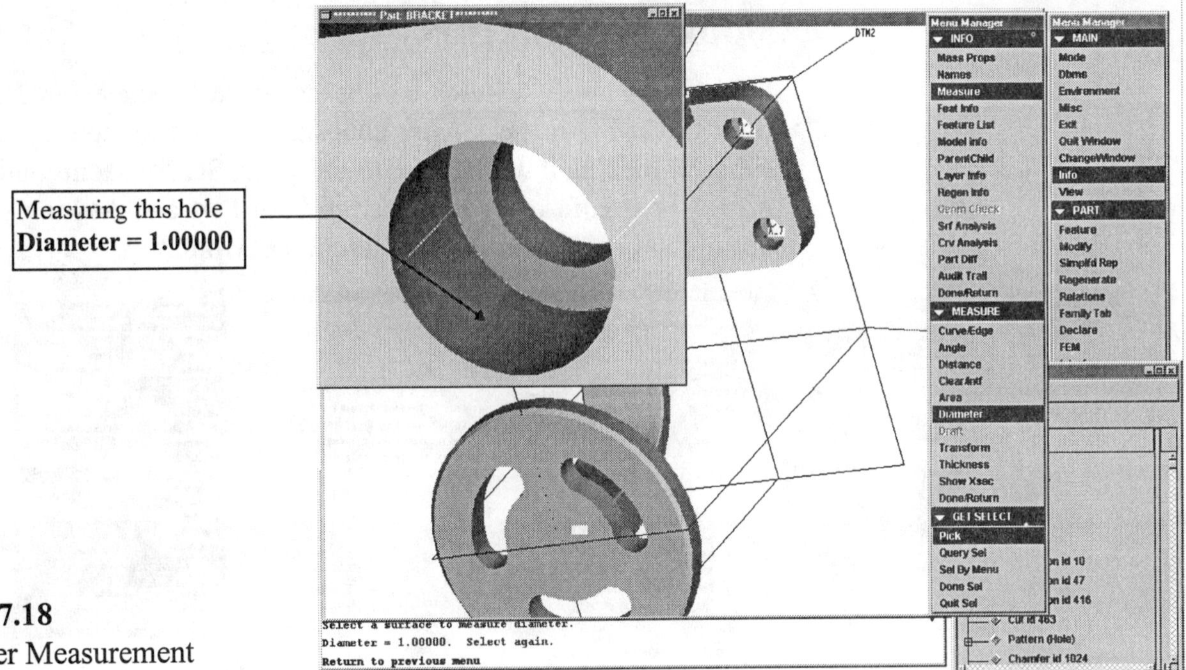

Figure 7.18
Diameter Measurement

Thickness

Thickness measures the thickness of a part to determine if a specified region has a thickness that is greater than or less than a user-specified maximum or minimum value. The region is displayed as a cross section (Fig. 7.19). If a region's thickness is over the maximum allowable, it will be displayed with a red border. If a region's thickness is below the minimum allowable, it will be displayed with a blue border.

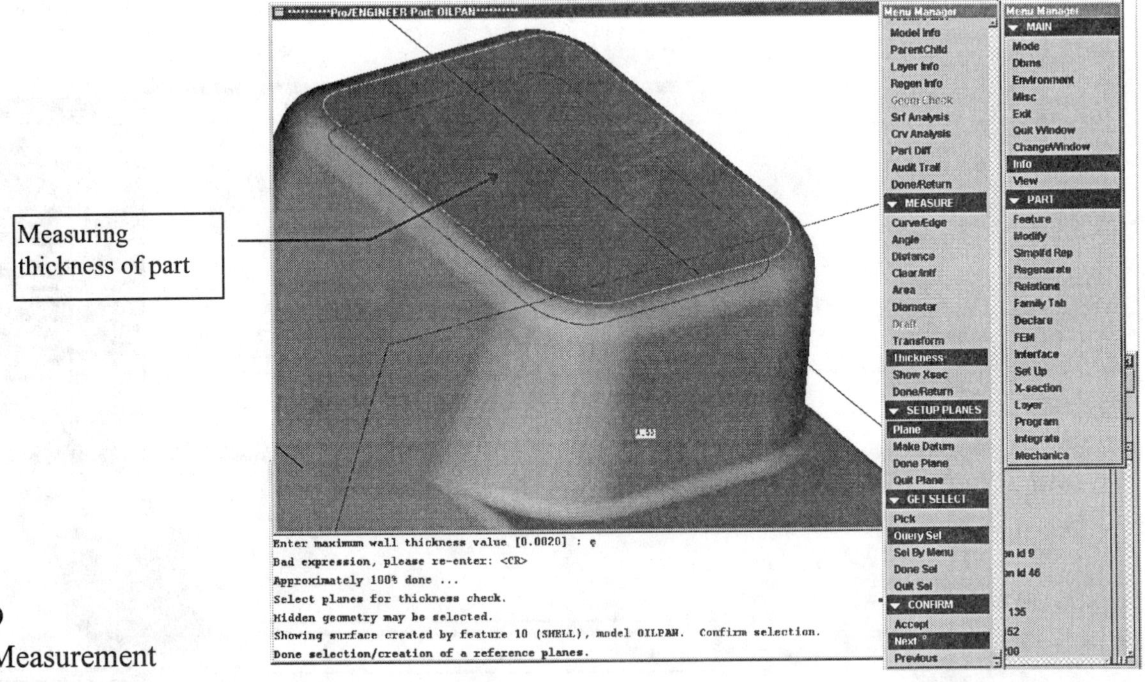

Figure 7.19
Thickness Measurement

MODEL INFORMATION

Information about features can be accessed by choosing **Feat Info** from the INFO menu. After choosing **Feat Info**, specify a feature either by picking it with mouse or using the **Sel By Menu** option (Fig. 7.20). **Feat Info** may be chosen in both Part and Assembly modes. Feature information is shown in the INFO window.

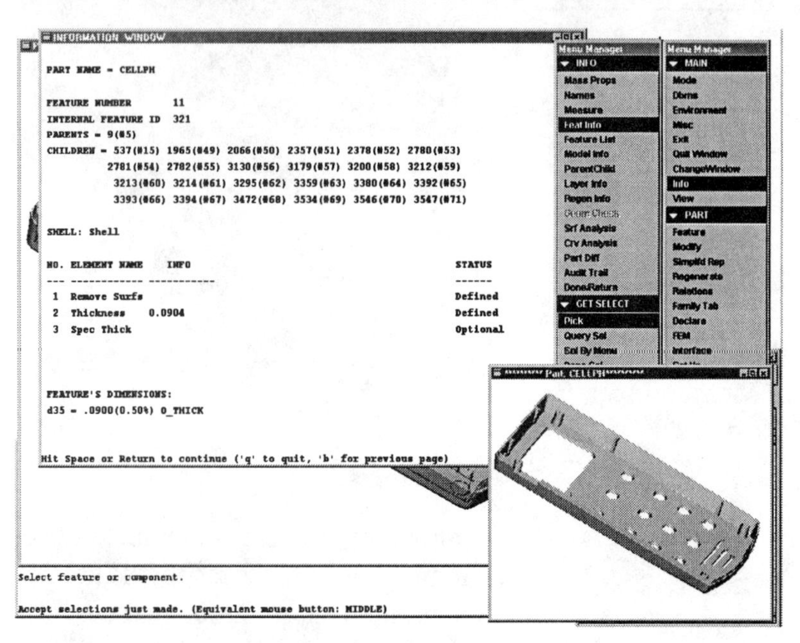

Figure 7.20
Feature Information

Feature List

When you choose **Feature List**, the Information Window appears with a table listing the features in order and giving information, such as Number, Name, Type, Suppression Order and Regeneration Status (Regenerated, Unregenerated, Failed etc.). An example is shown in (Fig. 7.21).

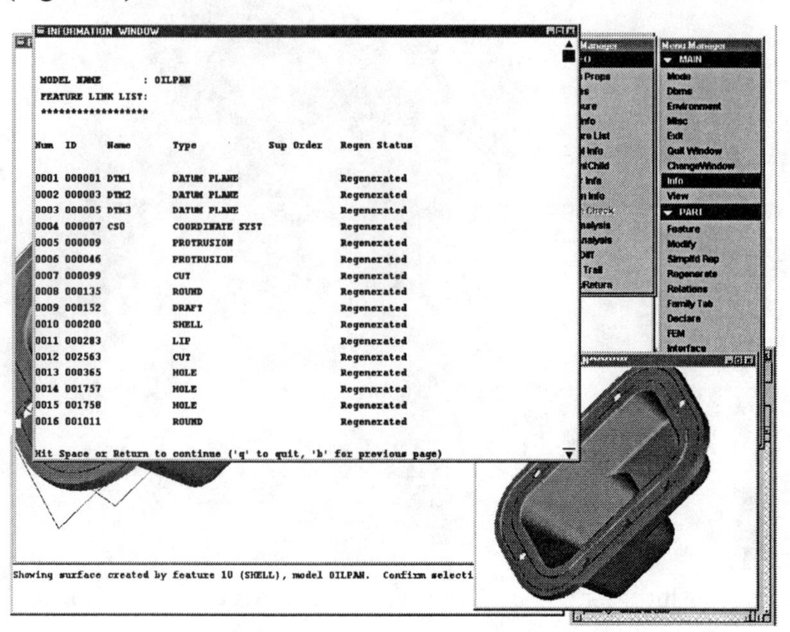

Figure 7.21
Feature List

Part Information

Information about every feature on a part can be accessed by choosing **Model Info**. In Part mode, the INFO window will appear immediately. In Assembly mode (Fig. 7.22), choose **Part** from the MODEL INFO menu and specify a part, either by picking a part with the mouse (**Pick**) or by entering the name of the part (**Name**).

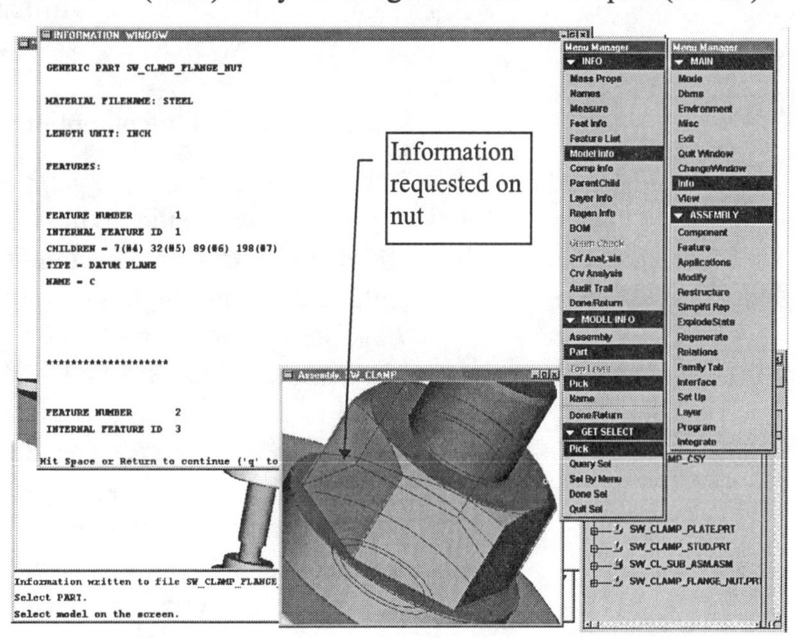

Figure 7.22
Part Model Information

Assembly Information

Assembly information can be accessed by choosing **Model Info** and then **Assembly** from the INFO menu, and then either picking on an assembly or specifying the name of the assembly. An information window displaying the assembly information will then appear. The names of the components in the assembly are displayed in a hierarchical structure to show how it was assembled (Fig. 7.23).

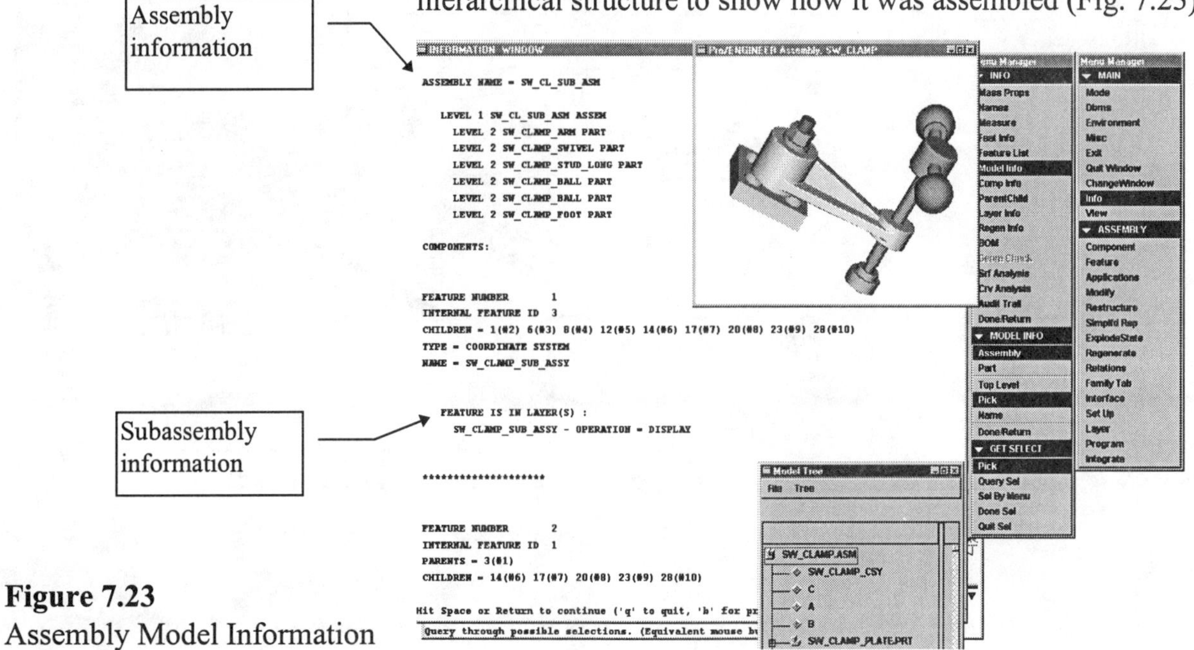

Figure 7.23
Assembly Model Information

Parent/Child Information

The **Parent/Child** option in the INFO menu is used to highlight the relationships between features. If you select either **Parents** or **Children**, you can then choose either to create an information file or to highlight the appropriate geometry on the screen. If the file option is chosen, the information is written to a file and displayed in the INFORMATION WINDOW.

The PARENT/CHILD menu options are as follows:

Parents Shows all the parents for the selected feature. The parent features are highlighted in the Reference color.

Children Shows all the children for the selected feature. The child features are highlighted in the Reference color. In Figure 7.24, the cylindrical protrusion was chosen and its children were listed in the INFORMATION WINDOW.

Child Ref Shows all the references for the each of the children of the selected feature. The child features are highlighted in the Surface Mesh color, and the surfaces, edges or points that they reference are highlighted in the Internal Quilt Edge color.

References All the references used to construct a feature are shown one at a time. These can be axes, datums, surfaces, edges, or other features.

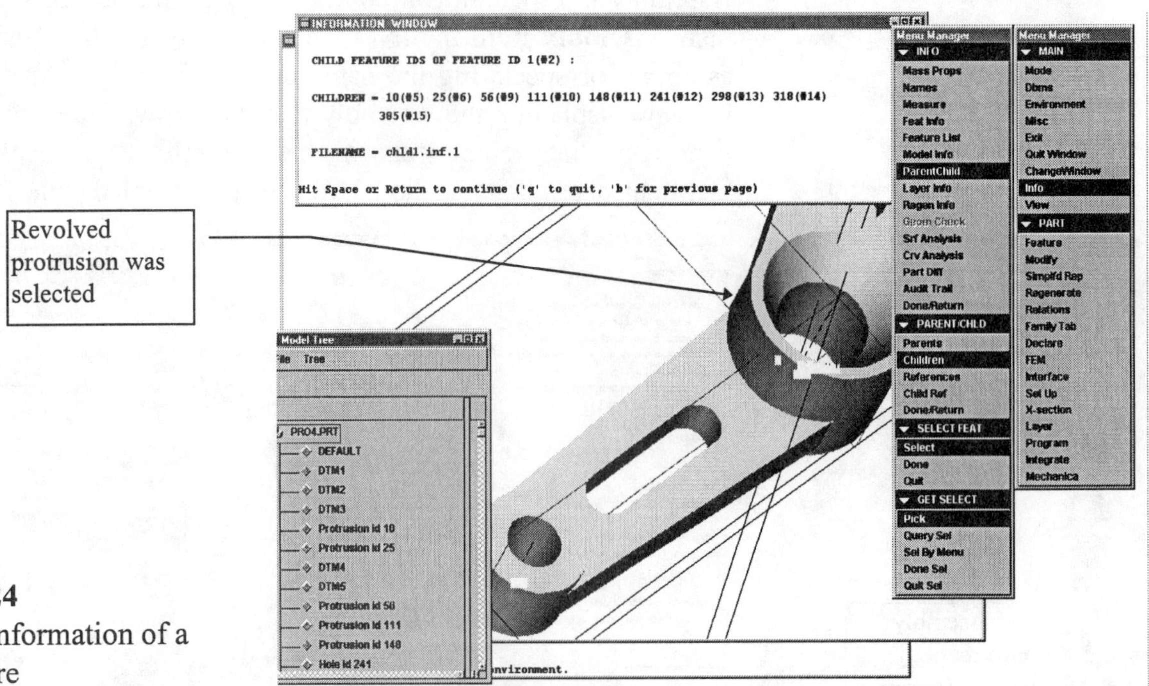

Figure 7.24
Children Information of a Part Feature

Displaying Parent/Child Information

This command is used to show parent/child relationships of a particular feature for a component or if the information is needed for an assembly:

1. Select **Parent/Child** from the INFO menu.
2. Choose one of the options from the PARENT/CHILD menu.
3. Chose **Parents** (Fig. 7.25) or **Children** (Fig. 7.24), and then select one of the options from the FILE/HILITE menu:
   File A text file with the corresponding feature IDs listed will be written to disk and displayed in the INFORMATION WINDOW. The file name is displayed (Fig. 7.25)
   Highlight The appropriate geometry will be highlighted with the color codes.
4. Select the feature. In Figure 7.25 the swivel component of the assembly was selected.

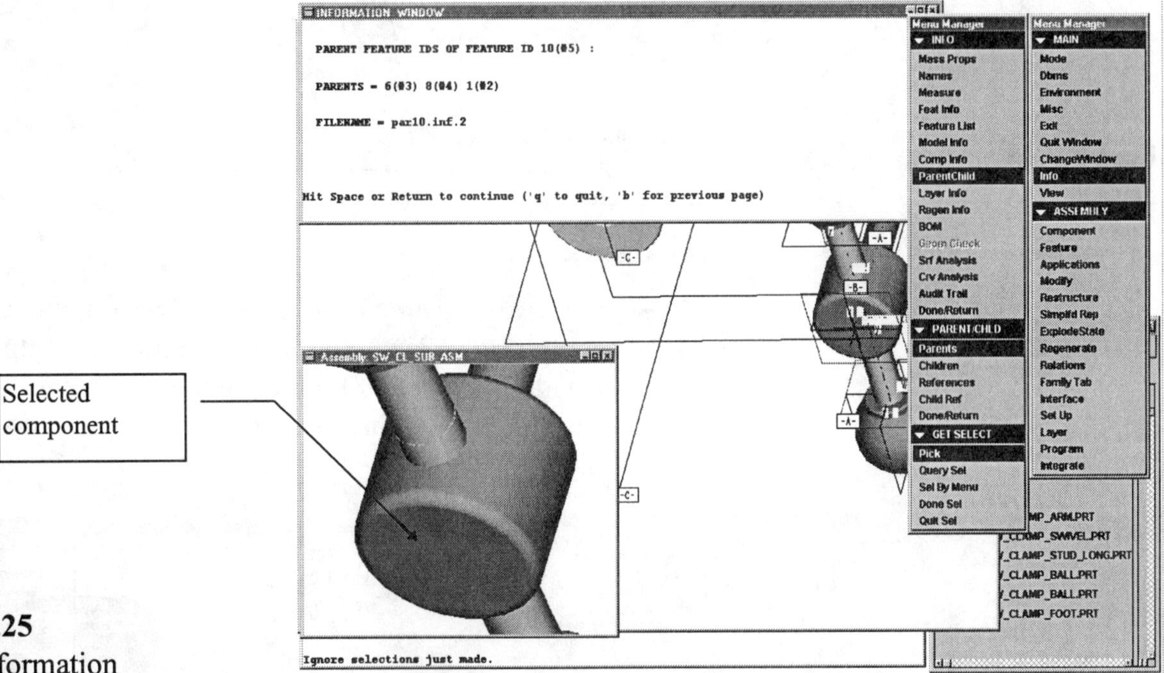

Figure 7.25
Parent Information

5. If **References** is chosen, the SHOW REF menu appears. Choose one of the following:
   Next or **Previous** Select the next or previous reference.
   Info Displays information about the highlighted reference in the INFORMATION WINDOW.
6. If **Child Ref** is chosen, the CHILD REFS menu appears. Choose one of the following:
   Next or **Previous** Select the next or previous reference.
   Ref Info Displays information about the highlighted reference.
   Child Info Displays information about the child feature(s).

File Name Listing

The INFO menu option **Names** displays a lists of all stored files in an INFORMATION WINDOW (Fig. 7.26). The first portion of the listing shows parts, assemblies, drawings, layouts, and sections in memory. The second portion gives a complete listing of all Pro/ENGINEER objects in the working directory.

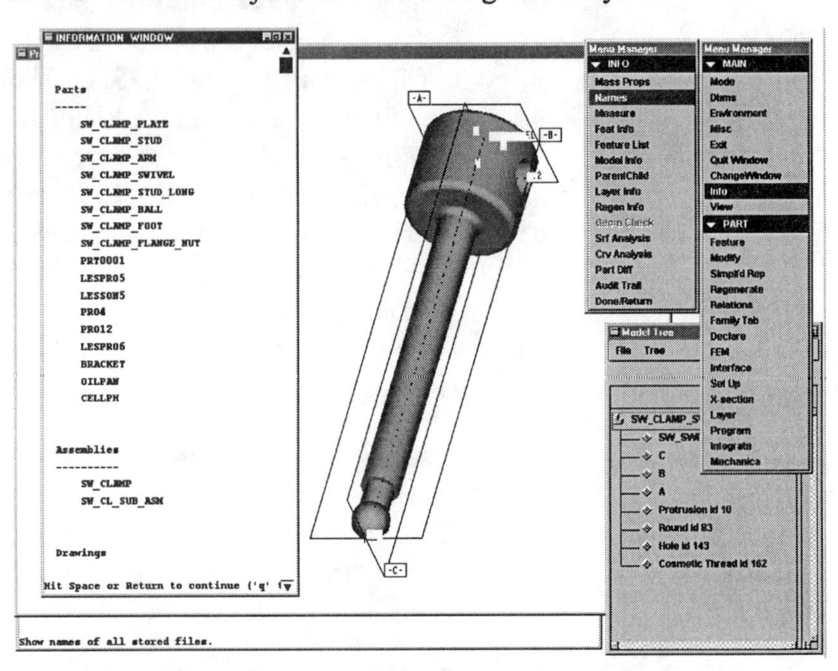

Figure 7.26
Names

Bill of Materials (BOM)

A *bill of materials* is a listing of all parts and parts parameters in the current assembly (Fig. 7.27). You can customize the output formats to produce a particular form of presentation and content. BOMs can be created for assemblies in Assembly mode or from assembly drawings in Drawing mode.

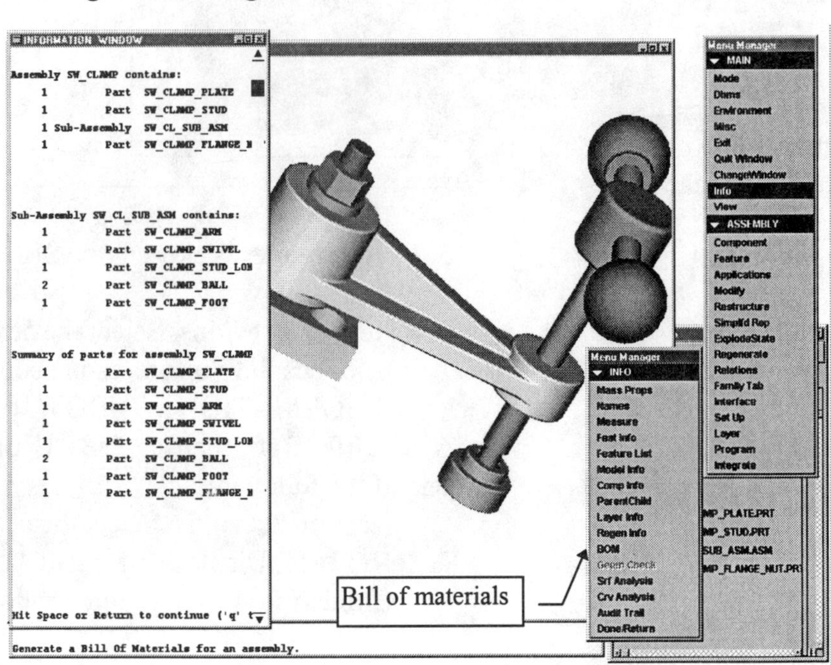

Figure 7.27
BOM

The information that follows explains how to create and format simple text BOMs, which are stored as text files. The optional module, Pro/REPORT, provides functionality for creating BOM reports: graphical BOMs with complex formatting and indexing.

The source of the bill of materials (BOM) output format can be configured by the configuration file. An example of the configuration file option for a user-defined formats is:

bom_format **bomcompany.fmt**

The default output format for the BOM is divided into two sections:

1. **Breakdown** Lists the name, type, and number of instances of each member and submember.
2. **Summary** Lists the total quantity of each part included in the assembly. It amounts to a "shopping list" of all the parts needed to build the assembly from the part level.

The BOM in Figure 7.28 contains the subassembly shown in Figure 7.27. Note that the **Model Tree** also displays information about an assembly.

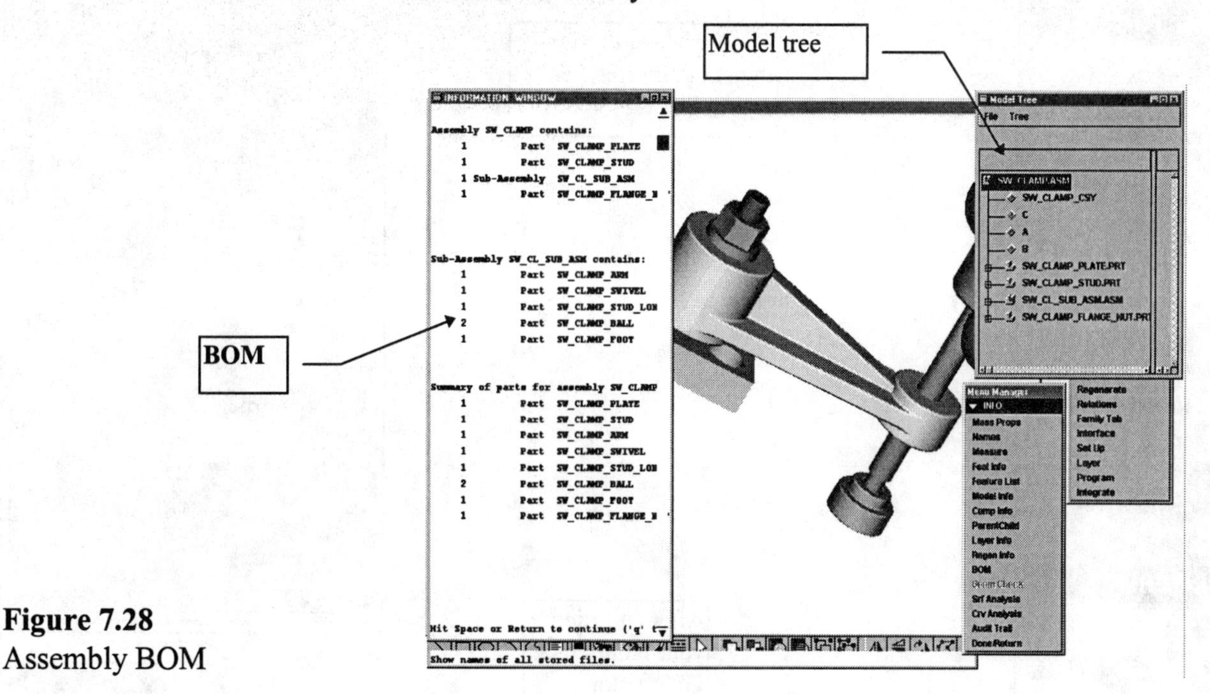

Figure 7.28
Assembly BOM

A user-defined BOM output format specifies separately the format of the breakdown section and of the summary section. You can include one or both sections, but you must specify the column titles, row content, and display format for each included section.

The Interface Menu

To import or export information through Pro/E, the INTERFACE menu is available (Fig. 7.29). The options that appear in this menu will change depending on what mode you are working in. Also, the Pro/E modules that are available will determine the options you have access to in this menu (i.e., Pro/INTERFACE, Pro/PLOT, Pro/ECAD, Pro/STEP, etc.). Pro/E can import and export data in a variety of formats, including: IGES, IGES Groups, STEP, DXF, SET, VDA, CGM, SLA, Plotter files, Neutral files, Render, Inventor, 3DPAINT, PATRAN Geom, COSMOS Geom, SUPRTB Geom, CatiaFacets, PDGS, ECAD, CGM, TIFF, PHOTORENDER, CATIA, CATIA IIF, CDRS, ENGEN, and VRML.

Plotting can be accomplished with the **Interface** command from the PART menu (ASSEMBLY menu, DRAWING menu, and other menus). Your system administrator will set the path to the plotting and printing devices on your system. Choose **Interface** ⇒ **Export** ⇒ **Plotter** ⇒ **OK** ⇒ **OK**.

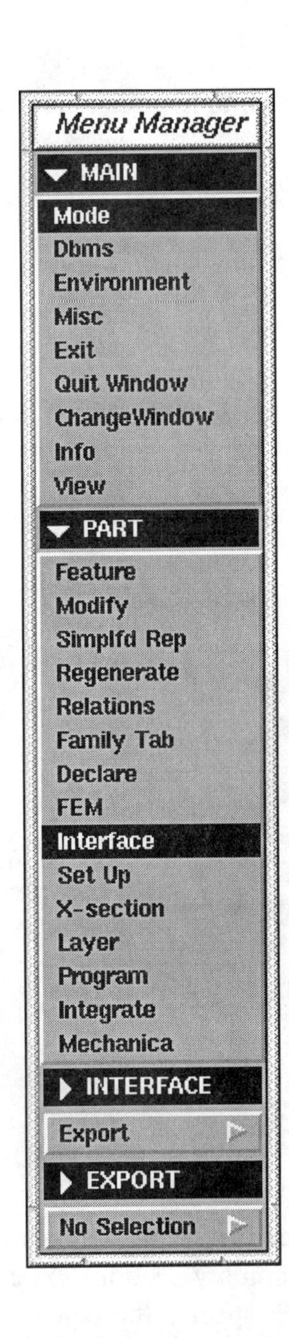

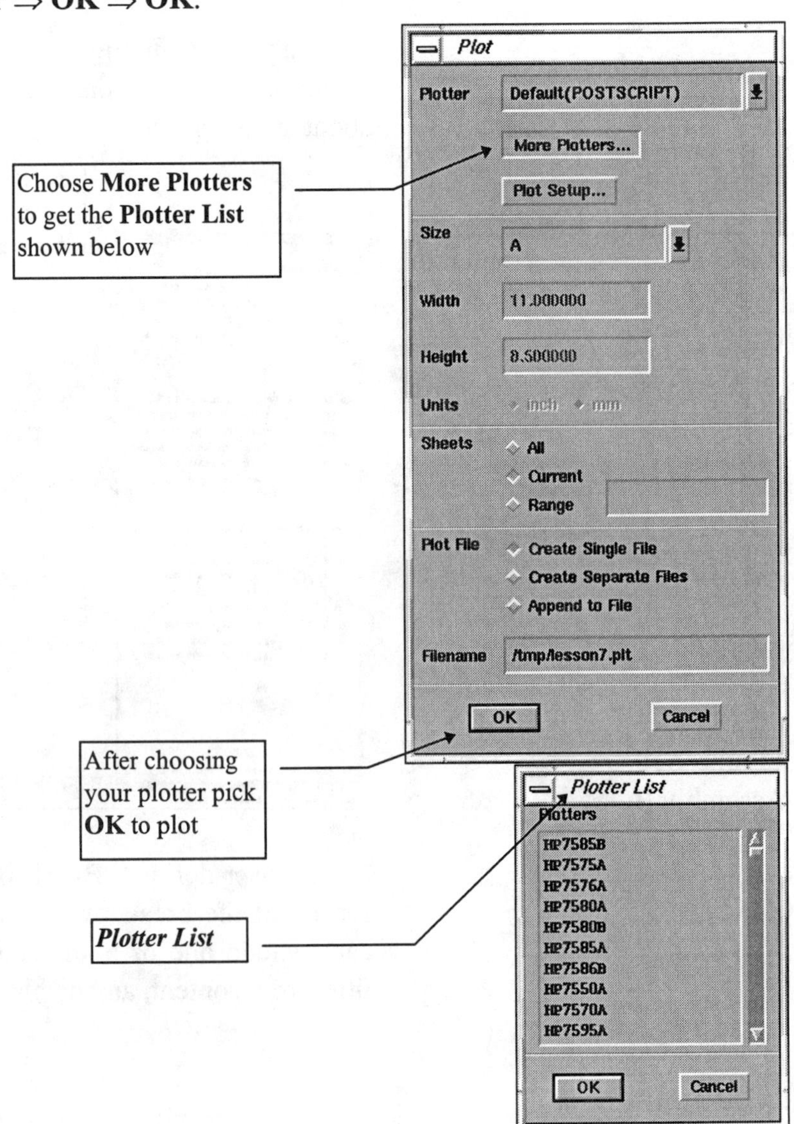

Figure 7.29
Interface

Section 8

Layers

Layers are an essential tool for grouping items and performing operations on them, such as selecting, displaying or blanking, plotting, and suppressing. Any number of layers can be created. User-defined names are available, so they can be easily recognized. Most companies have a layering scheme that serve as a *default standard* so that all projects follow the same naming conventions and objects/items are easily located by anyone with access. Layer information, such as display status, is stored with each individual part, assembly, or drawing.

Using Layers

The LAYERS menu (Fig. 8.1) contains five operations that can be performed on layers:

> **Setup layer** Performs operations on layers such as creating, deleting, and renaming; manipulating layers using the from file options; and adding parameters when exporting layers to another system.
> **Set Items** Adds or removes items from selected layers.
> **Set Display** Changes the display status (display or blank) of selected layers.
> **SetDefLayer** Accesses default layer editor.
> **Info** Displays layer information.

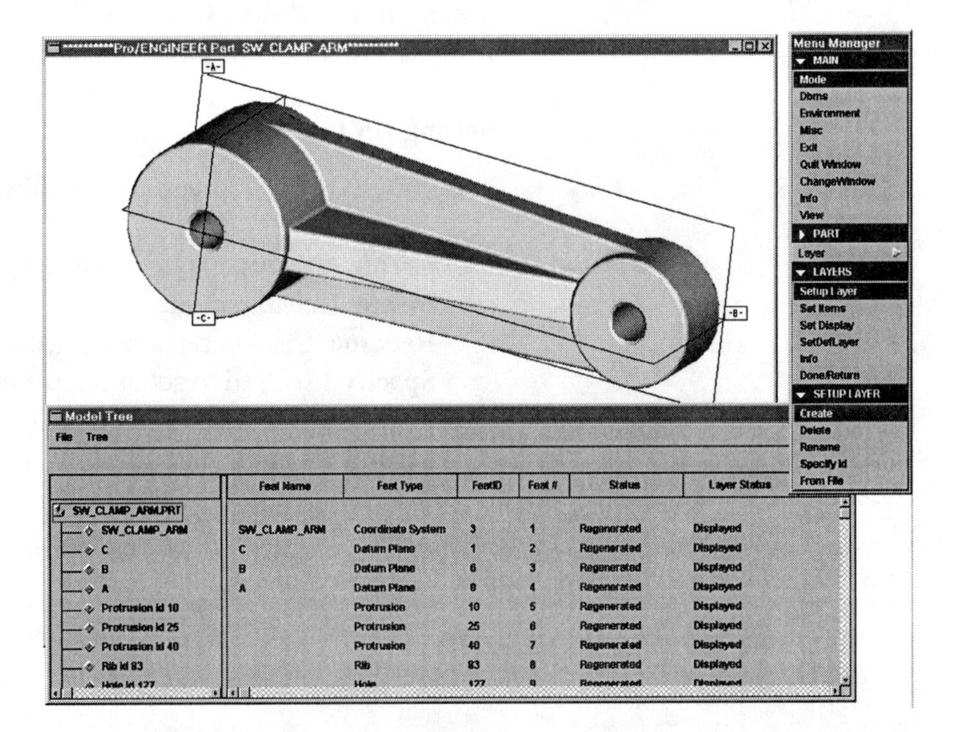

Figure 8.1
Layers Menu

To use layers:

1. Set up the layers to which items will be added.
2. Add items to the specified layers:
   * Some items can be automatically added to layers as they are created, using configuration file options.
   * An item can be added by selecting its type and then picking the items themselves.
   * Features are also added by selecting an option that adds all features of a particular type to the layer.
   * A range of features can be added to a layer.
   * One layer can be added onto another layer.
   * Items can be copied from one layer to an existing layer or to a new layer.
3. Set the display status of the layers:
   * A layer's name can be picked from a Namelist menu.
   * A layer can be retrieved from a file that contains the desired layer status, which automatically sets each layer to the status specified in the file.
   * The current layer file can be edited. All changed layers will reflect the new status once the file is saved.

Layers Names

Layers are identified by name. Layer names can be expressed in numeric or alphanumeric form, with a maximum of 31 characters per name. When layers are displayed using **Sel Menu,** numeric layer names are sorted first, then alphanumeric layer names. Layers names in alphabetic form are sorted alphabetically.

Setting Up Layers

The SETUP LAYER menu contains the following options:

Create Creates a layer by entering its name.
Delete Deletes selected layers.
Rename Used to rename a selected layer.
Specify Id Used to set or remove an interface layer ID.
From File Allows manipulation of a layer by modifying parameters in the layer file.

Creating a Layer

Before you can place items on a layer, you must first create the layer using the following method (Fig. 8.2):

1. Pick **Layer** from the PART or ASSEMBLY menu
(select the active model if you are in Assembly mode).
2. Choose **Setup Layer** from the LAYERS menu.
3. Select **Create** from the SETUP LAYER menu, and enter the layer name. Continue to create layers by entering new names at the prompt. Press **Enter** on the keyboard to quit the creating mode.

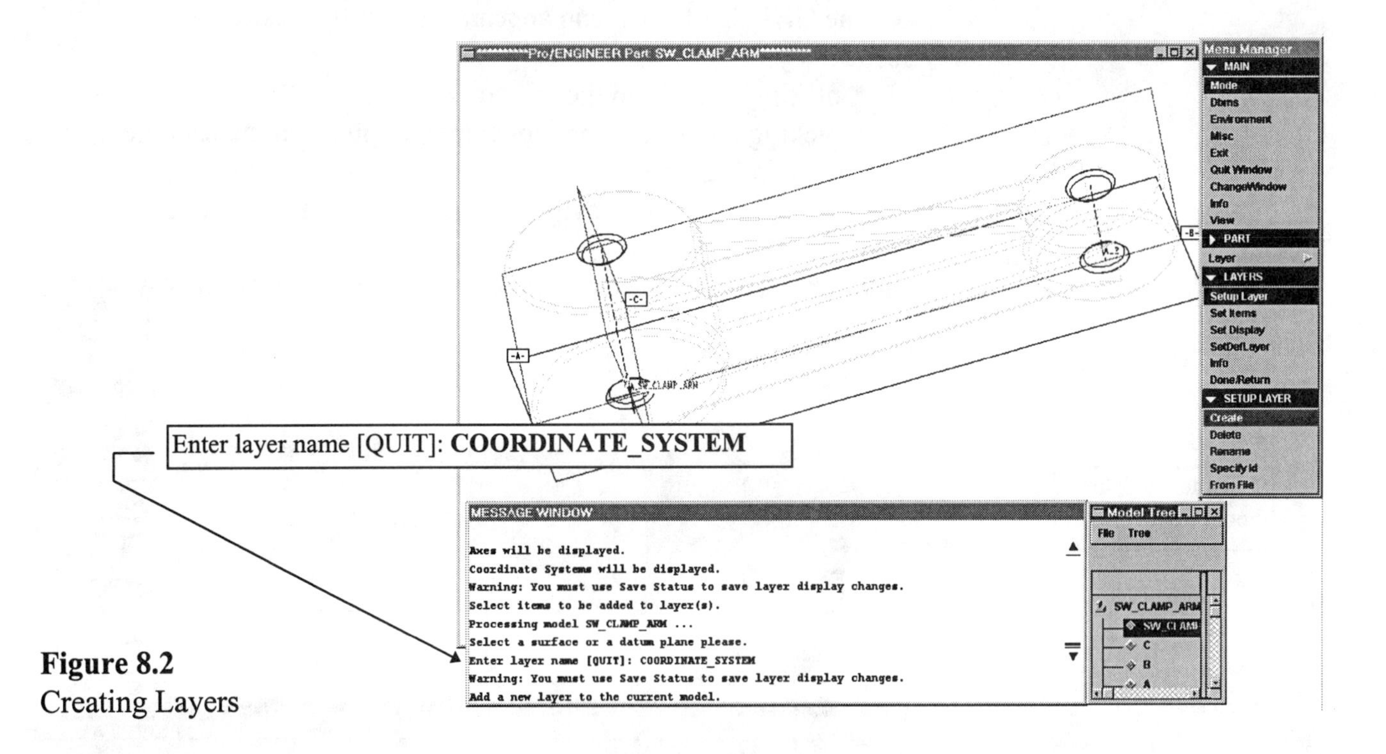

Figure 8.2
Creating Layers

In Figure 8.2, two layers were added. Note that the first attempt to enter a layer name called *coordinate system* was met with a system response of:

Illegal characters in COORDINATE SYSTEM. Reenter:

This message is saying that there can be no spaces in the layer name. Therefore, the next attempt included an underline character: **coordinate_system**. A layer called **datum_plane** was also added. These layers have been created for this part but still do not contain items. The next step is to set the items we want on each layer. We want to put the coordinate system on one layer and the three datum planes on the other.

Adding Items to layers

After you have created a layer, you can associate items to it by taking the following steps:

1. Choose **Layer** from the PART or ASSEMBLY menu (select the active model if you are in Assembly mode).
2. Choose **Set Items** from the LAYERS menu. The SEL NAMELIST menu appears.
3. Check (✓) one or more layers on which you are going to add the same items and then pick **Done**.
4. Select **Add Items** from the SET LAYER menu.
5. The LAYER OBJ menu appears with a list of possible item types.
6. The GET SELECT menu appears. Select the desired objects or features by:
   * picking them from the screen
   * picking them from the model tree by highlighting each item, as shown in Figure 8.3
   * selecting them by navigating through the menu structure

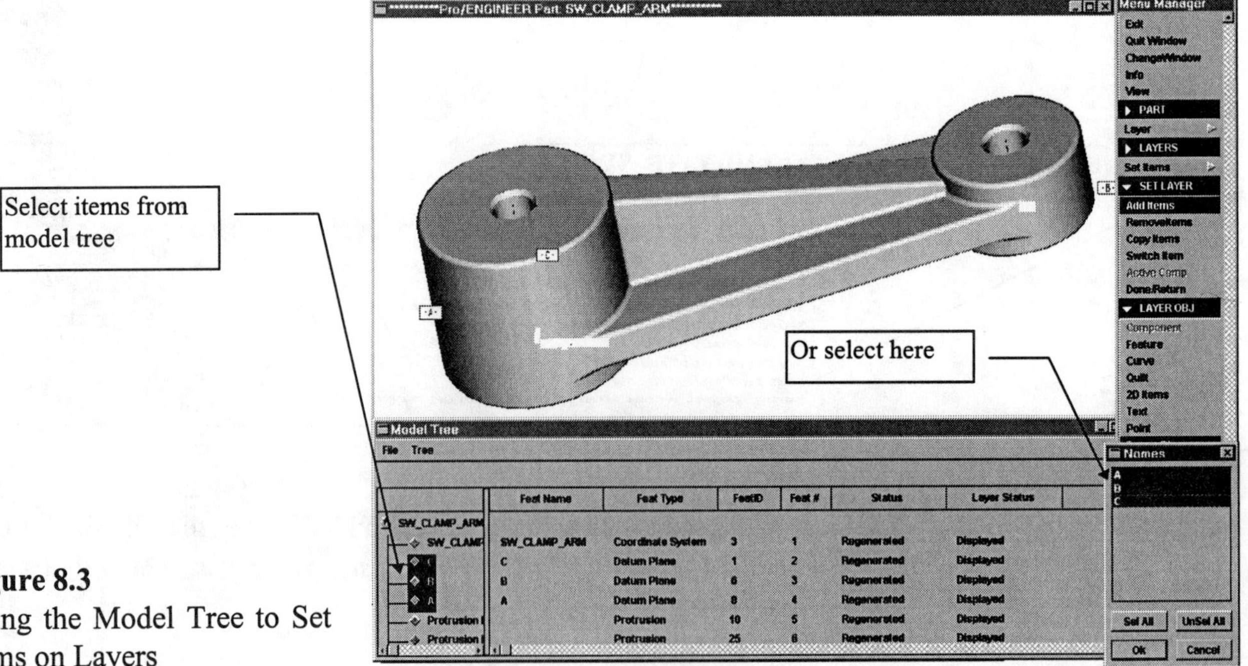

Figure 8.3
Using the Model Tree to Set Items on Layers

After the items have been set, you can verify that the items were placed by picking **Info** ⇒ **Layer** ⇒ **Disp Status** ⇒ **Select All** ⇒ **Done Sel**, as shown in Figure 8.4.

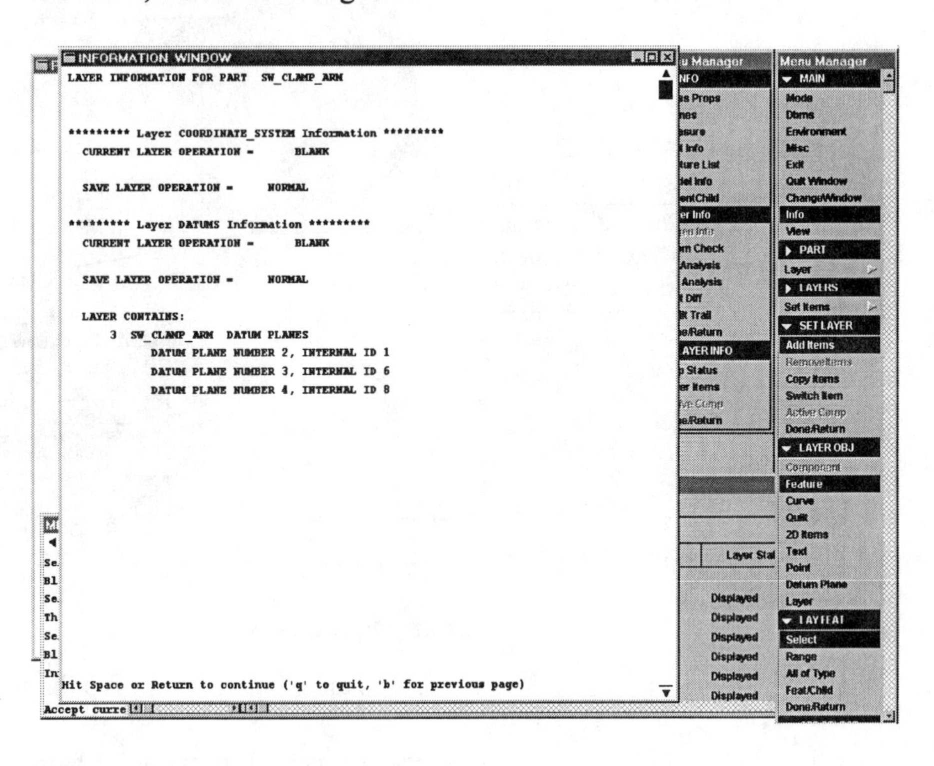

Figure 8.4
Layer Info

Removing Items from Layers

When you remove items from a layer in the active model, you disassociate them from the layer. In Figure 8.5, the **Remove Items** command has been given and the datum plane layer was chosen. The procedure is similar to the one for adding items. Use the following steps:

1. Pick **Layer** from the PART or ASSEMBLY menu
(select the active model if you are in Assembly mode).
2. Choose **Select Items** from the LAYERS MENU. The LAYER SEL NAMELIST menu appears.
3. Check the layers from which you wish to remove items, and then choose **Done**.
4. Select **Remove Items** from the SET LAYERS menu.
5. The DEL ITEM menu appears, with these choices:

   Specify Allows you to select the items you want to remove from the selected layer(s). If you choose this, the LAYER OBJ and GET SELECT menus appear, just as they do when you are adding items.
   Remove all Gives you the ability to remove all the items on the selected layer(s). The system prompts you to confirm your choice by entering [**y**].

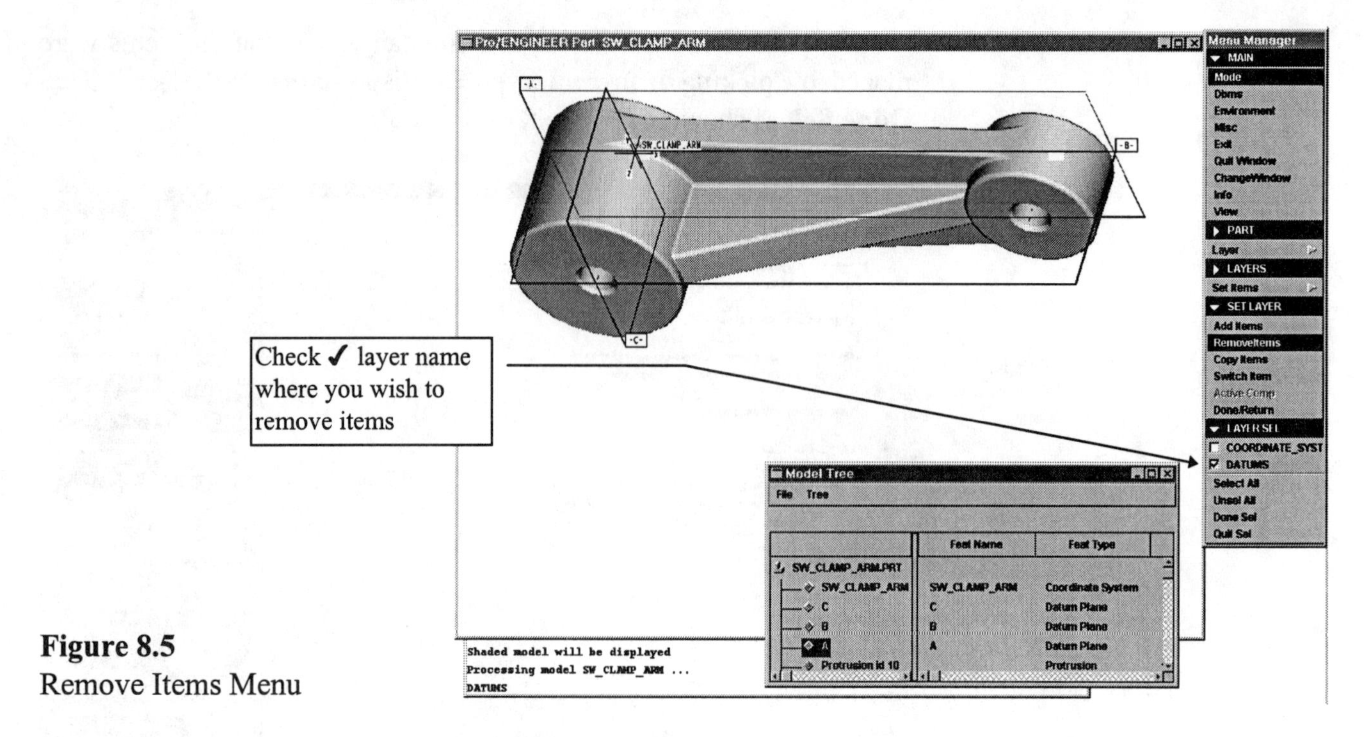

Figure 8.5
Remove Items Menu

Renaming a Layer

To rename an existing layer give the following command sequence:

1. Choose **Layer** from the PART or ASSEMBLY menu
(if you are in Assembly mode, select the active model).
2. Pick **Setup Layer** from the LAYERS menu.
3. Choose **Rename** from the SETUP LAYER menu.
4. Select the name of the layer you want to rename from the LAYER SEL menu.
5. Provide a new name for the layer.

Deleting a Layer

To delete an existing layer give the following command sequence:

1. Choose **Layer** from the PART or ASSEMBLY menu
(select the active model if you are in Assembly mode).
2. Pick **Setup Layer** from the LAYERS menu.
3. Choose **Delete** from the SETUP LAYER menu.
4. Select the names of the layers you want to delete from the LAYER SEL NAMELIST menu. A check mark (✓) to the left of the names indicates the ones you have selected.
5. Choose **Done Sel** and the layers will be deleted.

System Default Layering

Pro/E automatically places certain types of items on specified layers when they are created, using the configuration file option "**def_layer**". Use the following format:

Def_layer type-option layername

where *type-option* determines the item type and *layername* is the name that you assign to the layer. A few valid options are as follows:

Type option	Description
Layer_assem_member	Assembly members
Layer_feature	All features
Layer_axis	Features with axes
Layer_geom_feat	Features with geometry
Layer_nogeom_feat	Features without geometry
Layer_cosm_sketch	Cosmetic sketches
Layer_surface	Surface features
Layer_datum	Datum planes
Layer_datum_point	Datum point features
Layer_slot_feat	Slot
Layer_dgm_highway	Diagram highways
Layer_dgm_rail	Diagram rails
Layer_shell_feat	Shell
Layer_assy_cut_feat	Assembly cut
Layer_chamfer_feat	Chamfer
Layer_corn_chamf_feat	Corner chamfer
Layer_cut_feat	Cut
Layer_draft_feat	Draft
Layer_hole_feat	Hole
Layer_protrusion_feat	Protrusion
Layer_rib_feat	Rib
Layer_round_feat	Round

When you create an entity of one of these types, Pro/E will automatically add it to the specified default layer. If a feature (e.g., a hole) has an axis (other than a datum axis), then that axis can be automatically placed on two default layers, one for features with axes (option "**layer_axis**") and one for the particular type of feature (e.g., Hole - option "**layer_hole_feat**").

As an alternative to editing the configuration file, you can use the **SetDefLayer** option to edit the default layer table (Fig. 8.6). The layering options defined or changed here will *not* reflect back to the configuration file and are good only for *new* features created in the current session in *all* models. It is a good practice to set up the default layering *before* you start work on a part, drawing, or assembly.

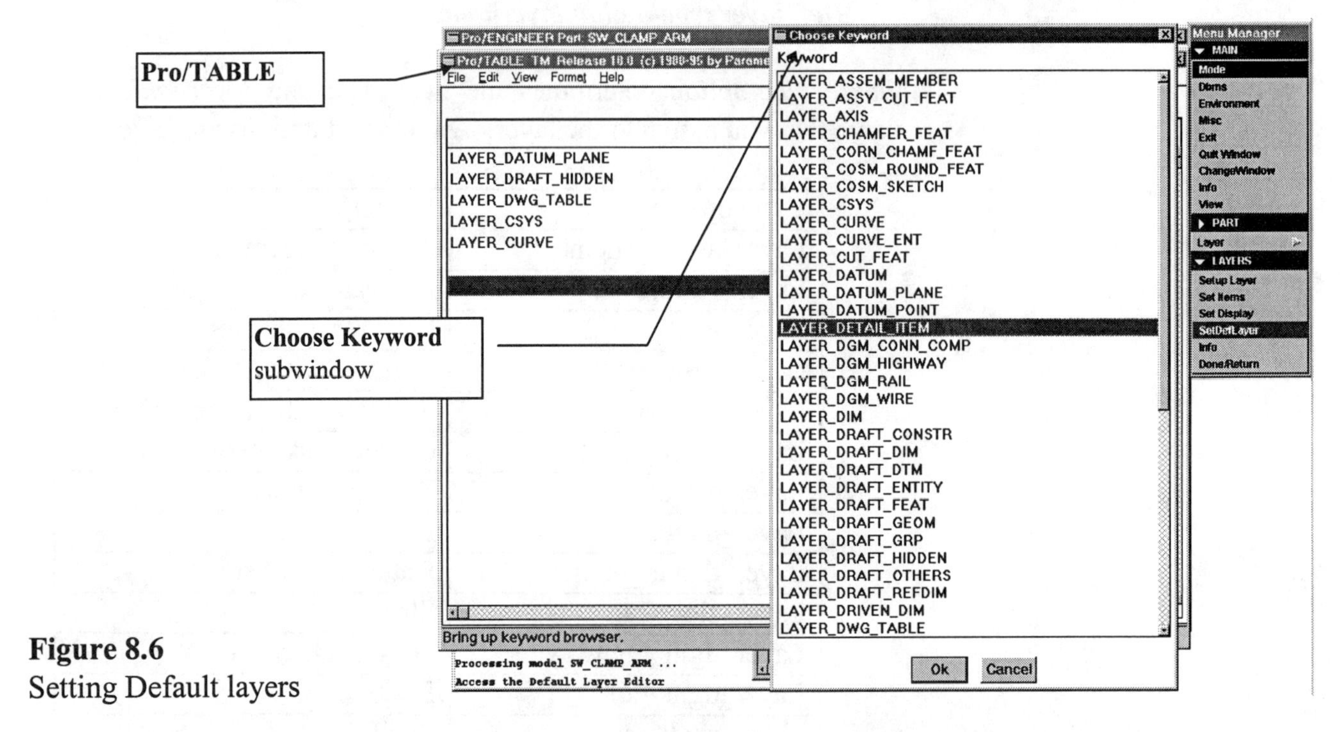

Figure 8.6
Setting Default layers

The left column lists layer items. The right column lists the layers to which the layer items are assigned.

When you are adding layer items using Pro/E, do the following:

1. Select an empty cell in the left-hand column.
2. Press the **F4** key.
3. Pick on an item (here the **LAYER_DETAIL_ITEM** keyword is highlighted) and the editor will copy it into the active cell (Fig. 8.6).

The system supplies a number of default layer names:

Layer_dim	Def_dims
Layer_corn_chamf_feat	Def_chamfers
Layer_protrusion_feat	Def_protrusions
Layer_axis	Def_axis
Layer_assem_member	Def_components
Layer_assy_cut_feat	Def_features
Layer_chamfer_feat	Def_chamfers

You can enter one of them into the right column by:

1. Selecting an empty cell in the left column.
2. Picking the **Choose Keywords** option in the EDIT menu. This will bring up the CHOOSE KEYWORD NAMELIST menu of default layer names.
3. Picking on a layer name. The editor will copy it into the active cell. Alternatively, you can type a layer name from the keyboard, as was done in the example. When finished, choose Exit.

In Figure 8.7 the **Info** command has been given, and the screen shows that a set of datums and a coordinate system have been created in the Part mode. The coordinate system has been *automatically* placed on a default layer.

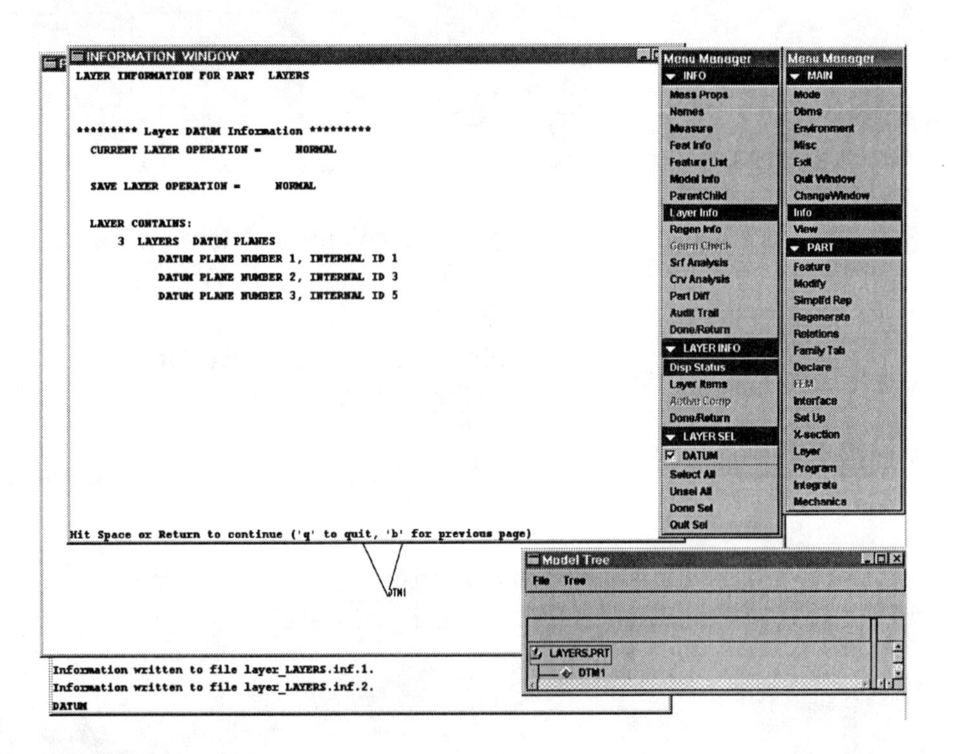

Figure 8.7
Layer Info

Displaying Layers

In the MAIN menu, we have a choice to turn on or off a variety of items, including datum planes, coordinate systems, and datum axes. Another way to control the display of the items on the screen is to put groups of items on layers and display or blank them as needed. Later you will learn that the datums can be set as default geometric tolerance features. Datum planes set in this way *cannot be turned off using environment settings*; therefore, when the object is displayed in Part, Assembly, or Drawing mode, the only way to control the datum display is to blank its layer. This will become obvious when you see assemblies with 5 to 2000 components, all with set datums displayed at once! Blanking layers becomes essential when this happens.

In Figure 8.8, the datums have been blanked.

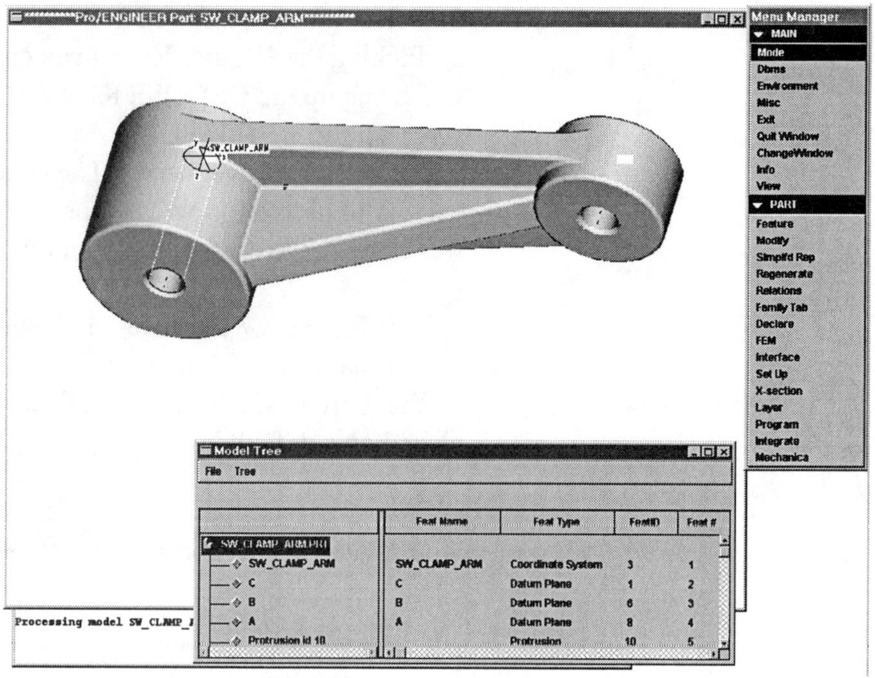

Figure 8.8
Blanked Layer

The command **View** ⇒ **Layer Disp** ⇒ **Blank** (the default) ⇒ **✓datum_plane** (checked) ⇒ **Done Sel** was given. The results show that the part is now displayed without datum planes. Note, if you pick **Environment**, the **Disp DtmPln** selection is still checked, as shown in Figure 8.9.

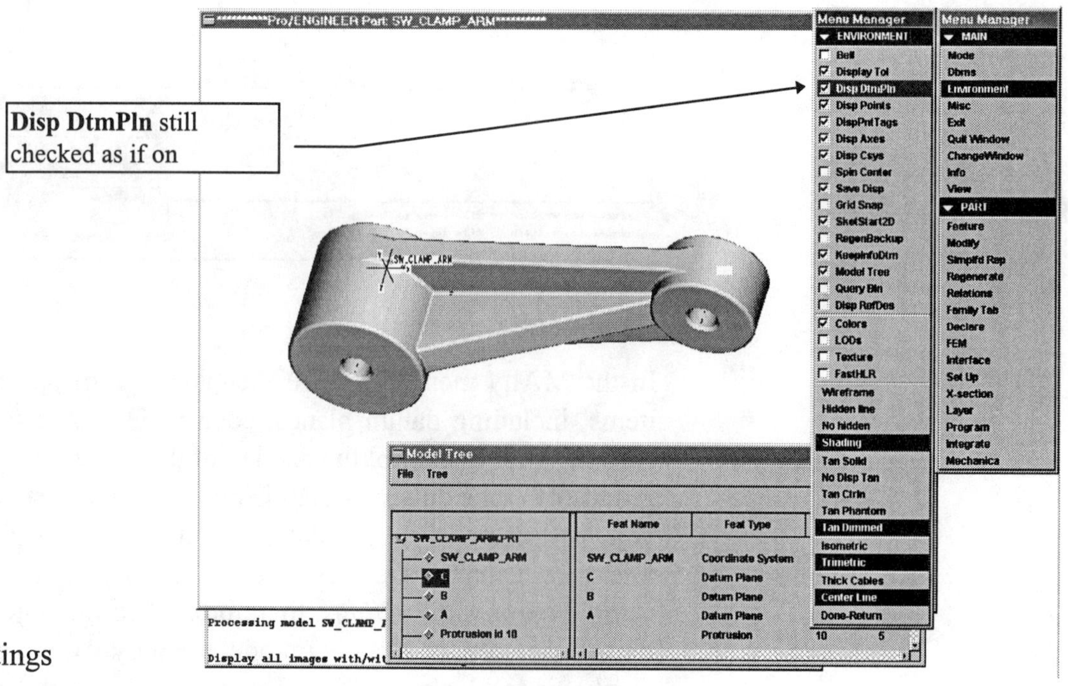

Figure 8.9
Environment Settings

Section 9

The Sketcher

In Pro/ENGINEER, *sketching* is done in the **Sketcher**. Almost all traditional lines, circles, arcs, and their variations can be accomplished on the screen without creating exact and perfectly constructed geometry. The system will assume a variety of conditions, such as tangency, similar sizes for same-type geometry, parallelism, perpendicularity, verticals, horizontals, coincident endpoints, tangent points, and symmetry.

The Sketcher can be entered by picking **Mode ⇒ Sketcher**, or you are automatically put in the Sketcher when creating most geometry in the **Part mode**. If you go directly into the Sketcher, you will be creating sections that can be recalled later during a part feature creation. The sections are like sheets of graph paper with sketched geometry on them. They can be saved for later use in any feature creation where a section is called for. The section in Figure 9.1 is an example of a section created in the Sketcher mode (**Pro/ENGINEER Section: S2D0001** is shown in the window bar). Note that you must give it a unique name or use the default, **S2D0001**.

A section created in Part mode or Assembly mode can be saved for later use as well as being used to create the geometry needed for the present feature.

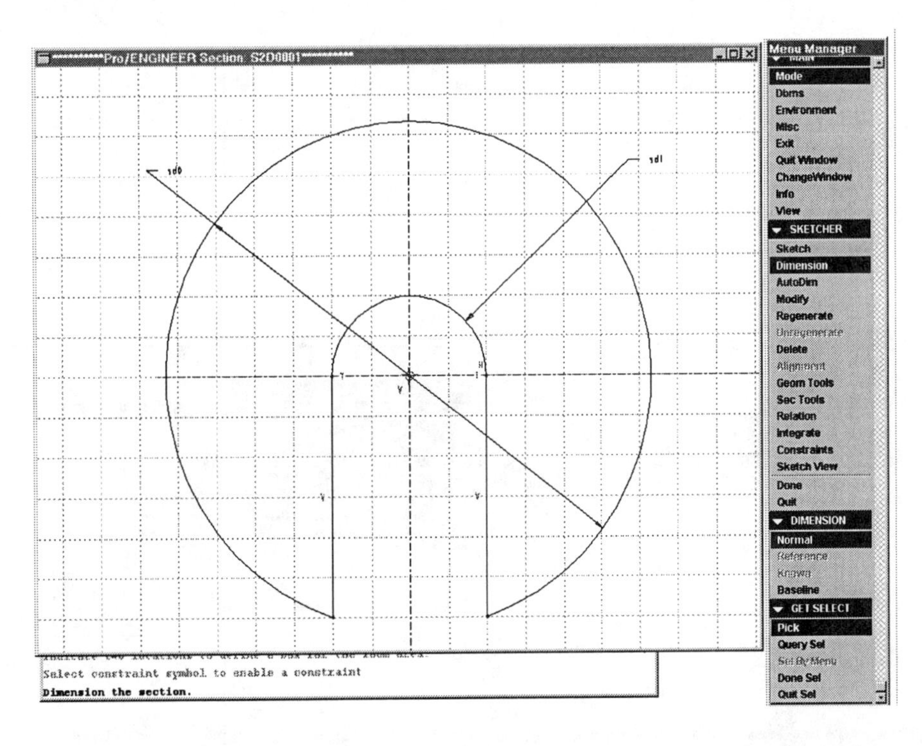

Figure 9.1
Sketcher Mode

Sketcher techniques are used in many areas of parametric design. The premise of the Sketcher, like that of parametric design, is to enable the quick and simple creation of geometry for your model. The Sketcher requires you to create and dimension this geometry, but during the sketching process you do not have to be concerned with dimensional sizes or the creation of perfect and accurate geometry.

Creating sections within the SKETCHER menu is not difficult. There are only a few steps to remember:

1. **Sketch** Sketch the section geometry. Use SKETCHER tools to create the section geometry (Fig. 9.2).
2. **Dimension** Dimension the section. Use a dimensioning scheme that you want to see in a drawing. Dimension to control the characteristics of the section geometry (Fig. 9.3).
3. **Alignment** Align the section geometry to a datum feature or to a part feature (Fig. 9.3).
4. **Regenerate** Regenerate the section. Regeneration solves the section sketch based on your dimensioning scheme (Fig. 9.4).
5. **Relations** Add section relations. Add relations to control the parametric behavior of your section.

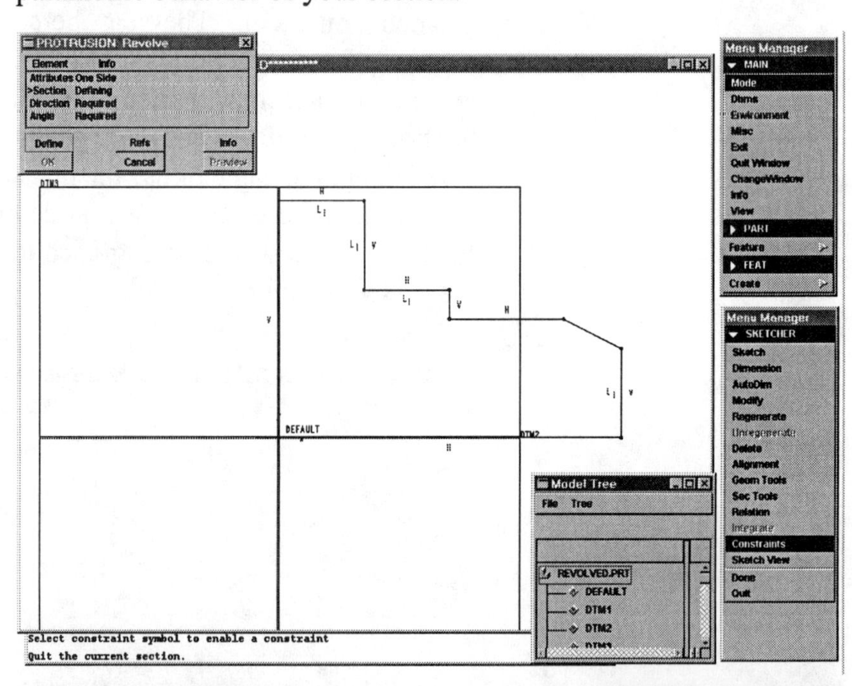

Figure 9.2
Sketch the Features

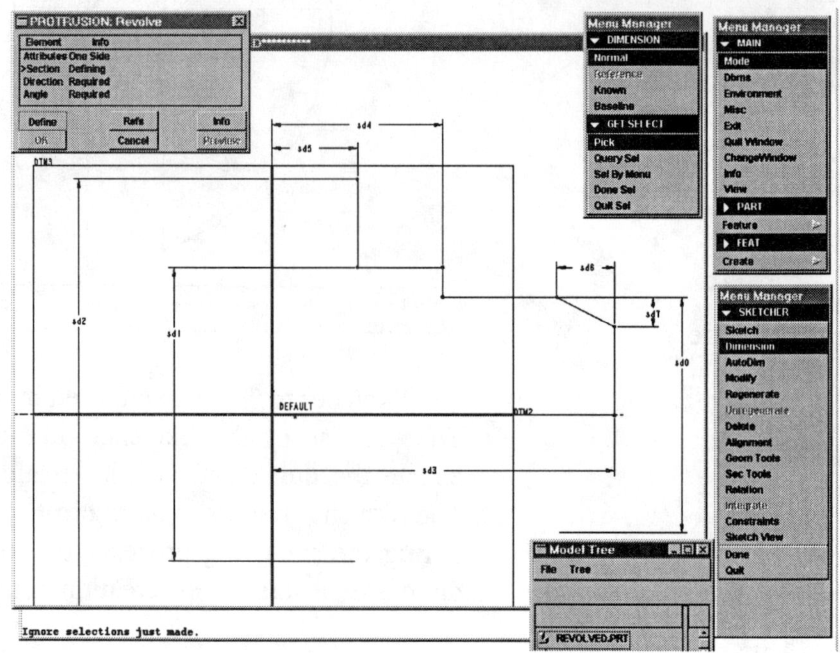

Figure 9.3
Dimension and
Align the Sketch

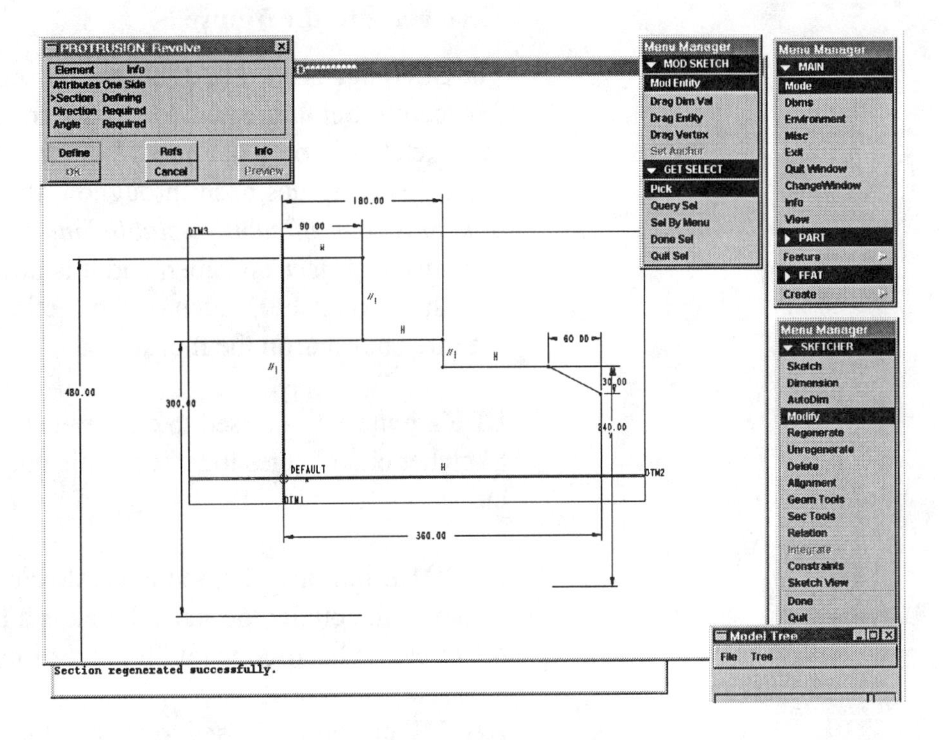

Figure 9.4
Regenerate the Section

After Pro/E regenerates the sketch, the feature can be completed, as shown in Figure 9.5. More features can then be added, using *sketched* features or *pick-and-place* features such as holes and chamfers.

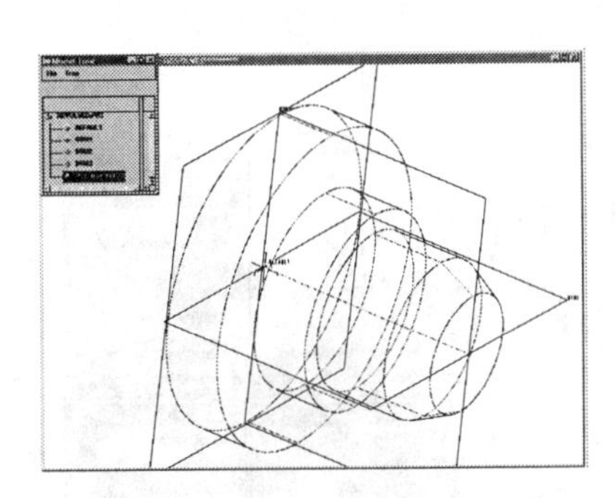

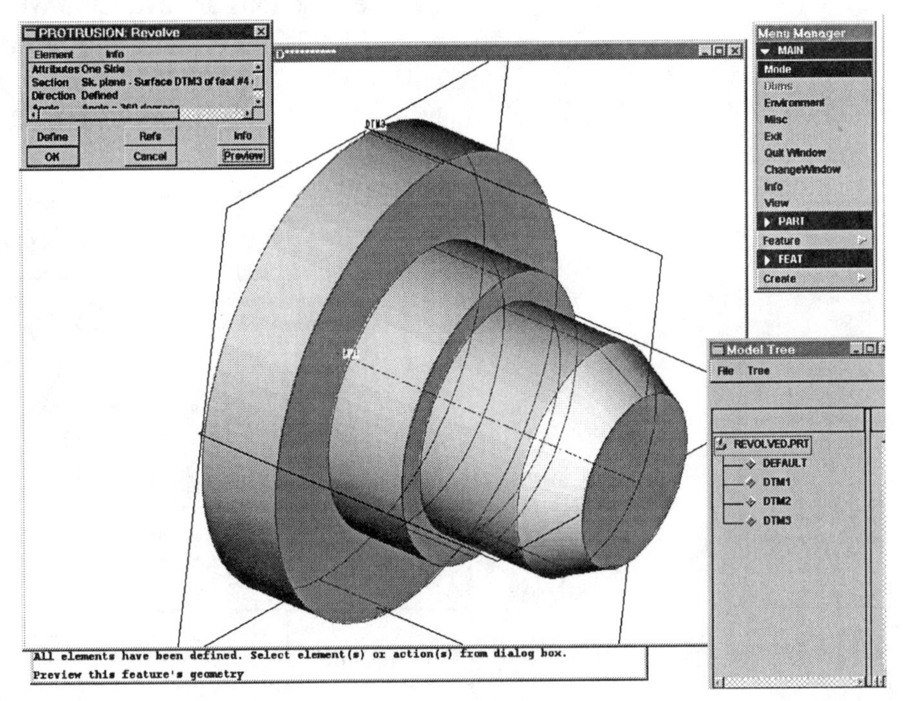

Figure 9.5
Revolved Feature Created from the Sketch

Sketcher and the Mouse

The **Sketcher** is used to establish 2D sections that are the basis for the 3D feature being created. In order to understand just how powerful the Sketcher is on a parametric design system, you need only look at what sketching has been throughout the ages: *Sketching is a process of simply and efficiently establishing the basic design and intent of a designer-engineer* on paper, and it is now possible with the **mouse.**

In the Sketcher, much of the section geometry can be created via the three buttons on the mouse.

LEFT button Used to create lines. The line command in the Sketcher chains lines together. This button also aborts the creation of circles.

MIDDLE button Used to create circles. The first pick is the center of the circle; the second pick is a location on the diameter of the circle. Also used to end line creation.

RIGHT button Used to create tangent arcs. *There must be an existing line, arc, or spline to reference for tangency.* Place cursor near the tangent entity, and press the right button to start the arc. Press the right button again to set the endpoint of the arc.

These functions are available when the **Sketch** button is selected from the SKETCHER menu and **Mouse Sketch** (Fig 9.6) is left as the default in the GEOMETRY menu. Also, there are other geometry functions found under the GEOMETRY menu.

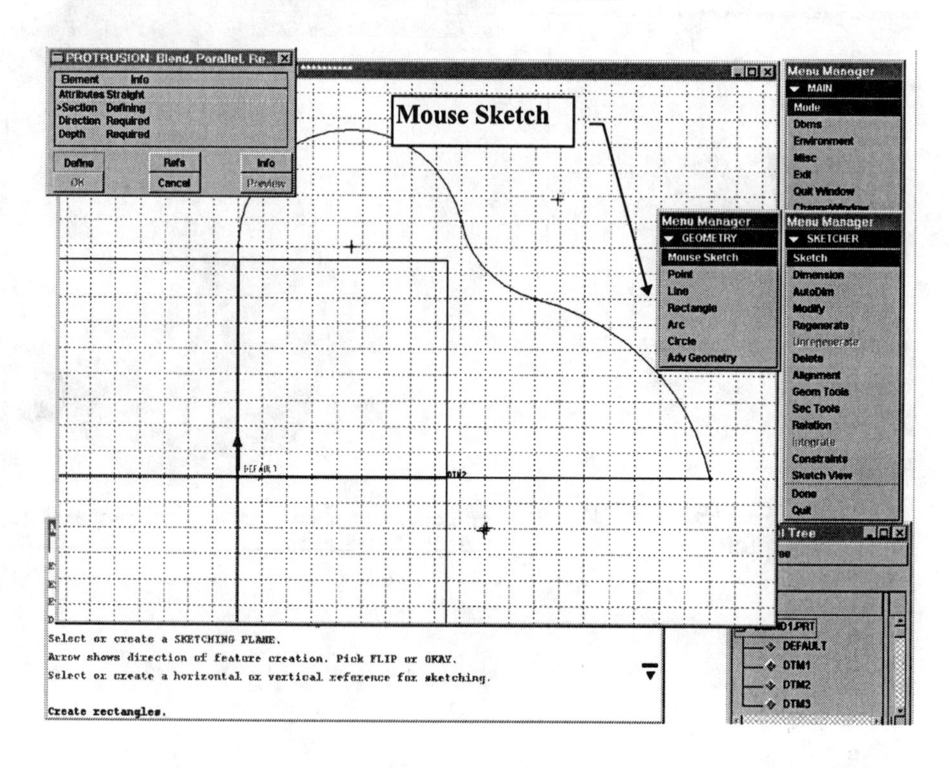

Figure 9.6
Mouse Sketch

Regenerating a Section Sketch

In the past, designers have sketched on paper, showing lines, arcs, circles, and other geometric forms in rough, simplified outline and internal forms. The sketched shapes are assumed to be what they *sort of* look like. Round shapes approximating a circle are assumed by the reader of the sketch to be circles, curved shapes are assumed to be arcs, and lines drawn straight up or down are assumed to be vertical. Lines drawn left to right are assumed to be horizontal. Lines sketched at an angle are straight lines that are angled. Dimensions roughly sketched on a less-than-perfect drawing of a part are assumed to represent the exact perfect shape desired by the person sketching. All this seems obvious to most people involved in engineering design. With the introduction of parametric design, we can now sketch on the screen and allow Pro/E to make all the assumptions that were traditionally made by a person creating a sketch or reading a sketch. These assumptions include, but are not limited to, the following: *symmetry*, *tangency*, *parallelism*, *perpendicularly*, *equal angles*, *same-size arcs* and *circles*, and *coincident centers*.

The sketch started in Figure 9.7 was completed without the **Grid Snap** activated. Note that the lines and arcs are not sketched perfectly. After regeneration, Pro/E will straighten the lines and align the features according to a set of assumptions or rules. Almost-vertical lines will be vertical, close-to-horizontal lines will become horizontal, etc. We suggest that you keep the grid snap "on" (**✓Grid Snap**) most of the time while sketching the first features of a part, experienced users normally keep it off and trust the Pro/E assumptions to clean up any sketching inconsistencies.

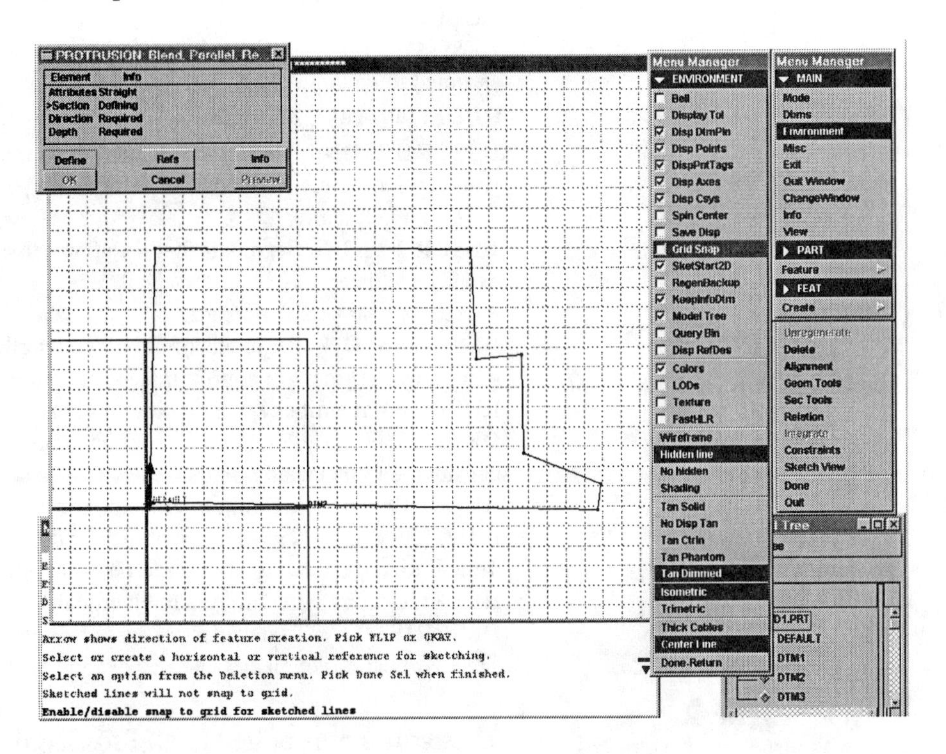

Figure 9.7
Sketching Without Grid Snap

During regeneration, Pro/E checks to make sure that it understands your dimensioning scheme and that you have created a complete and independent set of parameters. Pro/E analyzes your section based on the geometry that you have sketched and the dimensions you have created. In the absence of explicit dimensions, implicit information based on the sketch is used. You can quickly understand and control the assumptions made in solving the sketch. This streamlines the process of sketch creation and behavior diagnosis.

Modifications made in sketches are *animated* over a brief time. If a sketch fails, it changes shape up to the point of failure, allowing you to view the section at the point of failure with the option to restore the dimensions to the old values. This provides an understanding of how the sketch fails by showing the point of failure, and by displaying through animation how that point was achieved. Corrective actions can then be taken.

Here is a list of implicit information that Pro/ENGINEER uses to regenerate a section:

RULE: Equal radius/diameter
DESCRIPTION: If two or more arcs or circles are sketched with approximately the same radius, they are assigned the same radius value.
RULE: Symmetry
DESCRIPTION: Entities sketched symmetrically about a centerline are assigned equal values with respect to the centerline.
RULE: Horizontal and vertical lines
DESCRIPTION: Lines that are approximately horizontal or vertical are considered to be exactly horizontal or vertical.
RULE: Parallel and perpendicular lines
DESCRIPTION: Lines that are sketched approximately parallel or perpendicular are considered to be exactly horizontal or vertical.
RULE: Tangency
DESCRIPTION: Entities sketched approximately tangent to arcs or circles are assumed to be tangent.
RULE: 90°, 180°, 270° arcs
DESCRIPTION: Arcs are considered to be multiples of **90°** if they are sketched with approximately horizontal or vertical tangents at the endpoints.
RULE: Collinearity
DESCRIPTION: Segments that are approximately collinear are considered to be exactly collinear.
RULE: Equal segment lengths
DESCRIPTION: Segments of unknown length are assigned a length equal to that of a known segment of approximately the same length.
RULE: Point entities lying on other entities
DESCRIPTION: Point entities that lie approximately on lines, arcs, or circles are considered to be exactly on them.
RULE: Centers lying on the same horizontal
DESCRIPTION: Two centers of arcs or circles that lie approximately along the same horizontal direction are set to be exactly horizontally aligned.
RULE: Centers lying on the same vertical
DESCRIPTION: Two centers of arcs or circles that lie approximately along the same vertical direction are set to be exactly vertically aligned.

These rules are applied to all Pro/ENGINEER sketches.

Sketcher Mode and Constraints

Constraints used in solving a sketch display graphically on the sketch with the aid of *small symbols* that appear next to the entities to which they apply. You can turn off the display of these symbols, if desired. You can also click on the symbols to disable or enable the constraints and to obtain a brief explanation. Also, endpoints of section entities are highlighted with the aid of small dot symbols. This graphical display of symbols for constraints replaces the old user interface of a constraints dialog box.

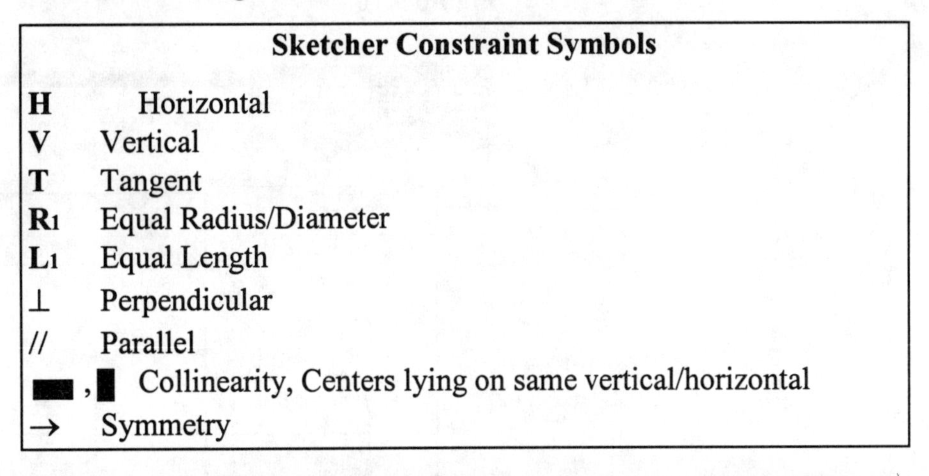

Sketcher Constraint Symbols

Symbol	Meaning
H	Horizontal
V	Vertical
T	Tangent
R1	Equal Radius/Diameter
L1	Equal Length
⊥	Perpendicular
//	Parallel
▬ , ▮	Collinearity, Centers lying on same vertical/horizontal
→	Symmetry

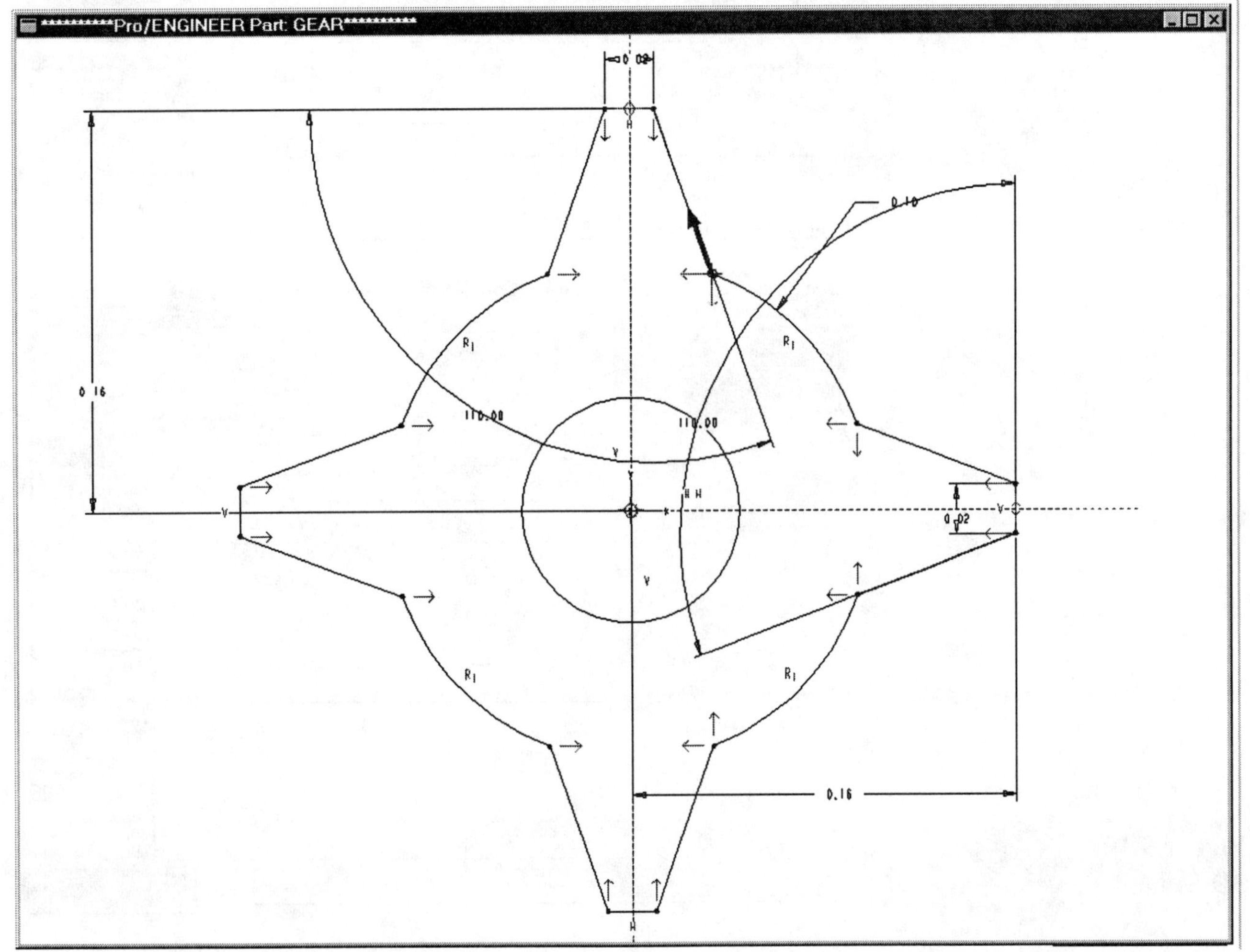

Figure 9.8
Sketch with Constraints Displayed

Sketching Lines

You can create two types of **lines** with Pro/E: geometry lines and centerlines. **Geometry lines** are used to create *feature geometry*. **Centerlines** are used to define the *axis of revolution* of a revolved feature, to define a line of symmetry within a section, or to create construction lines. In Figure 9.9, the section contains lines, centerlines, and arcs shown in 2D **Sketch View**. You can also reorient the sketch into a 3D **Default** view orientation, as shown in Figure 9.10.

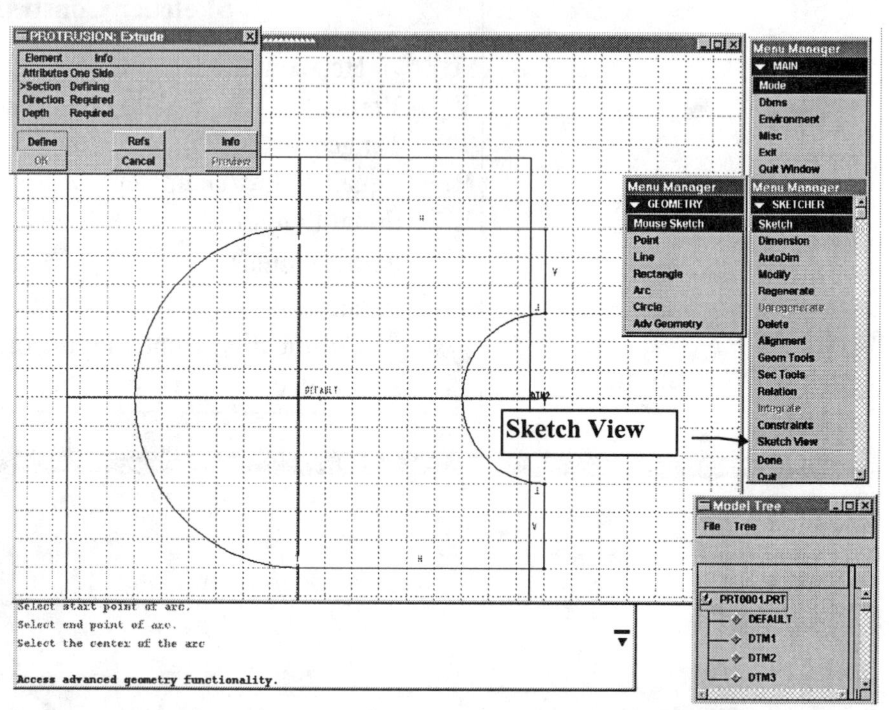

Figure 9.9
2D Sketch View

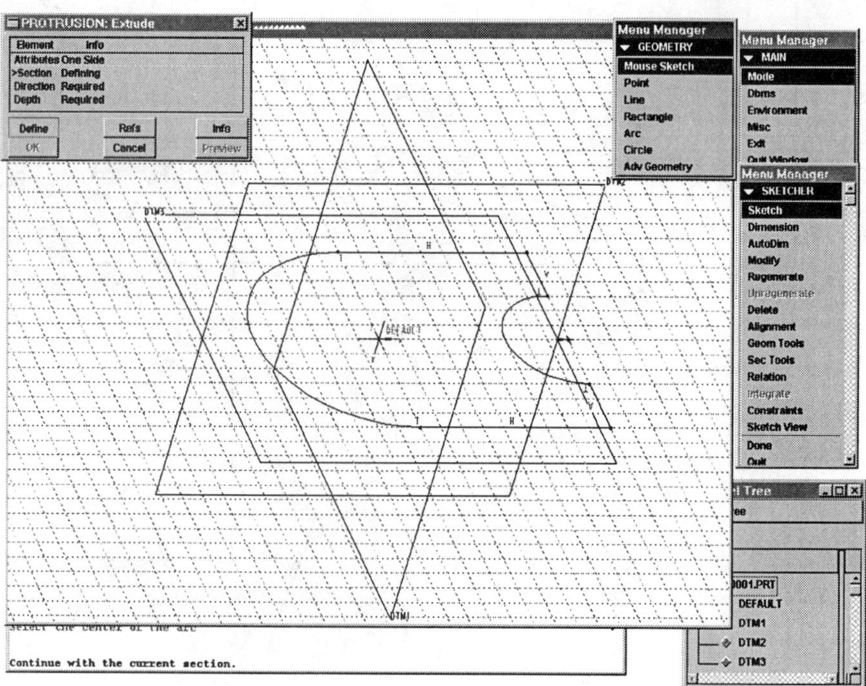

Figure 9.10
3D Default View

To sketch **geometry lines** or **centerlines** do the following:

1. Choose **Line** from the GEOMETRY menu. The LINE TYPE menu appears.
2. Choose **Geometry** or **Centerline** from the top portion of the menu to indicate the type of line that you want.
3. Choose an option from the bottom portion of the menu to indicate how to create the line:

2 Points Create a line by picking the start point and endpoint. Geometry lines created using this command will automatically be chained together.
Parallel Pick an existing line to determine the new line's direction, and then pick the start point and endpoint. For a centerline, only a single pick is needed to determine the parallel placement of the line, and the ends of the *centerline* will be chosen to fit model or section outlines.
Perpendicular Pick an existing line to determine the new line's direction, and then pick the start point and endpoint. For a centerline, only a single pick is needed to determine the perpendicular placement of the line, and the ends of the *centerline* will be chosen to fit model or section outlines.
Tangent Pick an endpoint of an arc or spline to start the new line and determine its direction, then pick the endpoint of the line. For a centerline, only a single pick is needed to determine the tangent placement of the line, and the ends of the *centerline* will be chosen to fit model or section outlines.
2 Tangent Pick two arcs, splines, or circles to determine the direction of the new line. The line is automatically created between the selected entities. A 2 Tangent line created to construction entities will not split the entity. A 2 Tangent *centerline*, created as a 2 Tangent line defined using two circles, will not split the circles.
Pnt/Tangent Pick a point anywhere in the current section, and then pick an arc, spline, or circle to which the line must be tangent.
Horizontal Create a line that is horizontal relative to the orientation of the section. For a geometry line, the endpoint is automatically the starting point of a chained vertical line. For a *centerline*, only a single pick is needed to determine the horizontal location of the line.
Vertical Creates a line that is vertical relative to the orientation of the section. For a geometry line, the endpoint is automatically the starting point of a chained horizontal line. For a *centerline*, only a single pick is needed to determine the vertical location of the line.

Circles

A variety of **geometry circles** and **construction circles** can be created in the Sketcher (Fig. 9.11). Geometry circles are used to create feature geometry, whereas construction circles serve as guides and references but do not create feature geometry. Construction circles are displayed in the same color as circles but with a *phantom line font* rather than the solid font that is used for circles and other geometry.

To sketch geometry and construction circles, give the following command picks:

1. Select **Circle** from the GEOMETRY MENU.
2. Pick **Geometry** or **Construction** from the top part of the menu.
3. Choose one of the following options from the bottom part of the menu:

Crt/Point Basically the same as creating a circle using **Mouse Sketch** and the middle button of the mouse, except here the left button is used for the picks.
Concentric Pick an existing circle or arc, and then pick on the radius of the new circle.
3 Tangent Create a circle between three existing reference entities. These can be centerlines, construction features, or geometry features.
Fillet Create a circle tangent to two existing entities.
3 Point Pick three points you wish to define the circle's circumference

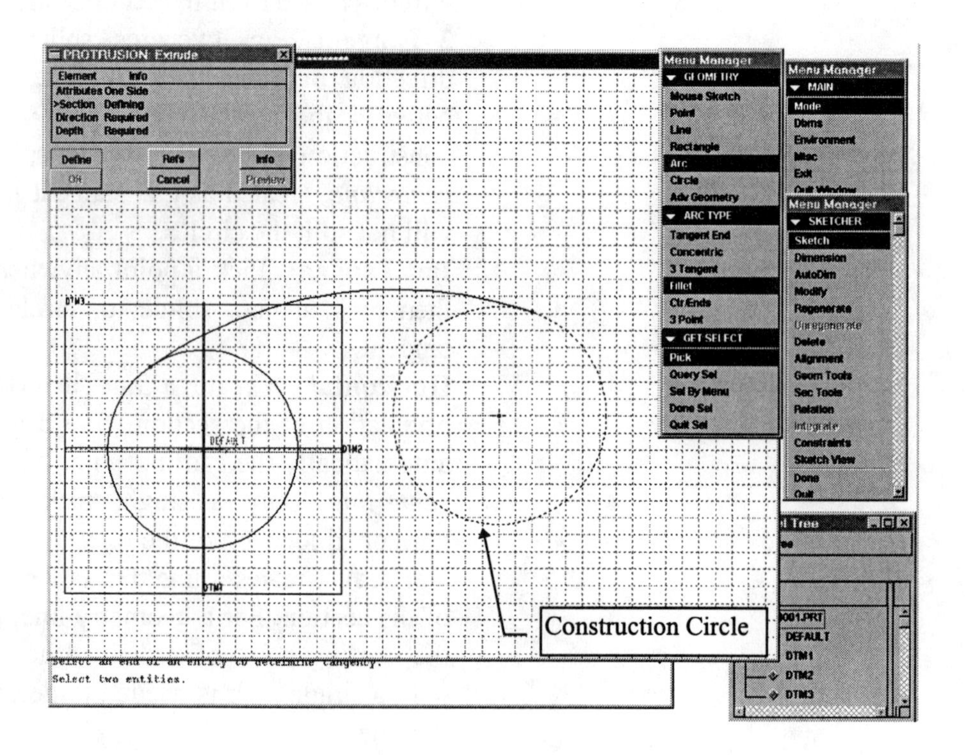

Figure 9.11
Geometry Circles, Arcs, and Construction Circles

Arcs

Arcs are sketched using the menu or the mouse. To sketch arcs:

1. Select **Arc** from the GEOMETRY menu. The ARC TYPE menu appears.
2. Pick one of the following options from the ARC TYPE menu:

   Tangent End This is the same as creating an arc using **Mouse Sketch**, except you must use the left mouse button. Pick an end of an entity to determine tangency, and then pick the endpoint of the arc.
   Concentric Pick an existing circle or arc as a reference, and then pick the endpoints of the new arc. As you create the arc, a radial line will appear through its center to assist you in aligning the endpoint.
   3 Tangent Select three entities for the new arc to be tangent to, and then create the arc in the same direction as the reference picks.
   Fillet Pick two entities to create a tangent arc between (Figure 9.12).
   Ctr/Ends Pick the center point of the arc, and then pick the arc's endpoints.
   3 Points Pick the endpoints of the arc, and then pick a point on the arc.

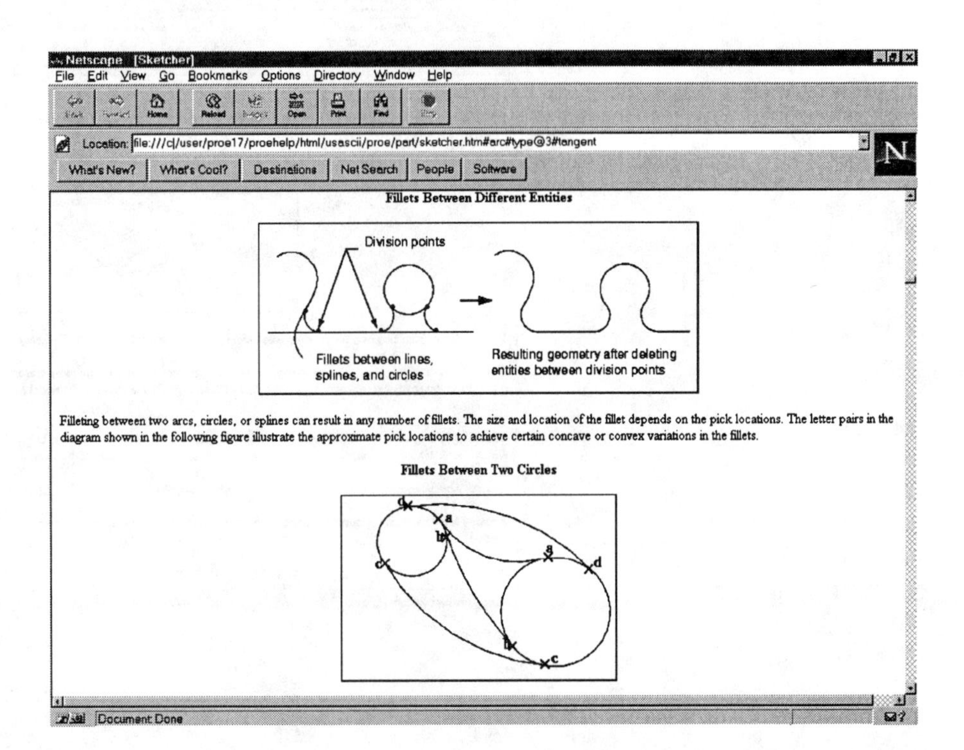

Figure 9.12
Arc Fillets and Fillets Between Circles

Advanced types of geometry, such as a conics (Fig. 9.13) and splines (Fig. 9.14), can also be created in the Sketcher.

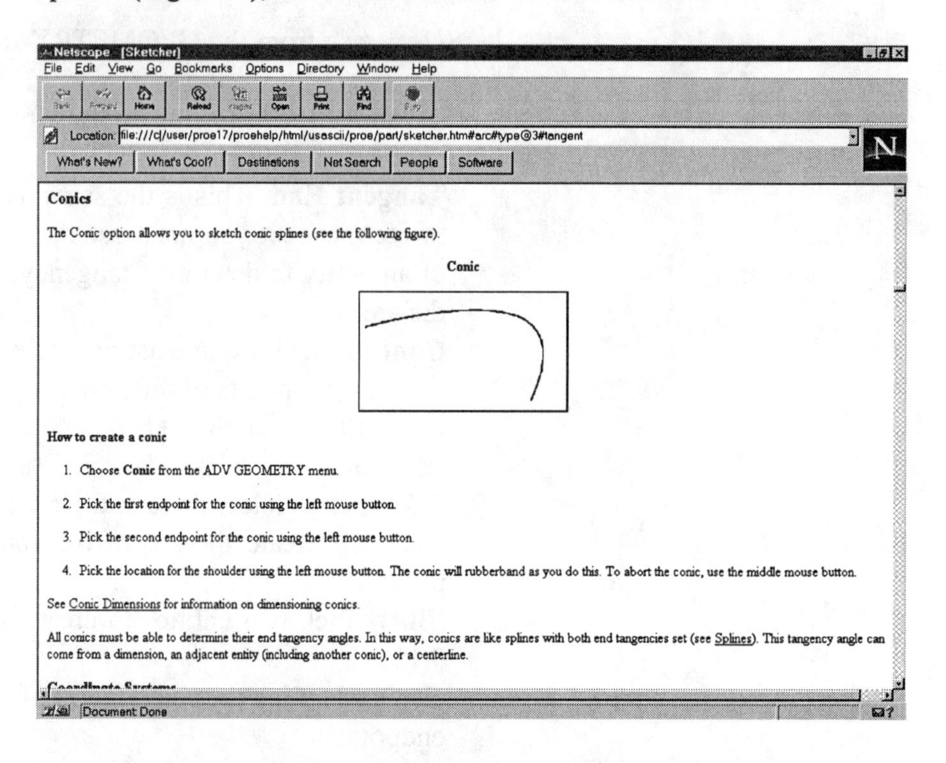

Conics

The Conic option allows you to sketch conic splines (see the following figure).

Conic

How to create a conic

1. Choose **Conic** from the ADV GEOMETRY menu.
2. Pick the first endpoint for the conic using the left mouse button.
3. Pick the second endpoint for the conic using the left mouse button.
4. Pick the location for the shoulder using the left mouse button. The conic will rubberband as you do this. To abort the conic, use the middle mouse button.

See Conic Dimensions for information on dimensioning conics.

All conics must be able to determine their end tangency angles. In this way, conics are like splines with both end tangencies set (see Splines). This tangency angle can come from a dimension, an adjacent entity (including another conic), or a centerline.

Figure 9.13
Sketching Conics

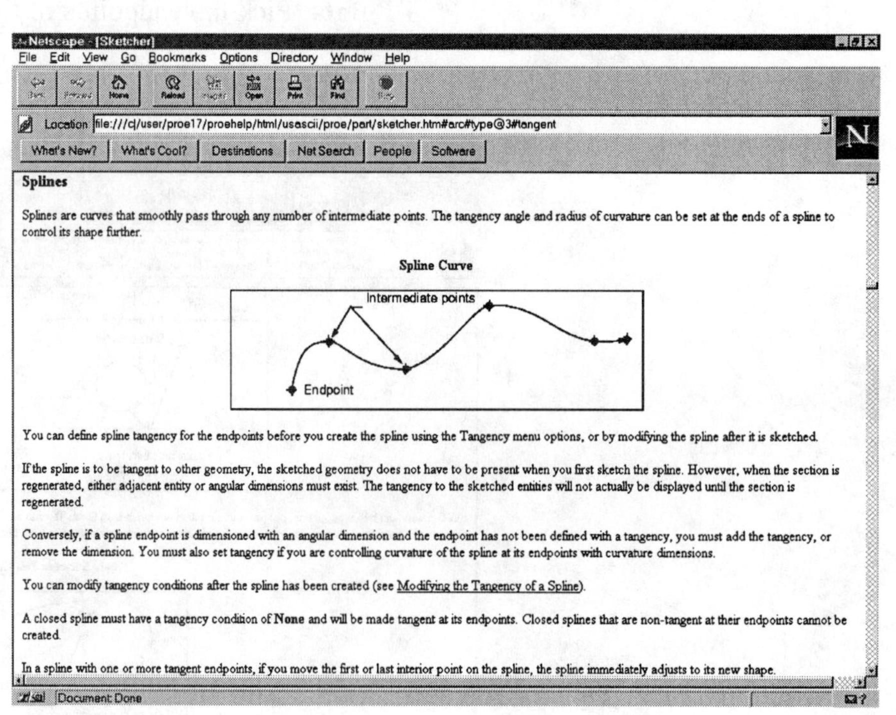

Splines

Splines are curves that smoothly pass through any number of intermediate points. The tangency angle and radius of curvature can be set at the ends of a spline to control its shape further.

Spline Curve

You can define spline tangency for the endpoints before you create the spline using the Tangency menu options, or by modifying the spline after it is sketched.

If the spline is to be tangent to other geometry, the sketched geometry does not have to be present when you first sketch the spline. However, when the section is regenerated, either adjacent entity or angular dimensions must exist. The tangency to the sketched entities will not actually be displayed until the section is regenerated.

Conversely, if a spline endpoint is dimensioned with an angular dimension and the endpoint has not been defined with a tangency, you must add the tangency, or remove the dimension. You must also set tangency if you are controlling curvature of the spline at its endpoints with curvature dimensions.

You can modify tangency conditions after the spline has been created (see Modifying the Tangency of a Spline).

A closed spline must have a tangency condition of **None** and will be made tangent at its endpoints. Closed splines that are non-tangent at their endpoints cannot be created.

In a spline with one or more tangent endpoints, if you move the first or last interior point on the spline, the spline immediately adjusts to its new shape.

Figure 9.14
Sketching Splines

DIMENSIONING SECTIONS

HINT

You can specify critical dimensions to identify key design intent, and the rest of the dimensioning is completed automatically.

In order to regenerate a sketch successfully, it must be properly dimensioned. The Sketcher provides the ability to dimension a sketch with just a push of a button. If any references are required to locate a section, you are prompted to select the desired references and then complete the dimensioning scheme There are two steps in dimensioning an entity: Pick the entity or entities with the ***left mouse button***, and then place the dimension at the desired location using the ***middle mouse button***.

Linear Dimensions

Linear dimensions indicate the length of a line segment or the distance between two entities. Only horizontal and vertical dimensions are allowed when creating a dimension between two arc or circle extents (tangency points). The dimension is created to the tangency point closest to the pick point. The dimension value (Fig. 9.15) is displayed as a symbol until the sketch is successfully regenerated.

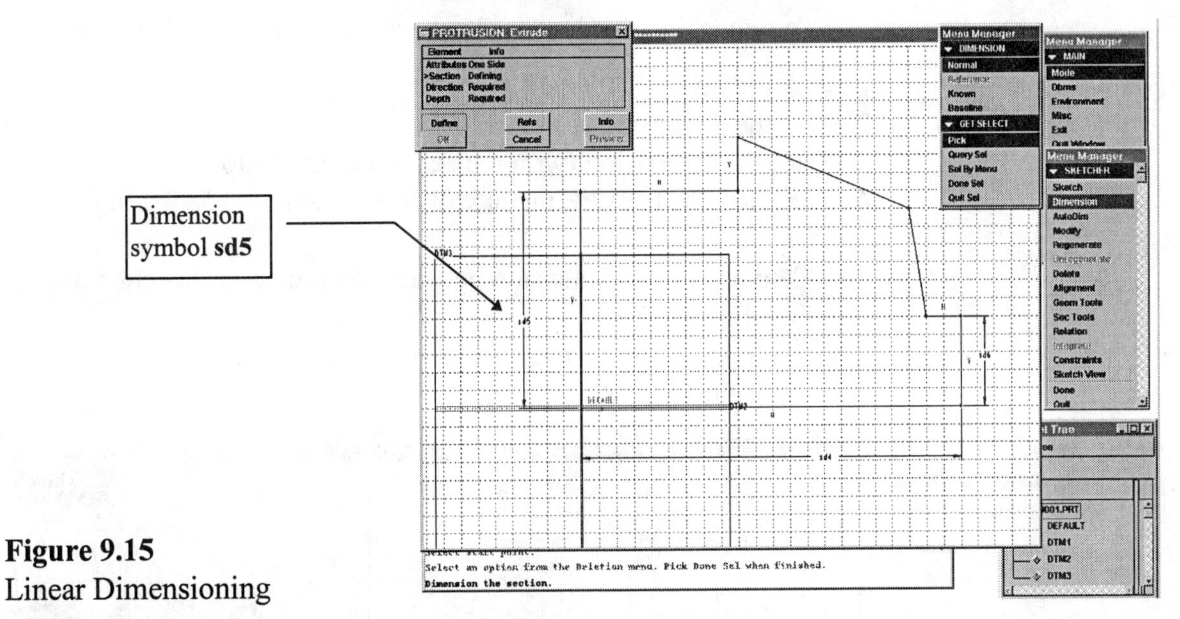

Figure 9.15
Linear Dimensioning

Linear dimensions (Fig. 9.15) let you do the following:

* Dimension the explicit length of a line: pick the line and then place the dimension.
* Dimension the distance between two parallel lines; pick the two lines and then place the dimension.
* Dimension the distance between a point and a line: pick the line, pick the point, and place the dimension.
* Create a dimension between two points (centerpoints and coordinate systems are included, but vertices are excluded): pick the points and location for the dimension. The DIM PNT menu appears:

Horizontal Horizontal distance between the points
Vertical Vertical distance between the points

Slanted Shortest distance between the points or, as in Figure 9.16, the shortest distance between a point and a circle

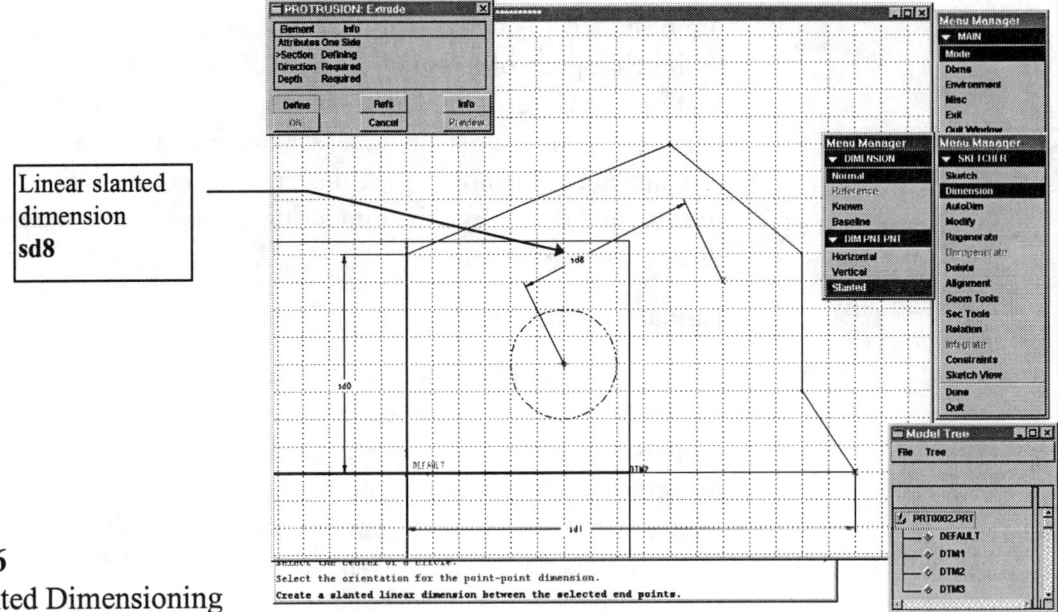

Figure 9.16
Linear Slanted Dimensioning

To dimension the distance between a line and a circle or arc:

1. Pick the line.
2. Pick the arc or circle.
3. Place the dimension (with the center mouse button).
4. The ARC PNT TYPE menu will appear, with the following options:

Center Use to dimension between the arc or circle center and the line (Fig. 9.17).

Tangent Use to dimension between a line and the point of nearest tangency on the arc or circle.

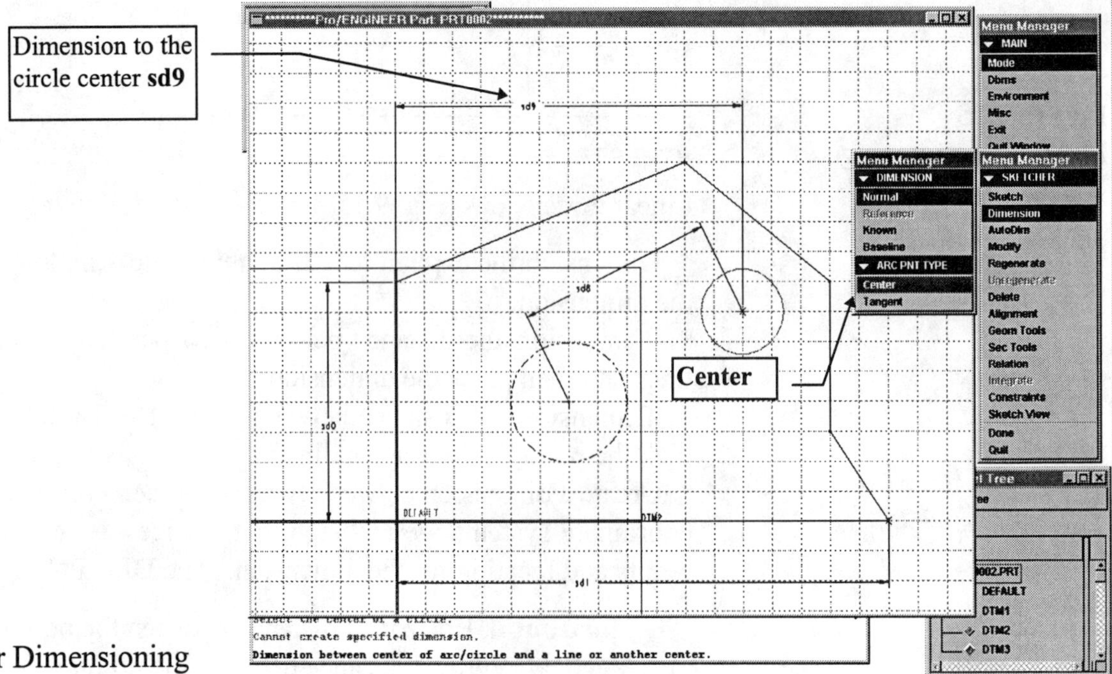

Figure 9.17
Linear Center Dimensioning

Figure 9.18 shows the dimension between a line and a circle at its tangency. To dimension between tangencies (Fig. 9.19):

1. Pick the first arc or circle.
2. Pick the second arc or circle.
3. Place the dimension.
4. Select **Tangent** from the ARC PNT TYPE menu.
5. Select either **Vert** or **Horiz** for the proper orientation.

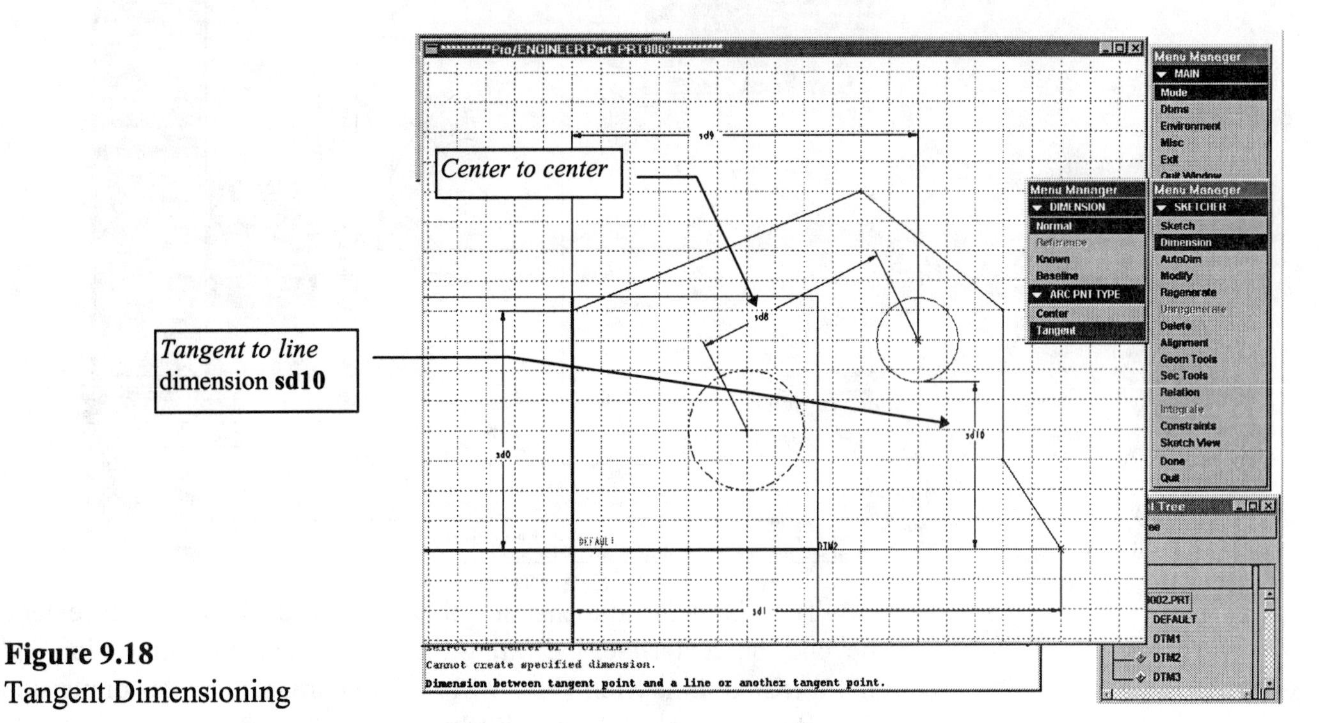

Figure 9.18
Tangent Dimensioning

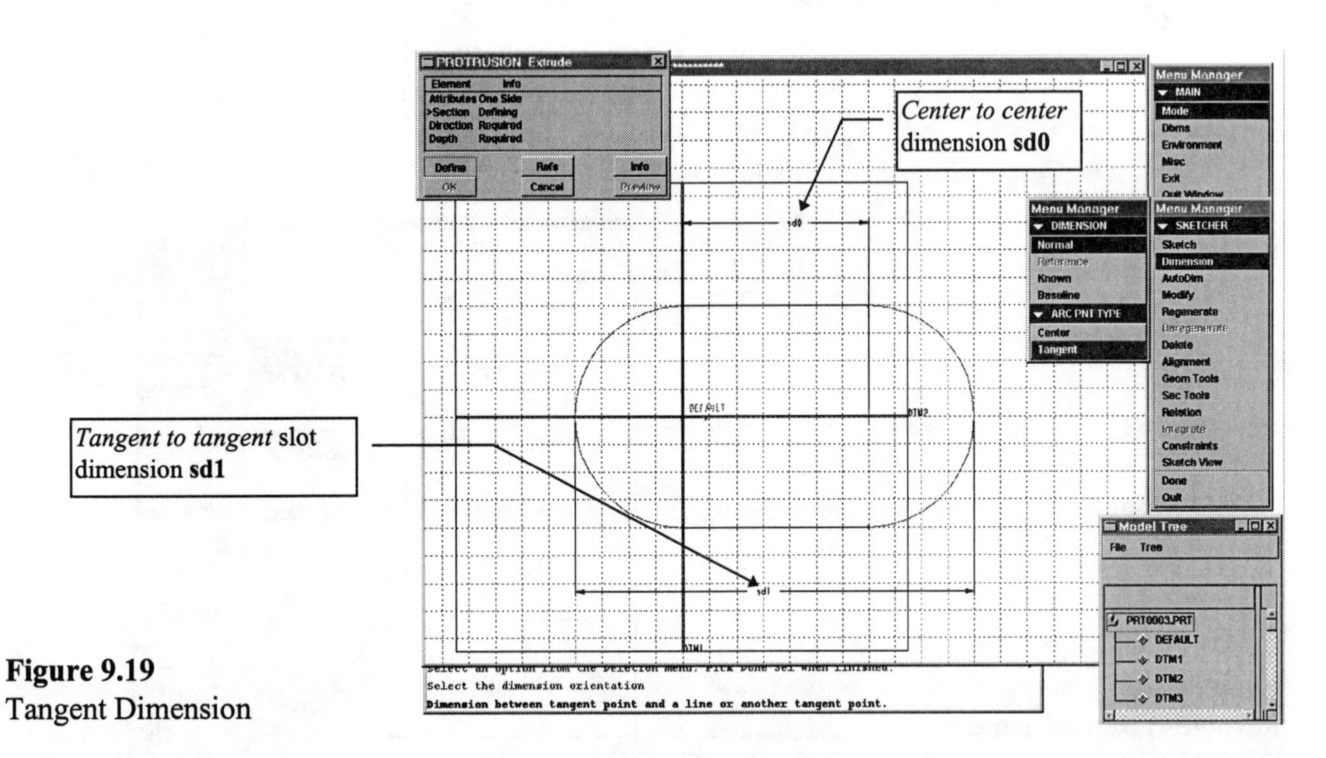

Figure 9.19
Tangent Dimension

Diameter Dimensions

Diameter dimensions measure the diameters of sketched circles and arcs or the diameters for sketching sections about the axis.

To create a diameter dimension for an arc or a circle, pick on the arc or circle twice, and then place the dimension (Fig. 9.20). Note that a polar grid was used in this section.

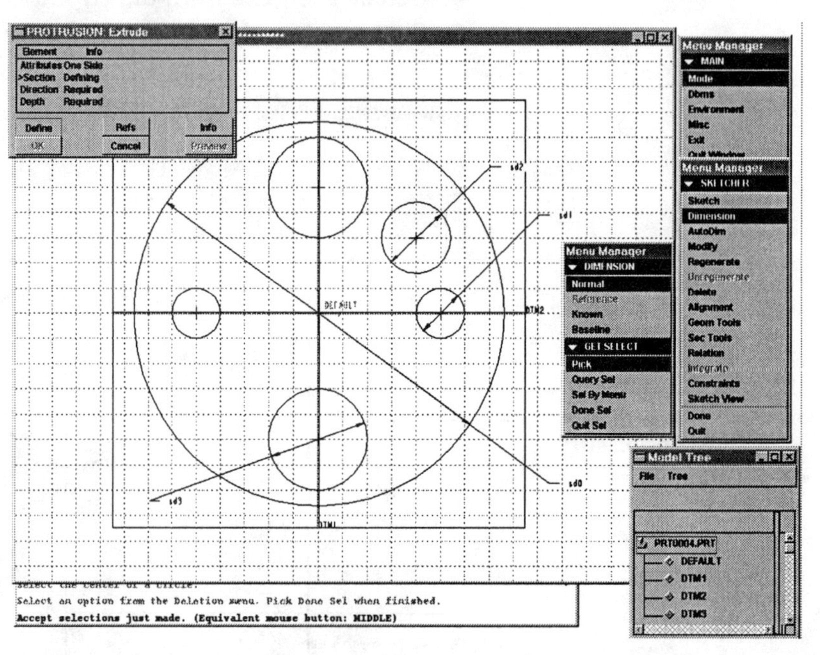

Figure 9.20
Diameter Dimensions

The diameter dimension for a revolved feature will extend beyond the centerline, indicating that it is a diameter dimension and not a radius dimension (Fig. 9.21). To create a diameter dimension for a section that will be revolved:

1. Select the entity to be dimensioned.
2. Pick the centerline that will be the axis of revolution.
3. Pick the entity again.
4. Place the dimension.

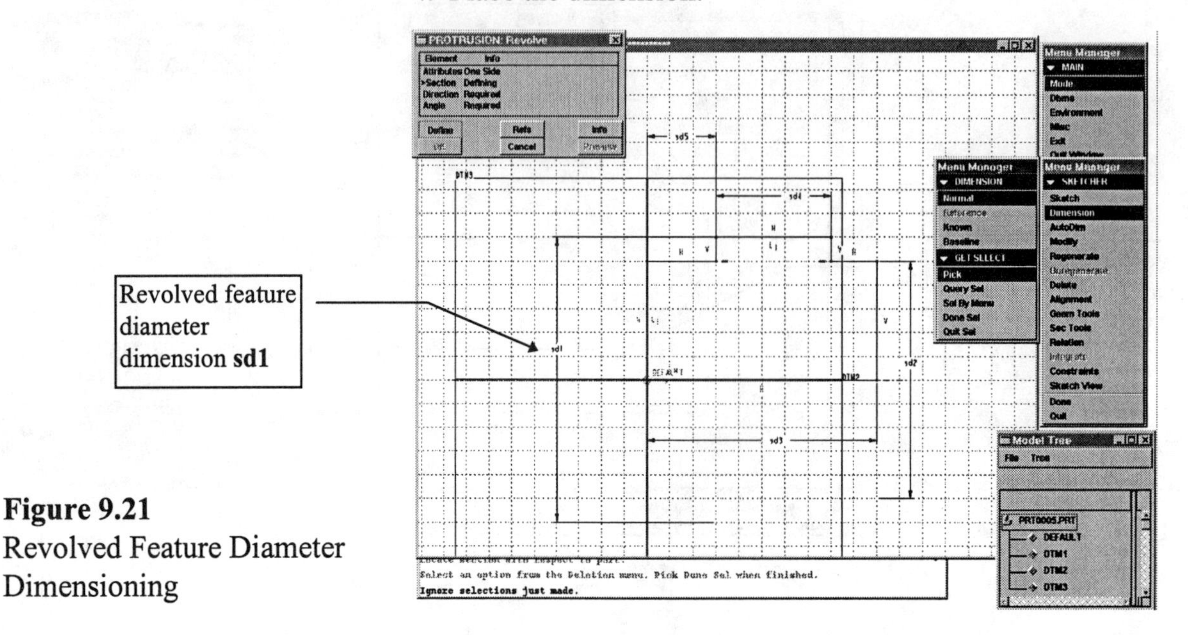

Figure 9.21
Revolved Feature Diameter Dimensioning

Radial Dimensions

Radial dimensions measure the radii of circles and arcs and the radii of circles and arcs created by revolving a section about an axis.

To create a radial dimension for an arc or circle, pick on the circle or arc and then place the dimension. In general, circles are dimensioned as diameters and arcs as radii (Fig. 9.22).

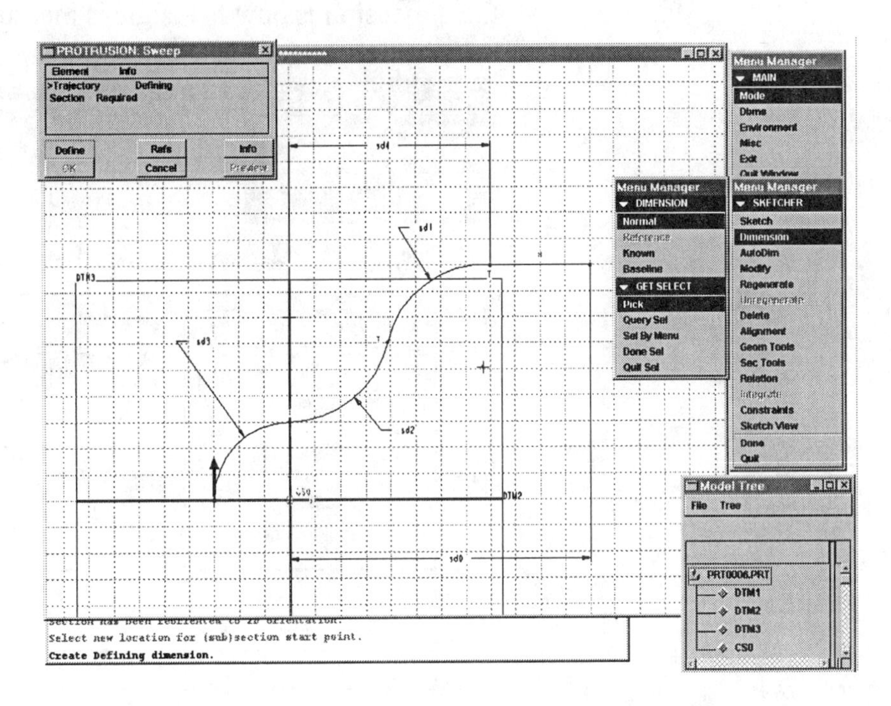

Figure 9.22
Arc Dimensioning

To create a diameter dimension for a section that will be revolved, pick on the entity, pick on the centerline axis, pick on the entity, and then place the dimension. The example in Figure 9.23 was dimensioned in the default 3D **Default** view instead of the 2D **Sketch View**.

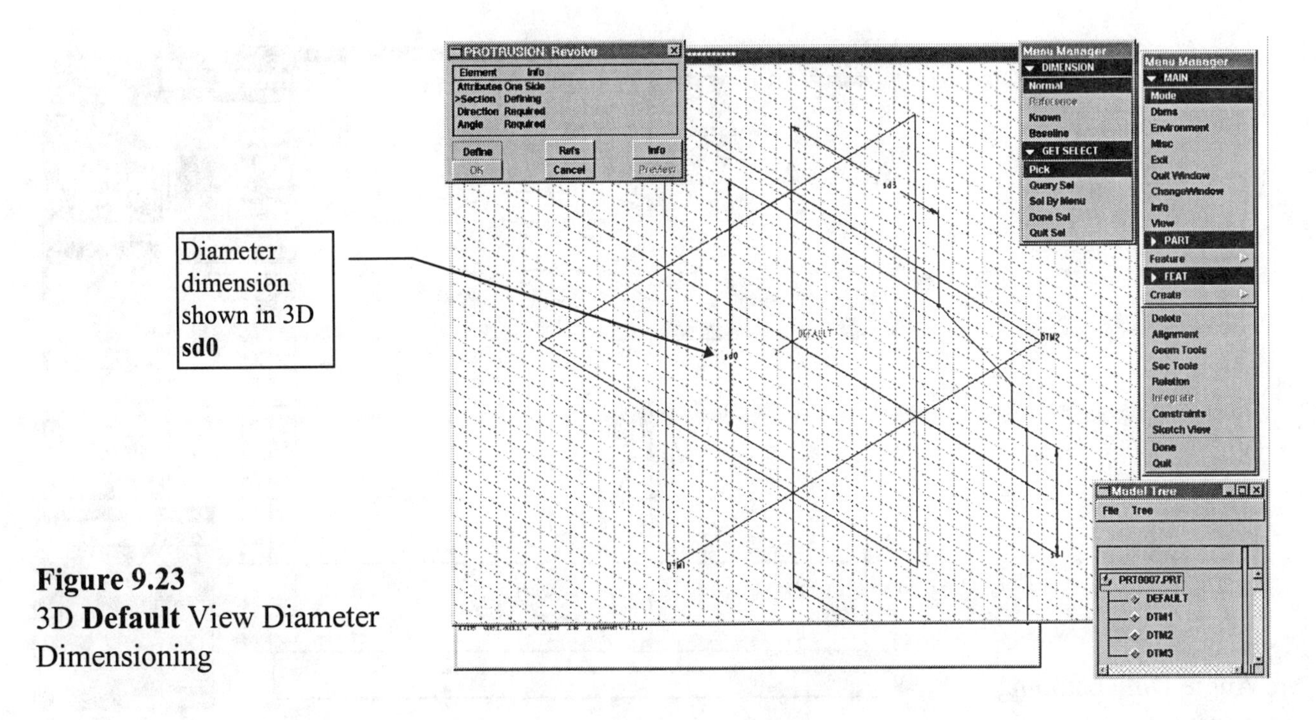

Figure 9.23
3D **Default** View Diameter Dimensioning

Angular Dimensions

Angular dimensions measure the angle between two lines (Fig. 9.24) or the angle of an arc between its endpoints.

To create an angular dimension between lines, pick the first line, pick the second line, and then place the dimension. Where you place the dimension is how the angle is measured (either acute or obtuse).

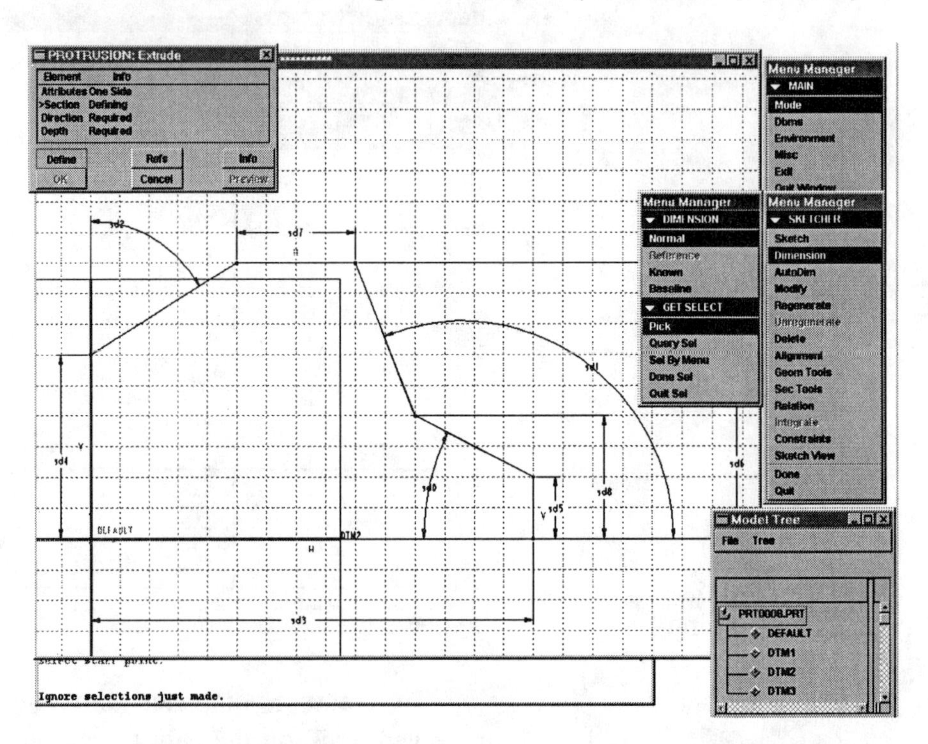

Figure 9.24
Angle Dimensioning

To create an arc angle dimension, pick one endpoint of the arc, pick the other endpoint of the arc, pick on the arc, and then place the dimension (Fig. 9.25).

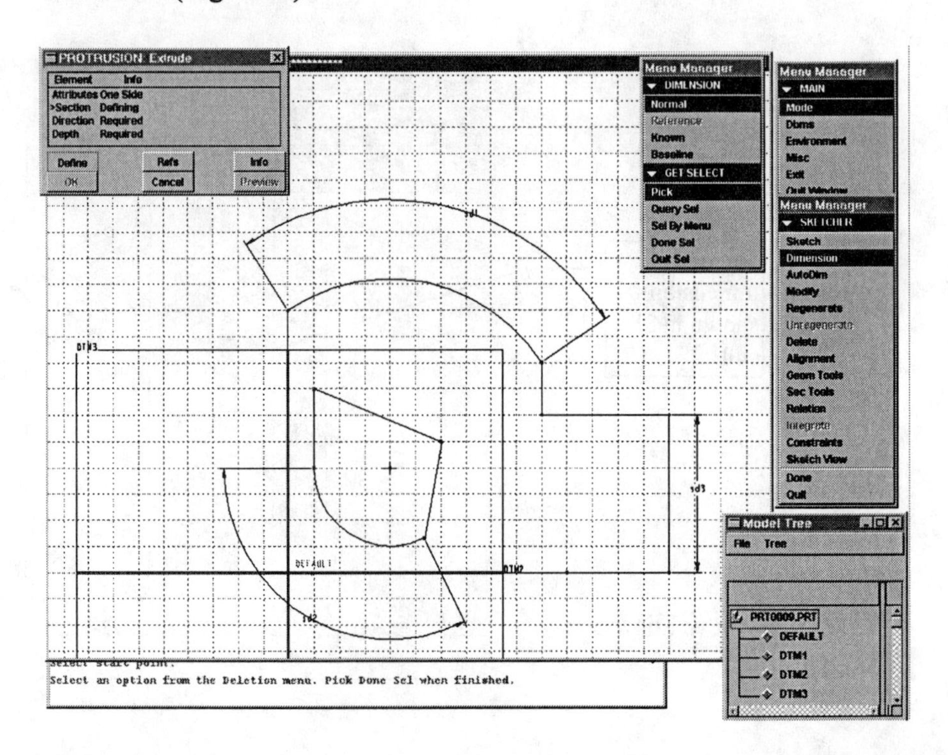

Figure 9.25
Arc Angle Dimensioning

Conclusion

The aim of the Sketcher in parametric feature-based design is to create quick and simple geometry for your model. The sketching process enables you to create and dimension the geometry for a feature or set of features based on your design. Remember, during the sketching process you need not concern yourself with creating perfect geometry or exact dimension sizes. You can modify your dimensions later in the design process (Figs. 9.26 and 9.27).

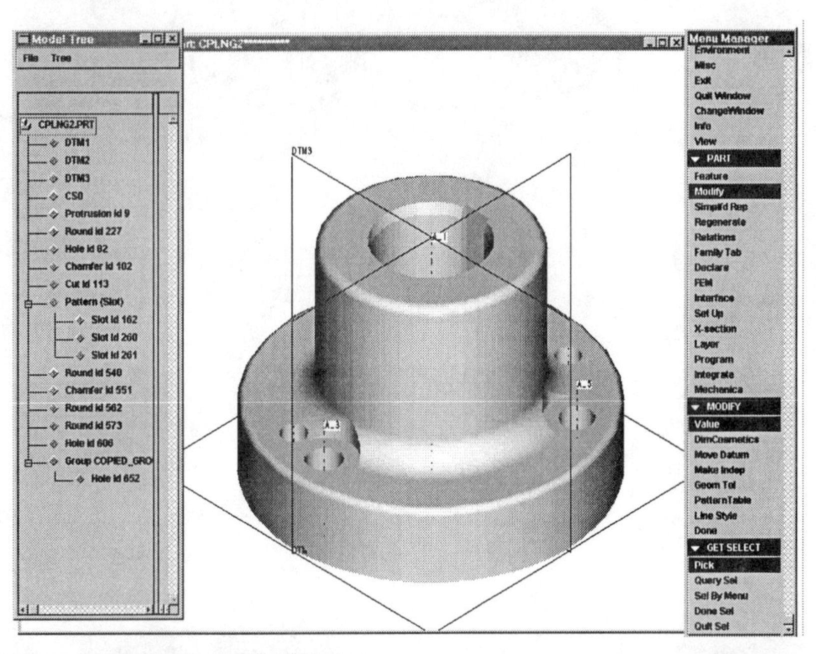

Figure 9.26
Original Part Design

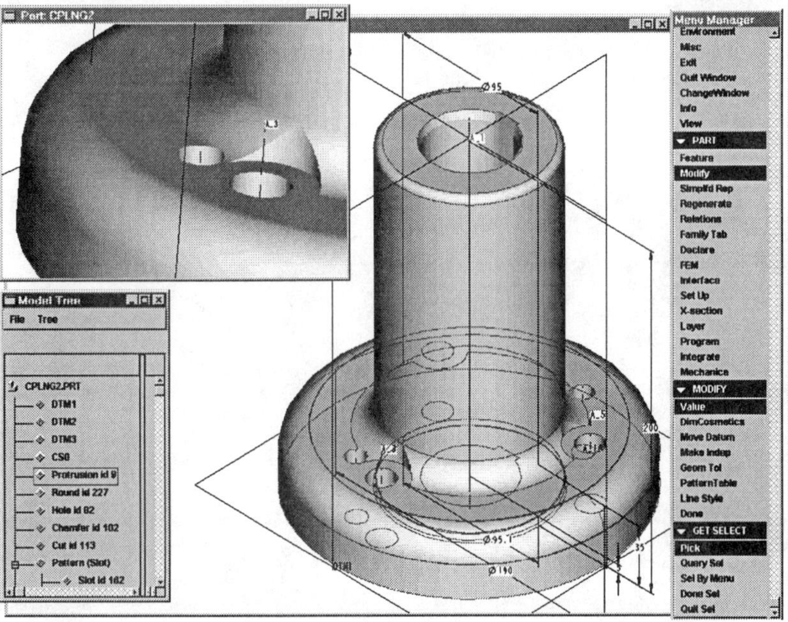

Figure 9.27
Modified Final Design

The remaining portion of this text is devoted to individual lessons. In each **lesson** you will be called upon to follow a set of steps to complete the **lesson part**, which in most cases includes the creation of a variety of sketches. After you complete the lesson part, the material that was introduced in that lesson will be applied to the subsequent **lesson project**.

Part One

Creating Parts

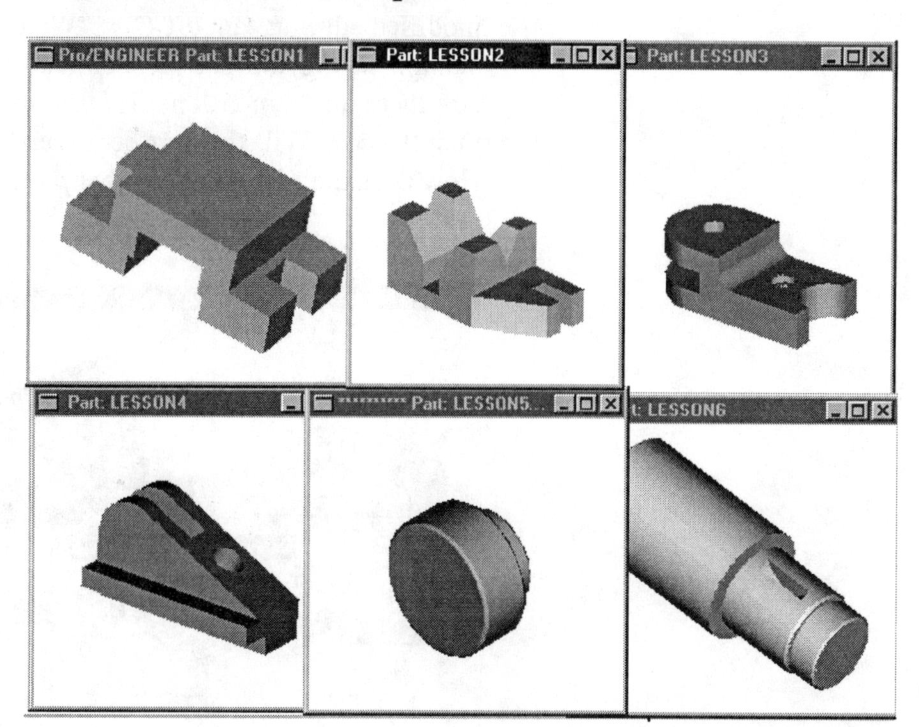

Lesson Parts 1-6

CREATING PARTS

The *design intent* of a feature, a part, or an assembly (and even a drawing) should be established before any work is done on the system. Skipping this step in the design process is a recipe for disaster. In industry, there are thousands of stories of how a designer created a *graphically correct* part or assembly that *"looked"* visually precise.

Upon closer examination, the model or assembly had too many or too few datum planes, parent-child relationships that were nothing but an example of the designers incorrect use of Pro/E, and massive feature failures that resulted when minor ECOs were introduced after the original design was complete. Pro/E is only as good as the person designing with it.

Without proper process planning, organization, and a well-defined design intent, the part model is useless. In most cases, such poor design habits result in the parts being remodeled, since it would take more time to reorder, modify, redefine, and reroute. In fact, most poor designs can't be fixed.

Part Design Philosophy: Design Intent

The **design intent** of a project must be understood before modeling geometry is started. Use the Design Intent Planning Sheets provided in Appendix D to sketch and analyze your part before modeling.

The *dimensioning scheme* will establish the dimensions that are critical for the design: What dimensions on the part might be modified during an *ECO*? What dimensions are required for *manufacturing* the part economically and to the correct *tolerances*? Are there any dimensional *relationships* that must be established and maintained? Will the part be a member of a *family of similar parts*? How does the part relate to other *parts in the assembly*?

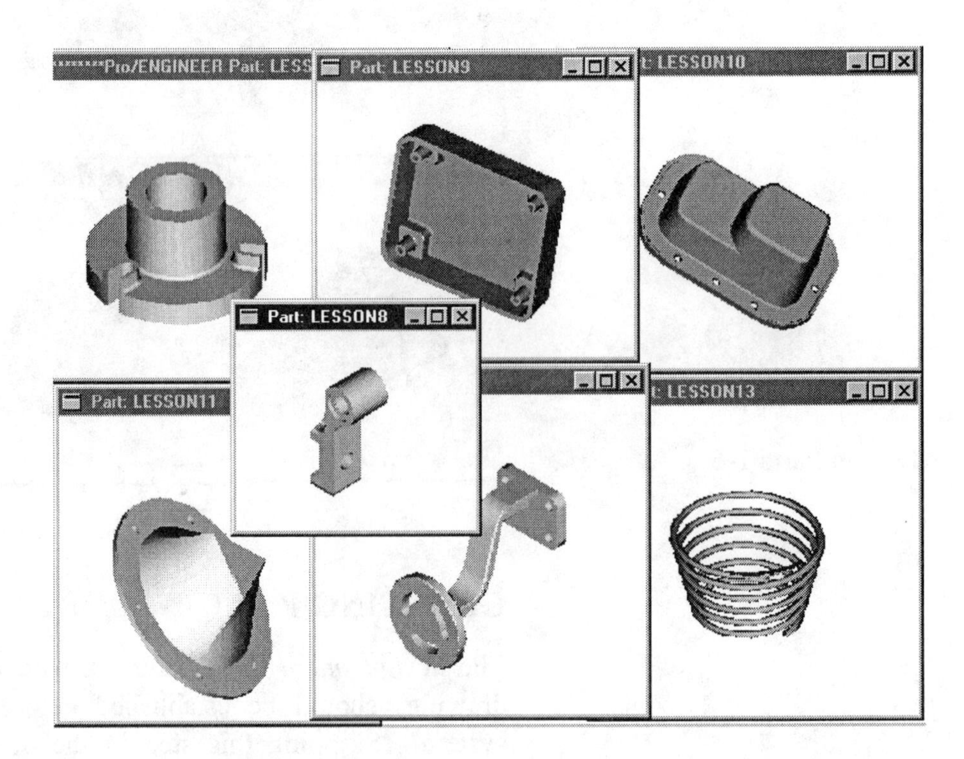

Lesson Parts 7-13

Use the following basic guideline to create a typical part:

1. Establish the system environment that you will work in for the project, including environment setting and *config.pro* settings
2. Use setup to establish the material and units.
3. Establish the datum planes and coordinate system.
4. Create a layering scheme, and set the layers and coordinate system on a layer.
5. Rename the datum planes per the part and geometric tolerance requirements.
6. Determine the base feature and protrusion type.
7. Sketch the base feature on the appropiate datum plane.
8. Establish the dimensioning scheme for the feature.
9. Determine what construction features should be used on the part.
10. Build a construction feature using a dimensioning scheme, keeping relationship requirements in mind.
11. Add relations to control the feature where desired, per the design intent.
12. Adjust dimension cosmetics as desired.
13. Create new layers, and establish a layering scheme for dimensions and features.
14. Add reference dimensions required for documentation.
15. Repeat steps 1-14 for each feature of the part until the part is completed.

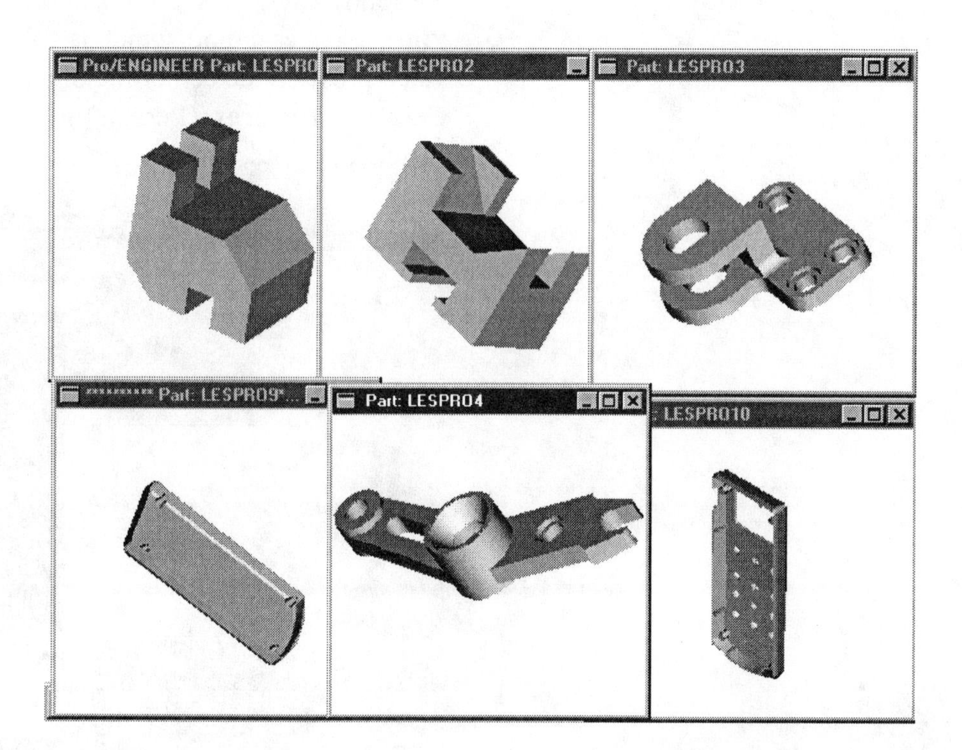

Lesson Project Parts

A wide variety of features can be created with Pro/ENGINEER. The following 13 part creation lessons incorporate most of the following capabilities:

Protrusions (Lesson 1) Part features that add material
Slots and Cuts (Lesson 1) Features that remove material from a part
Holes (Lesson 3) Creates different types of holes - through, counterbored (sketched), blind, etc.
Shafts Creates shafts
Rounds (Lesson 3) Creates many types of rounds
Chamfers (Lesson 6) Creates edge and corner chamfers, which remove flat sections of material to create a beveled surface
Necks (Lesson 5) Creates a neck, which is a special type of revolved slot that creates a groove around a revolved part or feature
Flanges Creates a flange, which is analogous to a neck, except it adds material to the revolved solid
Ribs (Lesson 8) Creates a rib, which creates a thin fin or web that is attached to a part
Tweak features Creates drafts (Lesson 9), local pushes, domes, ears, lips, patches, bends, and free-form features
Shells (Lesson 10) Creates a shell feature, which removes a surface or surfaces from the solid, then hollows out the inside of the solid, leaving a shell of a specified wall thickness
Pipes Creates a pipe, which is a three-dimensional centerline that represents the centerline of a pipe
Cosmetic features (Lesson 6) Creates cosmetic features, sketched, thread, groove, and user-defined

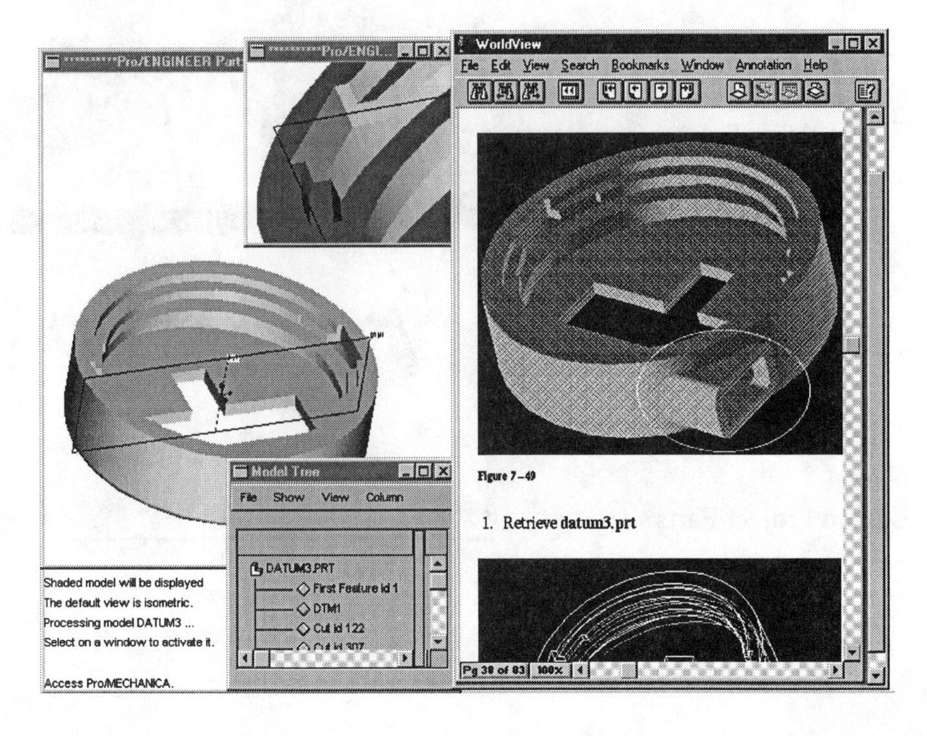

Lesson 1

Protrusions and Cuts

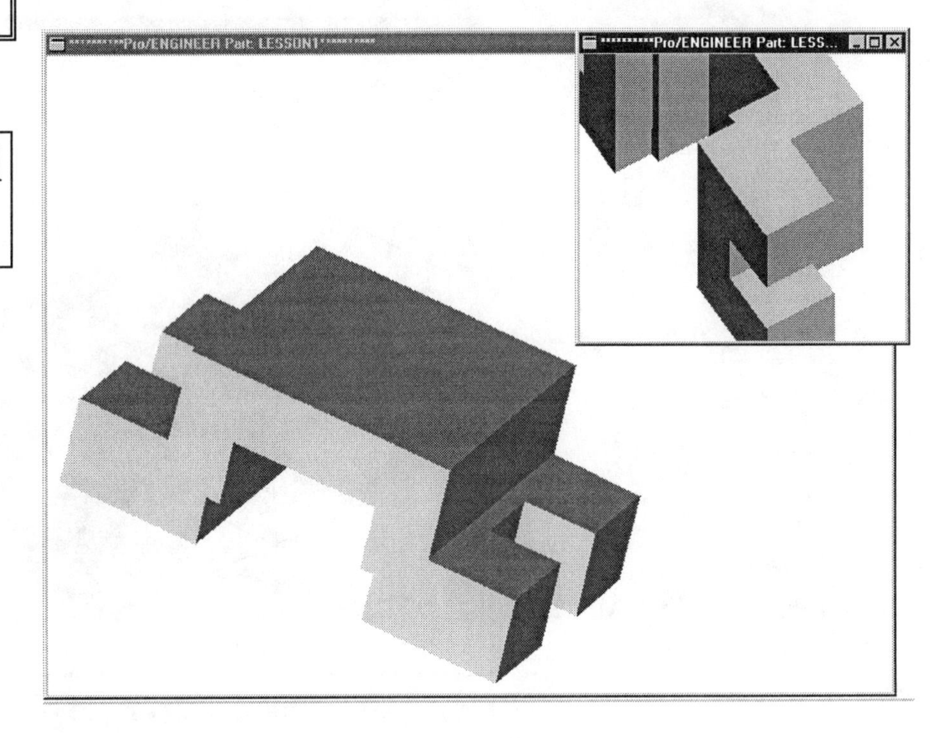

Figure 1.1
Clamp

OBJECTIVES

1. **Create a base feature using an extruded protrusion**
2. **Understand setup and environment settings**
3. **Define and set a material type**
4. **Create and use datums**
5. **Sketch a protrusion and a cut feature in the Sketcher**
6. **Understand the feature dialog box**
7. **Learn how to align sketch geometry**
8. **Shade a part**
9. **Copy a cut feature**
10. **Save and purge a part file**

EGD REFERENCE
Engineering Graphics and Design
by L. Lamit and K. Kitto
Read Chapters: 5, 10
See Pages: 111, 192-193, 302, 409

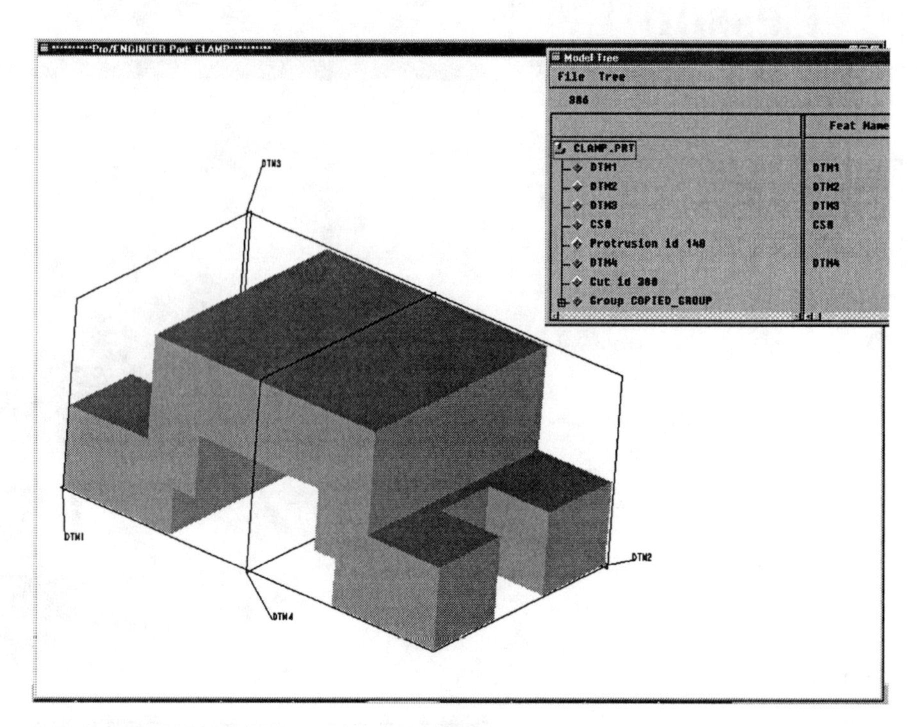

Figure 1.2
Clamp Part Showing the Datum Planes and the Model Tree

PROTRUSIONS AND CUTS

A **protrusion** is a part feature that adds material. You can sketch different geometry by combining a variety of form options and attributes during creation of the protrusion feature. **Cuts** and **slots** are used to remove material from existing solid features. Figures 1.1 and 1.2 show a simple protruded part with identical cuts on both sides.

Protrusions

A protrusion (Fig.1.3) is *always the first solid feature created.* This can be the **base feature** or the first feature created after a base feature of datum planes.

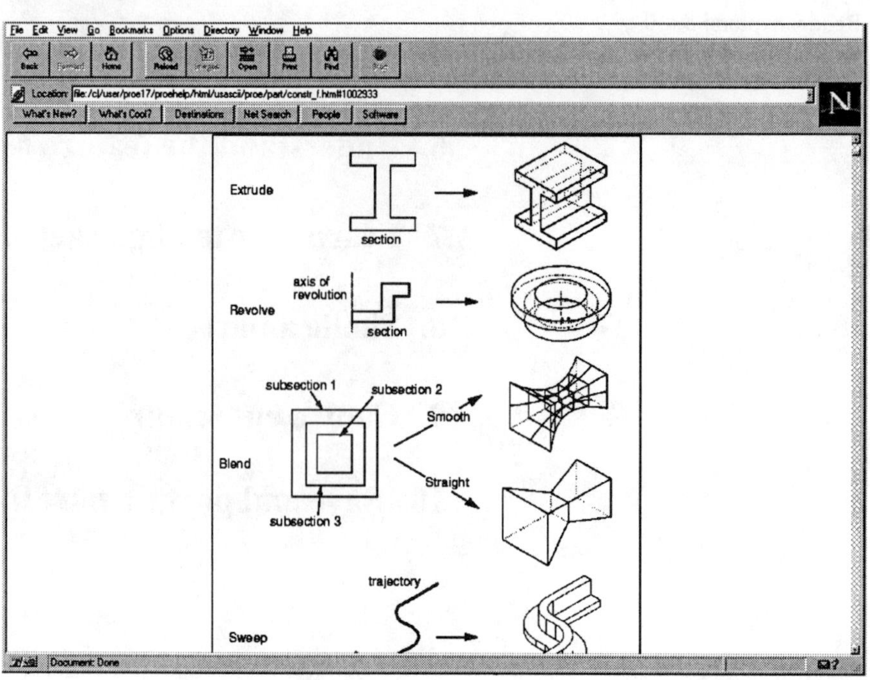

Figure 1.3
Online Documentation Protrusions

To create an extruded protrusion:

1. Choose **Feature** from the PART menu, then **Create** from the FEAT menu. Figure 1.4 is taken from **COAch for Pro/ENGINEER**. If you have this tutorial on your system, go to Creating a Sketch with a Circle, Segment 2 of the Creating a Basic Model in the Modeling module.
2. Choose **Protrusion** from the SOLID menu.
3. Choose **Extrude** ⇒ **Solid** ⇒ **Done** ⇒ from the SOLID OPTS menu.
4. Pro/E displays the PROTRUSION: Extrude dialog box, which lists the elements for creating protrusions.
5. Pro/E displays the ATTRIBUTES menu, which lists the following options:
   - **One Side** Creates the feature on one side of the sketching plane.
   - **Both Sides** Creates the feature on both sides of the sketching plane.
6. Choose **One Side** or **Both Sides** ⇒ **Done** from the ATTRIBUTES menu.
7. Select the sketching plane and the sketch orientation reference.
8. **Sketch** the protrusion.
9. Align (**Alignment**) and **Dimension** the section geometry
10. **Regenerate** the section.
11. **Modify** and **Regenerate** the section sketch.
12. Specify the depth of the protrusion and choose **OK**.

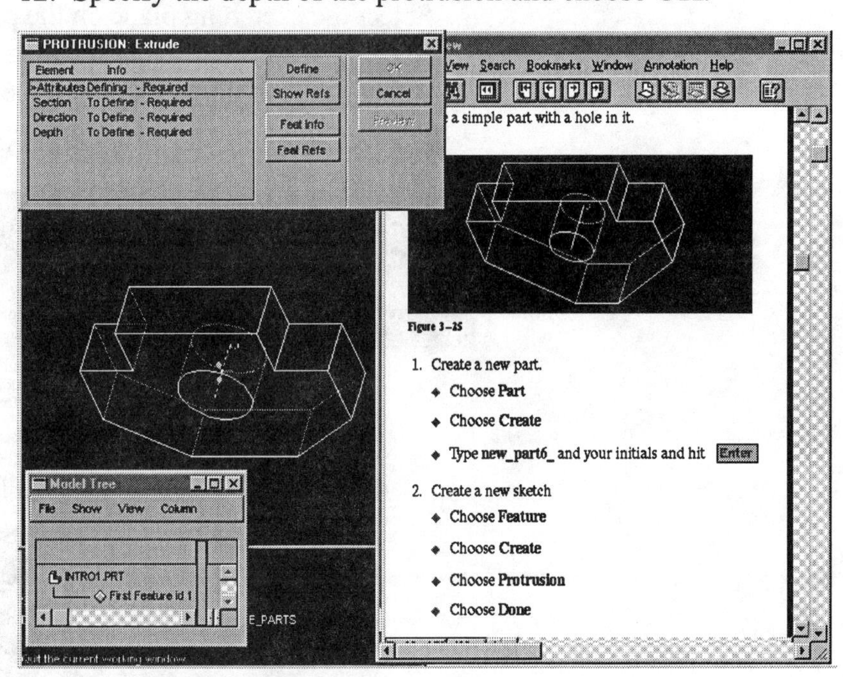

Figure 1.4
COAch for Pro/ENGINEER, Creating a Basic Model--Modeling (Segment 2: Creating a Sketch with a Circle)

Cuts (and Slots)

To remove material from a part, use one of the following features:

- **Cut** Removes material within a closed section (Fig. 1.5).
- **Slot** Removes material from a specified side.

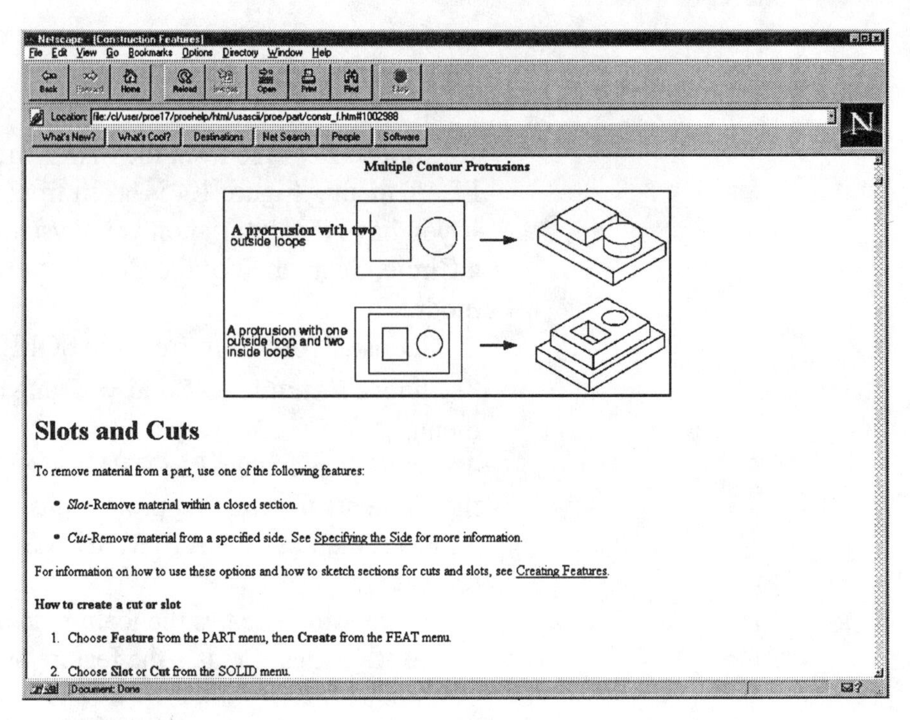

Figure 1.5
Online Documentation for Cuts and Slots

To create a cut (Fig. 1.6) or slot, do the following:

1. Choose **Feature** from the PART menu, then **Create** from the FEAT menu.
2. Choose **Slot** or **Cut** from the SOLID menu.
3. Choose **Extrude** ⇒ **Solid** ⇒ **Done** ⇒ (SOLID OPTS menu).
4. The appropriate dialog box is displayed.
5. Choose **One Side** or **Both Sides** ⇒ **Done** (ATTRIBUTES menu).
6. Select the sketching plane on the part and the part's orientation.
7. Sketch, **Alignment**, **Dimension**, and **Regenerate** the section.
8. **Accept** the cut direction or flip the arrow.
9. Determine the depth of the cut.

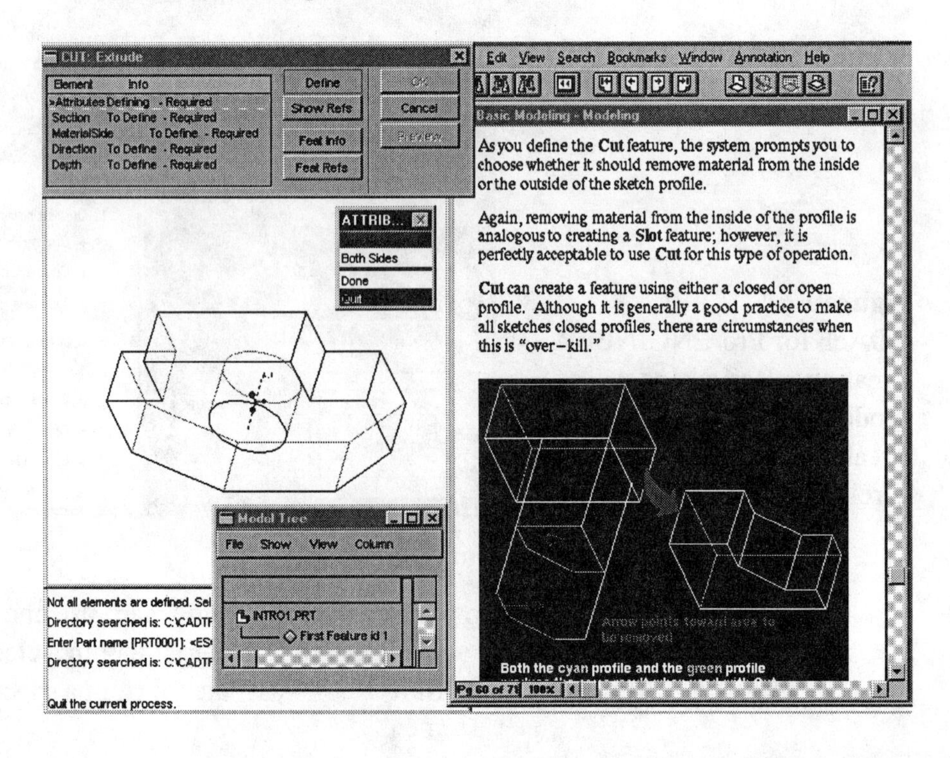

Figure 1.6
COAch for Pro/ENGINEER, Creating a Basic Model--Modeling (Segment 6: Cuts)

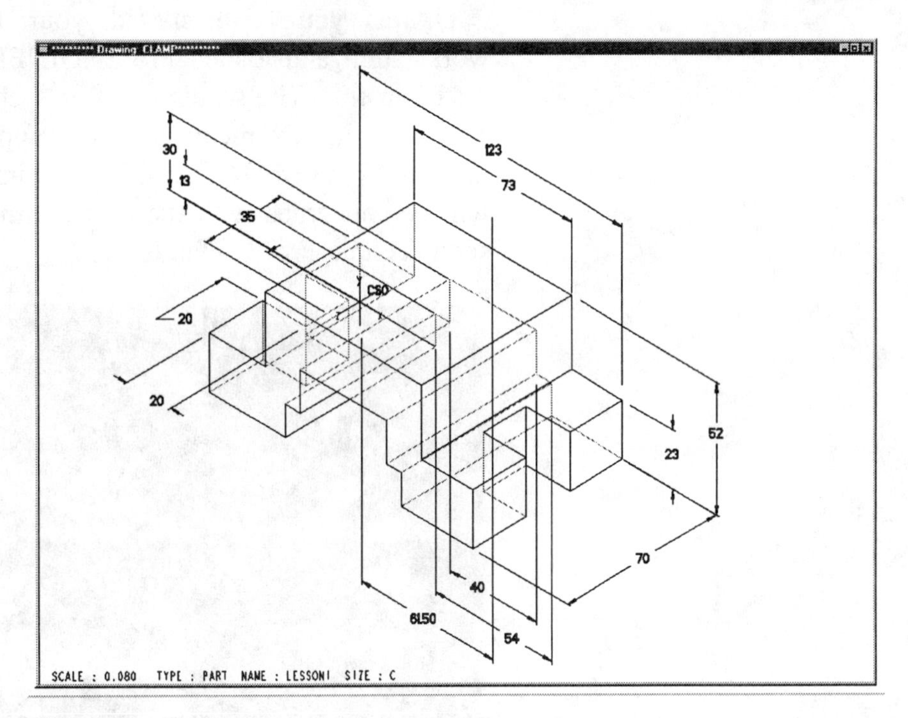

Figure 1.7
Clamp Dimensions

Clamp

The clamp (Fig. 1.7) will be our first ***lesson part***. It is composed of a simple protrusion and a cut. A number of things need to be established before you actually start modeling. These include setting up the *environment*, selecting the *units*, and establishing the *material* for the part.

Before you begin any part using Pro/E, you must plan the design. The **design intent** will depend on a number of things that are out of your control and a number that you can establish. Asking yourself a few questions will clear up the design intent you will follow: Is the part a component of an assembly? And if so, what surfaces or features are used to connect one part to another? Will geometric tolerancing be used on the part and assembly? What units are being used in the design, SI or decimal inches? What is the part's material? What is the primary part feature? How should I model the part? And what features are best used for the primary protrusion (the first solid mass)? On what datum plane should I sketch to model the first protrusion? These and many other questions will be answered as you follow the step-by-step lesson part. But you must answer many of the questions on your own when completing the ***lesson project***, which does not come with step-by-step instructions.

HINT

Before you start modeling with Pro/E, copy the **DIPS** from Appendix D to have them available for planning parts, feature sketches, assemblies, and drawings

Using the appropiate ***Design Intent Planning Sheet*** for each project will increase your chance of having a part, assembly, or drawing with the appropiate design intent and project sequence. For the lesson part step-by-step designs and the lesson projects found in Lessons 1-13 you should uses **DIPS 1, 2, 7,** and **8**. Block out some trial feature sequences to establish the parent-child relationships required by your design intent using **DIPS 1** and **2**. Use **DIPS 7** and **8** for your feature geometry sketches and dimensioning scheme.

After you have started your Unix, NT, or Windows 95 workstation and loaded Pro/ENGINEER, you can set up the Pro/E environment. The screen on your system will look similar to Figure 1.8. The MAIN menu is in the upper right-hand portion of your monitor. The MAIN menu has a number of choices available. We will be concerned with the **Environment** command first in order to set up certain Pro/E defaults.

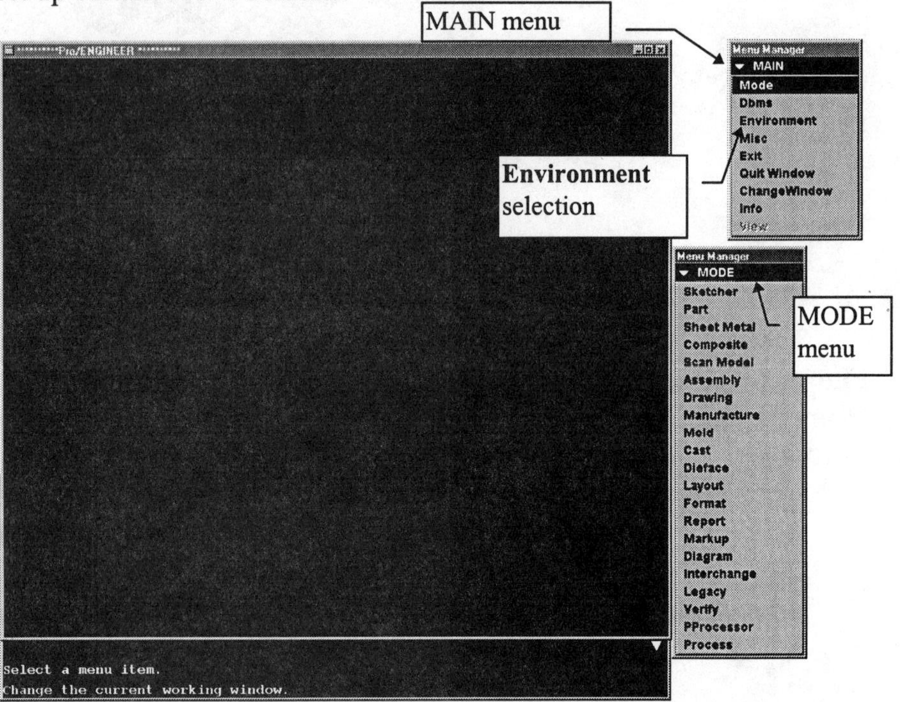

Figure 1.8
Starting a Pro/E Session

Your screen will have a *dark blue background* (by default). For the majority of this text, an alternate system color will be used for the background color. The text illustrations of screen captures show a white MAIN WINDOW color, not a solid black one as you see in Figure 1.8.

When starting a new part, assembly, or drawing, the first thing you should always do is check your environment settings. The following command block will appear at the beginning of each lesson in the text (you must do this without being prompted when starting the lesson projects). The command block will also prompt you to *give the* **Setup** *command after the part mode, assembly mode, or drawing mode has been activated.* You can change the environment settings before or after the MODE has been selected and at any time afterward during an active Pro/E session.

HINT
For setting material choose: **Material** ⇒ **Define** ⇒ (Type **Steel** and then press the **Enter** key) ⇒ *remember that the material properties window will appear on the screen and you must fill in the data, choose* **File** ⇒ **Exit** *and then* **assign** ⇒ (pick **Steel**) ⇒ **Accept** ⇒ **Done**

ENVIRONMENT AND SETUP

Environment ⇒ ✓**Grid Snap** ✓ **Model Tree**
Hidden Line Tan Dimmed
Setup ⇒ **Units** ⇒ **Millimeter** ⇒ **Done** ⇒ **Material** ⇒ **Define** ⇒ (Type **Steel** and then press the **enter** key) ⇒ **Assign** ⇒ (pick **Steel**) ⇒ **Accept** ⇒ **Done**

NOTE
Always use **Setup** to set your **Units** and **Material** at the beginning of every part.

If material files have not been generated, Pro/E will prompt you the enter the material properties into a file.

The first command you will give in this lesson will be to set the *environment* for the lesson part to be created. Using your mouse, highlight the **Environment** command. Choose it by clicking the left mouse button. This will bring up the ENVIRONMENT menu, as shown in Figure 1.9. To activate an environment setting, pick on the small box ☐ to the left of the selections. A ✓ means that the choice is activated. To deactivate a choice, pick on a box ☐ to remove the ✓. To set the model visibility, choose **Hidden Line** and **Tan Dimmed**, as shown in Figure 1.9. Also activate the **Grid Snap** if it is not already on (**✓Grid Snap**). When you are done with your selections, pick **Done-Return** from the ENVIRONMENT menu.

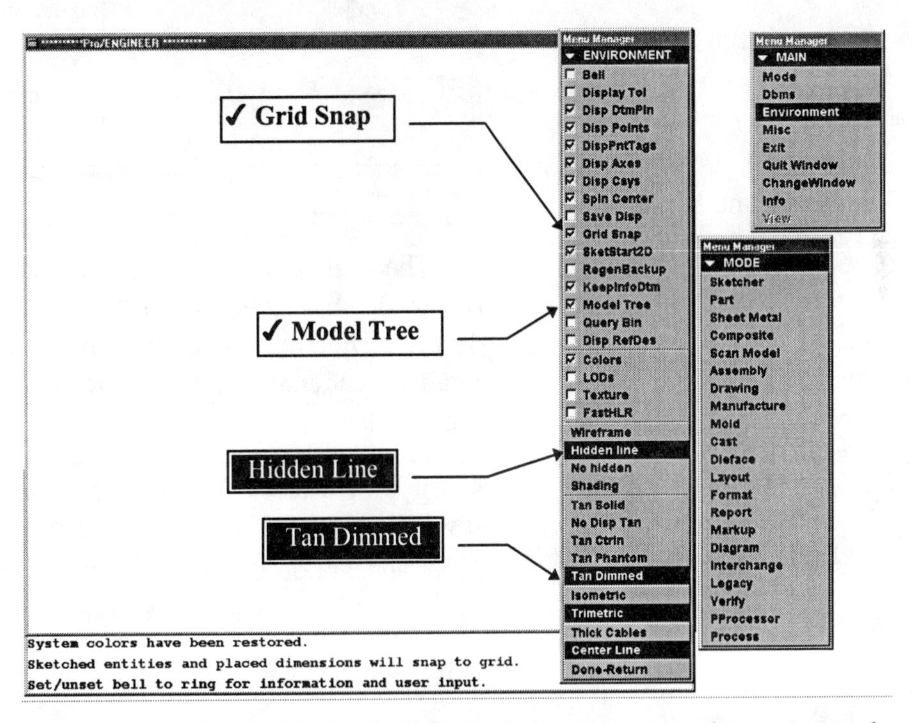

Figure 1.9
Setting the Environment

To start the actual part construction, you must choose a mode from the MODE menu. Since you will be creating a part, the **Part** mode will be selected. In Lessons 14 and 15 the **Assembly** mode will be used, and in Lessons 16-20 the **Drawing** mode is accessed. Give the following commands:

Part ⇒ Create (type the part name at the command prompt provided in the lower left-hand of the screen) **CLAMP** and press the **Enter** key on your keyboard (Fig 1.10).

The screen will show the **Model Tree** in the upper left side of the screen. You can move the Model Tree window to another spot, out of the way of the working area, the MAIN WINDOW of the screen. Using the mouse, pick on the upper bar of the menu and hold down the left mouse button while dragging the Model Tree window to a new screen location. Notice that the part name appears along the top of the MAIN WINDOW, as shown in Figure 1.10. The part name also shows in the Model Tree, which has been moved in Figure 1.11.

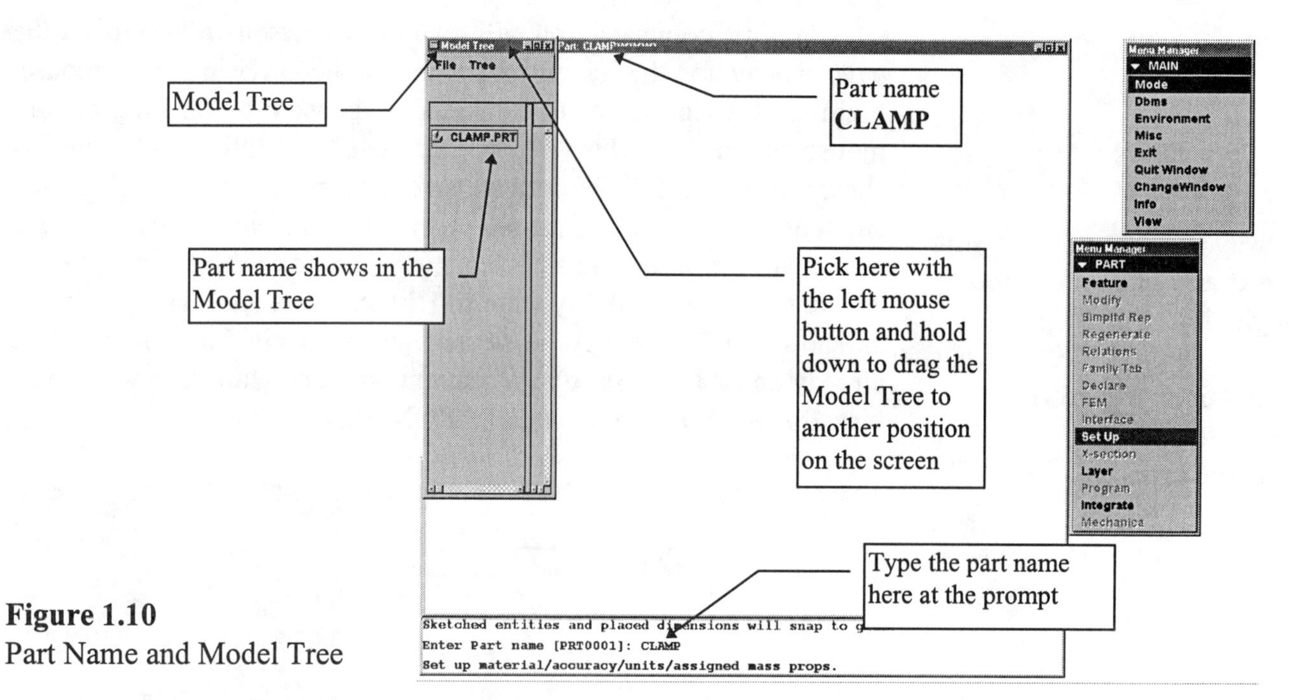

Figure 1.10
Part Name and Model Tree

The next step is to set up the part. Normally, the text will prompt you to do this step using the environment and setup command box shown earlier in this lesson. It is repeated here to provide a more detailed understanding of the process. Choose the following commands:

Setup ⇒ **Units** ⇒ **Millimeter** ⇒ **Done** ⇒ **Material** ⇒ **Define** ⇒ (Type **Steel** at the prompt and then press the **Enter** key. A material *Table* appears on the screen-- enter the values) ⇒ **File** (from material table) ⇒ **Save** ⇒ **Exit** ⇒ **Assign** ⇒ (pick **Steel)** ⇒ **Accept** ⇒ **Done**

This will set the part's units as millimeters (Fig. 1.11) and the material as steel.

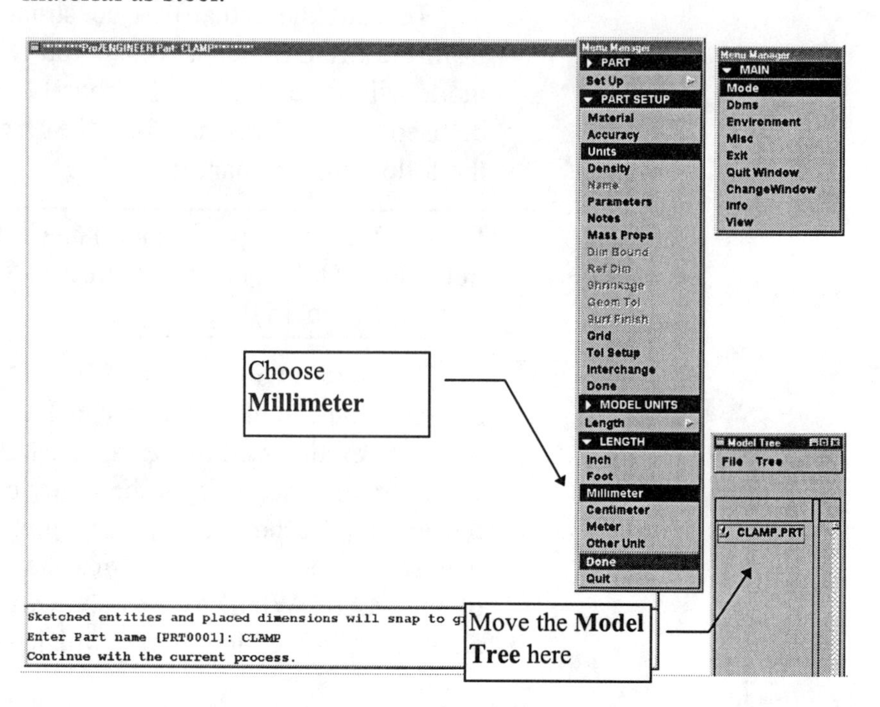

Figure 1.11
Setting Up the Part Units

The **Material** command brings up the Material Table using the default text editor, as shown in Figure 1.12. You can fill in the information if it is available from your instructor or just save and exit the table. The material is assigned to the part after exiting the table.

NOTE
Your text editor may differ.

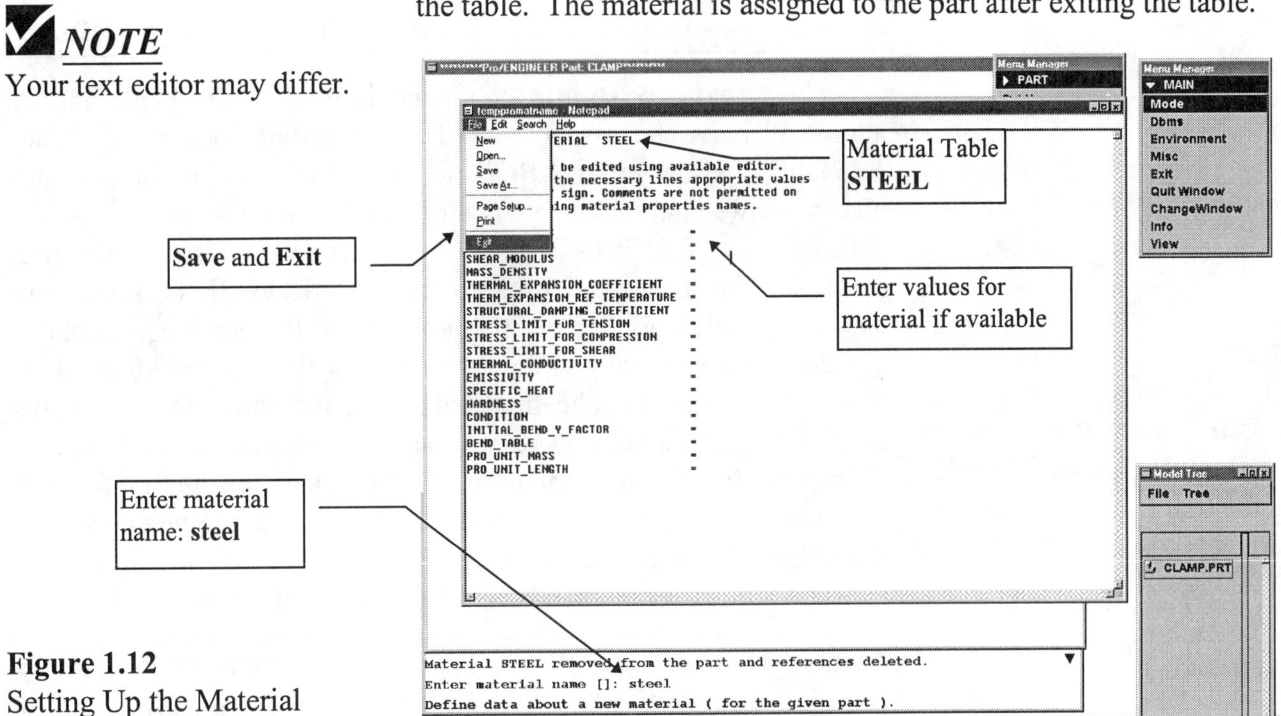

Figure 1.12
Setting Up the Material

At this point it may seem that there is a lot to do before any modeling takes place. You are building not just a solid model, but a *complete integrated database* that controls all aspects of the design, including the part's material, units, geometry, and other important parameters that capture the design intent of the project. The next step is to create three default datum planes and a default coordinate system (Fig. 1.13).

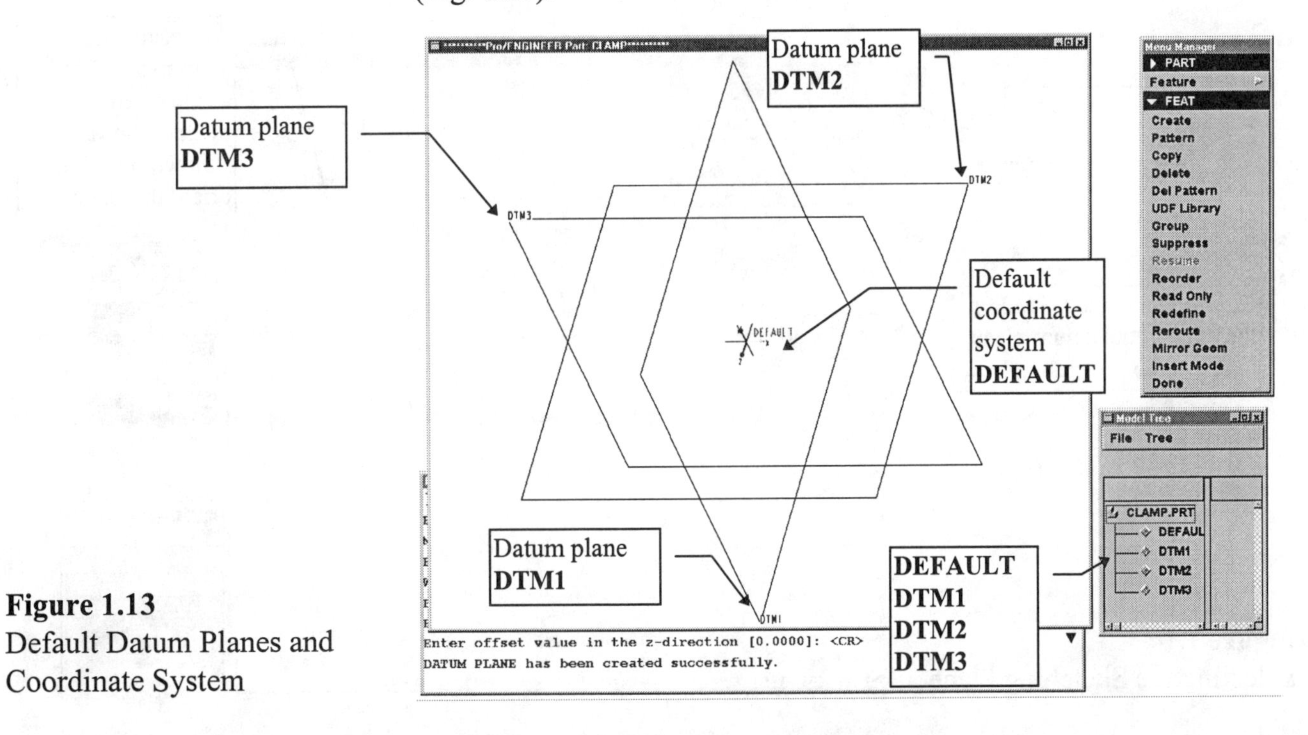

Figure 1.13
Default Datum Planes and Coordinate System

The datum planes will be used to sketch on and orient the part's section geometry. Give the following command picks:

> **Feature** ⇒ **Create** ⇒ **Datum** ⇒ **Plane** ⇒ **Offset** ⇒ **enter** (3 times at the prompt—the cursor must be inside the Pro/E window) ⇒ **Done**

The **Create** ⇒ **Datum** command is used at the beginning of almost all parts and assemblies. The screen will look as in Figure 1.13 after the last **enter**. Since an isometric view of three datum planes shows ambiguous geometry, the **Trimetric** setting is the default in the ENVIRONMENT menu. After you create the first protrusion for a part, go back to the ENVIRONMENT menu and choose **Isometric** for a more lifelike view of the part (or assembly). The coordinate system and the datum planes show in the Model Tree (Fig. 1.14). They are the first features of the part. These features become the *parents* of any feature tied to them through sketching or alignment. It is always good to have datum planes as the first features of a part instead of the first protrusion. This gives the modeler more flexibility later in the design.

HINT

Default Datum Planes and the Default Coordinate will be the first features on all parts *and* assemblies.

The following commands create the first protrusion (Fig. 1.14):

> **Feature** ⇒ **Create** ⇒ **Solid** ⇒ **Protrusion** ⇒ **Extrude** ⇒ **Solid** ⇒ **Done** ⇒ **One Side** (default from ATTRIBUTES menu) ⇒ **Done** (pick **DTM3** as the sketching plane) ⇒ **Okay** (for selecting the direction of protrusion projection- *red arrow*)

Pick on **DTM3** as shown in Figure 1.14. This is the plane on which you will be sketching the first protrusion's section geometry. The protrusion's feature dialog box appears on the screen at this time. Many of the actions you will be performing on this part will take place using the dialog box buttons.

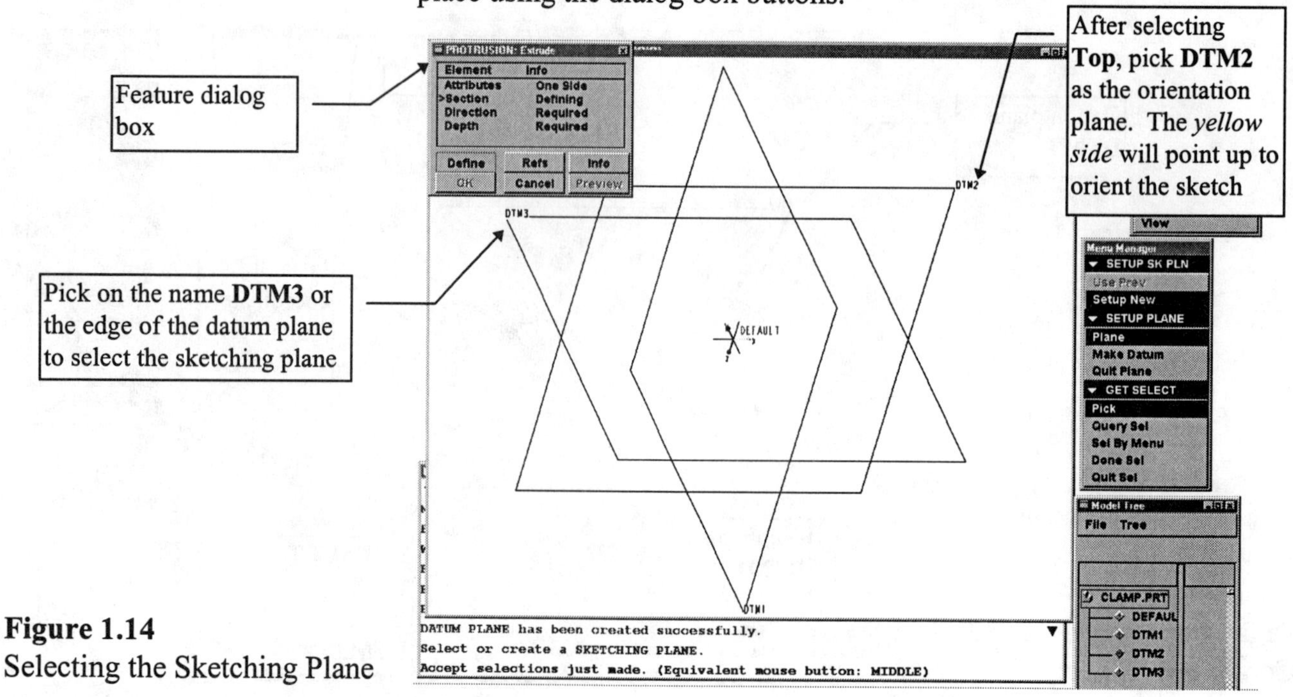

Figure 1.14
Selecting the Sketching Plane

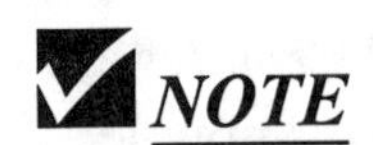

NOTE

Most sketches will be created on **DTM3** and aligned to **DTM1** and **DTM2**

You must still orient the sketch. Complete the process by choosing **Top** and picking **DTM2**. The sketch is now orientated as shown in Figure 1.15. The sketch is displayed in 2D, with the coordinate system at the middle of the sketch, where **DTM1** and **DTM2** intersect. The **X** coordinate arrow points to the right, and the **Y** coordinate arrow points up. The **Z** value is coming toward you (out from the screen). The square box you see is the limited display of **DTM3**, which is like a piece of graph paper you will be sketching on when you create the protrusion's geometry. The coordinate system is used for a few commands and is required for Pro/MANUFACTURING. Pro/E is not a coordinate-based-system, so you need not enter geometry with **X**, **Y**, and **Z** coordinates as on many other CAD systems. In reality, you don't need to turn on the coordinate system, but many modelers prefer to see the comforting axes displayed on the screen using the right-hand rule (similar to other systems they may have used in the past).

HINT

DTM3 is facing you and **DTM2** (*yellow side*) is facing the top of the screen (you are seeing it as an edge)

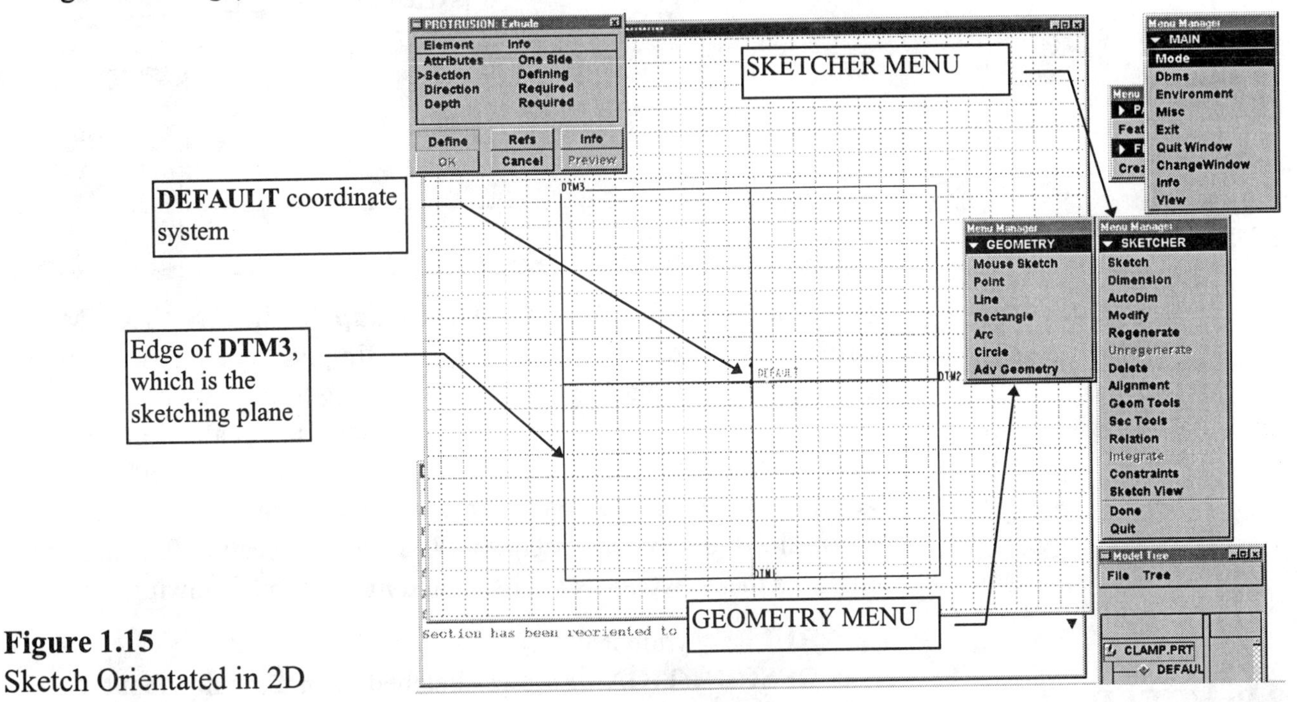

Figure 1.15
Sketch Orientated in 2D

You should reread Section 9, The Sketcher, at this time. It will help you understand some of the Sketcher's capabilities and what is required before a successful feature can be completed. In general, many of the part base protrusion features you will be modeling start with sketching in the first quadrant (Fig. 1.16). To make this more convenient and have a greater sketching area we need to change the position and size of the sketch, as shown in Figure 1.16. Here the sketch has been moved on the screen by placing the mouse cursor in the center of the screen and holding down the keyboard's control key (**Ctrl**) while simultaneously holding down the left mouse button; moving the mouse will enlarge or shrink (*zoom*) the display of the sketch.

HINT

If you double-click with the left mouse button (with the **Ctrl** key pressed), you start a zoom box. Move the cursor to the desired end location of the zoom box and pick once with the left mouse button.

Move the mouse up or down until the size of the screen and grid are correct. While you press the **Ctrl** key, the mouse button *pans* the sketch about the screen and the middle button *rotates* the sketch in 3D. You can sketch in 2D or 3D with Pro/E; for now let's stick with 2D sketching. If you rotate your sketch and wish to return to the 2D orientation, pick **Sketch View** from the SKETCHER menu.

You will be sketching in the first quadrant

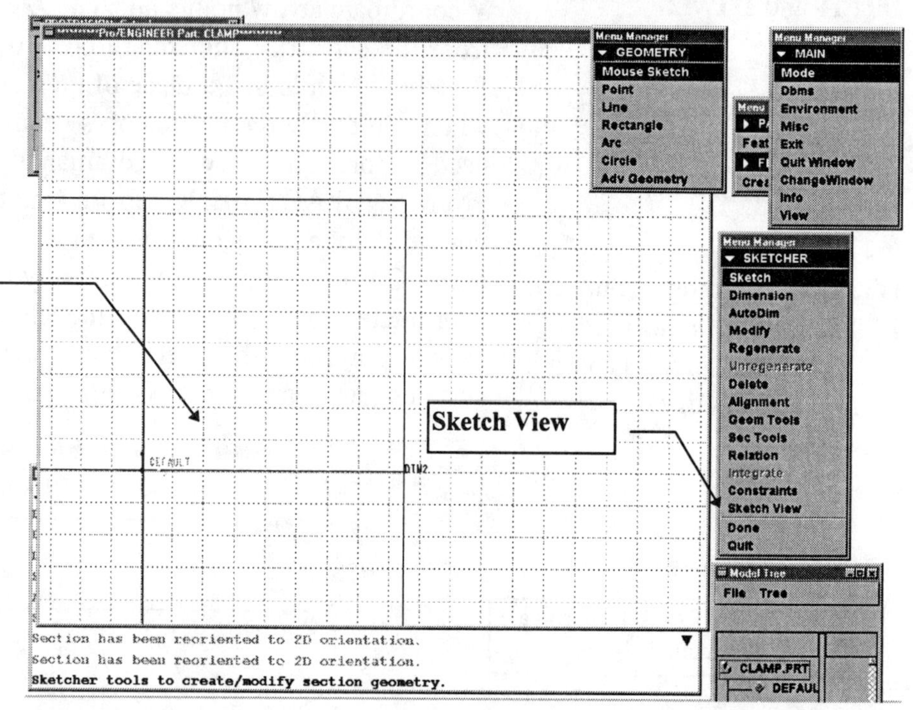

Figure 1.16
2D Sketch

Since you turned on the **Grid Snap** in the ENVIRONMENT MENU, you can now sketch by simply picking grid points representing the part's geometry (outline). Since this is a sketch in the true sense of the word, you only need to create geometry that *approximates* the shape of the feature; the sketch does not have to be accurate as far as size or dimensions are concerned. The system constrains the geometry according to rules. In Section 9, some of the rules for the **constraints** are listed and include the following:

? Pro/HELP

Highlight the **Constraint** command and press the right mouse button

RULE: Symmetry
DESCRIPTION: Entities sketched symmetrically about a centerline are assigned equal values with respect to the center line.
RULE: Horizontal and vertical lines
DESCRIPTION: Lines that are approximately horizontal or vertical are considered to be exactly horizontal or vertical.
RULE: Parallel and perpendicular lines
DESCRIPTION: Lines that are sketched approximately parallel or perpendicular are considered to be exactly parallel or perpendicular.
RULE: Tangency
DESCRIPTION: Entities sketched approximately tangent to arcs or circles are assumed to be tangent.

HINT

Do not draw lines or other entities over the top of each other. Pro/E will not regenerate the sketch, since it thinks you have two sections, not one. It is better to **Delete ⇒ Delete All** and start the sketch again. Later you will feel more comfortable trouble-shooting a failed sketch.

The **Mouse Sketch** default is active in the GEOMETRY menu, so we can begin the sketching process using the mouse buttons without selecting another command. The left button on the mouse creates lines, the middle button circles, and the right mouse button arcs that are tangent to the end of a line/arc. Place the mouse near the center of the coordinate system at the intersection of the **DTM1** and **DTM2** datum planes and click the left button. Continue picking until you have sketched an outline (Fig. 1.17) approximating the primary feature of the clamp part shown in Figure 1.2 and Figure 1.7.

The outline of the part's primary feature is sketched using a set of connected lines. The part's dimensions and general shape are provided in Figure 1.7. Since a majority of the part can be defined by sketching an outline similar to its front view, you can complete most of the part's geometry with one protrusion. The cuts on the sides will be the second feature created. The part will have its base (bottom horizontal edge) aligned with the edge of **DTM2**, its left edge aligned with **DTM1**, and, since you are sketching on **DTM3**, its back face aligned with that datum. If you are at all confused about this, reread Sections 3 and 9. It is important not to create any unintended constraints while sketching. Therefore, remember to exaggerate the sketch geometry and not to align edges that have no relationship. Pro/E is very smart: If you draw two lines at the same horizontal level, Pro/E thinks they are horizontally aligned even if you later dimension them differently! Figure 1.17 shows how to sketch vertical lines that *are not in line with each other.*

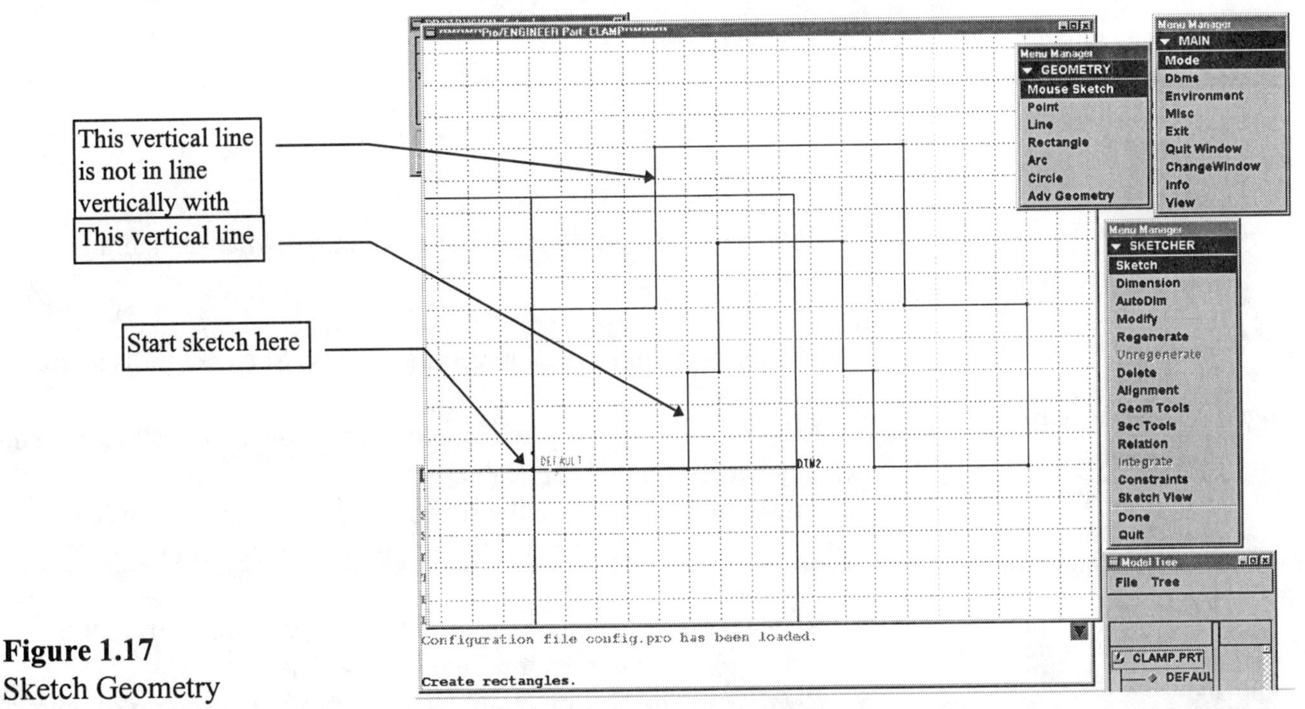

Figure 1.17
Sketch Geometry

After completing the sketch outline (section), select **Regenerate** from the SKETCHER menu. Notice that the endpoints of section entities are highlighted with the aid of small, colored dot symbols •.

Sketcher constraints can be turned off while sketching. Sketcher *Constraints Symbols* appear next to the entity that is controlled by that constraint. An **H** next to a line means horizontal, a **T** means tangent. Pick on the symbols to disable and enable the constraints and to obtain a brief explanation. The constraints can also be turned off. If the constraints are not displayed on the sketch, select **Constraints** from the SKETCHER menu and check **✓Display** in the CONSTRAINTS menu (Fig. 1.18). The constraints will appear on the section. They will also display when the sketch is oriented in 3D.

HINT

Choose **View** ⇒ **Default** ⇒ **Done/Return** to orient the sketch in 3D

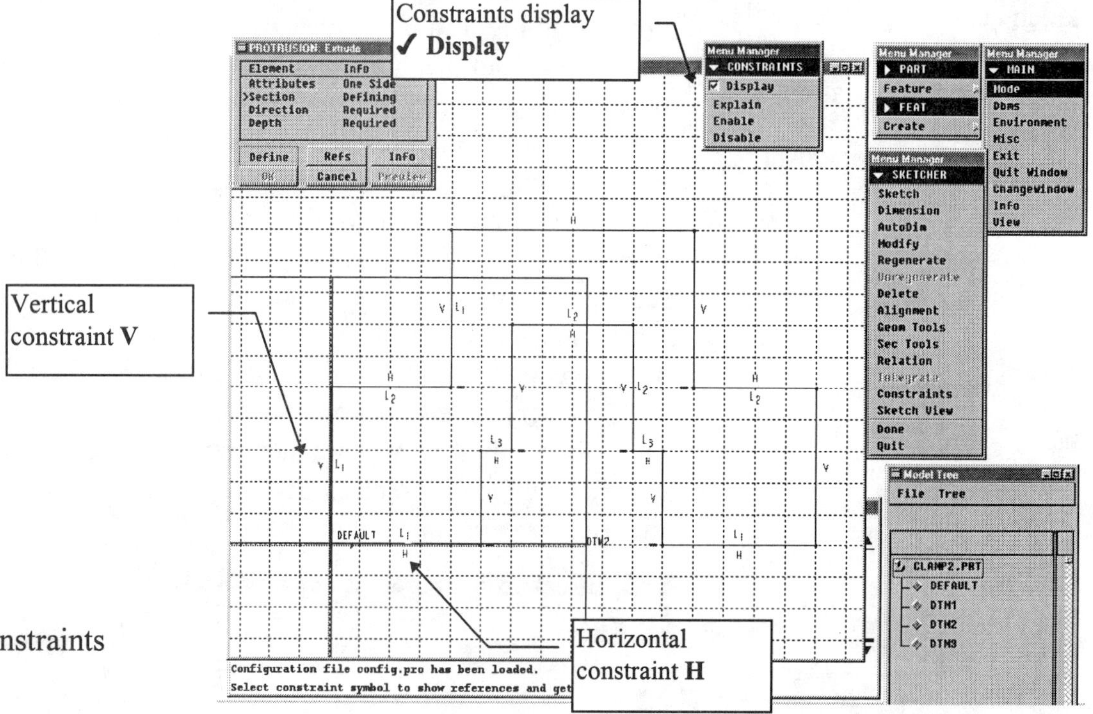

Figure 1.18
Sketch with Constraints Displayed.

HINT

Memorize this:
Sketch
Regenerate
Alignment
Regenerate
Dimension
Regenerate
Relations
Regenerate
Modify
Regenerate

In general, the following six steps are used when sketching a section:

1. **Sketch** Sketch the section geometry. Use Sketcher tools to create the section geometry.
2. **Alignment** Align the section geometry to a datum feature or to an existing part feature (if applicable).
3. **Dimension** Dimension the section. Use a dimensioning scheme that you want to see in a drawing. Dimension to control the characteristics of the section.
4. **Regenerate** Regenerate the section. Regeneration solves the section sketch based on your dimensioning scheme.
5. ***Relations*** Add section relations, to control the behavior of your section. *This step depends on the feature.*
6. **Modify** Modify the dimension values. Change the dimension values to reflect the design intent. **Regenerate** again after modifying.

Next, add a centerline to the sketched section:

Sketch ⇒ Line ⇒ Centerline ⇒ Vertical (pick on a vertical grid line running through the center of the sketch)

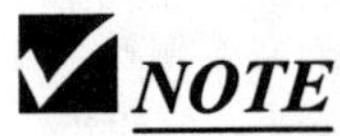

NOTE

You can change the size and type of grid in the Sketcher. You will learn how to do this in Lesson 2.

This is where your sketching skills come in handy. Did you sketch the section so it is symmetrical? Did you leave an even number of grid squares on each side of the section's center? It's not mandatory that you do so, but when you are starting Pro/E it helps to simplify the sketch and pay attention to details, since you do not have the necessary skills at this time to get the sketch to regenerate easily.

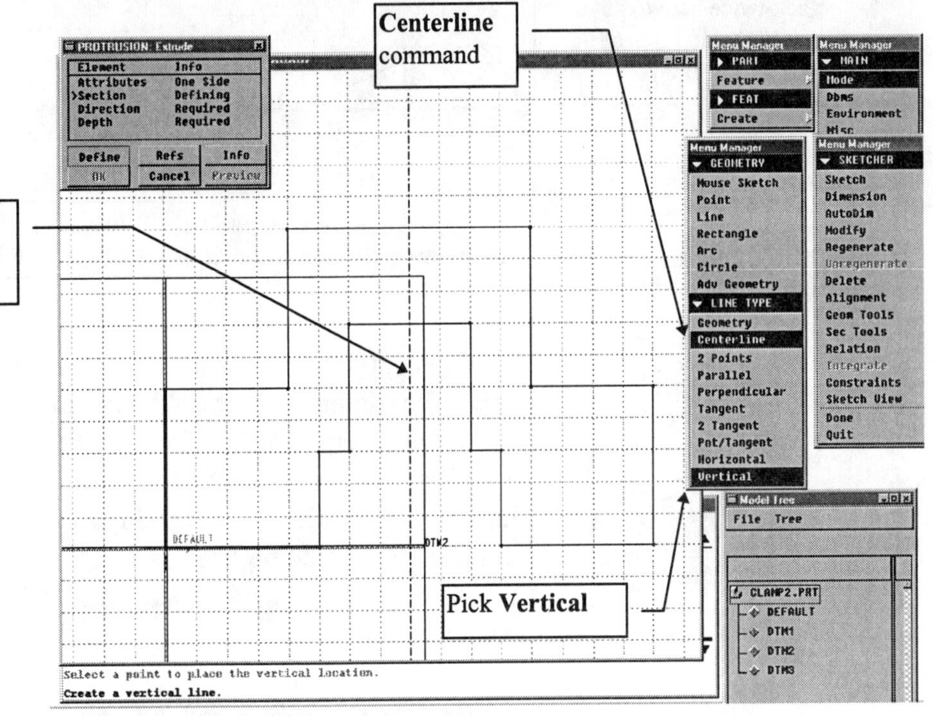

Figure 1.19
Sketching a Centerline

It is now time to *align* the sketch. This simple procedure sometimes confuses the beginner. Even though you have sketched the base and left edge against **DTM2** and **DTM1**, you still need to tell Pro/E what part of the sketch will be aligned with **DTM1** and **DTM2**. Choose **Alignment** from the SKETCHER menu, and pick twice with the left mouse button on **DTM2** and the horizontal line of the sketch (you need not move the cursor between the picks), as shown in Figure 1.20. Do the same with **DTM1** and the left edge vertical line of the sketch. The command line will say **ALIGNED**. Remember, since the line and edge of **DTM1** are coincident, you need only click twice with the left mouse button on the exact same spot. The alignment of the sketch tells Pro/E that the bottom edge of the part will be on **DTM2** and the left side of the part will be against **DTM1**.

In a later lesson you will find out how to set the datum planes according to **ASME Y14.5M 1994** geometric tolerance standards as default datums and rename the datum planes to **A**, **B**, and **C**.

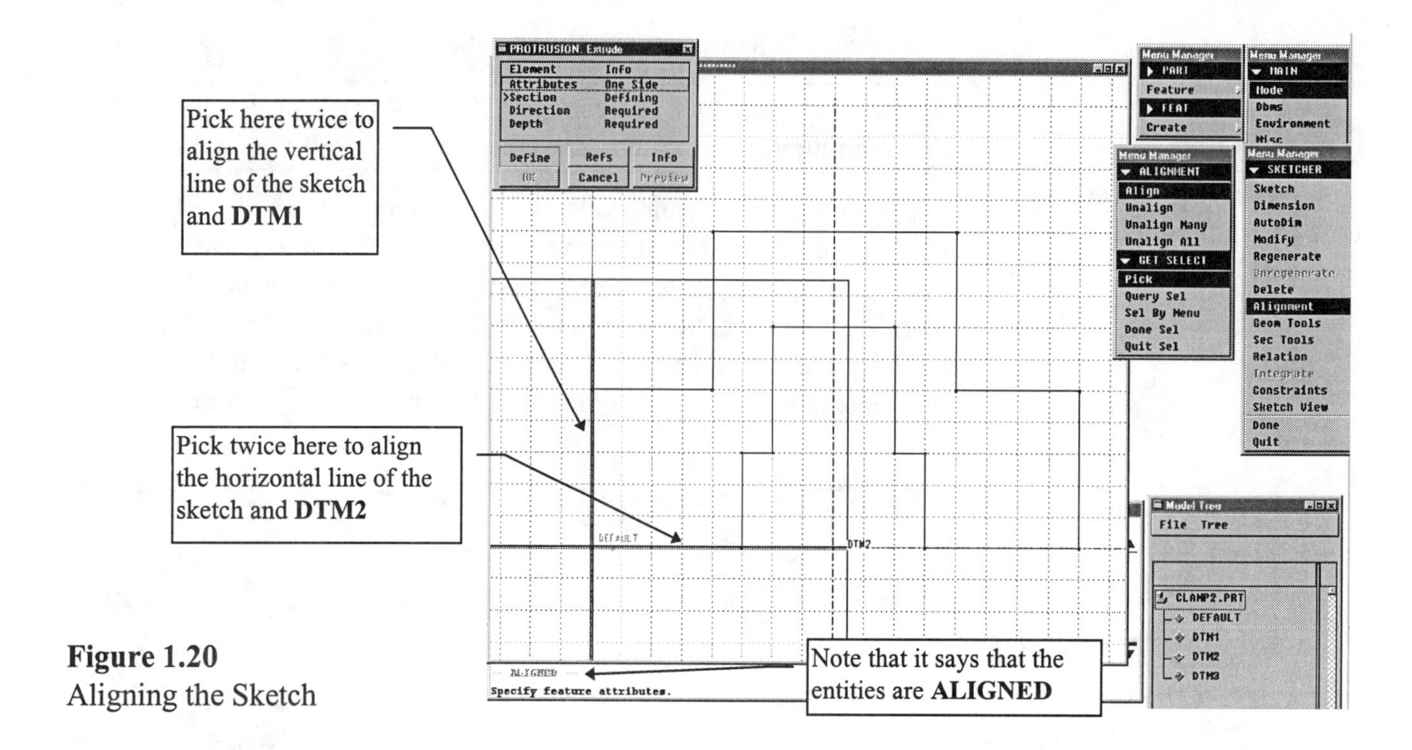

Figure 1.20
Aligning the Sketch

Look at the part again in Figures 1.1, 1.2, and 1.7. Datum **DTM2** could be datum **A**, **DTM3** datum **B**, and **DTM1** datum **C**, if geometric tolerancing was used on the part.

The sketch now needs to be dimensioned according to the design intent of the part. Look at Figure 1.7 and Figure 1.21 to see the dimensioning scheme used to create the part's geometry. Pro/E will automatically dimension a sketch, but the design intent might not be what you want.

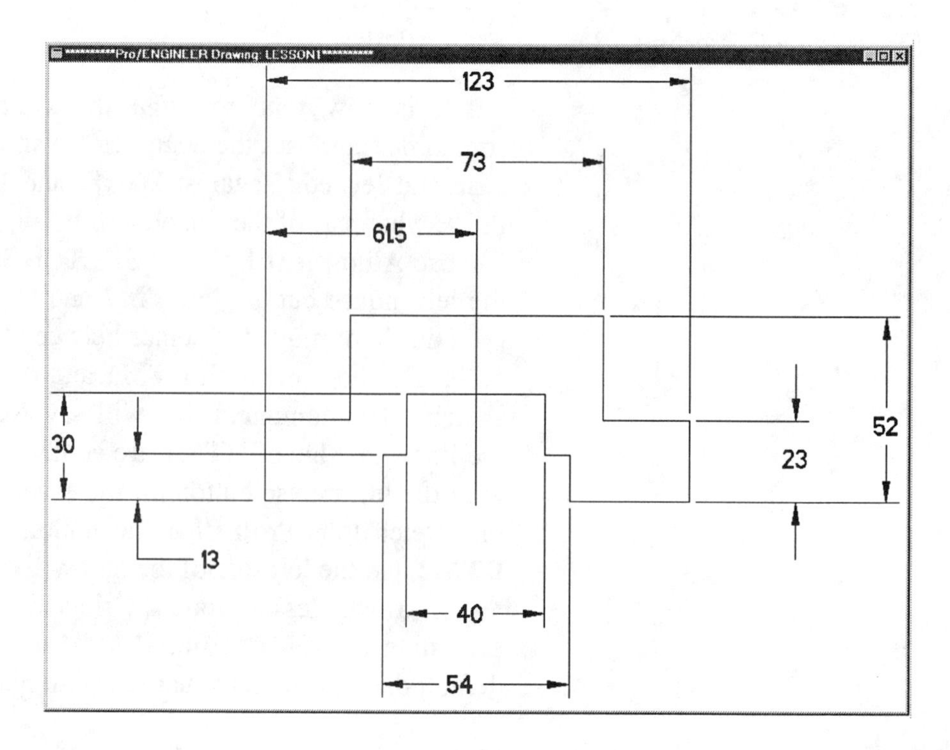

Figure 1.21
Part Dimensions

Place the dimensions as shown in Figure 1.22. Do not be concerned with the perfect positioning of the dimensions, but try to follow the spacing and positioning standards found in the **ASME Geometric Tolerancing and Dimensioning** standards. It will save you time when you create a drawing of the part.

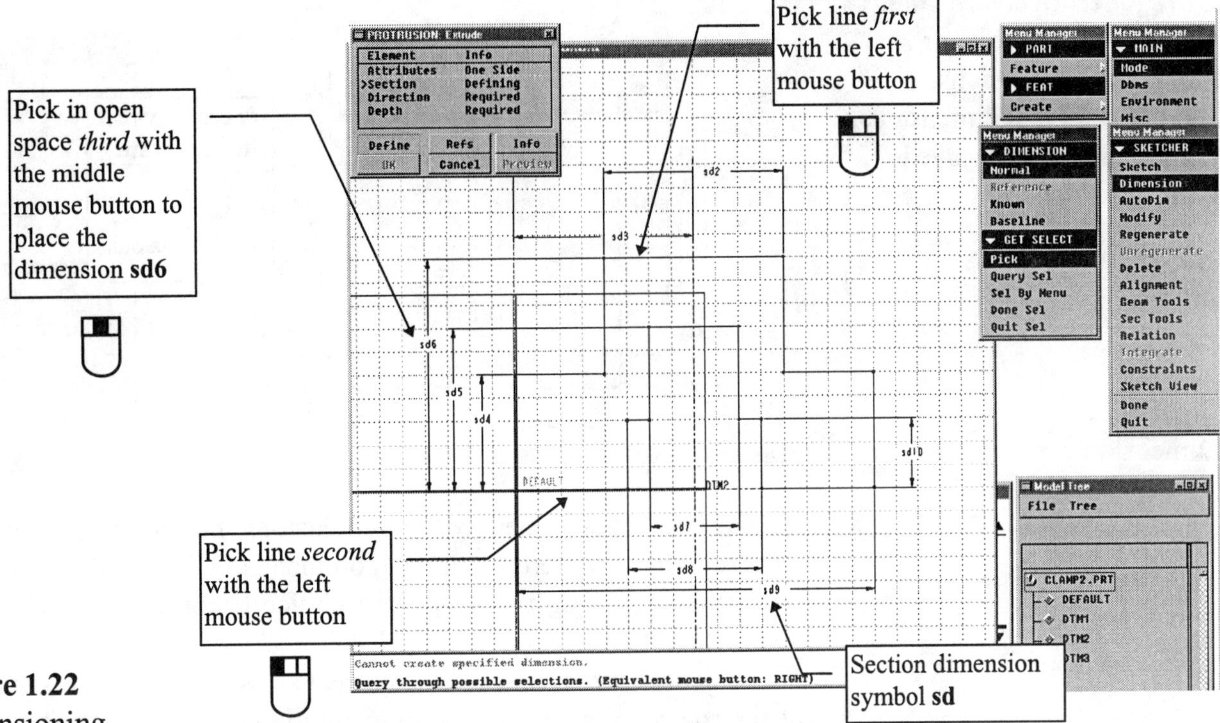

Figure 1.22
Dimensioning

The dimension you place at this stage of the design process will display on the drawing document by simply asking to show all the dimensions. The most important thing is to get enough dimensions on the sketch to regenerate. The dimensions should be ones required to manufacture the part. To dimension between two lines, simply pick the lines with the left mouse button and place the dimension value with the middle button. At this stage of the dimensioning process the dimensions are displayed as **sd** symbols: **sd1**, **sd2**, **sd3**, etc.

HINT

If you pick too many lines or choose an incorrect sequence of entities, simply pick the **Dimension** command to start and restart the dimensioning process.

It is now time to see if the sketch and dimensions will regenerate. Choose **Regenerate** from the SKETCHER menu (Fig. 1.23). The command line says: **Section Regenerated Successfully**. The **Modify** common is automatically initiated at the successful regeneration of the section. You can now change-modify the dimensions to the *design sizes*. Note that the regenerated dimensions are based on a grid size of **30** units, here **30** millimeters. All of the dimensions are too big. Pick on each dimension individually, and type the correct value at the prompt using the correct dimension from Figures 1.7 and 1.21. Make sure you change every dimension correctly. If you changed all but the center cut dimension and left it as a value greater than the part's width, the part will fail at the next regeneration. If this happens, immediately pick **Unregenerated** from the SKETCHER menu and correct your neglected dimension(s).

HINT

If the section fails, always **Unregenerate** *before* going to the next command or modifying dimensions. Modify only real dimensions, not dimensions displayed with **sd** values

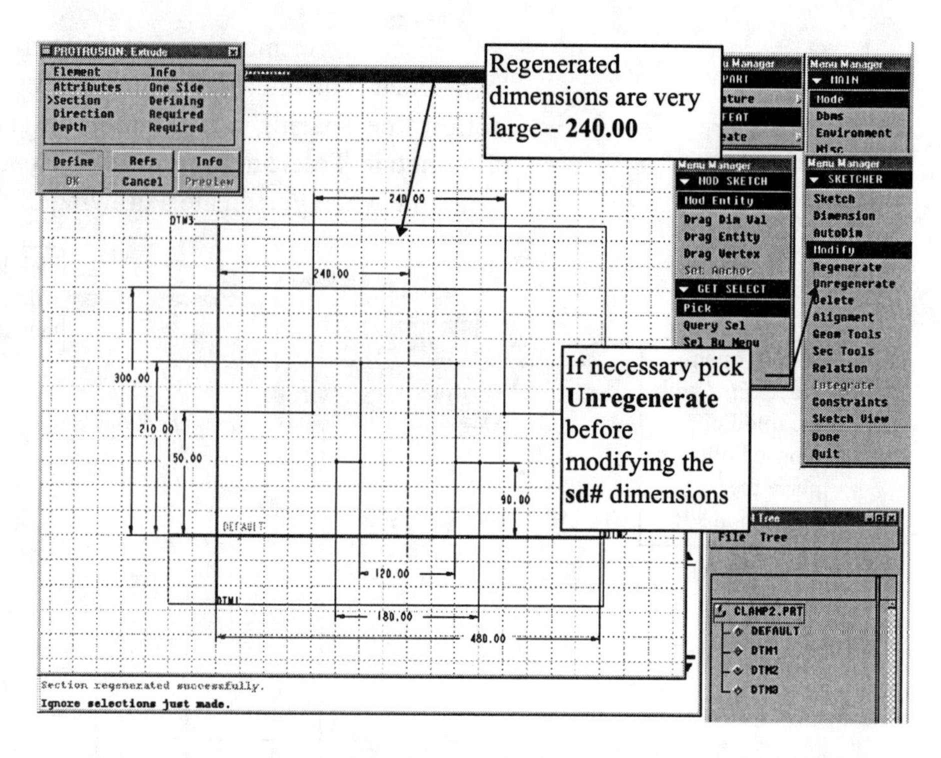

Figure 1.23
Dimensioning

Do not try to change the dimensions before unregenerating, because initiating certain Pro/E commands will block the use of **Unregenerate**. The **Unregenerate** choice becomes *dimmed* on the menu when it is unavailable (Fig. 1.24).

Regenerate the sketch. The correct sizes for the features are now displayed according to the design intent dimensions, and the section is complete. Turn off the constraint display to see the sketch more clearly (Fig 1.24).

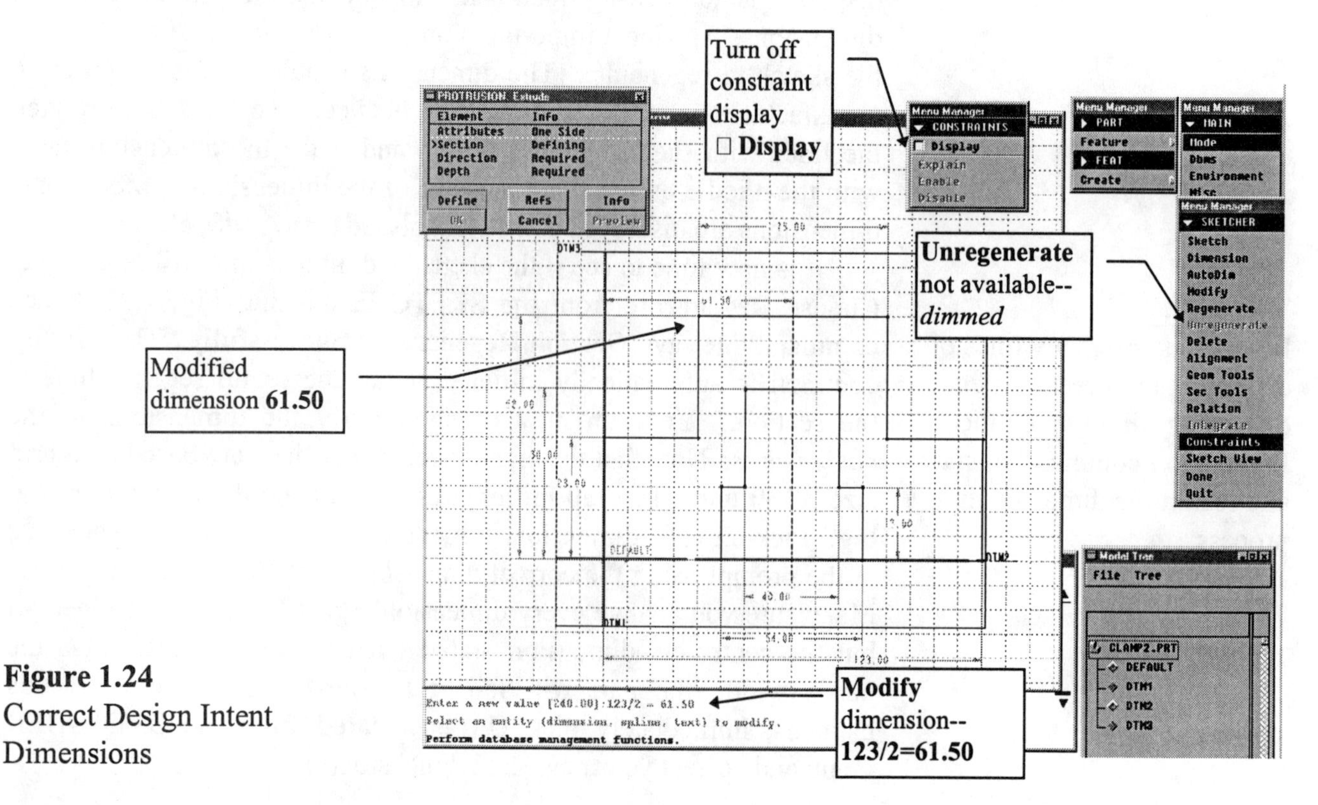

Figure 1.24
Correct Design Intent Dimensions

Complete the protrusion by choosing **Done** (in the Sketcher) ⇒ **View** (from the MAIN menu) ⇒ **Default** ⇒ **Done-Return** (from the ORIENTATION menu) to see the sketch in 3D trimetric. Pro/E now prompts for the depth in the *direction* of protrusion creation and displays an arrow → pointing in the direction previously set when **DTM3** was selected as the sketch plane (Fig. 1.25).

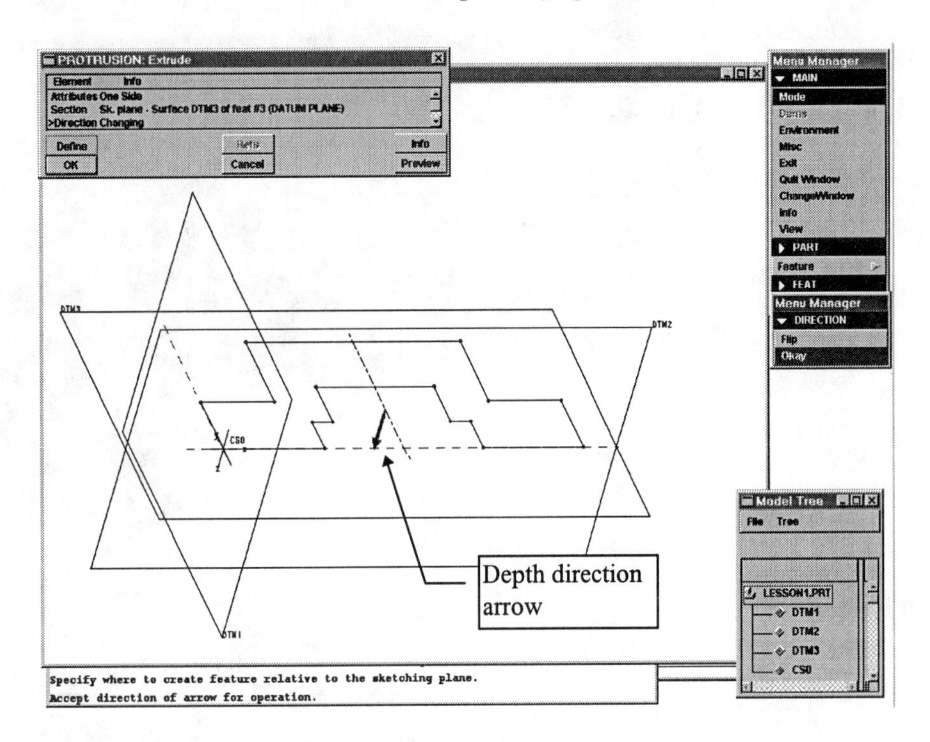

Figure 1.25
3D View of Sketch Showing the Arrow Direction

Choose **Blind** (from the SPEC TO menu) ⇒ **Done**, and at the prompt type the depth dimension of the part (**70**). Pick **Preview** from the dialog box to see the feature (Fig. 1.26).

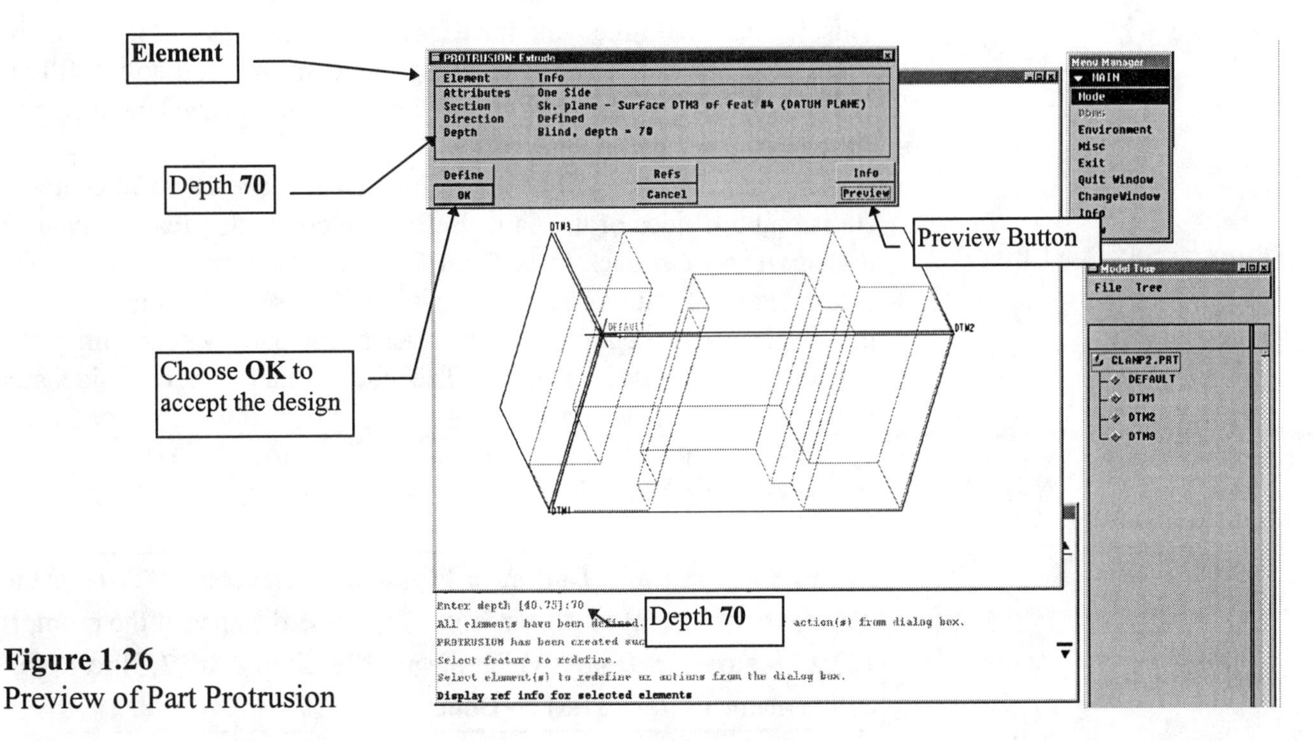

Figure 1.26
Preview of Part Protrusion

HINT

Use **Trimetric** before the first protrusion is created, and use **Isometric** as the view default after the first protrusion is completed.

At this point you could pick one of the elements from the dialog box and **Define** to change anything completed up to this point. Instead, pick **OK** from the dialog box. To see the model in a shaded state, choose **Environment** and turn off the spin center, datum planes, datum axis, etc. Make **Isometric** and **Shading** defaults, and then pick **Done-Return**. The part will be displayed as in Figure 1.27.

Shaded protrusion

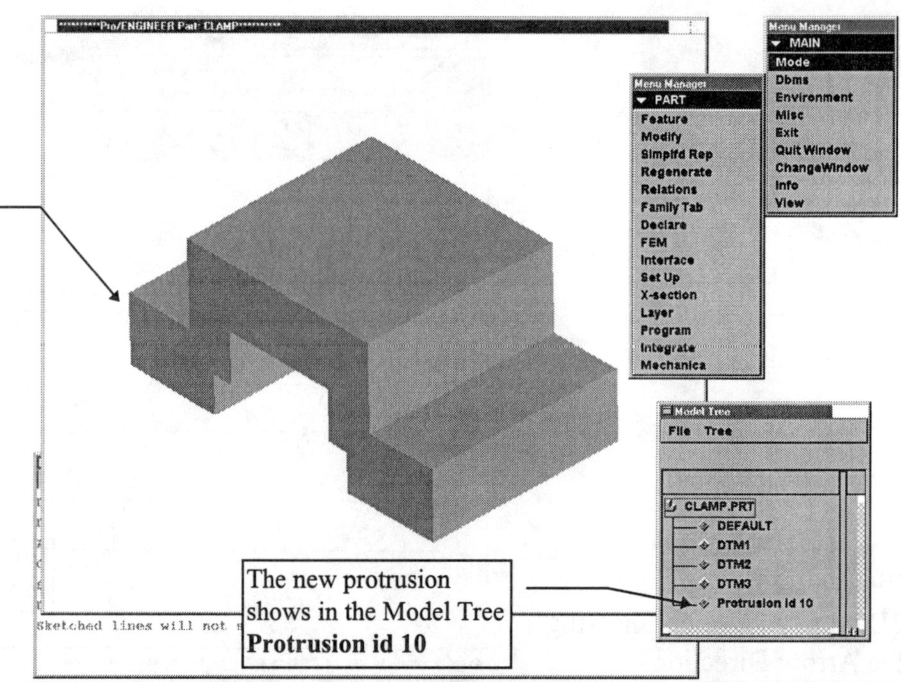

Figure 1.27
Shaded Part in Isometric View

Choose **Dbms** (from the MAIN menu) ⇒ **Save** ⇒ **Enter** ⇒ **Purge** ⇒ **Enter** ⇒ **Done-Return**. This will save the part at its present stage. Always use **Purge** after the **Save** command to eliminate previous versions of the part that may have been saved. This is the first time you have saved this part, so **Purge** is not necessary. But it is a good habit to get into so that you do not fill up your hard drive with versions of the same part saved at different stages in the design process.

Dbms ⇒ Save ⇒ Enter
Purge ⇒ enter ⇒
Done-Return

The next feature will be a cut (Fig. 1.28). The **20 X 20** centered cut is on both sides of the part. Before we create this feature, change the **Environment** back to **Hidden Line** and turn on the settings for datum planes, datum axes, etc. Leave the view set as isometric. Since the cut feature is identical on both sides of the part, we can mirror the cut after it has been created. The cut is made on one side and mirrored about a datum plane to the other side of the protrusion. Start by creating a new datum plane that is offset from **DTM1**. Give the following commands (Fig. 1.29):

Feature ⇒ **Create** ⇒ **Datum** ⇒ **Plane** ⇒ **Offset** (pick **DTM1** as the plane to offset from) ⇒ **Enter Value** (type the distance at the prompt) **123/2** ⇒ **enter** ⇒ **Done** (which means **123** divided by **2**; Pro/E will do the math: **123/2 = 61.5**) ⇒ **Done**

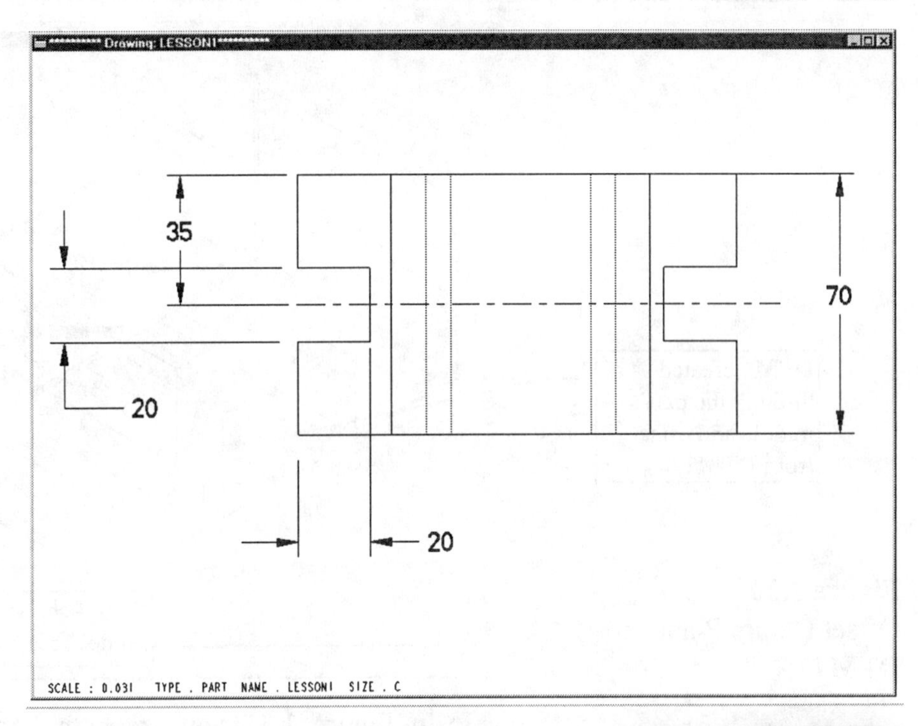

Figure 1.28
Cut Dimensions

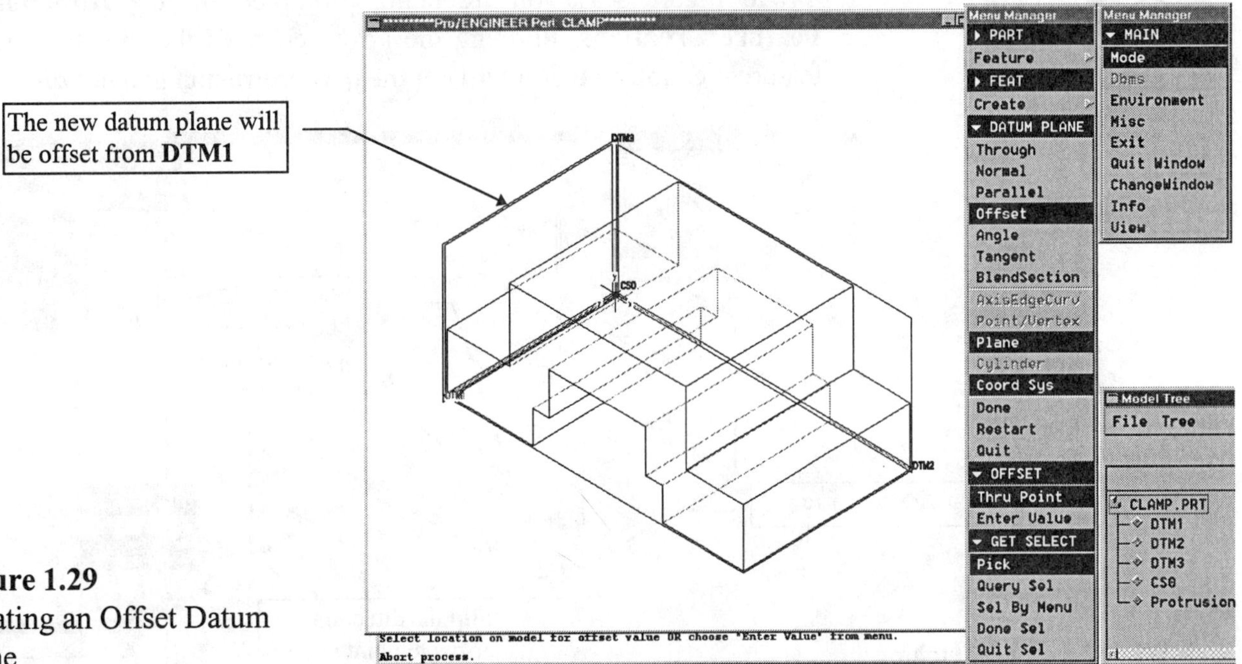

Figure 1.29
Creating an Offset Datum Plane

The datum plane is offset from **DTM1,** which is its *parent feature*. The new datum plane is shown on the screen passing through the center of the part and also appears in the Model Tree (Fig. 1.30). **DTM4** can now be used to construct other features as required by the design.

The cut will be a sketched feature similar to the first protrusion except that it will remove material. Give the following commands:

Feature ⇒ **Create** ⇒ **Cut** ⇒ **Extrude** (default) ⇒ **Solid** (default) ⇒ **Done** ⇒ **One side** (default) ⇒ **Done** (pick **DTM2** as the sketching plane) ⇒ **Flip** ⇒ **Okay** ⇒ **Top** (pick **DTM3** to orient the sketch)

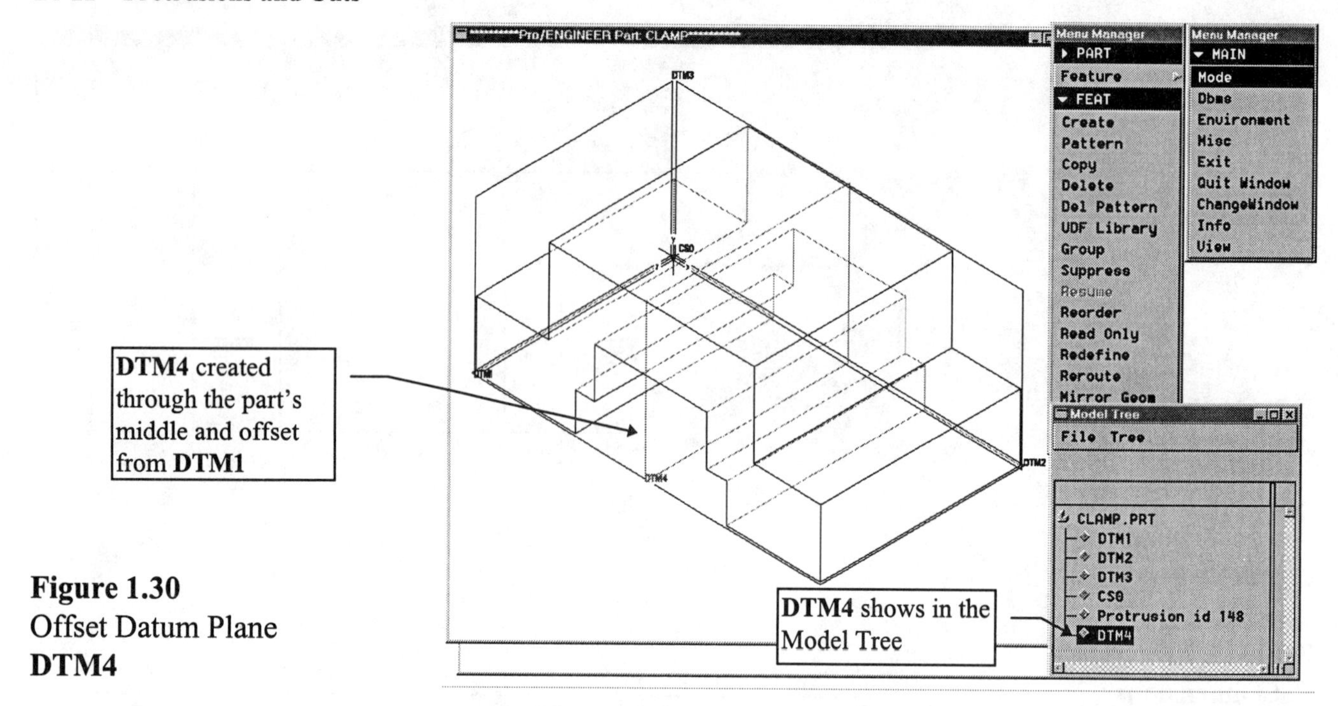

Figure 1.30
Offset Datum Plane
DTM4

In Figure 1.31 you are being prompted for the **Direction of Feature Creation**. Flipping the arrow changes the direction of the feature creation so that it will cut the part protrusion and not *air*!

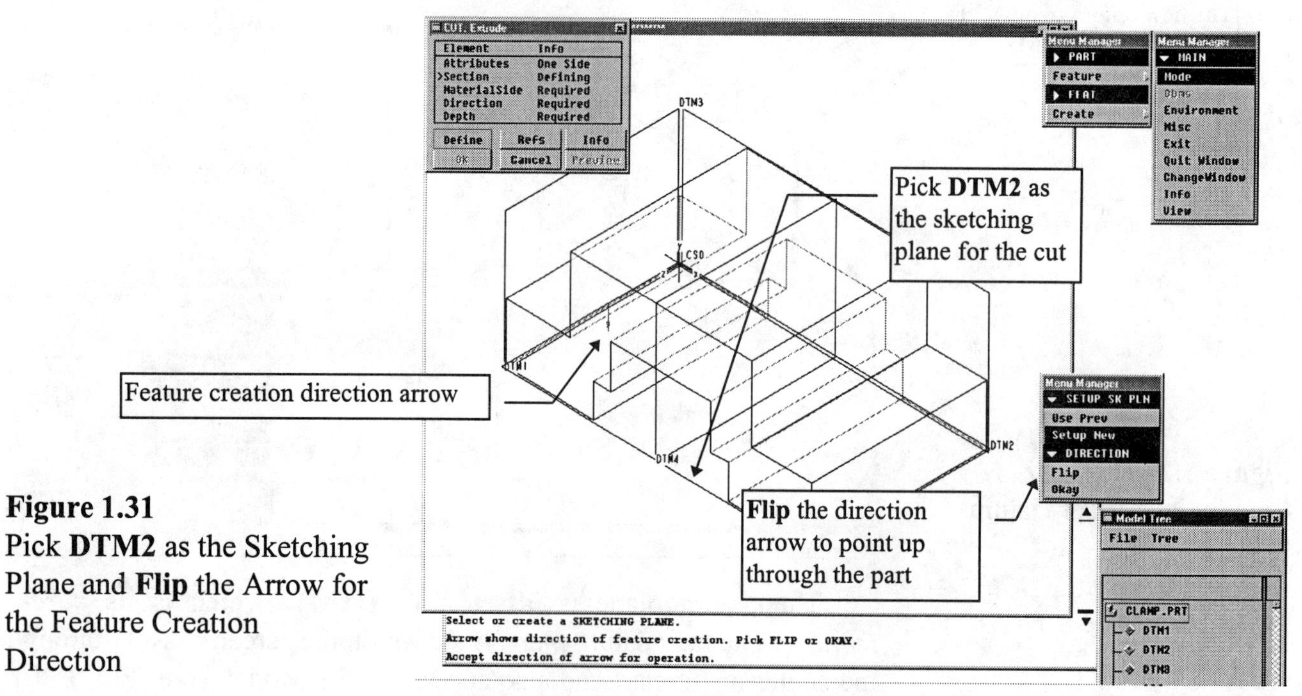

Figure 1.31
Pick **DTM2** as the Sketching Plane and **Flip** the Arrow for the Feature Creation Direction

In Figure 1.32 you are looking at the *bottom of the part*. At first it is hard to see the orientation of the part. To get visually oriented, note the location of the coordinate system. **DTM1** and **DTM3** show as edges and will be used to dimension the sketch. The sketch is composed of three lines forming an *open section*. The endpoints of the two lines that touch **DTM1** will be aligned with that datum plane. The dimensioning scheme and values will be the same as those shown in Figure 1.28. The edge of the cut is dimensioned from **DTM3**.

Turn the **Grid Snap** off in the ENVIRONMENT menu, and continue the commands (Fig. 1.32):

Environment ⇒ □ **Grid Snap** ⇒ **Done-Return** ⇒ **Sketch** ⇒ **Line** ⇒ **Horizontal** (pick on **DTM1** to start the line)

Complete the three continuous lines by picking four endpoints, as shown in Figure 1.32. After the second pick, the **Sketcher** toggles from horizontal to vertical line option automatically.

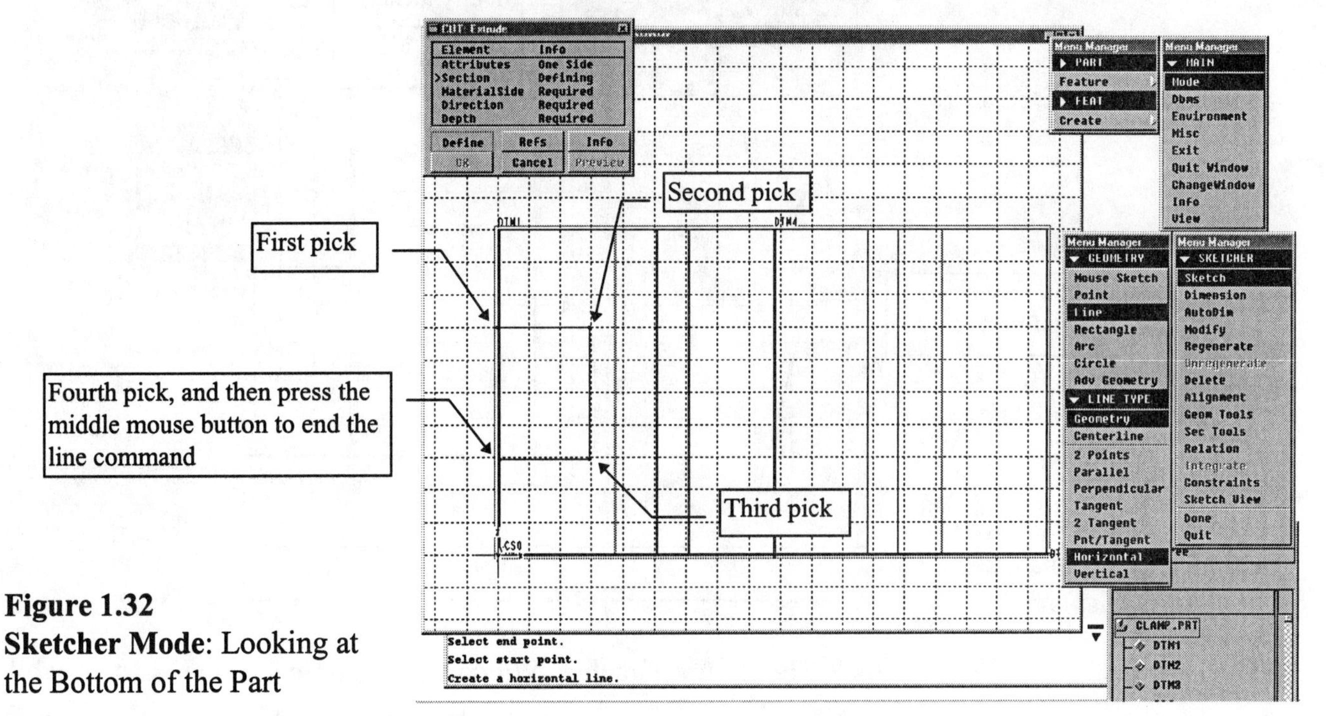

Figure 1.32
Sketcher Mode: Looking at the Bottom of the Part

Next, align the endpoints of the cut where they touch **DTM1** (Fig. 1.33). Three dimensions are needed to fix the feature's position on the part. The view shown in Figure 1.33 was panned and zoomed.

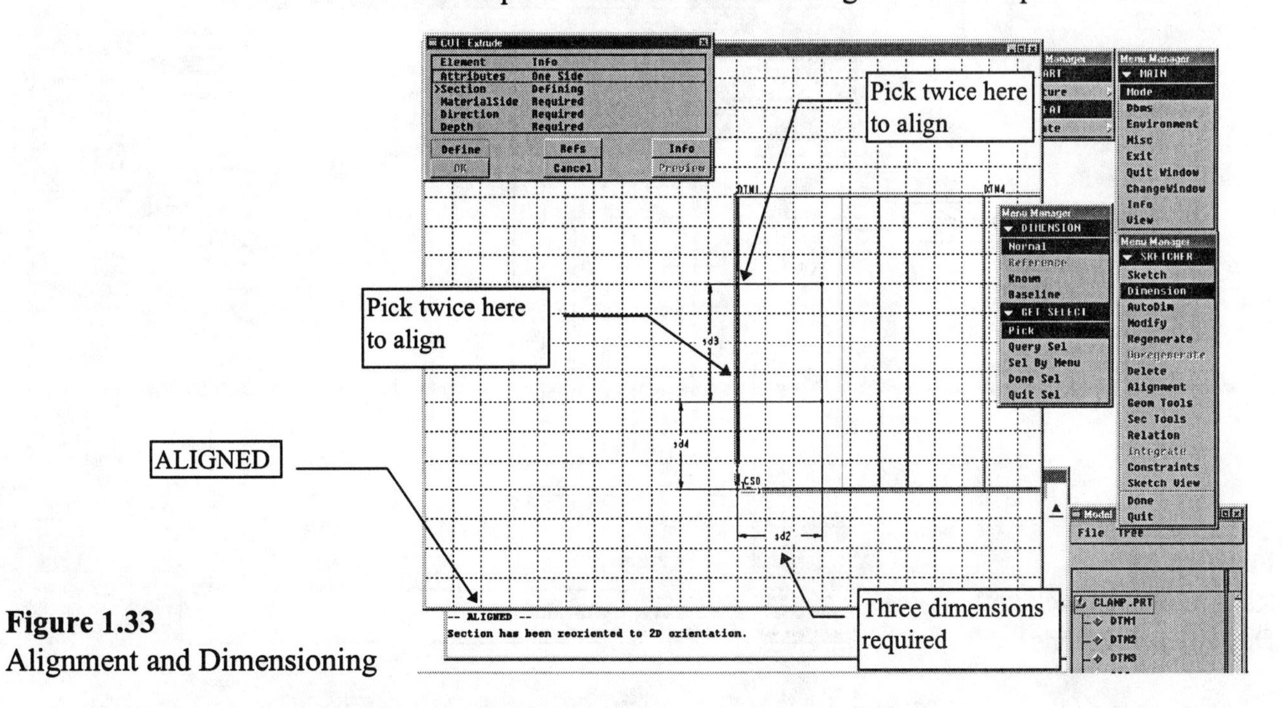

Figure 1.33
Alignment and Dimensioning

It is now time to **Regenerate** the sketch to see if you created, aligned, and dimensioned it correctly (Fig. 1.34). After the successful regeneration of the rough sketch, the sketch constraints will show on screen and the MOD SKETCH menu options are available. Pick on each dimension and type the design sizes for the dimensions. When done modifying the values, choose **Regeneration** again.

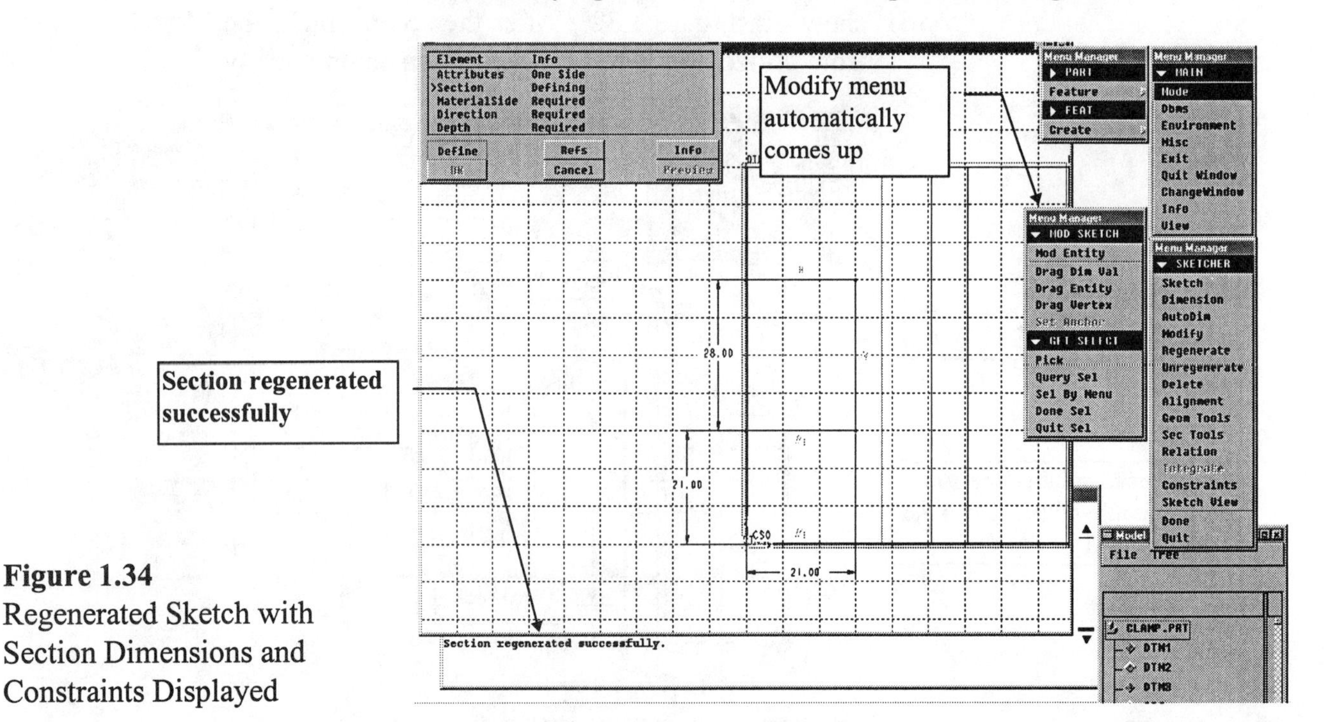

Figure 1.34
Regenerated Sketch with Section Dimensions and Constraints Displayed

The Main Graphic Window will *animate* the sketch while regenerating. The cut's section is now complete (Fig. 1.35). Choose **Done** from the SKETCHER menu.

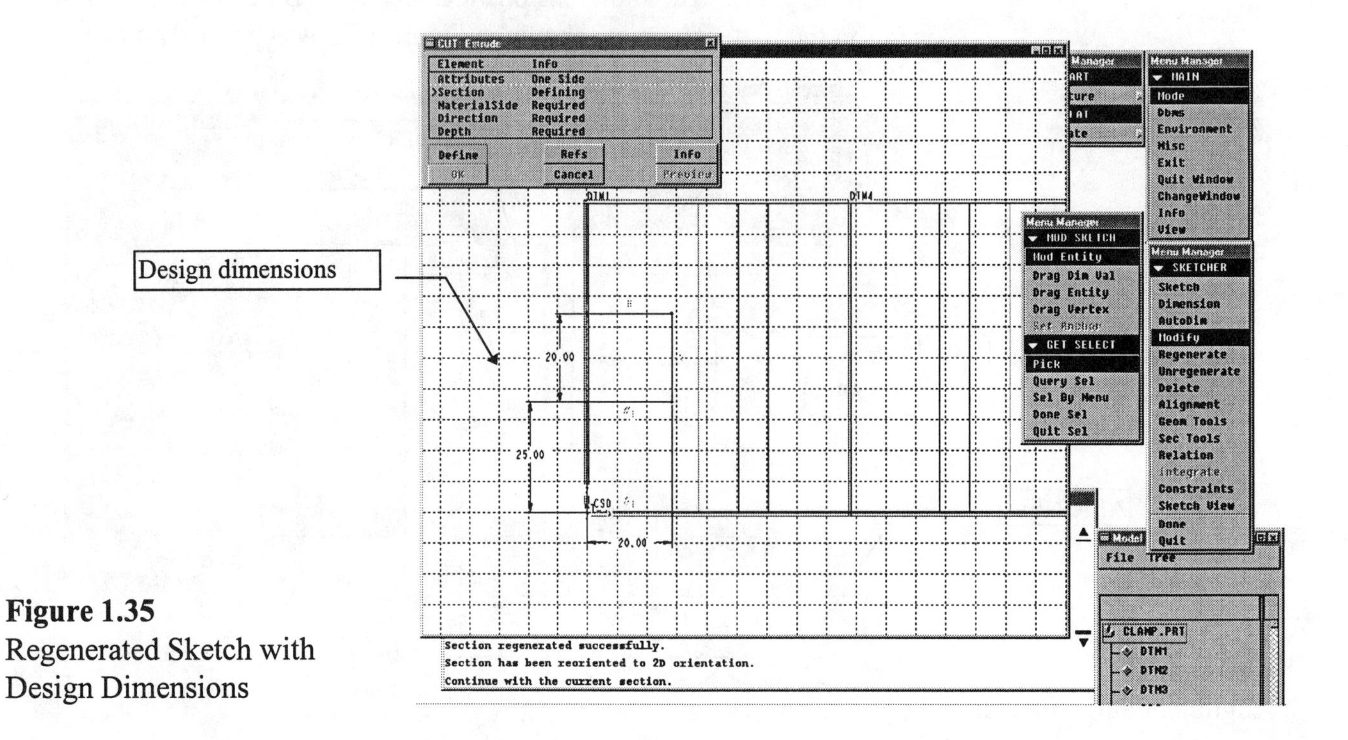

Figure 1.35
Regenerated Sketch with Design Dimensions

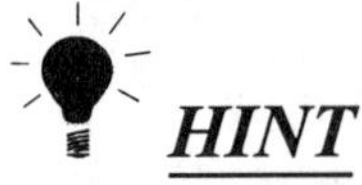

HINT

If the removal direction was pointing away from the cut and all that was left after the command was completed was a small square block of material, pick **Direction** and **Define** from the dialog box and redo the command.

The direction of material removal is required at this stage of the cut creation. An arrow is displayed showing the direction that the material will be removed (Fig. 1.36). Here, the direction arrow points toward the area to be removed for the cut; therefore pick **Okay** from the DIRECTION menu. If you select the incorrect direction for the arrow material removal direction, the whole part will be cut away and the small **20 X 20** square piece will be all that's left of your part!

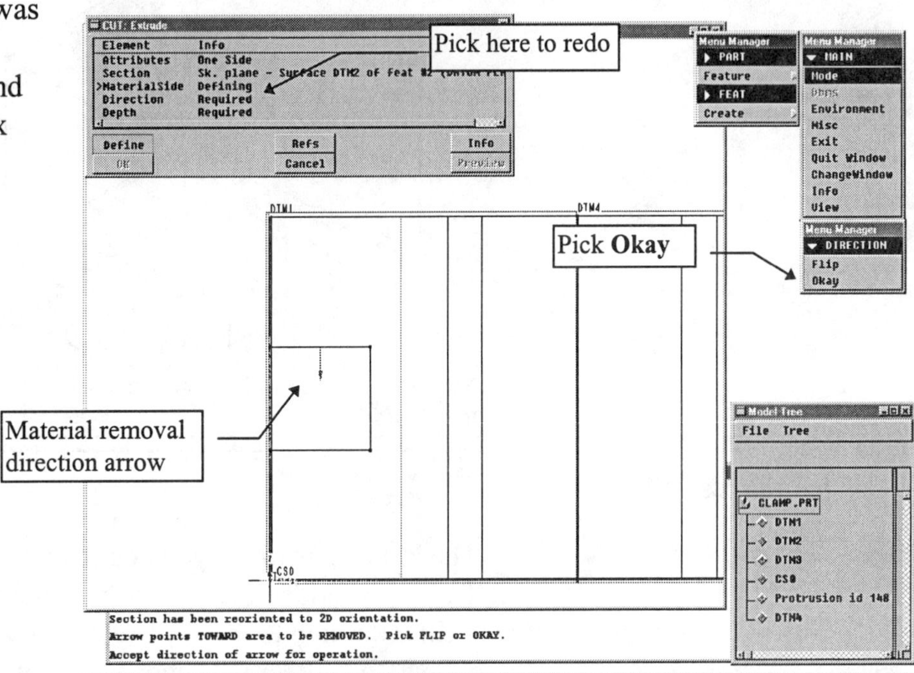

Figure 1.36
Direction of Material Removal

We recommend changing to a 3D view orientation in order to identify the correct cut-depth direction (Fig. 1.37).

Choose **View** ⇒ **Default** ⇒ **Done-Return** ⇒ **Okay** (make sure the cut arrow points into the model) ⇒ **Thru All** (from SPEC TO menu)

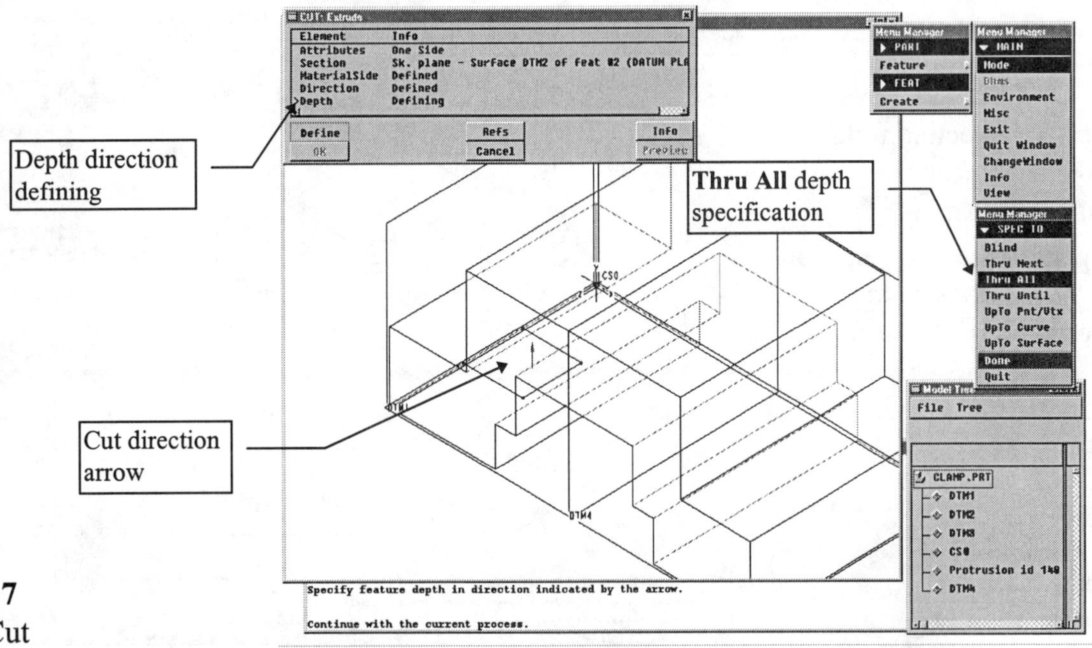

Figure 1.37
Depth of Cut

Choose **Done** from the SPEC TO menu and **Preview** from the dialog box; or if you don't wish to the see the feature previewed, choose **OK** from the dialog box (Fig. 1.38).

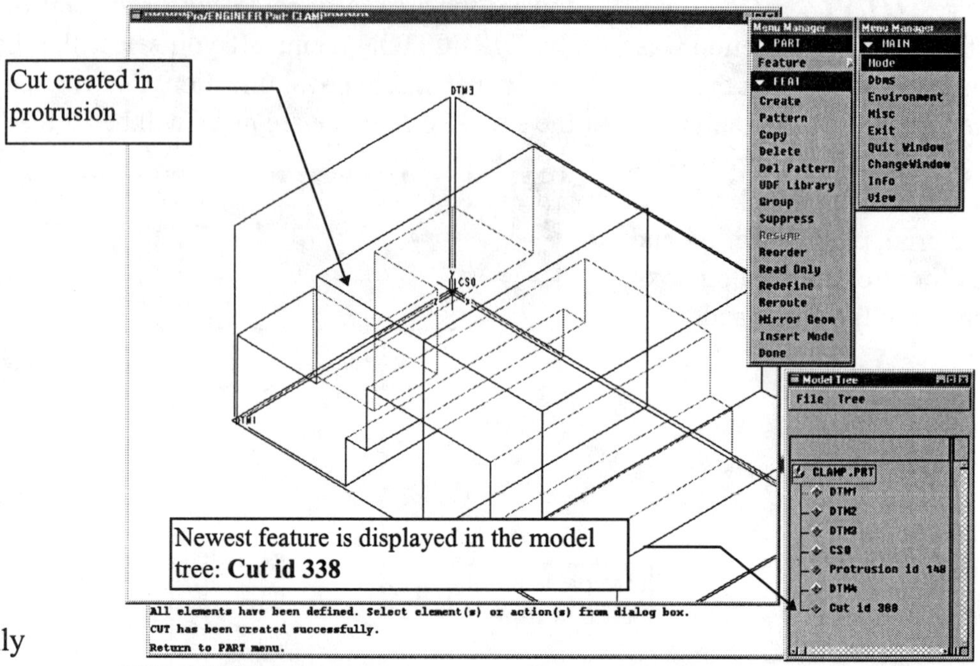

Figure 1.38
Cut Created Successfully

The last step is to copy the cut feature to the right side of the part. The feature is hidden behind the protrusion, so you must use **Query Sel** to filter through to the cut feature. It will highlight when selected. From the FEATURE menu choose the following (Fig. 1.39):

Copy ⇒ **Mirror** ⇒ **Independent** ⇒ **Done** ⇒ **Query Sel** (pick on the cut feature) ⇒ **Next** (from the CONFIRM menu) ⇒ **Accept** ⇒ **Done-Select** ⇒ **Done**

HINT
Instead of selecting from the CONFIRM menu use your *right mouse button* to query to the next selection and the *middle mouse button* to accept the selection

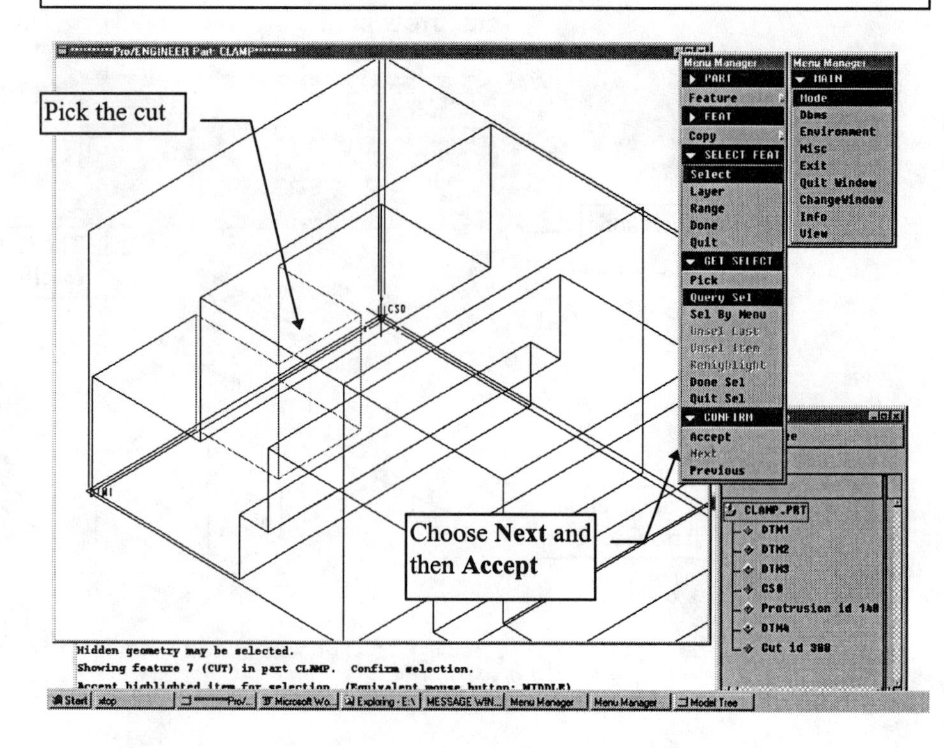

Figure 1.39
Copying the Cut

Pro/E now prompts you to select the plane or datum to use as the mirroring plane (Fig. 1.40). The cut will be mirrored about **DTM4**.

Dbms ⇒ Save ⇒ Enter
Purge ⇒ Enter ⇒
Done-Return

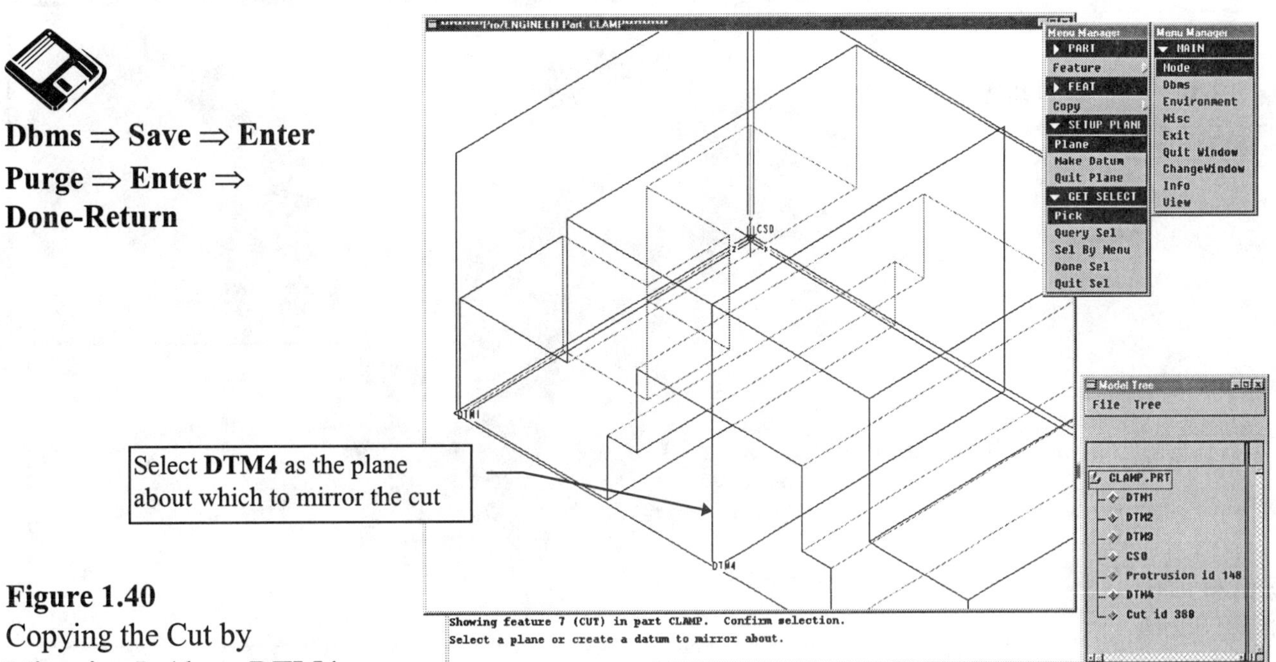

Figure 1.40
Copying the Cut by Mirroring It About **DTM4**

Pick on the edge or name of datum plane **DTM4**. You can use **Query Sel** here if you wish, or use your mouse buttons. After you select **Done Sel**, Pro/E completes the command and displays the part. You have completed your first Pro/E part! To see the protrusion and the cuts clearly, rotate and shade the part as shown the Figure 1.41:

Environment ⇒ Shading ⇒ Done-Return ⇒ View ⇒ Default ⇒ Done-Return ⇒ Dbms ⇒ Save ⇒ enter ⇒ Purge ⇒ enter

HINT
Hold down the keyboard **Ctrl** key and press the right (**Pan**), left (**Zoom**), or middle (**Rotate**) mouse button to orient the part on the screen

Pan **Zoom** **Rotate**

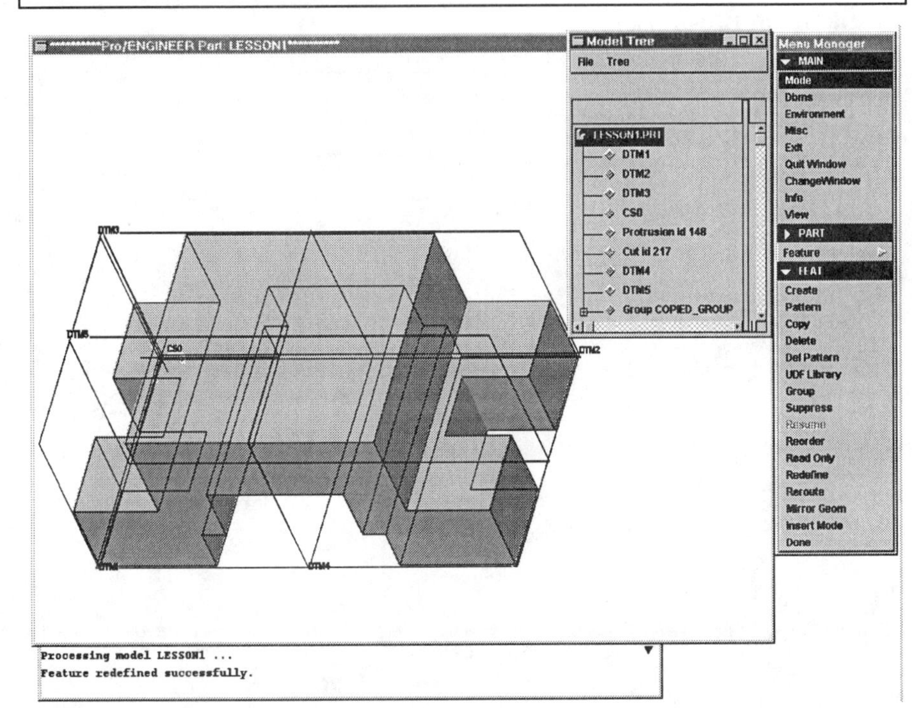

Figure 1.41
Rotated and Shaded Clamp

Lesson 1 Project

Angle Block

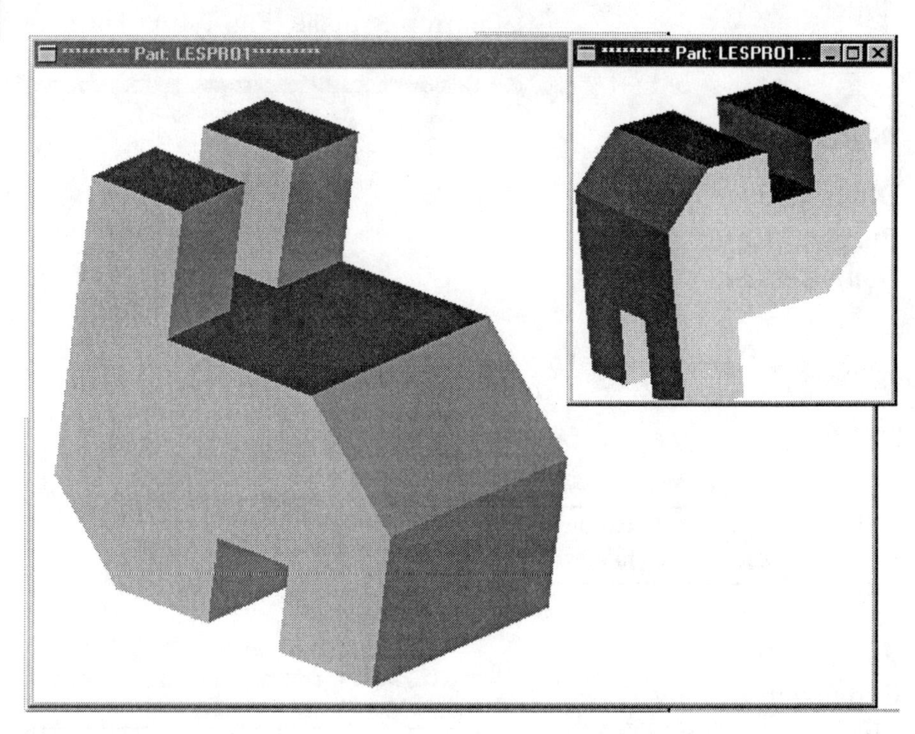

Figure 1.42
Angle Block

Angle Block

The first **lesson project** is a simple block that requires many of the same commands as the **Clamp**. Using the datums, create the part shown in Figures 1.42 through 1.46. First, create a default coordinate system and the default datum planes. Sketch the protrusion on **DTM3**. Sketch the cut on **DTM1**, and align it to the upper surface/plane of the protrusion, as shown in Figure 1.43.

? Pro/HELP
Remember to use the help available on Pro/E by highlighting a command and pressing the right mouse button 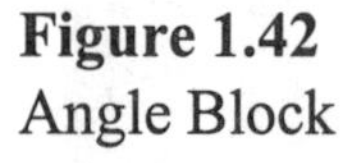 **? GetHelp**

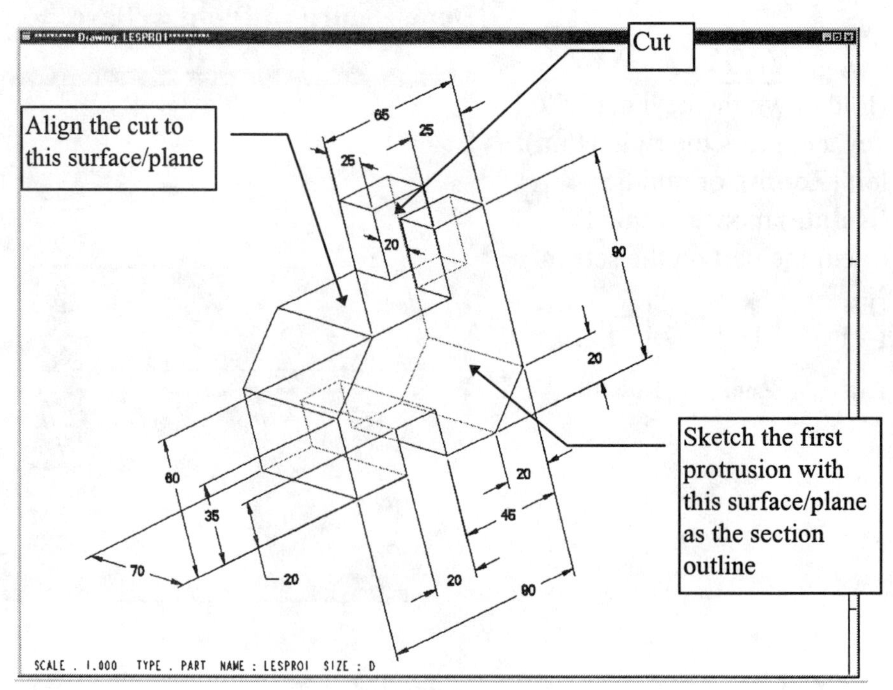

Figure 1.43
Angle Block Dimensions

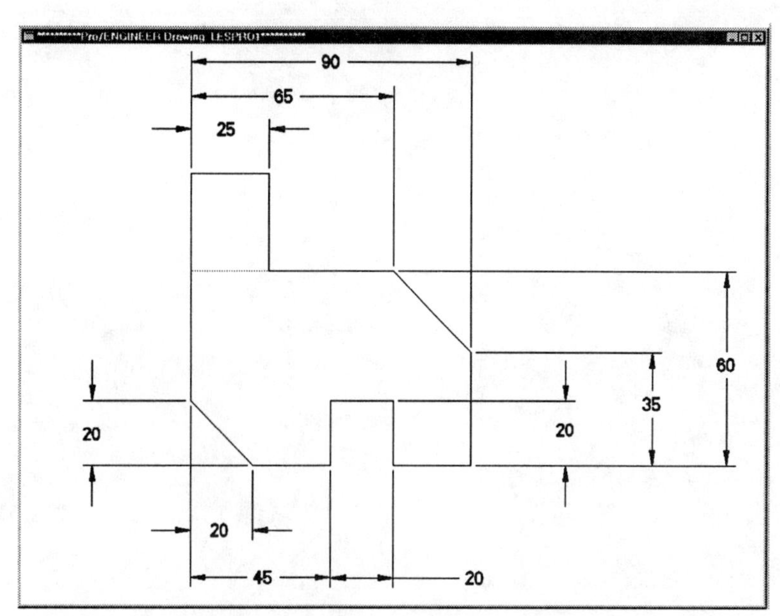

Figure 1.44
Angle Block Protrusion Sketch Dimensions

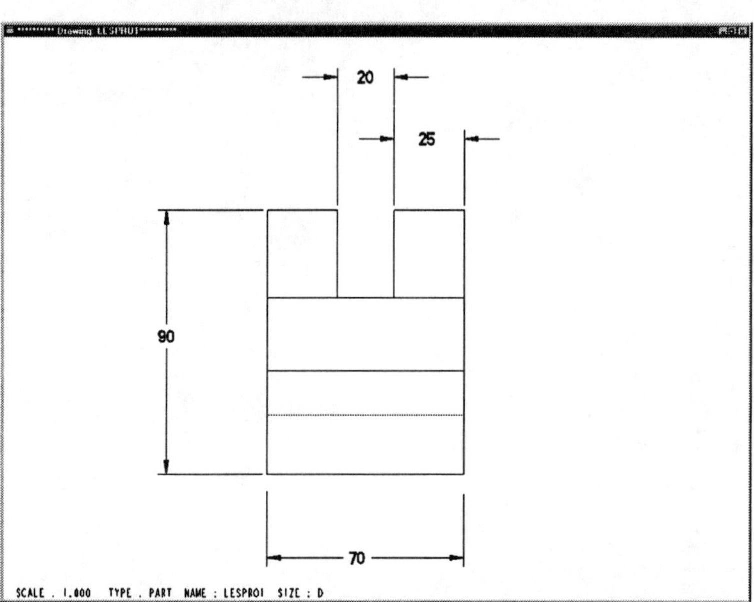

Figure 1.45
Angle Block Cut Sketch Dimensions

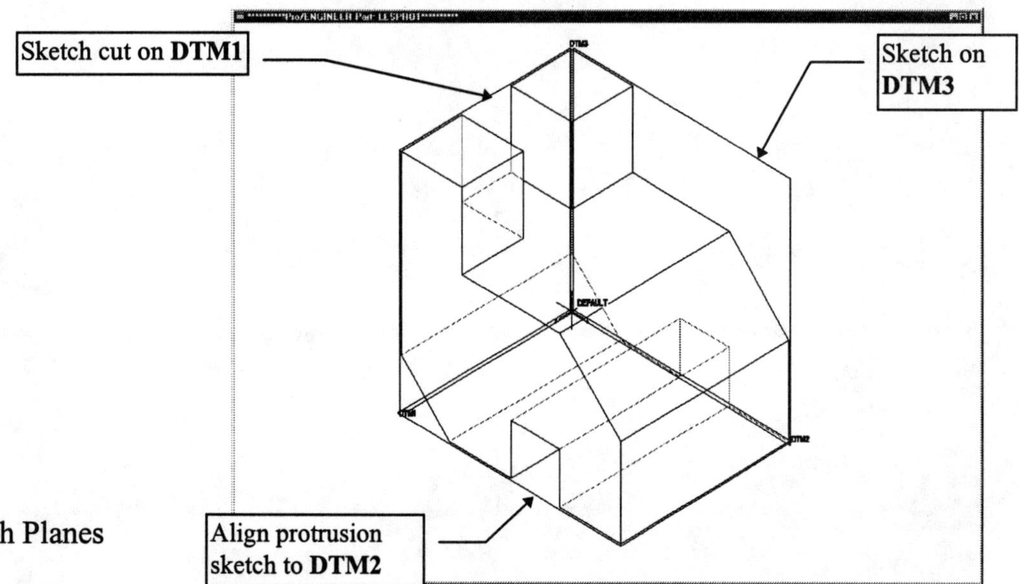

Figure 1.46
Angle Block Sketch Planes and Alignments

Lesson 2

Modify and Redefine

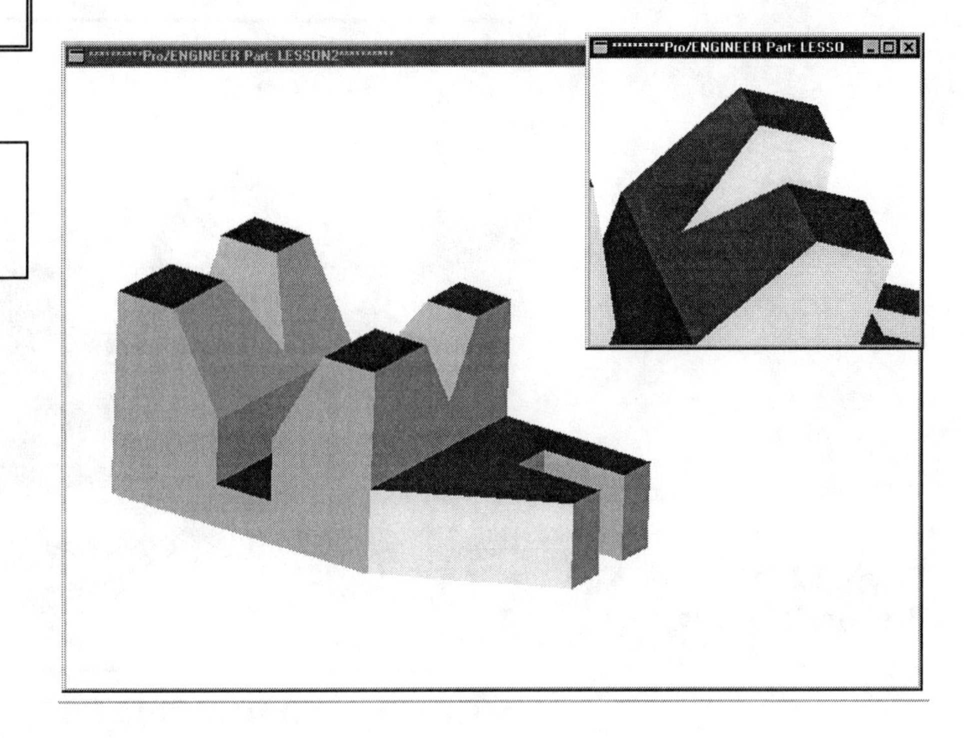

Figure 2.1
Base Angle

OBJECTIVES

1. **Change existing features by modifying dimensions**
2. **Alter the view consideration in dimension display**
3. **Redefine a part's features**
4. **Modify the number of decimal places of dimensions**
5. **Regenerate modified and redefined parts**
6. **Input a *config.pro* change**
7. **Use the Info command to get feature and model information**
8. **Use the Model Tree to modify, redefine, and get information**

EGD REFERENCE
Engineering Graphics and Design
by L. Lamit and K. Kitto
Read Chapters: 10
See Pages: 298, 399

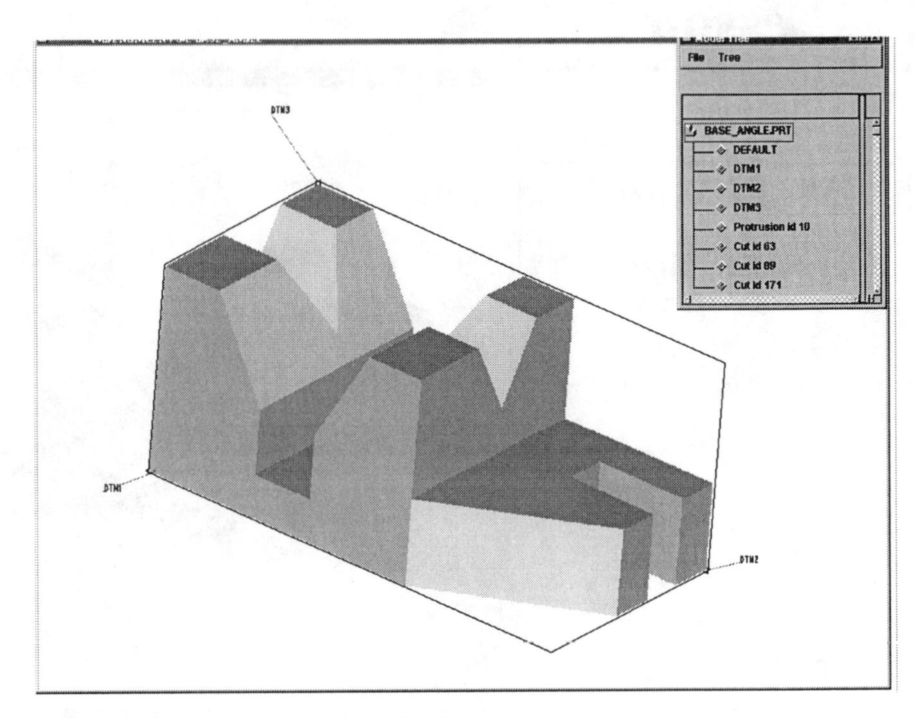

Figure 2.2
Base Angle Showing Datums, and Model Tree

MODIFY AND REDEFINE

Both **Modify** and **Redefine** are important capabilities in the design of parts (Fig. 2.1). With **Modify**, you can change any dimension used in the creation of a feature. With **Redefine**, the feature's attribute, the placement plane, the placement/orientation references, and the size and configuration of the feature can be redone. **Modify** is for simple dimensional changes, and **Redefine** is for more comprehensive changes to the model's features. You will be creating the **BASE ANGLE** shown in Figure 2.2 and then modifying and redefining some of its features. If you have **COAch for Pro/Engineer**, complete the appropiate segment (Fig. 2.3).

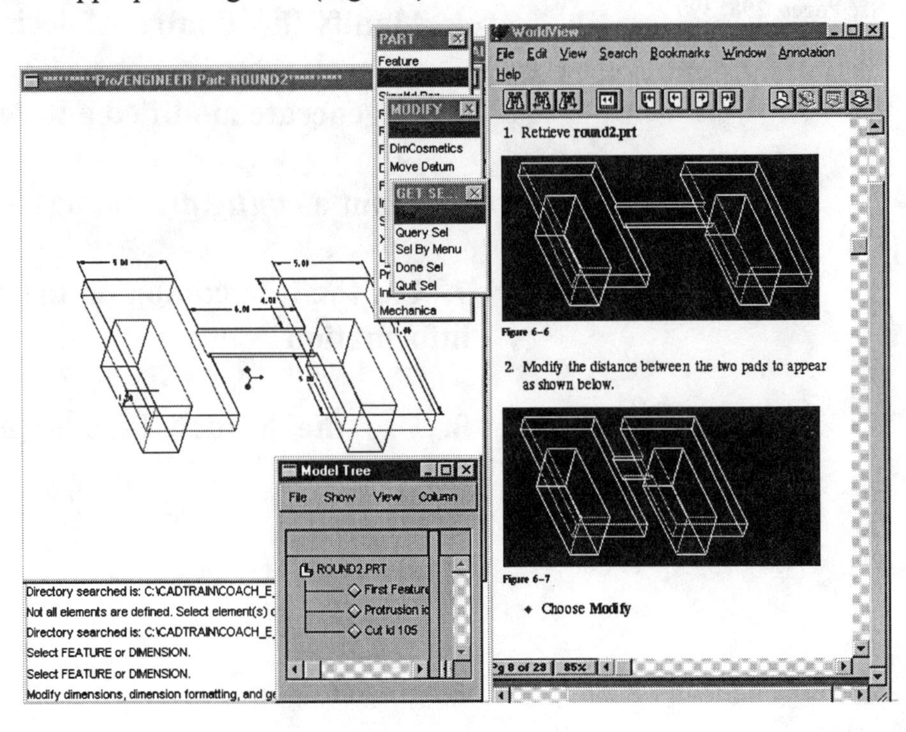

Figure 2.3
COAch for Pro/E, Basic Modifications to a Part--Modeling (Segment 1: Modify)

Modify

To modify dimensions, choose **Modify**, then **Value** (default) from the MODIFY menu, and select a feature. Pro/E displays all the dimensions associated with the selected feature. If you pick an edge that is shared by two features while you are using the **Query Sel** option, Pro/E highlights the associated features in turn. Pro/E displays the CONFIRM menu, which, after you pick **Next**, lets you step back and forth through the highlighted features to accept the ones you want.

View Orientation When Displaying Dimensions

The view orientation of a part can be adjusted to improve clarity in viewing dimensions. The need to change the view orientation becomes apparent when the dimensions overlap, the dimensions are in planes that are perpendicular to the current view, or Pro/E displays the section dimensions on the original sketching plane. Hold down the **Ctrl** key and press the middle button of the mouse to rotate the model to the desired position. Figure 2.4 shows the available help for the modify--view consideration.

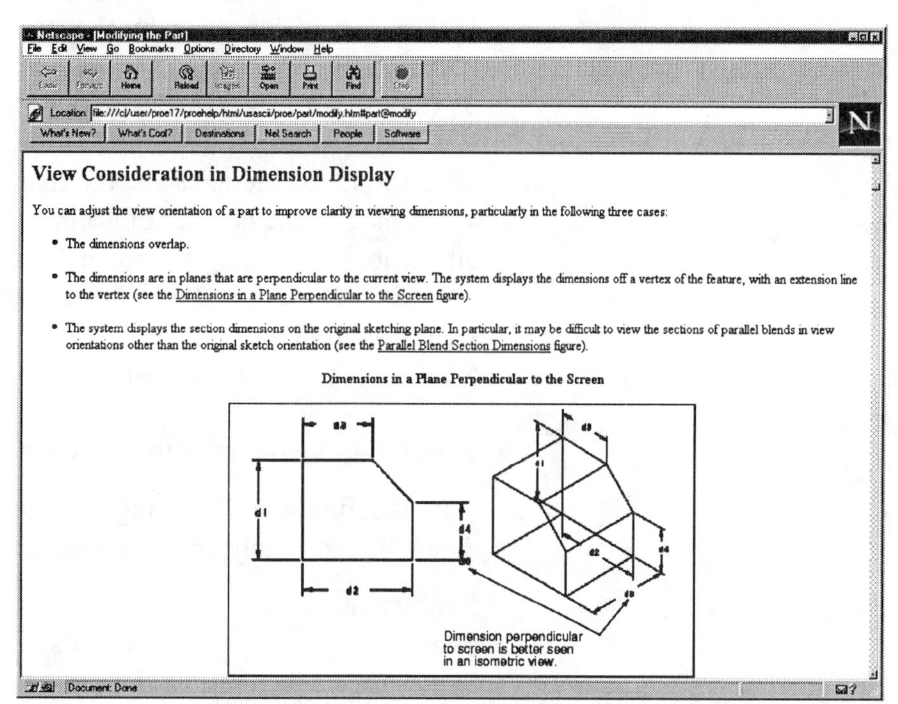

Figure 2.4
Online Documentation
Modify--View Consideration

Modifying Dimension Values

When you modify the value of a dimension, you can enter a new number or a **relation**. Pro/E supports the use of negative dimensions. The value entered depends on the displayed sign of the dimension. Pro/E displays all dimensions as positive values, and entering a negative value tells Pro/E to create the section geometry to the opposite side; but the *direction* of a feature creation cannot be changed by entering a negative number. Use the **Redefine** option to redefine the direction of the feature.

To modify a dimension value:

1. Choose **Modify** from the PART menu
2. Display the dimensions of a feature by picking on any surface of the desired feature.
3. Pick the dimension to change. The value highlights **(red)**, and Pro/E displays a prompt in the MESSAGE WINDOW.
4. Enter a new value, or accept the current value by pressing the **Enter** key. In many cases, this value can be negative. This new value (displayed in white) replaces the old value.
5. Modify other dimensions as required.
6. When you have completed all the changes, choose **Regenerate** to recalculate the part using the new dimension values.

Modifying the Number of Decimal Places for Dimensions

The default number of decimal places for dimensions is two. To increase the precision of a particular dimension, enter a new value with the desired precision. Modifying the number of decimal places for a dimension rounds the value of the dimension.

To decrease the precision of a particular dimension:

1. Turn the tolerances on by choosing **Display Tol** from the ENVIRONMENT menu.
2. Choose **Modify DimCosmetics Format.**
3. Choose **Nominal** from the DIM FORMAT menu, then pick a dimension. Its tolerance display changes to nominal.
4. To modify the number of decimal places to display for one or more dimensions (including reference dimensions), choose the DIM COSMETIC menu and pick **Num Digits**. Enter the number of significant digits.
5. Select the dimensions whose display is to be changed.

Redefining Features with Elements

You redefine the following features using the Feature Definition dialog box to change the elements with which they were created.

Protrusions	Cuts	Slots
Shells	Rounds	Holes
Some surface features	Drafts	Draft offsets
Some datum curves	Shafts	

To redefine a feature that has elements:

1. Choose the **Redefine** option from the FEAT menu, and pick the feature to be redefined.
2. Pro/E displays the Feature Definition dialog box. Each element and its current value are listed. Select the element to redefine, and then select the **DEFINE** button. Pro/E prompts for the information needed to redefine the element.

Redefining Features

The **Redefine** option in the FEAT menu allows you to change how a feature is created, including section geometry. The types of changes you can make depend on the selected feature. A cut has a *section*, therefore it can be redefined. If you have **COAch for Pro/ENGINEER** on your system, go to SEARCH and complete the appropiate Segment shown in Figure 2.5.

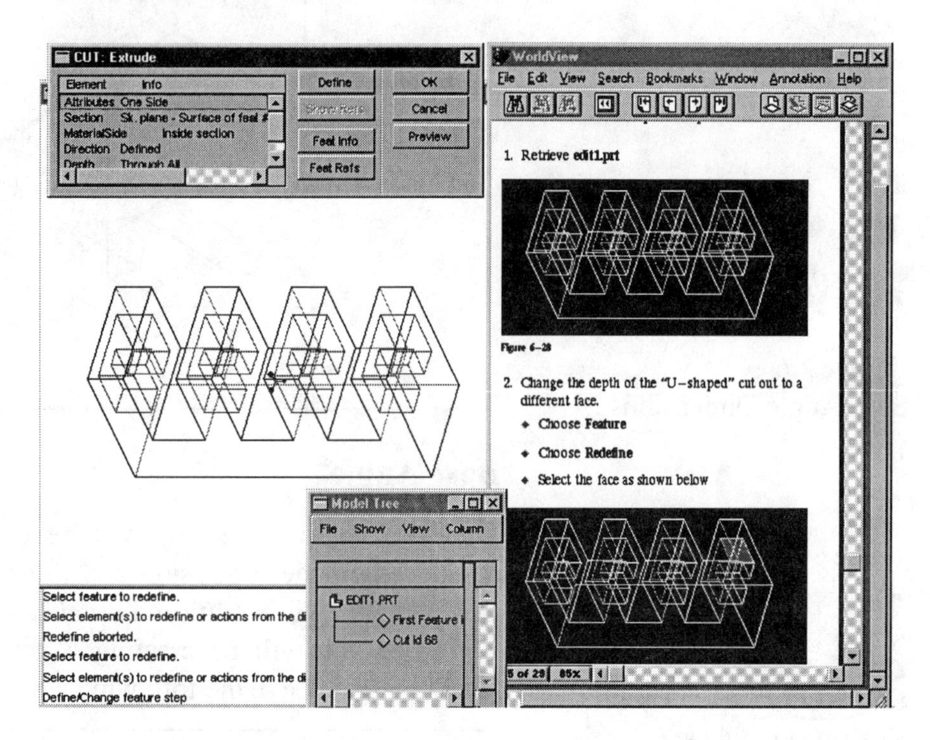

Figure 2.5
COAch for Pro/E, Basic Modifications to a Part--Modeling (Segment 3: Redefine)

Pro/E recreates the feature using the new feature definitions. When feature sections are redefined, you may need to redimension any child feature whose reference edge or surface was replaced. Redimension the child feature using the options **Redefine** and **Scheme**, or **Reroute**. If you make any changes to the feature that cause the feature creation to abort, you enter the **Resolve** environment.

When using the **Redefine** option for a feature that was created with the options **Copy**, **Mirror**, and **Dependent**, Pro/E issues a warning message stating that the selected feature is a dependent copy of the highlighted feature. If **Continue** is chosen from the WAITING menu, Pro/E will display the REDEFINE menu with the options **Attribute**, **Direction**, **Section**, and **Scheme**. For example, if you choose **Section** after you select the options to redefine, Pro/E asks for confirmation, because the section of the selected feature will become independent.

When you apply the redefinition, Pro/E removes the feature geometry and creates temporary geometry for your changes. When you exit from the user interface, *remember to regenerate the part.*

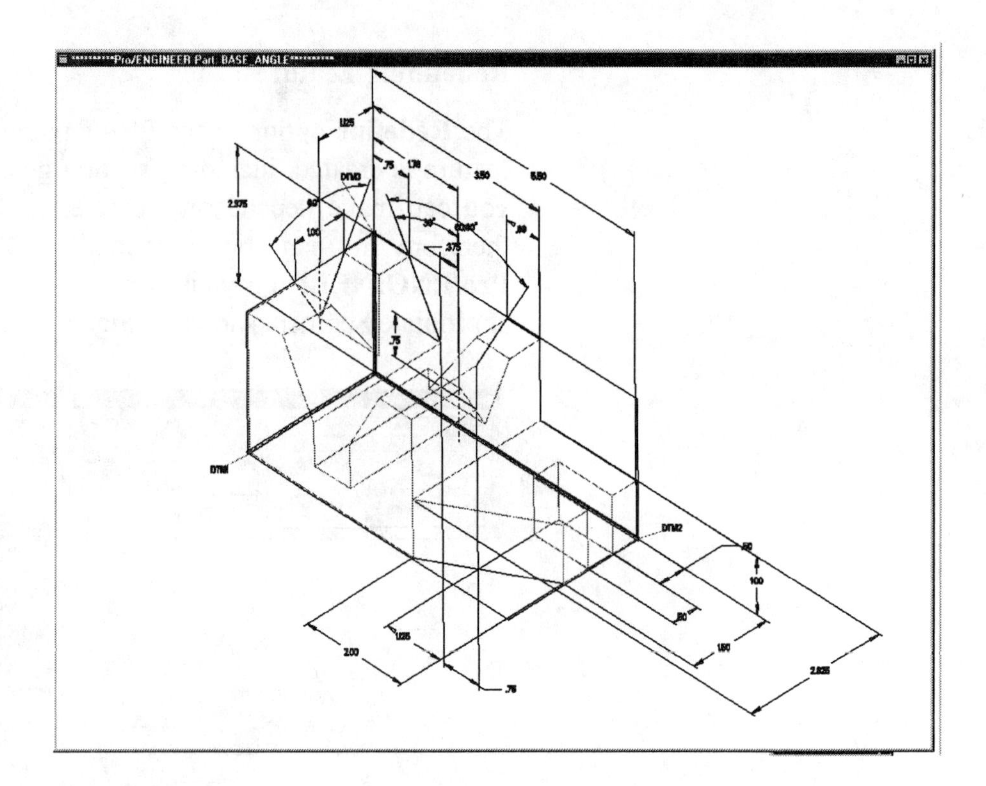

Figure 2.6
Base Angle Dimensions

Base Angle

The base angle (Fig. 2.6) will be our second ***lesson part***. It is composed of one protrusion and three cuts. Along with creating a new part, you will *modify* and *redefine* the part after it is completed.

Since you will be creating a new part, choose **Part** from the MODE menu. Give the following commands:

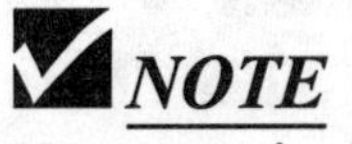

NOTE

You must include an underline character when creating file names that have a space. **BASE ANGLE** needs to be typed as **BASE_ANGLE**

Mode ⇒ **Part** ⇒ **Create** (type the part name at the command prompt) **BASE_ANGLE** and press the **Enter** key.

As in Lesson 1, the *environment*, *units*, and *material* for the part need to be established:

ENVIRONMENT AND SETUP

Environment ⇒ ✔ **Grid Snap** ✔ **Model Tree** **Hidden Line Tan Dimmed** **Setup** ⇒ **Units** ⇒ **Length** ⇒ **Inch** ⇒ **Done** ⇒ **Material** ⇒ **Define** ⇒ (type **Aluminum** and then press the **Enter** key; complete the table) ⇒ **Assign** ⇒ (pick **Aluminum**) ⇒ **Accept** ⇒ **Done**

Create a default coordinate system and the three default datum planes (Fig. 2.7). The datum planes are used to sketch on and orient the part's section geometry. Give the following commands:

Feature ⇒ **Create** ⇒ **Datum** ⇒ **Plane** ⇒ **Offset** ⇒ (press **enter** three times at the prompt) ⇒ **Done**

The next command creates the first protrusion for the **BASE_ANGLE** (Fig. 2.7):

Create ⇒ **Protrusion** ⇒ **Extrude** ⇒ **Solid** ⇒ **Done** ⇒ **One Side** ⇒ **Done** ⇒ (pick **DTM3** as the sketching plane) ⇒ **Okay** (for selecting the *direction* of protrusion projection) ⇒ **Top** ⇒ (pick **DTM2** as the *orientation* plane)

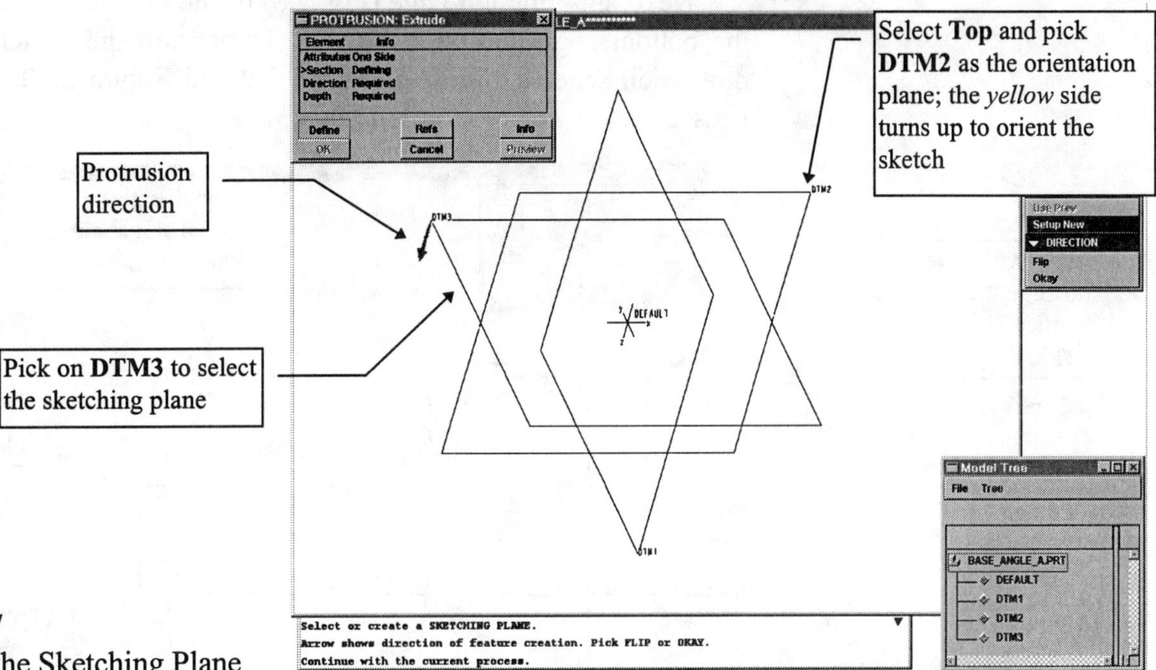

Figure 2.7
Selecting the Sketching Plane

The **Grid Snap** is on and the **Mouse Sketch** default is active in the GEOMETRY menu, so you can begin the sketching process using the mouse buttons. Sketch the outline of the part's primary feature using a set of connected lines as was done in Lesson 1. The part's dimensions and general shape are provided in Figures 2.1 and 2.6. A majority of the part can be defined by sketching an outline similar to its front view. Therefore, as with the CLAMP in Lesson 1, you can complete most of the part's geometry with one protrusion. The cuts on the top and side will be features created later. The part will have its base aligned with the edge of **DTM2**, its left edge aligned with **DTM1**, and, since you are sketching on **DTM3**, its back face aligned with that datum.

HINT

default_dec_places
sets the default number of decimal places to be displayed in all model modes; it does not affect the displayed number of digits.

sketcher_dec_places
controls the number of digits displayed when in the Sketcher.

The default number of digits for the part is **2**. You can set the number of digits using the configuration file option **"SKETCHER_DEC_PLACES"** (a value in the range **0** to **14**).

Misc ⇒ **Edit Config** ⇒ **enter** ⇒ (move down the menu to pick **SKETCHER_DEC_PLACES** and **3** as the new default) **enter** ⇒ **Save** and **Exit** the table ⇒ **Load Config** ⇒ **enter** ⇒ **Done/Return**

In the example that follows we have left the decimal places to be **2**. Pro/E will round three places to two (**1.125** becomes **1.13**).

Sketch the part's outline as shown in Figure 2.8. Add a vertical centerline down the middle of the slot. Do not create any constraints that you do not want as part of the design. Make sure lines not supposed to be at the same level (horizontal) are not sketched in line, features of differing size are sketched with different lengths, etc.

Make sure that lines at the same angle are sketched that way, that symetrical features are drawn equally about the centerline, etc. **Regenerate** after you complete the sketch outline.

Next, align the left edge (Fig. 2.8) of the sketch with **DTM1** and the bottom edge line with **DTM2**. Dimension the sketch with the dimension scheme shown in Figure 2.9 and Figure 2.10. (Refer to Lesson 1).

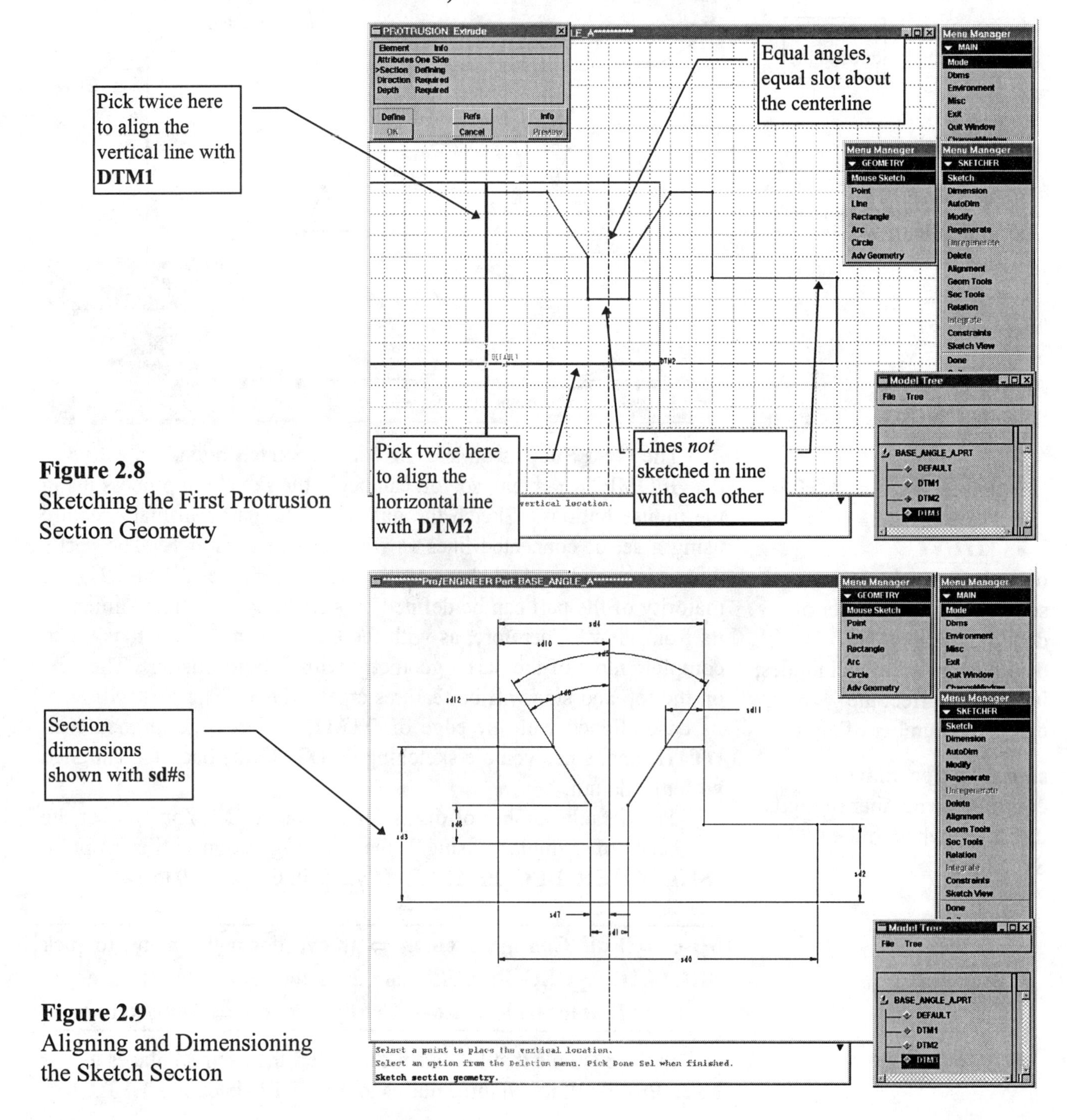

Figure 2.8
Sketching the First Protrusion Section Geometry

Figure 2.9
Aligning and Dimensioning the Sketch Section

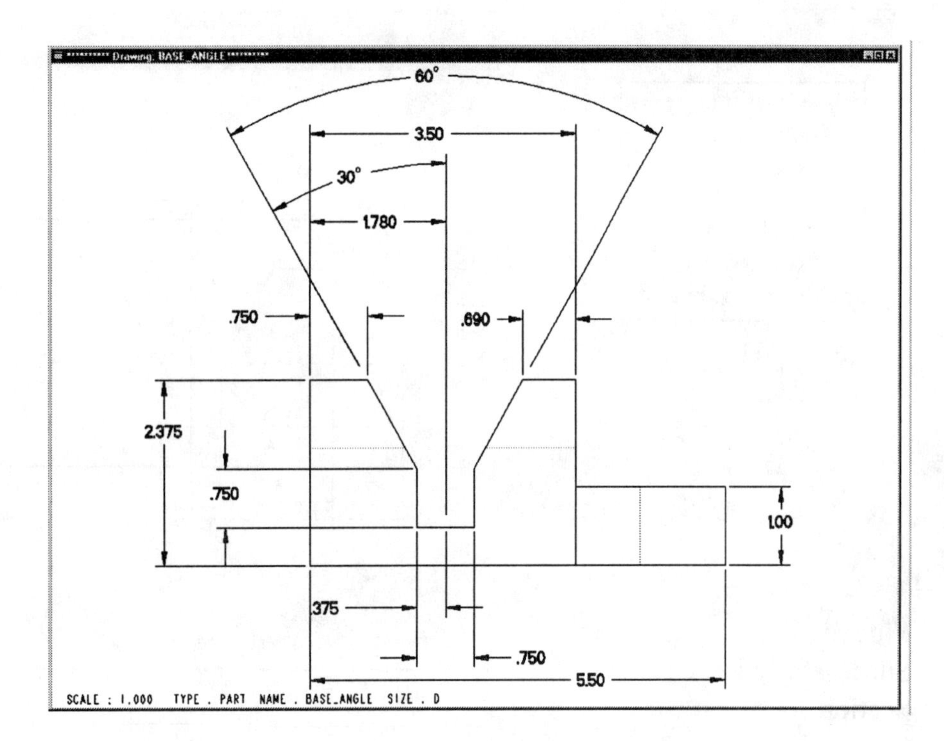

Figure 2.10
Part Dimensions

Figure 2.11 shows the sucessfully regenerated sketch. All of the dimension values are larger than the part at this point in the process. The sketch grid was **30** units and decimal inches were selected at the beginning of the part. Therefore the grid is **30** inches square, which accounts for the huge size of the sketch at this point in the process.

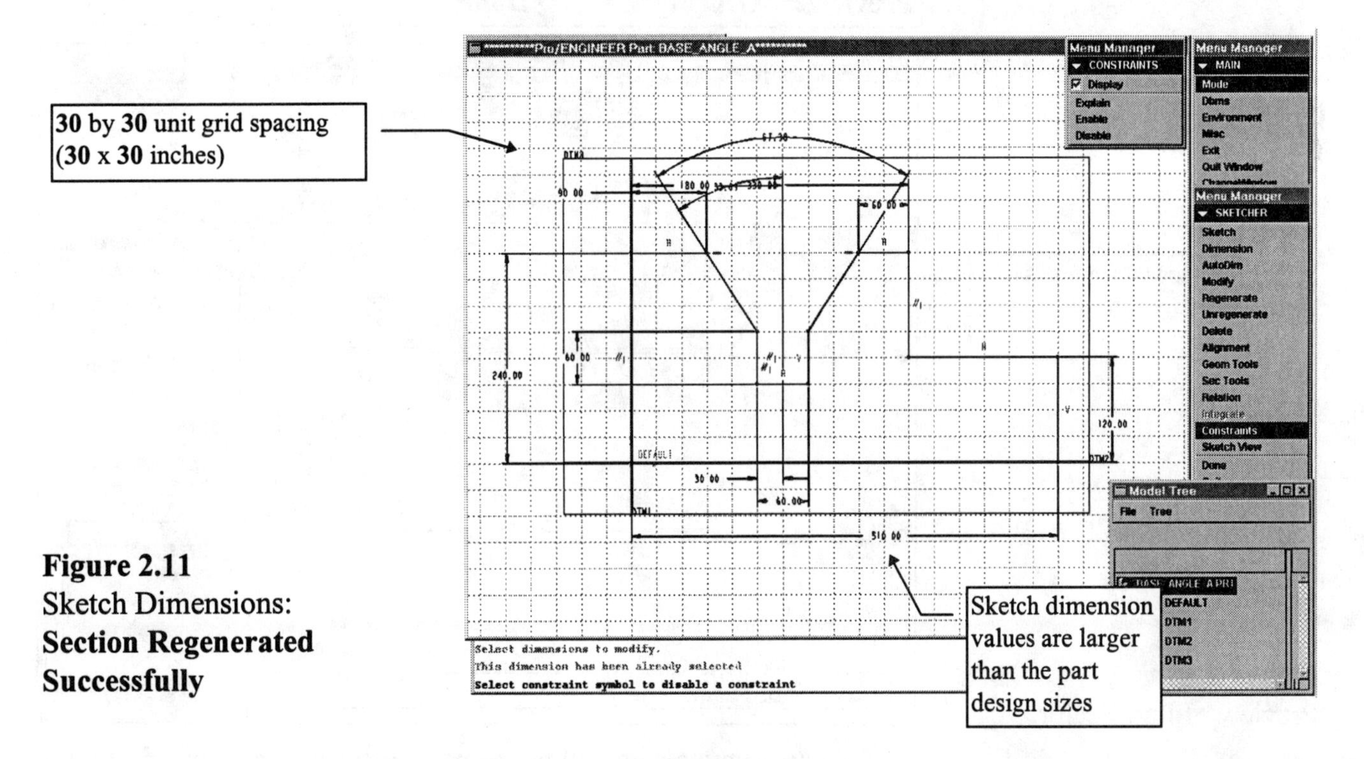

Figure 2.11
Sketch Dimensions:
Section Regenerated Successfully

Modify the dimensions one at a time, and then **Regenerate** the sketch as shown in Figure 2.12.

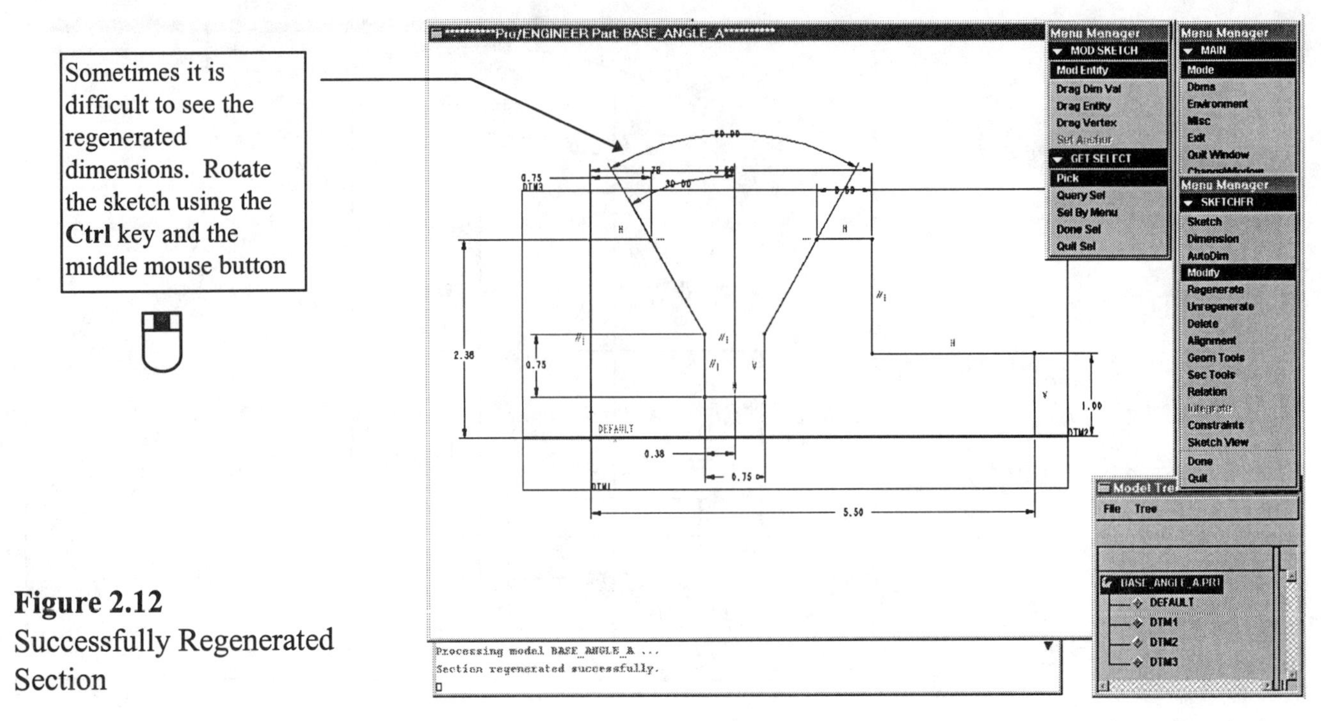

Figure 2.12
Successfully Regenerated Section

After the sketch is successfully regenerated with the design dimensions, give the following commands:

Done ⇒ **View** ⇒ **Default** ⇒ **Done-Return** ⇒ **Blind** ⇒ **Done** ⇒ (at the command prompt type **2.625** as the part's depth) ⇒ **enter** ⇒ **Preview** ⇒ **OK** (Fig. 2.13) ⇒ **Done**

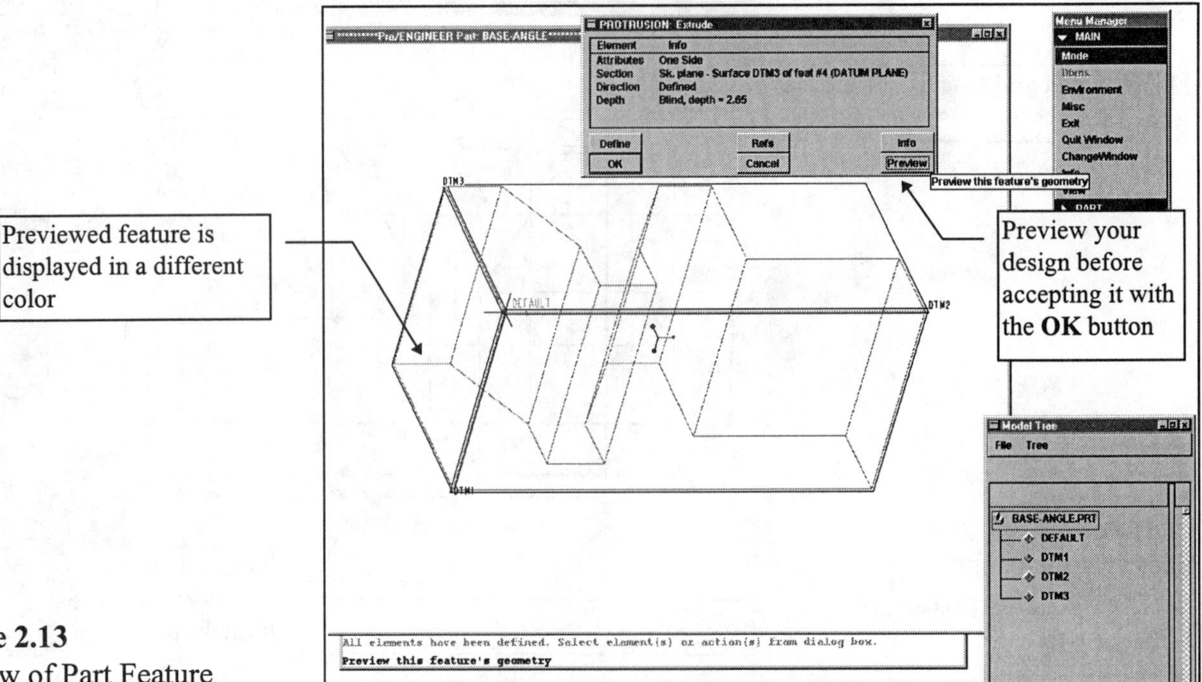

Figure 2.13
Preview of Part Feature

Three separate cuts will be the next features to be created. Figure 2.14 shows the right side view of the part with dimensions for the **V** cut, and Figure 2.15 shows the top view with dimensions for two more cuts.

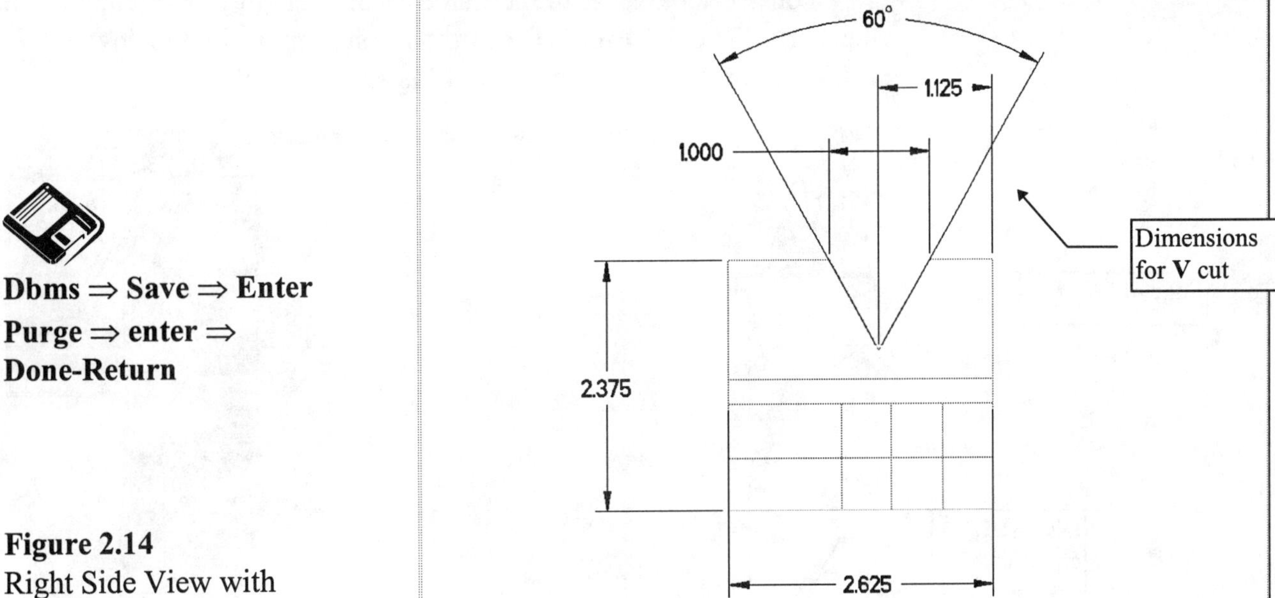

Dbms ⇒ Save ⇒ Enter Purge ⇒ enter ⇒ Done-Return

Figure 2.14
Right Side View with Dimensions

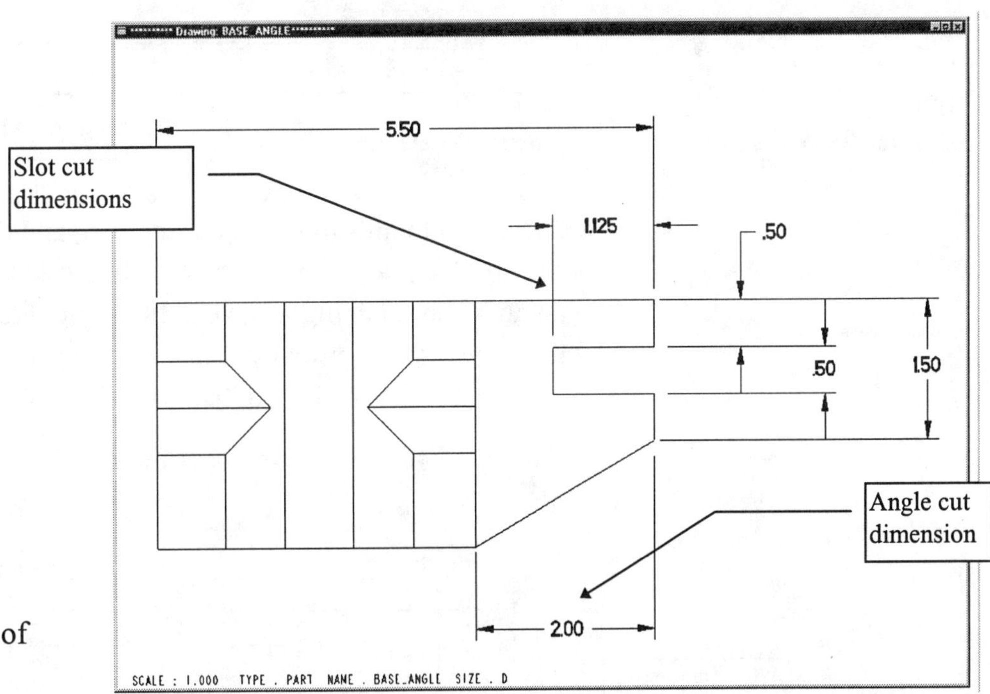

Figure 2.15
Dimensioned Top View of BASE_ANGLE

The next feature that will be created is the V-shaped cut. The dimensions for the cut are shown in Figure 2.14. The **V** cut is sketched similar to the first protrusion except that it removes material.

Give the following commands:

Feature ⇒ Create ⇒ Cut ⇒ Extrude (default) ⇒ **Solid** (default) ⇒ **Done ⇒ One side** (default) ⇒ **Done** (pick **DTM1** as the sketching plane) **Flip ⇒ Okay ⇒ Top** (pick **DTM2** to orient the sketch)

If necessary, use **Query Sel** to pick the datum planes. Pro/E now enters the **Sketcher** automatically and displays the part and datum planes as shown in Figure 2.16.

You are looking at the left side of the part in this orientation of the part. The **DEFAULT** coordinate system is in the lower left, **DTM2** is on the bottom, and **DTM3** is along the left.

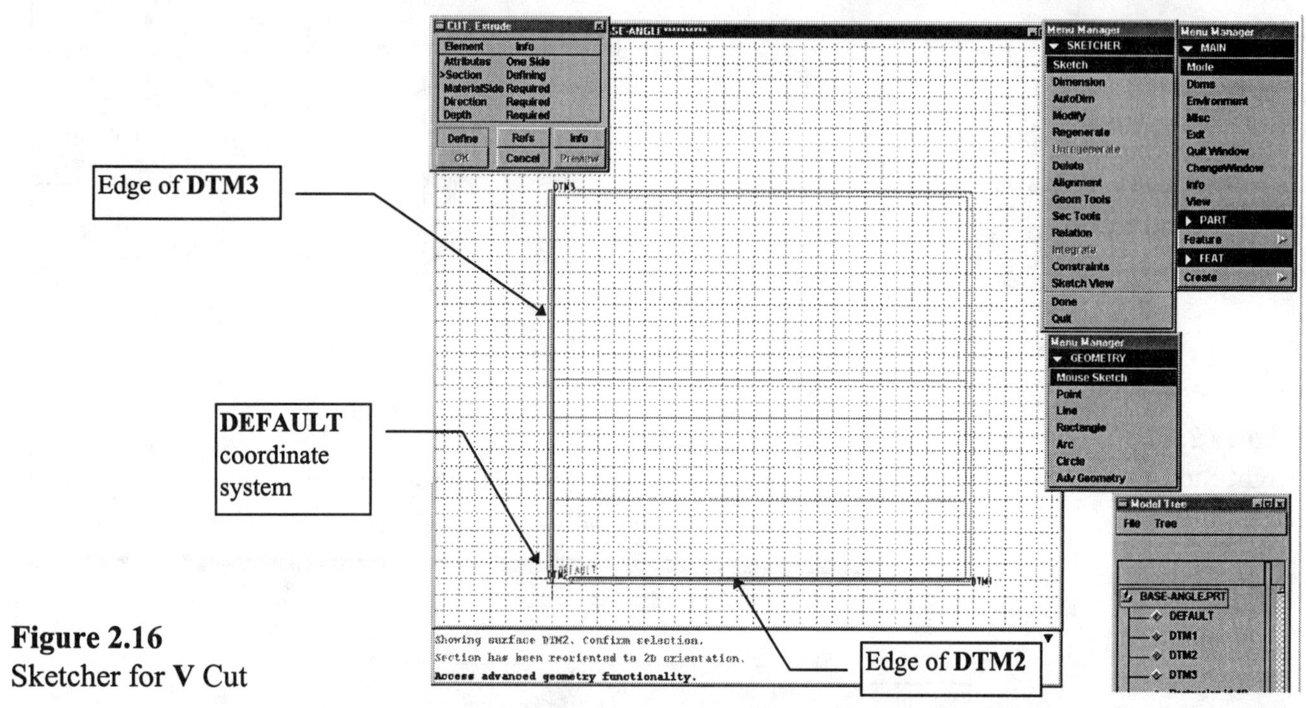

Figure 2.16
Sketcher for **V** Cut

Sketch the **V** cut symmetrically about a vertical centerline, align the two endpoints to the top of the part, and dimension the **V** shape as shown in Figure 2.14. The centerline establishes the symmetrical **V** angle without having to give a **30°** angle dimension. If the sketch is not drawn symmetrically, then you may need to add the **30°** angle, regenerate the sketch, and then delete the **30°** dimension.

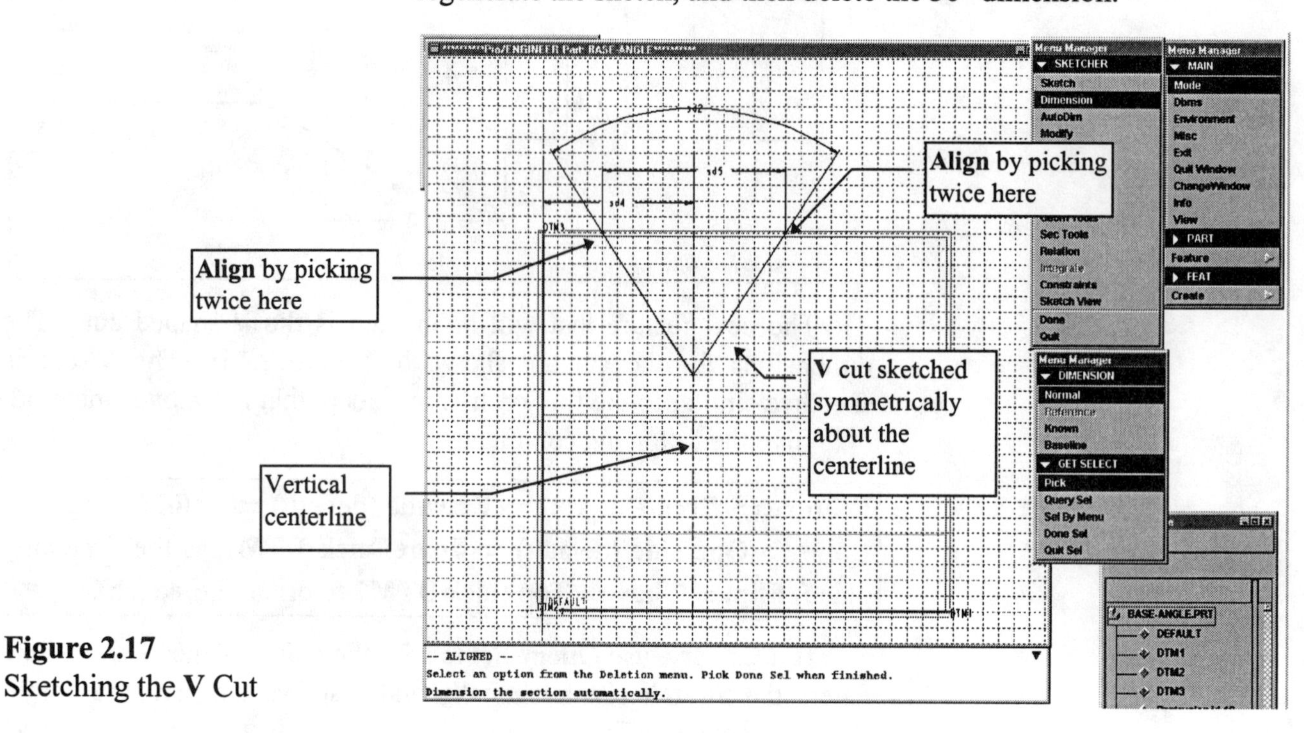

Figure 2.17
Sketching the **V** Cut

Regenerate the sketch as shown in Figure 2.18.

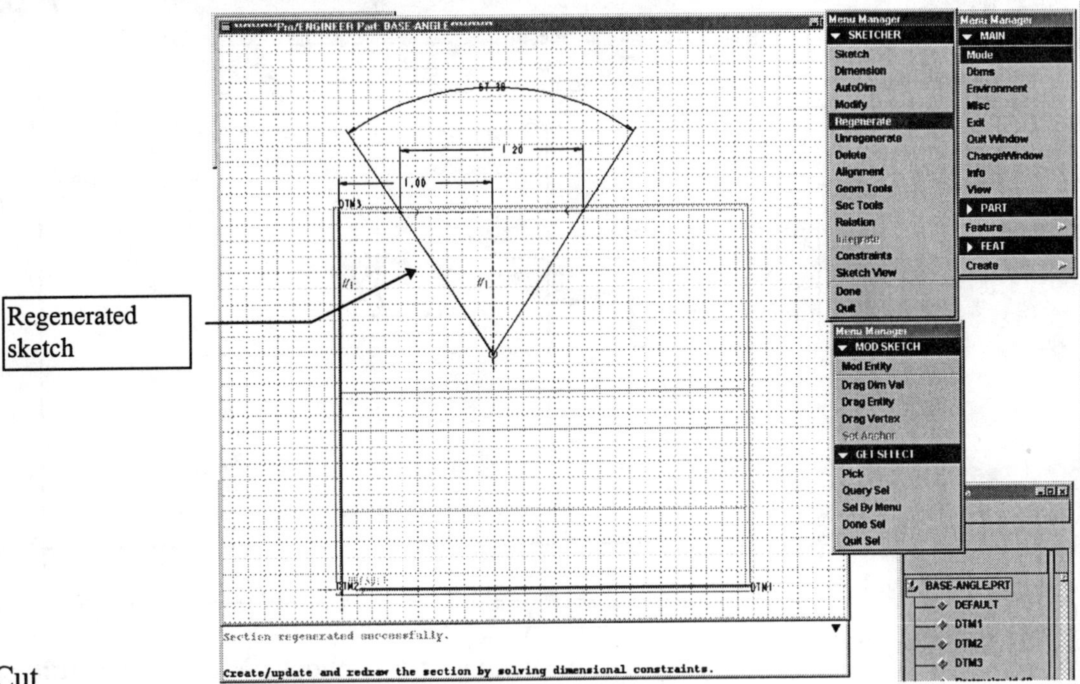

Figure 2.18
Regenerated **V** Cut

Modify the dimensions and **Regenerate** the sketch as shown Figure 2.19.

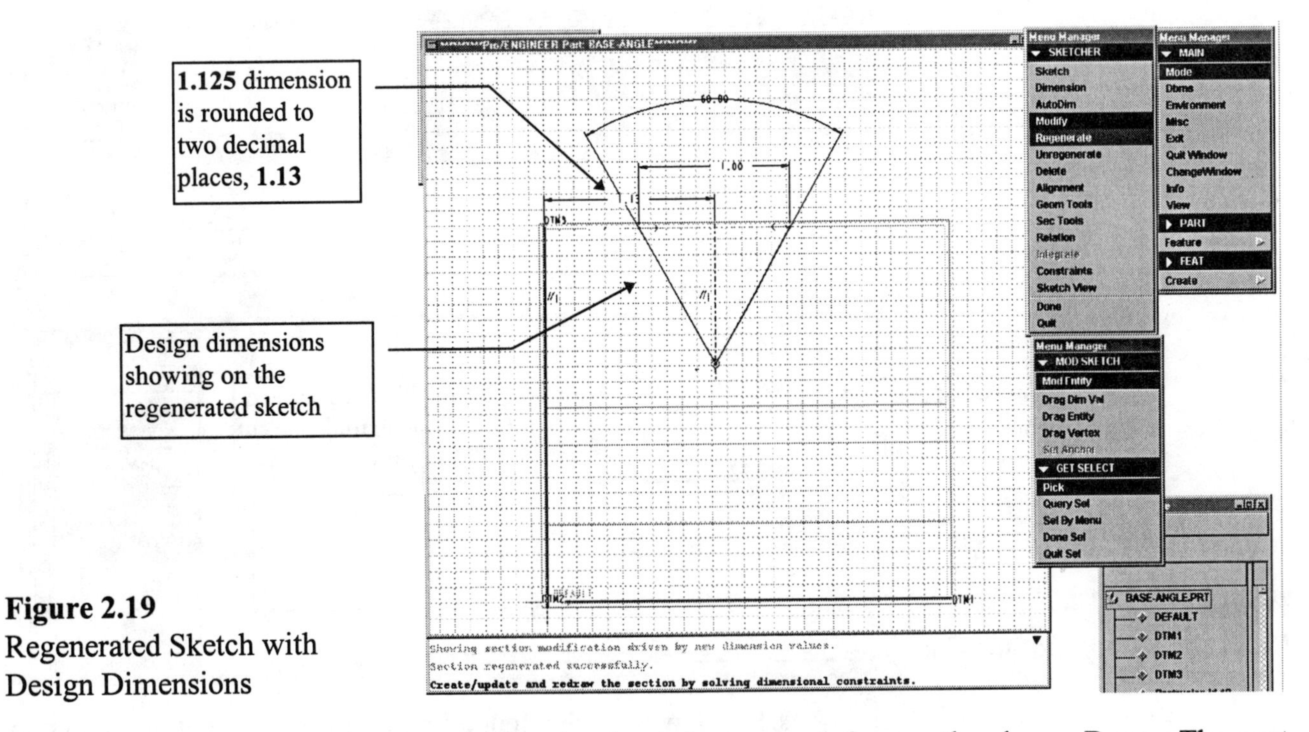

Figure 2.19
Regenerated Sketch with Design Dimensions

After the sketch is regenerated correctly, choose **Done**. The next **Element** that needs defining is the material removal direction. An arrow is displayed (Fig. 2.20), and you are prompted with: **Arrow points TOWARD the area to be REMOVED. Flip** or **Okay.** Choose **Okay**, since the arrow is correct.

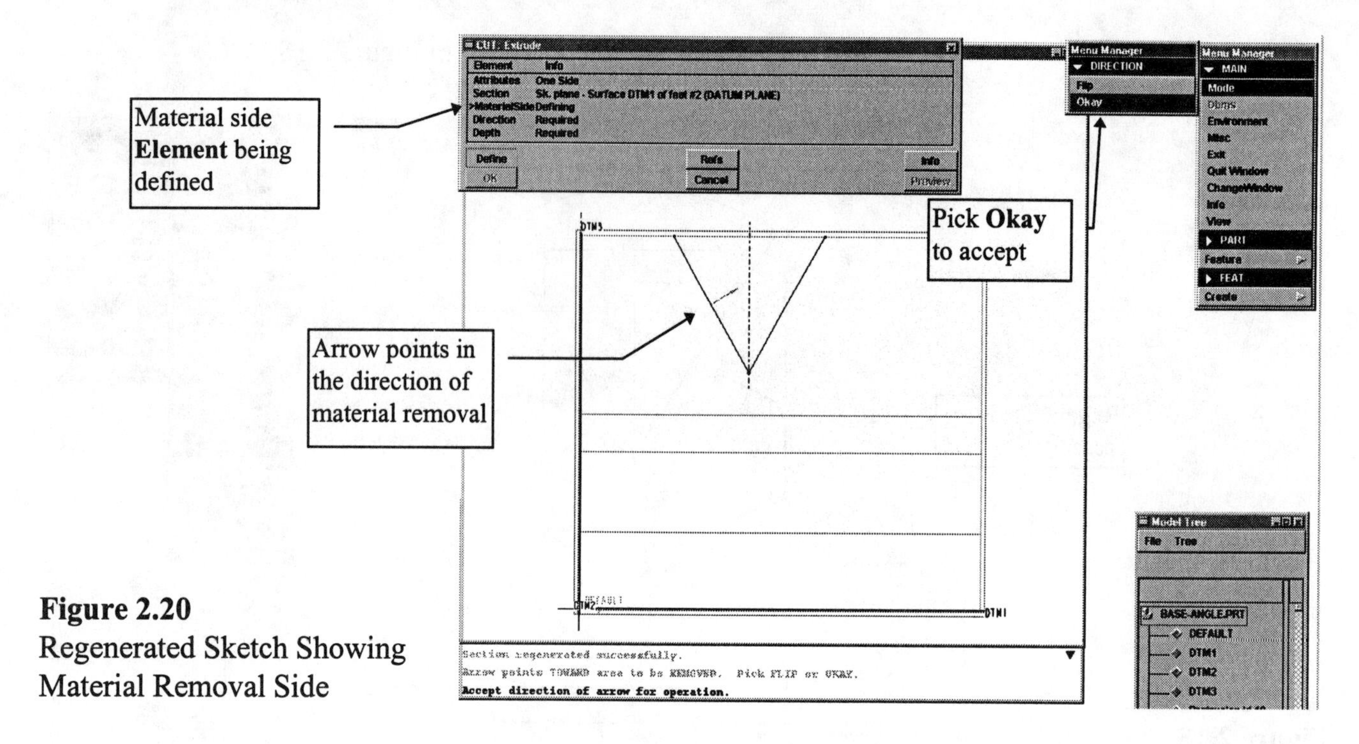

Figure 2.20
Regenerated Sketch Showing Material Removal Side

After picking **Okay**, choose **View** ⇒ **Default** ⇒ **Done-Return** ⇒ **Thru Until** ⇒ **Done** (pick the right upper plane shown in Fig. 2.21).

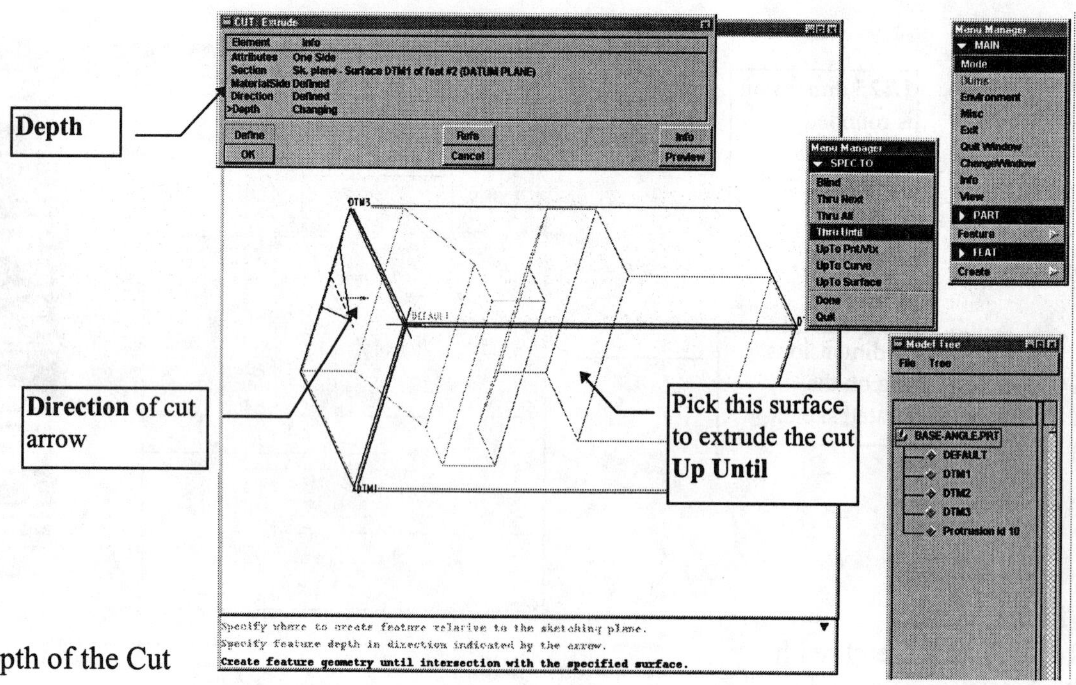

Figure 2.21
Defining the Depth of the Cut

Pick **Preview** from the dialog box to display the cut, as shown in Figure 2.22. Instead of accepting the design by picking the **OK** button, pick the **Depth** element in the dialog box and **Define**. Choose **Thru Next** from the SPEC TO menu (Fig. 2.23). Choose **Preview** from the dialog box.

The cut now goes up to the next feature it encounters and stops there. Pick **Depth** and **Define** from the dialog box again, select **Upto Surface**, and pick the same surface as in the first example when you selected **Thru Until** and then **Done** ⇒ **OK** (from the dialog box). The part will now look like Figure 2.24.

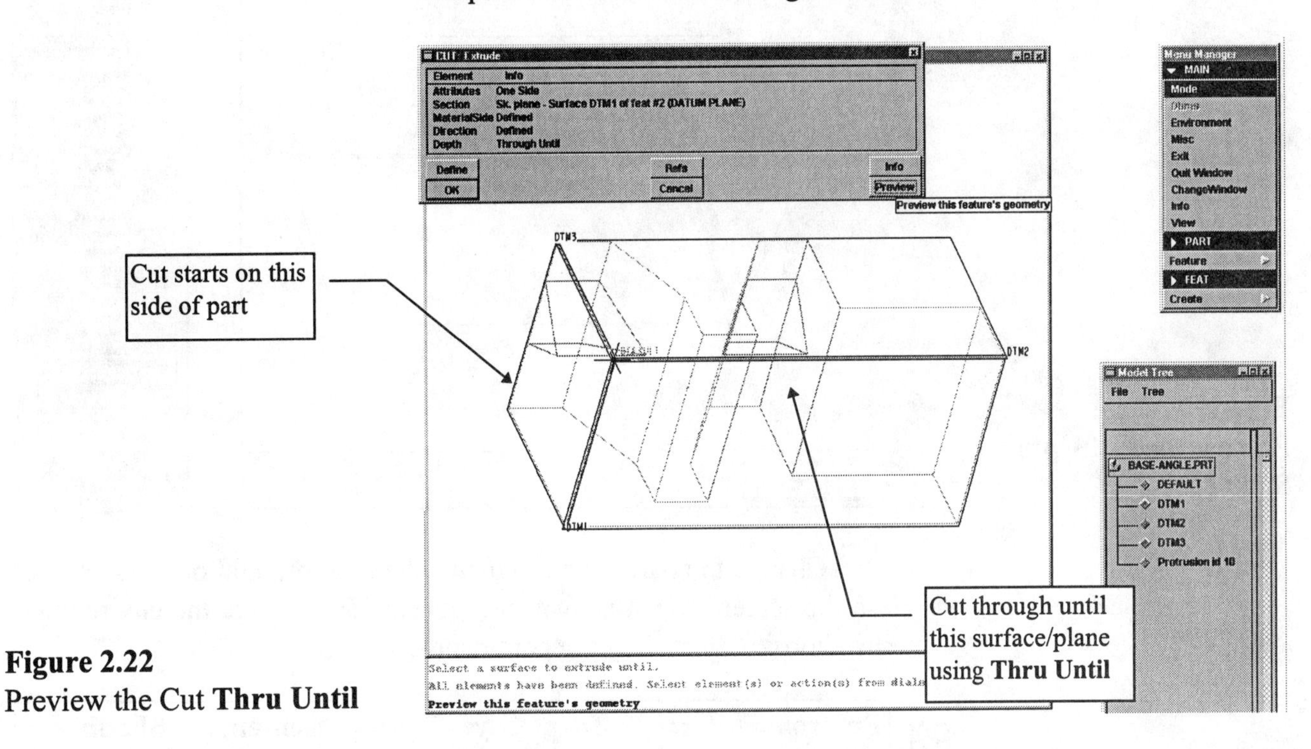

Figure 2.22
Preview the Cut **Thru Until**

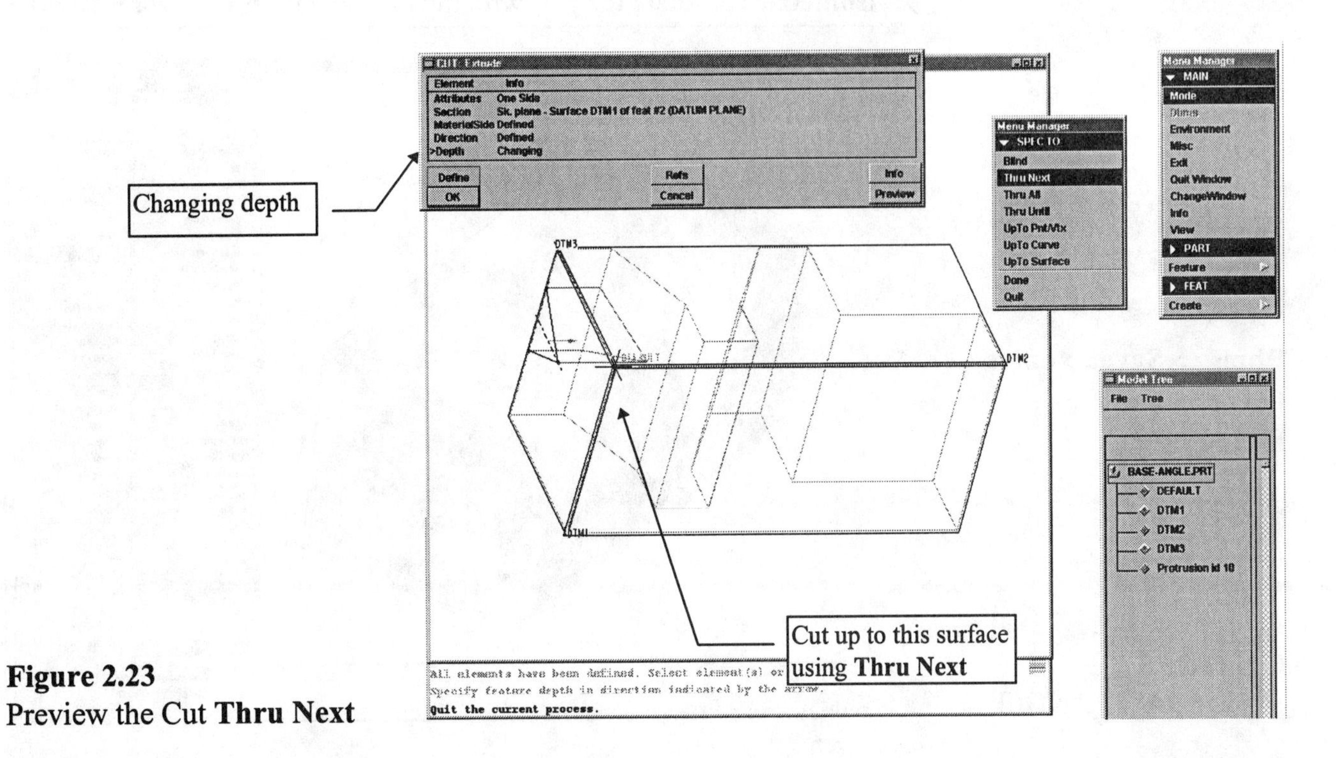

Figure 2.23
Preview the Cut **Thru Next**

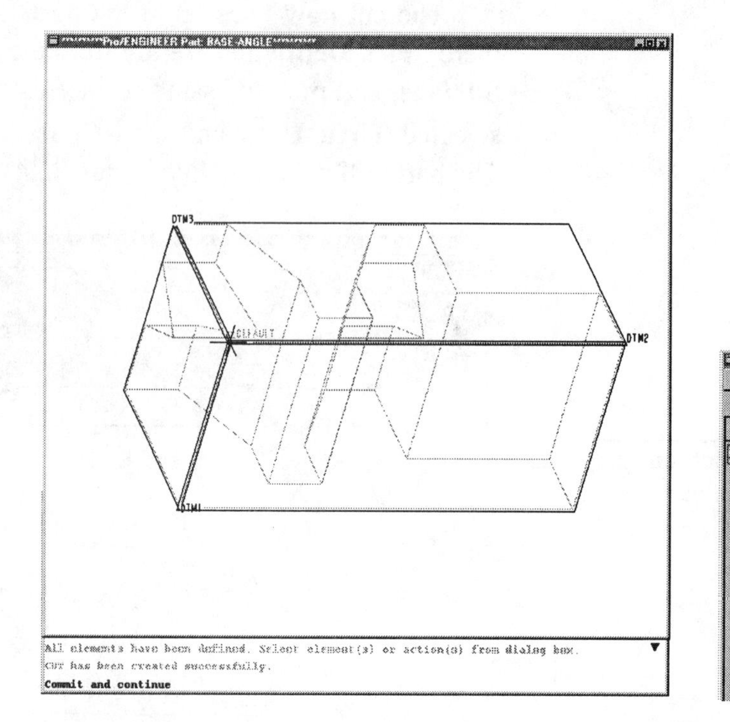

Figure 2.24
Preview the Cut **Upto Surface**

Change the part's environment, shading, orientation, and position on the screen to that shown in Figure 2.25. Change the environment by choosing the following commands:

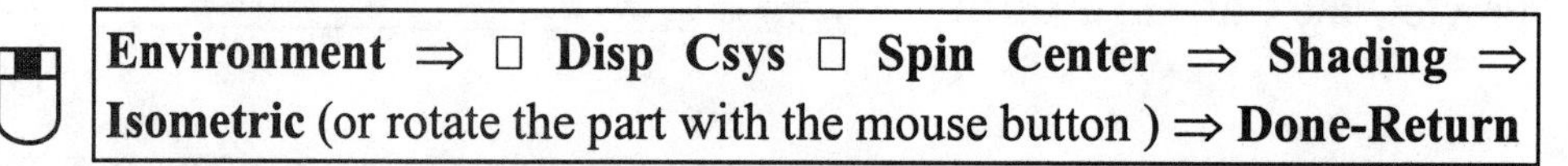

Environment ⇒ □ **Disp Csys** □ **Spin Center** ⇒ **Shading** ⇒ **Isometric** (or rotate the part with the mouse button) ⇒ **Done-Return**

Dbms ⇒ **Save** ⇒ **enter**
Purge ⇒ **enter** ⇒
Done-Return

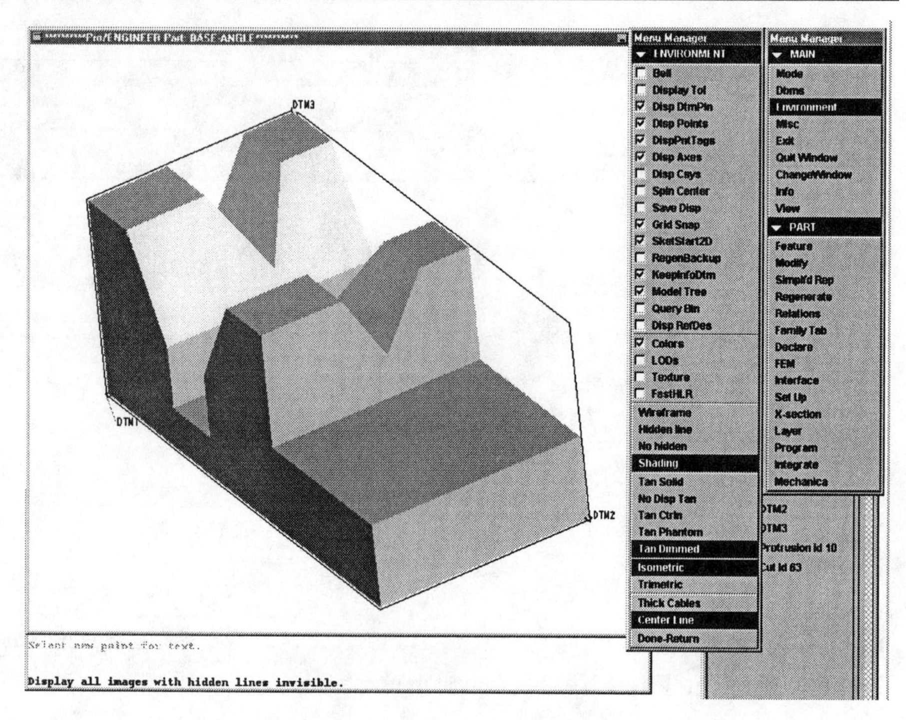

Figure 2.25
Shaded BASE_ANGLE

The cut is now complete. Next, the angle cut and the **U**-shaped cut need to be created.

Before creating the next cut, set the environment and view with the following commands:

HINT

You can make only *one open* section cut at a time. You can make *multiple closed* sections as one feature, but this is considered poor design practice.

Environment ⇒ Hidden Line ⇒ Isometric ⇒ Done-Return ⇒ View ⇒ Repaint ⇒ Done-Return ⇒ View ⇒ Default ⇒ Done-Return

The next feature will be another cut. There are two cuts still required for the completion of the part. They can be created together using closed sections or separately with one open section at a time. It is better design intent to make the cuts separately, since they do not have any particular relationship except that they cut the same direction and start on the same surface/plane. A *closed section* is a sketched set of entities that start and end at the same position, like a square □ **shape**. An *open section* does not form a closed figure, such as a ∩ or ∪ **shape**.

Start the angled cut by the using the following commands:

Feature ⇒ Create ⇒ Cut ⇒ Extrude ⇒ Solid ⇒ Done ⇒ One Side ⇒ Done ⇒ (pick **DTM2** as the sketching/placement plane) ⇒ **Flip** (change the direction of the cut to pass through the part, not out into space) ⇒ **Okay ⇒ Top ⇒** (pick **DTM3** as the orientation plane)

In Figure 2.26 the direction of cut is selected.

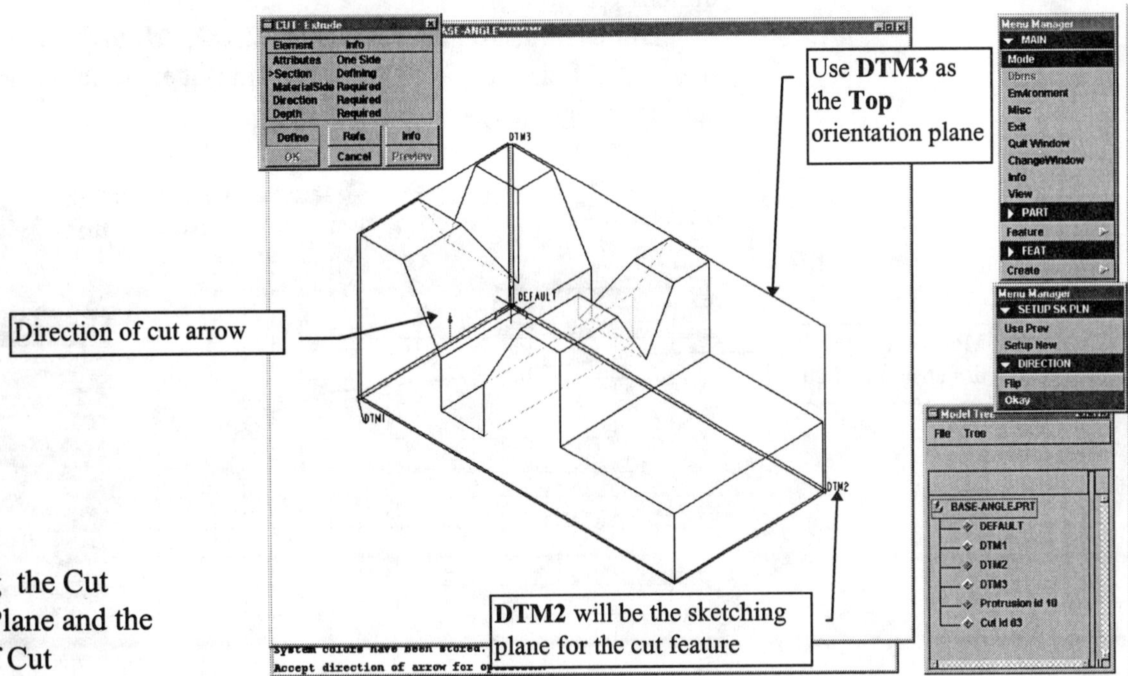

Figure 2.26
Establishing the Cut Placement Plane and the Direction of Cut

If already on, turn off the **Grid Snap** in the ENVIRONMENT menu: **Environment ⇒ □ Grid Snap ⇒ Done-Return**. Your screen should show the part in the Sketcher looking from the bottom, through the part, as in Figure 2.27. Because you are looking at the bottom, the cut will be sketched on the upper right of the part.

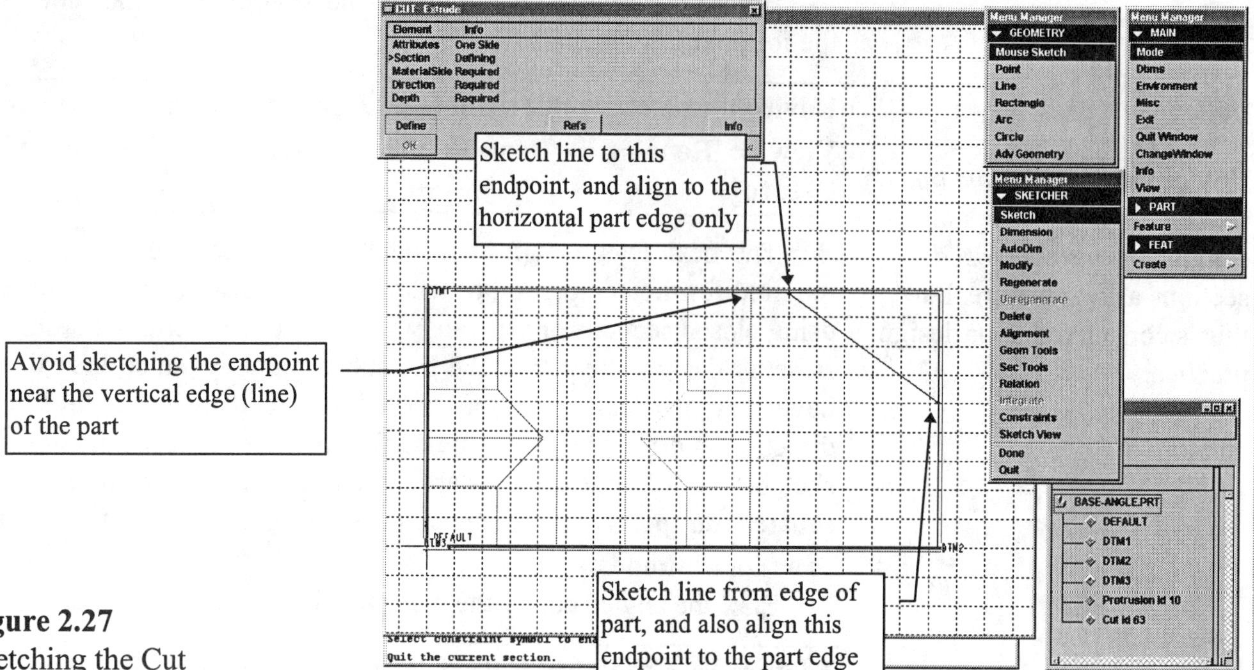

Figure 2.27
Sketching the Cut

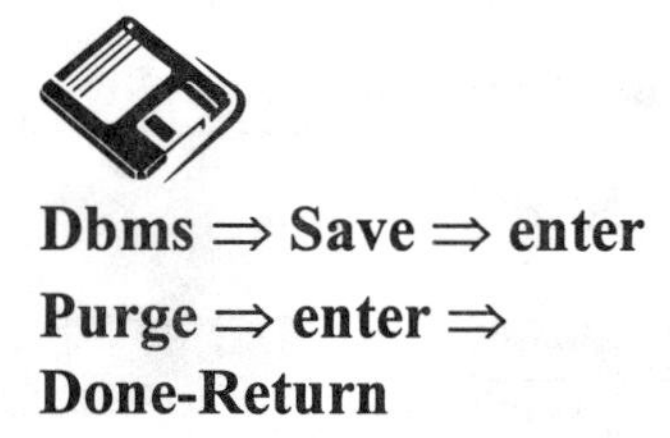

Dbms ⇒ Save ⇒ enter
Purge ⇒ enter ⇒ Done-Return

Do not sketch the endpoint of the line near the vertical edge (Fig. 2.27) of the part feature, in order to avoid an unwanted assumption that the point and the part edge are exactly at the same position. This will allow the cut to be modified later by changing the value of the cut dimension.

Dimension, **Regenerate** (Fig. 2.28), **Modify**, and **Regenerate** the cut. After you pick **Done**, the side of material removal must be selected (Fig. 2.29).

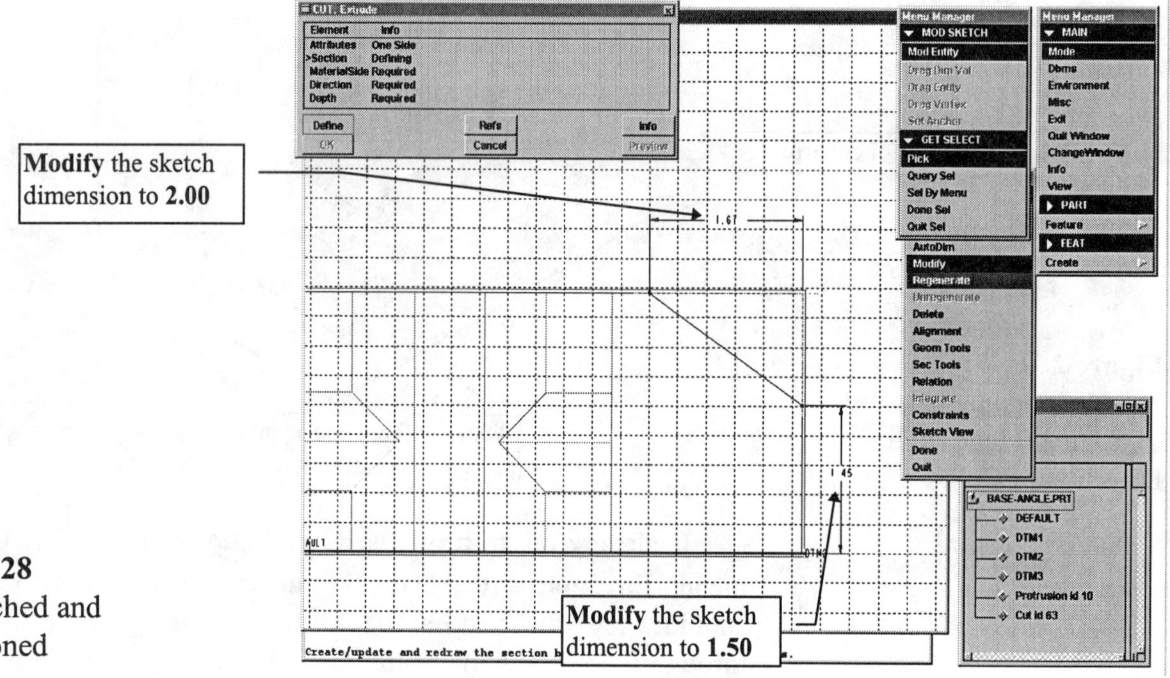

Figure 2.28
Cut Sketched and Dimensioned

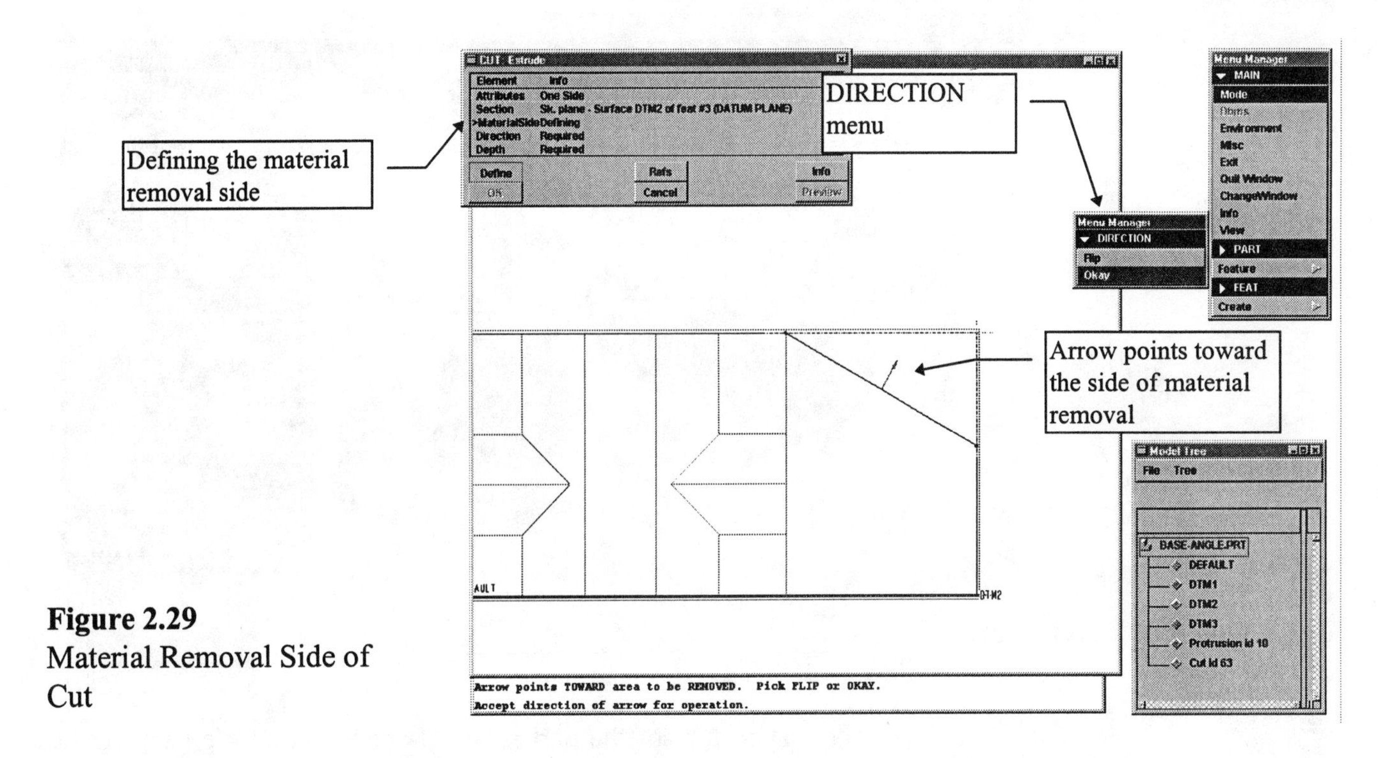

Figure 2.29
Material Removal Side of Cut

To complete the cut, give the following commands:

Okay (to accept the material removal side) ⇒ **View** ⇒ **Default** ⇒ **Done-Return** (rotates the model to see the cut direction) ⇒ **Thru All** ⇒ **Done** ⇒ **OK** (from the dialog box)

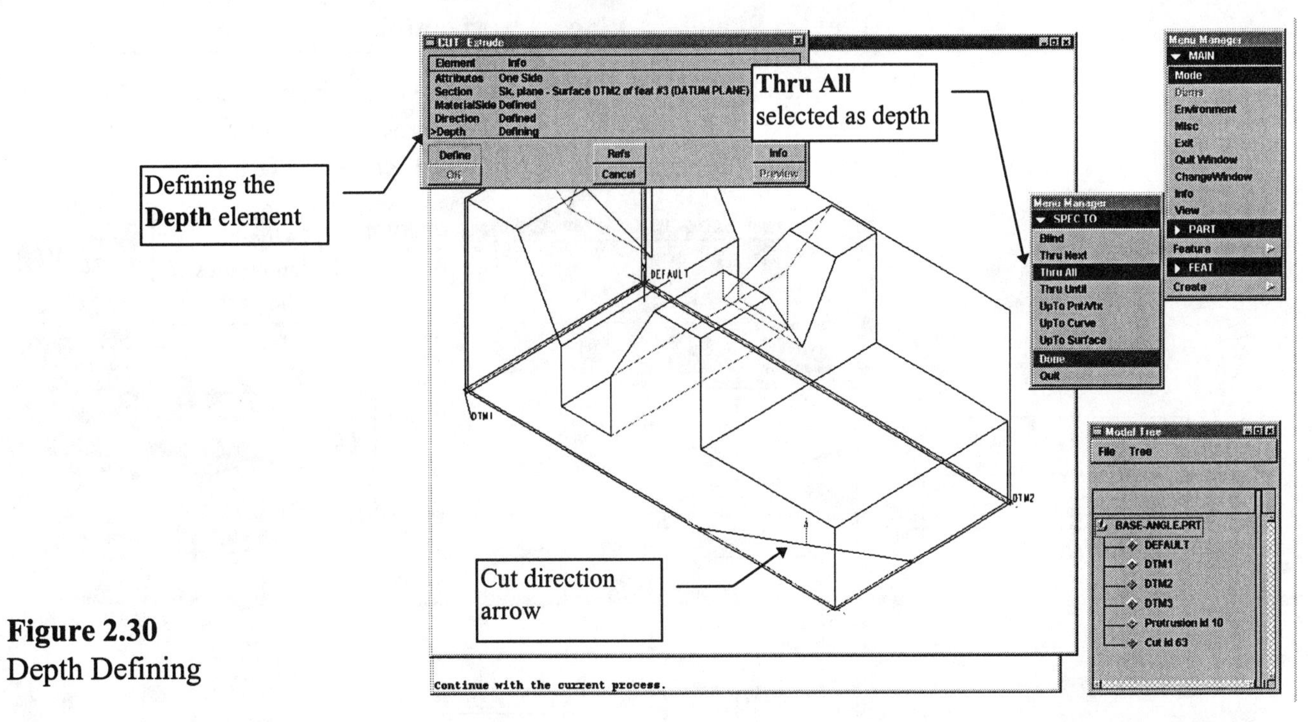

Figure 2.30
Depth Defining

Shade the part to see the angle part as shown in Figure 2.31:

View ⇒ **Cosmetic** ⇒ **Shade** ⇒ **Display**

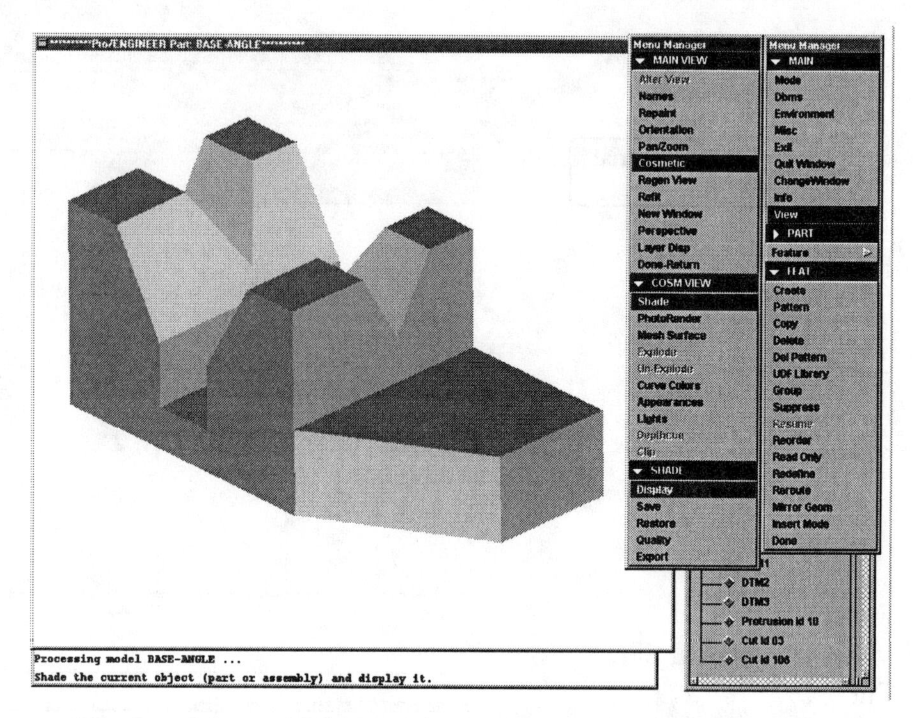

Figure 2.31
Completed Cut Shown as a Shaded Model

The last feature of the part is the **U-shaped** cut. We can use the previous cut's placement plane and orientation. Use the following commands:

Feature ⇒ Create ⇒ Cut ⇒ Extrude ⇒ Solid ⇒ Done ⇒ One Side ⇒ Done ⇒ Use Prev (sketching plane) **⇒ Flip** (change the direction of the cut to pass through the part, not out into space) **⇒ Okay ⇒ Sketch ⇒ Line ⇒ Horizontal**

Sketch the three lines as shown in Figure 2.32. Align the endpoints of the two horizontal lines with the right side of the part. **Dimension, Regenerate, Modify** and **Regenerate** the cut.

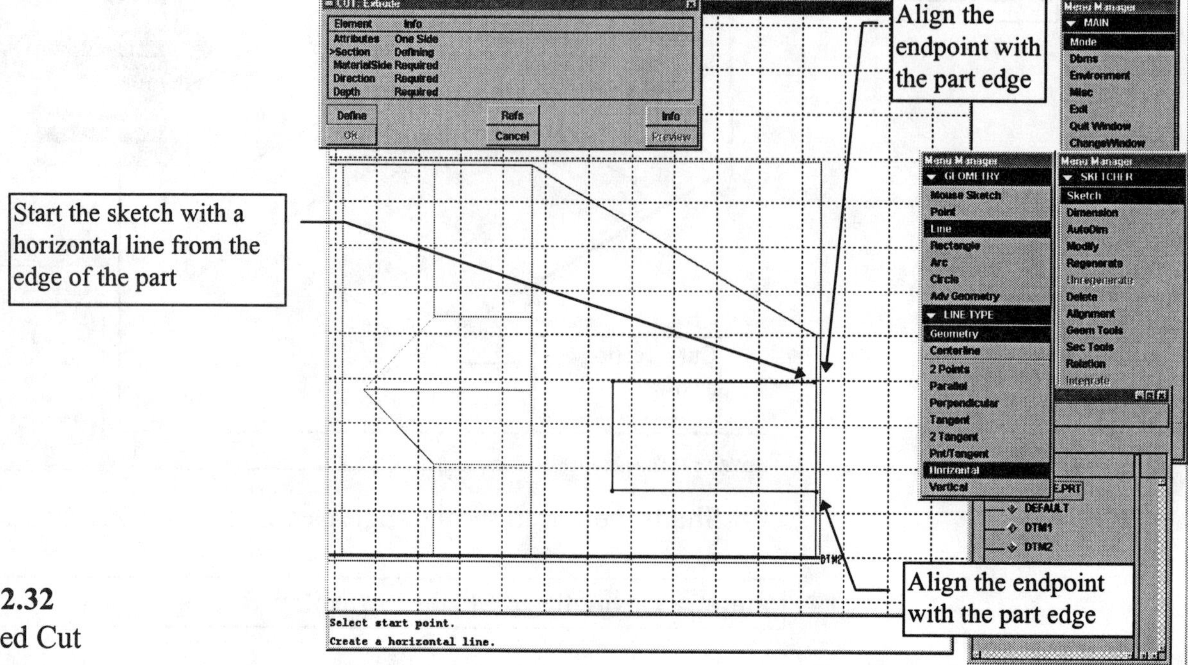

Figure 2.32
U-Shaped Cut

The cut is dimensioned, modified and regenerated in Figure 2.33.

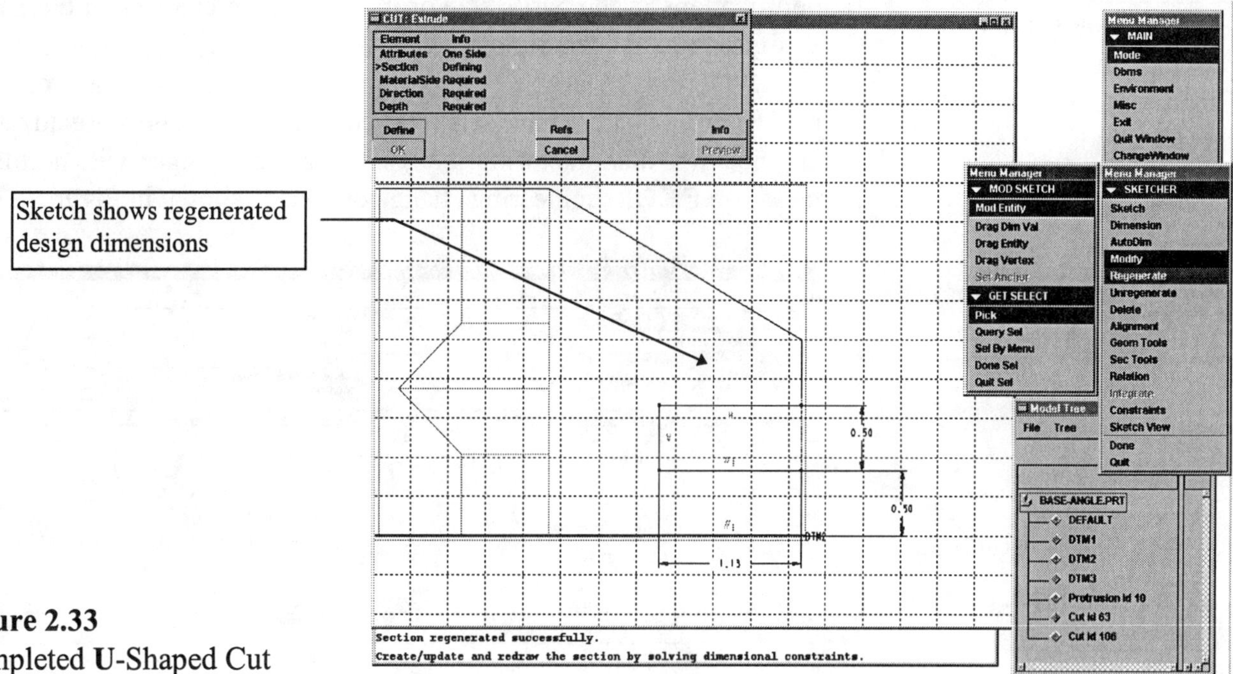

Figure 2.33
Completed **U**-Shaped Cut

Complete the part with the following commands (Fig. 2.34):

Okay (accepts material removal side) ⇒ **Thru All** ⇒ **Done** ⇒ **OK** (from dialog box) ⇒ **View** ⇒ **Default** ⇒ **Done-Return**

HINT

Save and **Purge** after every new feature is created:

Dbms ⇒ **Save** ⇒ **enter**

Purge ⇒ **enter** ⇒ **Done-Return**

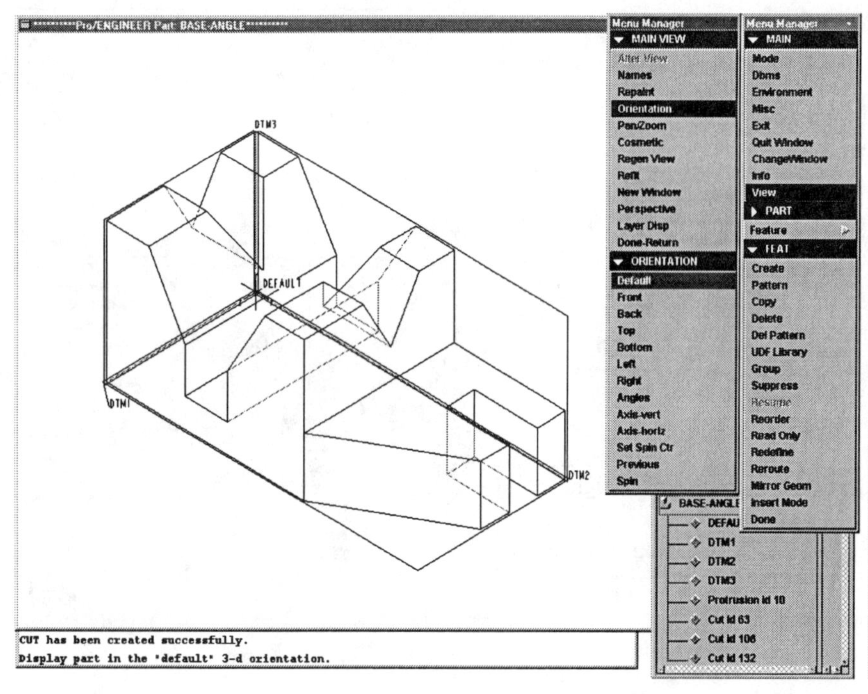

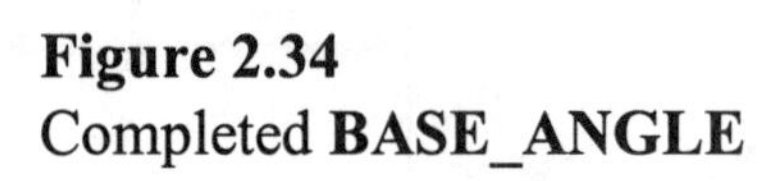

Figure 2.34
Completed **BASE_ANGLE**

Since you will be modifying and redefining the part, it might be a good idea to save this version under another name (ask your instructor):

Dbms ⇒ **Save As** ⇒ **enter** ⇒ **BASE_ANGLE_A** ⇒ **enter**.

NOTE

The **ECO** shown here is provided in the Pro/E **Format** mode. To see the format, choose:

Mode ⇒ **Format** (from the MODE menu) ⇒ **Search/Retr** ⇒ **Format Dir** ⇒ **ecofrom.frm**

Very few projects make it through the design, engineering, and manufacturing sequence without changes. The changes can be simple modifications in the part's size or more extreme changes in the part's configuration. When simple dimensional changes are requested, the **Modify** command is used; when configuration changes are needed, the **Redefine** command is called upon. The organization (engineering, manufacturing, etc.) issuing the changes will normally release an **ECO** (engineering change order), as shown in Figure 2.35.

ECO

New design requirements for the cuts

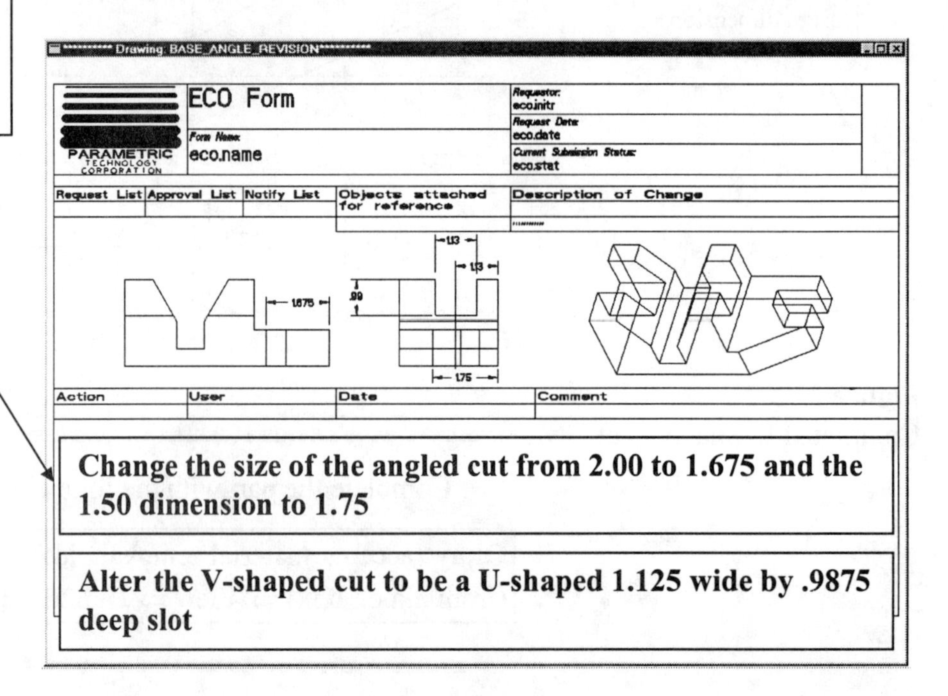

Figure 2.35
ECO Changing the Angled Cut Size and the V-shaped Cut

You can **Modify** and **Redefine** from the menu structure or from the Model Tree. Use the **Model Tree** for both ECOs (Fig. 2.36).

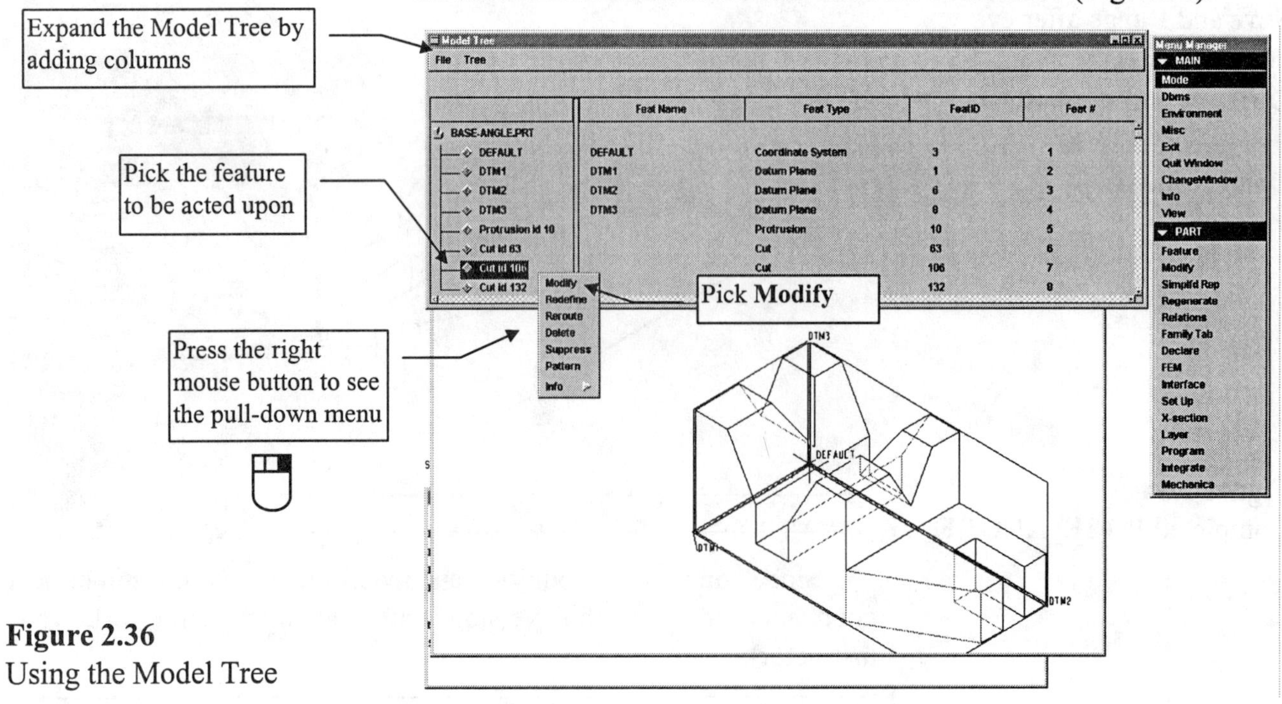

Figure 2.36
Using the Model Tree

After selecting the feature to be modified from the Model Tree, press and hold down the right mouse button to display the pull-down menu. Move the mouse to highlight the **Modify** command and then release the mouse button. Pick the **2.00** dimension, type the new value in the command line (**1.675**), and press **Enter**; do likewise to the **1.50** dimension to change it to **1.75**. **Regenerate** the part to see the changes. **Shade** and rotate the part to see the changes better (Fig. 2.38).

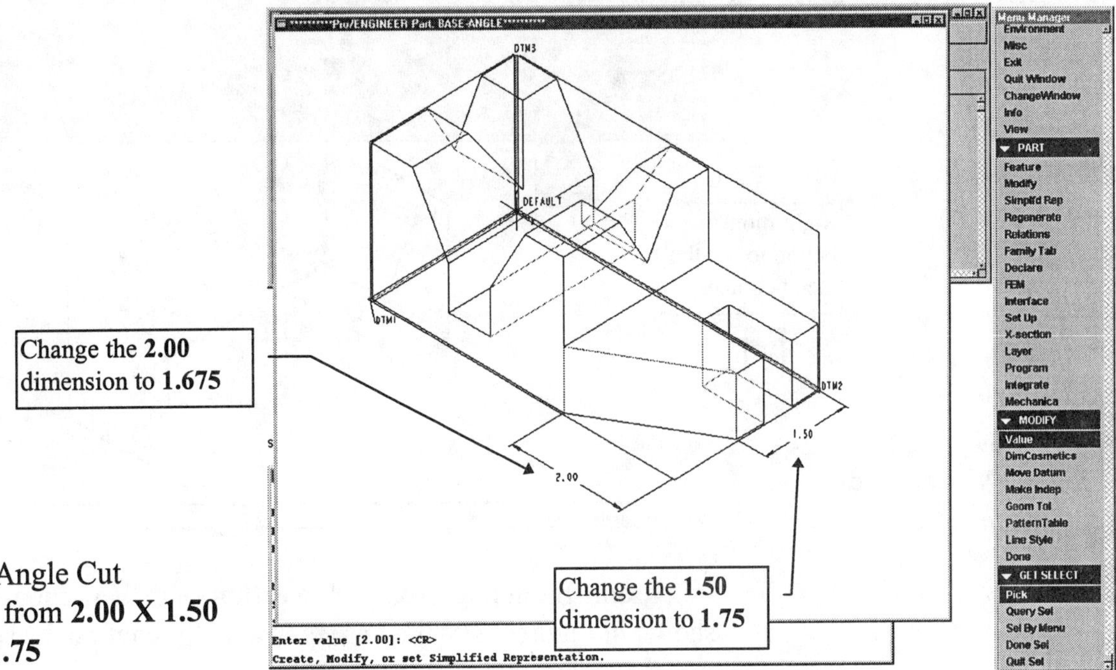

Figure 2.37
Modify the Angle Cut Dimensions from **2.00 X 1.50** to **1.675 X 1.75**

Dbms ⇒ Save ⇒ enter
Purge ⇒ enter ⇒
Done-Return

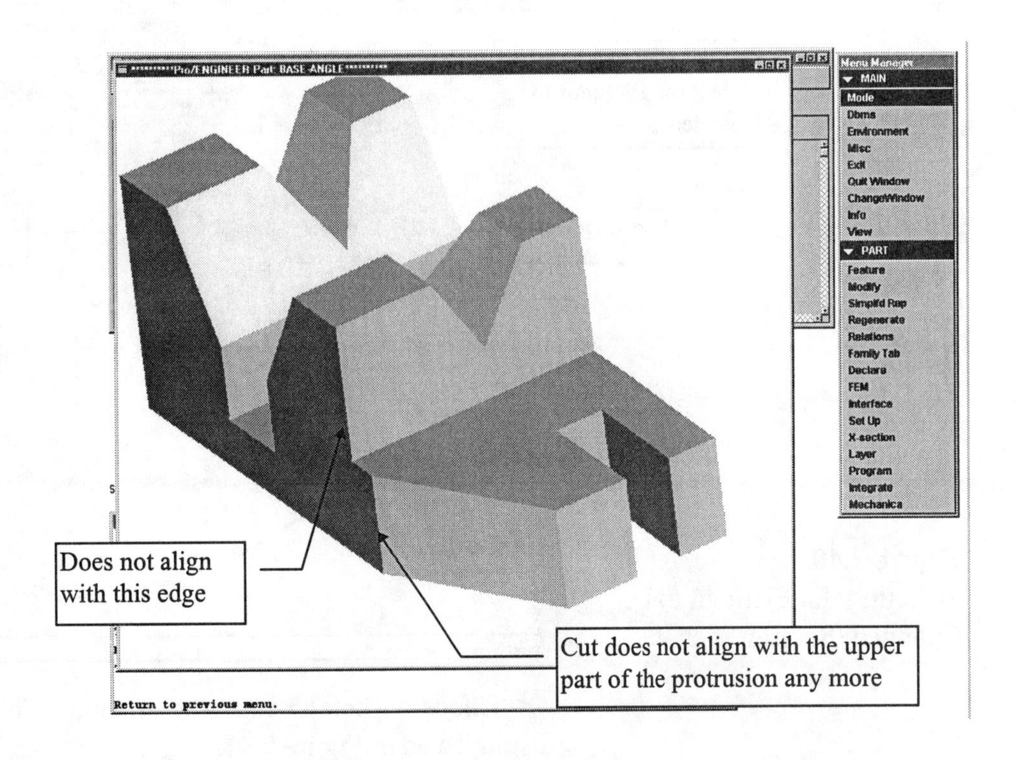

Figure 2.38
Modified Cut

The other requirement from the ECO was to change the shape and the size of the **V** cut. Alter the **V**-shaped cut to be a **U**-shaped slot **1.125** wide by **.9875** deep. Again let's use the Model Tree for the feature and the command selection (Fig. 2.39).

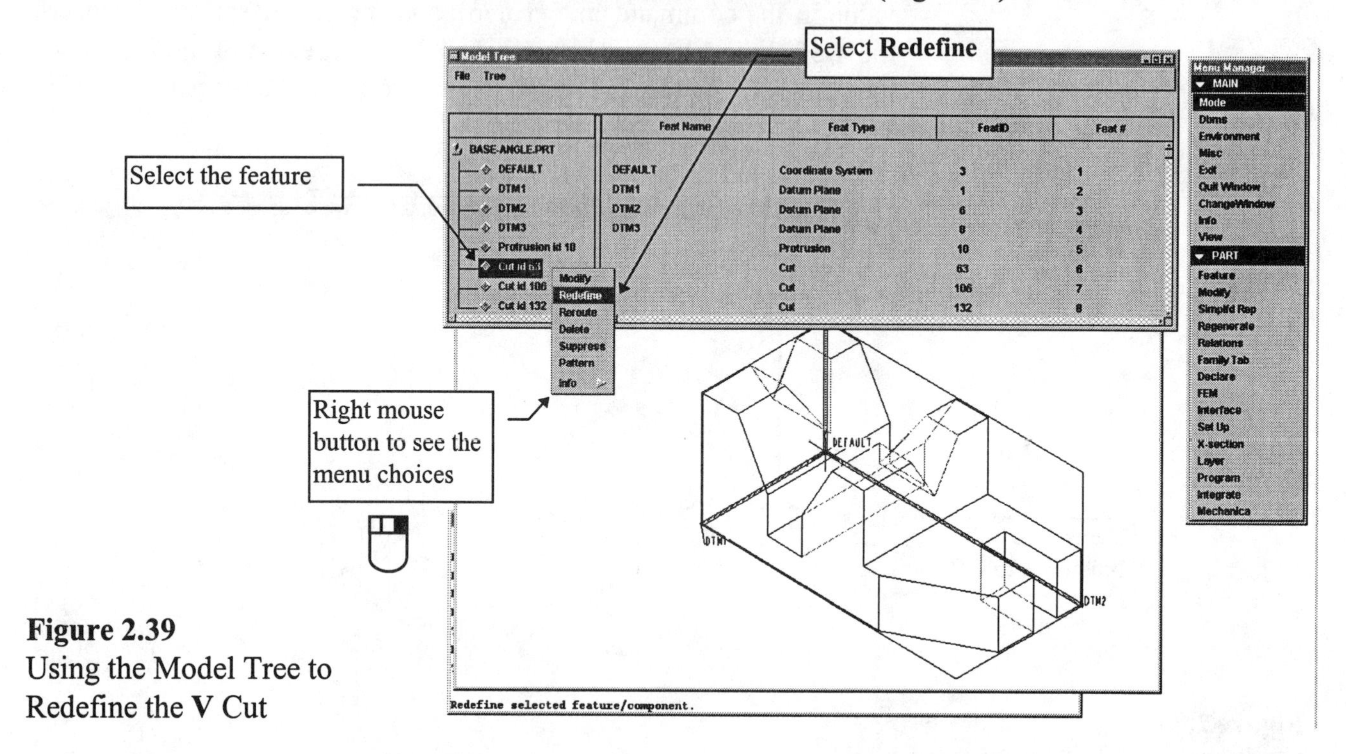

Figure 2.39
Using the Model Tree to Redefine the **V** Cut

From the dialog box, pick **Section** and then choose **Define**, as shown in Figure 2.40. You are now able to redefine the section.

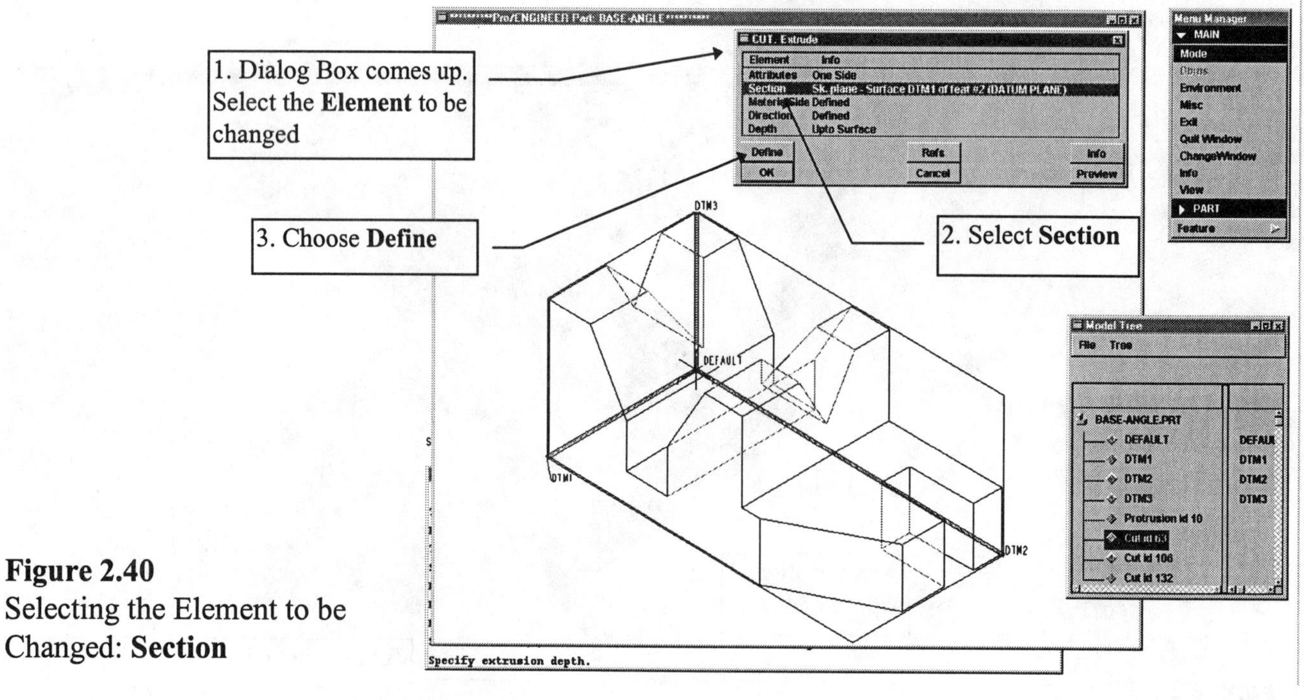

Figure 2.40
Selecting the Element to be Changed: **Section**

Choose **Sketch** from the menu. The section sketch will be displayed as in Figure 2.41.

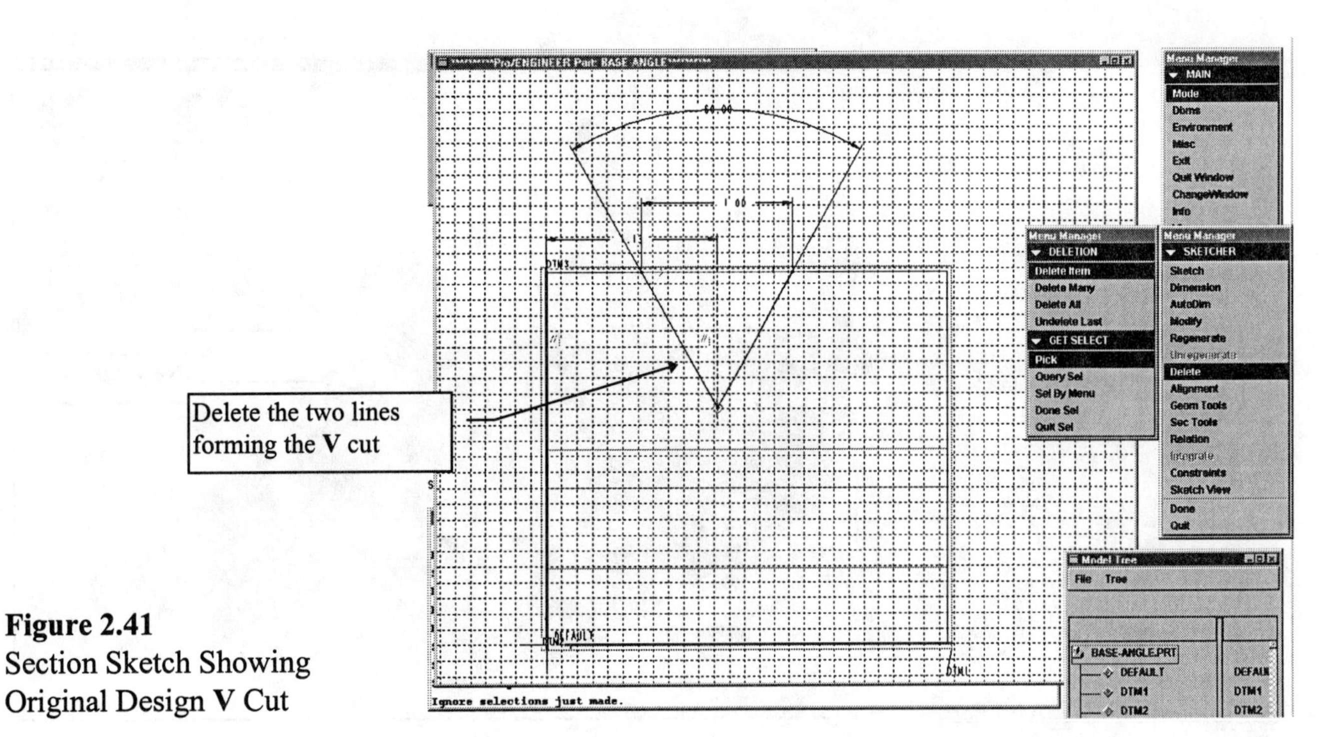

Figure 2.41
Section Sketch Showing Original Design **V** Cut

Use the same centerline for the redefined cut section. Delete the two lines forming the **V**, and sketch three lines of the slot as shown in Figure 2.42. Align the vertical sketch lines to the top of the part, and dimension using the **ECO** sizes of **1.125** wide by **.9875** deep (Fig. 2.42). The **.9875** dimension rounds to **.99**. Regenerate the sketch, and pick **Done** from the SKETCHER menu and **OK** from the dialog box. Rotate and shade the part (Fig. 2.43). Save the modified and redefined part.

Dbms ⇒ Save ⇒ enter
Purge ⇒ enter ⇒
Done-Return

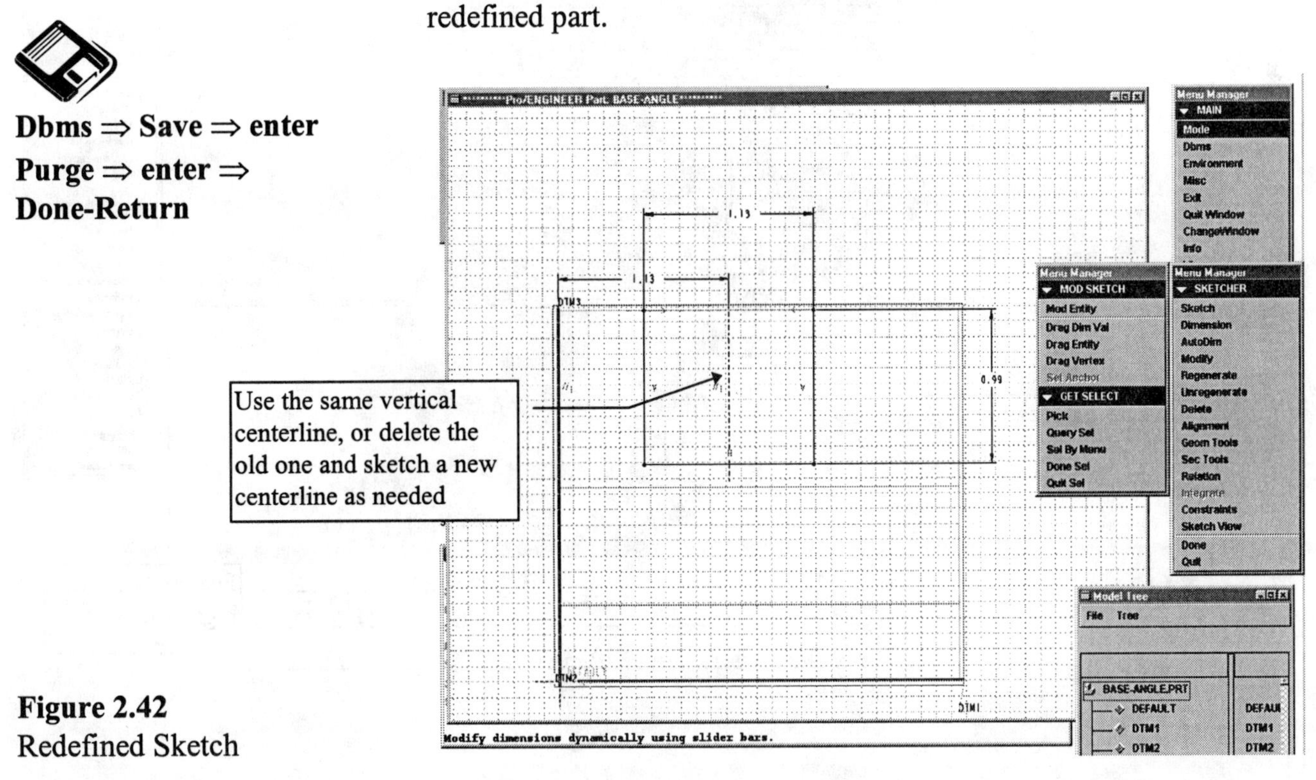

Figure 2.42
Redefined Sketch

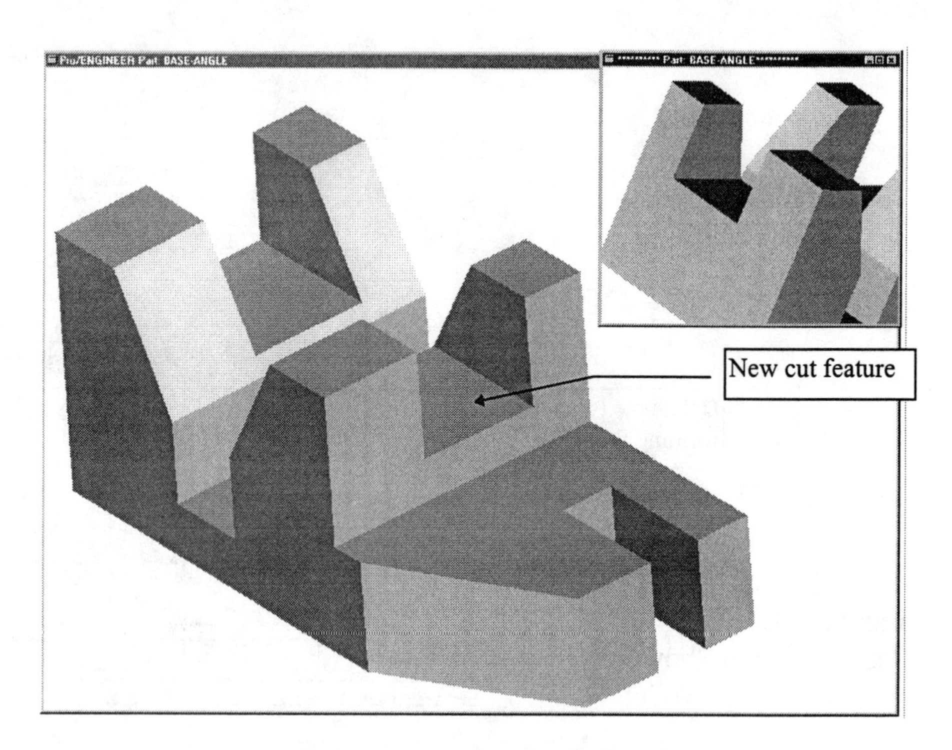

Figure 2.43
Completed BASE_ANGLE

You can also get information regarding your model at any time in the design or redesign process. Using the Model Tree, request information on the part feature just redefined (Fig. 2.44). A variety of information is displayed on the screen, including the **Feature's Dimensions**, **Feature Number**, **Internal Feature Id Number**, and **Parents** (Fig. 2.45).

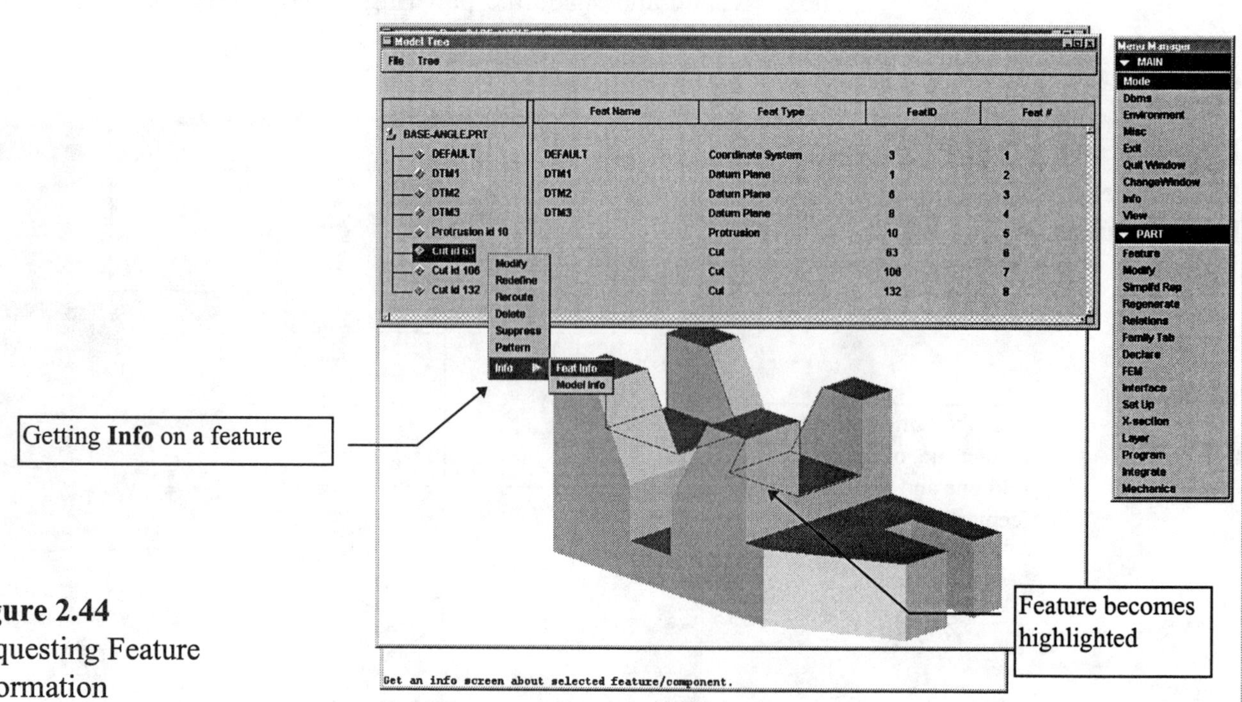

Figure 2.44
Requesting Feature Information

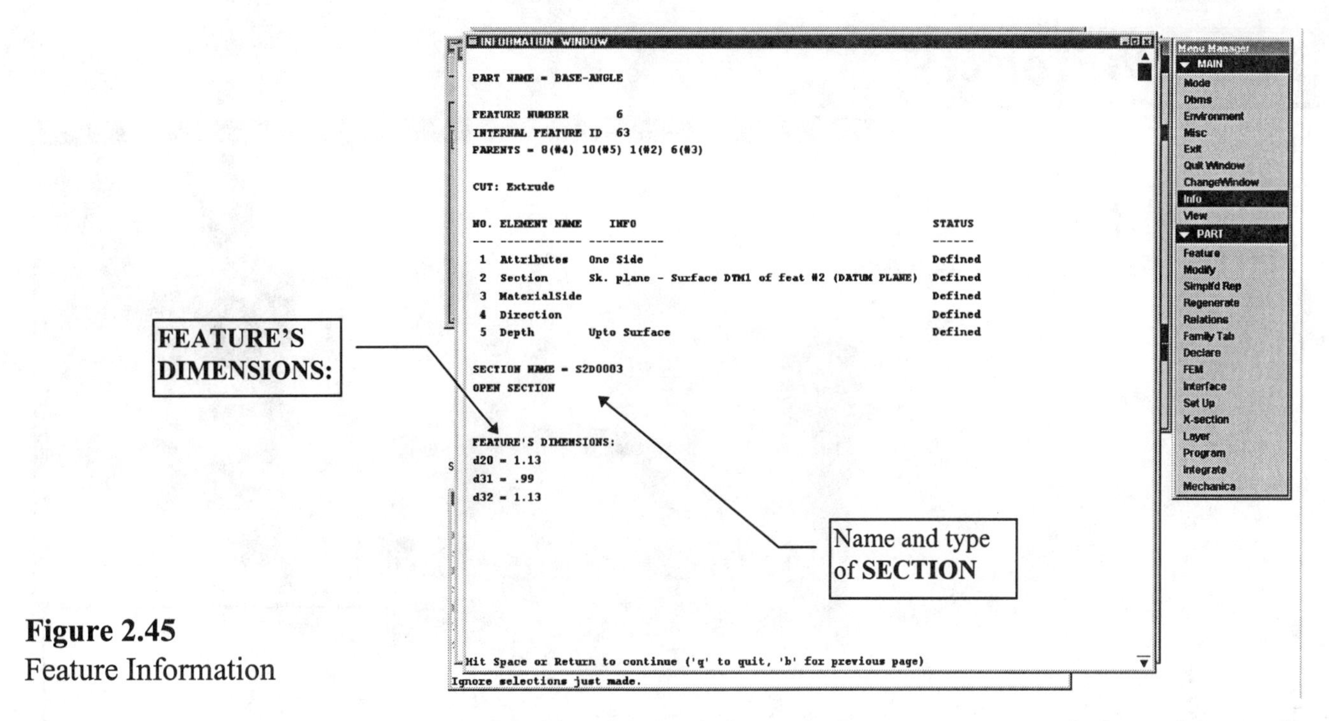

Figure 2.45
Feature Information

Besides feature information, you can extract information on the whole model by giving the following commands:

Info ⇒ **Model Info** (Fig. 2.46) ⇒ type **"q"** to quit, "**b**" for previous page, or press **Enter** or the ***Spacebar*** to continue

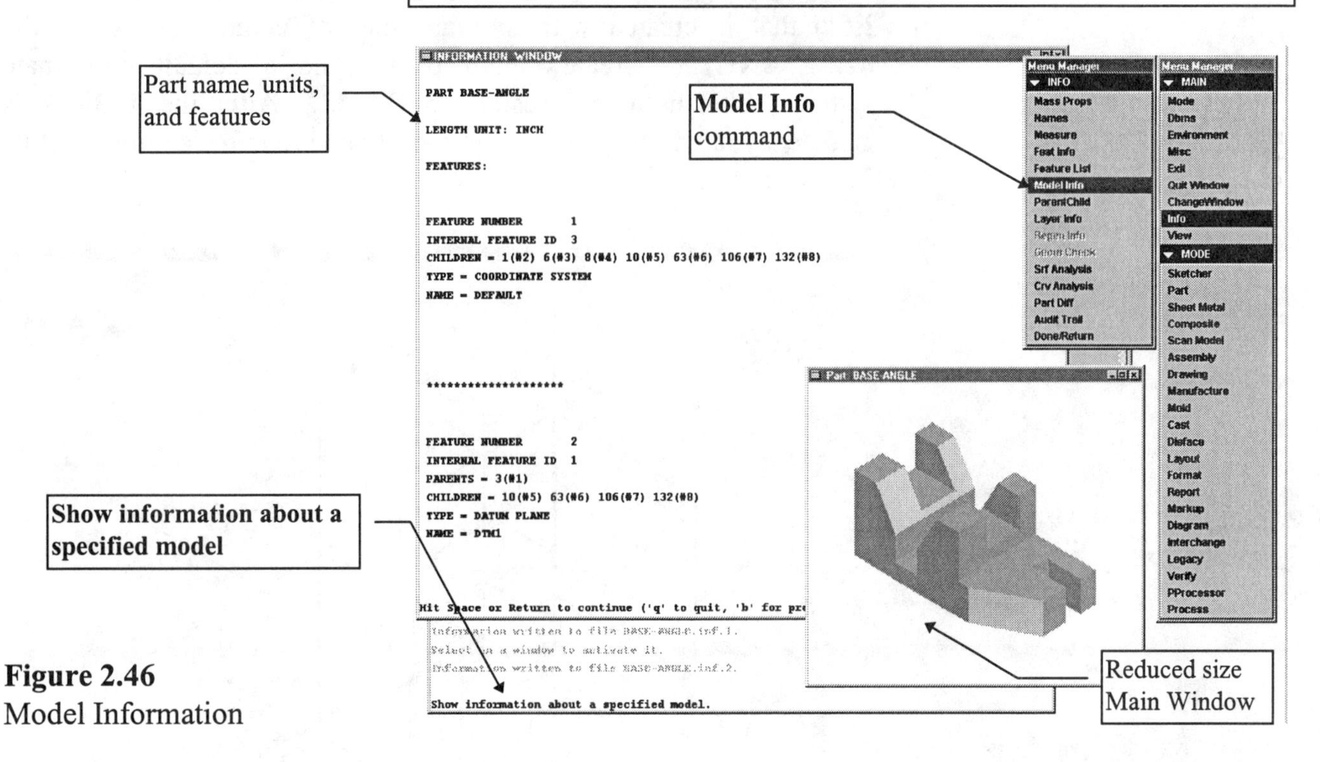

Figure 2.46
Model Information

Lesson 2 Project

T-Block

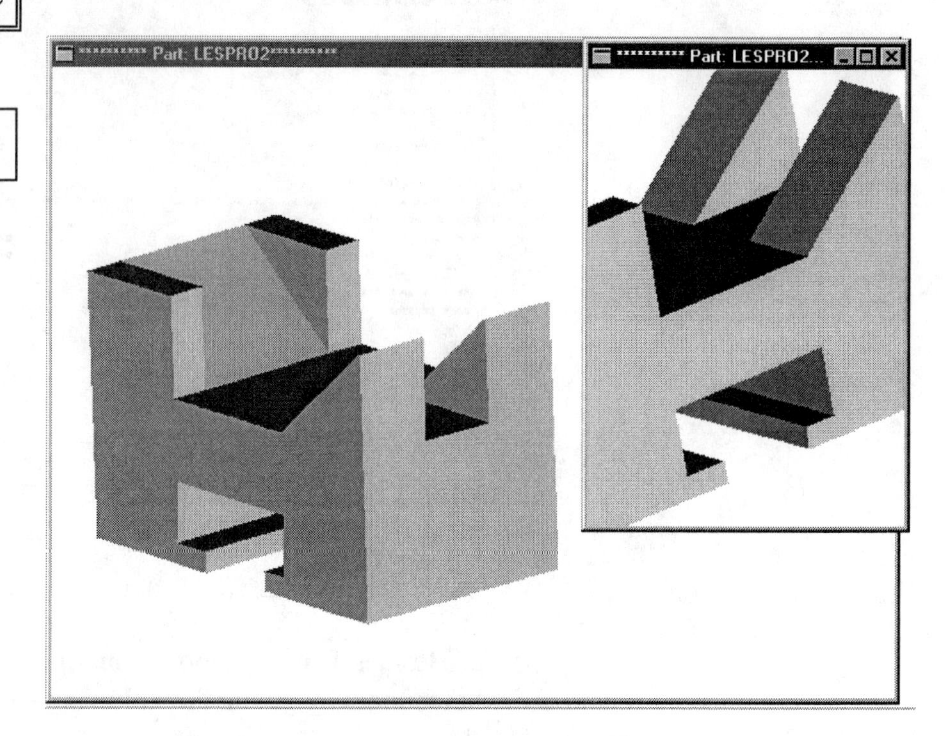

Figure 2.47
T-Block

T-Block

The second **lesson project** is a block (Figure 2.47 through Figure 2.51) that is created with similar types of commands as for the **BASE_ANGLE**. Create datum planes and a default coordinate system. Sketch the protrusion on **DTM3**. After the T-Block is modeled, you will be prompted to modify and redefine a number of its features from an **ECO**.

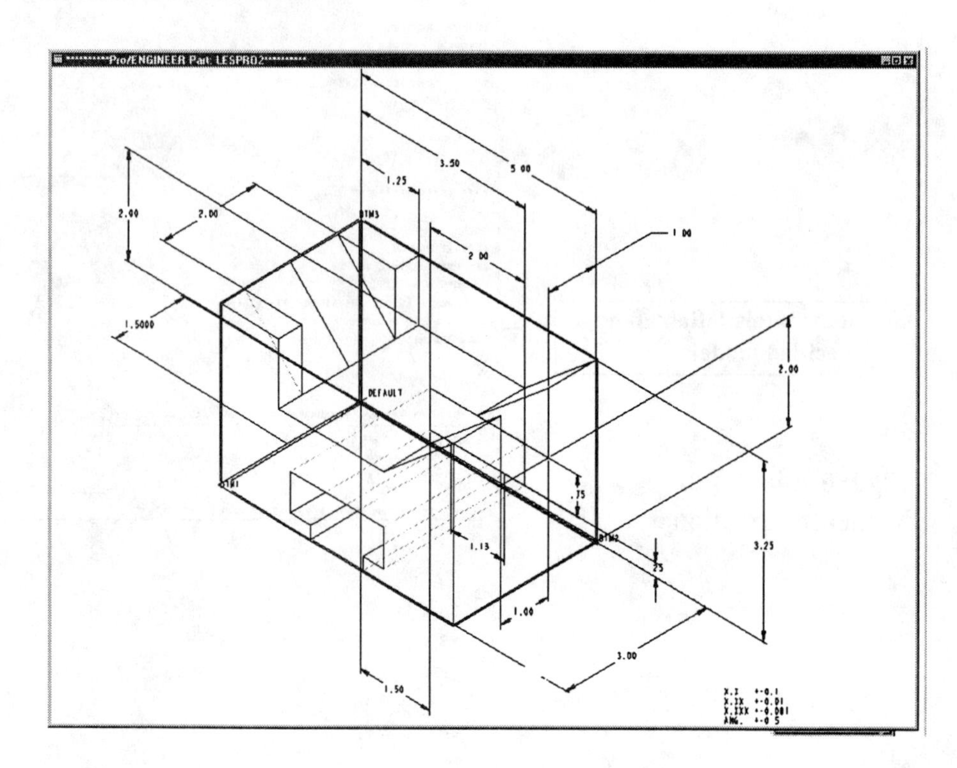

Figure 2.48
T-Block Dimensions

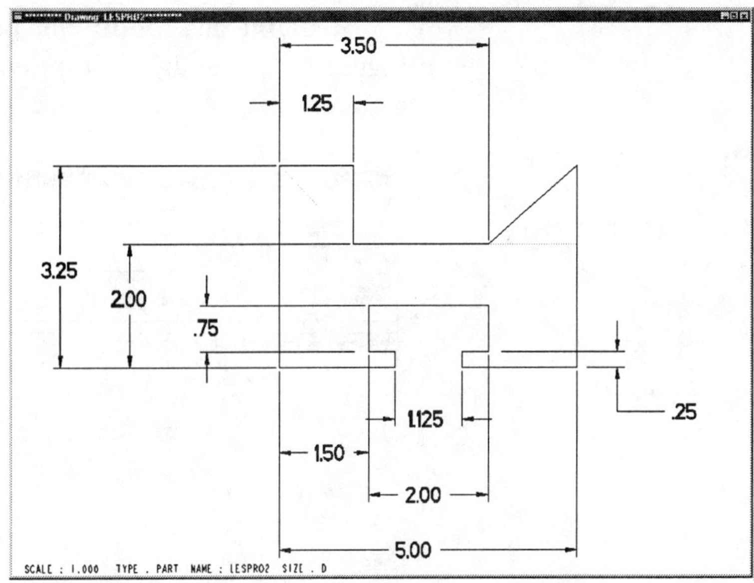

Figure 2.49
Front View

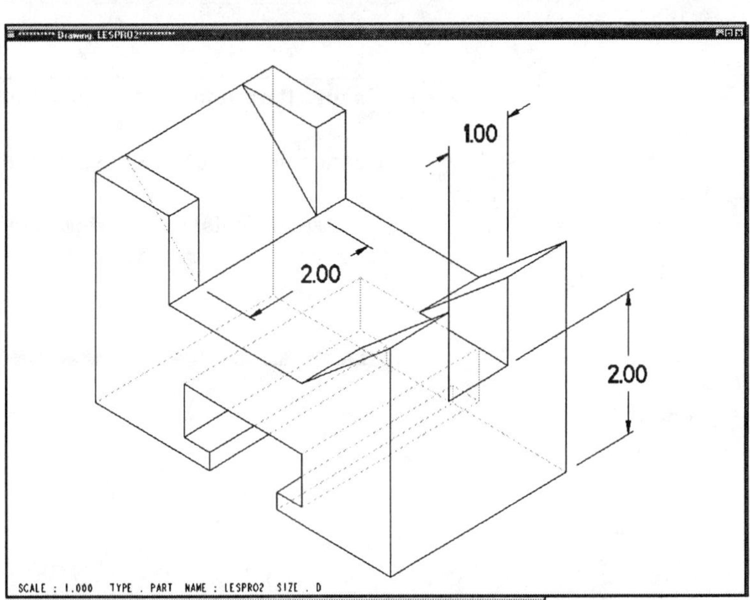

Figure 2.50
Pictorial View

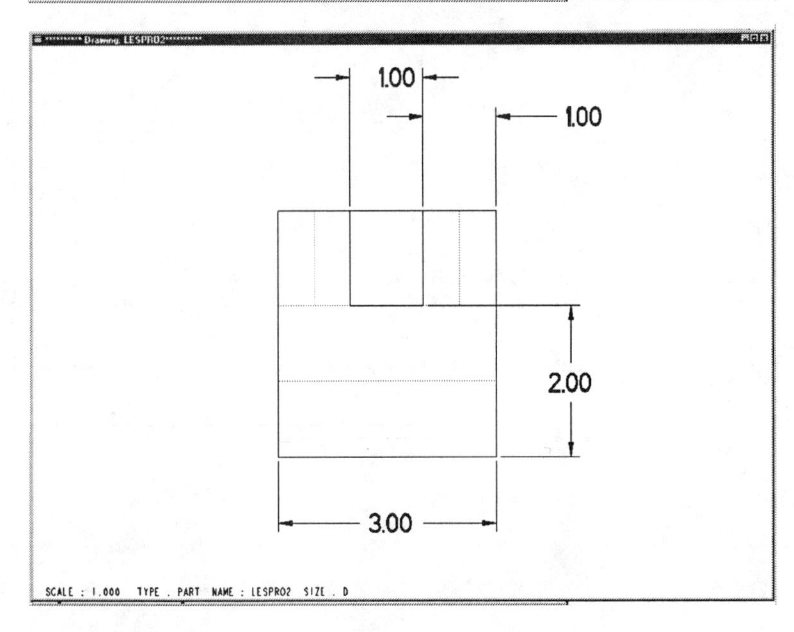

Figure 2.51
Right Side View

Redefine and **Modify** the part after you save it to another name. Figures 2.52 through 2.56 provide the ECO and feature redefinition requirements.

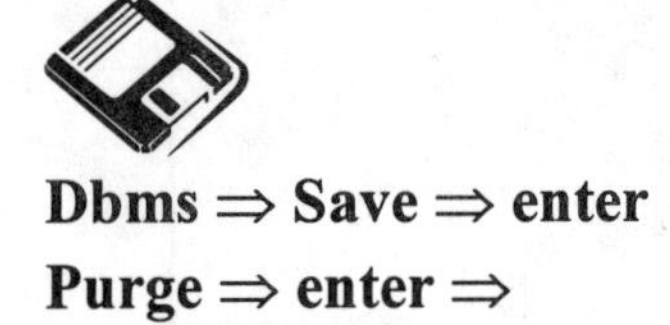

Dbms ⇒ Save ⇒ enter
Purge ⇒ enter ⇒
Done-Return

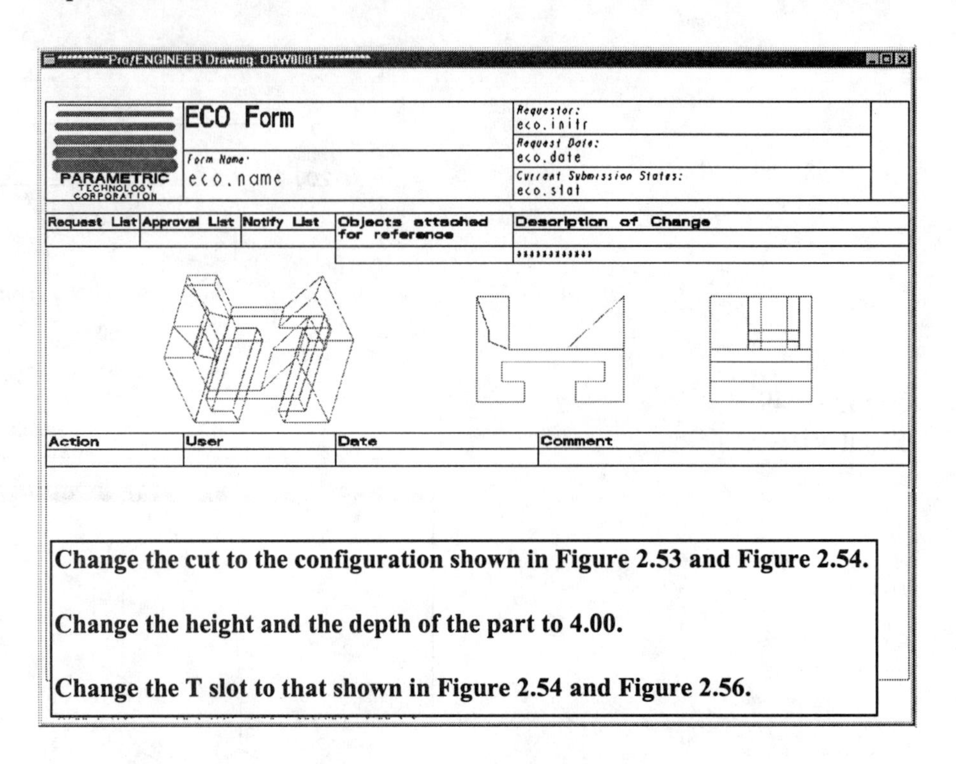

Pro/ENGINEER Drawing: DRW0001

PARAMETRIC TECHNOLOGY CORPORATION

ECO Form

Form Name: eco.name

Requestor: eco.initr

Request Date: eco.date

Current Submission States: eco.stat

Request List	Approval List	Notify List	Objects attached for reference	Description of Change

Action	User	Date	Comment

Change the cut to the configuration shown in Figure 2.53 and Figure 2.54.

Change the height and the depth of the part to 4.00.

Change the T slot to that shown in Figure 2.54 and Figure 2.56.

Figure 2.52
ECO

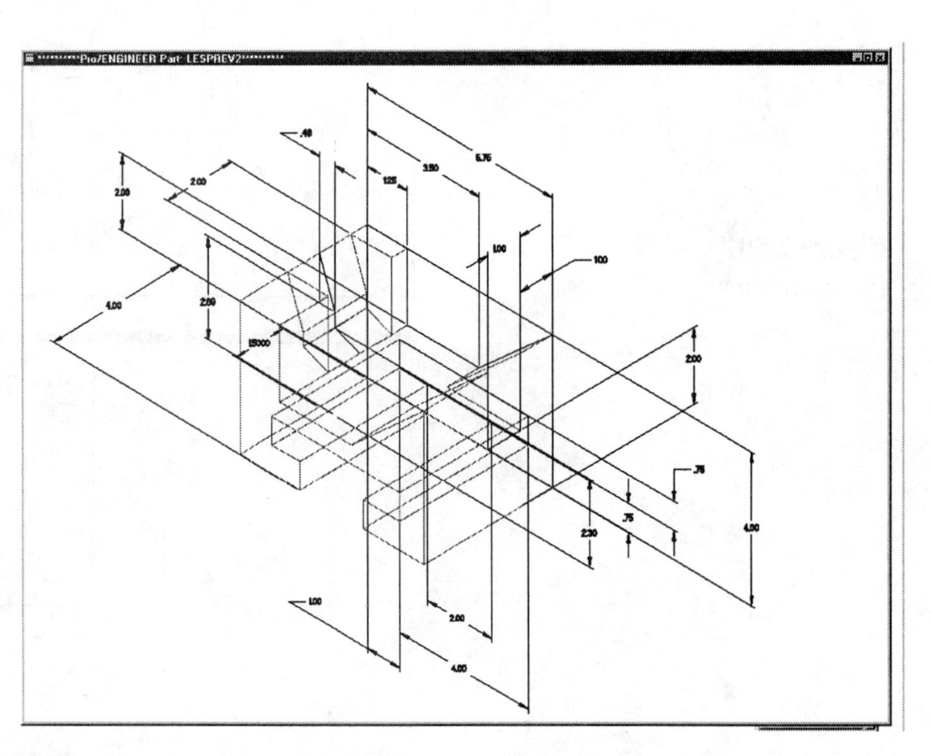

Figure 2.53
Modified Dimensions for T-Block

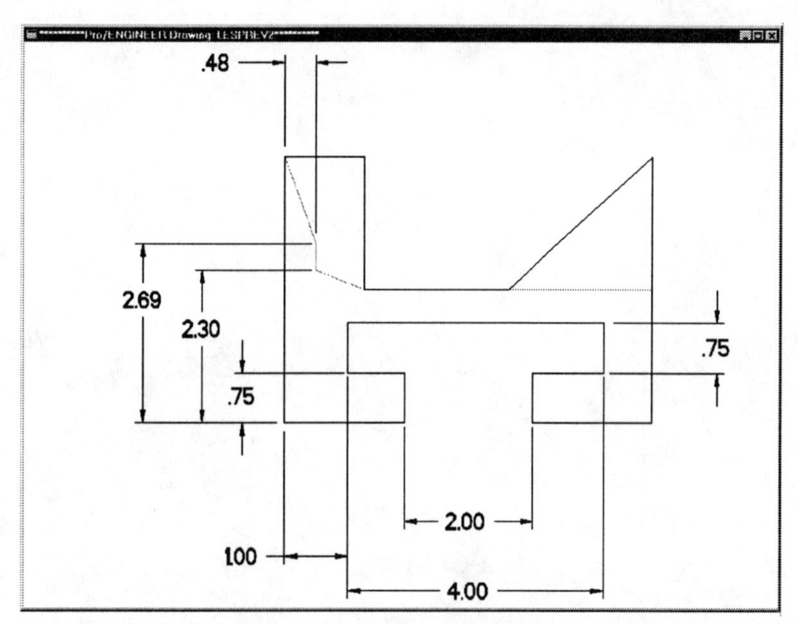

Figure 2.54
Front View

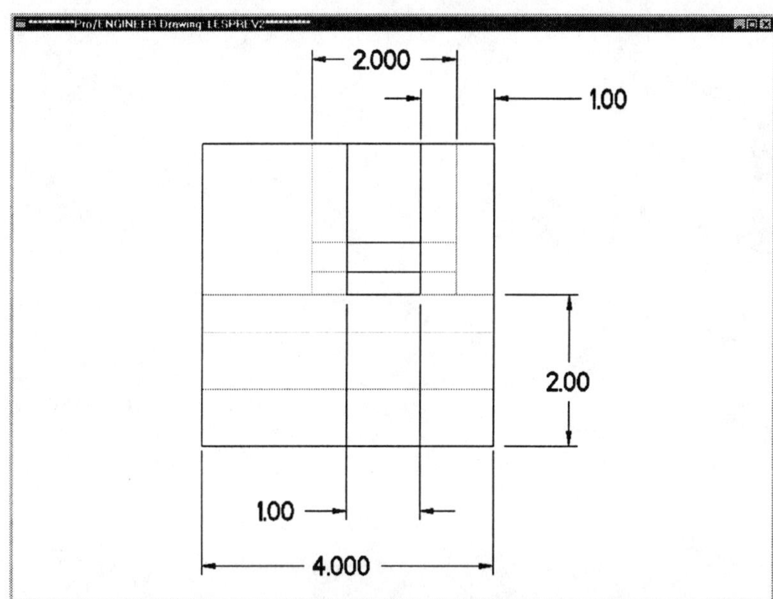

Figure 2.55
Side View

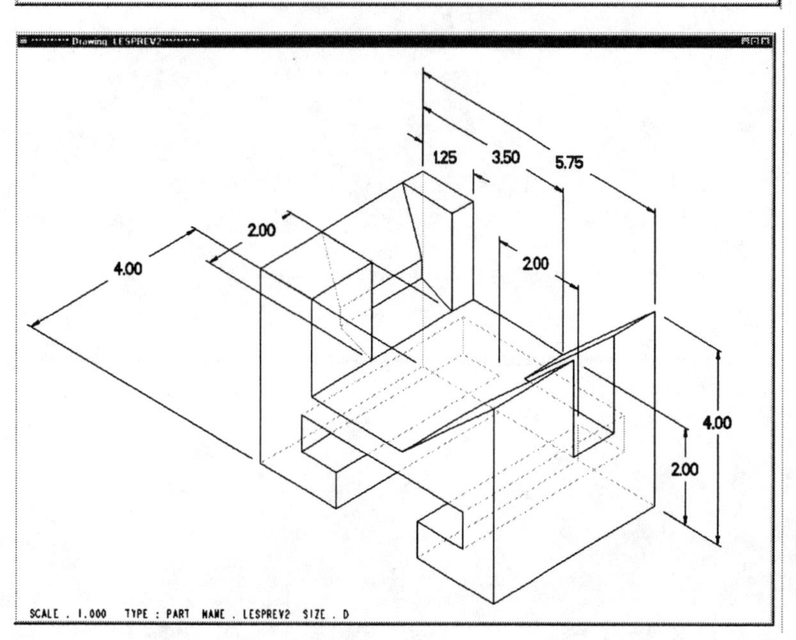

Figure 2.56
Pictorial View

Lesson 3

Holes and Rounds

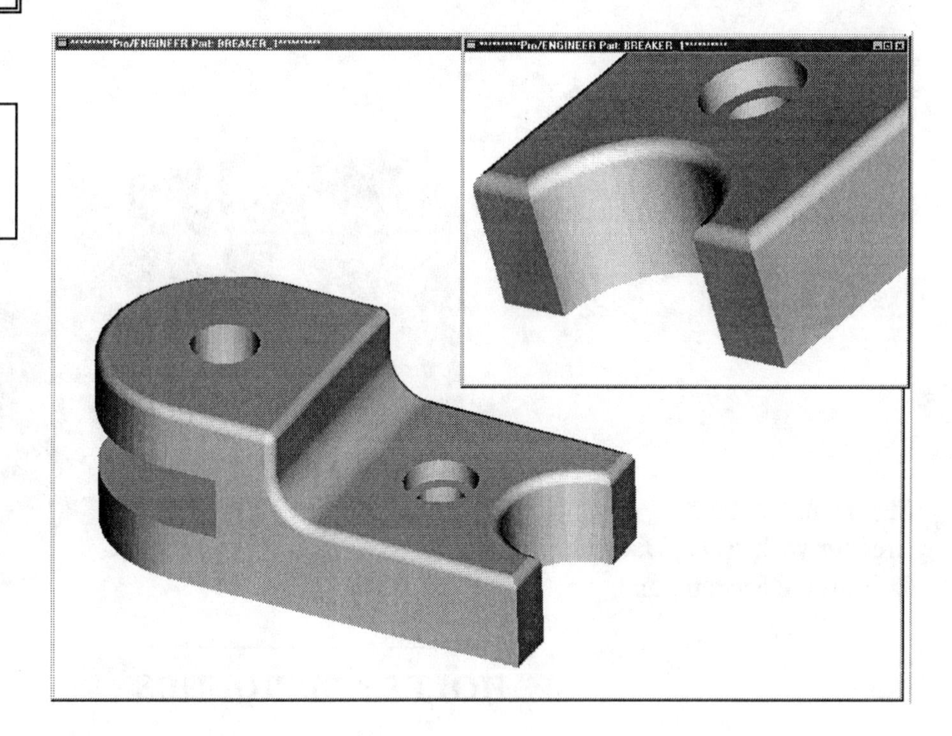

Figure 3.1
Breaker

OBJECTIVES

1. **Create simple rounds along model edges**

2. **Sketch arcs on sections**

3. **Create a straight hole through a part**

4. **Complete a sketched hole**

5. **Understand the difference between sketched and pick-and-place features**

6. **Understand the options for specifying hole depth, including Blind, Thru Next, Thru All, and Thru Until**

7. **Use the Info command to extract feature information**

8. **Understand the types of round creation options**

EGD REFERENCE
Engineering Graphics and Design
by L. Lamit and K. Kitto
Read Chapters: 10
See Pages: 286-341, 364, 570

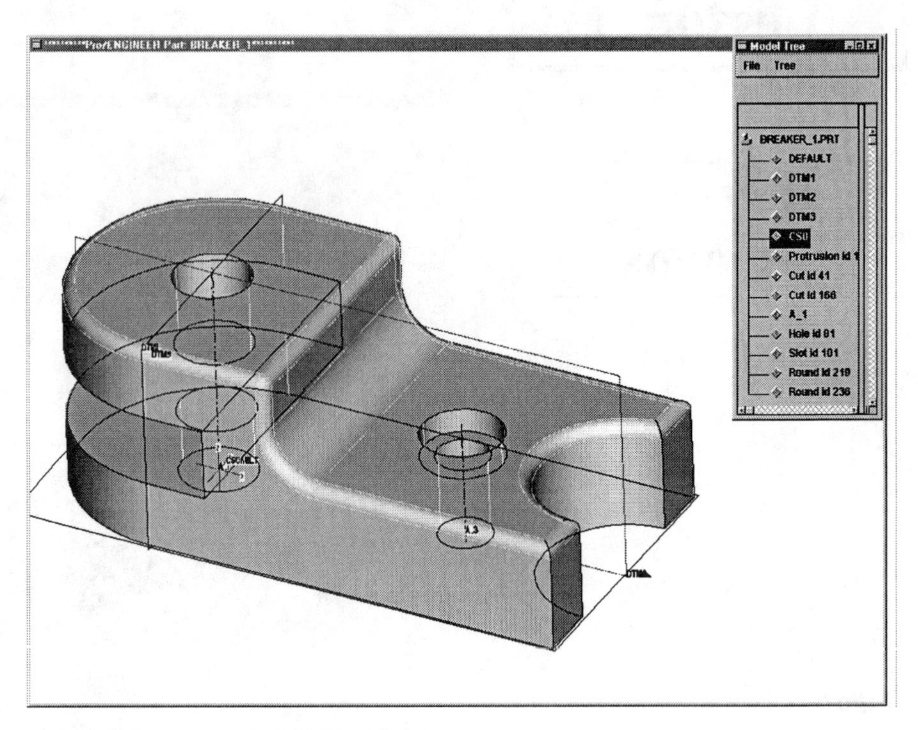

Figure 3.2
Breaker with Axes, Datums, Coordinate System, and Model Tree

HOLES AND ROUNDS

A variety of geometric shapes and constructions are accomplished automatically with Pro/E, including *holes* and *rounds*. These features are called *pick-and-place* features, since they are created automatically from your input and then placed according to prompts by Pro/E. A hole can also be created using **Cut,** but it must be sketched. In general, pick-and-place features are not sketched (except for the **Sketched** option when creating a complex hole shape such as a countersink hole or counterbore as in Fig. 3.1 and Fig. 3.2). **Round** creates a fillet, or a round on an edge, that is a smooth transition with a circular profile between two adjacent surfaces.

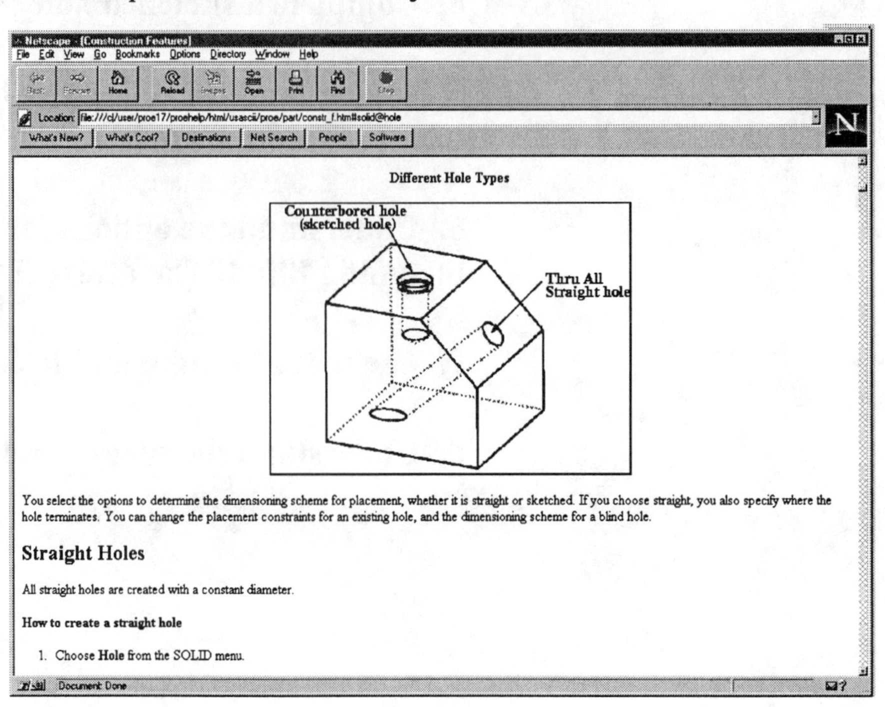

Figure 3.3
Online Documentation Holes

Holes

The **Hole** option creates many types of holes, including through, counterbore, and blind. Figure 3.3 shows the online help available for the hole command. All holes are based on two basic types of hole geometry:

> **Straight hole** An extruded slot with a circular section. This type passes from the placement surface to the specified end surface.
> **Sketched hole** A revolved feature defined by a sketched section. Counterbore and countersink holes, for example, are created as sketched holes.

Straight Holes

All straight holes are created with constant diameter. To create a straight hole:

1. Choose **Hole** from the SOLID menu.
2. Pro/E displays the HOLE OPTS menu. Choose **Straight** ⇒ **Done**.
3. Pro/E displays the Feature Creation dialog box and the PLACEMENT menu, which lists the options **Linear**, **Radial**, **Coaxial**, and **On Point**. Choose one of these options, then **Done**.
4. Select the placement plane.
5. Select the first reference (an edge, axis, planar surface, or datum).
6. Enter the distance from the first reference in the message area.
7. Select the second reference.
8. Enter the distance from the second reference in the message area.
9. Pro/E displays the SIDES menu. Choose **One Side** or **Both Sides**, then **Done**.
10. Select the extent to which the hole will be created, then choose **Done**. The SPEC TO menu options include:

> **Blind** Creates a hole with a flat bottom.
> **Thru Next** Creates a hole that continues until it reaches the next part surface.
> **Thru All** Creates a hole that intersects all the surfaces.
> **Thru Until** Creates a hole that goes through all the surfaces until it reaches intersection with the specified surface.
> **UpTo Pnt/Vtx** Creates a hole with a flat bottom that continues until it reaches the specified point or vertex.
> **UpTo Curve** Creates a hole with a flat bottom that continues until it reaches the specified curve that you draw in a plane parallel to the placement plane.
> **UpTo Surface** Extrudes the hole from material until the bottom of the hole conforms to the selected bounding surface.

11. Enter the depth of the hole at the prompt, if **Blind** was selected.
12. Enter the diameter of the hole at the prompt.
13. Select the **OK** button in the dialog box to create the hole.

Sketched Holes

A *sketched hole* is created by sketching a section for revolution and then placing the hole onto the part. Sketched holes are always blind and one-sided. Sketched holes must have a vertical centerline (*axis of revolution*), with at least one of the entities sketched normal to the axis centerline. Pro/E aligns the normal entity with the placement plane. The remainder of the sketched feature is cut from the part, as with a revolved cut. You can also use the revolved cut command to create holes. If you have **COAch for Pro/ENGINEER** on your system, go to SEARCH and do the sketched holes segment (Fig. 3.4).

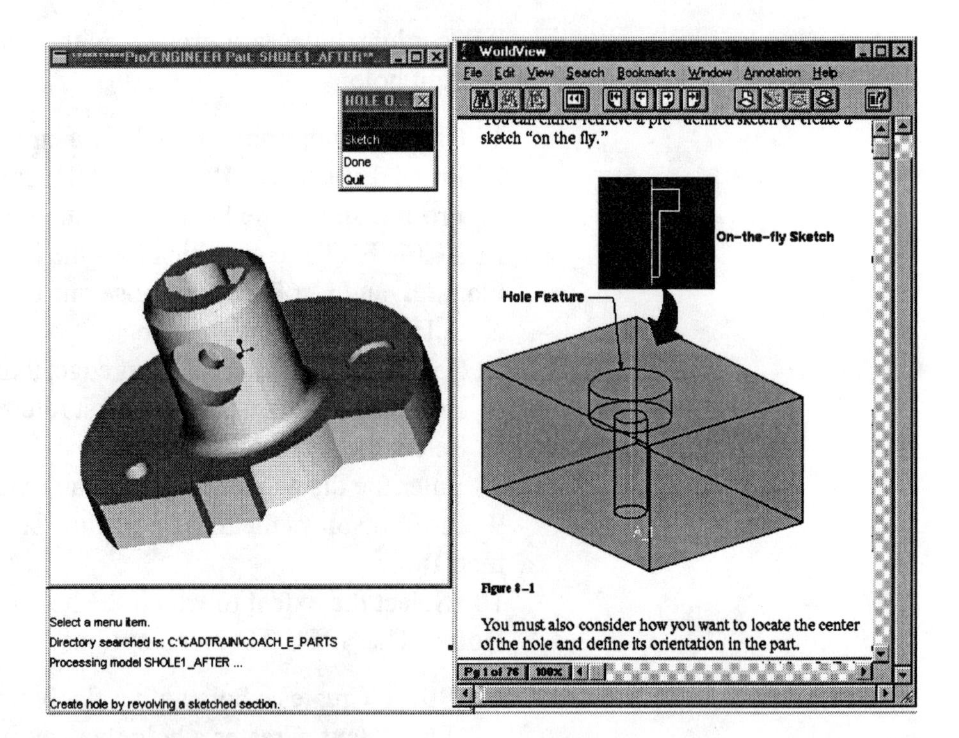

Figure 3.4
COAch for Pro/E, Holes--Modeling (Segment 2: Sketched Holes)

To create a sketched hole:

1. Choose **Hole** from the SOLID menu. Choose **Sketch** and **Done** from the HOLE OPTS menu.
2. Pro/E displays the feature creation dialog box.
3. Choose the dimensioning scheme for the hole using the PLACEMENT menu options. Choose **Done.**
4. Pro/E displays a grid in a subwindow. Sketch a vertical centerline and then the cross section of the hole. **Dimension** and **Regenerate** the section. **Modify** the dimensions. Choose **Done.**
5. Select the placement plane.
6. Select the first reference edge, planer surface, or axis.
7. Enter the distance from the first reference at the prompt.
8. Select **OK** from the dialog box.
9. Select the second reference, enter the distance, and choose **OK**.

Rounds

Rounds (Fig. 3.5) are created automatically at selected edges of the part. Tangent arcs are introduced as rounds between two adjacent surfaces of the solid model.

There are cases where rounds should be added early; but in general, wait until later in the design to add the rounds. Introducing rounds to a complex design early in the project can cause a series of failures later in the design. You might also choose to place all rounds on a layer and suppress that layer to speed up your working session. There are a number of basic rounds to consider, including: **Edge Chain, Surface-Surface, Edge-Surface, Edge Pair**.

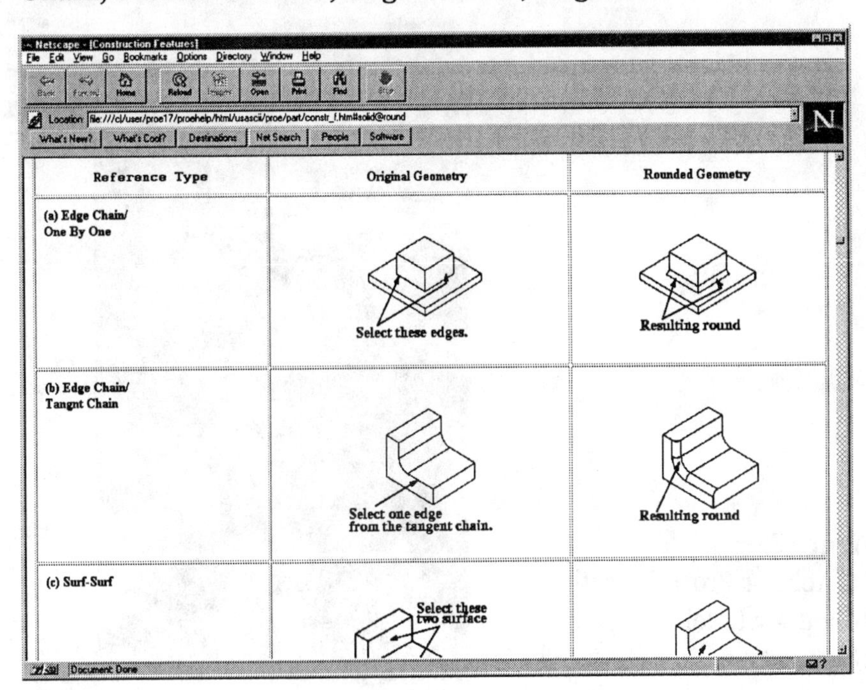

Figure 3.5
Online Documentation Rounds

Two categories of rounds are available: simple and advanced. Much of the time, you will create *simple* rounds. These rounds smooth the hard edges between two adjacent surfaces. If you have **COAch for Pro/ENGINEER** on your system, go to SEARCH and do the appropiate segment for rounds, shown in Figure 3.6.

Creating a Simple Round

Figures 3.7 (edge pair) and 3.8 (advanced variable) show some of the online documentation available on rounds. You should read your manual, or highlight the round command on the screen, press the right-hand mouse button, and choose **? GetHelp**.

The basic steps to create a simple round include:

1. Choose **Round** from the SOLID menu.
2. Choose **Simple** and **Done** from the ROUND TYPE menu.
3. A dialog box appears, listing elements of the round feature.
4. The Attributes element is selected by default. Use the RND SET ATTR menu options to specify the round's attributes.

Specify the type of round by selecting one of these options:

Constant Creates a round between two sets of surfaces with a constant radius.
Variable Creates a round between two sets of surfaces with variable radii. Specify radii at the ends of the chain of edges or at the ends of the spine (when the spine is required) and, optionally, at additional points along the edges or along the spine.
Full Round Creates a round by removing a surface; the consumed surface becomes a round.

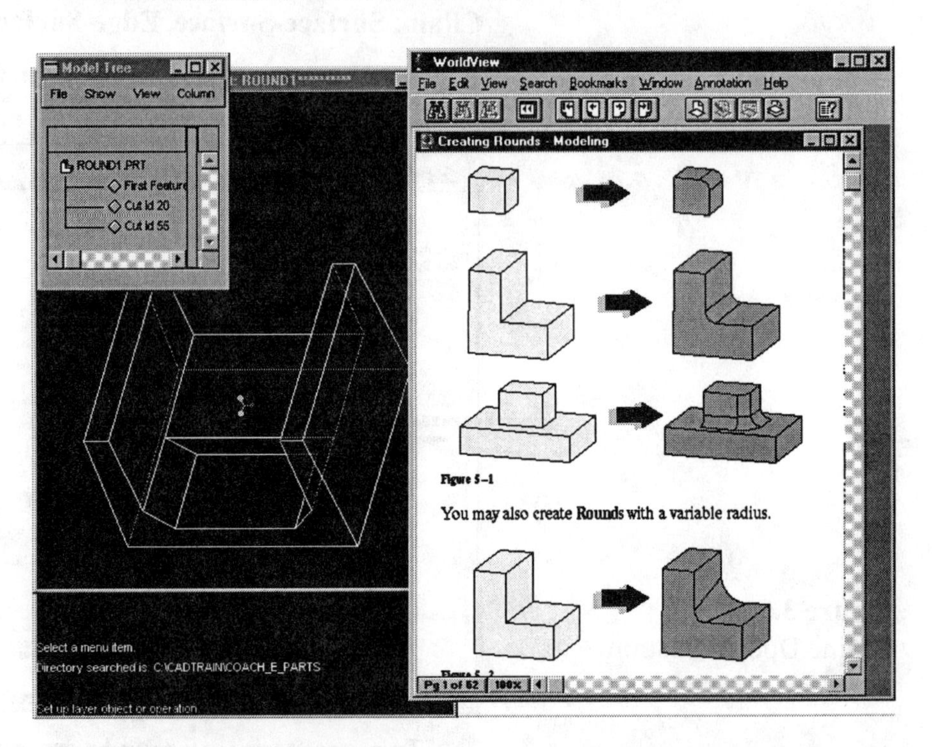

Figure 3.6
COAch for Pro/E, Creating Rounds--Modeling (Segment 1)

Select one of the following options to specify the type of references for placing the round:

Edge Chain Places a round by selecting a chain of edges. To select the chain, use options in the CHAIN menu.
Surf-Surf Places a round by selecting two adjacent surfaces.
Edge-Surf Places a round by specifying an edge and a tangent surface.
Edge Pair Places a full round by specifying a pair of edges.

5. Choose **Done** from the RND SET ATTR menu.
6. Pro/E prompts you to select the placement references.
7. For other than a full round, enter the radius for the round.
8. For other than a full round, define the extension boundaries of the round by specifying the **Round Extent** element (optional).
9. If required, define the **Attach Type** element.
10. Choose **OK** from the dialog box.

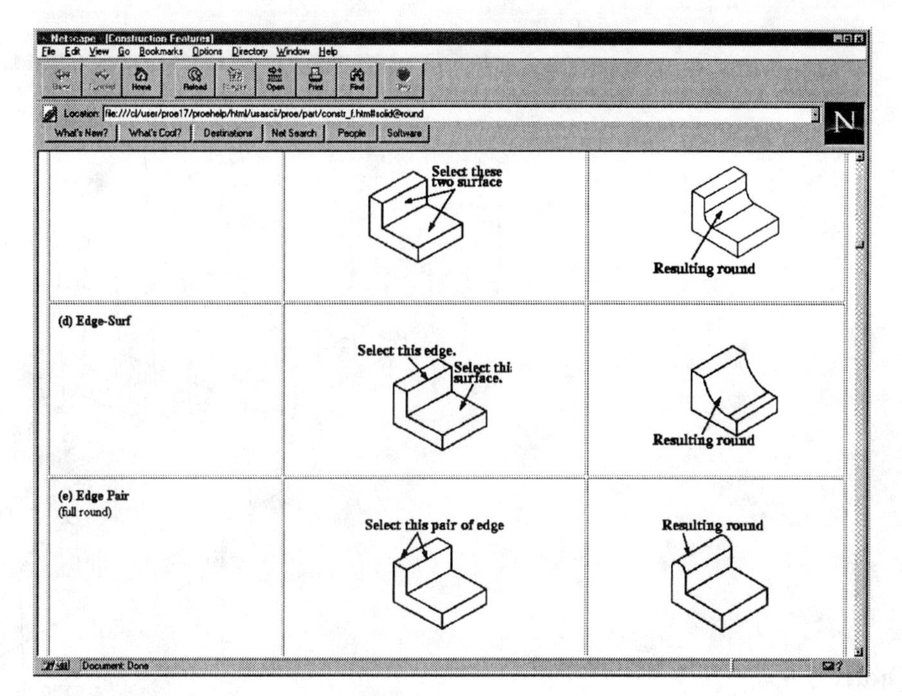

Figure 3.7
Online Documentation Rounds--**Edge-Surf** and **Edge Pair**

Using the Chain Menu Options

When you select reference edges with the **Edge Chain** option, Pro/E displays the CHAIN menu. Note that you can choose more than one option: Choose an option, select the references as prompted by Pro/E, and then choose the next option. The CHAIN menu lists the following options:

One By One Define a chain one at a time by selecting individual edges and curves.
Tangnt Chain Define a chain by selecting an edge. All tangent edges are included in the selection.
Surf Chain Define a chain of edges by selecting a surface.
Unselect Unselect references from one of the preceding options.

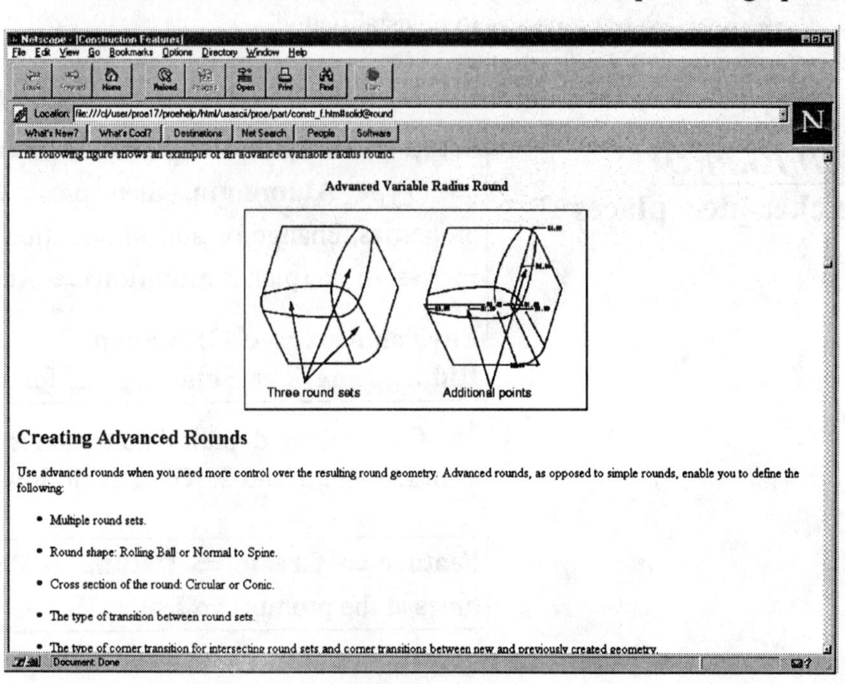

Figure 3.8
Online Documentation Rounds--Advanced Variable Radius Rounds

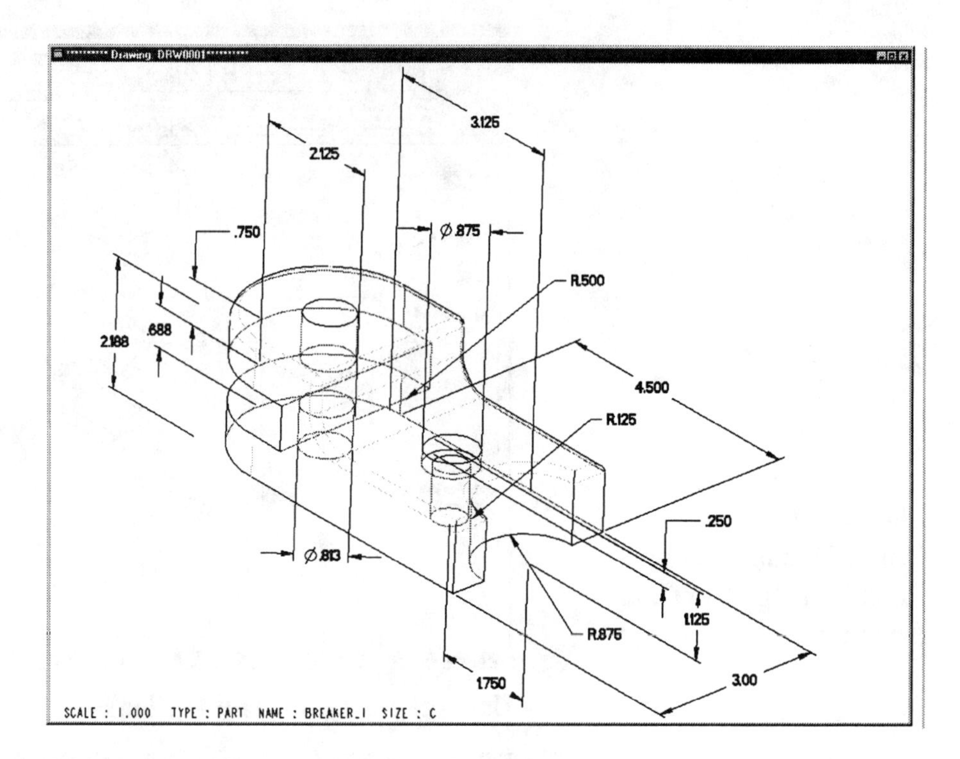

Figure 3.9
Breaker Dimensions

Breaker

The **Breaker** (Fig. 3.9) is the third lesson part. This part introduces two new features, *holes* and *rounds*. Also, in the Sketcher, the **Arc** command will be used to create the rounded end and the half circle cut of the Breaker. Choose **Part** from the MODE menu. Give the following commands:

Part ⇒ **Create** (type the part name **BREAKER**) ⇒ **enter.**

As always, the *environment*, *units*, and *material* for the part need to be established:

CONFIG.PRO
sketcher_dec_places 3

SETUP AND ENVIRONMENT

Setup ⇒ **Units** ⇒ **Length** ⇒ **Inch** ⇒ **Done** ⇒ **Material** ⇒ **Define** ⇒ (Type **Aluminum**, then press **Enter**) ⇒ (table of material properties, change or add information) **File** ⇒ **Save** ⇒ **File** ⇒ **Exit** ⇒ **Assign** ⇒ (pick **Aluminum**) ⇒ **Accept** ⇒ **Done**

Environment ⇒ ✓**Grid Snap**
Hidden Line Tan Solid (try this for a different look)

Create three default datum planes and a default coordinate system as in Lessons 1 and 2. Give the following commands:

Feature ⇒ **Create** ⇒ **Datum** ⇒ **Plane** ⇒ **Offset** ⇒ **enter** (three times at the prompt) ⇒ **Done**

The first protrusion for the Breaker is created using an extruded protrusion and sketching on **DTM3** the outline of the part as seen from its top. Give the following commands to set up the section for the sketch:

Feature ⇒ **Create** ⇒ **Protrusion** ⇒ **Extrude** ⇒ **Solid** ⇒ **Done** ⇒ **One Side** ⇒ **Done** ⇒ (pick **DTM3** as the sketching plane) ⇒ **Okay** (for selecting the *direction* of feature creation) ⇒ **Top** ⇒ (pick **DTM2** as the *orientation* plane)

Though not really necessary for the sketching of this section, sometimes the grid spacing needs to be altered to a different size. Change the size of the grid spacing by giving the following commands:

Sec Tools ⇒ **Sec Environ** ⇒ **Grid** ⇒ **Params** ⇒ **X&Y Spacing** ⇒ (type **15** at the prompt) ⇒ **enter** ⇒ **Done-Return**

Your screen should look similar to Figure 3.10. The grid will now be twice as dense.

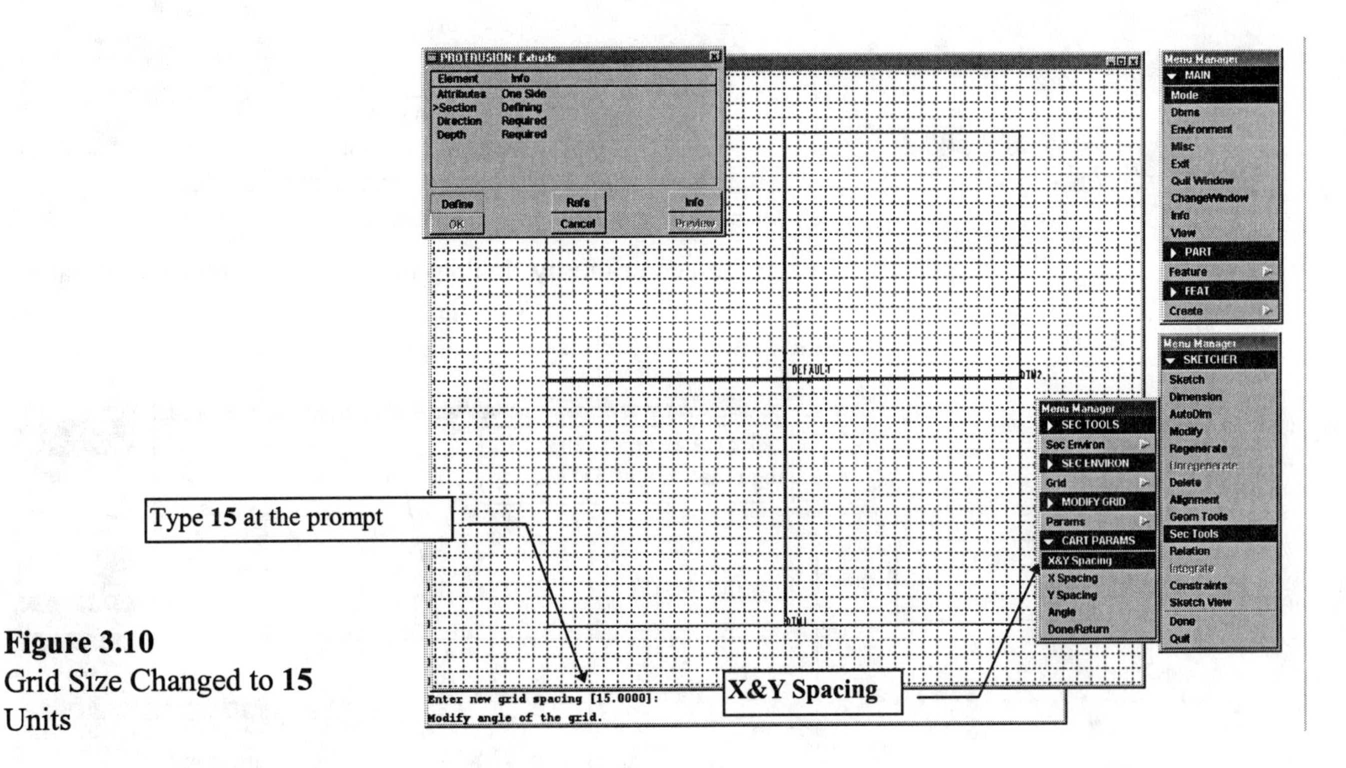

Figure 3.10
Grid Size Changed to **15** Units

Sketch the section by creating an arc representing the part's curved end. The origin of the arc will be at the intersection of **DTM1** and **DTM2**, as shown in Figure 3.11. The section will be composed of two arcs, four lines, and a horizontal centerline. Align the arcs with **DTM1** and **DTM2**.

Create the arcs first (Fig. 3.11), and then add the lines (Fig. 3.12). Give the following commands:

Sketch ⇒ **Arc** ⇒ **Ctr/Ends** (pick the intersection of **DTM1** and **DTM2** as the arc's center, and then pick the starting and ending points; repeat the process for the smaller arc) ⇒ **Sketch** ⇒ **Mouse Sketch** ⇒ (create the lines of the section geometry) ⇒ **Regenerate**

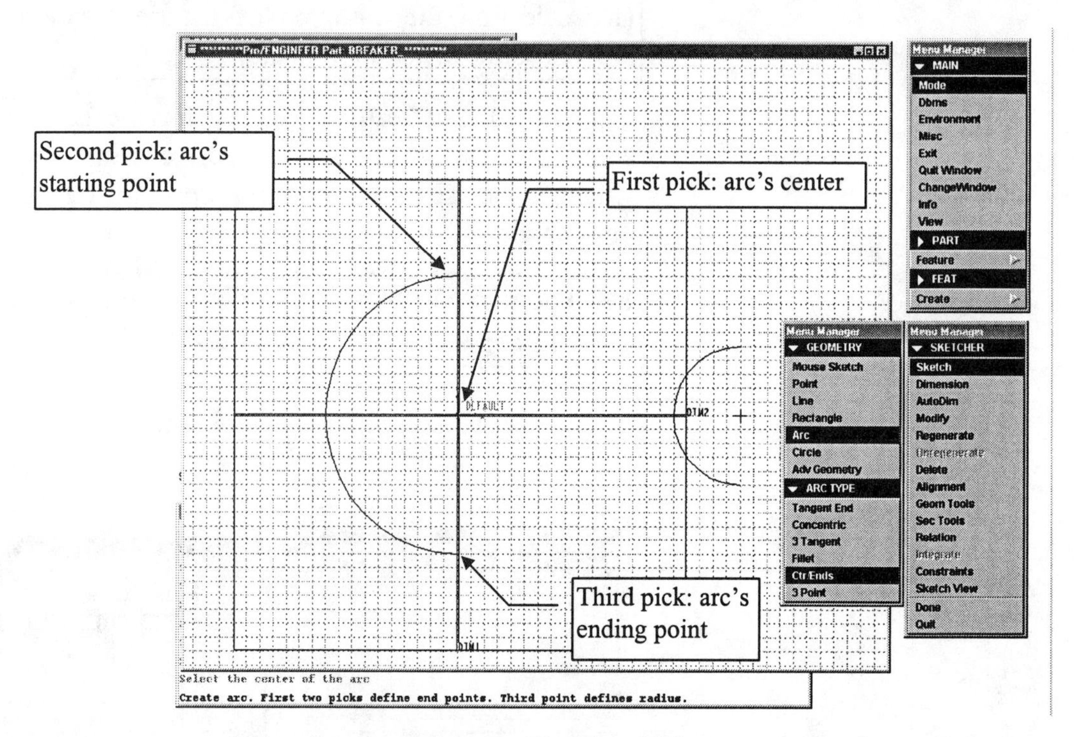

Figure 3.11
Sketched Arcs

Align the centerline by double-picking on the entity and datum plane. To align the arcs, pick on the arc itself (or its center point) and then the datum to which it will be aligned. **Regenerate** again.

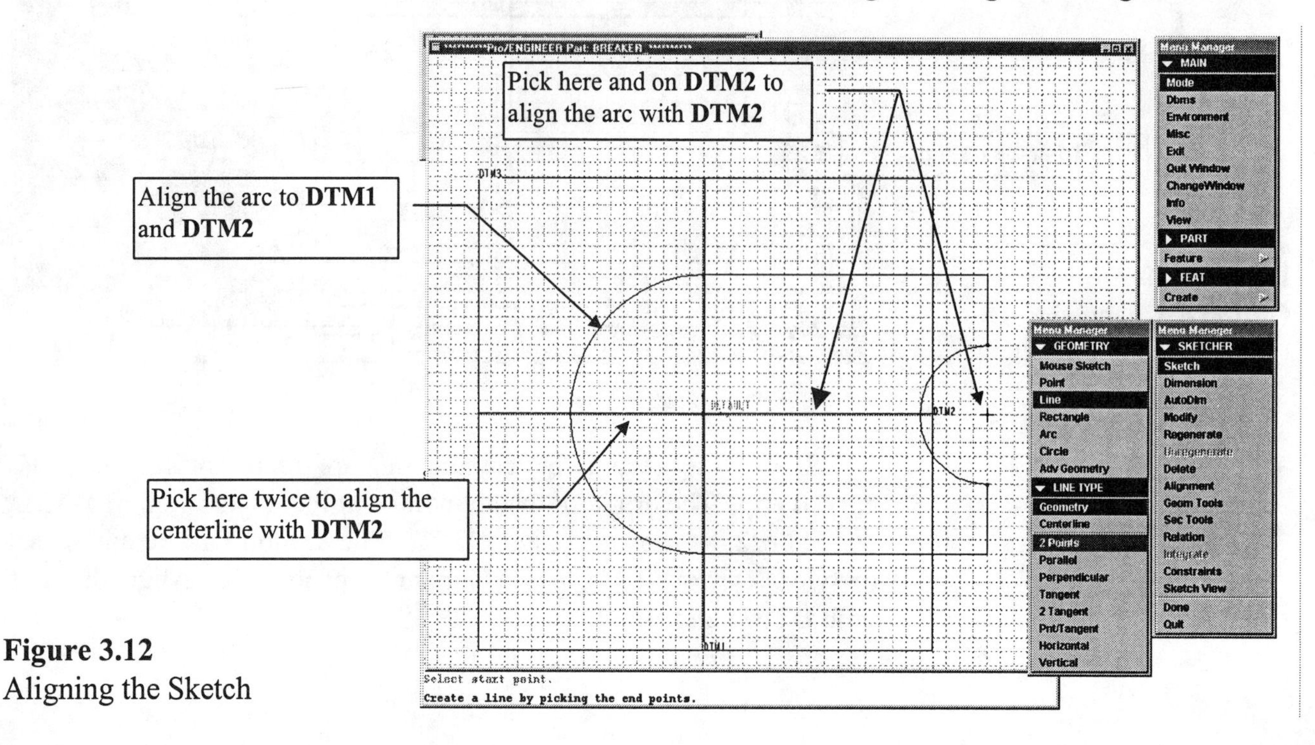

Figure 3.12
Aligning the Sketch

Use the dimensions provided in Figure 3.13 as the dimensioning scheme. **Dimension**, **Regenerate**, **Modify**, and **Regenerate** the sketch as shown in Figure 3.14. Only three dimensions are required for the successful regeneration of the sketch.

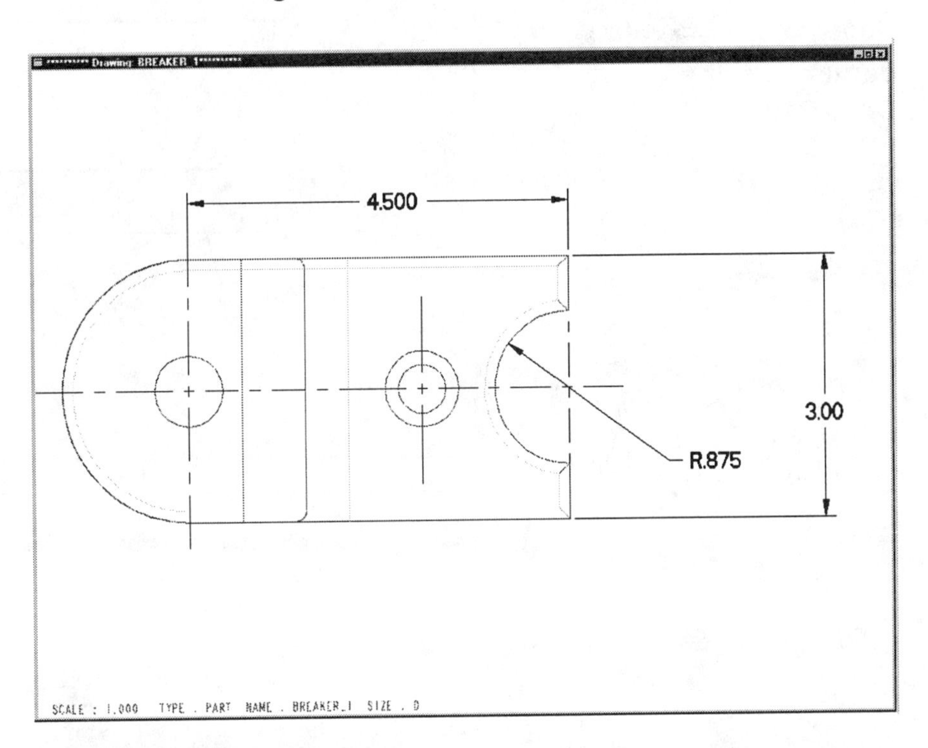

Figure 3.13
Design Intent Dimensions

HINT

You can change the grid size at any point in the sketching process. After the sketch is regenerated the grid disappears, since it is **15** inches square. Change the grid to **.25.**

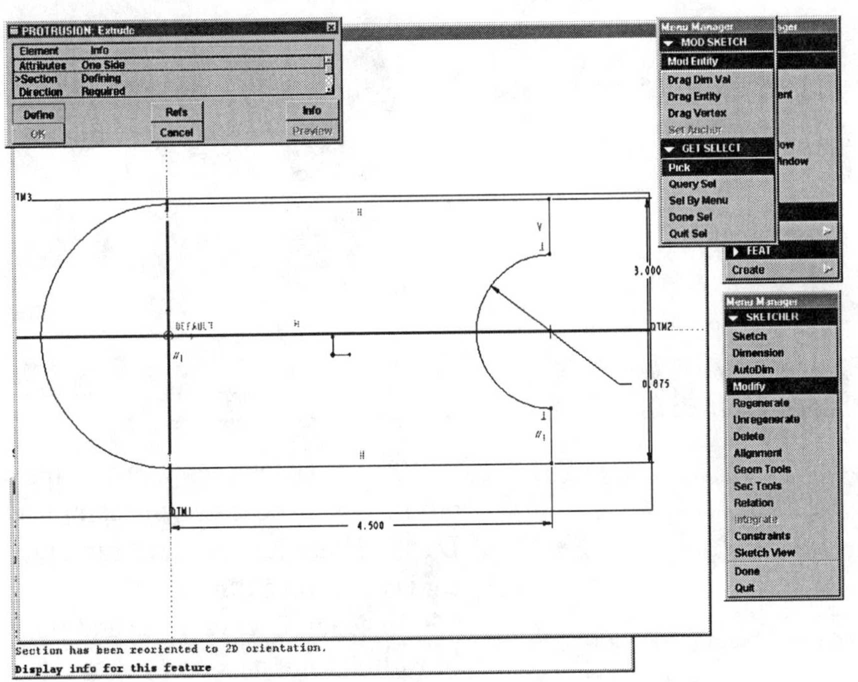

Figure 3.14
Modified and Regenerated Sketch

To complete the feature give the following commands:

Regenerate ⇒ Done ⇒ Blind ⇒ Done ⇒ 2.1875 (as the depth of protrusion) ⇒ **OK** (from the dialog box) ⇒ **View ⇒ Default ⇒ Done-Return** (to view the protrusion as in Figure 3.15) ⇒ **View ⇒ Cosmetic ⇒ Shade ⇒ Display** (shade the part as in Figure 3.16)

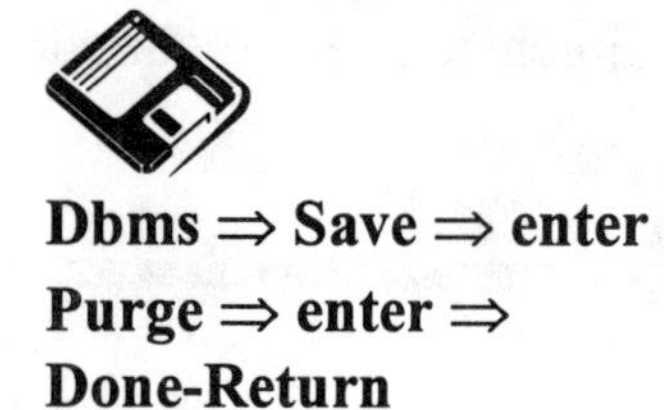

Dbms ⇒ Save ⇒ enter
Purge ⇒ enter ⇒
Done-Return

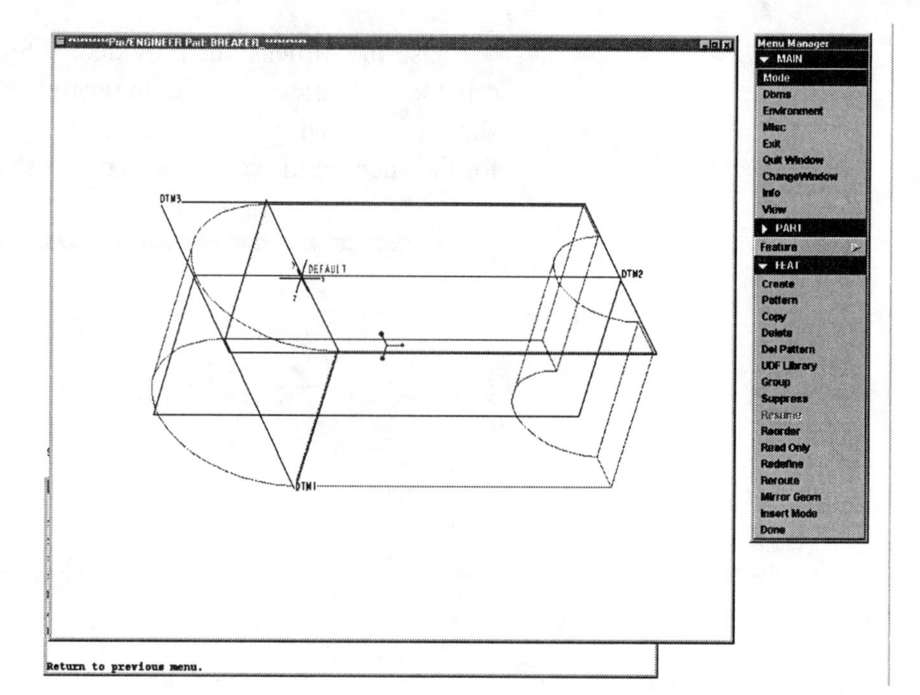

Figure 3.15
Completed Protrusion

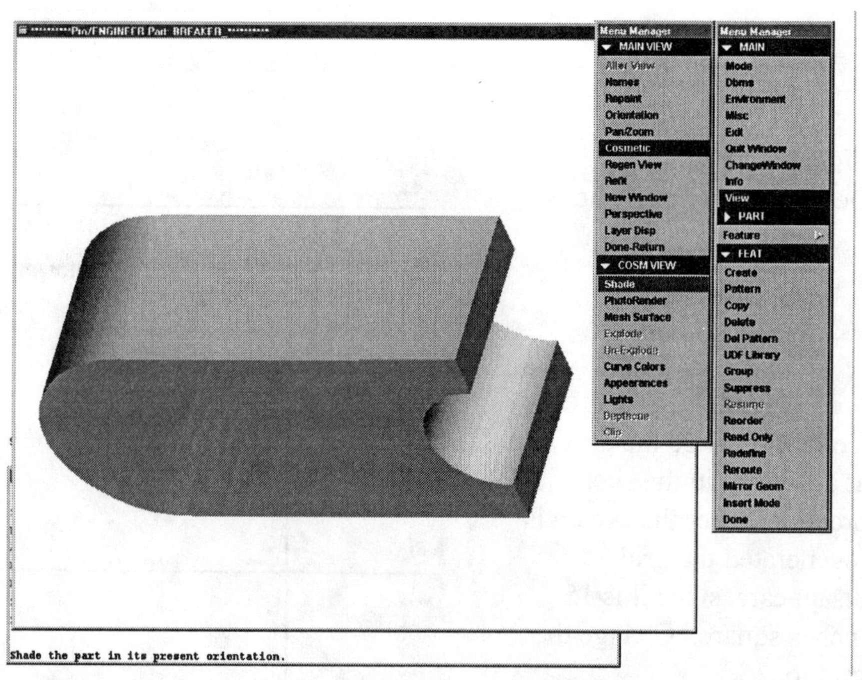

Figure 3.16
Shaded Protrusion

The next features will be cuts created to remove portions of the protrusion. The cuts will complete the primary features of the part. **DTM2** is used as the sketching plane for both cuts, and the extruded cut is made on **Both Sides**.

In general, leave *holes* and *rounds* as the final features of the part. Most holes are *pick-and-place* features that are added to the model at a similar step, like when they are drilled, reamed, or bored during actual manufacturing. In most cases, this means after a majority of the machining has been completed. Rounds are the very last features created. A good many model failures happen when creating a set of rounds. Leaving them for the final features reduces the effort needed to resolve problems.

The next feature that will be created is the cut on the top of the part. The dimensions for the cut are shown in Figure 3.17. Give the following commands:

View ⇒ **Repaint** ⇒ **Done-Return** ⇒ **Feature** ⇒ **Create** ⇒ **Cut** ⇒ **Extrude** ⇒ **Solid** ⇒ **Done** ⇒ **Both Sides** ⇒ **Done** (pick **DTM2** as the sketching plane) ⇒ **Flip** ⇒ **Okay** ⇒ **Top** (pick **DTM3** to orient the sketch)

If necessary use **Query Sel** to pick the datum planes (Fig. 3.18). Pro/E now enters the **Sketcher** and displays the part and datum planes, as shown in Figure 3.19.

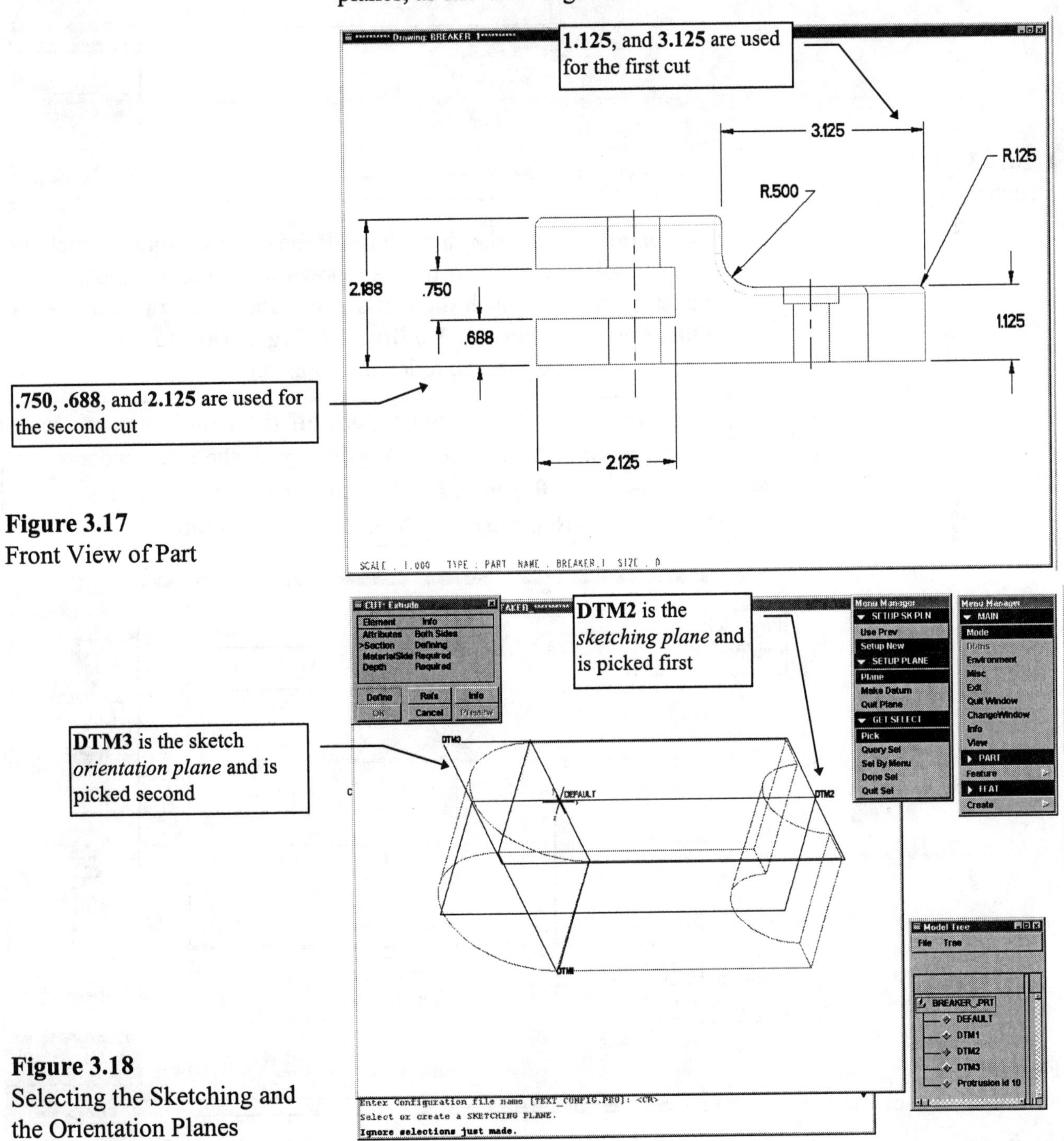

Figure 3.17
Front View of Part

Figure 3.18
Selecting the Sketching and the Orientation Planes

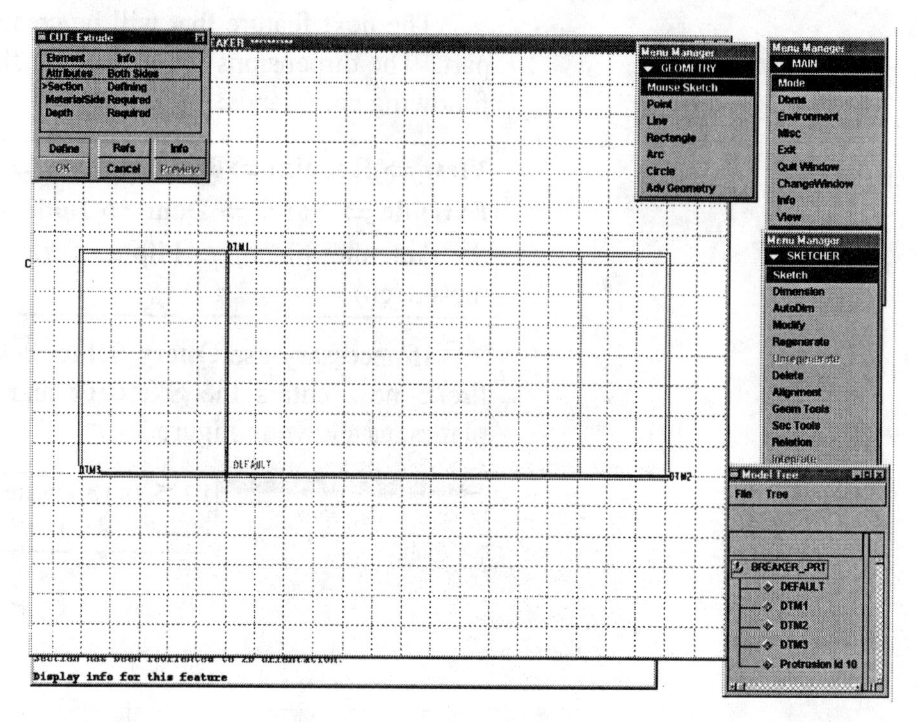

Figure 3.19
Sketcher

Before you start sketching, turn off the □ **Grid Snap**. Sketch the three endpoints of the two lines as shown in Figure 3.20. Align the two endpoints that touch the part's edges and regenerate the sketch. **Dimension, Regenerate, Modify** and **Regenerate** to complete the sketch (Fig. 3.21). Use the following commands:

Environment ⇒ □ **Grid Snap** (turn off the Grid Snap) ⇒ **Done-Return** ⇒ **Sketch** ⇒ **Line** ⇒ **Vertical** (pick the three endpoints of the lines) ⇒ **Regenerate** ⇒ **Alignment** ⇒ **Regenerate** ⇒ **Dimension** ⇒ **Regenerate** ⇒ **Modify** ⇒ **Regenerate**

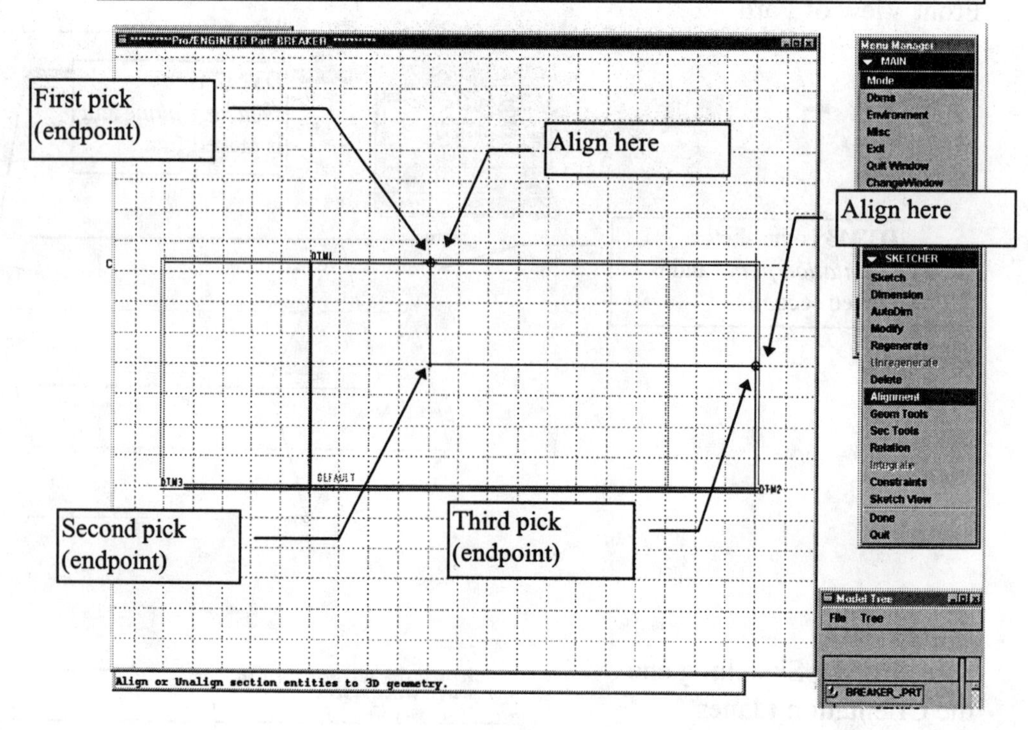

Figure 3.20
Section Sketch of Cut

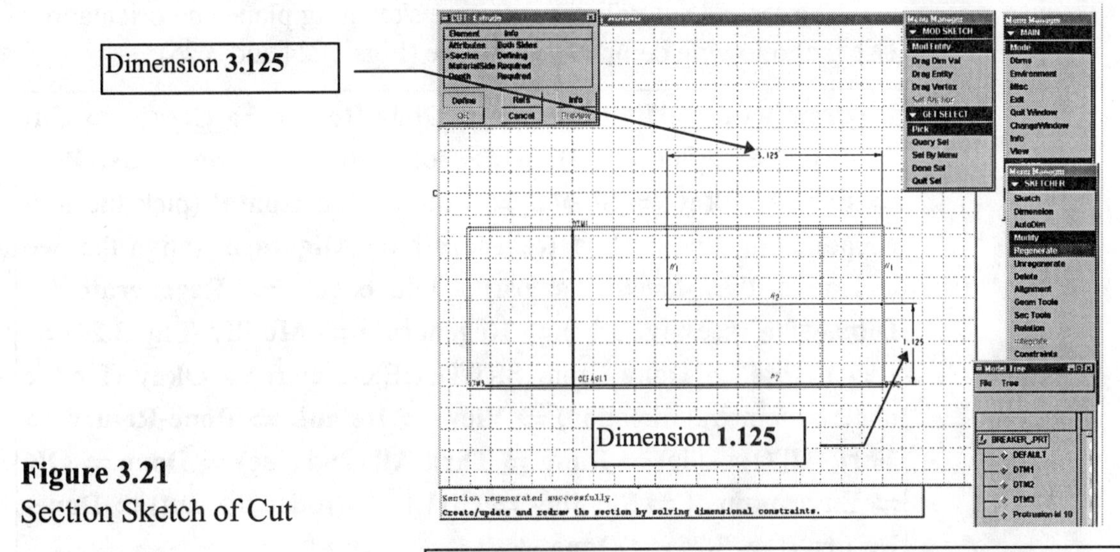

Figure 3.21
Section Sketch of Cut

Complete the cut: **Done** (from SKETCHER menu) ⇒ **Okay** (for the material removal direction, as shown in Figure 3.22) ⇒ **View** ⇒ **Default** ⇒ **Thru All** (first side) ⇒ **Done** ⇒ **Thru All** (second side) ⇒ **Done** ⇒ **Preview** ⇒ **OK** ⇒ **Done** ⇒ **Environment** ⇒ **Shading** ⇒ **Isometric** ⇒ **Done-Return** (Fig. 3.23)

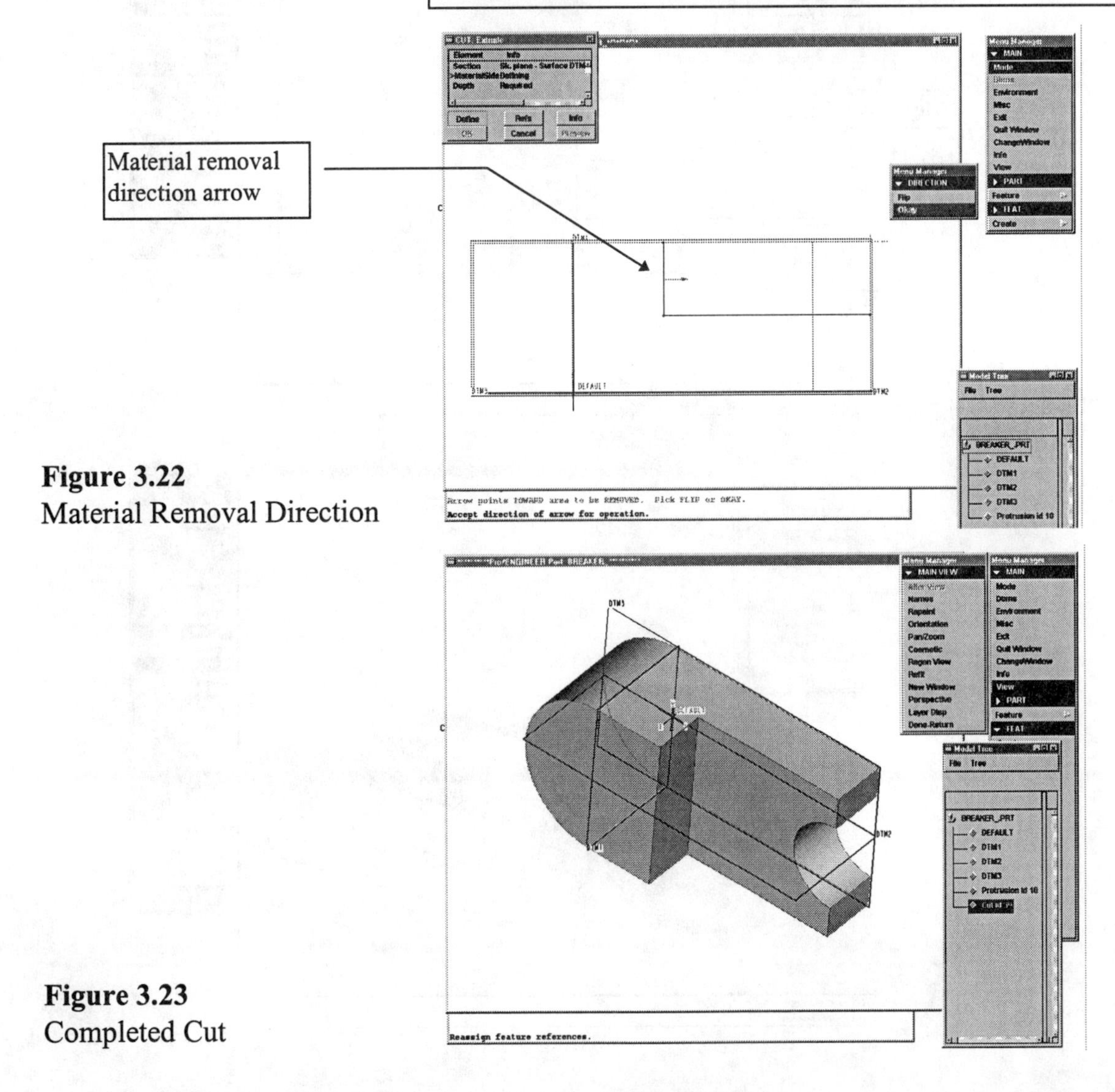

Figure 3.22
Material Removal Direction

Figure 3.23
Completed Cut

The second cut will use the same sketching plane and orientation. Use the following command sequence (Figs. 3.24 and 3.25):

Environment ⇒ **Hidden Line** ⇒ **Done-Return** ⇒ **Create** ⇒ **Cut** ⇒ **Extrude** ⇒ **Solid** ⇒ **Done** ⇒ **Both Sides** ⇒ **Done** ⇒ **Use Prev** ⇒ **Flip** ⇒ **Okay** ⇒ **Sketch** ⇒ **Line** ⇒ **Horizontal** (pick the four endpoints of the lines) ⇒ **Regenerate** ⇒ **Alignment** (align the two endpoints that touch the part's left edge) ⇒ **Regenerate** ⇒ **Dimension** (see Fig. 3.17) ⇒ **Regenerate** ⇒ **Modify** (Fig. 3.24) ⇒ **Regenerate** ⇒ **Done** (from SKETCHER menu) ⇒ **Okay** (for the material removal direction) ⇒ **View** ⇒ **Default** ⇒ **Done-Return** ⇒ **Thru All** (1st side) ⇒ **Done** ⇒ **Thru All** (2nd side) ⇒ **Done** ⇒ **OK** ⇒ **Environment** ⇒ **Shading** ⇒ **Ctrl** (rotate the part) ⇒ **Done-Return** (Fig. 3.25) ⇒ **Done**

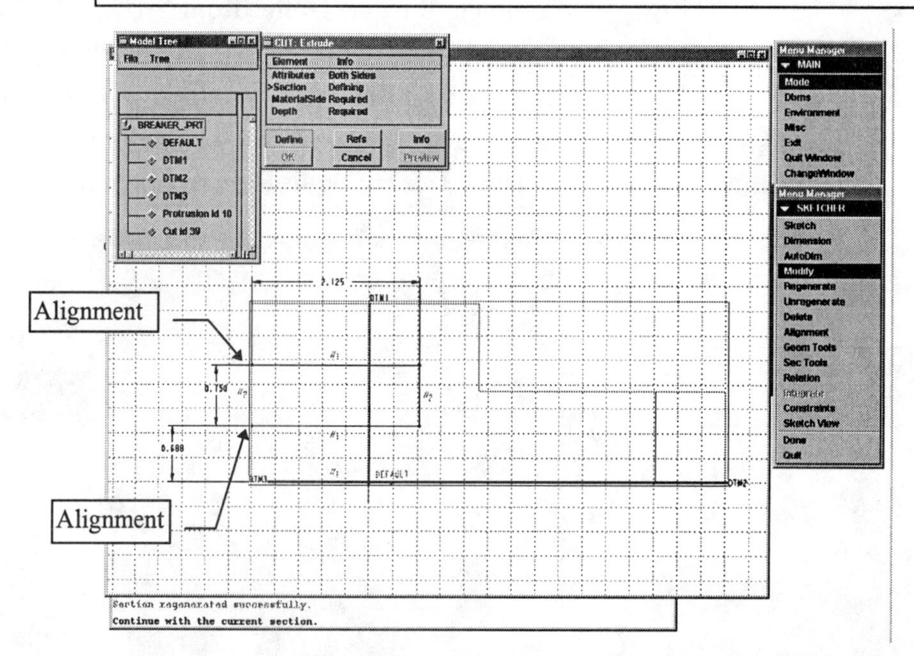

Figure 3.24
Second Cut Section

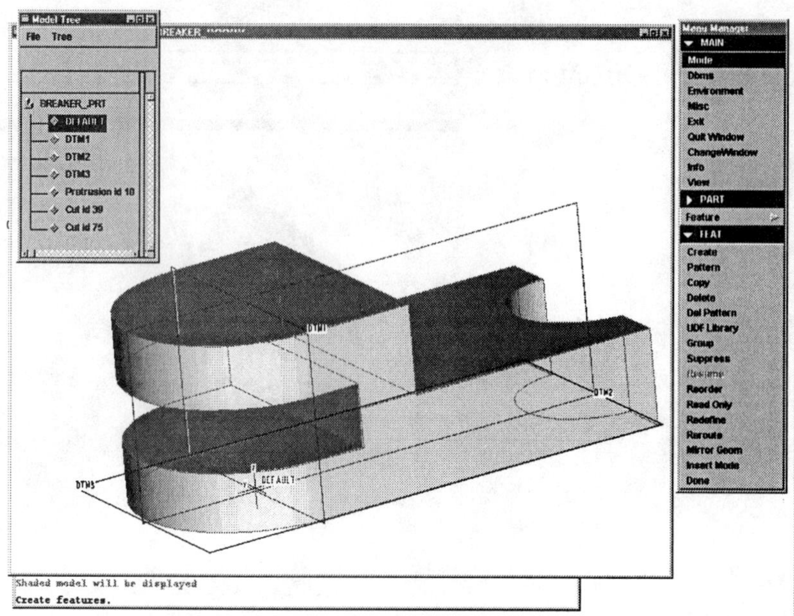

Figure 3.25
Second Cut Completed

The next feature to be created is a hole. This is a *pick-and-place* feature that does not require a sketch (Fig. 3.27). The placement plane is the part's top surface; the dimensioning edges/planes are **DTM1** and **DTM2**. The hole will be at the intersection of the two datum planes; therefore the distance from both will be **0** inches. Give the following commands:

? Pro/HELP

To get more information about holes, highlight the hole command and press the right mouse button and **? GetHelp**

Create ⇒ Hole ⇒ Straight ⇒ Done ⇒ Linear ⇒ Done ⇒ (pick the hole's placement plane) ⇒ (select the first edge to dimension from, which will be **DTM2**, and type **0** at the prompt) ⇒ (select the second edge to dimension from, which is **DTM1**, and type **0** at the prompt) ⇒ **One Side ⇒ Done ⇒ Thru All ⇒ Done ⇒** enter diameter **.8125** at the prompt ⇒ **enter ⇒ OK** (Fig. 3.27) ⇒ **Done**

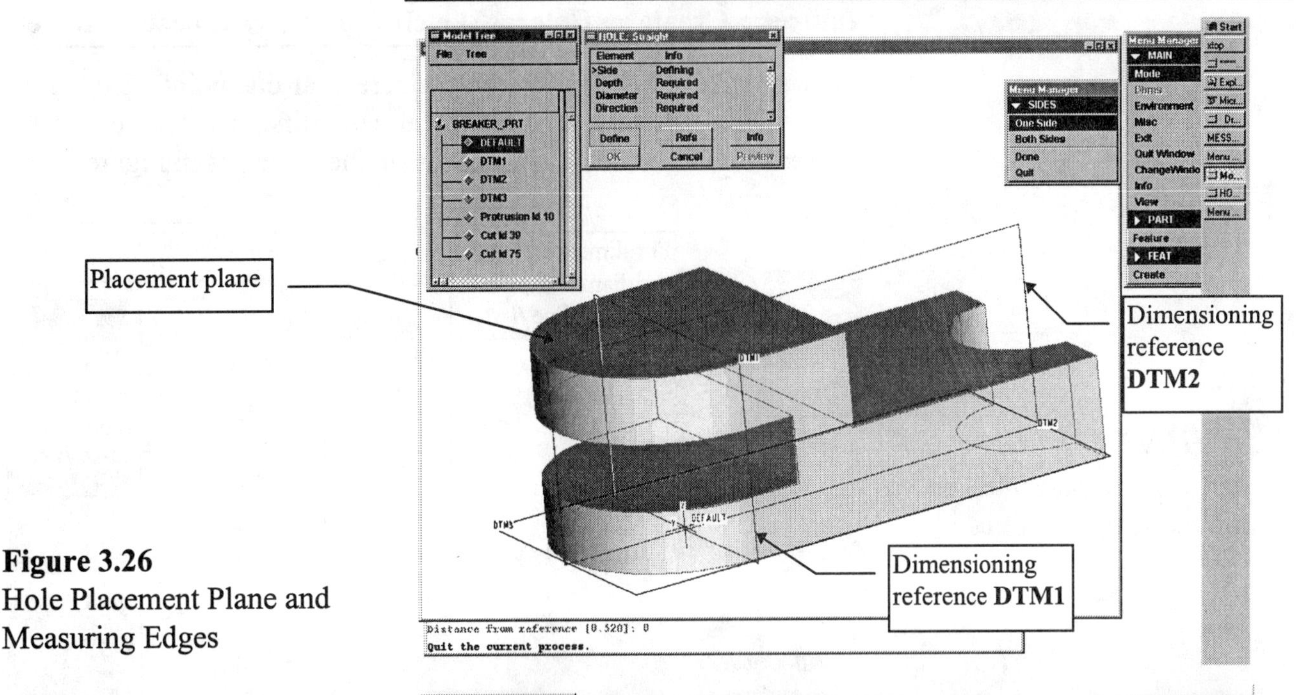

Figure 3.26
Hole Placement Plane and Measuring Edges

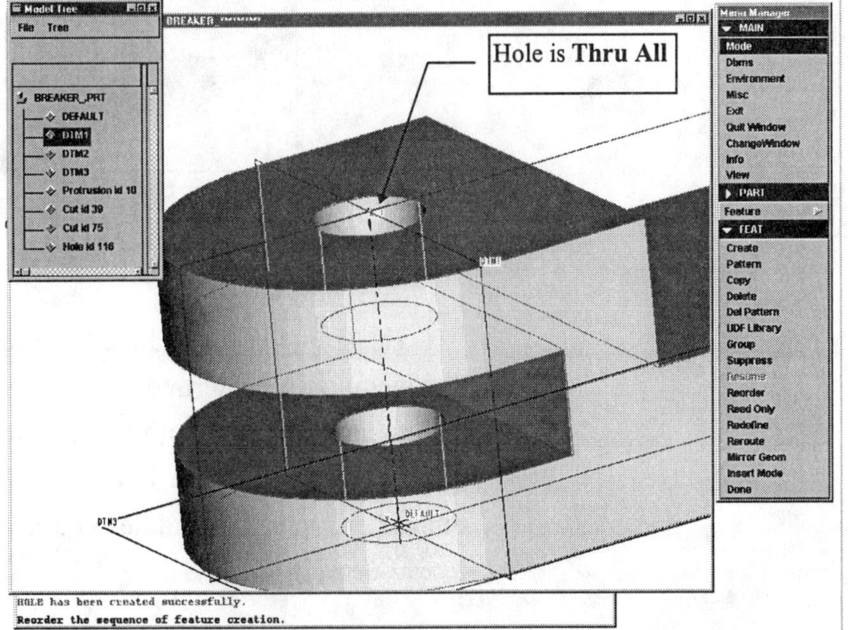

Figure 3.27
Completed Hole

Dbms ⇒ Save ⇒ enter Purge ⇒ enter ⇒ Done-Return

The next hole will be a *sketched hole*. A sketched hole can be almost any configuration. Sketched holes can be used to create counterbore holes, countersunk holes, spotfaced holes, etc. In this part of the lesson, you will create a counterbored hole. Sketched holes are really nothing more than revolved cuts.

Sketched holes are created with a section sketch, just like a cut. ***The section must have a vertical centerline, and all entities must be on one side of that centerline.*** Always start by sketching the vertical centerline first. Next, sketch the entities required to describe the hole's shape (half of the shape). No alignment is necessary, but dimensions are required. ***The section must be closed.*** Use the following commands:

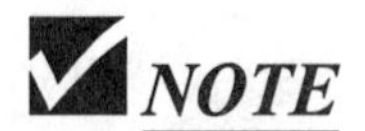

NOTE

All entities of a sketched hole must be on one side of a vertical centerline, and the section must be closed

Feature ⇒ Create ⇒ Hole ⇒ Sketch ⇒ Done ⇒ Linear ⇒ Done

A small window appears on the screen at this point. Sketch the hole's section, starting with a vertical centerline, and then create the required closed geometry and dimension the sketch as in Figure 3.28.

HINT

Sketch the vertical centerline before the section entities

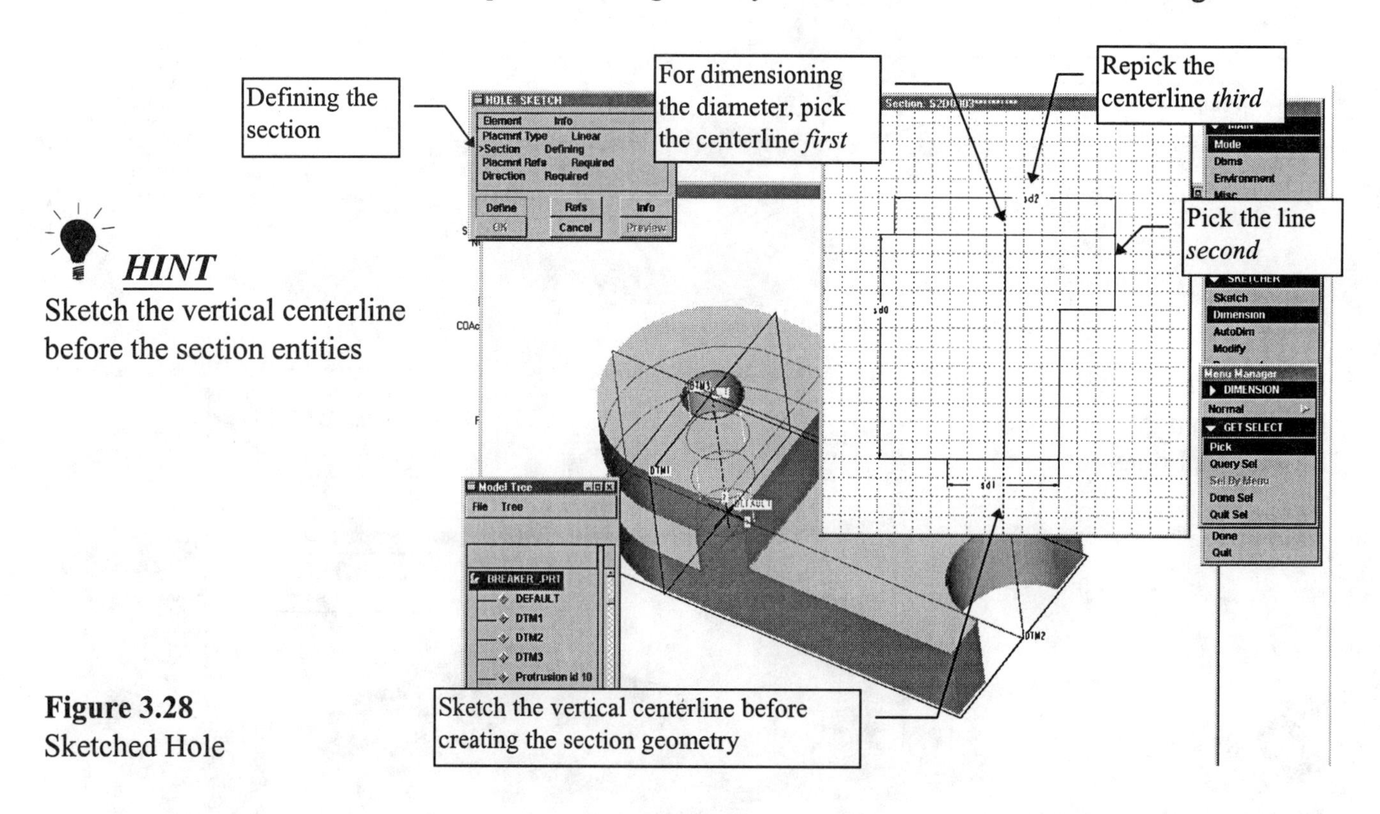

Figure 3.28
Sketched Hole

The diameter of the hole is dimensioned by choosing **Dimension** and picking the centerline with the left mouse button, the edge to be dimensioned, also with the left mouse button, and the centerline a second time with the left mouse button, and then placing the dimension with the middle button of the mouse. You should reread Section 9 at this time.

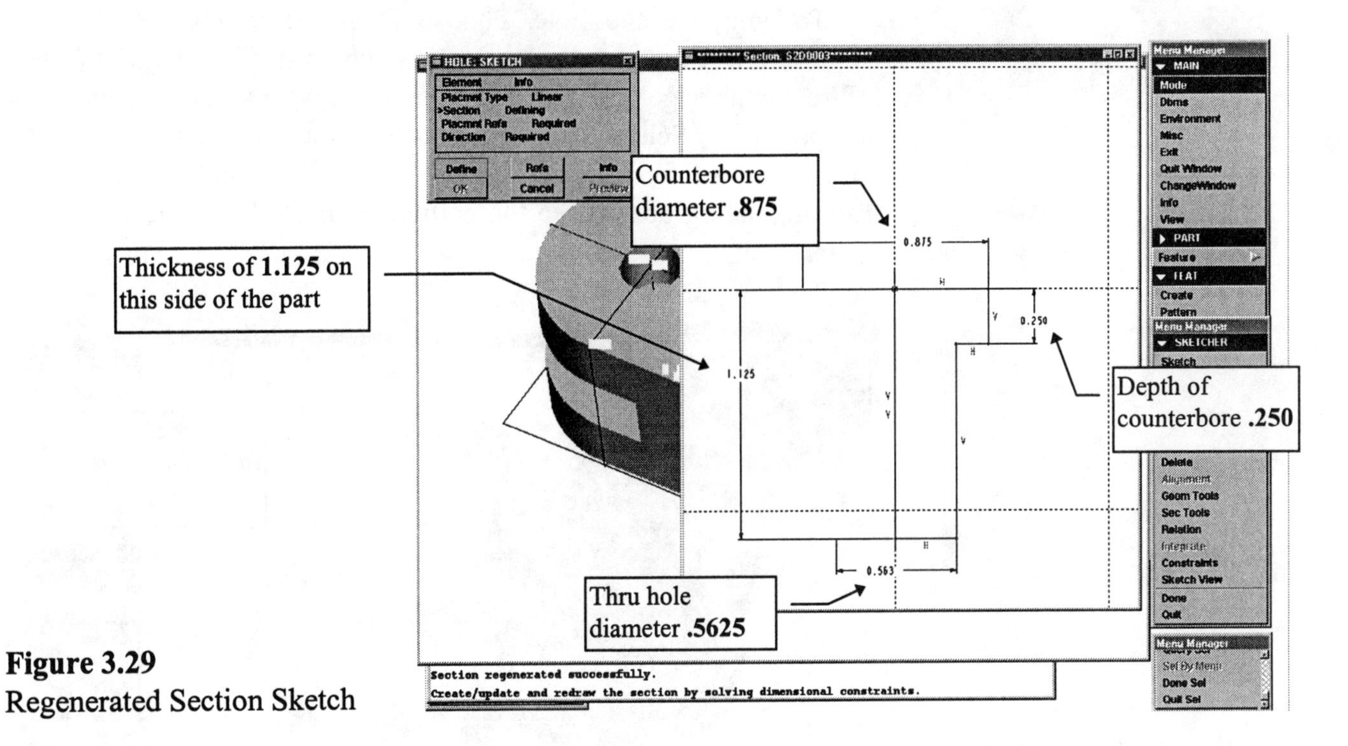

Figure 3.29
Regenerated Section Sketch

Regenerate the sketch, and **Modify** the dimensions (Fig. 3.29). The counterbore diameter is **.875**, and the thru hole diameter is **.5625** (Fig. 3.30). The depth of the hole is the same as the thickness of the part where the hole is placed, **1.125**. Choose **Done**.

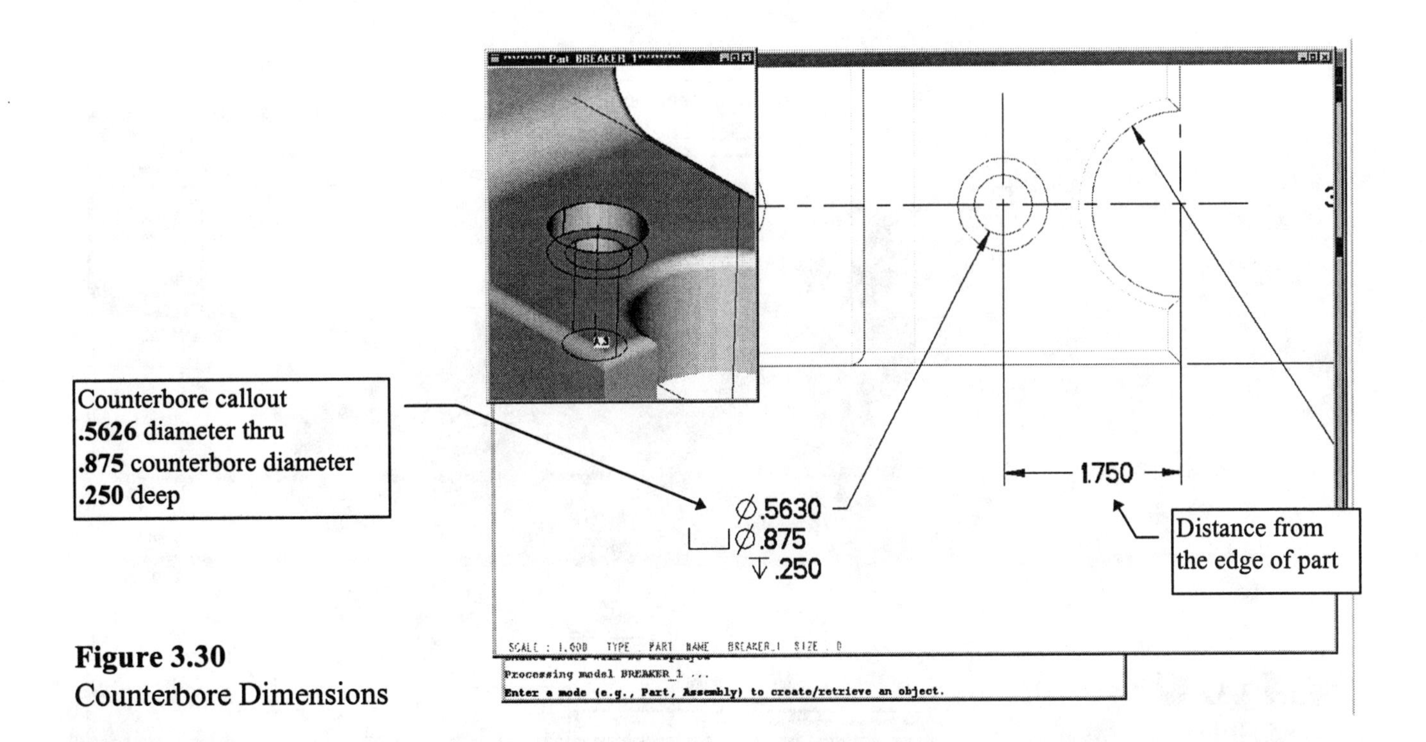

Figure 3.30
Counterbore Dimensions

To complete the hole, choose **Done** (from the SKETCHER menu). Select the placement plane as shown in Figure 3.31. Select **DTM2** as the edge to dimension from, and give a value of **0** inches at the prompt. Select the right side of the part for the second dimensioning reference, and type **1.75** at the prompt. Choose **OK** from the dialog box to see the completed hole (Fig. 3.32). Choose **Done**.

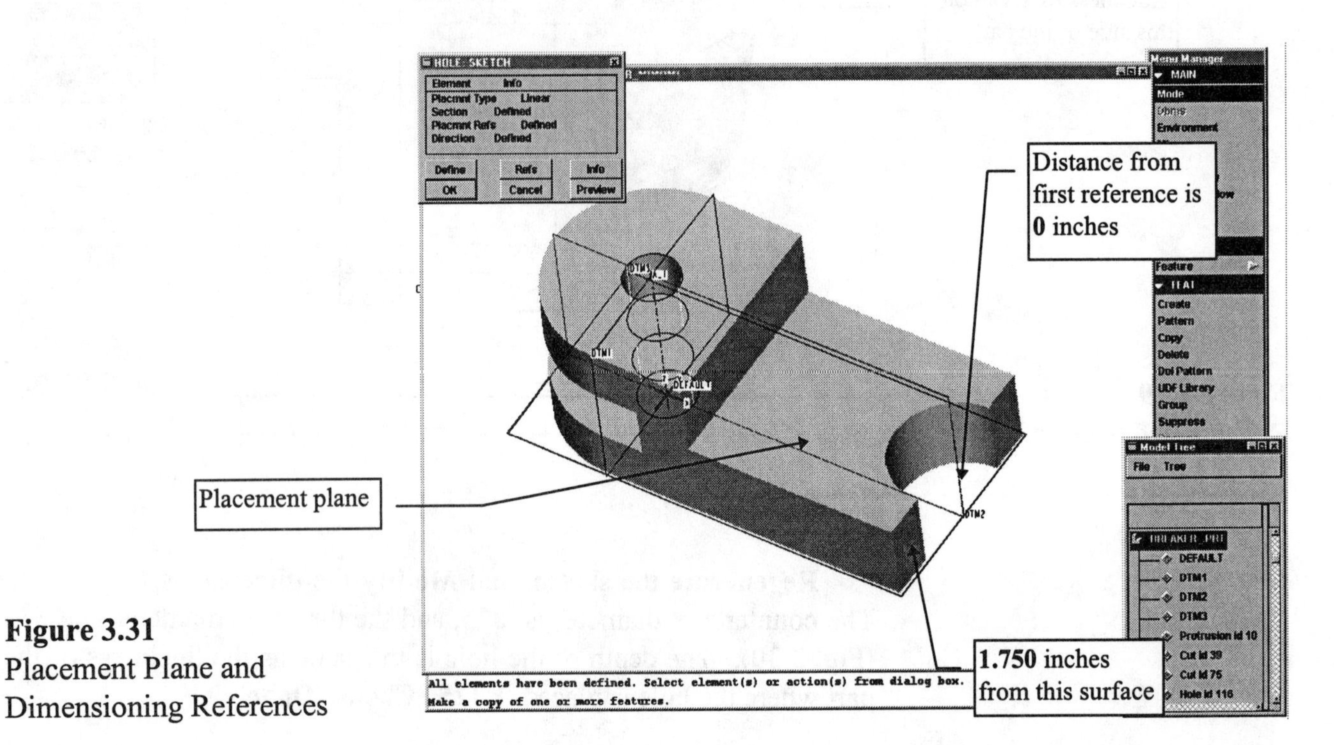

Figure 3.31
Placement Plane and Dimensioning References

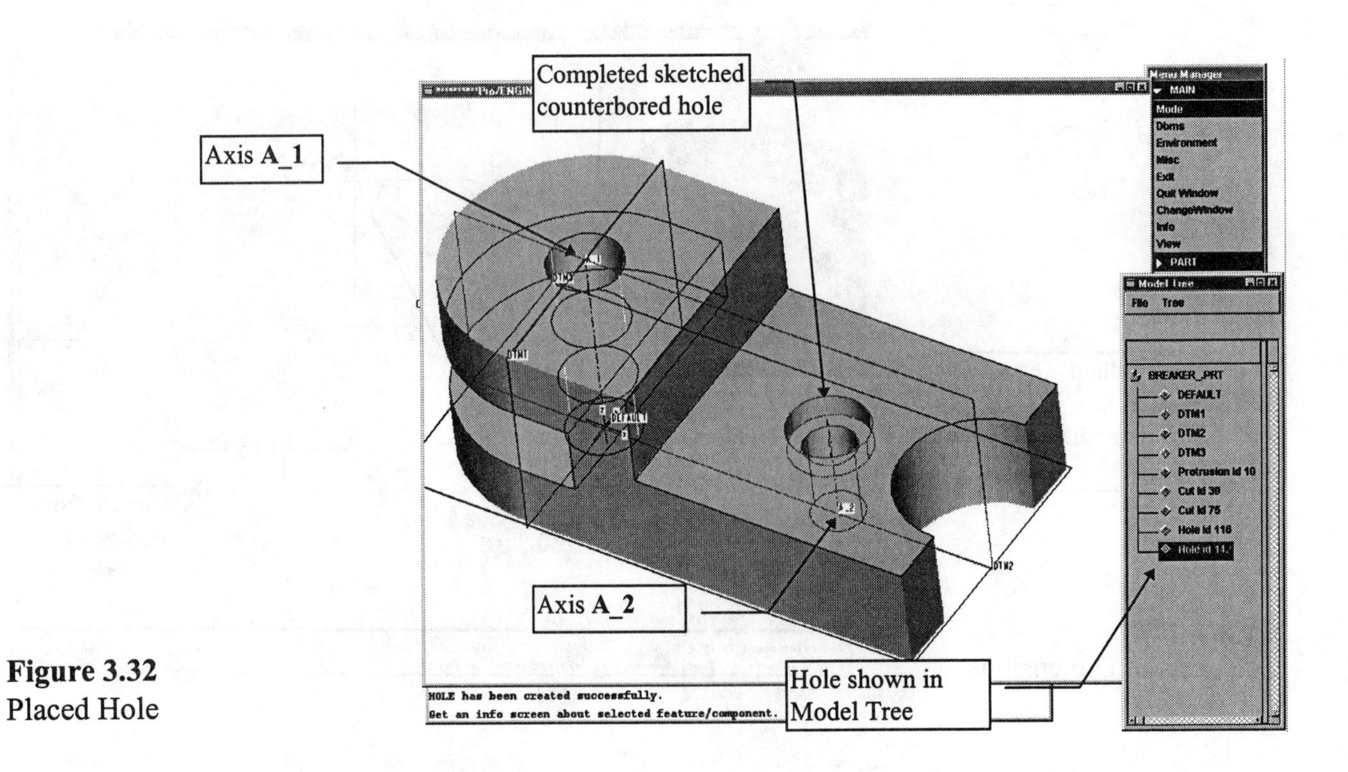

Figure 3.32
Placed Hole

Open a new window and zoom in on the counterbore (Fig. 3.33). Highlight the feature in the Model Tree and press the right mouse button. Select **Feat Info**. The INFORMATION WINDOW will appear on your screen, as in Figure 3.34.

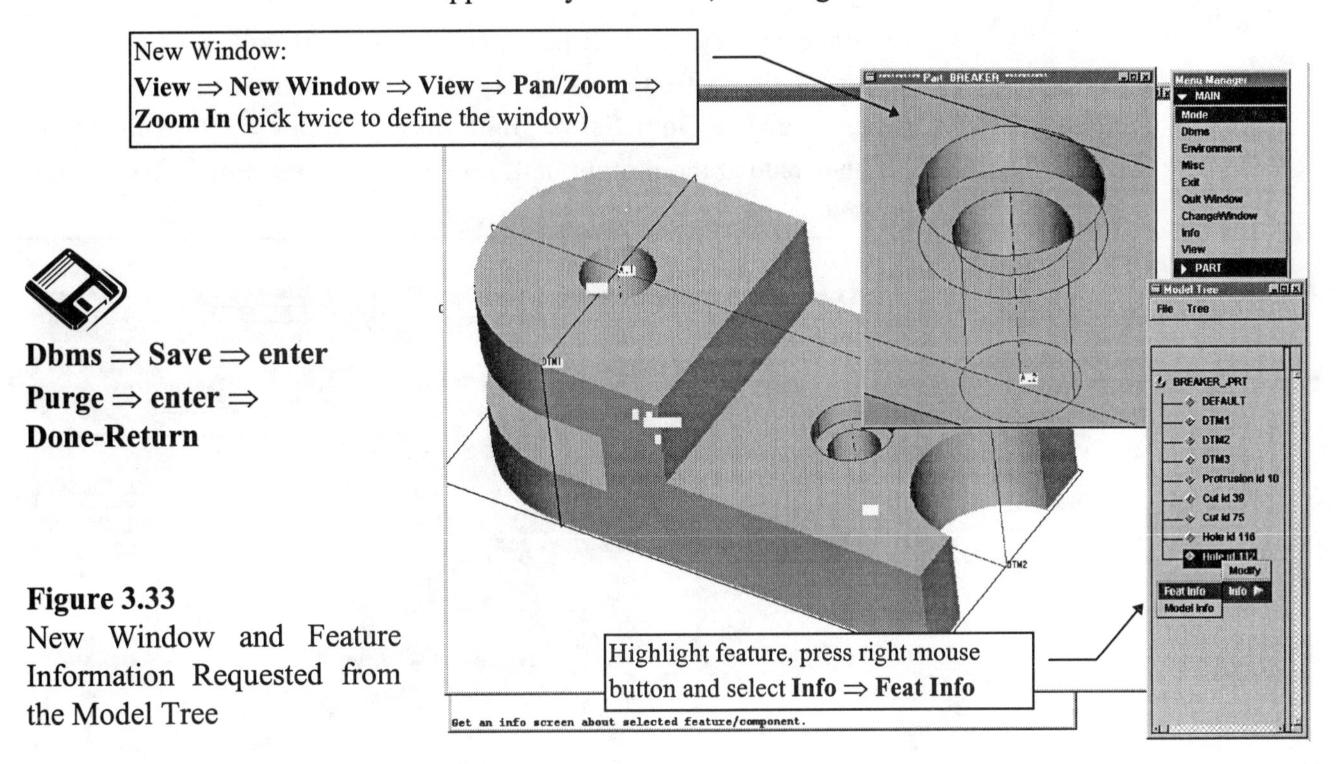

Dbms ⇒ Save ⇒ enter
Purge ⇒ enter ⇒
Done-Return

Figure 3.33
New Window and Feature Information Requested from the Model Tree

HINT

Information about a feature, part, assembly, etc. can be requested as needed during a Pro/E Session.

To close an Information Window type "**q**" to quit, or choose **Quit Window ⇒ Change Window** and pick in the Main Window.

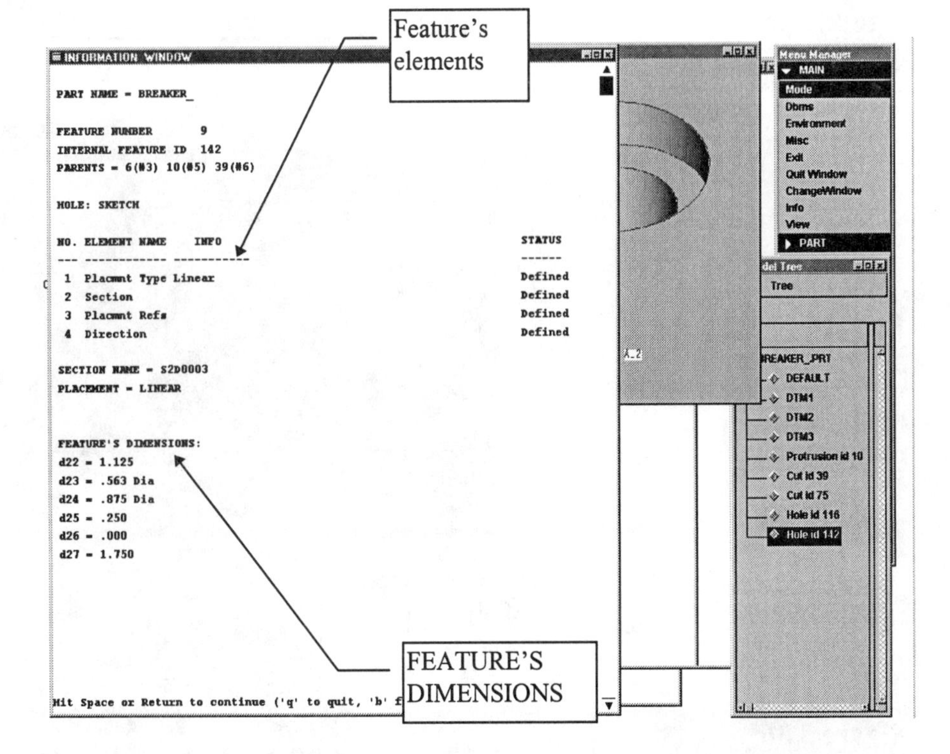

Figure 3.34
Feature Information

To complete the part, a number of rounds need to be created. The first round is an edge round between the vertical and horizontal faces of the first cut. Give the following commands:

Feature ⇒ **Create** ⇒ **Round** ⇒ **Simple** ⇒ **Done** ⇒ **Constant** ⇒ **Edge Chain** ⇒ **Done** ⇒ **One By One** ⇒ (pick the edge as shown in Figure 3.25) ⇒ **Done Sel** ⇒ **Done** ⇒ **New Value** ⇒ (type the round radius value at the prompt, **.50**) ⇒ **enter** ⇒ **Preview** ⇒ **OK** (round will appear as in Figure 3.36) ⇒ **Done**

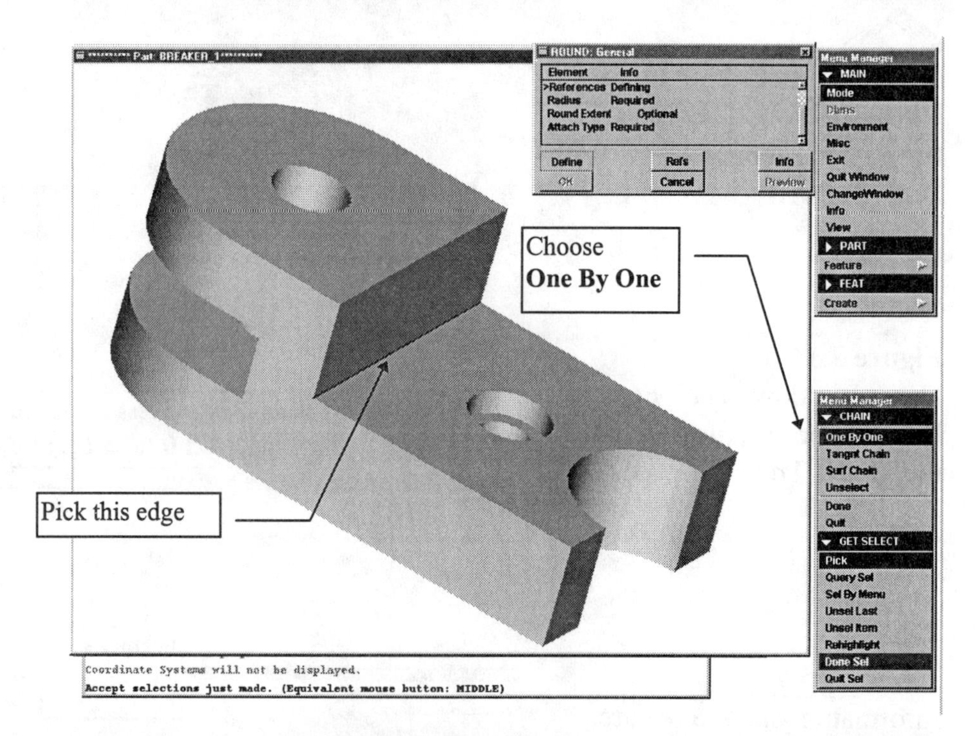

Figure 3.35
Edge to be Rounded

Dbms ⇒ **Save** ⇒ **enter**
Purge ⇒ **enter** ⇒
Done-Return

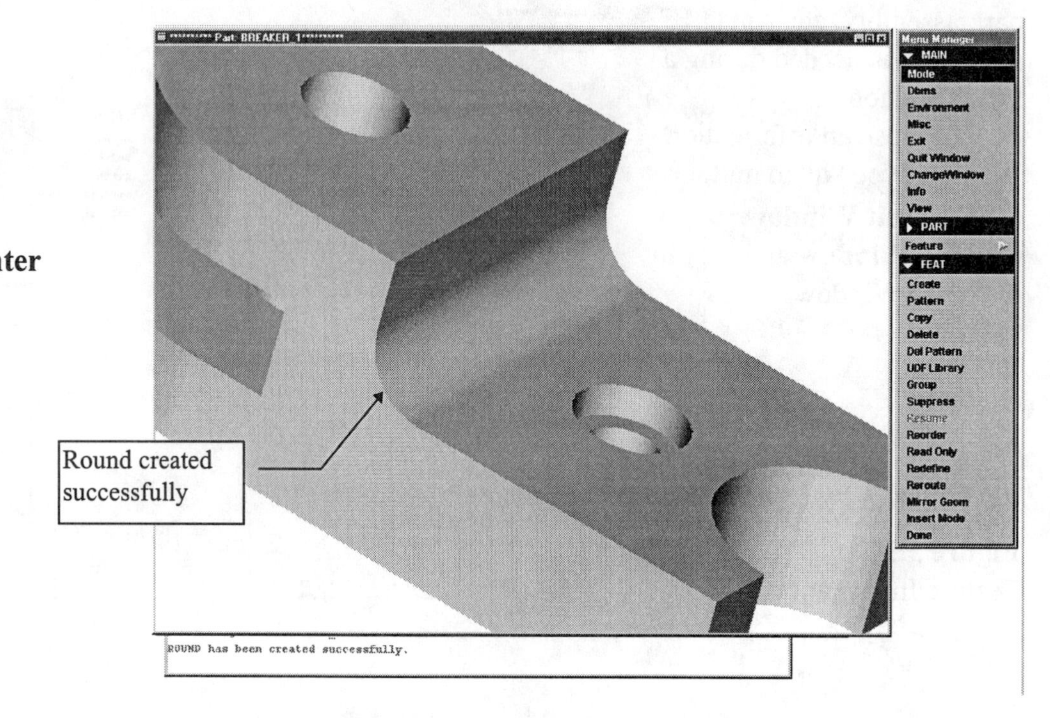

Figure 3.36
Round

Finally, the last round is a set that can be created all at the same time. Give the following commands:

Feature ⇒ Create ⇒ Round ⇒ Simple ⇒ Done ⇒ Constant ⇒ Edge Chain ⇒ Done ⇒ Tangnt Chain (pick the upper edge as shown in Figure 3.37) ⇒ **One By One** ⇒ (pick the remaining edges as shown in Figure 3.37 in sequence) ⇒ **Done Sel ⇒ Done ⇒ New Value** ⇒ (type the round radius value at the prompt, **.125**) ⇒ **enter ⇒ Preview ⇒ OK** (Figure 3.38)

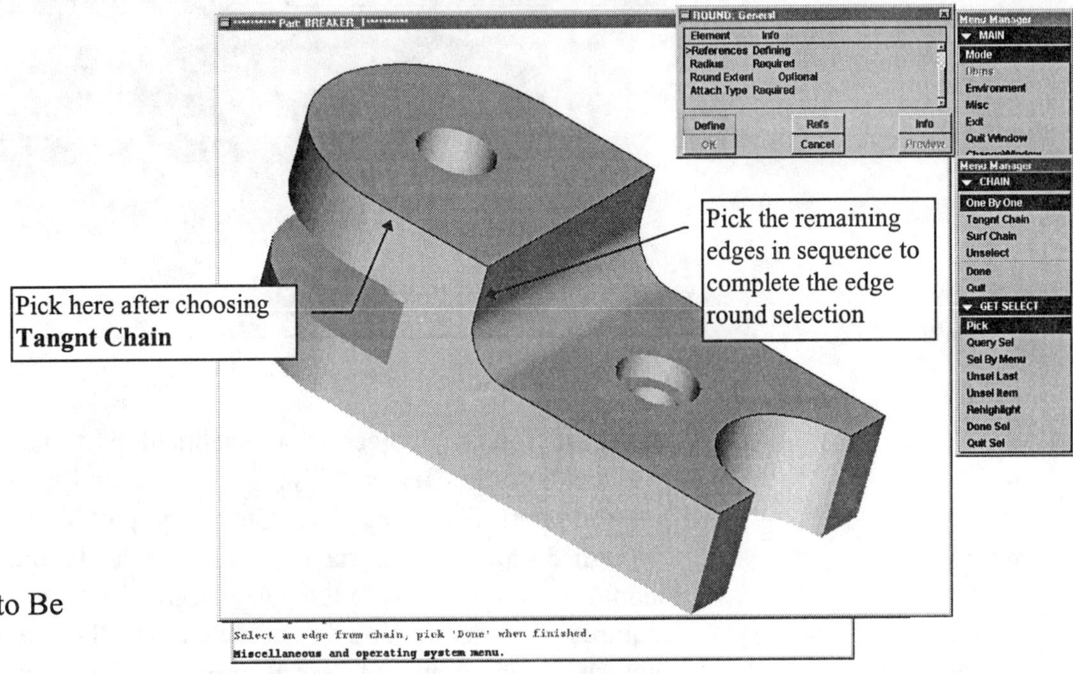

Figure 3.37
Pick the Edges to Be Rounded

Dbms ⇒ Save ⇒ enter Purge ⇒ enter ⇒ Done-Return

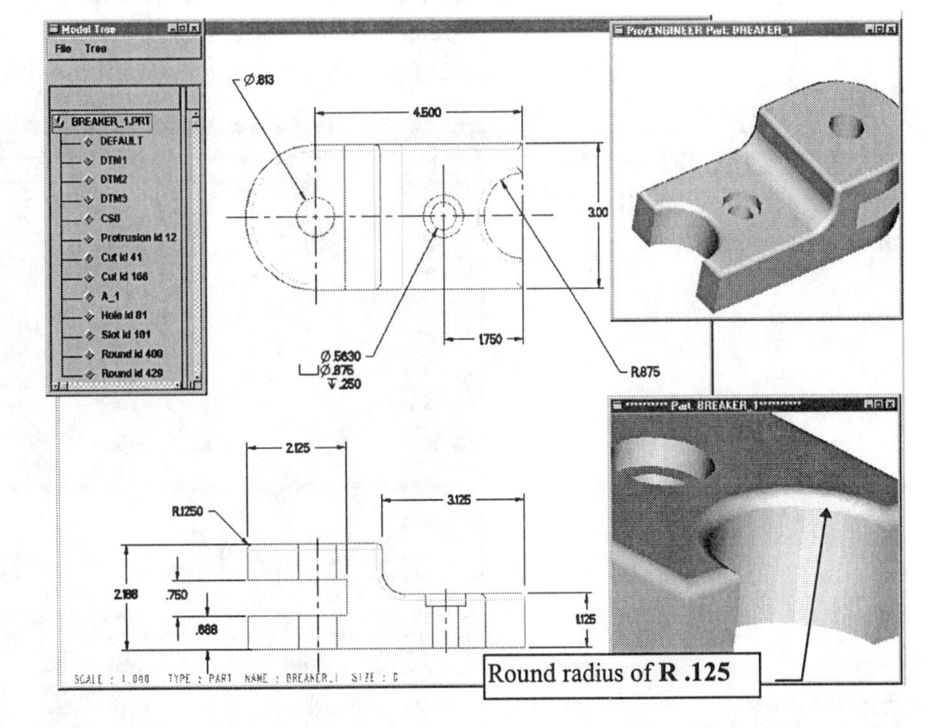

Figure 3.38
Completed Part

Lesson 3 Project

Guide Bracket

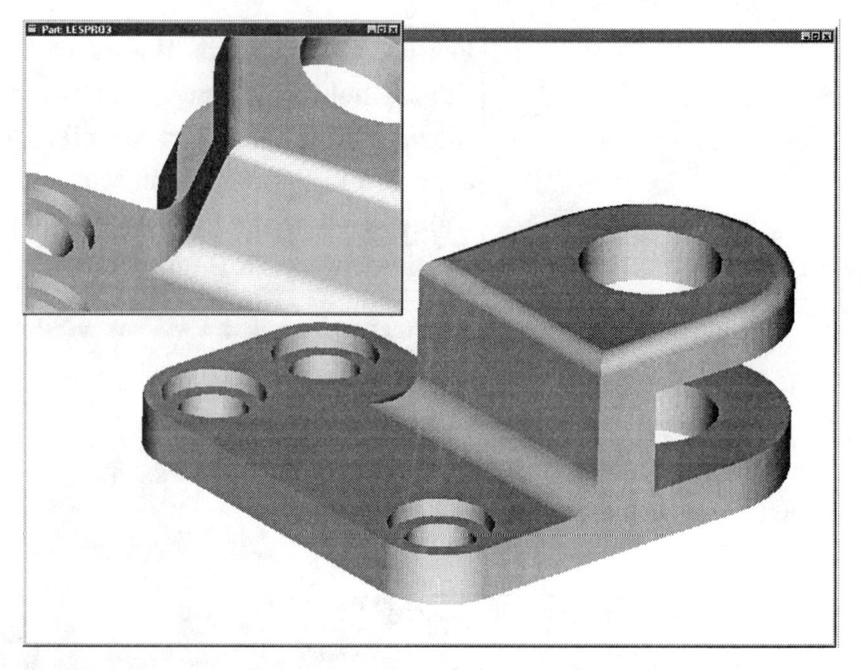

Figure 3.39
Guide Bracket

Guide Bracket

The third **lesson project** is a machined part that requires similar commands to the **Bracket**. Simple rounds and straight and sketched holes are part of the exercise. Create the part shown in Figures 3.39 through 3.45. At this stage in your understanding of Pro/E, you should be able to analyze the part and plan out the steps and features required to model it. You do not have to use the exact design intent as that shown here in the lesson project. You must use the same dimensions and dimension scheme, but the choice and quantity of datum planes and sequence of modeling can be different.

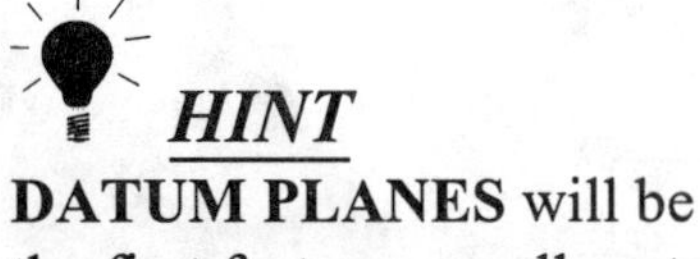

HINT
DATUM PLANES will be the first features on all parts and assemblies.

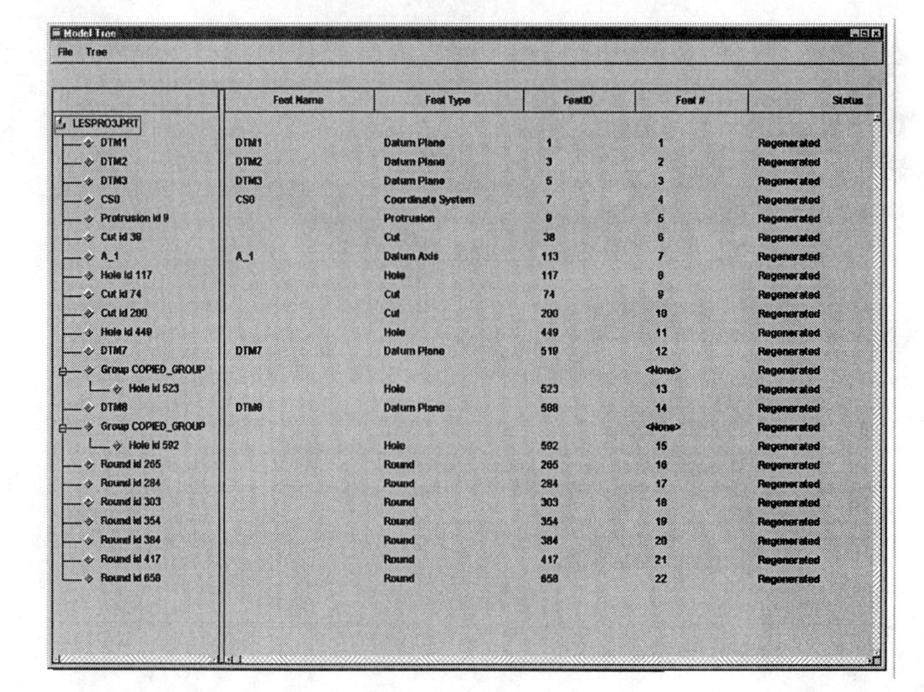

LESPRO3.PRT	Feat Name	Feat Type	FeatID	Feat #	Status
DTM1	DTM1	Datum Plane	1	1	Regenerated
DTM2	DTM2	Datum Plane	3	2	Regenerated
DTM3	DTM3	Datum Plane	5	3	Regenerated
CS0	CS0	Coordinate System	7	4	Regenerated
Protrusion id 9		Protrusion	9	5	Regenerated
Cut id 38		Cut	38	6	Regenerated
A_1	A_1	Datum Axis	113	7	Regenerated
Hole id 117		Hole	117	8	Regenerated
Cut id 74		Cut	74	9	Regenerated
Cut id 200		Cut	200	10	Regenerated
Hole id 449		Hole	449	11	Regenerated
DTM7	DTM7	Datum Plane	519	12	Regenerated
Group COPIED_GROUP				<None>	Regenerated
Hole id 523		Hole	523	13	Regenerated
DTM8	DTM8	Datum Plane	588	14	Regenerated
Group COPIED_GROUP				<None>	Regenerated
Hole id 592		Hole	592	15	Regenerated
Round id 265		Round	265	16	Regenerated
Round id 284		Round	284	17	Regenerated
Round id 303		Round	303	18	Regenerated
Round id 354		Round	354	19	Regenerated
Round id 384		Round	384	20	Regenerated
Round id 417		Round	417	21	Regenerated
Round id 658		Round	658	22	Regenerated

Figure 3.40
Guide Bracket Model Tree

Use the DIPS to plan out your feature creation sequence and the selection of datum sketching planes.

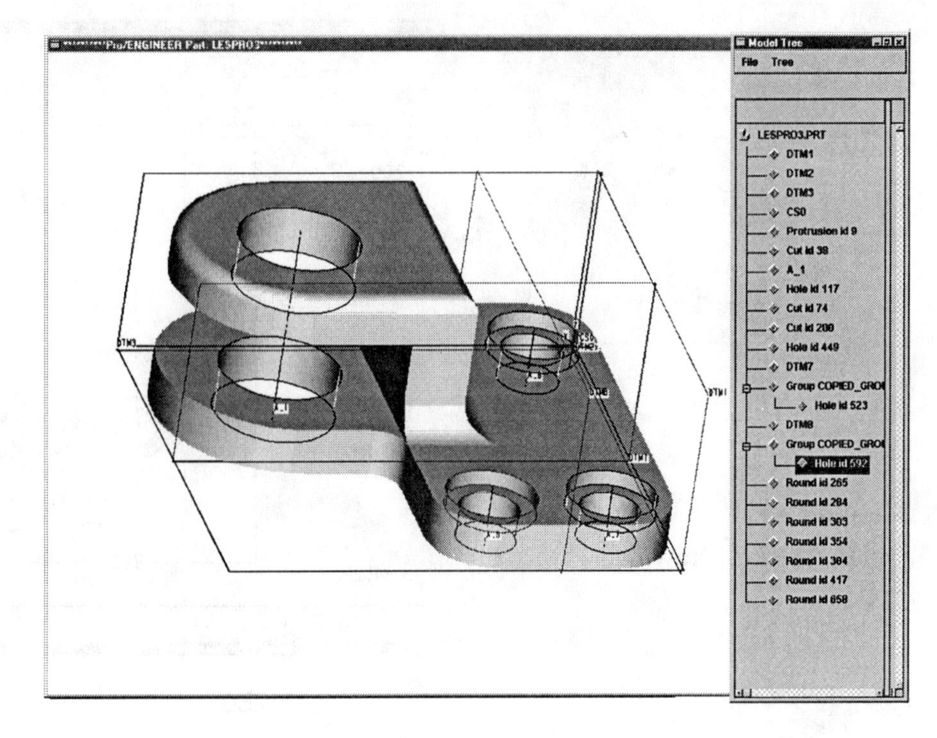

Figure 3.41
Guide Bracket Showing Datums and Coordinate System

Figure 3.42
Guide Bracket Drawing

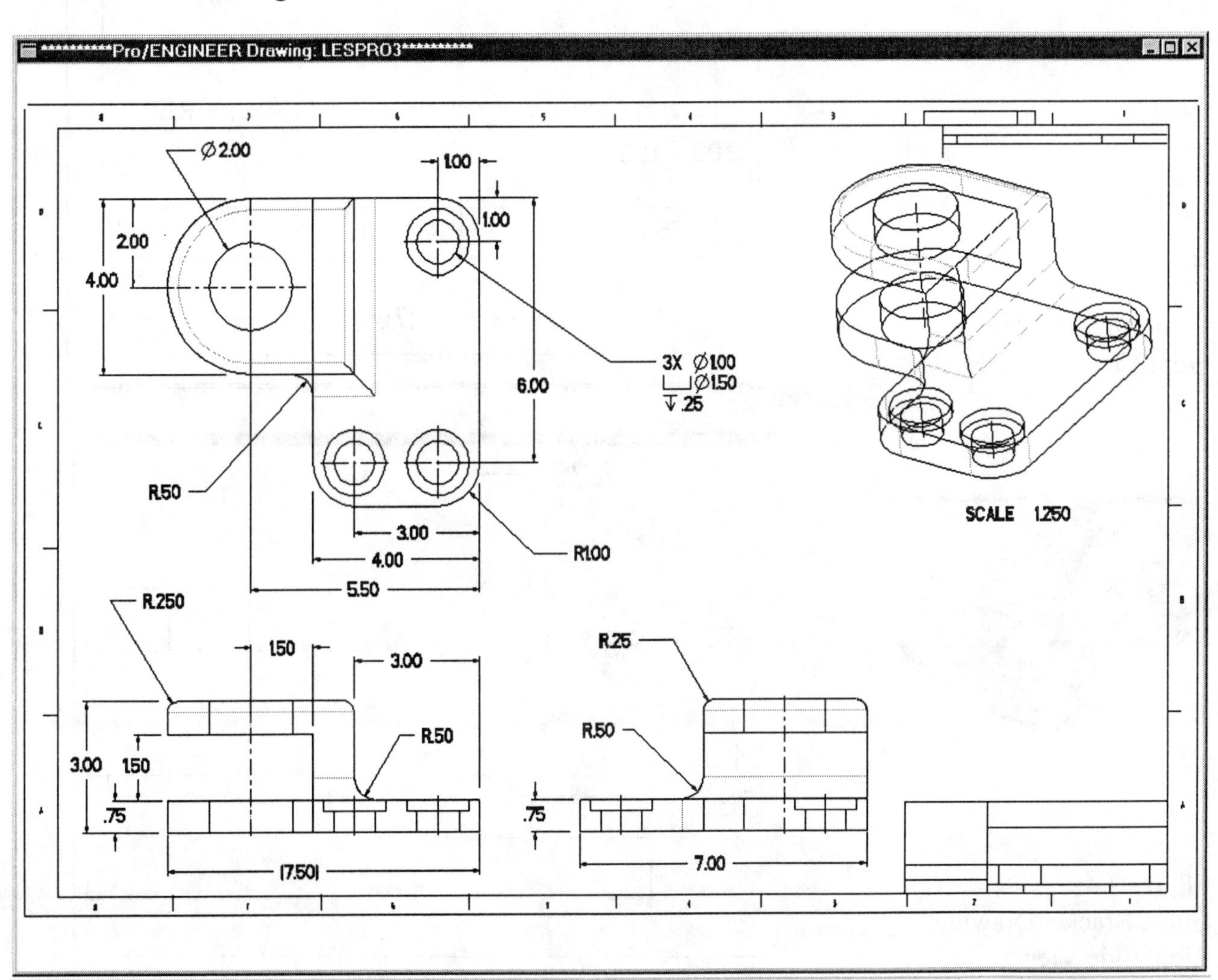

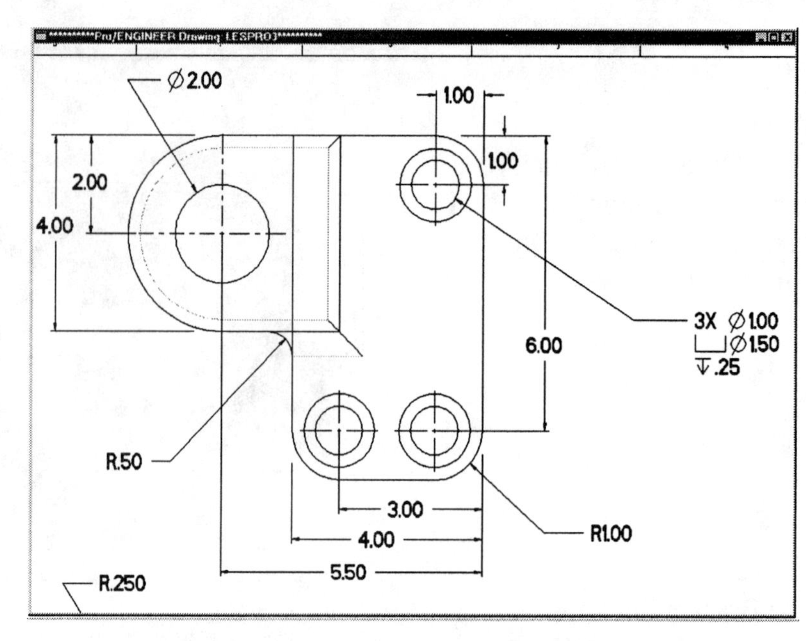

Figure 3.43
Guide Bracket Drawing,
Top View

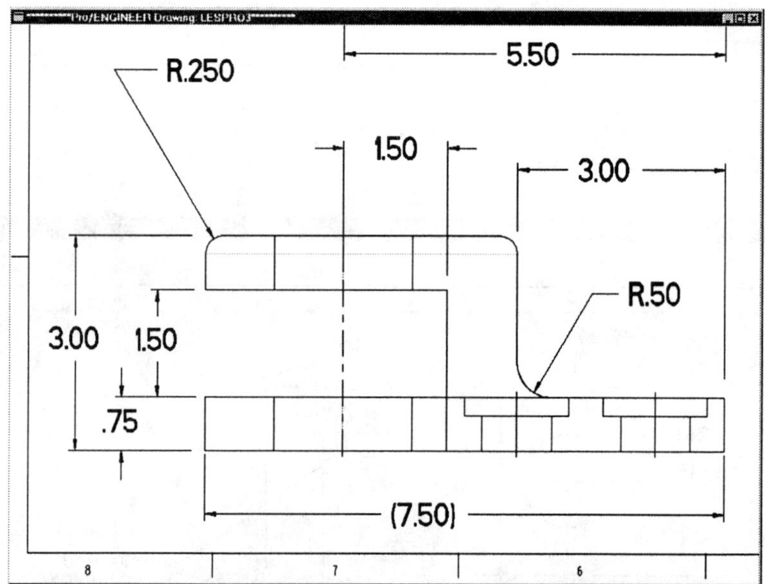

Figure 3.44
Guide Bracket Drawing,
Front View

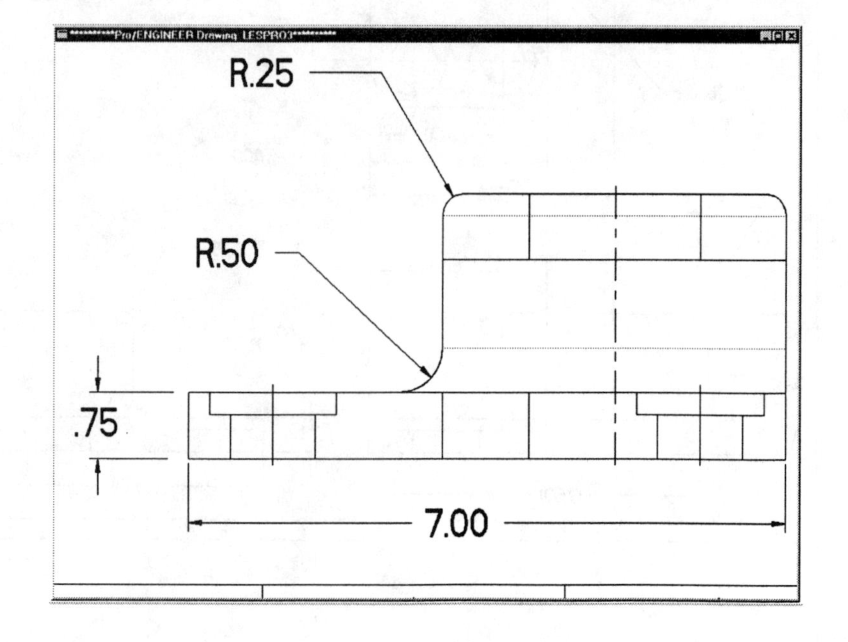

Figure 3.45
Guide Bracket Drawing,
Right Side View

Lesson 4

Datums and Layers

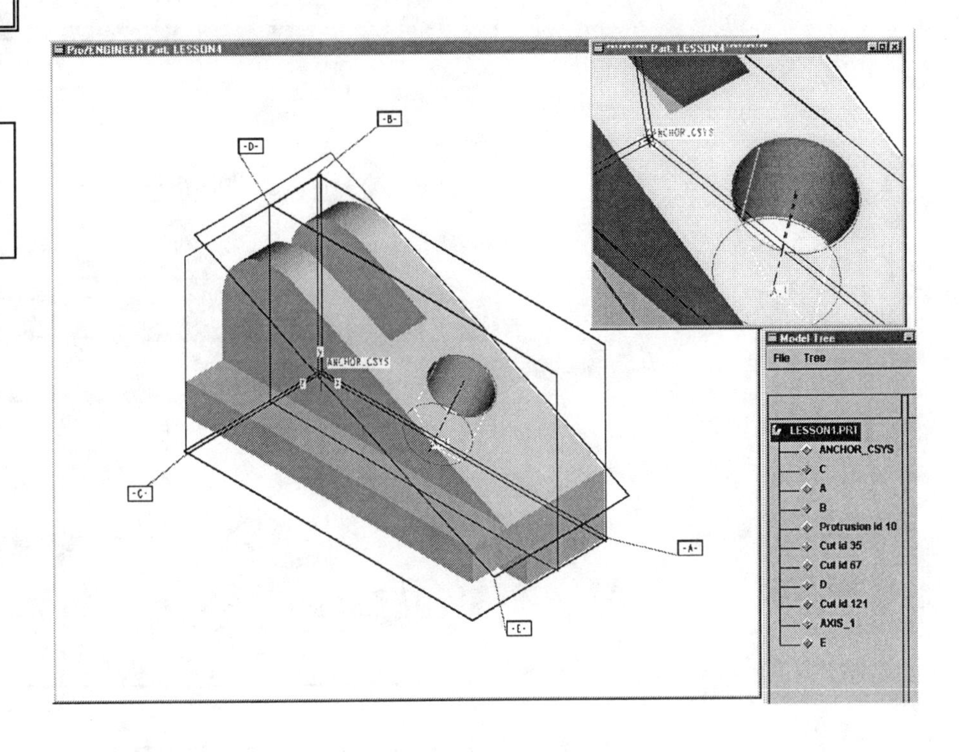

Figure 4.1
Anchor

EGD REFERENCE
Engineering Graphics and Design
by L. Lamit and K. Kitto
Read Chapters: 10, 25
See Pages: 388, 549, 929-932

OBJECTIVES

1. **Create datums to locate features**
2. **Use layers to organize part features**
3. **Set datum planes for geometric tolerancing**
4. **Rename datums**
5. **Add a simple relation to control a feature**
6. **Use datum planes to establish sections**
7. **Reroute a features references**
8. **Use Info command to get layer information**
9. **Learn how to change the color and shading of models**

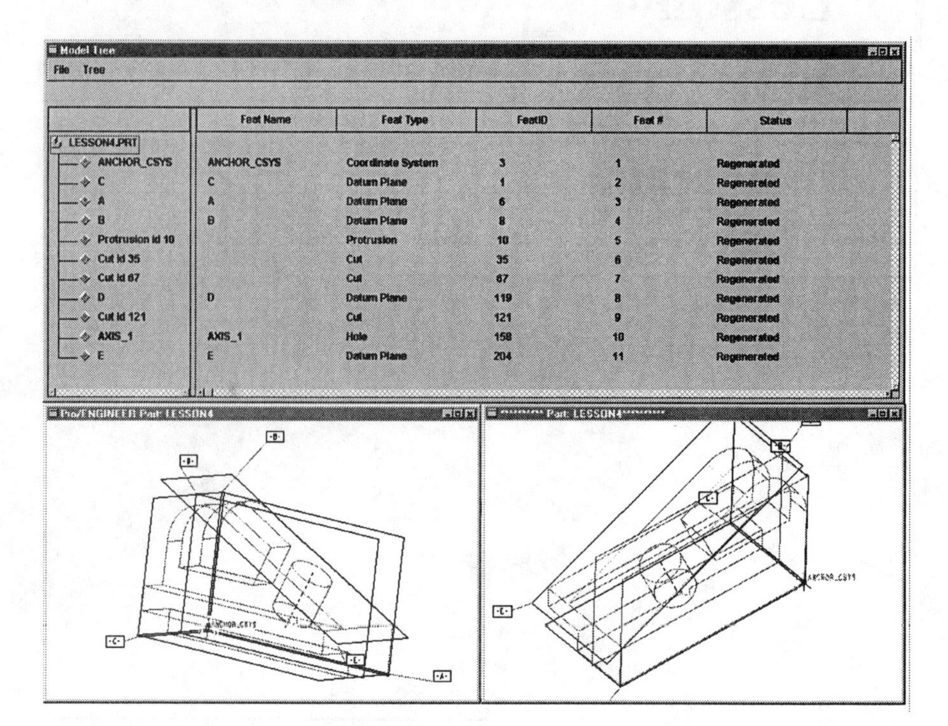

Figure 4.2
Anchor with Datums

DATUMS AND LAYERS

Datum planes and layers are two of the most useful mechanisms for creating and organizing your design (Figs. 4.1 and 4.2). **Layers** were covered in detail in Section 8 and that section should be reread at this point. Datum features such as *datum planes* and *datum axes* are essential for the creation of all parts, assemblies, and drawings using Pro/E. If you have **COAch for Pro/ENGINEER** on your system, do the appropiate segment, shown in Figure 4.3.

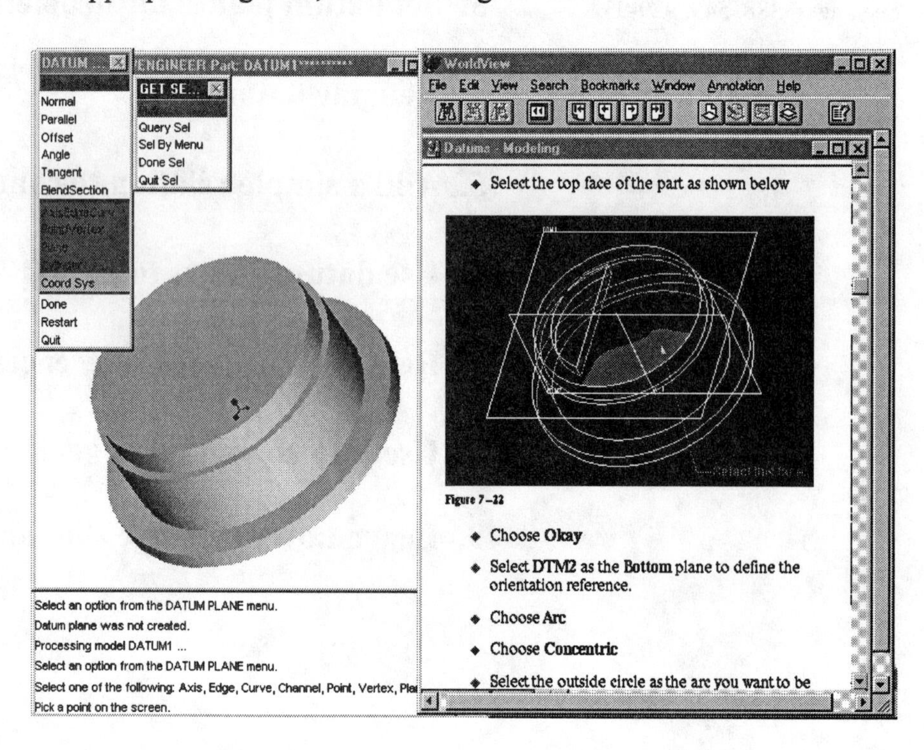

Figure 4.3
COAch for Pro/E, Datums-Modeling (Segment 1: Datum Planes)

Datum Planes

Datum planes are used to create a reference on a part where one does not already exist. For example, you can sketch or place features on a datum plane when there is no appropriate planar surface; you can also dimension to a datum plane as if it were an edge. When you are constructing an assembly, you can use datums with assembly commands. All datums have a red side and a yellow side so that you know on which side you are working.

Datum planes can be used as references, as sketching planes, and as parent features for a variety of nonsketched part features.

In most part designs, the first features created will be three default datums-- **DTM1**, **DTM2**, and **DTM3** (Fig. 4.4). The base construction feature (a protrusion) is created using these datums.

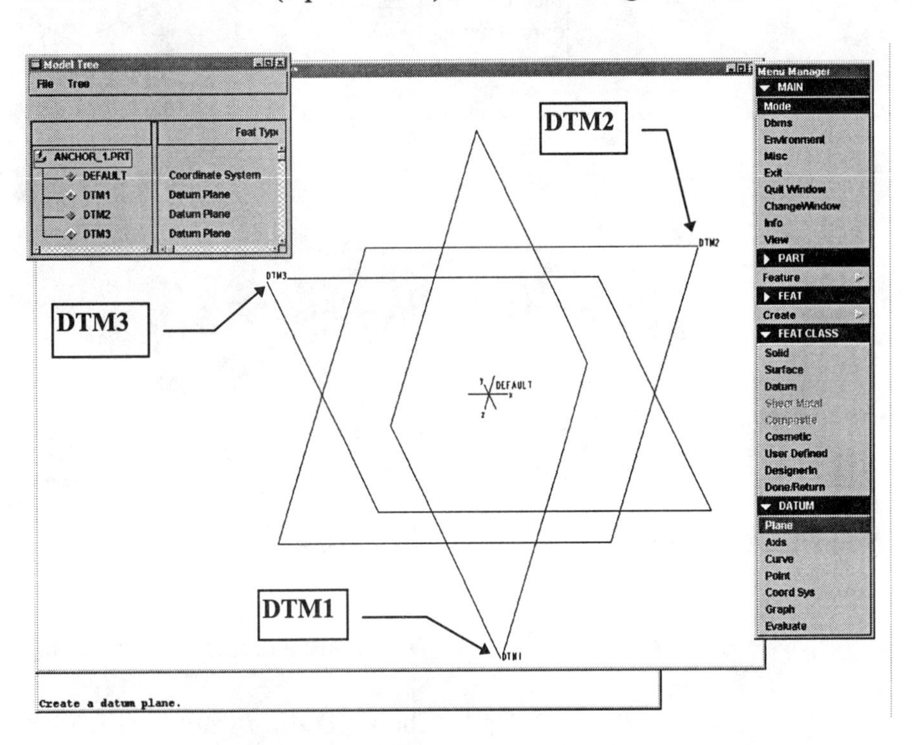

Figure 4.4
Default Datums

A nondefault datum is created by specifying **constraints** that locate it with respect to existing geometry. As an example, a datum plane might be created to pass tangent to a cylinder and parallel to a planar surface. Selected constraints must locate the datum plane relative to the model without ambiguity. Figure 4.5 shows an angled datum used to model the part. The following datum constraints are used alone, since each locates the datum plane completely:

Through/Plane Creates a datum plane coincident with a planar surface.

Offset/Plane Creates a datum plane that is parallel to a plane and offset from the plane by a specified distance.

Offset/Coord Sys Creates a datum plane that is normal to one of the coordinate system axes and offset from the origin of the coordinate system. When this option is selected, you are prompted to select which axis the plane will be normal to, then to enter the offset distance along this axis.
BlendSection Creates a datum plane through the section that was used to make a feature. When sections exist, as for a blend, you will be prompted for the section number.

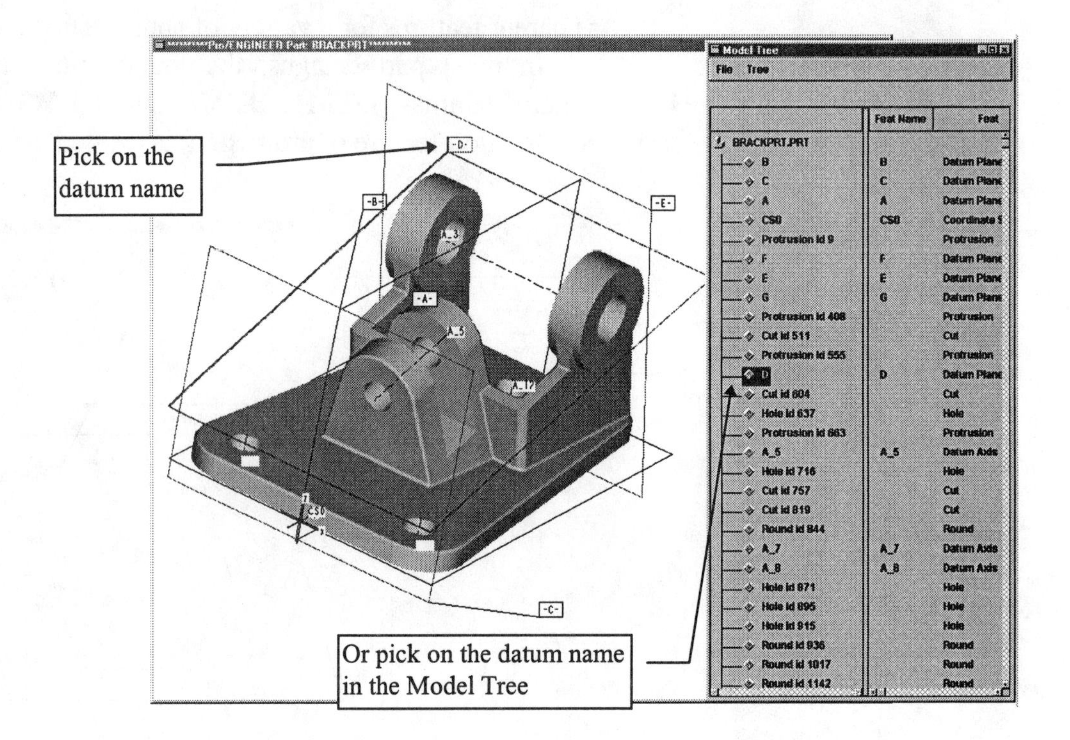

Figure 4.5
Nondefault Datums

To create a datum plane:

1. Choose **Datum** from the FEAT CLASS menu (or **Make Datum** from the SETUP PLANE menu), and then choose **Plane** from the DATUM menu.
2. Choose the desired constraint option from the DATUM PLANE menu. All appropriate geometry options in the lower section of the menu will be selected automatically. To limit the items to select, click on those highlighted menu options to unhighlight them.
3. Pick the necessary references.
4. Repeat Steps 2 and 3 until the required constraints have been established. When the maximum number of constraints have been specified, Pro/E notifies you by *dimming out* all options except **Done** and **Quit**. Although they are actually *infinite planes*, datum planes are displayed scaled to the model size. To select a datum plane, you can pick on its name or select one of its boundaries, or select it by picking on its name in the **Model Tree** (Fig. 4.5).

The size of a displayed datum plane changes with the dimensions of a part. All datum planes, except those made ***on-the-fly*** (within other commands using **Make Datum**), can be sized to specific geometry using the **Redefine** command. These allow you to make your datum plane as big as the model or as small as an edge or surface on the model. Figure 4.6 shows a page of the help available on datum planes.

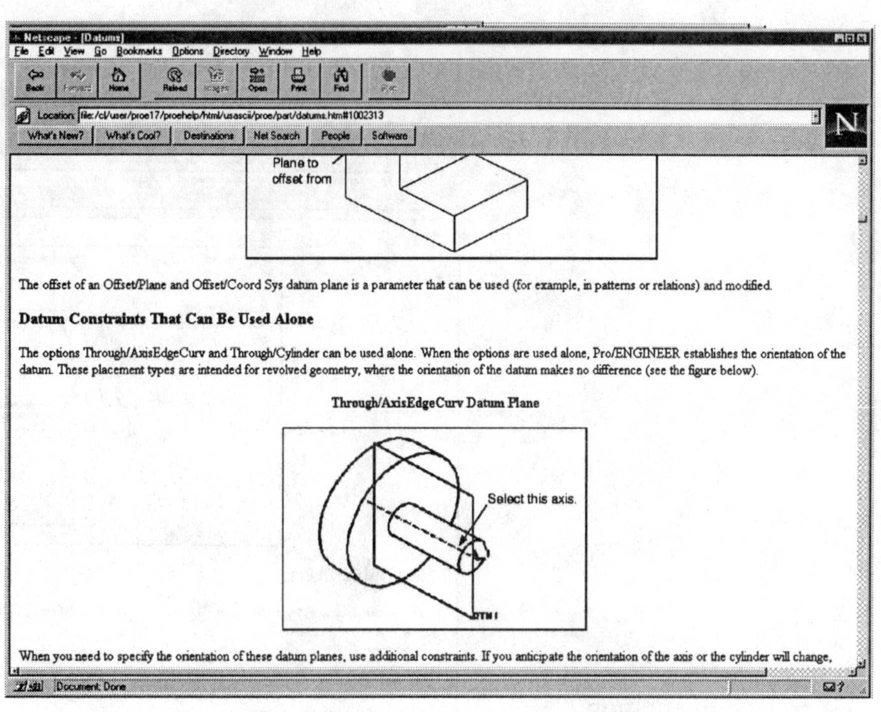

Figure 4.6
Online Documentation Datums

The options available for sizing the datum plane outline are:

Default The datum plane is sized to the model (part or assembly).
Fit Part Sizes the datum plane to a part in Assembly mode.
Fit Feature Sizes the datum plane to a part or assembly feature.
Fit Surface Sizes the datum plane to any surface.
Fit Edge Sizes the datum plane to fit an edge.
Fit Axis Sizes the datum plane to fit an axis.
Fit Radius Sizes the datum plane to fit a specified radius, centering itself within the constraints of the model.

Datum Axes

Datum axes (Fig. 4.7) can be used as references for feature creation, such as the coaxial placement of a hole. They are particularly useful for making datum planes, for placing items concentrically, and for creating radial patterns. Axes can be used to measure from, place coordinate systems, and place specific features. The angle between features and an axis, the distance between an axis and a feature, etc. can be determined using the **Info** command. Axes (appearing as centerlines) are automatically created for:

Revolved features All features whose geometry is revolved, including revolved base features, holes (Fig 4.8), shafts, revolved slots, cuts, and circular protrusions (Fig. 4.9).

Extruded circles An axis is created for every extruded circle in any extruded feature.

Extruded arcs An axis can be created automatically for extruded arcs only when you set the configuration options.

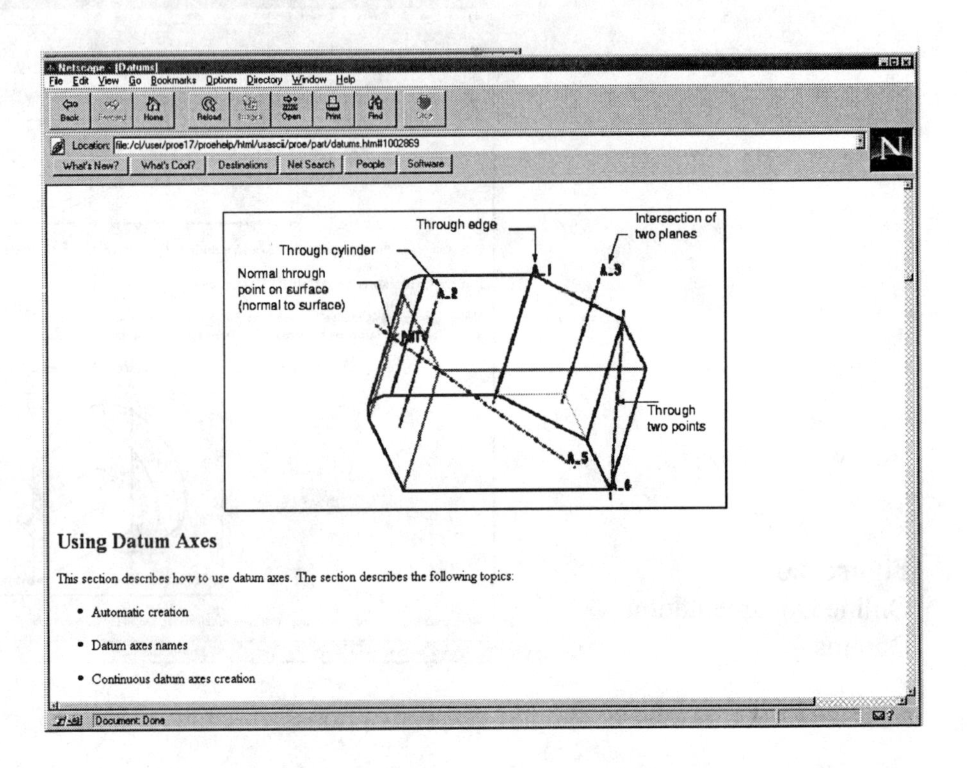

Figure 4.7
Online Documentation
Datum Axes

To create a datum axis:

1. Choose **Datum** from the FEAT Class menu and then the **Axis** from the DATUM menu.
2. Choose the desired constraint option from the DATUM AXIS menu:

Thru Edge Creates a datum axis through a straight edge. Select the edge.

Norm Pln Creates an axis that is normal to a surface, with linear dimensions locating it on that surface.

Pnt Norm Pln Creates an axis through a datum point and normal to a specified plane.

Thru Cyl Creates an axis through the "imaginary" axis of any surface of revolution (where an axis does not already exist). Select a cylindrical surface or a revolved surface.

Two Planes Creates a datum axis at the intersection of two planes (datum planes or surfaces). Select two planes; they cannot be parallel, but they need not be shown to intersect on the screen.

Two Pnt/Vtx Create an axis between two datum points or edge vertices. Select datum points or edge vertices.

Pnt on Surf Create an axis through any datum point located on a surface; the point does not need to have been created using **On Surface**. The axis will be normal to the surface at that point.

Tan Curve Create an axis that is tangent to a curve or edge at its endpoint. Select the curve/edge to be tangent to, then select an endpoint of the curve/edge.

3. Pick the necessary references for the selected option.

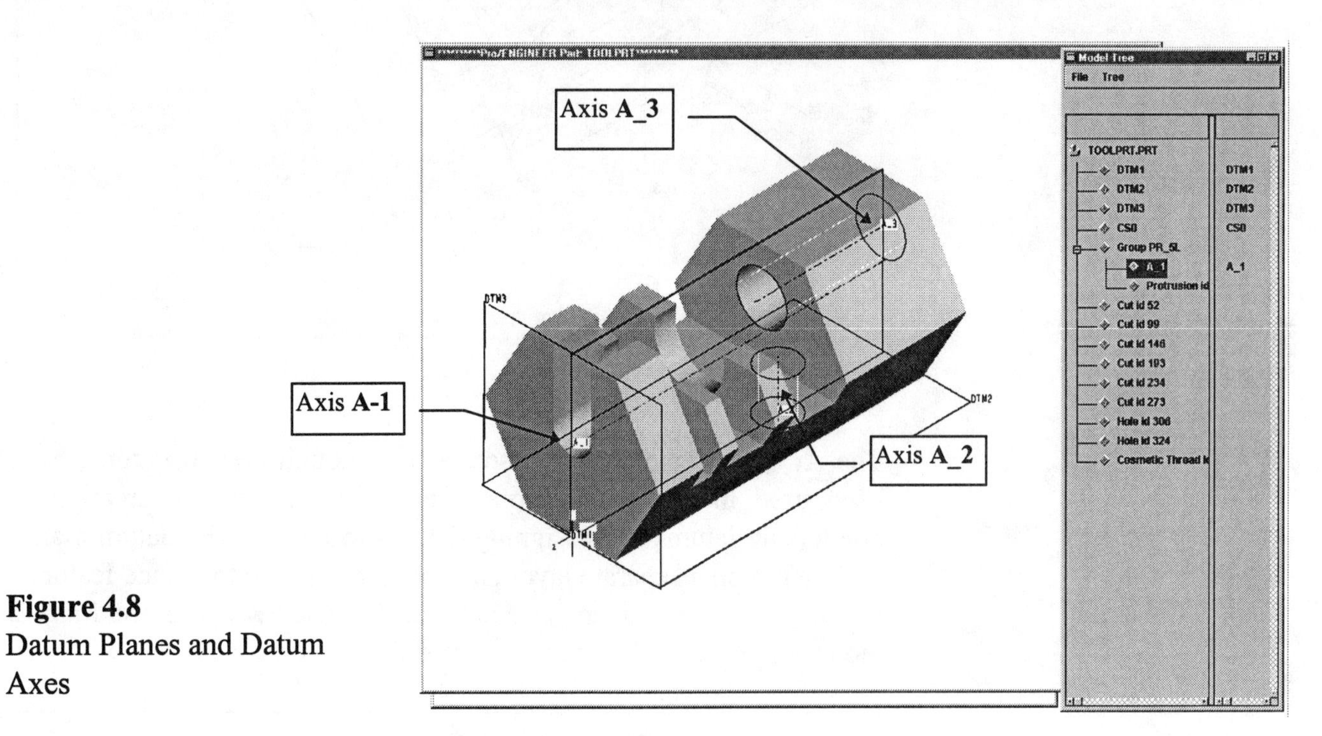

Figure 4.8
Datum Planes and Datum Axes

Axis for each patterned hole (**A_24**)

Axis for each drafted circular protrusion (**A_32**)

Figure 4.9
Datum Axes for Holes and Circular Protrusions

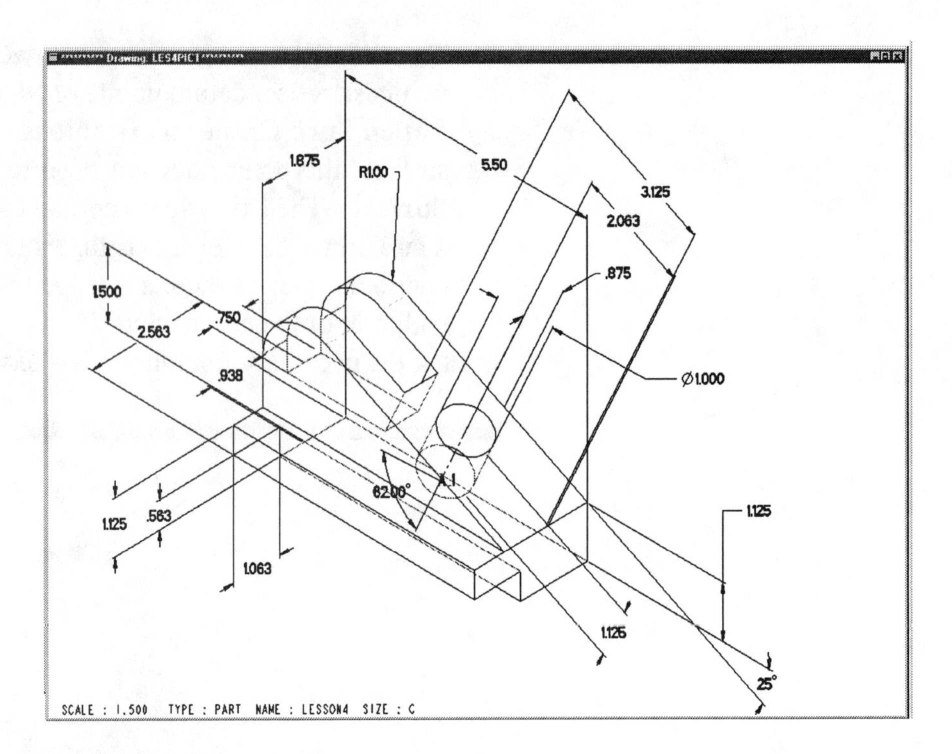

Figure 4.10
Anchor Dimensions

Anchor

The **Anchor** is the fourth lesson part. Though default datums have been used in all other lessons, Lesson 4 requires the creation of nondefault datums and assigning datums to layers. The datum planes will be put on a separate layer and set as geometric tolerance features. Reread Section 8, Layers, at this time. Choose **Part** from the MODE menu, and give the following commands:

Part ⇒ **Create** (type the part name **Anchor**) ⇒ **enter.**

Set up the units and the environment:

SETUP AND ENVIRONMENT

Setup ⇒ **Units** ⇒ **Inch** ⇒ **Material** ⇒ **Define** ⇒ (type **Steel**, then press **Enter** key) ⇒ **Assign** ⇒ (pick **Steel**) ⇒ **Accept**

Environment ⇒ ✓**Grid Snap**
Hidden Line Tan Phantom

Create a default coordinate system and three default datum planes as in Lessons 1 and 2. Give the following commands:

Feature ⇒ **Create** ⇒ **Datum** ⇒ **Plane** ⇒ **Offset** ⇒ **enter** (three times at the prompt) ⇒ **Done**

The first protrusion will be sketched on **DTM3,** as in previous lessons. Use Figure 4.11 for the protrusions dimensions, and sketch the outline in Figure 4.12. Only **5.50**, **R1.00**, **1.125**, and **25°** are needed. Give the following commands:

> **Feature ⇒ Create ⇒ Protrusion ⇒ Extrude ⇒ Solid ⇒ Done ⇒ One Side ⇒ Done ⇒** (pick **DTM3** as the *sketching* plane) **⇒ Okay** (for selecting the *direction* of feature creation) **⇒ Top ⇒** (pick **DTM2** as *orientation* plane) **⇒ Done ⇒ Sketch ⇒ Mouse Sketch**

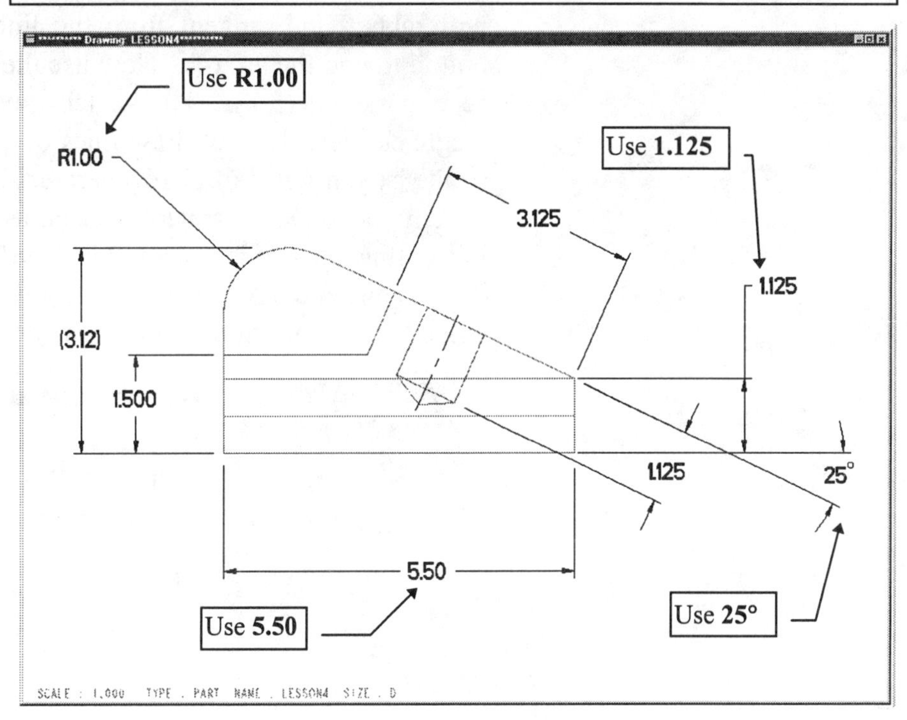

Figure 4.11
Protrusion Dimensions

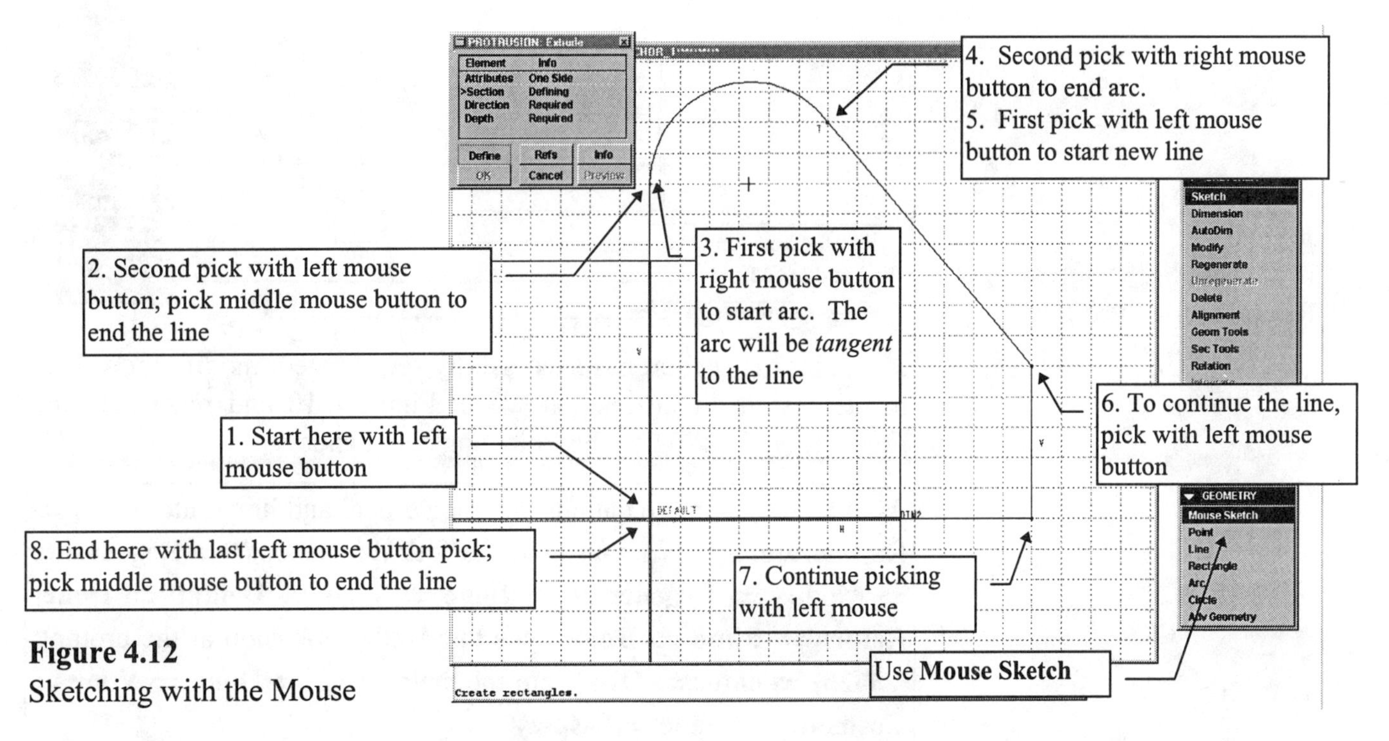

Figure 4.12
Sketching with the Mouse

Sketching with the mouse is a fast and efficient method of creating geometry. The **left mouse button** is used to create a continuous set of *lines*. The **middle mouse button** is used for creating *circles*, and the **right mouse button** creates *tangent arcs* when the first pick is near the end of an existing line or arc.

? Pro/HELP

Use online documentation to understand the use of mouse buttons better.

Start the sketch (Fig. 4.12) by picking the first endpoint of the vertical line with the left mouse button (1). Pick the second endpoint at a position needed to create a vertical line (2). Use the middle mouse button to end the line. Next, using the right mouse button pick near the last endpoint created for the line to start an arc (3). The arc will rubberband tangent from the line. Pick with the right mouse button to end the arc (4). Now use the left mouse button to finish the section sketch (5-8). The sketch must be closed and composed of single entities. Do not draw lines or arcs over the top of one another or the sketch will fail to regenerate. The sketch does not have to be exactly the same dimensional scale as the physical part. As long as the outline is similar, then Pro/E will correct the sketch when the dimensions are modified. After the sketch is complete, align the edges, add dimensions, and regenerate the section sketch (Fig. 4.13).

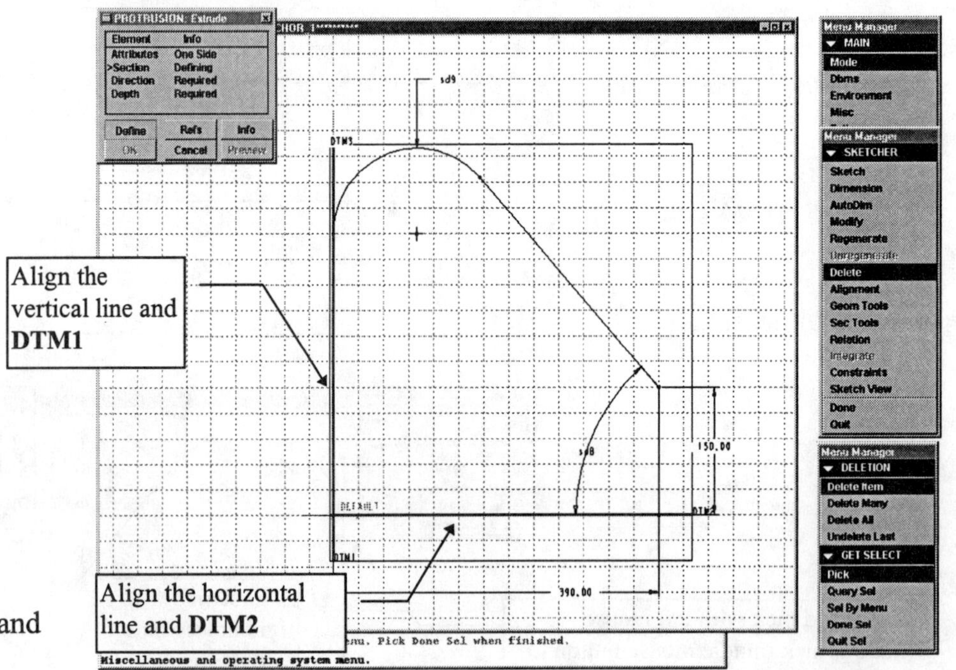

Figure 4.13
Dimensioning, Aligning, and Regenerating the Sketch

After the section has successfully regenerated, modify the values to the design dimensions shown in Figure 4.10 and regenerate the sketch (Fig. 4.14) using the following commands:

Alignment (align the datums and the vertical and horizontal lines) ⇒ **Regenerate**⇒ **Dimension** (add the four dimensions) ⇒ **Regenerate** ⇒ **Modify** ⇒ **Regenerate** ⇒ **Done** ⇒ **View** ⇒ **Default** ⇒ **Done-Return** ⇒ **Blind** ⇒ **Done** (type the depth dimension at the prompt: **2.5625**) ⇒ **enter** ⇒ **OK** (from the dialog box) ⇒ **Done** ⇒ **View** ⇒ **Cosmetic** ⇒ **Shade** ⇒ **Display**

Dbms ⇒ Save ⇒ enter Purge ⇒ enter ⇒ Done-Return

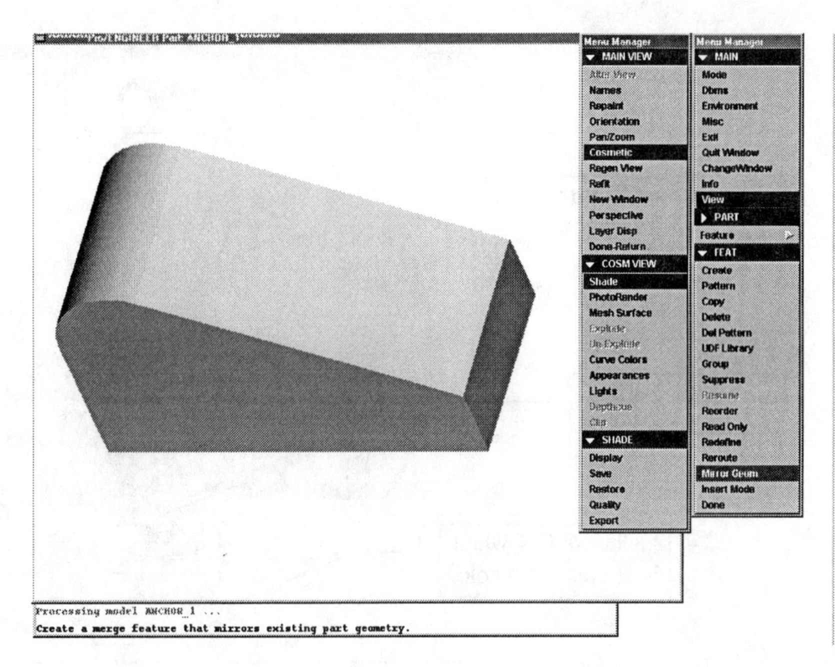

Figure 4.14
Shaded First Protrusion

At this point let's change the color of the model. In general, try to avoid the colors that are used as defaults. Red and colors close to red should be avoided since feature and entity highlighting is defaulted to red. Datum planes have red and yellow sides; therefore yellow should be avoided. Shades of blue and green work well. It is really up to you to choose colors that are pleasant to look at when gazing at the monitor for hours on end!

View ⇒ Cosmetic ⇒ Appearances ⇒ Define (will activate the **Appearance Editor** on your screen, as shown, in Figure 4.15. Pick on the color box to initiate the **Color Editor**, as shown in Figure 4.16)

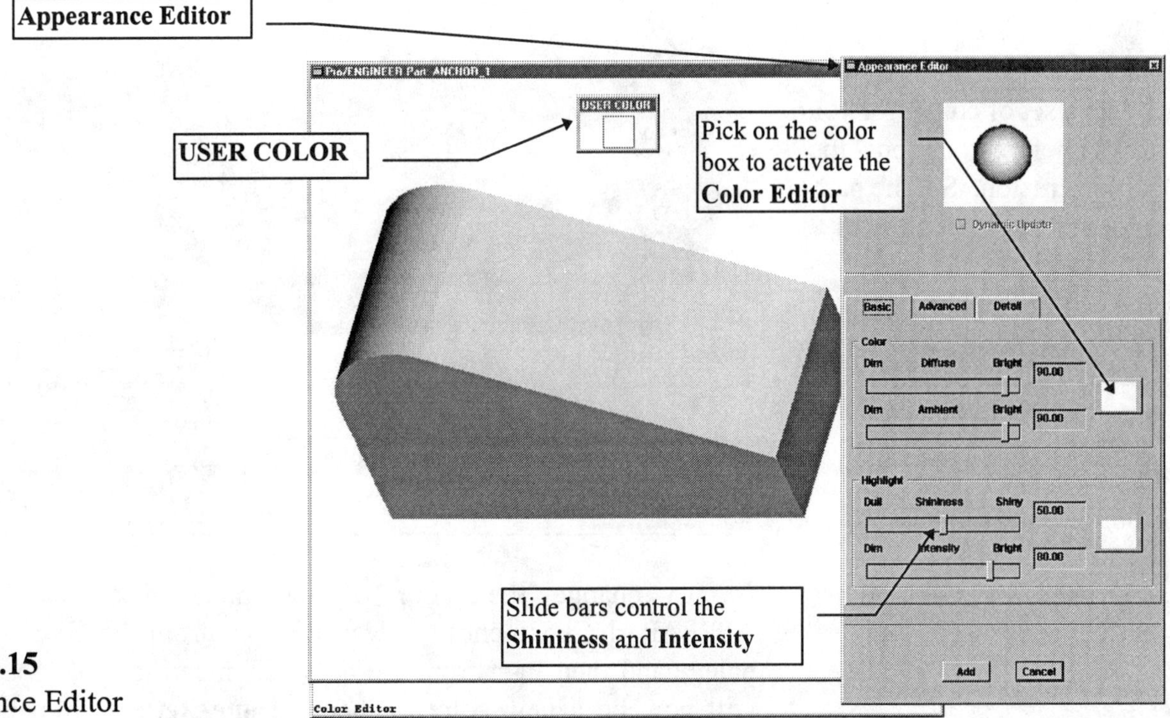

Figure 4.15
Appearance Editor

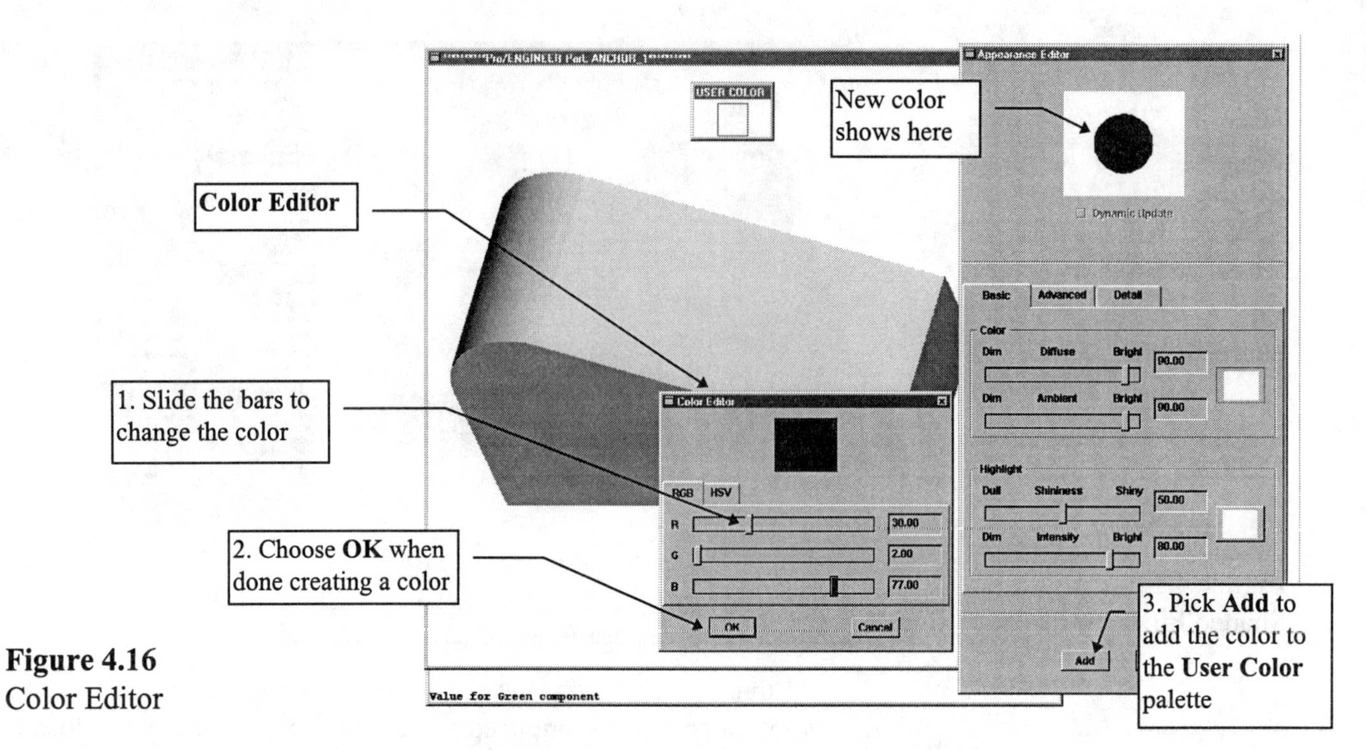

Figure 4.16
Color Editor

Slide the RGB bars to create useful colors. Add the color to the **USER COLOR** palette by choosing **OK** from the **Color Editor** and then **Add** from the **Appearance Editor** (Fig. 4.17). Continue making a few more colors, adding each to the palette. Select the **Cancel** button from the appearance editor.

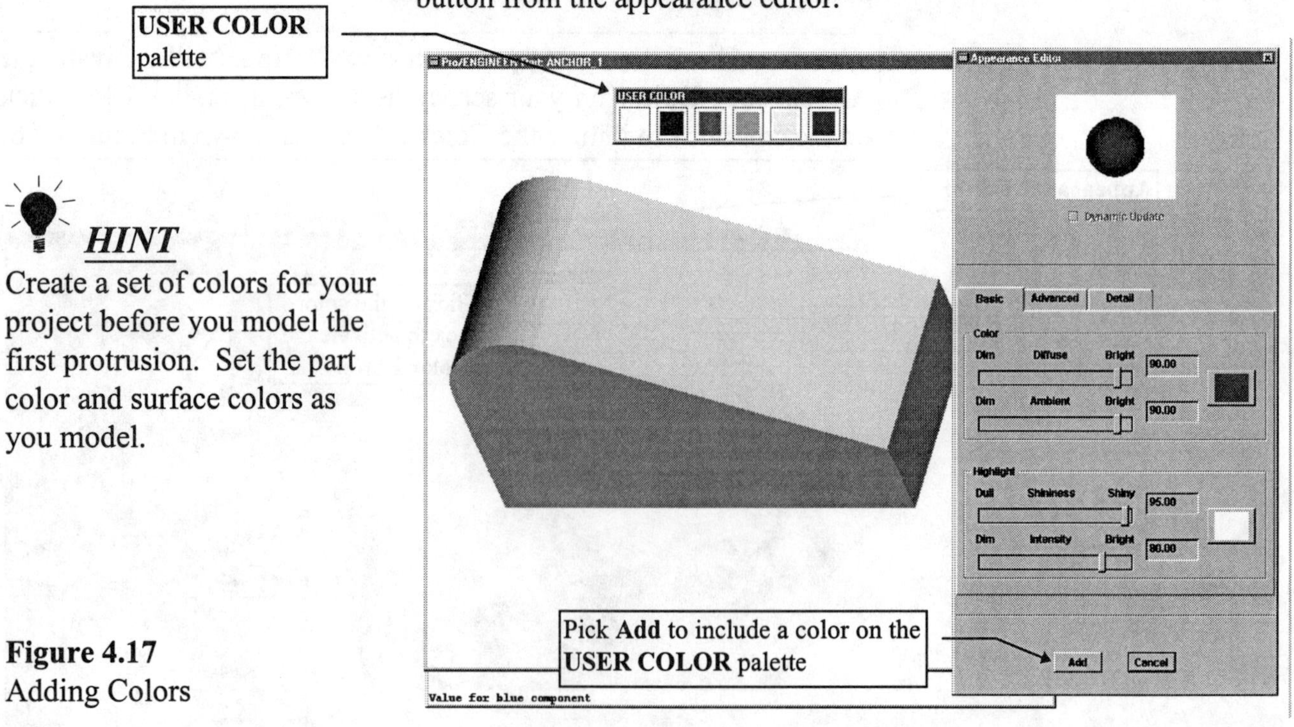

HINT
Create a set of colors for your project before you model the first protrusion. Set the part color and surface colors as you model.

Figure 4.17
Adding Colors

To change the color of the part, pick **Set** (from the APPEARANCES menu). Select the color from the **USER COLOR** palette and then **Part** from the OBJECT menu (Fig. 4.18). The part will now be a new color. Choose **Done-Return** from the MAIN menu to end the process.

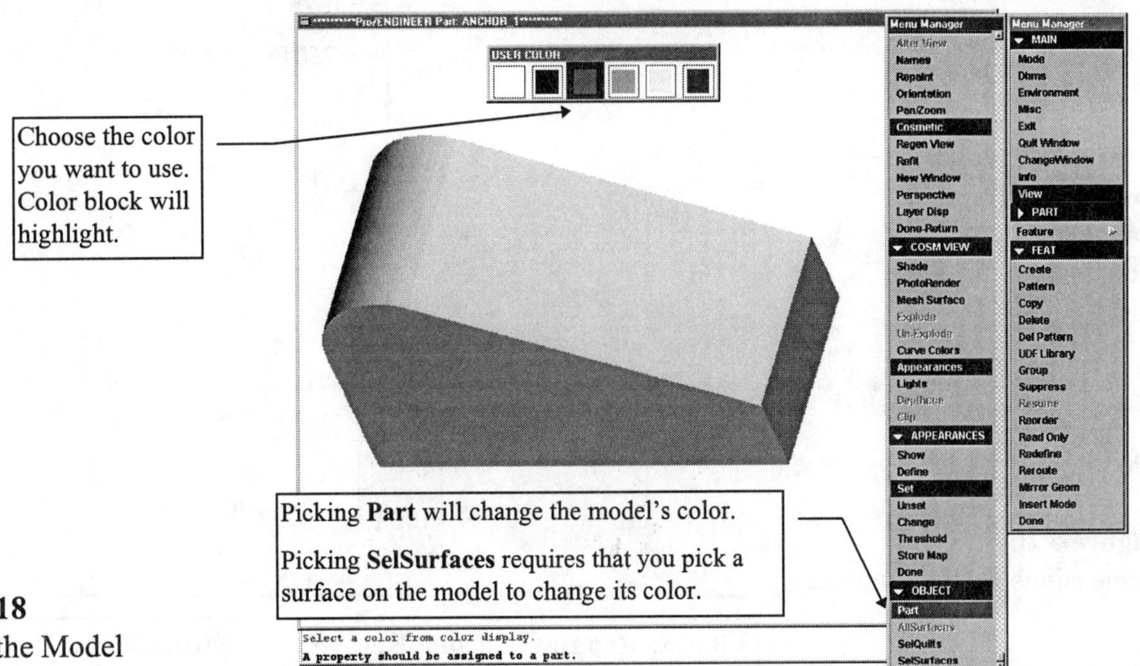

Figure 4.18
Coloring the Model

Dbms ⇒ Save ⇒ enter
Purge ⇒ enter ⇒
Done-Return

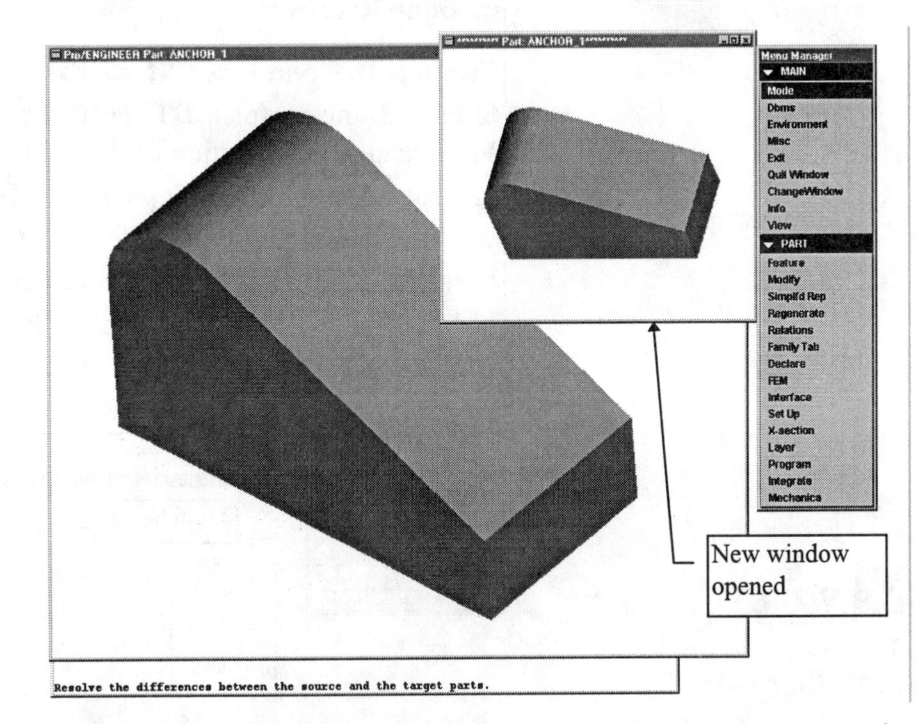

Figure 4.19
Model with New Color

Unfortunately, this text is not in color, so you cannot see the changes to the model (Fig. 4.19). Experimenting with the colors is most students' idea of fun. Enjoy yourself, but don't create too many colors; they really aren't necessary at this stage of the project.

The next two features will be cuts. For both of the cuts use **DTM1** as the sketching plane and give **Thru All** as the depth. Each cut requires just two lines, two dimensions, and two alignments. Use the first cut's sketching/placement plane and reference/orientation (**Use Prev**) for the second cut. Create the cuts separately, each as an open section. Figure 4.20 shows the dimensions for each cut. Give the following commands:

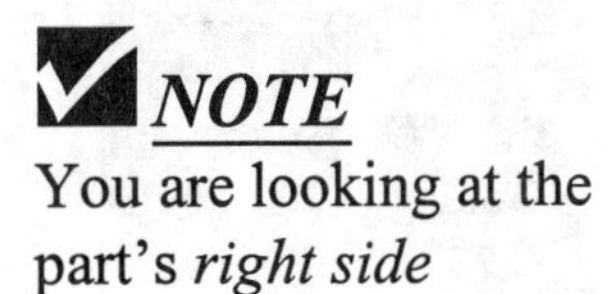

NOTE

You are looking at the part's *right side*

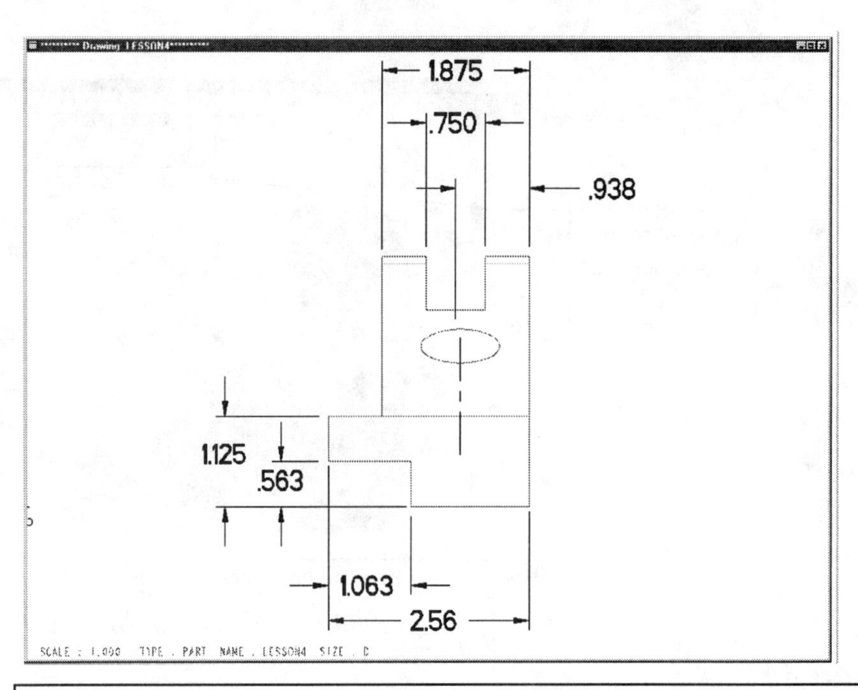

Figure 4.20
Dimensions for the Two Cuts

View ⇒ Repaint ⇒ Done-Return ⇒ Environment ⇒ □ Grid Snap ⇒ Done-Return ⇒

Feature ⇒ Create ⇒ Cut ⇒ Extrude ⇒ Solid ⇒ Done ⇒ One Side ⇒ Done ⇒ (pick **DTM1** as the sketching/placement plane) ⇒ **Flip** (change the direction of the cut to pass through the part, not out into space) ⇒ **Okay ⇒ Top ⇒** (pick **DTM2** as the orientation plane; you are looking at the left side) ⇒ **Line ⇒ Vertical** (sketch the two lines) ⇒ **Alignment** (align the vertical line with the top of the part and the horizontal line with the right side edge) ⇒ **Regenerate ⇒ Dimension** (add the dimensions) ⇒ **Regenerate ⇒ Modify** (change the dimensions to **1.125** and **1.875**) ⇒ **Regenerate** (Fig. 4.21)

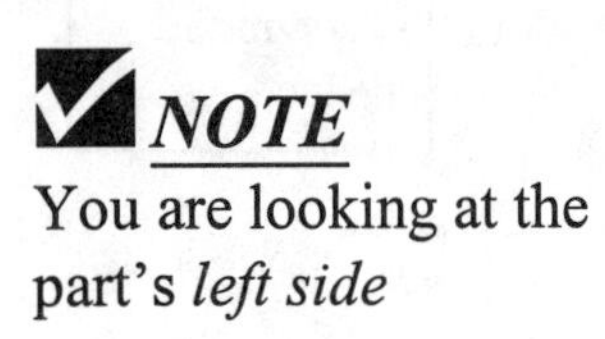

NOTE

You are looking at the part's *left side*

Figure 4.21
Cut Section Dimensions

Complete the cut (Fig. 4.22) using the following commands:

Done ⇒ OKAY ⇒ Thru All ⇒ Done ⇒ View ⇒ Default ⇒ OK ⇒ View ⇒ Cosmetic ⇒ Shade ⇒ Display ⇒ Done-Return ⇒ Done

Dbms ⇒ Save ⇒ enter Purge ⇒ enter ⇒ Done-Return

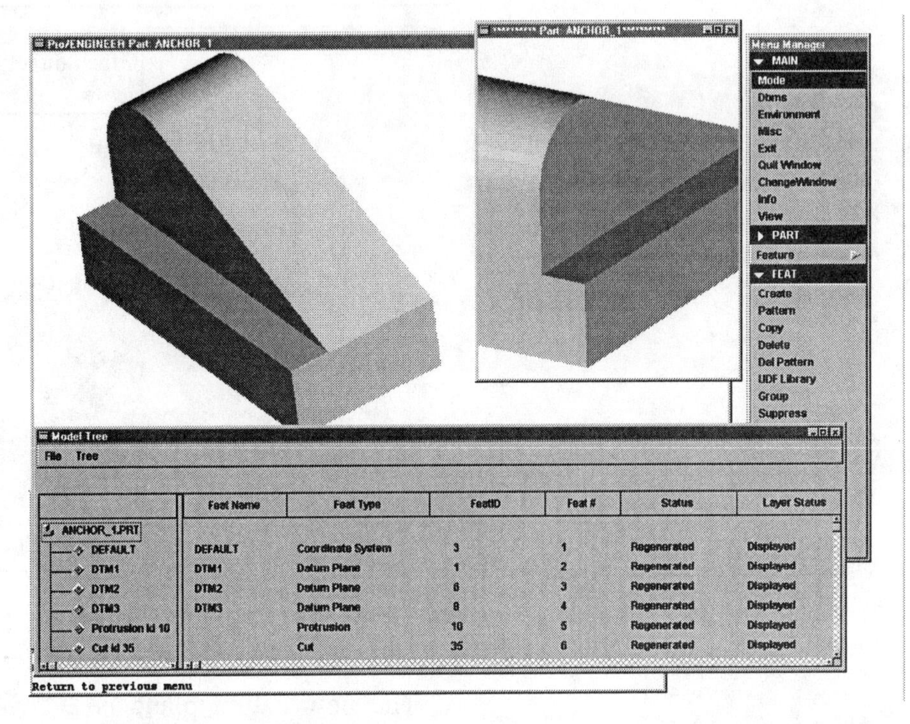

Figure 4.22
Completed Cut

Complete the second cut using the last two command blocks (Fig. 4.23). Everything will be the same except for the dimensions **.563** and **1.063**. You can expedite things by selecting **Use Prev** for the sketching and orientation planes.

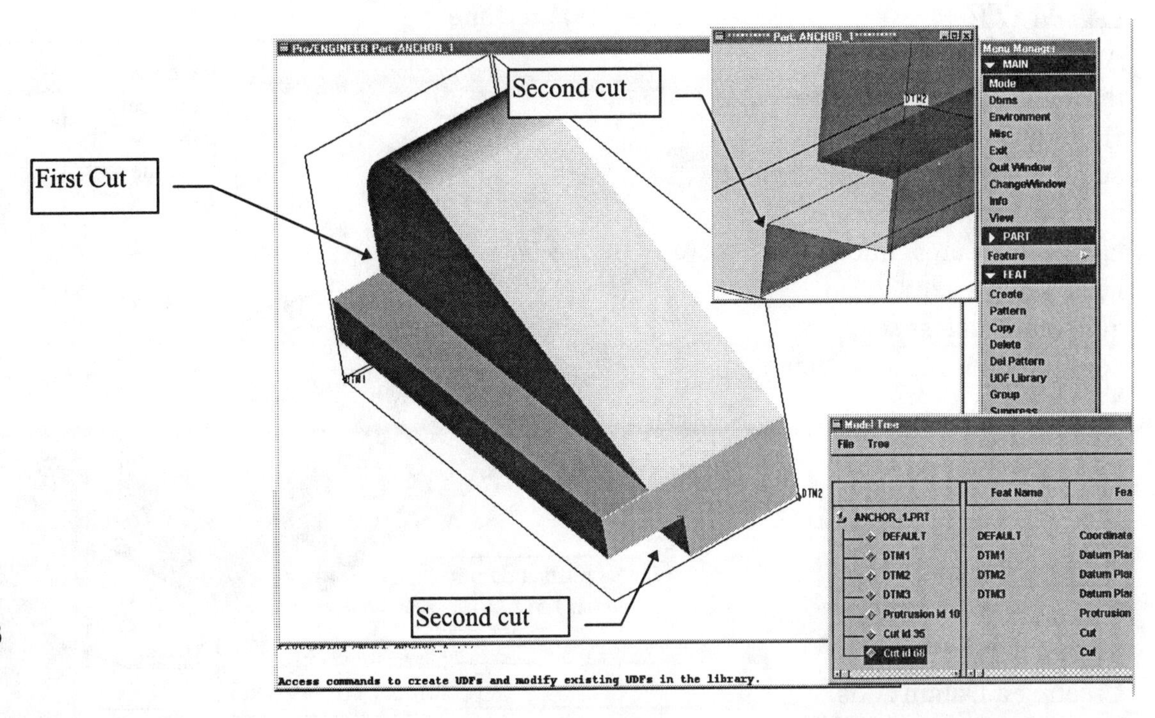

Figure 4.23
Second Cut

On the next feature we will create a new datum plane to sketch on. A datum axis will also be created at this time. Give the following commands:

View ⇒ Repaint ⇒ Done-Return ⇒ Feature ⇒ Create ⇒ Datum ⇒ Plane ⇒ Offset (pick **DTM3** to offset from) ⇒ **Enter Value** (type the offset distance of **.9375** at the prompt) ⇒ **Done** (Fig. 4.24)

Dbms ⇒ Save ⇒ enter Purge ⇒ enter ⇒ Done-Return

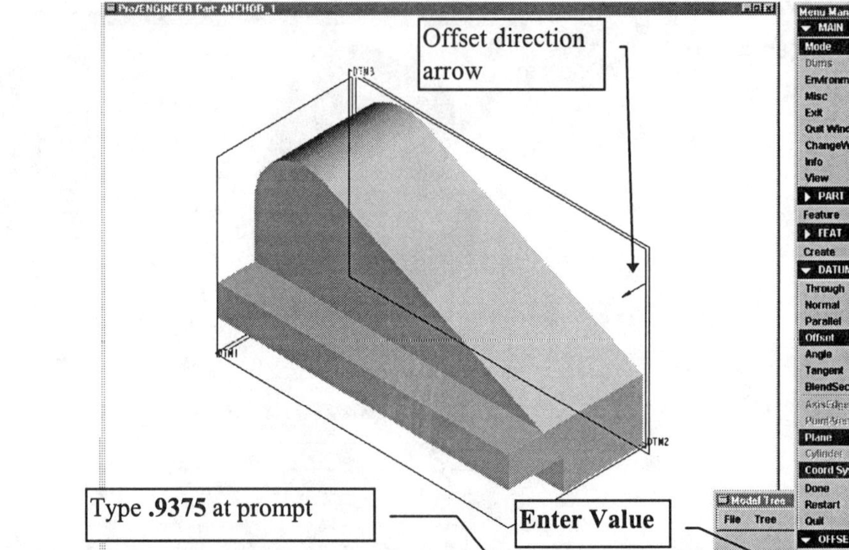

Figure 4.24
Creating an Offset Datum Plane

The new datum plane is **DTM4**. A datum axis will now be created through the curved top of the part using the following commands:

Feature ⇒ Create ⇒ Datum ⇒ Axis ⇒ Thru Cyl (pick as shown in Fig. 4.25) ⇒ **Done**

NOTE
An axis can be inserted through any curved feature. Holes and circular features automatically have axes when they are created. Features created with arcs, fillets, etc. need to have axes added afterwards unless set in the configuration file as a default

Pick on cylindrical surface to create the axis **Thru Cyl**

Offset from **DTM3 .9375**

Axis **A_1** created through the cylinder

Offset datum **DTM4**

Axis shows in Model Tree as **A_1**

Figure 4.25
Creating a Datum Axis

The cut on the top of the part is created by sketching on **DTM4** and projecting it toward both sides. Use the Model Tree to select the appropiate datum planes. Give the following commands:

Feature ⇒ **Create** ⇒ **Cut** ⇒ **Extrude** ⇒ **Solid** ⇒ **Done** ⇒ **Both Sides** ⇒ **Done** ⇒ (pick **DTM4** as the sketching/placement plane) ⇒ **Okay** ⇒ **Top** ⇒ (pick **DTM2** as the orientation plane) ⇒ **Line** ⇒ **Perpendicular** (pick the angled edge and then the starting and ending points of the line) **Line** ⇒ **Horizontal** ⇒ (pick the endpoints of the horizontal line as shown in Figure 4.26) ⇒ **Alignment** (align the angled line's endpoint with the angled edge of the part and the horizontal line with the left side edge) ⇒ **Regenerate** ⇒ **Dimension** (add the dimensions) ⇒ **Regenerate** ⇒ **Modify** (change the dimensions to **3.125** and **1.50**, as in Figure 4.11) ⇒ **Regenerate** (Fig. 4.27)

HINT

To get the slanted dimension of **3.125**, select the slanted line itself and then the end of the block. Place the dimension properly.

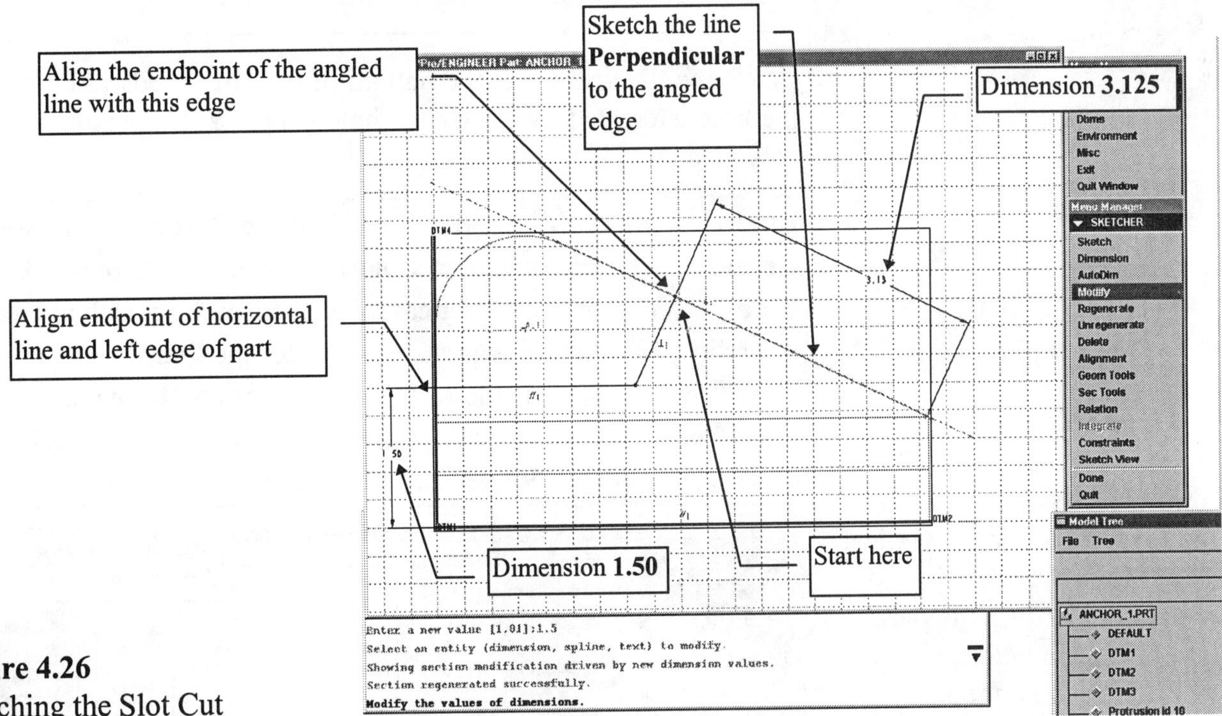

Figure 4.26
Sketching the Slot Cut

Complete the cut with the following commands:

Done ⇒ **Okay** (for the material removal side) ⇒ **Blind** ⇒ **Done** ⇒ (type the full width of **.750** for the slot at the prompt) ⇒ **OK** ⇒ **Done** ⇒ **View** ⇒ **Default** (shade the display as in Figure 4.27) ⇒ **Environment** ⇒ **Isometric** ⇒ **Done-Return**

The hole drilled in the angled surface appears to be aligned with **DTM4**. Upon closer inspection, it can be seen that the hole is a different distance from the edge (**.875** from **DTM3**) and is not in line with the slot and datum plane. Create the feature using a sketched hole (Fig. 4.28).

Dbms ⇒ Save ⇒ enter
Purge ⇒ enter ⇒
Done-Return

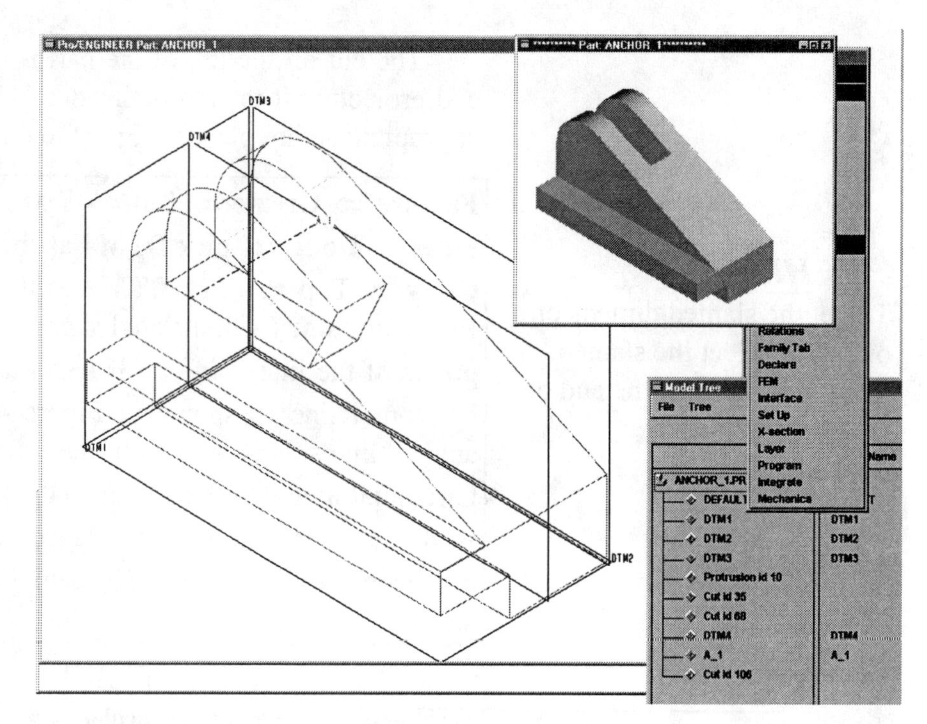

Figure 4.27
Cut Created on Both Sides of **DTM4**

Since the drill tip at the bottom of the hole needs to be modeled, the hole is created as a sketched hole (Fig. 4.28). Use the following commands:

Environment ⇒ ✓ Grid Snap ⇒ Done-Return ⇒ Feature ⇒ Create ⇒ Hole ⇒ Sketch ⇒ Done ⇒ Linear ⇒ Done ⇒ Sketch ⇒ Line ⇒ Centerline ⇒ Vertical (pick once on the sketch to create the centerline) ⇒ **Line** (pick five endpoints for the four lines to create the *closed section* on one side of the centerline) ⇒ **Dimension** (add the **1.125** depth, **62°** representing one-half of the drill tip angle, and a diameter of **1.00**) ⇒ **Regenerate ⇒ Modify**

HINT
Remember that the diameter dimension is created by picking the rightmost vertical line. Then pick the centerline, and finally pick the line again before placing the dimension on top.

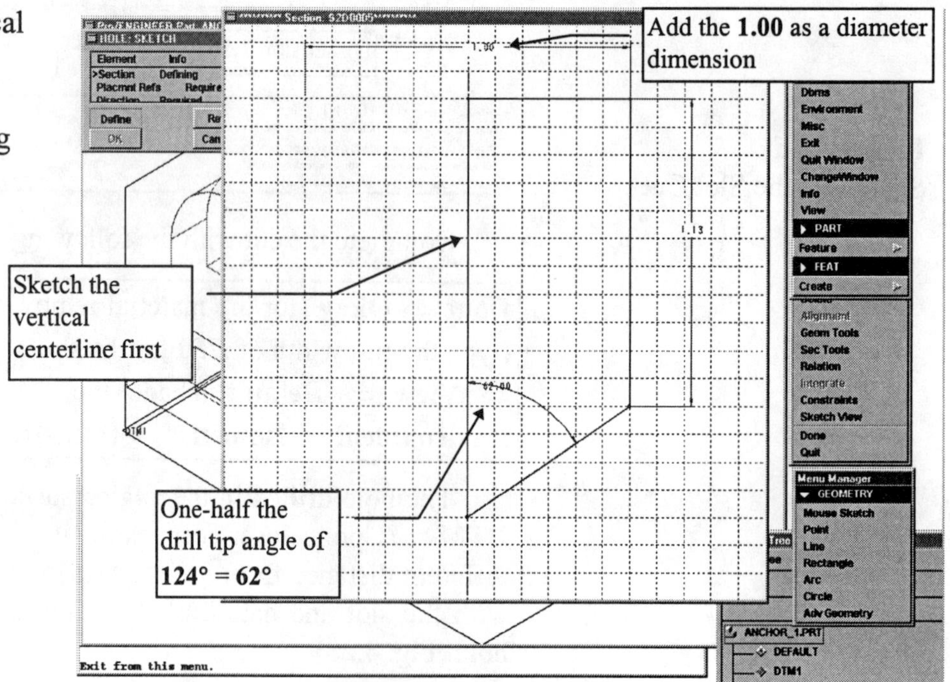

Figure 4.28
Sketching the Hole

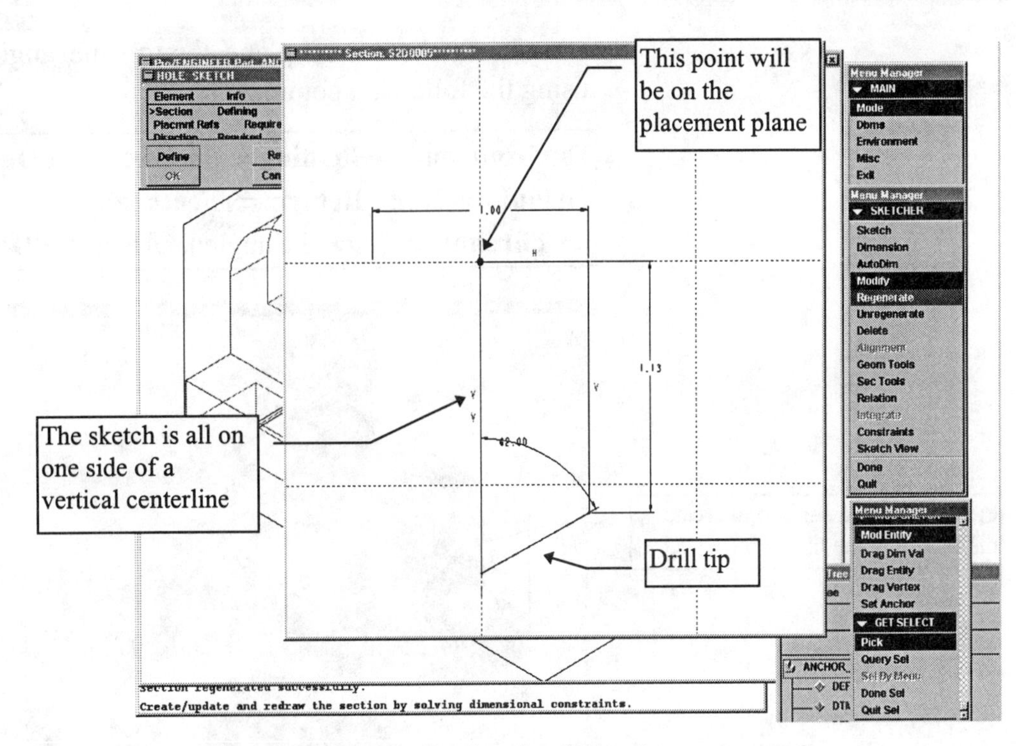

Figure 4.29
Regenerated Sketch

To complete the hole give the following commands:

Regenerate (Fig. 4.29) ⇒ **Done** ⇒ (pick the angled surface as the placement plane) ⇒ (select two edges to dimension from, pick the edge as shown in Figure 4.30, and type **2.0625** at the prompt) ⇒ **enter** ⇒ (next, pick **DTM3** and type **.875** at the prompt) ⇒ **enter** ⇒ **OK** (from the dialog box) ⇒ **Done**

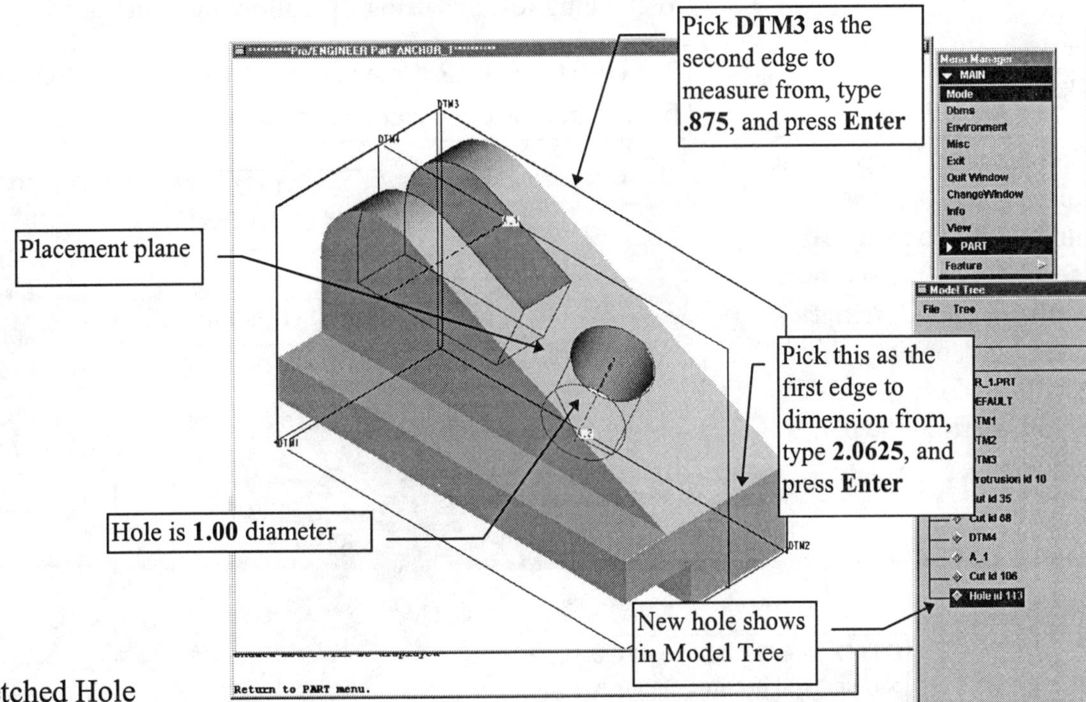

Figure 4.30
Completed Sketched Hole

A new datum plane will now be created that is through the angled surface. All datum planes are to be *set* as geometric tolerancing features and put on a layer.

Create the datum plane though the angled surface (Fig. 4.31) using the following commands:

Environment ⇒ **Shading** ⇒ **Isometric** ⇒ **Done-Return** ⇒ **View** ⇒ **Default** ⇒ **Done-Return** ⇒ **Feature** ⇒ **Create** ⇒ **Datum** ⇒ **Plane** ⇒ **Through** ⇒ (pick the angled surface) ⇒ **Done** ⇒ **Done**

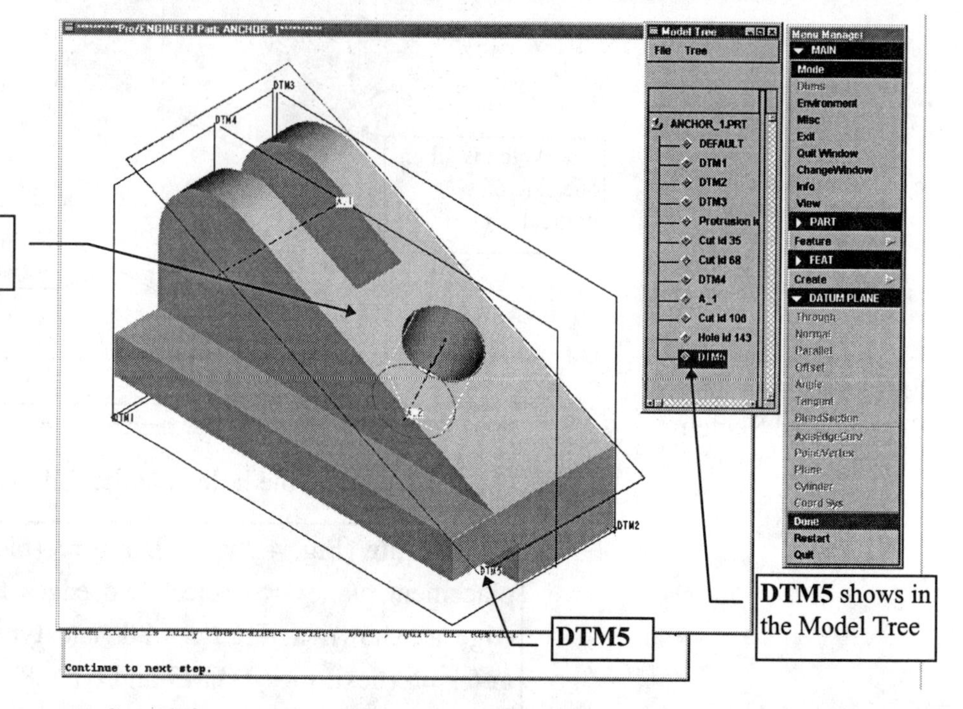

Figure 4.31
DTM5 Created Through the Angled Surface

Set (Fig. 4.32) and rename the datum planes as geometric tolerancing features using the following commands:

Set Up (from PART menu) ⇒ **Geom Tol** ⇒ **Set Datum** ⇒ (pick **DTM2** from Model Tree) ⇒ (change name from **DTM2** to **A**) ⇒ **OK**

HINT

Datums used in geometric tolerancing; **A** (primary--three-point contact), **B** (secondary—two-point contact), and **C** (tertiary--one-point contact), should be established on your ***DIPS*** before starting the part model!

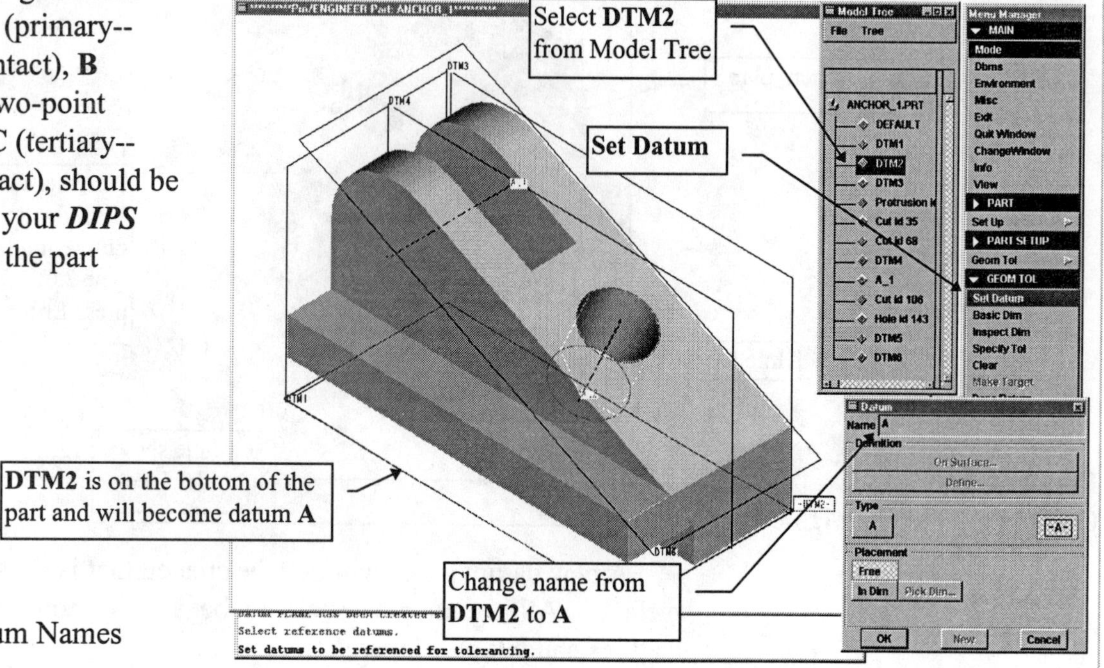

Figure 4.32
Changing Datum Names

Complete the renaming and setting of all five datum planes. Move the position of the names of the datums by giving the following commands (Fig. 4.33):

Done/Return ⇒ **Done** ⇒ **Modify** (from the PART menu) ⇒ **Move Datum** ⇒ (pick the datum name, edge, or name from the Model Tree) ⇒ (pick a new position on the screen for the name) ⇒ **Done**

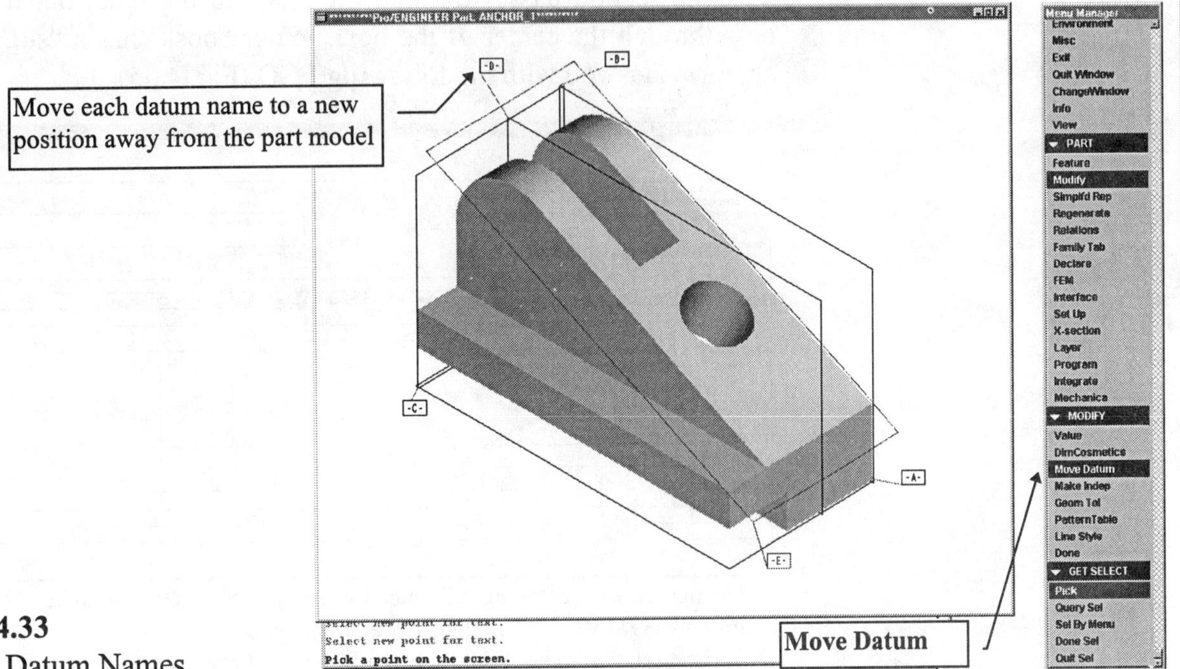

Figure 4.33
Moving Datum Names

Next you will create a section to be used when detailing the model in the drawing mode. The section will pass through the part lengthwise using datum plane **D**. The section will be named **A** and will show as **Section A-A** when detailing the view in the **Drawing Mode**. Figure 4.34 shows the section.

HINT

Create sections in the **Part Mode** to be used later in the **Drawing Mode** when detailing the part model.

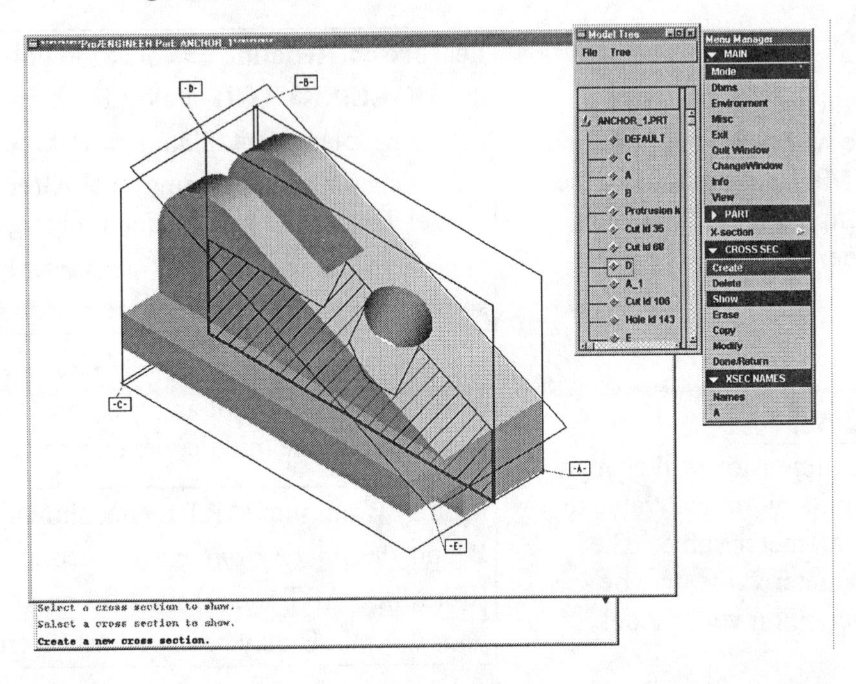

Figure 4.34
Creating a Section

Give the following commands to create the section shown in Figure 4.34:

> **X-section** (from the PART menu) ⇒ **Create** ⇒ **Planar** ⇒ **Single** ⇒ **Done** ⇒ (enter the **NAME** for the cross section at the prompt: **A)** ⇒ **enter** ⇒ (select the planar surface of datum plane: **D)** ⇒ **Done-Return**

The section passes through the slot and the hole, but it doesn't pass through the center of the hole. Your "boss" has a "suggestion" and provides you with the following ECO (Fig. 4.35).

ECO

Pro/ENGINEER Drawing: DRW0001

PARAMETRIC TECHNOLOGY CORPORATION

ECO Form

Form Name.
eco.name

Requestor:
eco.initr

Request Date:
eco.date

Current Submission Status:
eco.stat

Request List	Approval List	Notify List	Objects attached for reference	Description of Change

-D- -B- -C- -A- AULT

Action	User	Date	Comment

1. Change the dimensioning reference for the **1.00** diameter hole to be **.00** from datum plane **D** instead of **.875** from datum plane **B**.
2. Write a relation to keep datum plane **D** centered exactly on the upper portion of the part.
3. Add a **.250** diameter hole coaxially through the cylindrical surface.

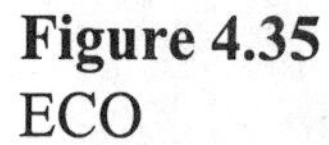

Figure 4.35
ECO

Give the following commands to change the dimensioning reference for the hole:

> **Feature** ⇒ **Redefine** ⇒ (pick the hole) ⇒ (pick **PlacemntRefs** from the **HOLE:SKETCH** dialog box) ⇒ **Define** ⇒ **Same Ref** (to keep the same placement plane) ⇒ **Same Ref** (to keep the same edge reference for dimensioning) ⇒ **Alternate** (pick datum **D** from the Model Tree or the part model to change the second reference-- if asked, "Align feature to reference? [N]" press **enter** key**)** ⇒ (type **0** at the prompt, as shown in Figure 4.36) ⇒ **enter** ⇒ **OK** ⇒ **Done**

HINT
If you answer yes (**Y**) the hole will be aligned to **DTMD** and you will not be prompted for a dimension value.

The hole (and the slot) is now a ***child*** of datum **D**. If datum **D** moves, so will the slot and the hole. In order to ensure that the plane stays through the middle, create a relation to control its location:

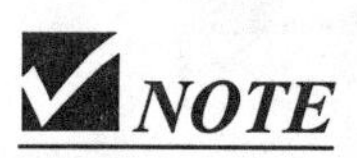

NOTE
The dimension will change from its numerical value to its parameter value. The parameter value may be different on your model.

> **Modify** (from the PART menu, shown in Figure 4.37) ⇒ (pick datum **D** and the *upper front cut* to display the dimensions) ⇒ **Relations** (from the PART menu) ⇒ **Add** ⇒ (type **d18=d12/2** at the prompt-- your **d** symbols may be different) ⇒ **enter** ⇒ **enter** ⇒ **Done**

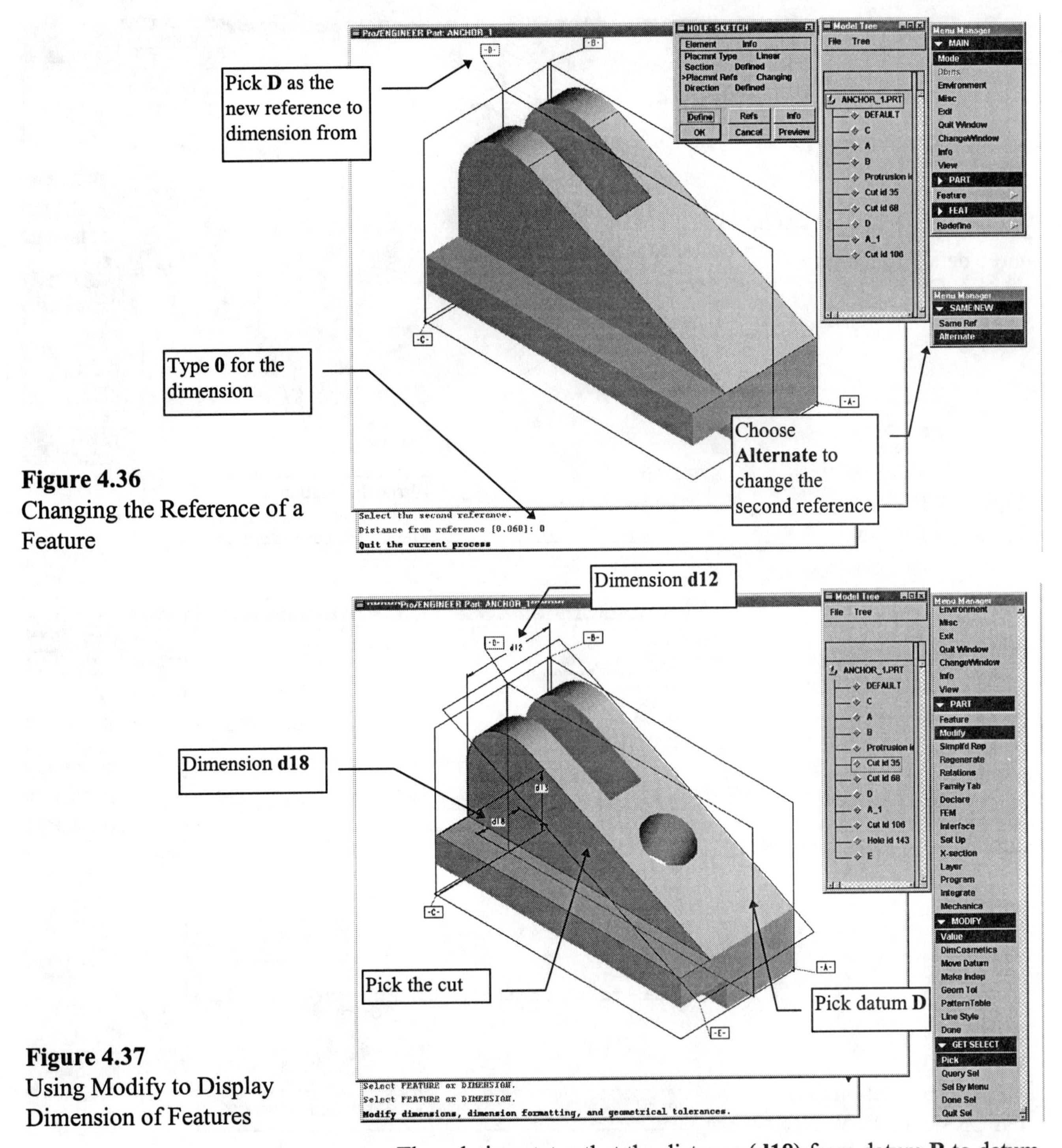

Figure 4.36
Changing the Reference of a Feature

Figure 4.37
Using Modify to Display Dimension of Features

The relation states that the distance **(d18)** from datum **B** to datum **D** will be one-half the value of the distance from datum **B** to the cut surface **(d12)**. If the thickness of the upper portion of the part **1.875** changes, then datum **D** will remain centered, as will the slot cut and the **1.00** diameter hole. If your *dimension symbols* do not show on the screen after selecting them during the **Modify** command (Fig. 4.37), pick **Switch Dim** under the **Relations** command (Fig. 4.38). To see the new relation, select **Show Rel** before completing the command (Fig. 4.39) or choose **Relations** ⇒ **Show Rel** ⇒ (type **q** to quit the INFORMATION WINDOW) ⇒ **Done**.

NOTE
Relations will be covered in more detail in Lesson 8.

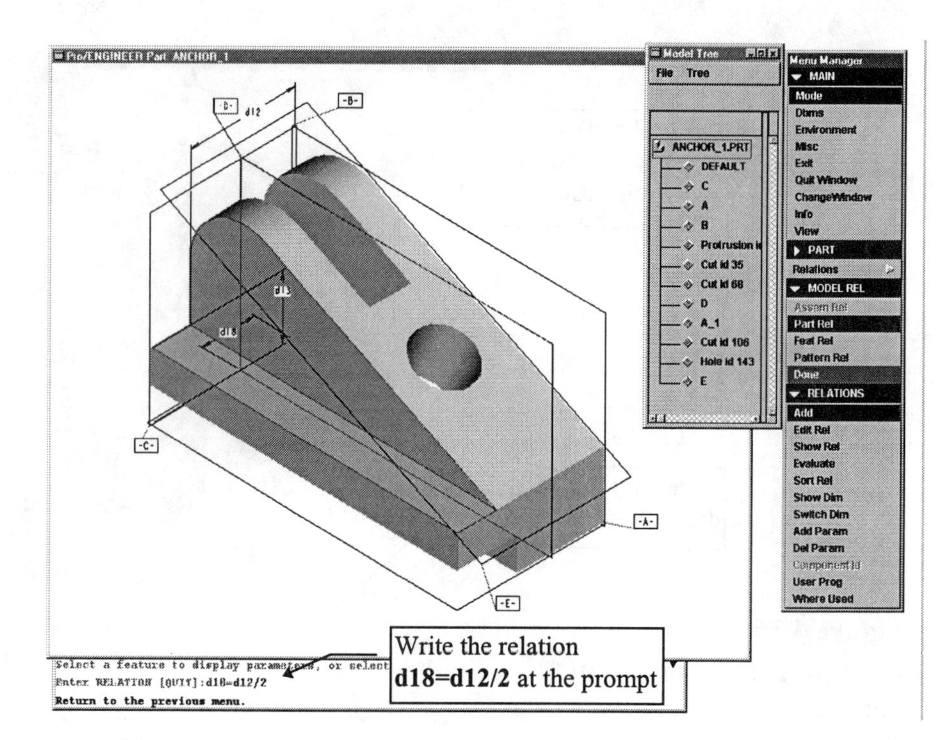

Figure 4.38
Writing a Relation

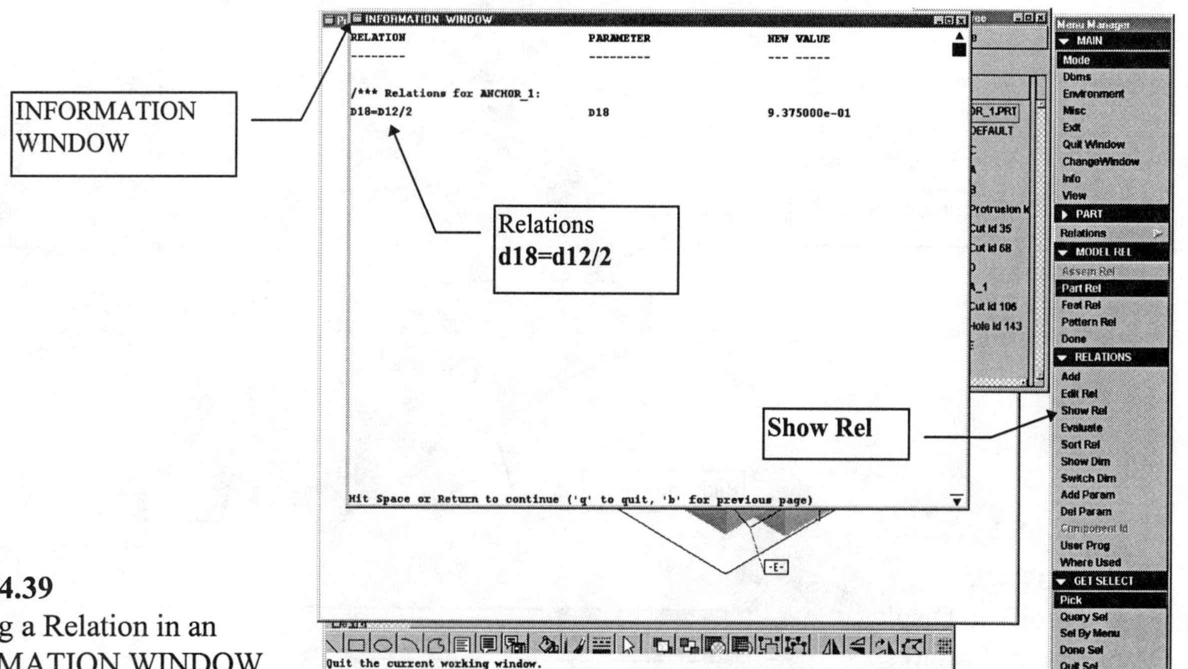

Figure 4.39
Showing a Relation in an INFORMATION WINDOW

The last feature to create is a **.250** diameter hole to be placed coaxially with **A_1**. Give the following commands:

Feature ⇒ **Create** ⇒ **Hole** ⇒ **Straight** ⇒ **Done** ⇒ **Coaxial** ⇒ **Done** ⇒ (pick on the axis line **A_1**) ⇒ (select the placement plane- datum **D**) ⇒ **Okay** ⇒ **Both Sides** ⇒ **Done** ⇒ **Thru All** ⇒ **Done** ⇒ **Thru All** ⇒ **Done** ⇒ (enter the diameter of **.250** at the prompt) ⇒ **enter** ⇒ **Preview** (Fig. 4.40) ⇒ **OK** ⇒ **Done**

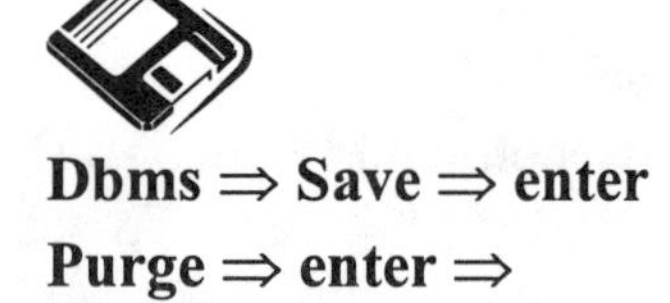

Dbms ⇒ **Save** ⇒ **enter**
Purge ⇒ **enter** ⇒
Done-Return

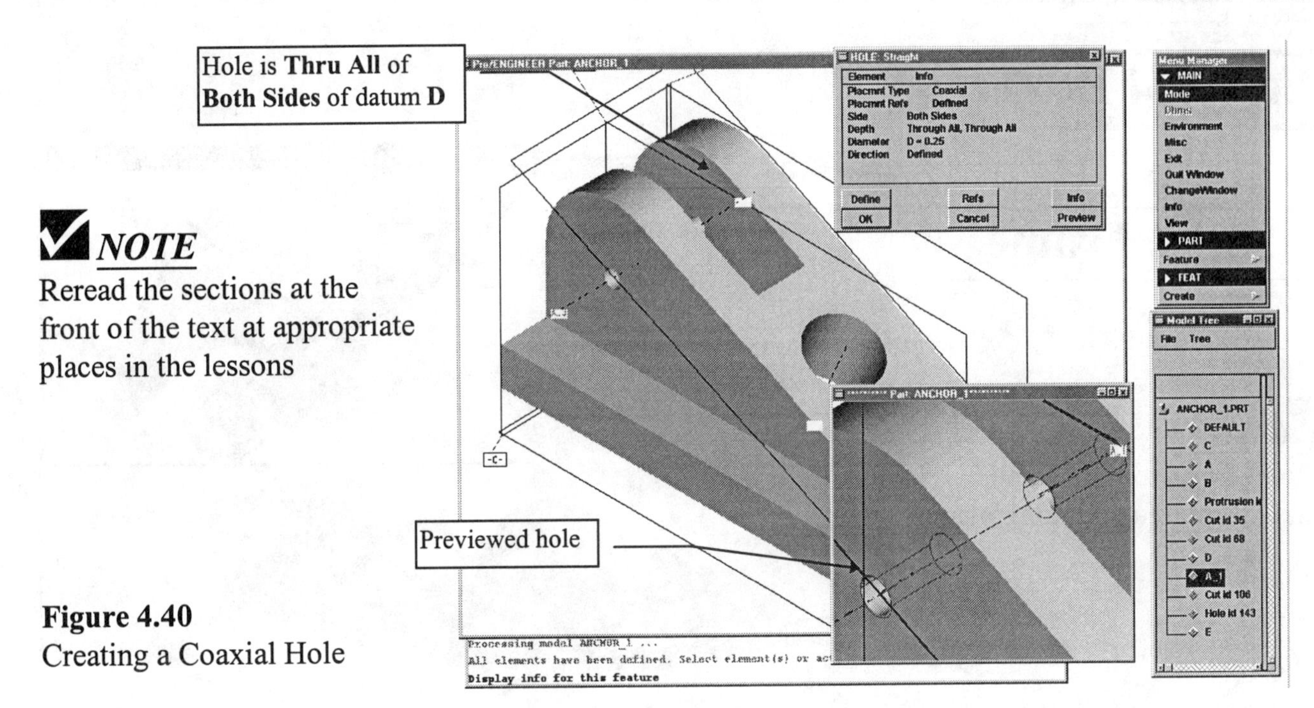

NOTE

Reread the sections at the front of the text at appropriate places in the lessons

Figure 4.40
Creating a Coaxial Hole

Create a layer and add the datum planes to it (reread Section 8, Layers, at this time) using the following commands:

Dbms ⇒ **Save** ⇒ **enter**
Purge ⇒ **enter** ⇒
Done-Return

Layer (from the **PART** menu) ⇒ **Setup Layer** ⇒ **Create** ⇒ (type name of new layer at prompt: **DATUM_LAYER)** ⇒ **enter** ⇒ **enter** ⇒ **Set Items** ⇒ **Add Items** ⇒ (check **✓DATUM_LAYER)** ⇒ **Done Sel** ⇒ **Datum Plane** (from LAYER OBJECT menu) ⇒ **Sel By Menu** ⇒ **Name** ⇒ (highlight all, or pick **Sel All-- A, B, C, D, E)** ⇒ **OK** ⇒ **Done Sel** ⇒ **Done-Return** ⇒ **Info** (from MAIN menu) ⇒ **Layer Info** ⇒ **Disp Status** ⇒ (check **✓DATUM_LAYER)** ⇒ **Done Sel** (Fig. 4.41) ⇒ (type **q)** ⇒ **Done-Return** (three times)

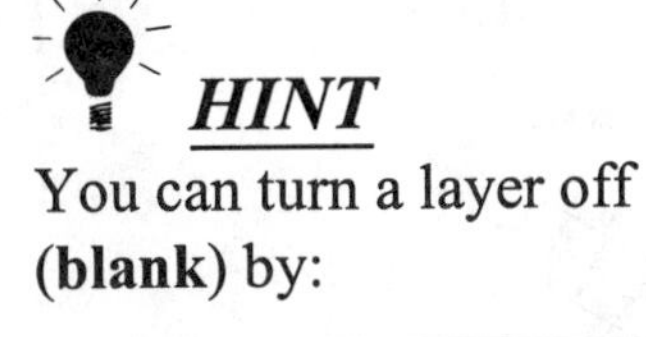

HINT

You can turn a layer off (**blank**) by:

Set Display ⇒ **Blank** ⇒
✓ Default Dat ⇒ **Done Sel**
⇒ **View** ⇒ **Repaint** ⇒
Done-Return

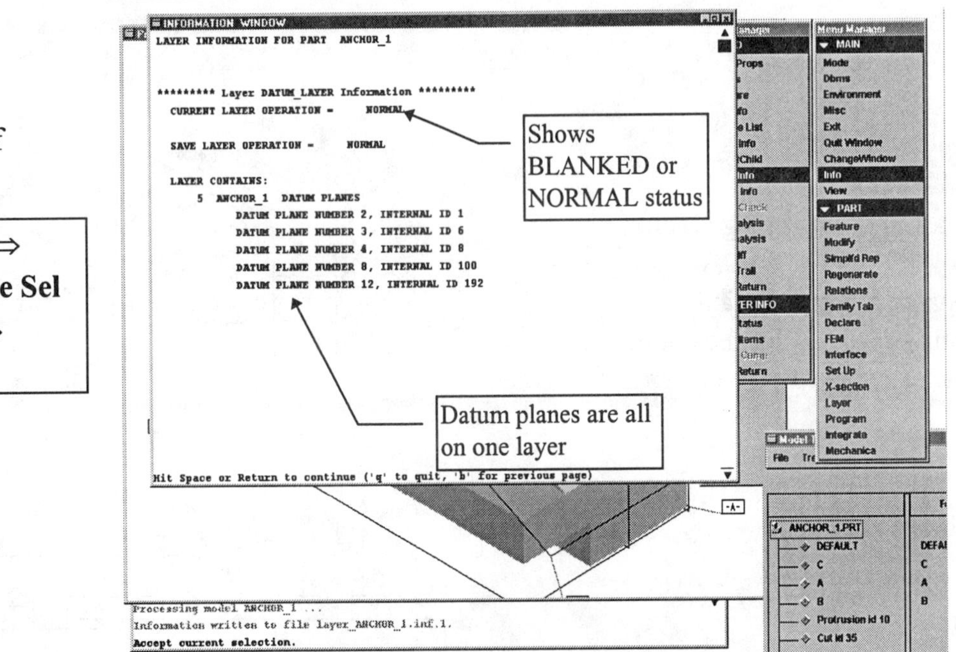

Figure 4.41
Layer Information

Lesson 4 Project

Angle Frame

NOTE

Don't forget to set the units and material (aluminum) for the part.

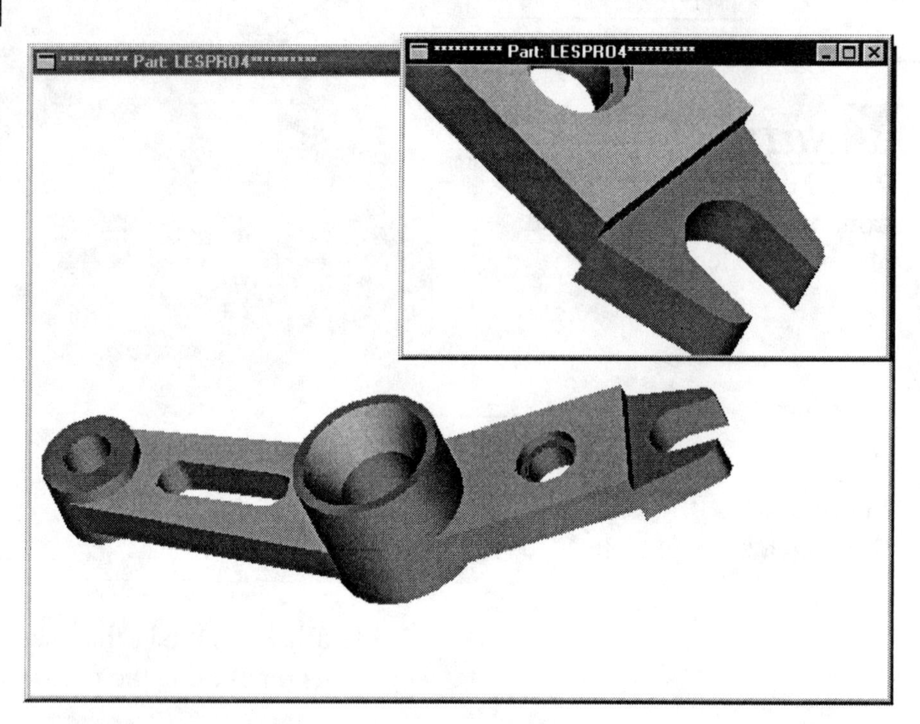

Figure 4.42
Angle Frame

EGD REFERENCE

Engineering Graphics and Design
By L. Lamit and K. Kitto
See Pages: 311, 461, 548, and 929

Angle Frame

The fourth **lesson project** is a machined part that requires the use of a variety of datum planes and a layering scheme. You will also add a relation to control the depth of the large countersink hole at the part's center. Analyze the part and plan out the steps and features required to model it. Use the **DIPS** in Appendix D to establish a feature creation sequence before you start modeling. Create the part shown in Figures 4.42 through 4.51.

NOTE

Set the datums using **Geom Tol**, rename **A**, **B**, **C**, etc. to rename all three default datum planes. Put the datum planes on a separate layer.

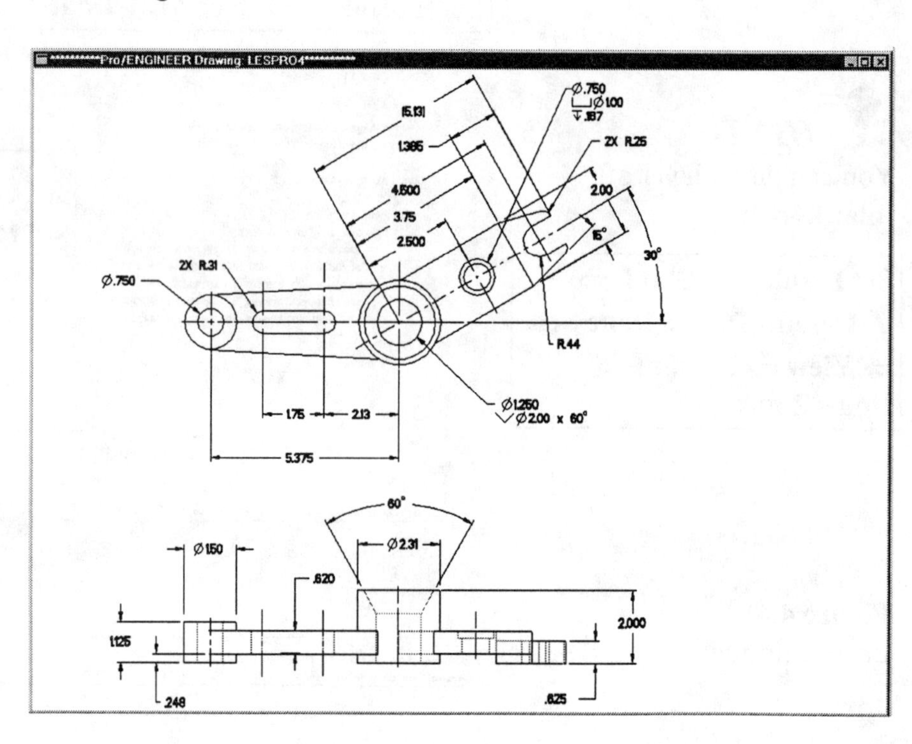

Figure 4.43
Angle Frame Drawing

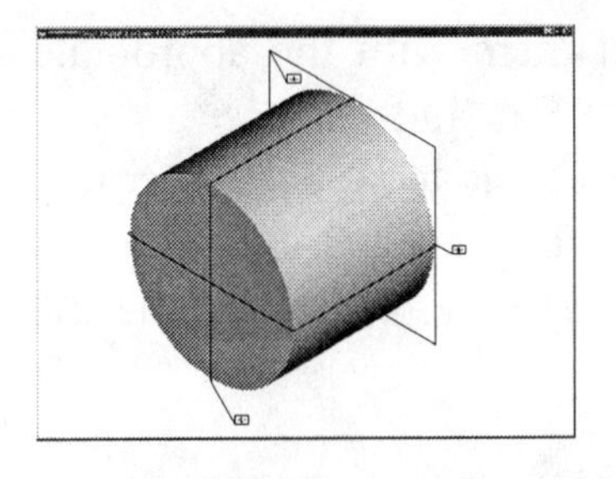

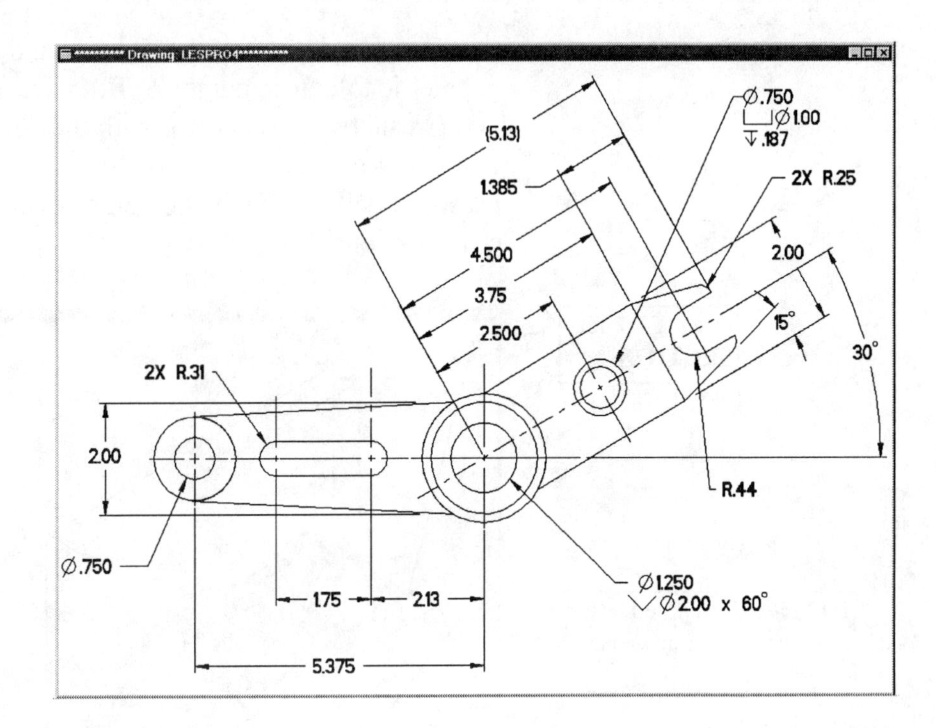

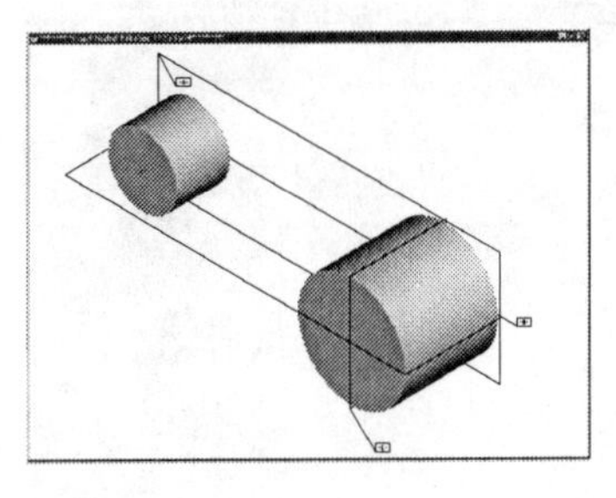

Figure 4.44
Angle Frame Top View

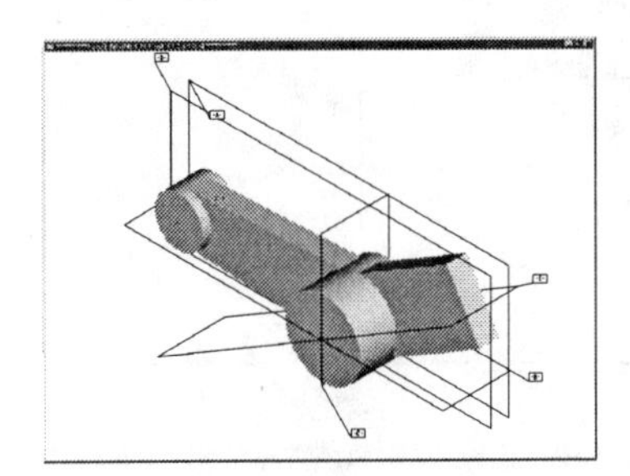

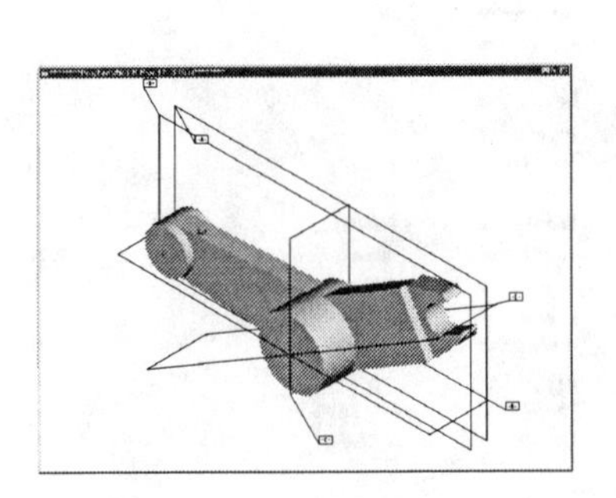

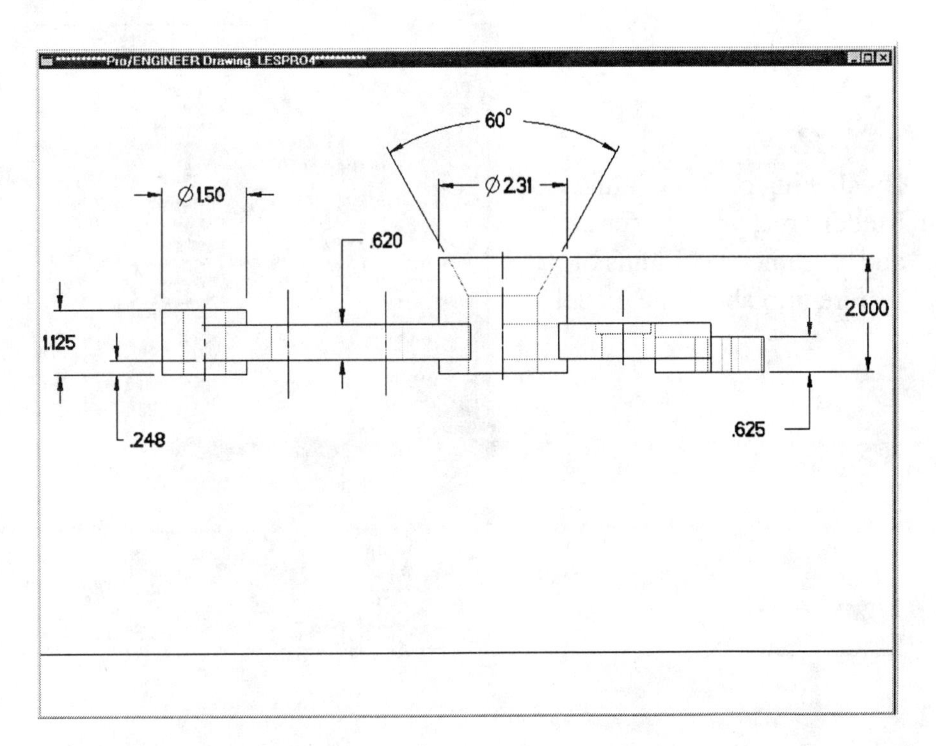

Figure 4.45
Angle Frame Front View

Create a layer for the datums and set them with the appropiate geometric tolerance name: **A**, **B**, **C**, etc. (see Fig. 4.49).

Create two *sections* through the Angle Frame to be used later in a Drawing Lesson. For the sections, use datum planes **B** and **E** which pass vertically through the center of the part (Fig. 4.51). Name the cross sections **A** and **B**.

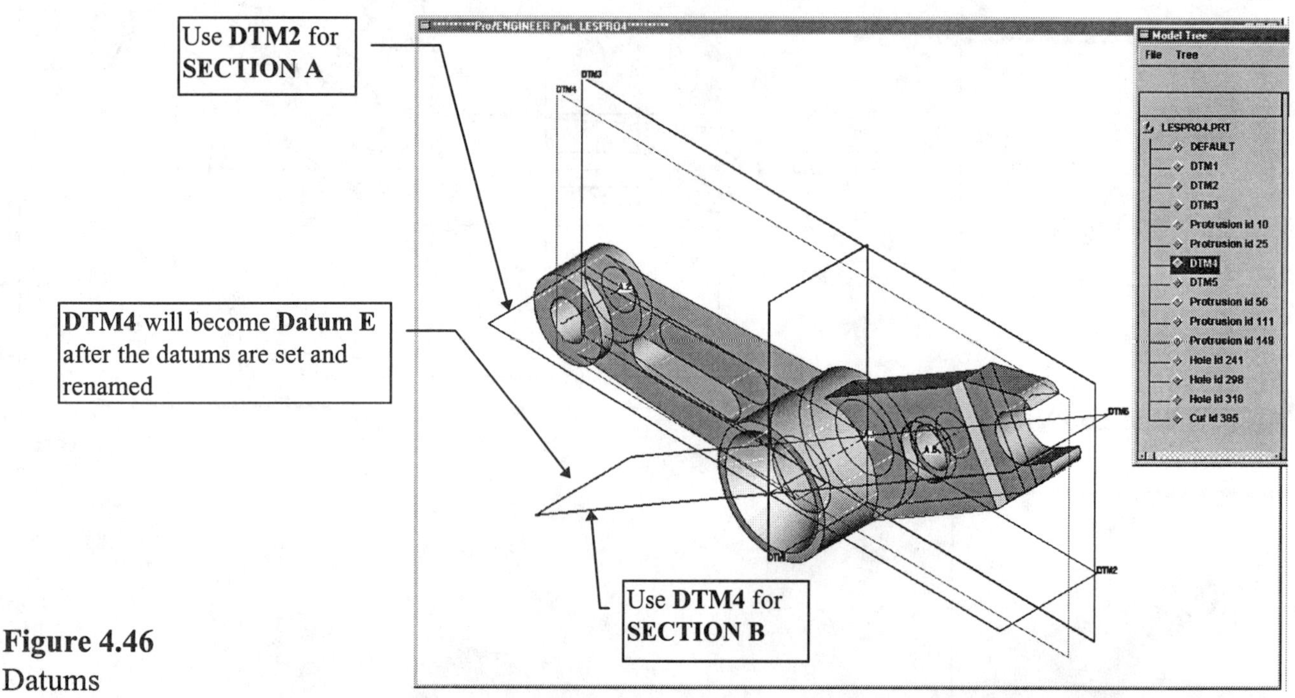

Figure 4.46
Datums

HINT
By clicking on **Tree** in the Model Tree, you can format, add, or remove columns of feature info about the model

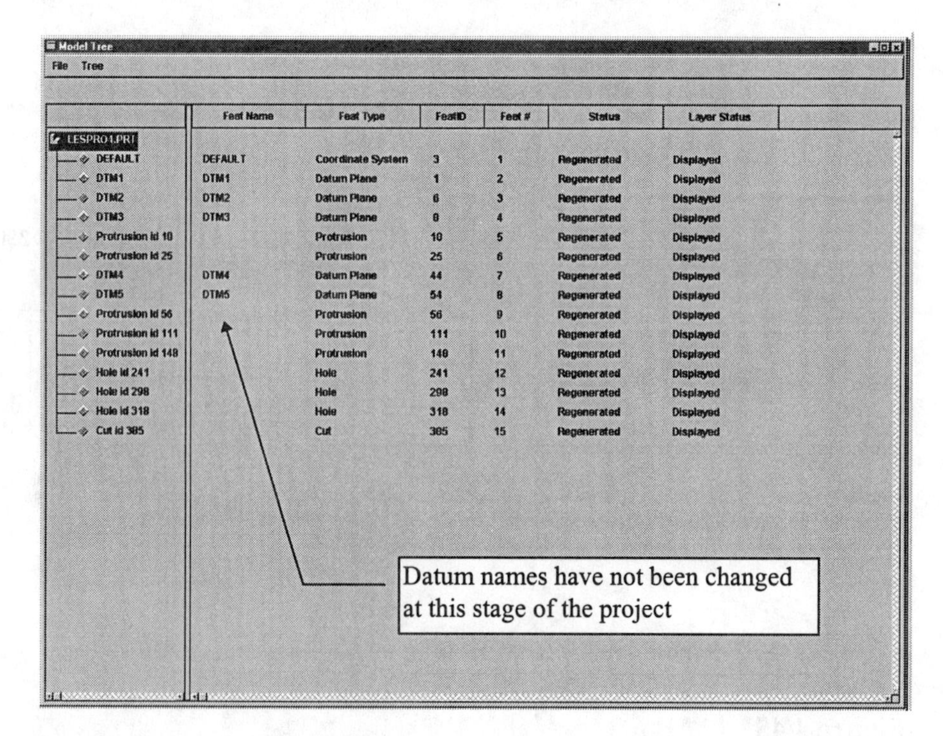

Figure 4.47
Model Tree

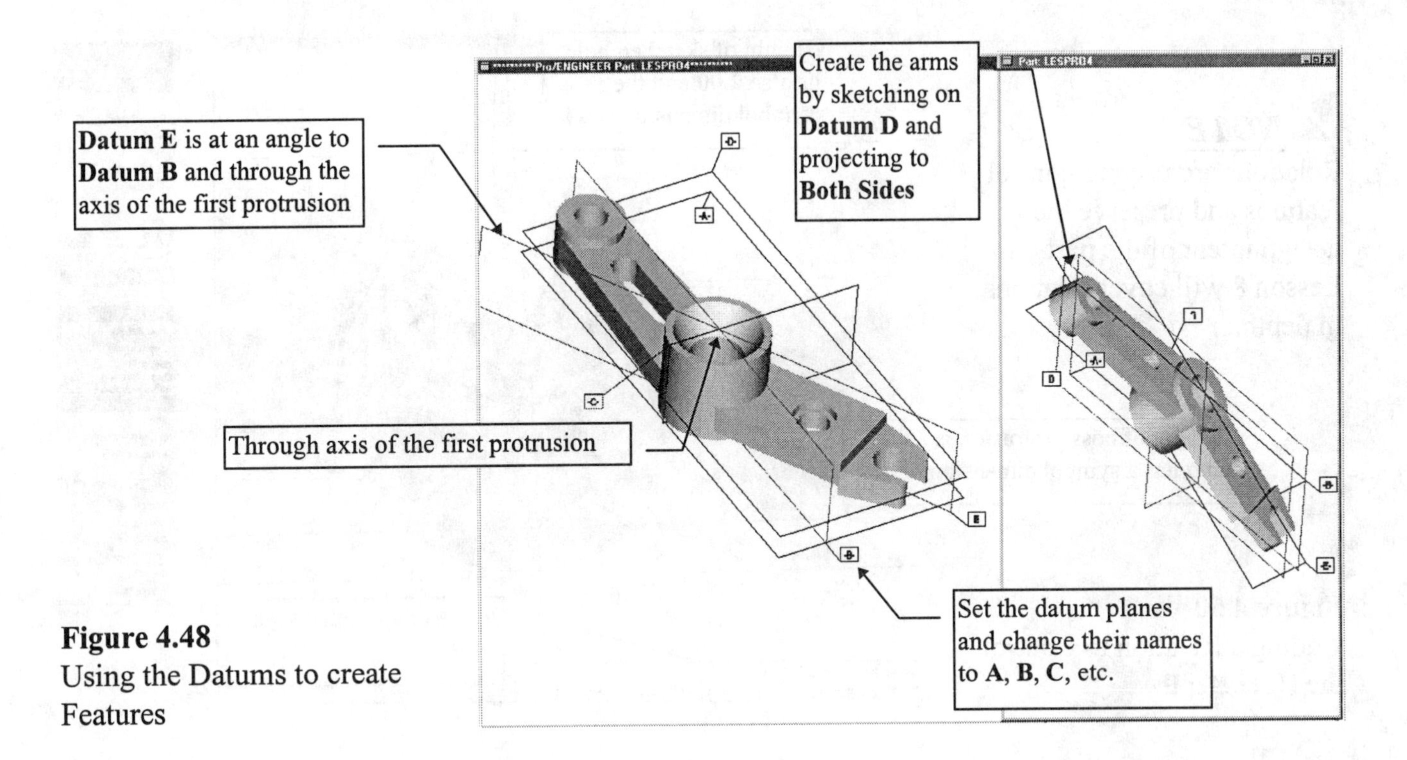

Figure 4.48
Using the Datums to create Features

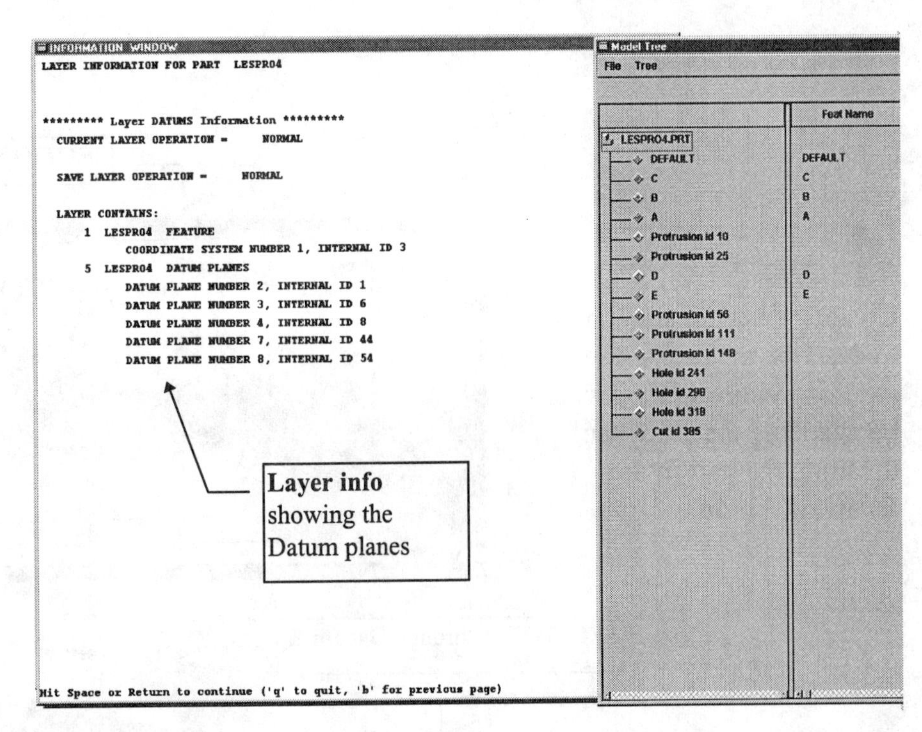

Save as you model!
Dbms ⇒ Save ⇒ enter
Purge ⇒ enter ⇒ Done-Return

Figure 4.49
Layer Info

Modify the thickness of the boss from **2.00** to **2.50**. Note that the hole does not go through the part. Modify the boss back to the original design dimension of **2.00**. Add a relation to the hole that says the depth of the hole should be equal to the thickness of the boss (**d43=d6**), as shown in Figure 4.50).

Now change the thickness of the boss (original protrusion) to see that the hole still goes through the part. No matter what the boss thickness dimension changes to, the hole will always go completely through it. This relation controls the *design intent* of the hole.

HINT
Your **d#** symbols will probably be different than the ones shown here.

NOTE

Relations are used to control features and preserve the design intent of the part. Lesson 8 will cover relations in depth.

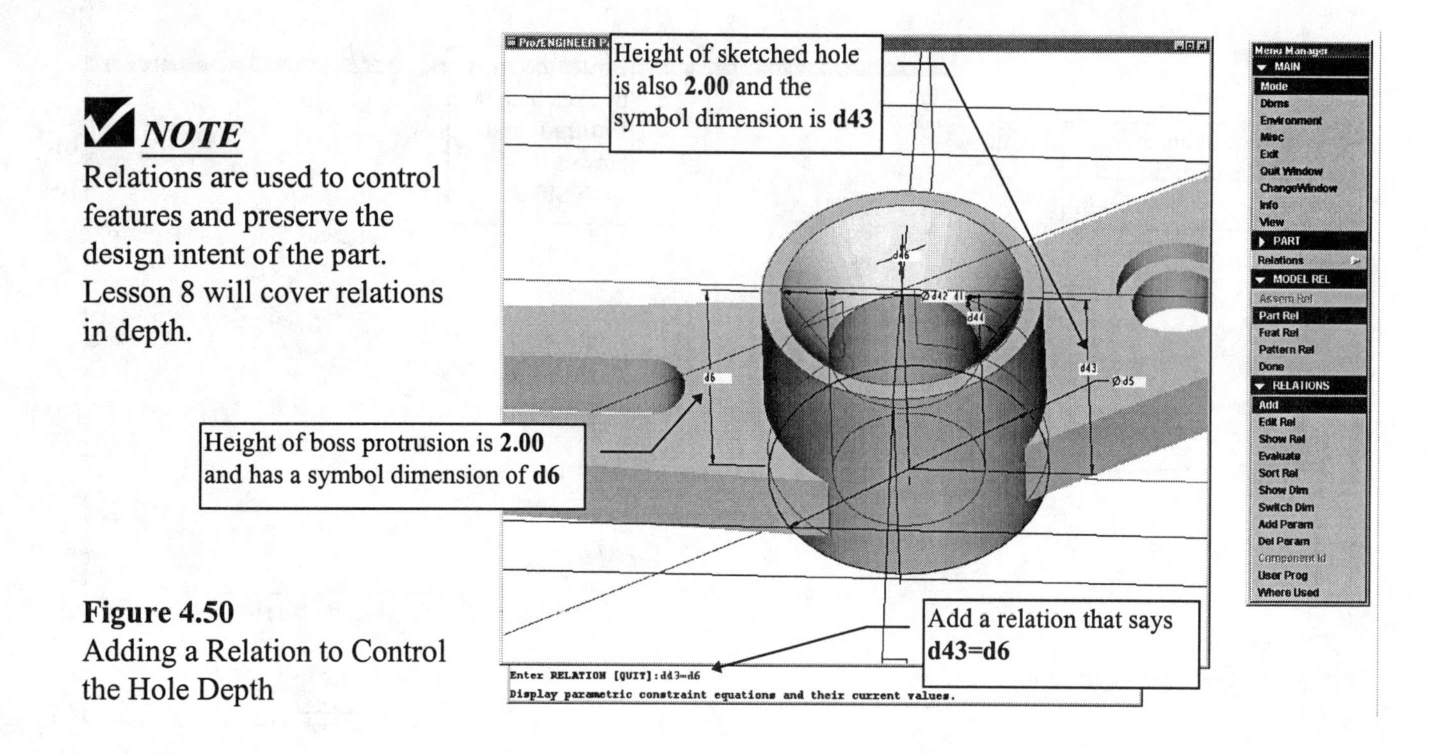

Figure 4.50
Adding a Relation to Control the Hole Depth

HINT

Create the sections required to describe the part while in the **Part Mode** so they will be available for use when detailing the part in the **Drawing Mode**.

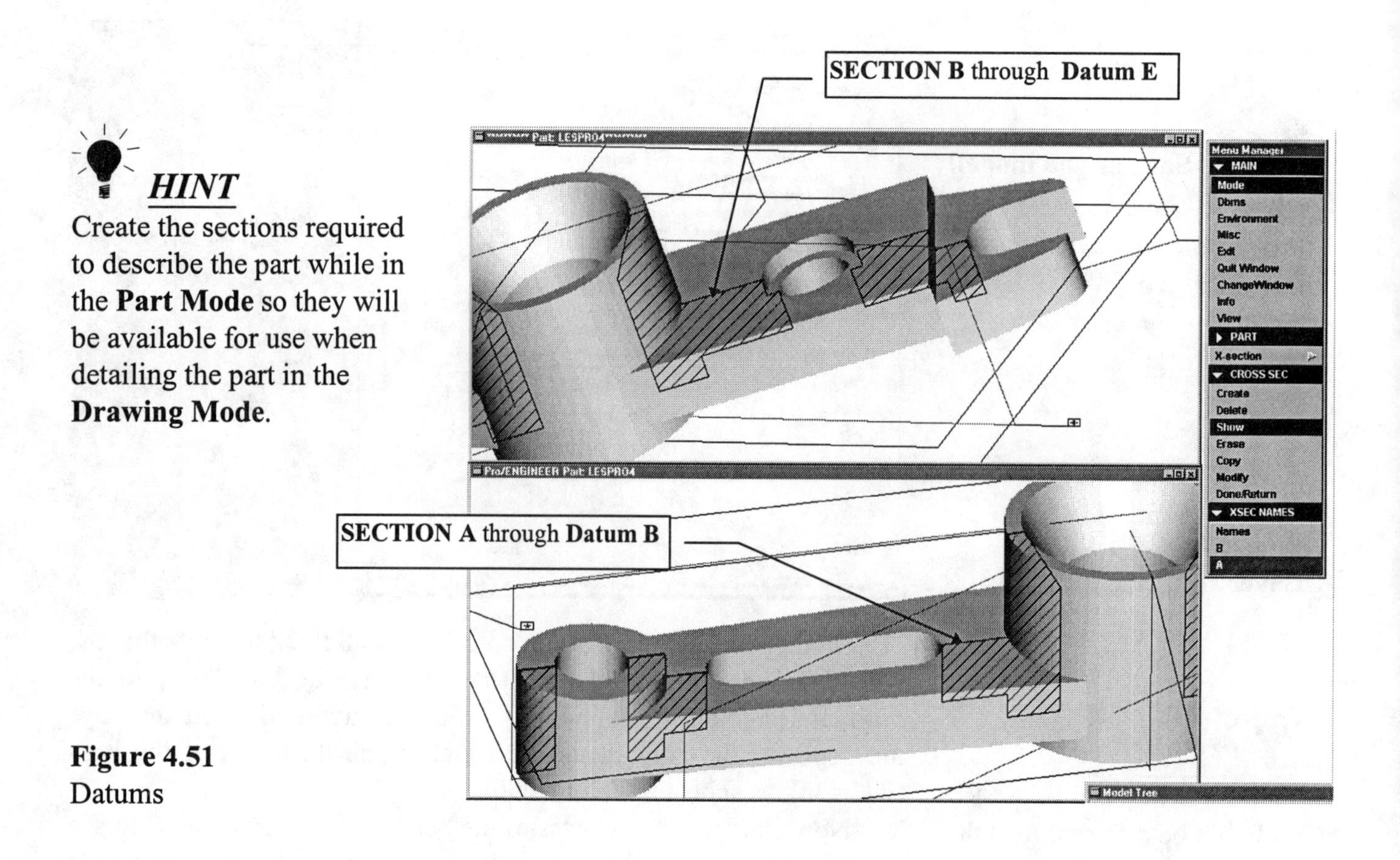

Figure 4.51
Datums

Lesson 5

Revolved Protrusions and Neck Cuts

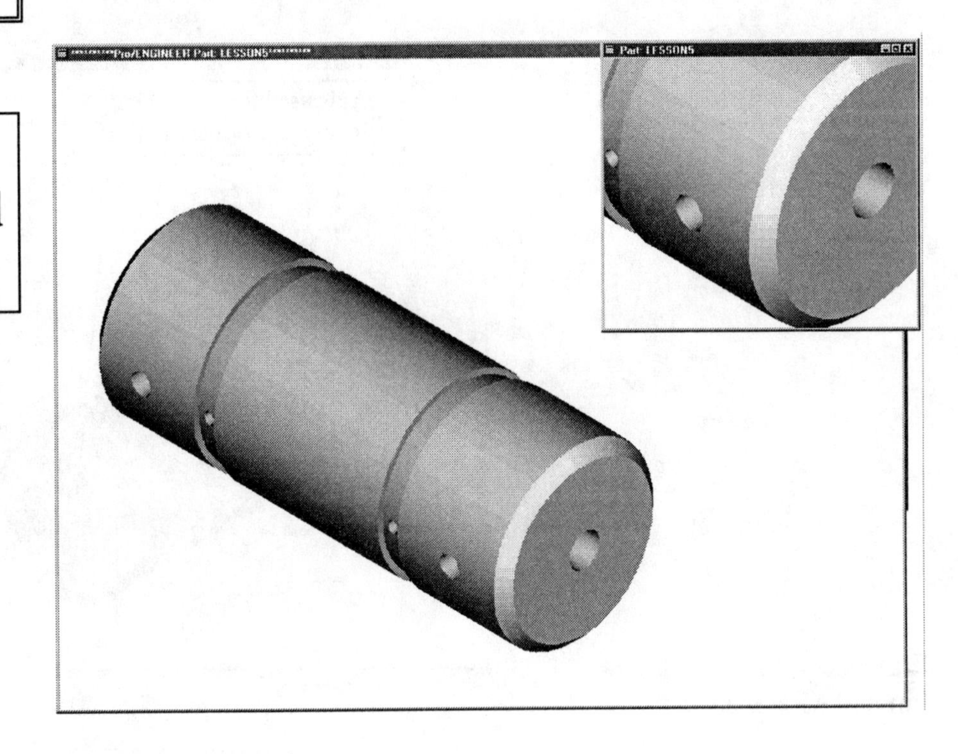

Figure 5.1
Pin

OBJECTIVES

1. **Create a simple revolved protrusion**

2. **Understand the angle options used to create revolved features**

3. **Use datums to locate holes**

4. **Cut necks in revolved protrusions**

5. **Create a conical revolved cut**

6. **Use the Info command to measure a revolved feature**

7. **Get a hard copy using the Interface command**

EGD REFERENCE
Engineering Graphics and Design
by L. Lamit and K. Kitto
Read Chapters: 8, 14, and 26
See Pages: 239, 486, and 956-957

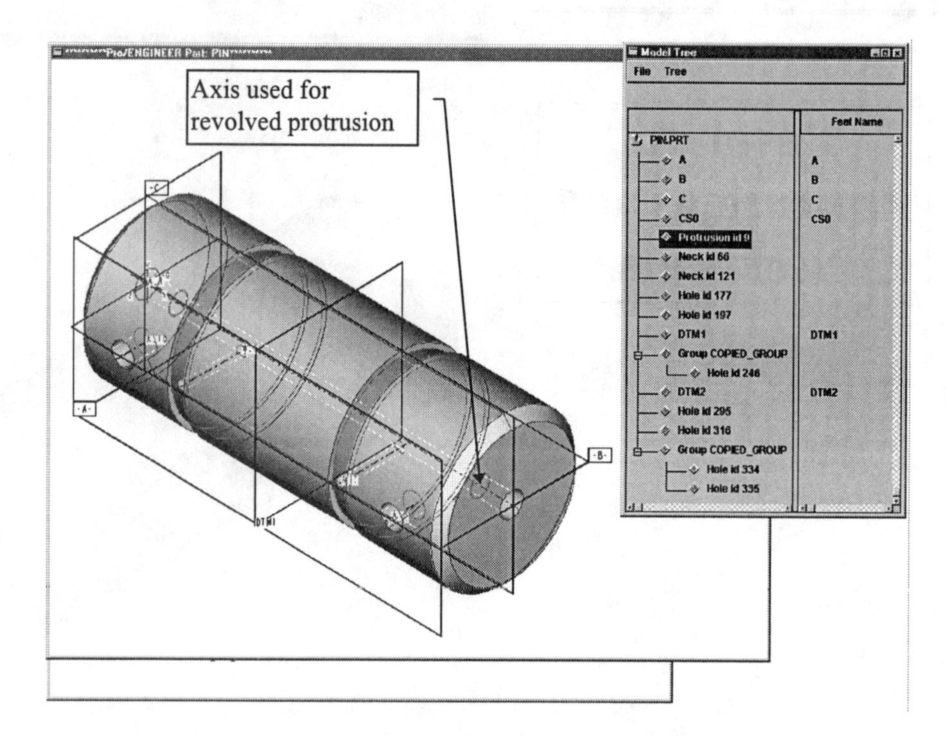

Figure 5.2
Pin and Model Tree

REVOLVED PROTRUSIONS AND NECK CUTS

The **revolve** option creates a feature by revolving the sketched section around a centerline from the sketching plane into the part (Fig. 5.1). You may have any number of centerlines in your sketch, but the first centerline will be the one used to rotate your section geometry.

When sketching the feature to be revolved, the first centerline sketched is the *axis of revolution* (Fig. 5.2). The section geometry must lie completely on one side of this centerline and must be closed (Fig. 5.3).

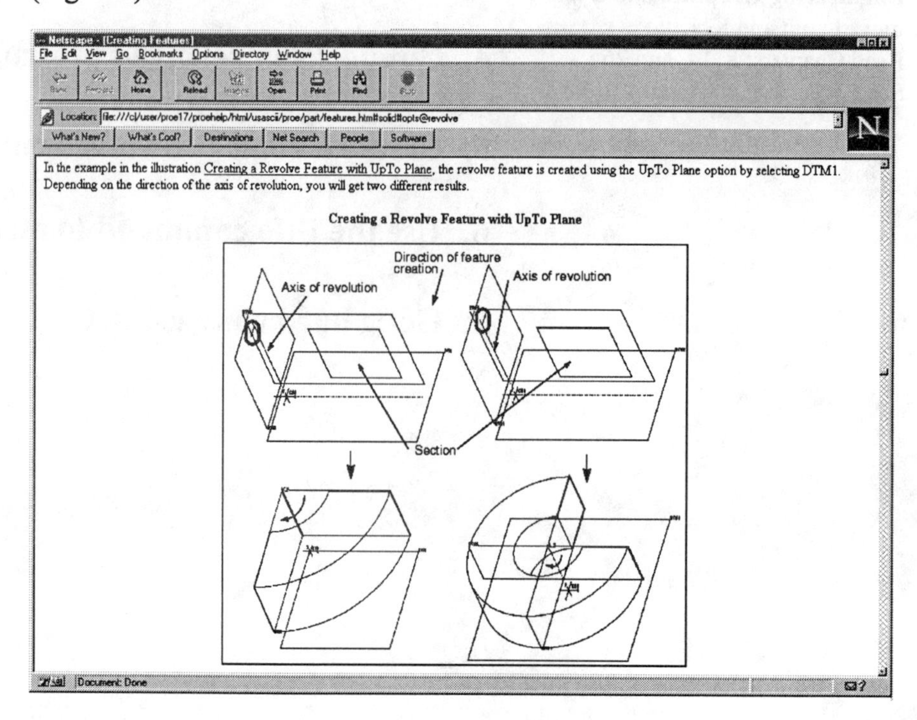

Figure 5.3
Online Documentation
Revolved Protrusions

A revolved feature can be created either entirely on one side of the sketching plane or symmetrically to both sides of the sketching plane. The **One Side** and **Both Sides** options are available for any but the first feature. If you choose **Both Sides**, the feature will be revolved symmetrically in each direction for one-half of the angle specified in the REV TO menu or the variable angle of revolution that you enter when prompted.

After successfully regenerating the revolved section, select **Done** and the REV TO menu appears. This menu allows you to specify the feature's angle of revolution and whether that angle is to be measured entirely on one side of the sketching plane or symmetrically on both sides of the sketching plane. You can choose the **Variable** option for a user-defined angle of revolution, or you can choose from one of four preset angles: **90**, **180**, **270**, and **360**.

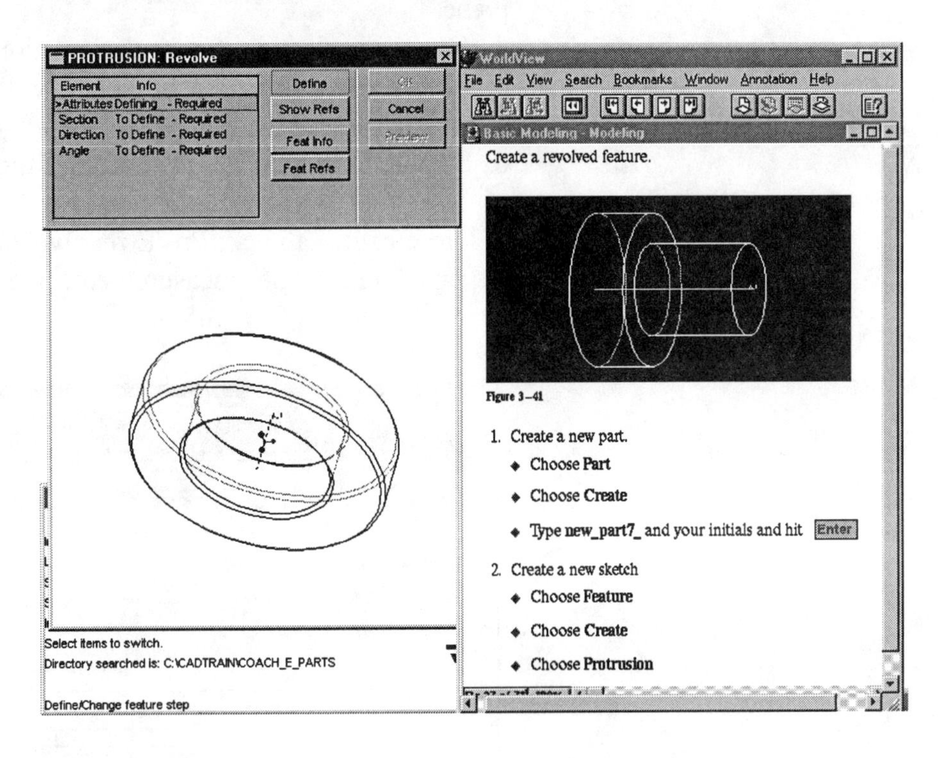

Figure 5.4
COAch for Pro/E, Creating a Basic Model-- Modeling (Segment 4: Revolved Protrusions)

If you choose **Variable**, the angle may be specified and modified after the section is created. This angle must be greater than **0°** and less than **360°**. The angle is controlled by a dimension that appears when modifying the part and in drawings. A corresponding dimension will not appear if a preset angle is chosen. The base feature of the pin was created with a **360°** revolved section.

If you have **COAch for Pro/ENGINEER** on your system, go to SEARCH and do the appropiate segment, shown in Figure 5.4.

Neck Cuts

A **Neck** is a special type of *revolved slot* that creates a groove around a revolved part or feature. You always create a neck on a **Through/Axis** datum plane, and the sketch is inside the part. You might align both ends of the section to the revolved surface of the parent feature.

To create a neck:

1. Choose **Neck** from the SOLID menu.
2. Choose an option from the OPTIONS menu to specify the number of degrees in the revolution.
3. Create or select a **Through/Axis** datum plane as the sketching plane.
4. Create or select a reference/orientation plane
5. Sketch the neck crossection open, with the ends aligned to the silhouette edge of the part or feature.
6. Sketch the centerline that becomes the axis of revolution.

In creating a neck, Pro/E revolves the section around the part to the specified angle measurement, thereby removing the material inside the section (Fig. 5.5).

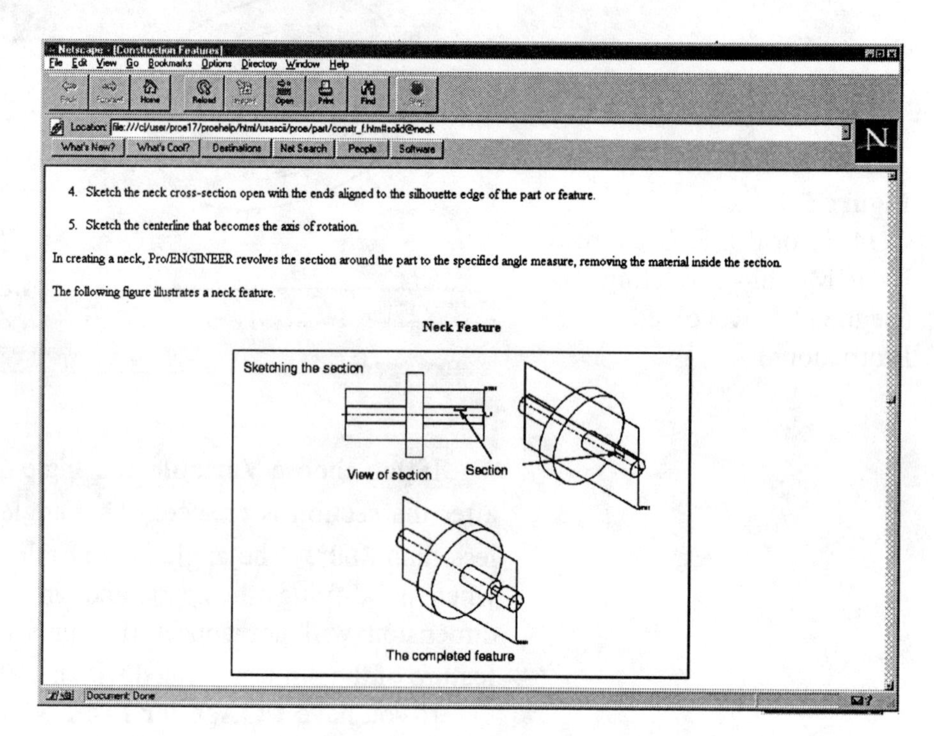

Figure 5.5
Online Documentation
Necks

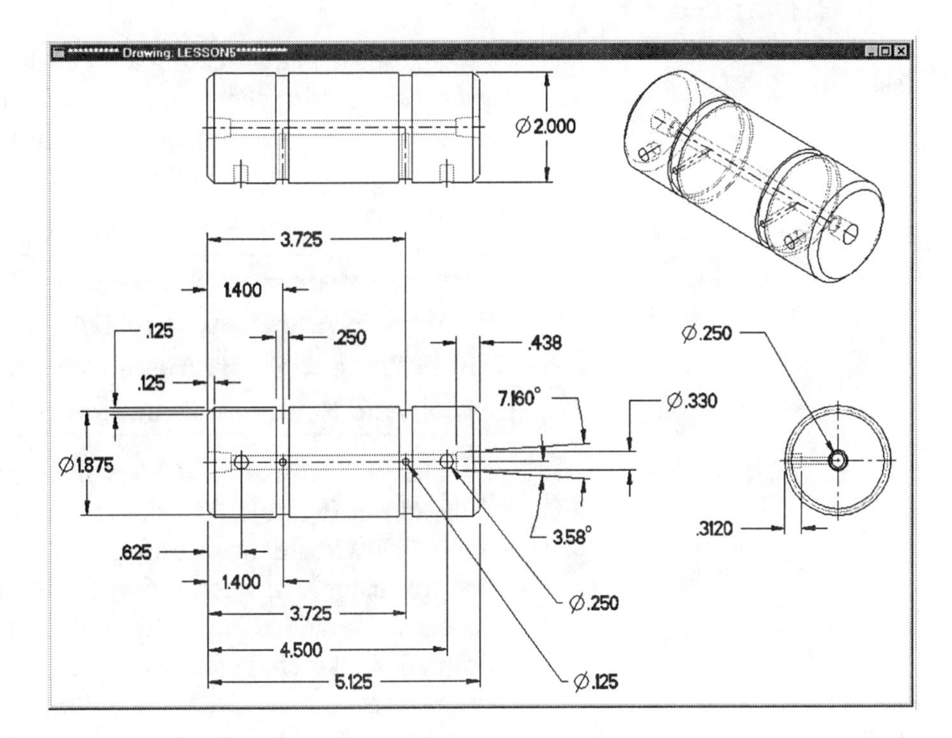

Figure 5.6
Pin Dimensions

Pin

The pin is an example of a part created by revolving one section about a centerline (Fig. 5.6). The pin was created as a **revolved protrusion**. The chamfers are created on the first revolved protrusion. The grooves were cut with the **Neck** command. The holes were added using datum axes and a new datum plane.

The pin's complete geometry (with the exception of the holes) could have been created with one revolved protrusion. In general, this is poor design practice since it limits the flexibility of modifying the geometry later in the design process. For most parts the basic revolved shape should be the first protrusion, followed by the most important secondary features (cuts, protrusions, etc.). The holes required for the part are then created. Lastly, the rounds and chamfers are created where required. Choose **Part** from the MODE menu and give the following commands:

Part ⇒ **Create** (type the part name, **Pin**) ⇒ **enter**

SETUP AND ENVIRONMENT

Setup ⇒ **Units** ⇒ **Inch** ⇒ **Material** ⇒ **Define** ⇒ (Type **Steel**, then press **Enter** key) ⇒ **Assign** ⇒ (pick **Steel**) ⇒ **Accept** ⇒ **Done**

Environment ⇒ ✓**Grid Snap**
Hidden Line Tan Dimmed

Feature ⇒ **Create** ⇒ **Datum** ⇒ **Plane** ⇒ **Default** ⇒ **Create** ⇒ **Datum** ⇒ **Coord Sys** ⇒ **Default** ⇒ **Done** ⇒ **Done**

Start the Pin by creating a set of datum planes and a coordinate system and putting them on a separate layer. Note that the datum and coordinate system command are slightly different than when using the **Offset** option. Give the following commands to create and set datums on the layer:

Layer ⇒ **Create** ⇒ (type **DATUM_LAYER**) ⇒ **enter** ⇒ **enter** ⇒ **Set Items** ⇒ **Add Items** ⇒ (**✓DATUM_LAYER**) ⇒ **Done Sel** ⇒ **Datum Plane** ⇒ **Sel By Menu** ⇒ **Name** ⇒ **Sel All** ⇒ **OK** ⇒ **Done Sel** ⇒ **Done-Return** ⇒ **Done/Return**

The first protrusion is a revolved protrusion. Use the front view of the pin in Figure 5.6 to sketch the revolved protrusion's section in Figure 5.7. Create the chamfers and the main body of the pin with the first protrusion. In Lesson 6 you will see that chamfers, like rounds, are normally added near the end of the modeling sequence. The commands, sketch, references, etc. are very similar to the previous lessons:

Feature ⇒ **Create** ⇒ **Solid** ⇒ **Protrusion** ⇒ **Revolve** ⇒ **Done** ⇒ **One Side** ⇒ **Done** ⇒ (pick **DTM3** as sketching plane) ⇒ **Okay** ⇒ **Top** (pick **DTM2** as the horizontal reference) ⇒ **View** ⇒ **Pan/Zoom** ⇒ **Pan** (pan the screen as shown) ⇒ **Done-Return** ⇒ **Line** ⇒ **Centerline** ⇒ **Horizontal** (sketch the horizontal centerline *first*) ⇒ **Mouse Sketch** (sketch the section geometry as shown in Figure 5.7) ⇒ **Alignment** (align the horizontal centerline with **DTM2**, the horizontal line with **DTM2**, and the vertical left side line with **DTM1**) ⇒ **Regenerate** (Fig. 5.7) ⇒ **Dimension** (Fig. 5.8) ⇒ **Regenerate**

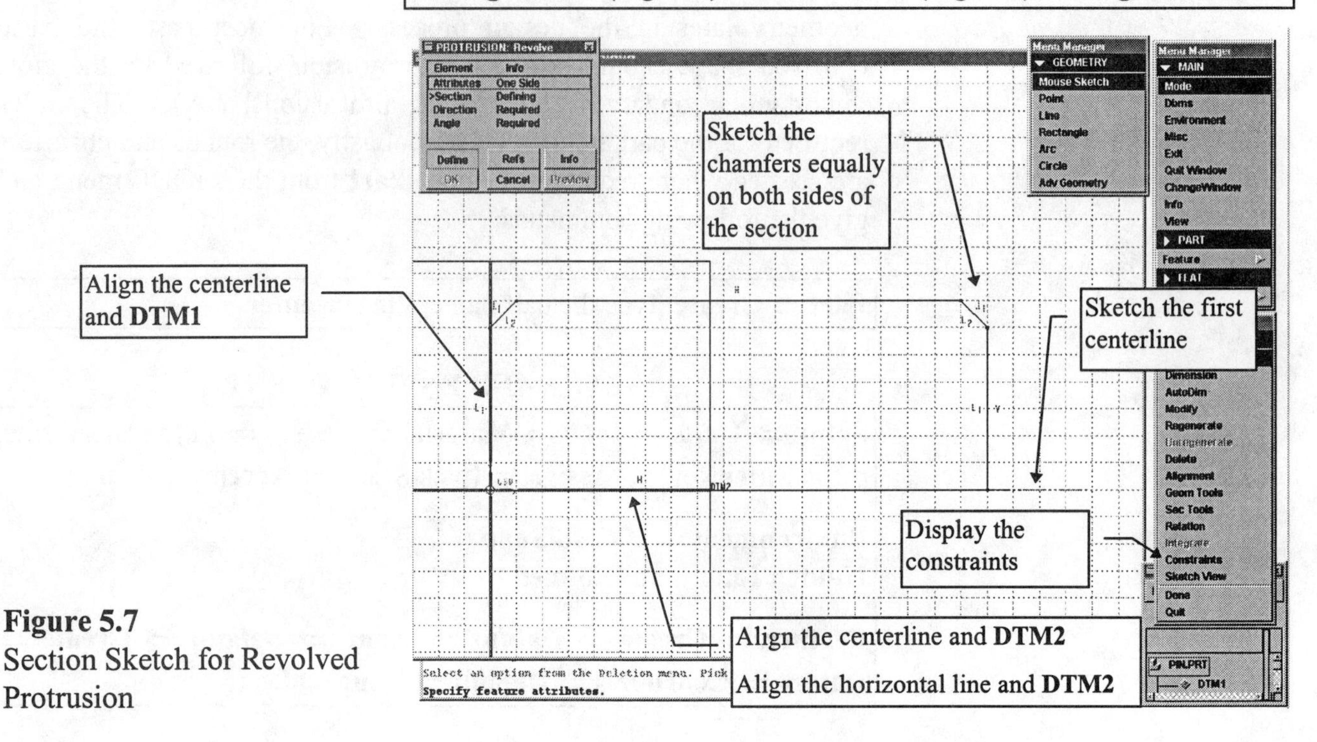

Figure 5.7
Section Sketch for Revolved Protrusion

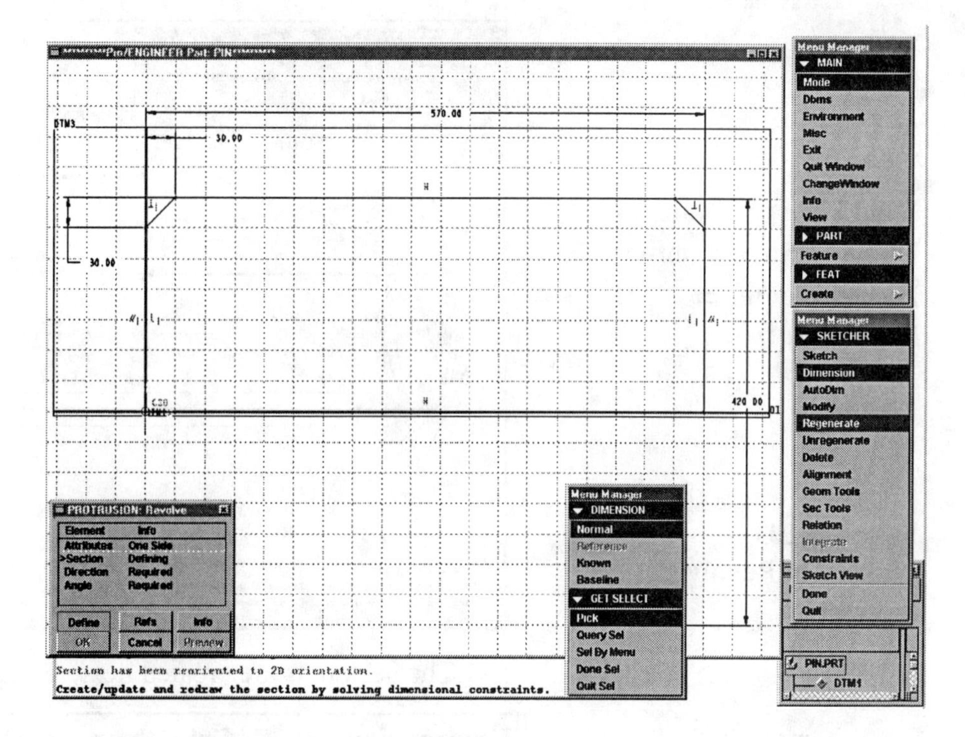

Figure 5.8
Dimensioned and Regenerated Section Sketch

Complete the sketch by choosing: **Modify** (change the dimensions to the design values shown in Figure 5.6) ⇒ **Regenerate**. If the section fails, **Unregenerate** immediately, then add and **Modify** the dimensions for the chamfer on the other side of the sketch and **Regenerate** again as in Figure 5.9).

HINT
If a section fails after you modify the dimensions, **Unregenerate** immediately.

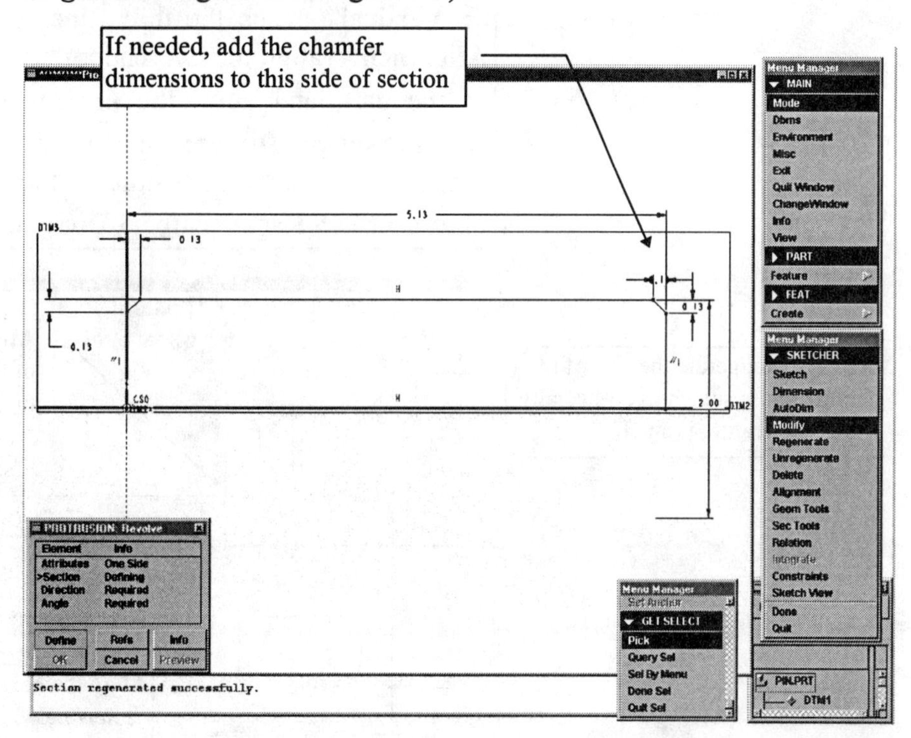

Figure 5.9
Regenerated Design Dimensions

Done ⇒ **360** (from the REV TO menu) ⇒ **Done** ⇒ **View** ⇒ **Default** ⇒ **Done-Return** ⇒ **Preview** (Fig. 5.10) ⇒ **OK** ⇒ **Done**

Dbms ⇒ Save ⇒ enter Purge ⇒ enter ⇒ Done-Return

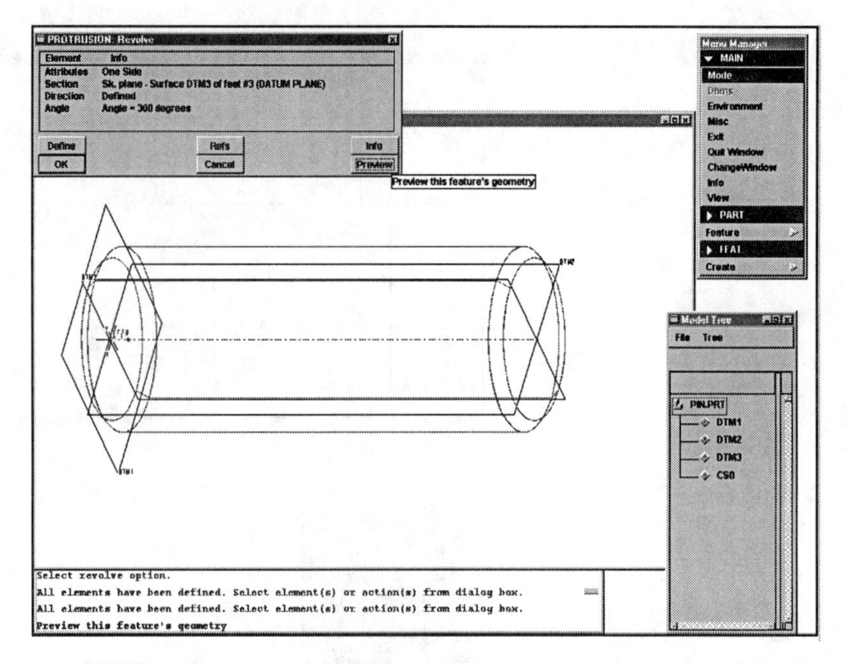

Figure 5.10
Preview of Revolved Protrusion

The neck cuts are created next. Exaggerate the neck's size:

Environment ⇒ □ **Grid Snap** ⇒ **Done-Return** ⇒ **Feature** ⇒ **Create** ⇒ **Neck** ⇒ **360** ⇒ **One Side** ⇒ **Done** ⇒ **Use Prev** ⇒ **Okay** ⇒ **Sketch** ⇒ **Line** ⇒ **Centerline** ⇒ **Horizontal** (sketch the centerline used to revolve the neck) ⇒ **Vertical** (sketch the vertical centerline to establish the middle of the neck cut) ⇒ **Sketch** ⇒ **Line** ⇒ **Vertical** (sketch the three lines of the neck) ⇒ **Regenerate** ⇒ **Alignment** (align the two endpoints of the vertical lines with the top of the part, and align the *horizontal* centerline with **DTM2**) ⇒ **Regenerate** ⇒ **Dimension** (Fig. 5.11) ⇒ **Regenerate** ⇒ **Modify** (change the sketch dimensions to the design dimensions as shown in Figure 5.6) ⇒ **Regenerate** ⇒ **Done** ⇒ **Done**

Always exaggerate the size of the feature when sketching, especially when the feature is small

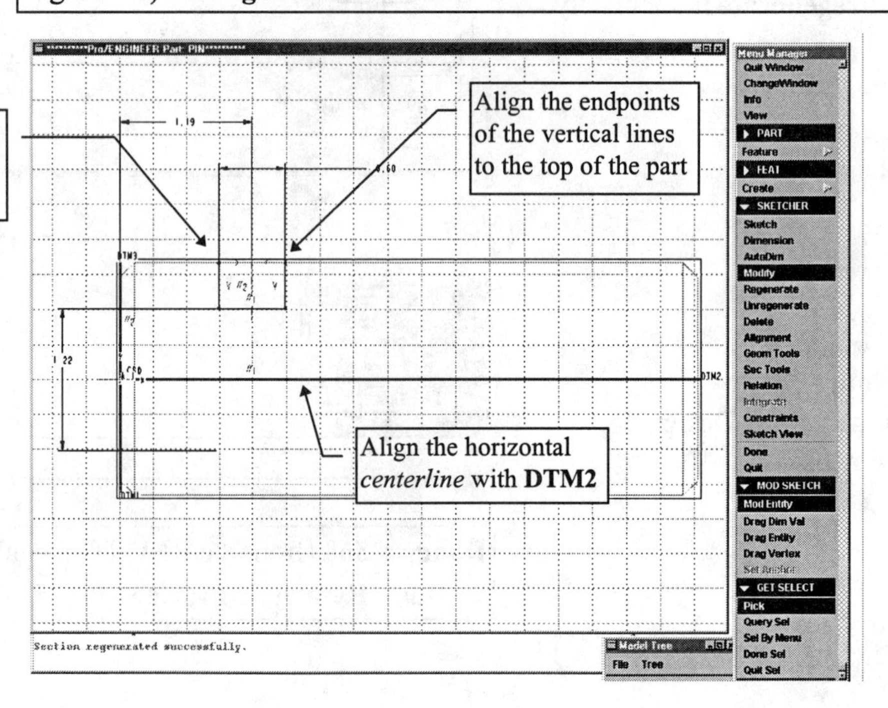

Figure 5.11
Neck Cut Sketch

The neck is now complete. Note that the neck command does not have a dialog box and that it is quicker to complete than a revolved protrusion (or a *revolved cut*, which is the same thing as a *neck*). Change the view, and shade and color the part (Fig. 5.12).

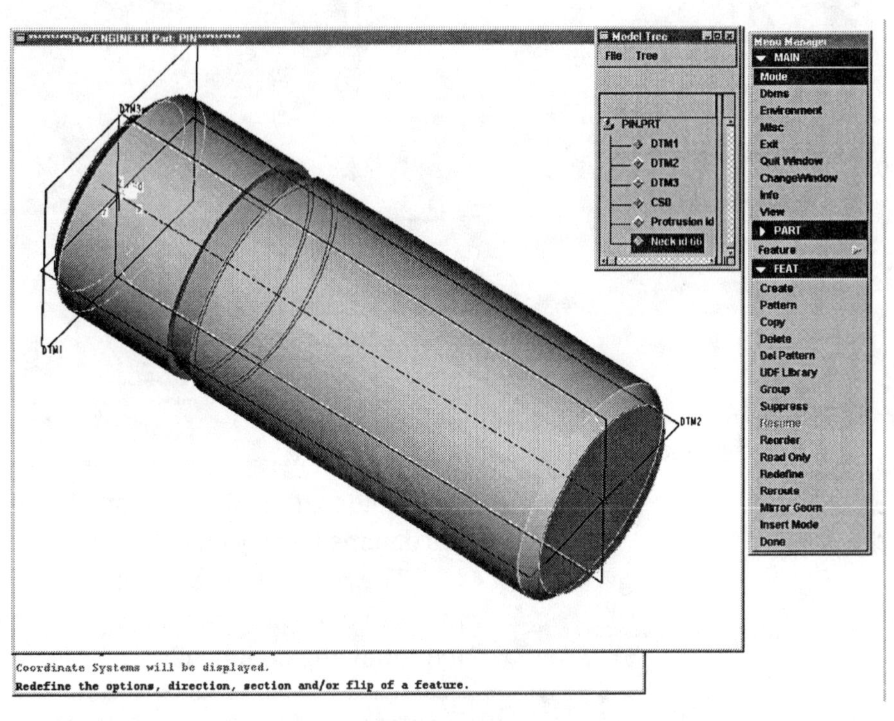

Figure 5.12
Completed Neck

Create the other neck cut using the same steps (Fig. 5.13). Try sketching with the shading on, as shown in Figure 5.14. Rotate the part to see the necks, as in Figure 5.14. Shade the neck cuts differently from the base part. In industry, the neck would have been copied instead of created again, but it is good to practice repeating commands at this stage of your understanding of Pro/E.

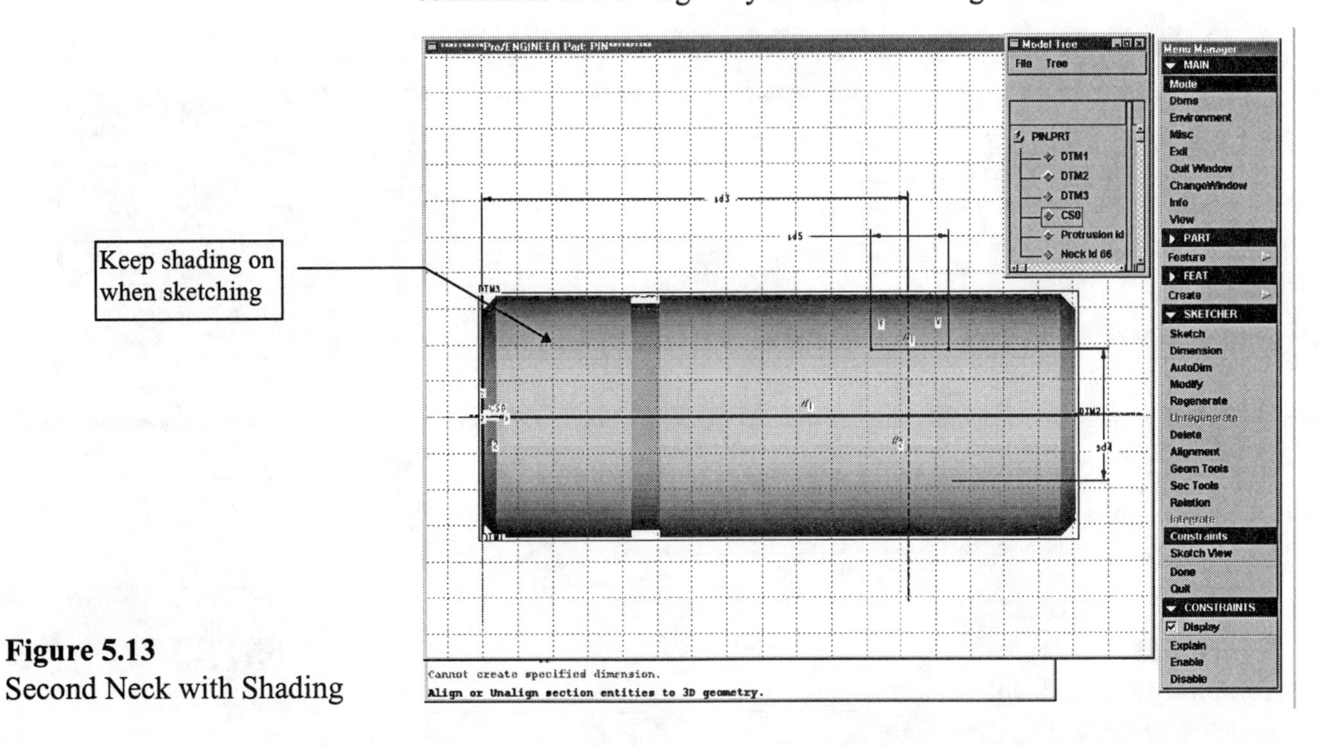

Figure 5.13
Second Neck with Shading

NOTE

In the example, we have renamed the datum planes:

DTM1 = datum **A**
DTM2 = datum **B**
DTM3 = datum **C**

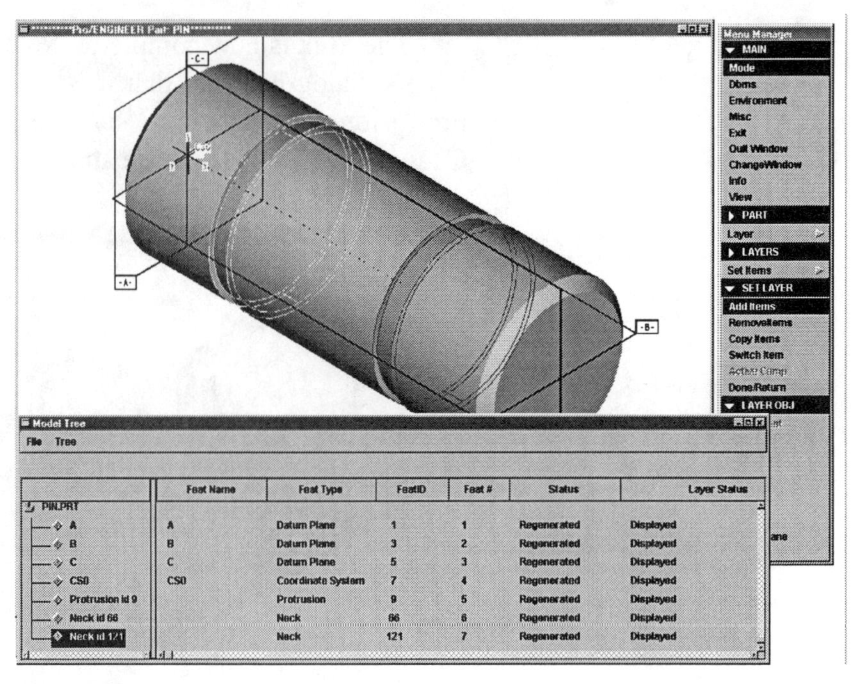

Figure 5.14
Completed Neck

Note that the datums have been *set* and the coordinate system and the datums have been put on a layer.

The remaining features are all holes. The first hole to create is the **.250** diameter hole through the center of the pin (Fig. 5.15). Give the following commands:

> **Feature** ⇒ **Create** ⇒ **Hole** ⇒ **Straight** ⇒ **Done** ⇒ **Coaxial** ⇒ **Done** ⇒ **Query Sel** (pick on axis **A_1** with left mouse button and accept with center mouse button; pick the placement plane datum **A**, which was **DTM1**) ⇒ **Flip** (if necessary pick **Flip** a couple of times to see the direction of the arrow; it must point toward the part) ⇒ **Okay** ⇒ **One Side** ⇒ **Done** ⇒ **Thru All** ⇒ **Done** ⇒ (type the diameter at the prompt: **.250**) ⇒ **enter** ⇒ **OK** ⇒ **Done**

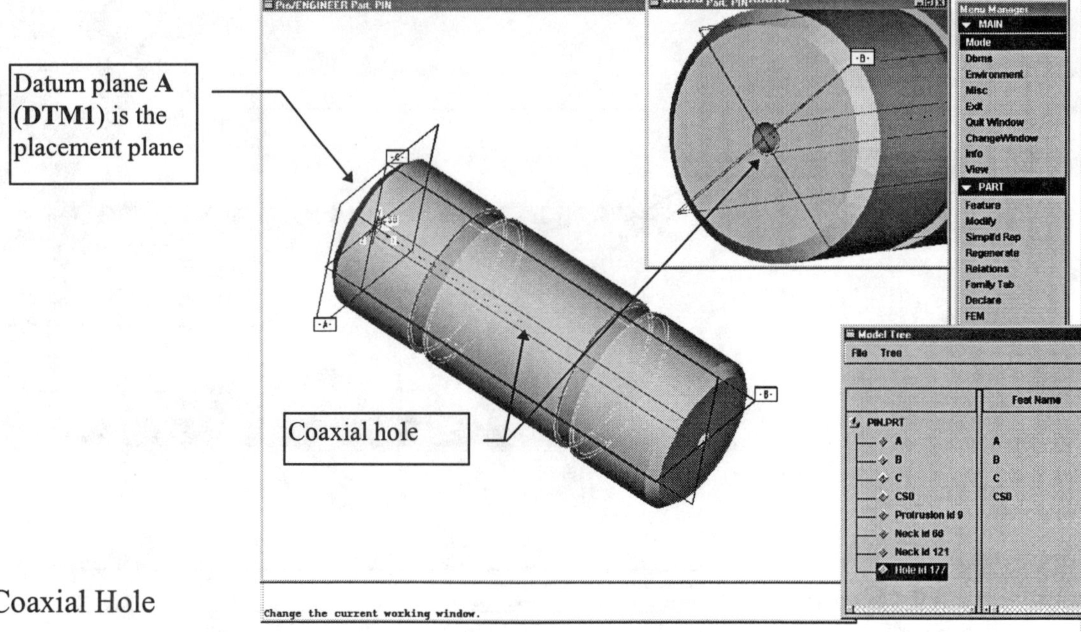

Figure 5.15
Completed Coaxial Hole

The pin has a conical cut (hole) at both ends that is coaxial with the **.250** hole and axis **A_1**. Make the conical feature with a *sketched hole* placed onto the datum plane **A** side of the pin (Fig. 5.16):

Environment ⇒ **✓Grid Snap** ⇒ **Done-Return** ⇒ **Feature** ⇒ **Create** ⇒ **Hole** ⇒ **Sketch** ⇒ **Done** ⇒ **Coaxial** ⇒ **Done** ⇒ **Sketch** ⇒ **Line** ⇒ **Centerline** ⇒ **Vertical** (pick once) ⇒ **Sketch Line** ⇒ **2Points** ⇒ (sketch the four lines of the *closed* section on one side of the centerline) ⇒ **Regenerate** ⇒ **Dimension** ⇒ **Regenerate** ⇒ **Modify** ⇒ **Regenerate** (Fig. 5.16) ⇒ **Done** ⇒ (pick axis **A_1**) ⇒ (pick datum **A** as the placement plane) ⇒ **Flip** ⇒ **Okay** ⇒ **Preview** (Fig. 5.17) ⇒ **OK** ⇒ **Done**

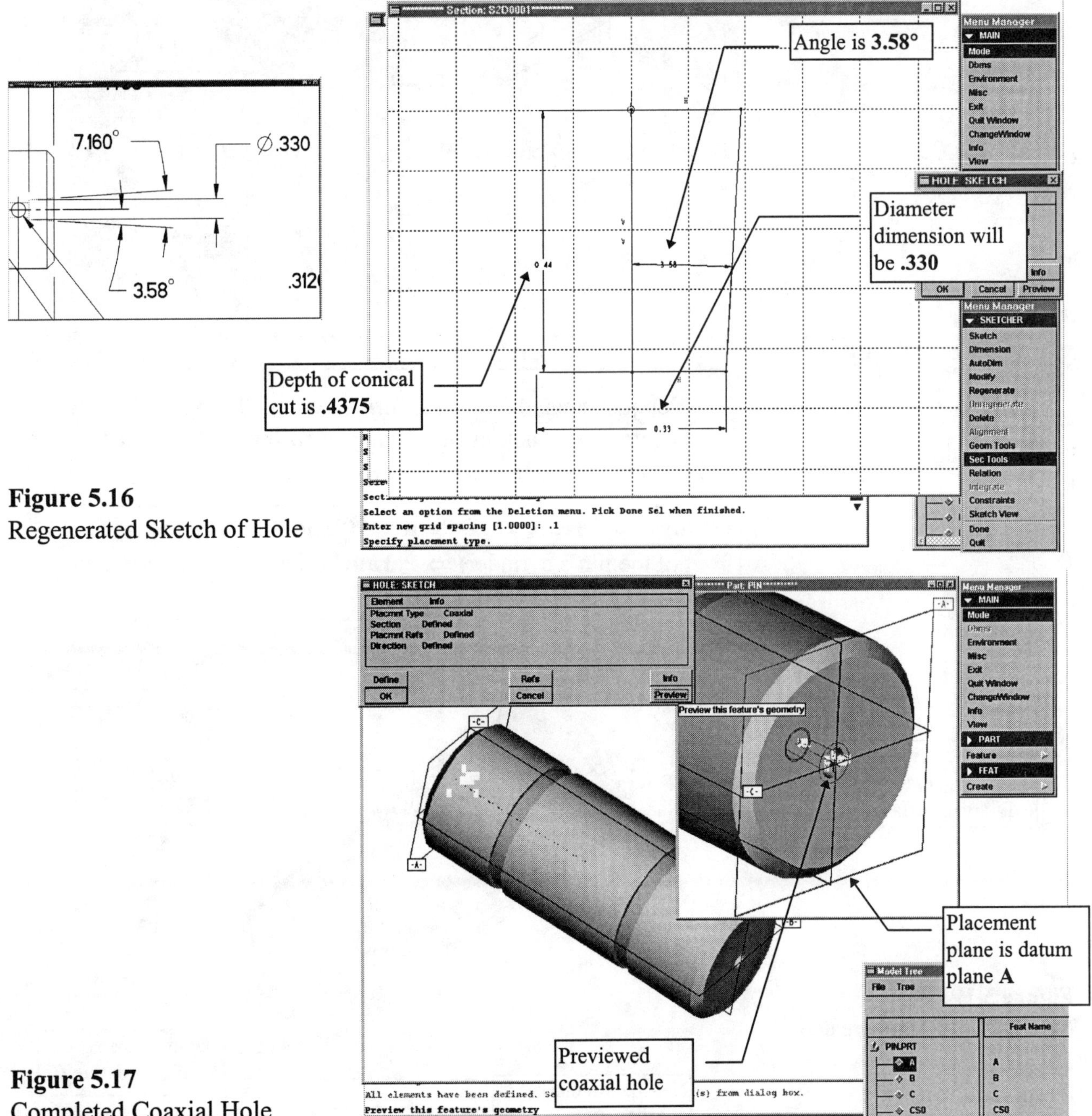

Figure 5.16
Regenerated Sketch of Hole

Figure 5.17
Completed Coaxial Hole

The conical hole is on both sides of the pin. Copy and mirror the hole using the following commands:

Feature ⇒ **Copy** ⇒ **Mirror** ⇒ **Dependent** ⇒ **Done** ⇒ (select the conical hole to be mirrored; use **Query Sel** or rotate the model to see the hole) ⇒ **Accept** ⇒ **Done Sel** ⇒ **Done** ⇒ **Make Datum** ⇒ **Offset** ⇒ (pick datum plane **A** to offset from) ⇒ **Enter Value** (type **5.125/2** at the prompt) ⇒ **enter** ⇒ **Done** (Fig. 5.18) ⇒ **Accept**

Dbms ⇒ **Save** ⇒ **enter**
Purge ⇒ **enter** ⇒
Done-Return

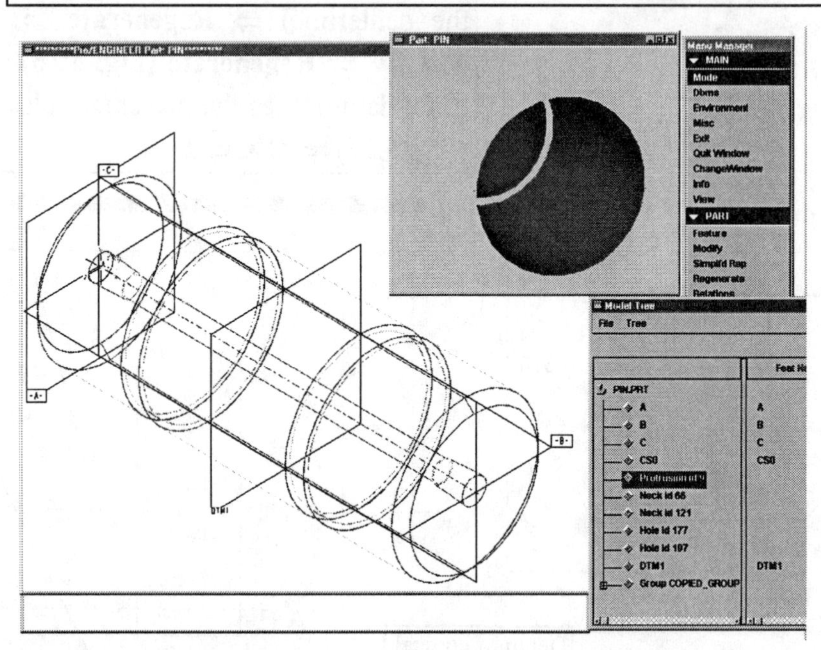

Figure 5.18
Mirrored Conical Hole

You just created your first ***datum-on-the-fly*** (**Make Datum**).

Create a new datum plane tangent to the pin's outside diameter and parallel to datum plane **C**:

Feature ⇒ **Create** ⇒ **Datum** ⇒ **Plane** ⇒ **Tangent** (pick the left front of the pin's cylinder) ⇒ **Parallel** (pick datum plane **C**, as shown in Figure 5.19) ⇒ **Done** ⇒ **Done**

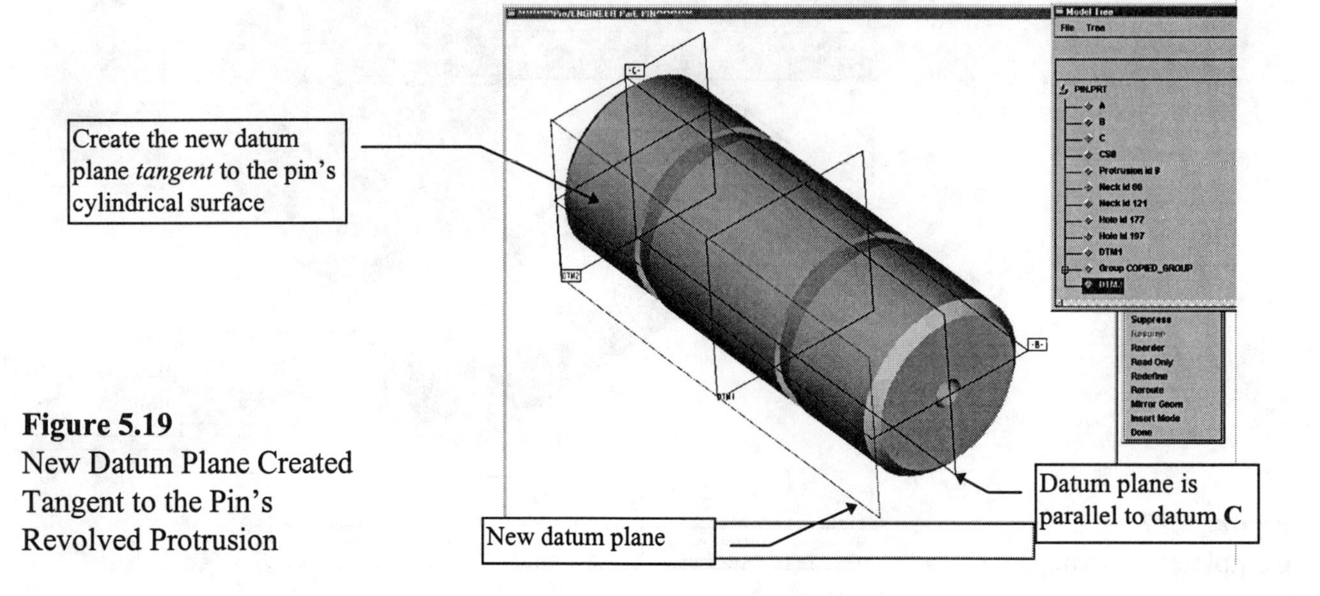

Figure 5.19
New Datum Plane Created Tangent to the Pin's Revolved Protrusion

Using this new datum plane, create the two **.125** diameter holes using the datum plane as the placement plane and the **.250** diameter coaxial hole running through the pin as the ending surface.

> **Feature ⇒ Create ⇒ Hole ⇒ Straight ⇒ Done ⇒ Linear ⇒ Done** ⇒ (select the new datum plane as the placement plane) ⇒ (select the feature placement location on the datum plane; pick anywhere on the plane) ⇒ **Okay** ⇒ (to accept the feature creation direction-- see the arrow) ⇒ (select two edges to dimension from; pick datum plane **B** and type **0.00** inches at the prompt) ⇒ (select datum plane **A** as the second reference and type **3.725** at the prompt) ⇒ **One Side ⇒ Done** ⇒ **UpTo Surface ⇒ Done** (pick the **.250** diameter hole by using **Query Sel** to filter through the part to get to the hole) ⇒ **Accept** ⇒ (type **.125** as the hole's diameter at the prompt) ⇒ **enter ⇒ OK** (from the dialog box) ⇒ **Done**

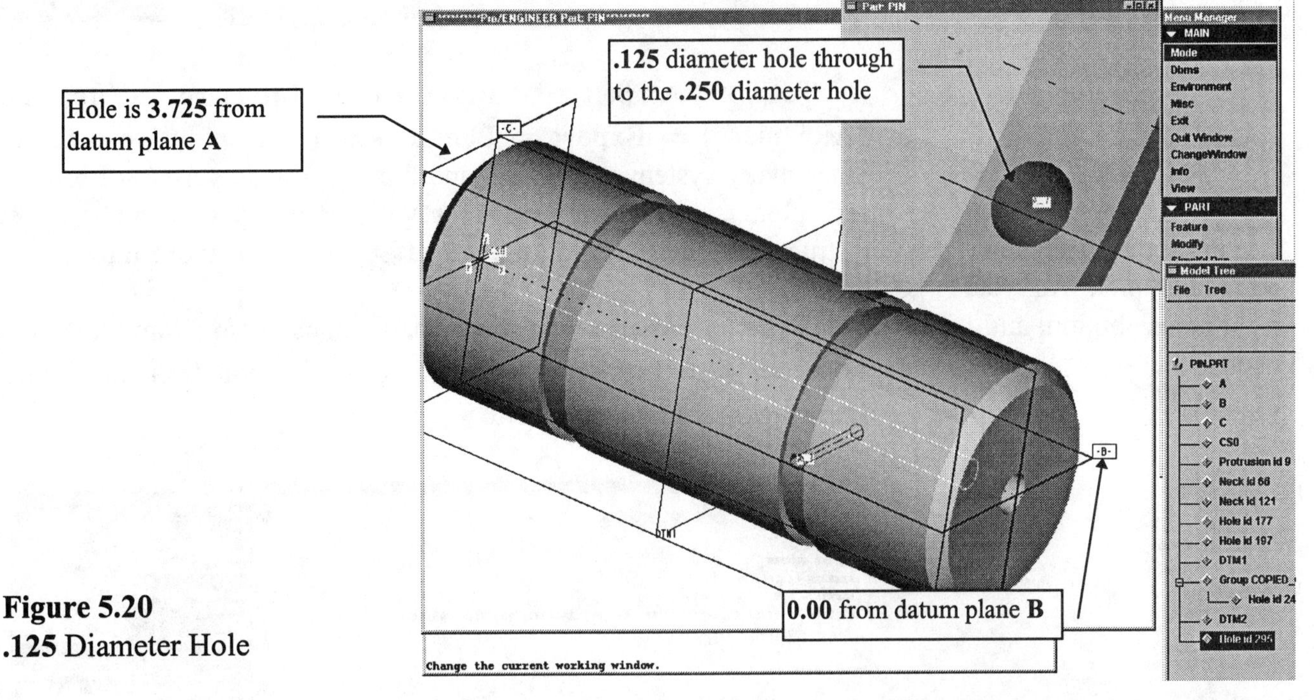

Figure 5.20
.125 Diameter Hole

Now create the **.250** diameter hole on the part's side (you will mirror and copy both holes later). The command is exactly the same as the previous command, for the **.125** diameter hole, except the diameter is **.250**, its distance from datum plane **A** is **4.50**, and it is a *blind* hole with a depth of **.3120** (Fig. 5.21).

The holes can now be copied and mirrored about the same datum plane as the conical hole was used to mirror about (Fig. 5.20). Use the following commands:

> **Feature ⇒ Copy ⇒ Mirror ⇒ Dependent ⇒ Done** ⇒ (select both the **.125** and the **.250** holes just created) ⇒ **Done Sel ⇒ Done ⇒** (pick **DTM1** as the plane to mirror about) ⇒ **Done**

Dbms ⇒ Save ⇒ enter Purge ⇒ enter ⇒ Done-Return

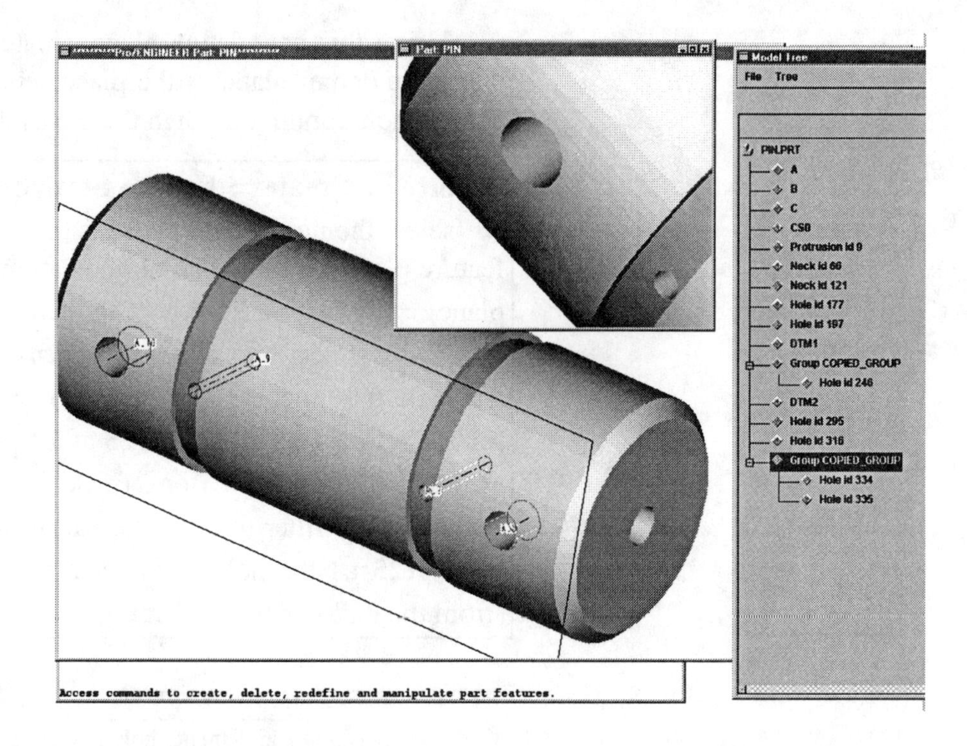

Figure 5.21
Mirrored Holes

To plot your part or drawing, choose **Interface** ⇒ (from the PART menu) ⇒ **Export** ⇒ **Plotter** (select your plotter or printer). Since every system is different and there is a wide variety of printers and plotters available, you must ask your system manager or instructor for help in plotting. Read Section 7 for more information on plotting.

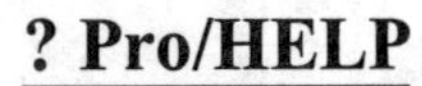

? Pro/HELP
Highlight the **Interface** command, pick with your right mouse button, and choose **?GetHelp**

To get information about your part, choose **Info** (from the MAIN menu) ⇒ **Feat Info** (pick the revolved protrusion from the Model Tree (Fig. 5.22) ⇒ **Done/Return**.

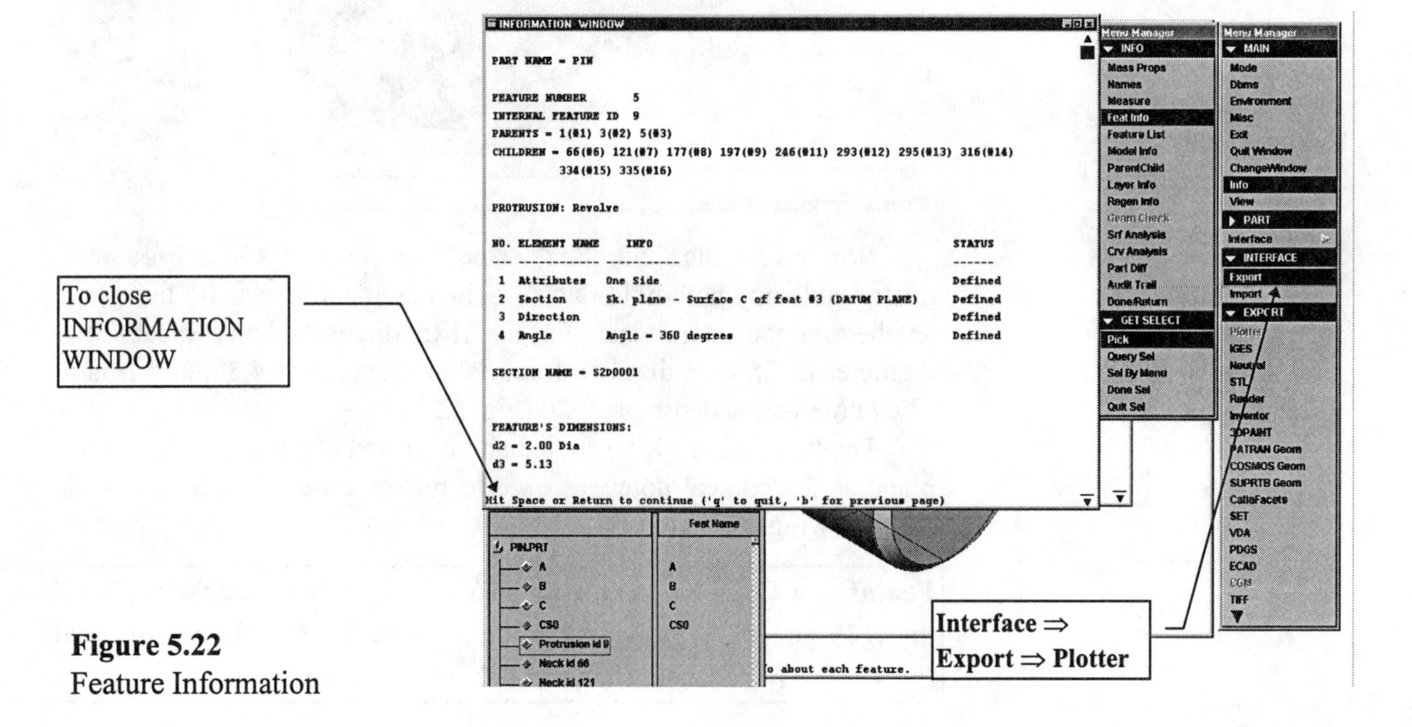

Figure 5.22
Feature Information

Lesson 5 Project

Clamp Foot and Clamp Swivel

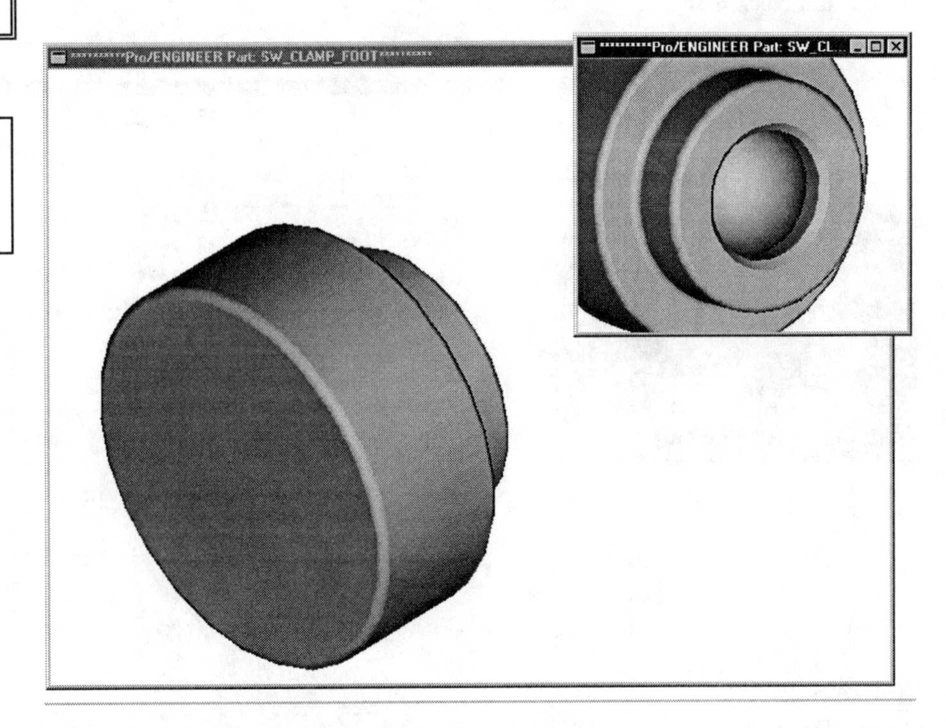

Figure 5.23
Clamp Foot

Clamp Foot and Clamp Swivel

Two lesson projects are provided in Lesson 5. You will use both of these two parts in Lessons 14 and 15 when creating an assembly. Both the clamp and the swivel are simple revolved protrusions. The Foot is nylon and the Swivel is steel. The Swivel fits inside the Foot.

Analyze each part, and plan out the steps and features required to model it. Use the **DIPS** in Appendix D to establish a feature creation sequence before the start of modeling. Remember to set up the environment, establish datum planes, and set them on layers.

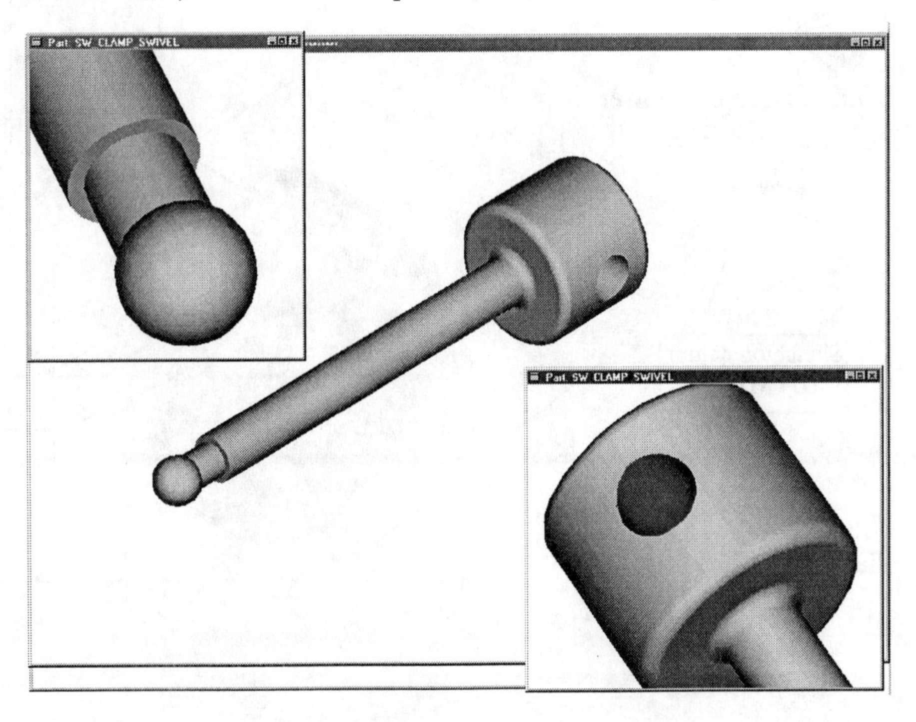

Figure 5.24
Clamp Swivel

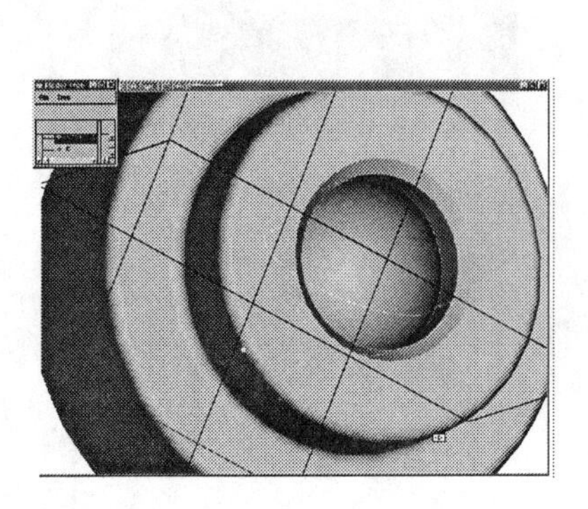

Figure 5.25
Clamp Foot Dimensions

Create the two parts (Figure 5.23 through Figure 5.28) with revolved protrusions using datum **C (DTM1)** as the sketching plane. Create the internal cut on the **Foot** with a *revolved cut*. Add the rounds on both parts at the end of the modeling process; do not include them on the first revolved protrusions.

Dbms ⇒ Save ⇒ enter
Purge ⇒ enter ⇒
Done-Return

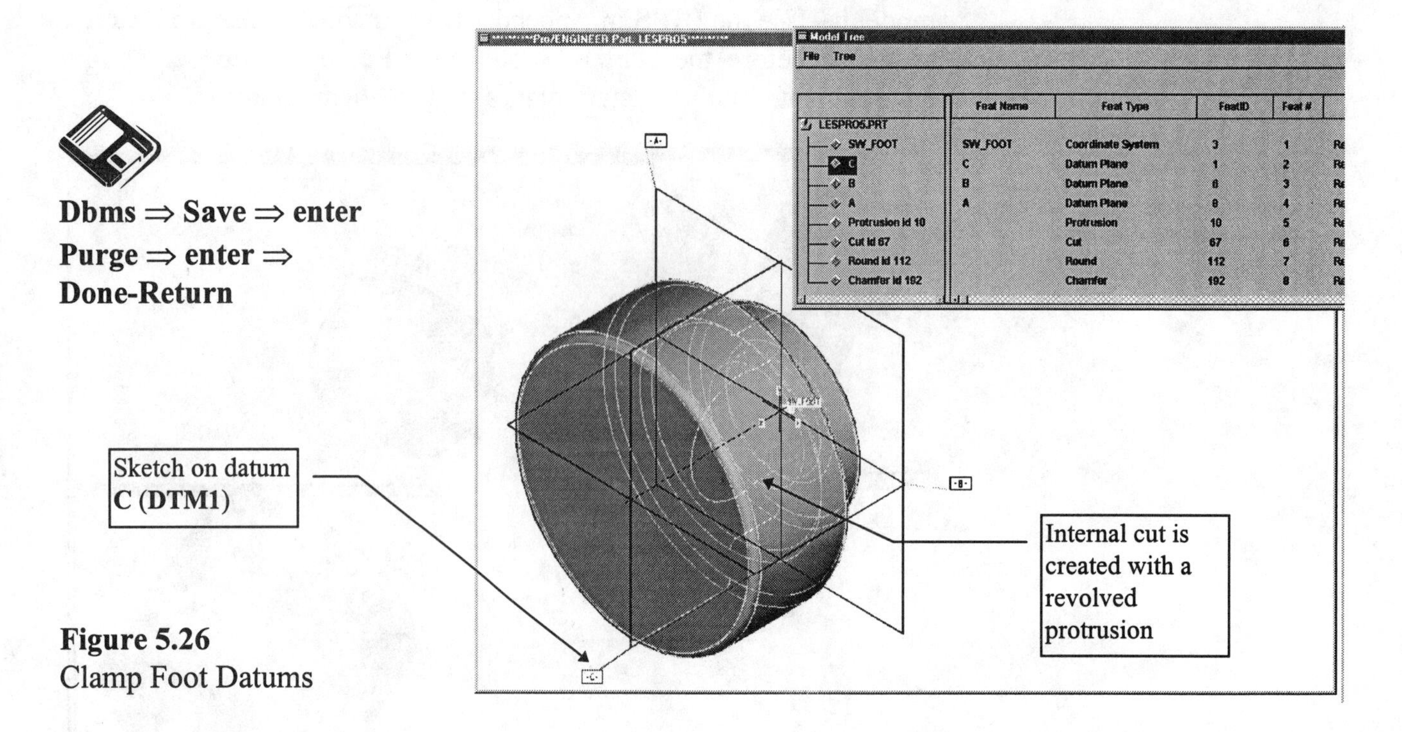

Figure 5.26
Clamp Foot Datums

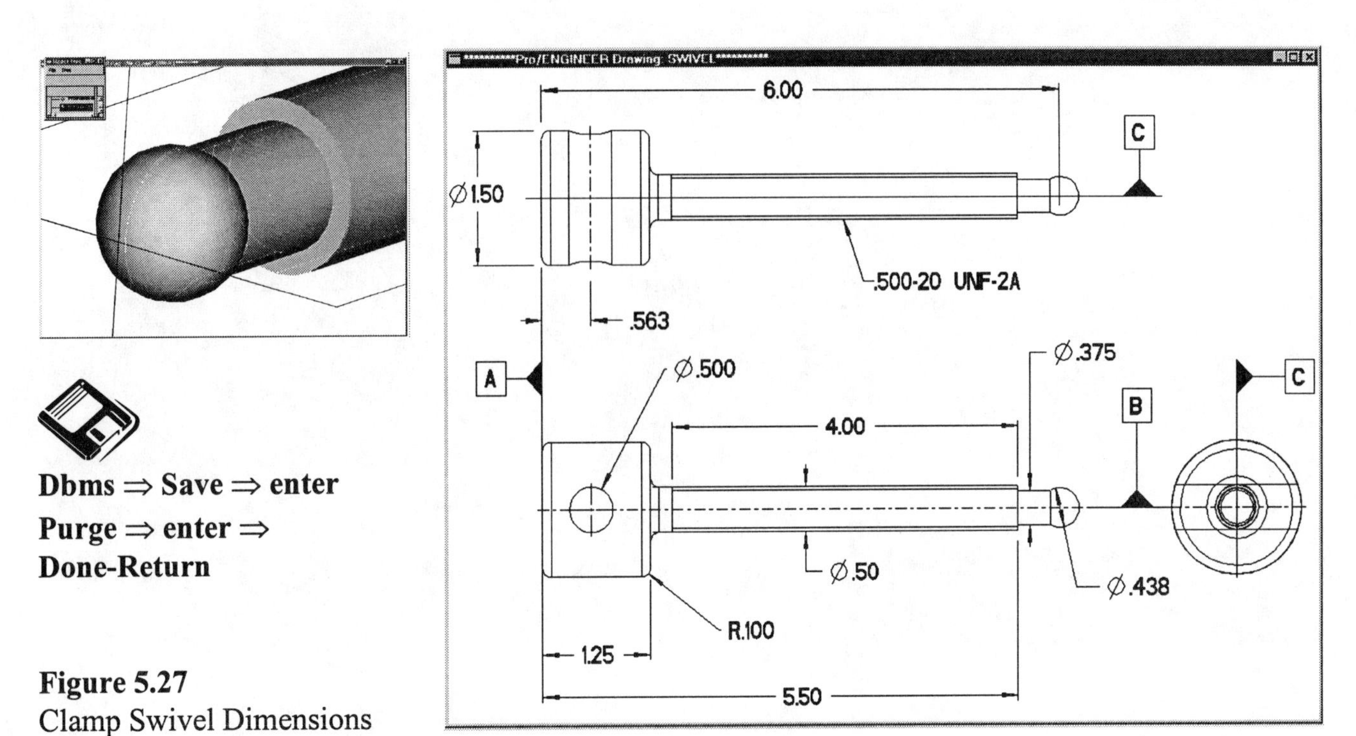

Dbms ⇒ Save ⇒ enter
Purge ⇒ enter ⇒
Done-Return

Figure 5.27
Clamp Swivel Dimensions

Figure 5.28
Clamp Swivel Datums

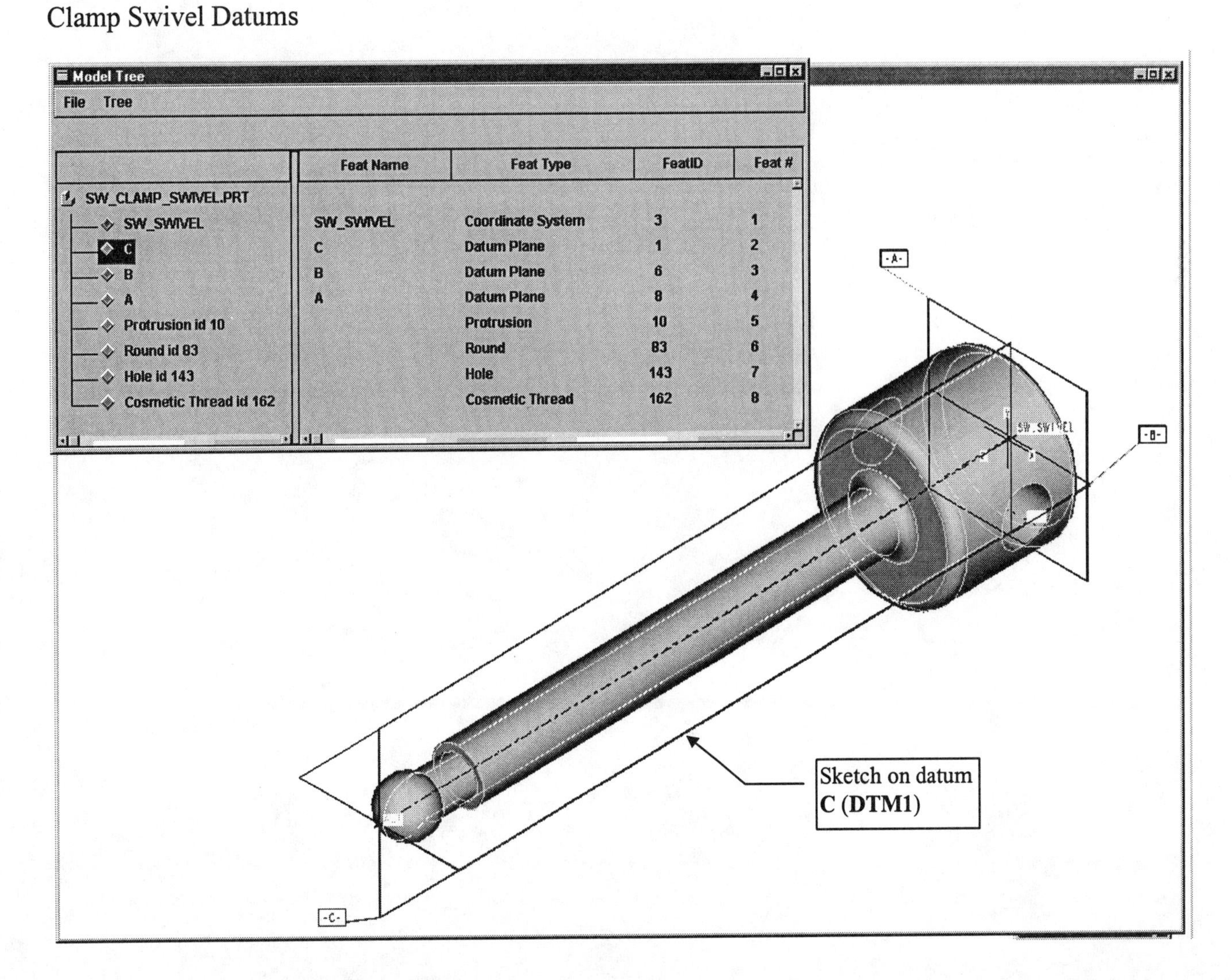

Lesson 6

Chamfers and Cosmetic Threads

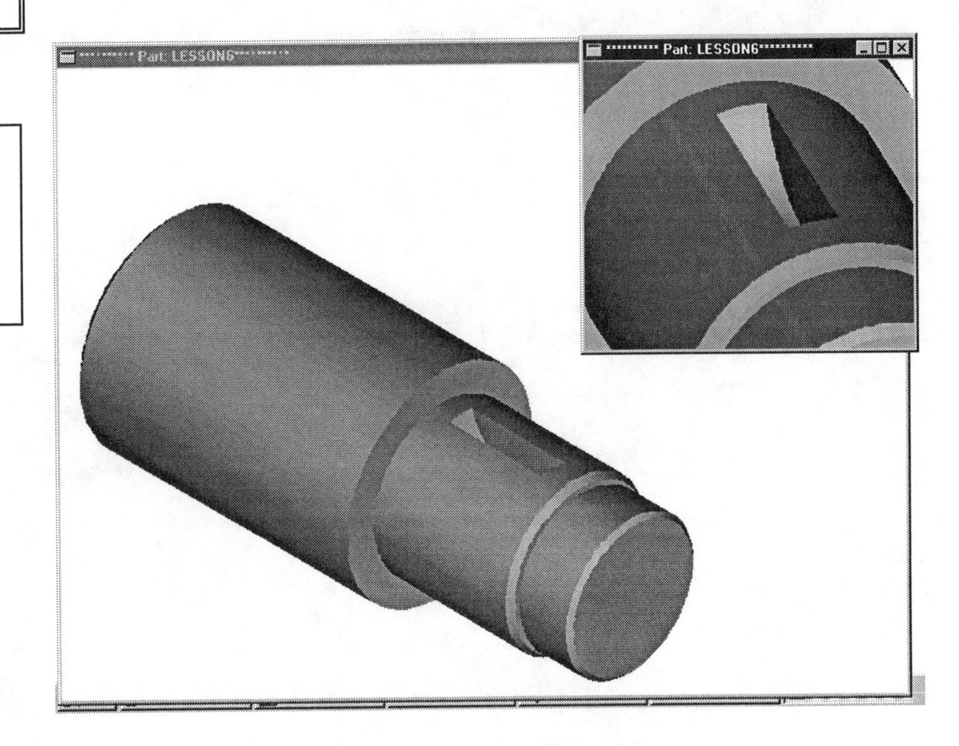

Figure 6.1
Cylinder Rod

OBJECTIVES

1. **Create simple chamfers along part edges**
2. **Learn how to sketch in 3D**
3. **Create cosmetic threads**
4. **Complete tabular information for threads**
5. **Get information on existing cosmetic threads**
6. **Dynamically modify sketch dimension values**

EGD REFERENCE
Engineering Graphics and Design
by L. Lamit and K. Kitto
Read Chapters: 14 and 17
See Pages: 497, 540-541, and 674-678

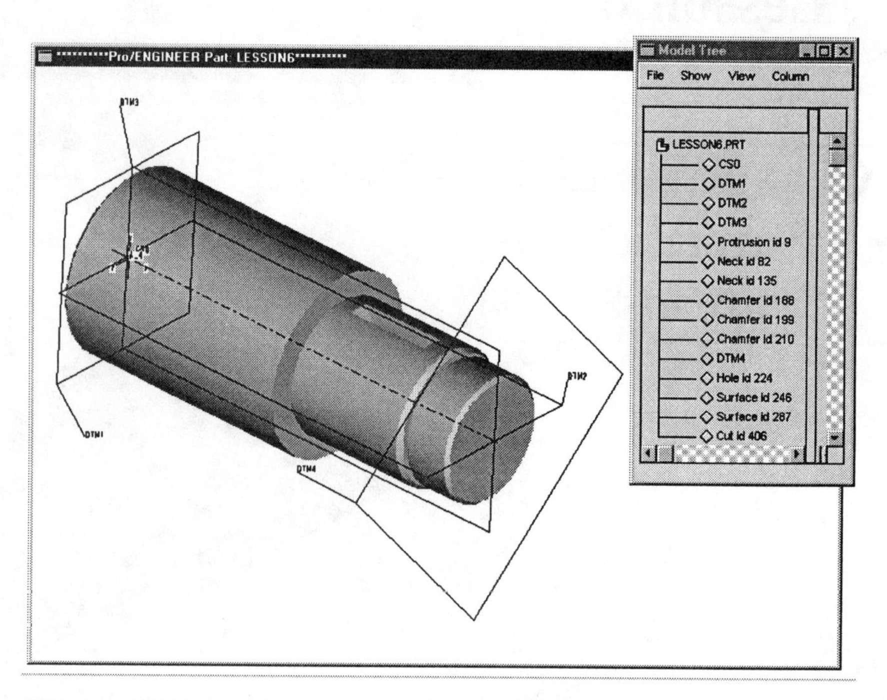

Figure 6.2
Cylinder Rod with Datums and Model Tree

CHAMFERS AND COSMETIC THREADS

A variety of geometric shapes and constructions are accomplished automatically with a CAD system using parametric modeling. For instance, **chamfers** are created automatically at selected edges of the part (Figs. 6.1 and 6.2). Chamfers are *pick-and-place* features (Fig. 6.3).

Threads are usually a *cosmetic feature* representing the *nominal diameter* or the *root diameter* of the thread. Information can be embedded into the feature. Threads show as a unique color (magenta). By putting cosmetic threads on a separate layer, you can display, blank, or suppress them as required.

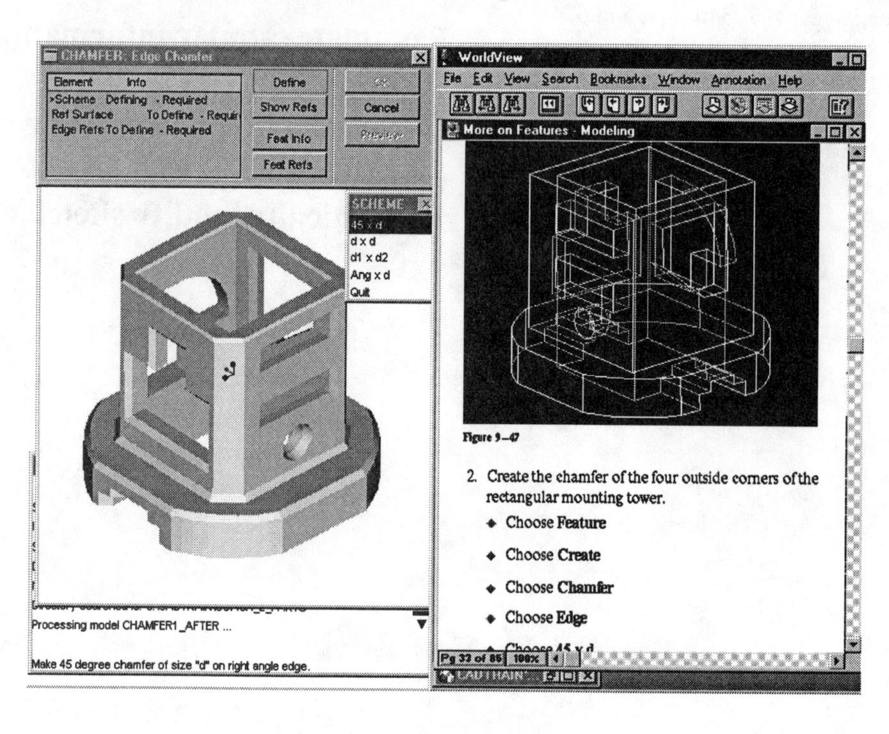

Figure 6.3
COAch for Pro/E, More on Features--Modeling (Segment 2: Chamfers)

Chamfers

Chamfers are created between abutting edges of two surfaces on the solid model. An **edge chamfer** removes a flat section of material from a selected edge to create a beveled surface between the two original surfaces common to that edge (Fig. 6.4). Multiple edges may be selected.

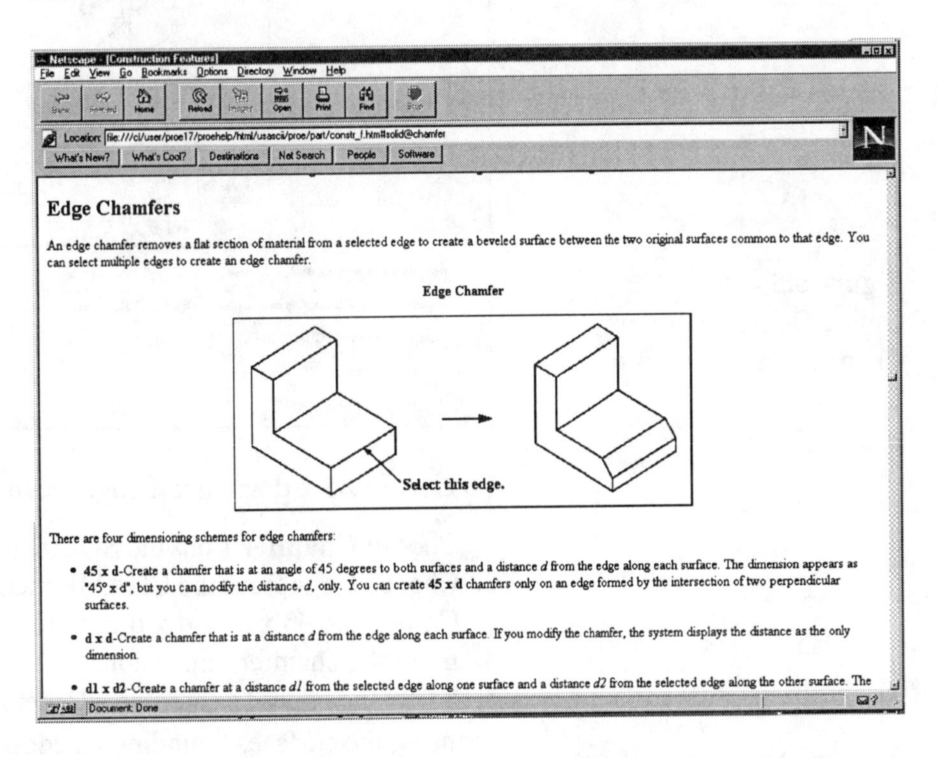

Figure 6.4
Online Documentation
Edge Chamfers

There are four dimensioning schemes for edge chamfers, as shown in Figure 6.5:

45 x d Creates a chamfer that is at an angle of **45°** to both surfaces and a distance **d** from the edge along each surface. The distance is the only dimension to appear when modified. **45 x d** chamfers can be created only on a edge formed by the intersection of two *perpendicular* surfaces.
d x d Creates a chamfer that is a distance **d** from the edge along each surface. The distance is the only dimension to appear when modified.
d1 x d2 Creates a chamfer at a distance **d1** from the selected edge along one surface and a distance **d2** from the selected edge along the other surface. Both distances appear along their respective surfaces when modified.
Ang x d Creates a chamfer at a distance **d** from the selected edge along one adjacent surface at a specified angle to that surface.

The dimensioning schemes appear as options under the SCHEME menu. The SCHEME menu appears after **Edge** is chosen from the CHAMF menu.

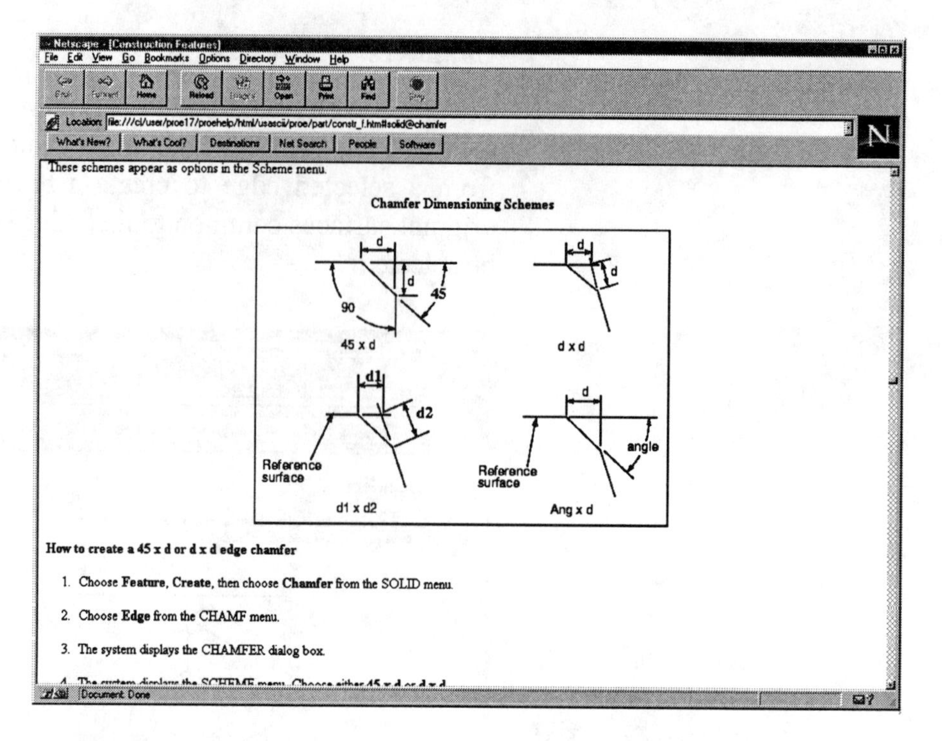

Figure 6.5
Online Documentation Chamfer Dimensioning Schemes

To create a **45 x d** and **d x d** edge chamfer:

1. Choose **Chamfer** from the SOLID menu.
2. Choose **Edge** from the CHAMF menu.
3. Choose the **45 x d** or **d x d** option.
4. Enter the chamfer dimension.
5. Select the edges to chamfer. Remember that for a **45 x d** edge chamfer, the surfaces bounding an edge must be **90°** to each other.

To create a **d1 x d2** chamfer:

1. Choose **Chamfer** from the SOLID menu.
2. Choose **Edge** from the CHAMF menu.
3. Choose the **d1 x d2** option.
4. Input a distance along a surface to be selected.
5. Input a second distance
6. Pick the surface along which the first distance will be measured, and pick the edge to chamfer.

To create a **Ang x d** chamfer:

1. Choose **Chamfer** from the SOLID menu.
2. Choose **Edge** from the CHAMF menu.
3. Choose the **Ang x d** option.
4. Input distance.
5. Input an angle from a surface to be selected.
6. Select the reference surface from which the values will be measured.
7. Pick the edge to chamfer and the dimensioning references.

To create a corner chamfer (Fig. 6.6):

1. Choose **Chamfer** from the SOLID menu, then choose **Corner** from the CHAMF menu.
2. Select the corner you want to chamfer.
3. Pro/E displays the PICK/ENTER menu, which allows you to specify the location of the chamfer vertex on the highlighted edge. The PICK/ENTER menu options are as follows:

   Pick Point Pick a point on the highlighted edge to define the chamfer distance along that edge.

   Enter input Type in a value for the chamfer distance along the highlighted edge.

4. Pick or enter values to describe the chamfer lengths along the edge. After you have selected the first vertex, Pro/E highlights the other edges, one at a time, so you can place the other two vertices
5. Select the **OK** button in the dialog box.

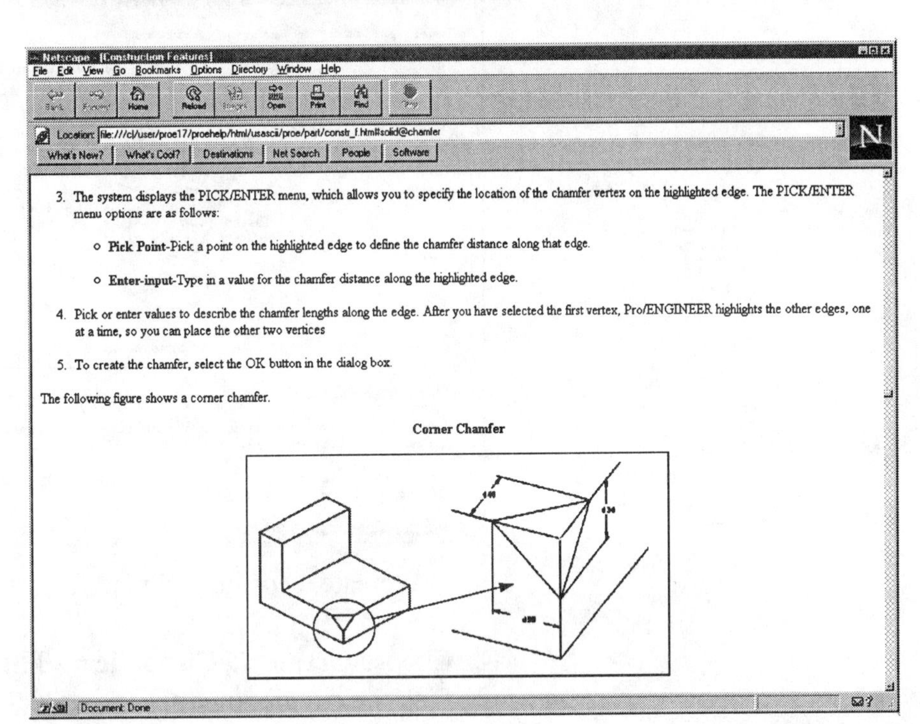

Figure 6.6
Online Documentation
Corner Chamfer

Threads

Cosmetic threads (Fig. 6.7) are displayed with magenta lines and circles. Cosmetic threads can be external or internal, blind or through. In the Rod part, one end has external blind threads and the opposite end has internal blind threads.

A thread has a set of supported parameters that can be defined at its creation or later when the thread is added.

The following parameters can be defined for a thread:

PARAMETER DESCRIPTION	PARAMETER NAME	PARAMETER VALUE
Thread major diameter	MAJOR_DIAMETER	Number
Threads per inch (pitch)	THREADS_PER_INCH	Number
Thread form	THREAD_FORM	String
Thread class	CLASS	Number
Thread placement (A-external, B-internal)	PLACEMENT	A or B
Thread is Metric	METRIC	TRUE/FALSE

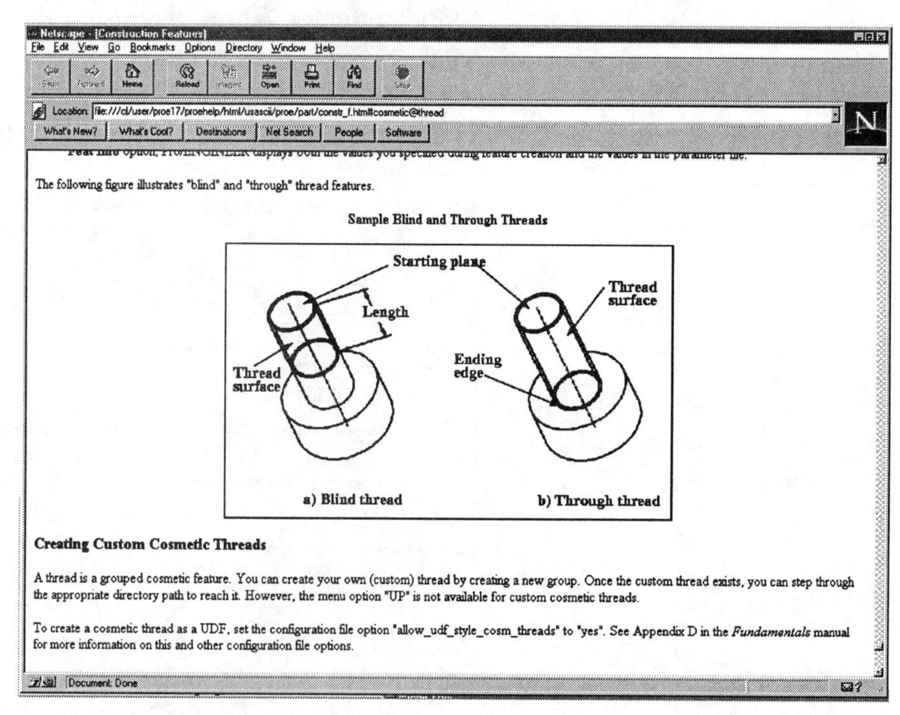

Figure 6.7
Online Documentation Cosmetic Threads

To create a cosmetic thread feature:

1. Choose **Create ⇒ Cosmetic ⇒ Thread**
2. Pick the circular internal or external thread surface at the prompt. The system automatically knows if the threads are internal or external, based on the feature selected.
3. Select the thread start surface, and **Flip** or **Okay** the direction as needed.
4. When finished, choose **Done-Return** to continue.
5. From the SPEC TO menu pick **Blind**, **UpTo Pnt/Vtx**, **UpTo Curve**, or **UpTo Surface** and then **Done.**
6. Follow the prompts, which differ depending on Step 5.
7. Enter the diameter at the prompt.
8. From the FEAT PARAM menu, select one of the options. In general you will be picking the **Mod Params** and completing the Pro/TABLE information. Select from the FEAT PARAM menu:

Retrieve Retrieves a previously created and saved thread file.
Save For completing the table, to save your thread file for use later.
Mod Params Modifies thread parameters in Pro/TABLE environment (Fig. 6.8).
Show Displays a set of thread parameters in the INFORMATION WINDOW.
Done-Return To complete the process, exit from this menu.

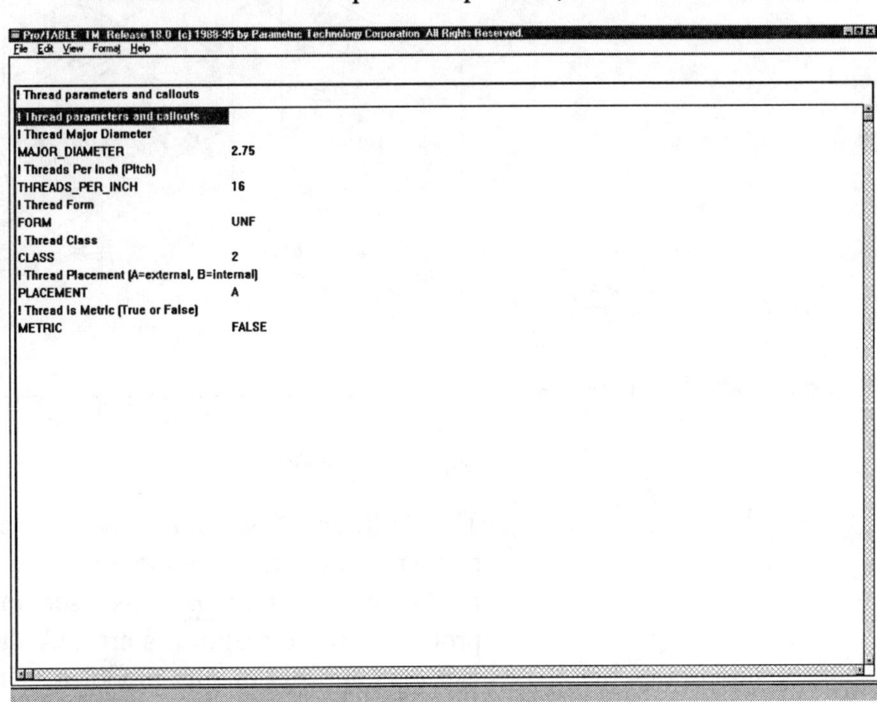

Figure 6.8
Using Pro/TABLE to Input Thread Parameters

Thread parameters can be manipulated like other user-defined parameters: they can be added, modified, deleted, or displayed using menu options.

The following information was extracted using the commands **Info ⇒ Feature Info ⇒** (select the thread). Information similar to the following appears in a window:

PART NAME = ROD
FEATURE NUMBER 13
INTERNAL FEATURE ID 246
PARENTS = 9(#5) 224(#12)

TYPE THREAD

FORM = 360 DEG. REVOLVED
SECTION NAME = S2D0003
OPEN SECTION

FEATURE DIMENSIONS:

d28(d11) = 1.25
d29(d12) = .63 Dia.
MAJOR_DIAMETER .625
THREADS_PER_INCH 11
FORM UNC
CLASS 2A

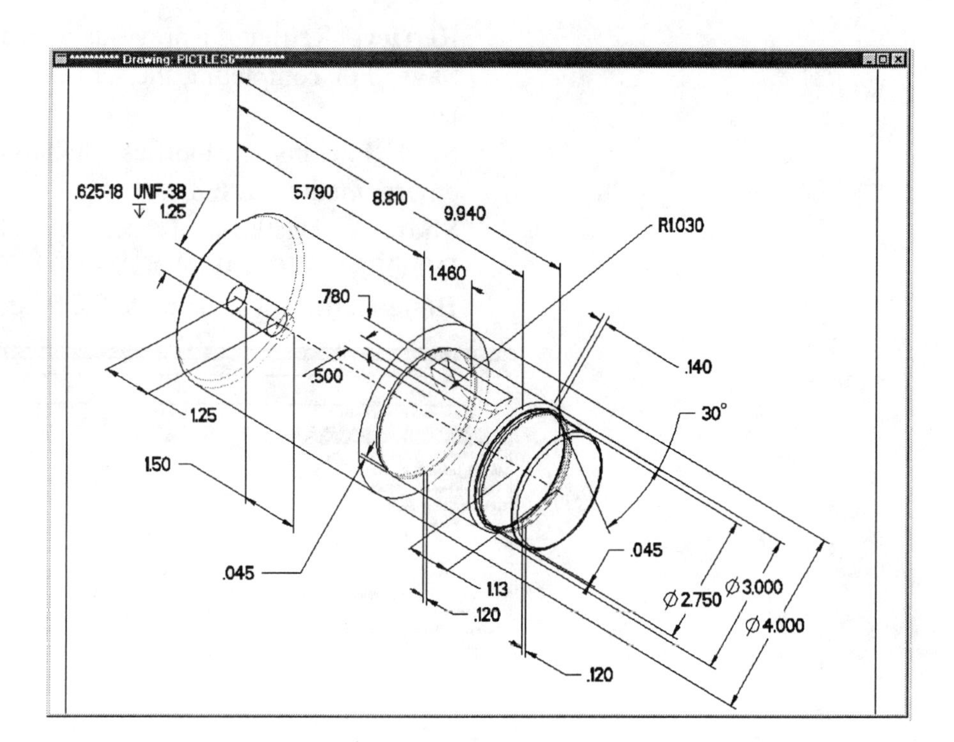

Figure 6.9
Cylinder Rod Dimensions

Cylinder Rod

The **Cylinder Rod** is modeled by creating a revolved protrusion, similar to the Pin in Lesson 5. The geometry of the revolved feature is shown in Figure 6.9 (also see Fig. 6.38). After the revolved protrusion (base feature) is created, the necks (reliefs), the chamfers, and the tap drill hole are modeled. In this lesson we will create our first revolved protrusion by sketching in 3D.

Three chamfers are required for this part. The **45 x d** option was used to chamfer the left side (**4.00** diameter) of the part (**45° x .125**). A **45° x .09** chamfer is added to the right side (**2.75** diameter) of the Rod and a **30° x .14** chamfer is used on the **3.00** diameter step of the Rod at the relief. Two necks are required, both are **.125 x .045 DEEP**.

The cosmetic threads for the external threaded shaft end and the internal hole threads are added last. They are created by specifying the minor or major diameter (for external or internal threads, respectively), starting plane (for the external threads, a **DTM4** was used for the external threads starting plane), and thread length or ending edge. The internal threaded hole is **.625-16 UNF-3B** by **1.25 DEEP**. The threaded shaft is **2.75-16 UN-2A**.

CONFIG.PRO

LINEAR_TOL_0.000

SKETCHER_DEC_PLACES 3

SKETCHER_DISPLAY_ CONSTRAINTS YES

DEFAULT_DEC_PLACES 3

NOTE

SETUP AND ENVIRONMENT blocks are for your information only; they are not meant to be input at this time.

SETUP AND ENVIRONMENT

Setup ⇒ Units ⇒ Inch ⇒ Material ⇒ Define ⇒ Aluminum ⇒ Assign ⇒ Aluminum ⇒ Accept
Environment ⇒ ✓Grid Snap
SketStart2D (we will be sketching in 3D)
Hidden Line Tan Dimmed

Start the part with the usual commands (Fig. 6.10):

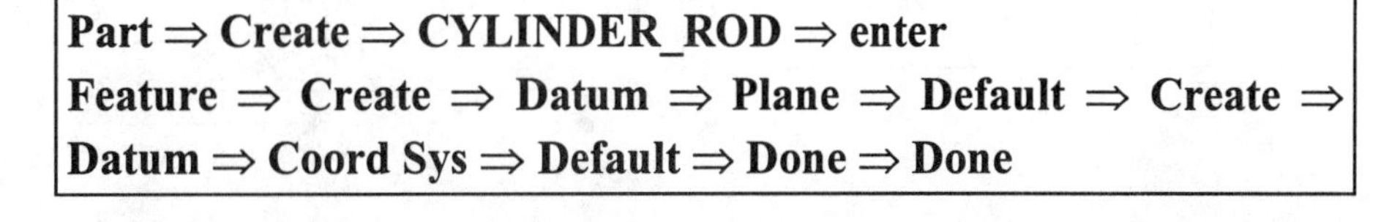

Part ⇒ Create ⇒ CYLINDER_ROD ⇒ enter
Feature ⇒ Create ⇒ Datum ⇒ Plane ⇒ Default ⇒ Create ⇒
Datum ⇒ Coord Sys ⇒ Default ⇒ Done ⇒ Done

HINT
Setup and **Environment** are set *after* the Part or Assembly mode has been entered.

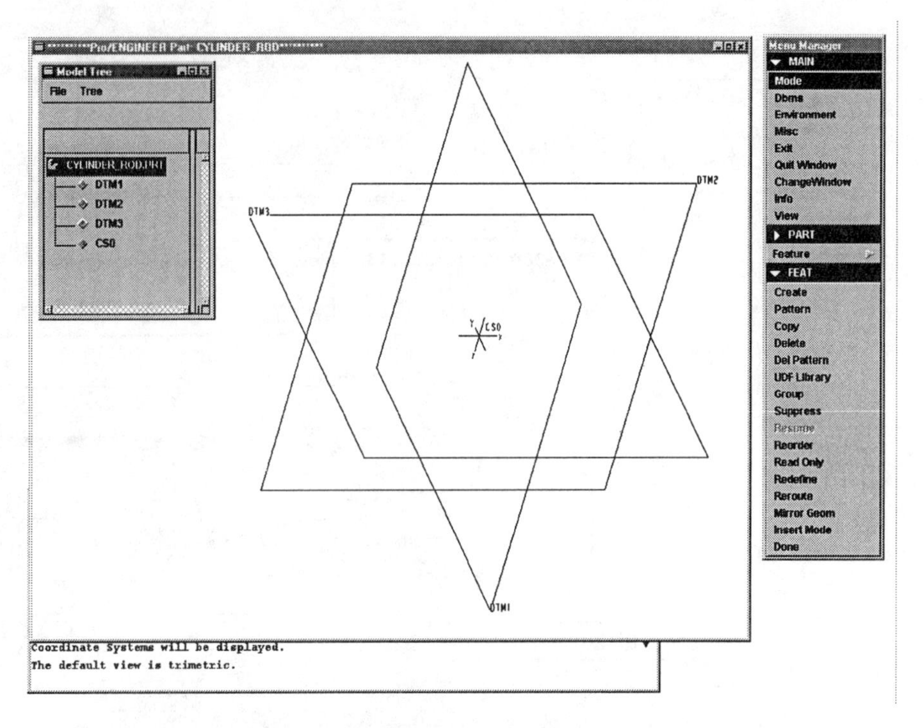

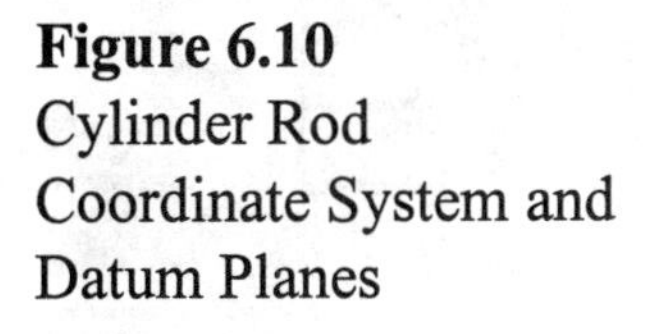

Figure 6.10
Cylinder Rod Coordinate System and Datum Planes

Create a layer for the datum planes and the coordinate system:

Layer ⇒ Create ⇒ (type **DATUM_LAYER**) **⇒ enter ⇒ enter ⇒ Set Items ⇒ Add Items ⇒** (✔**DATUM_LAYER**) **⇒ Done Sel ⇒ Feature ⇒ Sel By Menu ⇒ Name ⇒ Sel All ⇒ OK ⇒ Done Sel ⇒ Done/Return ⇒ Done/Return ⇒ Done/Return**

The first protrusion is a revolved feature, and you will be sketching the section in 3D for the first time:

Feature ⇒ Create ⇒ Solid ⇒ Protrusion ⇒ Revolve ⇒ Done ⇒ One Side ⇒ Done ⇒ (pick **DTM3** as sketching plane) **⇒ Okay ⇒ Top** (pick **DTM2** as horizontal reference) **⇒ View ⇒ Pan/Zoom ⇒ Pan** (pan the screen as in Figure 6.11) **⇒ Done-Return ⇒ Line ⇒ Centerline ⇒ Horizontal** (sketch the horizontal centerline *first*) **⇒ Sketch ⇒ Mouse Sketch** (sketch the section geometry as in Figure 6.11) **⇒ Alignment** (align the horizontal centerline with **DTM2**, the horizontal line with **DTM2**, and the vertical left side line with **DTM1**, as in Figure 6.12) **⇒ Regenerate ⇒ Dimension** (use dimension scheme shown in Figure 6.13) **⇒ Regenerate ⇒ Modify** (modify to design dimensions shown in Figure 6.9) **⇒ Regenerate**

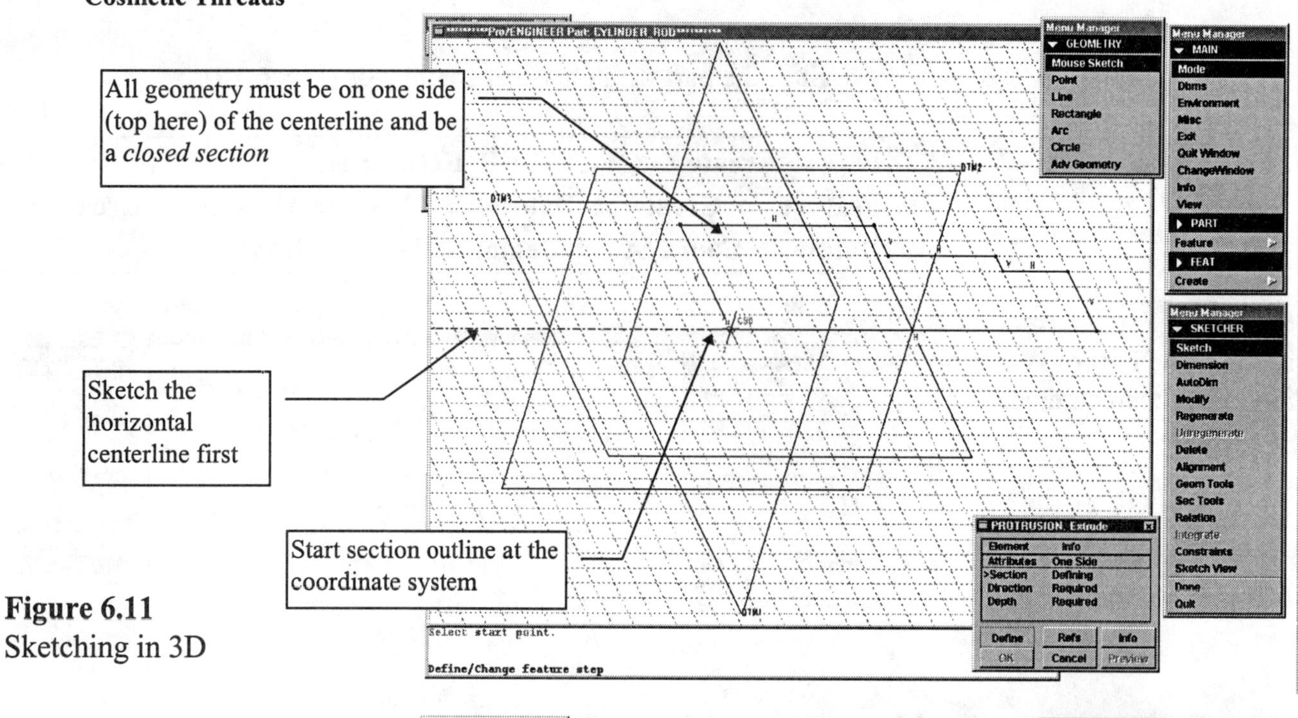

Figure 6.11
Sketching in 3D

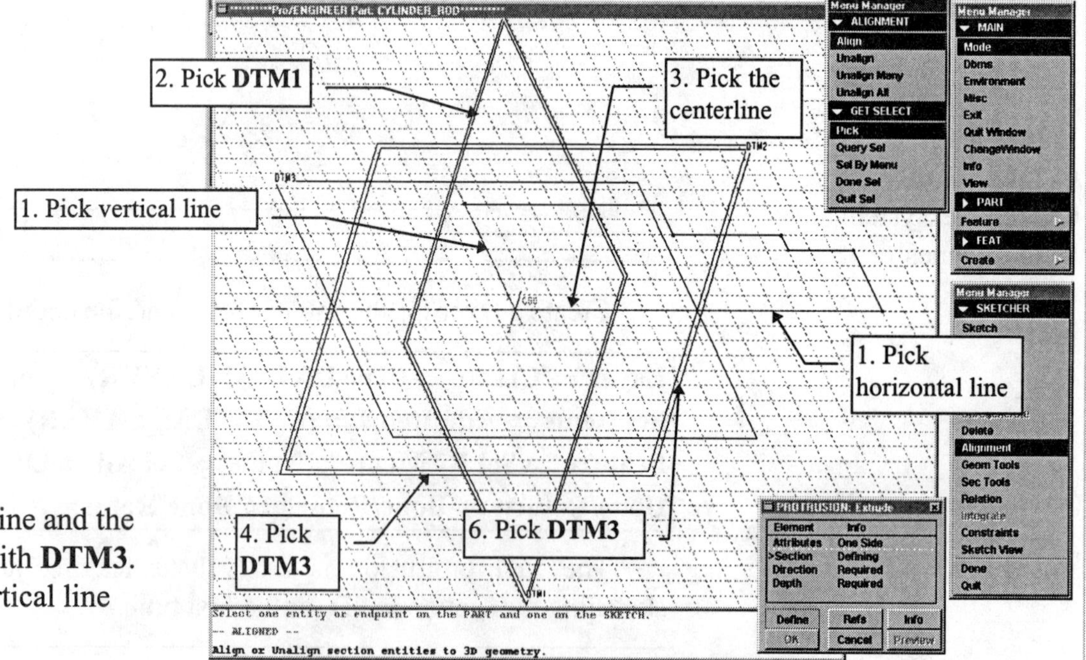

Figure 6.12
Align the centerline and the horizontal line with **DTM3**. Align the left vertical line with **DTM1**

HINT
You can toggle between sketching in 3D and 2D. If you want to go to 2D, choose **Sketch View**.

You can dynamically change any dimension on the sketch. This capability is available for 3D and 2D sketching. After the sketch has been regenerated with the design dimensions (Fig. 6.9), give the following commands:

Modify ⇒ **Drag Dim Val** ⇒ (pick the **9.94** dimension) ⇒ **Done Sel** ⇒ (the **Modify Dims** Scale Menu will appear on the screen ⇒ (move the mouse pointer into the region under the title **"Linear,"** pick once with the left mouse button, slide the linear indicator to the right until the dimension is about **12.50**, as shown in Figure 6.14, then pick once more with the left mouse button to finish)

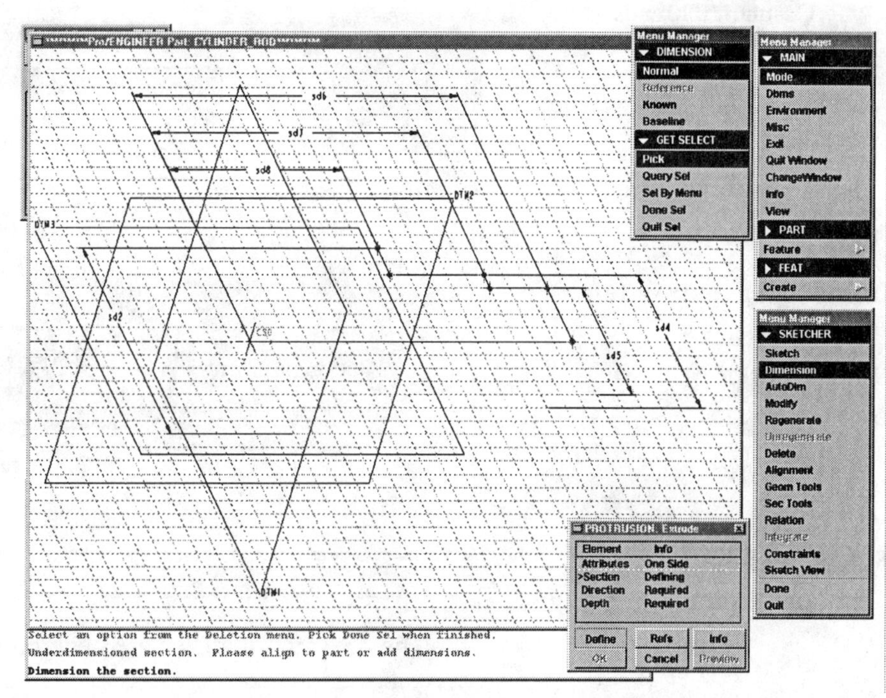

Figure 6.13
Section with Sketch Dimensions

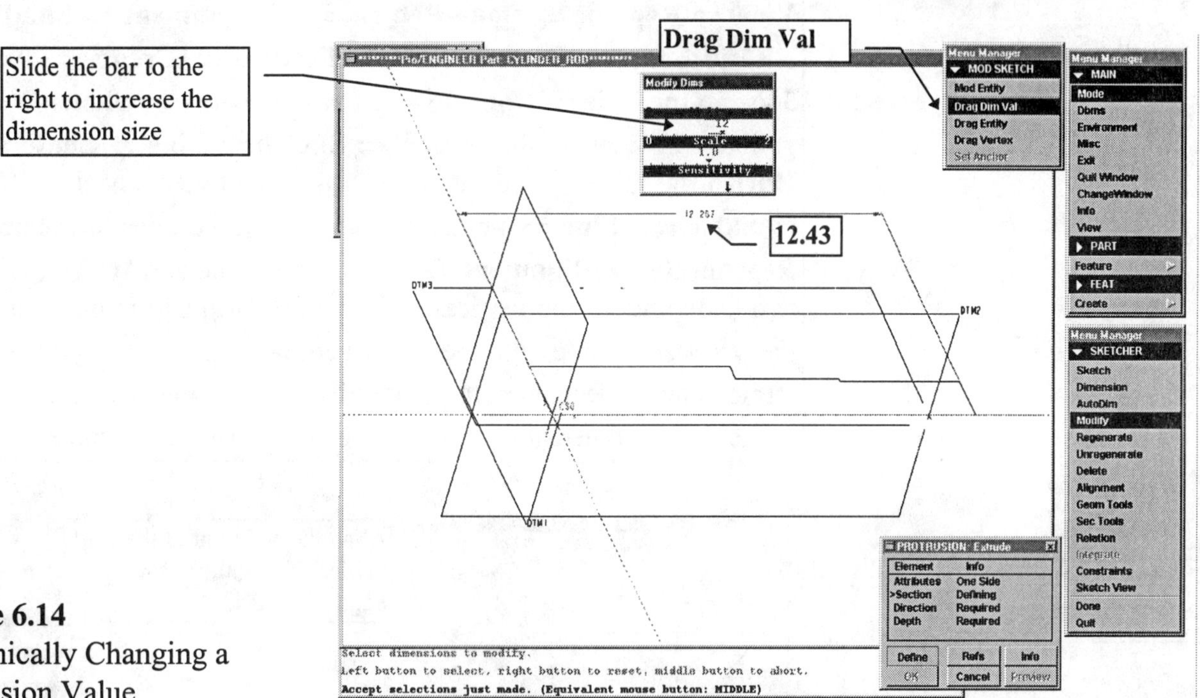

Figure 6.14
Dynamically Changing a Dimension Value

The sketch will dynamically change with the length value. After experimenting with this capability, **Regenerate** the sketch again, **Modify** the length back to the design dimension of **9.94**, and **Regenerate** once more. Complete the protrusion by giving the following commands:

Done ⇒ **360** (from the REV TO menu) ⇒ **Done** ⇒ **OK** ⇒ **View** ⇒ **Cosmetic** ⇒ **Shade** ⇒ **Display** ⇒ **Done-Return** (Fig. 6.15) ⇒ **Done**

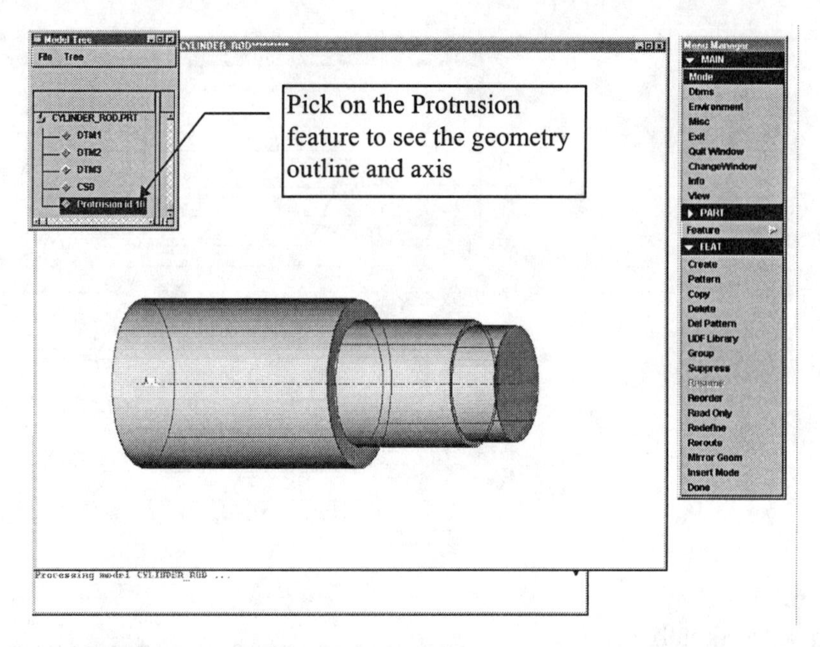

Dbms ⇒ Save ⇒ enter
Purge ⇒ enter ⇒ Done-Return

Figure 6.15
Revolved Protrusion

The next two features are neck cuts similar to the Pin in Lesson 5:

View ⇒ Repaint ⇒ Done-Return ⇒ Environment ⇒ Shading ⇒ □ **Grid Snap ⇒ Done-Return ⇒ Feature ⇒ Create ⇒ Neck ⇒ 360 ⇒ One Side ⇒ Done ⇒ Use Prev ⇒ Okay ⇒ Sketch View** (to go into a 2D view of the section) ⇒ **Sketch ⇒ Line ⇒ Centerline ⇒ Horizontal** (sketch the centerline used to revolve the neck on **DTM2**) ⇒ **Sketch ⇒ Line ⇒ Vertical** (sketch the three lines of the neck) ⇒ **Regenerate ⇒ Alignment** (align the centerline and **DTM2**, align the two endpoints of the vertical lines with the top of the part, and align the left *vertical* line with the part's vertical step, as in Figure 6.16) ⇒ **Dimension ⇒ Regenerate ⇒ Modify** (change the *sketch* dimensions to the *design* dimensions) ⇒ **Regenerate ⇒ Done ⇒ Done**

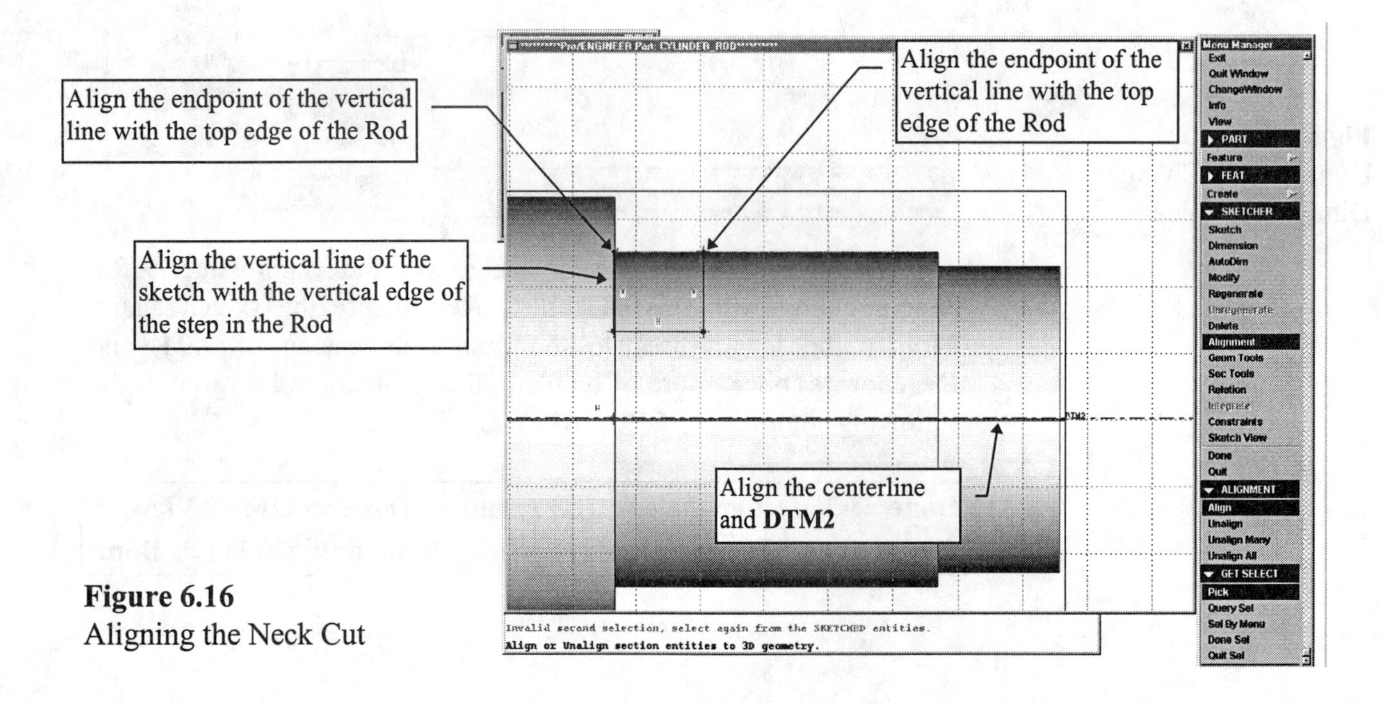

Figure 6.16
Aligning the Neck Cut

HINT

Always exaggerate the size of the sketch features. **Dimension** the sketch and **Regenerate**, then **Modify** to the design sizes. ***Never*** modify the sketch dimensions before you get a successful regeneration!

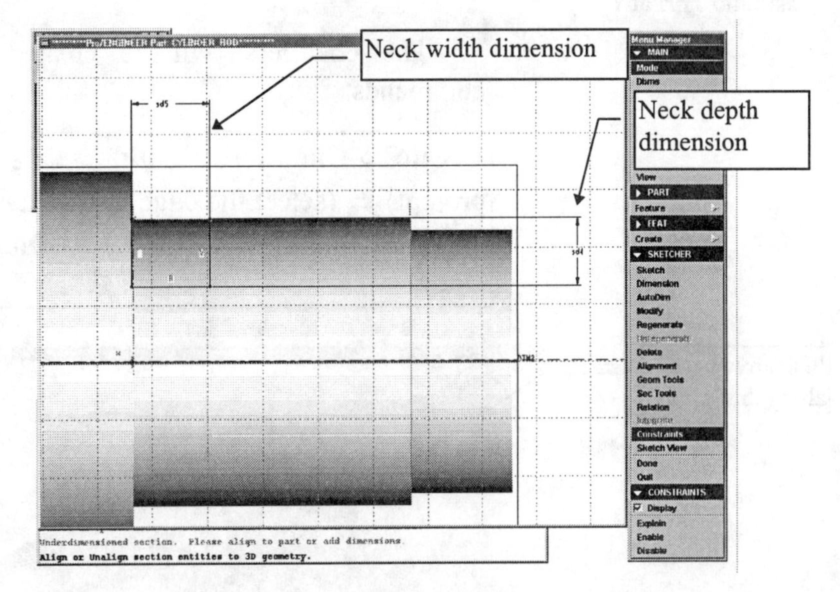

Figure 6.17
Dimensioning the **.14 X .045** Neck

After the first neck is successfully created (Fig. 6.18), model the second neck using the same procedure (Fig. 6.19).

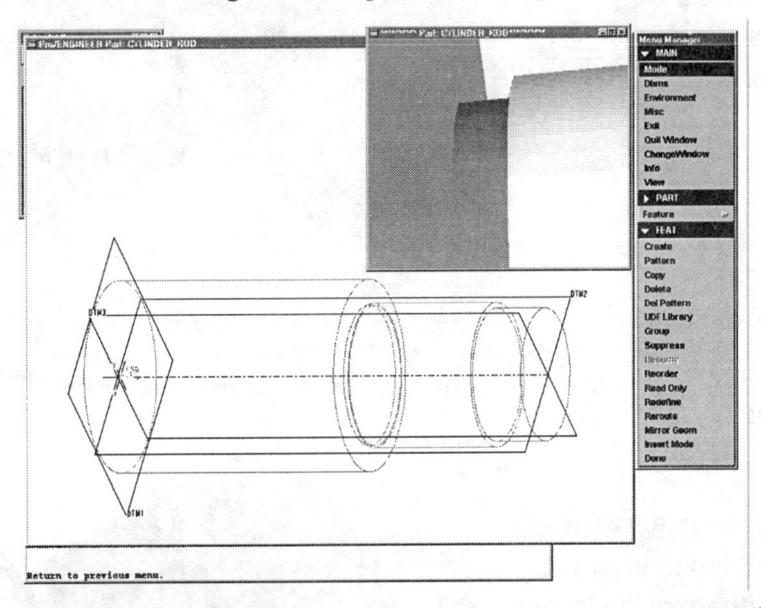

Figure 6.18
First Neck

Dbms ⇒ Save ⇒ enter
Purge ⇒ enter ⇒ Done-Return

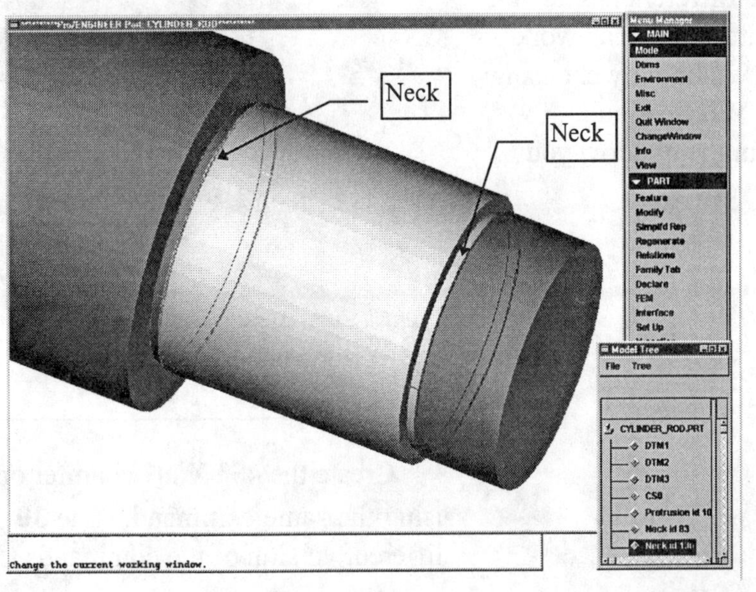

Figure 6.19
Both Necks Modeled Using the Same Dimensions

The chamfers will be created next. Give the following commands:

Create ⇒ **Chamfer** ⇒ **Solid** ⇒ **Edge** ⇒ **45 x d** ⇒ (type **.125** at the prompt) ⇒ (select the edge to be chamfered as shown in Figure 6.20 and Figure 6.21) ⇒ **Done Sel** ⇒ **Done Refs** ⇒ **Preview** ⇒ **OK** ⇒ **Done**

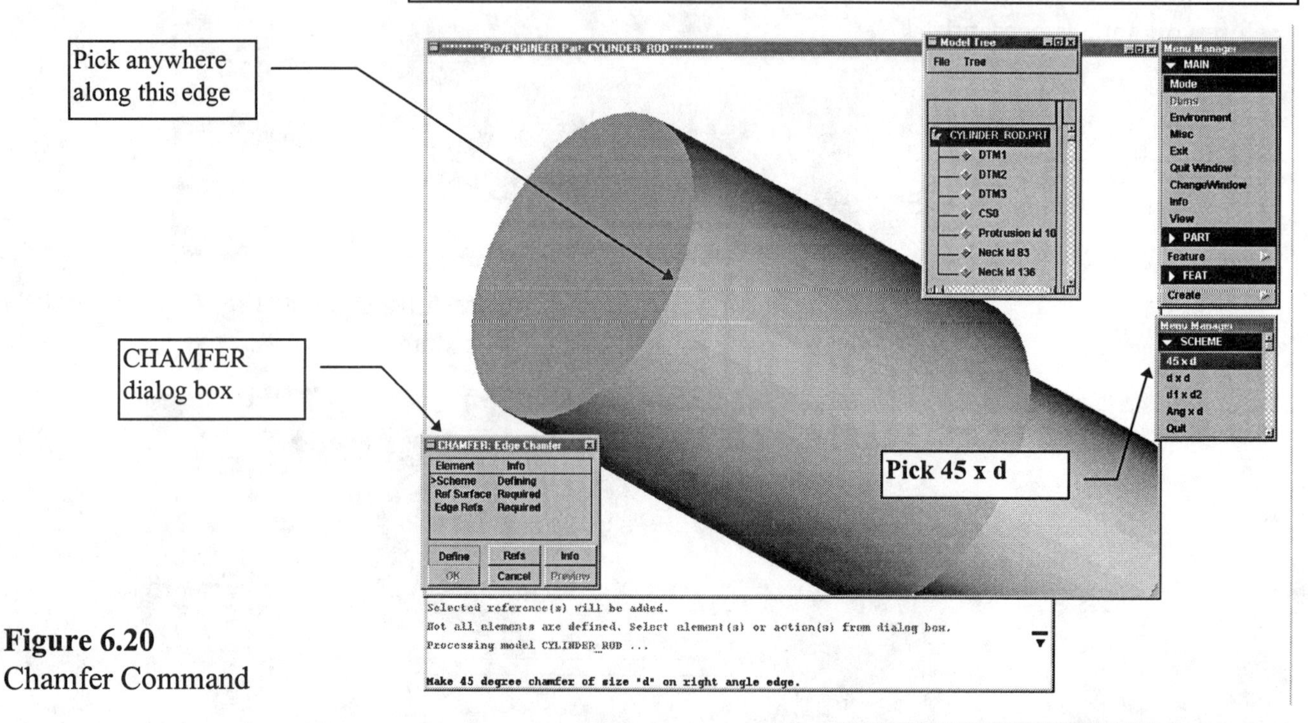

Figure 6.20
Chamfer Command

HINT

Open new window (**View** ⇒ **New Window**) to view your part from different angles and zoom states. You can work in either window by choosing **Change Window** and picking in the window you wish to work.

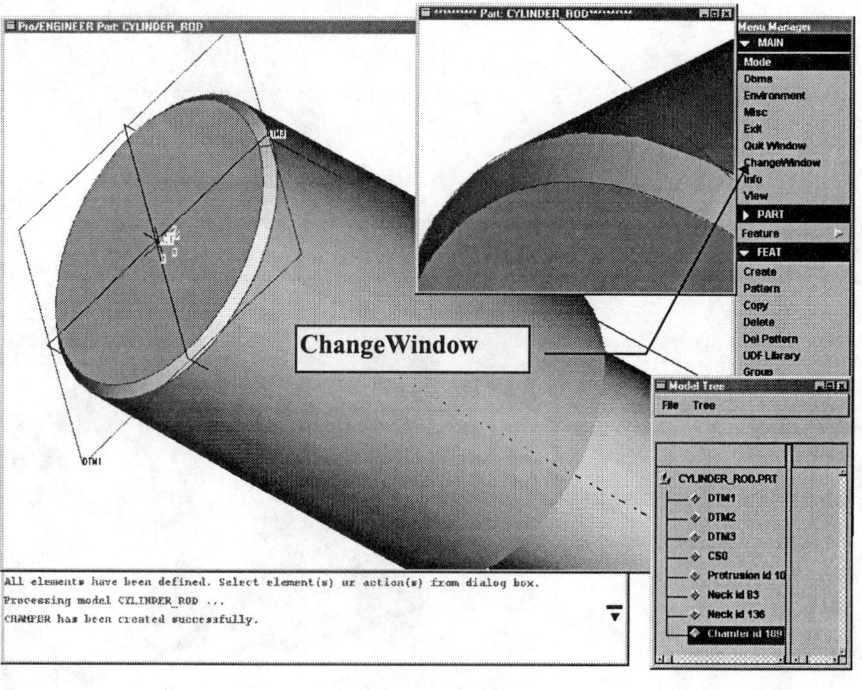

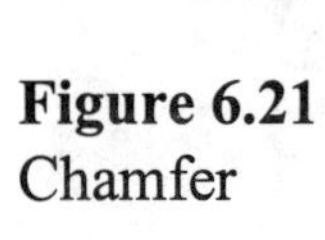

Figure 6.21
Chamfer

Create the **45° X .09** chamfer on the **2.75** diameter side of the part using the same command. The **30° X .14** chamfer requires a slightly different version of the commands (Fig. 6.22):

Feature ⇒ Create ⇒ Chamfer ⇒ Solid ⇒ Edge ⇒ Ang x d ⇒ (type **.14** when prompted for the distance) ⇒ (type **30** when prompted for the angle) ⇒ (select the cylinder's revolved surface as the reference surface) ⇒ (select the edge to be chamfered as shown in Figure 6.22) ⇒ **Done Sel** ⇒ **Done Refs** ⇒ **OK** (Fig. 6.23) ⇒ **Done**

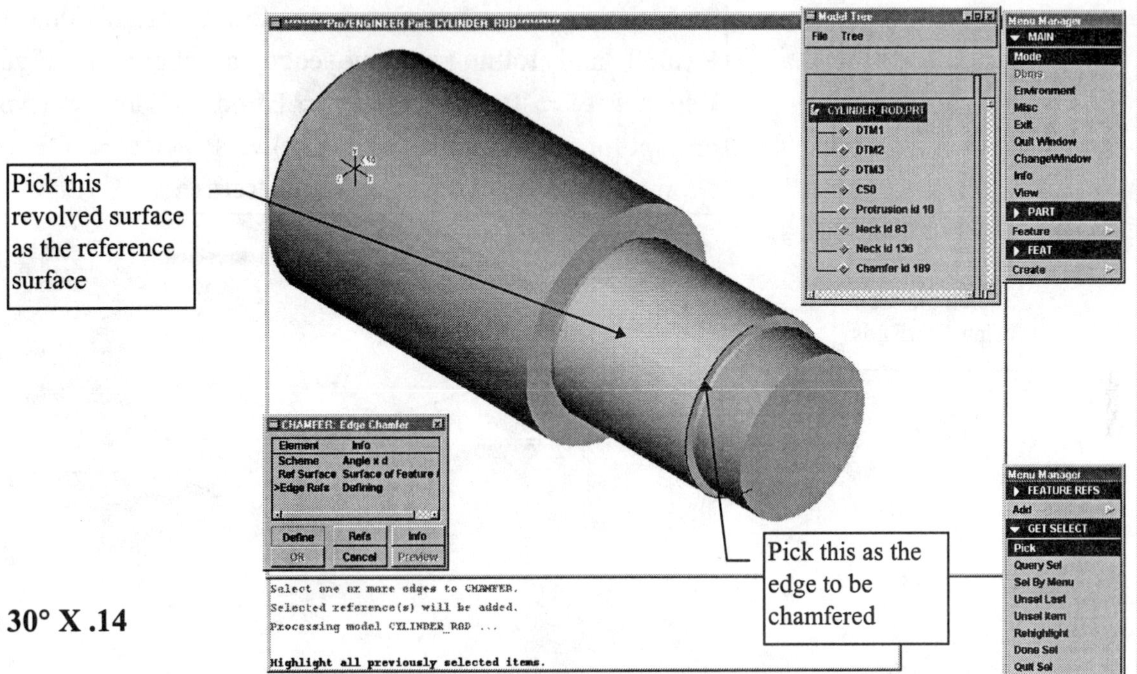

Figure 6.22
Creating the **30° X .14** Chamfer

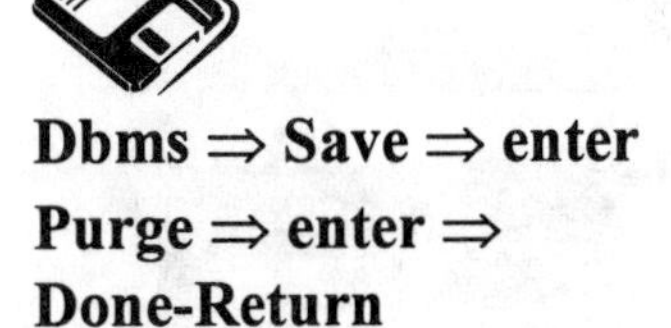

Dbms ⇒ Save ⇒ enter
Purge ⇒ enter ⇒
Done-Return

30° X .14
Chamfer

Figure 6.23
Chamfer **30° X .14**

Create the key seat using an extruded cut:

Feature ⇒ **Create** ⇒ **Cut** ⇒ **Extrude** ⇒ **Solid** ⇒ **Done** ⇒ **Both Sides** ⇒ **Done** ⇒ **Use Prev** ⇒ **OKAY** ⇒ **Sketch View** ⇒ **Arc** ⇒ **Ctr/Ends** ⇒ (sketch the arc as in Figure 6.24) ⇒ **Sketch** ⇒ **Line** ⇒ **Centerline** ⇒ **Vertical** (sketch the vertical centerline through the arc's center point, as in Figure 6.25) ⇒ **Regenerate** ⇒ **Alignment** (align the endpoints of the arc to the upper edge of the part, as in Figure 6.25) ⇒ **Dimension** (add the three dimensions) ⇒ **Regenerate** ⇒ **Modify** (change the dimensions to the design values of **.780** deep, **R1.030**, and **1.480** from the edge, as shown in Figure 6.26) ⇒ **Regenerate** ⇒ **Done** ⇒ **Okay** ⇒ **Blind** ⇒ **Done** ⇒ (type **.500** at the prompt for the width of the keyseat) ⇒ **Preview** ⇒ **OK** (Fig. 6.27) ⇒ **Done** ⇒ **View** ⇒ **Default** ⇒ **Done-Return**

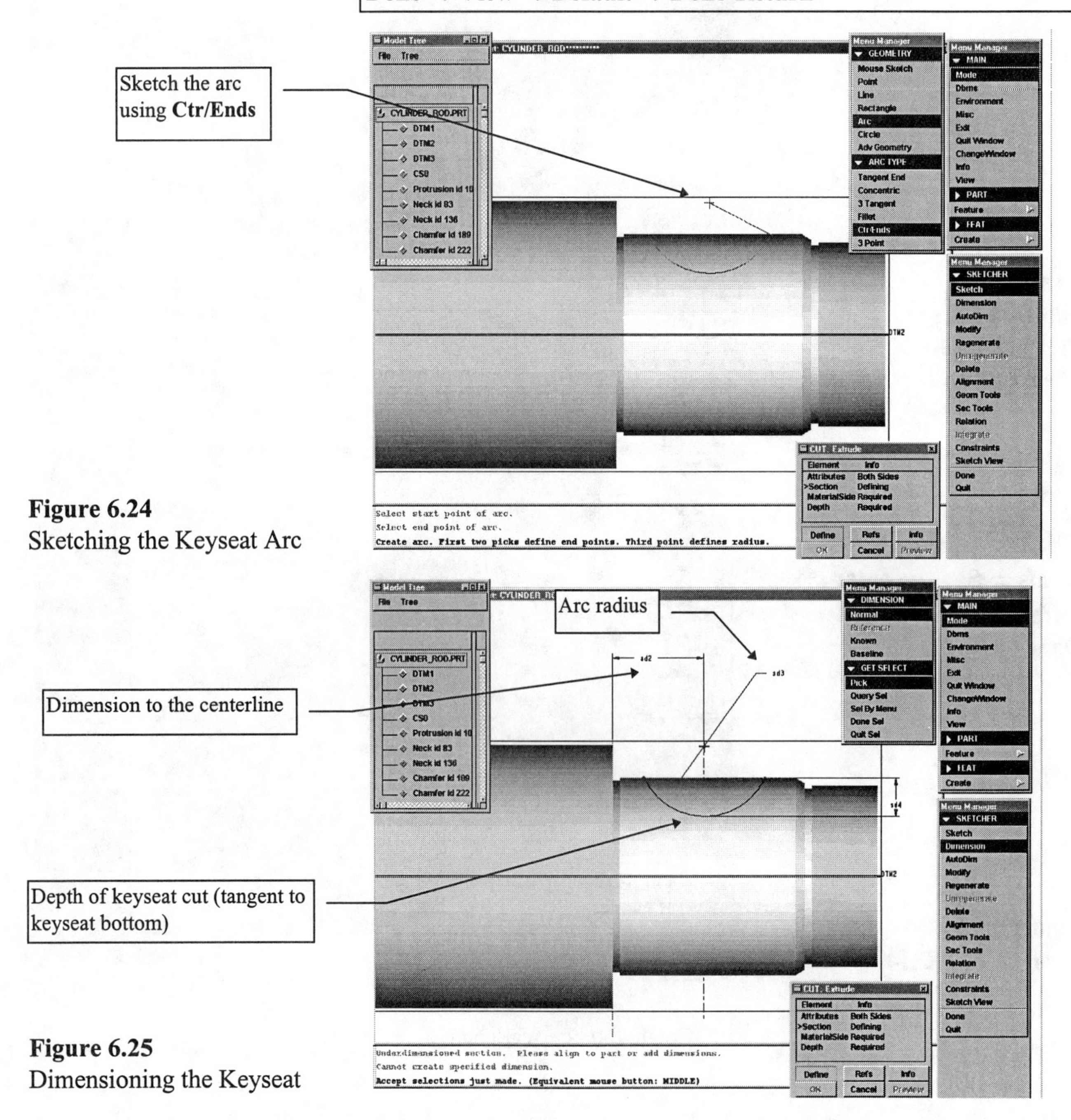

Figure 6.24
Sketching the Keyseat Arc

Figure 6.25
Dimensioning the Keyseat

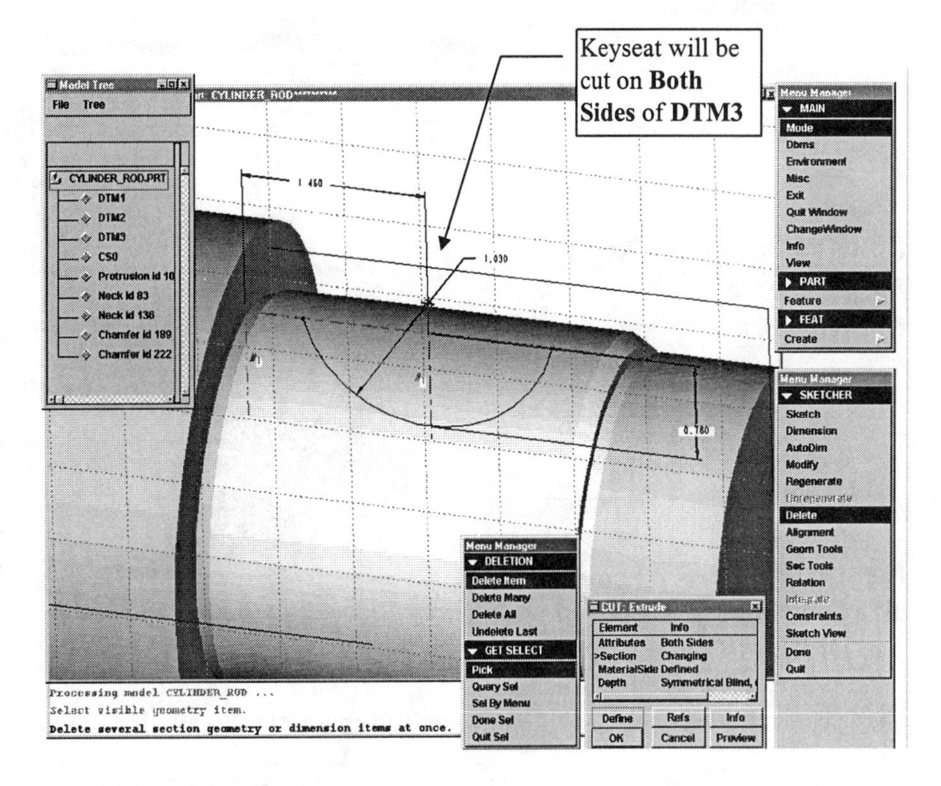

Figure 6.26
Keyseat Dimensions

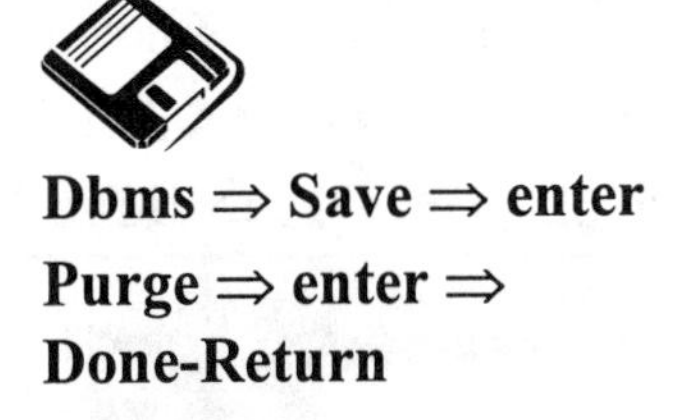

Dbms ⇒ Save ⇒ enter
Purge ⇒ enter ⇒
Done-Return

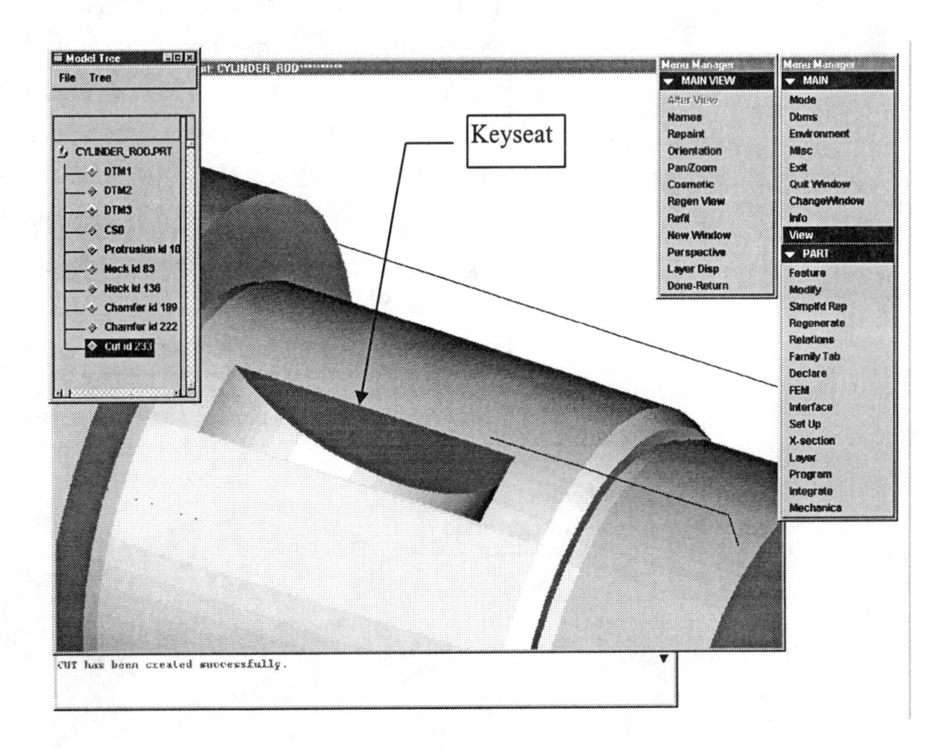

Figure 6.27
Keyseat

At this stage of your understanding of Pro/E, you should be able to add the tap drill hole of **.578 X 1.50 DEEP** to the **4.00** diameter side of the part. Use coaxial for the hole placement (Fig. 6.28).

The last features to be added to the Cylinder Rod will be cosmetic threads that will establish the thread size, type, class, etc. for the internal tap drill hole and the external thread that needs to be created on the **2.75** diameter side of the rod.

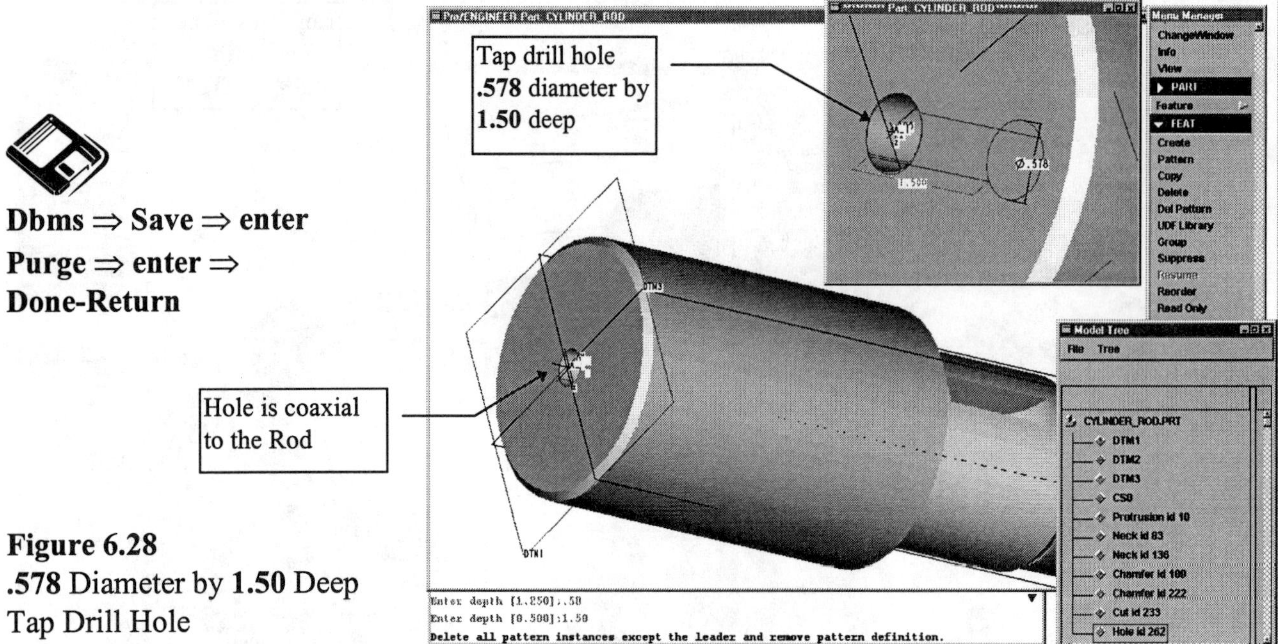

Dbms ⇒ Save ⇒ enter
Purge ⇒ enter ⇒
Done-Return

Figure 6.28
.578 Diameter by **1.50** Deep Tap Drill Hole

Create the cosmetic threads for the hole with the following command:

HINT
You can select feature geometry directly from the screen or from the Model Tree.

Feature ⇒ Create ⇒ Cosmetic ⇒ Thread ⇒ (pick the hole or select from the Model Tree) ⇒ (pick the starting surface for the thread feature as in Figure 6.29) ⇒ **Okay** (for the direction of feature creation) ⇒ **Blind ⇒ Done ⇒** (type **1.25** as the depth of the cosmetic thread) ⇒ **enter ⇒** (type **.625** as the thread diameter) ⇒ **enter ⇒ Mod Params ⇒** (Pro/TABLE displays the thread options as shown in Figure 6.30; fill in the table information) ⇒ **File ⇒ Save ⇒ File ⇒ Exit ⇒ Show** (Fig. 6.31) ⇒ type **Q** (to quit the INFORMATION WINDOW) ⇒ **Done/Return ⇒ Preview** (Fig. 6.32) ⇒ **OK ⇒ Done**

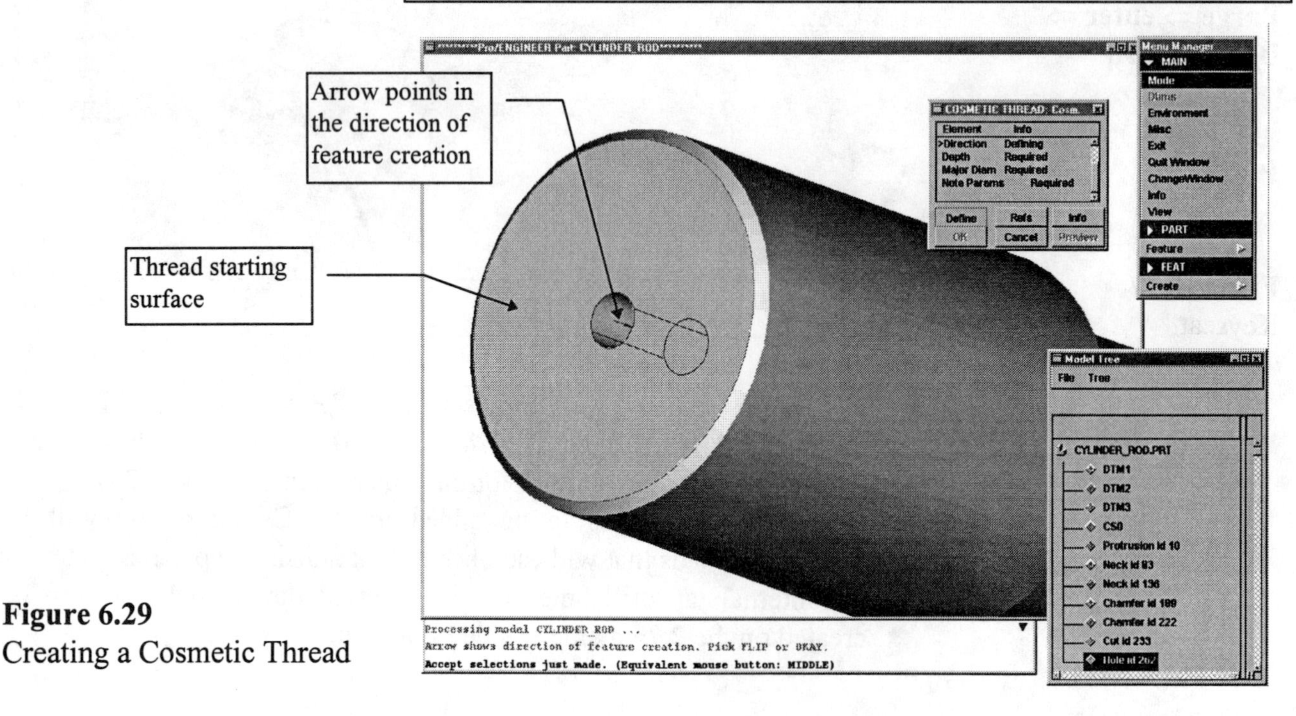

Figure 6.29
Creating a Cosmetic Thread

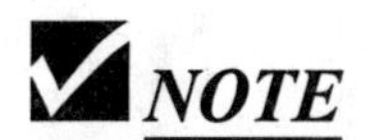

NOTE

Always **Save** the table information before exiting.

Pro/TABLE TM Release 18.0 (c) 1988-95 by Parametric Technology Corporation All Rights Reserved

File Edit View Format Help

! Thread Major Diameter

! Thread parameters and callouts
! Thread Major Diameter
MAJOR_DIAMETER .625
! Threads Per Inch (Pitch)
THREADS_PER_INCH 18
! Thread Form
FORM UNF
! Thread Class
CLASS 3
! Thread Placement (A=external, B=internal)
PLACEMENT B
! Thread Is Metric (True or False)
METRIC FALSE

MAJOR_DIAMETER	.625
THREADS_PER_INCH	18
FORM	UNF
CLASS	3
PLACEMENT	B
METRIC	FALSE

Hit Space or Return to continue ('q' to quit, 'b' for previous page)

FEAT PARAM
Retrieve
Save
Mod Params
Show
Done/Return

Figure 6.30
Thread Table

HINT

Turn off the shading and set Hidden Line as the default, otherwise you will not be able to see the cosmetic threads.

INFORMATION WINDOW

```
! Thread parameters and callouts
! Thread Major Diameter
MAJOR_DIAMETER 0.625
! Threads Per Inch (Pitch)
THREADS_PER_INCH 18
! Thread Form
FORM UNF
! Thread Class
CLASS 3
! Thread Placement (A=external, B=internal)
PLACEMENT B
! Thread Is Metric (True or False)
METRIC FALSE

Hit Space or Return to continue ('q' to quit, 'b' for previous page)
```

Processing model LESSON6 ...
The default view is isometric.
Modify parameters using the editor

Menu Manager
PART: Feature, Modify, Simplfd Rep, Regenerate, Relations, Family Tab, Declare, FEM, Interface, Set Up, X-section, Layer, Program, Integrate, Mechanica
MODIFY: Value, DimCosmetics, Move Datum, Make Indep, Geom Tol, PatternTable, Line Style, Done
GET SELECT: Pick, Query Sel, Sel By Menu, Done Sel, Quit Sel
FEAT PARAM: Retrieve, Save, Mod Params, Show, Done/Return

Figure 6.31
Showing the Feature Parameters

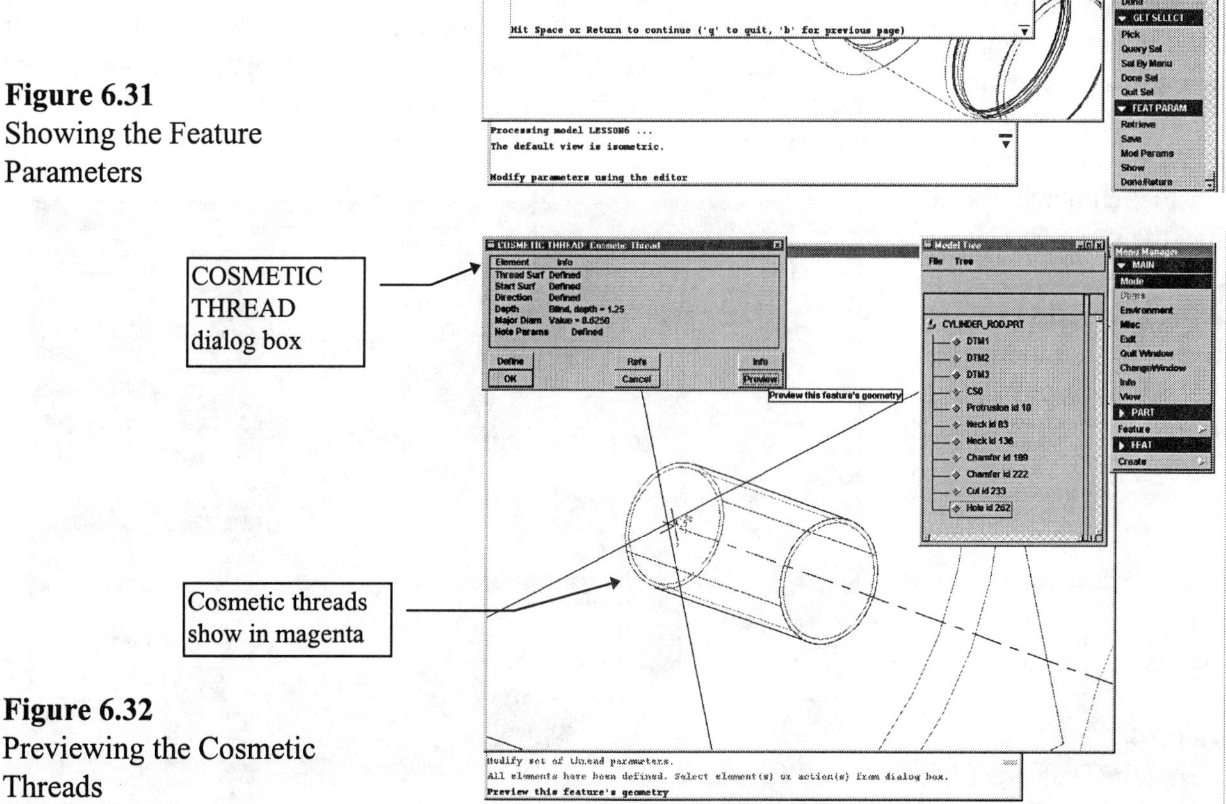

Figure 6.32
Previewing the Cosmetic Threads

Create **DTM4** offset from the end of the Rod a distance of **.09** so that it passes through the chamfer (Fig. 6.33)

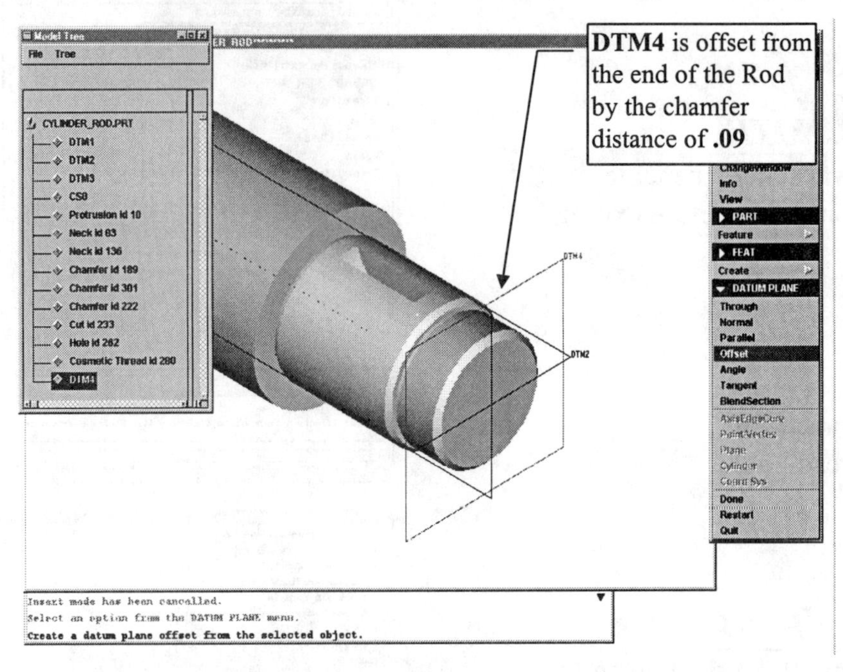

Figure 6.33
DTM4 Offset from the Rod End by **.09**

HINT

Internal cosmetic threads represent the nominal diameter. *External cosmetic* threads represent the root diameter. After creating the cosmetic thread, you must edit the thread table using Pro/TABLE. The diameter for a cosmetic internal thread will stay the same on the table, but the thread size of an external thread must be changed to the nominal size from the shown root diameter.

Measure the distance from **DTM4** to the lip of the neck cut before creating the cosmetic thread, as shown in Figure 6.34. Now create the other cosmetic thread, on the **2.75** diameter surface. The thread will start on **DTM4** and have a depth of **.920000** (you may need to type **-.920000** if the arrow points away from the part). The Thread Table information (Fig. 6.35) will show the diameter as **2.75**, but the magenta-colored cylinder representing the thread on your screen is **2.6875** in diameter (Fig. 6.36). The thread diameter will be smaller than the nominal thread size of **2.75** (Fig. 6.37) since you are representing the *root diameter* of the thread **(2.6875)**.

Distance measurement

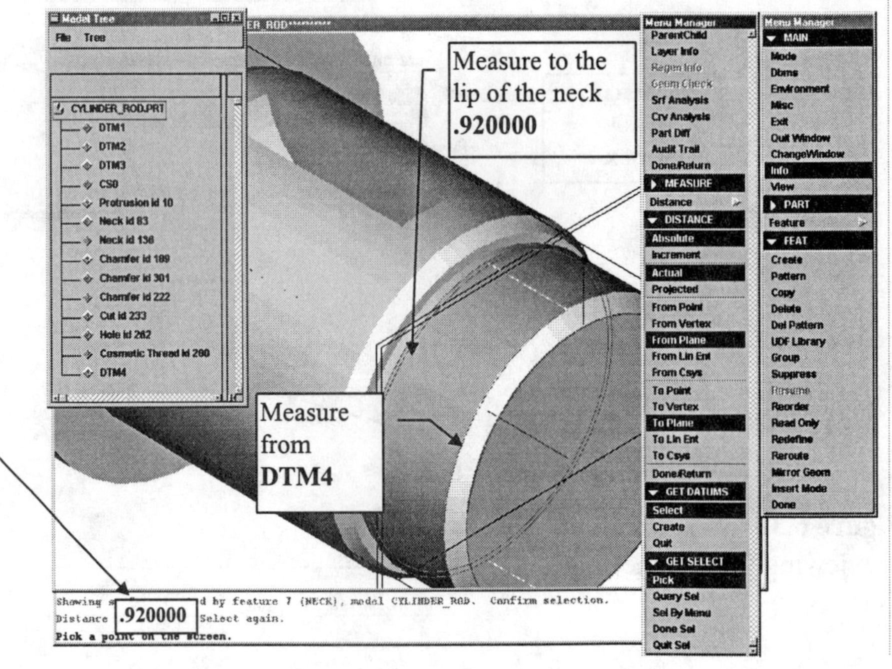

Figure 6.34
Measuring the Distance from **DTM4** to the Lip of the Neck

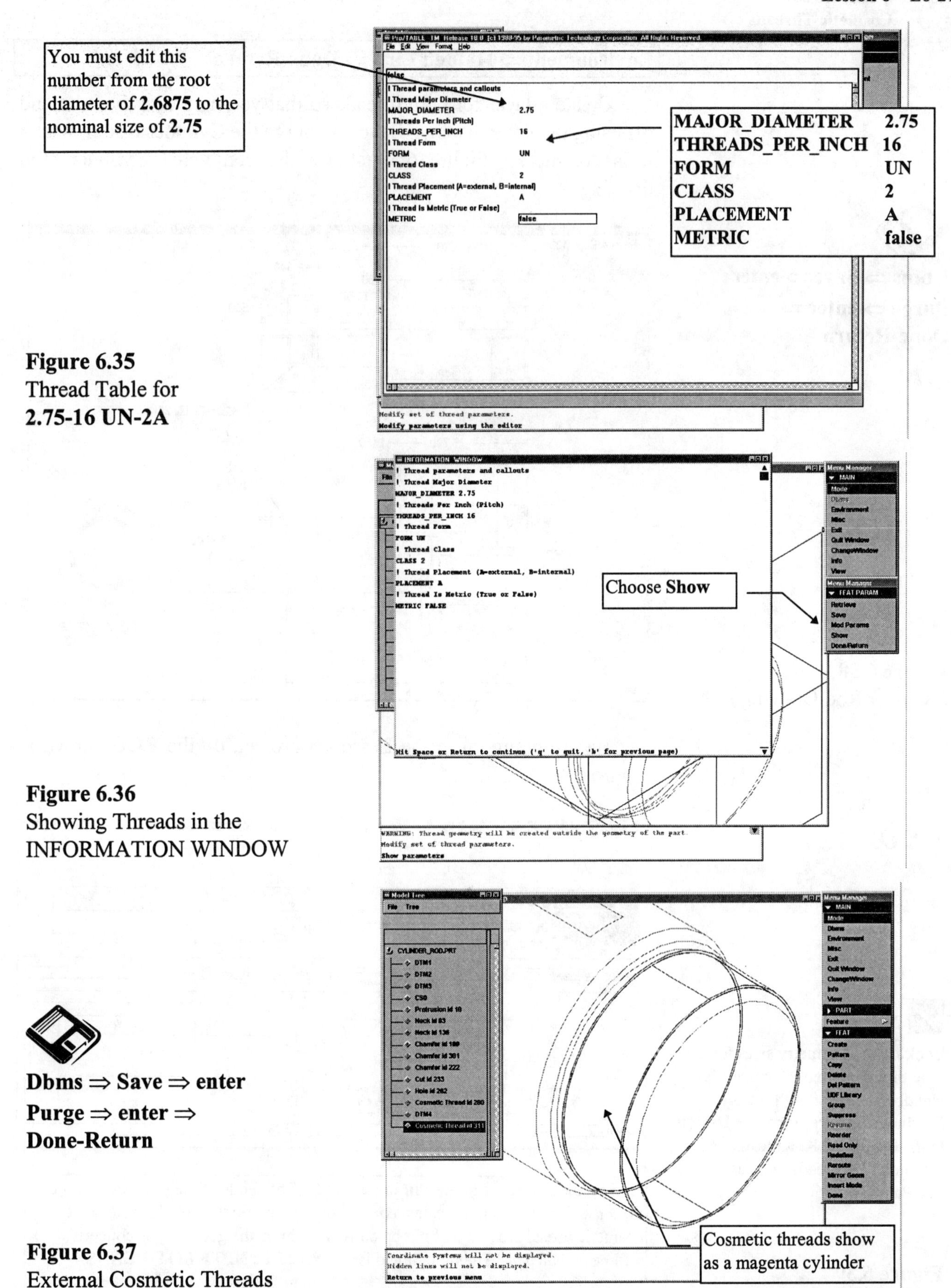

Figure 6.35
Thread Table for
2.75-16 UN-2A

Figure 6.36
Showing Threads in the
INFORMATION WINDOW

Dbms ⇒ Save ⇒ enter
Purge ⇒ enter ⇒
Done-Return

Figure 6.37
External Cosmetic Threads

Environment ⇒ Hidden Line ⇒ Done-Return

Create a layer for both threads so that you may turn them on and off using the layer display. Save and purge the Cylinder Rod now that it is complete. Figure 6.38 shows the completed Cylinder Rod drawing.

Dbms ⇒ Save ⇒ enter

Purge ⇒ enter ⇒ Done-Return

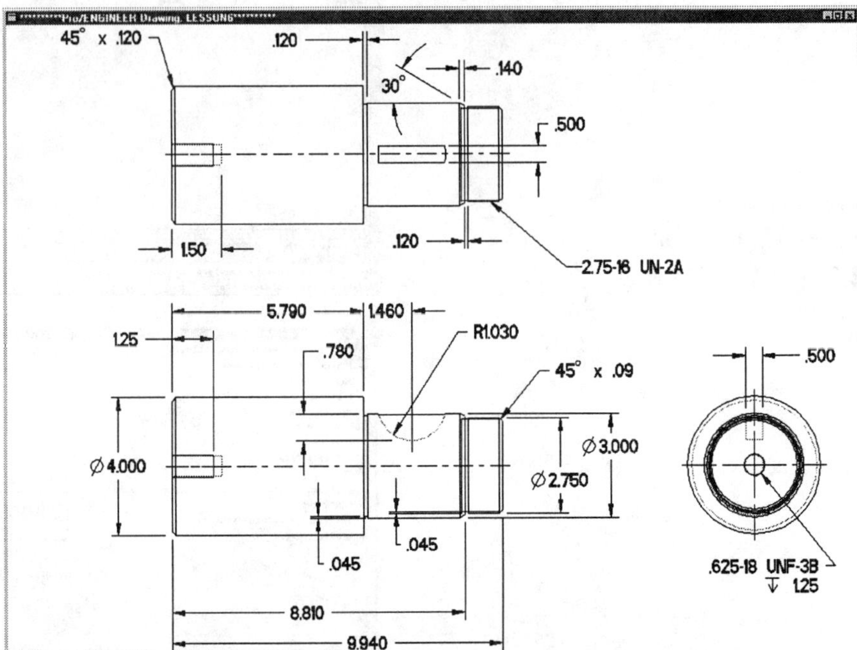

Figure 6.38
Cylinder Rod Drawing

Before going on to the Lesson 6 Project, do the **ECO** shown in Figure 6.39.

ECO

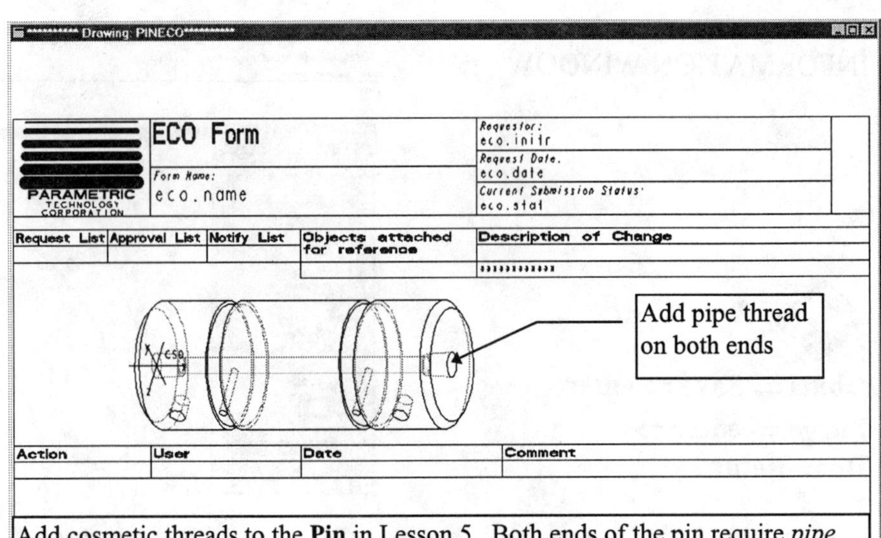

Drawing: PINECO

PARAMETRIC TECHNOLOGY CORPORATION	ECO Form	Requestor: eco.initr
		Request Date: eco.date
	Form Name: eco.name	Current Submission Status: eco.stat

Request List	Approval List	Notify List	Objects attached for reference	Description of Change

Action	User	Date	Comment

Add cosmetic threads to the **Pin** in Lesson 5. Both ends of the pin require *pipe threads*. Pro/E does not create a *conical* pipe cosmetic thread. Therefore the cosmetic thread created will not be correct as far as the geometry is concerned-- it will be *cylindrical*. The pipe thread is: **.125-27 NPTF BOTH ENDS .44 DEEP**

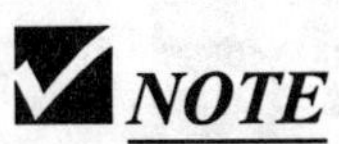

NOTE
Look up the geometry sizes for a pipe thread in your Machinery's Handbook, a drafting text, or **Engineering Graphics and Design** by L. Lamit and K. Kitto Chapter 17, Threads and Fasteners See Pages: 661-662

Figure 6.39
ECO for the Pin in Lesson 5

Lesson 6 Project

Clamp Ball and Coupling Shaft

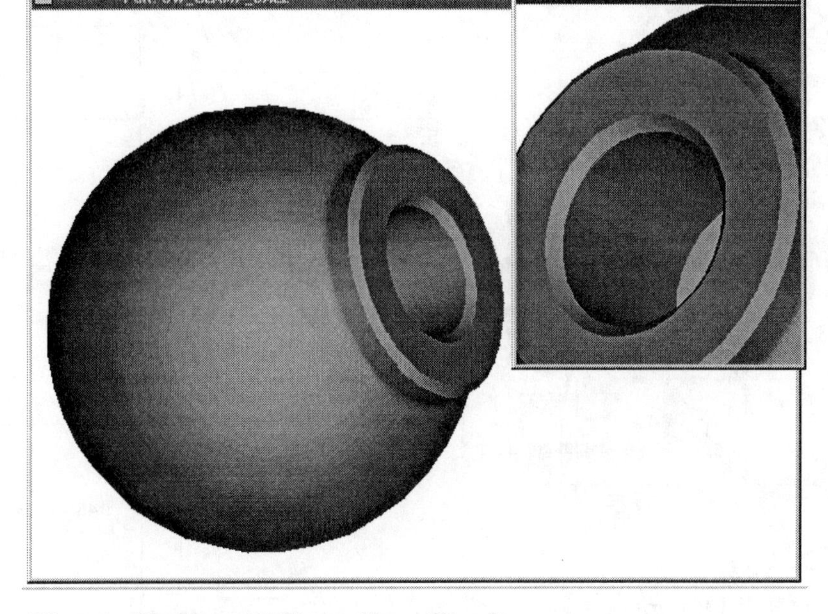

Figure 6.40
Clamp Ball

Clamp Ball and Coupling Shaft

As with Lesson 5, two lesson projects are provided here in Lesson 6 (Figure 6.40 through Figure 6.51). You will use both parts in Lessons 14 and 15 when creating different assemblies. Both the **Clamp Ball** (decimal inches) and the **Coupling Shaft** (SI units) are revolved protrusions. The Clamp Ball is simpler and easier to complete. The Ball is *black plastic* and the Shaft is *steel*. Create all cosmetic threads required on each part. The two parts are used on different assemblies.

Analyze the parts and plan out the steps and features required to model them. Use the **DIPS** in Appendix D to establish a feature creation sequence before the start of modeling. Remember to set up the environment, establish datum planes, and set them on layers.

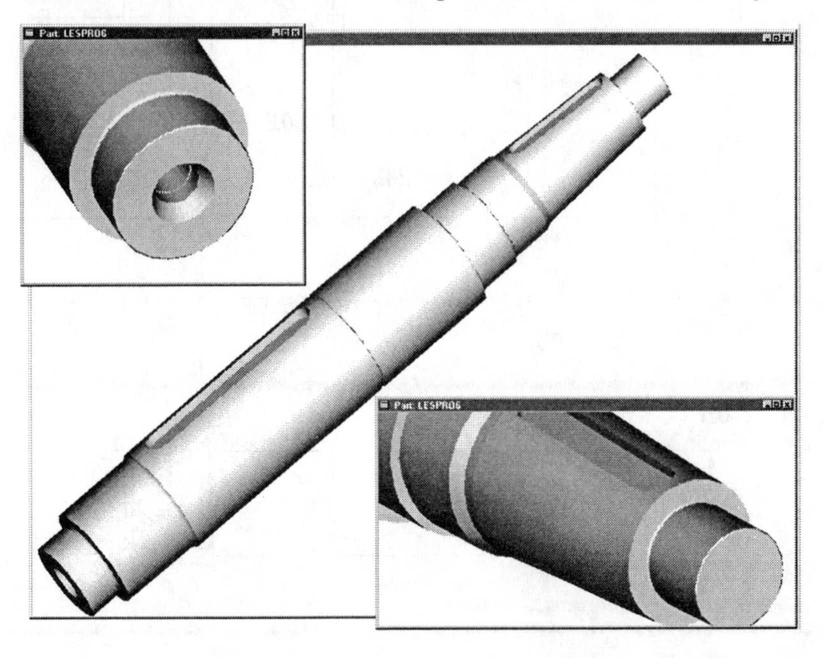

Figure 6.41
Coupling Shaft

Dbms ⇒ Save ⇒ enter
Purge ⇒ enter ⇒ Done-Return

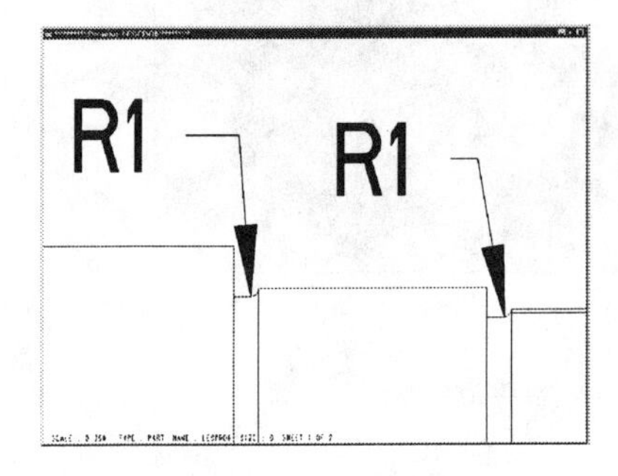

Figure 6.42
Clamp Ball Dimensions

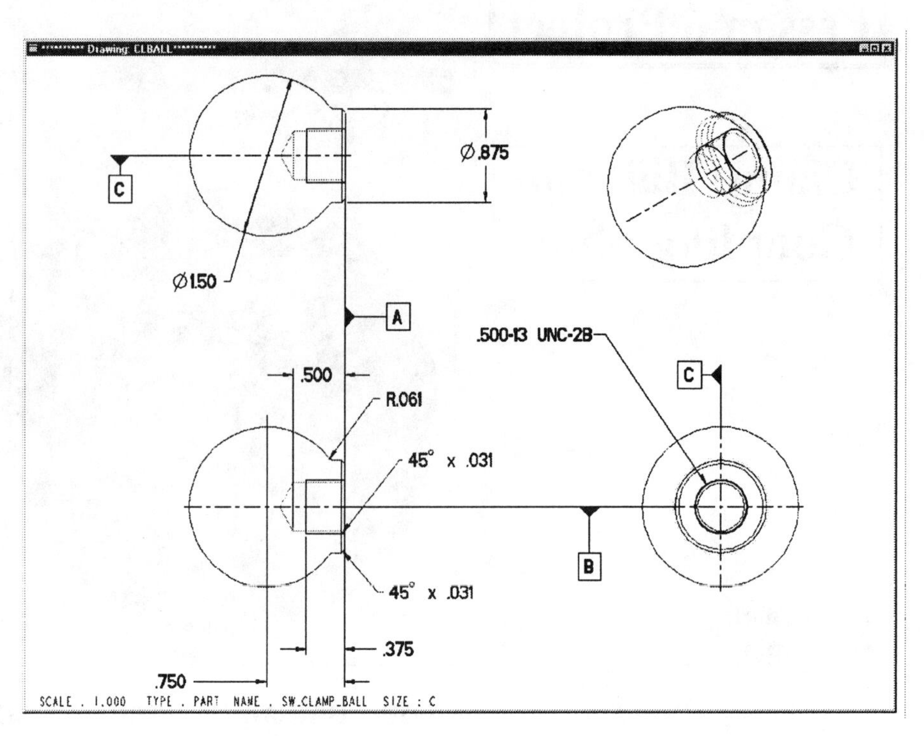

Figure 6.43
Coupling Shaft Drawing, Sheet One

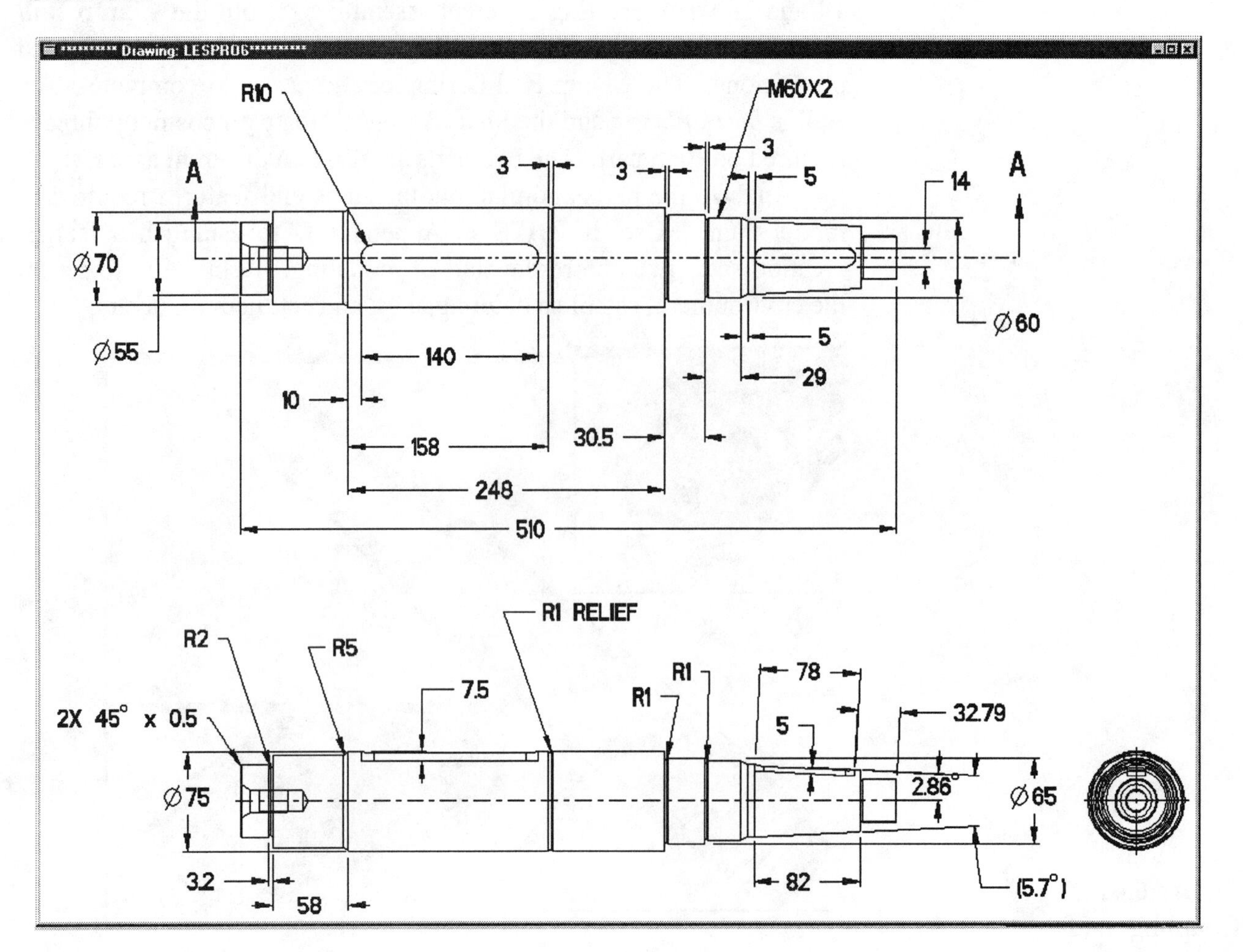

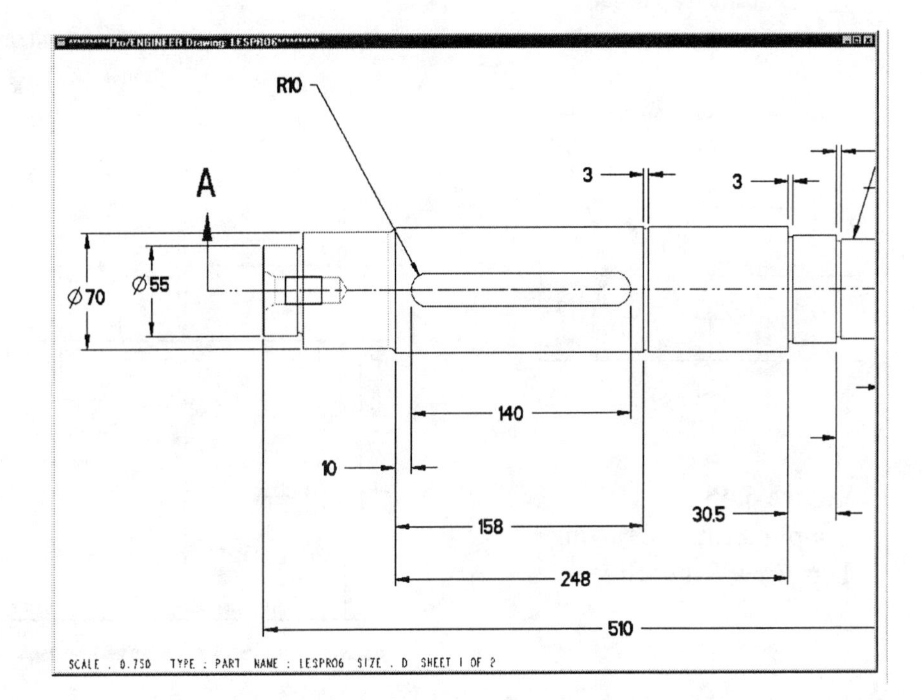

Figure 6.44
Coupling Shaft Drawing, Top View Left Side

Figure 6.45
Coupling Shaft Drawing, Sheet Two

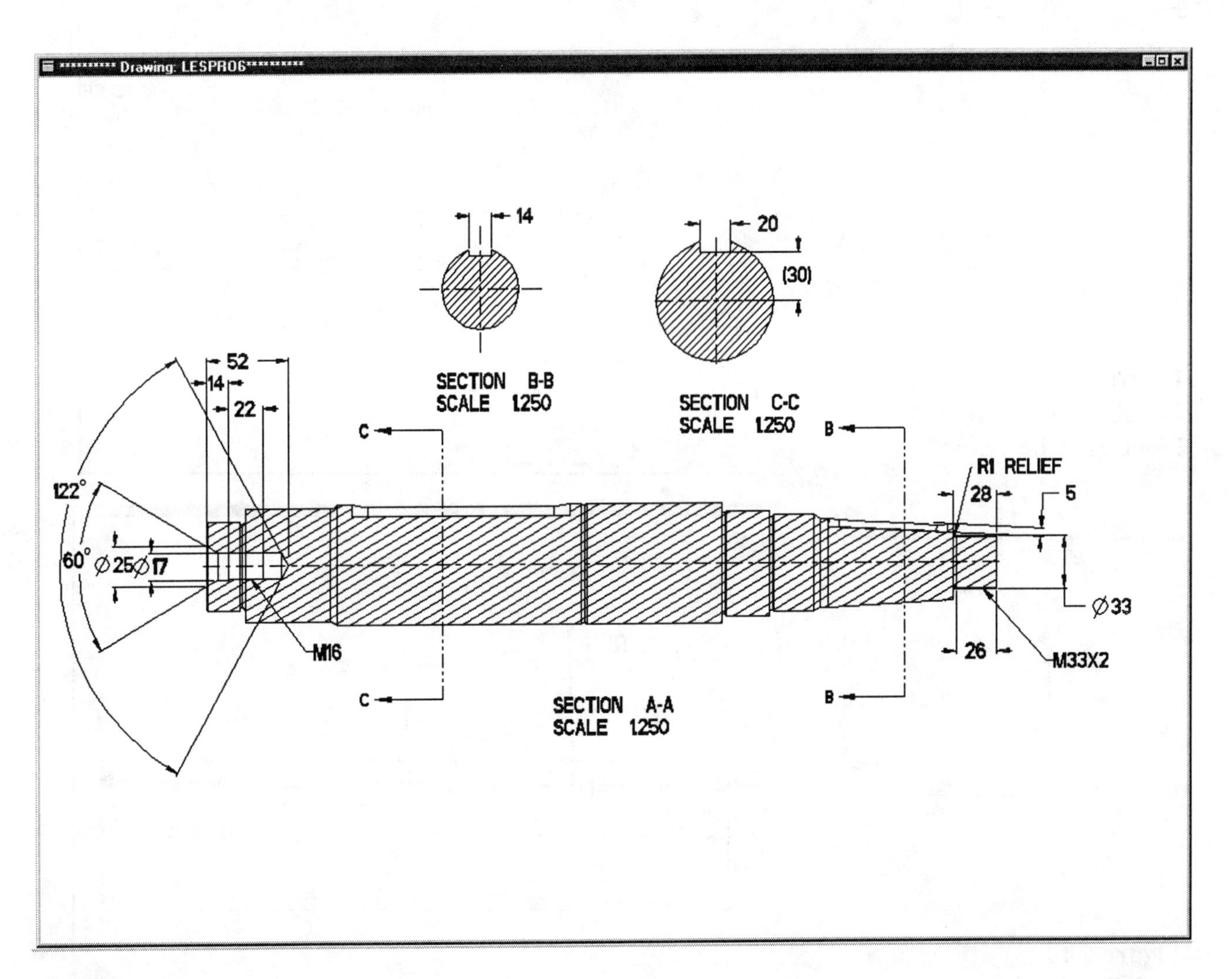

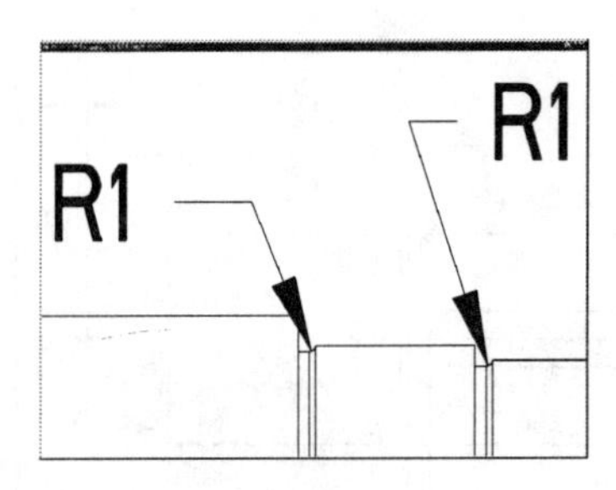

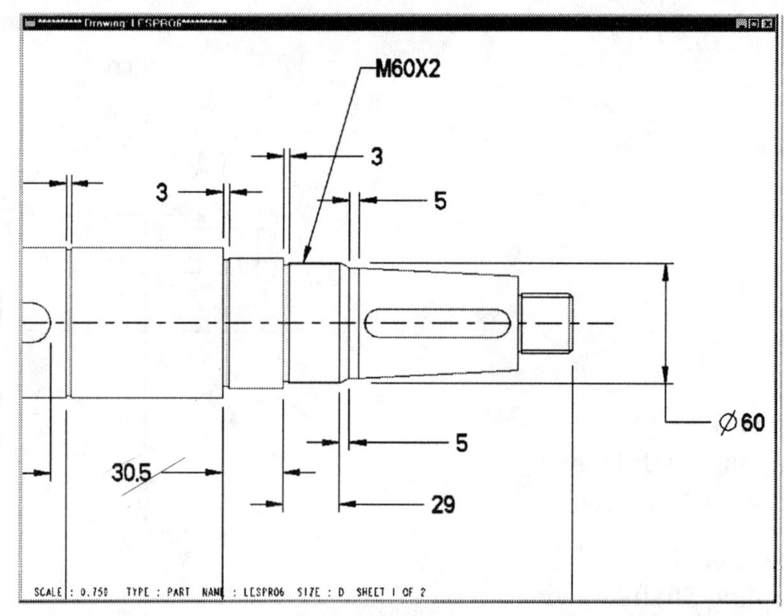

Figure 6.46
Coupling Shaft Drawing, Top View Right Side

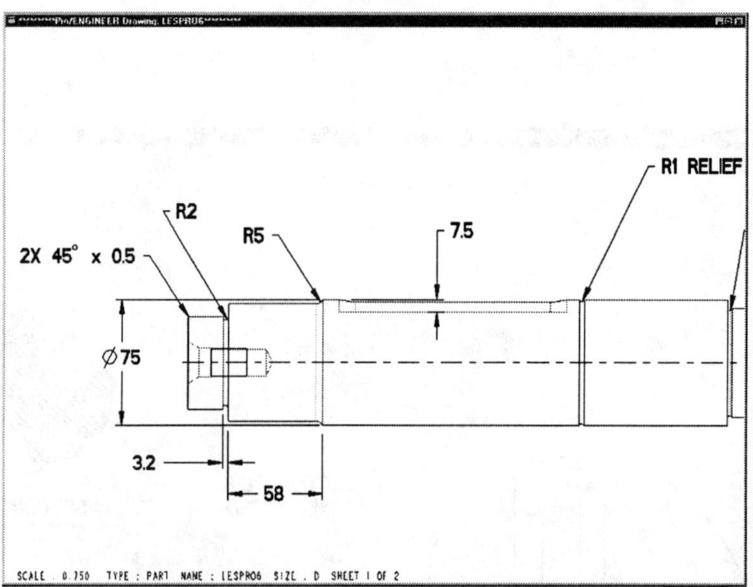

Figure 6.47
Coupling Shaft Drawing, Front View Left Side

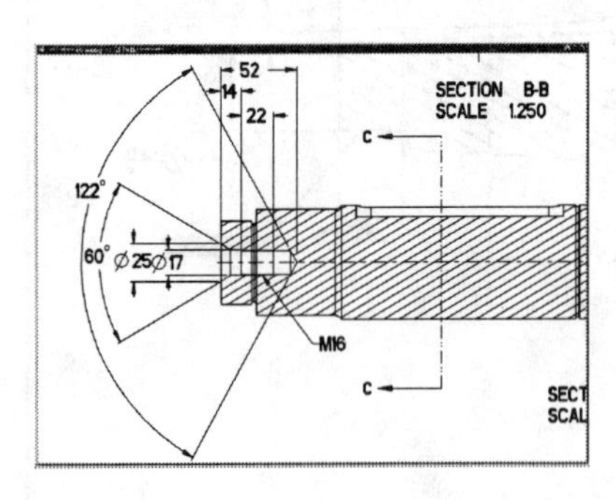

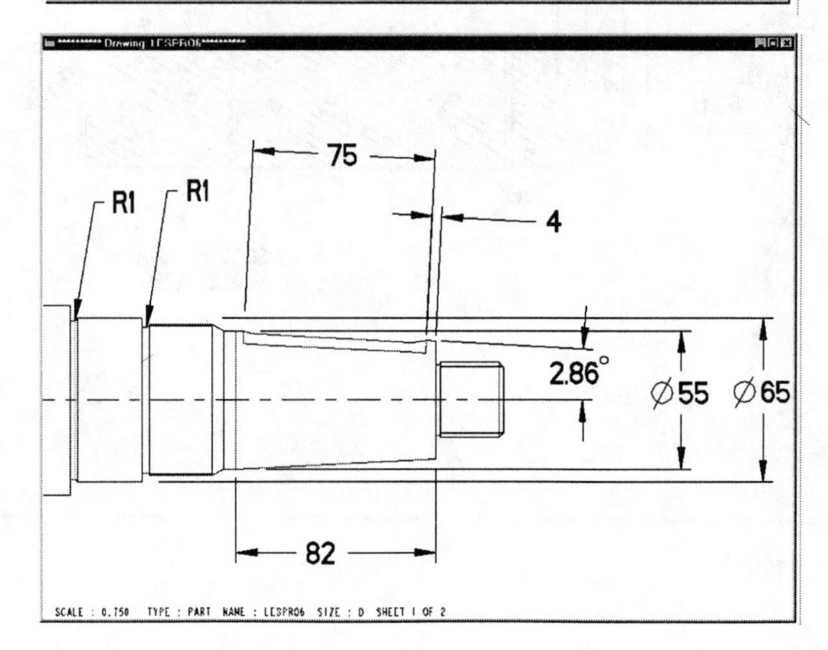

Figure 6.48
Coupling Shaft Drawing, Front View Right Side

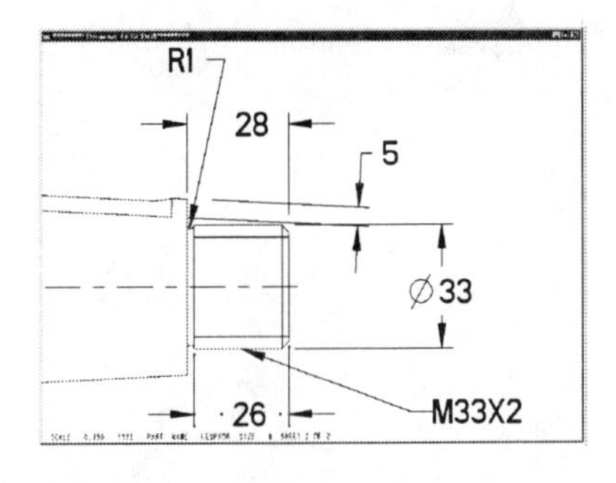

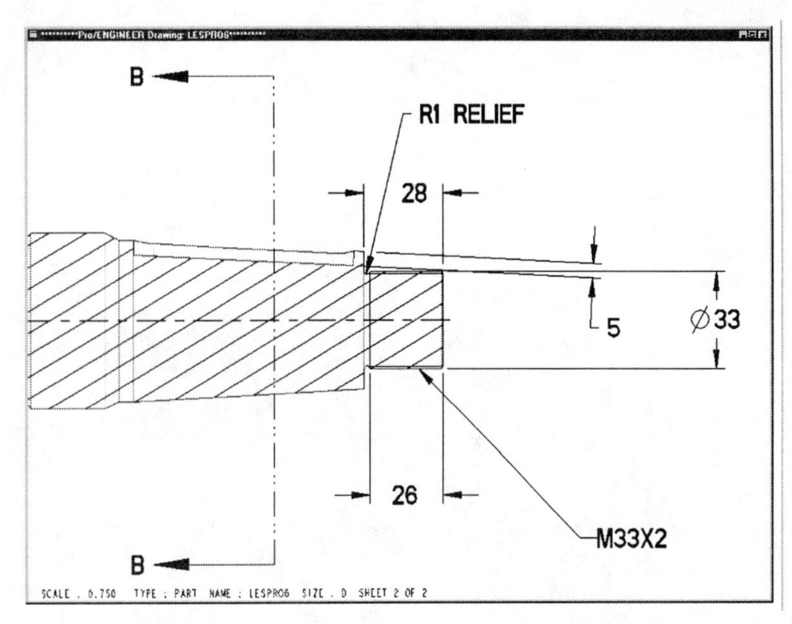

Figure 6.49
Coupling Shaft Drawing, Sheet Two, **SECTION A-A** Right Side

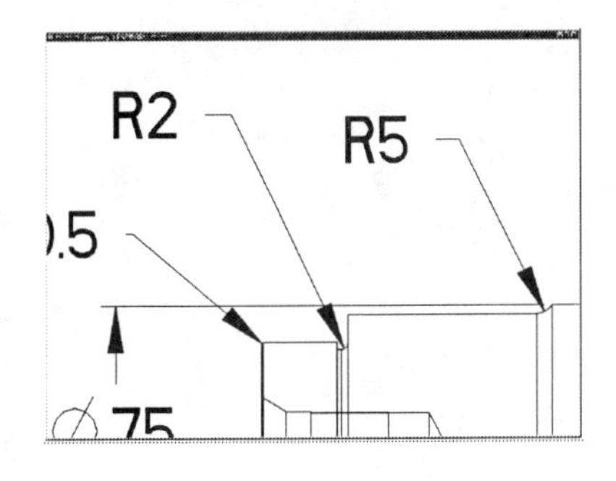

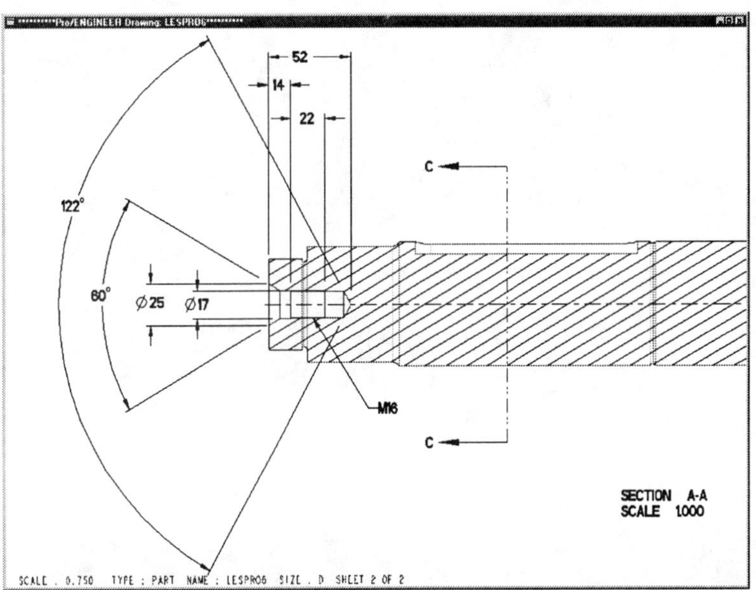

Figure 6.50
Coupling Shaft Drawing, Sheet Two, **SECTION A-A** Left Side

Dbms ⇒ Save ⇒ enter
Purge ⇒ enter ⇒
Done-Return

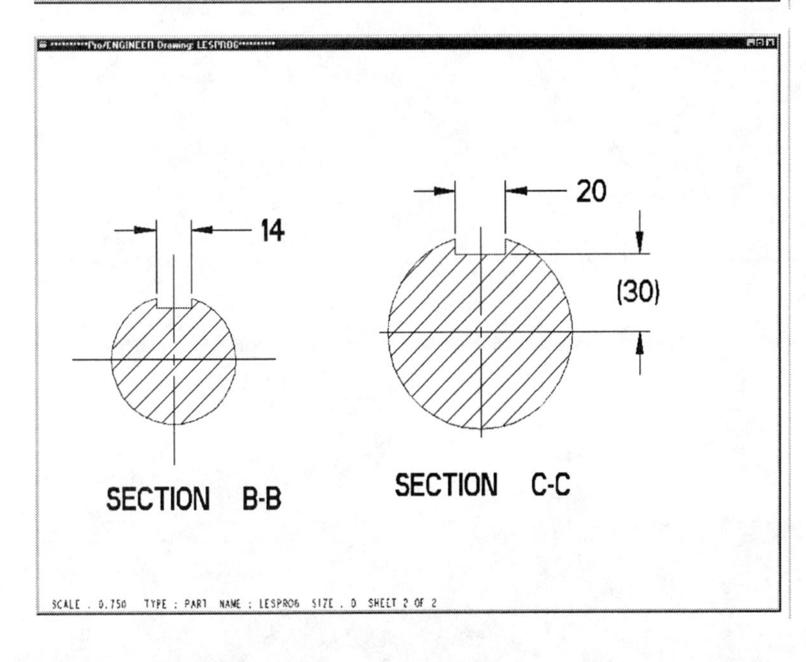

Figure 6.51
Coupling Shaft Drawing, Sheet Two, **SECTION B-B** and **SECTION C-C**

Lesson 7

Patterns and Groups

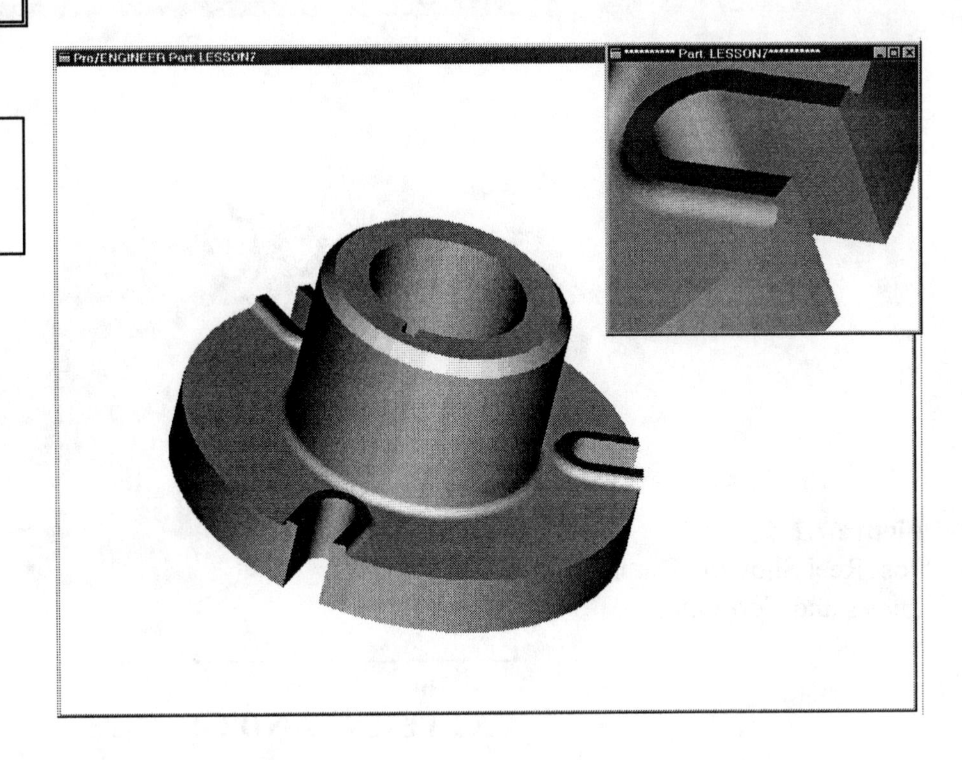

Figure 7.1
Post Reel

OBJECTIVES

1. Change dimension cosmetics and move part dimensions

2. Use offset edge to create sketch entities from existing geometry

3. Pattern features

4. Create and manipulate grouped features

5. Understand how to pattern groups

EGD REFERENCE
Engineering Graphics and Design
by L. Lamit and K. Kitto
Read Chapter: 10
See Pages: 296 and 392

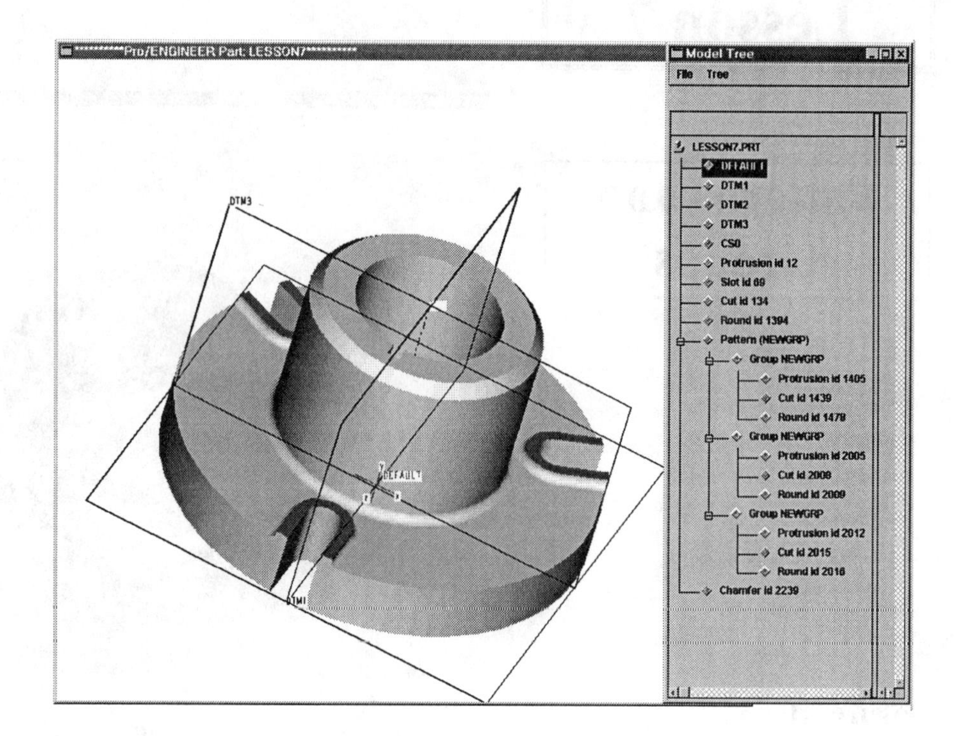

Figure 7.2
Post Reel Showing Datum Planes and Coordinate System

PATTERNS AND GROUPS

Patterns are multiple features created from a single feature (or group of features). After it is created, a pattern behaves as if it were a single feature. When you create a pattern, you create instances (copies) of the selected feature. **Patterns** and **Groups** may be used in many ways. **Groups** are features that have been united.

Creating a pattern is a quick way to reproduce a feature or a set of features that are related and grouped for easy manipulation. Manipulating a pattern may be more advantageous for you than operating on individual features.

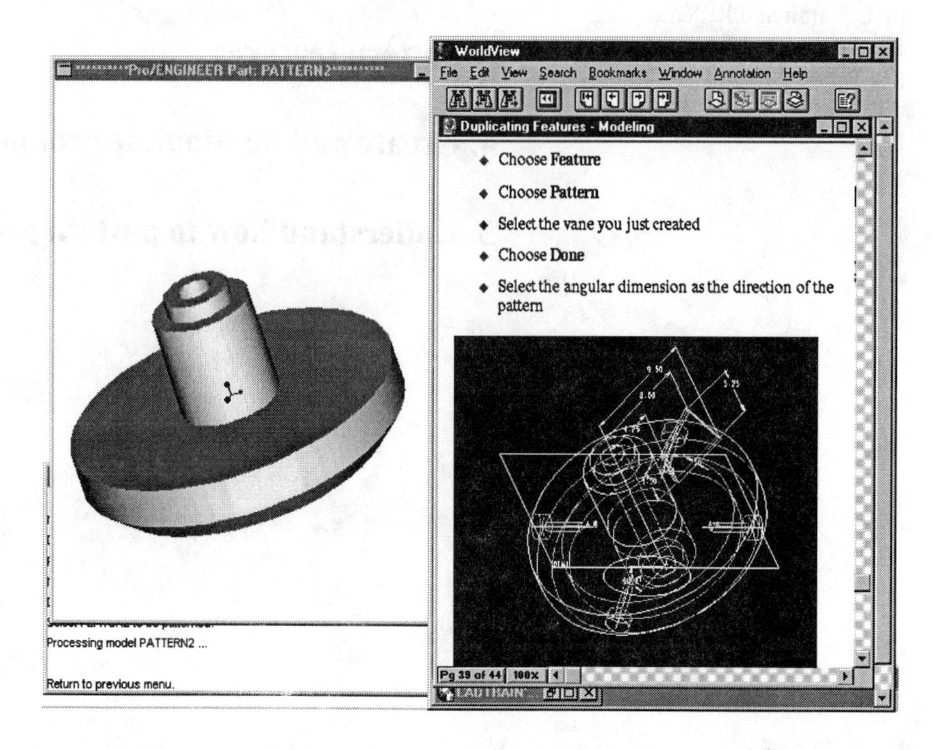

Figure 7.3
COAch for Pro/ENGINEER, Duplicating Features--Modeling (Segment 2: Circular Patterns)

Groups

When you create a *local group*, you must select the features in the consecutive sequential order of the regeneration list. A quick way to do this is to select the intended group by range. If there are features between the specified features in the regeneration list, Pro/E asks if you want to group all the features in between.

Features that are already in other groups cannot be grouped again. To create a local group, do the following:

1. Choose **Create** from the GROUP menu and **local group** from the CREATE GROUP menu.
2. Select each feature to include in the local group
   or
3. Answer "yes" to the prompt asking if you want to group all the features in the between. If you answer "no" to this prompt, Pro/E does not create the local group.

Patterns

Modifying patterns is more efficient than modifying individual features. In a pattern, when you change the dimensions of the original feature, the whole pattern will be updated automatically. A pattern is parametrically controlled. Therefore, a pattern can be modified by changing pattern parameters, such as the number of instances, spacing between instances, and the original feature dimensions. When you create a pattern, you create *instances* of the selected feature.

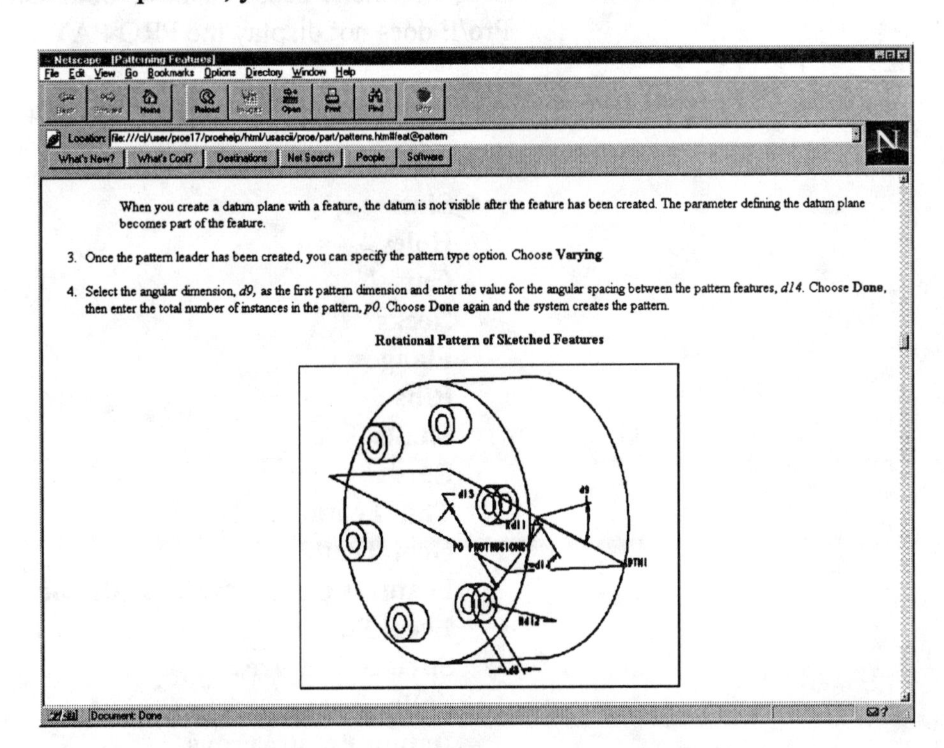

Figure 7.4
Online Documentation
Patterns

When you create a pattern, Pro/E assumes it is a "single" feature. Creating a pattern is a quick way to reproduce a feature. Patterning is an easier and more effective way to perform single operations on the multiple features contained in a pattern, rather than on the individual features. For example, you can easily suppress a pattern or add it to a layer.

You can pattern most features using the FEAT menu **Pattern** option. A thin feature "remembers" the surface to which it is attached and patterns to that surface.

Pro/E allows you to pattern a single feature only. However, you can pattern several features as if they were a single feature by arranging them in a **"local group,"** then patterning the group. After the pattern is created, you can unpattern and ungroup the instances, then make them independently modifiable using the option **Make Indep**. There are two ways to pattern a feature using the PRO PAT TYPE menu:

Dim Pattern Controls the pattern using driving dimensions to determine the incremental changes to the pattern. This is the case for all of the pattern.
Ref Pattern Controls the pattern by referencing another pattern. The dimension pattern must exist before you can create the next pattern type. For example, counterbore one hole in the pattern, then tell Pro/E to pattern the counterbore. It automatically makes a reference pattern.

When you are working with features or components for which it does not make sense to have both **Dim Pattern** and **Ref Pattern**, Pro/E does not display the PRO PAT TYPE menu.

The variety of features can be patterned including the following:

Protrusions
Slots
Holes
Cuts
Necks
Flanges
Ribs
Shafts
Ears
Thin Features
Gear Teeth
Features Copied by Translation
Local Pushes
Surface Features
Spline Teeth
Datum Point Arrays
Datum Planes
Cosmetics

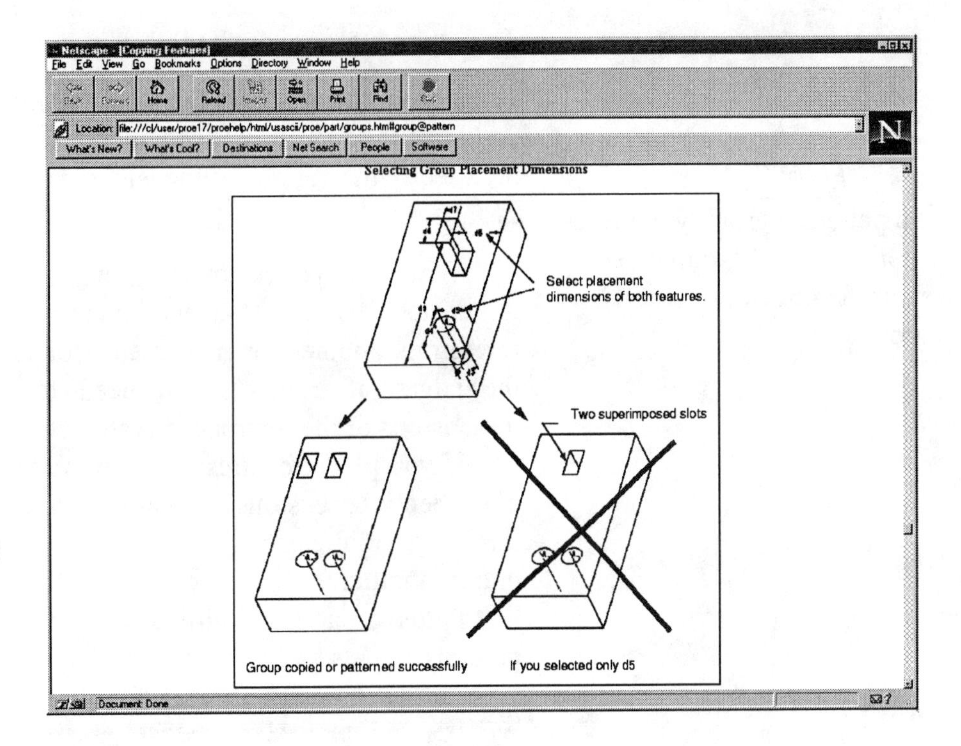

Figure 7.5
Online Documentation
Patterns

Patterning Groups

You can pattern groups created from UDFs (user-defined features) and local groups using the GROUP menu option **Pattern**. This option differs from the FEAT menu option **Pattern** in that the GROUP menu option treats an entire group as a single entity. You can use the FEAT menu option to pattern one feature at a time.

You can select all the dimensions in the selected group, except those used to create a feature within the group, as incremental dimensions. When you create a patterned group, one member represents the whole group. When regenerating, however, Pro/E regenerates all the features individually.

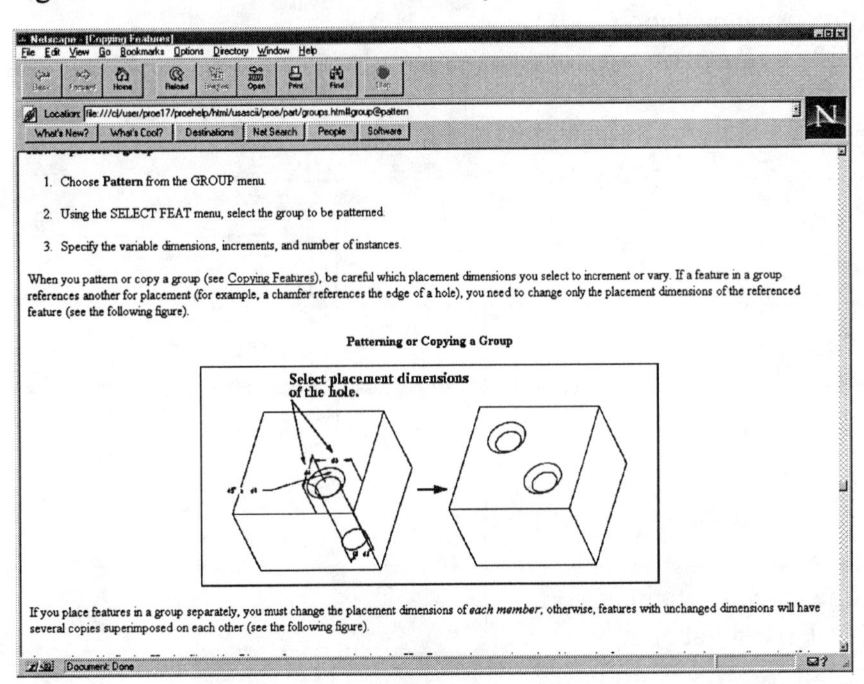

Figure 7.6
Online Documentation
Patterning Groups

To pattern a group, do the following:

1. Choose **Pattern** from the GROUP menu.
2. Using the SELECT FEAT menu, select the group to be patterned.
3. Specify the variable dimensions, increments, and number of instances.

HINT

To pattern a group you must first name and group two or more features into a local group.

When you pattern or copy a group, be careful which placement dimensions you select to increment or vary. If a feature in a group references another for placement (for example, a chamfer references the edges of a hole), you need to change only the placement dimensions of the referenced feature.

If you place features in a group separately, you must change the placement dimensions of ***each member***; otherwise, features with unchanged dimensions will have several copies superimposed onto one an another.

Patterns can be modified as shown in Figure 7.7. Table-driven patterns can also be created (Fig. 7.8).

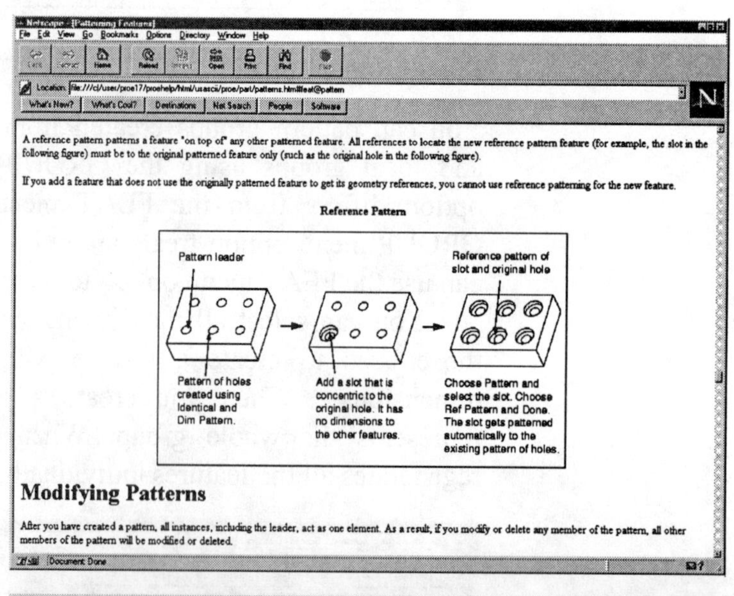

Figure 7.7
Online Documentation
Modifying Patterns

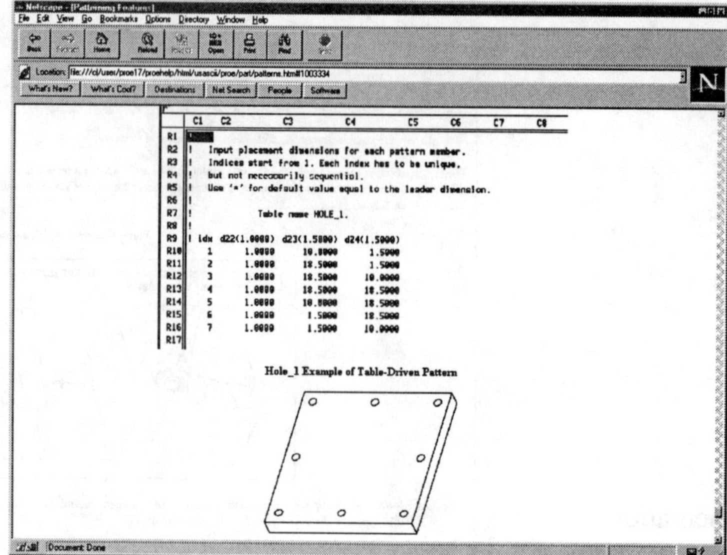

Figure 7.8
Online Documentation
Table-Driven Patterns

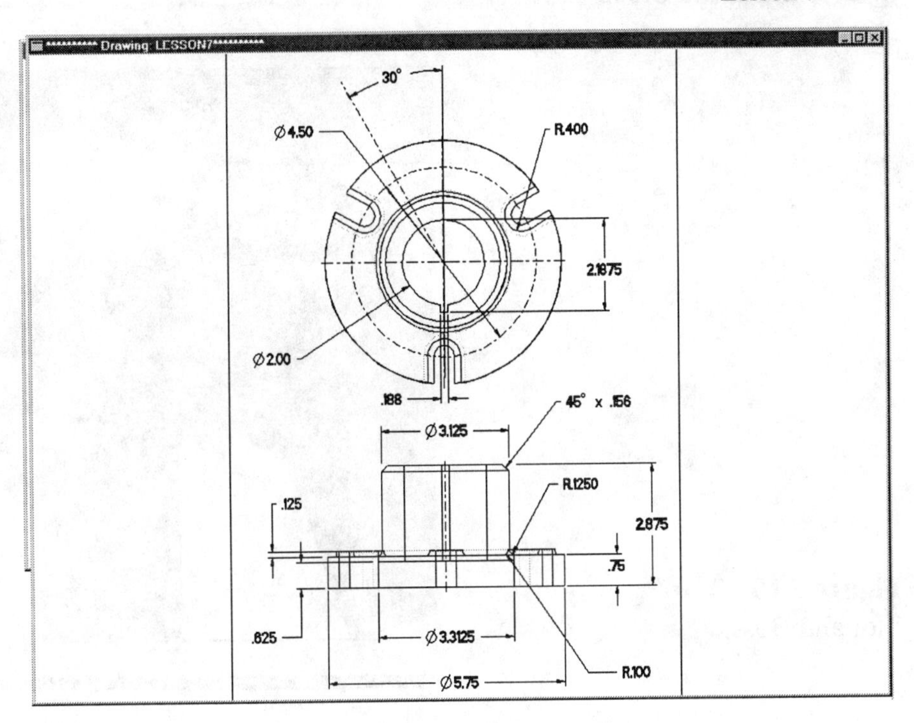

Figure 7.9
Post Reel Detail

Post Reel

The **Post Reel** (Fig. 7.9) is created with a *revolved protrusion*, as in Lesson 5 and Lesson 6. The internal geometry of the Post Reel can be created with a *revolved cut* instead of two holes of differing diameters or a sketched hole. The chamfers and the rounds are simple pick-and-place features. One boss and one slot (Fig. 7.10) are created using a datum on-the-fly (**Make Datum**), grouped and patterned to complete the part. A detailed set of instructions will be supplied only for the boss and slot, since the other geometry is similar to previous lessons. The dimensions for the part are provided in Figure 7.9 through Figure 7.12.

Set up the part using the following commands:

Part ⇒ Create ⇒ POST_REEL ⇒ enter
Feature ⇒ Create ⇒ Datum ⇒ Plane ⇒ Default ⇒ Create ⇒ Datum ⇒ Coord Sys ⇒ Default ⇒ Done ⇒ Done

Layer ⇒ Create ⇒ (type **DATUM_LAYER**) **⇒ enter ⇒ enter ⇒ Set Items ⇒ Add Items ⇒ (✓DATUM_LAYER) ⇒ Done Sel ⇒ Datum Plane ⇒ Sel By Menu ⇒ Name ⇒ Sel All ⇒ OK ⇒ Done Sel ⇒ Done/Return ⇒ Done/Return**

SETUP AND ENVIRONMENT

Set Up ⇒ Units ⇒ Inch ⇒ Material ⇒ Define ⇒ Steel_1020 ⇒ Assign ⇒ Steel_1020 ⇒ Accept
Environment ⇒ ✓Grid Snap
☐ **SketStart2D** (sketch in 3D)
Hidden Line Tan Dimmed

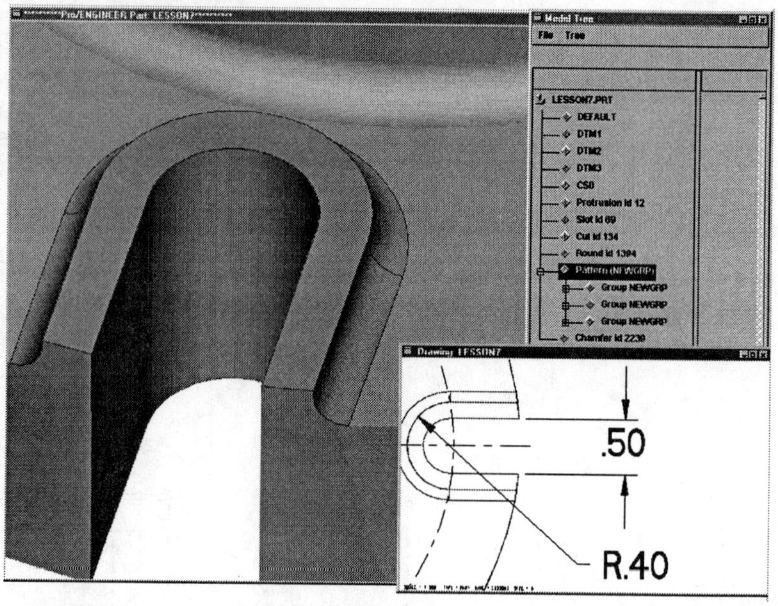

Figure 7.10
Slot and Boss

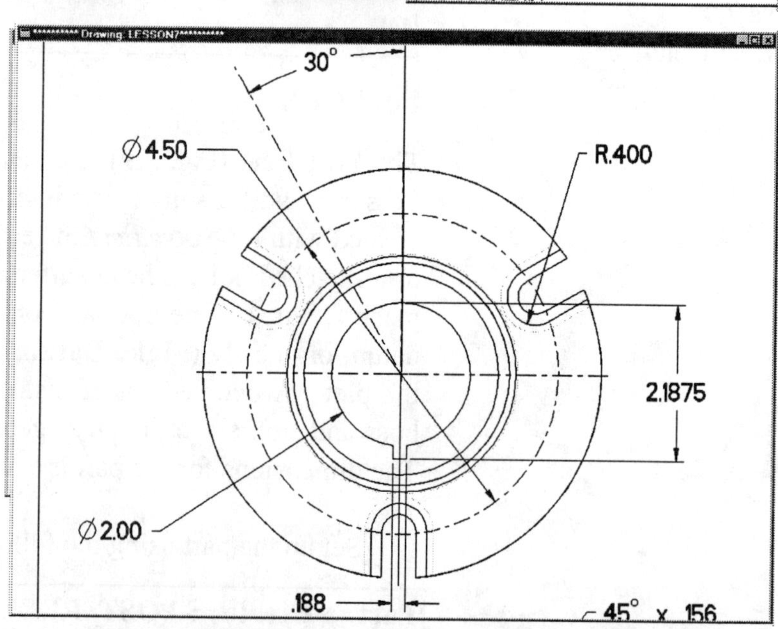

Figure 7.11
Post Reel Drawing, Top View

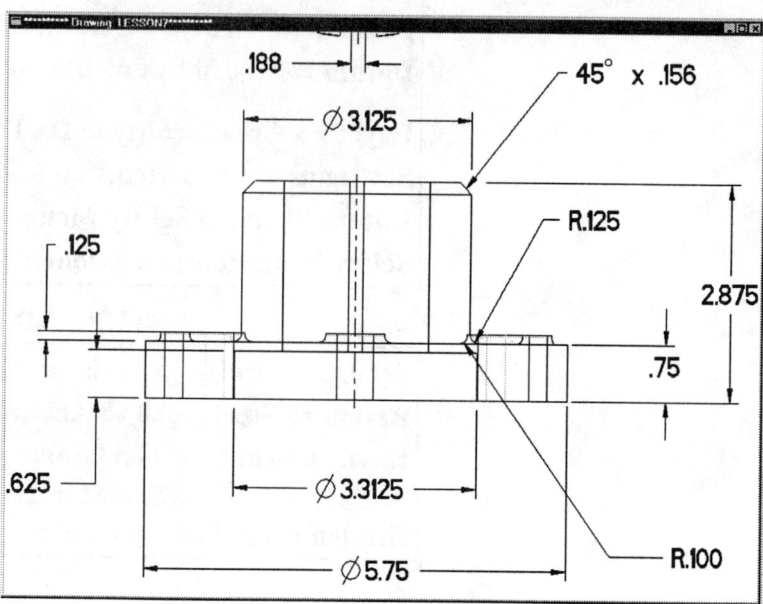

Figure 7.12
Post Reel Drawing, Front View

The first protrusion is revolved and consists of the flangelike geometry shown in Figure 7.13. Sketch on **DTM3**, and revolve the section **360°** about a *vertical axis*.

HINT

Remember to create a palette of colors and use them on the model as required.

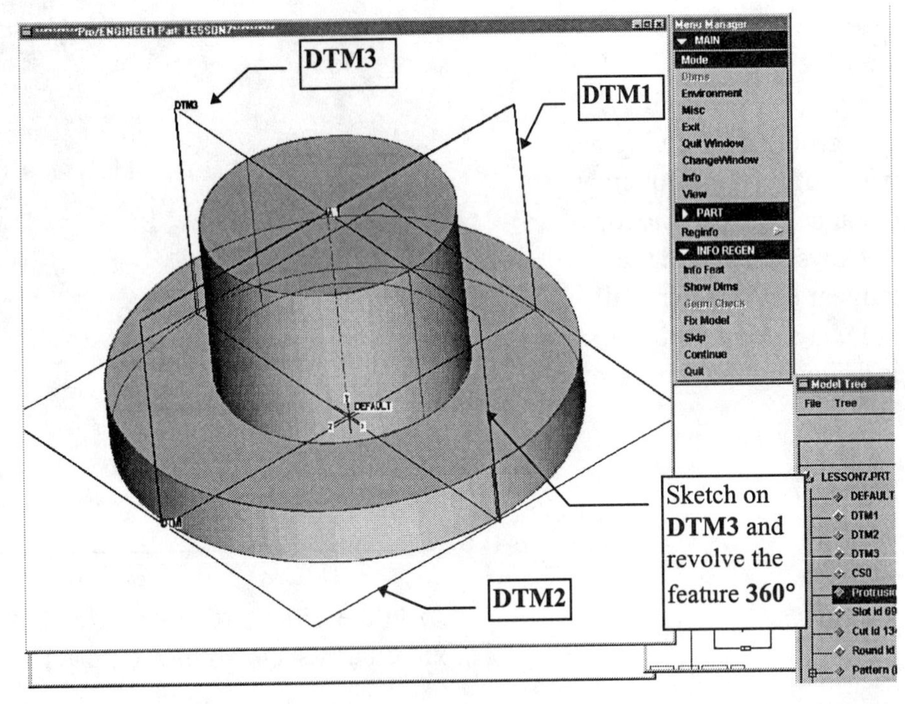

Figure 7.13
Revolved Protrusion

The second feature will be the internal cut shown in Figure 7.14. You can use the previous sketching plane and references.

Create revolved cut feature using the previous placement and reference planes

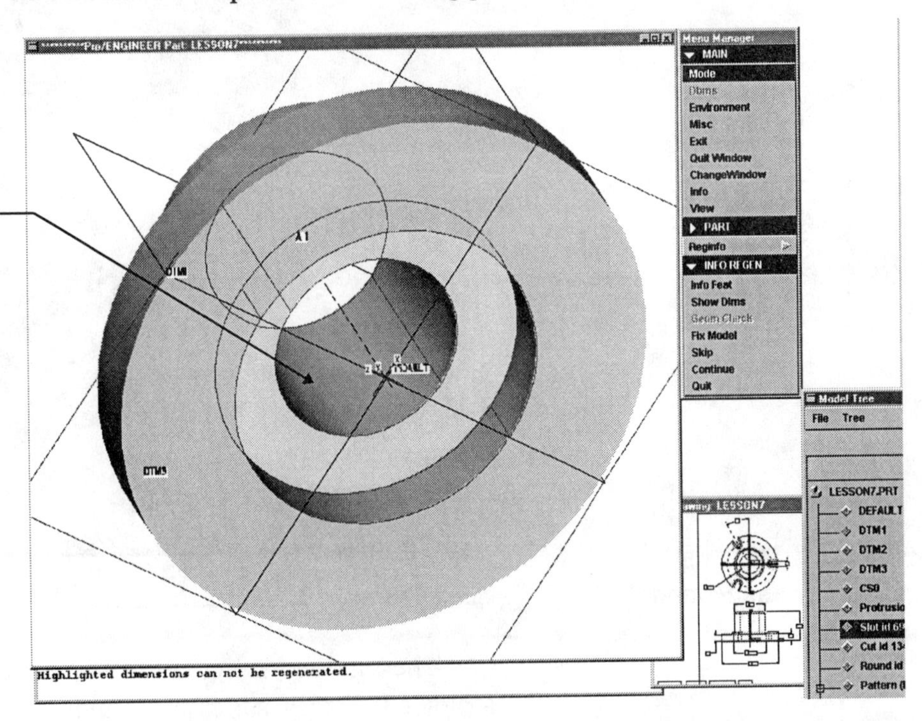

Dbms ⇒ Save ⇒ enter
Purge ⇒ enter ⇒
Done-Return

Figure 7.14
Revolved Cut

The keyseat can be created with a cut from the end of the Reel using three lines sketched on **DTM2**, (or the top face), or can be created with one line from the side, sketching on **DTM1** and projecting to both sides (Fig. 7.15). Be careful to use the proper dimensioning scheme.

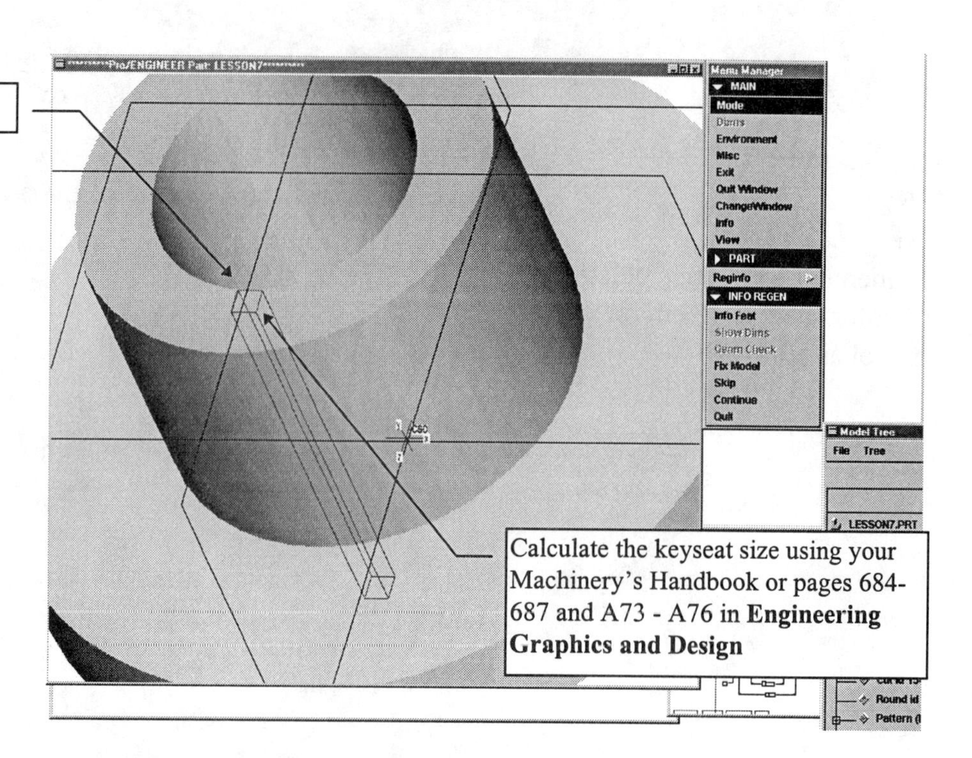

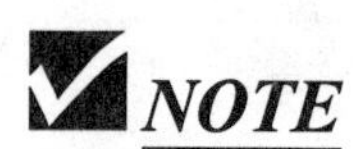

NOTE

Keyseats are normally dimensioned by giving the width of the key slot and the distance between the top of the keyseat cut and the tangent edge of the shaft hole. The size of the keyseat is driven by the shaft size.

Figure 7.15
Keyseat

Add the **45° X .1563** chamfer to the top of the Post Reel and the **R.125** round, as shown in Figure 7.16.

Dbms ⇒ Save ⇒ enter
Purge ⇒ enter ⇒
Done-Return

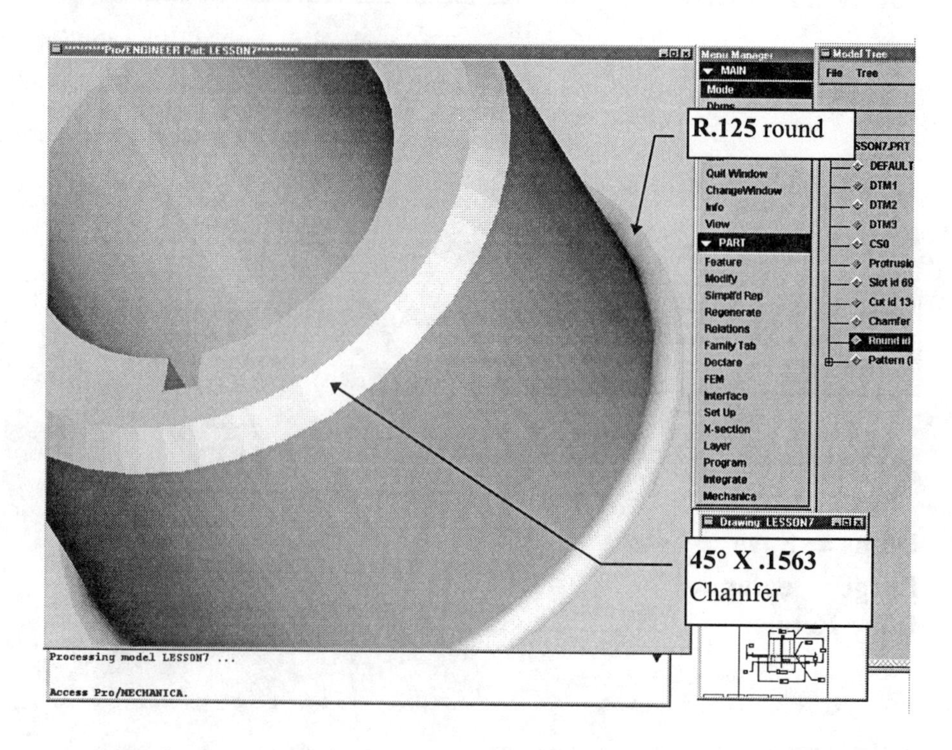

Figure 7.16
Chamfer and Round

The final feature creation sequence consists of a boss protrusion, a slot cut, and a series of rounds. The three types of features are grouped and the group is then patterned. The protrusion and the slot must be created with a *datum on-the-fly* (**Make Datum** within the SETUP PLANE menu).

Create the boss protrusion using the following commands:

Create ⇒ **Protrusion** ⇒ **Extrude** ⇒ **Solid** ⇒ **Done** ⇒ **One Side** ⇒ **Done** ⇒ (pick top of the flange as the placement surface, as in Figure 7.17) ⇒ **Okay** (for the direction of feature creation) ⇒ **Bottom** ⇒ **Make Datum** ⇒ **Through** (pick axis **A_1**) ⇒ **Angle** (pick **DTM3**) ⇒ **Done** ⇒ **Enter Value** (type **30** at the prompt) ⇒ **enter** (Fig. 7.18)

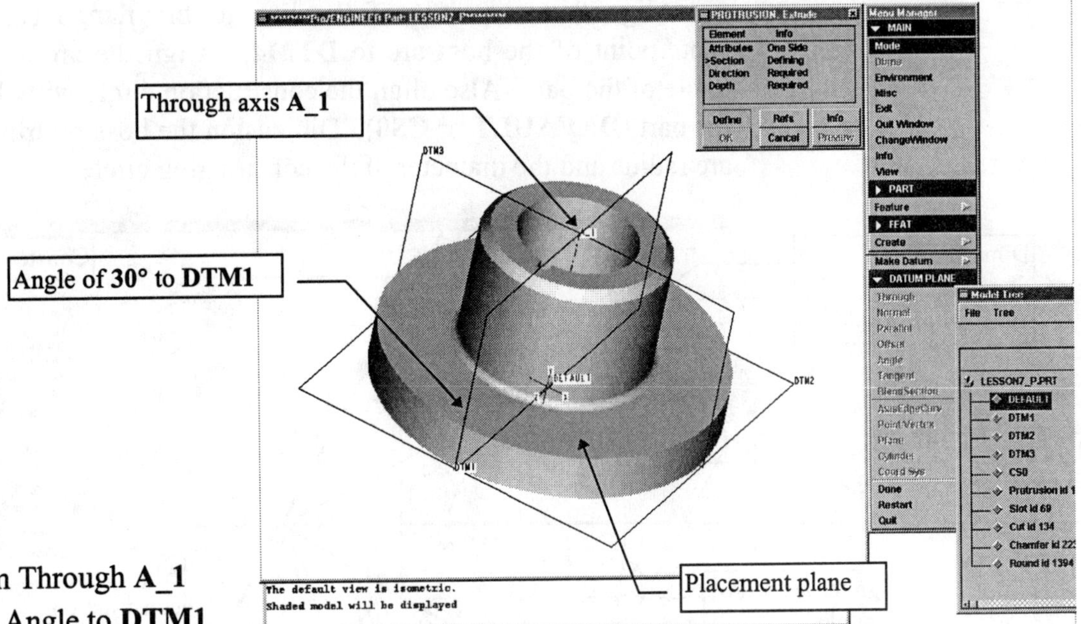

Figure 7.17
Make Datum Through **A_1** and at a **30°** Angle to **DTM1**

Turn on the grid snap and change the grid spacing:

Environment ⇒ ✓ **Grid Snap** ⇒ **Done-Return** ⇒ **Sec Tools** ⇒ **Sec Environ** ⇒ **Grid** ⇒ **Params** ⇒ **X&Y Spacing** ⇒ (type **.20** at the prompt) ⇒ (zoom and pan the sketch as in Figure 7.19)

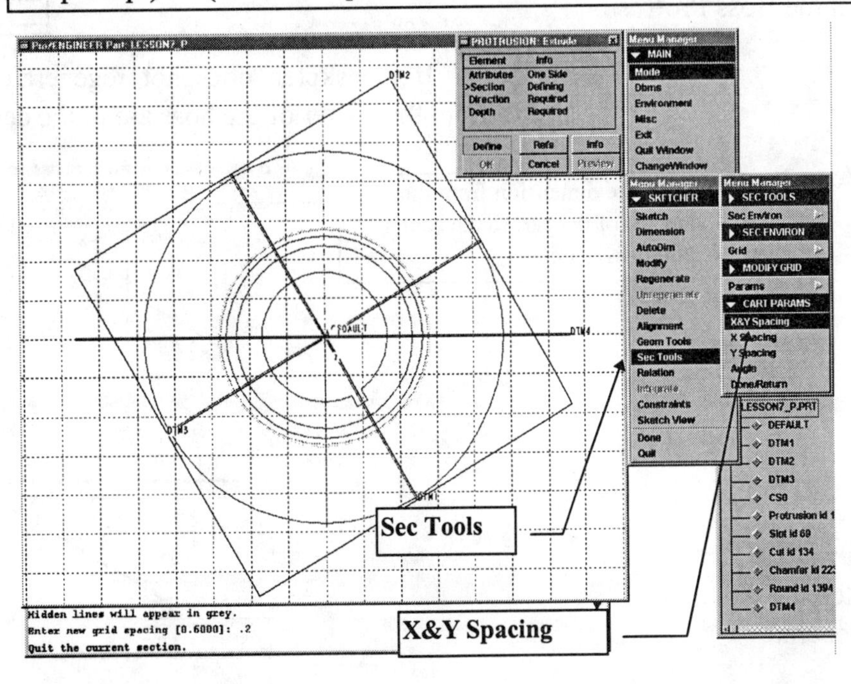

Figure 7.18
Sec Tools

Sketch the two horizontal lines and the two arcs. The section must be *closed*. Create a construction circle to locate the center of the boss arc as shown in Figure 7.19:

Sketch ⇒ **Circle** ⇒ **Construction** ⇒ **Concentric** ⇒ (pick any of the circles composing the part and then pick the center of the boss arc) ⇒ (sketch the remaining geometry) ⇒ **Regenerate**

Align the endpoints of the line to the flange curve and the centerpoint of the boss arc to **DTM4**. Align the arc to the outside circle of the part. Also align the construction circle with the center of the part (**DEFAULT** or **CS0**). Dimension the boss protrusion with an arc radius and the diameter of the construction circle.

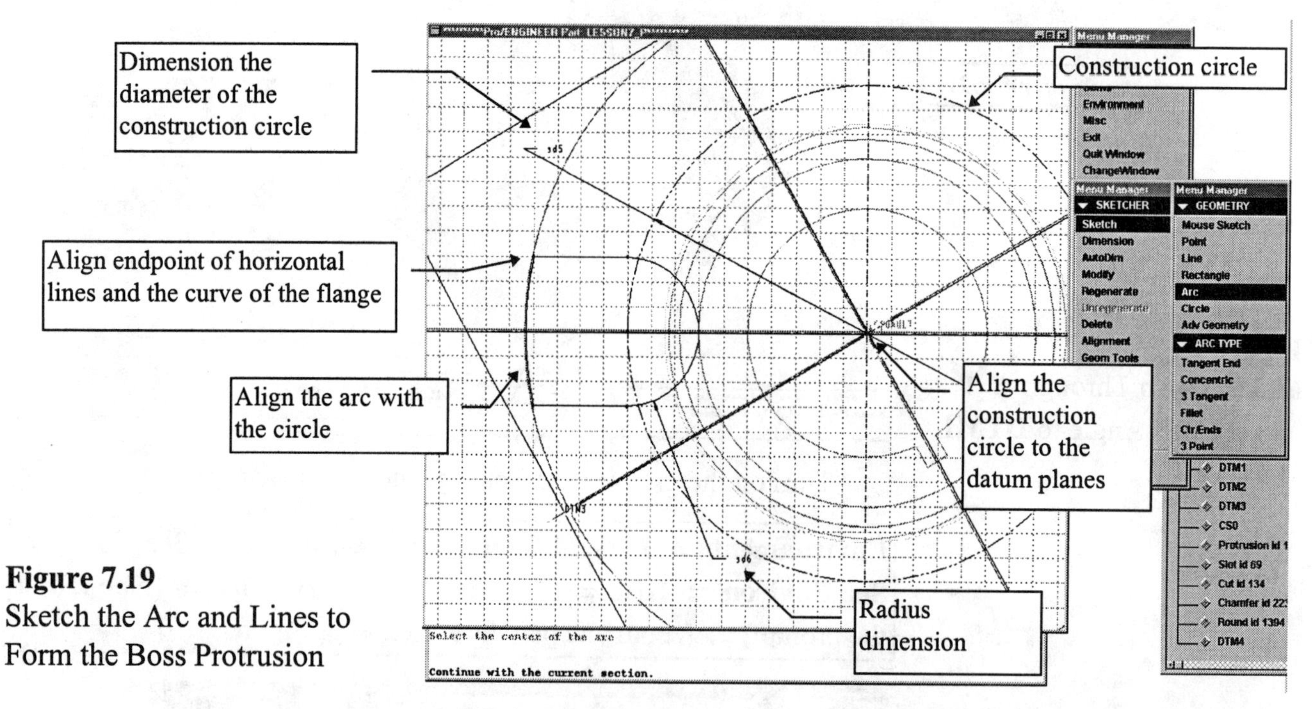

Figure 7.19
Sketch the Arc and Lines to Form the Boss Protrusion

If the sketch does not regenerate, add a horizontal dimension from the center of the boss arc to the center of the part (Fig. 7.20).

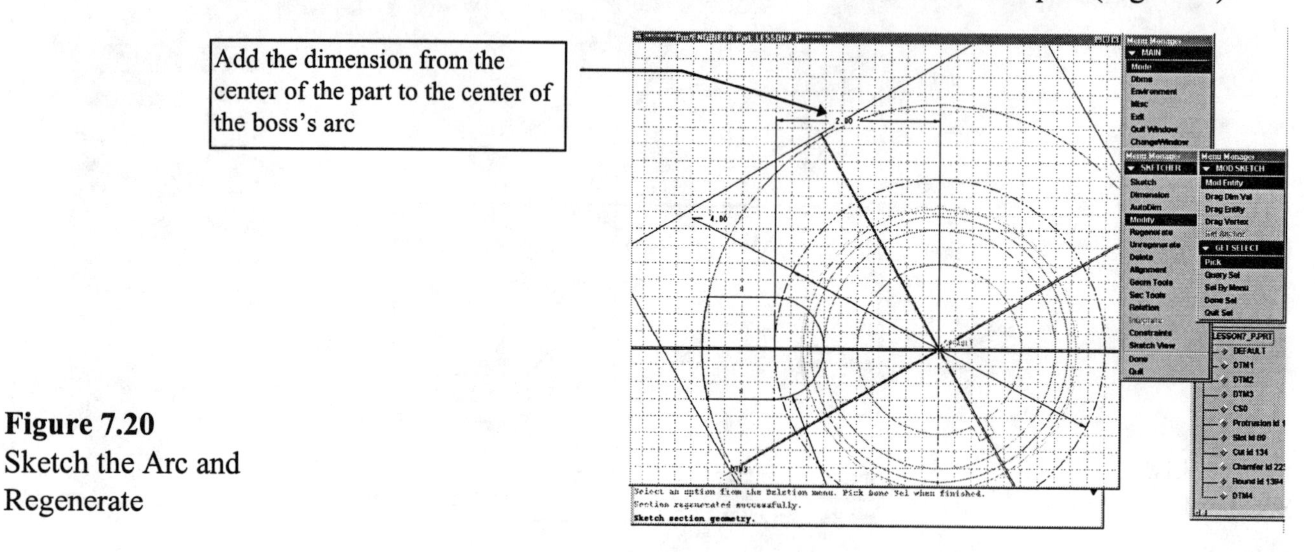

Figure 7.20
Sketch the Arc and Regenerate

After the sketch regenerates successfully, modify the dimensions and regenerate the sketch as shown in Figure 7.21.

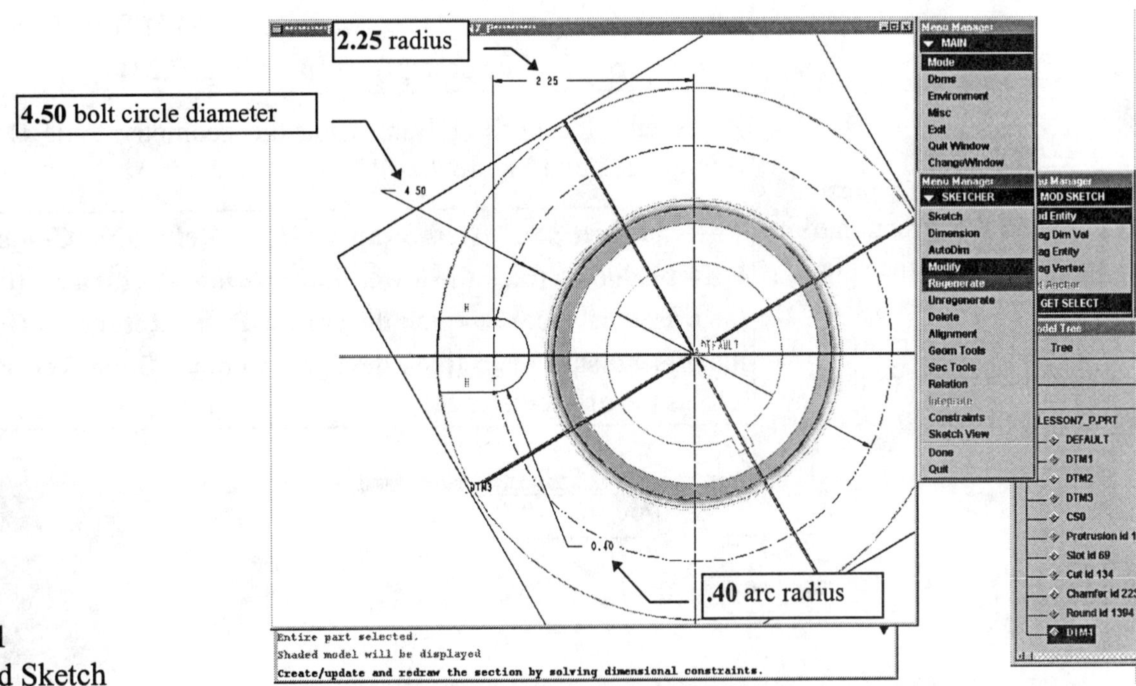

Figure 7.21
Regenerated Sketch

Complete the protrusion (Fig. 7.22) with the following command:

Done ⇒ **Blind** ⇒ **Done** ⇒ (enter the depth of **.125**) ⇒ **enter** ⇒ **View** ⇒ **Default** ⇒ **Done-Return** ⇒ **Preview** ⇒ **OK** ⇒ **Done**

Dbms ⇒ **Save** ⇒ **enter**
Purge ⇒ **enter** ⇒
Done-Return

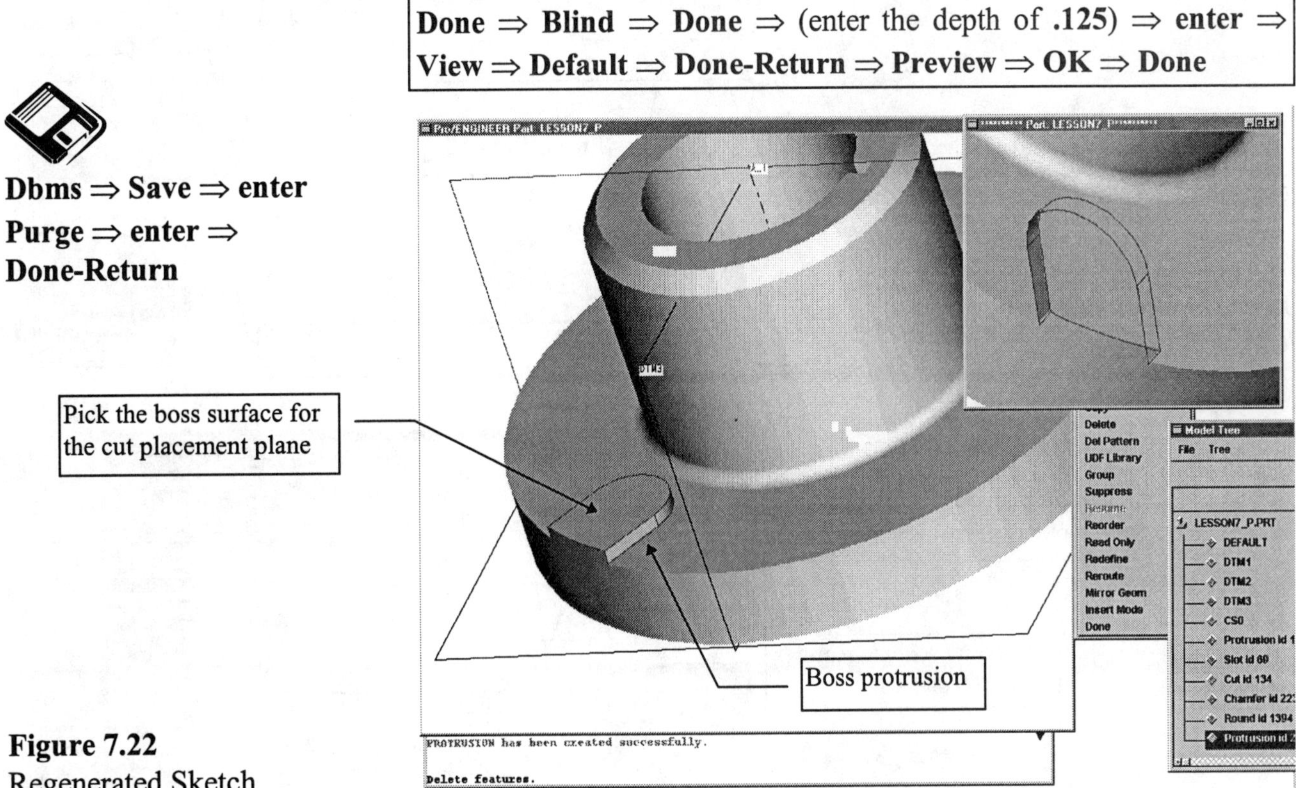

Figure 7.22
Regenerated Sketch

The next feature is a cut through the boss. The boss surface will be the placement plane (Fig. 7.22). The reference/orientation plane is created using **Make Datum**:

Feature ⇒ **Create** ⇒ **Cut** ⇒ **Done** ⇒ **Done** ⇒ (pick the top of the Boss's surface; see Figure 7.22) ⇒ **Okay** ⇒ **Make Datum** ⇒ **Through** (pick axis **A_1**) ⇒ **Angle** (pick **DTM3**) ⇒ **Done** ⇒ **Enter Value** (type **30** at the prompt) ⇒ **enter** (Fig. 7.23)

Sketch the lines and arc using the geometry tools **Offset Edge** (Figure 7.23 and Figure 7.24):

Environment ⇒ □ **Grid Snap** ⇒ **Done-Return** ⇒ **Geom Tools** ⇒ **Offset Edge** ⇒ (pick the lower line) ⇒ **Done-Return** ⇒ (type **-.15** as the offset distance) ⇒ (pick the arc) ⇒ **Done-Return** ⇒ (type **-.15** as the offset distance) ⇒ (pick the upper line) ⇒ **Done-Return** ⇒ (type **-.15** as the offset distance)

HINT

Type a *plus* **.15** or a *minus* **.15** based on the direction of the arrow showing the side on which the geometry will be created. You want to draw the lines and the arc to the inside of the boss.

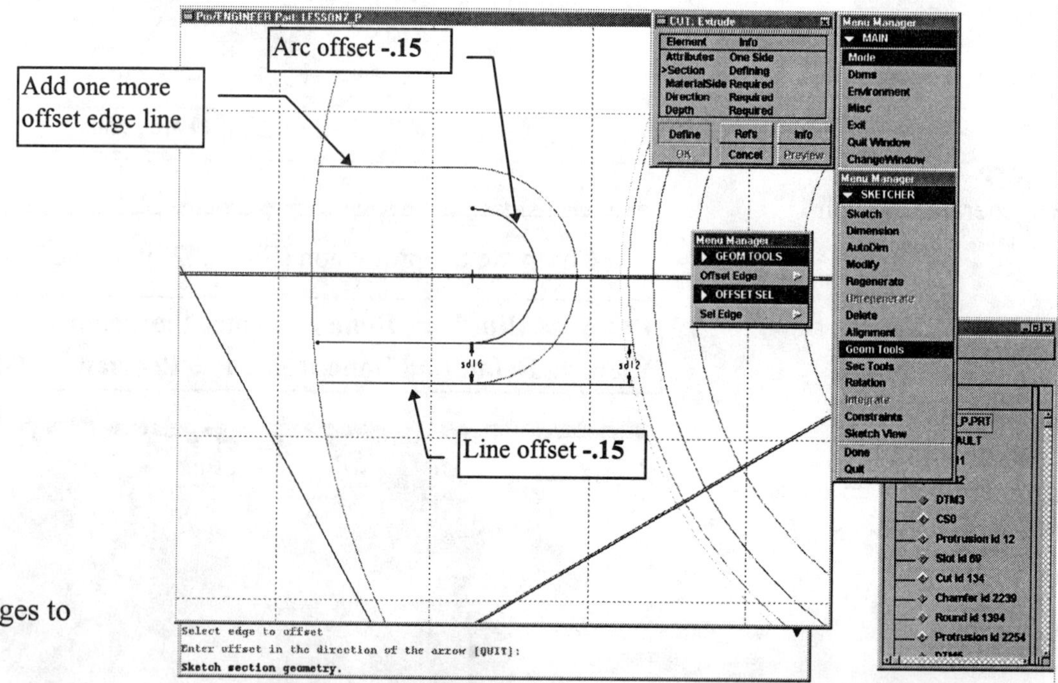

Figure 7.23
Offset the Boss Edges to Form the Slot

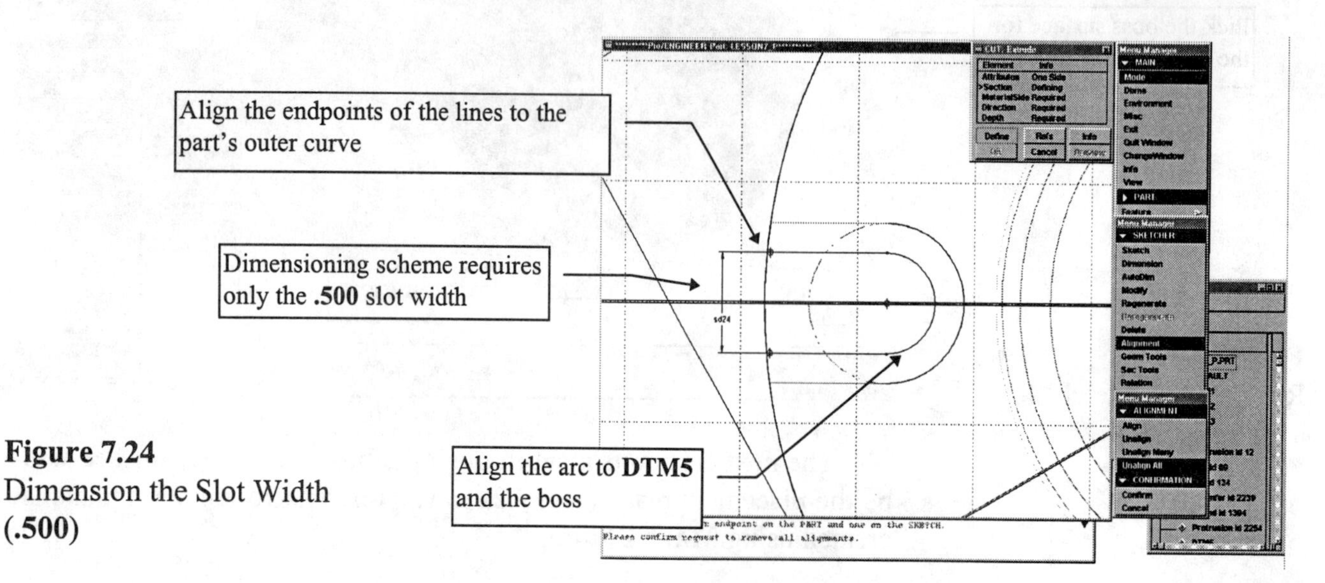

Figure 7.24
Dimension the Slot Width (.500)

Delete all three offset dimensions (**sd**), and repaint the screen to see the slot sketch geometry (Fig. 7.24 and Fig. 7.25). Align the endpoint of the horizontal lines to the outer curve of the part. Align the arc to the boss arc and to **DTM5**. Add the dimension for the slot width (**.500**).

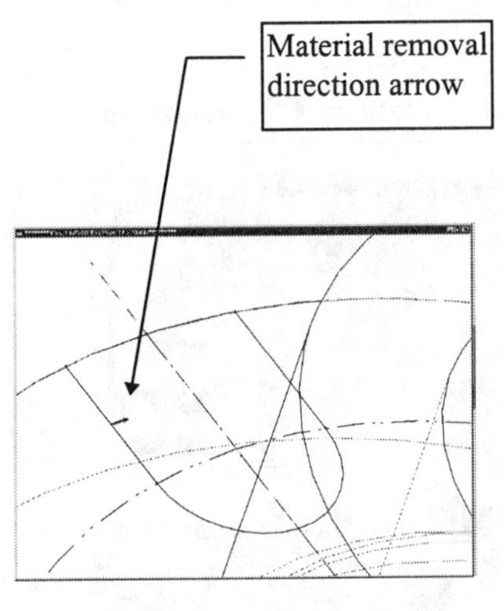

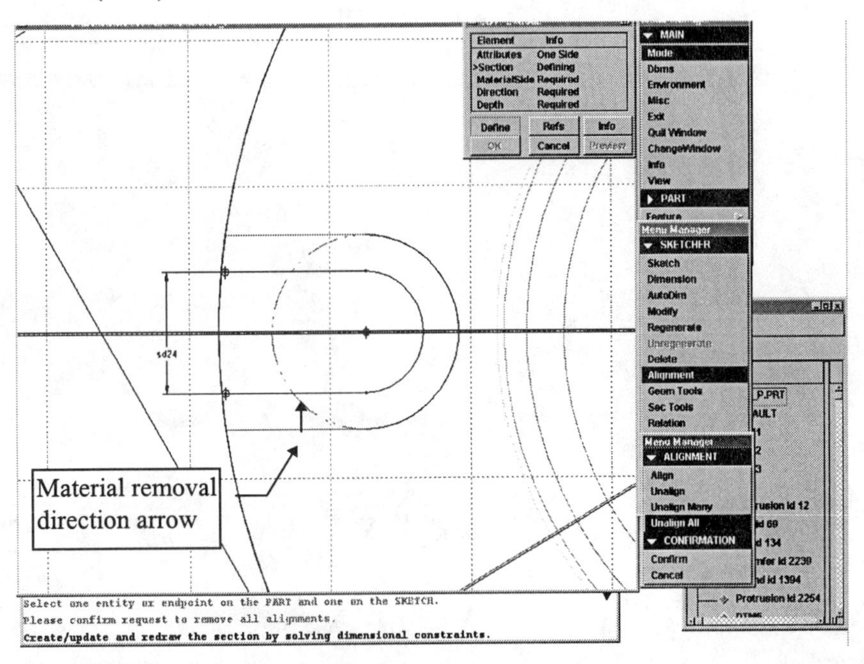

Figure 7.25
Regenerated Section Showing the Material Removal Side of Cut

Regenerate the sketch, and choose as **Okay** the material removal direction. *There is no need to modify the value if it was created at the correct design size.* Make sure the direction of cut is toward the part, not out into space. The depth will be **Thru All**. Change the sketch orientation by rotating or using the default view. If the cut does not intersect the part, pick **Direction** and **Define** from the dialog box and flip the arrow. Figure 7.26 shows the completed cut.

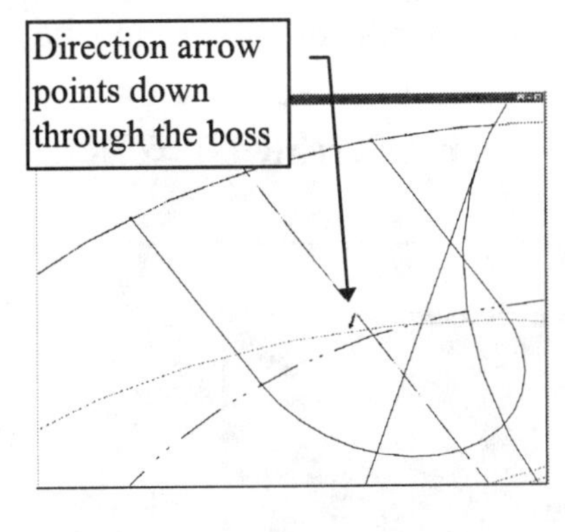

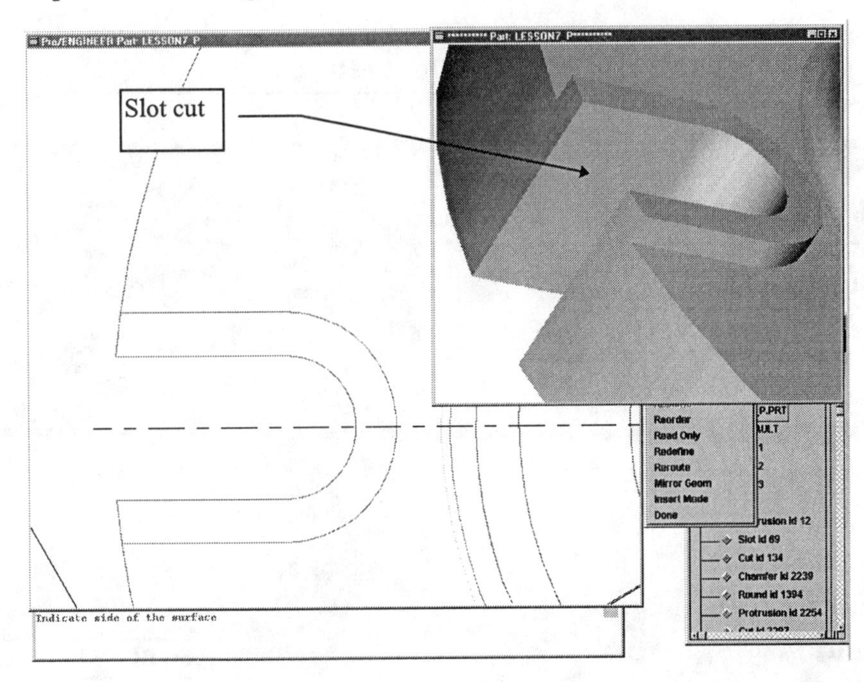

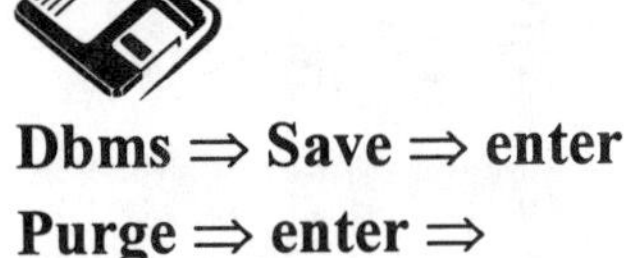

Dbms ⇒ Save ⇒ enter
Purge ⇒ enter ⇒
Done-Return

Figure 7.26
Slot Cut

Create a set of rounds around the boss, as show in Figure 7.27:

Feature ⇒ **Create** ⇒ **Round** ⇒ **Simple** ⇒ **Done** ⇒ **Done** ⇒ **Tangent Chain** ⇒ (pick the line at the edge of the boss; both lines and the arc should highlight) ⇒ **Done Sel** ⇒ **Done** ⇒ (type the radius value of **.10** at the prompt) ⇒ **enter** ⇒ **Preview** ⇒ **OK** ⇒ **Done**

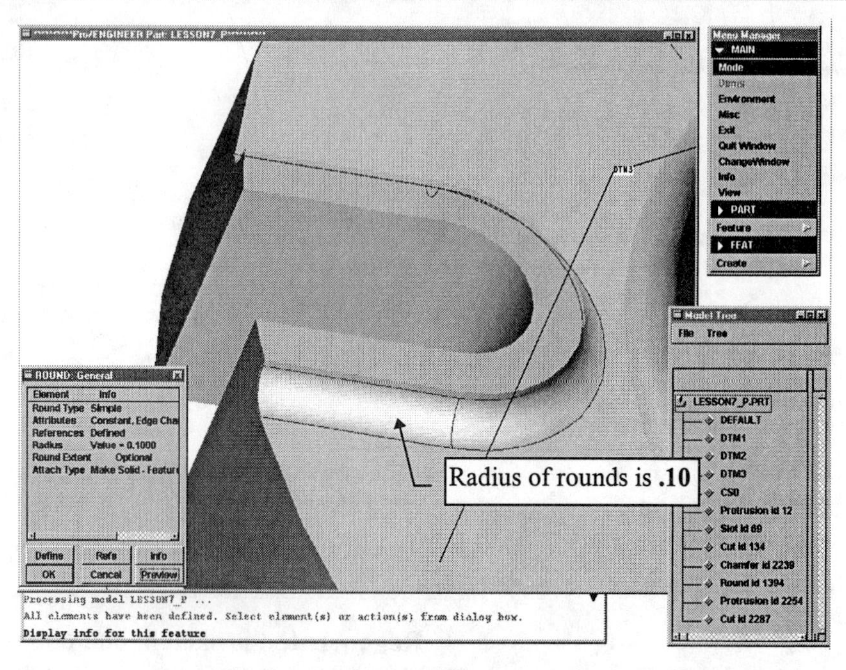

Figure 7.27
Rounds

Next we will group (Fig. 7.28) the three features together using the following commands:

Feature ⇒ **Group** ⇒ **Create** ⇒ **Local Group** ⇒ (enter the group name at the prompt: **BOSS**) ⇒ (select the features for the group: the *slot*, the *boss protrusion*, and the *rounds*) ⇒ **Done Sel** ⇒ **Done** ⇒ (Pro/E will respond with **"Group BOSS has been created"**) ⇒ **Done** ⇒ **Done/return**

Dbms ⇒ **Save** ⇒ **enter**
Purge ⇒ **enter** ⇒
Done-Return

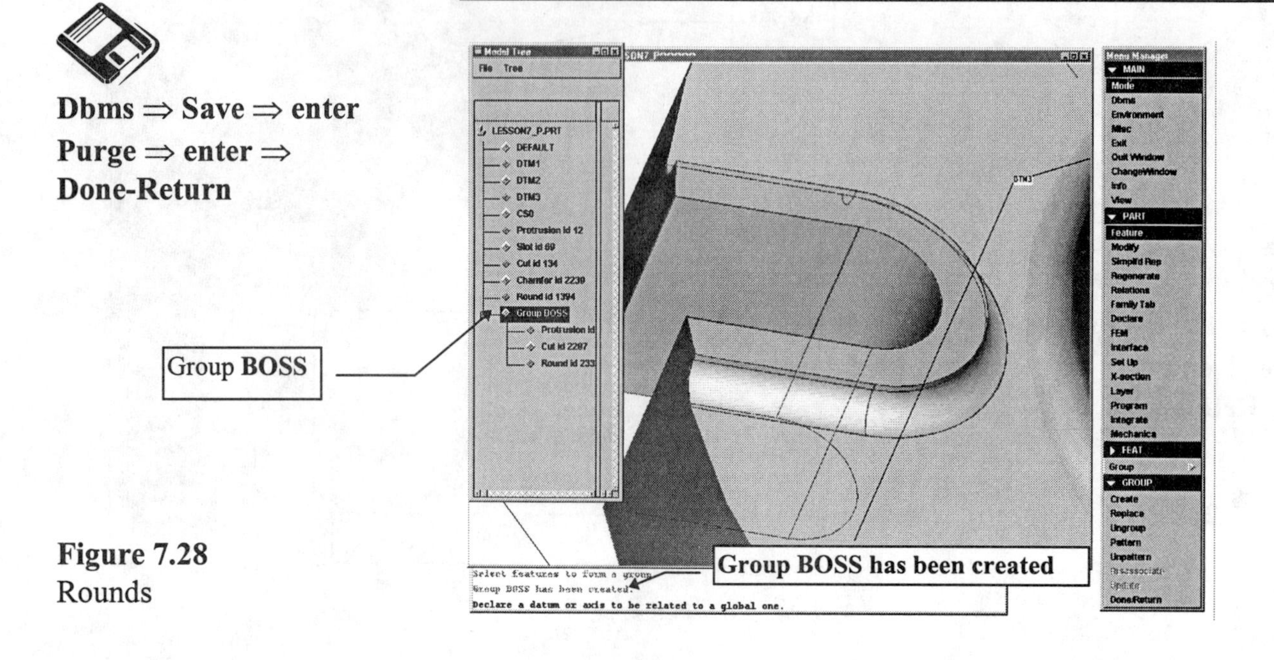

Figure 7.28
Rounds

Since the dimensions for the features that you will be using to pattern the group are the same for the angle of the boss protrusion and the angle of the cut, you need to move the dimension. This must be done before the group is patterned (Fig. 7.29). Use the following commands:

Dbms ⇒ **Save** ⇒ **enter**
Purge ⇒ **enter** ⇒
Done-Return

Modify ⇒ (pick the boss) ⇒ **DimCosmetics** ⇒ **Move Dim** ⇒ (pick the original position of the **30°** degree angle) ⇒ (pick the new position for the angle dimension)

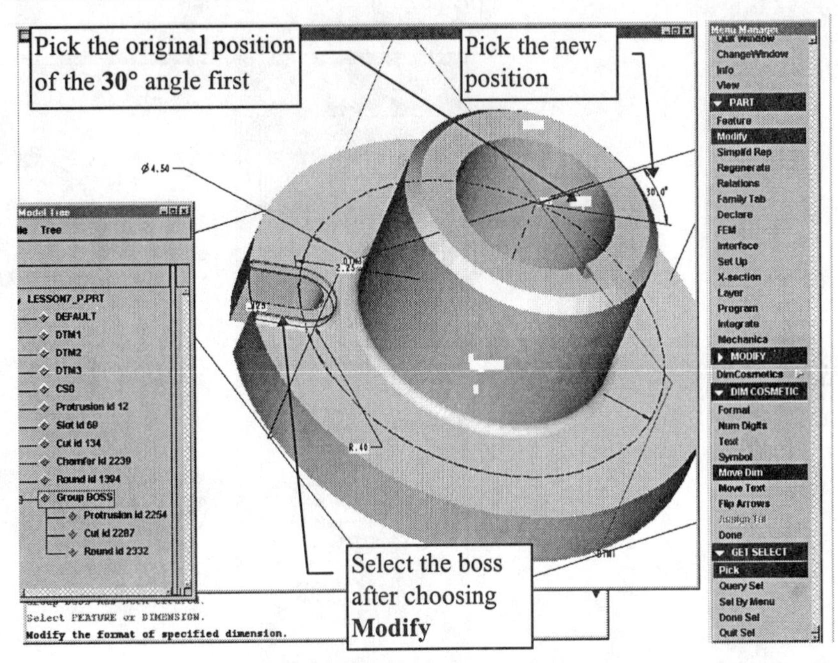

Figure 7.29
Modifying the Dimension Cosmetics

Finally, the BOSS group needs to patterned. To pattern a group you need to choose **Pattern** from the GROUP menu choices, not from the FEATURE menu. Give the following commands (Fig. 7.30):

Group ⇒ **Pattern** ⇒ (select the **BOSS** group from the Model Tree; notice that two **30°** angle dimensions show on the screen, one for the slot and another for the protrusion)

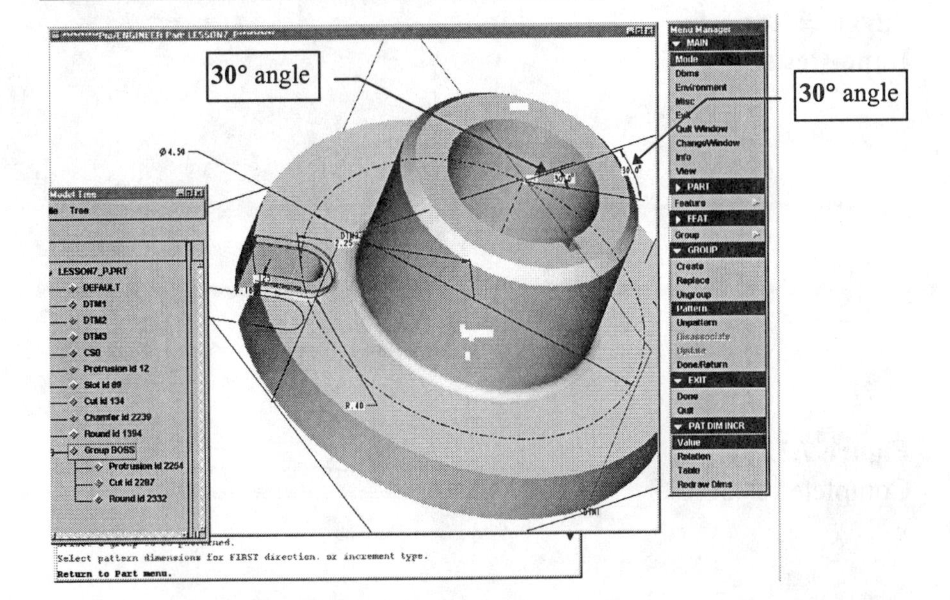

Figure 7.30
Modifying the Dimension Cosmetics

Pro/E will respond with "**Select pattern dimensions for the FIRST direction, or increment type**." Continue with the following commands (Fig. 7.30):

(pick one of the **30°** angles) ⇒ (type **120**) ⇒ **enter** ⇒ (pick the other **30°** angle) ⇒ (type **120**) ⇒ **Done** ⇒ (Pro/E responds with "**Enter TOTAL number of instances in this direction (including original):**") ⇒ (type **3**) ⇒ **enter** ⇒ **Done** (Figure 7.31 and Figure 7.32)

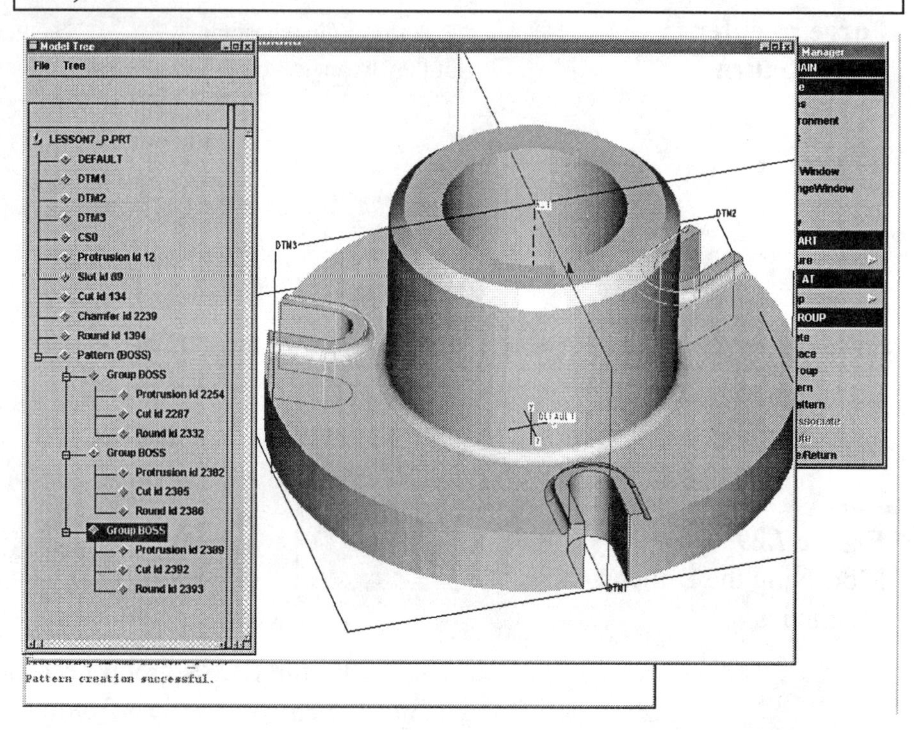

Figure 7.31
Patterning the Group

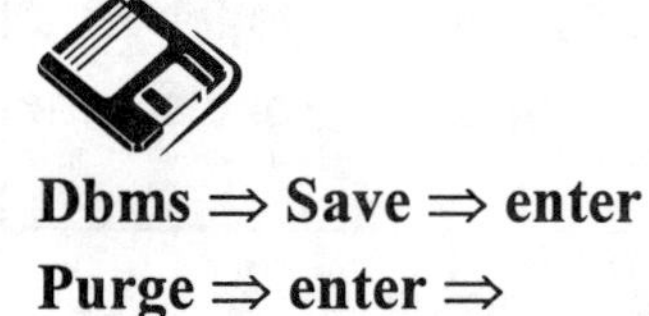

Dbms ⇒ Save ⇒ enter
Purge ⇒ enter ⇒
Done-Return

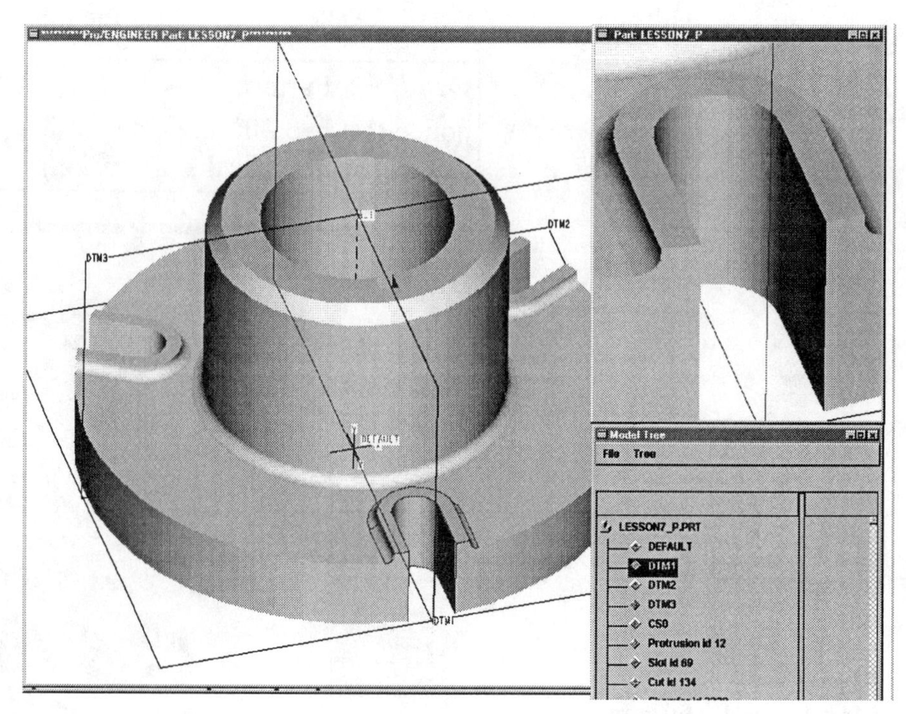

Figure 7.32
Completed Post Reel

Lesson 7 Project

Taper Coupling

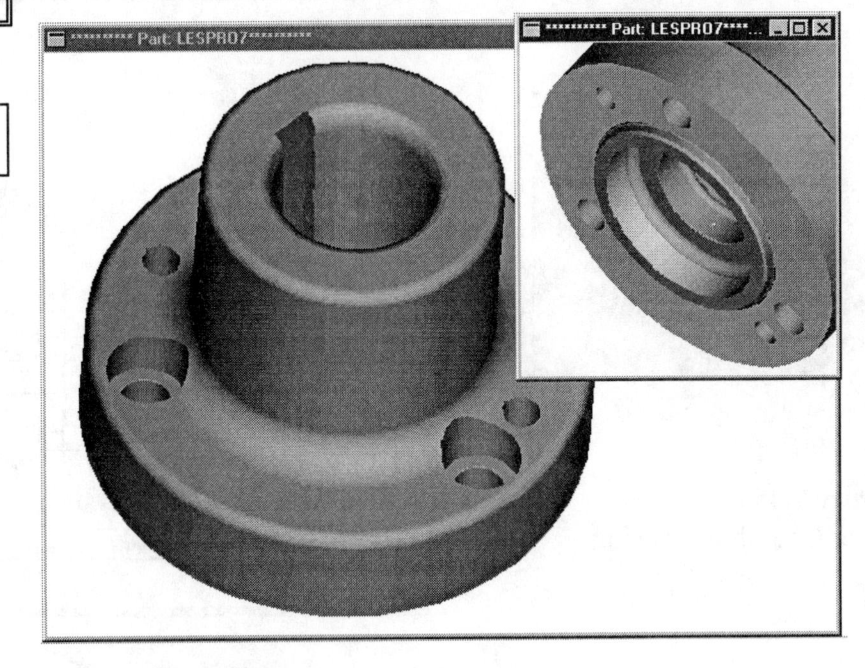

Figure 7.33
Taper Coupling

Taper Coupling

The seventh **lesson project** is a machined part that requires similar commands to the **Post Reel**. Create the part shown in Figure 7.33 through Figure 7.46. At this stage in your understanding of Pro/E you should be able to analyze the part and plan out the steps and features required to model it. Use the DIPS in Appendix D to plan out the feature creation and the parent-child relationships for the part. The coupling will be used for an assembly in Lessons 14 and 15. The machined face of the coupling mates with and is fastened to a similar surface when assembled. Plan your geometric tolerancing requirements accordingly, and set the datums to anticipate the mating surfaces.

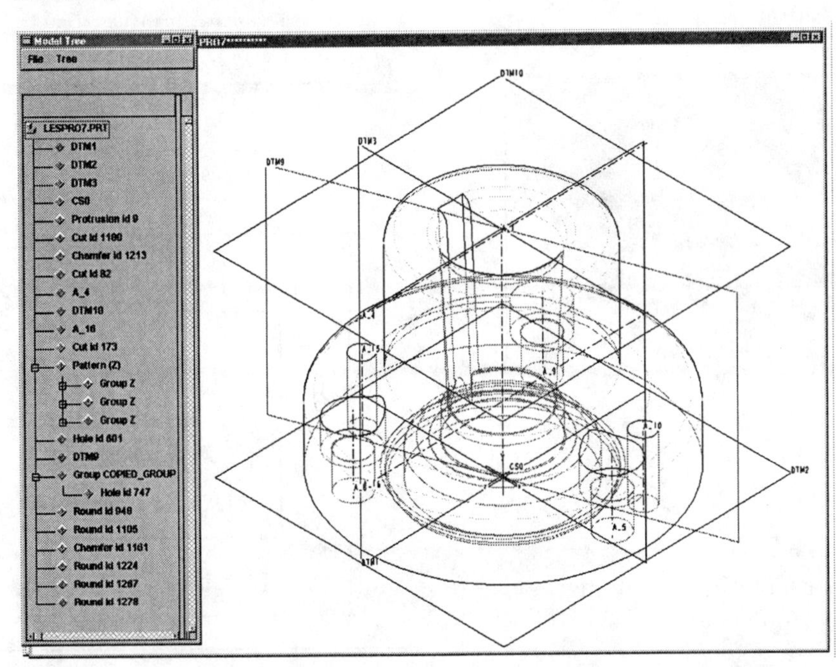

Figure 7.34
Taper Coupling Model with Datum Planes

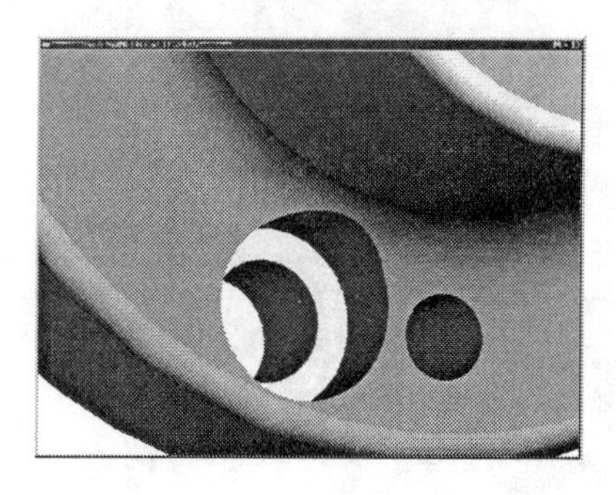

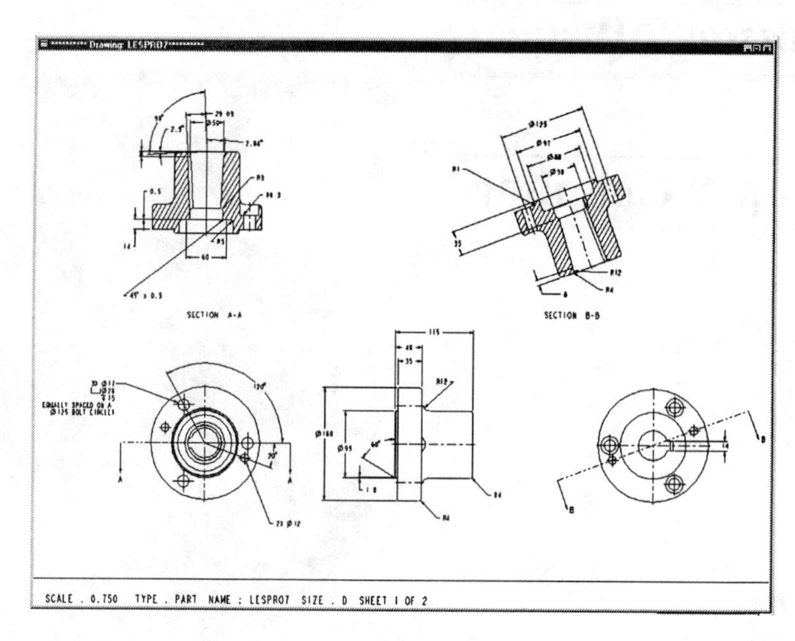

Figure 7.35
Taper Coupling Drawing

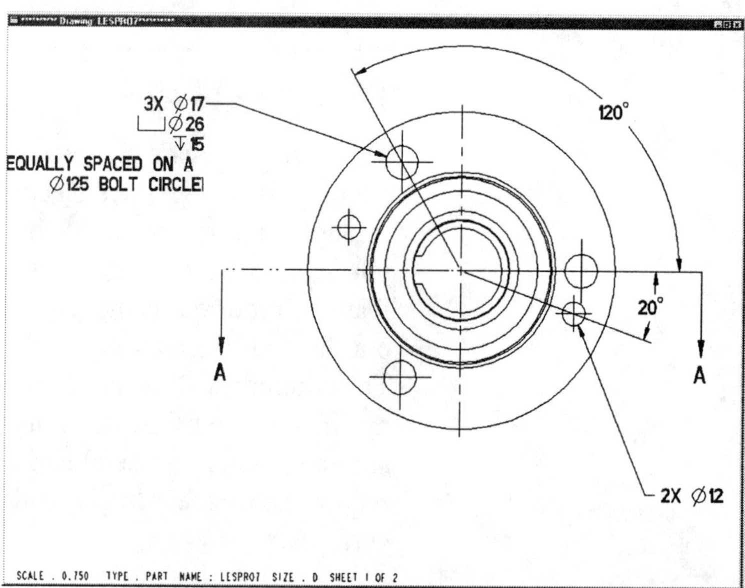

Figure 7.36
Taper Coupling Drawing, Bottom

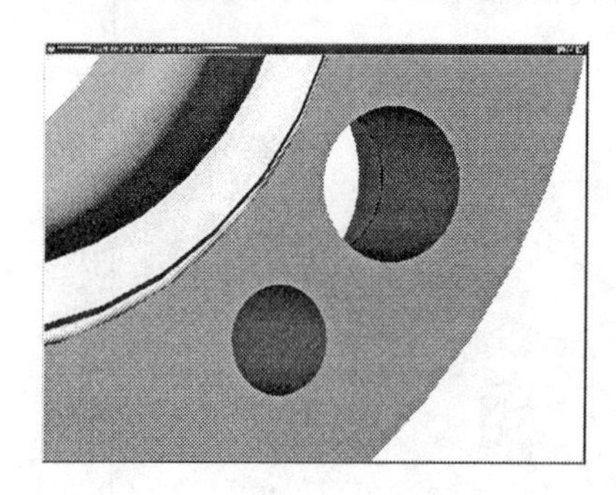

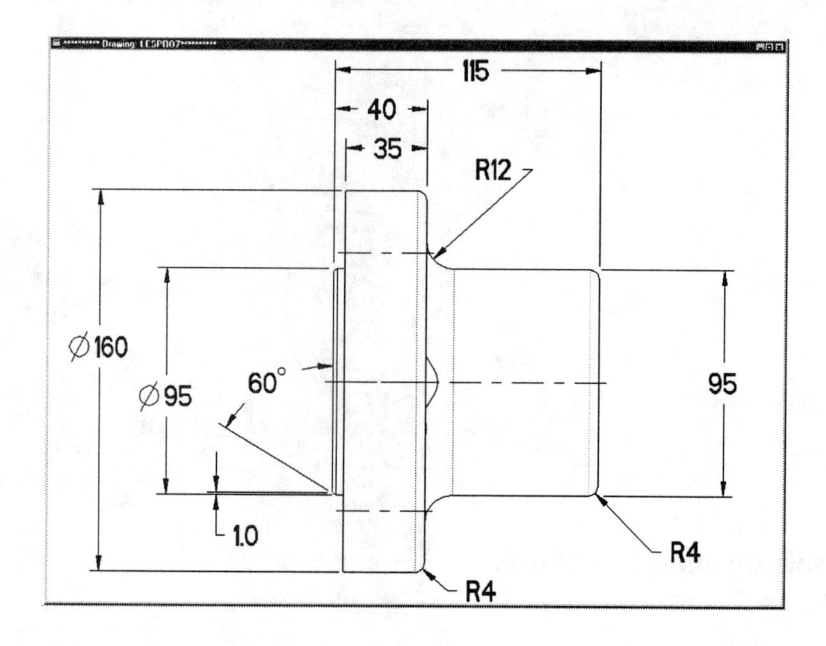

Figure 7.37
Taper Coupling Drawing, Side View

After completing Lesson 8, write a relation that will keep this dimension the depth of the counterbore plus the radius of the large round (**R12**).

Dim=15+Radius
d18=d9+d6

Your dim symbol values will differ.

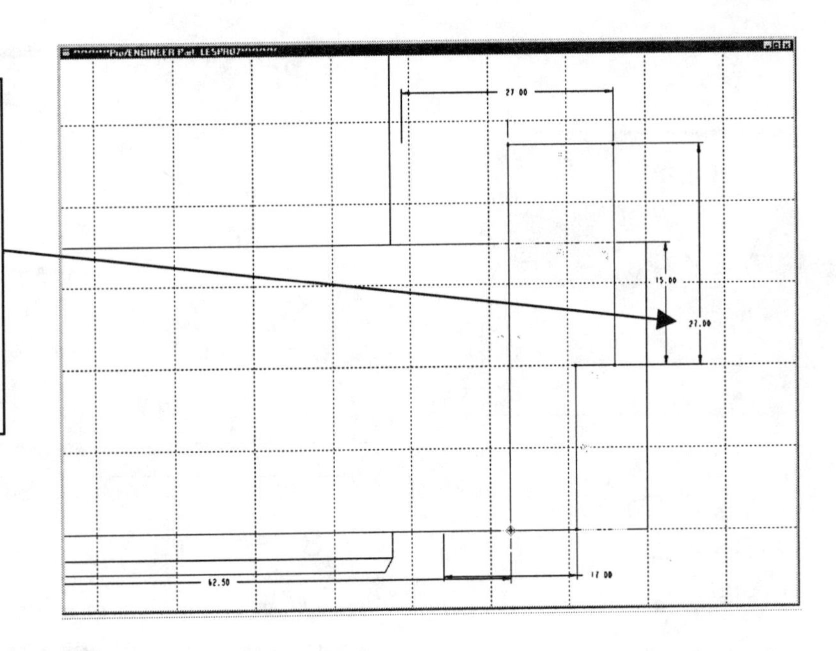

Figure 7.38
Taper Coupling Section, Counterbore

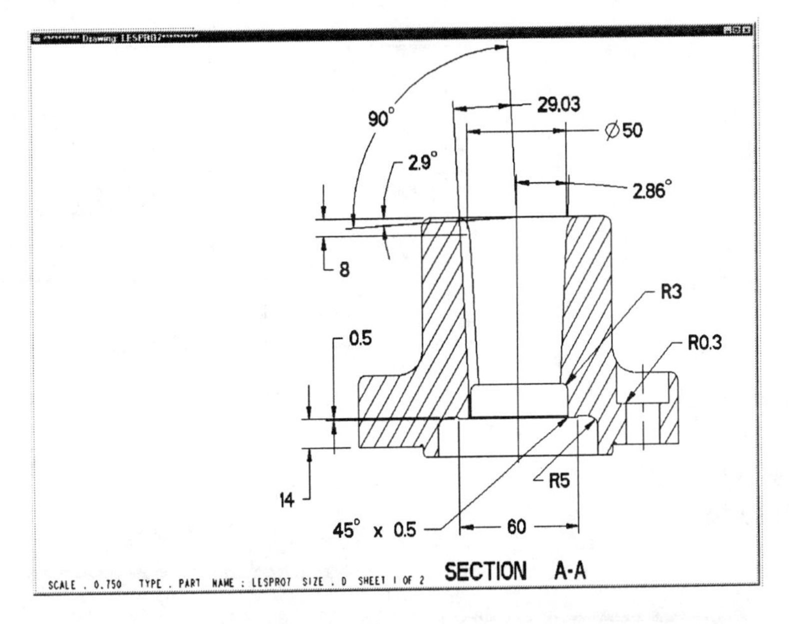

Figure 7.39
Taper Coupling Drawing,
SECTION A-A

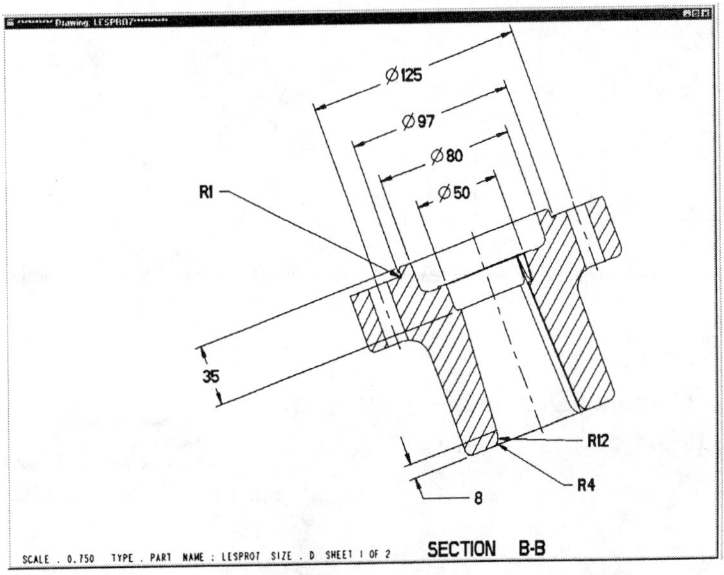

Figure 7.40
Taper Coupling Drawing,
SECTION B-B

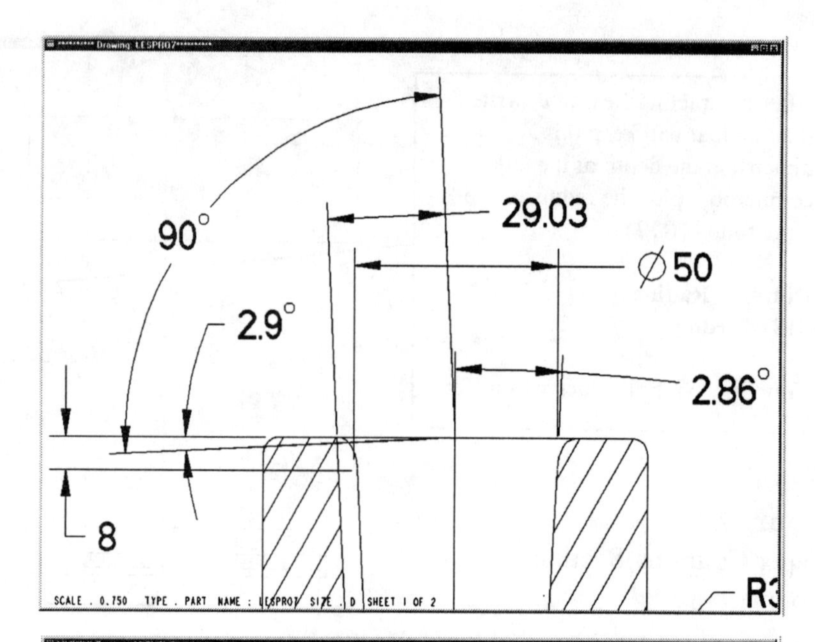

Figure 7.41
Taper Coupling Drawing,
SECTION A-A,
Taper Angle

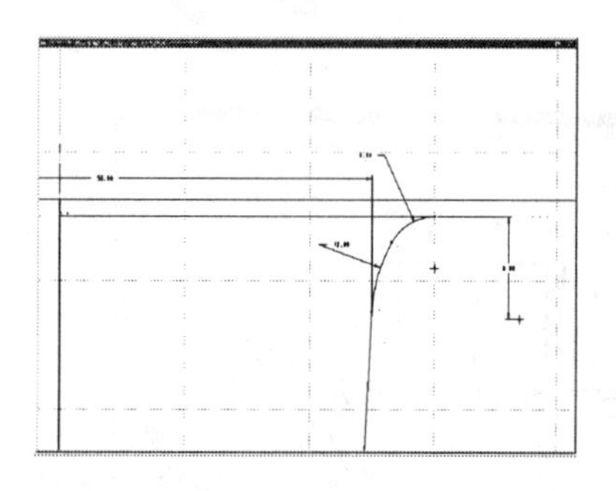

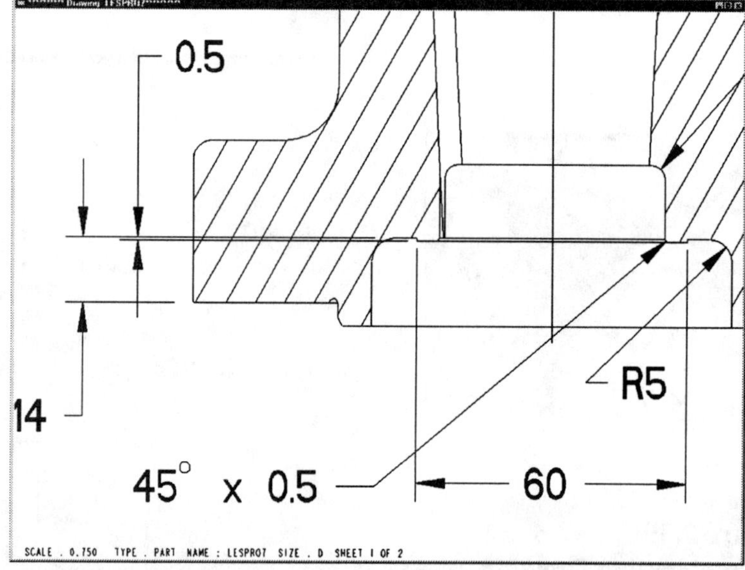

Figure 7.42
Taper Coupling Drawing,
SECTION A-A,
Close-up

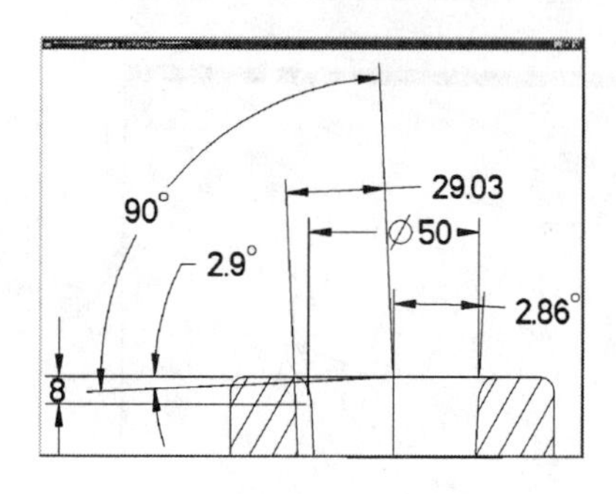

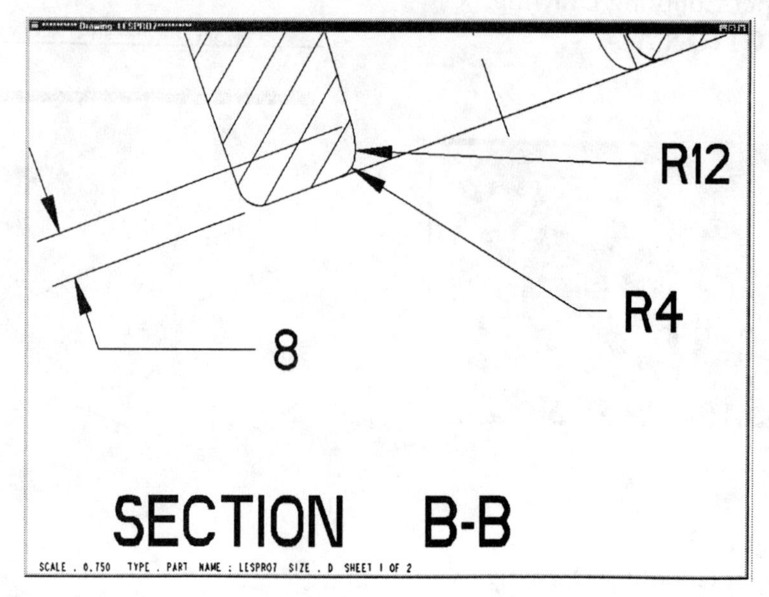

Figure 7.43
Taper Coupling Drawing,
SECTION B-B,
Close-up

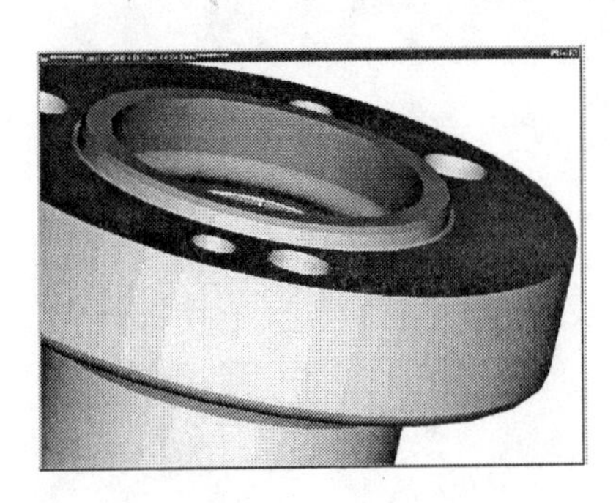

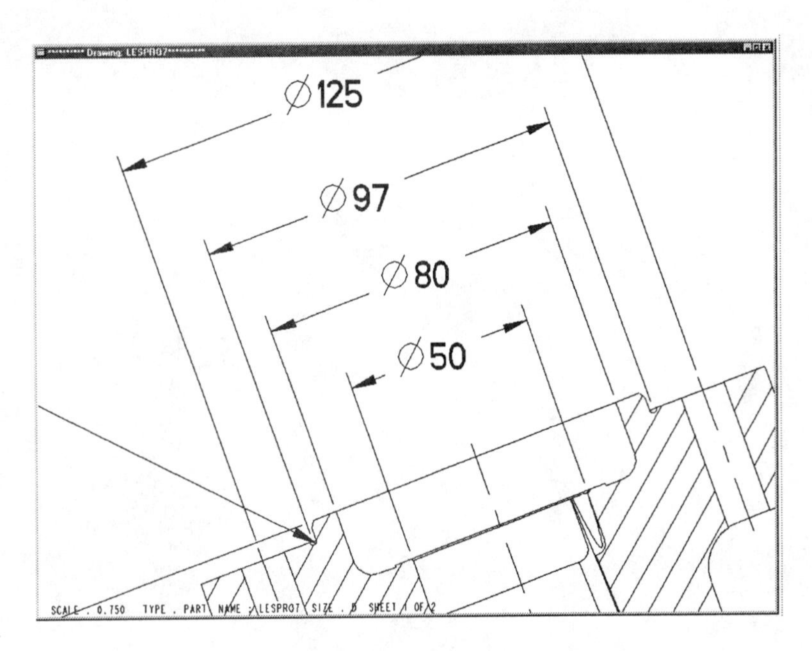

Figure 7.44
Taper Coupling Drawing,
SECTION B-B,
Mating Diameters

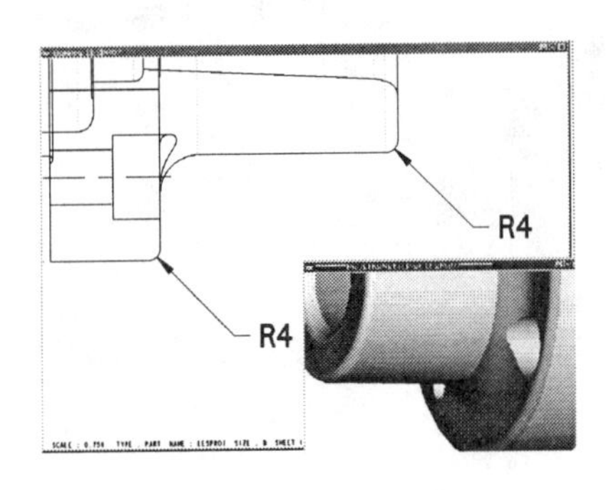

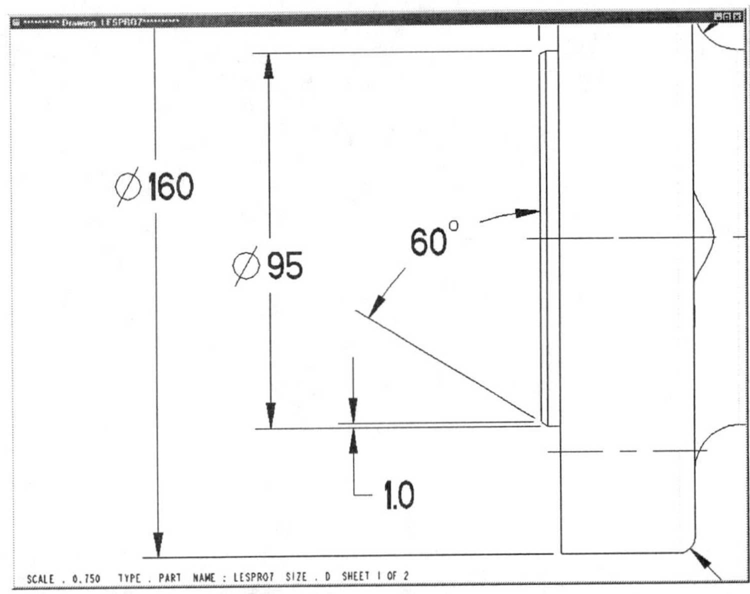

Figure 7.45
Taper Coupling Drawing,
Side View, Close-up

Dbms ⇒ Save ⇒ enter
Purge ⇒ enter ⇒
Done-Return

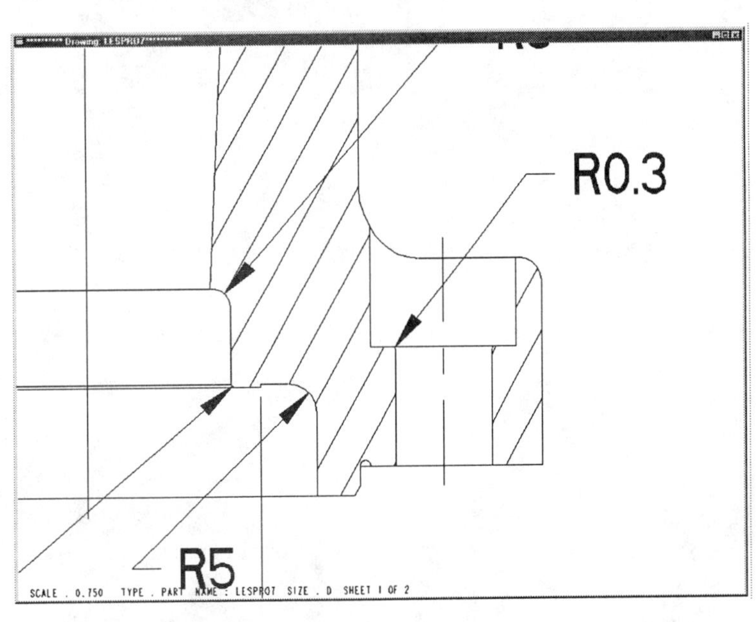

Figure 7.46
Taper Coupling Drawing,
SECTION A-A,
Close-up of Radii

Lesson 8

Ribs, Relations, and Failures

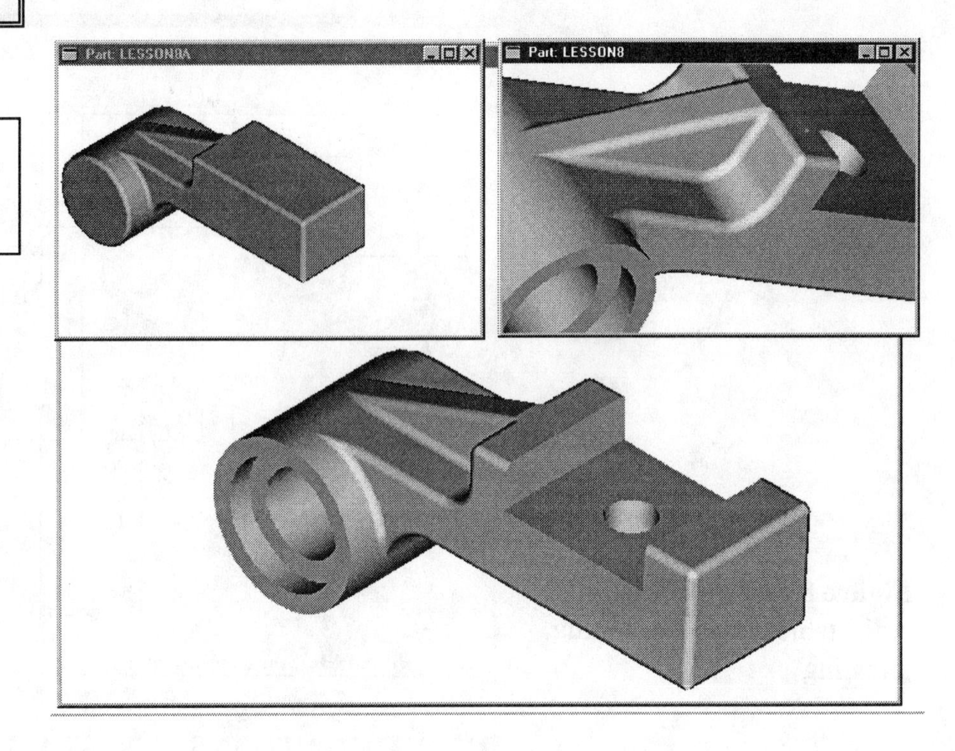

Figure 8.1
Adjustable Guide,
Casting and Machine Part

EGD REFERENCE
Engineering Graphics and Design
by L. Lamit and K. Kitto
Read Chapters: 11, and 14
See Pages: 359, 498

OBJECTIVES

1. **Understand parameters and relations**

2. **Create straight ribs**

3. **Troubleshoot and resolve failures**

4. **Write relations to control features**

5. **Create a manufacturing model**

6. **Model a workpiece and a design part to be used in Pro/MANUFACTURING**

7. **Establish parameters for a part**

8. **Create equality and comparison relations**

9. **Understand the four types of parameter symbols**

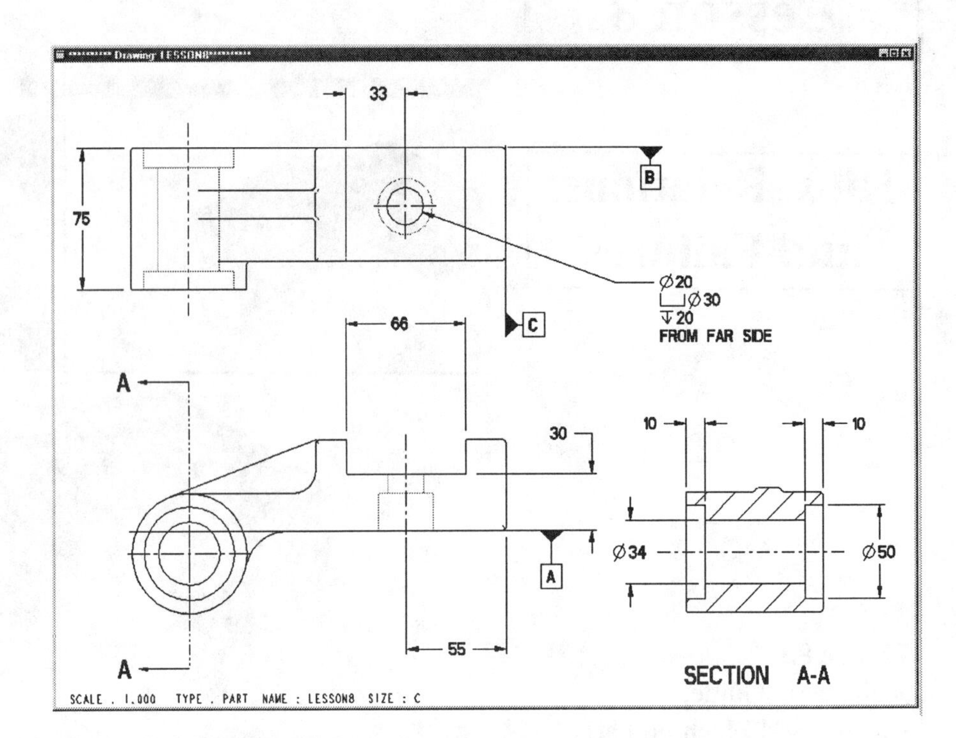

Figure 8.2
Adjustable Guide Machining Drawing

RIBS, RELATIONS, AND FAILURES

A **Rib** is a special type of protrusion designed to create a thin fin or web that is attached to a part (Fig. 8.1 through 8.3).

Relations are equations written between symbolic dimensions and parameters. By writing relations between dimensions in a part or an assembly, the effect of modifications can be controlled.

Failures happen when the model cannot be regenerated. You need to know how to avoid and how to resolve part and assembly failures.

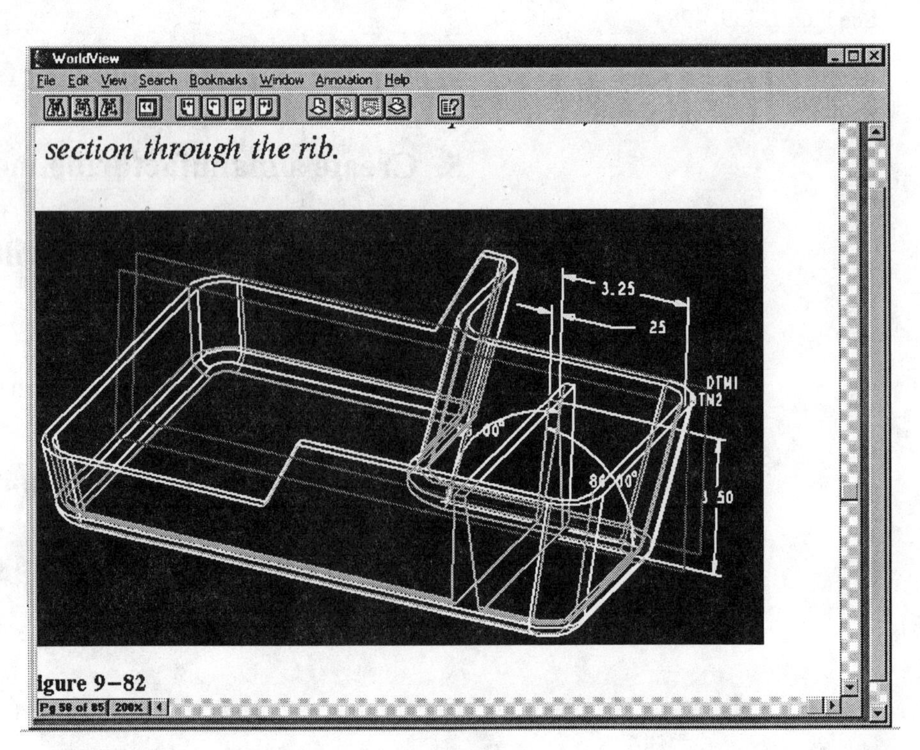

Figure 8.3
COAch for Pro/E, More on Modeling-- Modeling

You always sketch a rib from a side view, and it grows symmetrically about the sketching plane. Because of the way ribs are attached to the parent geometry, they are always sketched as open sections. A rib must "see" material everywhere it attaches to the part; otherwise, it becomes an unattached feature. There are two types of ribs, *straight* and *rotational*. We will confine our discussion to straight ribs. For more information on ribs and in particular rotational ribs, use online documentation (see Fig. 8.4).

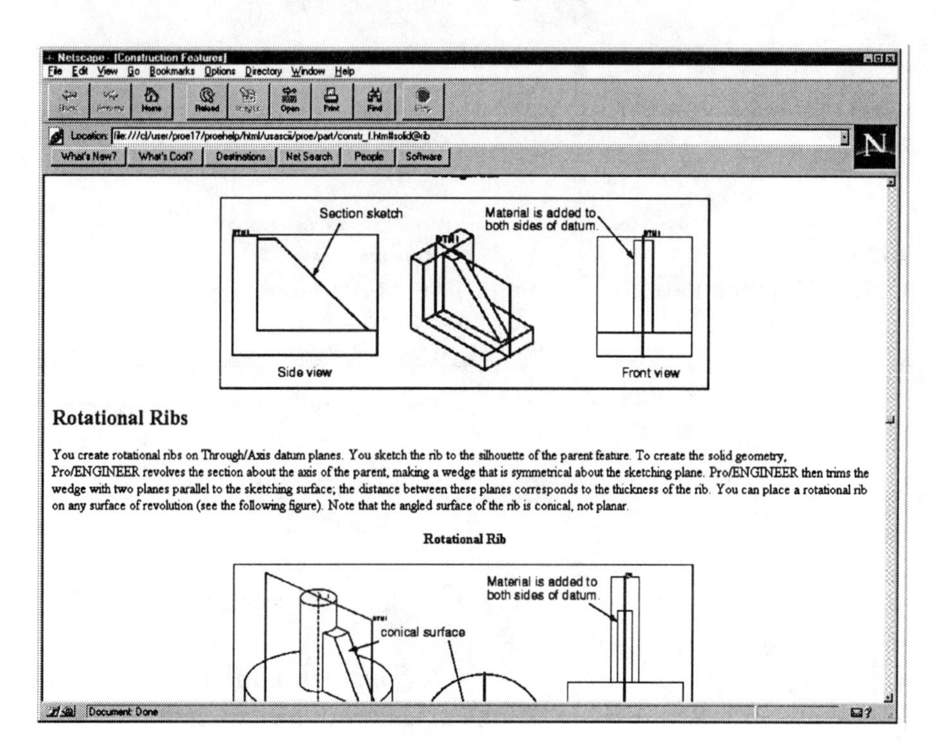

Figure 8.4
Online Documentation Ribs

Straight Ribs

Ribs that are not created on **Through/Axis** datum planes are extruded symmetrically about the sketching plane. You must still sketch the ribs as open sections. Because you are sketching an open section, Pro/E may be uncertain about which side to add the rib. Pro/E displays the DIRECTION menu after the rib section has been regenerated. Pro/E adds all material in the direction of the arrow. If the incorrect choice is made, correct the direction and finish with OK.

Relations

Relations (Fig. 8.5 and 8.6) can be used to provide a value for a dimension. But they can also be used to notify you when a condition has been violated, such as when a dimension exceeds a certain value.

There are two basic types of relations, *equality* and *comparison*. An *equality relation* equates a parameter on the left side of the equation to an expression on the right side. This type of relation is used for assigning values to dimensions and parameters.

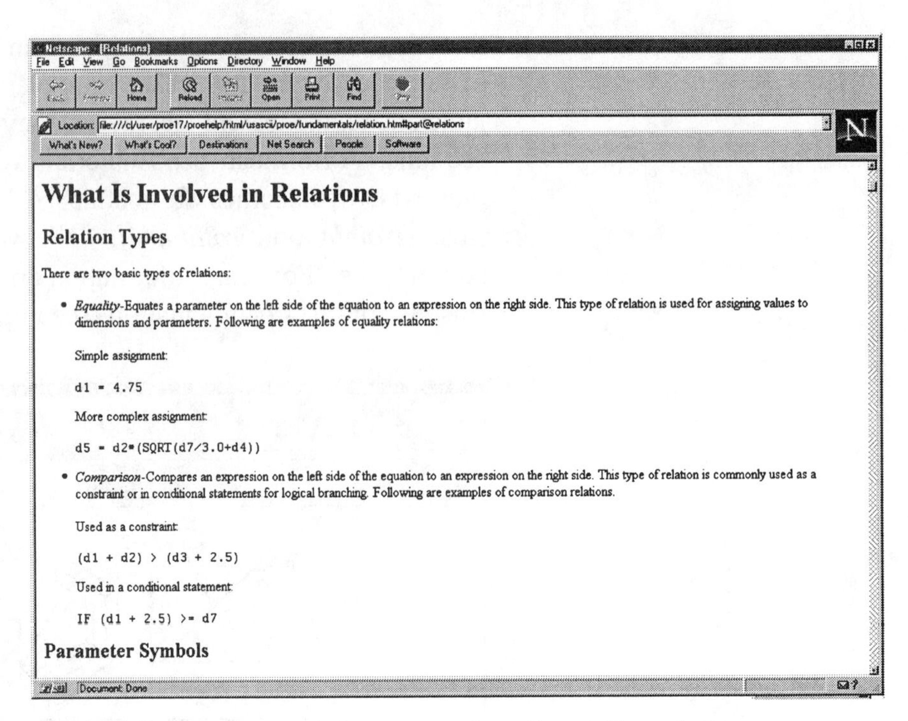

Figure 8.5
Online Documentation Relations

The following are a few examples of equality relations:

d2 = 25.500
d8 = d4/2
d7 = d1+d6/2
d6 = d2*(sqrt(d7/3.0+d4))

A *comparison relation* compares an expression on the left side of the equation to an expression on the right side. This type of relation is commonly used as a constraint or in conditional statements for logical branching.

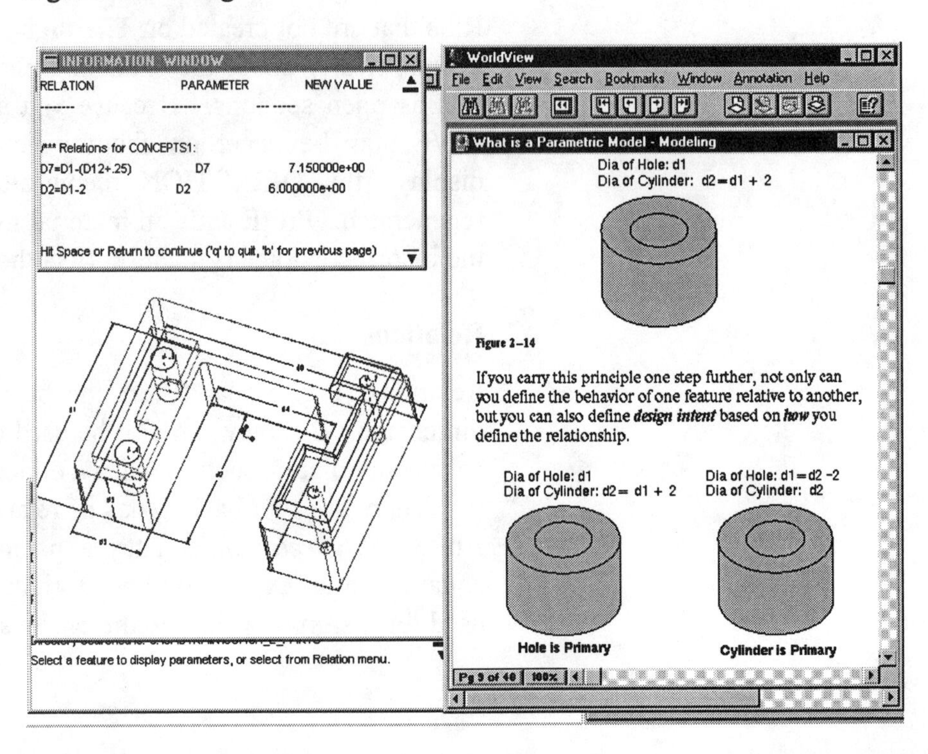

Figure 8.6
COAch for Pro/E, What Is a Parametric Model?--Modeling

The following are examples of comparison relations:

d1 + d2 > (d3 + 2.5)	Used as a constraint
IF (d1 + 2.5) >= d7	Used in a conditional statement

Parameter Symbols

There are four types of parameter symbols used in relations:

Dimensions These are dimension symbols, such as **d8, d12**.
Tolerances These are parameters associated with +- symmetrical and plus-minus tolerance formats. These symbols appear when dimensions are switched from numeric to symbolic.
Number of Instances These are integer parameters for the number of instances in a direction of a pattern.
User Parameter These can be parameters defined by adding a parameter or a relation (e.g., **Volume = d3 * d4 * d5**).

Operators and Functions

The following operators and functions can be used in equations and conditional statements.

Arithmetic Operators

+ Plus, for addition
- Minus, for subtraction
/ Divided by, for division
* Times, for multiplication
^ Exponentiation
( ) Parentheses, for grouping, for example, **d10 = (d5-d6)*d7**

Assignment Operator

The "equals" sign is an assignment operator that equates the two sides of an equation. Obviously, when it is used, the equation can have only a single parameter on the left side.

= Equal to

Comparison Operators

Comparison operators are used wherever a TRUE/FALSE value can be returned. For example, the equation **d1 >= .625** will return TRUE whenever **d1** is greater than or equal to **.625**. It will return FALSE whenever **d1** is less than **.625**.

The following comparison operators can be used:

== Equal to
> Greater than
>= Greater than or equal to
!=, < >, ~= Not equal to
< Less than
<= Less than or equal to
| Or
& And
~, ! Not

The "|", "&", "!", and "~" operators extend the use of comparison relations by allowing several conditions to be set in a single statement.

Functions

Mathematical functions-- Relations may also include the following mathematical functions:

cos ( ) cosine
tan ( ) tangent
sin ( ) sine
sqrt ( ) square root
asin ( ) arcsine
acos ( ) arccosine
atan ( ) arctangent
sinh ( ) hyperbolic sine
cosh ( ) hyperbolic cosine
tanh ( ) hyperbolic tangent

Failures

Sometimes model geometry cannot be constructed because features that have been modified or created conflict with or invalidate other features. For example, this can happen when the following occurs:

* A protrusion is created that is unattached and has a one-sided edge.
* New features are created that are unattached and have one-sided edges.
* A feature is resumed that now conflicts with another feature (such as having two chamfers on the same edge).
* The intersection of features is no longer valid because dimensional changes have moved the intersecting surfaces.
* A relation constraint has been violated.

Resolving Feature Failures During Creation/Redefinition

Depending on the type of environment used to create a feature (i.e., whether the feature uses the dialog box interface-- see Fig. 8.7), Pro/E handles feature failures that may occur during feature creation or redefinition in two different ways:

* *For features that use the dialog box interface*-- If the feature fails after you press OK or Preview, the Resolve button appears in the feature creation dialog box. You can either stay in the dialog box environment and redefine feature elements with the **Define** button, or click on the **Resolve** button to access the Resolve environment so you can obtain diagnostics or make changes to other parts of the model.

* *For features that do not use the dialog box interface-- If the* feature fails, Pro/E brings up the FEAT FAILED menu.

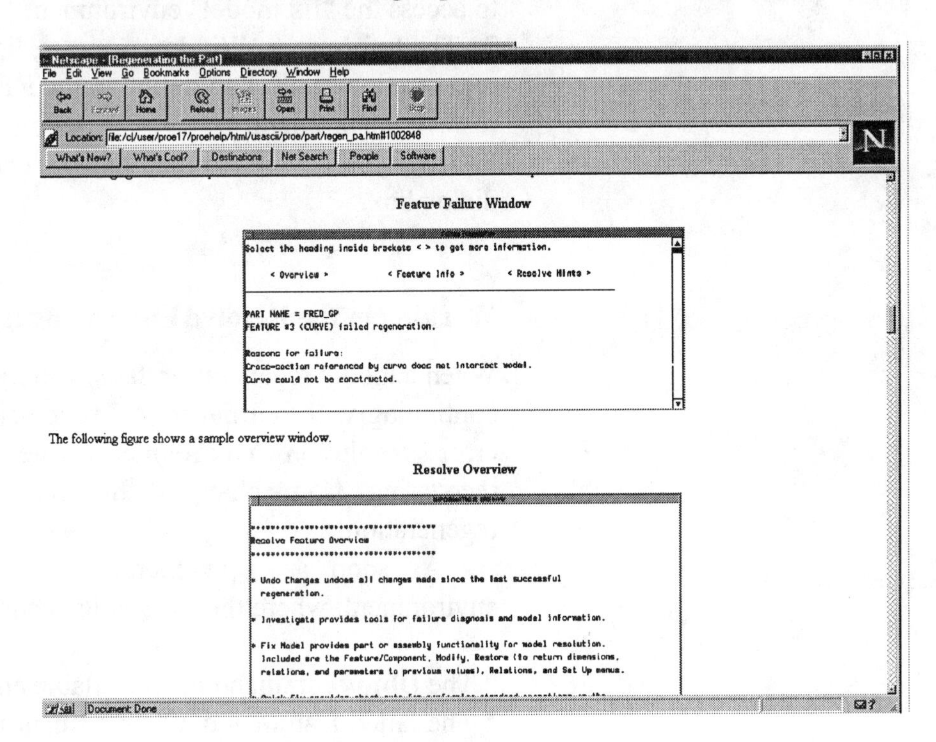

Figure 8.7
Online Documentation
Feature Failure Window

Using the FEAT FAILED Menu

If a feature fails during creation and it does not use the dialog box interface, Pro/E displays the FEAT FAILED menu, with the following options:

Redefine the feature.
Show Ref Display the SHOW REF menu so you can see the references of the failed feature. Pro/E displays the reference number in the MESSAGE WINDOW.
Geom Check for problems with overlapping geometry, misalignment, and so on. This command may be *dimmed.* If a shell, offset surface, or thickened surface fails, Pro/E stores information about the surfaces that could not be offset. The GEOM CHECK menu displays a list of features with failed geometry and a **Restore** command.
Feat Info Get information about the feature.

Using the "Resolve" Button in the Dialog Box

If a feature fails after you choose Preview or **OK** from a dialog box, the Resolve button appears in the dialog box, enabling you to enter the "fix model" environment. To resolve a feature failure by using the Resolve environment, follow these steps:

1. After a feature creation fails, choose **Resolve** from the dialog box to access the "fix model" environment.
2. The Information Window appears, listing features that failed regeneration. Select an option in the RESOLVE FEAT menu to resolve the problem.
3. After you have fixed the problem, choose Preview or **OK** in the dialog box.

Working in the Resolve Environment

When a model regeneration fails, you must resolve the problem before continuing with normal model processing. Pro/E provides a special error resolution environment (the Resolve environment) for recovering from changes that have caused the model to fail regeneration.

As soon as a regeneration fails, Pro/E enters the Resolve environment, where the following occurs:

* The **Dbms** command is unavailable and the model cannot be saved.
* The failed feature and all subsequent features remain unregenerated. The current model displays only the regenerated features as they were at the last successful regeneration. Pro/E displays a message that indicates the problem in the MESSAGE WINDOW.
* Pro/E displays the RESOLVE FEAT menu and the failed-feature diagnostic window.

The Resolve environment (Fig. 8.8) allows you to do the following:

* Undo all the changes made since the last successful regeneration.
* Diagnose the cause of the model failure.
* Fix the problems within this special environment while using standard part or assembly functionality.
* Attempt a quick fix of the problems using shortcuts for performing standard operations on the failed feature, including redefine, reroute, suppress (for parts), freeze (for assemblies), clip suppress, and delete.

For both diagnosing and fixing the problem, you can choose to work on the current (failed) model or on the backup model. The backup model shows all features in their preregenerated state, and can be used to modify or restore dimensions of the features that are not displayed in the current (failed) model.

Using the RESOLVE FEAT Menu

The RESOLVE FEAT menu options are as follows:

Undo Changes Undo the changes that caused the failed regeneration attempt, and return to the last successfully regenerated model. Pro/E displays the CONFIRMATION menu.
Investigate Investigate the cause of the regeneration failure using the INVESTIGATE submenu.
Fix Model Enter the environment to fix the cause of the regeneration failure.
Quick Fix Use the QUICK FIX menu to immediately perform the specified option on the current model. Use options:
- **Redefine** Redefine the failed feature.
- **Reroute** Reroute the failed feature.
- **Suppress** Suppress the failed feature and its children.
- **Clip Supp** Suppress the failed feature and all the features after it.
- **Delete** Delete the failed feature and its children.

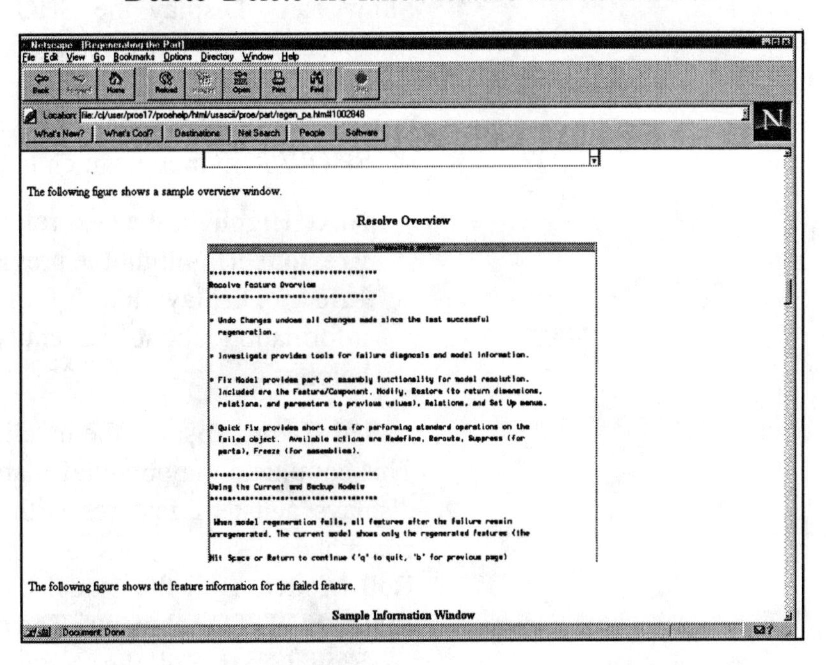

Figure 8.8
Online Documentation
Resolve Overview

If you choose the **Investigate** option, Pro/E displays the option INVESTIGATE menu, which has the following options:

Current Modl Perform operations on the current (failed) model.
Backup Modl Perform operations on the backup model, displayed in a separate window (current model displayed).

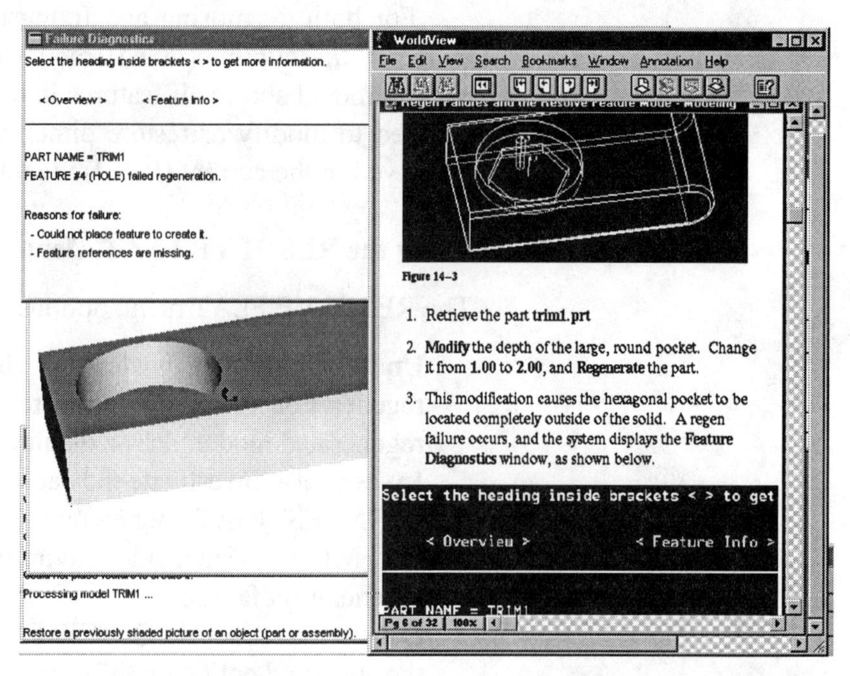

Figure 8.9
COAch for Pro/E, Regen Failures and the Resolve Feature Mode-- Modeling (Segment 1: Undoing Changes)

Diagnostics Toggle on or off the display of the failed-feature diagnostic window (Fig. 8.9).

List Changes Show the modified dimensions in the Main Window and in a preregenerated model window (Review Window), if available. Also, a table is displayed that lists all the modifications and changes.

Show Ref Display the SHOW REF menu to show all the references for the failed feature in the models, in both the Review Window and the Main Window. Pro/E highlights the first reference in the reference color (such as magenta), and displays the SHOW REF menu, which lists the following options:

> **Next** Highlight the next reference.
> **Previous** Highlight the previous reference.
> **Info** Display an Information Window that provides information about the entity and the feature to which it belongs.

Failed Geom Display the invalid geometry of the failed feature. This command may be unavailable. The FAILED GEOM menu displays a list of features with failed geometry and a Restore command.

Roll Model Roll the model back to the option selected in the ROLL MDL TO submenu. The options are as follows:

> **Failed Feat** Roll the model back to the failed feature (for the backup model only).
> **Before Fail** Roll the model back to the feature just before the failed feature.
> **Last Success** Roll the model back to the state it was in at the end of the last successful feature regeneration.
> **Specify** Roll the model back to the specified feature.

If you choose the **Fix Model** option, Pro/E displays the option FIX MODEL menu, which includes the following options:

Current Modl Perform operations on the current active (failed) model.
Backup Modl Perform operations on the backup model, displayed in a separate window from the current model in the active window.
Feature Perform feature operations on the model using the standard FEAT menu. Pro/E displays the CONFIRMATION menu, so you can confirm or cancel the request only if the **Undo Changes** option is not possible. However, the **Undo Changes** option is always possible if you used the **Regen Backup** option in the ENVIRONMENT menu.
Modify Modify dimensions using the standard MODIFY menu.
Regenerate Regenerate the model.
Switch Dim Switch the dimension display from symbols to values, or vice versa.
Restore Display the RESTORE menu so you can restore dimensions, parameters, relations, or all of these to their values prior to the failure. The RESTORE menu options include the following:

All Changes Restore all the changed items.
Dimensions Restore the dimensions.
Parameters Restore the parameters.
Relations Restore the relations.

Relations Add, delete, or modify relations, as necessary, to be able to regenerate the model, using the MODEL REL and RELATIONS menus.
Set Up Display the standard PART SETUP menu to perform additional part setup procedures.
X-Section Create, modify, or delete a cross-sectional view using the CROSS SEC menu.
Program-Access Pro/PROGRAM capabilities using the PROGRAM menu.

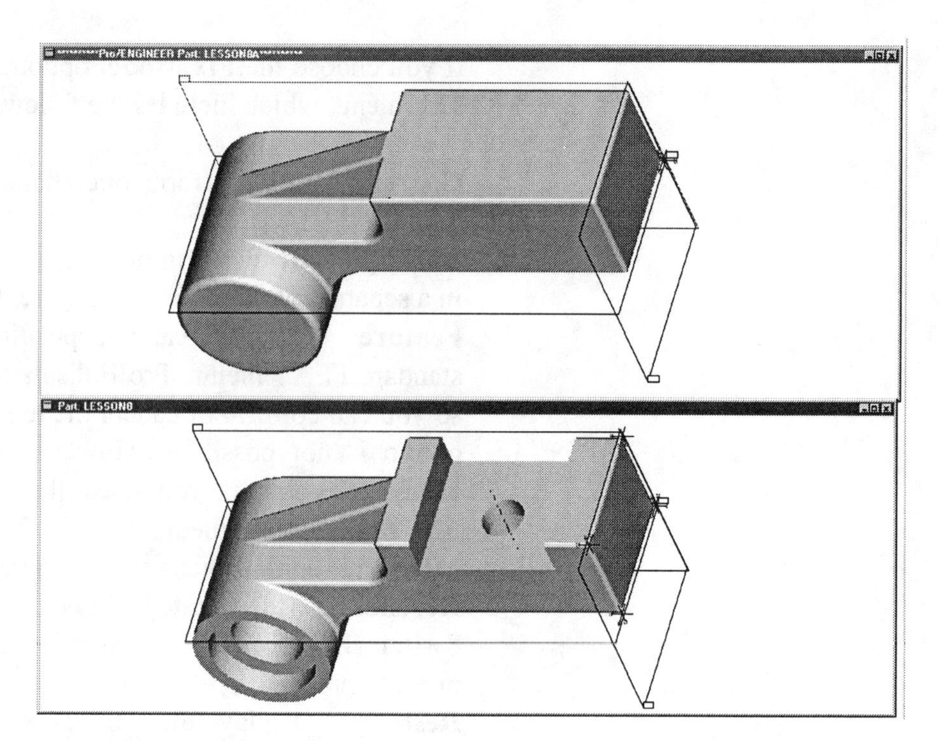

Figure 8.10
Adjustable Guide, Casting and Machined Part

Adjustable Guide

Model the casting first (Fig. 8.10, top) and save it when completed. After saving the casting under a different name, use the part model to create the machined (Fig. 8.10, bottom) **Adjustable Guide**. By having a *casting* (which is called a ***workpiece*** in **Pro/MANUFACTURING**) and a separate but almost identical *machined part* (which is called a ***design part*** in Pro/MANUFACTURING), you can create an operation for machining and an NC sequence. During the manufacturing process you merge the workpiece into the design part and create a ***manufacturing model*** (Fig. 8.11). The difference between the two files is the difference between the volume of the casting minus the volume of the machined part. The removed volume can be seen as *material removal* when doing an **NC Check** operation on the manufacturing model. The manufacturing model is *green*, the cuts completed in **NC Check** show as *yellow* until the machine tool reaches the design size, and then it shows *magenta*. If the machining process gouges the part, the gouge will display as *cyan*. The cutter location can also be displayed as an animated machining process, as shown in Figure 8.12.

The rib created in the casting model will have a **relation** added to it to control its location. The relation will keep the rib centered on the rectangular side of the part.

The Adjustable Guide is a simple part, so the process of describing step-by-step procedures will start with the creation of the rib. The rounds are added late in the modeling process but will cause the model to fail in most cases. The process of fixing the part so that the rounds do not make the regeneration fail is also described. The machined version of the part can be finished with two cuts, a sketched counterbore hole, and a thru hole with two counterbores.

NOTE
From this lesson on you will be required to set up your parts without step-by-step commands.

Set up the Adjustable Guide:

- Material = Steel
- Units = Millimeters
- Default Datum Planes
- Default Coordinate System
- Layers = **DATUM_PLANES**
- Set datums with Geo Tol
- Rename datums **A**, **B**, and **C** as shown in the figures
- Hidden Line
- Tan Dimmed
- ✓Grid Snap

HINT

* Create the part with **ADJ_GUIDE_MACHINE** as the part name.
* Model only the casting dimensions.
* Save the part using **Save As** with a name of **ADJ_GUIDE_CAST**.
* Continue modeling the machined features.
* Save the part under its original name.

Figure 8.11
Manufacturing Model

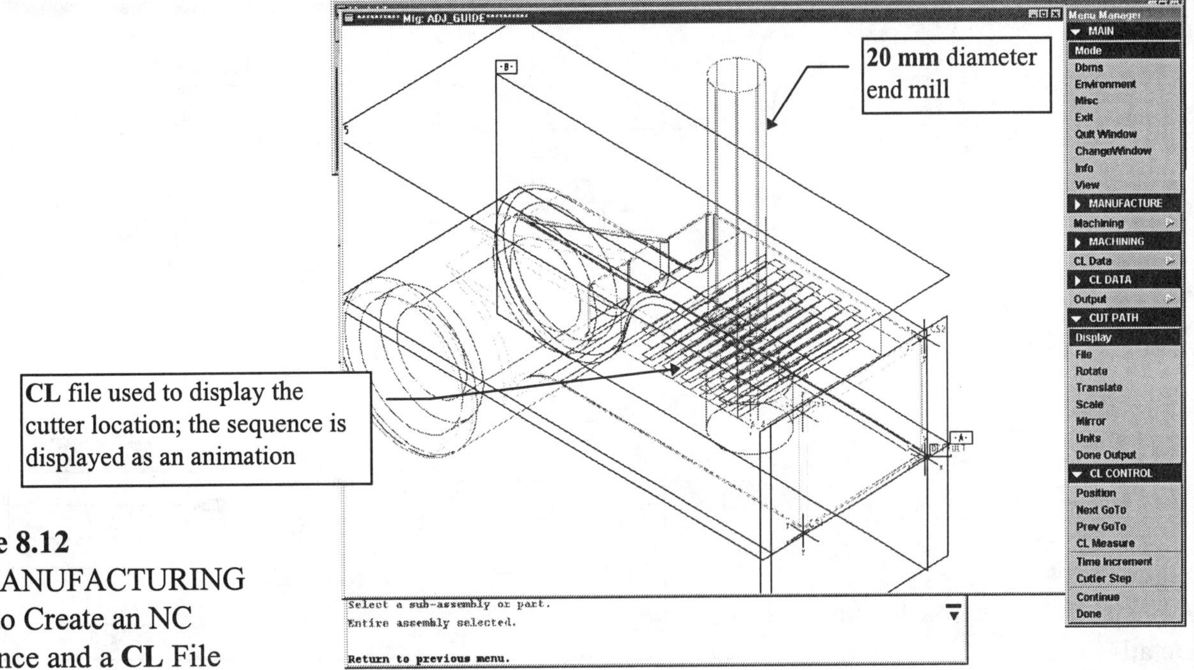

Figure 8.12
Pro/MANUFACTURING Used to Create an NC Sequence and a **CL** File

Start the process by creating a part called **ADJ_GUIDE_MACHINE**. Use Figure 8.13 for the modeling of the Adjustable Guide casting and Figure 8.14 for the Adjustable Guide machined part. Remember to create the casting first and save it using **Dbms** ⇒ **Save As** ⇒ **enter** ⇒ (type a new name, such as **ADJ_GUIDE_CAST)** ⇒ **enter** (to save). Continue working on the part by adding the machined features as required by the design. In the first edition of this text, we will not be covering Pro/MANUFACTURING, but if your company or school has the manufacturing module, you may wish to create a manufacturing model and machine it with the assistance of your instructor.

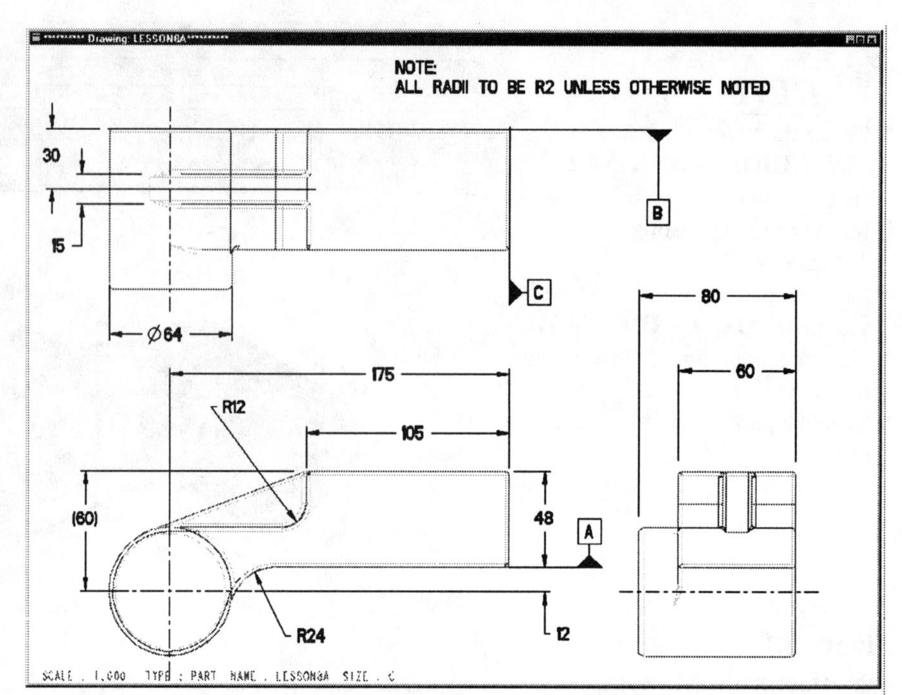

Figure 8.13
Adjustable Guide, Casting Detail

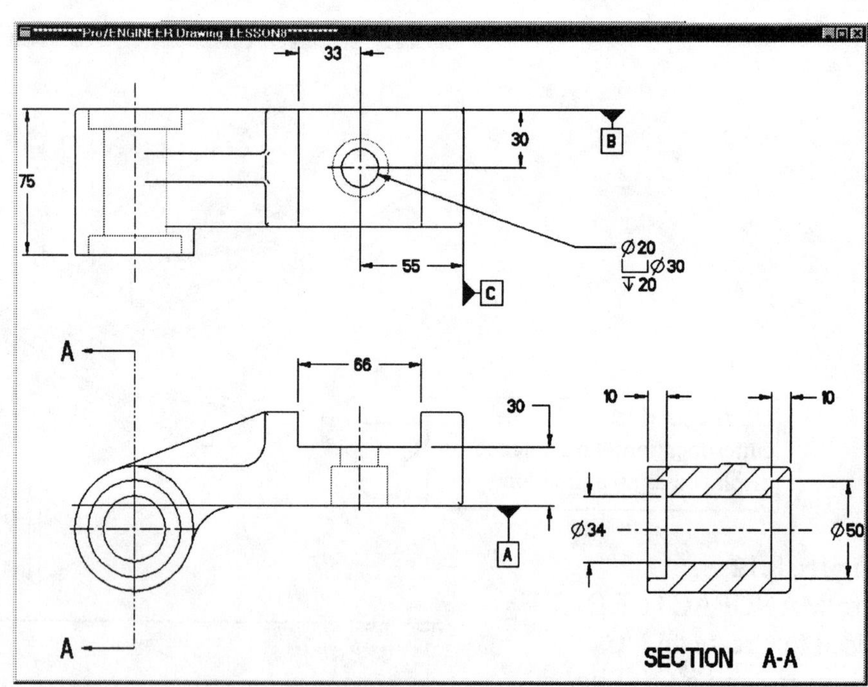

Figure 8.14
Adjustable Guide, Machining Detail

Use the same datum planes as shown in the following figures. The sketching plane for the part is **DTM3** and is on the back side of the part (datum **B** in Fig. 8.15). Datum **A** was originally **DTM2** and runs along the bottom of the part. A *jig and fixture* assembly will be used to hold, locate, and clamp the part for machining. Datum **C** was originally **DTM1** and runs along the right side of the part.

The first protrusion is shown in Figure 8.15. The circular protrusion is extended in Figure 8.16. The first round set is created in Figure 8.17. A round is needed to create a tangent rib. Normally, all rounds would be added last. Create the round by giving the following commands:

Feature ⇒ Create ⇒ Round ⇒ Simple ⇒ Done ⇒ Constant ⇒ Edge Chain ⇒ Done ⇒ Surf Chain ⇒ (pick the top surface of the part as shown in Figure 8.17) **⇒ Select All ⇒ Done ⇒** (type 2) **⇒ enter ⇒ OK ⇒ Done**

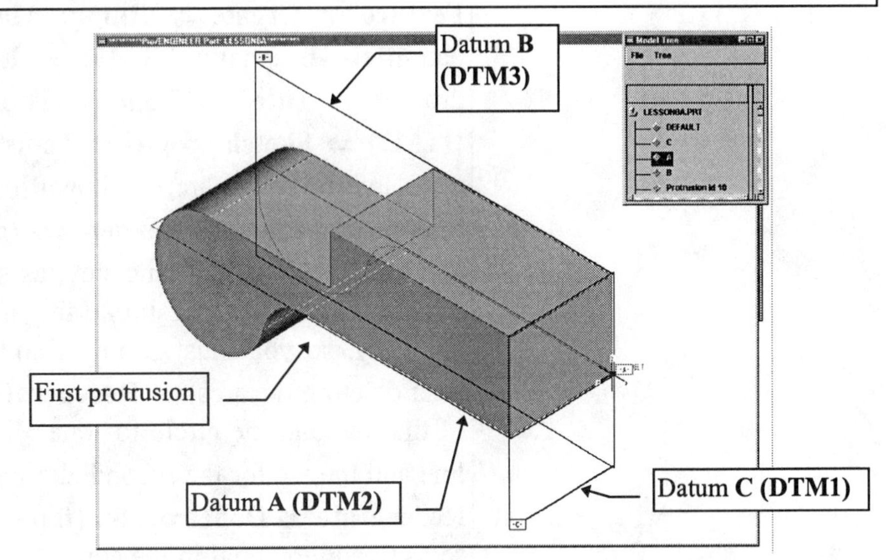

Figure 8.15
First Protrusion

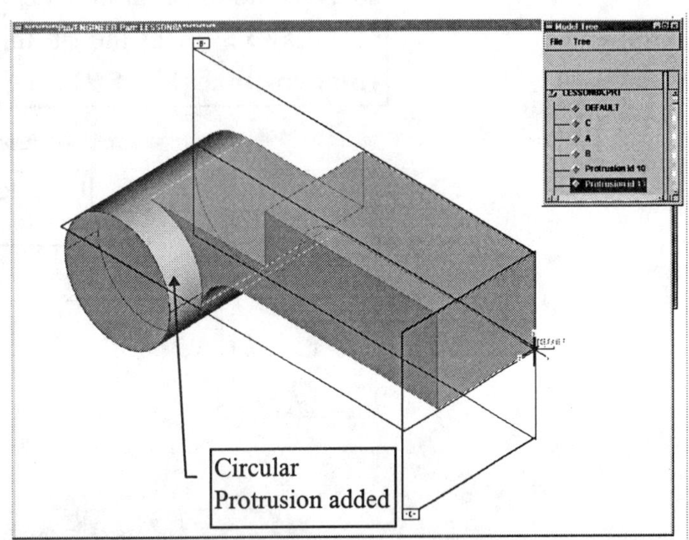

Figure 8.16
Round Protrusion Added to the Model

Dbms ⇒ Save ⇒ enter
Purge ⇒ enter ⇒
Done-Return

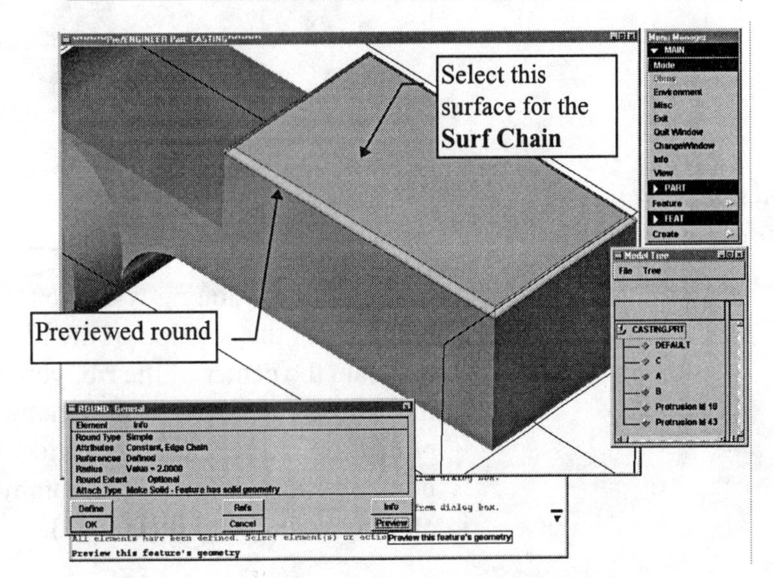

Figure 8.17
R2 Rounds Added to the Top Surface of the Part

The rib will be constructed using a datum on the fly (**Make Datum**). The sketch of the rib will require just one tangent line between the large circular protrusion and the round just created. Use the following commands:

Feature ⇒ **Create** ⇒ **Rib** ⇒ **Make Datum** ⇒ **Offset** ⇒ (pick datum **B**-- originally **DTM3**) ⇒ **Enter Value** ⇒ (type **30** at the prompt) ⇒ **enter** ⇒ **Done** ⇒ **Top** ⇒ (pick **datum A**-- originally **DTM2**) ⇒ **Sketch** ⇒ **Geom Tools** ⇒ **Use Edge** (pick the circular edge and the round arc, as shown in Figure 8.18) ⇒ **Done Select** ⇒ **Sketch** ⇒ **Line** ⇒ **2 Tangent** ⇒ (pick the circle and the arc at the approximate points of tangency, as shown in Figure 8.19) ⇒ **Delete** (delete the arcs-- Pro/E broke the arc and the circle at the points of tangency, so you must zoom in and make sure only the line is left after deleting the arcs) ⇒ **Regenerate** ⇒ **Alignment** ⇒ (pick the end of the line and the circle for one alignment and the other end of the line and the arc for the second alignment, as shown in Figure 8.19) ⇒ **Regenerate** ⇒ **Done** ⇒ **Flip** (flip the arrow to point toward the part so as to add material in the correct direction, as shown in Figure 8.20) ⇒ **Okay** ⇒ (input the rib thickness by typing **15** at the prompt) ⇒ **enter** ⇒ **Done** (Fig. 8.21)

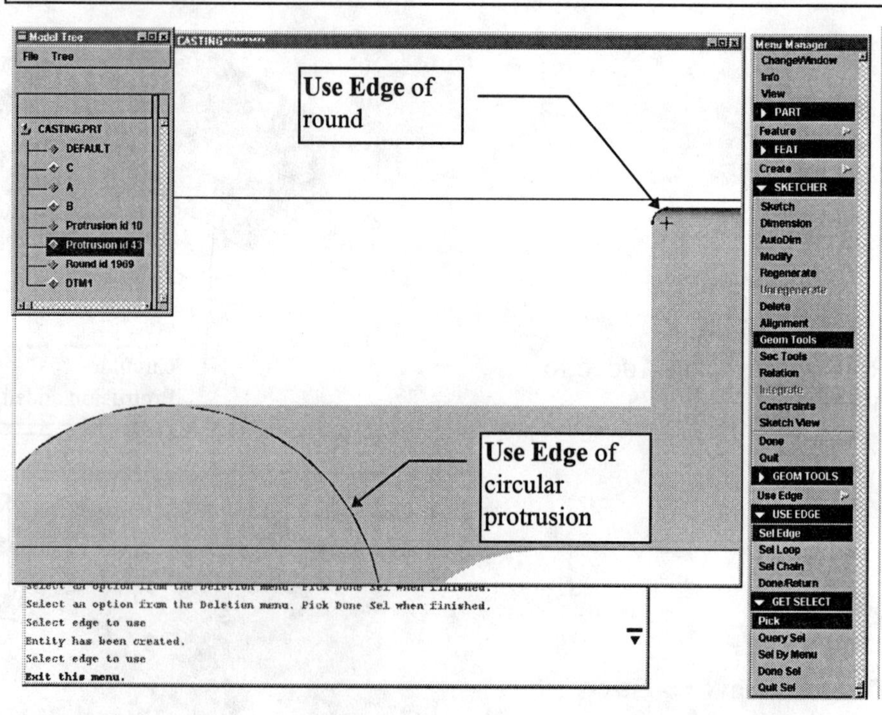

Figure 8.18
Use Edge

Before continuing with the modeling process write a relation that will control the position of the rib. Since you used a datum on the fly to locate the center of the rib, you need to write a relation that will say that the rib's center will remain at the center of the rectangular protrusion no matter how wide the protrusion becomes during a possible **ECO** change. The datum plane used to sketch on was offset **30 mm** from datum **B (DTM3)**.

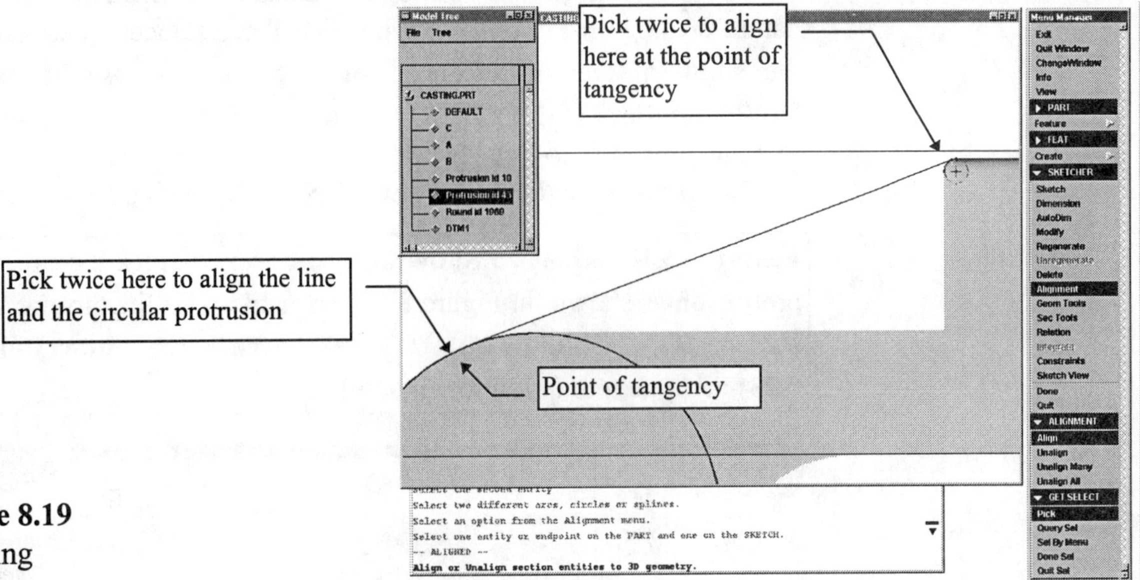

Figure 8.19
Aligning

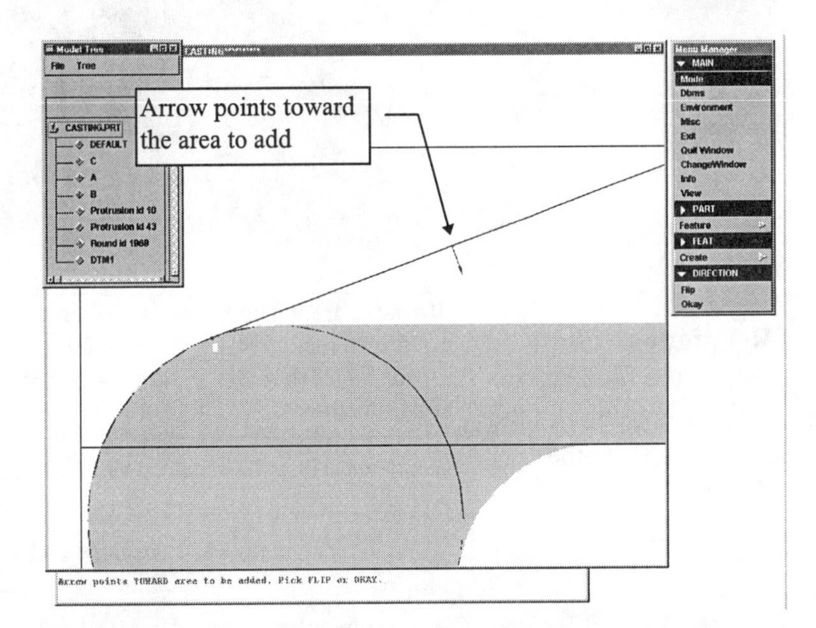

Figure 8.20
Arrow Points Toward the Area to Be Added

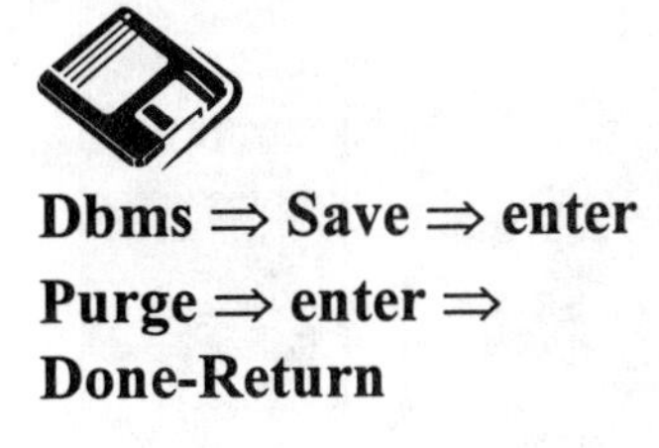

Dbms ⇒ Save ⇒ enter
Purge ⇒ enter ⇒
Done-Return

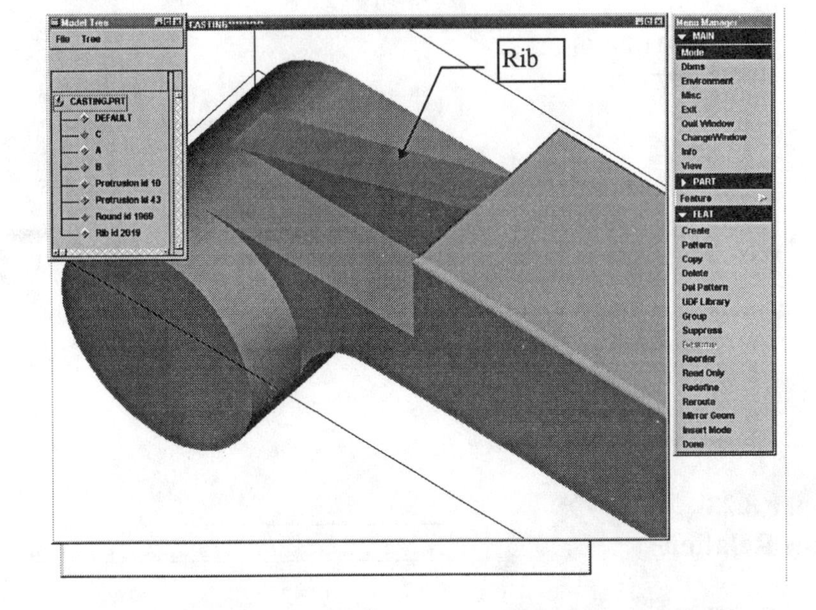

Figure 8.21
Completed Rib

The relation should control the distance from datum **B** to the datum created on the fly by saying that the distance equals one-half the width of the protrusion. You may need to use **Modify** ⇒ **DimCosmetics** ⇒ **Move Dim** ⇒ (select the desired feature and move to new position) to see the dimensions clearly. ***Your dimension symbol numbers will be different.*** Give the following commands:

Feature ⇒ **Relations** ⇒ **Show Dim** ⇒ **enter** ⇒ (pick the rib and the protrusion as shown in Figure 8.22) ⇒ **Add** ⇒ (at the prompt, **Enter RELATION: d100=d11/2**) ⇒ **enter** ⇒ **enter** ⇒ **Show Rel** (Fig. 8.23) ⇒ (type **q** to quit info window)

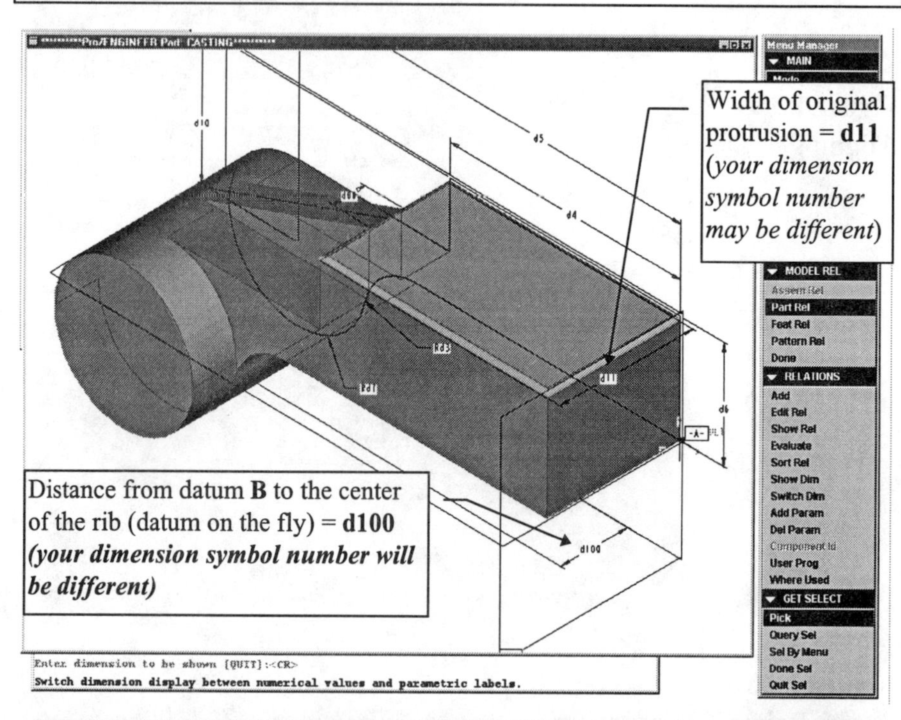

Figure 8.22
Using **Relations** ⇒ **Show Dim** to View the Dimension Symbols for the Center of the Rib and the Width of the Protrusion

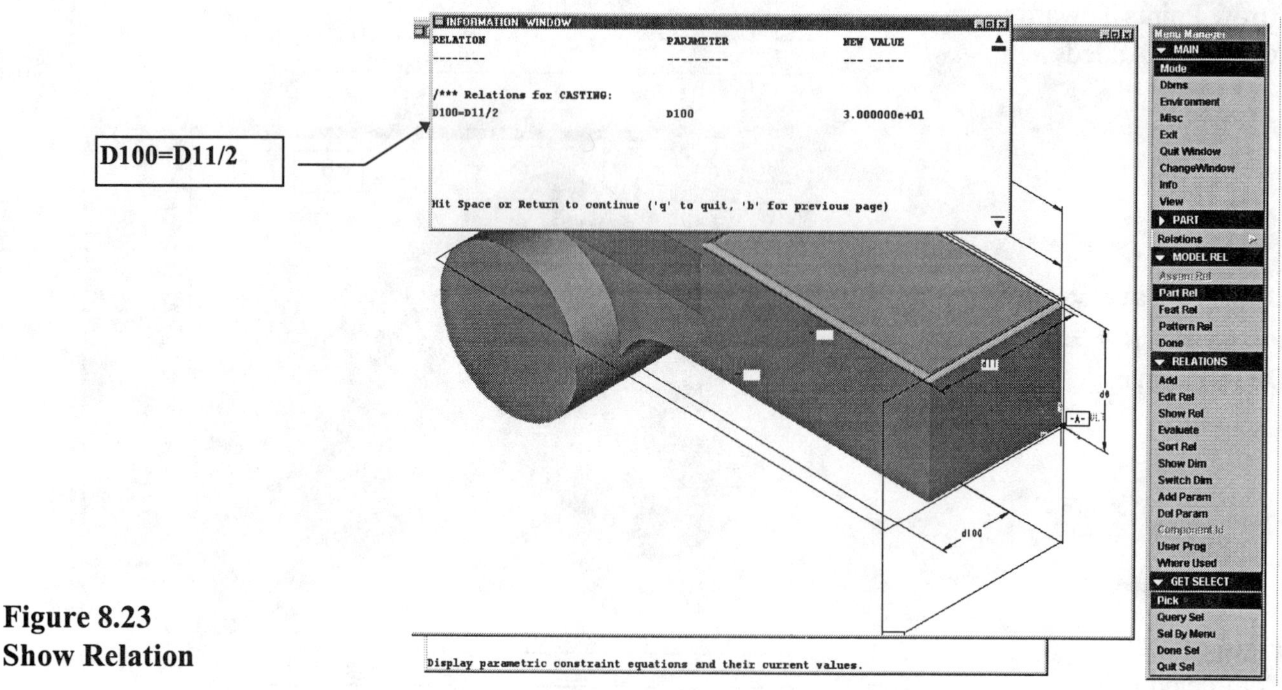

Figure 8.23
Show Relation

Create the rounds for the rest of the part. If you pick too many edges using **One By One**, you may get a failure, as shown in Figure 8.24, where the **Feat Info** choice was selected after the failure to explain the error. Redo the round, working one small set of connected edges at a time using **One By One** and **Tangent Chain**, as shown in Figure 8.25.

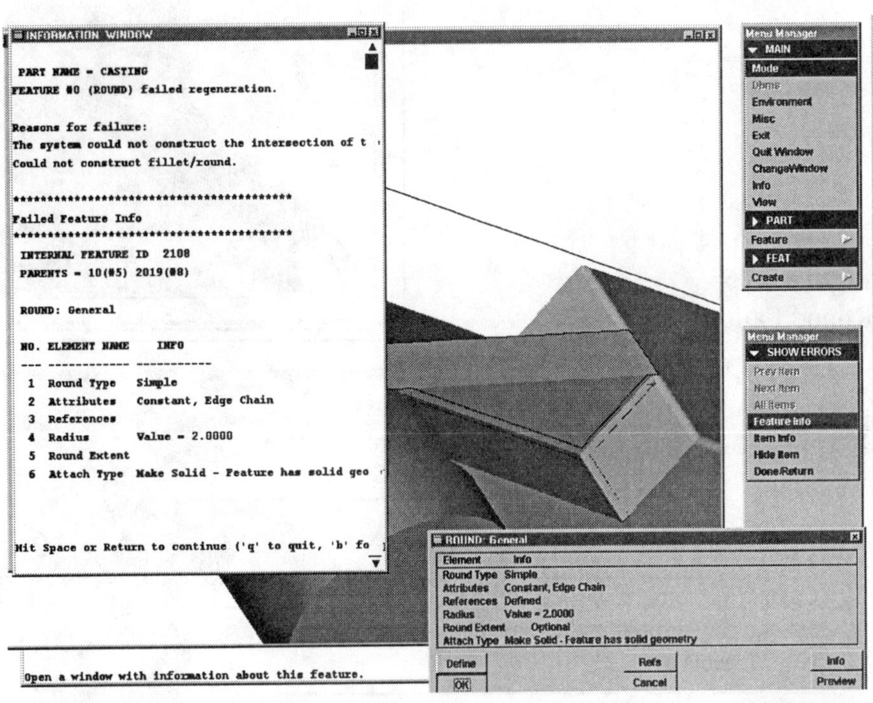

Figure 8.24
Feature Info During Failed Round

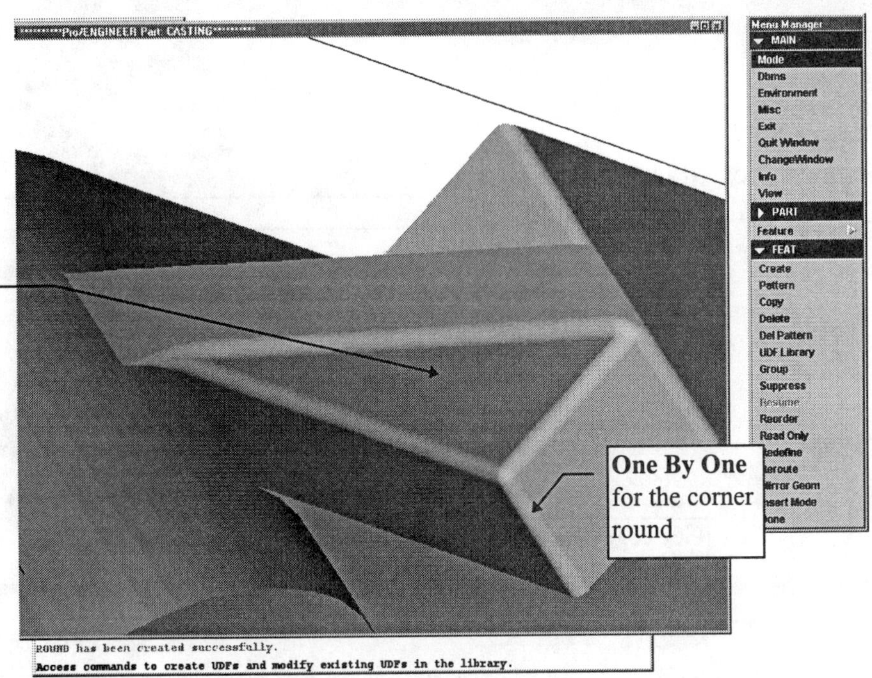

Figure 8.25
Rounds Created Successfully Using **One By One** and **Tangent Chain**

If you get a failure as shown in Figure 8.26, choose **Undo Changes ⇒ Confirm** and try another sequence of picks for the round selection. If you use **Quick Fix ⇒ Delete ⇒ Delete All** instead, you may remove other rounds that you wanted to keep.

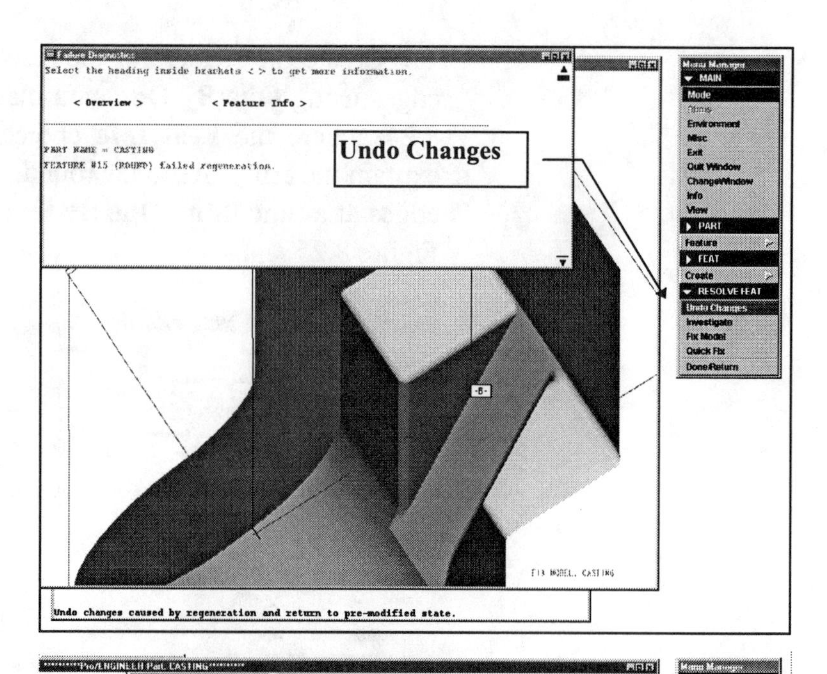

Figure 8.26
Round Failure

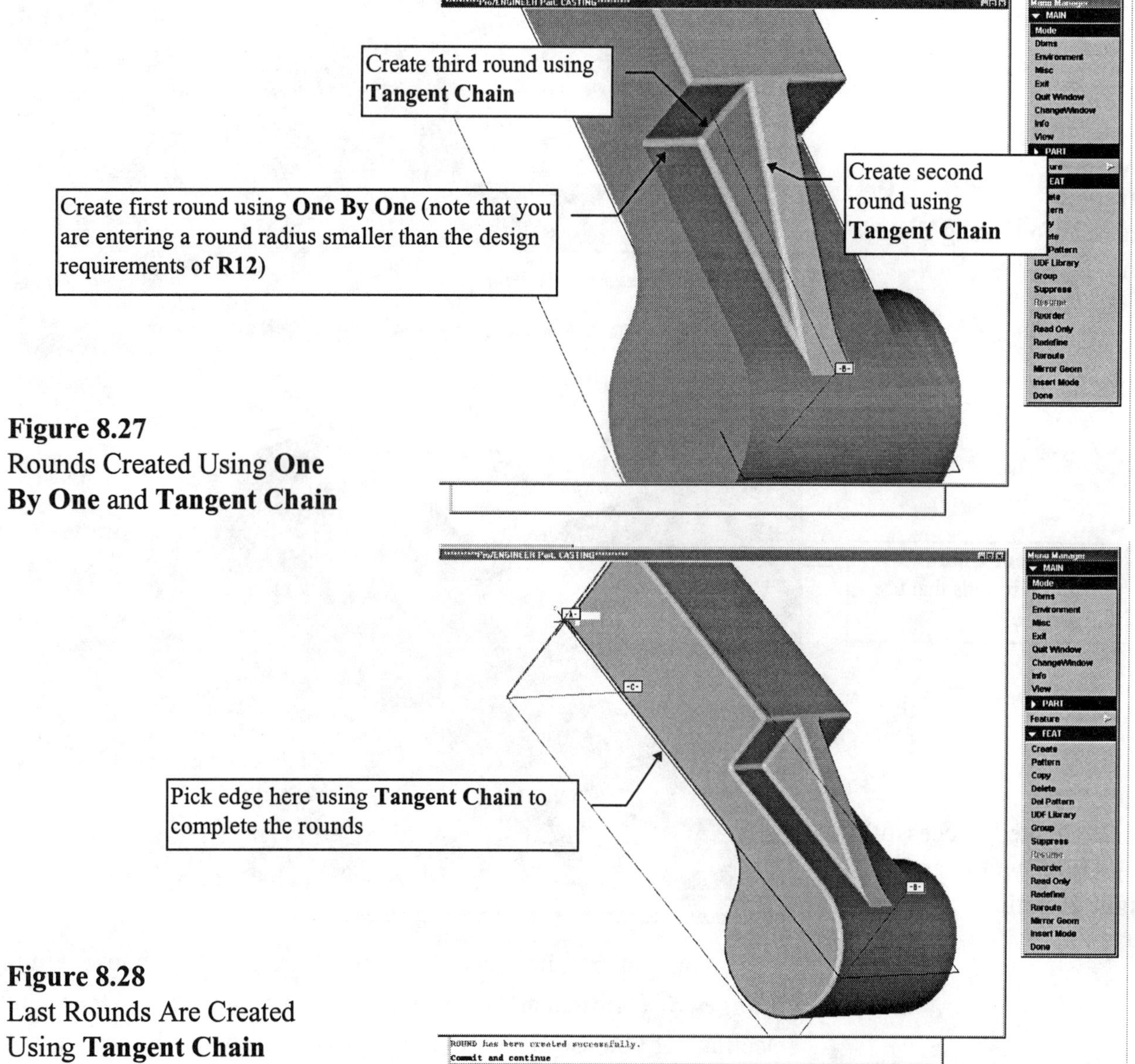

Figure 8.27
Rounds Created Using **One By One** and **Tangent Chain**

Figure 8.28
Last Rounds Are Created Using **Tangent Chain**

It will take some trial and error (Fig. 8.27 and 8.28), but it is possible to get all the rounds done correctly, as shown in Figure 8.29.

Dbms ⇒ Save ⇒ enter Purge ⇒ enter ⇒ Done-Return

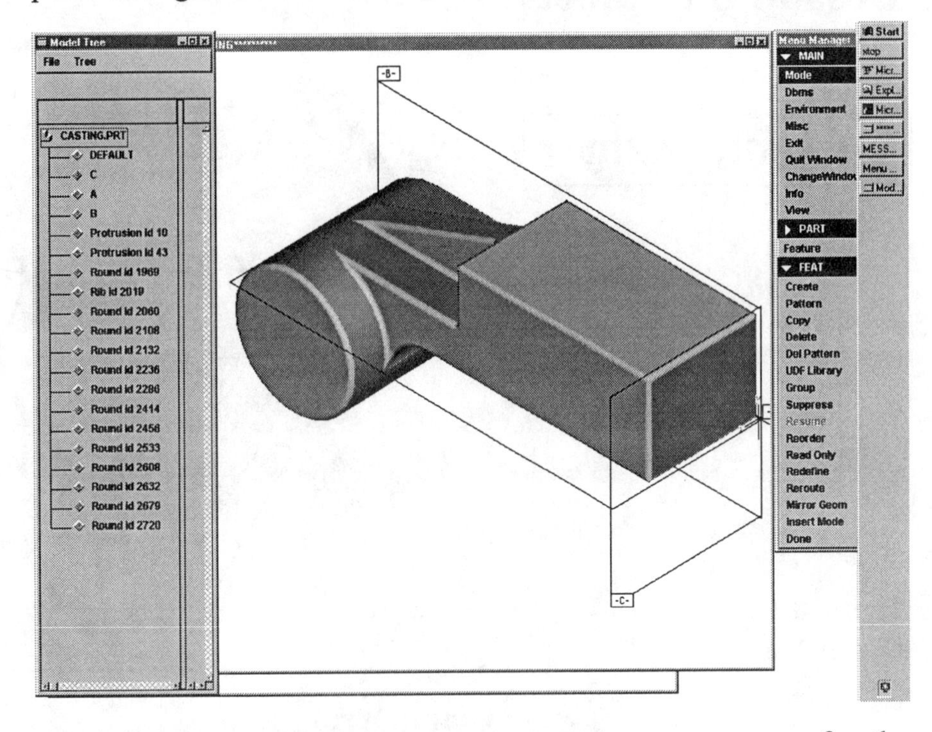

Figure 8.29
Completed Casting Part (***Workpiece***)

Use **Dbms** ⇒ **Save As** ⇒ **enter** ⇒ (type new name for the casting **ADJ_GUIDE_CAST**) ⇒ **enter** ⇒ **Done-Return** to save the casting separate from the part file on which you will continue to model.

Dbms ⇒ Save ⇒ enter Purge ⇒ enter ⇒ Done-Return

Modify the **R2** round to the design size of **R12** for the feature shown in Figure 8.30. The radius for the round must be updated on the casting and machine part to reflect the design requirements. Complete the machined part using the dimensions shown in Figure 8.14.

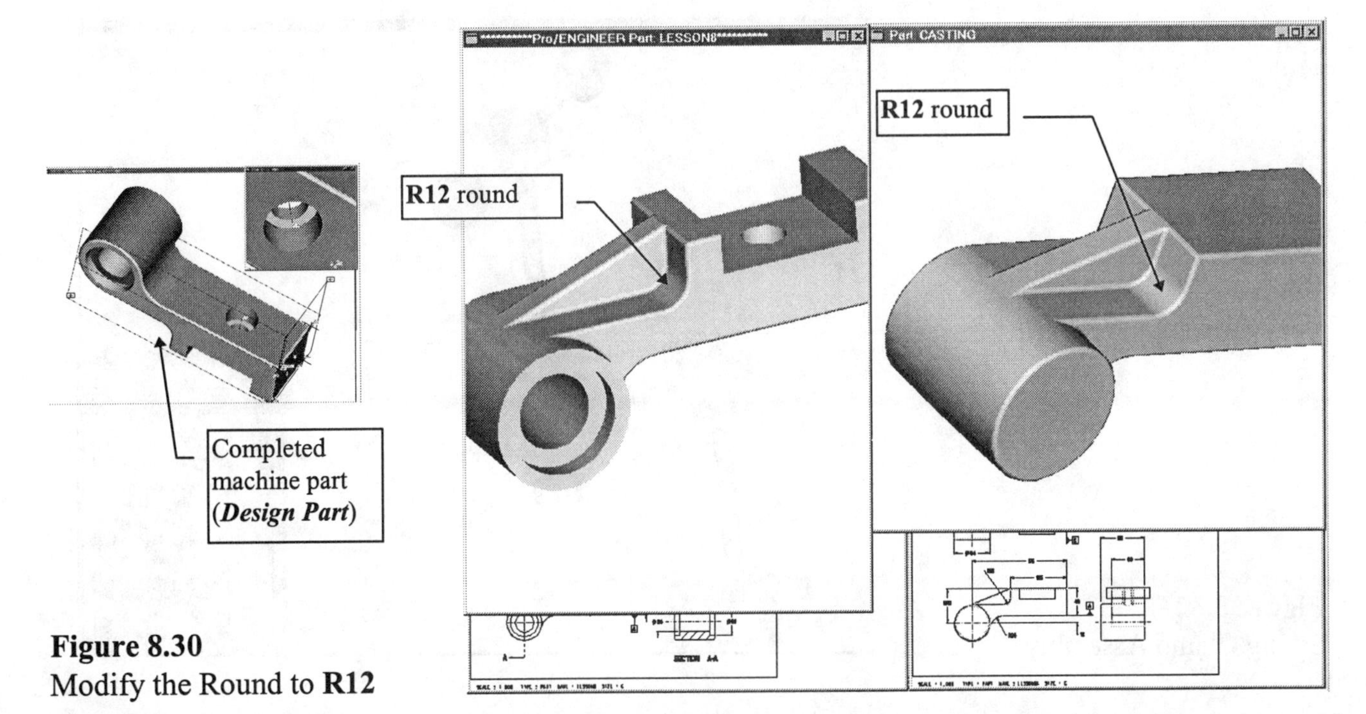

Figure 8.30
Modify the Round to **R12**

Lesson 8 Project

Clamp Arm

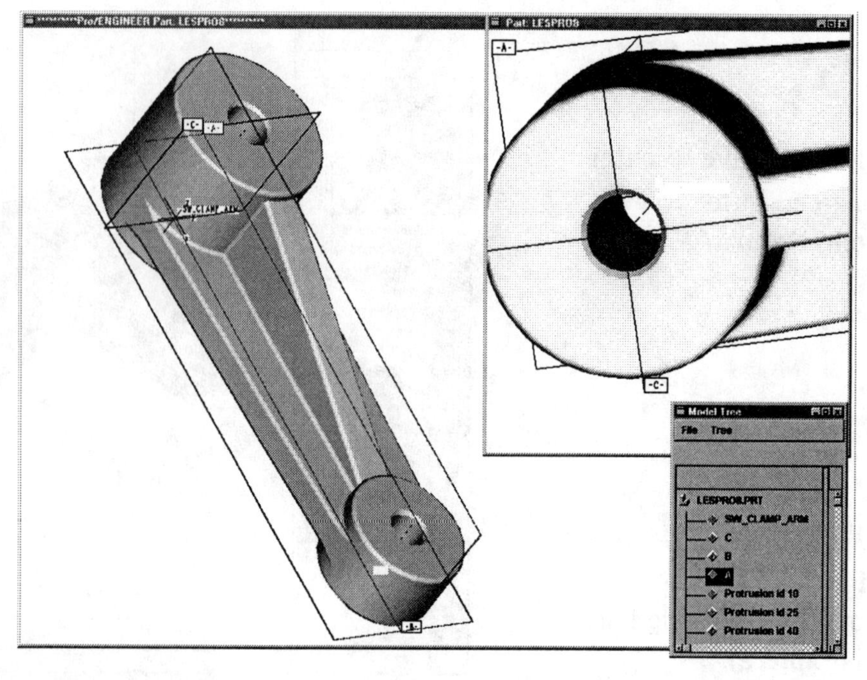

Figure 8.31
Clamp Arm

Clamp Arm

The eighth **lesson project** is a cast part that requires similar commands as the **Adjustable Guide**. Create the part shown in Figure 8.31 through Figure 8.37. Analyze the part and plan out the steps and features required to model it. Use the DIPS in Appendix D to plan out the feature creation sequence and the parent-child relationships for the part. Note that the Clamp Arm is used for an assembly (Fig. 8.32) in Lessons 14 and 15. Create two versions of the Arm, one with all cast surfaces and one with machined ends.

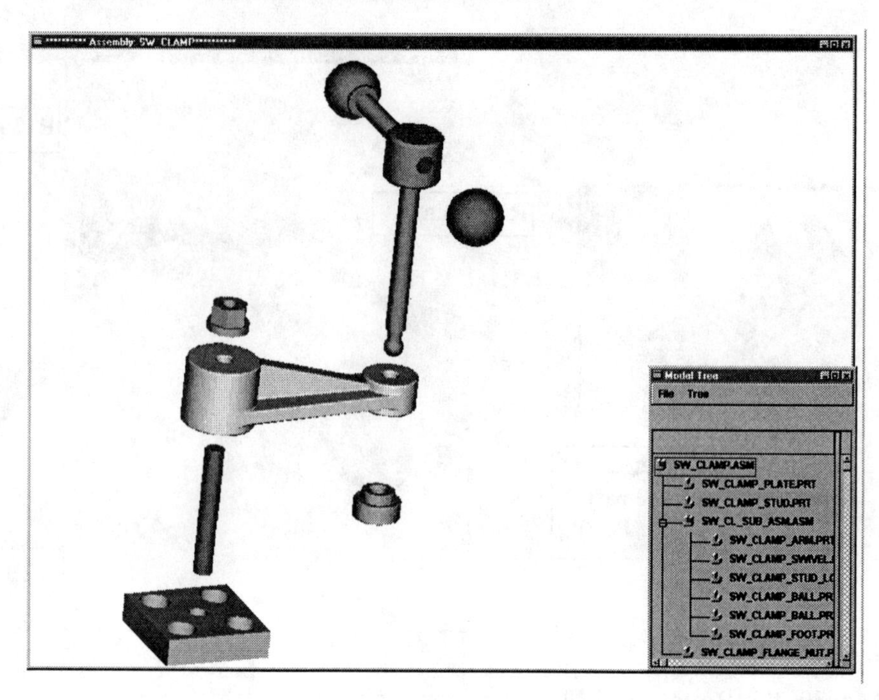

Figure 8.32
Swing Clamp Assembly

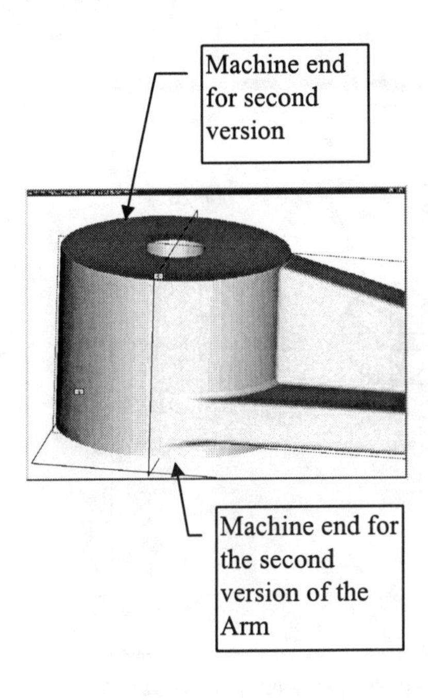

Machine the top and bottom of both round protrusions **.10** off either end. Save a casting version of the part (no holes or machined surfaces) and a machined version. Include cosmetic threads where required.

Write a relation to control the position of the horizontal ribs (Fig 8.33). They must remain at the center of the smaller circular end.

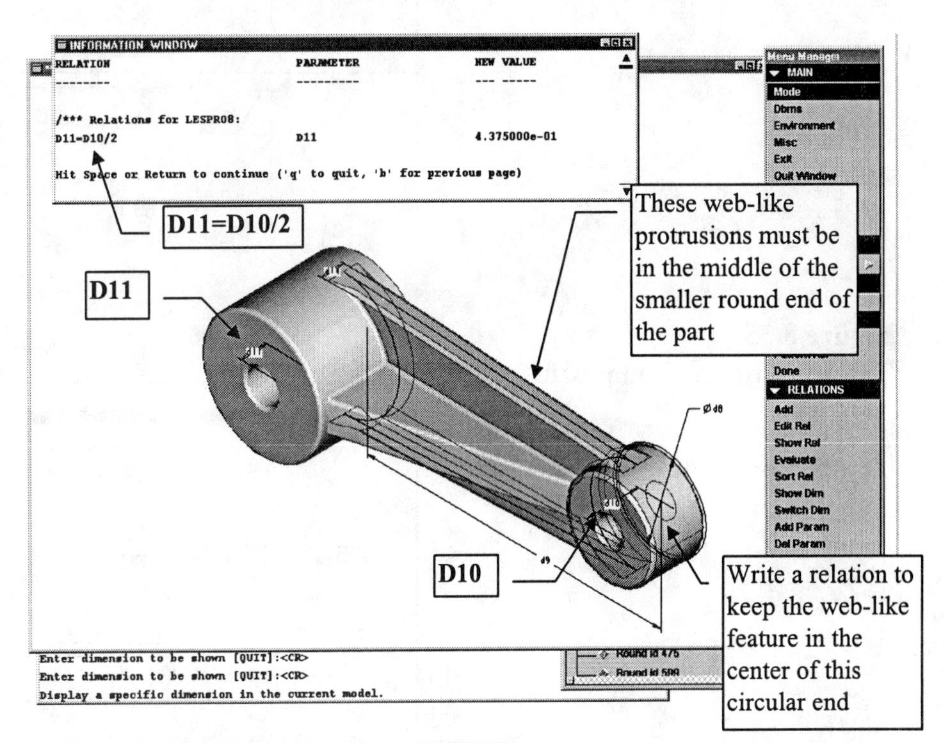

Figure 8.33
Relation

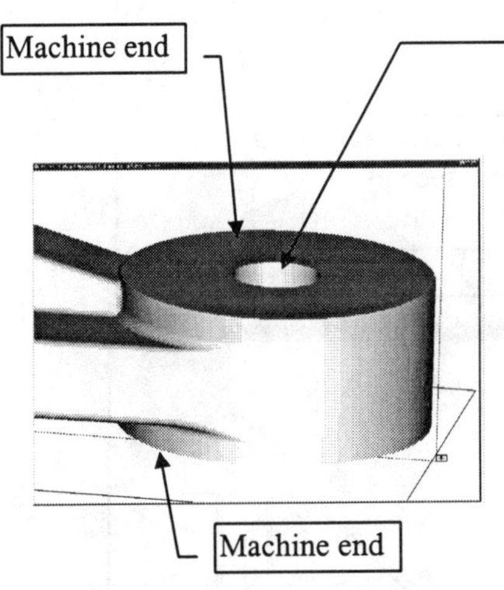

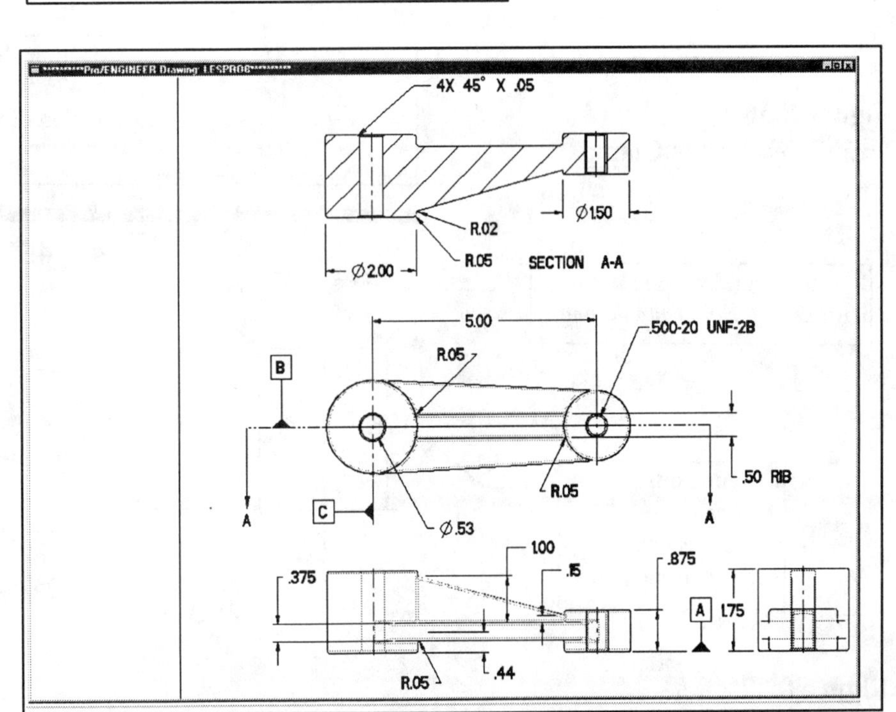

Figure 8.34
Detail of Clamp Arm
(ROUNDS ARE R.02 UNLESS OTHERWISE NOTED)

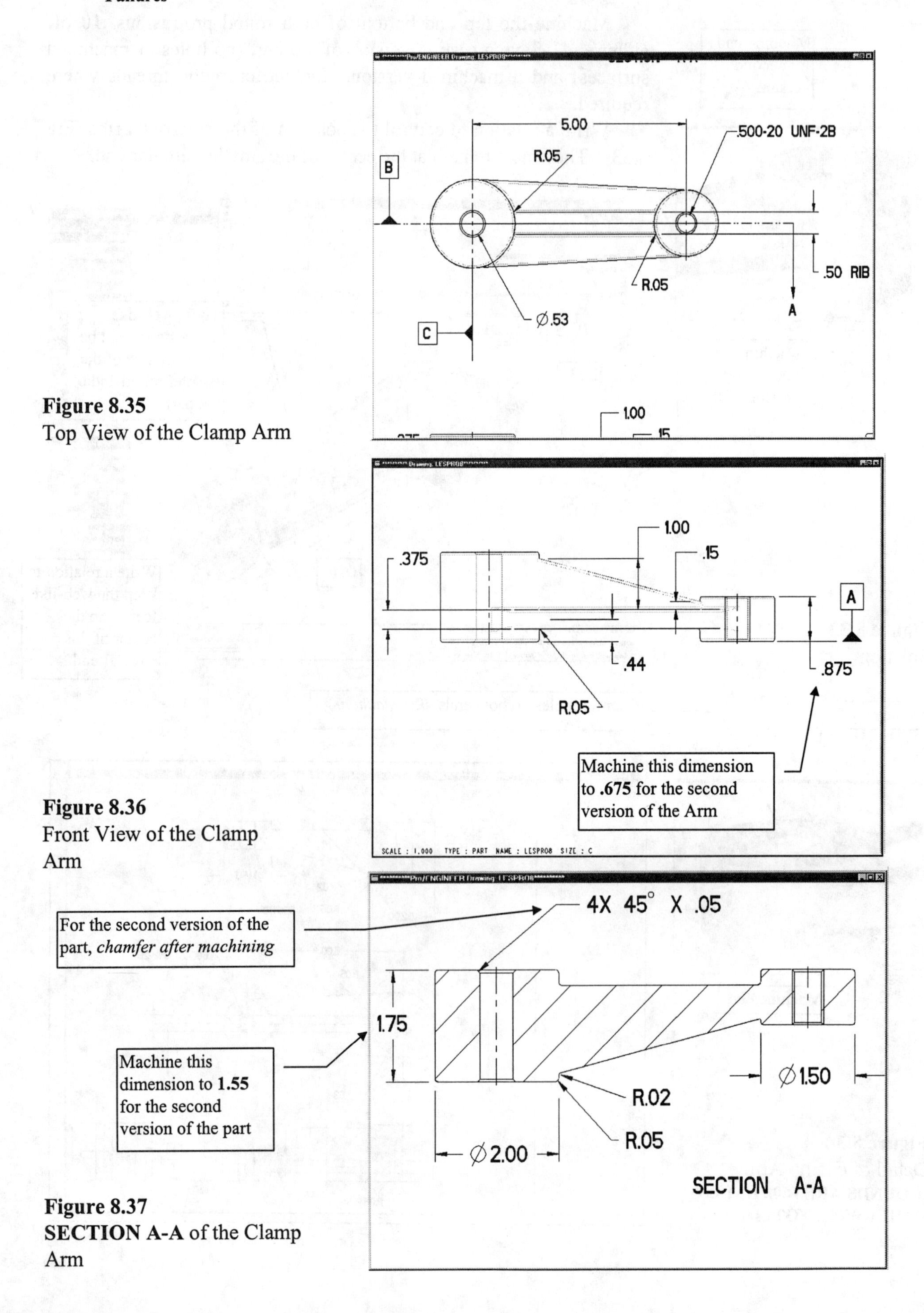

Figure 8.35
Top View of the Clamp Arm

Figure 8.36
Front View of the Clamp Arm

Figure 8.37
SECTION A-A of the Clamp Arm

Lesson 9

Drafts, Suppress, and Text Protrusions

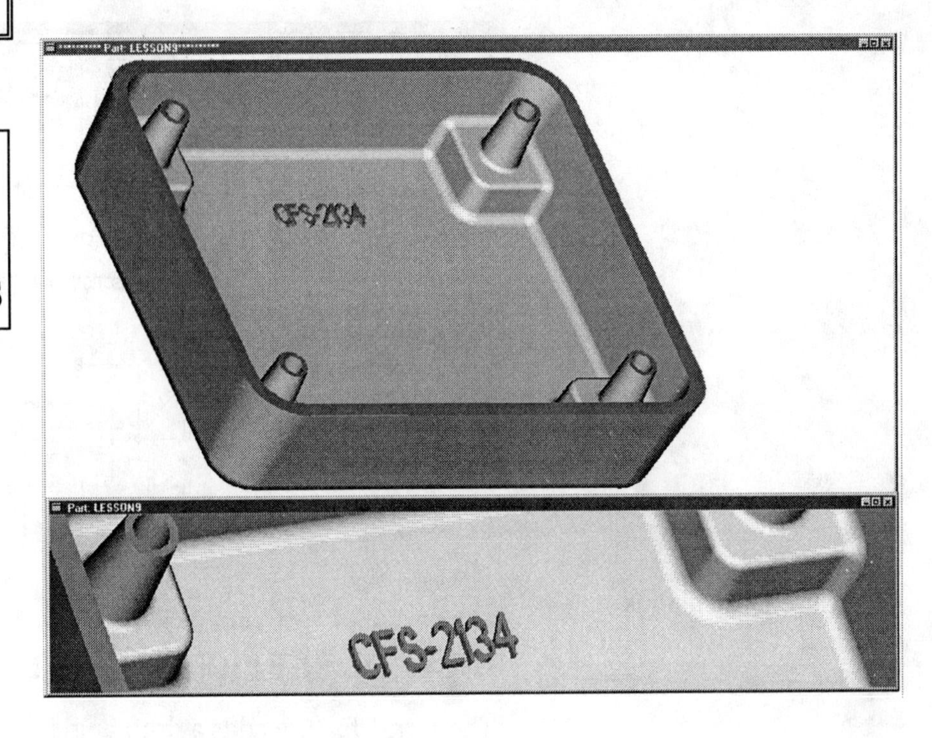

Figure 9.1
Enclosure

OBJECTIVES

1. **Create draft features**
2. **Create text protrusions on parts**
3. **Shell a part**
4. **Suppress features to decrease regeneration time**
5. **Resume a set of suppressed features**

EGD REFERENCE
Engineering Graphics and Design
by L. Lamit and K. Kitto
Read Chapters: 14
See Pages: 493-498

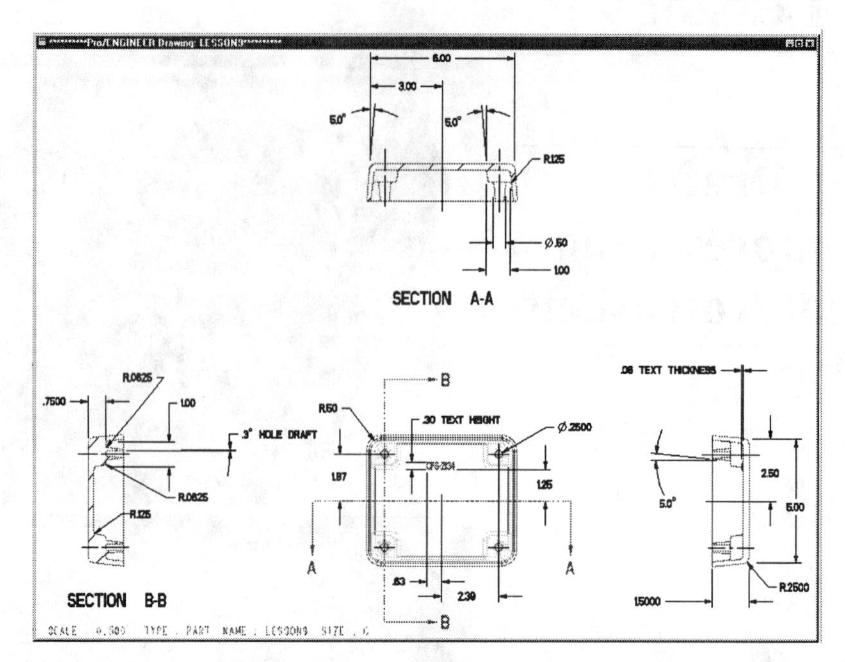

Figure 9.2
Enclosure Dimensions

DRAFTS, SUPPRESS, AND TEXT PROTRUSIONS

The **Draft** feature adds a draft angle between surfaces. A wide variety of parts incorporate drafts in their design. Casting, injection mold, and die parts normally have drafted surfaces. The ENCLOSURE in Figure 9.1 and Figure 9.2 is a plastic injection mold part.

Text can be included in a sketch for extruded protrusions and cuts, trimming surfaces, and cosmetic features. The text can be suppressed after it is created in order to decrease regeneration time of the model. Text can also be drafted.

Suppressing features using the **Suppress** command temporarily removes them from regeneration. Suppressed features can be "unsuppressed" (**Resume**) at any time. It is sometimes convenient to suppress text protrusions and rounds to speed up regeneration of the model.

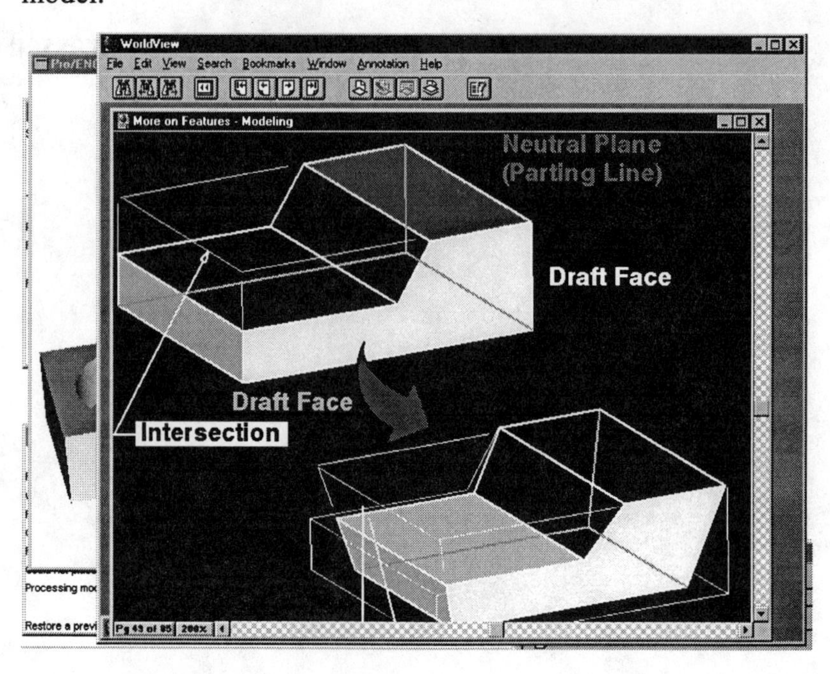

Figure 9.3
COAch for Pro/E, More on Features-- Modeling (Segment 3: Draft Features)

Drafts

The **Draft** feature adds a draft angle between **-15** degrees and **+15** degrees to individual surfaces or to a series of selected planar surfaces (Fig. 9.3). The following terminology is used in draft creation:

> **Draft surfaces** are the selected surfaces of the model designated for drafting.
> **Neutral plane** defines the pivot plane (Fig. 9.4) where the draft angle is **0°** (zero). The intersection of the neutral plane and drafted surfaces defines the *axis of rotation*.
> **Neutral curve** defines the curve on the surfaces to be drafted where the draft angle is **0°** (zero). Drafted surfaces are rotated about the neutral curve.

Draft direction, or **reference direction**, indicates the direction in which material is added. The draft direction is defined as a normal to the reference plane that you specify; it is shown on the screen as a *red arrow*. The normal also indicates the direction from which the draft angle is measured. **Draft angle** is measured between the draft direction and the resulting drafted surfaces. If the draft surfaces are split, you can define two independent angles for each portion of the draft. Draft angles must be within the range of **-15** to **+15** degrees. **Direction of rotation** defines how surfaces are rotated with respect to the neutral plane or neutral curve. **Split areas** of the draft allow you to divide draft surfaces so different draft angles can be applied.

When creating drafts, you can draft only surfaces that are formed by tabulated cylinders or planes. The reference plane used to establish the draft direction must be parallel to the neutral plane.

Surfaces with fillets around the edge boundary cannot be drafted. Therefore, draft the surfaces first, and then fillet the edges as required.

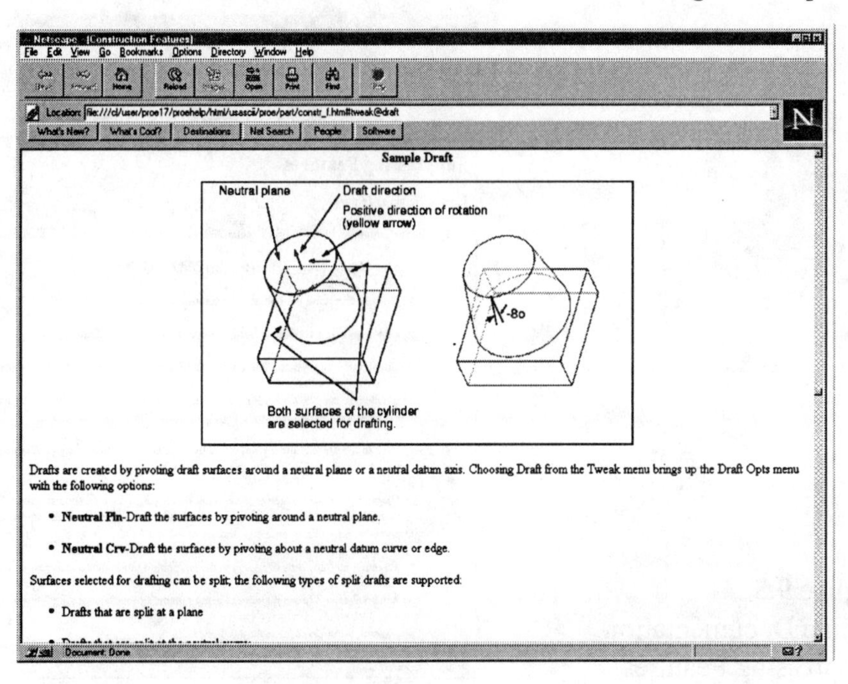

Figure 9.4
Online Documentation
Drafts

Suppressing and Resuming Features

Suppressing features is similar to removing the feature from regeneration temporarily (Fig. 9.5). You can "unsuppress" (**Resume**) suppressed features at any time. Features on a part can be suppressed to simplify the part model and decrease regeneration time. For example, while you work on one end of a shaft, it may be desirable to suppress features on the other end of the shaft. Similarly, while working on a complex assembly, you can suppress some of the features and components for which the detail is not essential to the current assembly process.

Unlike other features, the base feature cannot be suppressed. If you are not satisfied with your base feature, you can redefine the section of the feature, or delete it and start over again.

Suppressing Features

To suppress features, do the following:

1. Choose **Suppress** from the FEAT menu. Pro/E displays the SELECT FEAT and GET SELECT menus.
2. Choose one of the following options from the DELETE/SUPP menu:

   Normal Suppress the selected feature and all its children.
   Clip Suppress the selected feature and all the features that follow.
   Unrelated Suppress any feature that is not a child or parent of the selected feature.
3. Select a feature to suppress by picking on it, selecting from the Model Tree, specifying a *range*, entering its *feature number* or *identifier*, or using *layers*.

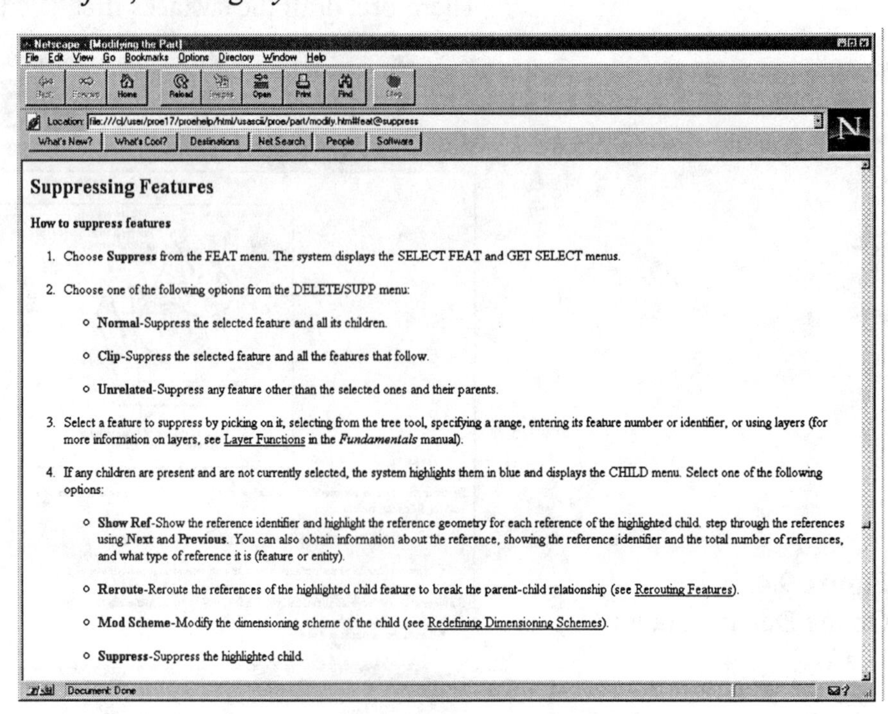

Suppressing Features

How to suppress features

1. Choose **Suppress** from the FEAT menu. The system displays the SELECT FEAT and GET SELECT menus.
2. Choose one of the following options from the DELETE/SUPP menu:
   - **Normal**-Suppress the selected feature and all its children.
   - **Clip**-Suppress the selected feature and all the features that follow.
   - **Unrelated**-Suppress any feature other than the selected ones and their parents.
3. Select a feature to suppress by picking on it, selecting from the tree tool, specifying a range, entering its feature number or identifier, or using layers (for more information on layers, see Layer Functions in the *Fundamentals* manual).
4. If any children are present and are not currently selected, the system highlights them in blue and displays the CHILD menu. Select one of the following options:
   - **Show Ref**-Show the reference identifier and highlight the reference geometry for each reference of the highlighted child. step through the references using **Next** and **Previous**. You can also obtain information about the reference, showing the reference identifier and the total number of references, and what type of reference it is (feature or entity).
   - **Reroute**-Reroute the references of the highlighted child feature to break the parent-child relationship (see Rerouting Features).
   - **Mod Scheme**-Modify the dimensioning scheme of the child (see Redefining Dimensioning Schemes).
   - **Suppress**-Suppress the highlighted child.

Figure 9.5
Online Documentation Suppressing Features

You can use suppress and resume to simplify the part before inserting features such as text protrusions. Also you may wish to suppress the text protrusion if there is other work to be done on the part. Text protrusions take time to regenerate since they increase the file size considerably.

Text Protrusions

When modeling, **text** can be included in a sketch for extruded protrusions and cuts, trimming surfaces, and cosmetic features (Fig. 9.6). The characters that are in an extruded feature must use the font "**font3d**" for Pro/E. For cosmetic features, any font may be used, and this is done by modifying the text after creating the sketch.

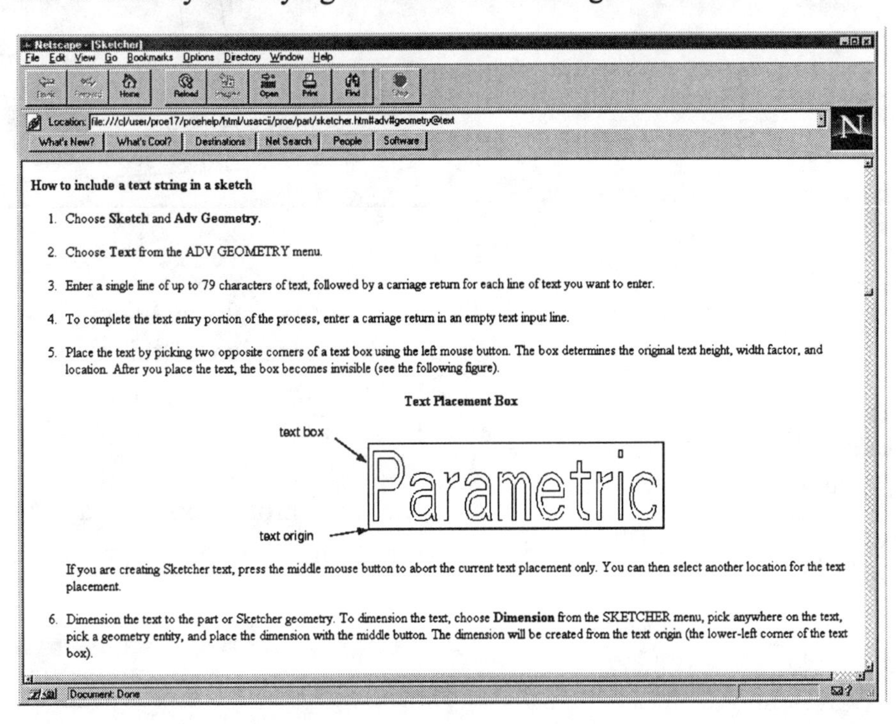

Figure 9.6
Online Documentation
Text Protrusions

To include a text entry in a sketch:

1. Choose **Sketch ⇒ Adv Geometry**
2. Choose **Text** from the ADV GEOMETRY menu.
3. Enter a single line of up to 79 characters of text. For the Enclosure, the part lot number, "**CFS-2134**", was added to the model, as shown in Figure 9.1 and Figure 9.2.
4. Place the text by picking two opposite corners of a text box. The box determines the original text height, width factor, and location. After the text is placed, the box becomes invisible.

Dimension the text to the part or Sketcher geometry. To dimension the text, choose **Dimension** from the SKETCHER menu, pick anywhere on the text, pick a geometry entity, and place the dimension. The dimension will be created from the text origin (the lower left corner of the text box).

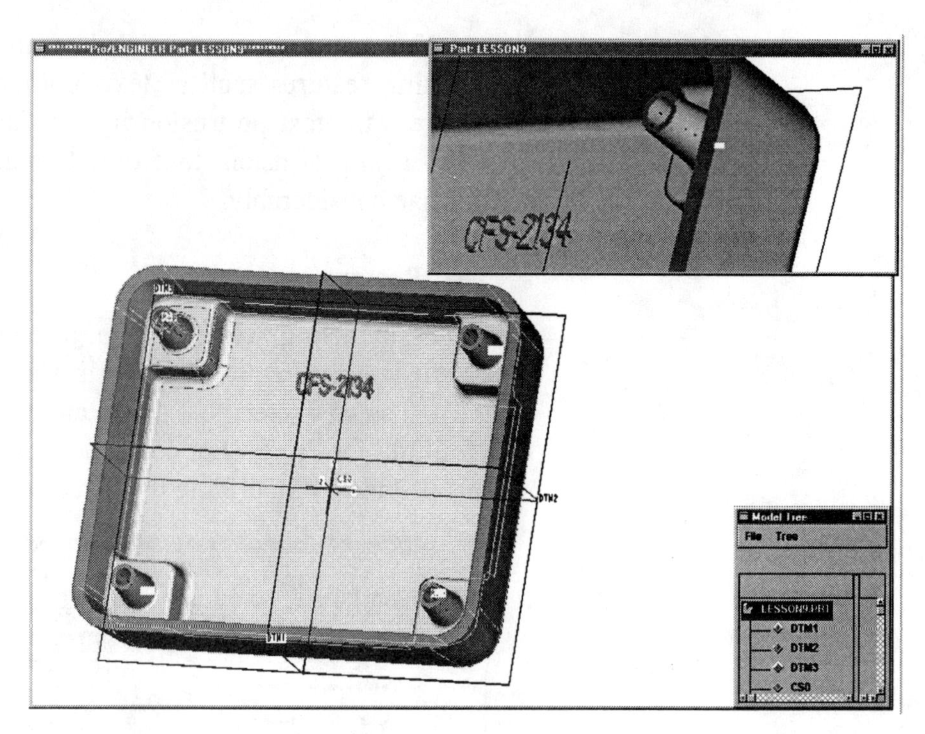

Figure 9.7
Enclosure with Datum Planes

Enclosure

The Enclosure is a plastic injection mold part. A variety of drafts will be used in the design of this part. A *raised text protrusion* will be modeled on the inside of the Enclosure, as shown in Figure 9.7. The dimensions for the part are provided in Figure 9.8 through Figure 9.13.

The first protrusion has a draft angle of **5°**, as do the pad protrusions and cylindrical protrusions. The holes have a **.3°** draft angle.

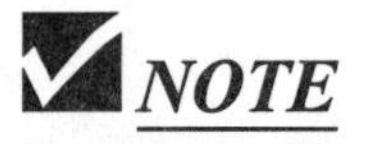

NOTE

Set up the Enclosure:

- Material = Plastic
- Units = Inches
- Default Datum Planes
- Default Coordinate System
- Layers = **DATUM_PLANES**
- Hidden Line
- Tan Dimmed
- ✓Grid Snap

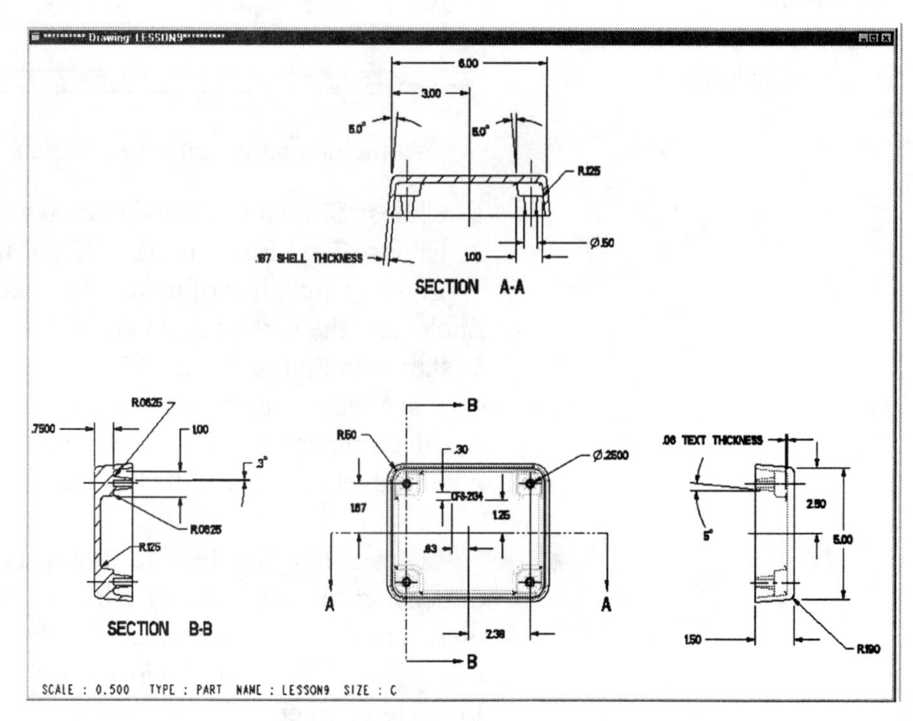

Figure 9.8
Enclosure Detail Drawing

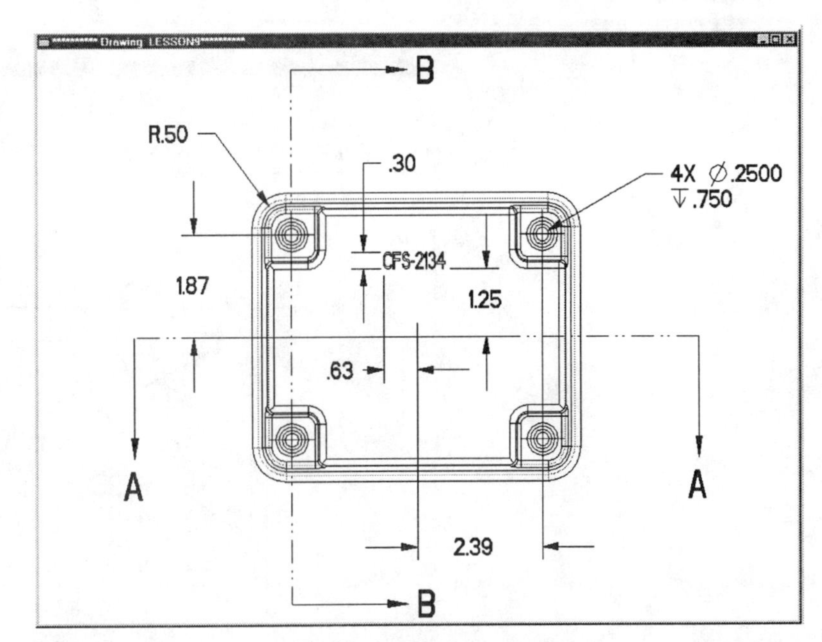

Figure 9.9
Enclosure Drawing,
Front View

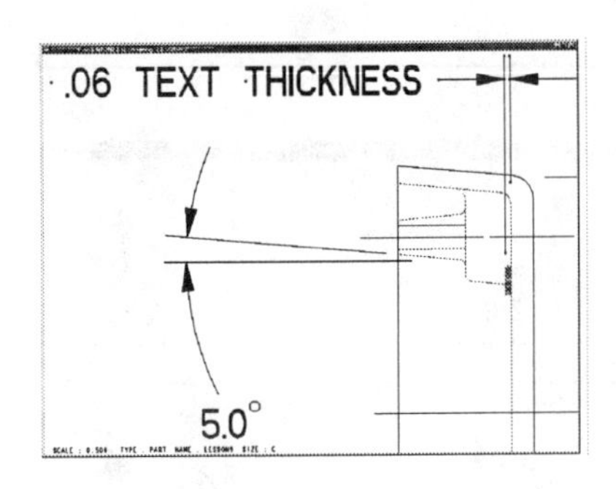

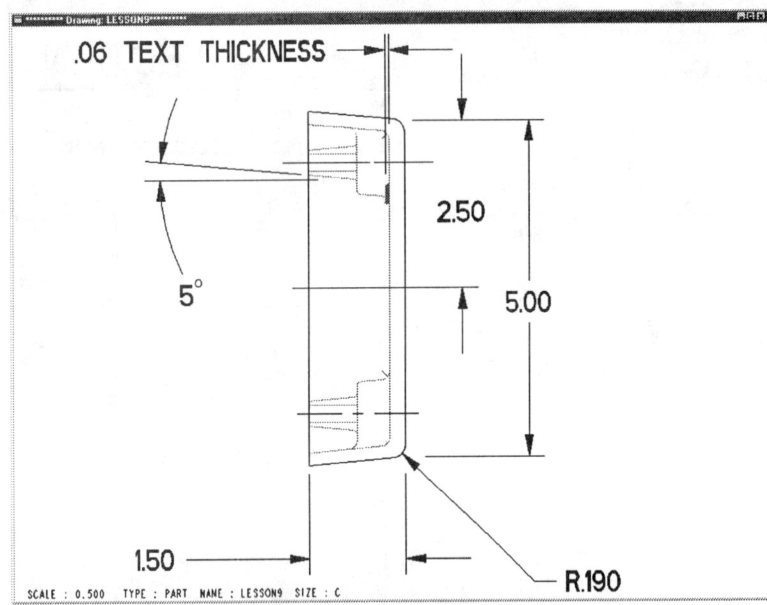

Figure 9.10
Enclosure Drawing,
Right Side View

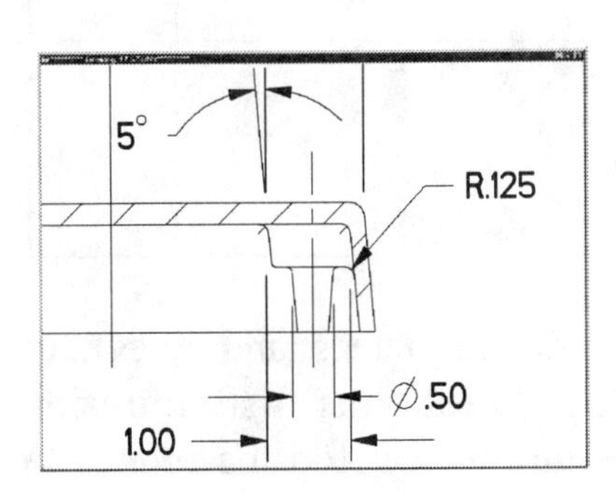

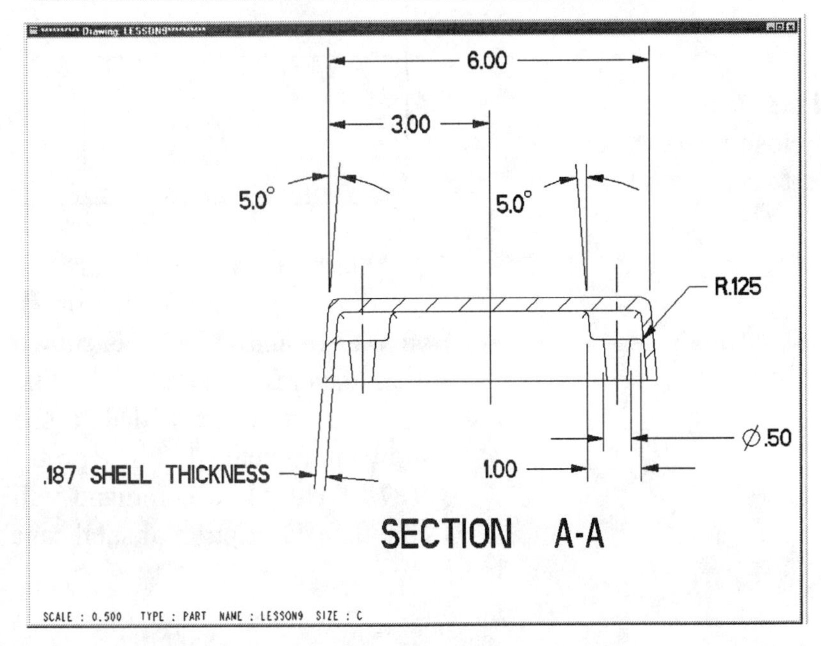

Figure 9.11
Enclosure Drawing,
SECTION A-A

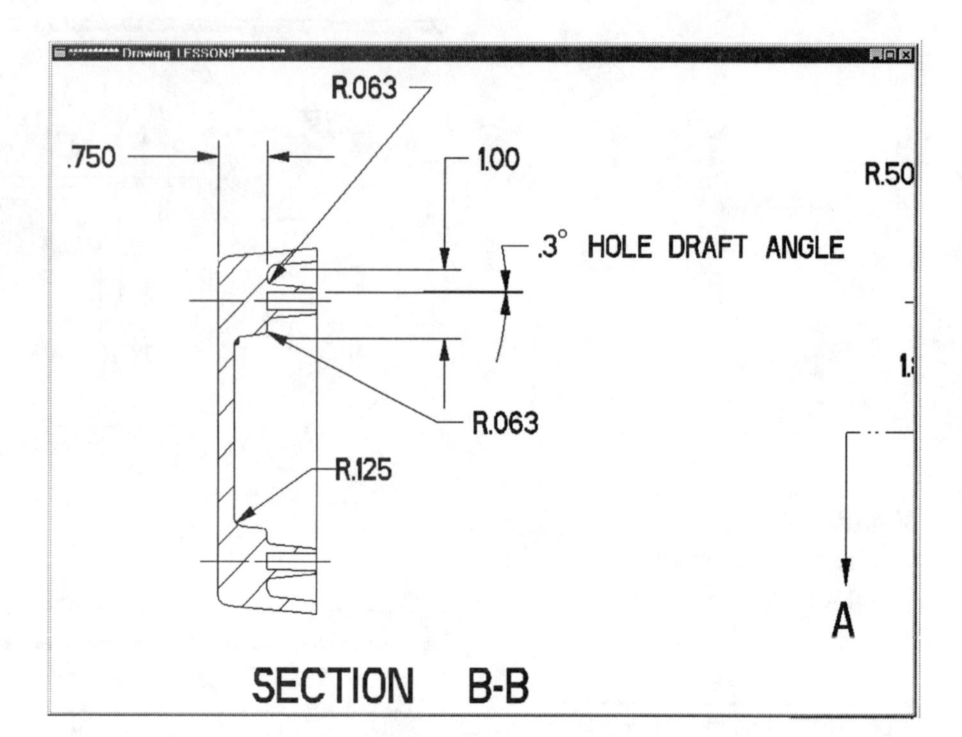

Figure 9.12
Enclosure Drawing,
SECTION B-B

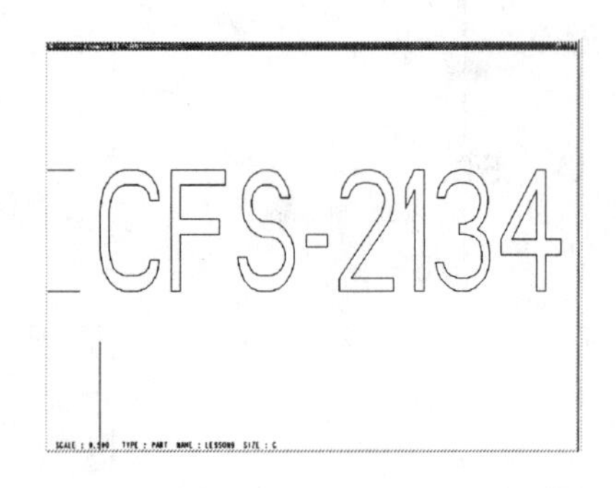

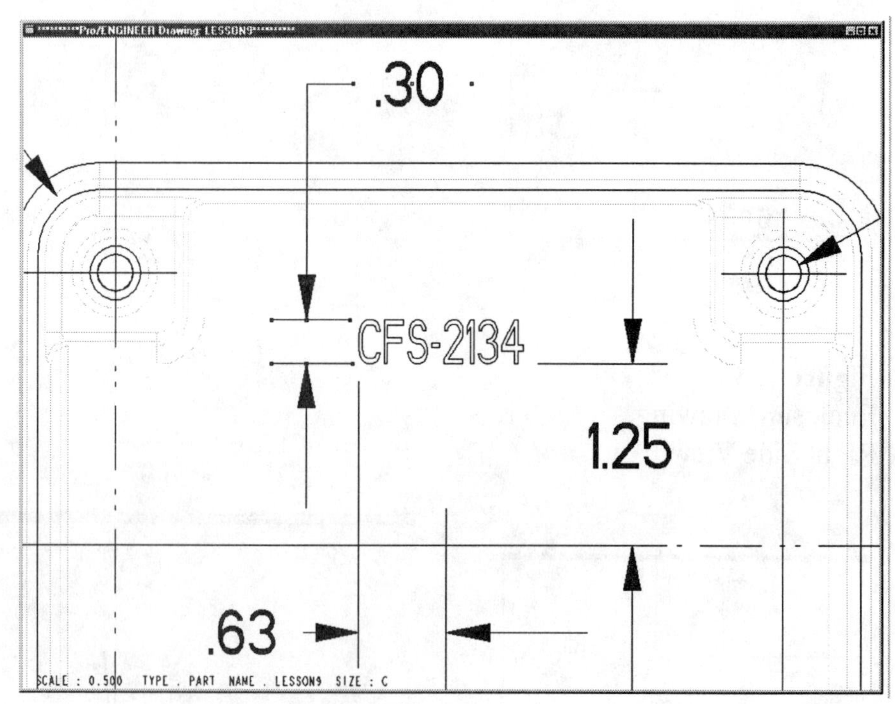

Figure 9.13
Enclosure Drawing,
Text Protrusion Location

Start the part with three default datum planes and a default coordinate system. Sketch on **DTM3**, and center the first protrusion on **DTM1** and **DTM2**, as shown in Figure 9.14. Add the rounds to the sketch instead of after the first protrusion is complete.

The drafts are added to the protrusion before it is shelled, as shown in Figure 9.15. The protrusion will be shelled to a thickness of **.1875**. The **Shell** command will be introduced in this lesson, but a complete description of shell capabilities is provided in Lesson 10.

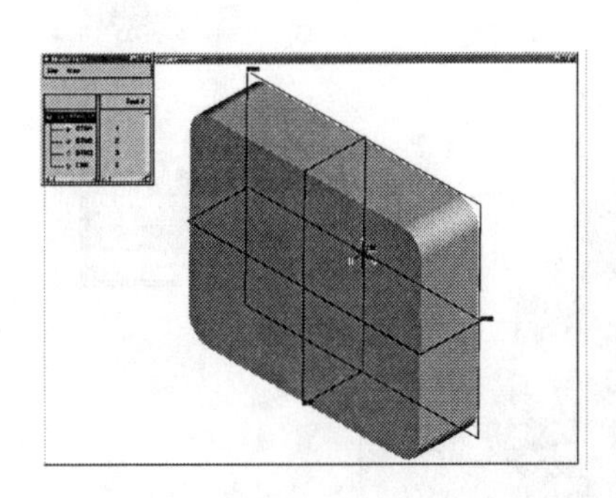

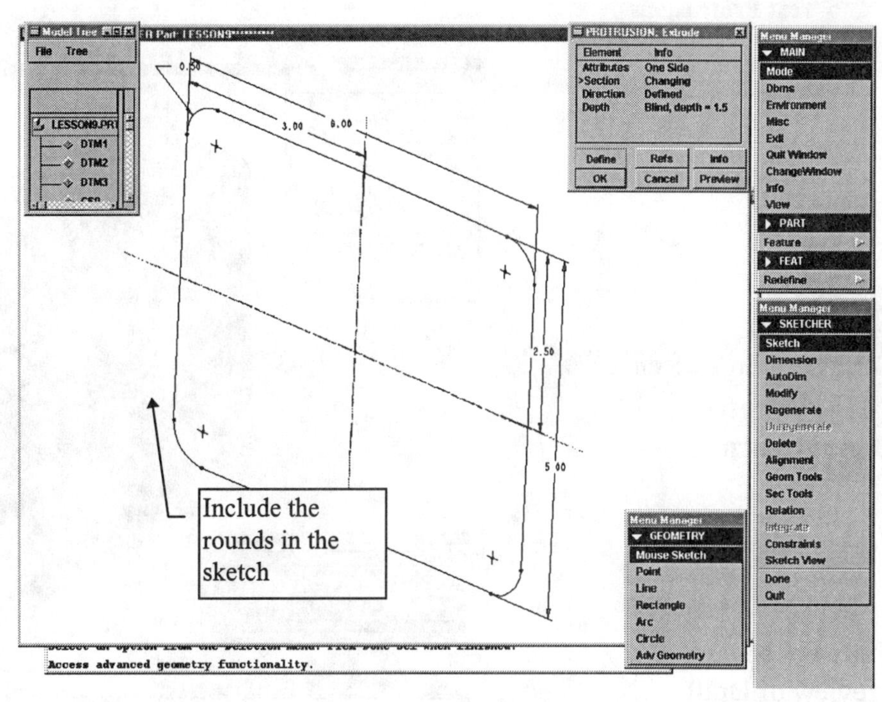

Figure 9.14
First Protrusion

Create the draft for the lateral surfaces of the protrusion using the following commands:

Create ⇒ **Tweak** ⇒ **Draft** ⇒ **Neutral Pln** ⇒ **Done** ⇒ **No Split** ⇒ **Constant** ⇒ **Done** ⇒ **Include** ⇒ **Indiv Surfs** ⇒ (pick all *eight lateral surfaces* of the protrusion, as shown in Figure 9.15) ⇒ **Done Sel** ⇒ **Done** ⇒ (select **DTM3** as the neutral plane) ⇒ (select **DTM3** as the plane the direction will be perpendicular to, as shown in Figure 9.15) ⇒ (specify the draft angle by typing **5** at the prompt) ⇒ **enter** ⇒ **Preview** (Fig. 9.16) ⇒ **OK** ⇒ **Done**

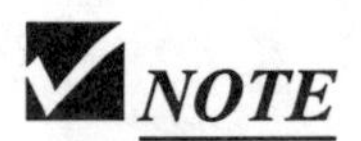

NOTE

The *neutral plane* and the *plane the direction will be perpendicular to* are many times the same plane. For the draft created here, **DTM3** is used for both.

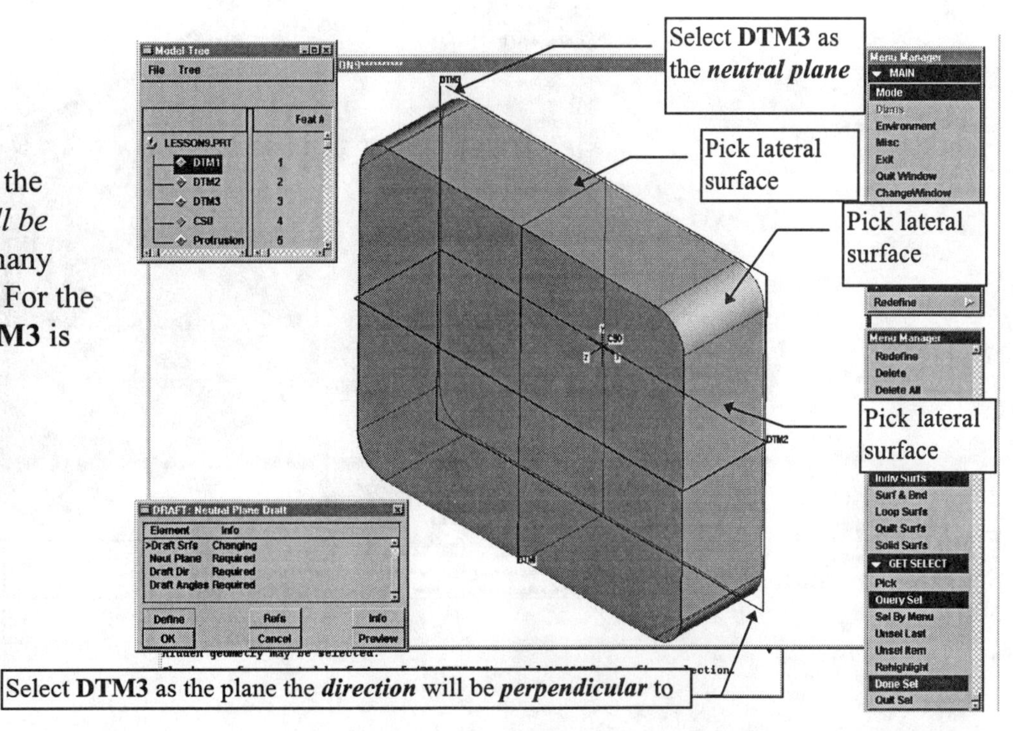

Figure 9.15
Draft

Dbms ⇒ Save ⇒ enter Purge ⇒ enter ⇒ Done-Return

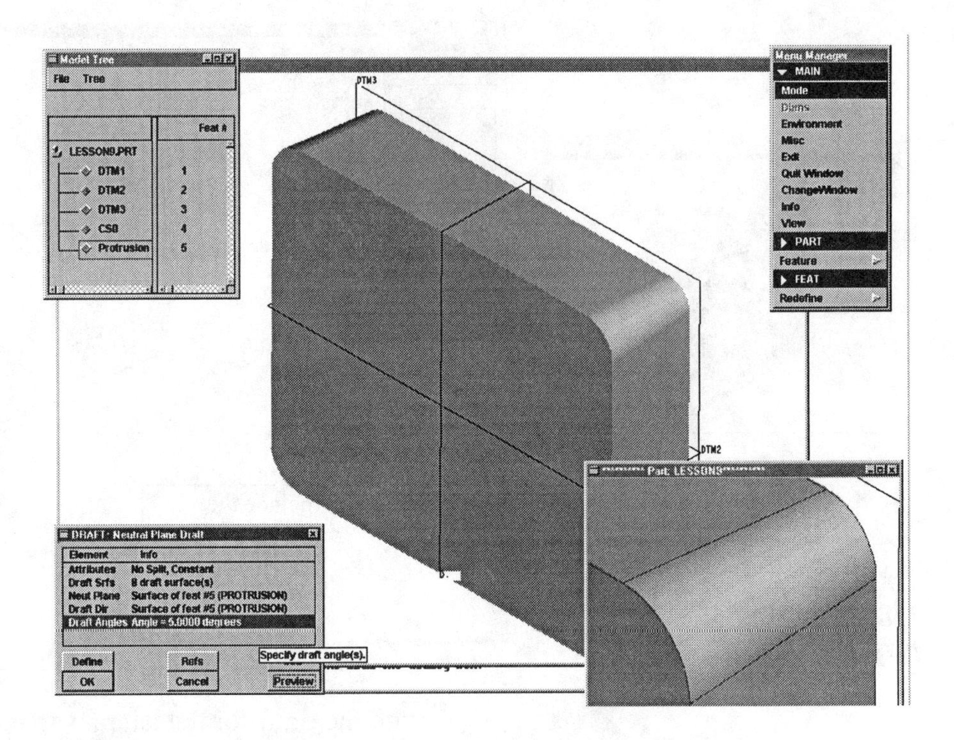

Figure 9.16
Preview of Draft

Next you will shell the part using the following commands:

Feature ⇒ **Create** ⇒ **Shell** ⇒ (select the surface to remove by picking the vertical face, as shown in Figure 9.17) ⇒ **Done Sel** ⇒ **Done Refs** ⇒ (enter the thickness by typing **.1875** at the prompt) ⇒ **enter** ⇒ **Preview** ⇒ **OK** (Fig. 9.18) ⇒ **Done**

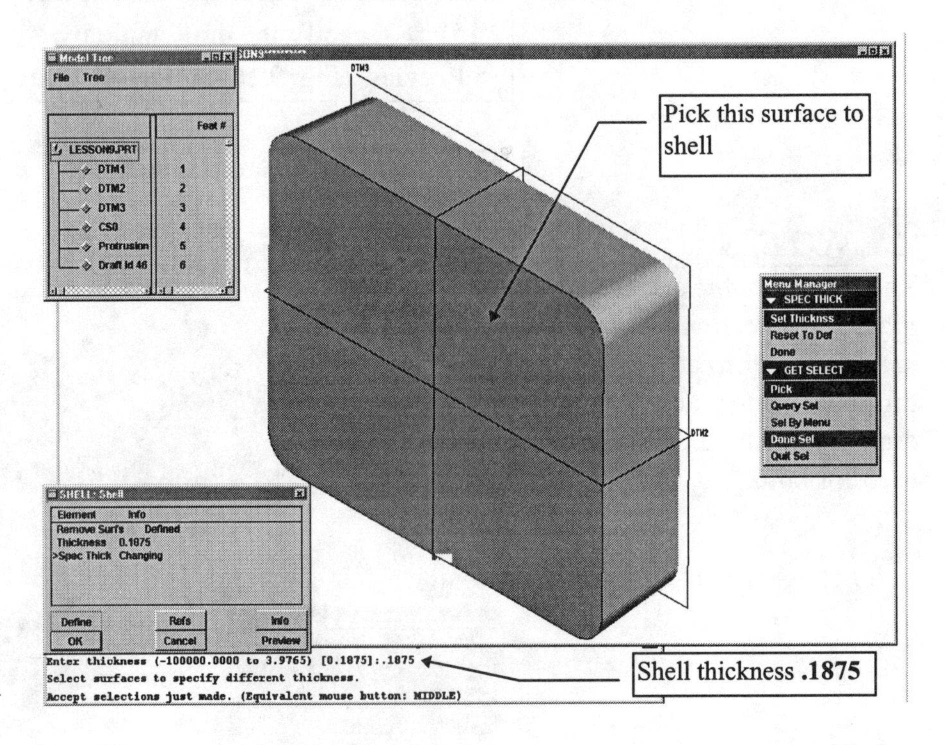

Figure 9.17
Selecting the Surface to Shell

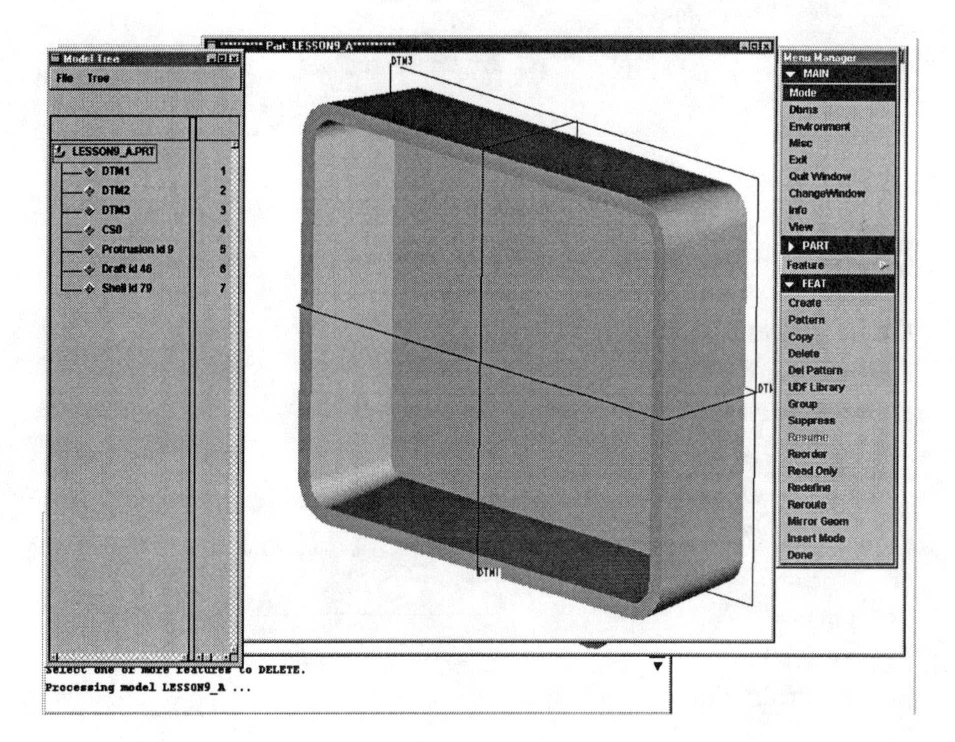

Figure 9.18
Shelled Part

Sketch the raised pedestal-like protrusion on **DTM3**, as shown in Figure 9.19. In most cases, the protrusion would be sketched on the inside of the shelled surface, but for this design we will project from the datum plane. Use the existing *internal* edge and the dimensioning scheme shown in the figure and in the detail drawings provided earlier. Figure 9.20 shows the completed protrusion.

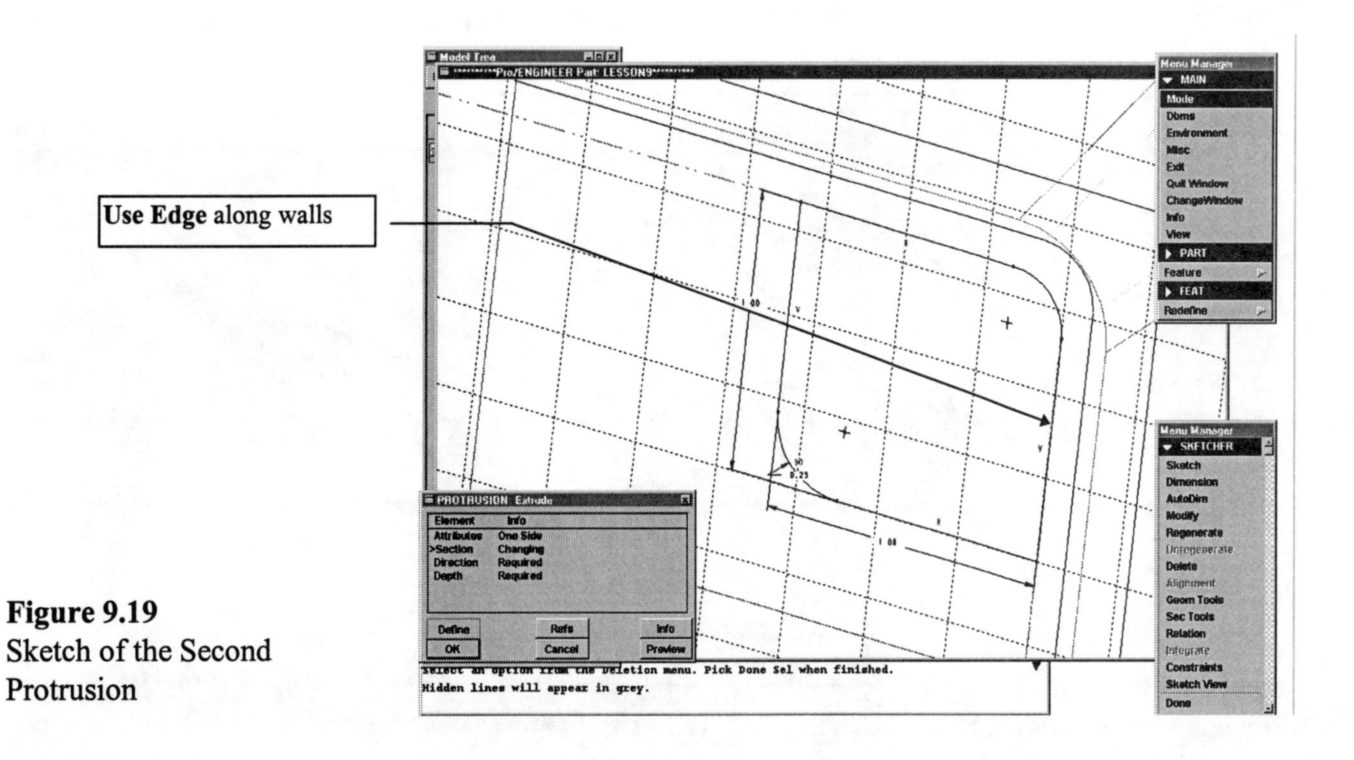

Figure 9.19
Sketch of the Second Protrusion

Dbms ⇒ Save ⇒ enter Purge ⇒ enter ⇒ Done-Return

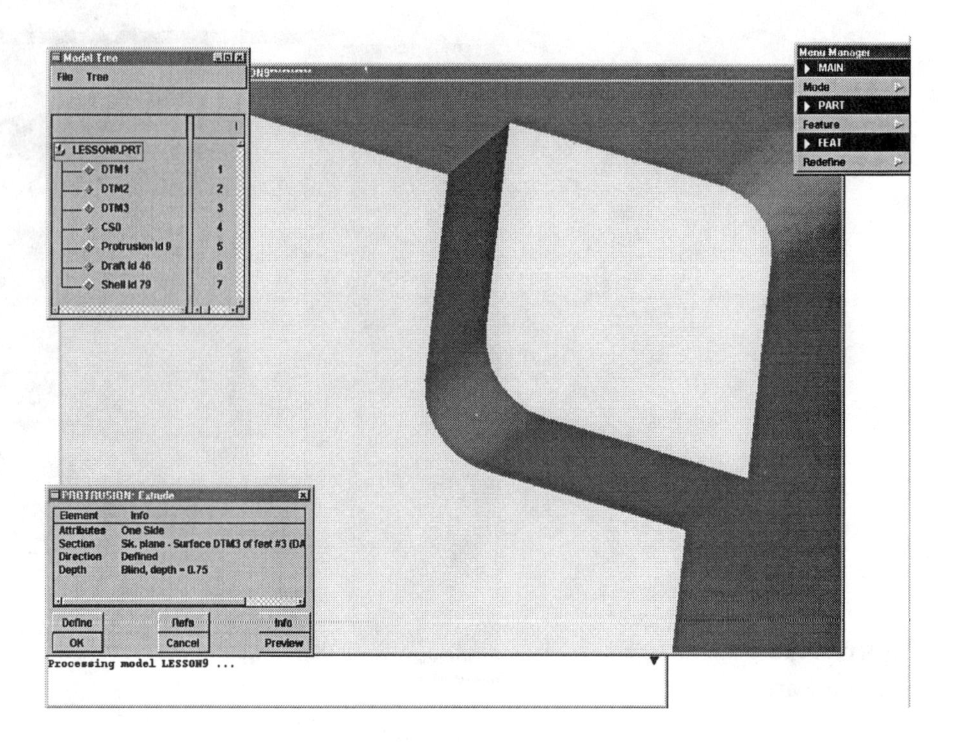

Figure 9.20
Second Protrusion

Use the following commands to draft the lateral surfaces of the second protrusion:

Create ⇒ Tweak ⇒ Draft ⇒ Neutral Pln ⇒ Done ⇒ No Split ⇒ Constant ⇒ Done ⇒ Include ⇒ Indiv Surfs ⇒ (pick all *three lateral surfaces* of the protrusion, as shown in Figure 9.21) **⇒ Done Sel ⇒ Done ⇒** (select **DTM3** as the neutral plane) ⇒ (select **DTM3** as the plane the direction will be perpendicular to) ⇒ (specify the draft angle by typing **-5** at the prompt) ⇒ **enter** ⇒ **Preview** (Fig. 9.21) ⇒ **OK**

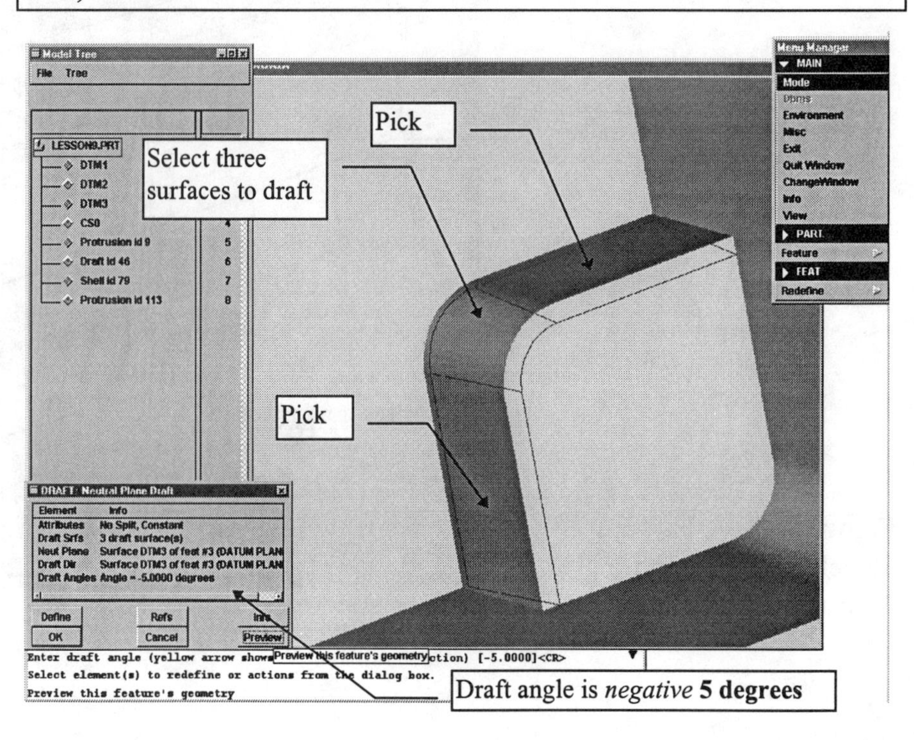

Figure 9.21
Preview of the Draft on the Second Protrusion

Model the circular protrusion (Fig. 9.22). Draft the new protrusion at **-5°** (Fig. 9.23). Create a hole (Fig. 9.24), and add an internal draft of -**.3°**.

Dbms ⇒ Save ⇒ enter
Purge ⇒ enter ⇒
Done-Return

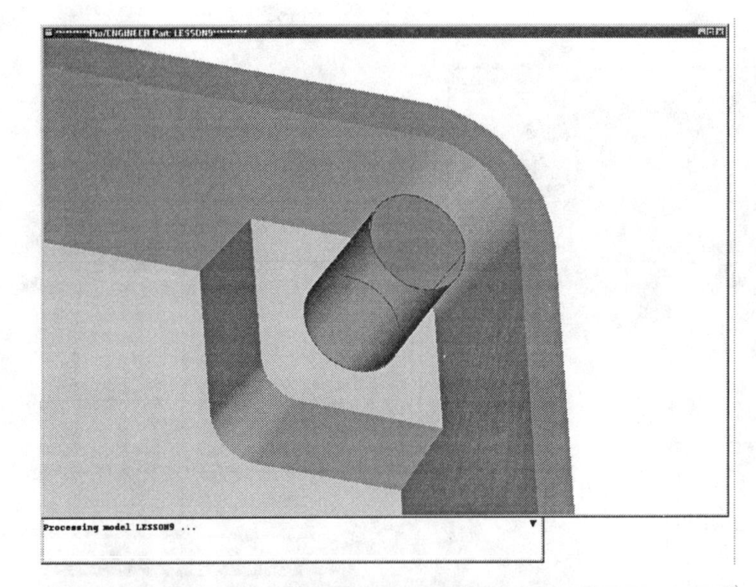

Figure 9.22
Circular Protrusion

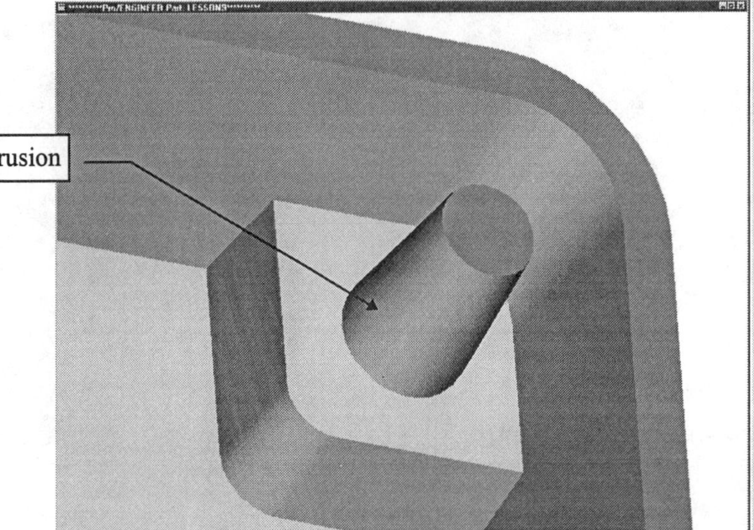

Figure 9.23
Draft

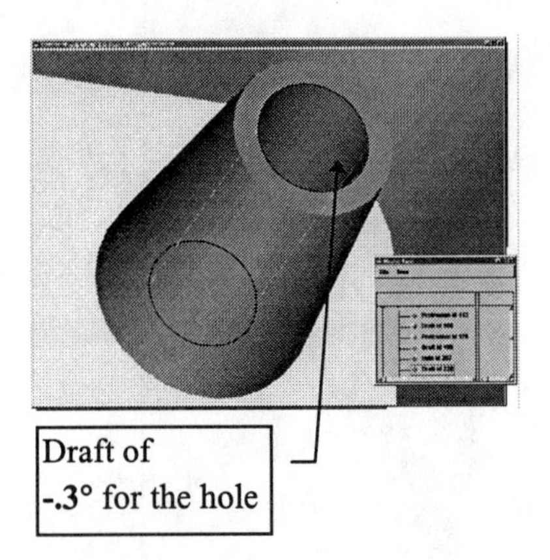

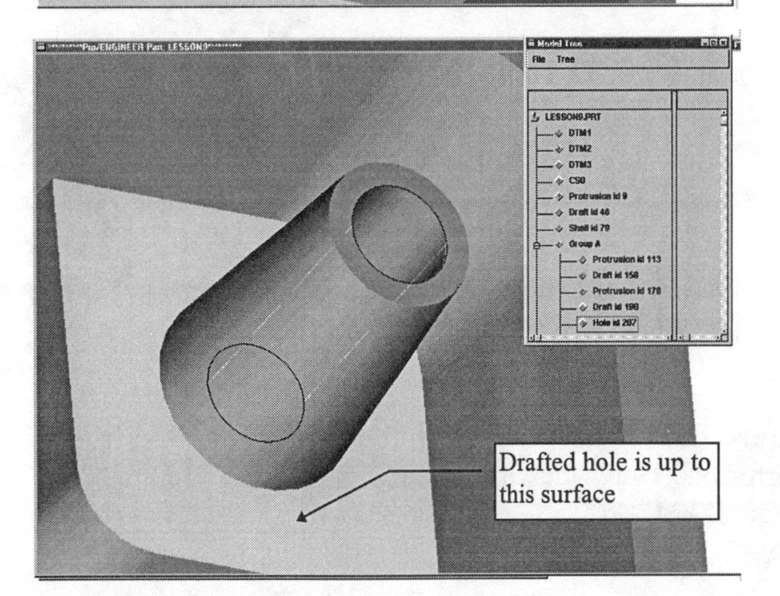

Figure 9.24
Hole

Create the rounds required for the part as shown in Figure 9.25. **Group** the protrusions, hole, and rounds and **Copy ⇒ Mirror** in both directions, as in Figure 9.26 and Figure 9.27.

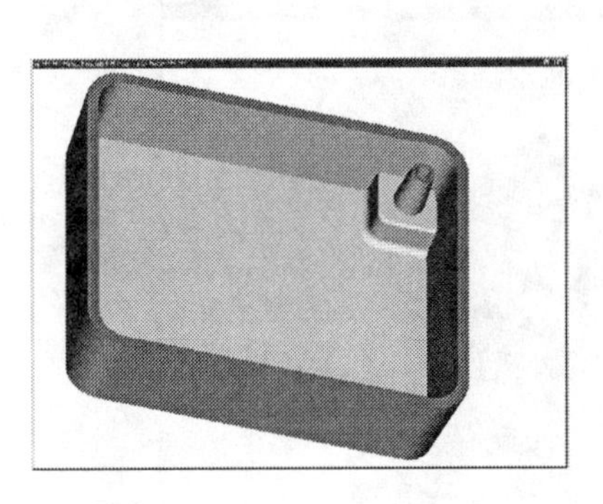

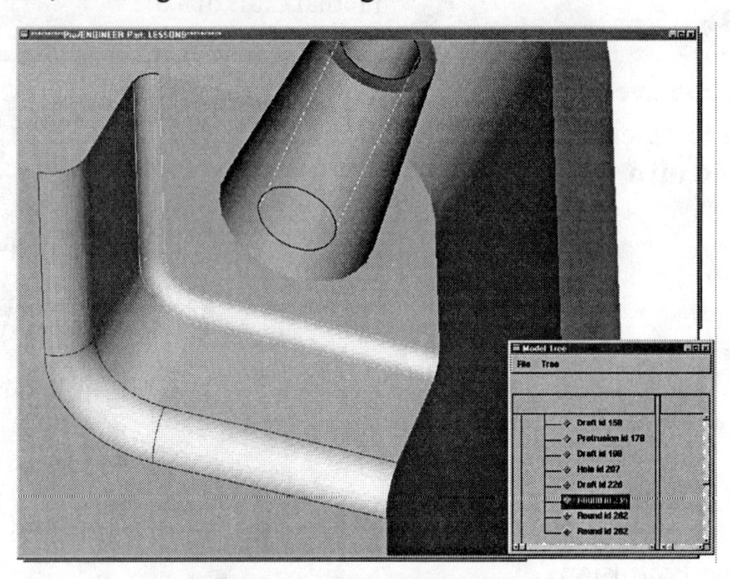

Figure 9.25
Group the Hole and Protrusions

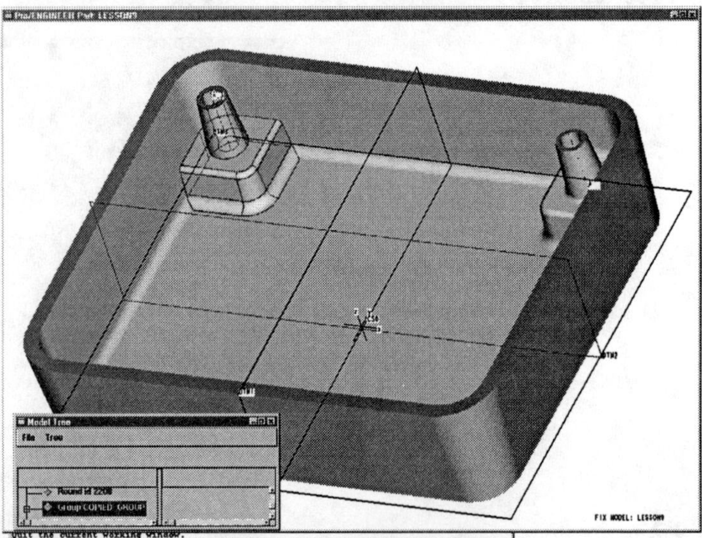

Figure 9.26
Copy ⇒ Mirror the **Group** About **DTM1**

Some surface-to-surface rounds still need to be created

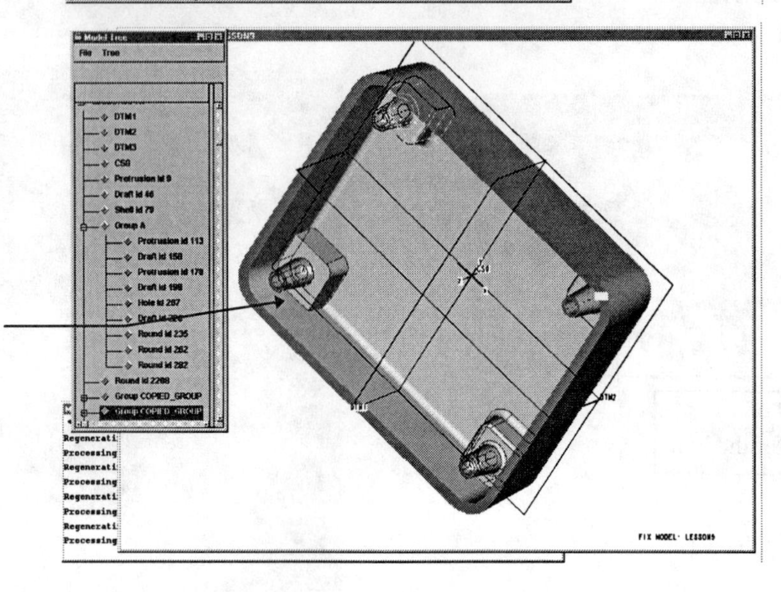

Figure 9.27
Mirror and Copy the Groups About **DTM2**

Create the missing rounds (Fig. 9.27) by using the following commands:

Create ⇒ **Round** ⇒ **Done** ⇒ **Surf-Surf** ⇒ **Done** (pick the first surface) ⇒ (pick the second surface) ⇒ (enter the radius **.125** at the prompt) ⇒ **enter** ⇒ **OK** (as shown in Figure 9.28)

Select the first surface

Select the second surface

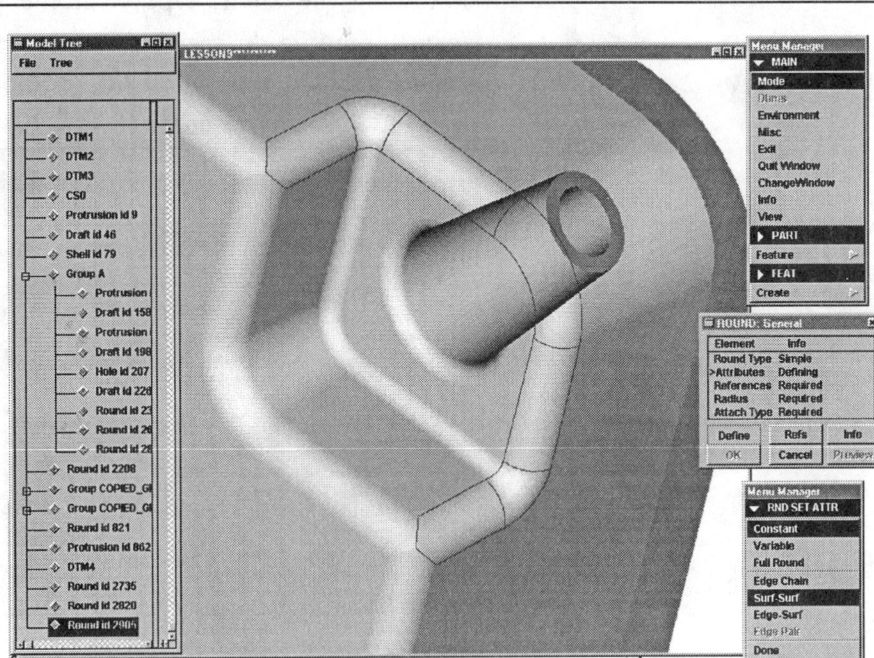

Figure 9.28
Create the Last Internal Rounds

Create the external round on the part, as shown in Figure 9.29. Try to **Reorder** the round to come before the shell. If the model fails, **Undo Changes**.

Dbms ⇒ **Save** ⇒ **enter**
Purge ⇒ **enter** ⇒ **Done-Return**

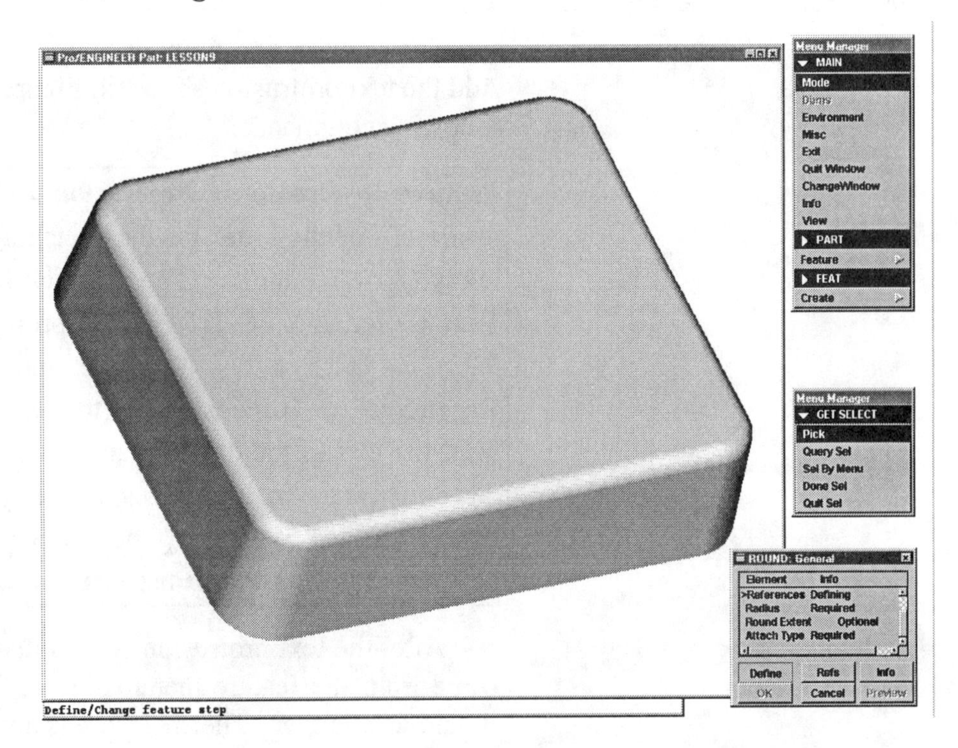

Figure 9.29
Create the Outside Round

Before creating the text protrusion, **Suppress** all of the features after the shell command. Expand the Model Tree to include the feature number and status. Use the following commands to suppress the features:

> **Feature** ⇒ **Suppress** ⇒ **Range** ⇒ (enter the number of the feature following the shell; type **8** at the prompt-- *your number for this feature may be different*) ⇒ **enter** ⇒ (enter regeneration number of ending feature; type **50** at the prompt-- *your number for this feature may be different*) ⇒ **enter** (Fig. 9.30) **Done Sel** ⇒ **Done** ⇒ **Done**

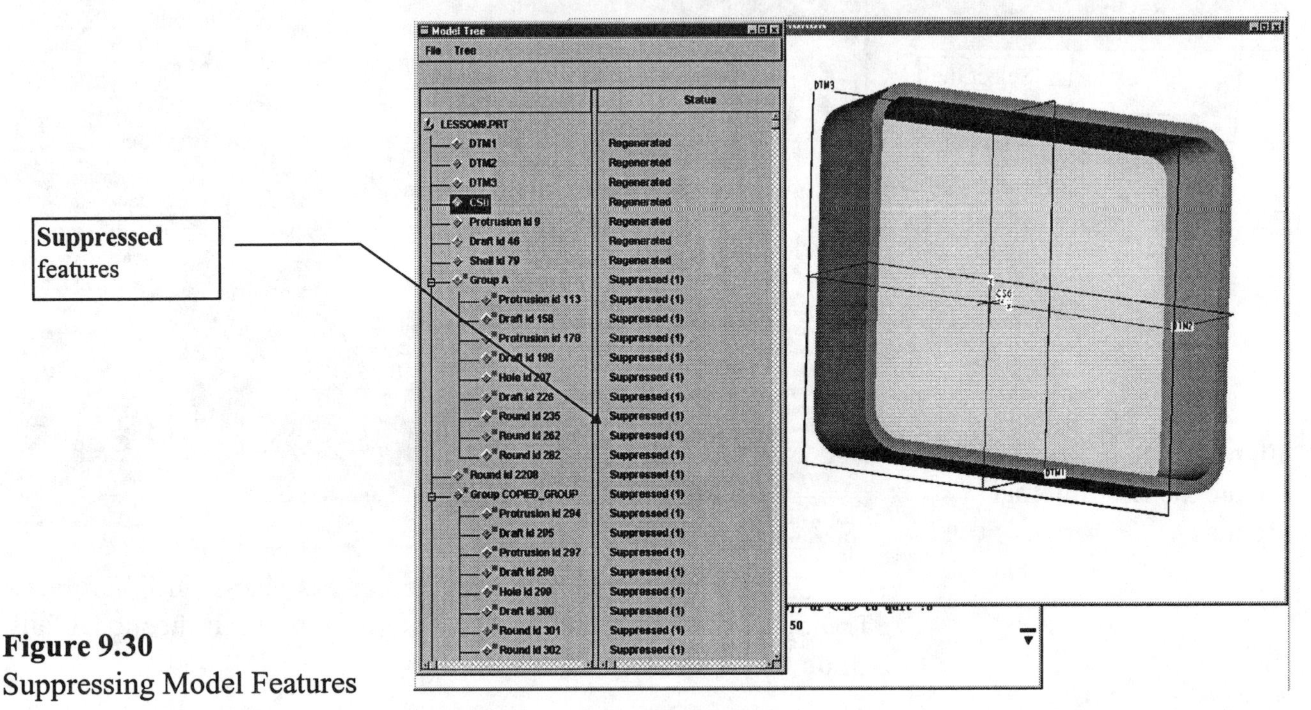

Figure 9.30
Suppressing Model Features

The regeneration time for your model will now be much shorter. Add the text protrusion shown in Figure 9.9 and Figure 9.13 using the following commands:

> **Feature** ⇒ **Create** ⇒ **Protrusion** ⇒ **Done** ⇒ **Done** ⇒ (pick the inside of the enclosure for the sketching plane as in Figure 9.31) ⇒ **Okay** (for direction) ⇒ **Top** ⇒ (pick **DTM2**) ⇒ **Adv Geometry** ⇒ **Text** ⇒ (enter **CFS-2134** at the prompt) ⇒ **enter** (Figs. 9.32 and 9.33) ⇒ (pick a box to place and size the text) ⇒ **enter** ⇒ **Regenerate** ⇒ **Dimension** ⇒ (to locate the text, add the dimensions from the text to the datum planes as in Figure 9.34) ⇒ **Modify** (use the dimensions in Figure 9.9 and Figure 9.34) ⇒ **Regenerate** ⇒ **Done** ⇒ **Blind** ⇒ **Done** (type **.0625** at the prompt) ⇒ **enter** ⇒ **OK** ⇒ **View** ⇒ **Default** ⇒ **Done-Return** ⇒ **New Window** (Fig. 9.35)

After the text protrusion is completed, choose **Resume** ⇒ **All** ⇒ **Done** from the feature menu (Fig. 9.36). Try regenerating now. It should take longer. The enclosure is now complete.

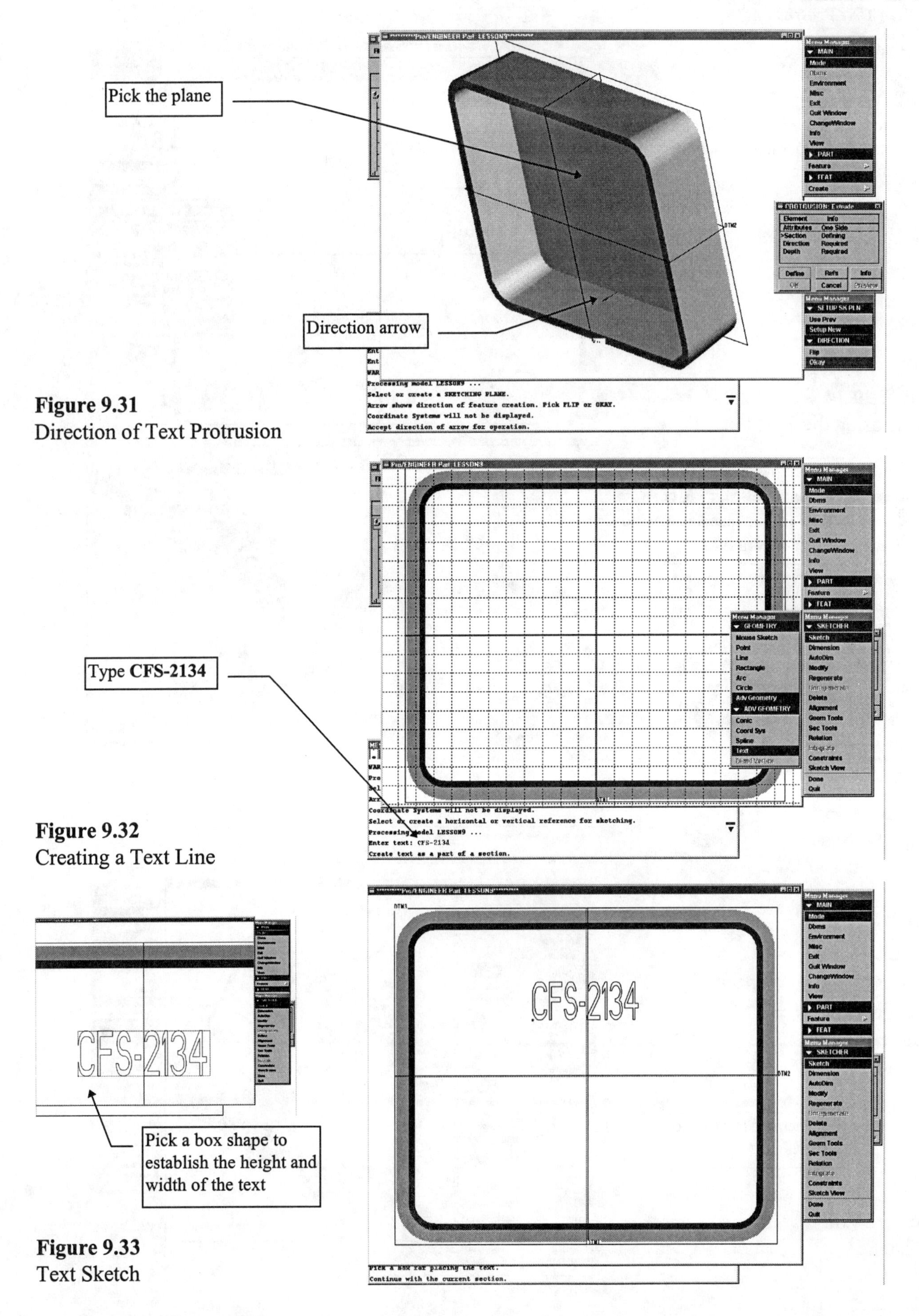

Figure 9.31
Direction of Text Protrusion

Figure 9.32
Creating a Text Line

Figure 9.33
Text Sketch

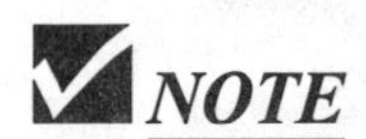

NOTE

So as to view the text clearly, it has been sketched larger here than in the detail drawings.

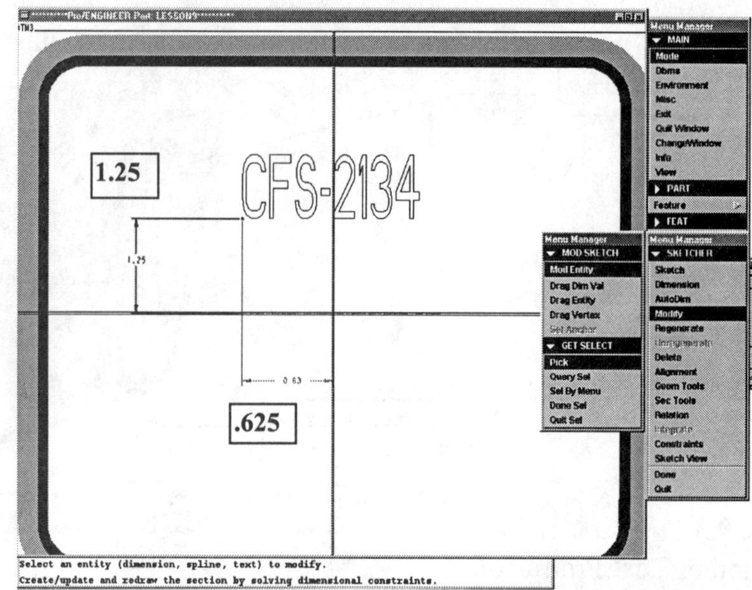

Figure 9.34
Dimension the Text Sketch

Try modifying the thickness and location of the text protrusion: **Feature ⇒ Modify ⇒** (pick the text) ⇒ (change the thickness to **.25**) ⇒ **Regenerate ⇒ Resume All**

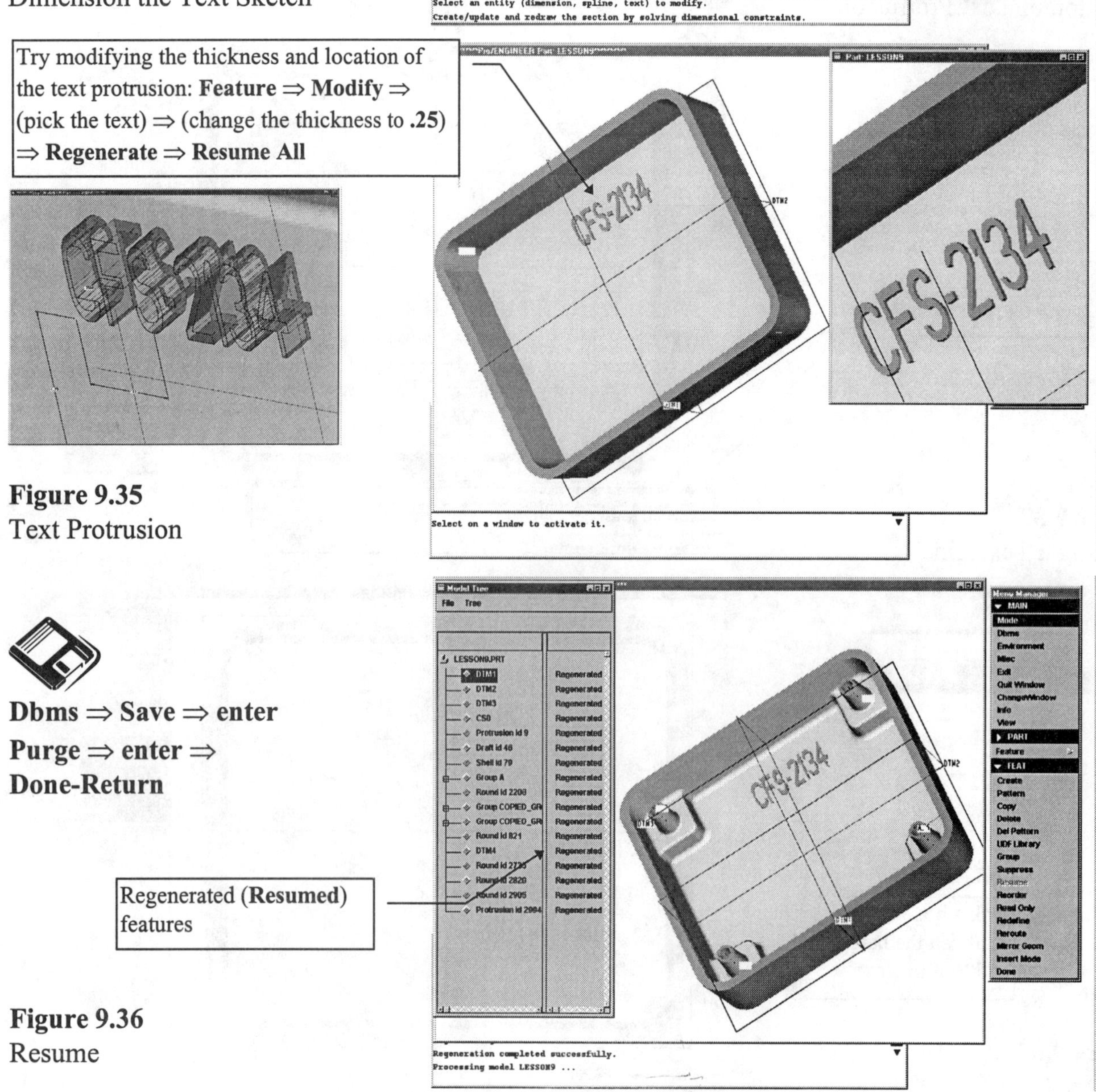

Figure 9.35
Text Protrusion

Dbms ⇒ Save ⇒ enter
Purge ⇒ enter ⇒
Done-Return

Regenerated (**Resumed**) features

Figure 9.36
Resume

Lesson 9 Project

Cellular Phone Bottom

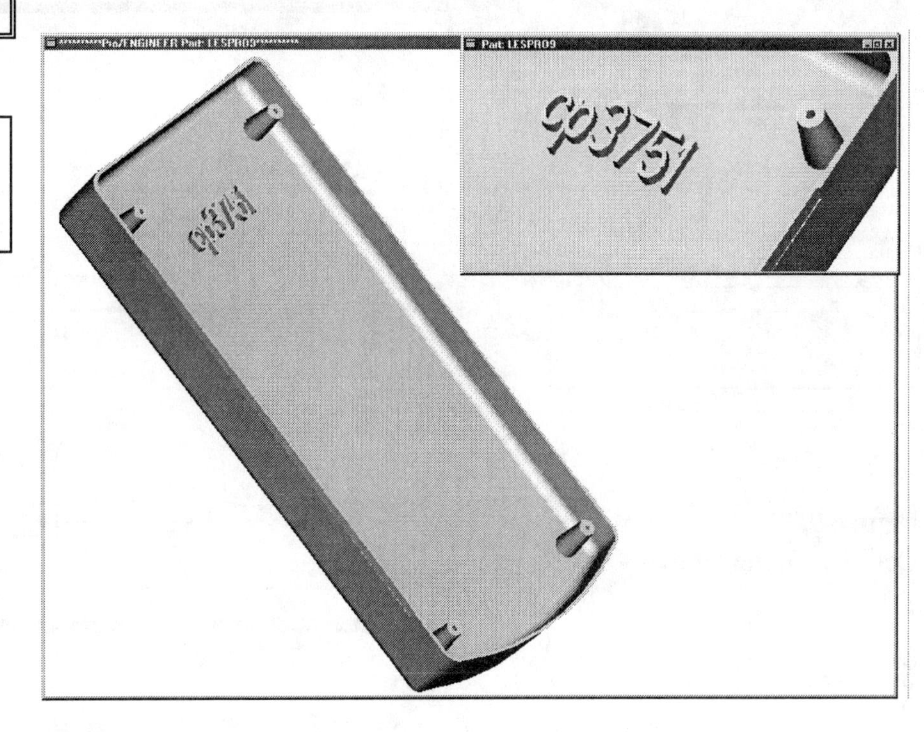

Figure 9.37
Cellular Phone Bottom

Cellular Phone Bottom

The ninth **lesson project** is a die-cast plastic part that requires similar commands as the **Enclosure**. Create the part shown in Figure 9.37 through Figure 9.47. Analyze the part and plan out the steps and features required to model it. Use the DIPS in Appendix D to plan out the feature creation sequence and the parent-child relationships for the part. The top of the cellular phone is created in the Lesson 10 Project. Do not shell the Cellular Phone Bottom.

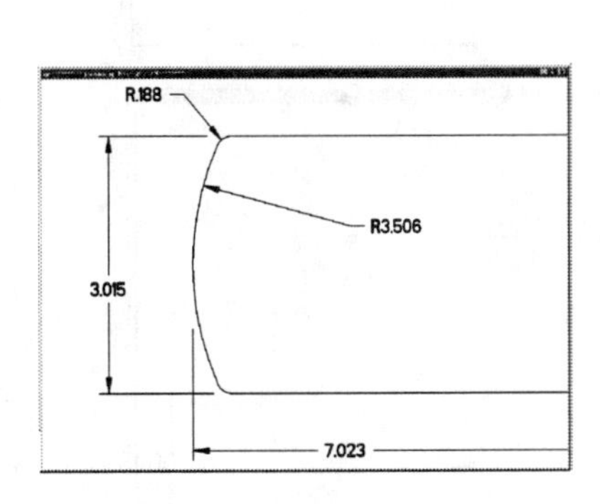

Figure 9.38
Cellular Phone Bottom
Showing Datum Planes

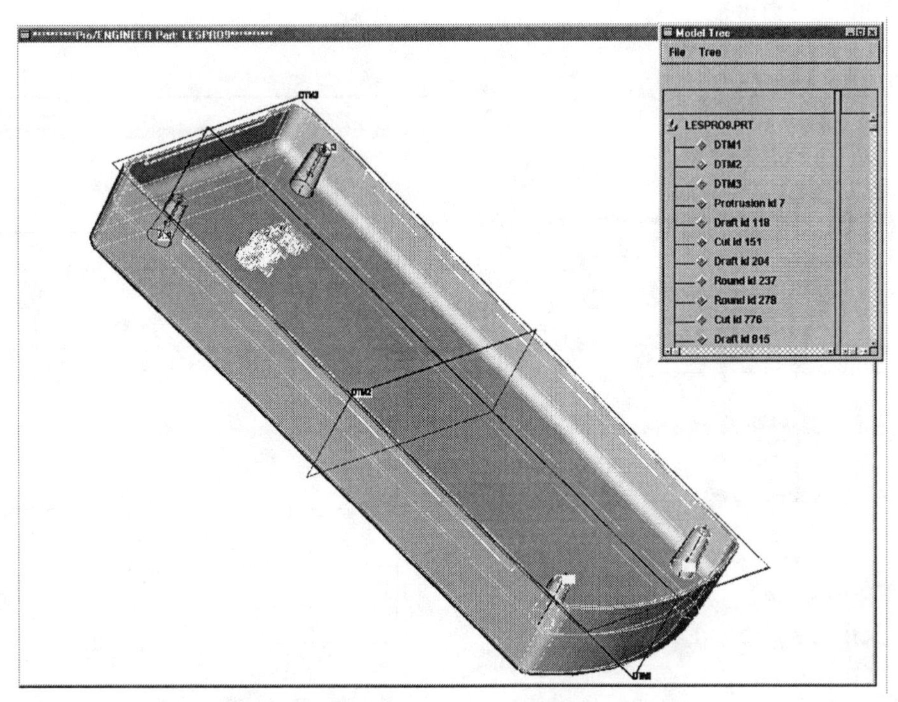

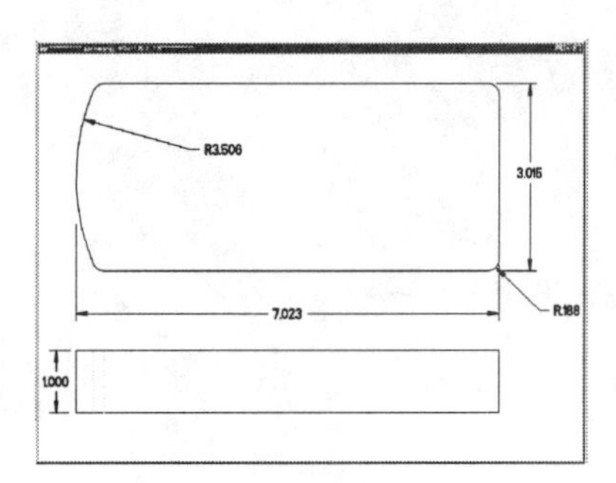

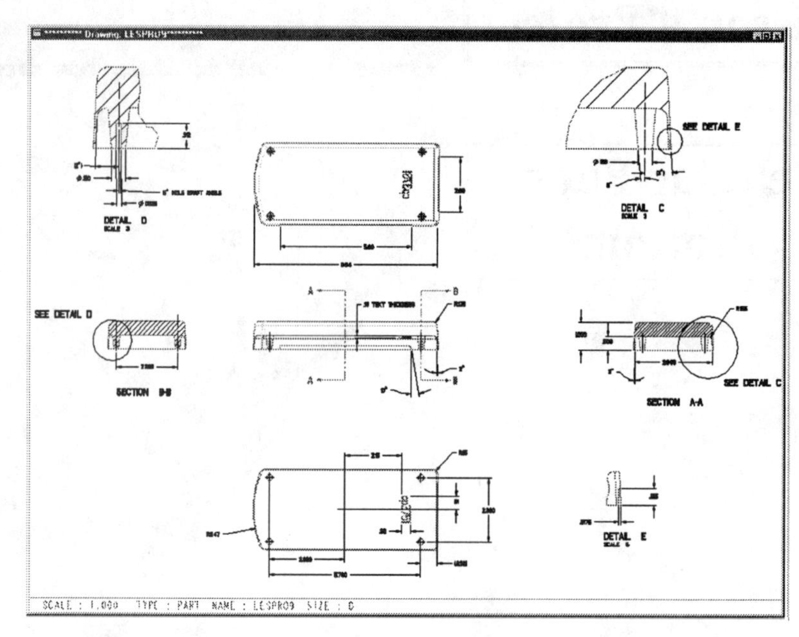

Figure 9.39
Cellular Phone Drawing,

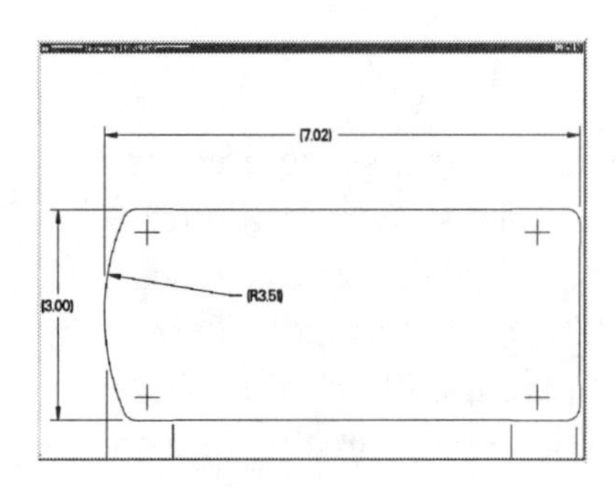

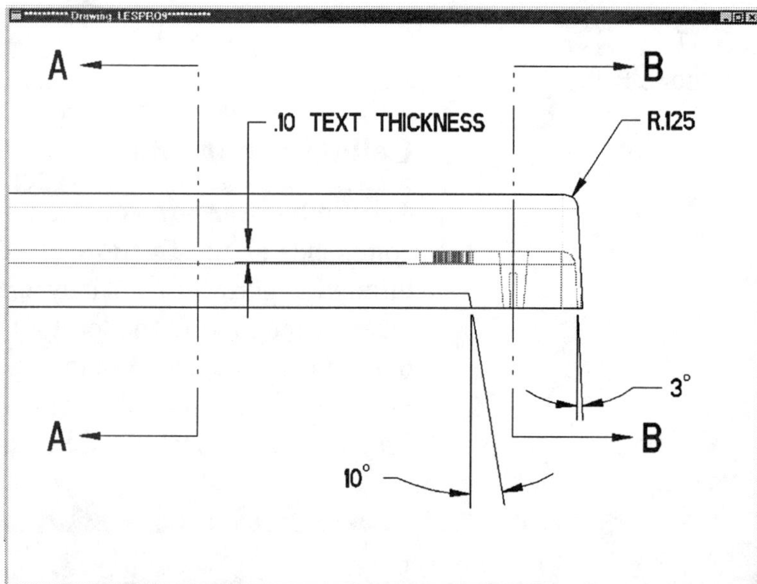

Figure 9.40
Cellular Phone Drawing,
Front View

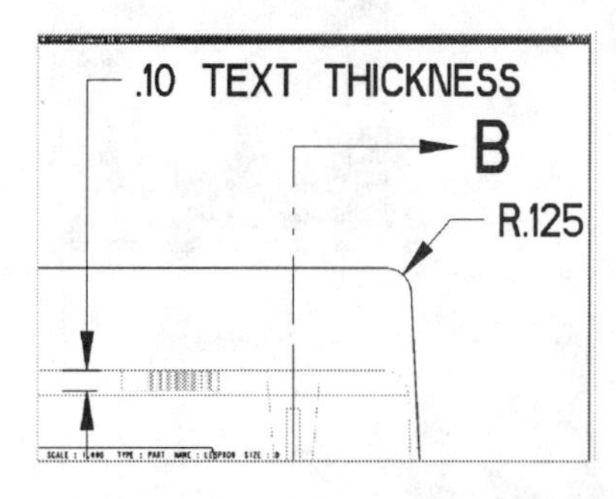

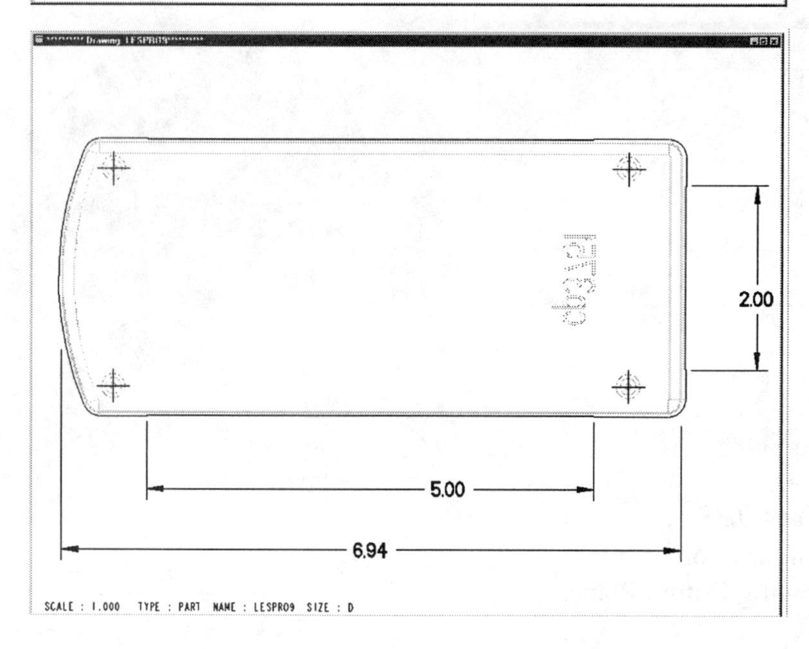

Figure 9.41
Cellular Phone Drawing,
Top View

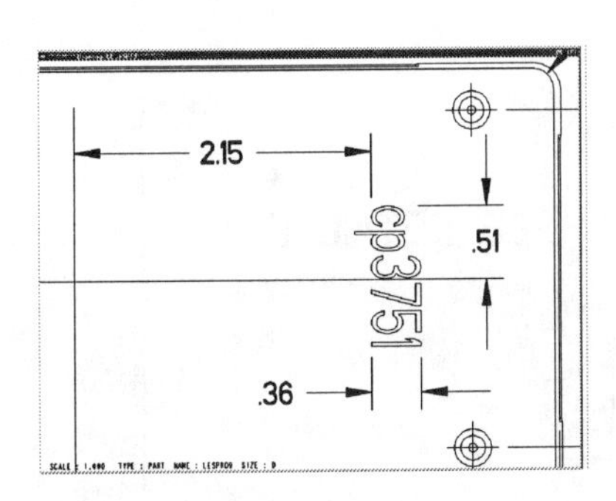

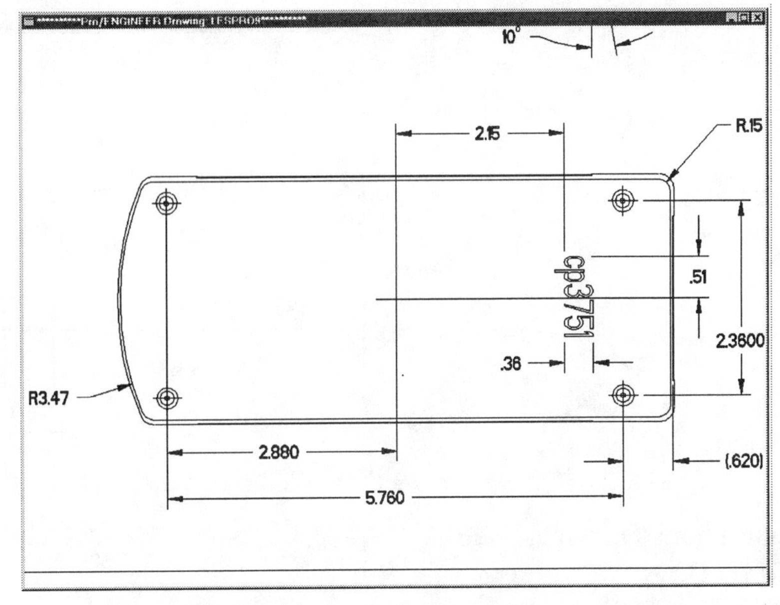

Figure 9.42
Cellular Phone Drawing,
Bottom View

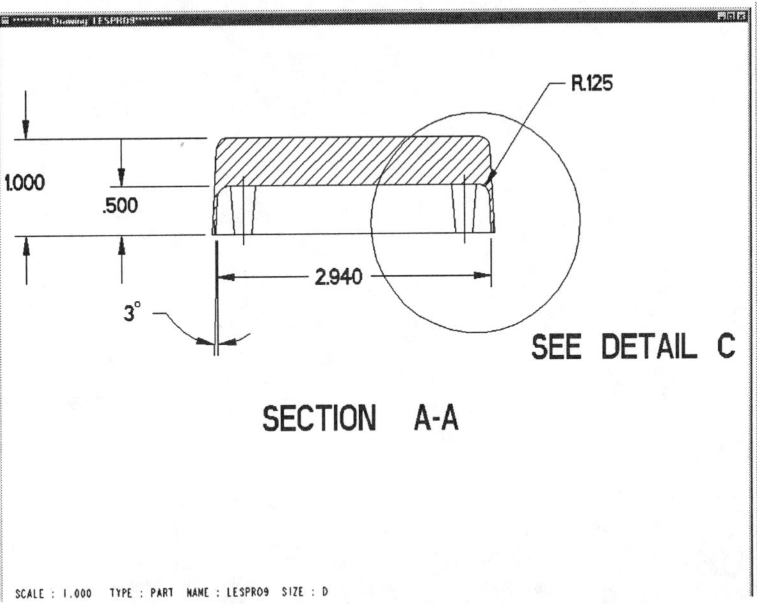

Figure 9.43
Cellular Phone Drawing,
SECTION A-A

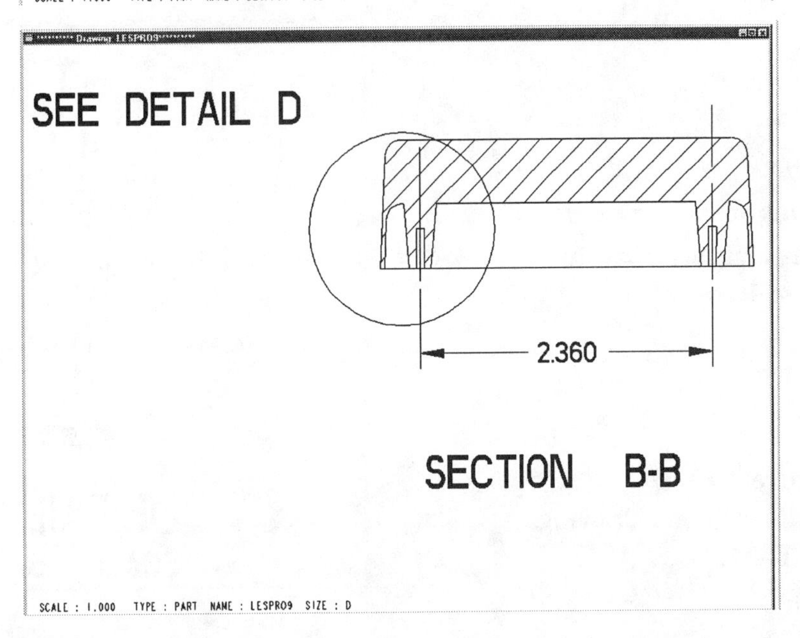

Figure 9.44
Cellular Phone Drawing,
SECTION B-B

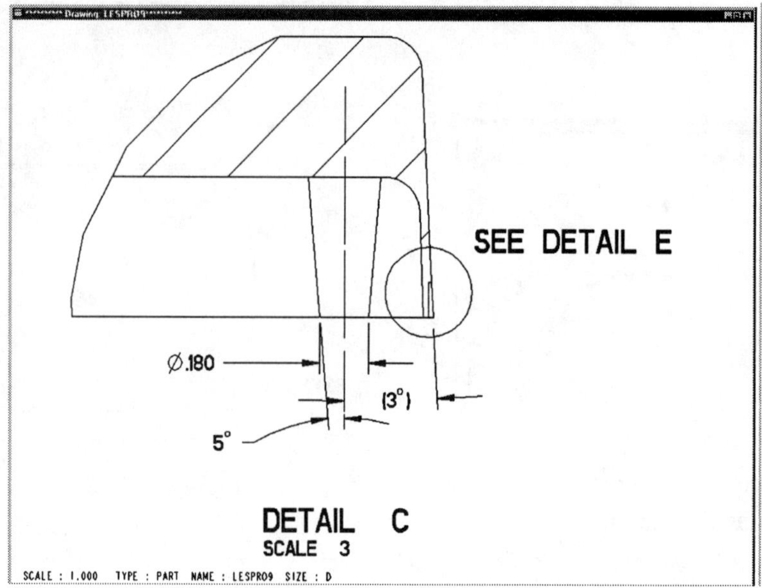

Figure 9.45
Cellular Phone Drawing, **DETAIL C**

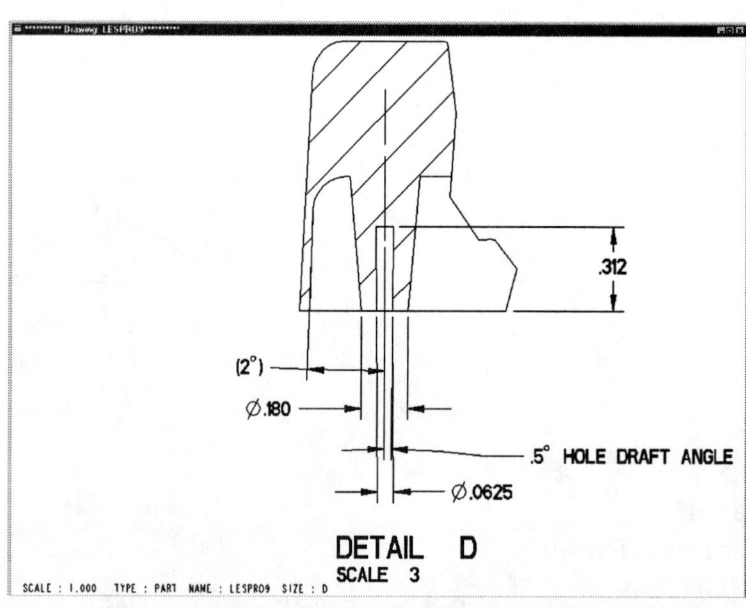

Figure 9.46
Cellular Phone Drawing, **DETAIL D**

Dbms ⇒ Save ⇒ enter
Purge ⇒ enter ⇒
Done-Return

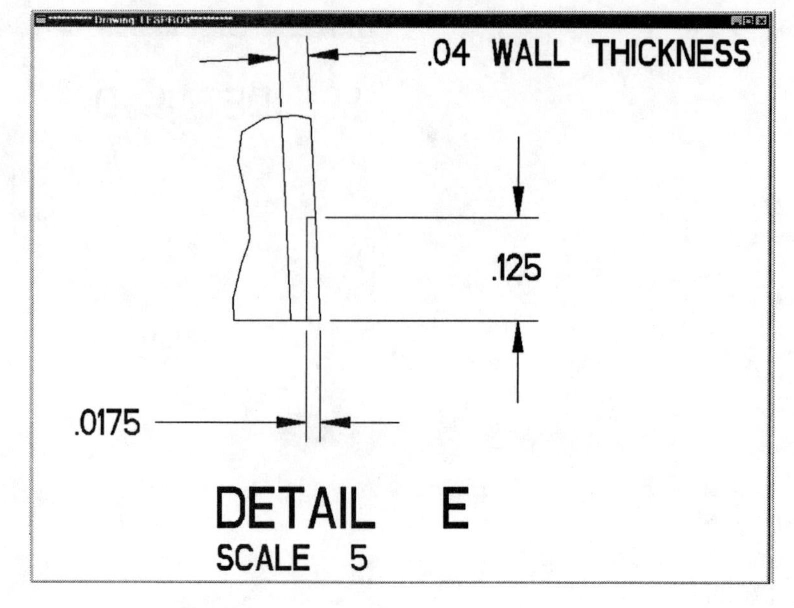

Figure 9.47
Cellular Phone Drawing, **DETAIL E**

Lesson 10

Shell, Reorder, and Insert Mode

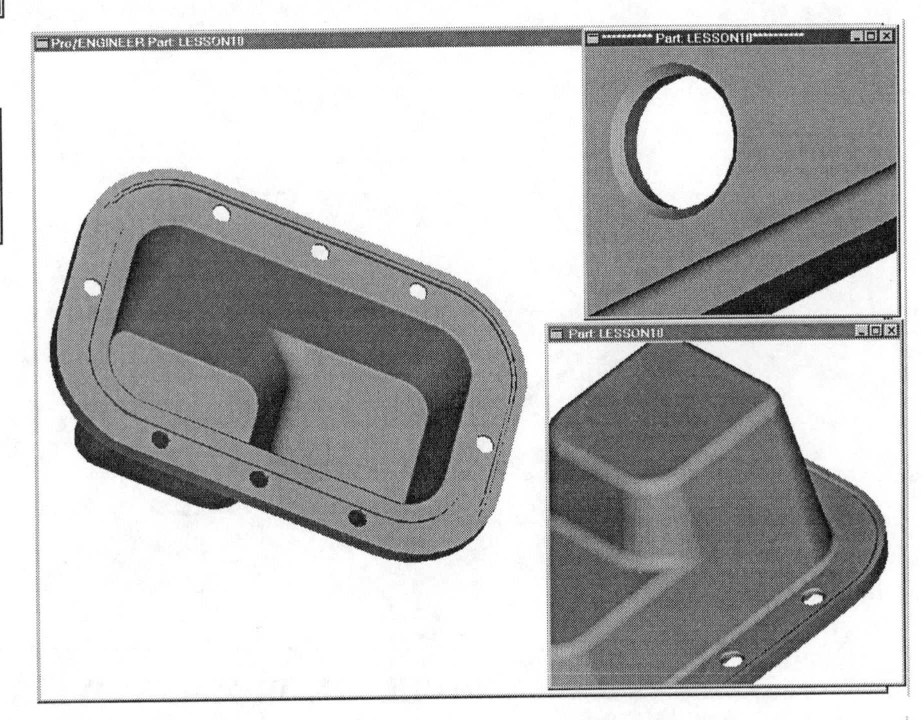

Figure 10.1
Oil Sink

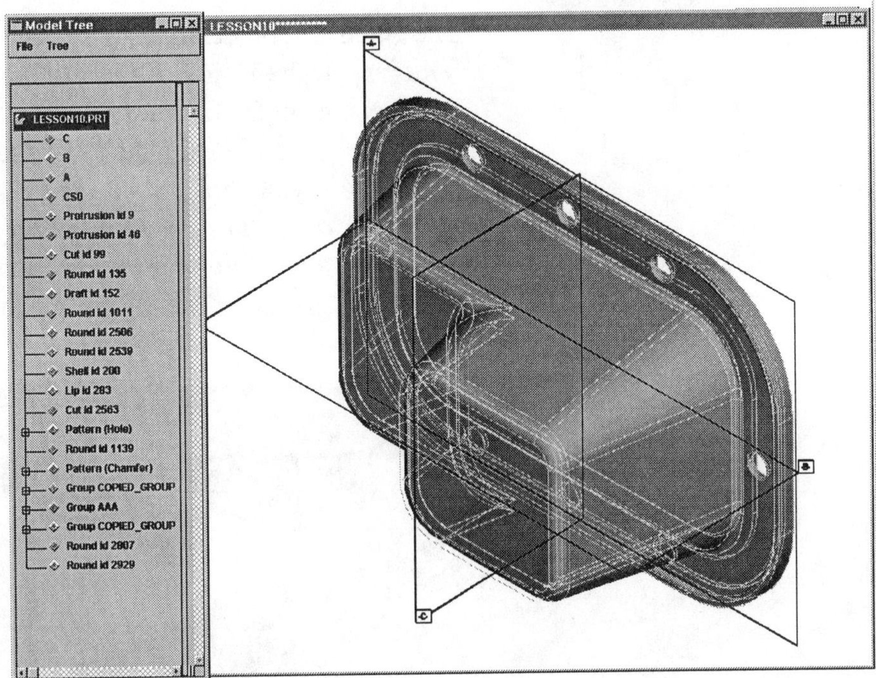

Figure 10.2
Oil Sink with Model Tree

OBJECTIVES

1. **Shell out a part**
2. **Alter the creation order of a feature**
3. **Insert a feature at a specific point in the design order**
4. **Create a lip on a part model**

EGD REFERENCE
Engineering Graphics and Design
by L. Lamit and K. Kitto
Read Chapter: 14

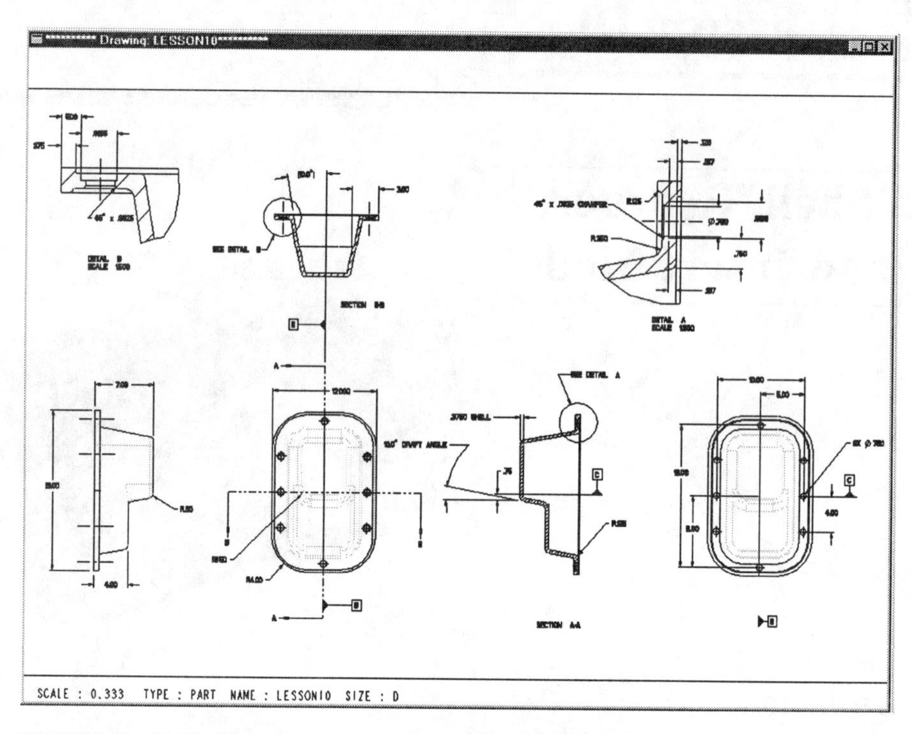

Figure 10.3
Oil Sink Dimensions

SHELL, REORDER, AND INSERT MODE

The **Shell** option removes a surface or surfaces from the solid, then hollows out the inside of the solid, leaving a shell of a specified wall thickness, as in the oil sink shown in Figure 10.1 through Figure 10.3. When Pro/E makes the shell, all the features that were added to the solid before you chose **Shell** are hollowed out (Fig. 10.4). Therefore, the *order of feature creation* is very important when you use shell. You can alter the feature creation order by using the **Reorder** command. Another method of placing a feature at a specific place in the feature/design creation order is to use the **Insert Mode** option.

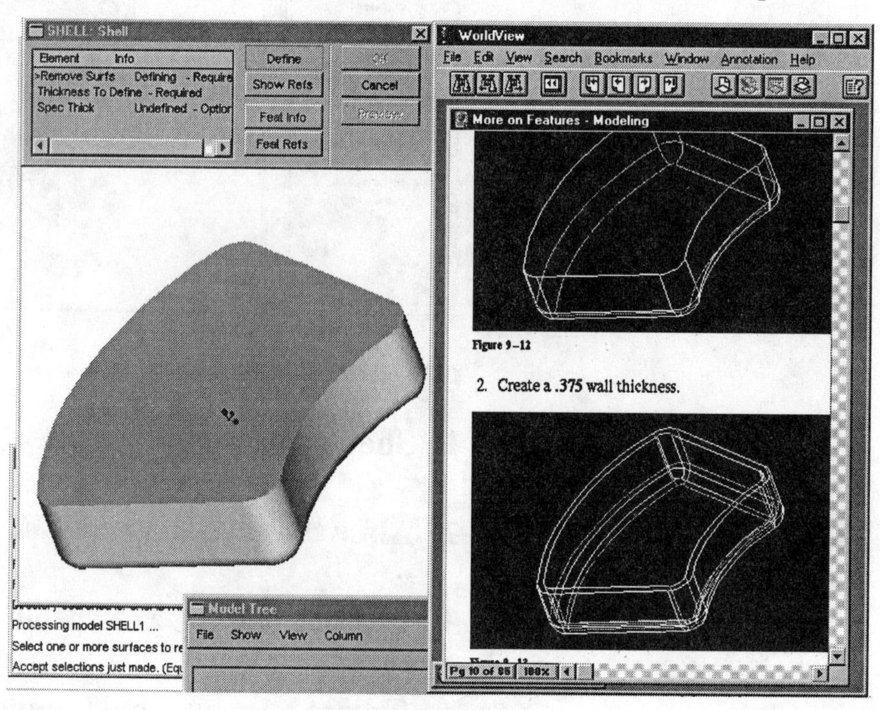

Figure 10.4
COAch for Pro/E, More on Features-- Modeling (Segment 1: Shell)

Creating Shells

To create a shell (Fig. 10.5), do the following:

1. Choose **Shell** from the SOLID menu.
2. Pro/E displays the feature creation dialog box. If desired, select the optional element **Spec Thick** to specify thickness individually. Select the Define button from the dialog box.
3. Select a surface or surfaces to be removed. When you have finished, choose **Done Refs** from the FEATURE REFS menu.
4. Enter the thickness of the wall. This thickness applies to all surfaces except those to which you assign a different thickness.
5. If you chose the Spec Thick element, Pro/E displays the SPEC THICK menu, which list the following options:

> **Set Thickness** Set thickness for the individual surface.
> **Reset to Def** Reset the surfaces to the default thickness.

Choose **Set Thickness**. Select a surface and enter the thickness. Continue this process until you have specified all the surfaces you want. When you have finished, choose **Done** from the SPEC THICK menu.

6. To create the shell, select **OK** from the dialog box. If you entered a positive value for the thickness, material will be removed, leaving the shell thickness "inside" the part. However, if you entered a negative value, the shell thickness is added to the "outside" of the part.

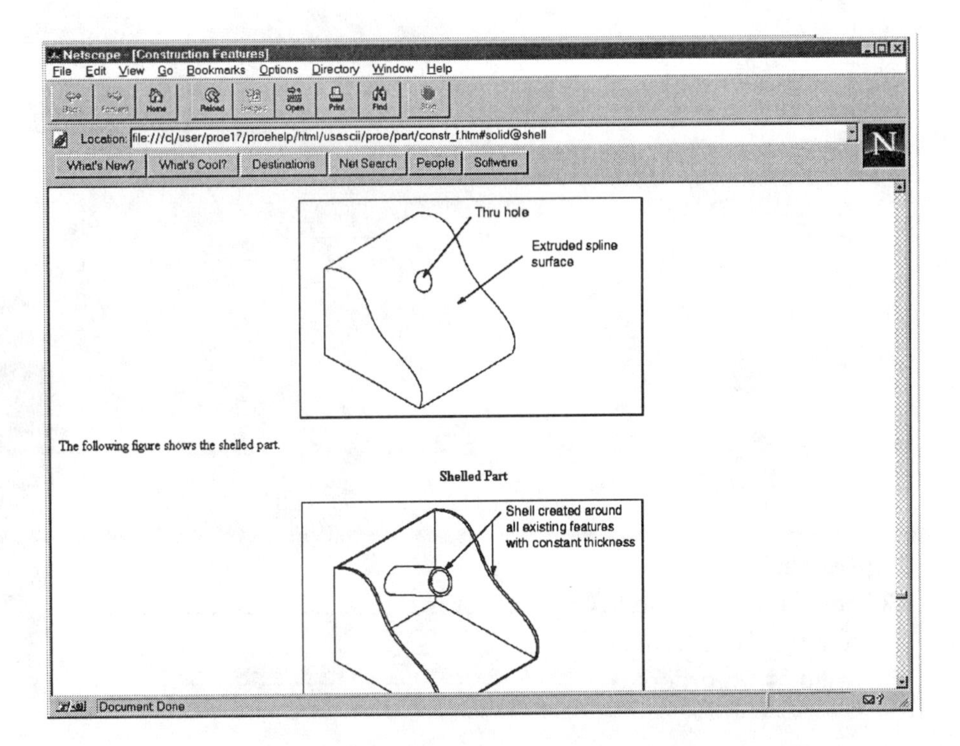

Figure 10.5
Online Documentation Shell

Reordering Features

You can move features forward or backward in the feature creation (regeneration) order list, thus changing the order in which they are regenerated (Fig. 10.6). You can reorder multiple features in one operation, as long as these features appear in ***consecutive*** order.

Feature reorder *cannot* occur under the following conditions:

Parents cannot be moved so that their regeneration occurs after the regeneration of their children.
Children cannot be moved so that their regeneration occurs before the regeneration of their parents.

To reorder a feature, do the following:

1. Use the command sequence **Part, Feature, Reorder.**
2. Specify the selection method by choosing an option from the SELECT FEAT menu:

Select Select features to reorder by picking on the screen and/or from the tree tool. You can also choose **Sel By Menu** to enter the feature number. Choose **Done Sel** when finished selecting.
Layer Select all features from a layer by selecting the layer. Choose **Done Sel** from the LAYER SEL menu when finished.
Range Specify the range of features by entering the regeneration number of the starting and ending feature.

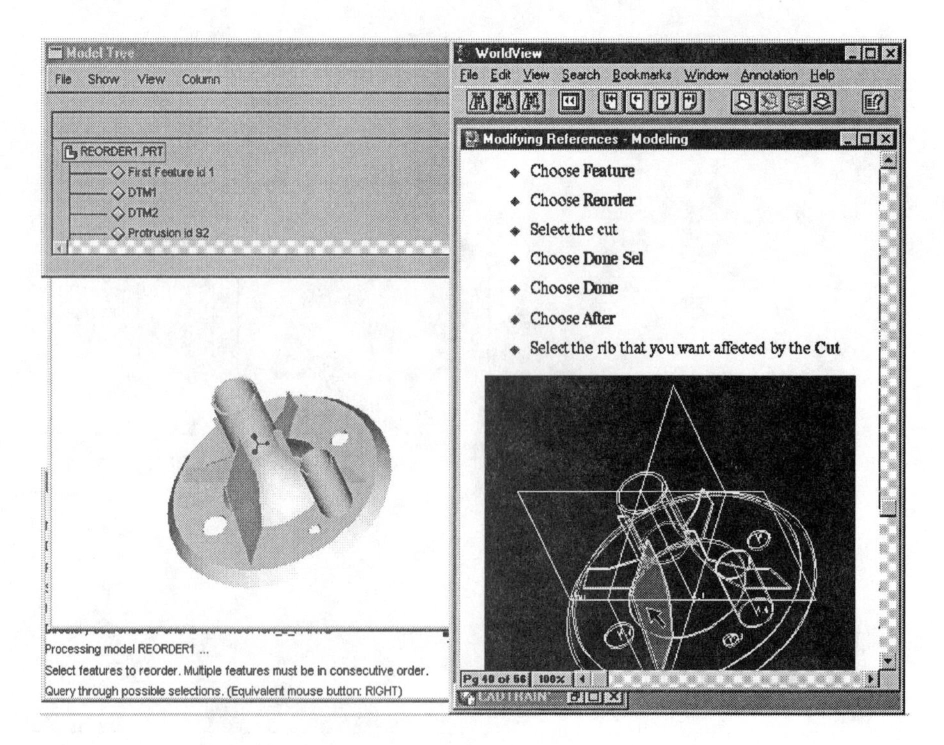

Figure 10.6
COAch for Pro/E, Modifying References-- Modeling (Segment 4: Reorder)

3. A system message lists the selected features for reorder and states the valid ranges for the new insertion point.
4. Choose **Done** from the SELECT FEAT menu.
5. Choose one of the options in the REORDER menu:
   Before Insert the feature before the insertion point feature.
   After Insert the feature after the insertion point feature.

Pick a feature indicating the insertion point, or choose Sel By Menu to enter the feature number.

Inserting Features

Normally, Pro/E adds a new feature after the last existing feature in the part, including suppressed features. Insert mode (Fig. 10.7) allows you to add new features at any point in the feature sequence, other than before the base feature or after the last feature.

To insert features, do the following:

1. Choose **Insert Mode** from the FEAT menu, then choose **Activate**.
2. Select a feature after which the new features will be inserted. All features after the selected one will be automatically suppressed.
3. Choose **Create** and create the new features as usual.
4. Cancel **Insert Mode** by choosing **Resume** from the FEAT menu and selecting to resume the features that were suppressed when you activated Insert mode, or choose **Cancel** from the INSERT MODE menu. Pro/E asks you whether to resume the features that were suppressed when you activated Insert mode, then automatically regenerates the part.

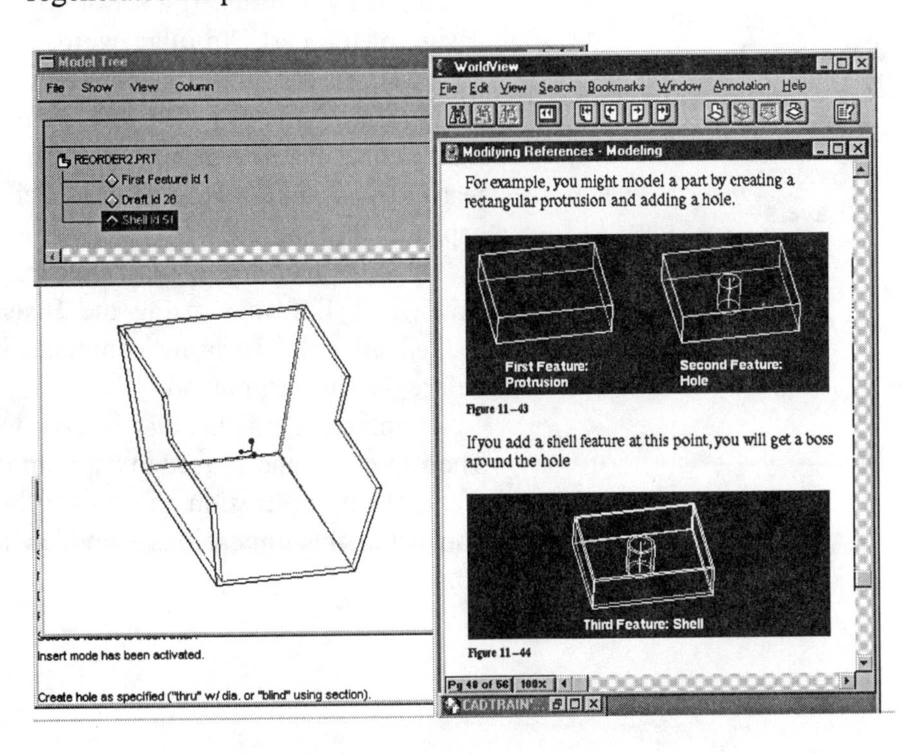

Figure 10.7
COAch for Pro/E, Modifying References-- Modeling (Segment 5: Insert Mode)

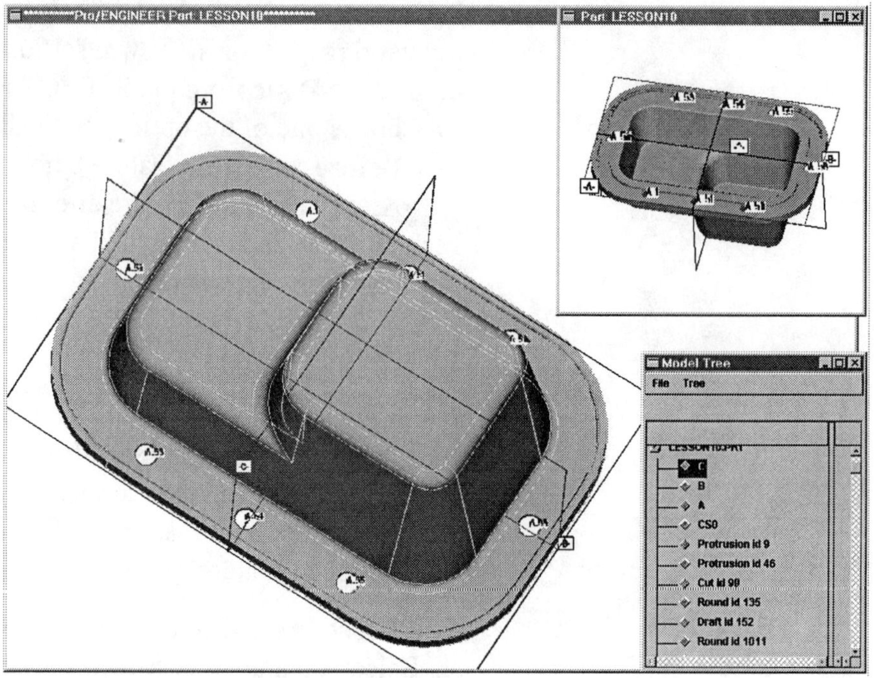

Figure 10.8
Oil Sink Showing Datum Planes and Coordinate System and Model Tree

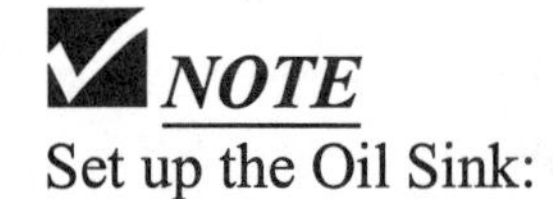

NOTE

Set up the Oil Sink:

- Material = Steel
- Units = Inches
- Default Datum Planes
- Default Coordinate System
- Layers = **DATUM_PLANES**
- Hidden Line
- ✓Grid Snap

Oil Sink

The **Oil Sink** (Fig. 10.8) requires the use of the **Shell** command. The shelling of a part should be done after every protrusion and round has been modeled. This lesson part will have you create a protrusion, a cut, and a set of rounds. Some of the required rounds will be left off the part model on purpose. Pro/E has a command called **Insert Mode** that allows you to insert a set of commands at a previous state in the design of the part. In other words, you can create a feature after or before a selected existing feature even if the whole model has been completed. You can also *move the order in which a feature was created* and therefore have subsequent commands affect the reordered feature. A round created after a shell command can be reordered to appear before the shell.

In this lesson you will also insert a round or two before the existing shell feature using the **Insert Mode**. The round will be shelled after the **Resume** command is given, since it now appears before the shell command.

Another command that is new to you will also be introduced. The **Lip** command is used to add/remove material along an edge.

The first protrusion and cut for the Oil Sink must be modeled by you using the dimensions provided in Figure 10.9 through Figure 10.16.

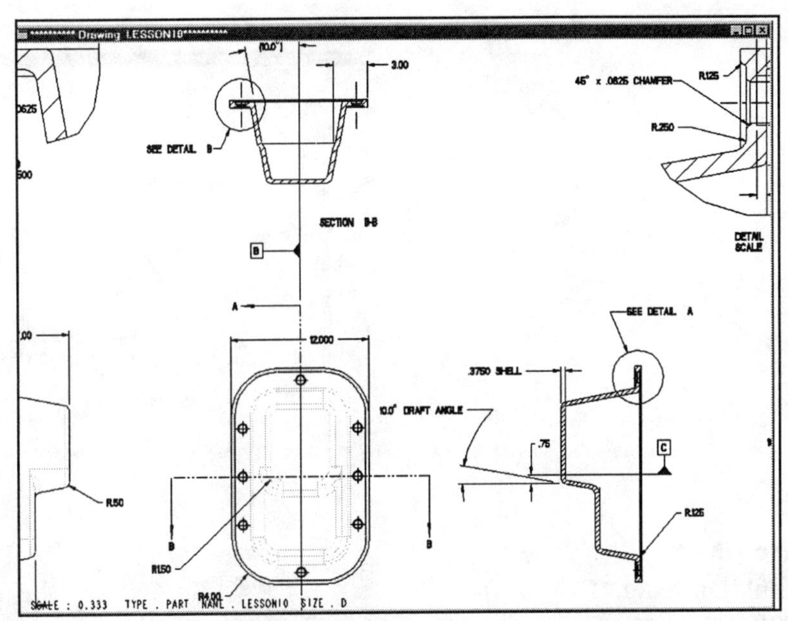

Figure 10.9
Oil Sink Detail

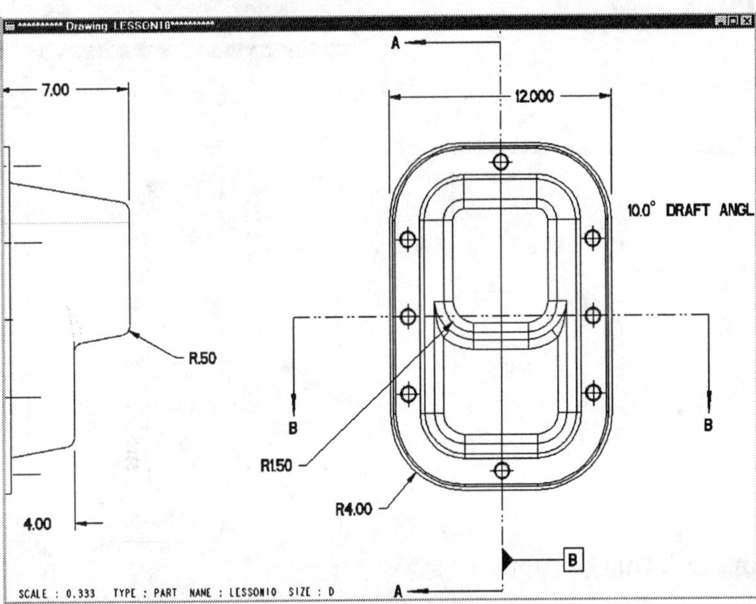

Figure 10.10
Oil Sink Drawing, Front View

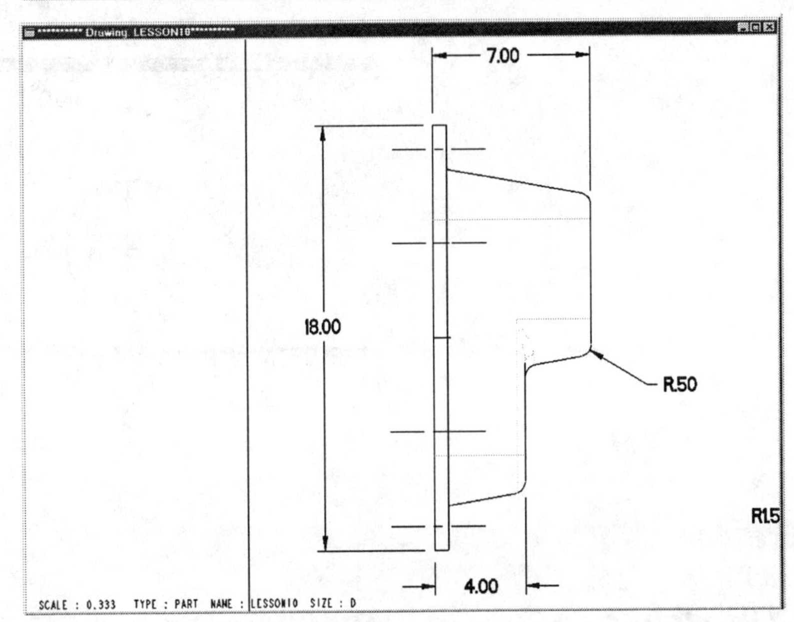

Figure 10.11
Oil Sink Drawing, Left Side View

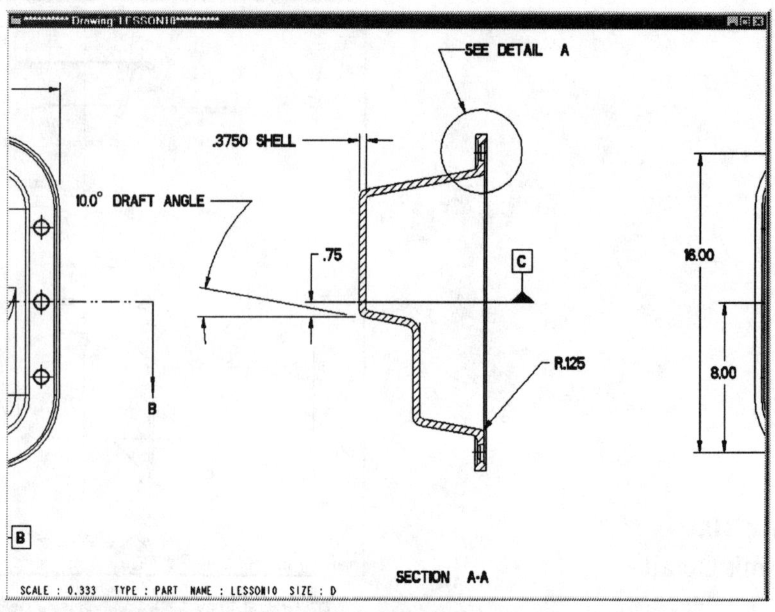

Figure 10.12
Oil Sink Drawing,
SECTION A-A

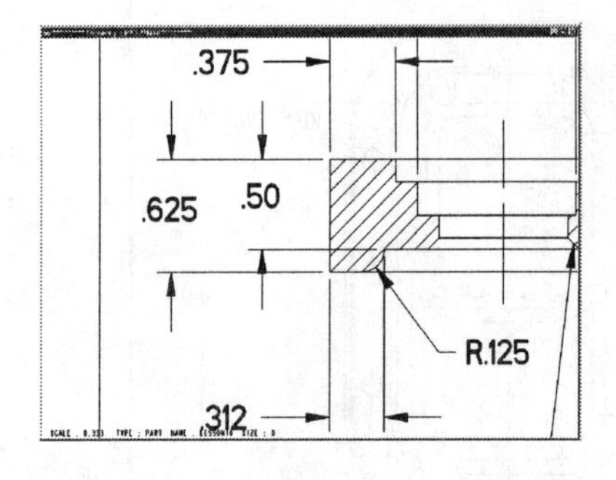

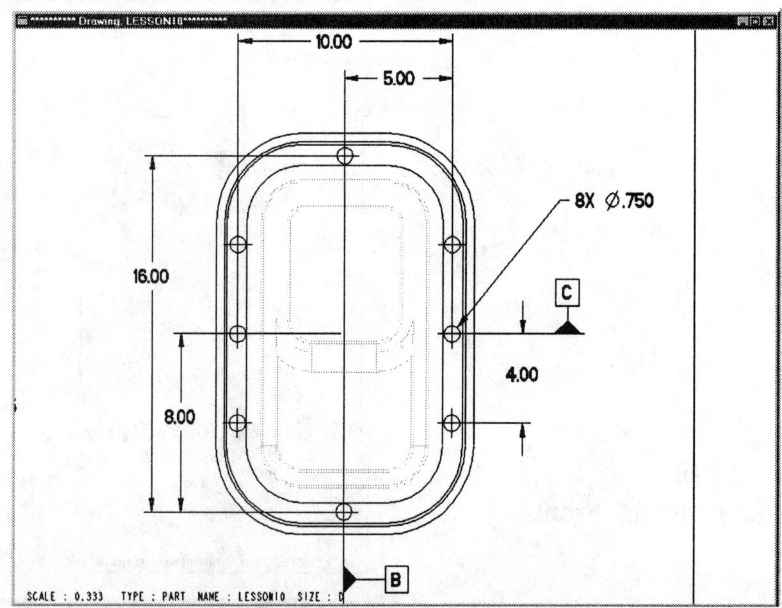

Figure 10.13
Oil Sink Drawing, Bottom View

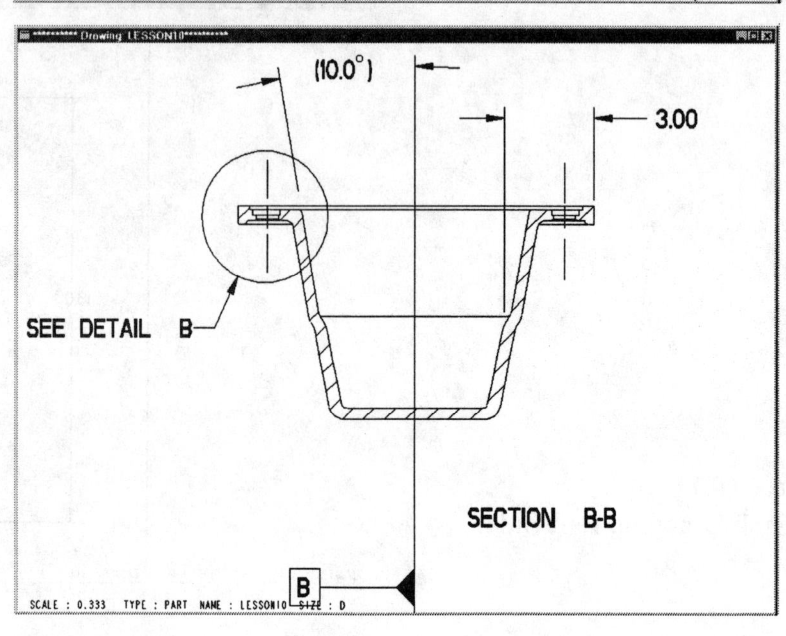

Figure 10.14
Oil Sink,
SECTION B-B

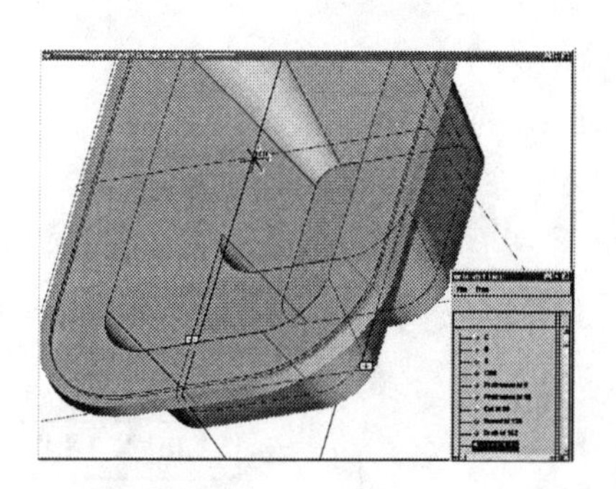

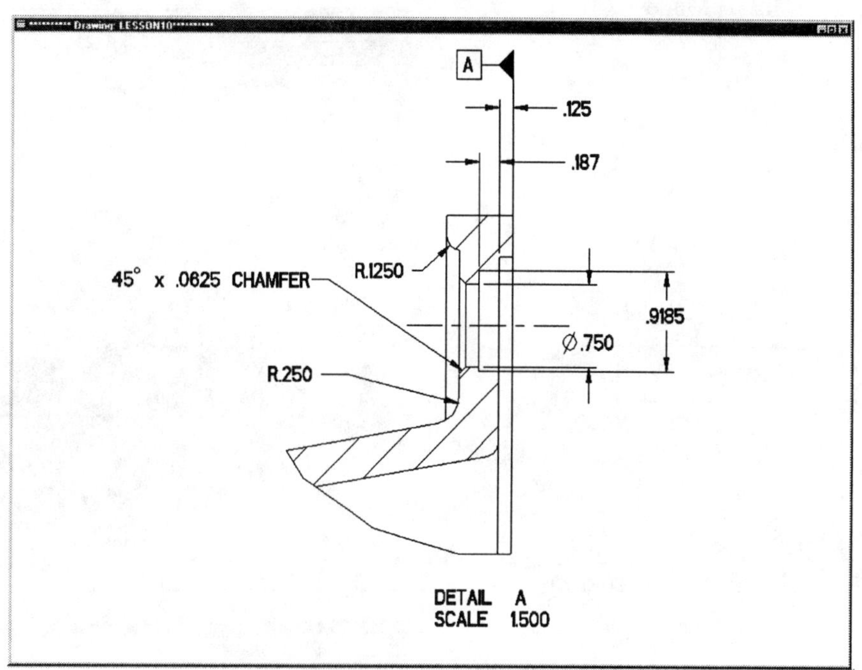

Figure 10.15
Oil Sink,
DETAIL A

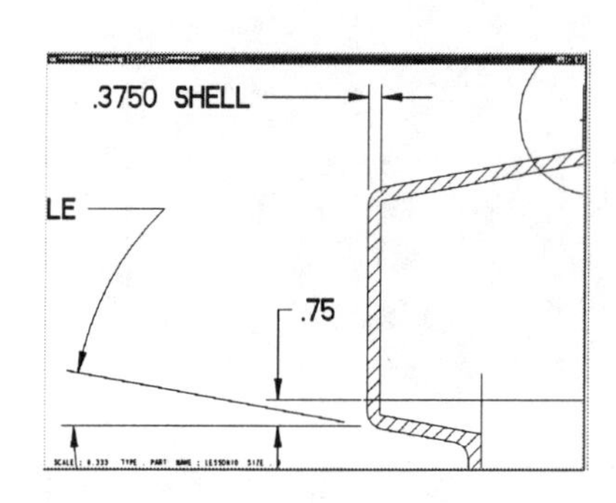

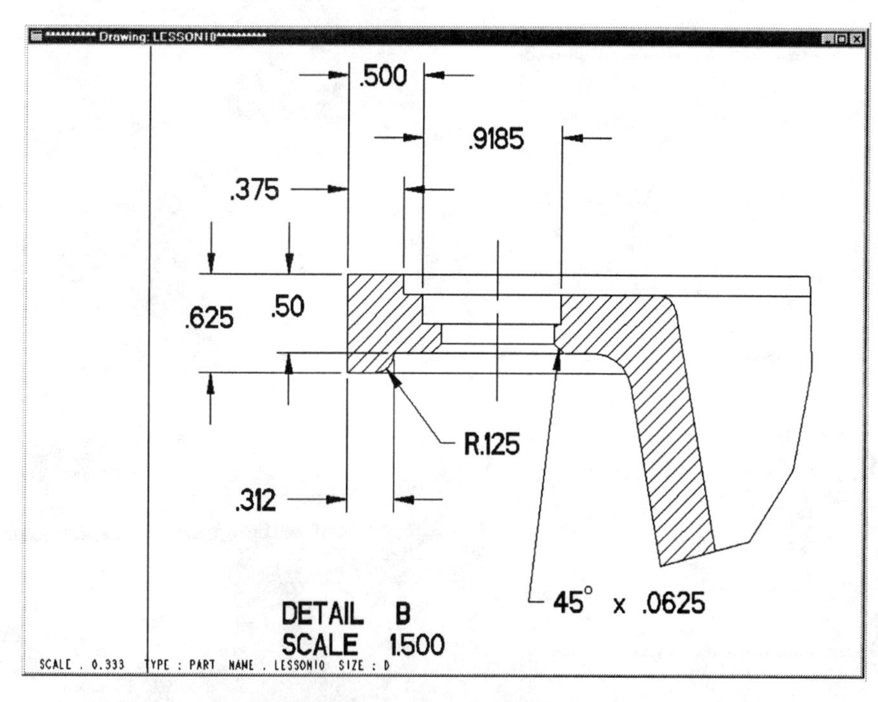

Figure 10.16
Oil Sink,
DETAIL B

After creating the default datum planes and coordinate system, model the first protrusion shown in Figure 10.17. The large protrusion forming the Oil Sink's tublike shape is modeled next (Fig. 10.18). In Figure 10.19 a cut is added to the model. The **R1.50** rounds are added next (Fig. 10.20). The lateral sides of the protrusion are drafted at a angle of **10°**, as shown in Figure 10.21. Shell the part with the following commands (Fig. 10.22):

Feature ⇒ **Create** ⇒ **Shell** ⇒ (select one or more surfaces-- pick the bottom surface ⇒ **Done Sel** ⇒ **Done Refs** ⇒ (enter thickness of **.375** at the prompt) ⇒ **enter** ⇒ **OK** ⇒ **Done**

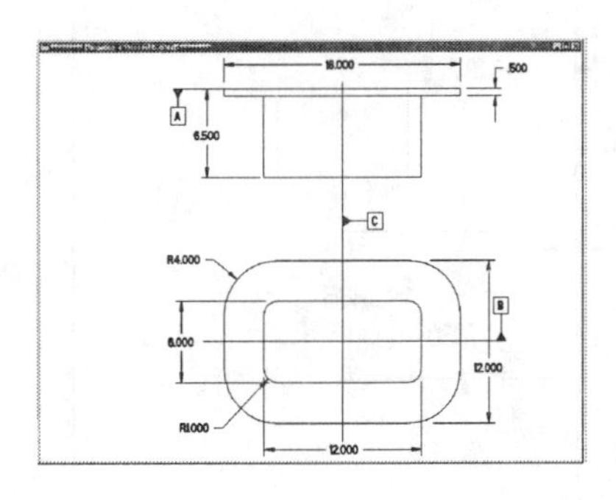

Figure 10.17
Oil Sink's First Protrusion

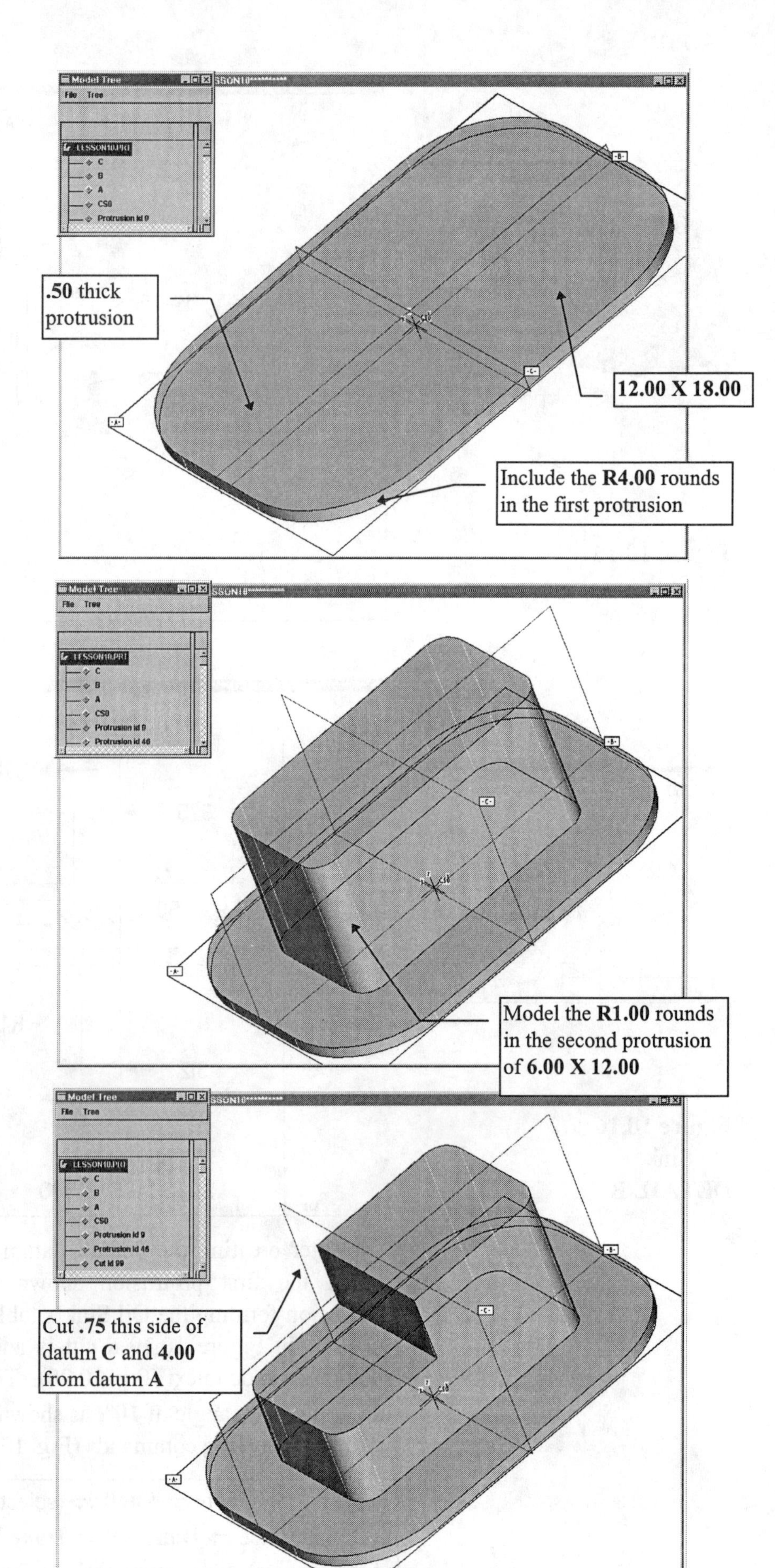

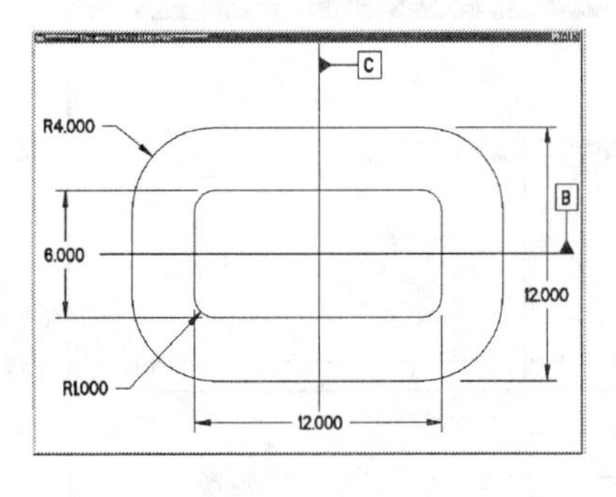

Figure 10.18
Oil Sink's Second Protrusion

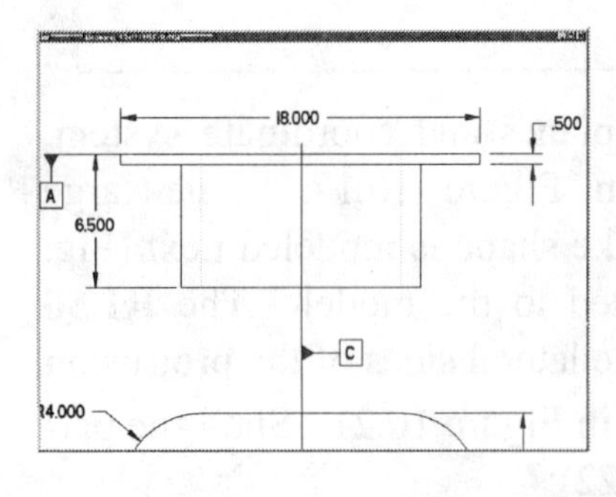

Figure 10.19
Oil Sink's First Cut

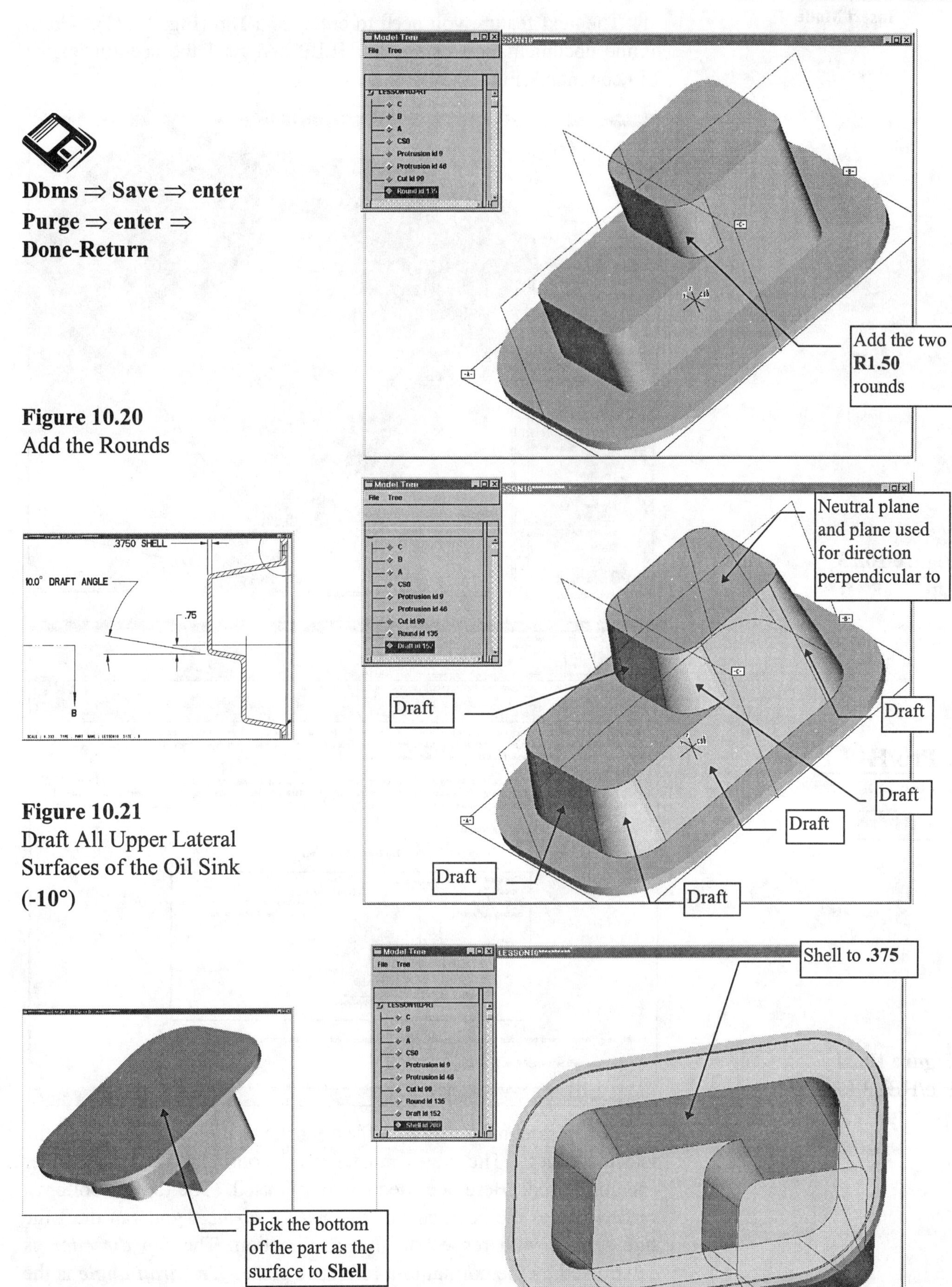

Dbms ⇒ Save ⇒ enter Purge ⇒ enter ⇒ Done-Return

Figure 10.20
Add the Rounds

Figure 10.21
Draft All Upper Lateral Surfaces of the Oil Sink (-10°)

Figure 10.22
Shell the Part

The next feature you need to create is a **Lip** (Fig. 10.23). Go to online documentation using Pro/HELP and read the section on the **Lip** command (Fig. 10.24).

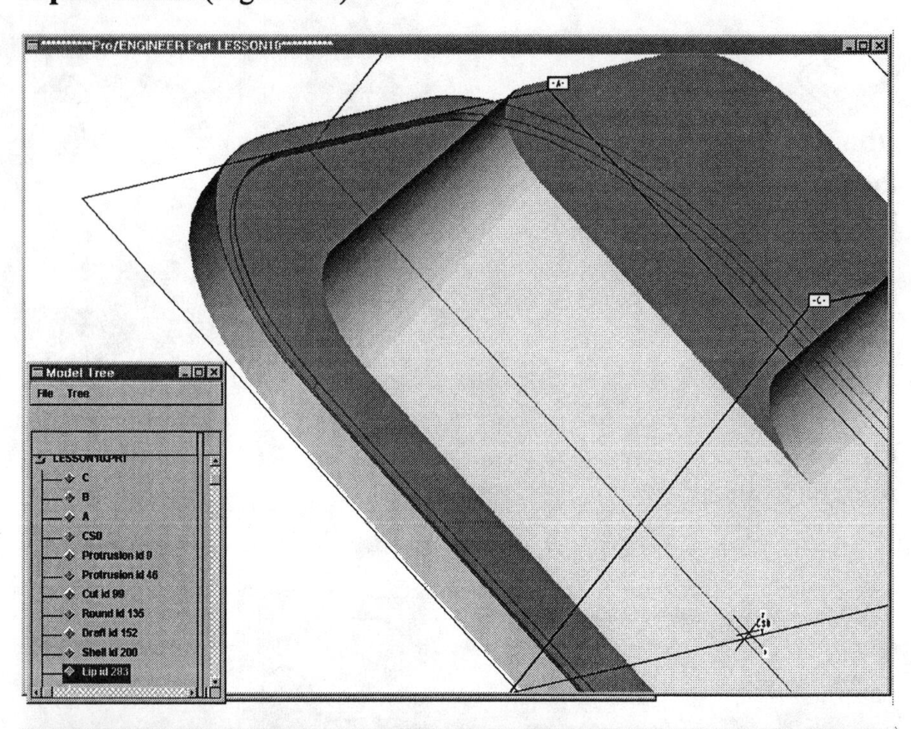

Figure 10.23
Lip

? Pro/HELP
Choose **Lip**

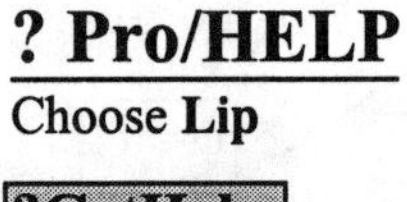

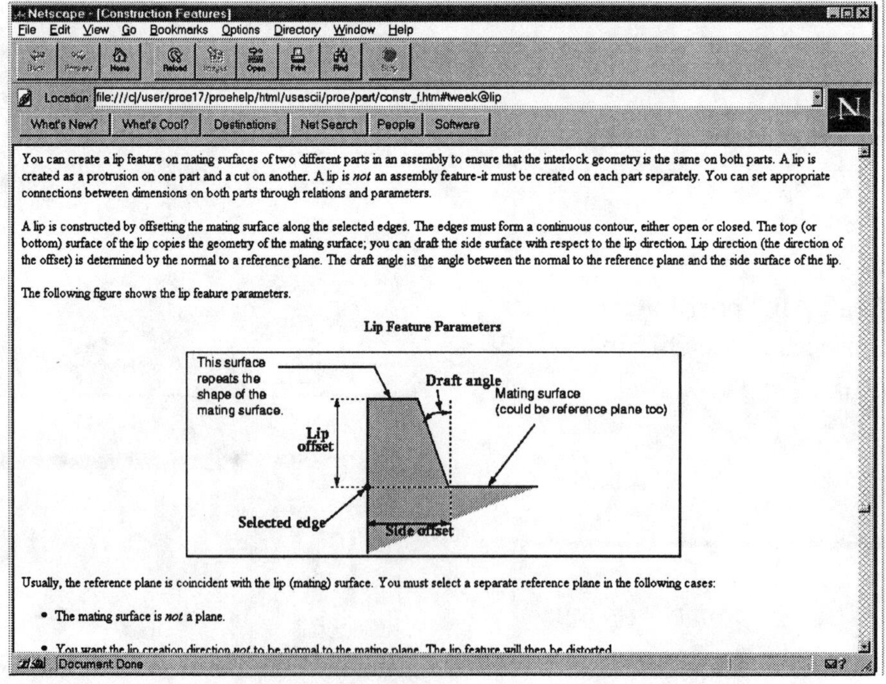

Figure 10.24
Pro/HELP for **Lip**

A **Lip** is constructed by offsetting the mating surface along the selected edges. The edges must form a continuous contour, either open or closed. Here, a closed contour is used. The top (or bottom) surface copies the geometry of the *mating surface*: you can draft the side surface with respect to the lip direction. The *Lip direction* is determined by the normal to a reference plane. The *draft angle* is the angle between the normal to the reference plane and the side surface of the Lip.

In most cases, the reference plane is coincident with the lip (mating) surface. You must select a separate reference plane in the following cases:

* The mating surface is not a plane
* You want the lip creation direction not to be normal to the mating plane. The lip feature will then be distorted.

Create the lip with the following commands (Fig. 10.25):

Feature ⇒ **Create** ⇒ **Tweak** ⇒ **Lip** ⇒ **Chain** (pick the edge to select the mating surface as in Figure 10.25) ⇒ **Done Sel** ⇒ **Done** ⇒ (pick the surface to be offset, in this case the mating surface) ⇒ (enter the lip offset from the mating surface: **.125**) ⇒ (enter the *side offset distance* of **.3125** from the selected edges to the *draft surfaces*) ⇒ (select the *draft reference surface*-- same as mating surface) ⇒ (enter the *draft angle*) ⇒ **enter** (for zero) ⇒ **Done** (Fig. 10.26)

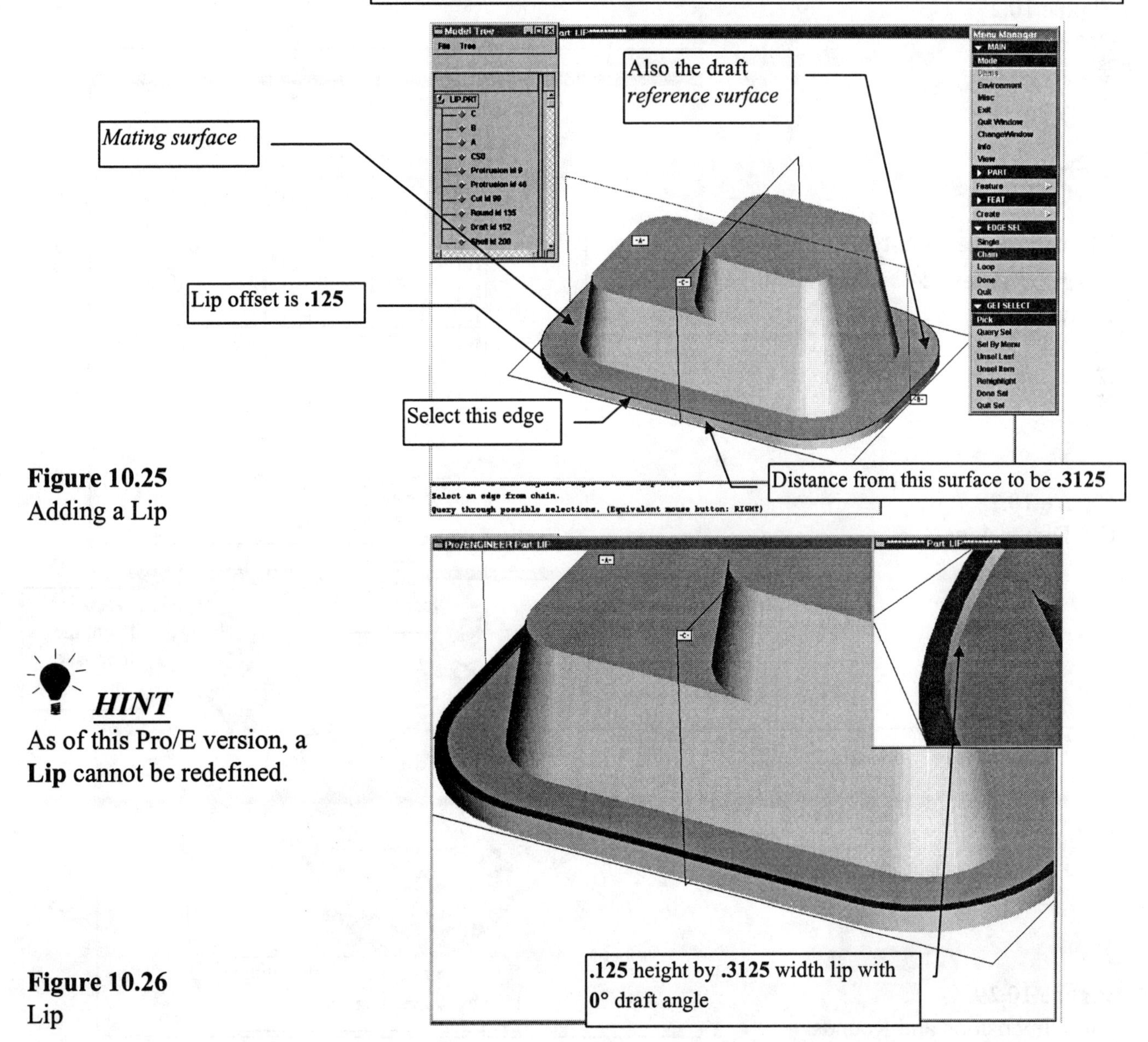

Figure 10.25
Adding a Lip

HINT
As of this Pro/E version, a **Lip** cannot be redefined.

Figure 10.26
Lip

The next feature is a cut measuring **.9185** wide by **.1875** deep (see Figure 10.15 and Figure 10.16). Figure 10.27 and Figure 10.28 show the cut. Add the countersunk holes and the rounds (Fig. 10.29).

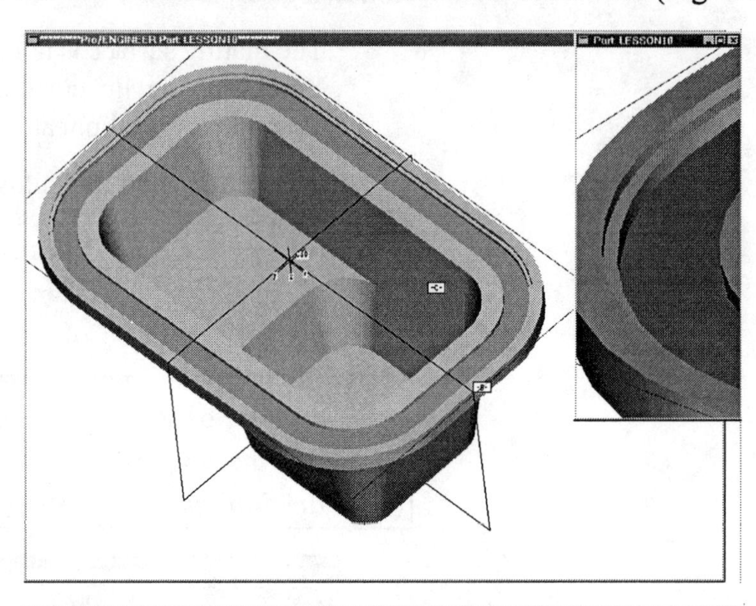

Figure 10.27
Cut

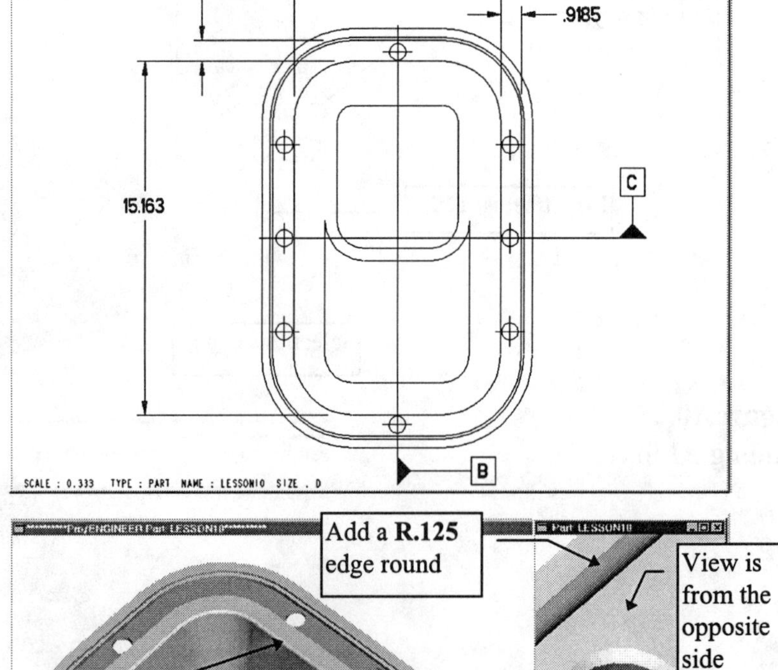

Figure 10.28
Cut Dimensions

Dbms ⇒ Save ⇒ enter
Purge ⇒ enter ⇒
Done-Return

Add a **R.125** edge round

View is from the opposite side

R.125 round

Eight **.750** diameter holes with **45°** by **.0625** chamfers

Figure 10.29
Cut Dimensions and Rounds

The next series of features will purposely be created at the wrong stage of the project. Create a **R.50** round as shown in Figure 10.30. Since the design intent should have been to have a constant thickness for the part, shouldn't the round have been created before the shell? Use the **Reorder** command to change the position of this round in the design sequence. Your prompts will have different feature numbers. Give the following commands:

Feature ⇒ **Reorder** ⇒ (pick the feature to reorder, the round) ⇒ **Done Sel** ⇒ **Done** ⇒ (system will prompt with: ***Feature can be inserted before feature [10-26] or [30-34] Please select features:*** ⇒ (from the Model Tree select the **Shell id**) ⇒ **Done** (Fig. 10.31)

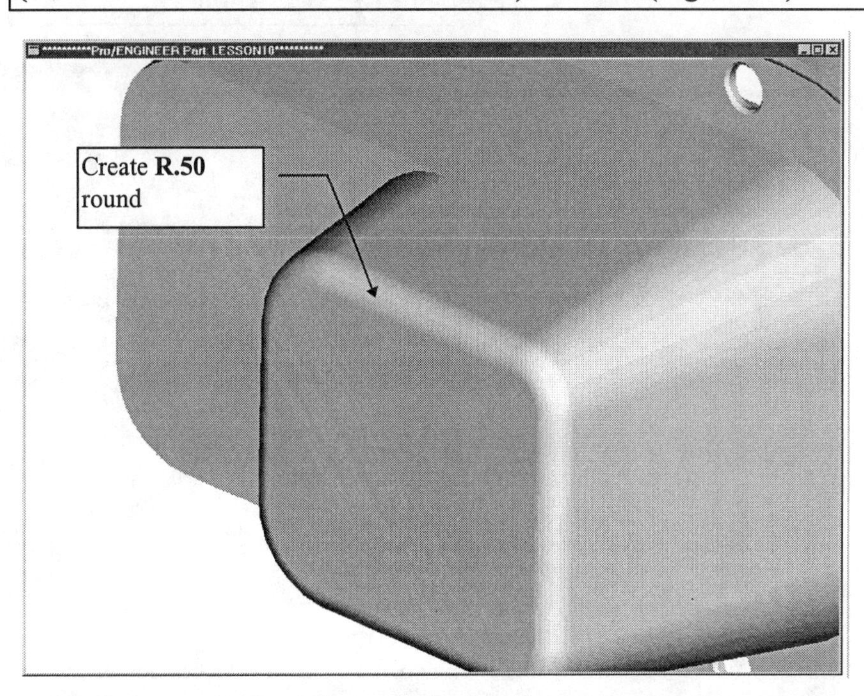

Figure 10.30
R.50 Round

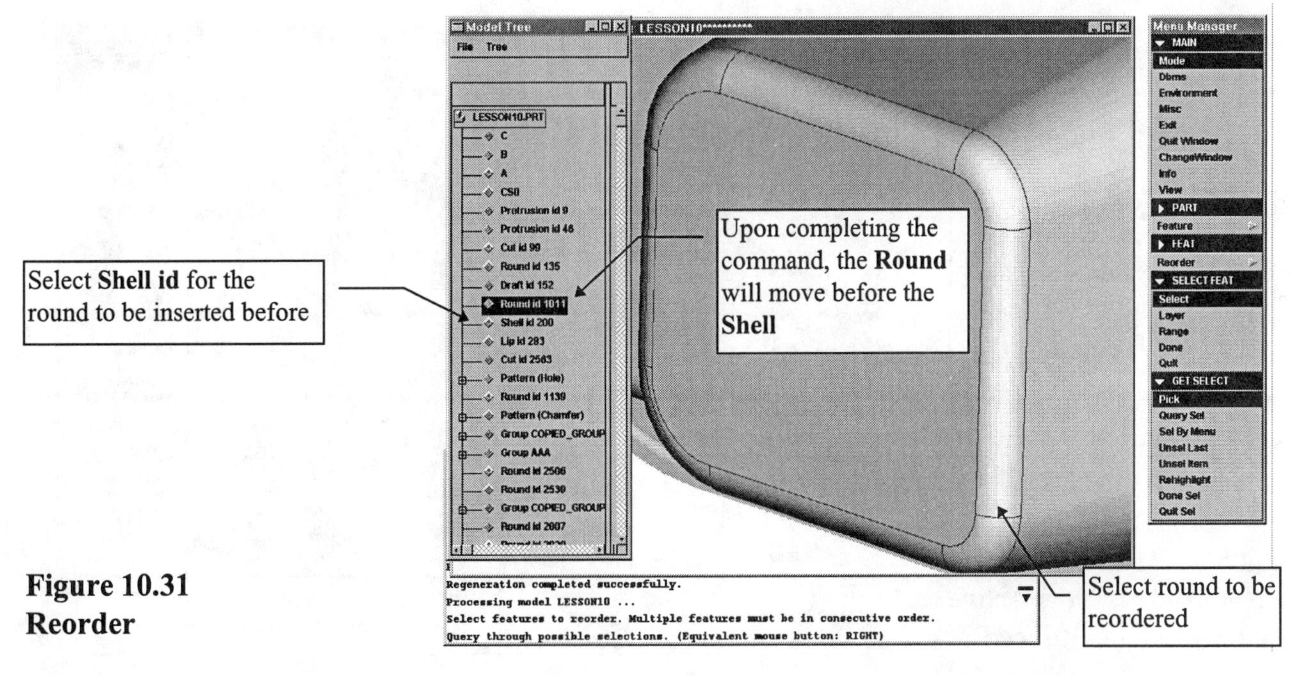

Figure 10.31
Reorder

The previous round was created at the wrong position in the design sequence and then reordered. To eliminate the reordering of a feature, the next round will be created using the **Insert Mode**. **Insert Mode** allows you to insert a feature at a previous stage of the design sequence. This is kind of like going back to the past and fixing or doing something you wish you would have done before-- not possible with life, but with Pro/E less of a problem. Give the following commands to enter **Insert Mode,** create the round, and then use **Resume** to return to the previous stage in the design (Fig. 10.32):

Feature ⇒ **Insert mode** ⇒ **Activate** ⇒ (select feature to insert after; from the Model Tree, pick the round you just reordered, as shown in Figure 10.33) ⇒ **Done**

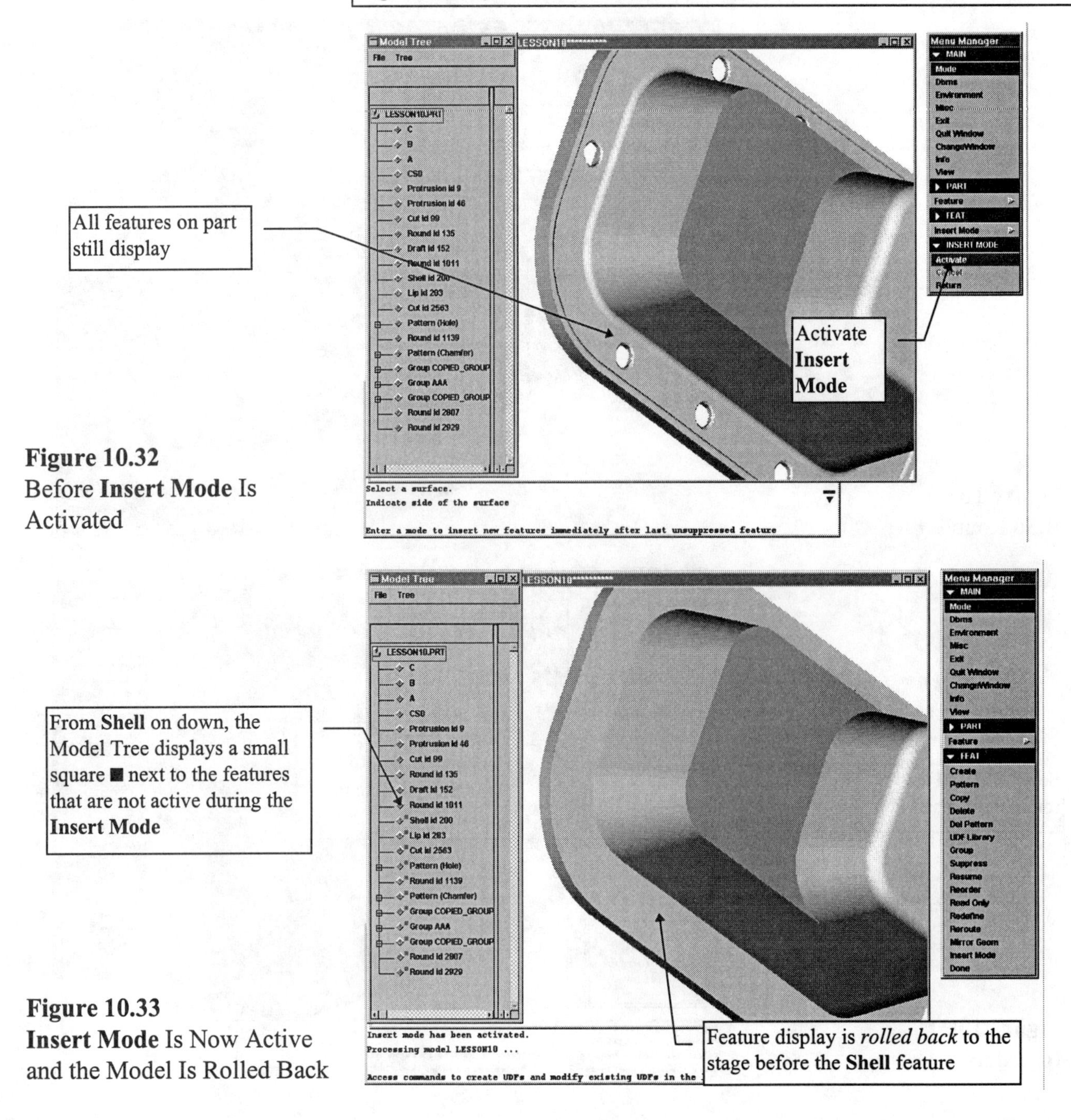

Figure 10.32
Before **Insert Mode** Is Activated

Figure 10.33
Insert Mode Is Now Active and the Model Is Rolled Back

The rounds can now be created using the following commands:

Feature ⇒ **Create** ⇒ **Round** ⇒ **Simple** ⇒ **Done** ⇒ **Constant** ⇒ **Edge Chain** ⇒ **Done** ⇒ **Tangent Chain** ⇒ (pick the edge shown in Figure 10.34) ⇒ **Done Sel** ⇒ **Done** ⇒ (enter the radius value of **.50** at the prompt) ⇒ **enter** ⇒ **OK** ⇒ **Done** ⇒ (*repeat the same commands* and pick the edge shown in Figure 10.35) ⇒ **Feature** ⇒ **Resume** ⇒ **All** (Fig. 10.36) ⇒ **Done** (Fig. 10.37)

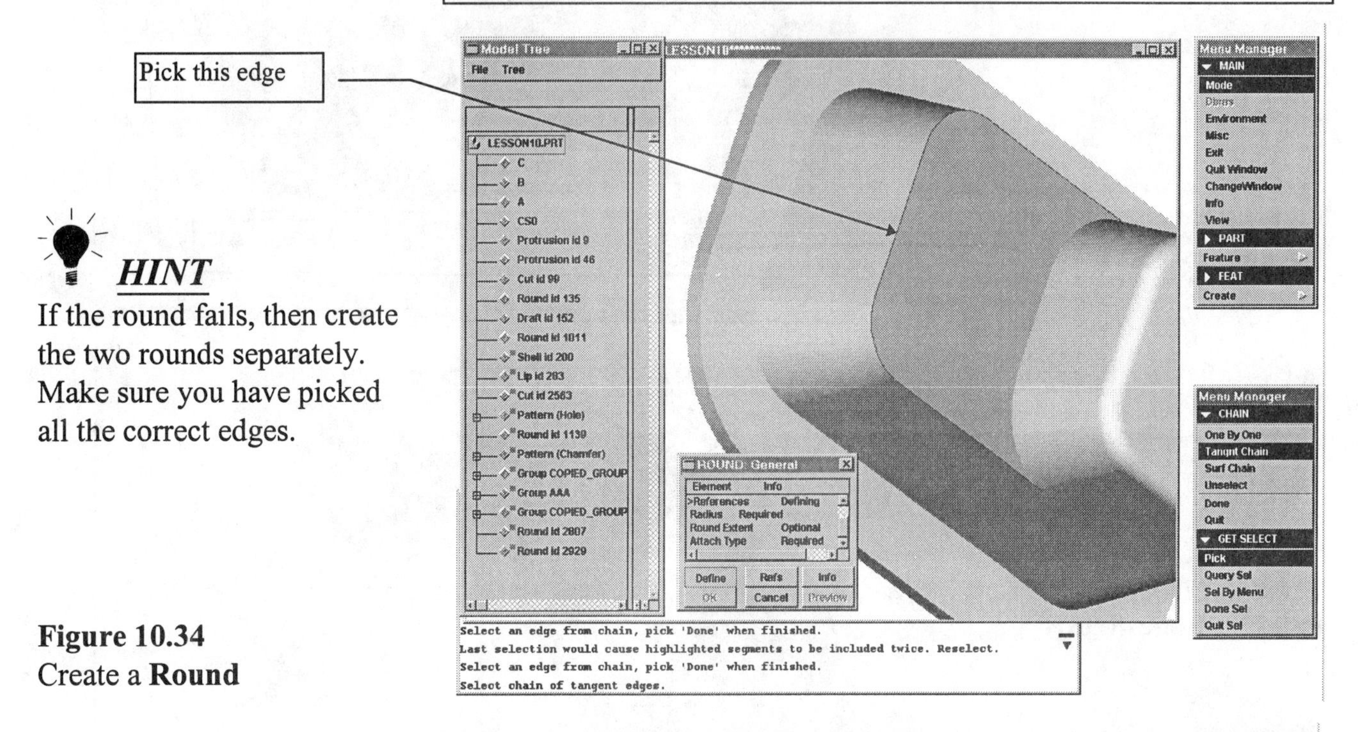

HINT
If the round fails, then create the two rounds separately. Make sure you have picked all the correct edges.

Figure 10.34
Create a **Round**

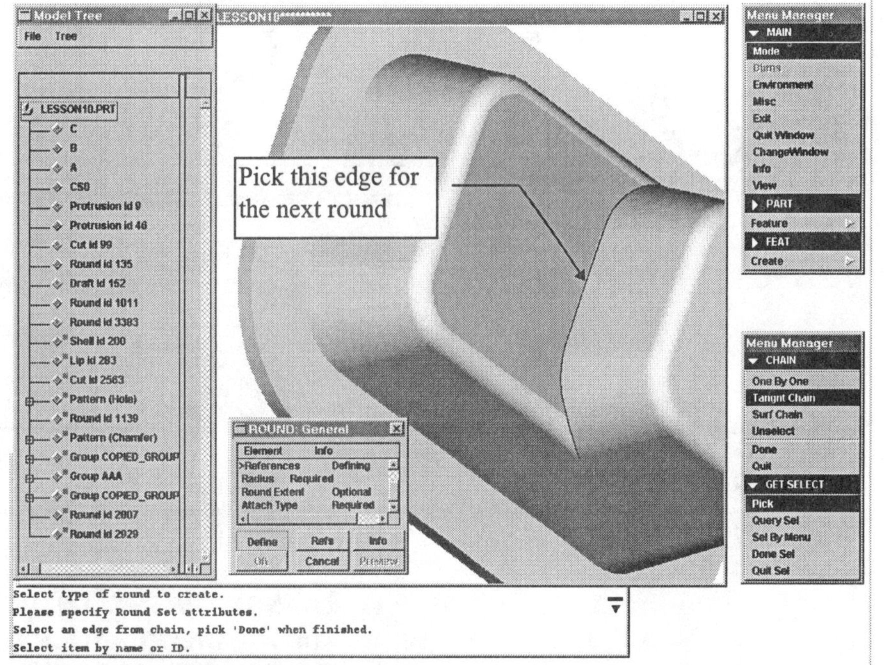

Dbms ⇒ **Save** ⇒ **enter Purge** ⇒ **enter** ⇒ **Done-Return**

Figure 10.35
Create Another **Round**

The rounds are now in the proper design sequence for the shell feature to create a constant thickness. After the part is completed, save it under a new name and do the **ECO** (Fig. 10.38)

Figure 10.36
Resume ⇒ All

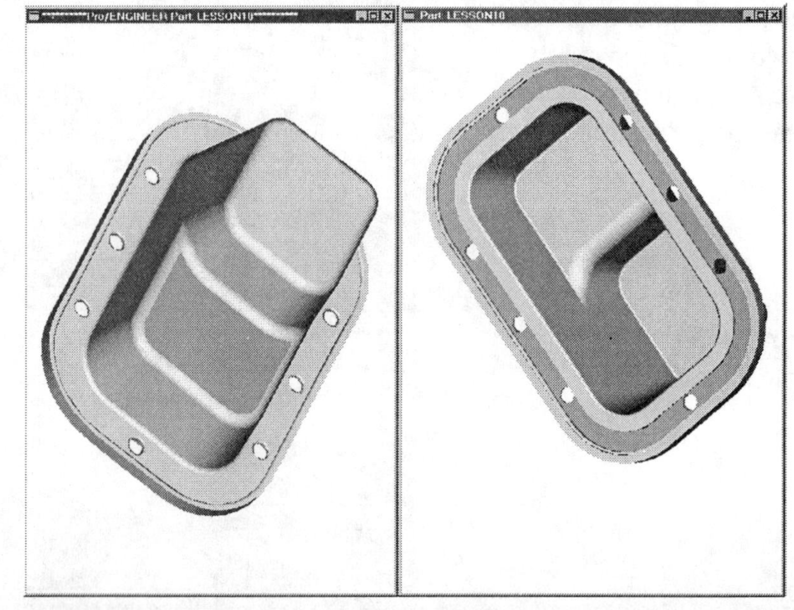

Dbms ⇒ Save As ⇒ enter
⇒ (give a new name)
enter ⇒ Done-Return

Figure 10.37
Resumed Part

ECO

Dbms ⇒ Save ⇒ enter
Purge ⇒ enter ⇒
Done-Return

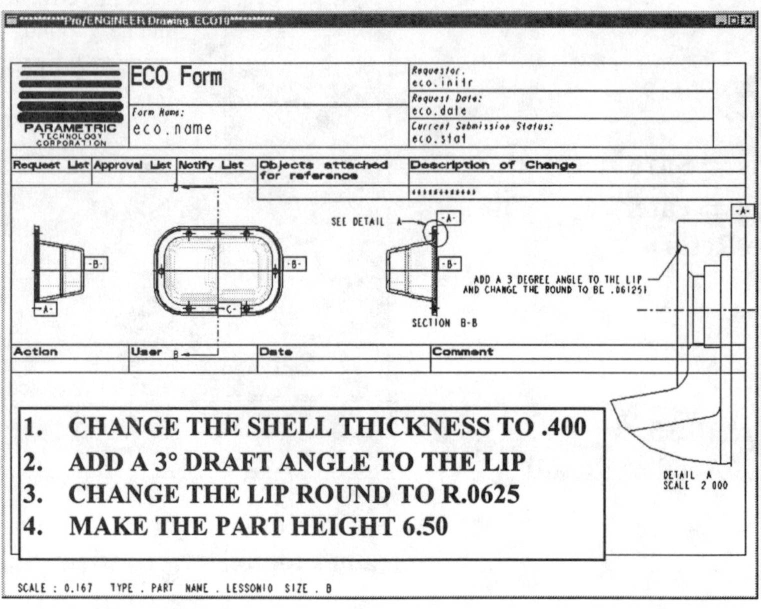

Figure 10.38
ECO

Lesson 10 Project

Cellular Phone Top

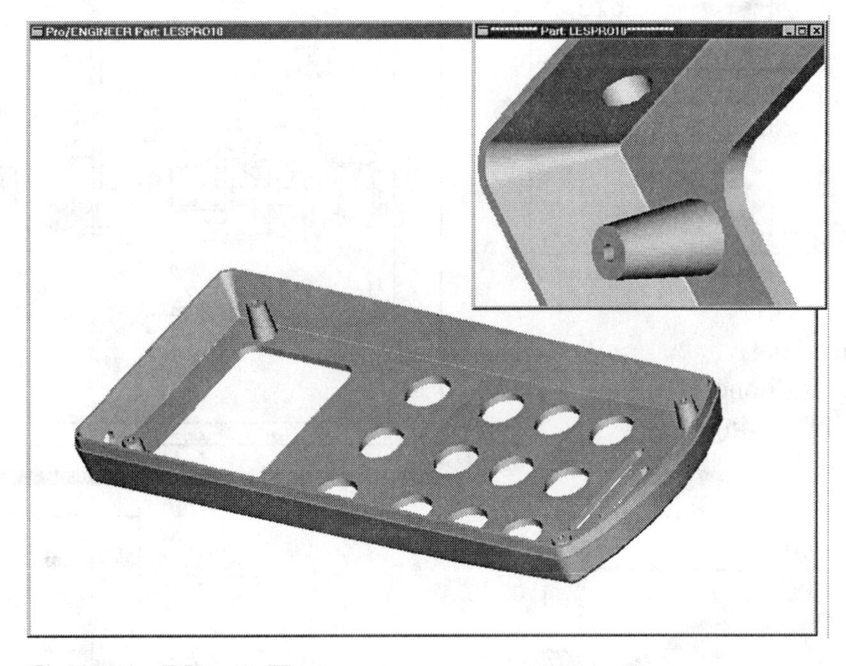

Figure 10.39
Cellular Phone Top

Cellular Phone Top

The Cellular Phone Top (Figure 10.39 through Figure 10.49) is one of two major components for a cellular phone. The other you created as a project in Lesson 9. The part is made of the same plastic as the Cellular Phone Bottom. If time permits, try and assemble the two pieces after completing Lesson 14, Assembly Constraints.

Analyze each part and plan out the steps and features required to model it. Use the **DIPS** in Appendix D to establish a feature creation sequence before the start of modeling. Remember to set up the environment, establish datum planes, and set them on layers.

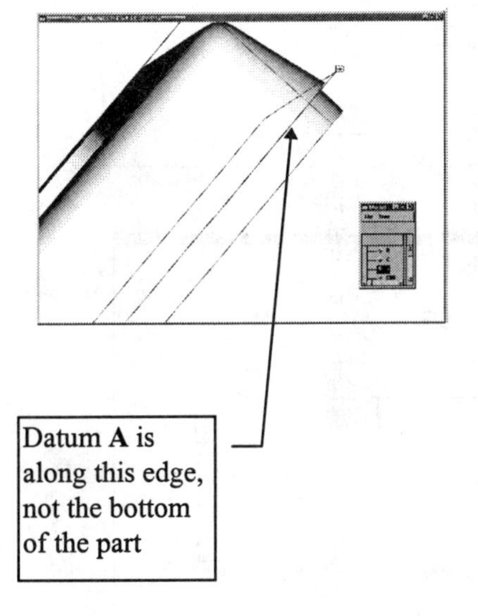

Figure 10.40
Cellular Phone Top Showing Datum Planes and Model Tree

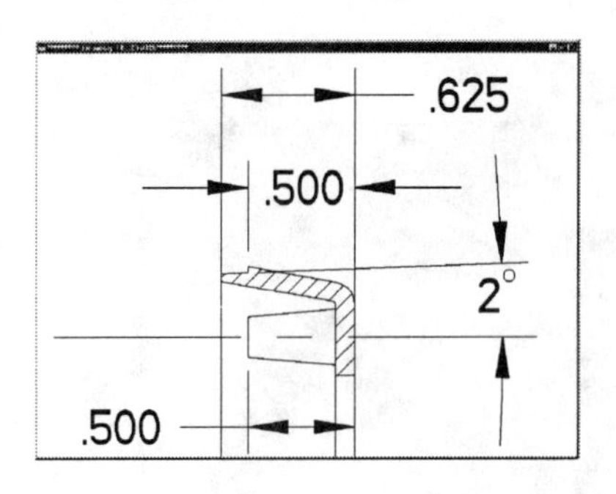

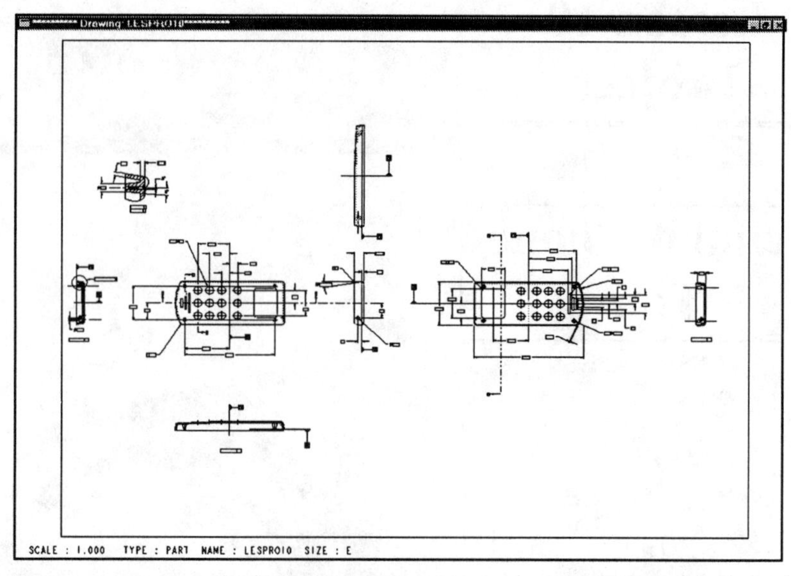

Figure 10.41
Cellular Phone Top, Detail Drawing

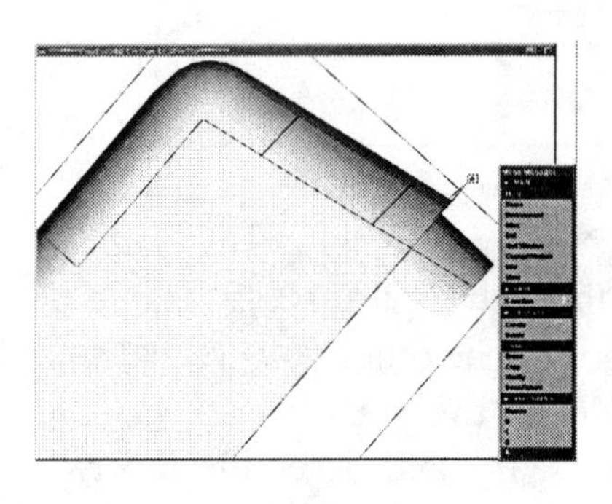

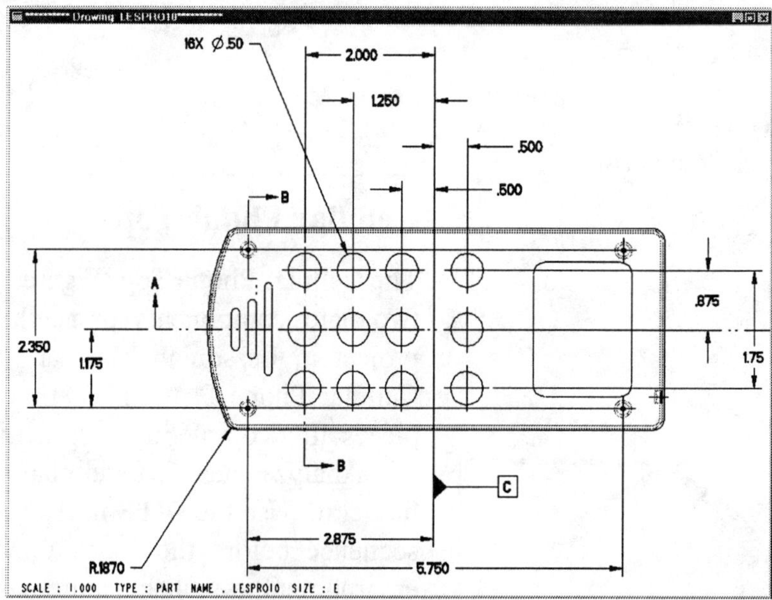

Figure 10.42
Cellular Phone Top, Front View

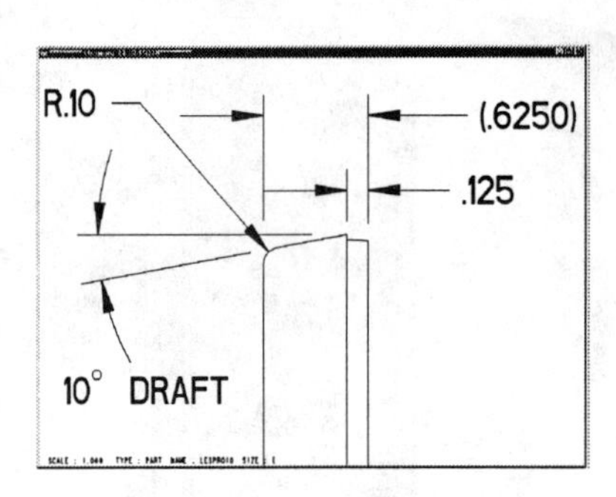

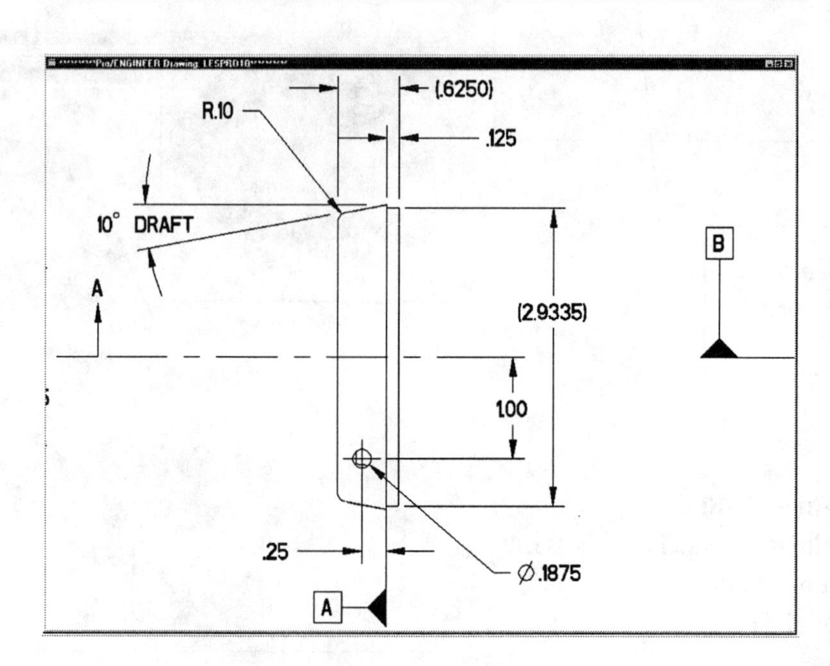

Figure 10.43
Cellular Phone Top, Right Side View

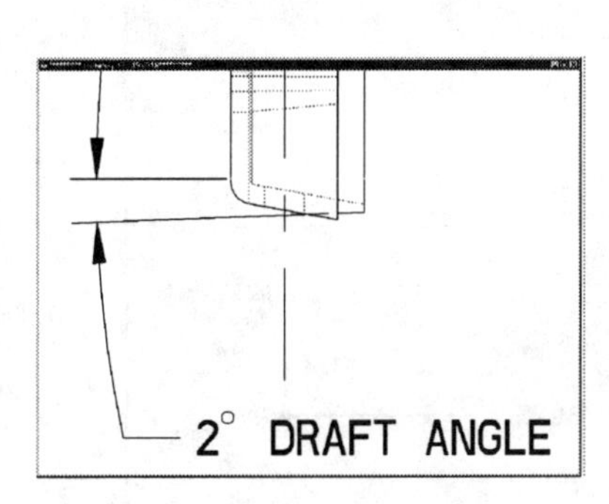

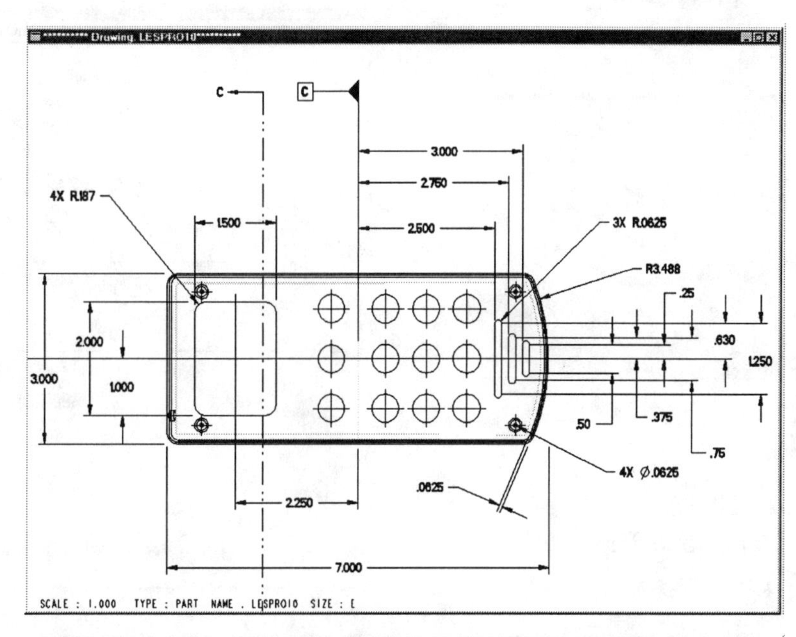

Figure 10.44
Cellular Phone Top,
Bottom View

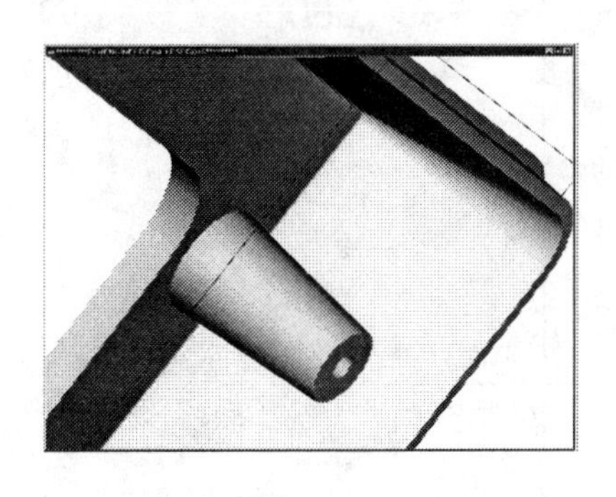

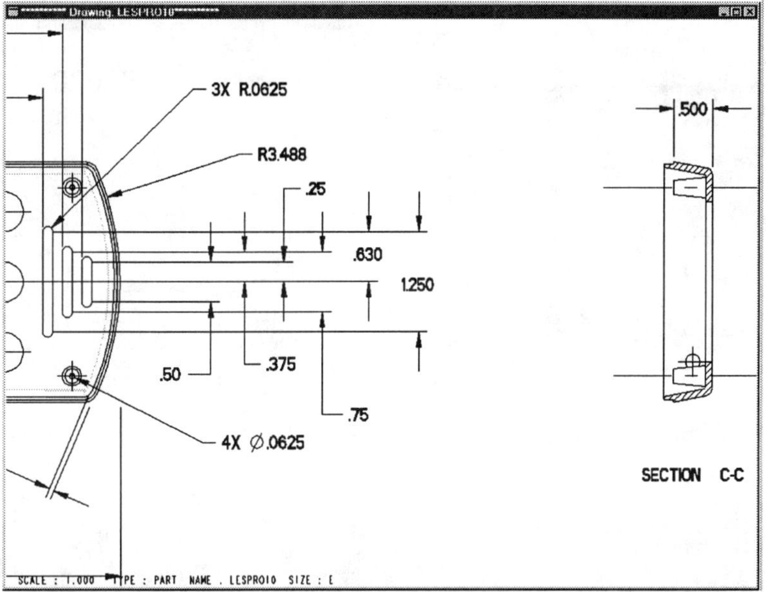

Figure 10.45
Cellular Phone Top,
SECTION C-C

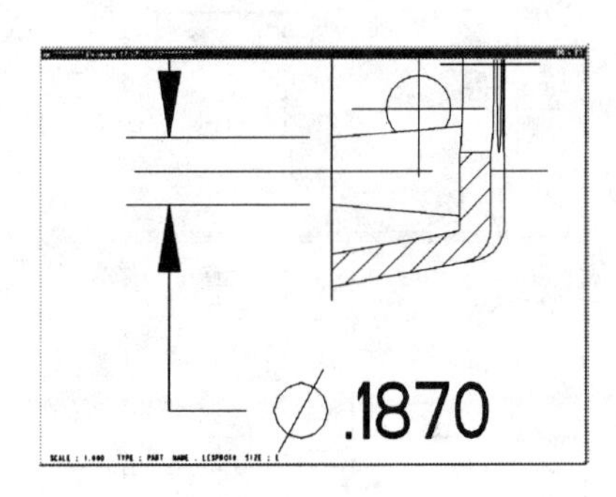

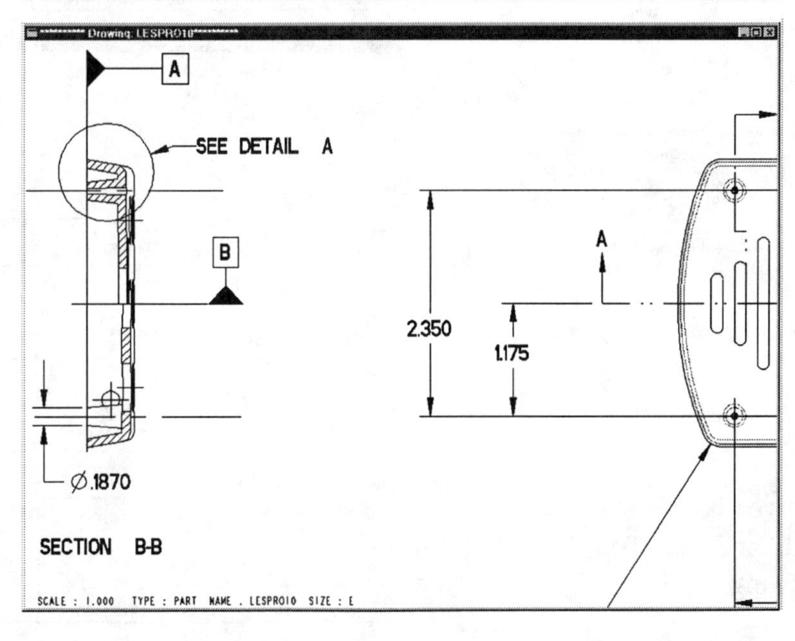

Figure 10.46
Cellular Phone Top,
SECTION B-B

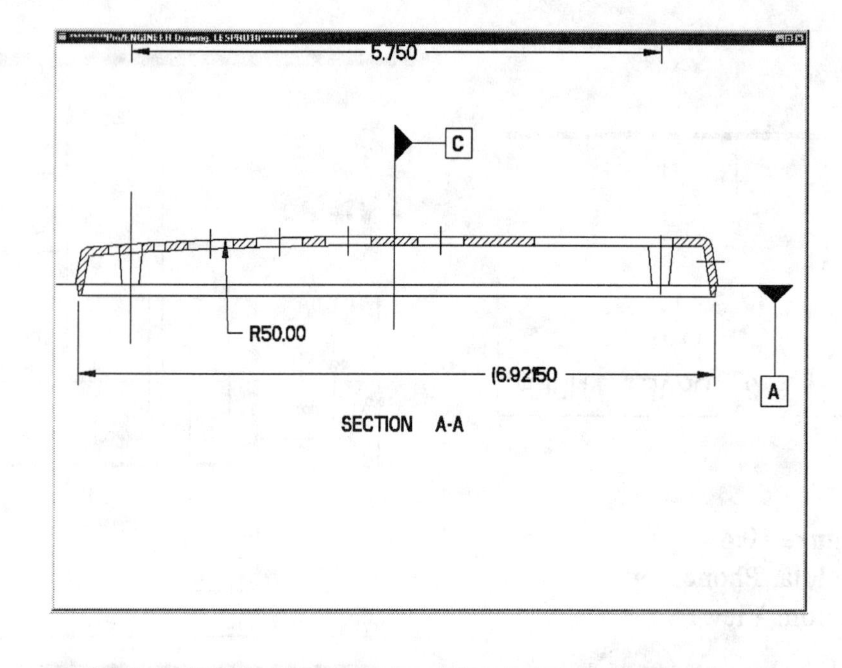

Figure 10.47
Cellular Phone Top,
SECTION A-A

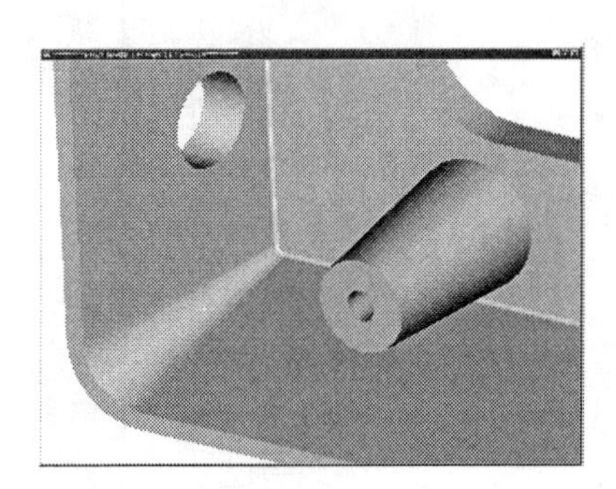

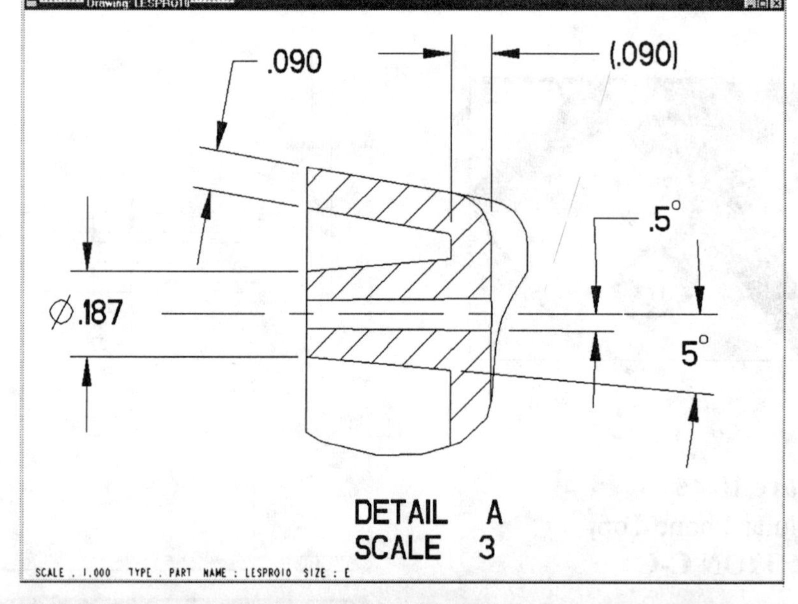

Figure 10.48
Cellular Phone Top,
DETAIL A

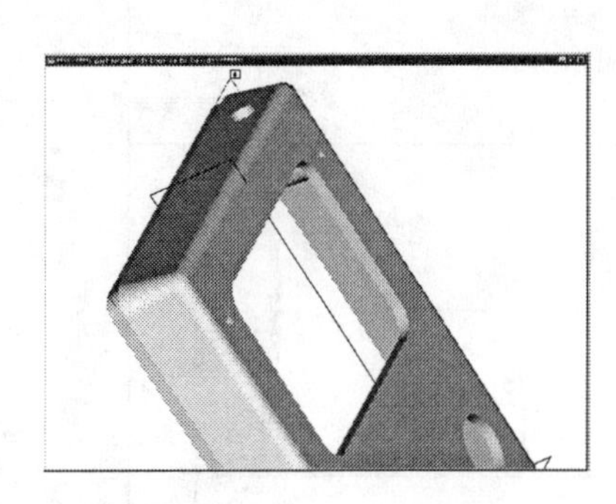

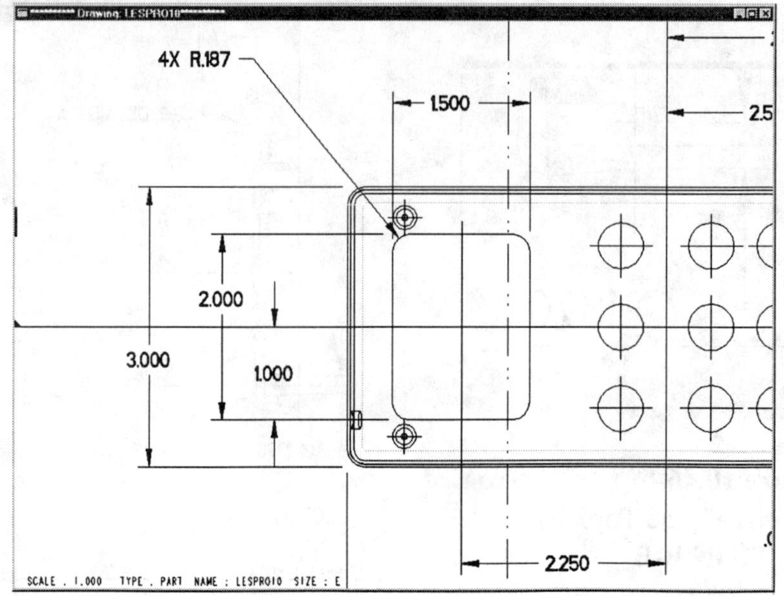

Figure 10.49
Cellular Phone Top,
Opening

Lesson 11

Sweeps

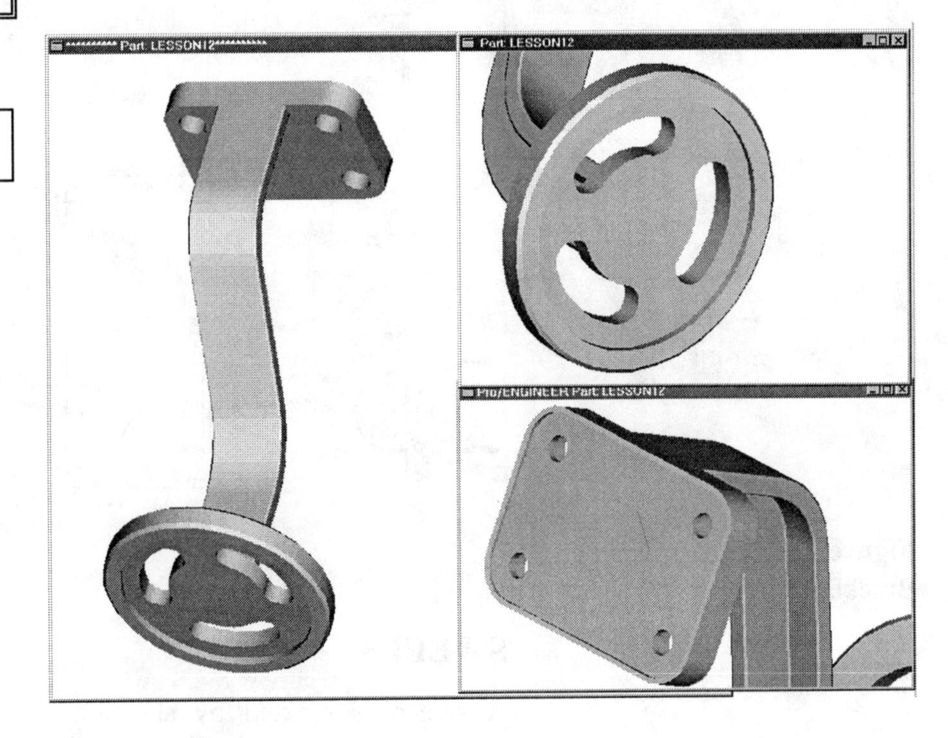

Figure 11.1
Bracket

OBJECTIVES

1. Create a constant section swept feature

2. Sketch a trajectory for a sweep

3. Sketch and locate a sweep section

4. Understand the difference between adding and not adding inner faces

5. Be able to redefine a sweep

6. Understand the difference between a sketched and a selected trajectory

7. Create variable sweeps

EGD REFERENCE
Engineering Graphics and Design
by L. Lamit and K. Kitto
Read Chapter: 20
See Pages: 382, 378

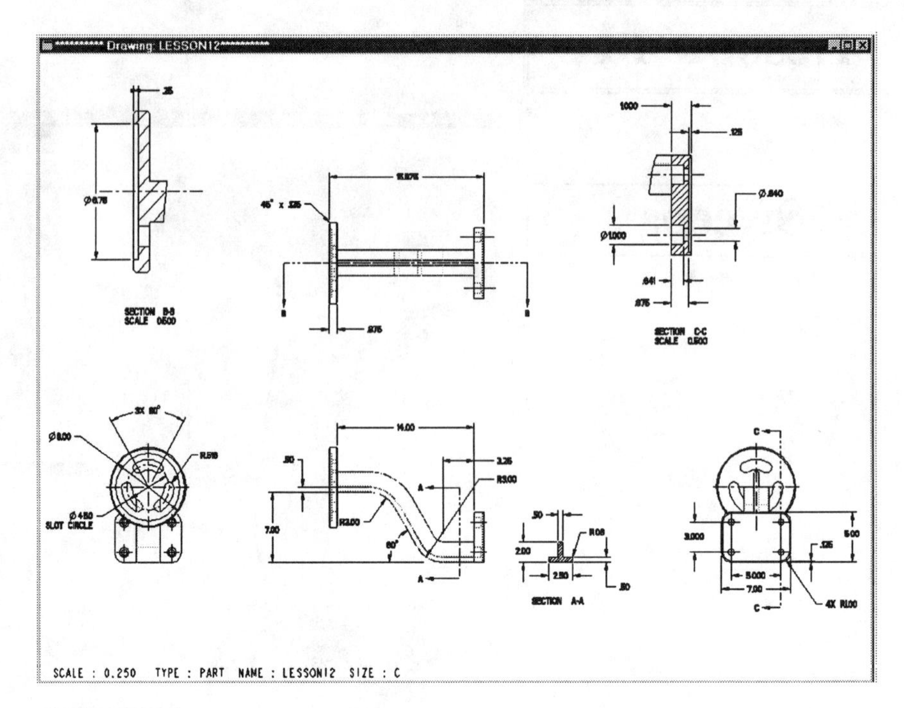

Figure 11.2
Bracket Dctail

SWEEPS

A sweep is created by sketching or selecting a *trajectory*, then sketching a *section* to follow along it. The **Bracket** in Figure 11.1 and Figure 11.2 uses a simple sweep in its design.

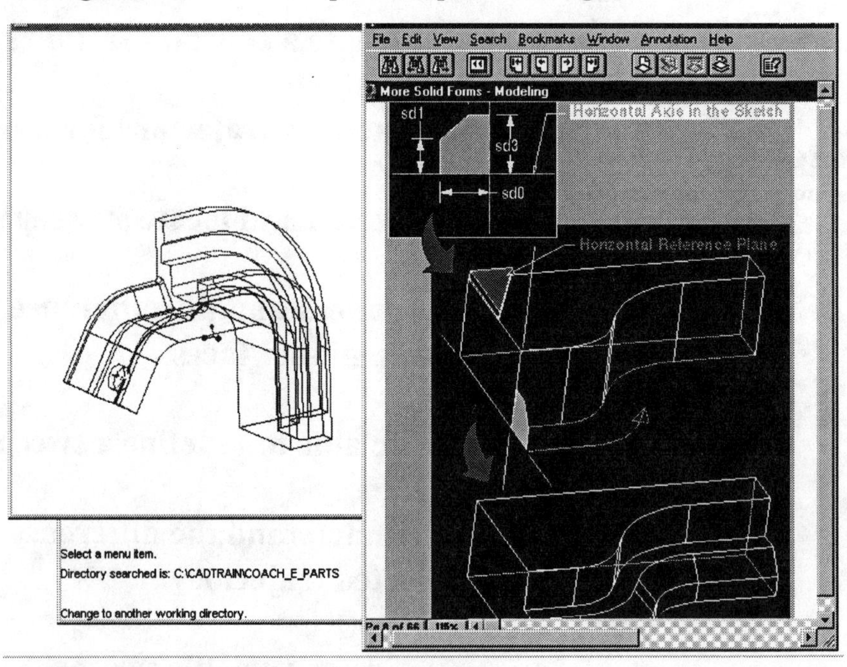

Figure 11.3
COAch for Pro/E, More Solid Features-- Modeling (Segment 1: Sweeps)

Defining a Trajectory

A *constant section sweep* can use either a trajectory sketched at the time of feature creation or a trajectory (Fig. 11.3) made up of selected datum curves or edges. The trajectory must have adjacent reference surfaces or be planar (Fig. 11.4). When defining a sweep (Fig. 11.5), Pro/E checks the specified trajectory for validity and establishes normal surfaces. When ambiguity exists, Pro/E prompts you to select a normal surface.

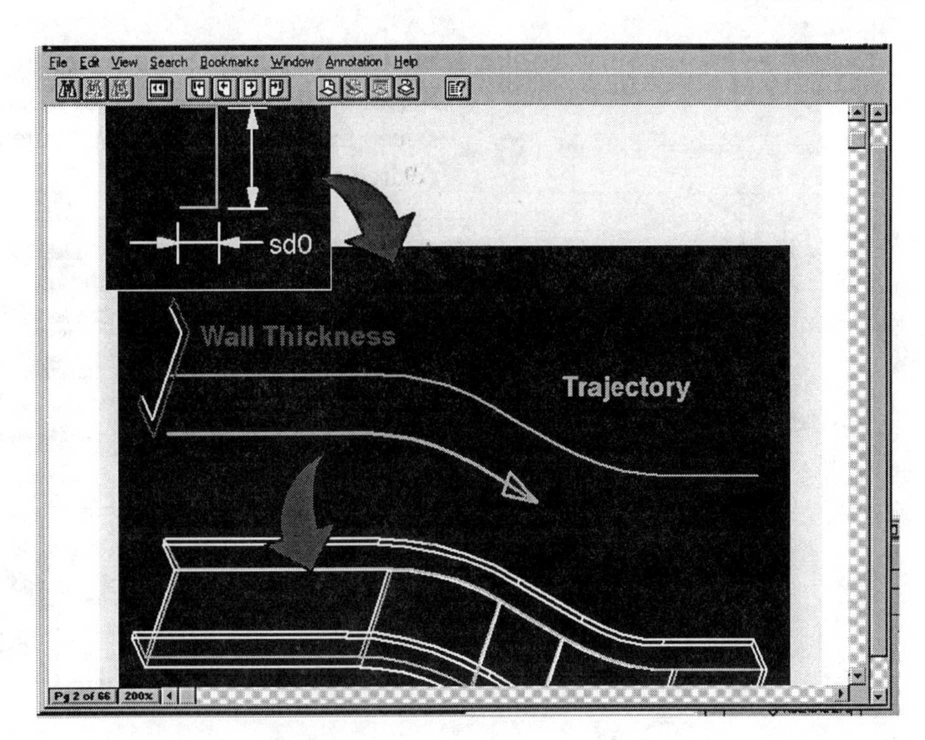

Figure 11.4
COAch for Pro/E, More Solid Features-- Modeling (Segment 1: Sweeps)

Creating a Swept Feature

To create a sweep, do the following:

1. **Feature ⇒ Create ⇒ Solid ⇒ Protrusion.**
2. Choose **Sweep** and **Done** from the SOLID OPTS menu.
3. The feature creation dialog box for sweeps is displayed.

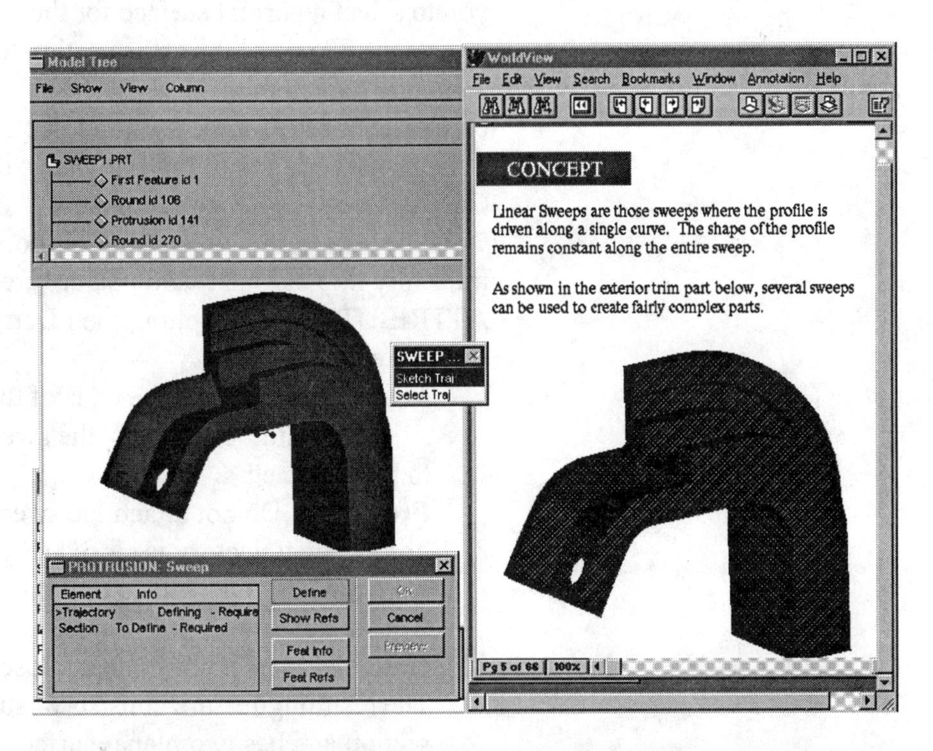

Figure 11.5
COAch for Pro/E, More Solid Features-- Modeling (Segment 1: Sweeps)

4. Sketch or select an open or closed trajectory (Fig. 11.6) using a SWEEP TRAJ menu option. The following options are available:

Sketch Traj Sketch the sweep trajectory using Sketcher mode.
Select Traj Select a chain of existing curves or edges as the sweep trajectory. The CHAIN menu allows you to select the desired trajector.

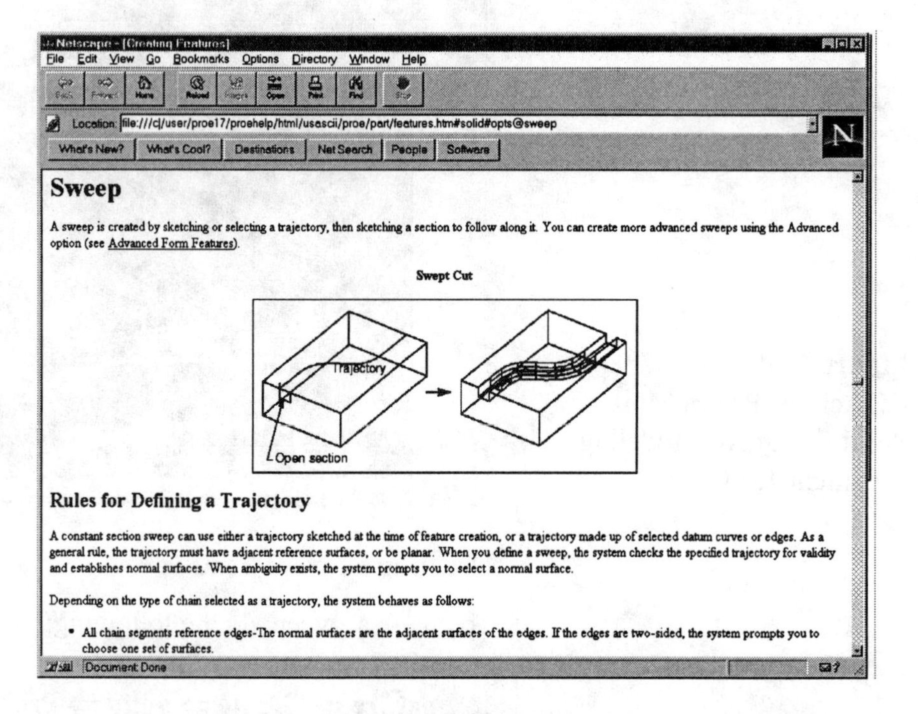

Figure 11.6
Online Documentation
Sweeps

5. If the trajectory lies in more than one surface, such as a trajectory defined by a datum curve created using **Intr Surfs**, Pro/E prompts you to select a normal surface for the sweep cross section. Pro/E orients the **Y** axis of the cross section to be normal to this surface along the trajectory.
6. Create or retrieve the section to be swept along the trajectory, and dimension it relative to the *crosshairs* displayed on the trajectory. Choose **Done**.
7. If the trajectory is open (the startpoint and endpoint of the trajectory do not touch) and you are creating a solid sweep, choose an ATTRIBUTES menu option, then **Done**. The possible options are as follows:

Merge Ends Merge the ends of the sweep, if possible, into the adjacent solid. To do this, the sweep endpoint must be attached to part geometry.
Free Ends Do not attach the sweep end to adjacent geometry.

8. If the sweep trajectory is closed (Fig. 11.7), choose one of the following SWEEP OPT menu options and then **Done**:

Add Inn Fcs For open sections, add top and bottom faces to close the swept solid (planar, closed trajectory, and open section). The resulting feature consists of surfaces created by sweeping the section and has two planar surfaces that cap the open ends.
No Inn Fcs Do not add top and bottom faces.

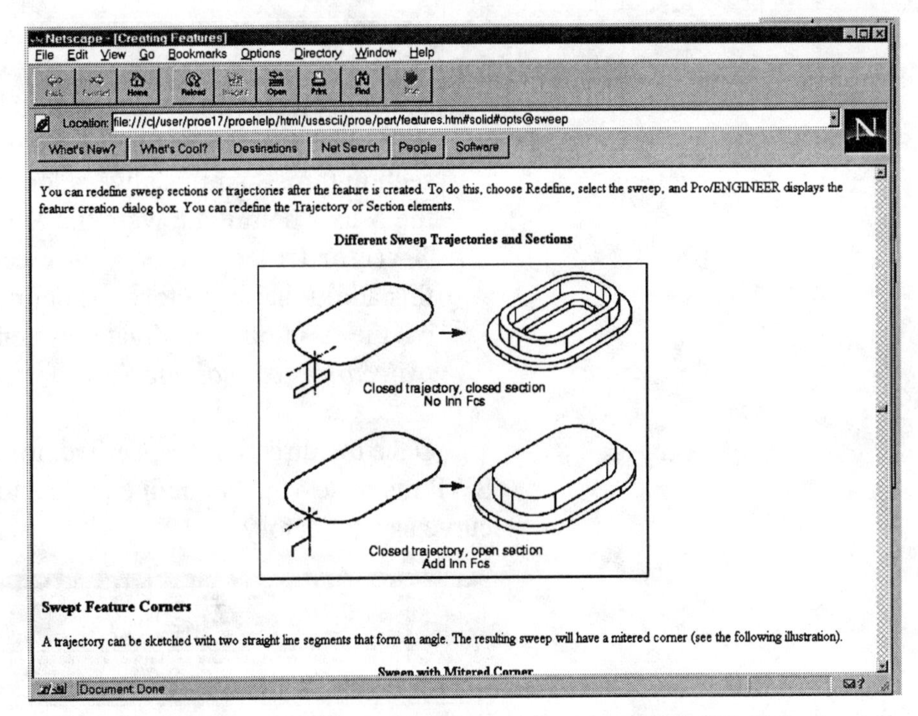

Figure 11.7
Online Documentation
Sweep Trajectories and Sections-- **Add Inn Fcs** and **No Inn Fcs**

9. Choose **Flip**, if desired, then **Okay** from the DIRECTION menu to select the side on which to remove material for swept cuts.
10. Pro/E issues a message stating that all the elements have been defined. If desired, select one of the buttons in the dialog box.
11. Select the **OK** button in the dialog box to create the sweep.

To redefine sweep sections or trajectories after the feature is created, choose **Redefine** and then select the sweep. Pro/E will display the feature creation dialog box. You can redefine the trajectory or section elements.

Variable Section Sweeps

A solid sweep feature using one or more longitudinal trajectories and a single variable section can also be created (Fig. 11.8). The parameters of the section can vary as the section moves along the sweep trajectories. These sweeps are called **variable section sweeps**.

Every variable section sweep requires one *longitudinal "spine" trajectory*. You can define a sweep for which the **X** axis of the section follows the **X**-vector trajectory while remaining normal to the spine at all times as it sweeps along the spine (**Nrm to Spine**). To do so, you must also specify an **X**-vector trajectory to orient the section as it sweeps along the spine The section plane is always normal to the spine trajectory at the point of their intersection. The **X** axis of each section's coordinate system is defined by the intersection of the plane and the spine to the point of intersection of the plane and the **X**-vector trajectory for that section. You can also define a variable section sweep for which the **Y** axis of the section remains constant. The section will follow the spine such that it is normal to the selected pivot plane. The **X** axis and **Z** axis will still follow the spine and **X**-vector trajectories.

Variable section sweeps need the following trajectories:

Spine trajectory The trajectory along which the section is swept (Fig. 11.9). If you choose **Nrm To Spine**, the origin of the section (crosshairs) is always located on the spine trajectory, with the **X** axis pointing toward the **X**-vector trajectory.
X-vector trajectory Sweeps created using **Nrm To Spine** need this additional trajectory. It defines the orientation of the **X** axis of the section coordinate system. *The X-vector and spine trajectories cannot intersect.*

Once the direction is specified, the system displays the VAR SEC SWP menu so you can define a trajectory. You can use a composites curve as a trajectory.

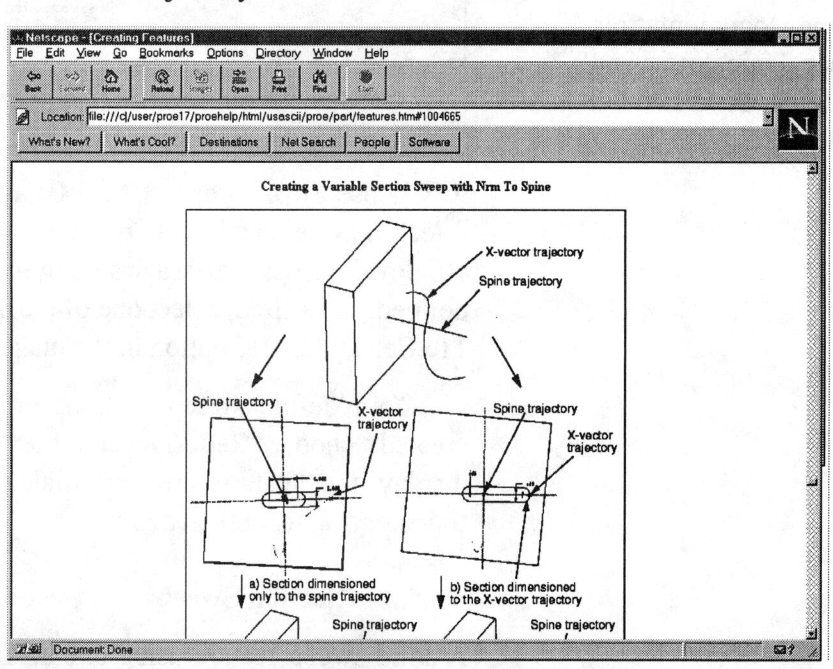

Figure 11.8
Online Documentation
Variable Section Sweeps

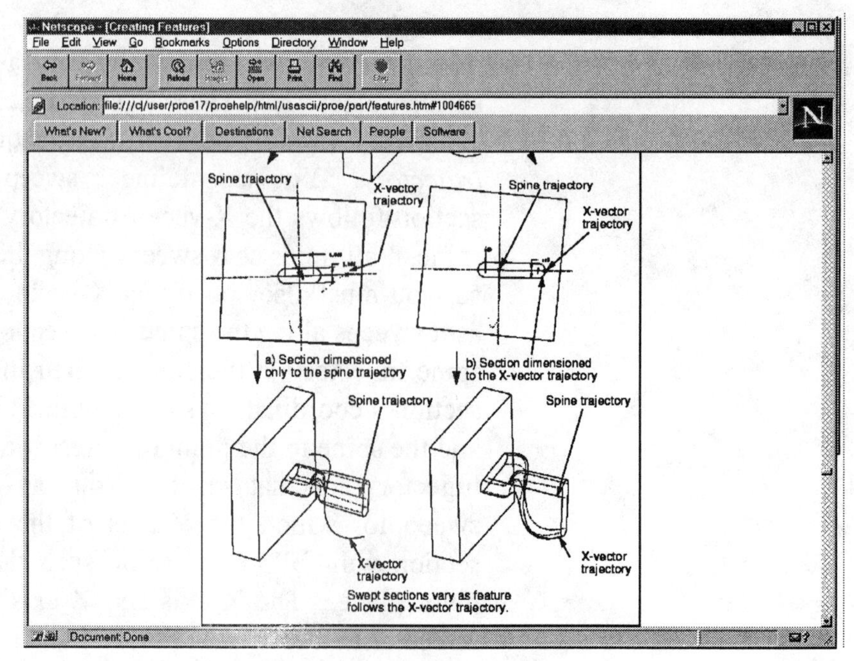

Figure 11.9
Online Documentation
Spine Trajectories

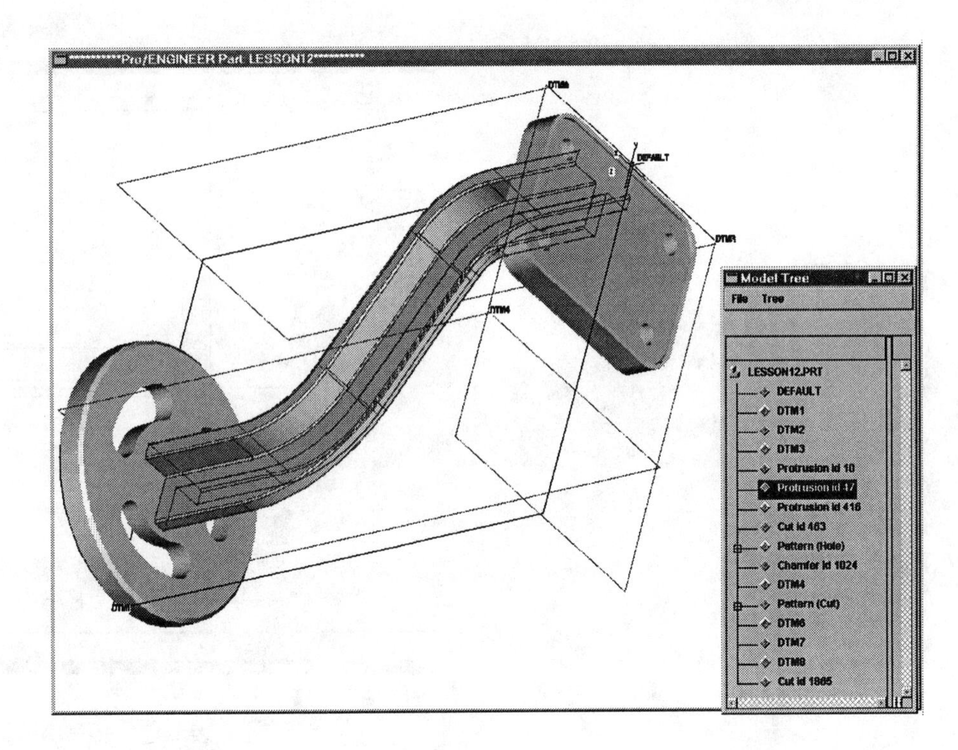

Figure 11.10
Bracket Showing Datum Planes and Coordinate System and Model Tree

Bracket

The **Bracket** (Figure 11.10 through Figure 11.39) requires the use of the **Sweep** command. The T-shaped section is swept along the selected *trajectory*. The protrusions on both sides of the swept feature are to be created with the dimensions given in Figures 11.11 through 11.17. Step-by-step commands are provided only for the sweep trajectory and the its crosssection.

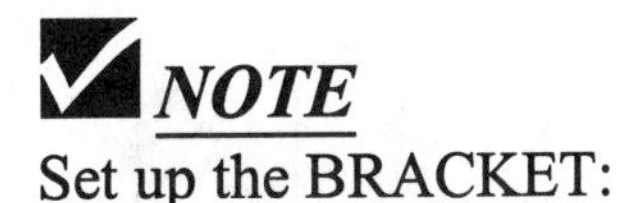

NOTE

Set up the BRACKET:

- Material = Cast Iron
- Units = Inches
- Default Datum Planes
- Default Coordinate System
- Layers = **DATUM_PLANES**
- Hidden Line
- ✓Grid Snap

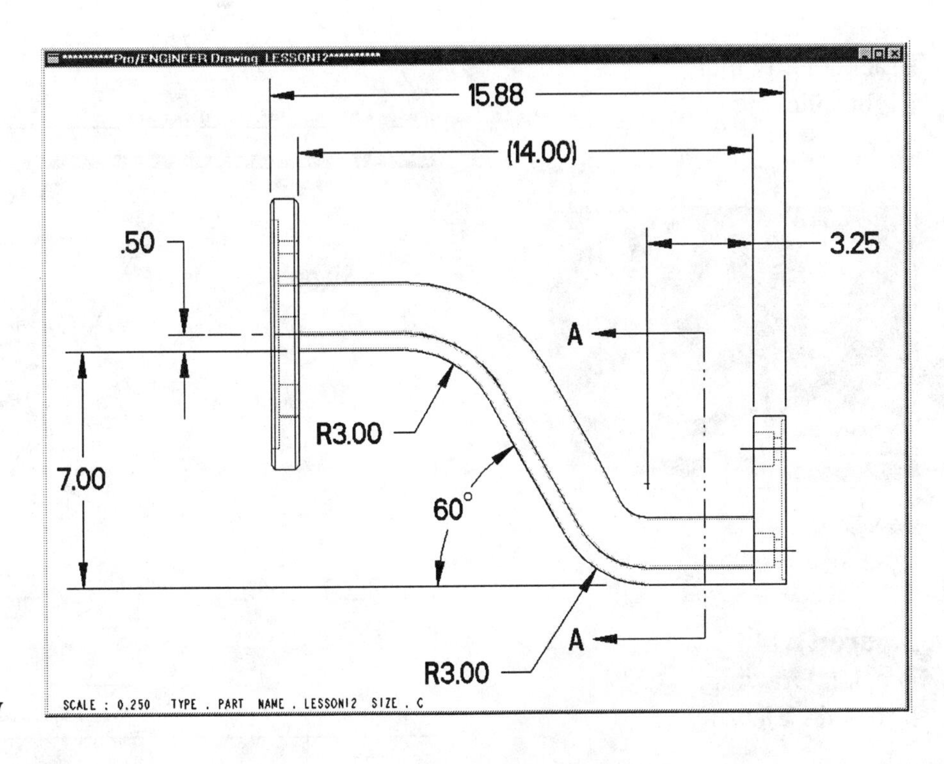

Figure 11.11
Bracket Drawing, Front View

Pro/ENGINEER Drawing: LESSON12

15.875

45° X .125

.875

Figure 11.12
Bracket Drawing,
Top View

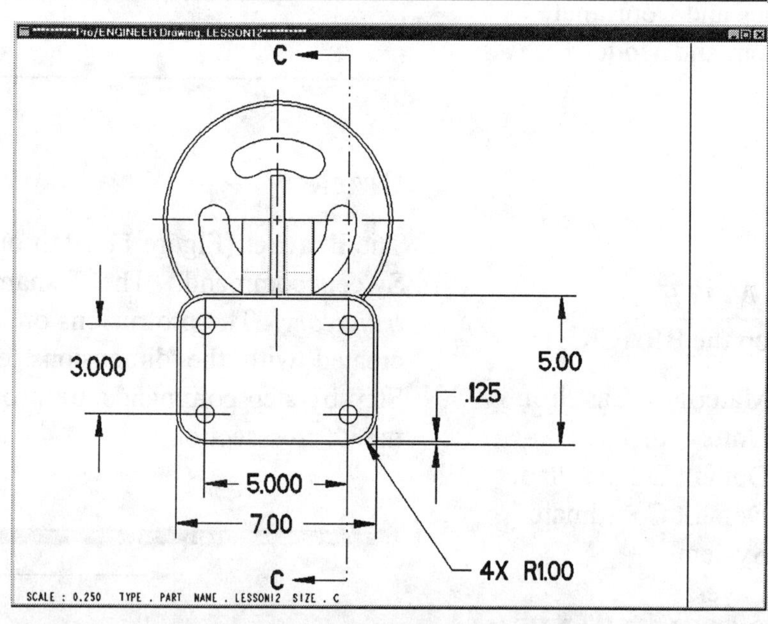

Figure 11.13
Bracket Drawing,
Right Side View

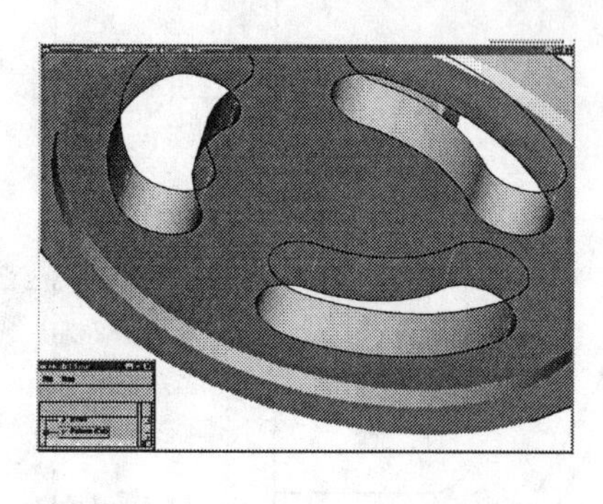

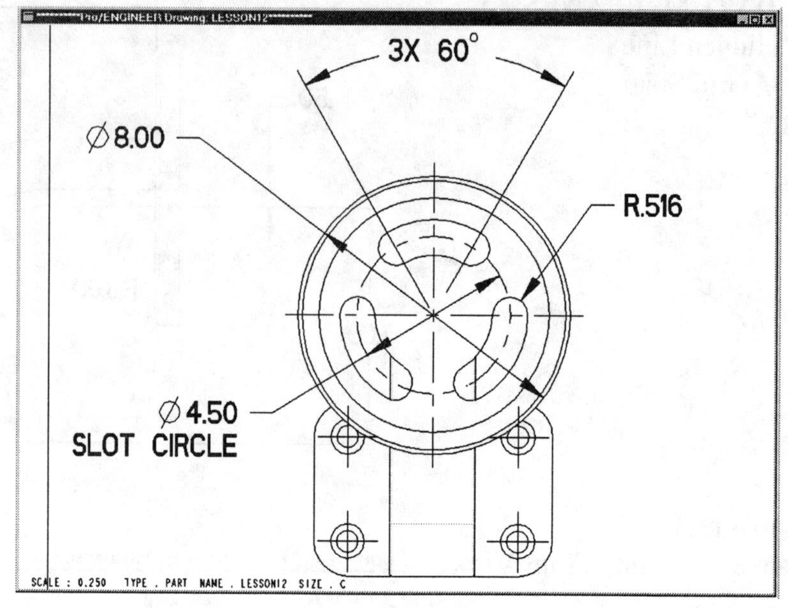

Figure 11.14
Bracket Drawing,
Left Side View

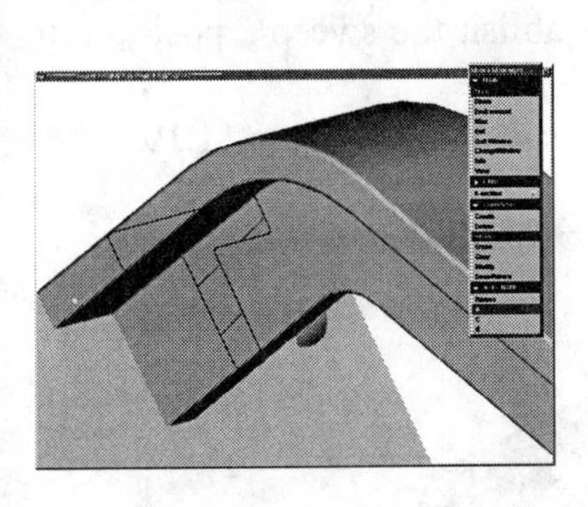

**********Pro/ENGINEER Drawing LESSON12**********

.50

R.06

2.00

2.50

.50

SECTION A-A

SCALE : 0.250 TYPE . PART NAME . LESSON12 SIZE . C

Figure 11.15
Bracket Drawing,
SECTION A-A

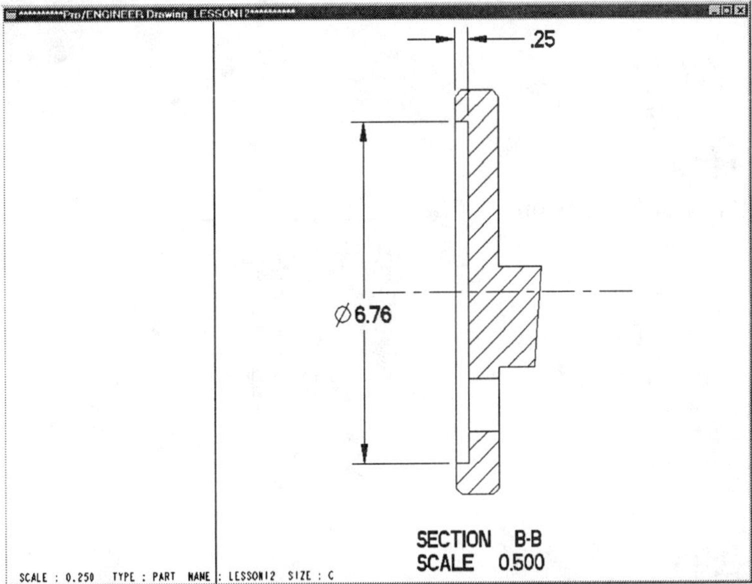

Figure 11.16
Bracket Drawing,
SECTION B-B

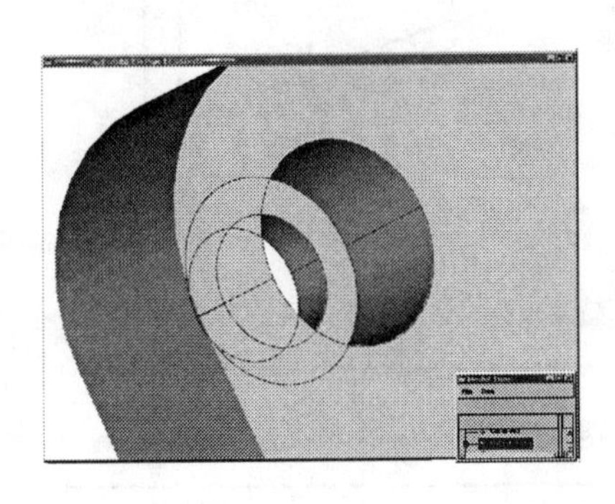

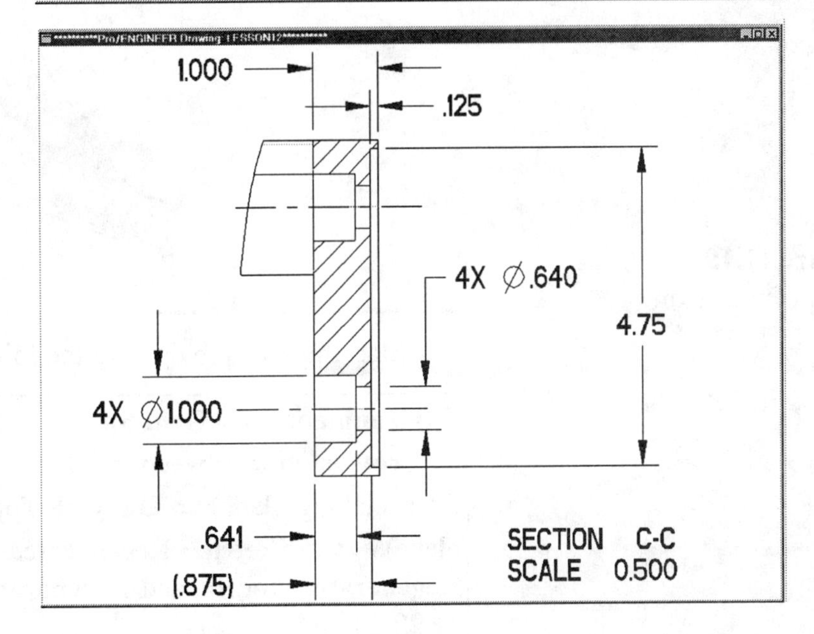

Figure 11.17
Bracket Drawing,
SECTION C-C

The Bracket is started by modeling the protrusion shown in Figure 11.18. This protrusion will be used to establish the sweep's position in space. Sketch the protrusion on **DTM3**.

The second protrusion is a sweep and is shown in Figure 11.19.

Dbms ⇒ Save ⇒ enter
Purge ⇒ enter ⇒
Done-Return

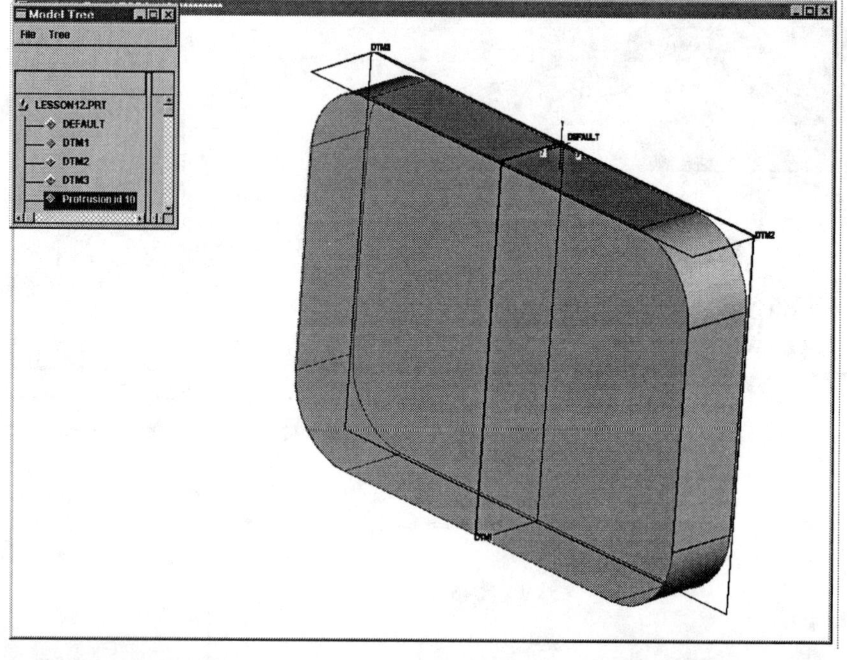

Figure 11.18
Bracket's First Protrusion

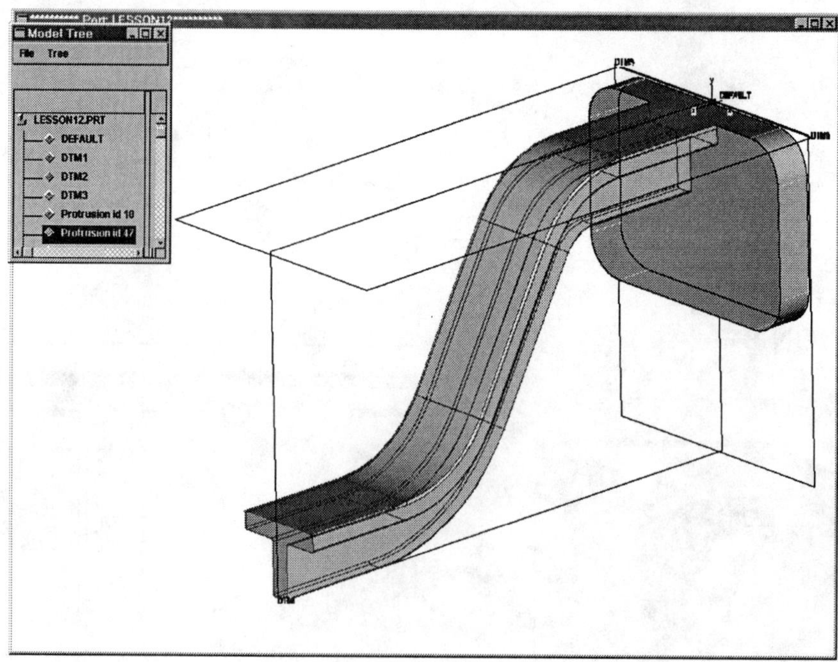

Figure 11.19
Swept Protrusion

Start the sweep by giving the following commands:

Environment ⇒ □ **Grid Snap** ⇒ **Done-Return** ⇒ **Feature** ⇒ **Create** ⇒ **Protrusion** ⇒ **Sweep** ⇒ **Done** ⇒ **Sketch Traj** ⇒ (pick **DTM1** as the sketching plane) ⇒ **Okay** ⇒ **Top** ⇒ (pick **DTM2** as the orientation plane) ⇒ (**Sketch**, **Regenerate, Dimension**, **Regenerate, Align**, **Regenerate**, **Modify**, and **Regenerate** the trajectory as shown in Figures 11.20 through 11.25)

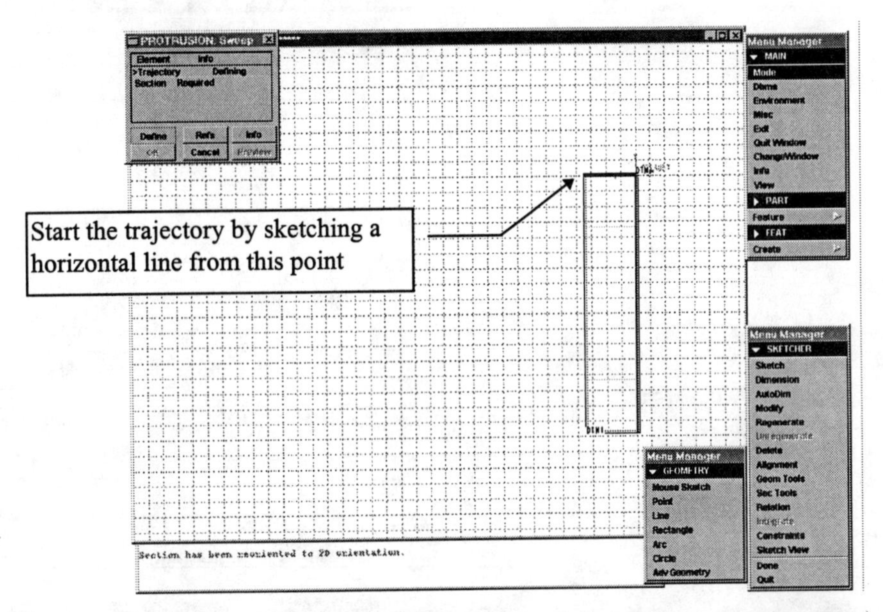

Figure 11.20
Starting the Sweep Trajectory Sketch

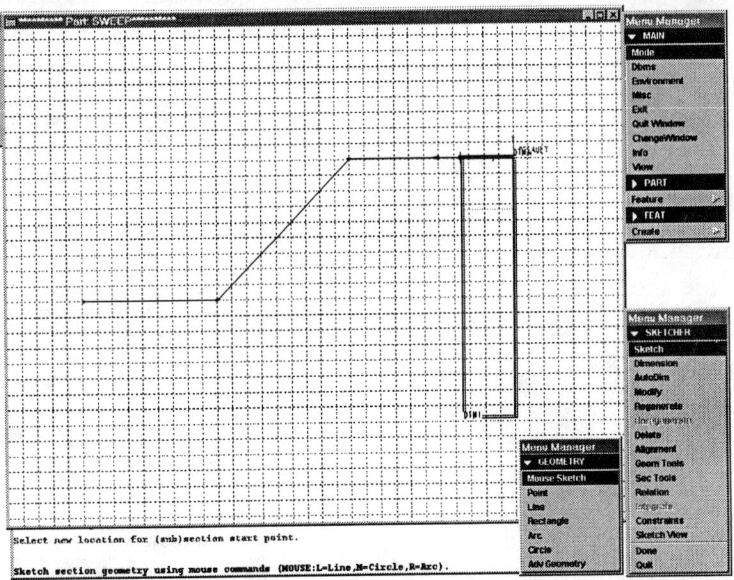

Figure 11.21
Sketch the Three Lines

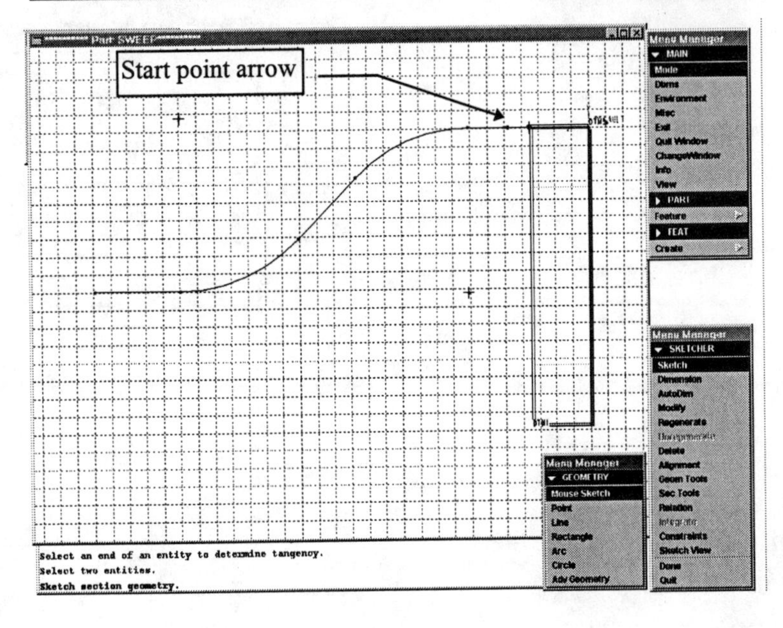

Figure 11.22
Add the Arc Fillets

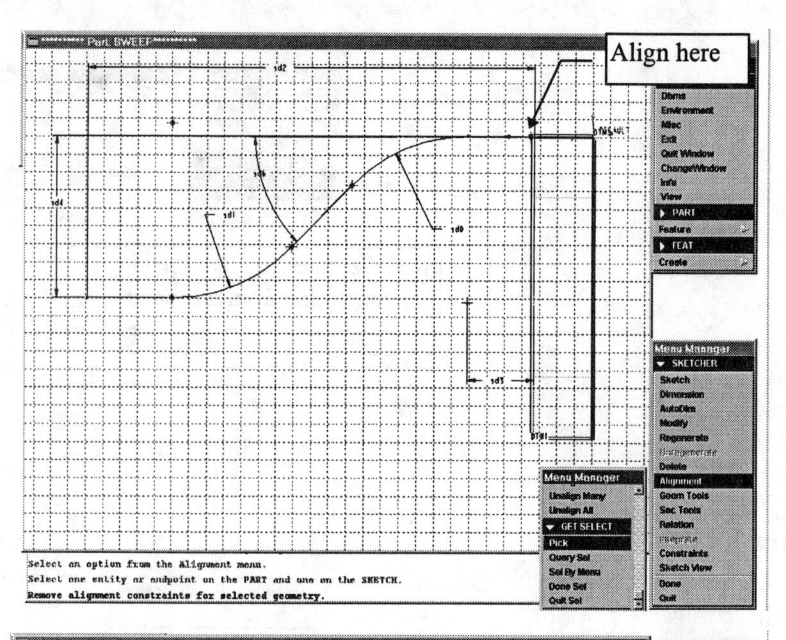

Figure 11.23
Align and **Dimension** the Sketch

HINT

Zoom out and regenerate again if the sketch fails to regenerate the first time.

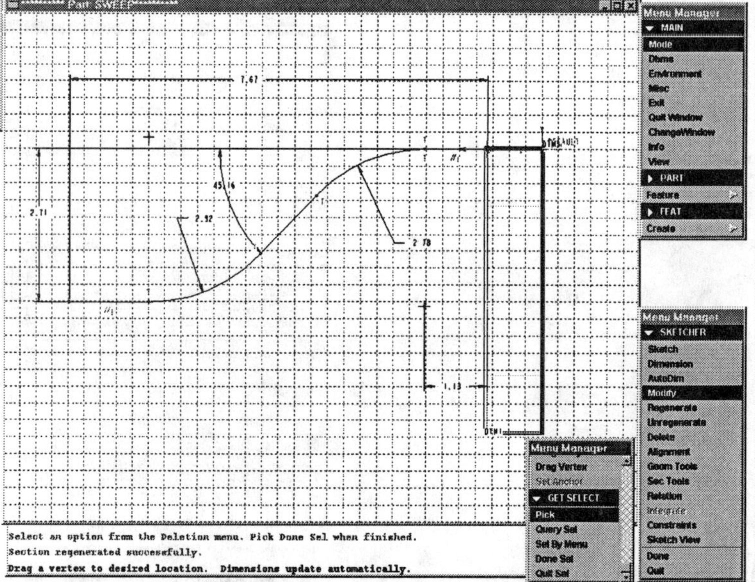

Figure 11.24
Regenerate the Sketch

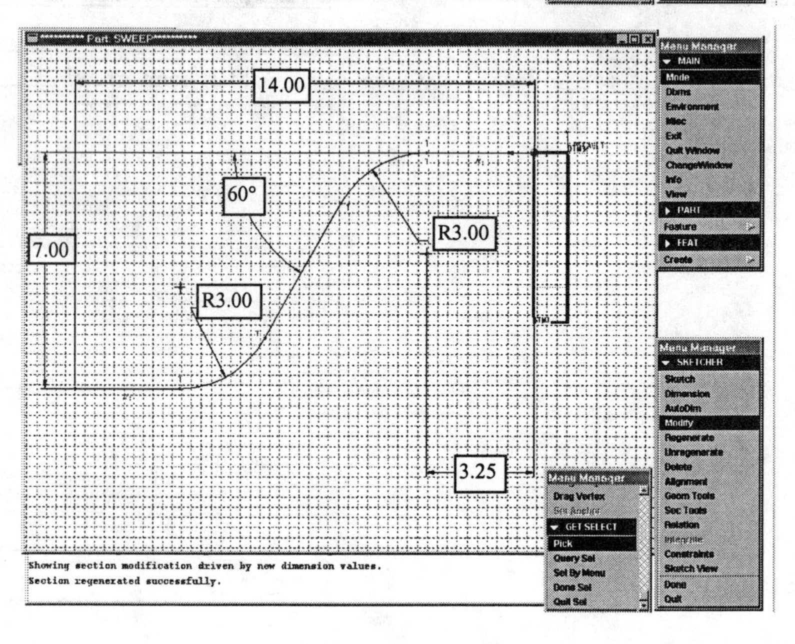

Figure 11.25
Modify and **Regenerate** the Sketch

After the trajectory has regenerated, complete the sweep with the following commands, as shown in Figure 11.28 through Figure 11.33:

Regenerate ⇒ Done ⇒ Free Ends ⇒ Done ⇒ (now sketch a section as shown in Figures 11.26 and 11.27) ⇒ (add the *eight* fillets to the sketch) ⇒ **Alignment** (align the vertical line of the sketch with **DTM2** and align the horizontal centerline to **DTM1)** ⇒ **Dimension ⇒ Regenerate ⇒ Modify ⇒ Regenerate ⇒ Done ⇒ Preview ⇒ Okay**

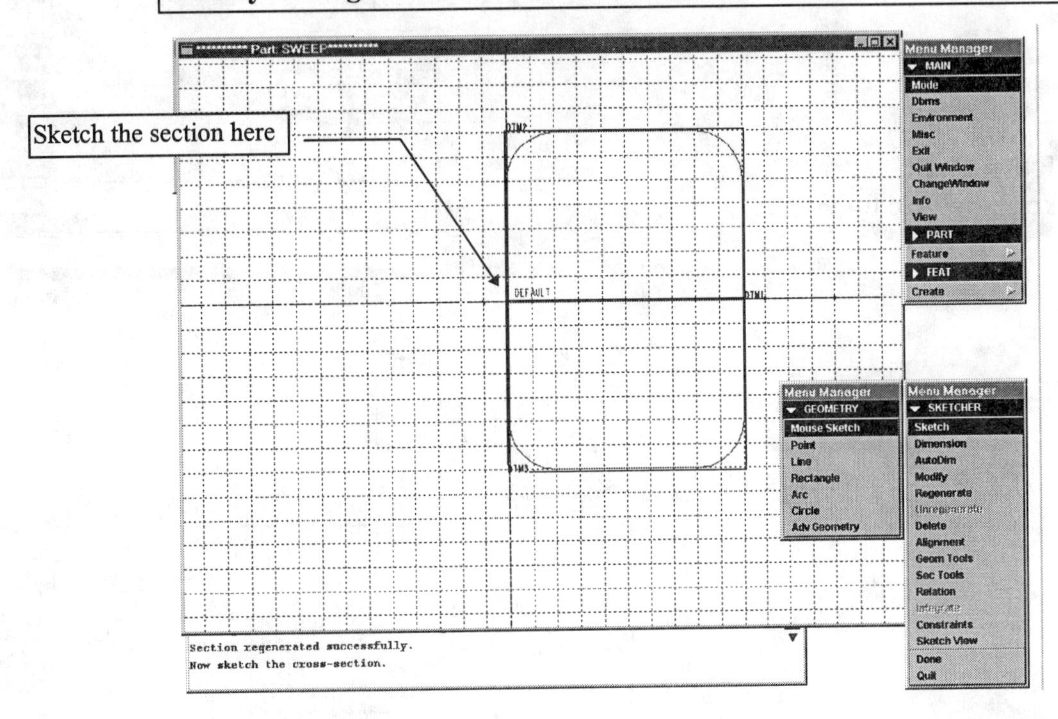

Figure 11.26
Sketch the Section

HINT
Change the grid size
Sketch ⇒ Sect Tools ⇒ Sec Environ ⇒ Grid ⇒ Params ⇒ X&Y Spacing ⇒ (type **.25** at prompt) ⇒ **enter**

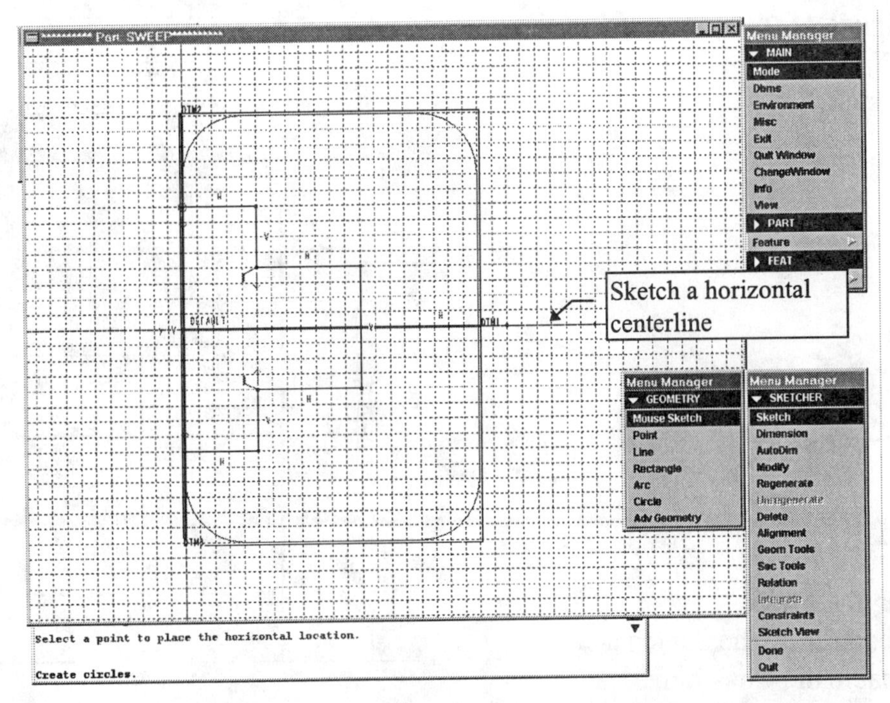

Figure 11.27
Sketch the Eight Lines of the Closed Section and a Centerline

Figure 11.28
Sketch the Arc Fillets

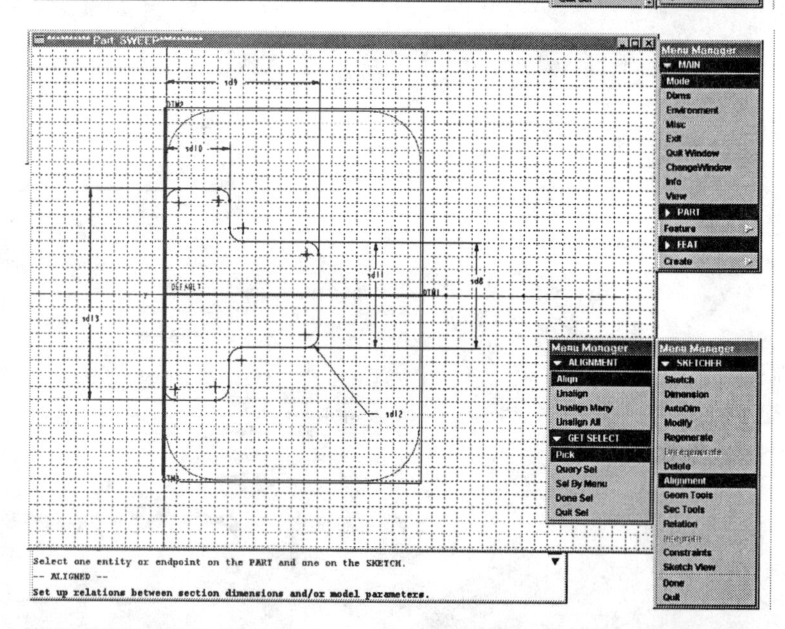

Figure 11.29
Align and **Dimension** the Sketch

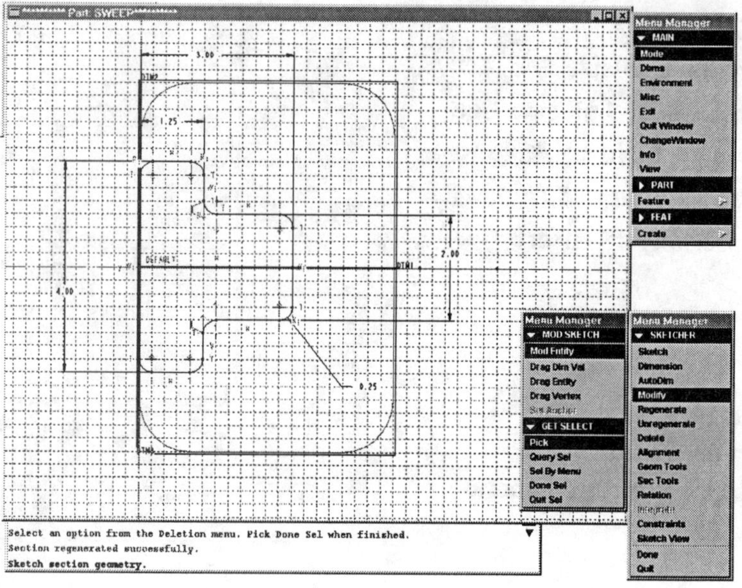

Figure 11.30
Regenerate the Sketch and Enable the Constraints

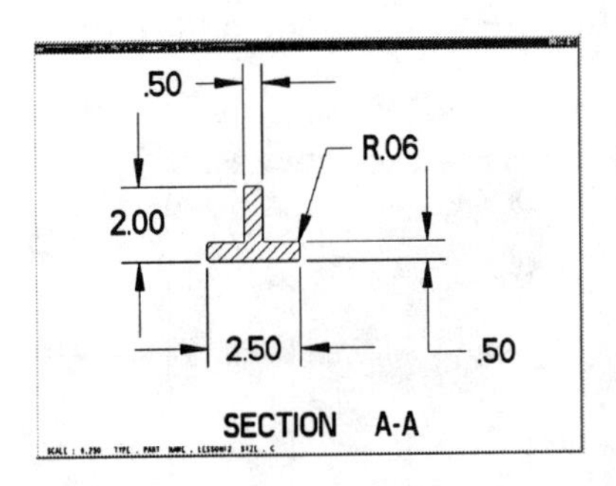

Figure 11.31
Modify and **Regenerate**

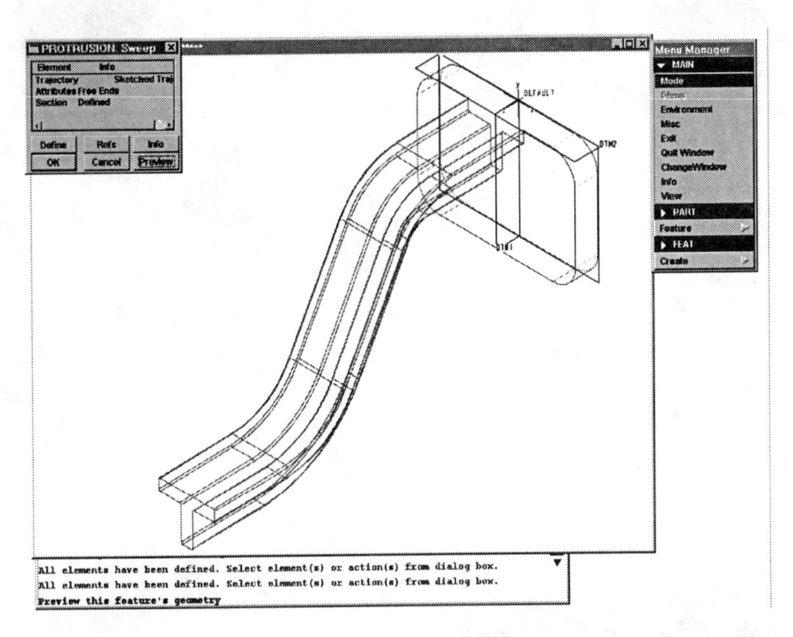

Figure 11.32
Preview

Dbms ⇒ Save ⇒ enter
Purge ⇒ enter ⇒
Done-Return

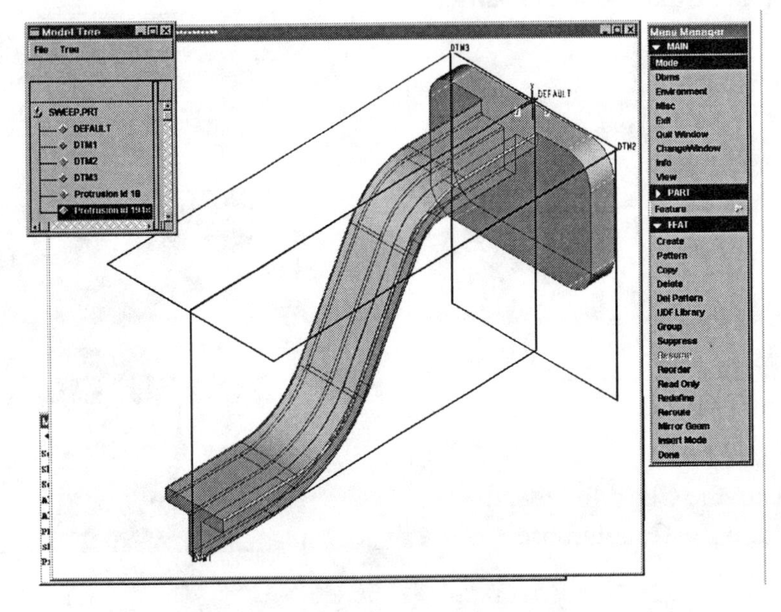

Figure 11.33
Completed Sweep

Model the third protrusion (Fig. 11.34), make the cuts (Figure 11.35 and Figure 11.36), and create and pattern the counterbore holes (Fig. 11.36). Complete the part by modeling the chamfers and the slots (Figure 11.37 through Figure 11.39).

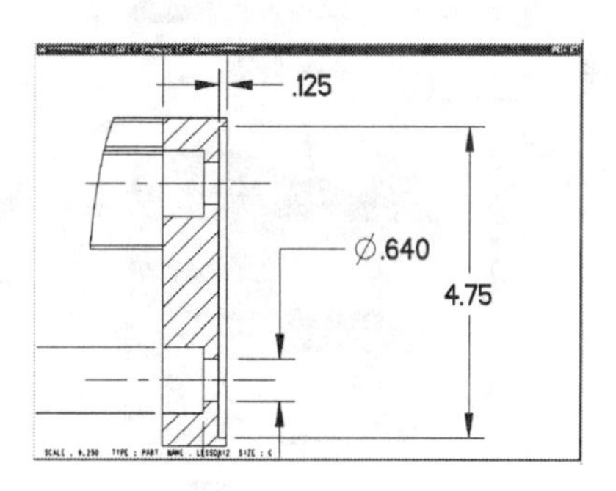

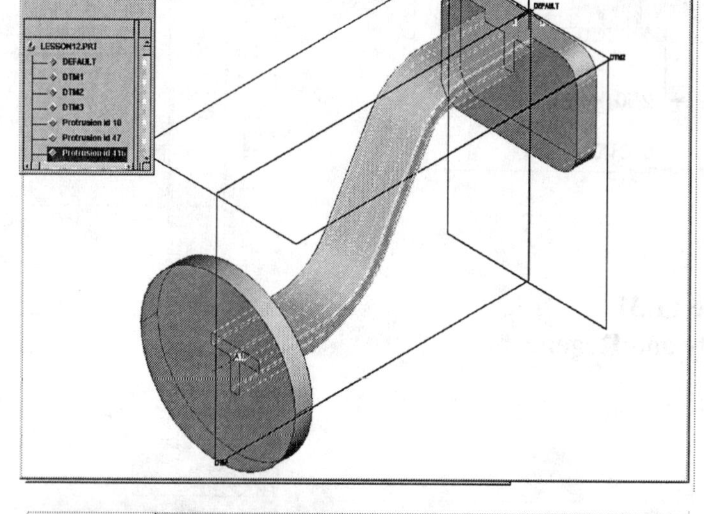

Figure 11.34
Third Protrusion

Dbms ⇒ Save ⇒ enter
Purge ⇒ enter ⇒
Done-Return

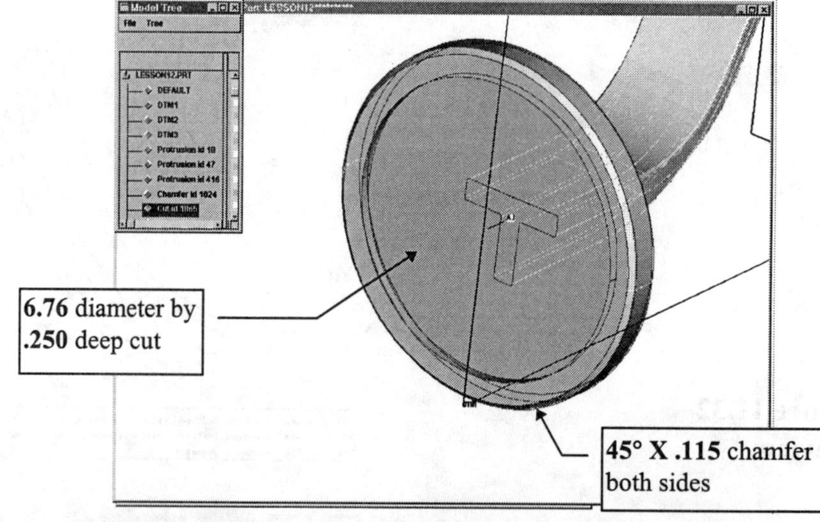

Figure 11.35
Add the Cuts and the Chamfers

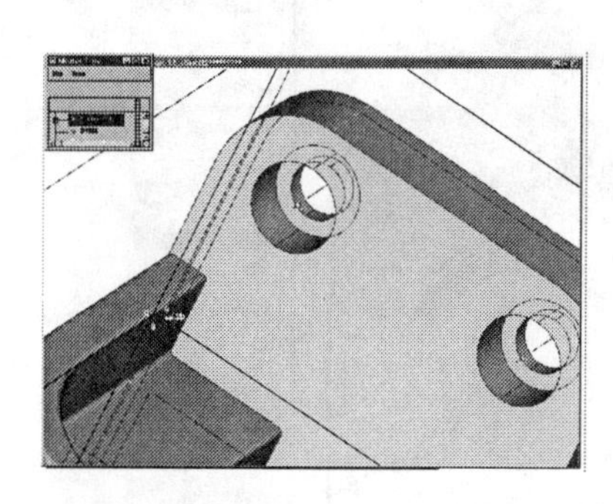

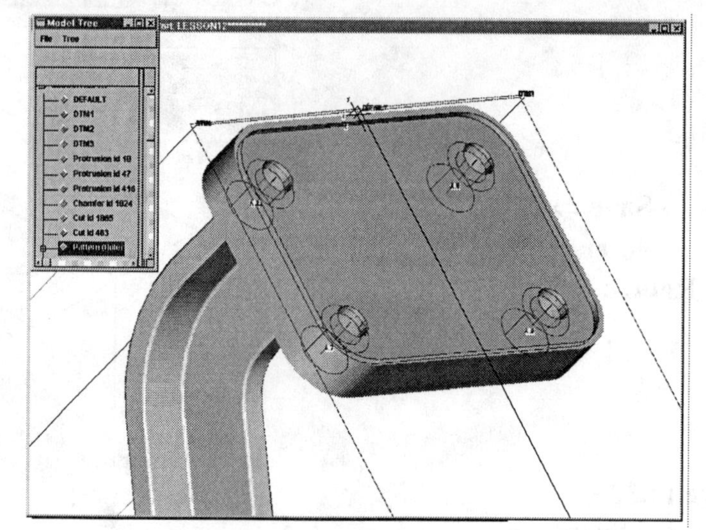

Figure 11.36
Create the Cut, Model and
Pattern the Counterbores

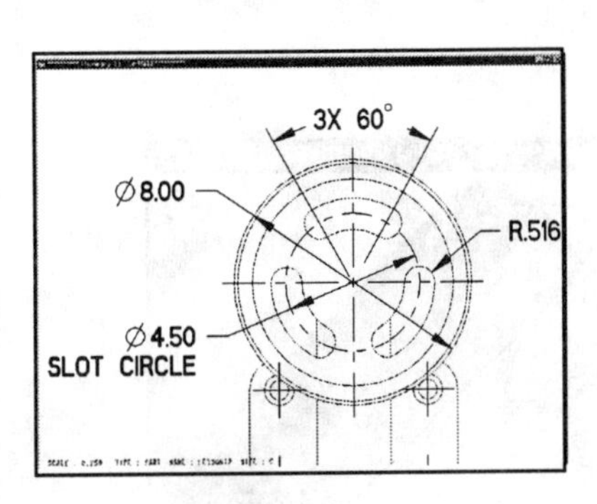

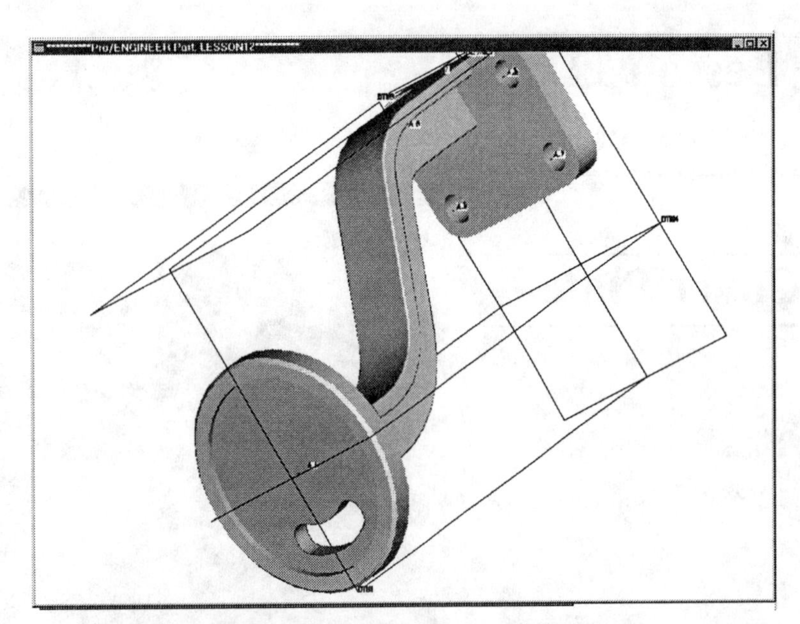

Figure 11.37
Create the Slot

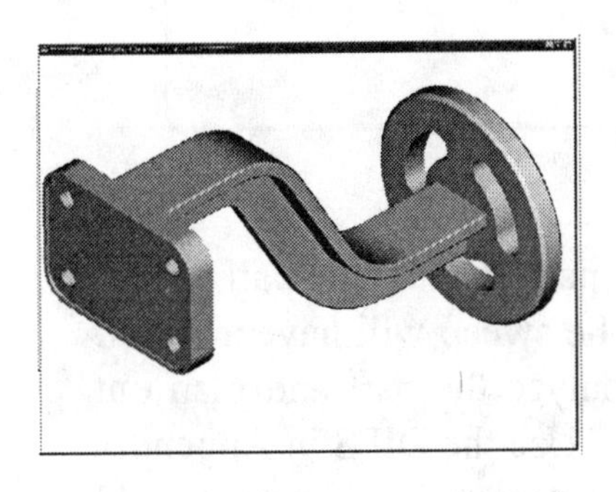

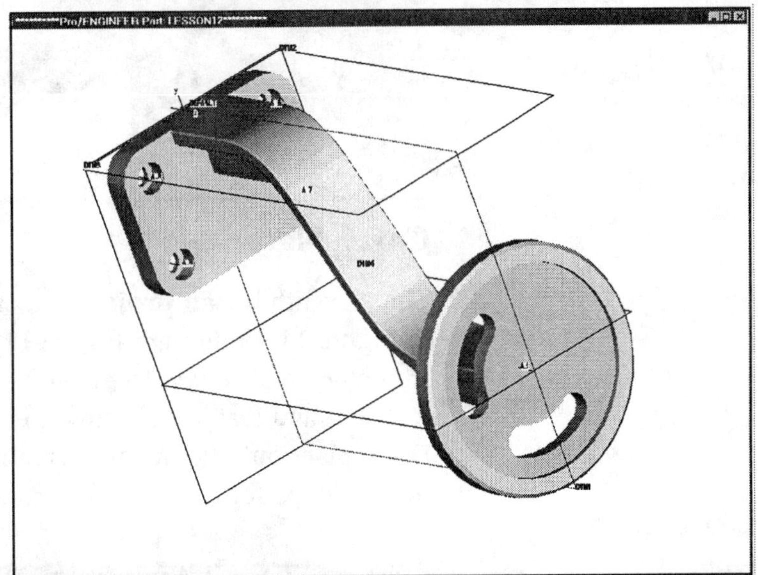

Figure 11.38
Pattern the Slots

Dbms ⇒ Save ⇒ enter
Purge ⇒ enter ⇒
Done-Return

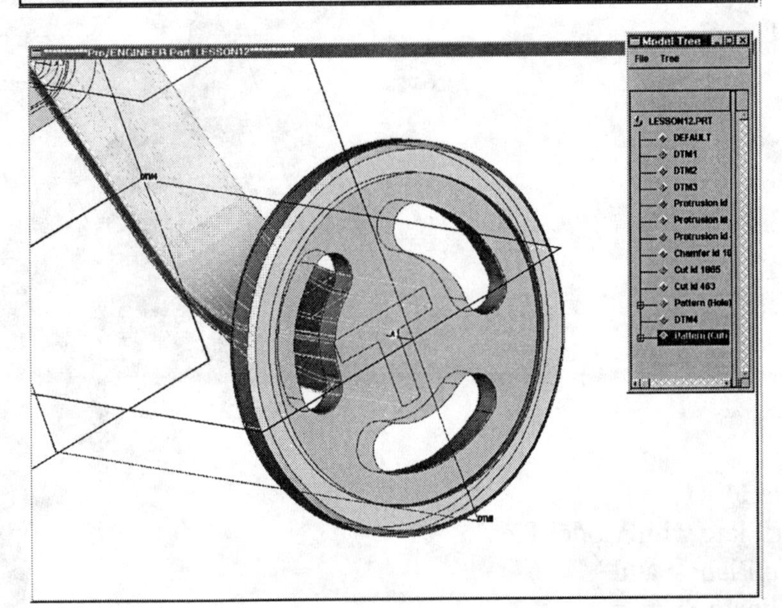

Figure 11.39
Completed Pattern and Part

Lesson 11 Project

Cover Plate

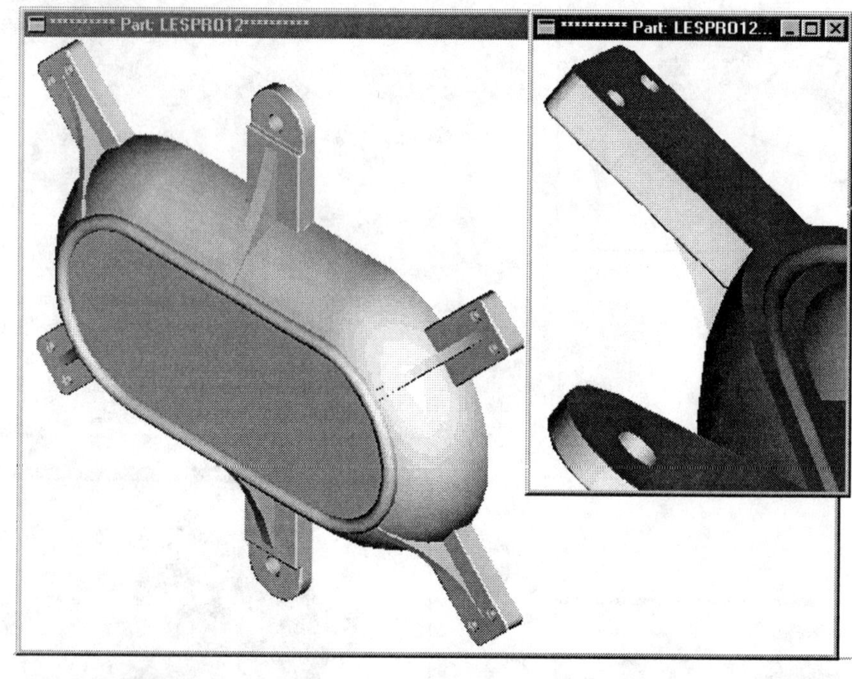

Figure 11.40
Cover Plate

Cover Plate

The eleventh **lesson project** is a cast iron part. Create the part shown in Figure 11.40 through Figure 11.53. The sweep will have a *closed trajectory* with *inner faces included.* Analyze the part and plan out the steps and features required to model it. Use the DIPS in Appendix D to plan out the feature creation sequence and the parent-child relationships for the part. Add rounds on all nonmachined edges.

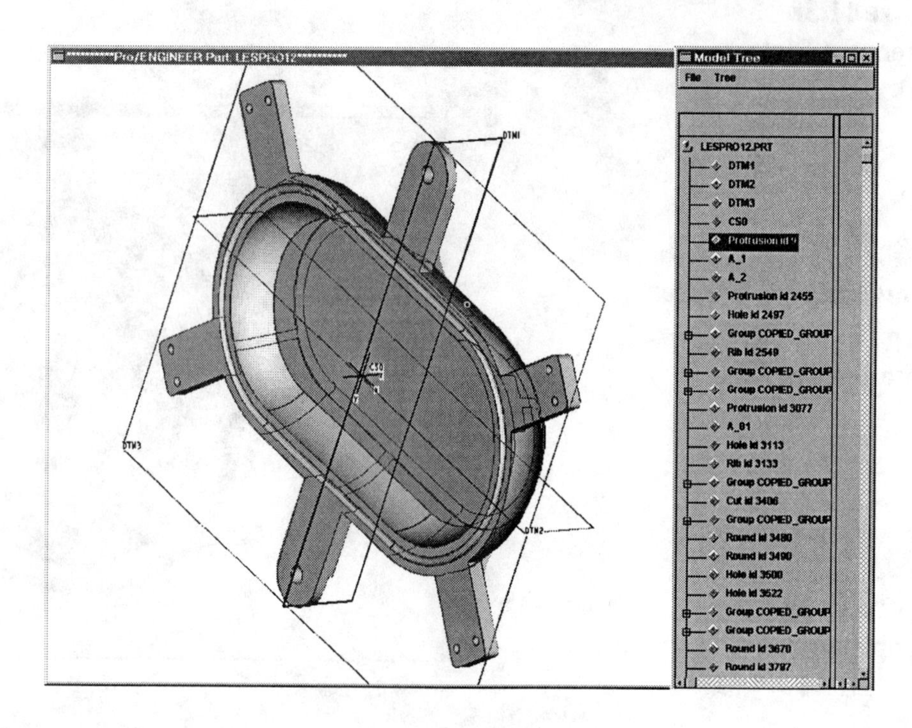

Figure 11.41
Cover Plate with Model Tree, Datum Planes, and Coordinate System

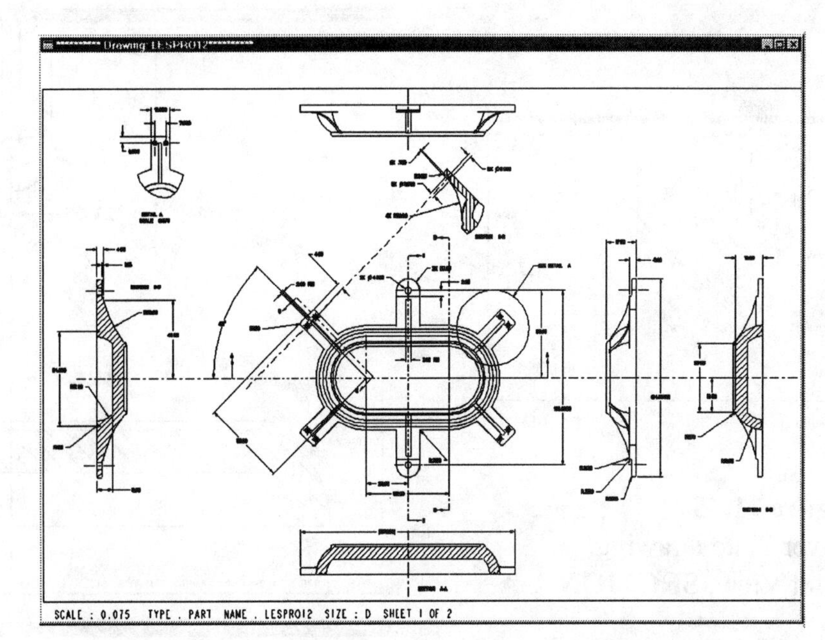

Figure 11.42
Cover Plate Detail Drawing, Sheet One

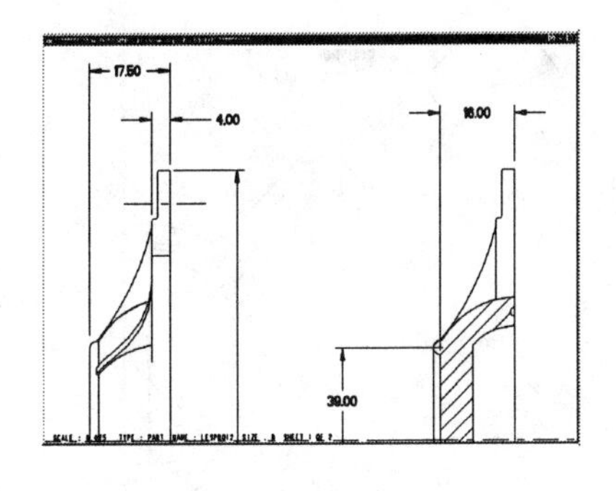

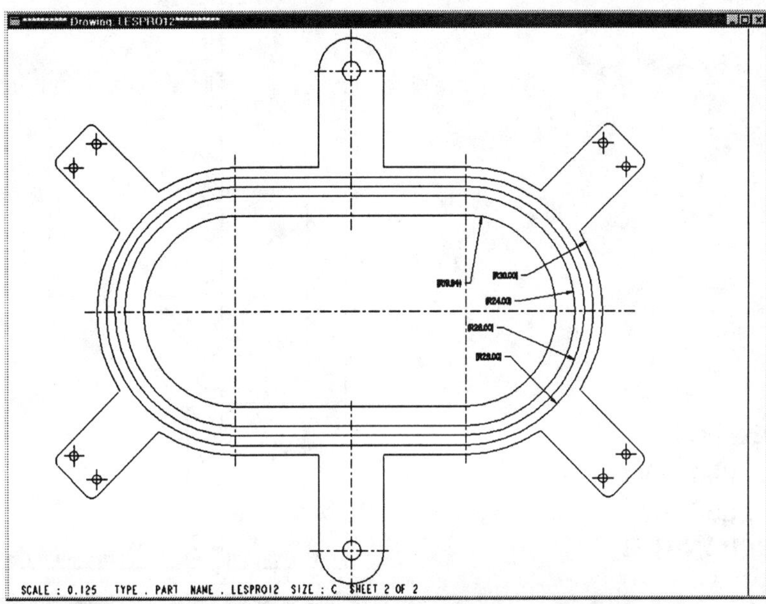

Figure 11.43
Cover Plate Detail Drawing, Sheet Two, Bottom View

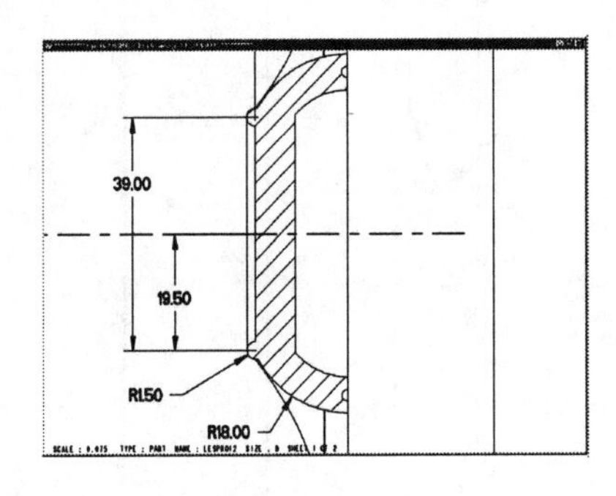

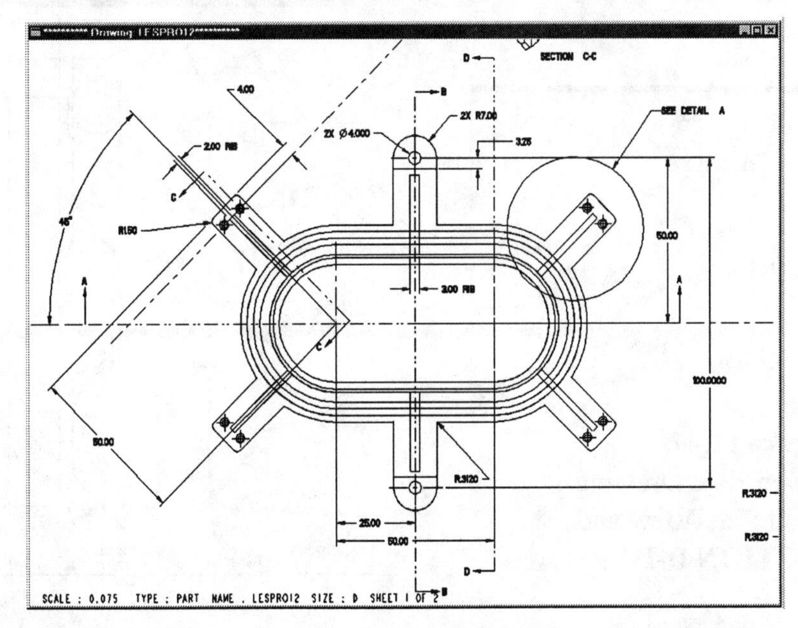

Figure 11.44
Cover Plate Drawing, Top View

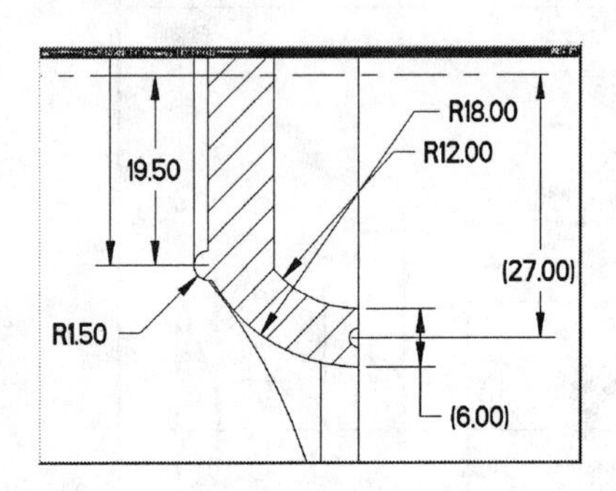

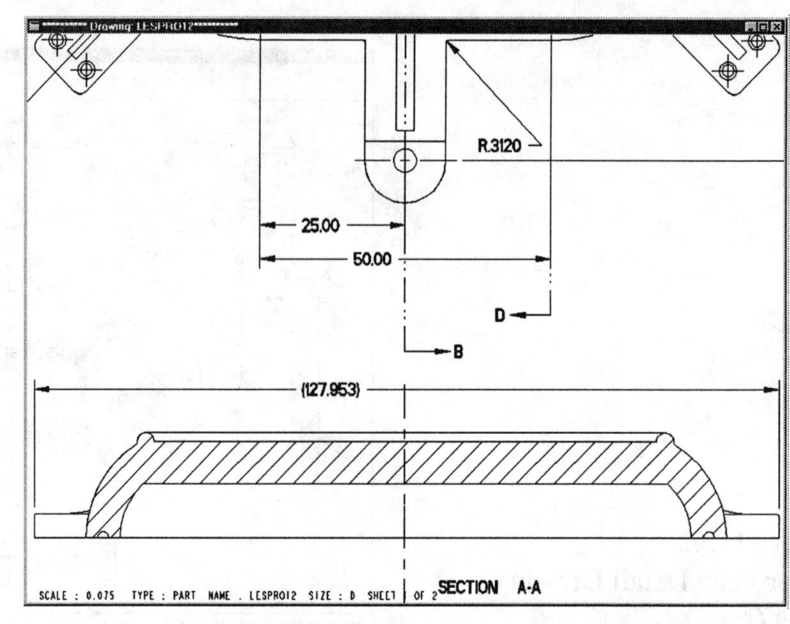

Figure 11.45
Cover Plate Drawing,
Front View, **SECTION A-A**

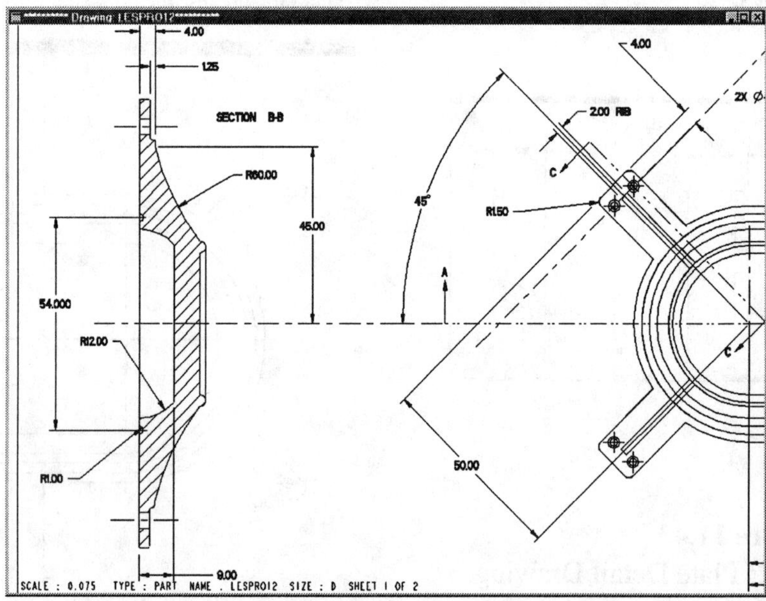

Figure 11.46
Cover Plate Drawing,
Left Side View,
SECTION B-B

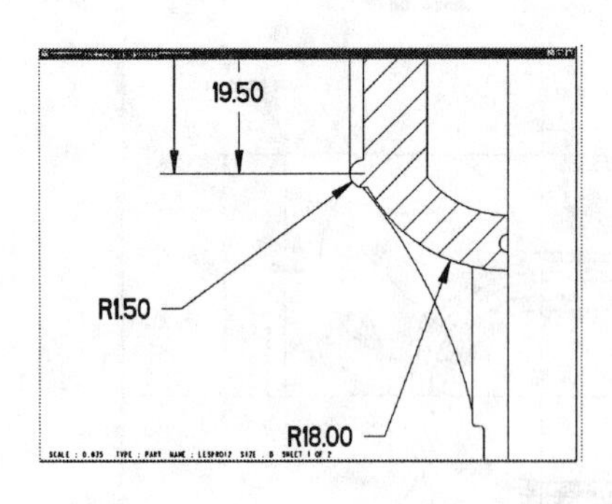

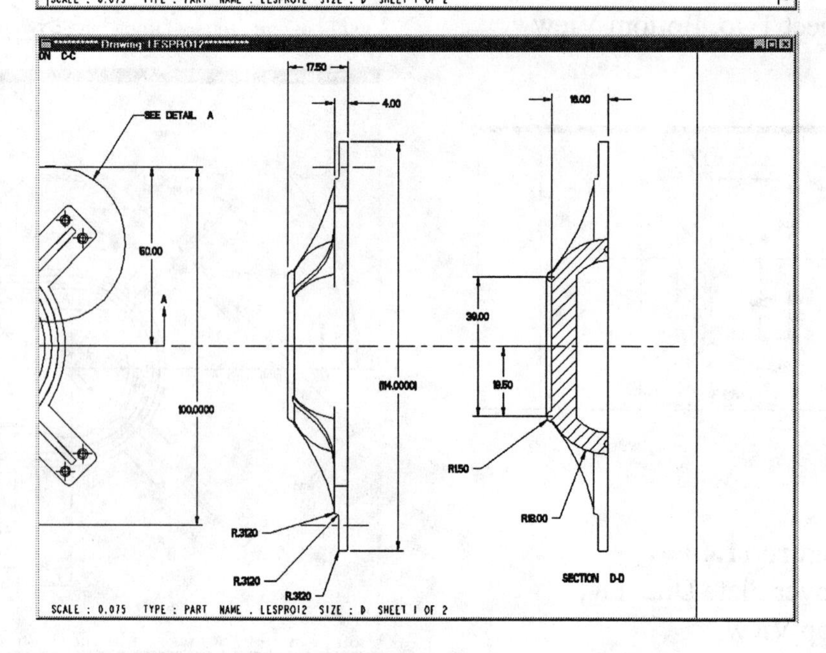

Figure 11.47
Cover Plate Drawing,
Right Side View and
SECTION D-D

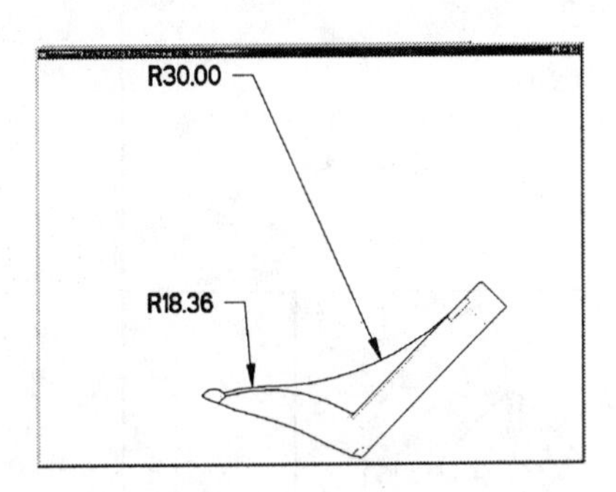

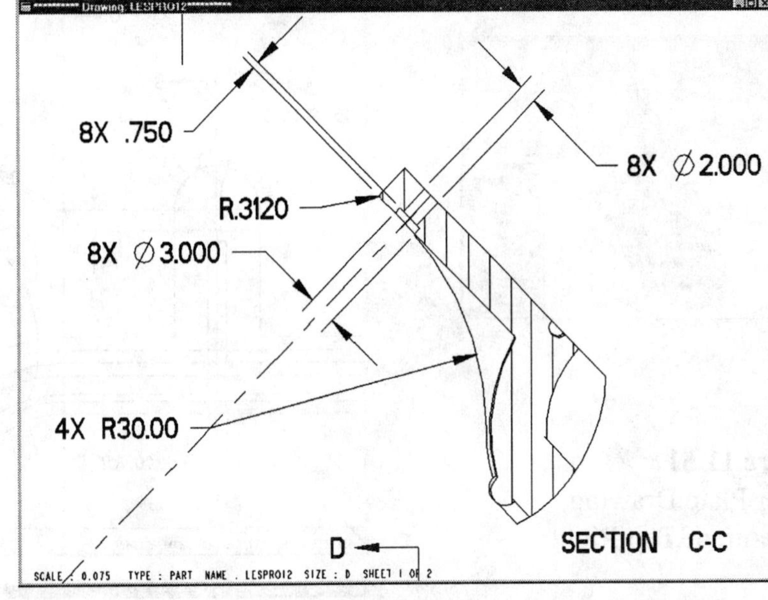

Figure 11.48
Cover Plate Drawing,
SECTION C-C

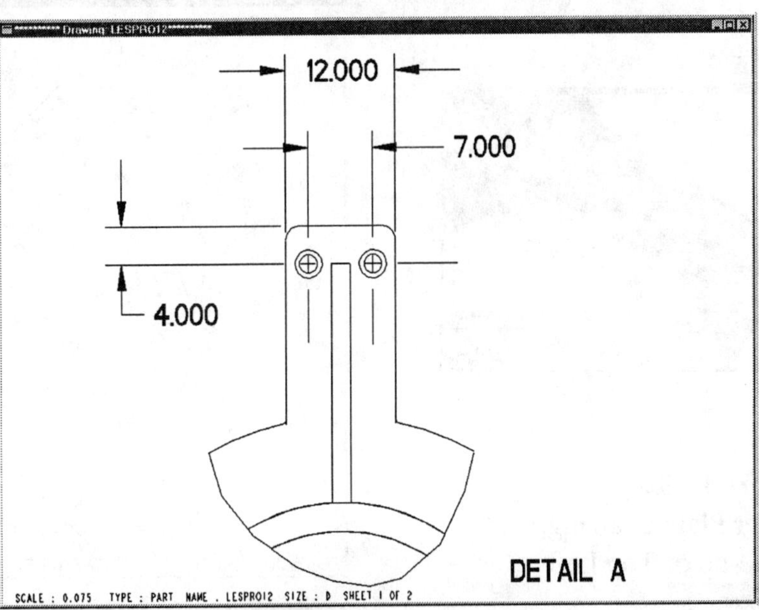

Figure 11.49
Cover Plate Drawing,
DETAIL A

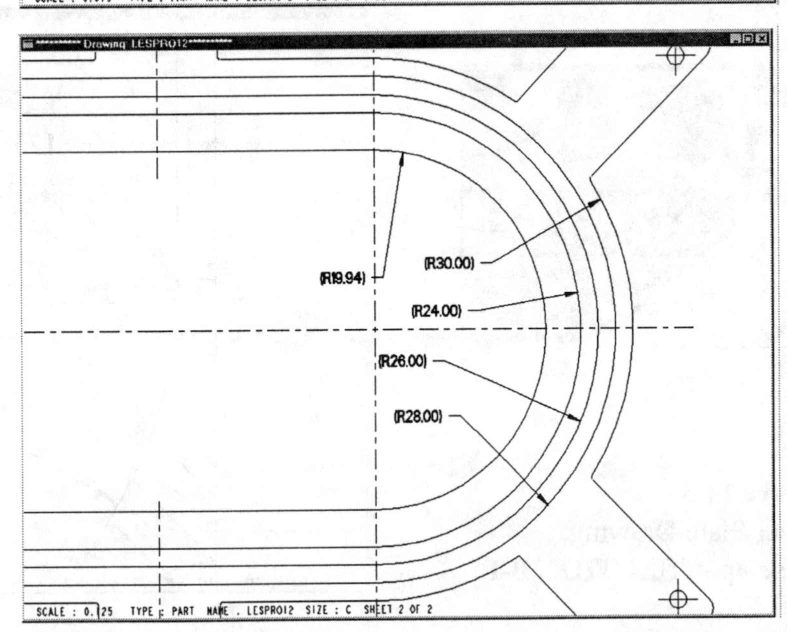

Figure 11.50
Cover Plate Drawing,
Second Sheet, Bottom View

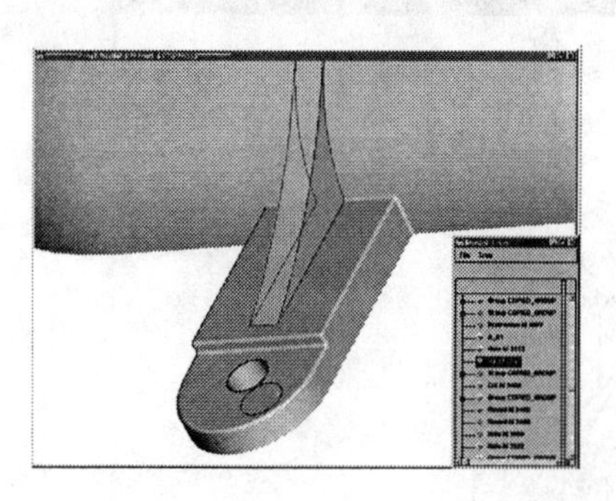

Figure 11.51
Cover Plate Drawing,
Close-up of Top Right

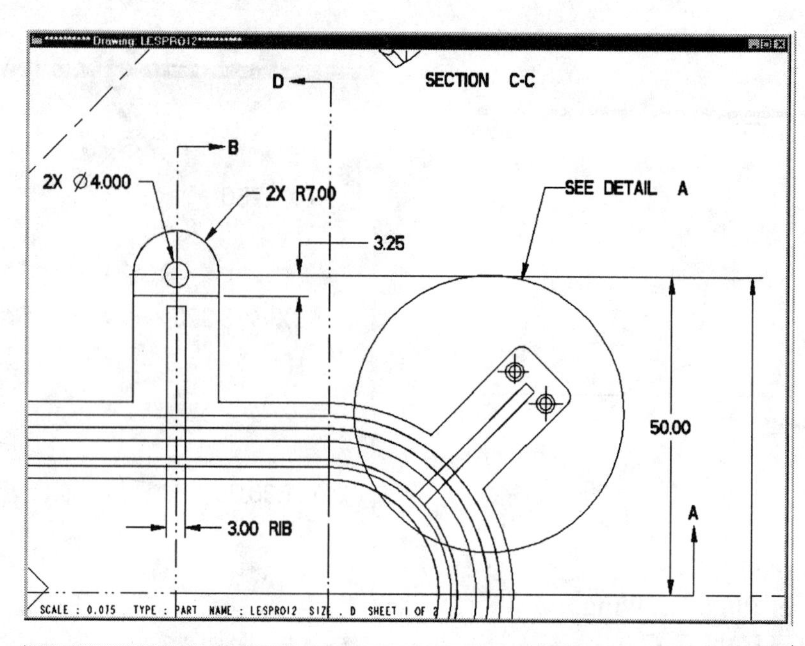

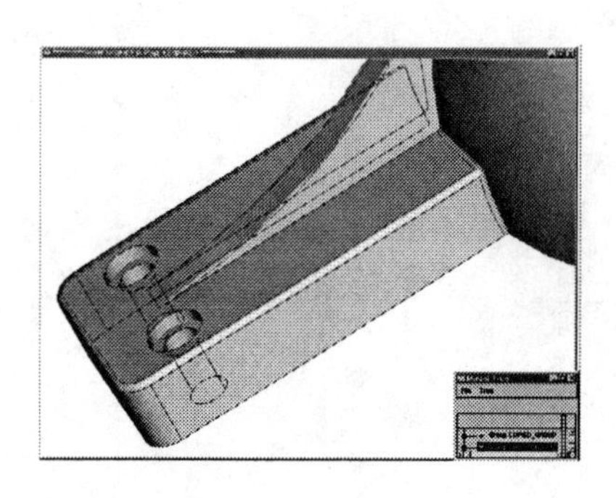

Figure 11.52
Cover Plate Drawing,
Close-up of Top Left

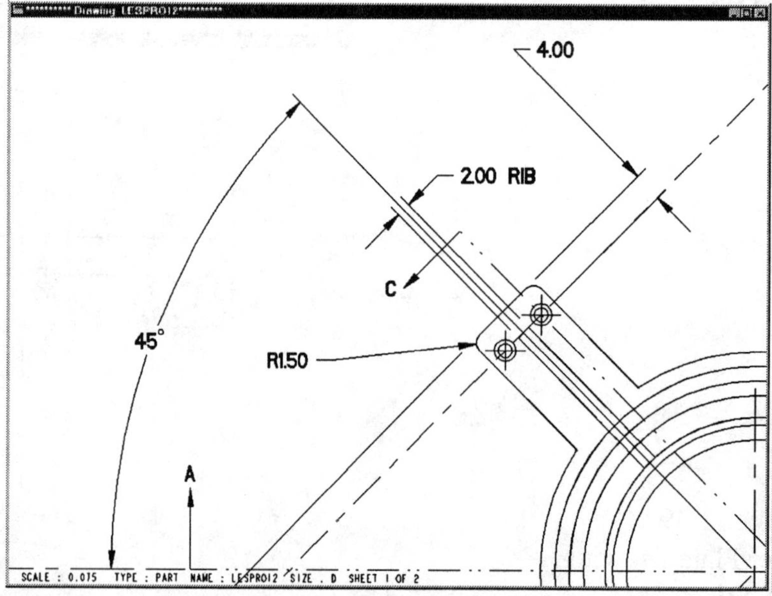

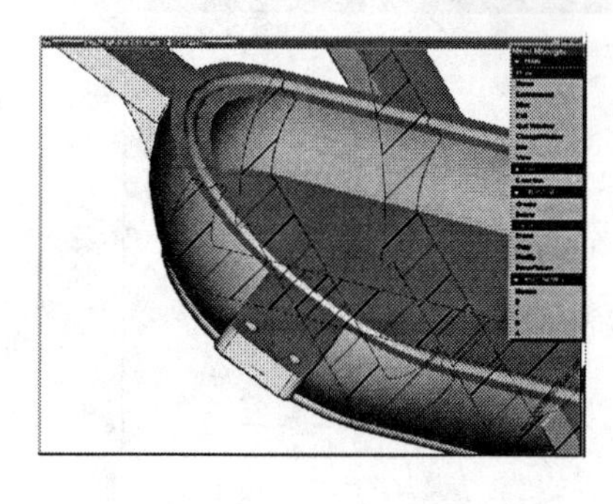

Figure 11.53
Cover Plate Drawing,
Close-up of **SECTION B-B**

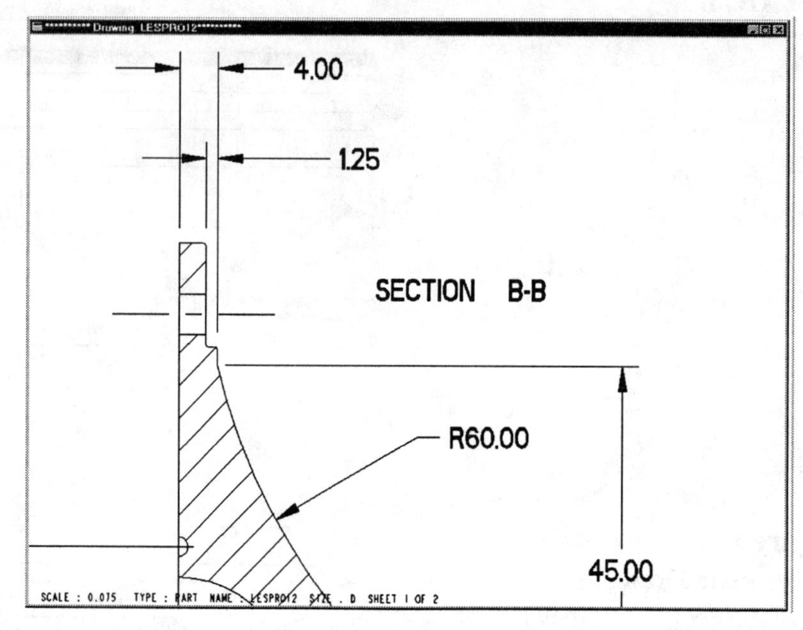

Lesson 12

Blends and Splines

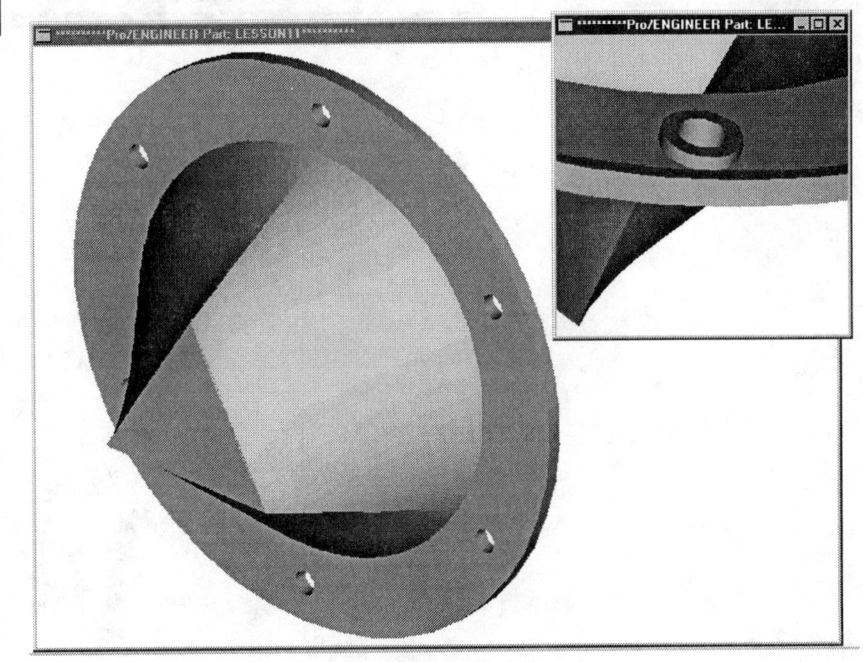

Figure 12.1
Cap

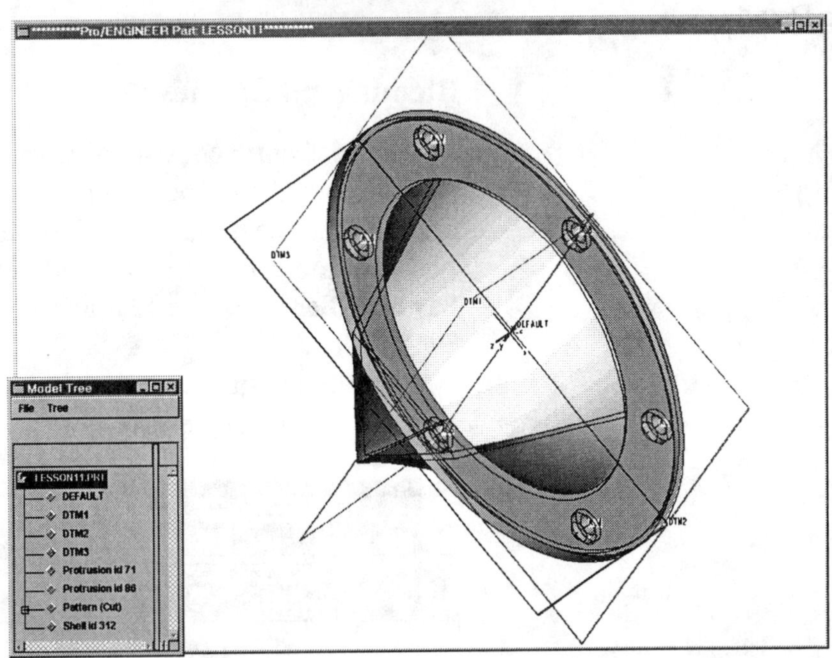

Figure 12.2
Cap with Datum Planes Displayed

OBJECTIVES

1. **Create a parallel blend feature**

2. **Shell a blend feature**

3. **Create a spline and use it in a swept blend**

4. **Create a swept blend feature**

5. **Create sections in Part mode**

EGD REFERENCE
Engineering Graphics and Design
by L. Lamit and K. Kitto
Read Chapter: 12
See Pages: 364-366

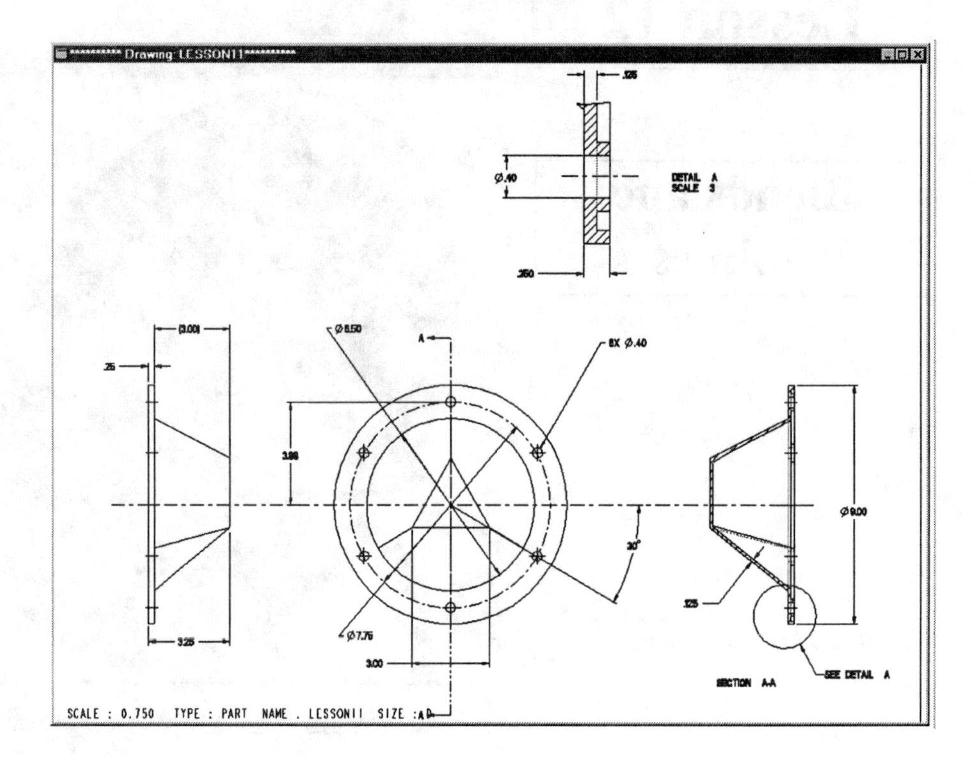

Figure 12.3
Cap Detail

Blends And Splines

A blended feature consists of a series of at least two planar sections that are joined together at their edges with transitional surfaces to form a continuous feature. The Cap in Figure 12.1 through Figure 12.3 uses a blend feature in its design. A **Blend** can be created as a **Parallel Blend** (Figure 12.4 and Figure 12.5), or you can construct a **Swept Blend**.

A **spline** is similar to drawing an **irregular** curve and is used in a variety of industrial designs.

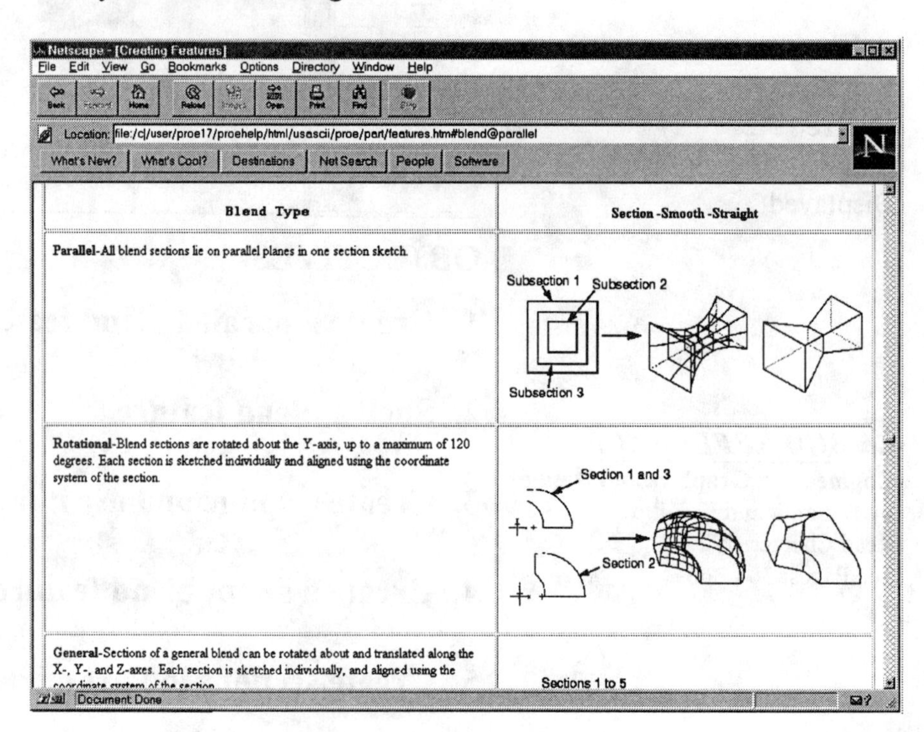

Figure 12.4
Online Documentation
Blends

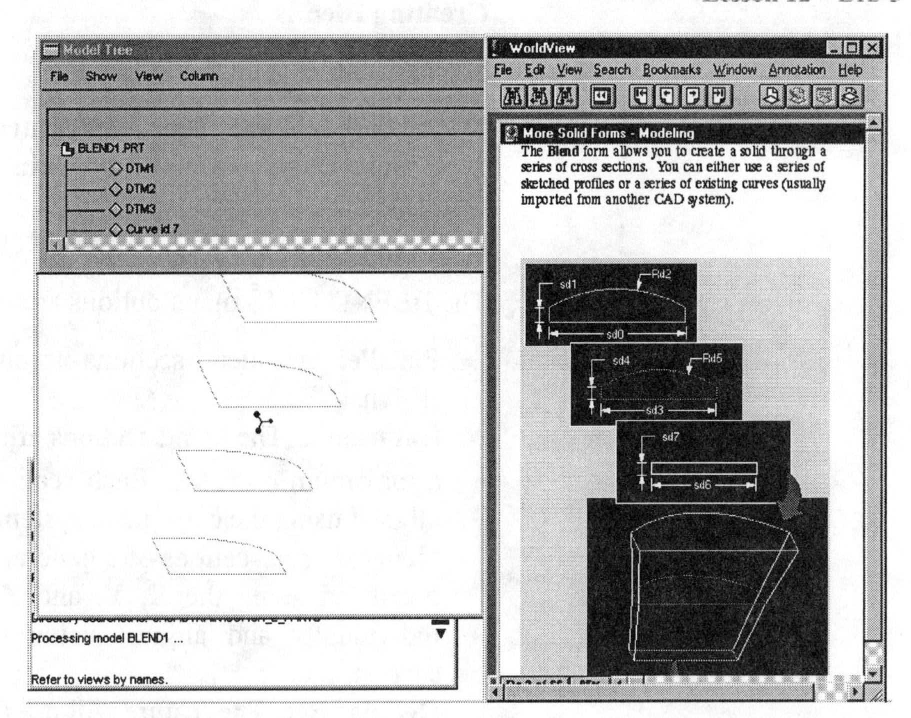

Figure 12.5
COAch for Pro/E, More Solid Forms-- Modeling (Segment 3: Parallel Blends)

Blend Sections

Figure 12.5 shows a parallel blend for which the *section* consists of four *subsections*. Each segment in the subsection is matched with a segment in the following subsection; the blended surfaces are created between the corresponding segments. With the exception of capping a blend, blends must always have the same number of entities in each section.

HINT
For the most part, blends must have the same number of entities in each section. The only exception is a **Blend Vertex Cap** blend

Starting Point of a Section

To create the transitional surfaces, Pro/E connects the *starting points* of the sections and continues to connect the vertices of the sections in a clockwise manner. By changing the starting point of a blend section, you can create blended surfaces that twist between the sections.

The default starting point is the first point sketched in the subsection. You can place the starting point at the endpoint of another segment by choosing the option **Start Point** from the SEC TOOLS menu and selecting the point.

Smooth and Straight Attributes

Blends use one of the following transitional surface ATTRIBUTES menu options:

Straight Create a straight blend by connecting vertices of different subsections with straight lines. Edges of the sections are connected with ruled surfaces.

Smooth Create a smooth blend by connecting vertices of different subsections with smooth curves. Edges of the sections are connected with ruled surfaces.

Creating Blends

To create a blend, do the following:

1. Use the command sequence **Feature, Create, Solid, Protrusion.**
2. Choose **Blend** and **Solid** or **Thin** from the SOLID OPTS menu, then **Done.**
3. Choose options from the BLEND OPTS menu, then **Done.**

The BLEND OPTS menu options are as follows:

Parallel All blend sections lie on parallel planes in one section sketch.
Rotational The blend sections are rotated about the **Y** axis, up to a maximum of **120°**. Each section is sketched individually and aligned using the coordinate system of the section.
General The sections of a general blend can be rotated about and translated along the **X, Y,** and **Z** axes. Sections are sketched individually and aligned using the coordinate system of the section.
Regular Sec The feature will use the regular sketching plane.
Project Sec The feature will use the projection of the section on the selected surface. This is used for parallel blends only.
Select Sec Select section entities (not available for parallel blends).
Sketch Sec Sketch section entities.

Parallel Blends

You create parallel blends (Fig. 12.6) using the **Parallel** option in the BLENDS OPTS menu. A parallel blend is created from a single section (Fig. 12.7) that contains multiple sketches, called *subsections*. First and last subsections can be defined as a point or a blend vertex.

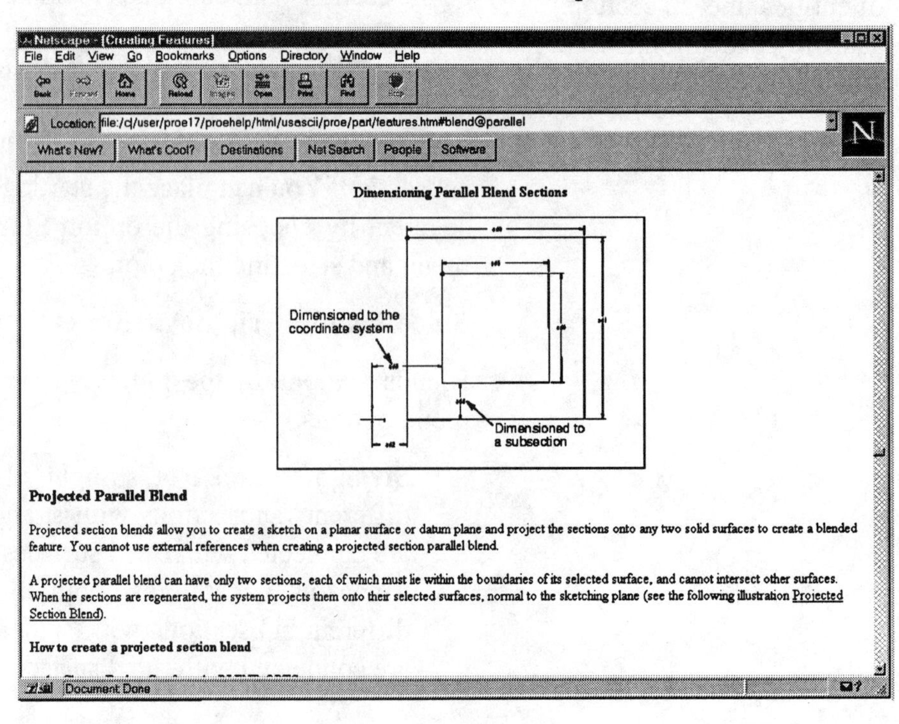

Figure 12.6
Online Documentation
Parallel Blends

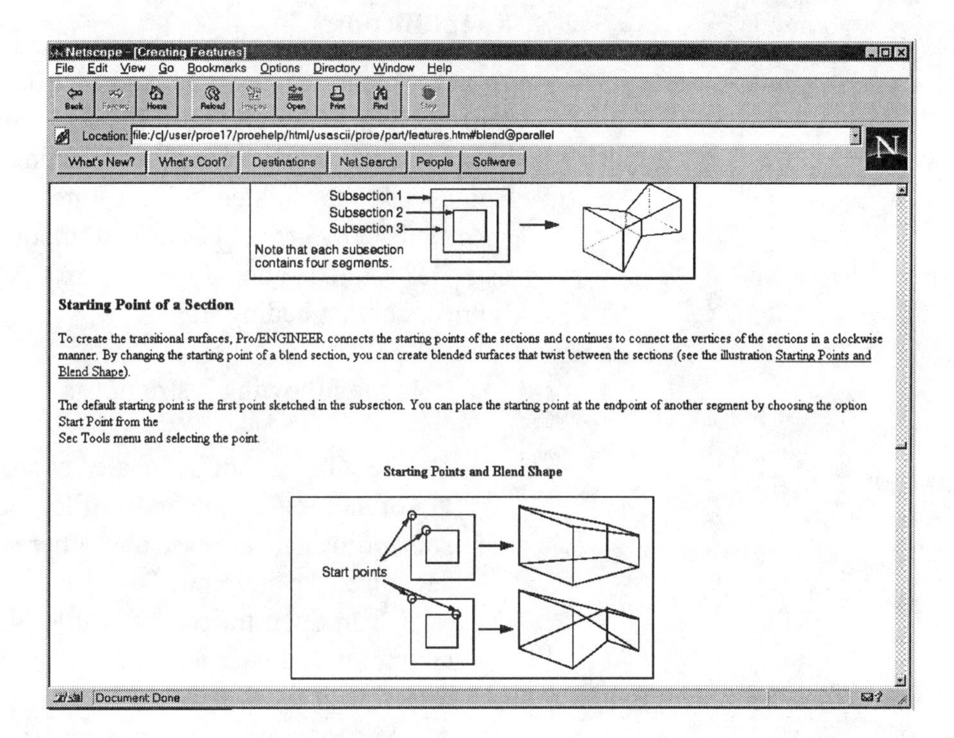

Figure 12.7
Online Documentation
Blend Sections

Creating a Parallel Blend

To create a parallel blend, do the following:

1. When you choose **Done** from the BLEND OPTS menu, Pro/E displays the feature creation dialog box and the ATTRIBUTES menu. Choose either **Straight** or **Smooth**.
2. Create the first subsection using the Sketcher. You determine the direction of feature creation as you set up the sketching plane. Dimension and regenerate each subsection sketch to ensure the validity of the dimensioning scheme. A parallel blend requires more than one subsection, so after successfully regenerating this section, choose **Sec Tools** from the SKETCHER menu.
3. Choose **Toggle** from the SEC TOOLS menu. The first subsection turns gray and becomes inactive.
4. Choose **Sketch** and sketch the second subsection. Make sure its starting point corresponds to the starting point of the first subsection in the manner that you intend. Dimension and regenerate it.
5. If you are sketching more than two subsections, choose **Toggle** repeatedly until all the current geometry is gray, then sketch the subsection. Repeat this step until all subsections are sketched.
6. To modify an existing subsection, toggle through until the subsection you want is active. While you can place or move the starting point of a subsection only when it is active, you can modify the dimensions of any subsection at any time.
7. When you have sketched all the subsections, choose **Done** from the SKETCHER menu. When prompted, enter the distances between each subsection.
8. Select the **OK** button to create the feature.

Swept Blends

A swept blend (Figure 12.8 and Figure 12.9) is created using a single trajectory (a spine) and multiple sections. You create the spine of the swept blend by sketching or selecting a datum curve or an edge. Spines can be created with splines. You sketch the sections at specified segment vertices or datum points on the spine. Each section can be rotated about the **Z** axis with respect to the section immediately preceding it.

Note the following restrictions:

* A section cannot be located at a sharp corner in the spine.
* For a closed trajectory profile, sections must be sketched at the start point and at least one other location. Pro/E uses the first section at the endpoint.
* For an open trajectory profile, you must create sections at the startpoint and endpoint. There is no option to skip placement of a section at those points.
* Sections cannot be dimensioned to the model, because modifying the trajectory would invalidate those dimensions.
* A composite datum curve cannot be selected for defining sections of a swept blend **(Select Sec)**. Instead, you must select one of the underlying datum curves or edges for which a composite curve is determined.
* If you choose **Pivot Dir** and **Select Sec**, all selected sections must lie in planes that are parallel to the pivot direction.
* You cannot use a nonplanar datum curve from an equation as a swept blend trajectory.

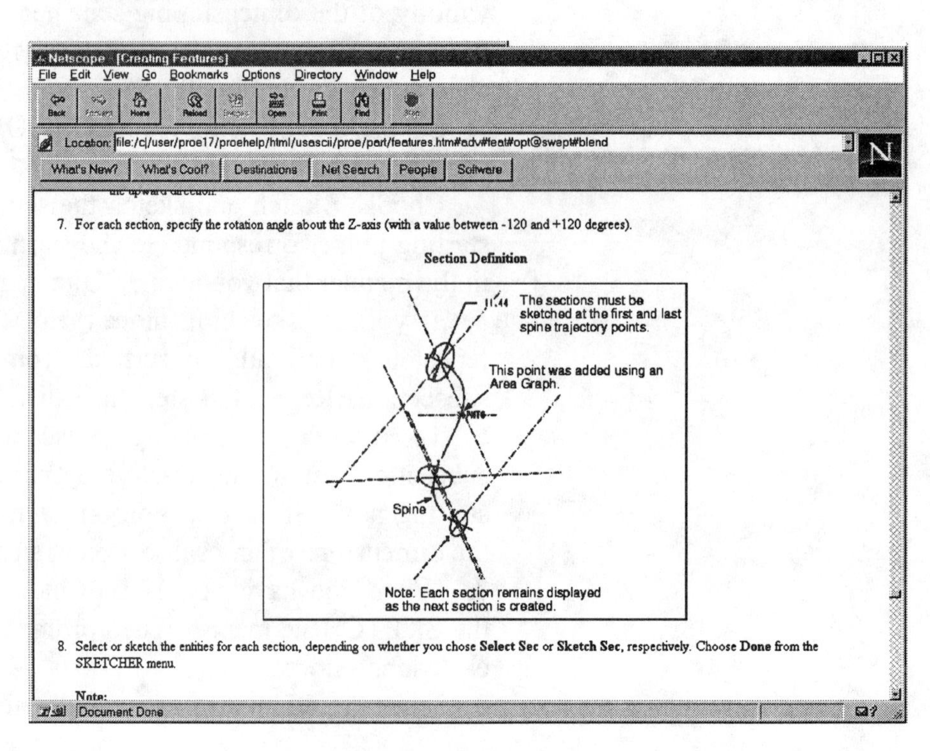

Figure 12.8
Online Documentation
Creating a Swept Blend

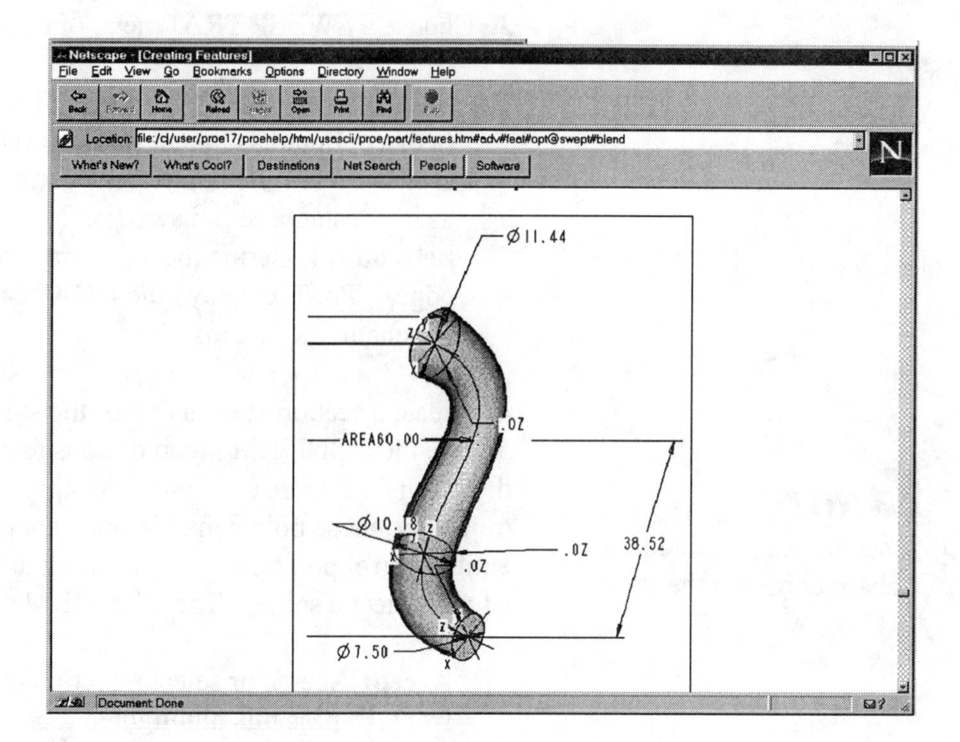

Figure 12.9
Online Documentation
Swept Blends

Creating a Swept Blend

To create a **Swept Blend** (Figure 12.9 and Figure 12.10), you can define the trajectory by sketching a trajectory or by selecting existing curves and edges and extending or trimming the first and last entity in the trajectory. Use the following procedure:

1. Choose **Advanced** from the SOLID OPTS menu and **Swept Blend** and **Done** from the ADV FEAT OPTS menu.
2. Choose the desired options from the BLEND OPTS menu mutually exclusive pairs, then choose **Done** from the BLEND OPTS menu. The possible options are as follows:

Select Sec Select existing curves or edges to define each section using the CRV SKETCHER menu.
Sketch Sec Sketch new section entities to define each section.
Nrm To Spine The section plane remains normal to the spine trajectory.
Pivot Dir The section plane will remain normal to the spine trajectory when viewed from a pivot direction, as the section plane sweeps along the trajectory, it always remains parallel to the pivot direction.

3. Choose a SWEEP TRAJ menu option. The options include:

   Sketch Traj Sketch a spine. The spine can have sharp corners (a discontinuous tangent to the curve), except at the endpoint of a closed curve. At nontangent vertices Pro/E mitres the geometry as in constant section sweeps.
   Select Traj Define the spine trajectory using existing curves and edges. Pro/E displays the CHAIN menu. Choose **Select**, define the chain, then choose **Done**.

4. Create a section at each end of the spine.
5. Use the CONFIRM menu options to choose the points at which to define any additional sections. As appropriate to the defined trajectory, those points may include endpoints of spine entities, spline entity control points, and any existing datum points on spine entities (if you select a spine). The CONFIRM menu options are as follows:

NOTE
A section cannot be located at a sharp corner in the spine.

   Accept Sketch or select a section at this highlighted location.
   Next Bypass this highlighted location and go to the next point.
   Previous Bypass this highlighted location and return to the previous point.

6. For each section, specify the rotation angle about the **Z** axis (with a value between **-120°** and **+120°**).
7. Select or sketch the entities for each section, depending on whether you choose **Select Sec** or **Sketch Sec**, respectively. Choose **Done** from the SKETCHER menu.
8. When all sections are sketched or selected, unless you chose **Area Graph** or **Blend Control** element, select the **OK** button in the dialog box to generate the swept blend feature.

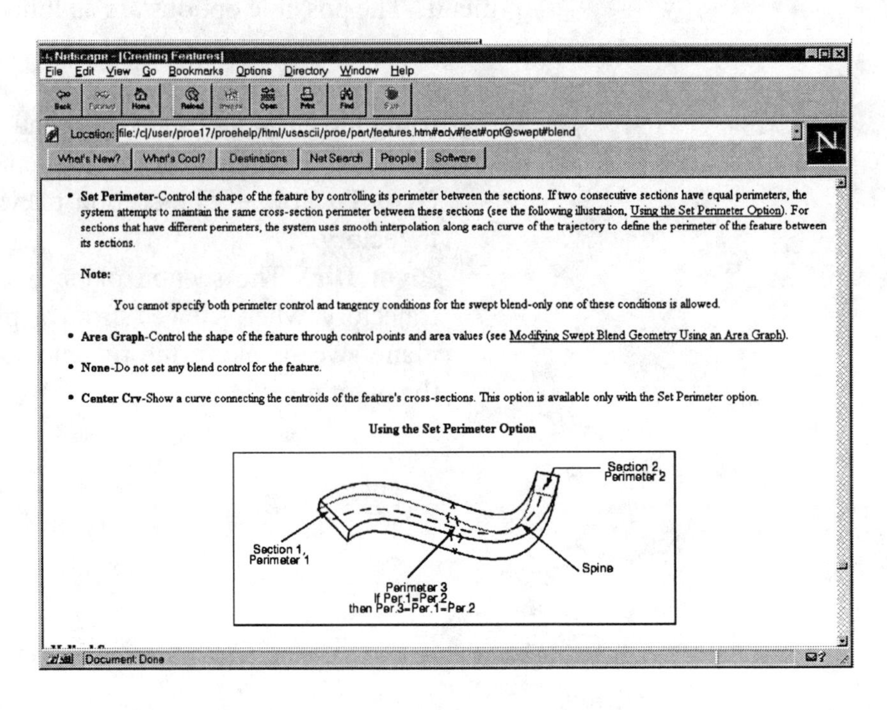

Figure 12.10
Online Documentation
Spines, Set Perimeter Option

Splines

Sketching a **spline** is similar to drawing an **irregular** curve (Fig. 12.11). Splines (Fig. 12.12) are created by picking or sketching a series of specific points.

To create a spline:

1. Choose **Spline** from the GEOMETRY menu.
2. The SPLINE MODE menu appears with the following commands:

    Sketch Points Create a spline by picking screen points for the spline to pass through.
    Select Points Create a spline by selecting existing Sketcher points. Once the point has been selected, there is no further link between the point and the spline.

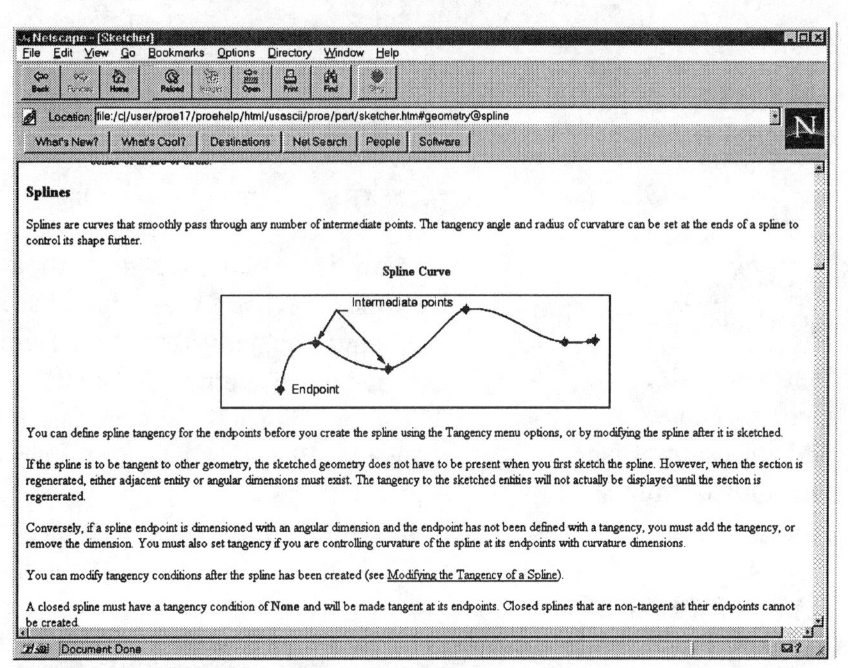
Netscape - [Sketcher]

File Edit View Go Bookmarks Options Directory Window Help

Location: file:/c|/user/proe17/proehelp/html/usascii/proe/part/sketcher.htm#geometry@spline

What's New? What's Cool? Destinations Net Search People Software

Splines

Splines are curves that smoothly pass through any number of intermediate points. The tangency angle and radius of curvature can be set at the ends of a spline to control its shape further.

Spline Curve

You can define spline tangency for the endpoints before you create the spline using the Tangency menu options, or by modifying the spline after it is sketched.

If the spline is to be tangent to other geometry, the sketched geometry does not have to be present when you first sketch the spline. However, when the section is regenerated, either adjacent entity or angular dimensions must exist. The tangency to the sketched entities will not actually be displayed until the section is regenerated.

Conversely, if a spline endpoint is dimensioned with an angular dimension and the endpoint has not been defined with a tangency, you must add the tangency, or remove the dimension. You must also set tangency if you are controlling curvature of the spline at its endpoints with curvature dimensions.

You can modify tangency conditions after the spline has been created (see Modifying the Tangency of a Spline).

A closed spline must have a tangency condition of **None** and will be made tangent at its endpoints. Closed splines that are non-tangent at their endpoints cannot be created.

Document: Done

Figure 12.11
Online Documentation Splines

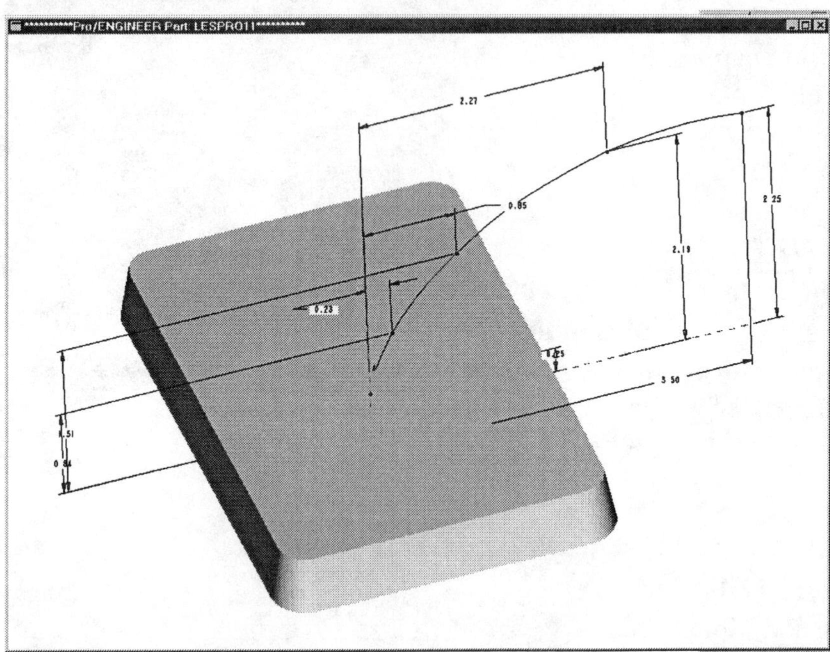

Figure 12.12
Spline

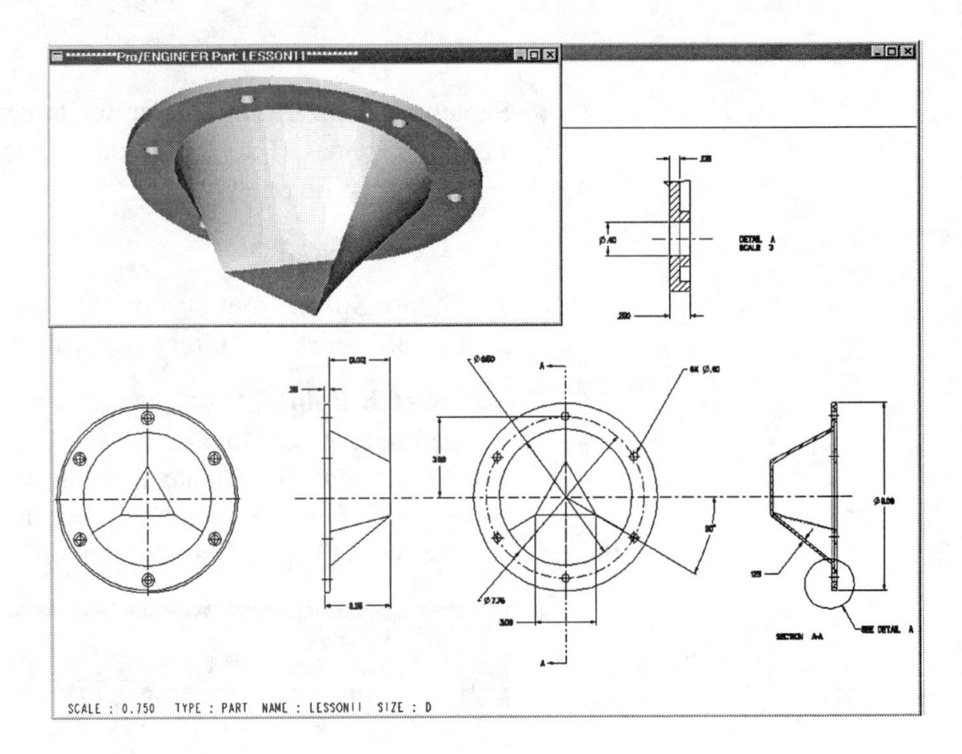

Figure 12.13
Cap Part and Detail

Cap

The **Cap** is a part created with a **Parallel Blend** (Figure 12.13 through Figure 12.19). The blend sections are a circle and a triangle. The circle is actually three equal arcs, since the sections of a blend must have equal segments.

The part is shelled as the last feature in its creation. The **Shell** command will create *bosses* around each hole as it hollows out the part.

For this part, a section will be created in the Part mode to be used when detailing the Cap in the Draw mode.

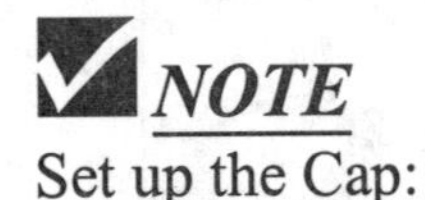

NOTE

Set up the Cap:

- Material = Plastic
- Units = Inches
- Default Datum Planes
- Default Coordinate System
- Layers = **DATUM_PLANES**
- Hidden Line
- Tan Dimmed

✓Grid Snap
✓Tol Display

CONFIG.PRO

angular_tol	**0**
tol_mode	**plusminus**
sketcher_dec_places	**3**
default_dec_places	**3**

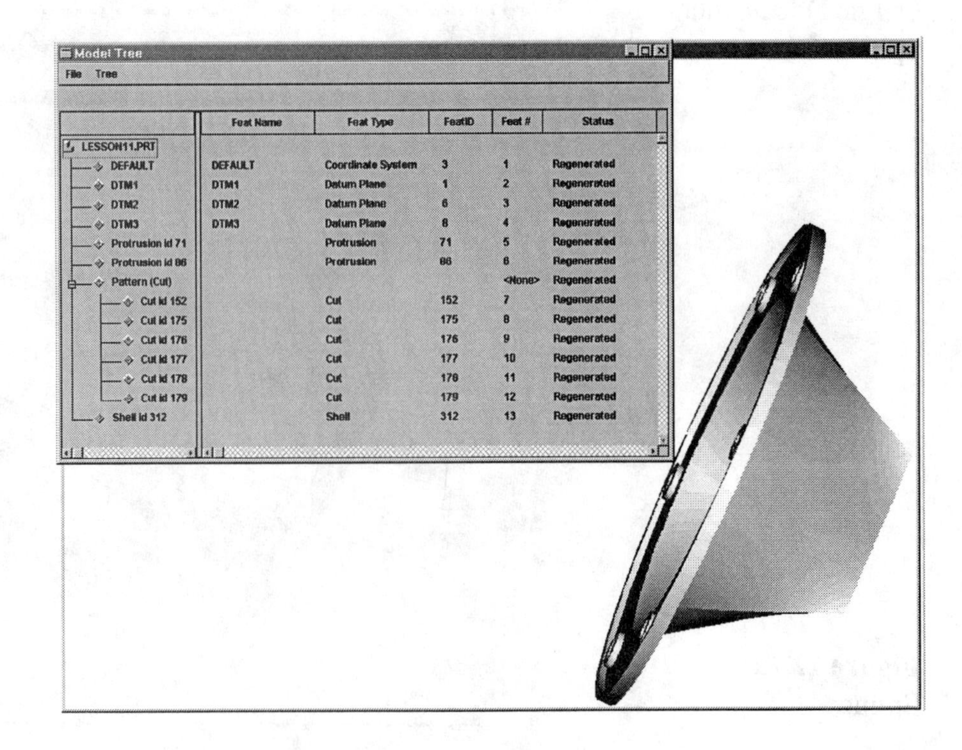

Model Tree

LESSON11.PRT	Feat Name	Feat Type	FeatID	Feat #	Status
DEFAULT	DEFAULT	Coordinate System	3	1	Regenerated
DTM1	DTM1	Datum Plane	1	2	Regenerated
DTM2	DTM2	Datum Plane	6	3	Regenerated
DTM3	DTM3	Datum Plane	8	4	Regenerated
Protrusion id 71		Protrusion	71	5	Regenerated
Protrusion id 86		Protrusion	86	6	Regenerated
Pattern (Cut)				<None>	Regenerated
Cut id 152		Cut	152	7	Regenerated
Cut id 175		Cut	175	8	Regenerated
Cut id 176		Cut	176	9	Regenerated
Cut id 177		Cut	177	10	Regenerated
Cut id 178		Cut	178	11	Regenerated
Cut id 179		Cut	179	12	Regenerated
Shell id 312		Shell	312	13	Regenerated

Figure 12.14
Cap and Model Tree

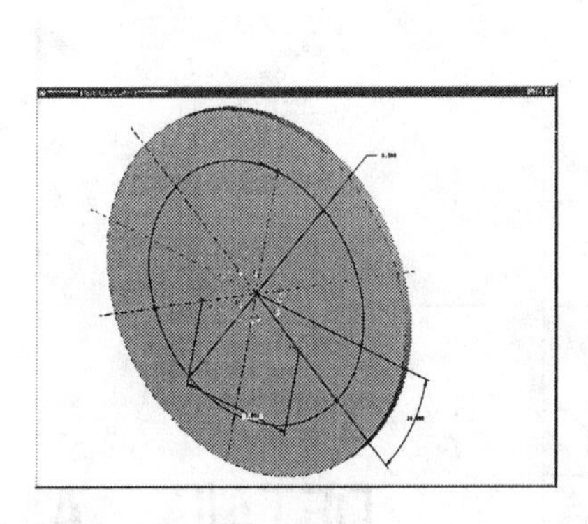

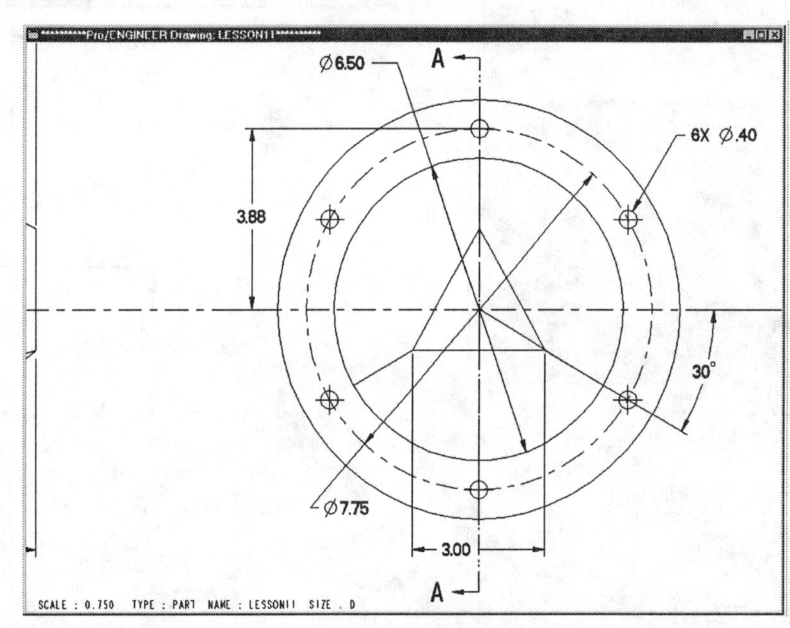

Figure 12.15
Cap, Top View

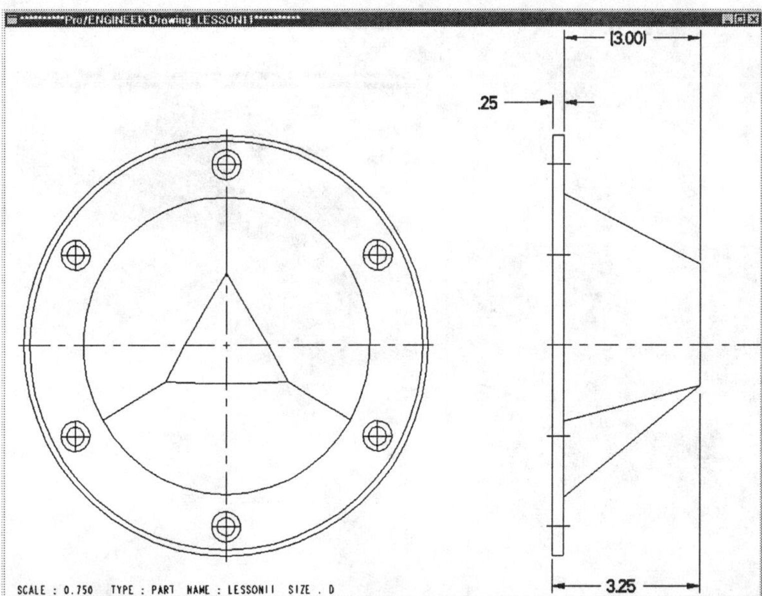

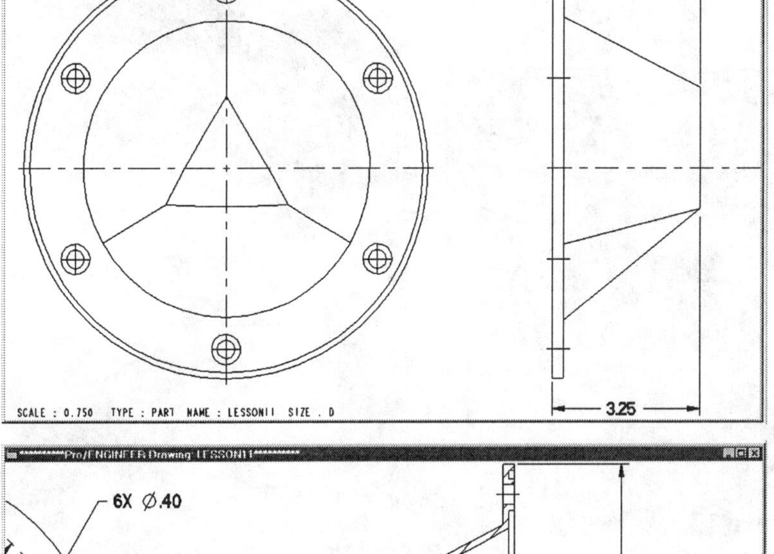

Figure 12.16
Cap, Left Side View and Back View

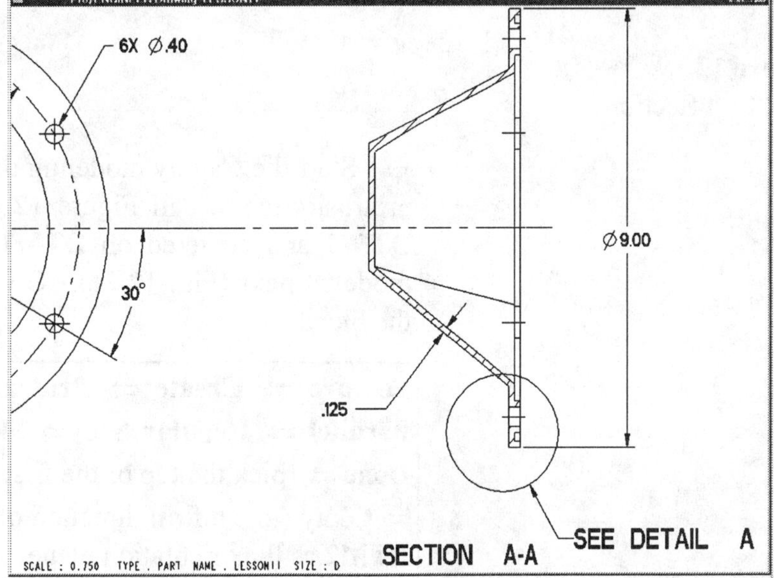

Figure 12.17
Cap, **SECTION A-A**

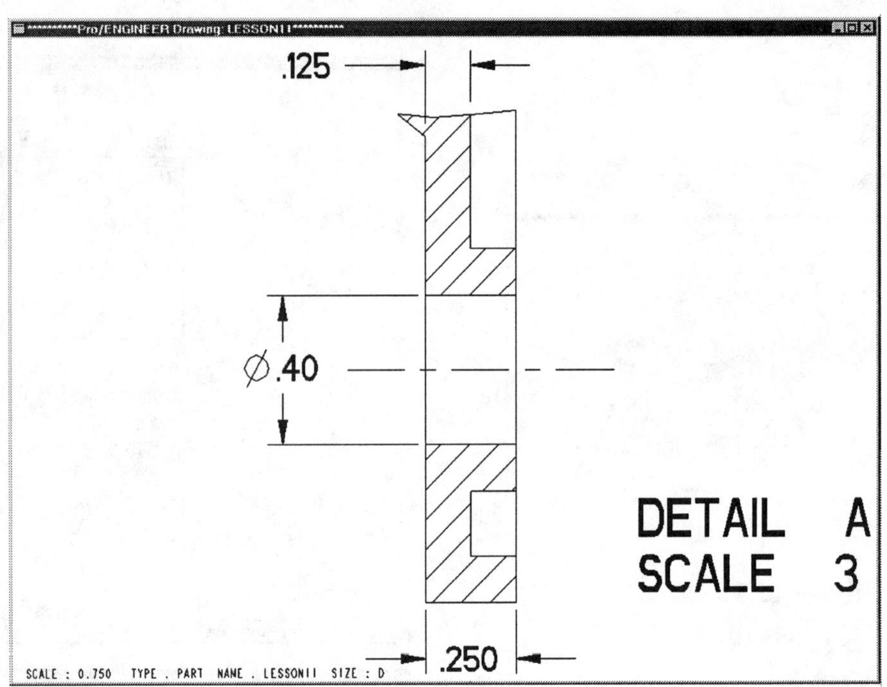

Figure 12.18
Cap, **DETAIL A**

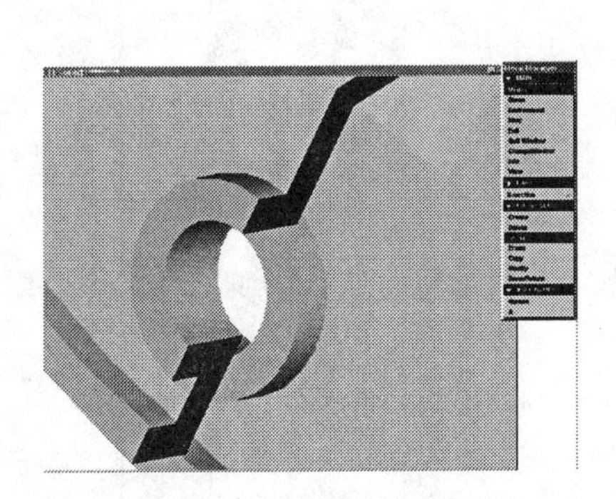

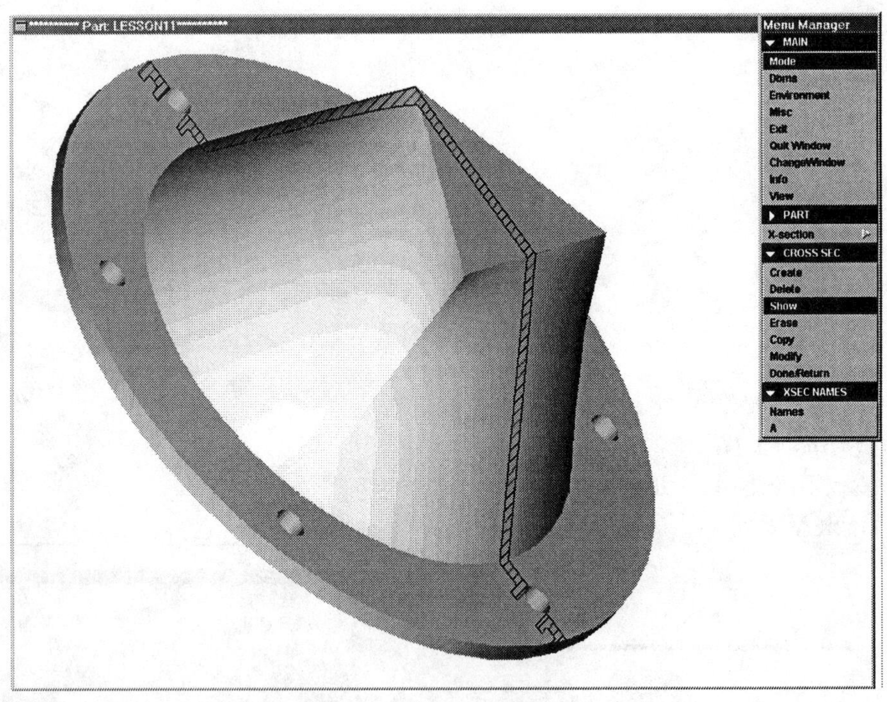

Figure 12.19
Cap, Part Section

Start the Cap by modeling the **9.00** diameter by **.25** thick circular protrusion shown in Figure 12.20. Sketch the first protrusion on **DTM3** and centered on **DTM1** and **DTM2**. The **Blend** feature is modeled next (Fig. 12.21). Give the following commands to create the blend:

Feature ⇒ **Create** ⇒ **Protrusion** ⇒ **Blend** ⇒ **Solid** ⇒ **Done Parallel** ⇒ **Regular Sec** ⇒ **Sketch Sec** ⇒ **Done** ⇒ **Straight** ⇒ **Done** ⇒ (pick the top of the first protrusion as shown in Figure 12.22) ⇒ **Okay** (to confirm direction of feature creation) ⇒ **Top** ⇒ (choose **DTM2** as the orientation plane)

Dbms ⇒ Save ⇒ enter Purge ⇒ enter ⇒ Done-Return

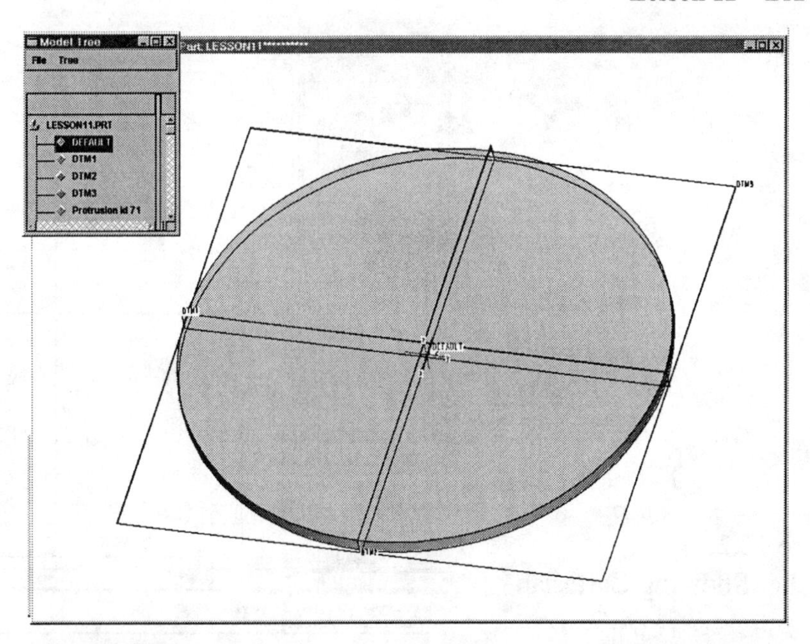

Figure 12.20
First Protrusion

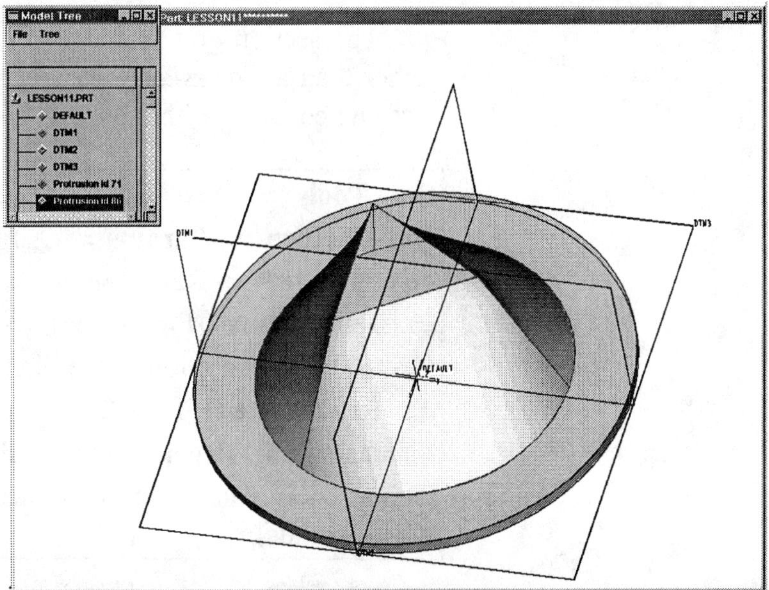

Figure 12.21
Blend Feature

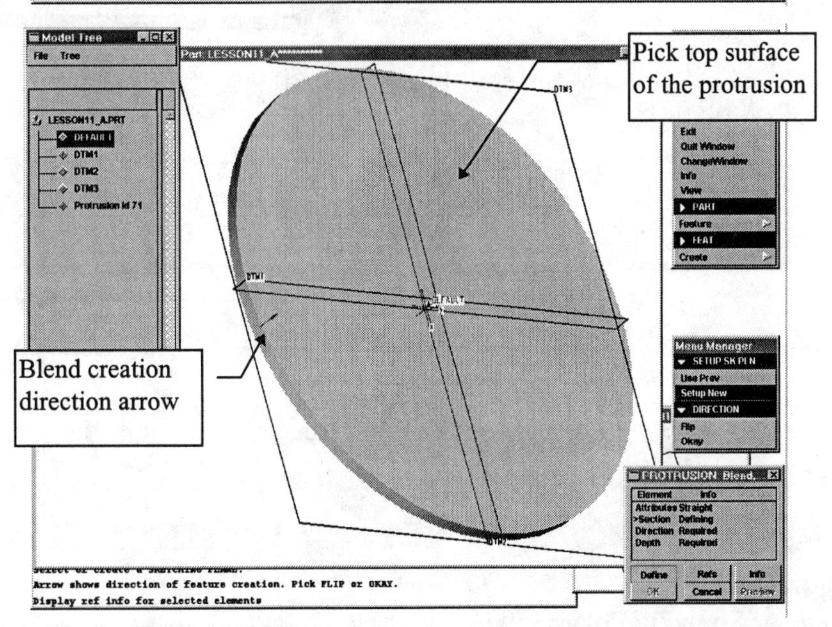

Figure 12.22
Blend Feature Starting Surface and Direction of Creation

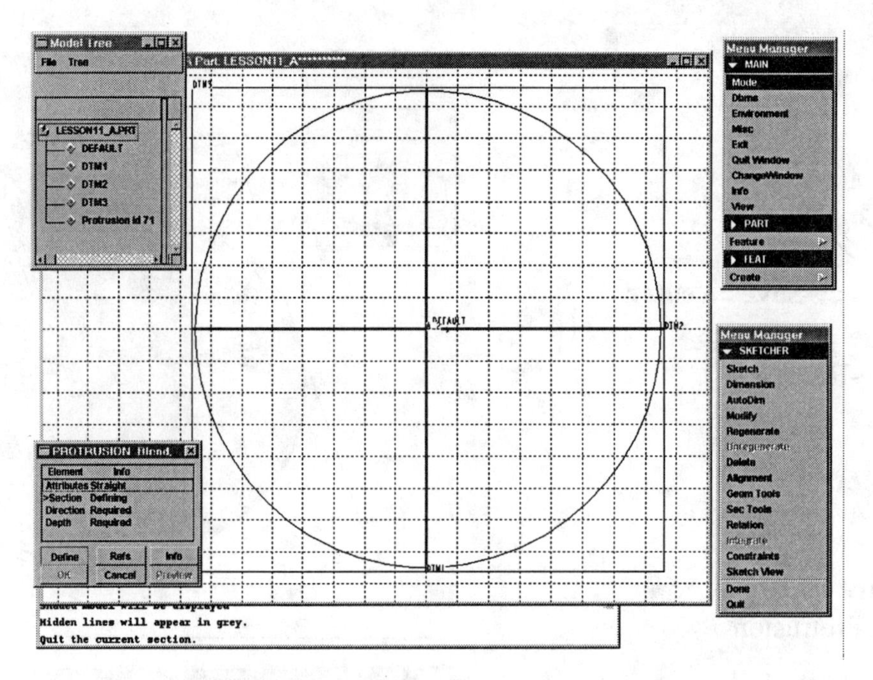

Figure 12.23
Sketcher Showing Cartesian Grid

The section grid would be better utilized if it were a *polar* grid rather than a *Cartesian* grid (Fig. 12.23). Change the grid type and size and continue with the blend commands:

Sec Tools ⇒ **Sec Environ** ⇒ **Grid** ⇒ **Type** ⇒ **Polar** ⇒ **Done/Return** ⇒ **Params** ⇒ **Ang Spacing** ⇒ **enter** (to accept the default of **30°**) ⇒ **Rad Spacing** ⇒ (type **.50** at the prompt) ⇒ **enter** ⇒ **Done/Return** (Fig. 12.24) ⇒ **Sketch** ⇒ **Arc** ⇒ **Ctr/Ends** ⇒ (sketch the first section of the blend by creating three equal **120°** arcs) ⇒ **Regenerate** (Fig. 12.25) ⇒ **Dimension** (add the diameter dimension) ⇒ **Regenerate** ⇒ **Alignment** (select each arc and the default coordinate system) ⇒ **Sketch** (add two centerlines to locate the arcs' ends)

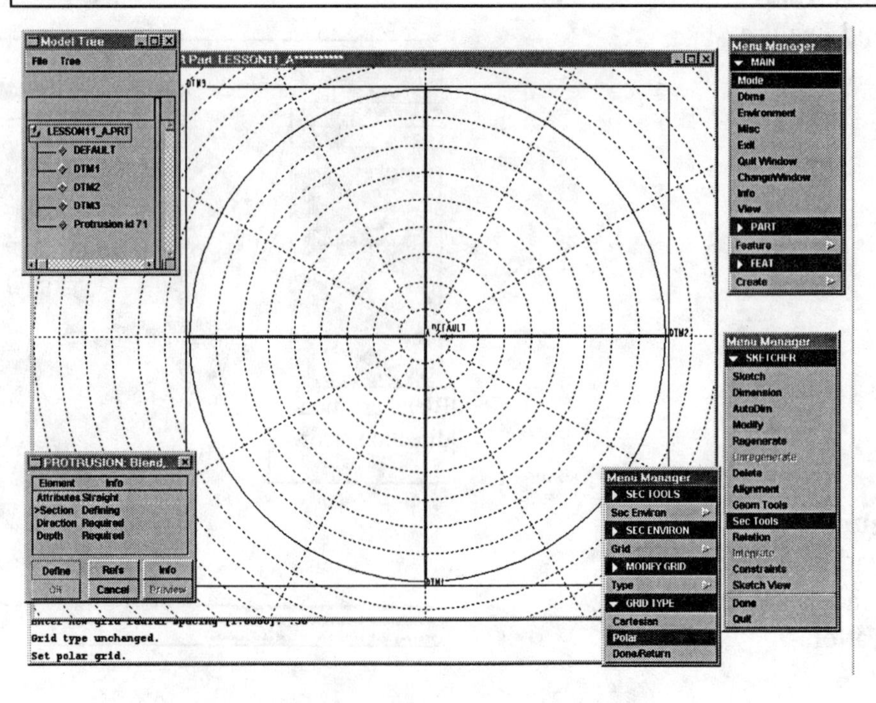

Figure 12.24
Sketcher Showing Polar Grid

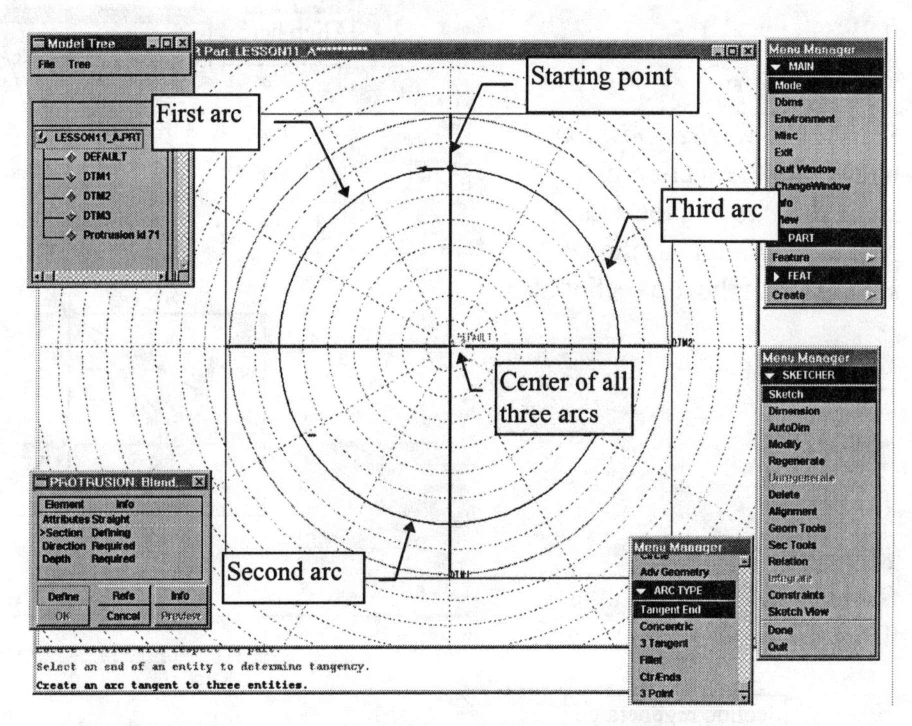

Figure 12.25
Sketch the Three Equal **120°** Arcs from the Center of the Part

The centerlines locating the arcs' ends are needed to get the sketch to regenerate. Continue with the command:

Line ⇒ **Centerline** ⇒ **Dimension** (dimension the centerlines from the vertical datum plane so as to create a **120°** angle as shown in Figure 12.26) ⇒ **Regenerate** ⇒ **Sec Tools** ⇒ **Toggle** (to sketch the second parallel section) ⇒ **Sketch** (the first section is *grayed out*) ⇒ **Line** ⇒ (sketch the three lines of the triangle starting at **DTM1** and picking points *in the same direction* as the arcs were created) ⇒ **Alignment** ⇒ **Regenerate** ⇒ **Dimension** ⇒ **Regenerate** (Fig. 12.27)

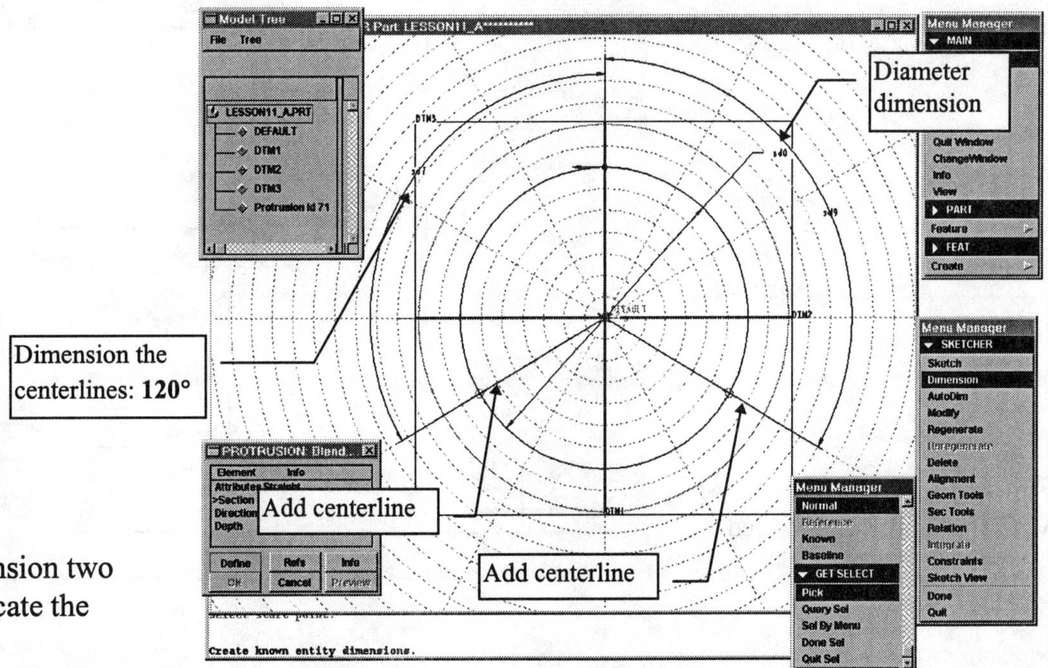

Figure 12.26
Sketch and Dimension two Centerlines to Locate the Ends of Each Arc

HINT

Start at the *vertical position* and create the section in the *same direction* as the arcs. Make the triangle section *smaller* than the arc section.

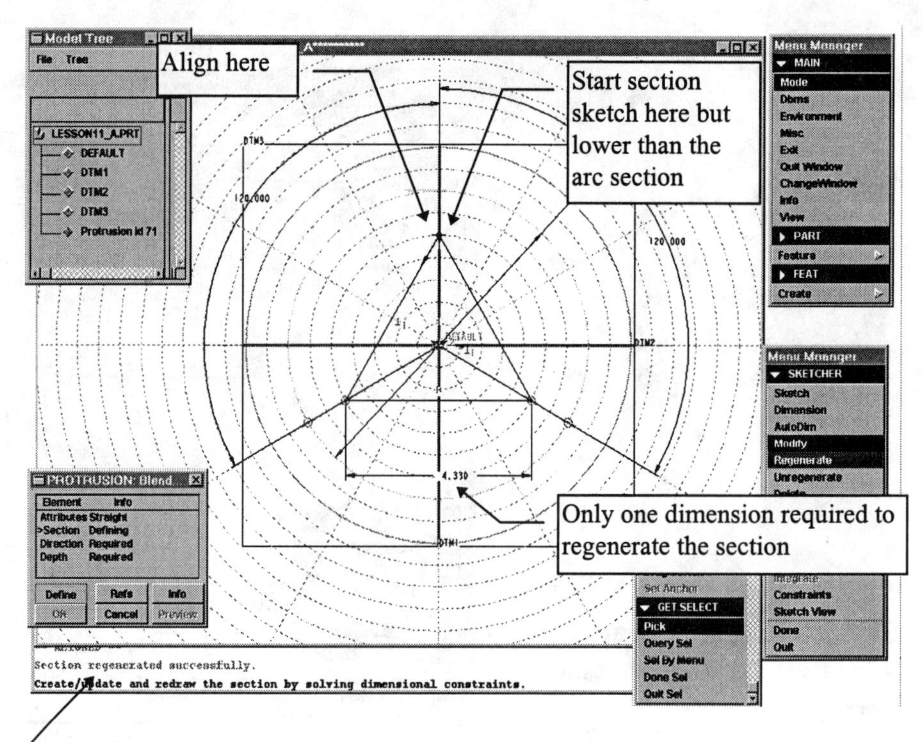

Figure 12.27
Sketch and Dimension the Three Lines

Modify and regenerate the sketch:

Modify (change the diameter dimension to **7.75**, the leg of the triangle dimension to **3.00** and the two angles to 120°) ⇒ **Regenerate** (Fig. 12.28) ⇒ **Done** ⇒ **View** ⇒ **Default** ⇒ **Done-Return** ⇒ **Blind** ⇒ **Done** (type **3.00** at the prompt) ⇒ **enter** ⇒ **Preview** ⇒ **View** ⇒ **Cosmetic** ⇒ **Shade** ⇒ **Display** ⇒ **Done-Return** (Fig. 12.29) ⇒ **OK** ⇒ **Done**

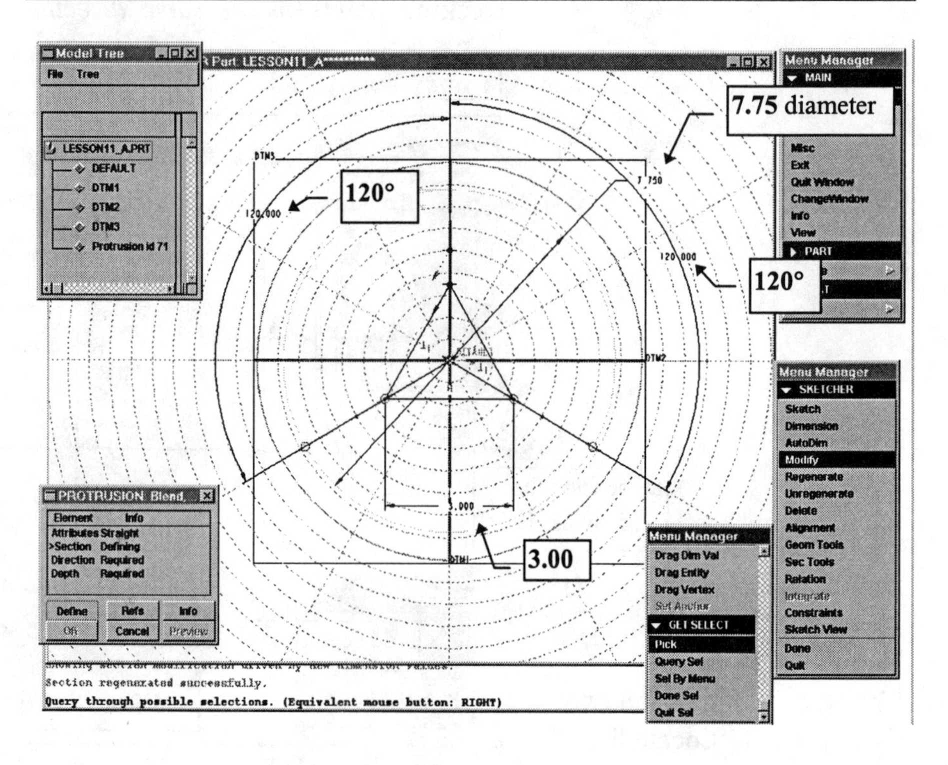

Figure 12.28
Modified and Regenerated Sketch

Dbms ⇒ Save ⇒ enter
Purge ⇒ enter ⇒
Done-Return

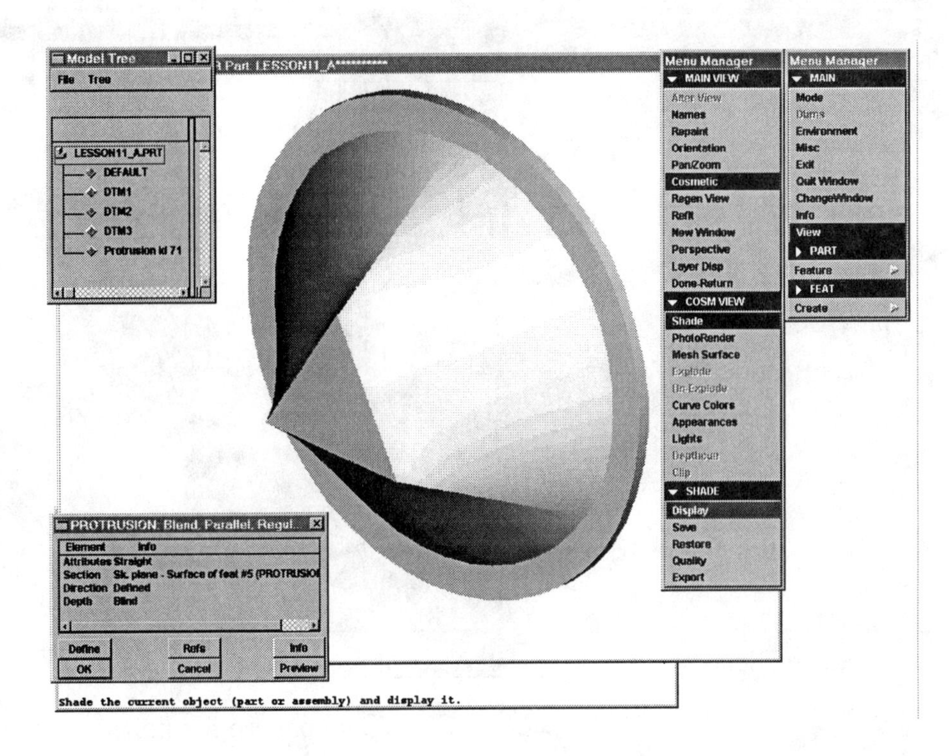

Figure 12.29
Shaded Preview of the Blend

You made a mistake (*we made a mistake*). The diameter of the blend was suppose to be **6.50,** not **7.75**, which is the diameter of the bolt circle.

Modify ⇒ (pick the blend protrusion from the Model Tree, as shown in Figure 21.30) ⇒ (pick the **7.75** dimension) ⇒ (change the value to **6.50**; see Figure 12.15) ⇒ **enter** ⇒ **Regenerate** (Fig. 12.31)

Dbms ⇒ Save ⇒ enter
Purge ⇒ enter ⇒
Done-Return

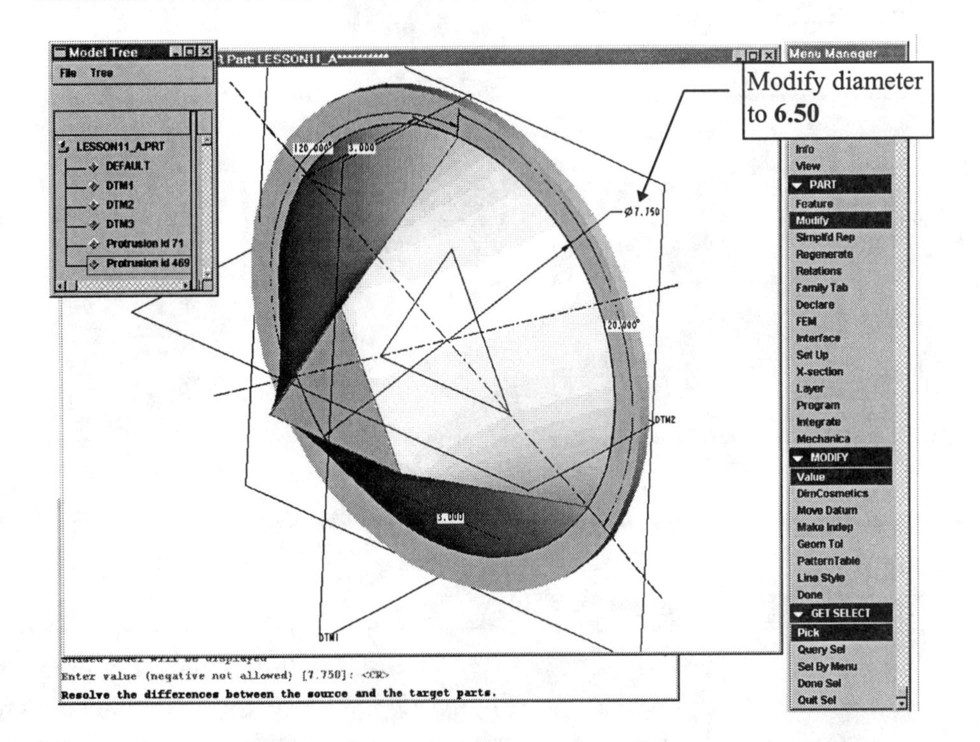

Figure 12.30
Modify the **7.75** diameter to **6.50**

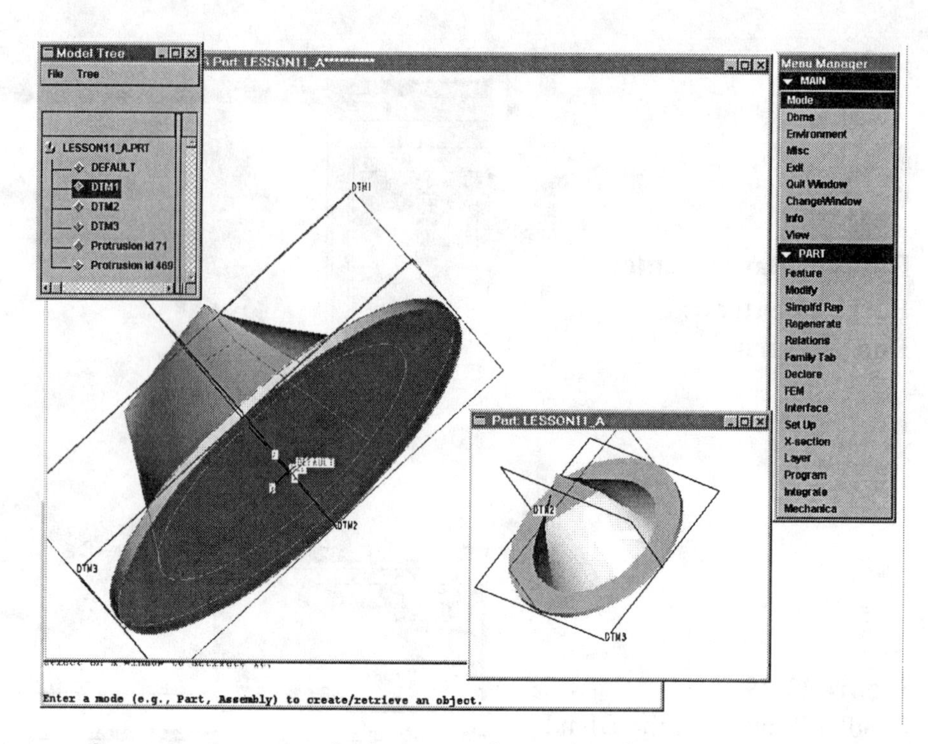

Figure 12.31
Completed Blend

Create and pattern the holes as shown in Figure 12.32. Shell the part using the following commands:

Feature ⇒ **Create** ⇒ **Shell** ⇒ (pick the bottom surface of the part as the surface to remove, as shown in Figure 12.33) ⇒ **Done Sel** ⇒ **Done Refs** ⇒ (type the thickness of **.125** at the prompt) ⇒ **enter** ⇒ **OK** (Fig. 12.33) ⇒ **Done**

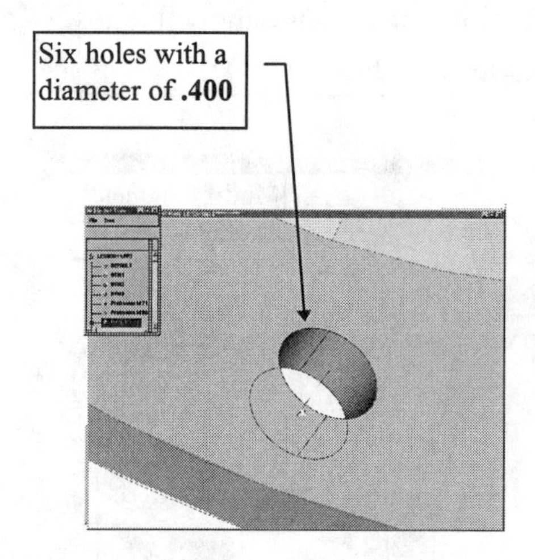

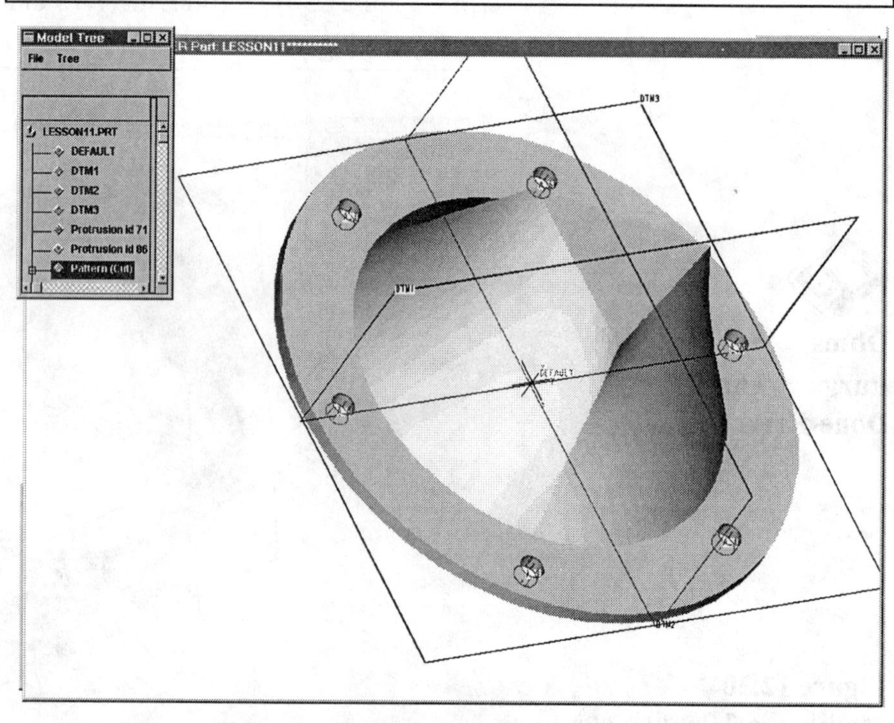

Figure 12.32
Create the Hole Pattern

The **Shell** will automatically create the bosses (Figure 12.34 and Figure 12.35) since the **.125** thickness is left around all previously created features. If the bosses were not desired, then you would simply **Reorder** the holes to come after the shell feature.

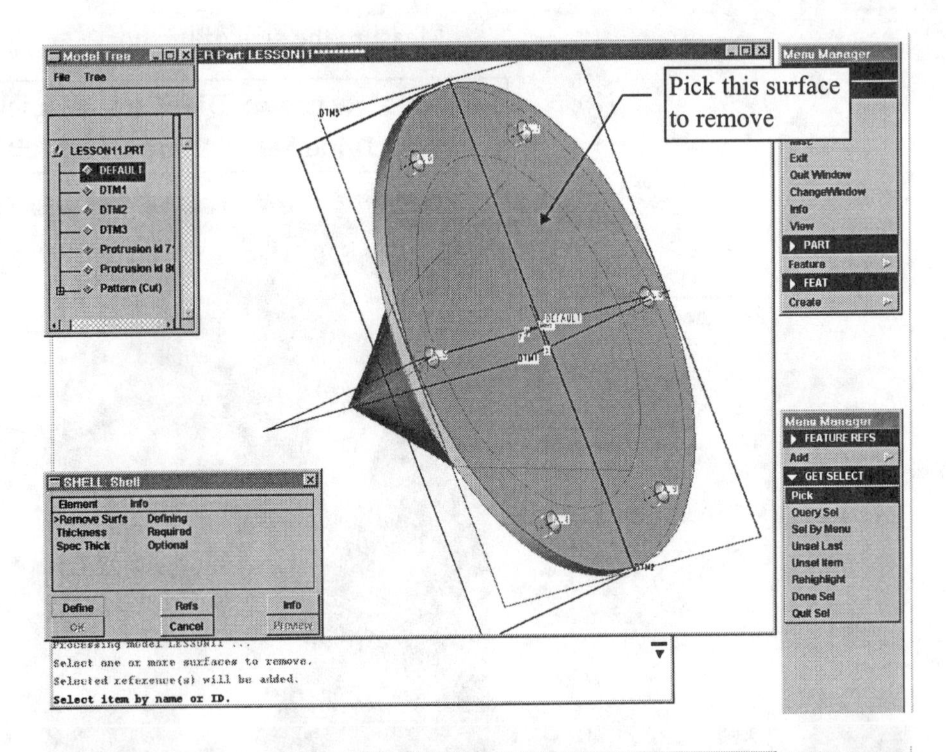

Figure 12.33
Shelling the Part

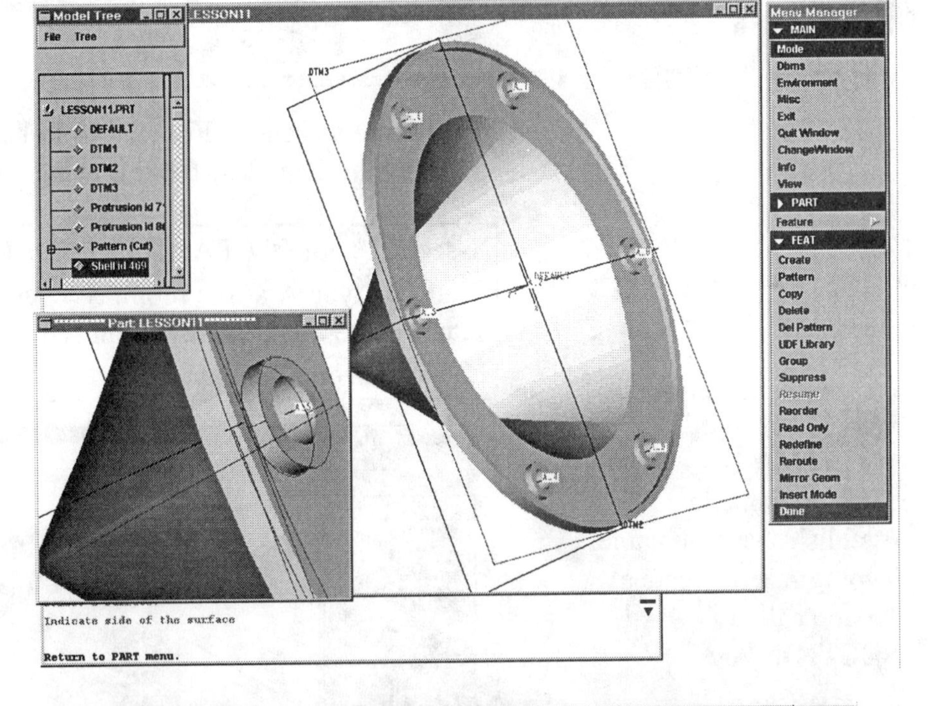

Figure 12.34
Shelled Part

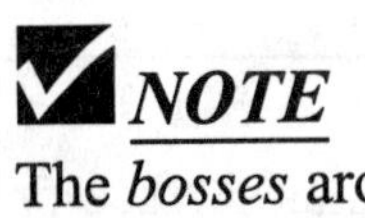

NOTE

The *bosses* around the holes are created automatically at **.250** larger than the holes (**.250** + **.400** = **.65** diameter)

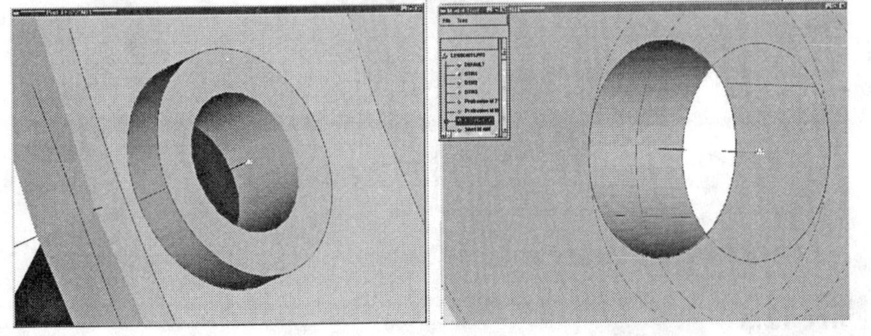

Figure 12.35
Hole and Boss

Measure the size of the boss:

Info ⇒ **Measure** ⇒ **Diameter** ⇒ (pick the boss as shown in Figure 12.36) ⇒ **Done Sel** ⇒ **Done/Return** ⇒ **Done/Return**

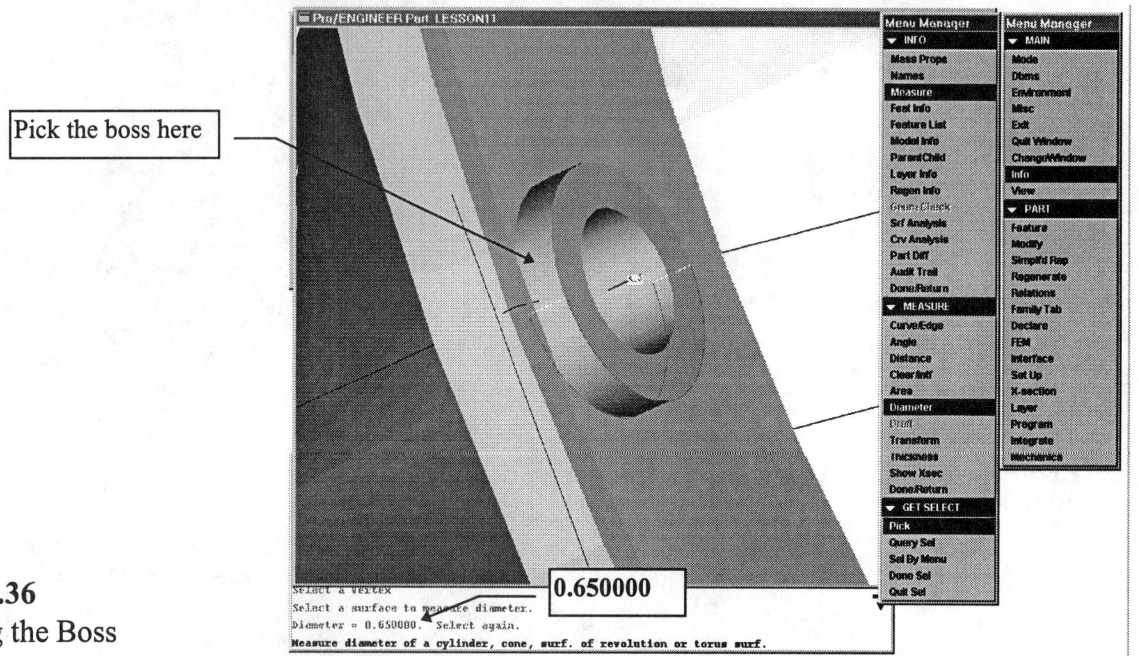

Figure 12.36
Measuring the Boss

Create a section (Fig. 12.37) to be used in the Draw mode when detailing the Cap. Use the following commands:

X-section (from the PART menu) ⇒ **Create** ⇒ **Planar** ⇒ **Single** ⇒ **Done** ⇒ (type **A** at the prompt) ⇒ **enter** ⇒ (select planar surface or datum plane; pick **DTM1** from the Model Tree, as shown in Figure 12.37) ⇒ **Done-Return**

HINT
Type only one letter to establish a section name. Typing **A** will create a section called **SECTION A-A**.

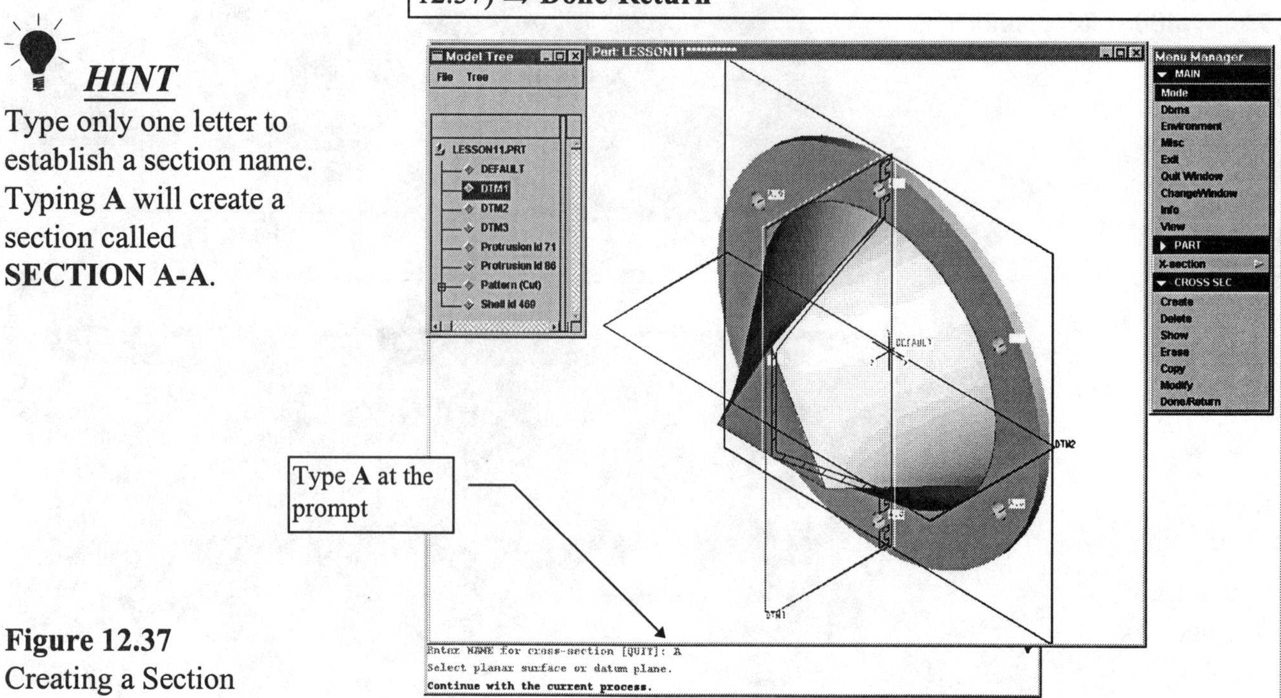

Figure 12.37
Creating a Section

Save the Cap under a new name, and complete the **ECO** to establish a second part with different sizes and features.

ECO

Dbms ⇒ Save As ⇒ enter ⇒ CAP_PART_ONE
Purge ⇒ enter ⇒ Done-Return

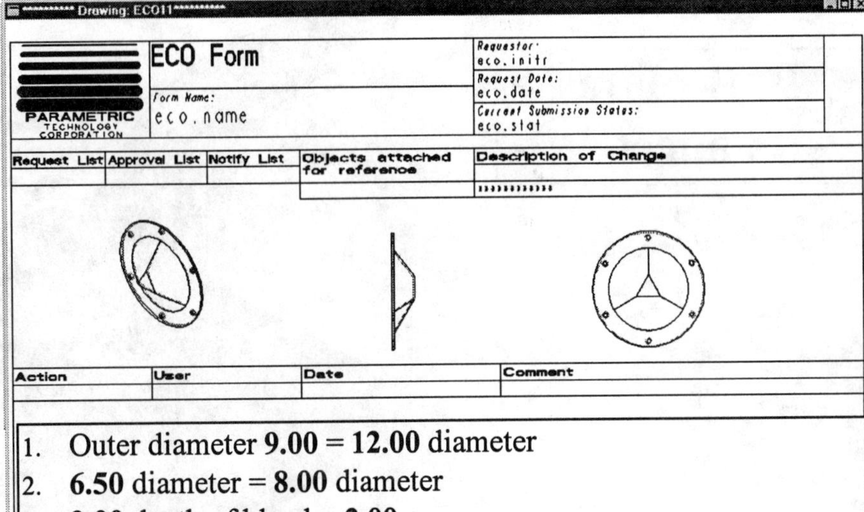

1. Outer diameter **9.00** = **12.00** diameter
2. **6.50** diameter = **8.00** diameter
3. **3.00** depth of blend = **2.00**
4. Triangle leg **3.00** = **2.00**
5. Bolt circle diameter **7.75** = **10.00** diameter
6. *Add rounds to all edges*-- design size just below failure

Figure 12.38
ECO

Dbms ⇒ Save ⇒ enter
Purge ⇒ enter ⇒ Done-Return

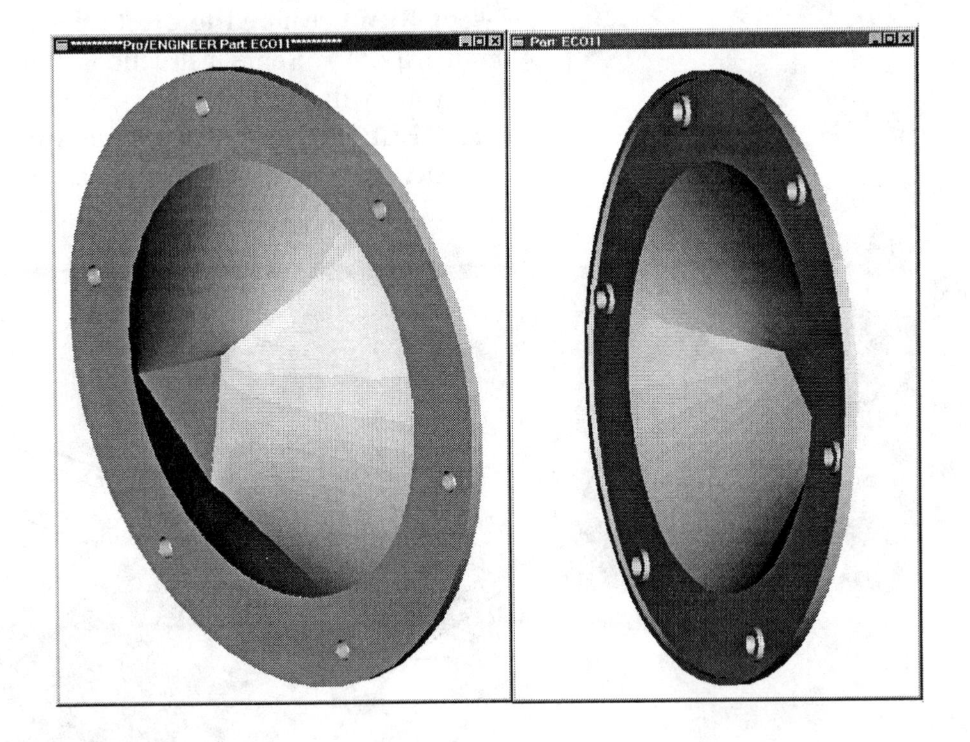

Figure 12.39
ECO Changes to Model

Lesson 12 Project

Bathroom Faucet

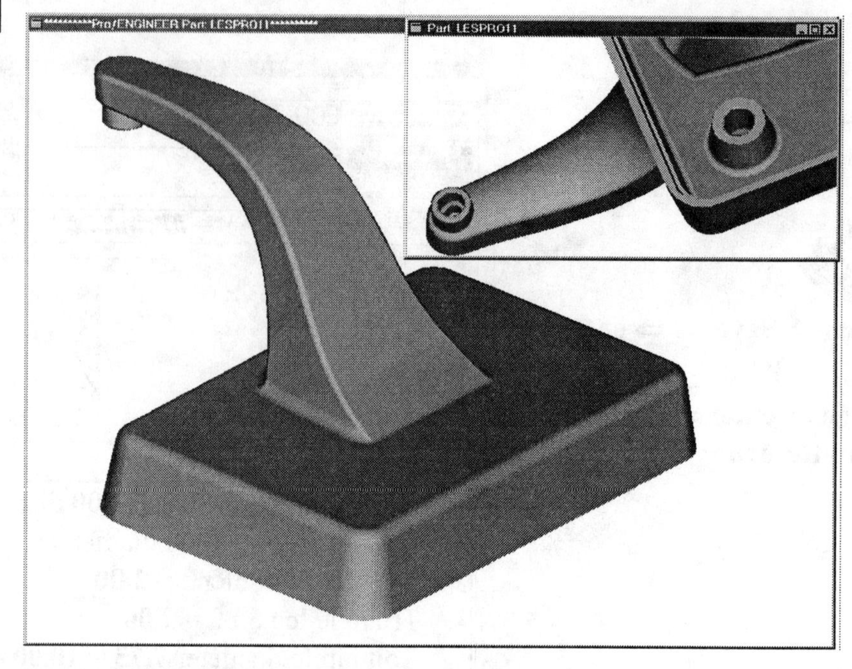

Figure 12.40
Bathroom Faucet

Bathroom Faucet

This is an advanced project. Since you have created over twenty parts using Pro/E you should be able to use that knowledge and model the **Swept Blend** required to create the **Bathroom Faucet** (Figs. 12.40 through 12.68). Some instructions will accompany the lesson project, but you will be required to research online documentation **(Pro/HELP)** and learn more about **Splines** and **Swept Blends**.

After the model is complete, create a number of sections that can be used in the Drawing mode when detailing the Faucet.

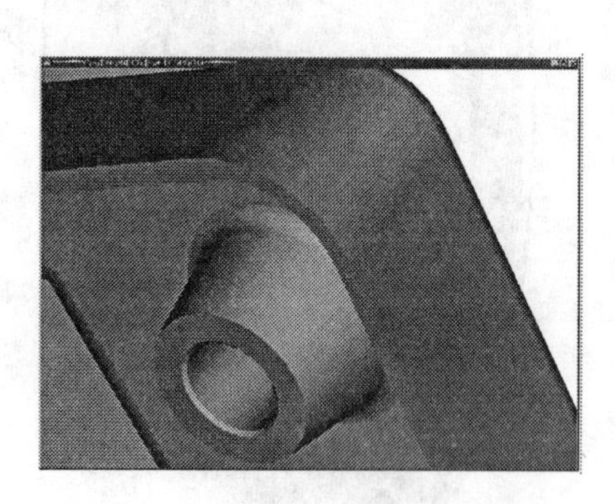

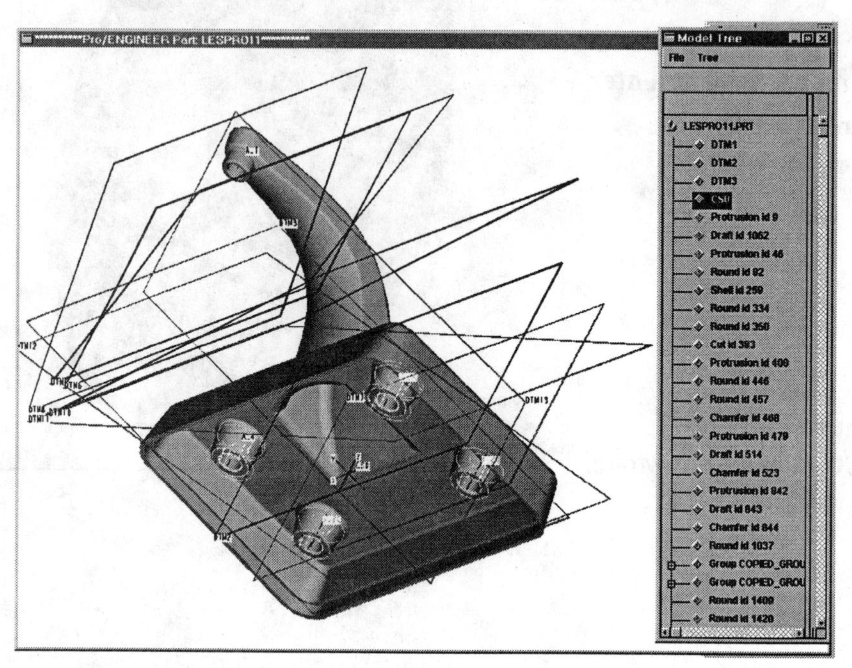

Figure 12.41
Bathroom Faucet with Model Tree and Datum Planes

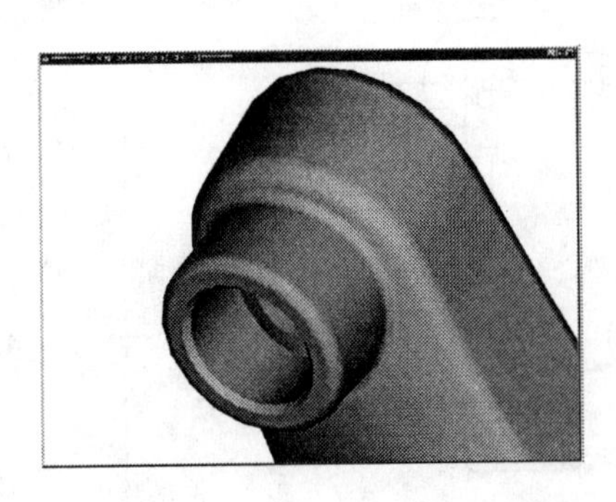

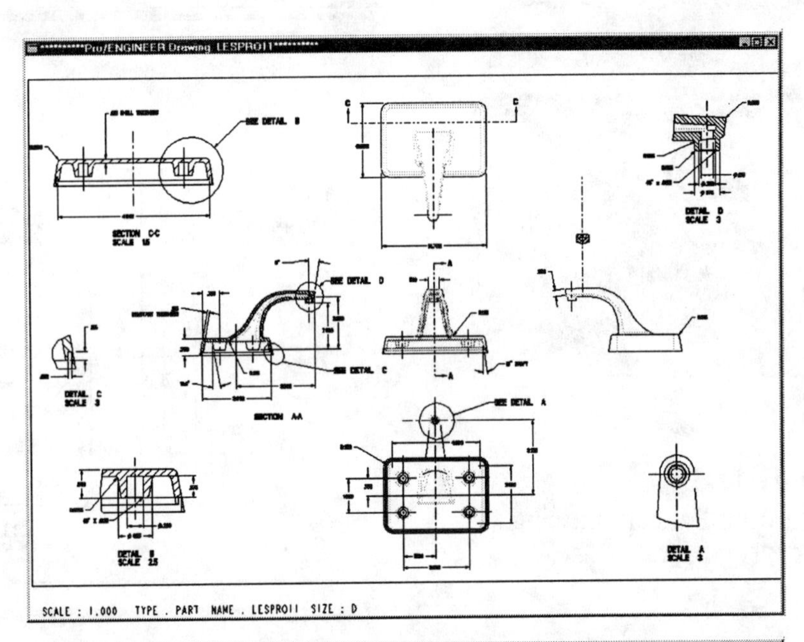

Figure 12.42
Bathroom Faucet, Detail Drawing

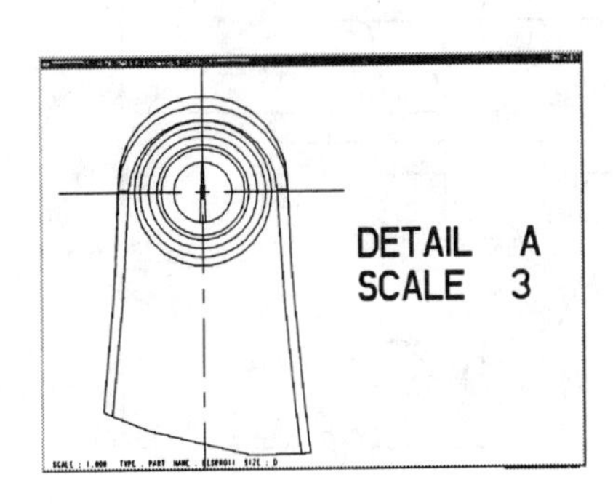

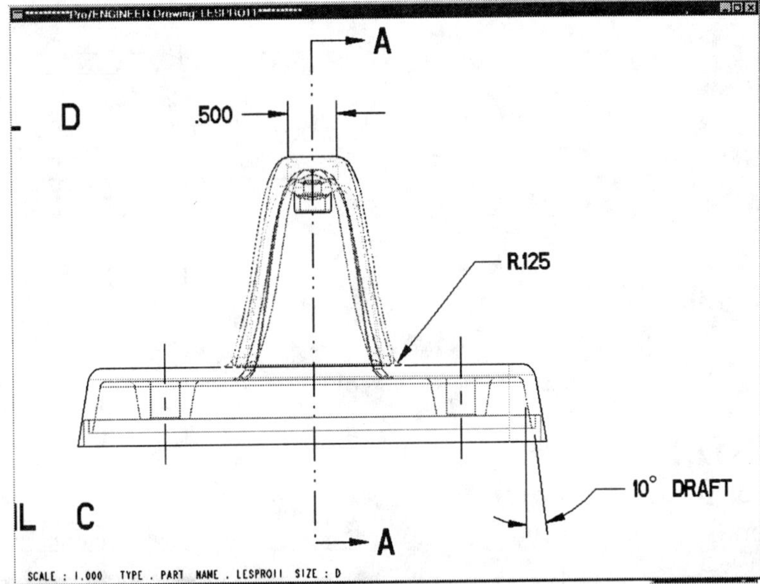

Figure 12.43
Bathroom Faucet Drawing, Front View

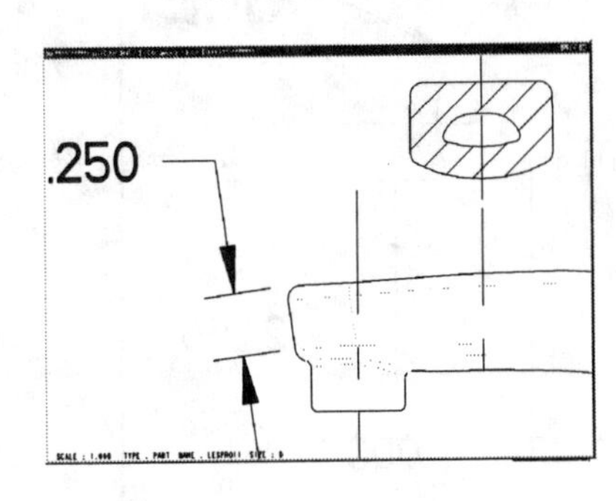

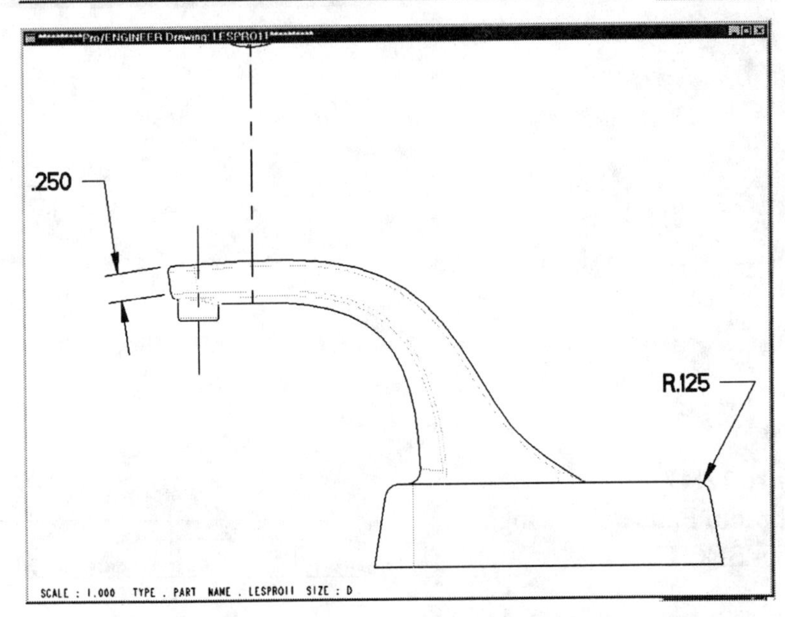

Figure 12.44
Bathroom Faucet Drawing, Right Side View

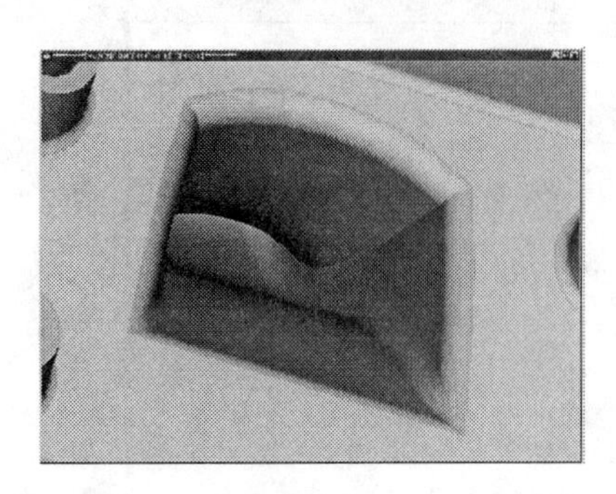

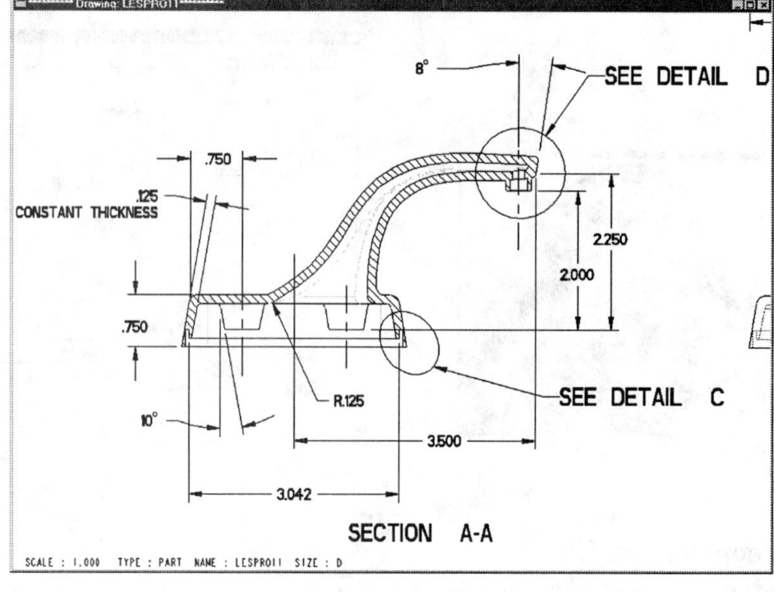

Figure 12.45
Bathroom Faucet Drawing,
SECTION A-A

Figure 12.46
Bathroom Faucet Drawing,
Lower **SECTION A-A**

Figure 12.47
Bathroom Faucet Drawing,
Upper **SECTION A-A**

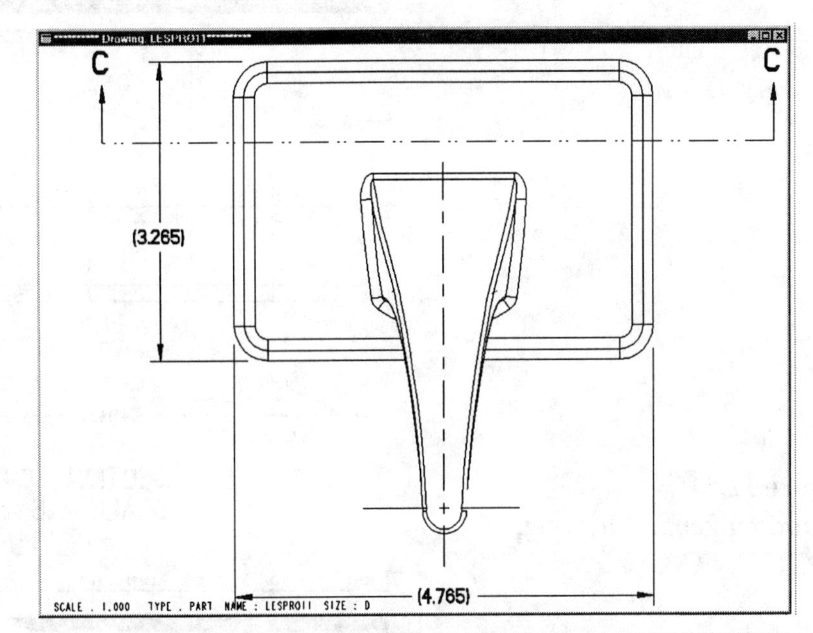

Figure 12.48
Bathroom Faucet Drawing,
Top View

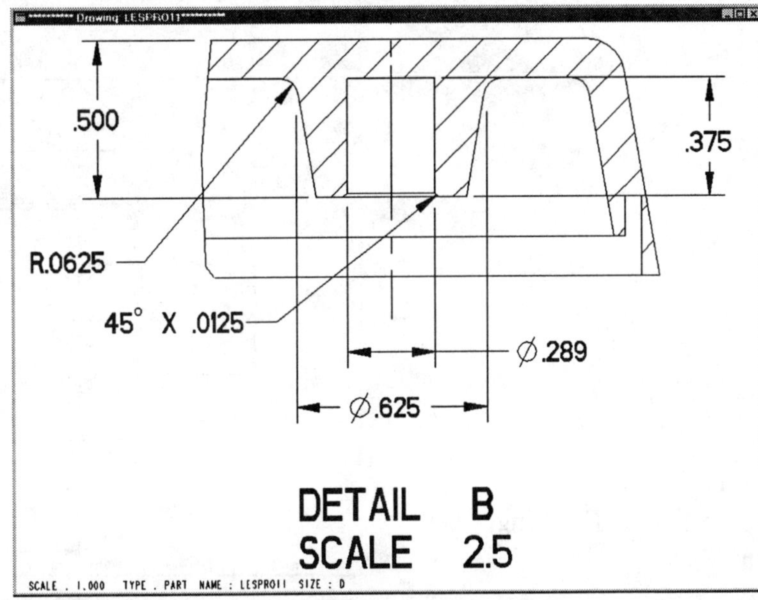

Figure 12.49
Bathroom Faucet Drawing,
DETAIL B

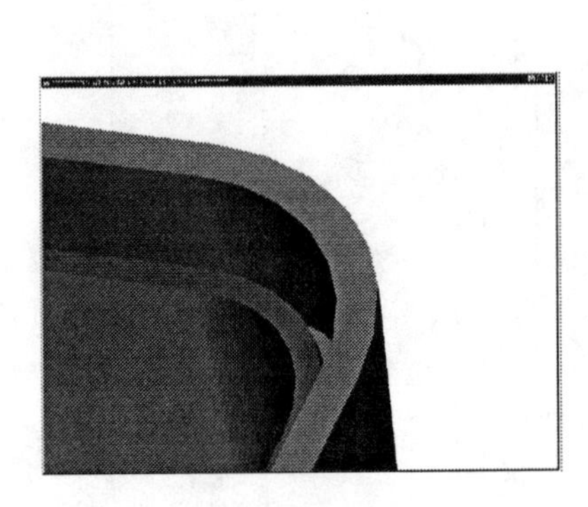

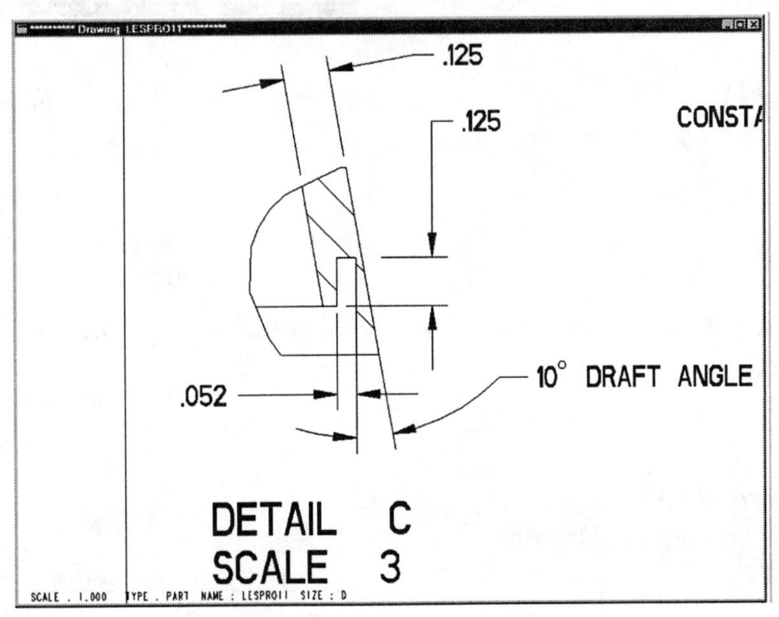

Figure 12.50
Bathroom Faucet Drawing,
DETAIL C

Figure 12.51
Bathroom Faucet Drawing,
SECTION C-C

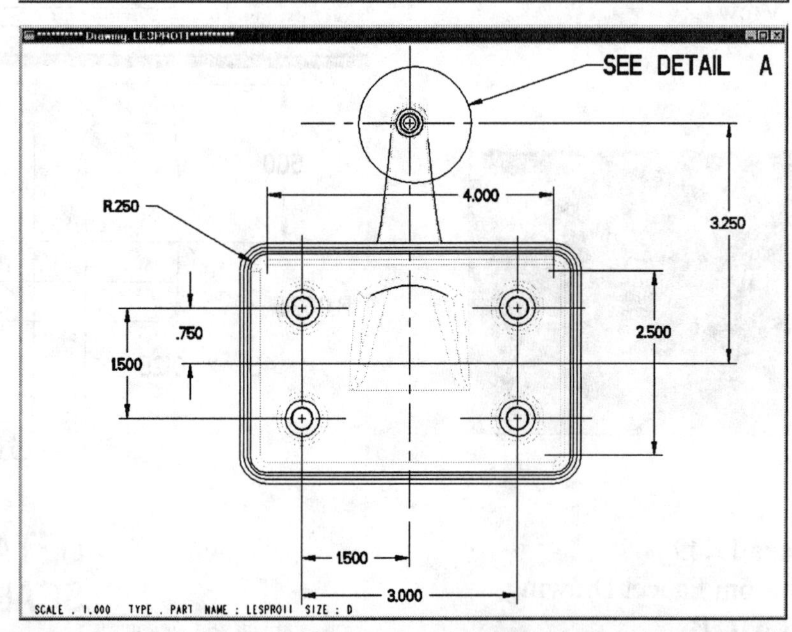

Figure 12.52
Bathroom Faucet Drawing,
Bottom View

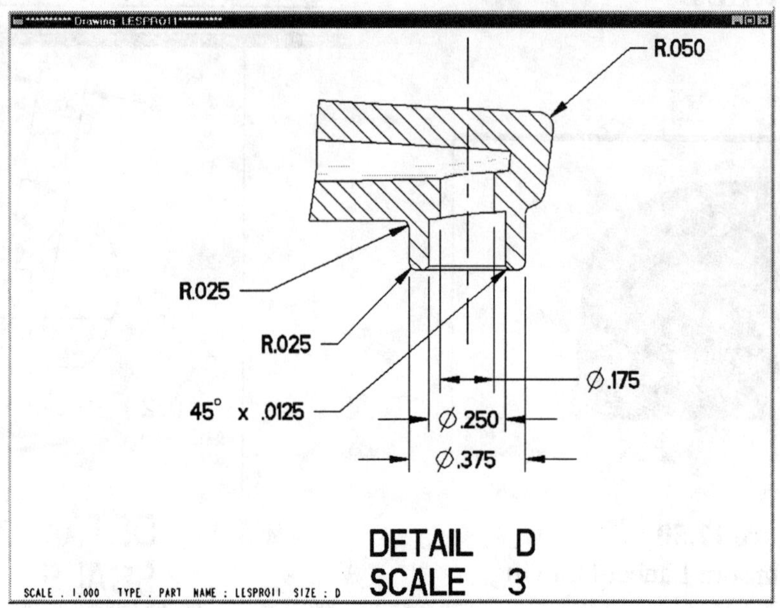

Figure 12.53
Bathroom Faucet Drawing,
DETAIL D

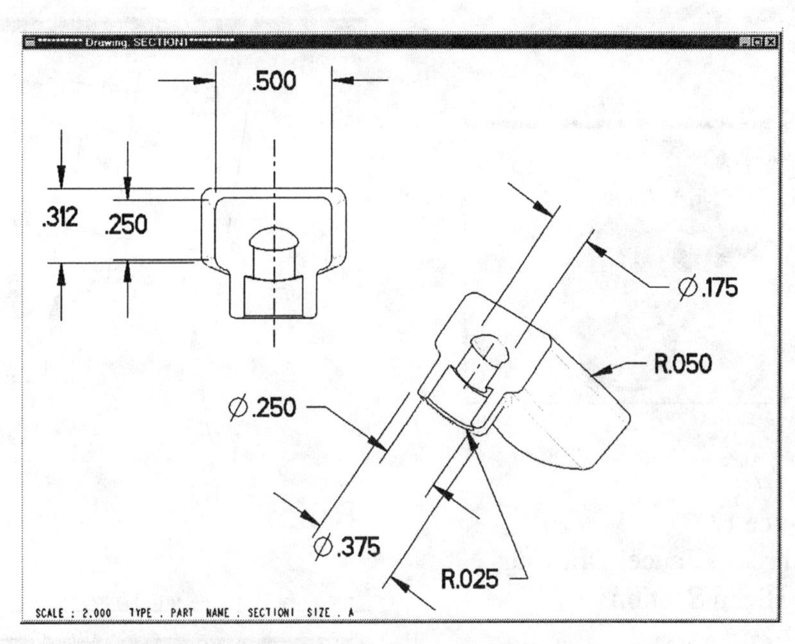

Figure 12.54
Bathroom Faucet Drawing,
Spout Section

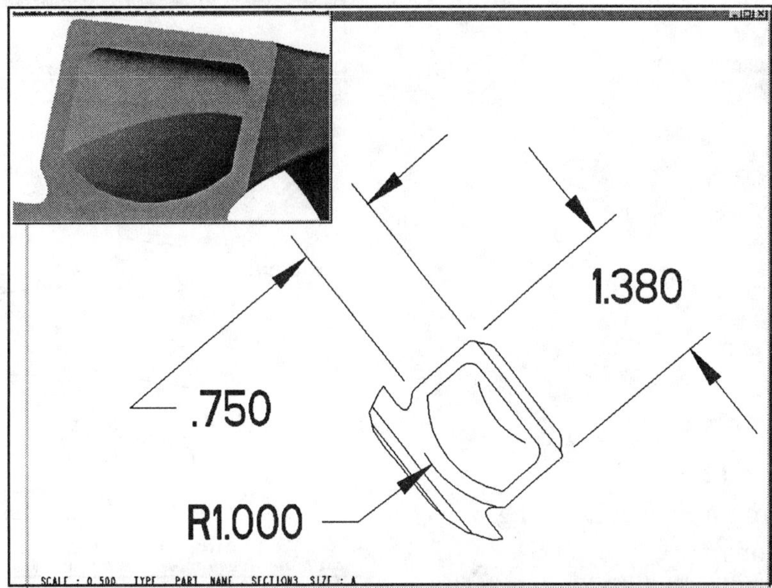

Figure 12.55
Bathroom Faucet Drawing,
2nd Blend Section

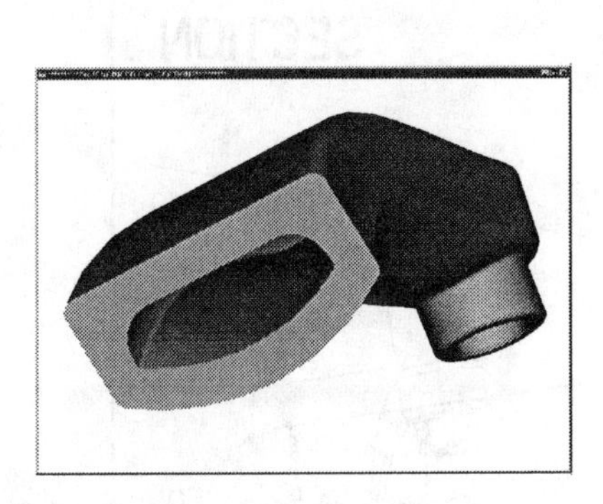

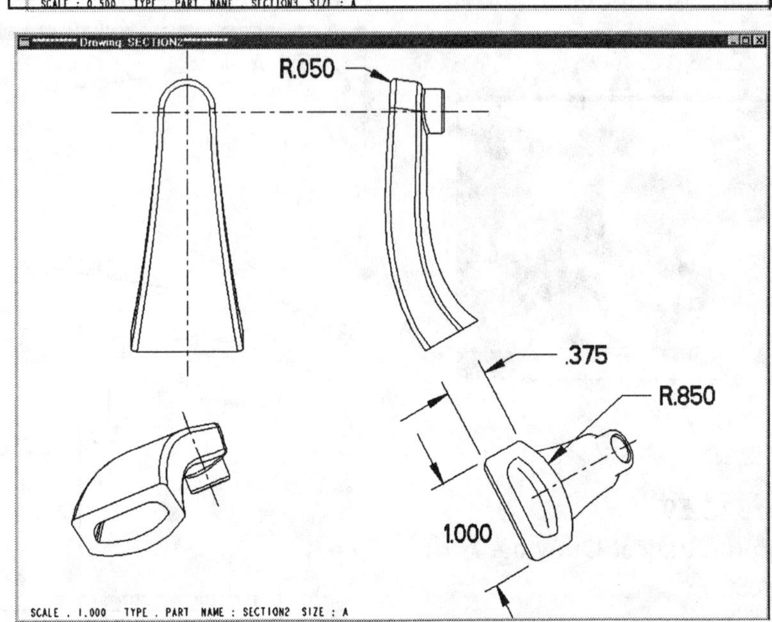

Figure 12.56
Bathroom Faucet Drawing,
3rd Blend Section

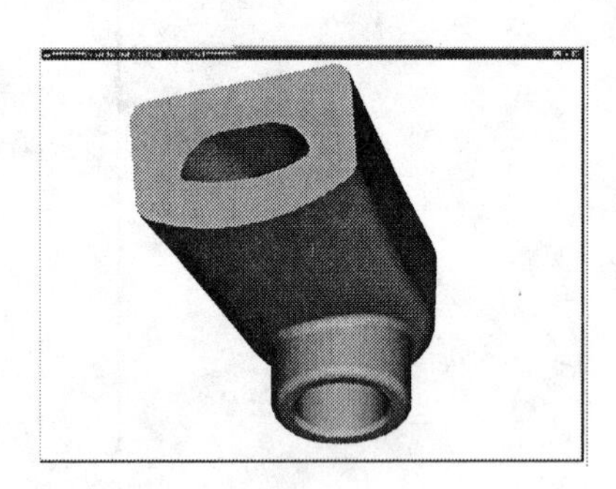

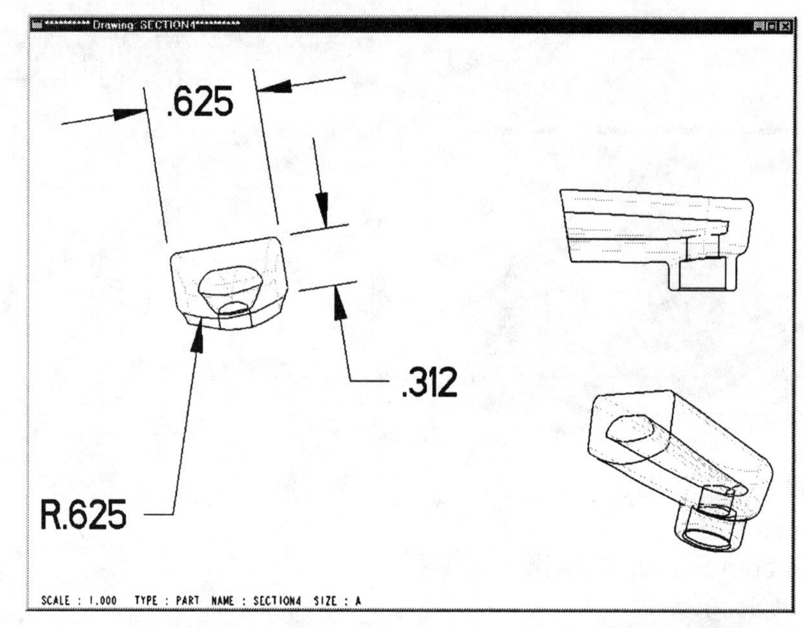

Figure 12.57
Bathroom Faucet Drawing,
4th Blend Section

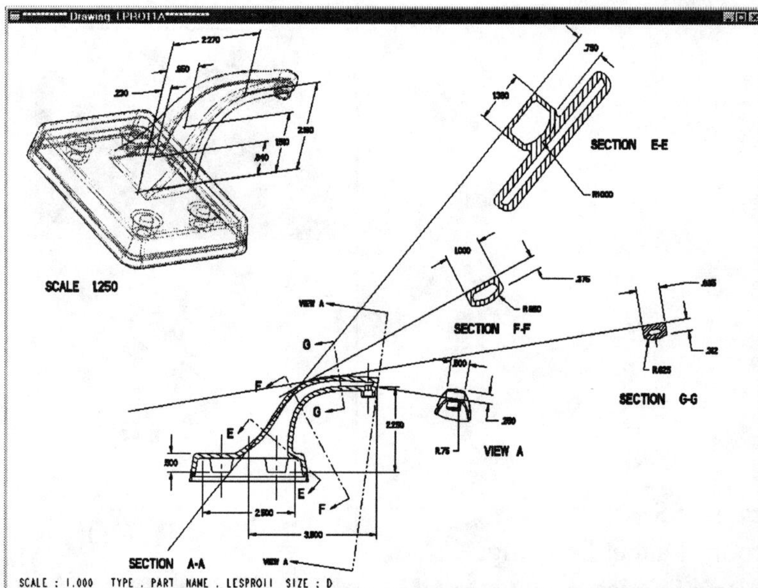

Figure 12.58
Bathroom Faucet Drawing,
Sections

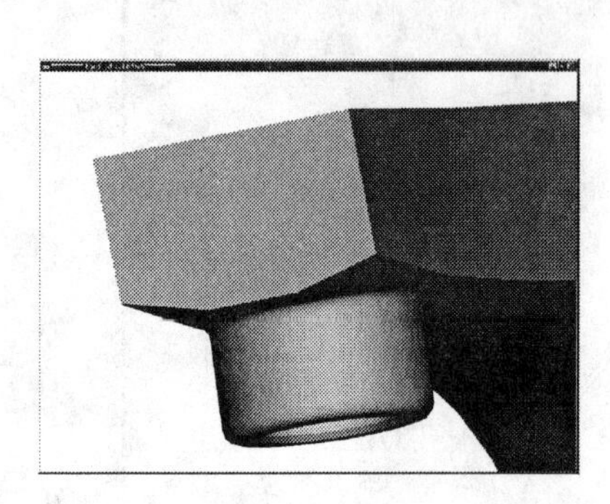

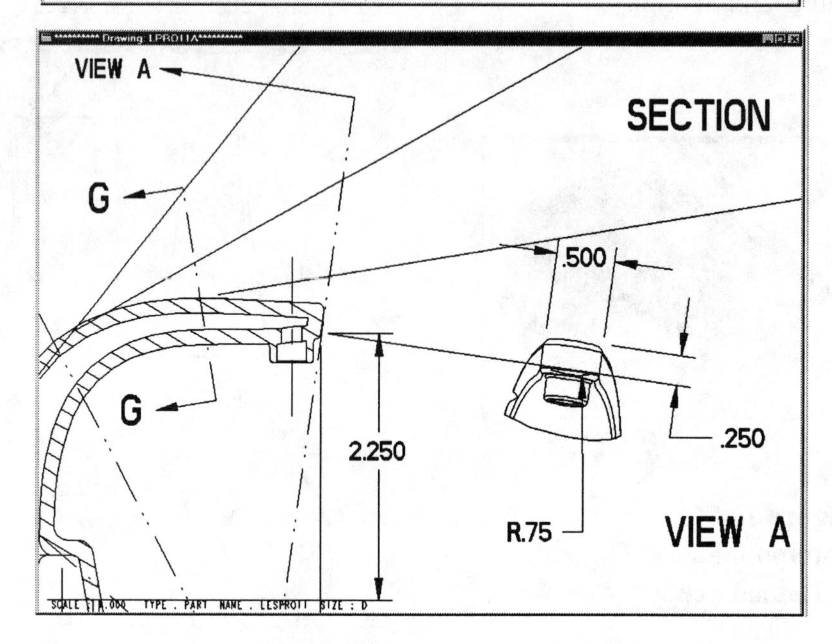

Figure 12.59
Bathroom Faucet Drawing,
5th Section

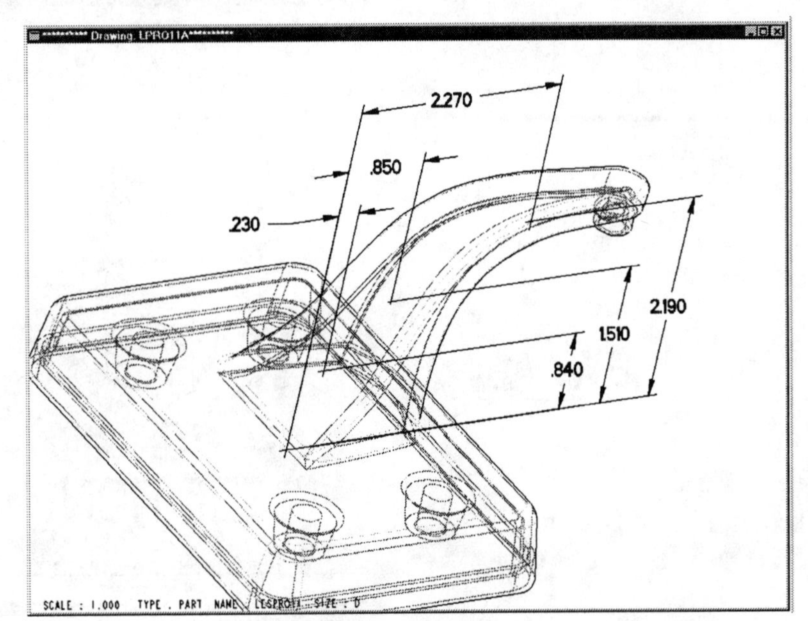

Figure 12.60
Bathroom Faucet Drawing, First Three Blend Section Locations

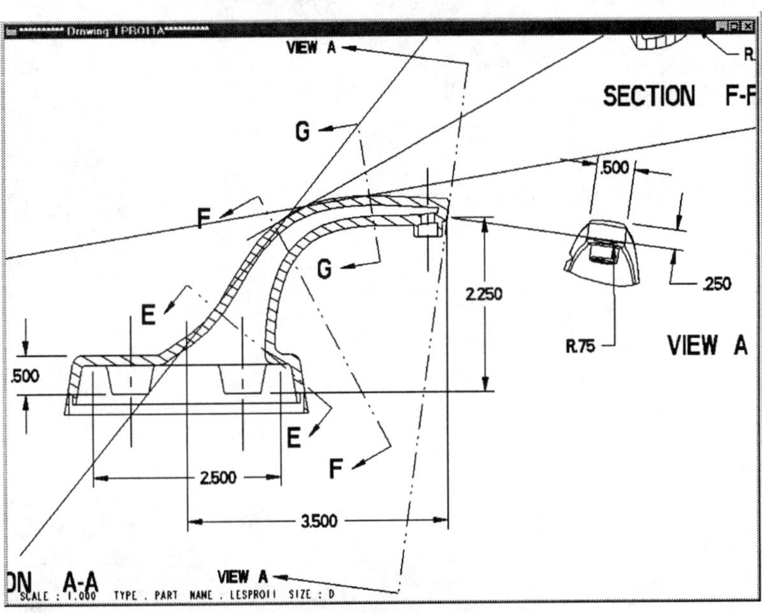

Figure 12.61
Bathroom Faucet Drawing,
SECTION A-A

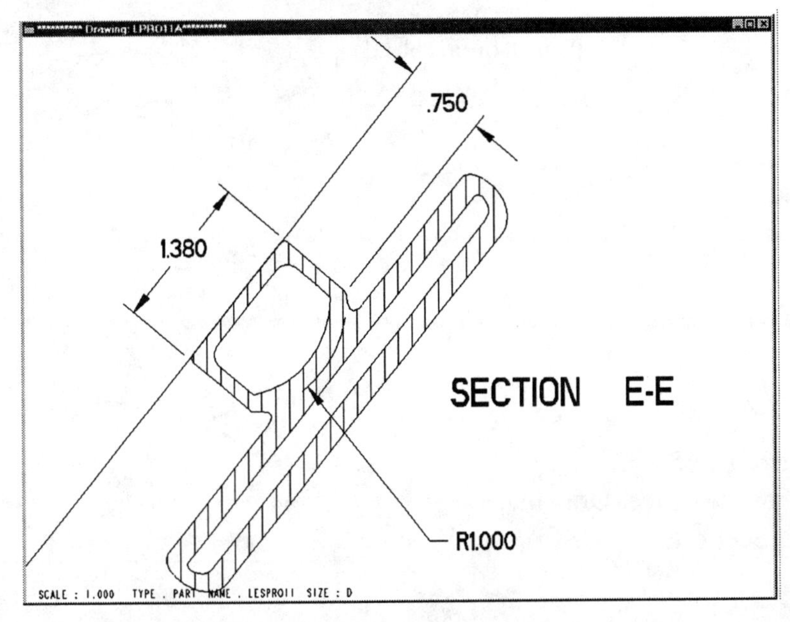

Figure 12.62
Bathroom Faucet Drawing,
SECTION E-E

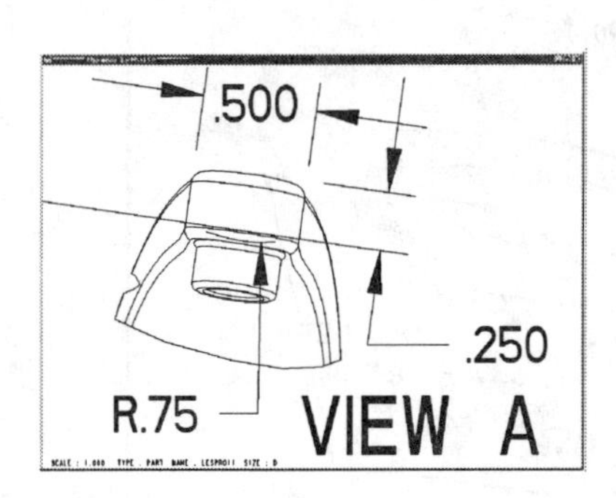

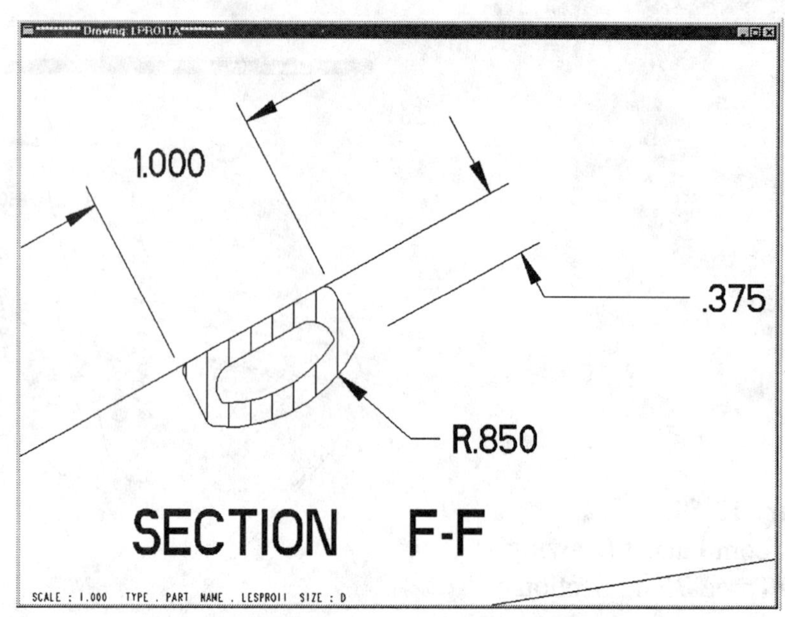

Figure 12.63
Bathroom Faucet Drawing,
SECTION F-F

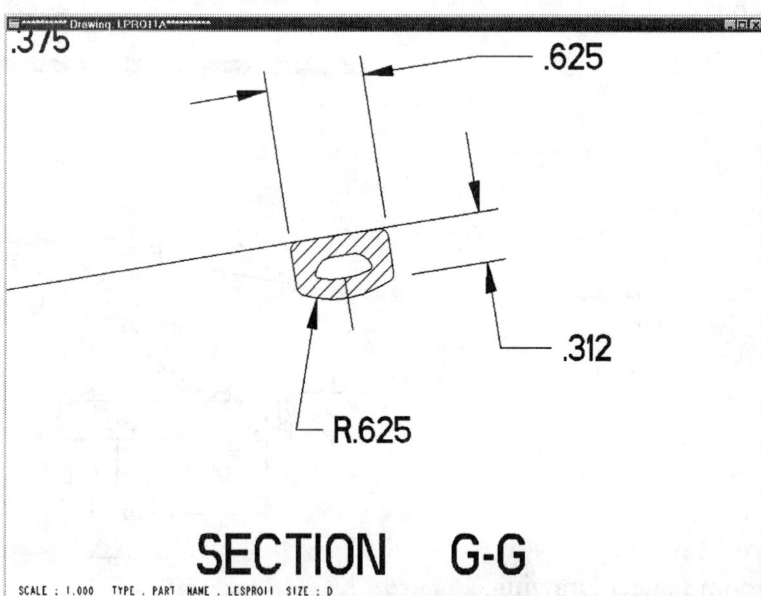

Figure 12.64
Bathroom Faucet Drawing,
SECTION G-G

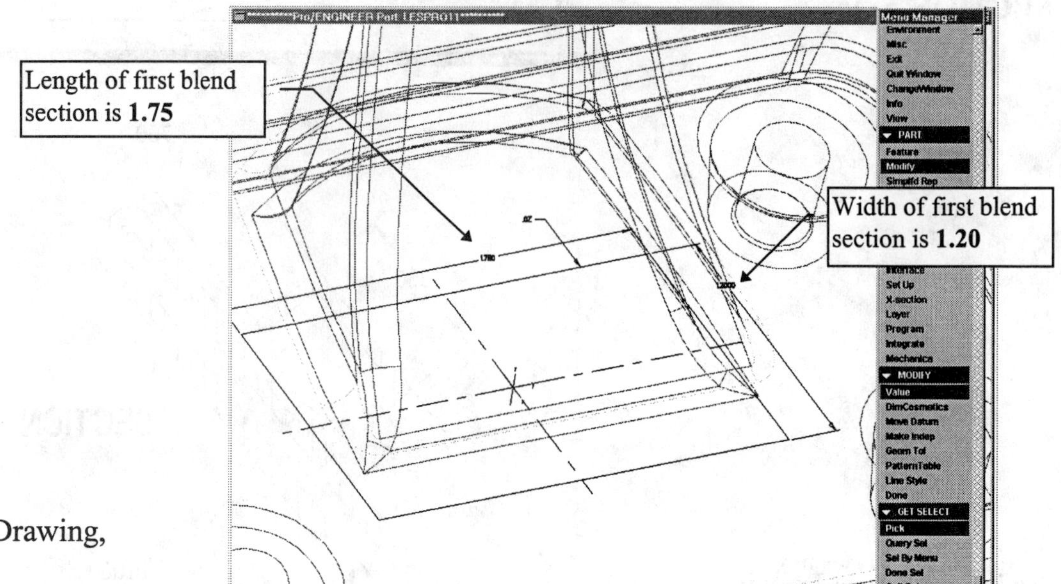

Figure 12.65
Bathroom Faucet Drawing,
First Section

Figure 12.66
Bathroom Faucet, First Protrusion and Draft

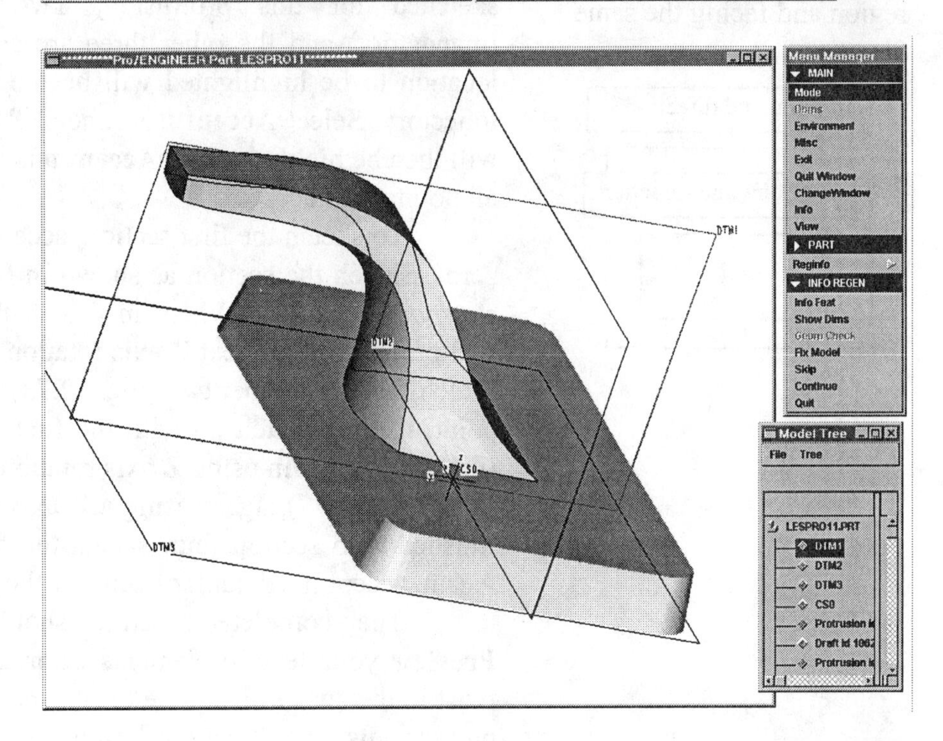

Figure 12.67
Bathroom Faucet, Swept Protrusion

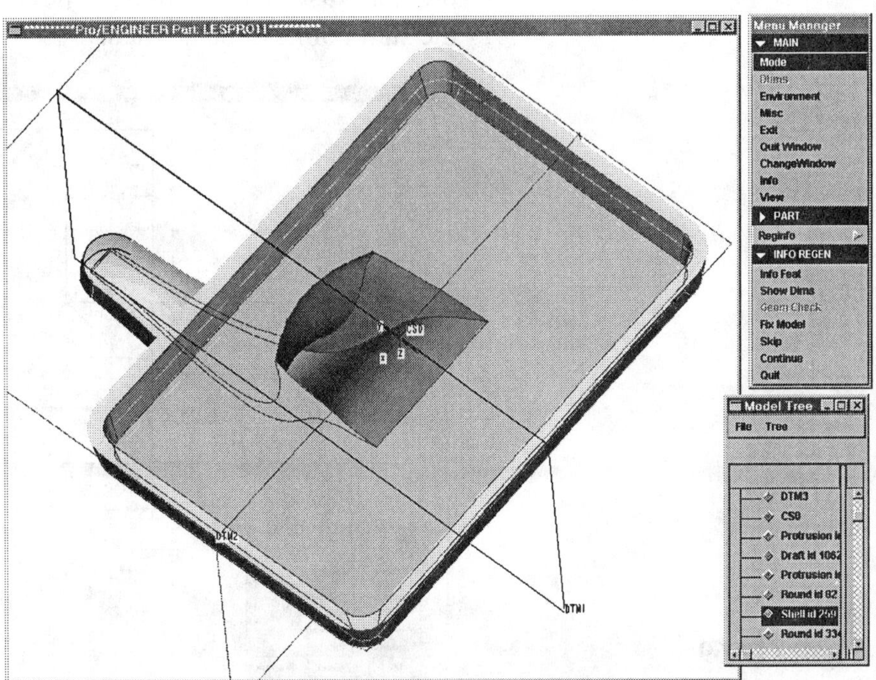

Figure 12.68
Bathroom Faucet, Shell

HINT

Each section in the **Swept Blend** has four entities. Keep the start point direction arrow of each section in the same region and facing the same direction.

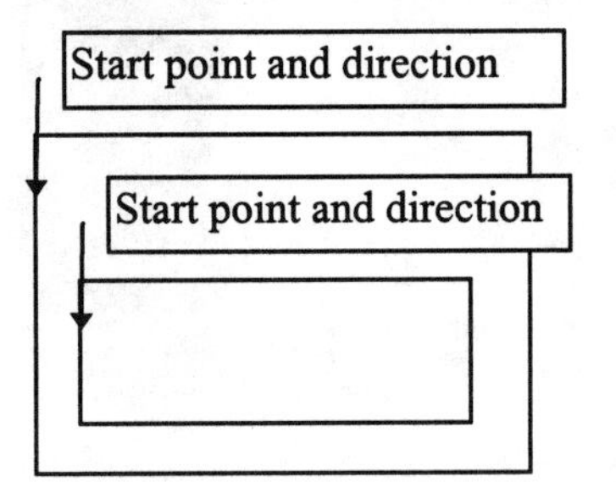

Create the first protrusion and draft as shown in Figure 12.66. Next, a **Swept Blend** will be used to create the geometry for the **Bathroom Faucet** as shown in Figure 12.69 through Figure 12.81. The default options of **Sketch Sec** and **Nrm To Spline** will be used. When prompted for the trajectory, choose **Sketch Traj**. Sketch a Spline using **Adv Geometry**. Create a total of five points along the trajectory. These will locate the five sections of the blend. Before you begin to sketch any sections, you are prompted for where the sections are to be located. There will be a total of five sections sketched for this protrusion. Two will be at the endpoints (mandatory) and the other three are at the datum points. The first location to be highlighted will be the second point of the sketched trajectory. Select **Accept** from the CONFIRM menu. The next point will then be highlighted so **Accept** this location. Choose **Accept** until all points are accepted.

To sketch the first section, accept the default **Z** axis rotation of zero. Sketch the section as shown in Figure 12.72. Be aware of the *startpoint* location. When finished with the first section, the system will prompt for the next **Z** axis rotation. Accept the default value and sketch section number two (Fig. 12.73). Again, keep track of the start point, it must match up with the first section. Sketch section three (Fig. 12.75), again using **Z** axis rotation value of zero. Sketch section four (Fig. 12.77), again using a **Z** axis rotation value of zero. When finished with section four, sketch the final section five (Fig. 12.79). Again, accept the default of zero for the **Z** axis rotation.

The completed feature should look like Figure 12.81. **Preview** your feature elements prior to choosing **OK**, to check for twist in the swept blend due to start points not lining up. If there is a problem, use the "Sections" option from the dialog box and change the start points to be aligned.

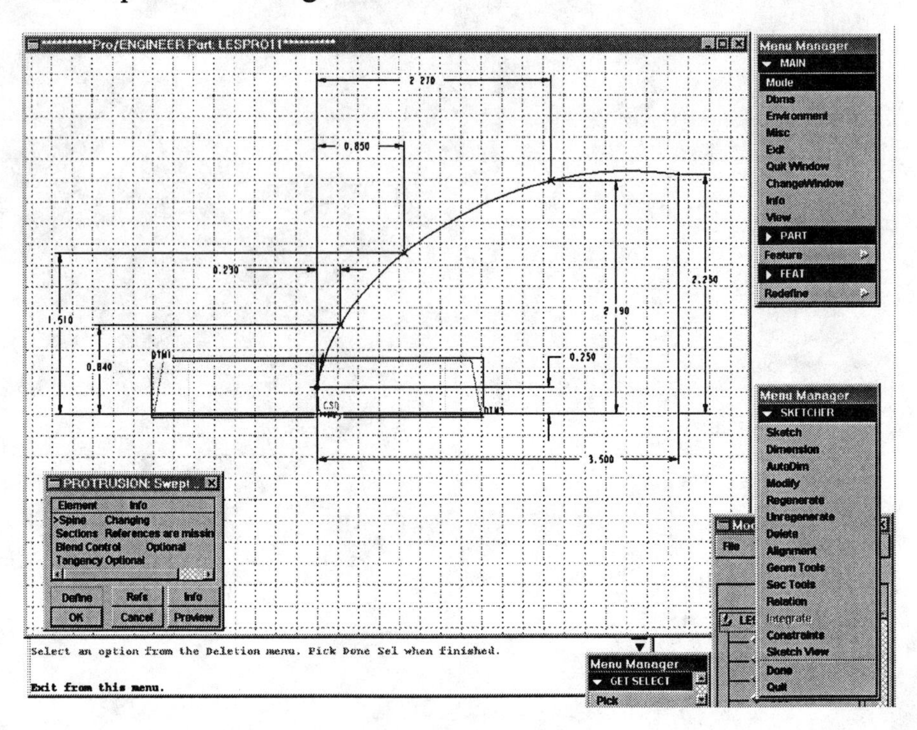

Figure 12.69
Bathroom Faucet, Trajectory Dimensions

Create the **Swept Blend** with the following commands:

Feature ⇒ Create ⇒ Protrusion ⇒ Advanced ⇒ Solid ⇒ Done ⇒ Swept Blend ⇒ Done ⇒ Sketch Sec ⇒ Nrm To Spine ⇒ Done Sketch Traj ⇒ (pick **DTM1**) **⇒ Okay ⇒ Right ⇒** (pick **DTM2**) **⇒ (sketch the three points) Adv Geometry ⇒ Spline ⇒ Sketch Points ⇒** (sketch the trajectory as shown in Figure 12.69) **⇒ Regenerate ⇒ Alignment** (align endpoint to **DTM1**) **⇒ Regenerate ⇒ Dimension** (pick the spline to display the points, dimension as shown in Figure 12.69) **⇒ Regenerate ⇒ Modify ⇒ Regenerate ⇒ Done ⇒ Accept** (All three middle points will be highlighted one at a time as you accept them. The first and last points are considered accepted already) **⇒ enter** (accept the default of **0°**) **⇒** (sketch section one at the center of the light blue crosshairs; starting point--each section will have one) **⇒ Dimension ⇒ Modify ⇒ Regenerate ⇒ Done ⇒** (continue creating all five sections) **⇒ Done ⇒ Preview ⇒ OK**

HINT

Create the first section approximately centered about its respective light blue crosshairs-- starting point. Use centerlines when possible.

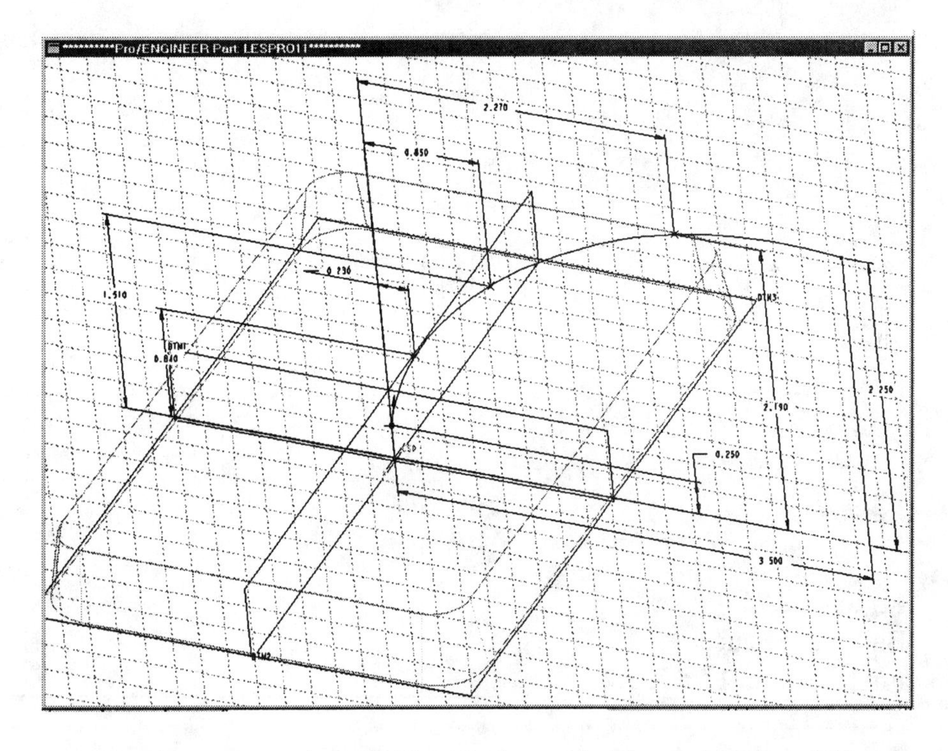

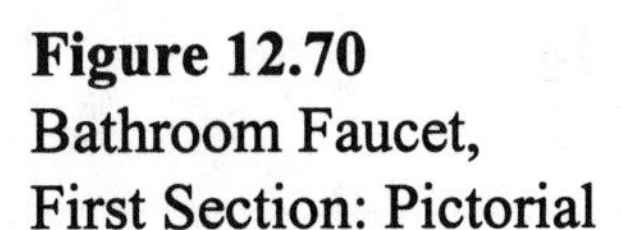

Figure 12.70
Bathroom Faucet,
First Section: Pictorial

The *first section* is **1.75 X 1.20** shown in Figure 12.71 and Figure 12.72. The *second section* is **.75 X 1.380 X R1.00** shown in Figure 12.55, Figure 12.73 and Figure 12.74. The *third section* is **1.00 X .375 X R.850** shown in Figure 12.56, Figure 12.75 and Figure 12.76. The *fourth section* is **.625 X .312 X R.625** shown in Figure 12.57, Figure 12.77 and Figure 12.78. The *fifth section* is **.50 X .25 X R.75** shown in Figure 12.59, Figure 12.79 and Figure 12.80.

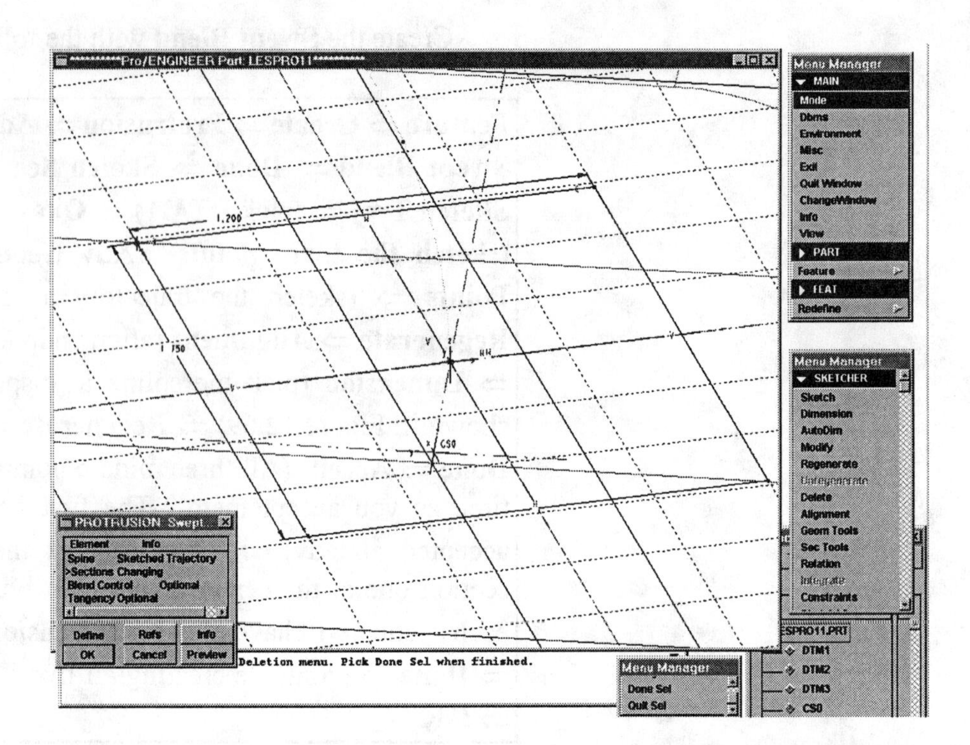

Figure 12.71
Bathroom Faucet,
Trajectory Dimensions

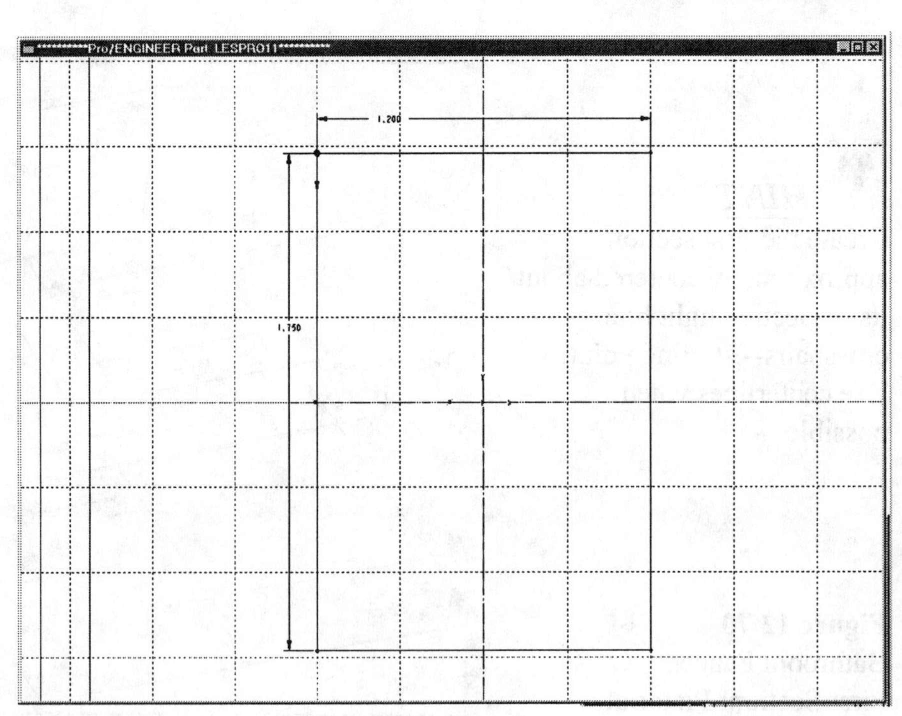

Figure 12.72
Bathroom Faucet,
First Section: **1.75 X 1.20**

HINT

For the second section, align the arc center to the light blue crosshairs-- starting point.

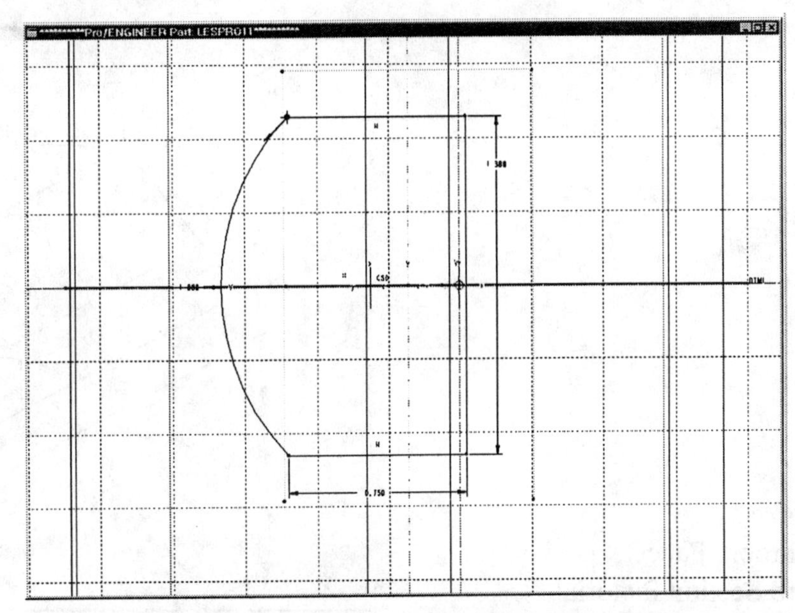

Figure 12.73
Bathroom Faucet,
Second Section:
.75 X 1.380 X R1.00

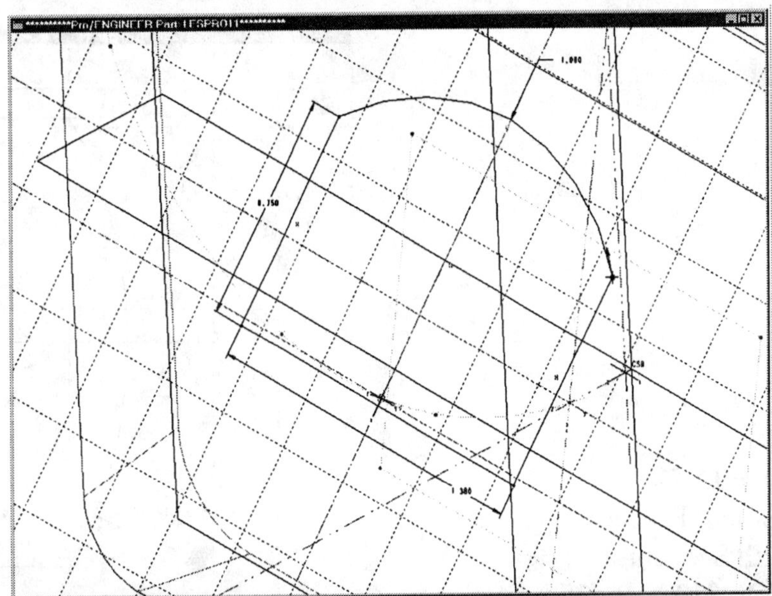

Figure 12.74
Bathroom Faucet,
Second Section Pictorial

HINT

Note where all endpoints are in relation to the light blue crosshairs-- starting point.

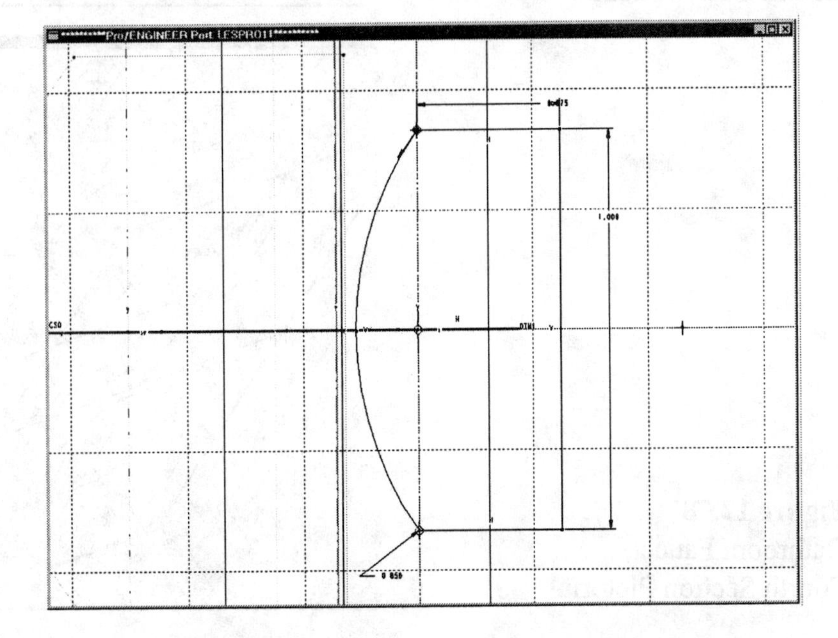

Figure 12.75
Bathroom Faucet,
Third Section:
1.00 X .375 X R.850

Figure 12.76
Bathroom Faucet,
Third Section Pictorial

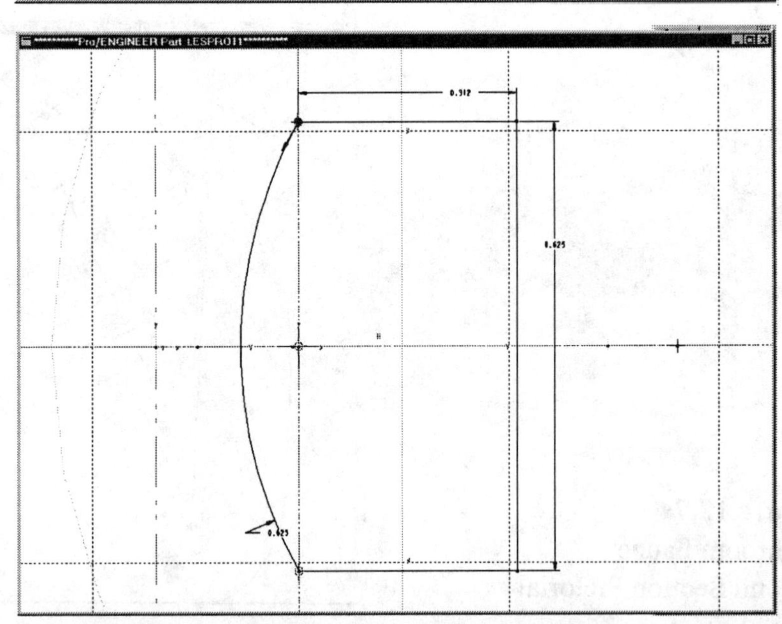

Figure 12.77
Bathroom Faucet
Fourth Section:
.625 X .312 X R.625

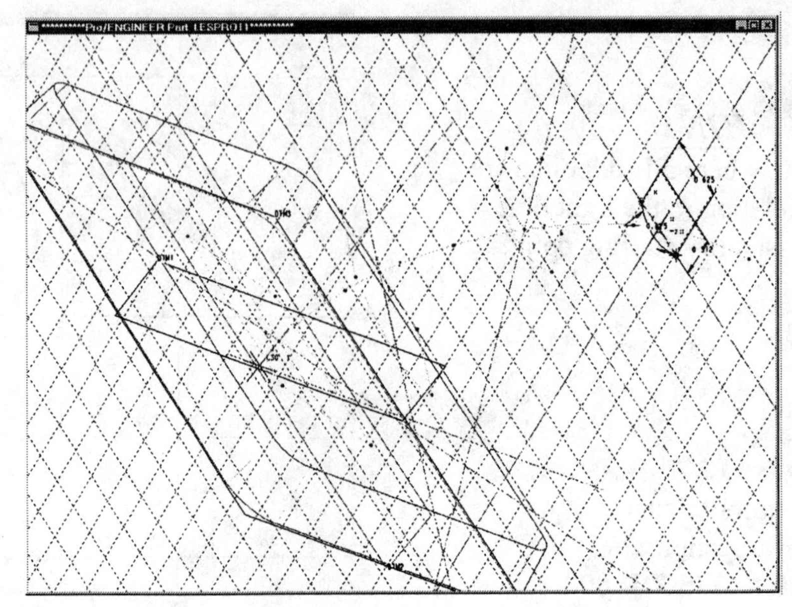

Figure 12.78
Bathroom Faucet,
Fourth Section Pictorial

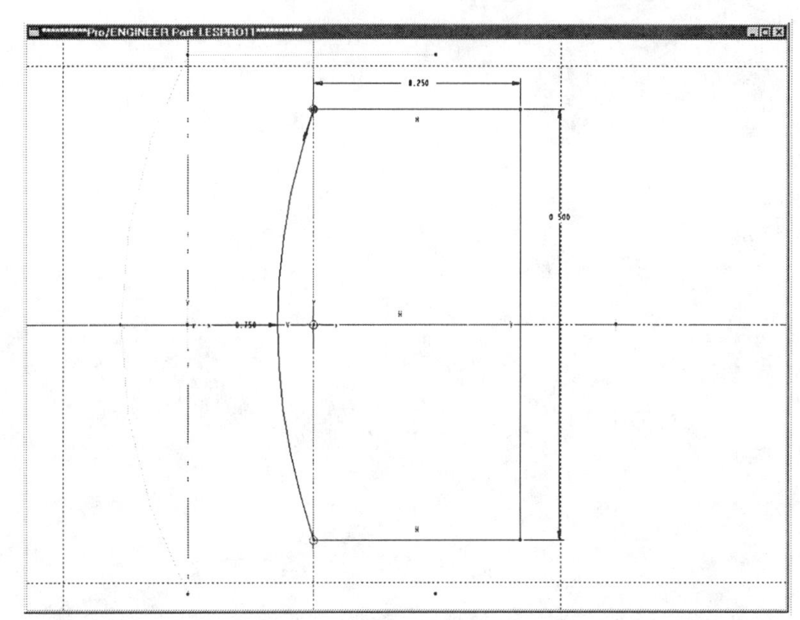

Figure 12.79
Bathroom Faucet,
Fifth Section:
.500 X .25 X R.75

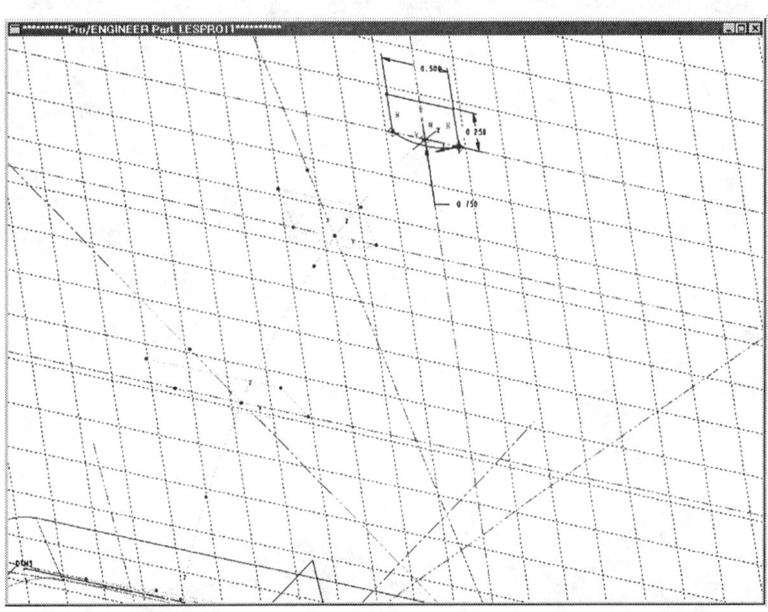

Figure 12.80
Bathroom Faucet,
Fifth Section Pictorial

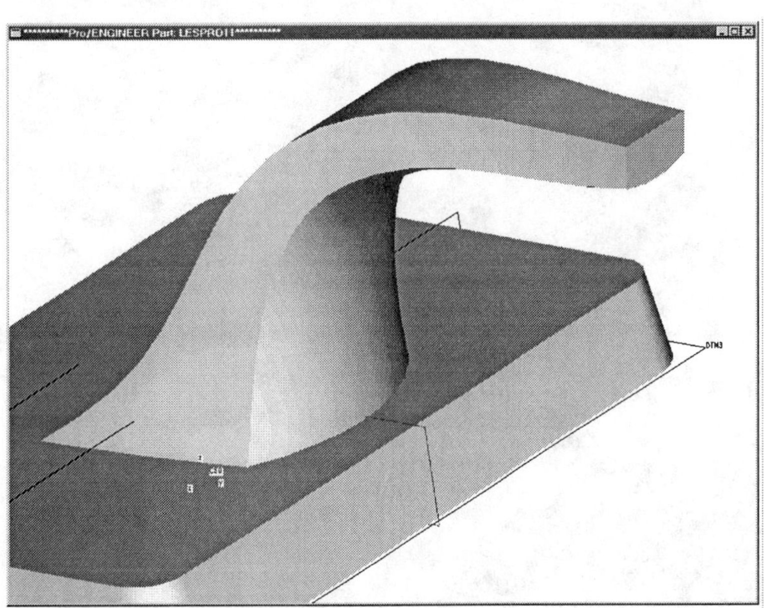

Figure 12.81
Bathroom Faucet,
Completed Swept Blend

Lesson 13

Helical Sweeps

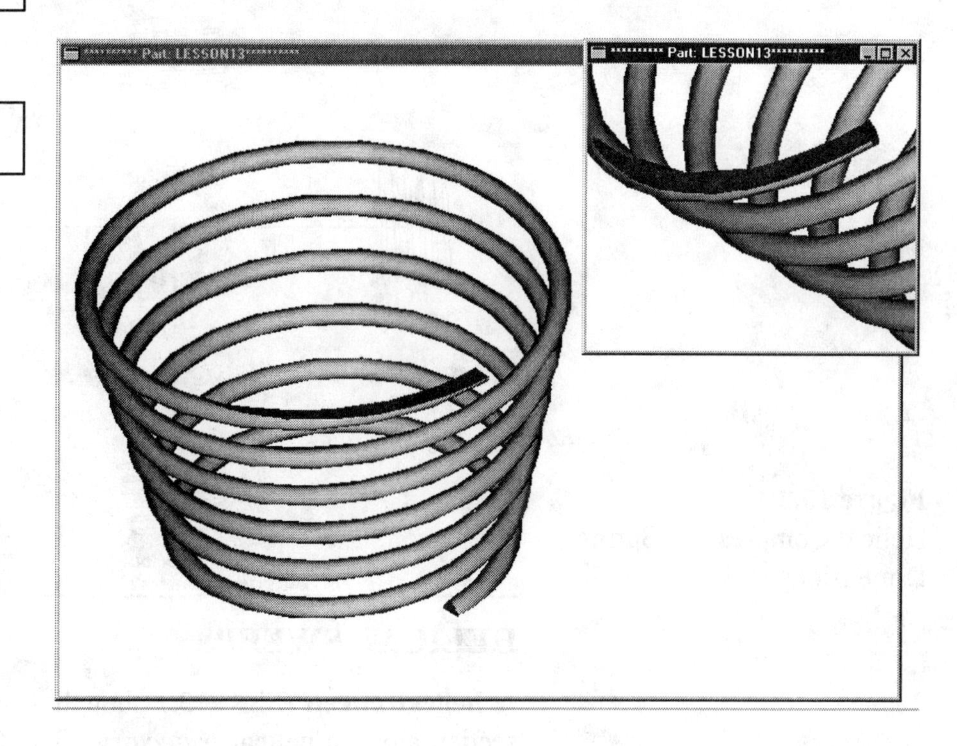

Figure 13.1
Helical Compression Spring

☑ ***EGD REFERENCE***
Engineering Graphics and Design
by L. Lamit and K. Kitto
Read Chapter: 18
See Pages: 702-704

OBJECTIVES

1. **Create springs with a helical sweep**
2. **Model a helical compression spring**
3. **Use sweeps to create hooks on extension springs**
4. **Design an extension spring with a machine hook**
5. **Create plain ground ends on a spring**
6. **Model a convex spring**

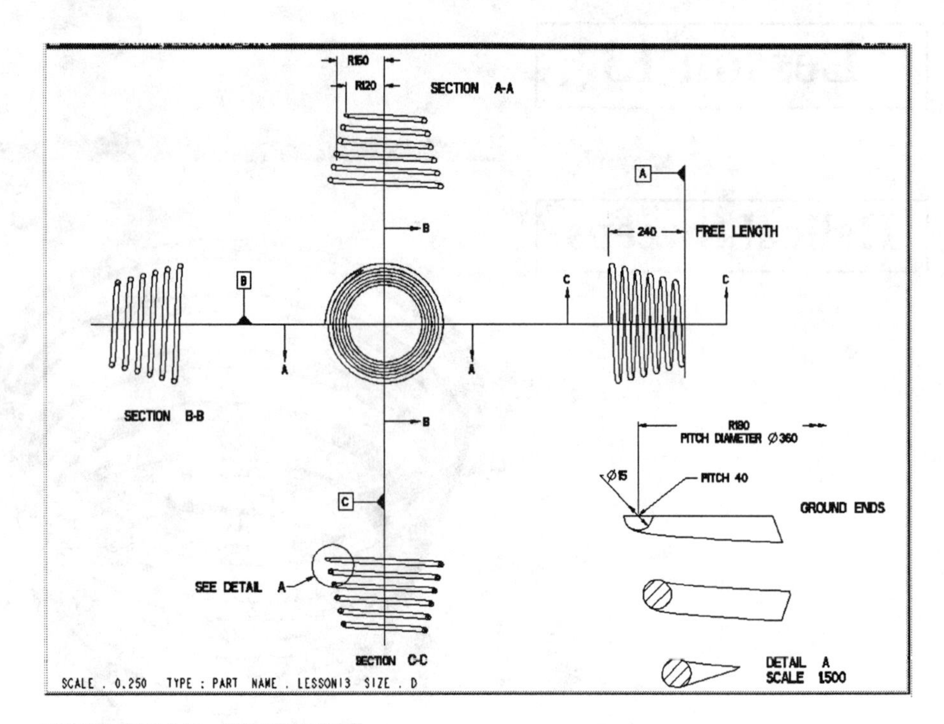

Figure 13.2
Helical Compression Spring Dimensions

HELICAL SWEEPS

A **helical sweep** (Figs. 13.1 though 13.4) is created by sweeping a section along a helical *trajectory*. The trajectory is defined by both the "*profile*" of the *surface of revolution* (which defines the distance from the section origin of the helical feature to its *axis of revolution*) and the "*pitch*" (the distance between coils). The trajectory and the surface of revolution are construction tools and do not appear in the resulting geometry.

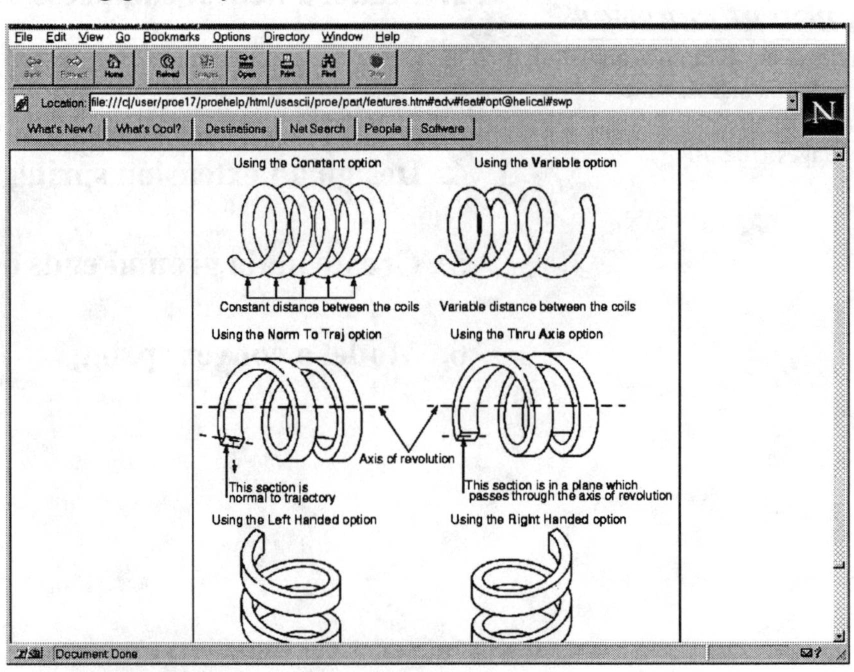

Figure 13.3
Online Documentation Helical Sweeps

The **Helical Sweep** option in the **ADV FEAT OPT** menu is available for both solid and surface features. Use the following **ATTRIBUTES** menu options in mutually exclusive pairs to define the helical sweep feature:

Constant The pitch is constant.
Variable The pitch is variable and defined by a graph.
Thru Axis The section lies in a plane that passes through the axis of revolution.
Norm To Traj The section is oriented normal to the trajectory or surface of revolution.
Right Handed The trajectory is defined by the right-hand rule.
Left Handed The trajectory is defined by the left-hand rule.

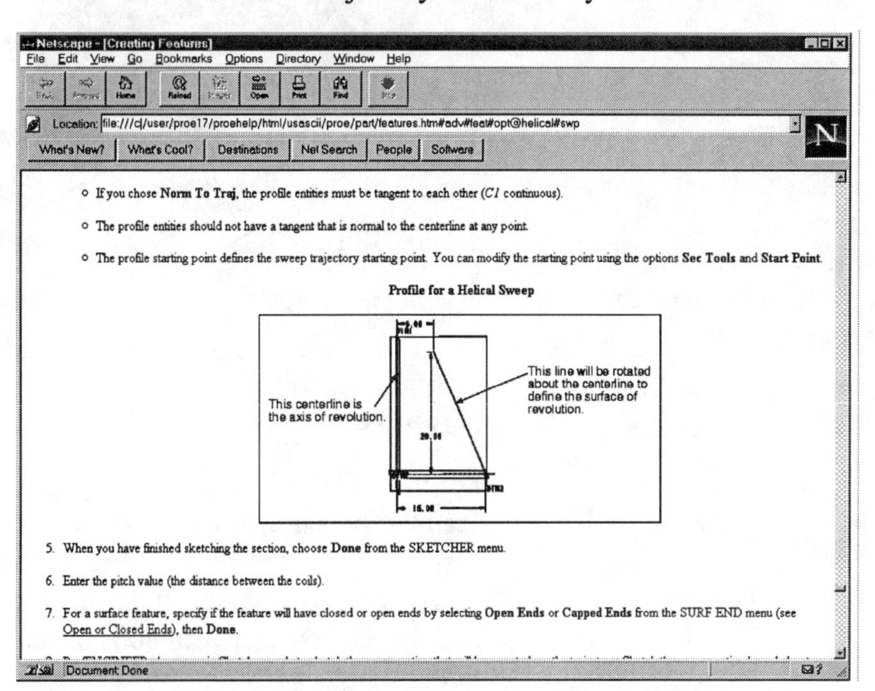

Figure 13.4
Online Documentation
Profile for a Helical Sweep

Constant Pitch Helical Sweeps

To create a helical sweep with a constant pitch value, give the following command sequence:

1. Choose **Advanced** ⇒ **Done** from the **SOLID OPTS** menu, then **Helical Swp** ⇒ **Done.** Pro/E displays the feature creation dialog box.
2. Define the feature by selecting from the **ATTRIBUTES** menu, then choose **Done.**
3. Pro/E places you in Sketcher mode. Specify the sketching plane and its orientation, and the axis of revolution. Sketch, dimension, and regenerate the surface of revolution profile. The sketched entities must form an *open loop*. As with a revolved feature, you must *sketch a centerline* to define the **axis of revolution**.
4. Pro/E places you in another Sketcher orientation. Sketch, dimension, and regenerate the section profile to be swept along the trajectory.

If you choose **Norm To Traj,** the profile entities must be tangent to each other (continuous). The profile entities should not have a tangent that is normal to the centerline at any point. The profile starting point defines the sweep trajectory starting point. Modify the starting point using the options **Sec Tools** and **Start Point**.

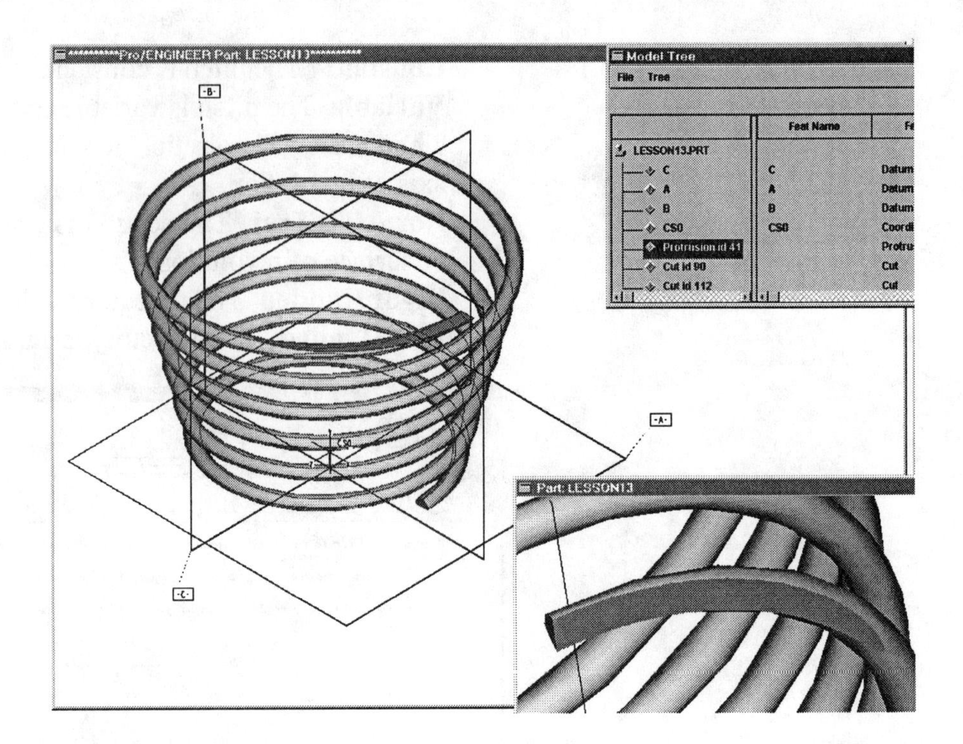

Figure 13.5
Helical Compression Spring with Datum Planes and Model Tree

Springs

Springs (Fig. 13.5) and other helical features are created with the **Helical Sweep** command. A helical sweep is created by sweeping a *section* along a *trajectory* that lies in the *surface of revolution*: the trajectory is defined by both the *profile* of the surface of revolution and the distance between coils. The part for this lesson is a *constant-pitch, right-handed, helical compression spring with ground ends, a pitch of **40 mm**, and wire diameter of **15 mm*** (Figure 13.6 through Figure 13.9).

NOTE
Set up the Helical Compression Spring:

- Material = Spring Steel
- Units = Millimeters
- Default Datum Planes
- Default Coordinate System
- Layers = **DATUM_PLANES**
- Hidden Line
- No Disp Tan
- ✓Grid Snap

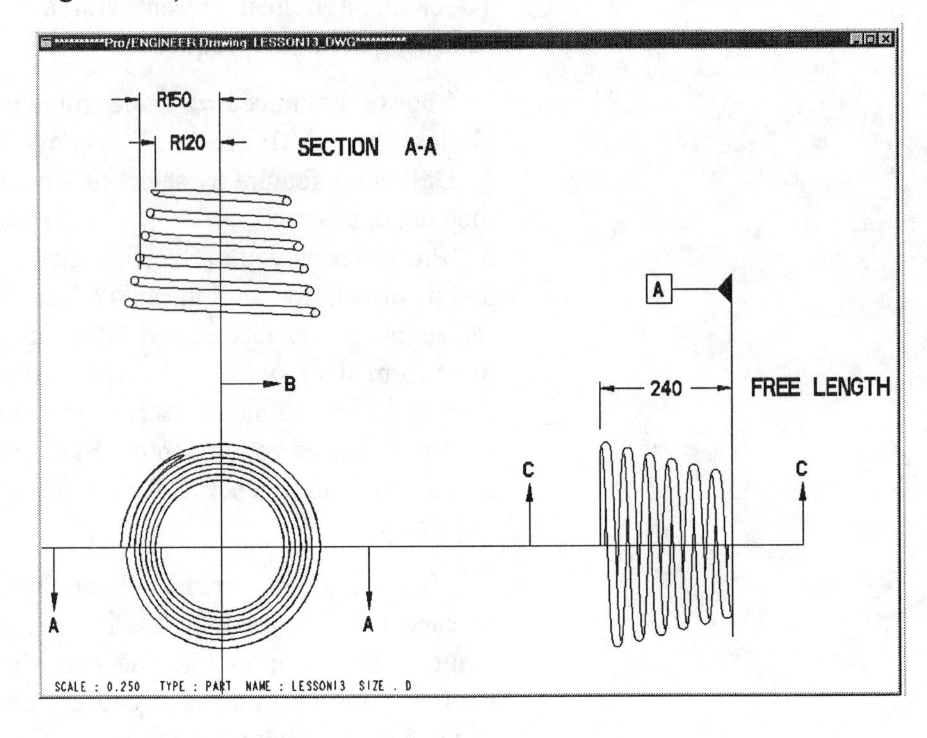

Figure 13.6
Helical Compression Spring

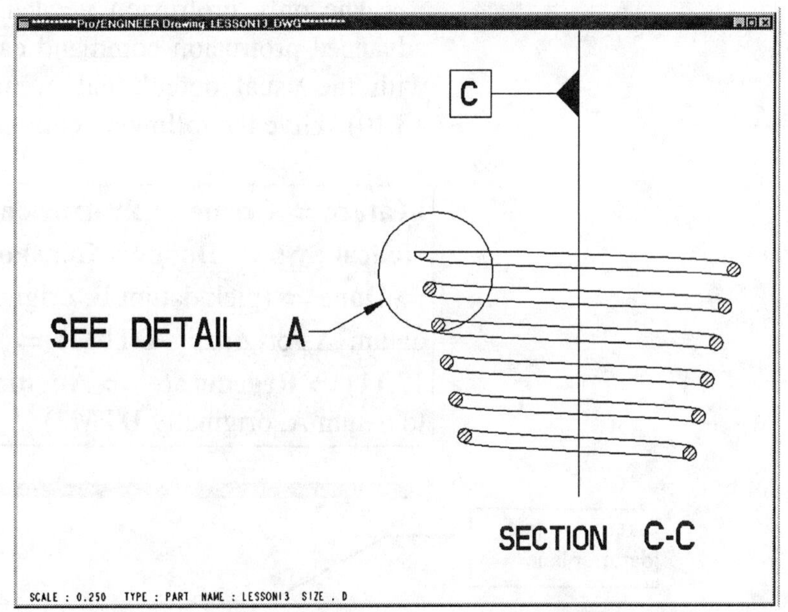

Figure 13.7
Helical Compression Spring Drawing, **SECTION C-C**

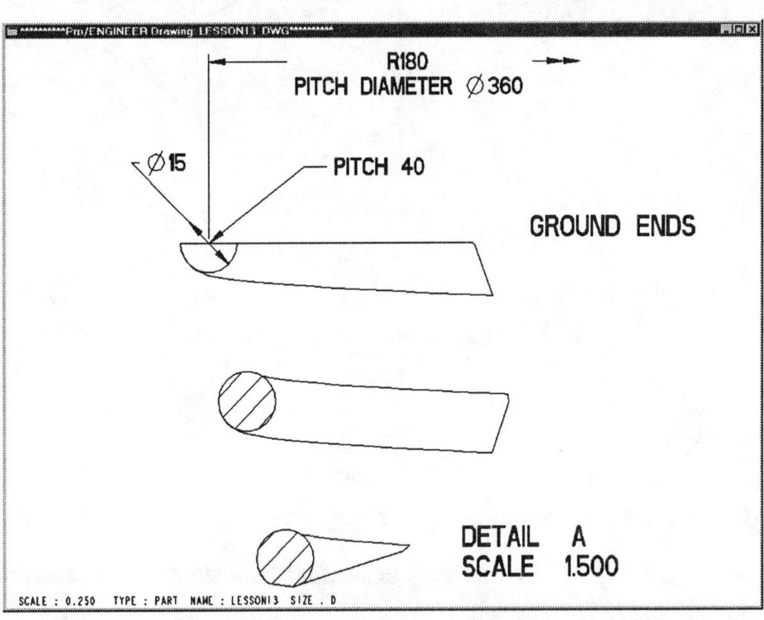

Figure 13.8
Helical Compression Spring Drawing, **DETAIL A**

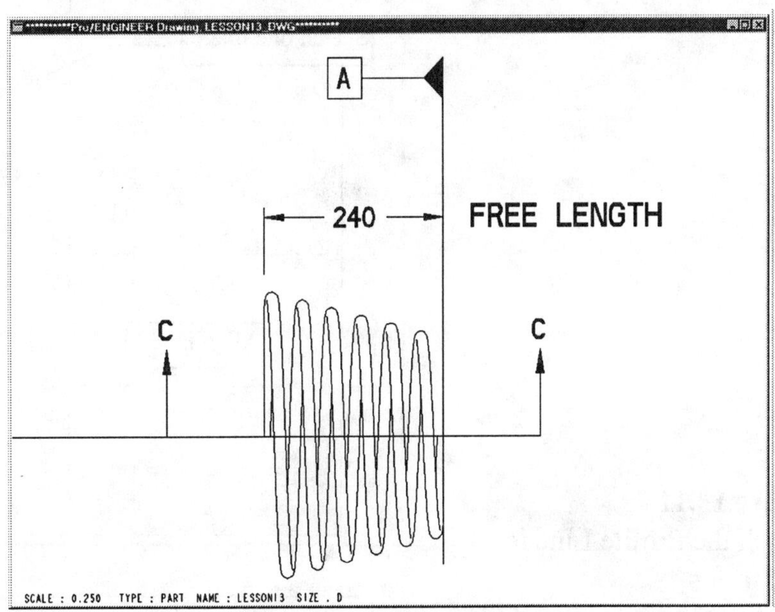

Figure 13.9
Helical Compression Spring Drawing, Right Side View

The only protrusion needed for the spring is created with an advanced protrusion command called **Helical Sweep**. Start the part with the usual default datum planes and coordinate system (Fig. 13.10). Give the following commands to create the first protrusion:

Feature ⇒ **Create** ⇒ **Protrusion** ⇒ **Advanced** ⇒ **Solid** ⇒ **Done** ⇒ **Helical Swp** ⇒ **Done** ⇒ **Constant** ⇒ **Thru Axis** ⇒ **Right Handed** ⇒ **Done** ⇒ (pick datum **B**, originally **DTM3**) ⇒ **Okay** ⇒ **Top** (pick datum **A**, originally **DTM2**) ⇒ (sketch a line as shown in Figure 13.11) ⇒ **Regenerate** ⇒ **Alignment** (align the lower end of the line to datum **A**, originally **DTM2**)

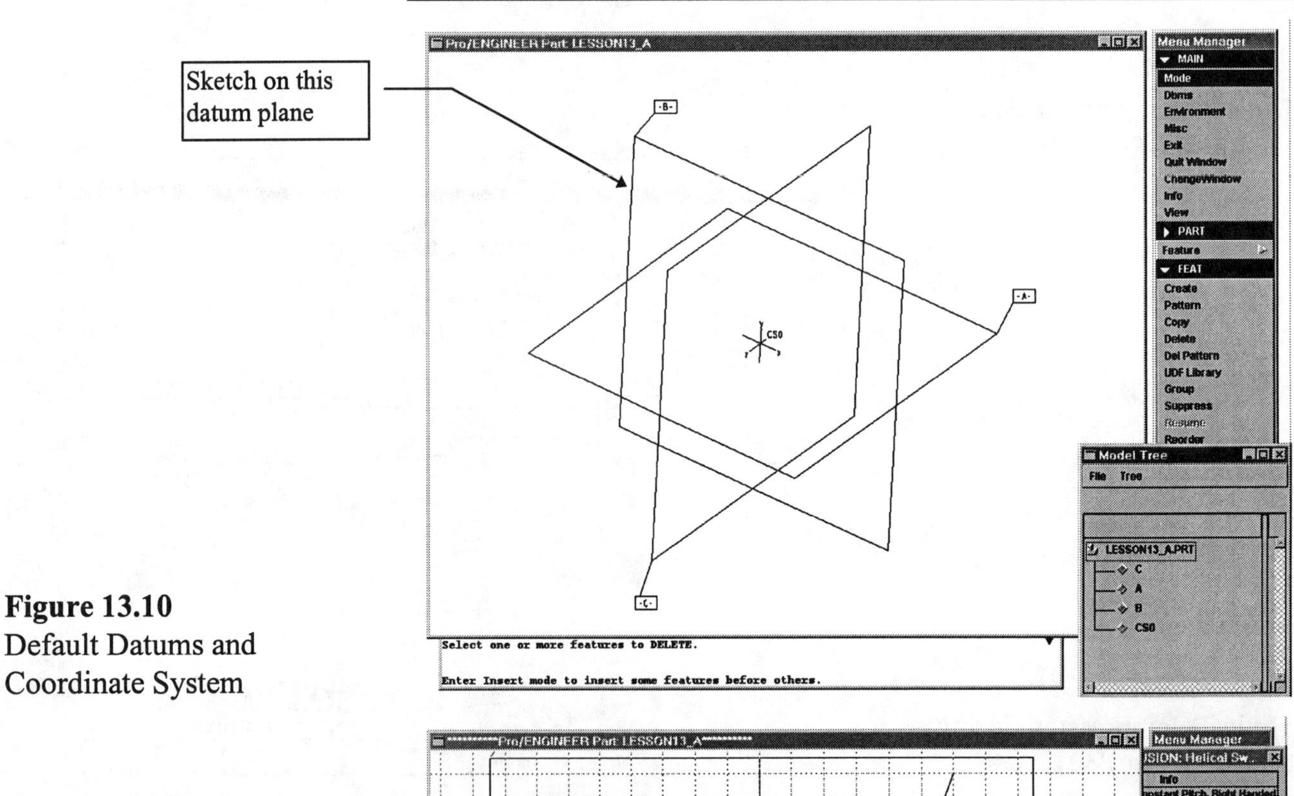

Figure 13.10
Default Datums and Coordinate System

Figure 13.11
Sketch the Profile Line for the Spring

> **Regenerate** ⇒ **Dimension** (add dimensions as shown in Figure 13.12) ⇒ **Regenerate** ⇒ **Sketch** ⇒ **Line** ⇒ **Centerline** ⇒ **Vertical** ⇒ (add a vertical axis line along datum **C**, **DTM1**) ⇒ **Alignment** (align the axis line to datum **C**, **DTM1**) ⇒ **Regenerate** ⇒ **Modify** (change the values to the design sizes) ⇒ **Regenerate** ⇒ **Done** ⇒ (enter the pitch value **40** at the prompt, as shown in Figure 13.13) ⇒ **enter**

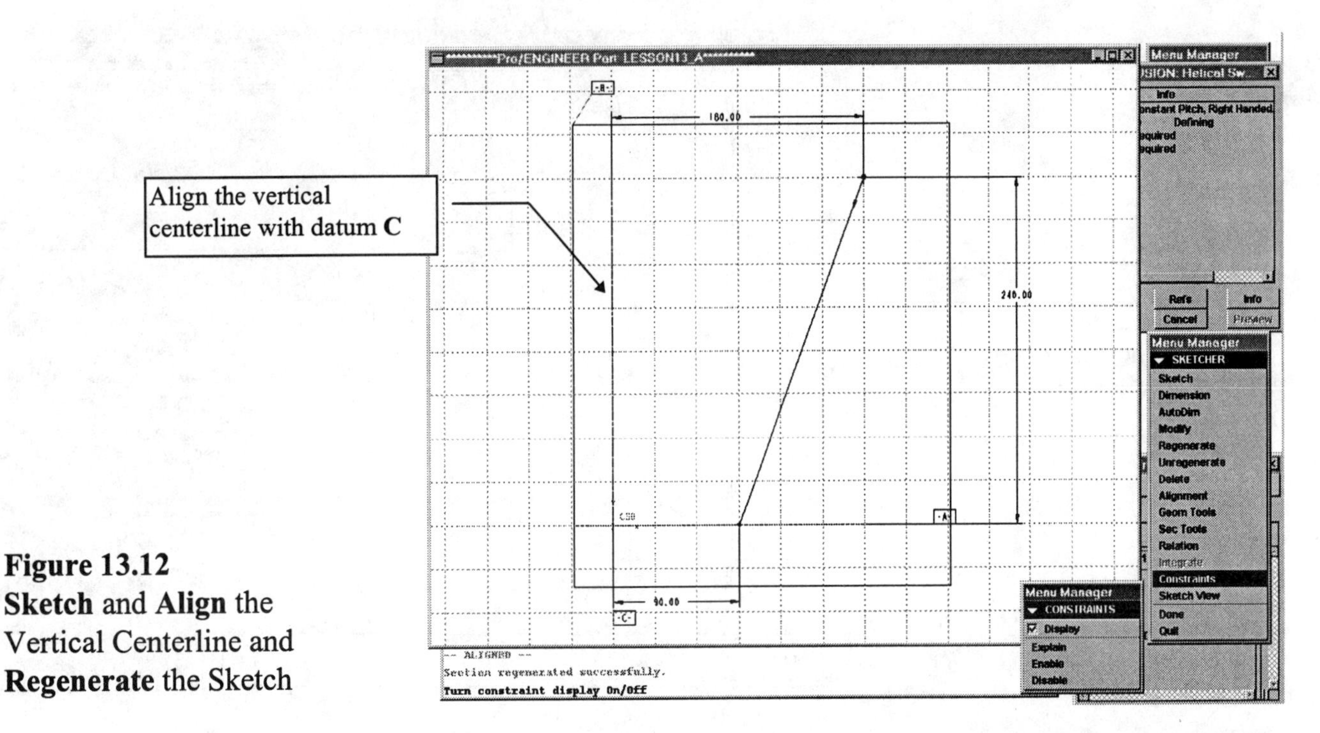

Figure 13.12
Sketch and **Align** the Vertical Centerline and **Regenerate** the Sketch

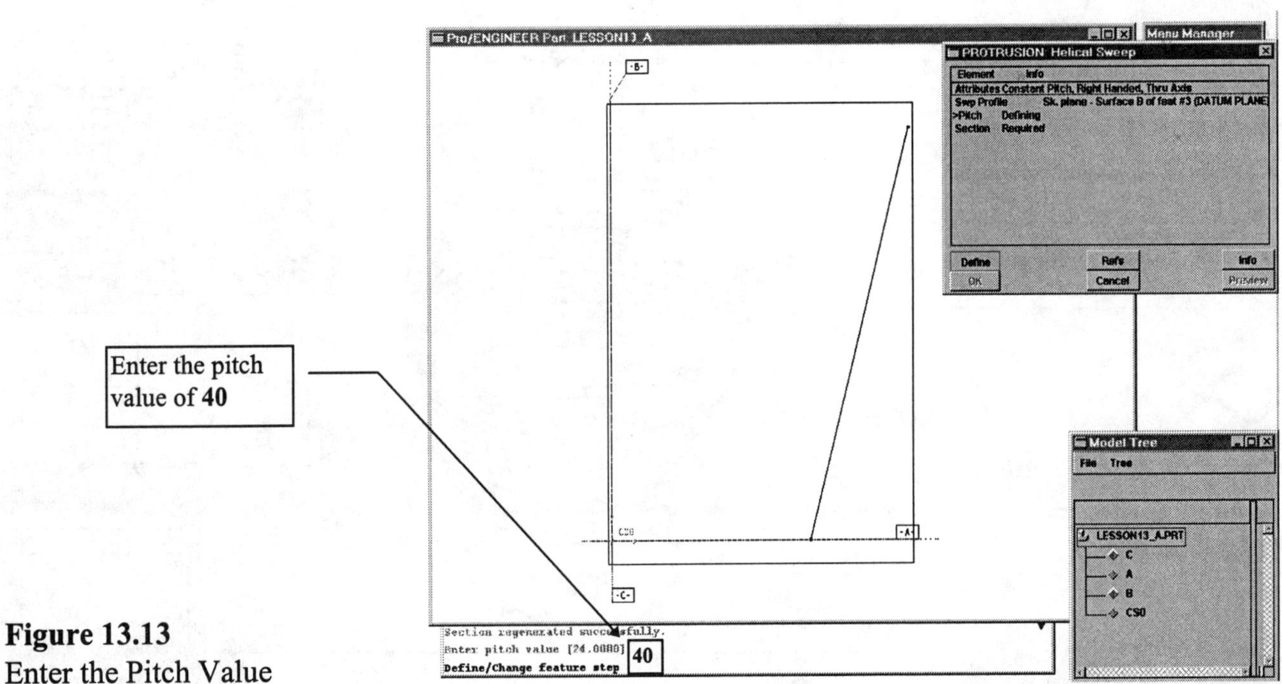

Figure 13.13
Enter the Pitch Value

Sketch the section geometry of the spring (here it is a circle):

Sketch ⇒ **Circle** ⇒ **Ctr/Point** (Fig. 13.14) ⇒ **Alignment** (align center of circle with the datum plane) ⇒ **Regenerate** ⇒ **Dimension** (add diameter dimension) ⇒ **Regenerate** ⇒ **Modify** (enter **15**) ⇒ **Regenerate** ⇒ **Done** ⇒ **OK** ⇒ **View** ⇒ **Default** ⇒ **Cosmetic** ⇒ **Shade** ⇒ **Display** (Fig. 13.15)

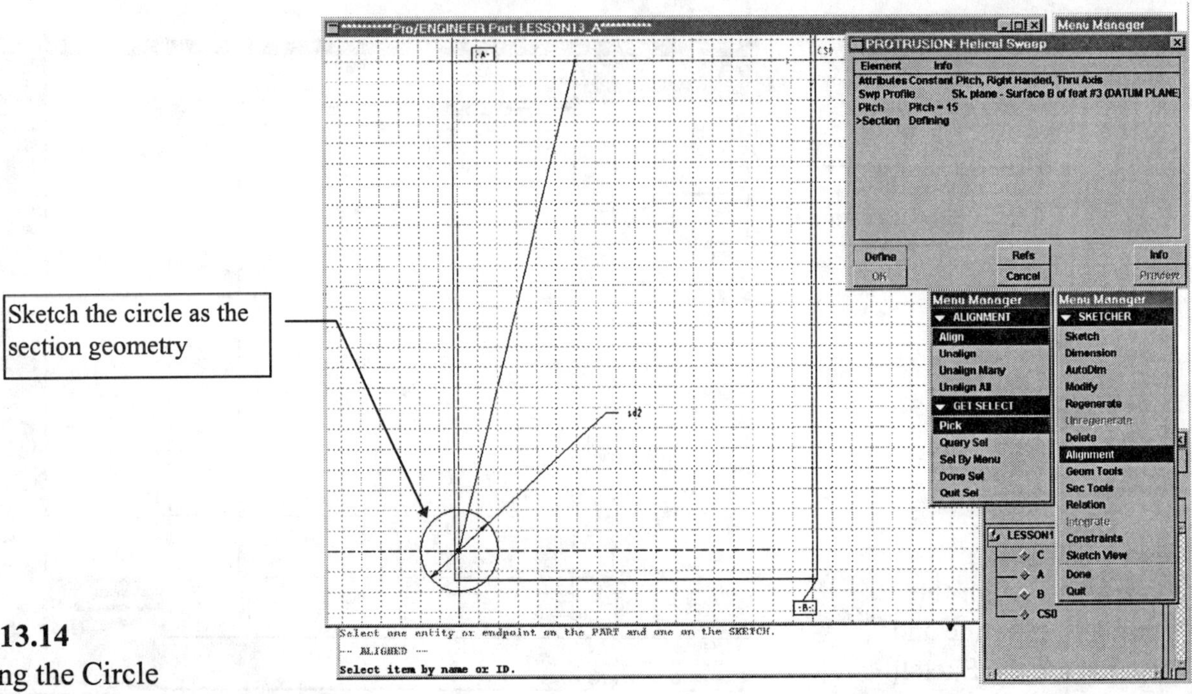

Figure 13.14
Sketching the Circle

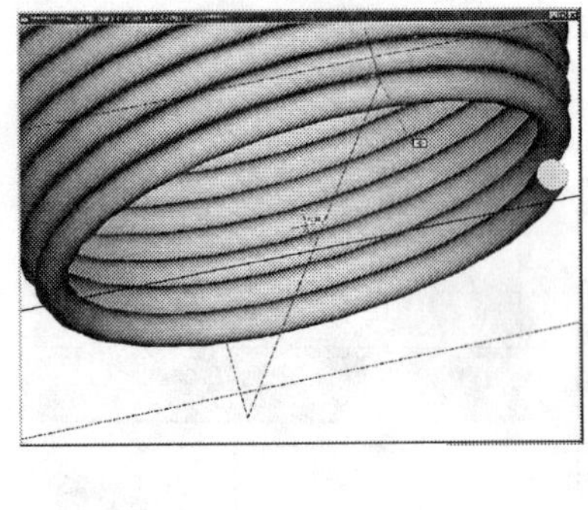

Dbms ⇒ **Save** ⇒ **enter**
Purge ⇒ **enter** ⇒
Done-Return

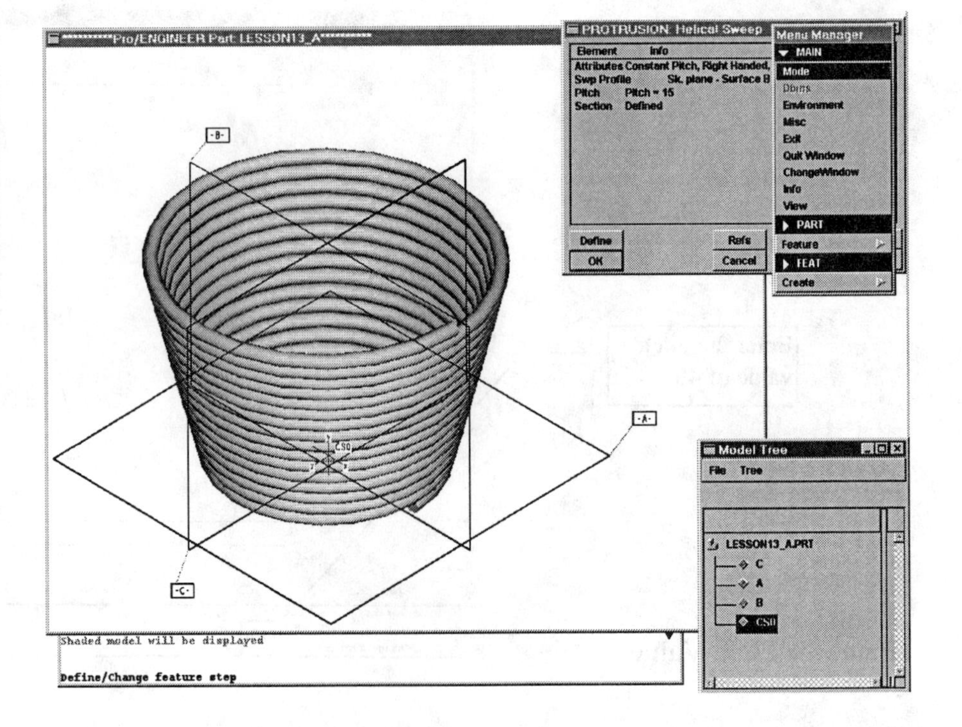

Figure 13.15
Completed Helical Sweep

The spring is almost complete (Fig. 13.16). Add the cut line to create the ground end as shown in Figure 13.17:

Create ⇒ Cut ⇒ Extrude ⇒ Solid ⇒ Done ⇒ Both Sides ⇒ Done ⇒ Plane (pick datum **C, DTM3) ⇒ Okay ⇒ Bottom** (pick datum **A, DTM2 ⇒ Sketch ⇒ Line ⇒ Horizontal** (on **DTM2**) **⇒ Alignment** (align the line to datum **A, DTM2**, and the end of the line to datum **B, DTM1**) **⇒ Regenerate Dimension ⇒ Regenerate** (there is no need to modify the dimension) **⇒ Done ⇒ Flip ⇒ Okay ⇒ Thru All ⇒ Done ⇒ Thru All ⇒ Done ⇒ OK** (Fig. 13.18)

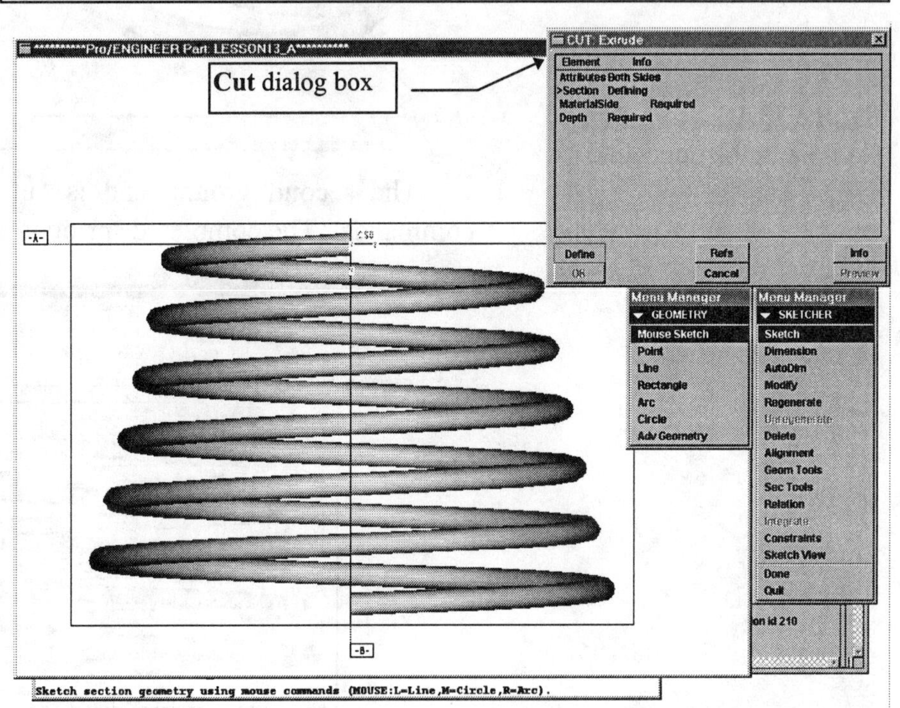

Figure 13.16
Creating a Cut

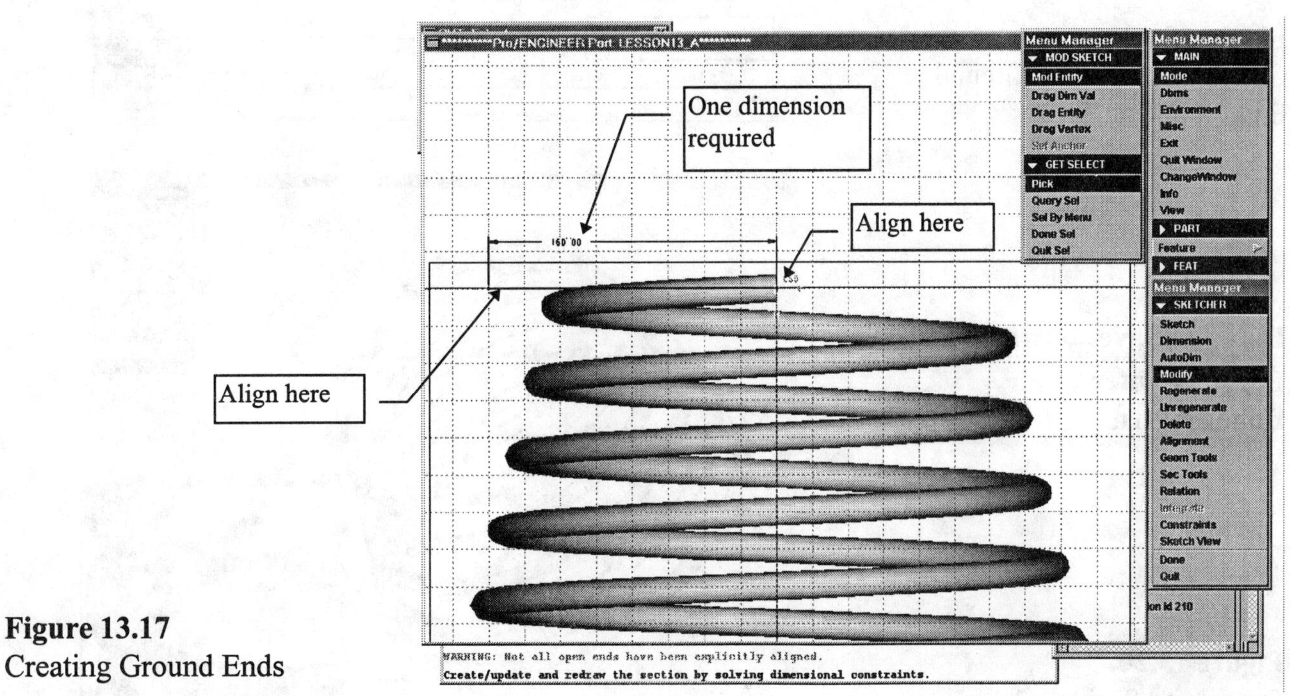

Figure 13.17
Creating Ground Ends

Dbms ⇒ Save ⇒ enter
Purge ⇒ enter ⇒
Done-Return

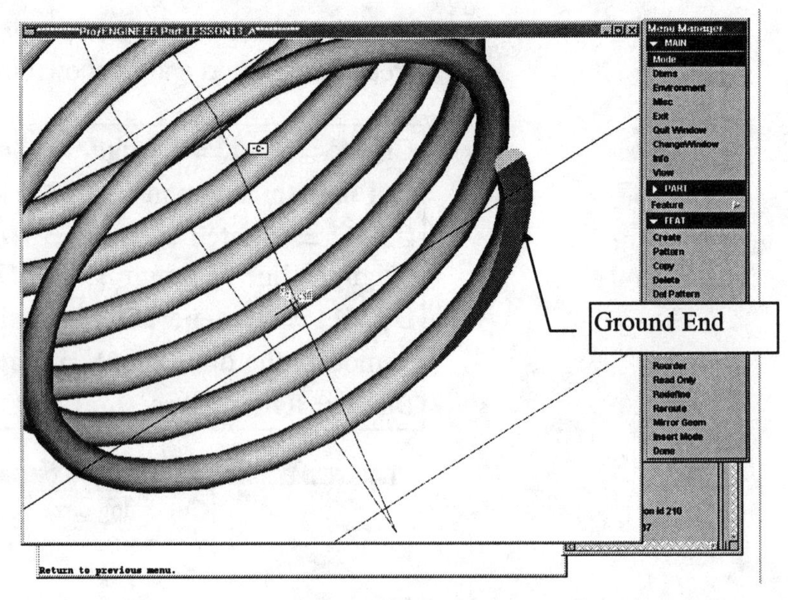

Figure 13.18
Completed Ground End

The second ground end is cut in Figure 13.19 using similar commands. The completed spring is shown in Figure 13.20.

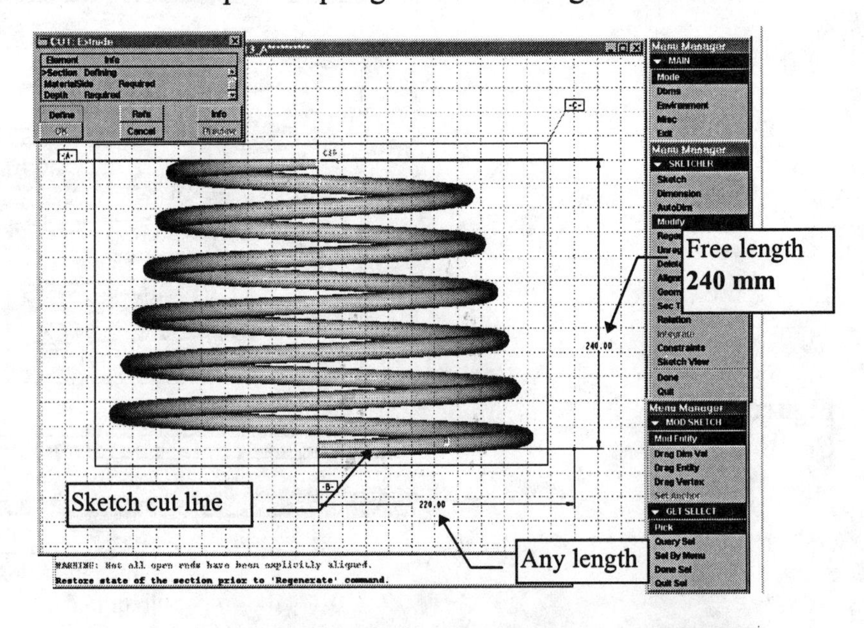

Figure 13.19
Creating the Second Ground End

Dbms ⇒ Save ⇒ enter
Purge ⇒ enter ⇒
Done-Return

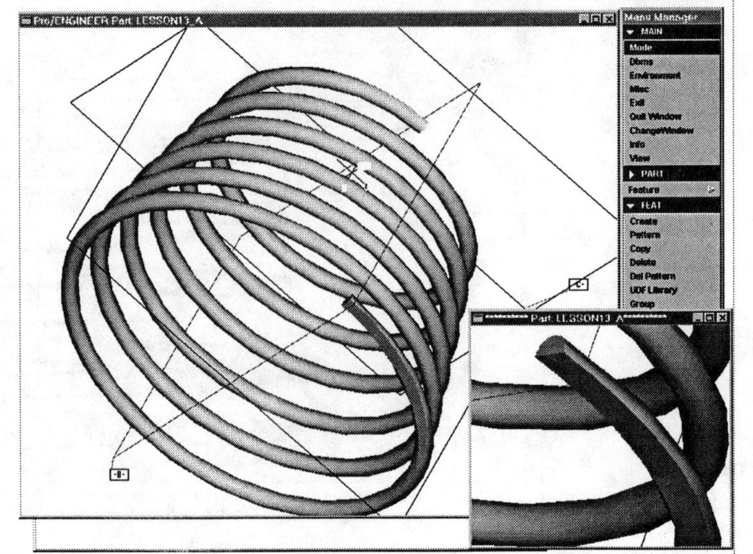

Figure 13.20
Completed Spring

ECO

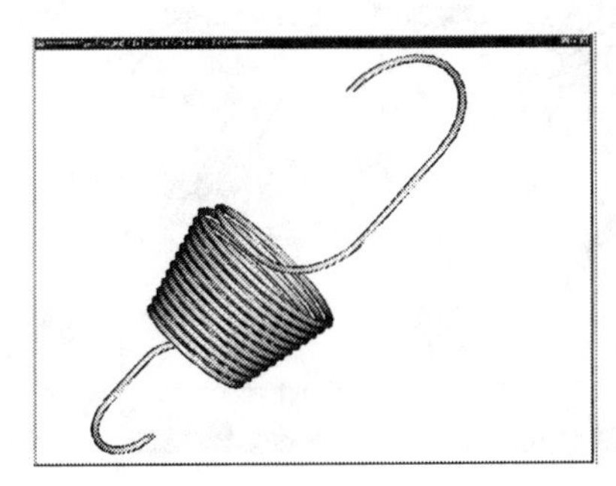

Save the Helical Compression Spring by giving the command: **Dbms ⇒ Save As,** and also giving it a different name, for example, **HEL_COMP_SPR_GRND_ENDS**. Rename the file you are working on by using **Dbms ⇒ Rename**, and give it a name such as **HEL_EXT_SPR_MACH_ENDS**. Delete the existing ground ends and modify the pitch to **10**. Figure 13.21 provides an **ECO** for the new spring. Figure 13.22 shows the wire diameter changed to **7.5** and the pitch changed to **10**. The ground ends have been deleted.

Complete the spring as shown in Figure 13.23 through Figure 13.31. The free length is to be **120 mm**, the large radius is now **90 mm**, and the small radius is **60 mm**.

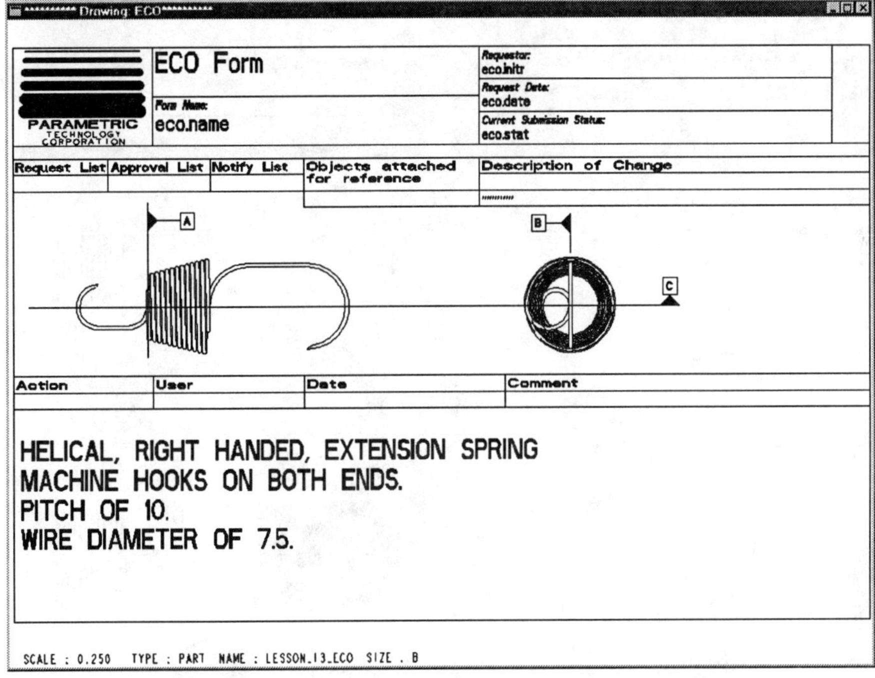

Drawing: ECO

PARAMETRIC TECHNOLOGY CORPORATION

ECO Form	Requestor: eco.initr
	Request Date: eco.date
Form Name: eco.name	Current Submission Status: eco.stat

Request List	Approval List	Notify List	Objects attached for reference	Description of Change

Action	User	Date	Comment

HELICAL, RIGHT HANDED, EXTENSION SPRING
MACHINE HOOKS ON BOTH ENDS.
PITCH OF 10.
WIRE DIAMETER OF 7.5.

SCALE : 0.250 TYPE : PART NAME : LESSON_13_ECO SIZE : B

Figure 13.21
ECO to Create a Helical Extension Spring

Dbms ⇒ Save ⇒ enter
Purge ⇒ enter ⇒
Done-Return

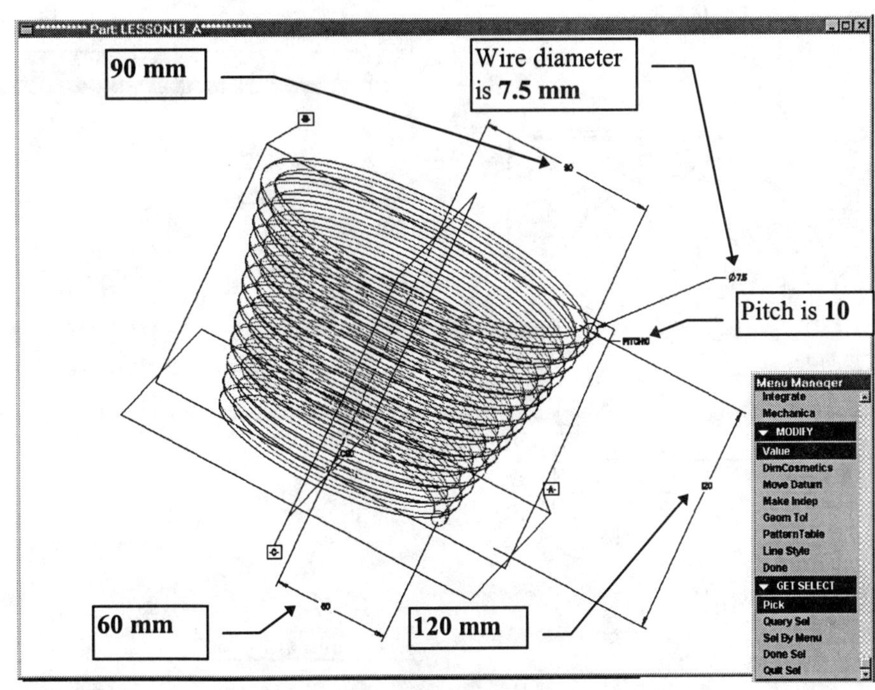

Figure 13.22
ECO Changes

Figure 13.23
Helical Extension Spring with Machine Hook Ends

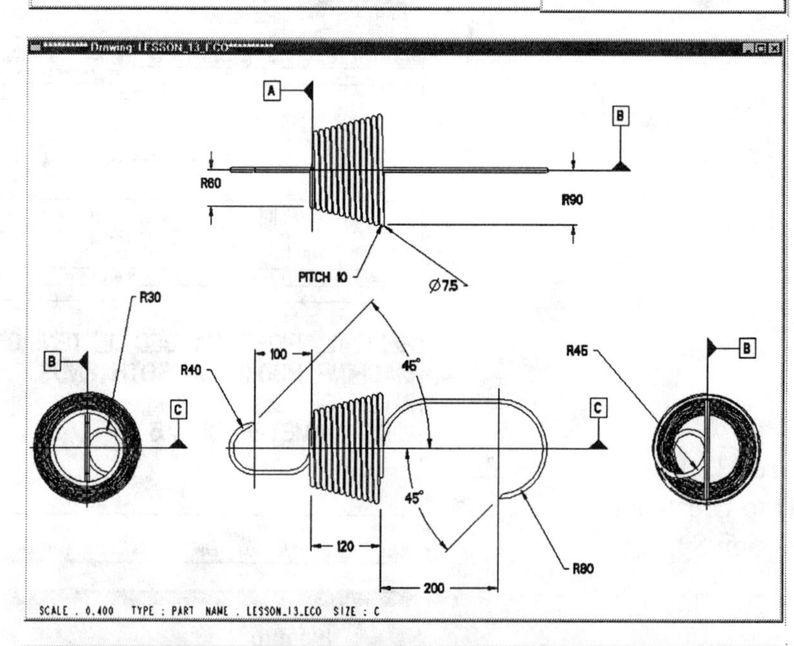

Figure 13.24
Detail Drawing of Helical Extension Spring with Machine Hook Ends

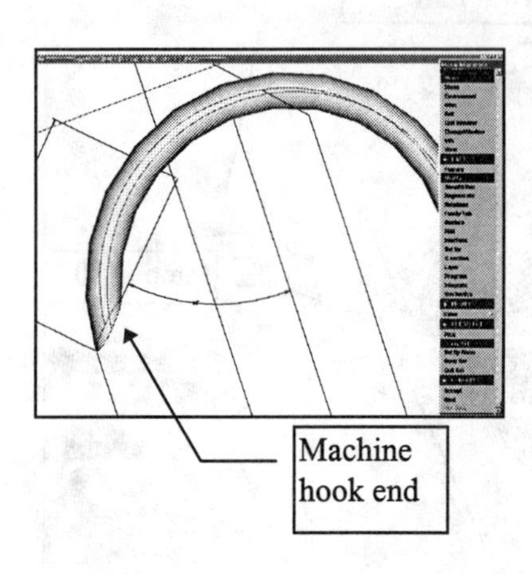

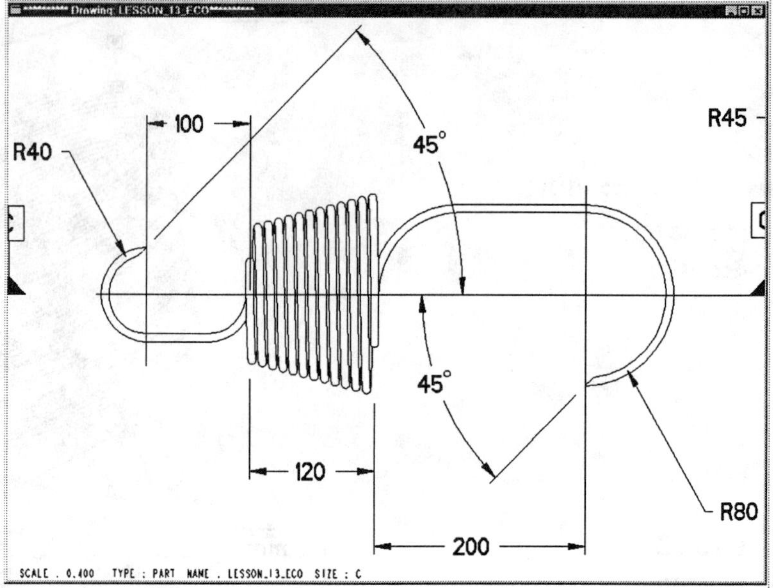

Figure 13.25
Front View

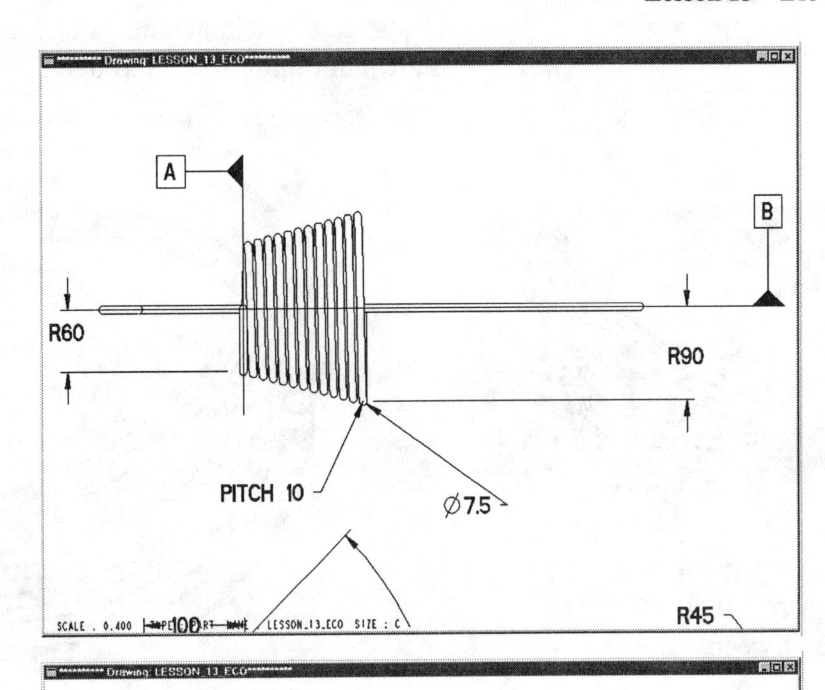

Figure 13.26
Top View

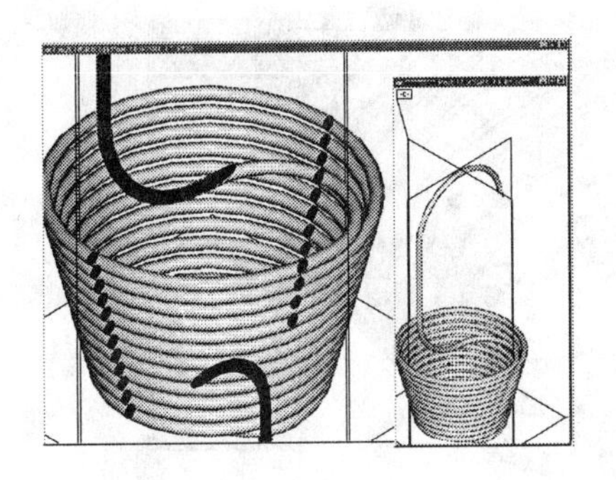

Figure 13.27
Right Side View

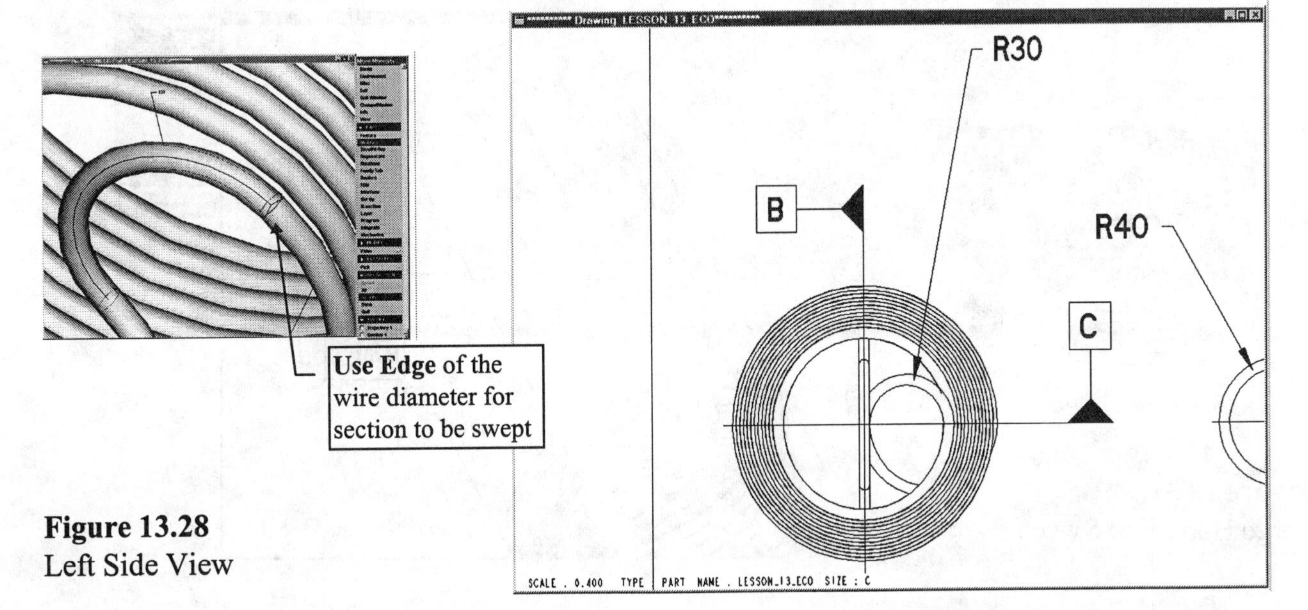

Figure 13.28
Left Side View

Create the machine hooks using simple sweeps and cuts, as shown in Figure 13.29 through Figure 13.31.

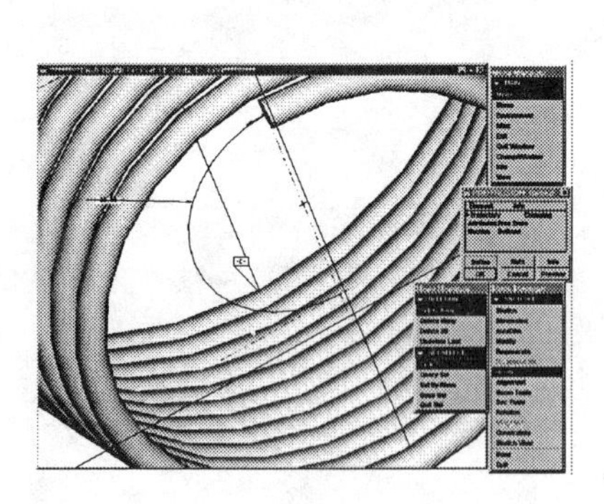

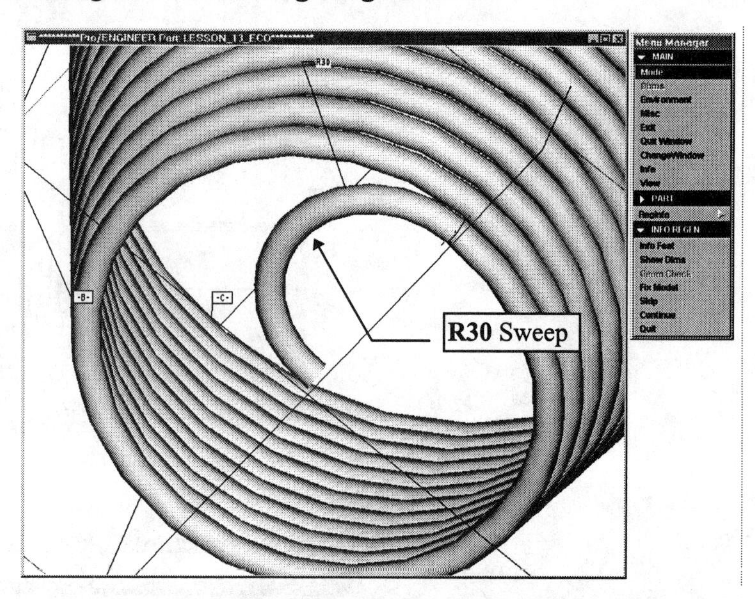

Figure 13.29
R30 Sweep

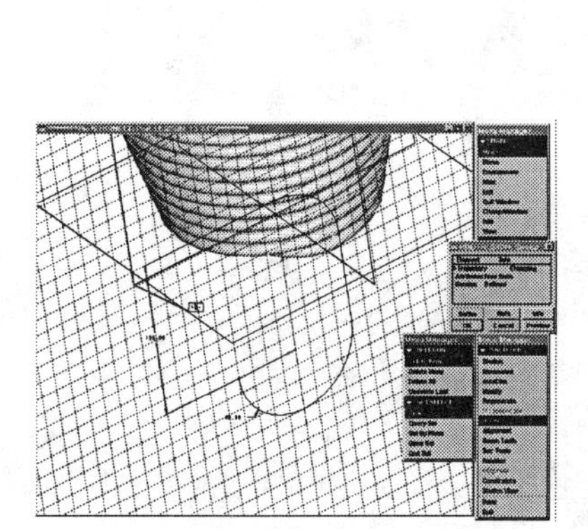

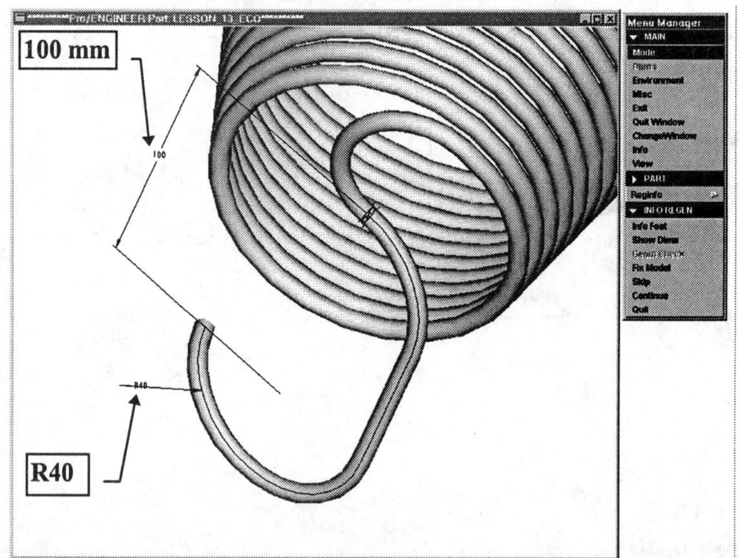

Figure 13.30
Small Hook End Sweep

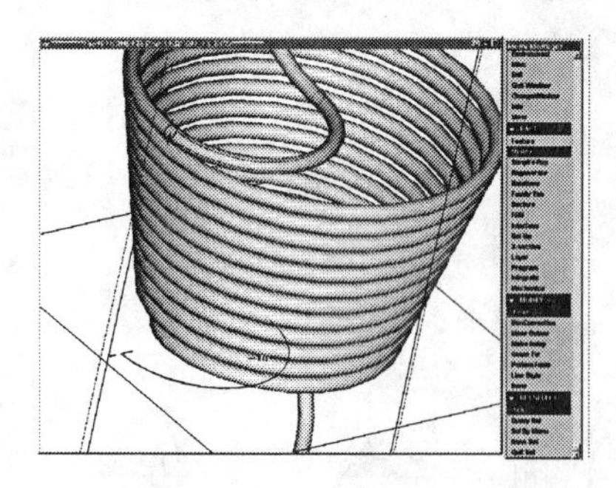

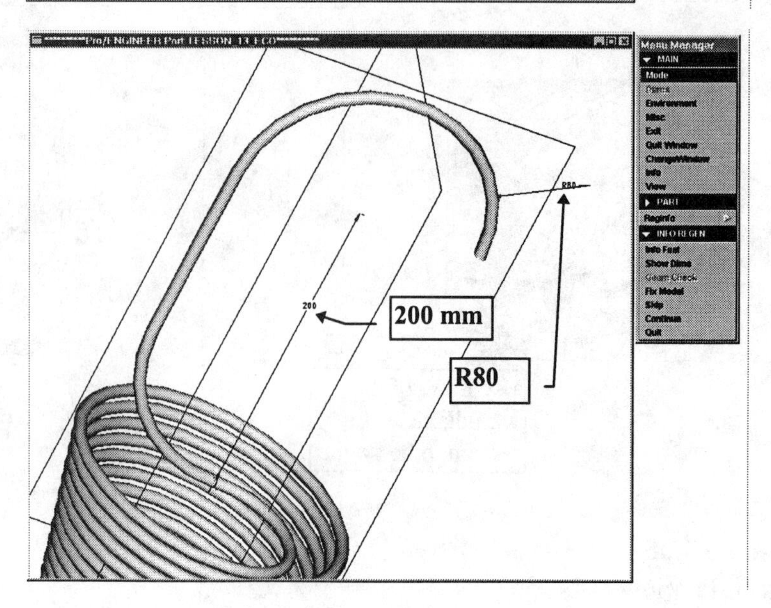

Figure 13.31
Large Hook End Sweep

Lesson 13 Project

Convex Compression Spring

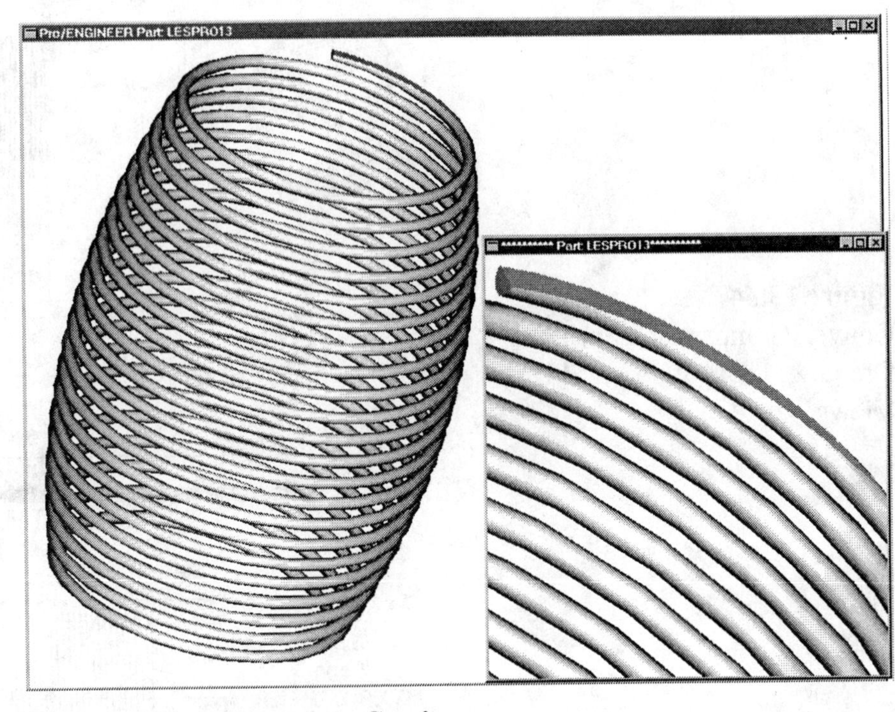

Figure 13.32
Convex Compression Spring

Convex Compression Spring

The last **lesson project** is a **Convex Compression Spring** that requires similar commands as the **Helical Spring**. Create the part shown in Figure 13.32 through Figure 13.37. Analyze the part and plan out the steps and features required to model it. Use the DIPS in Appendix D to plan out the feature creation sequence and the parent-child relationships for the part. The spring is made of *spring steel*.

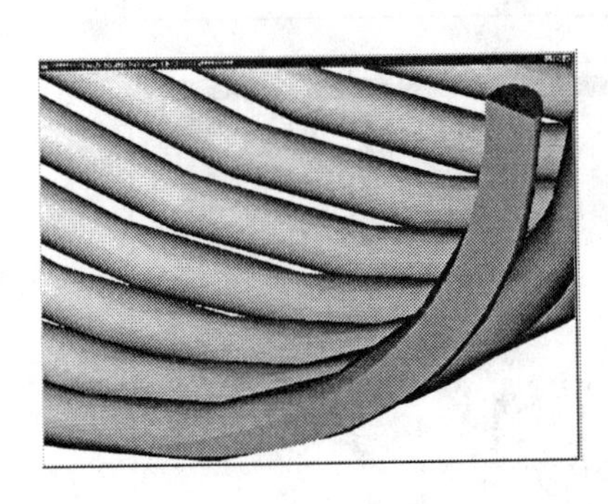

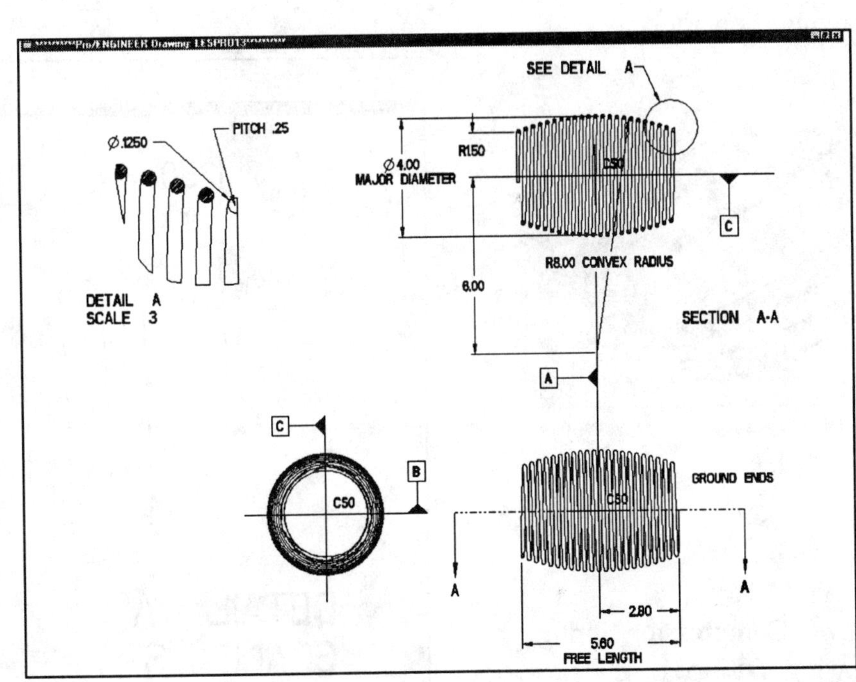

Figure 13.33
Convex Compression Spring, Detail Drawing

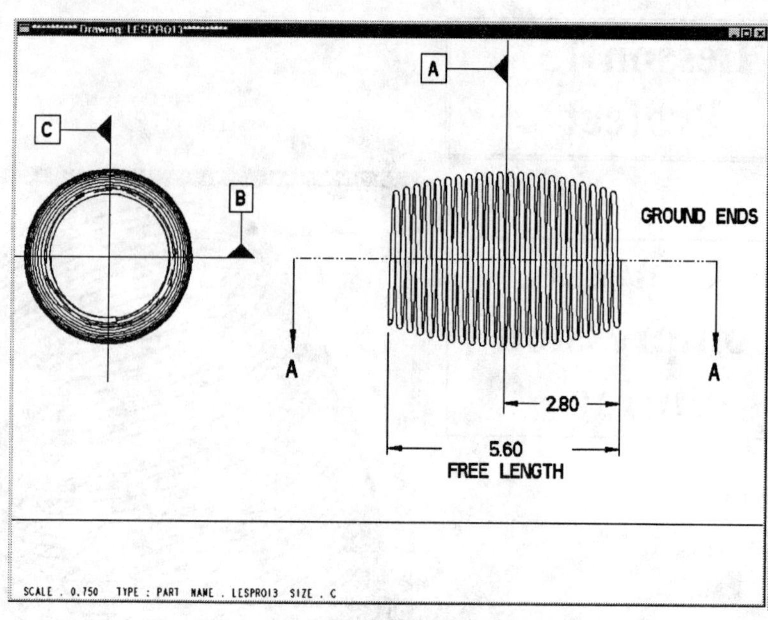

Figure 13.34
Convex Compression Spring Drawing, Front and Left Side Views

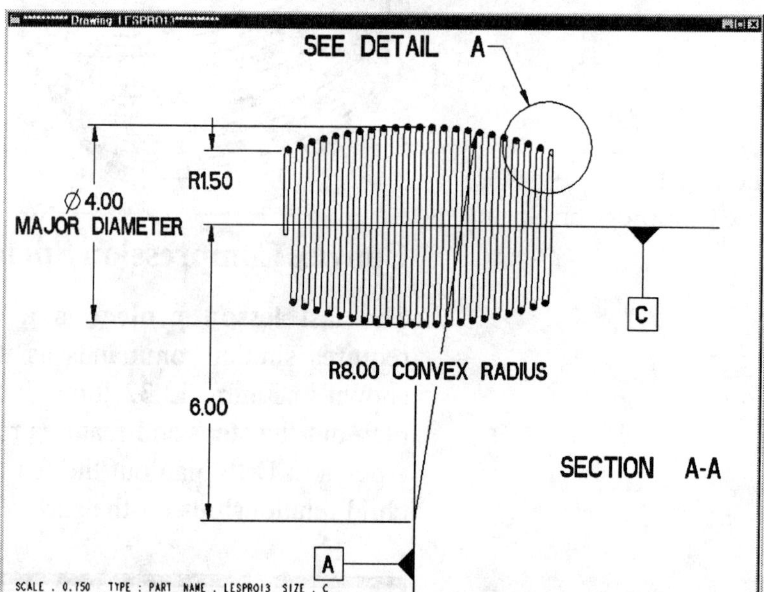

Figure 13.35
Convex Compression Spring Drawing, Top View

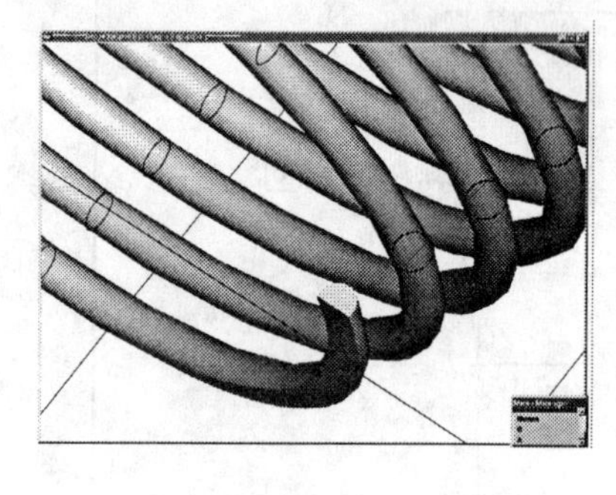

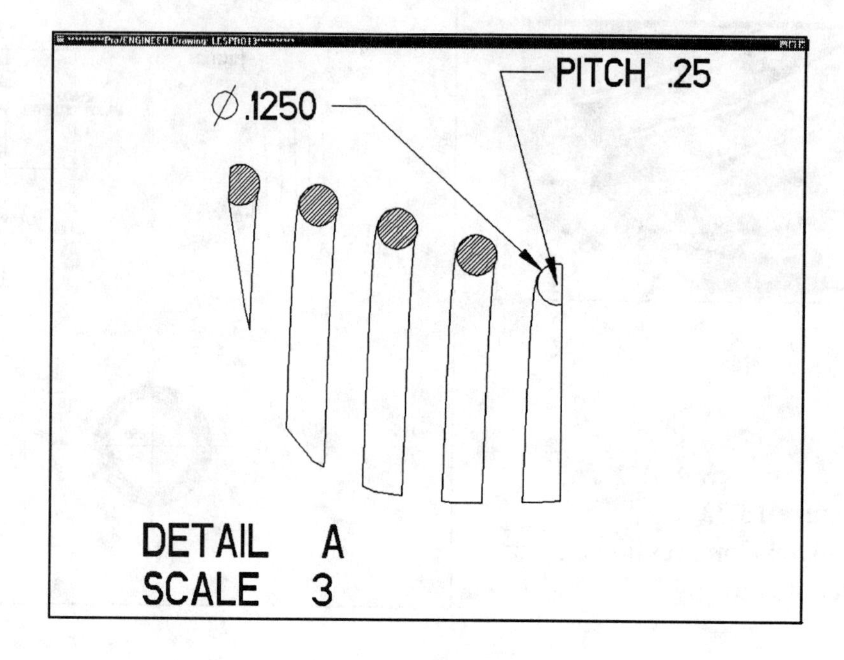

Figure 13.36
Convex Compression Spring Drawing, **DETAIL A**

Dbms ⇒ Save ⇒ enter
Purge ⇒ enter ⇒
Done-Return

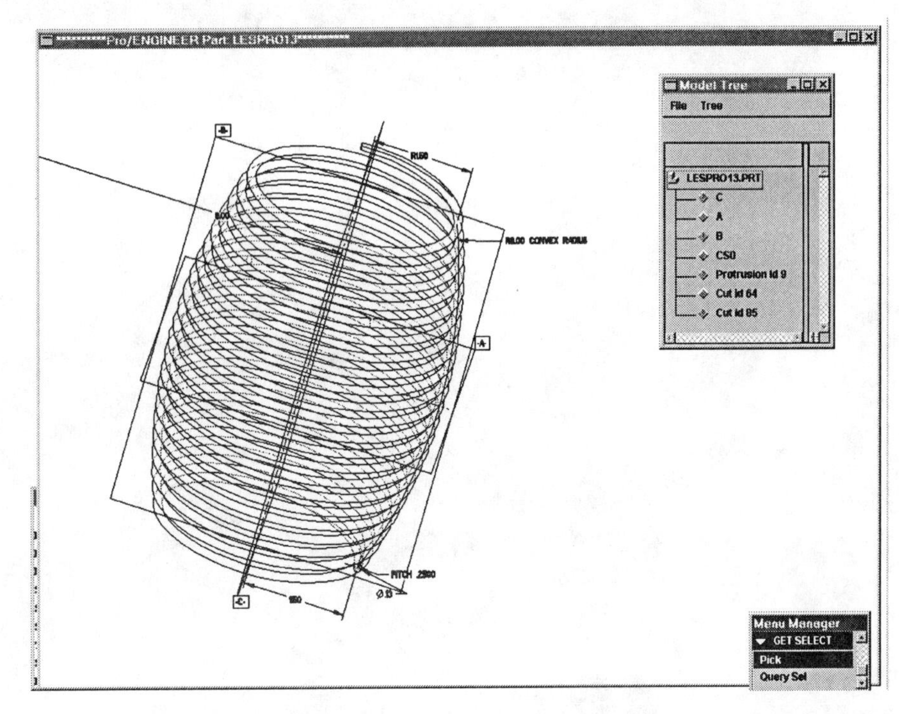

Figure 13.37
Convex Compression Spring Showing Datum Planes and Model Tree

The ECO in Figure 13.38 is not for altering or changing the convex compression spring you just created. The ECO requests that a new *extension* spring be designed with hook ends instead of ground ends. The same size and dimensions are to be used for the new spring as were required in the Convex Spring.

Save the Convex Spring under a new name, for example, **CONVEX_COM_SPR_GRND_ENDS**. Rename the active part to something like **CON_EXT_SPR_HOOK_ENDS**. Delete the ground ends and design machine hooks for both ends of the extension spring. Refer to a Machinery's Handbook, or your engineering graphics text (**EGD**, Chapter 18, Springs) for acceptable design options.

ECO

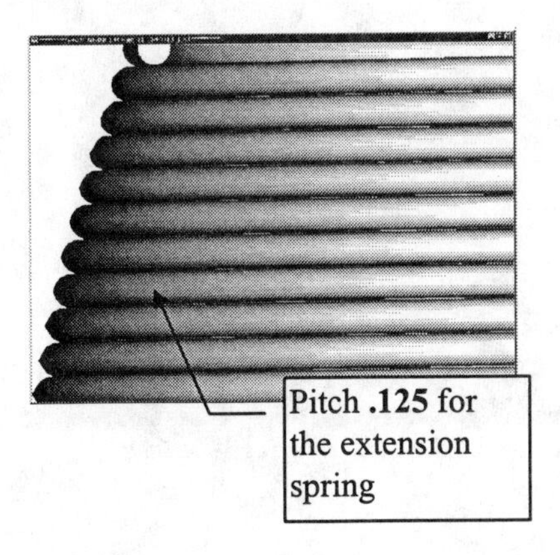

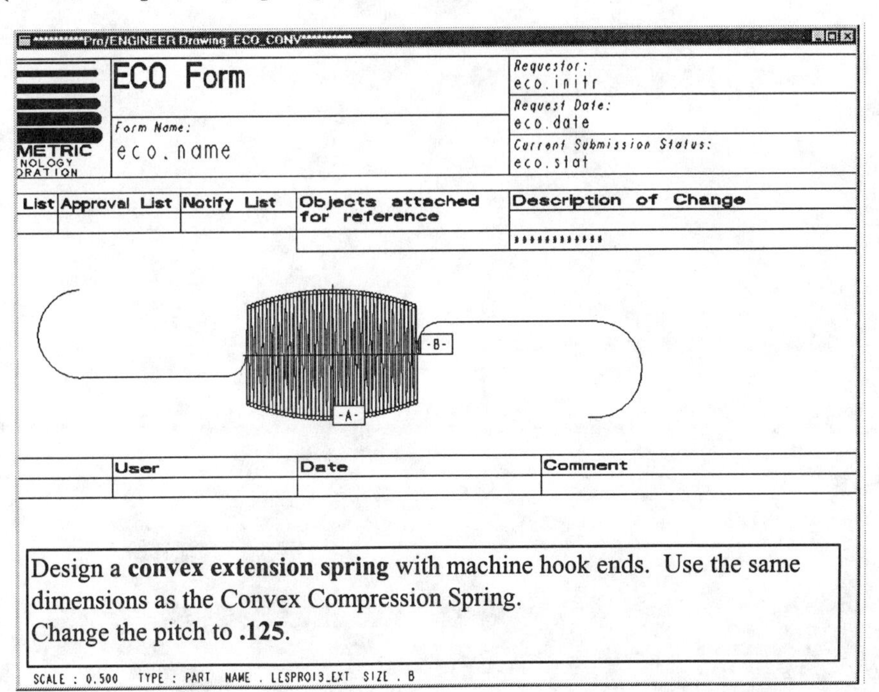

Figure 13.38
ECO for New Version of the Convex Spring

Part Two

Assemblies

Lesson 14 Assembly Constraints
Lesson 15 Exploded Assemblies

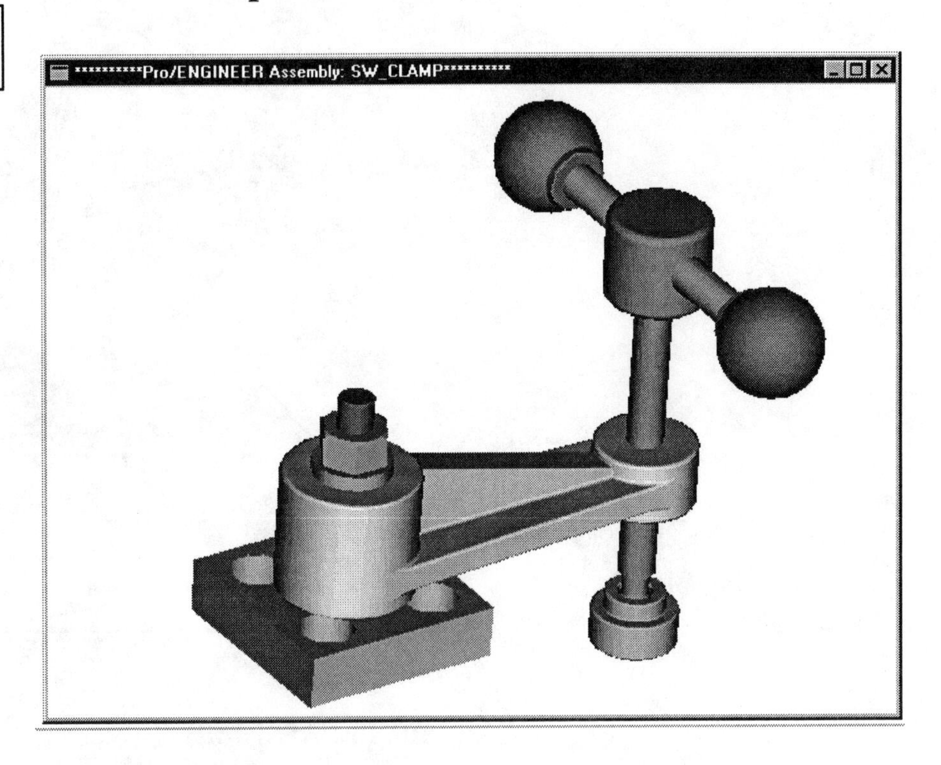

Swing Clamp Assembly

ASSEMBLIES

The **Assembly mode** allows you to place together component parts and subassemblies to form assemblies. Assemblies can then be modified, documented, analyzed, or reoriented. Assembly mode is used for the following functions:

* Placing components into assemblies-- Bottom-up assembly design
* Exploding views of assemblies
* Altering the display settings for individual components
* Designing in Assembly Mode-- Top-down assembly design
* Part modification, including feature construction
* Analysis of assemblies

With **Pro/ENGINEER** you can:

* Assemble-- place together component parts and subassemblies to form assemblies
* Delete or Replace-- remove or replace assembly components.
* Modify assembly placement offsets, and create and modify assembly datum planes, coordinate systems, and sectional views.
* Modify parts directly in assembly mode

* Get assembly engineering information, perform viewing and layer operations, create reference dimensions, and work with interfaces.

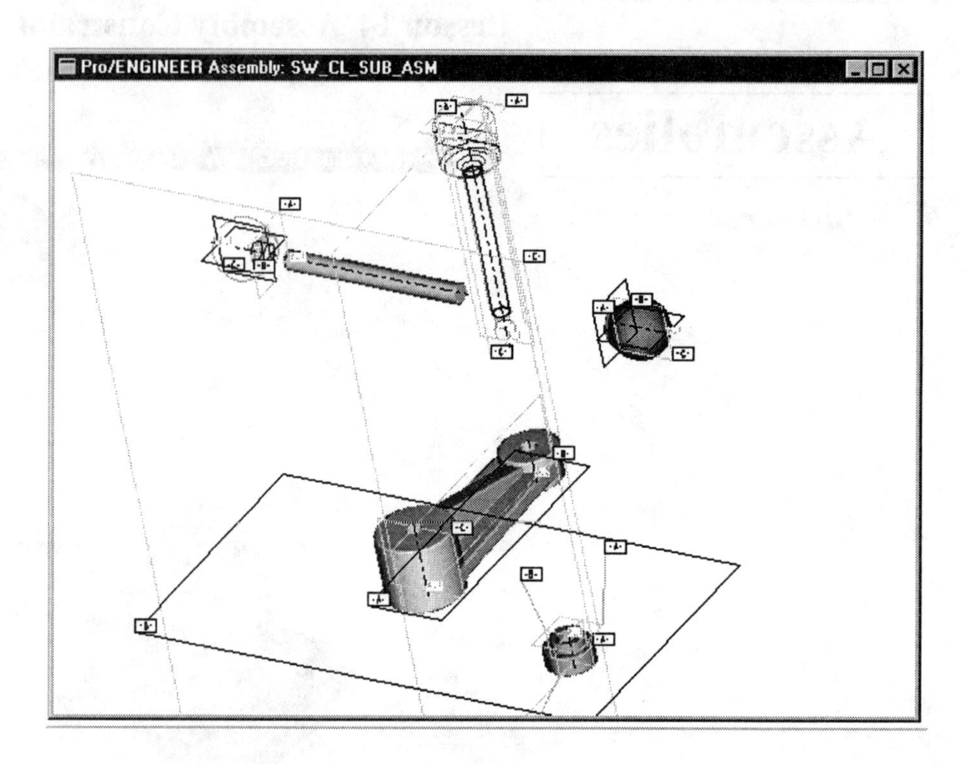

Exploded Swing Clamp Subassembly

With **Pro/ASSEMBLY** you can:

* Create new parts in Assembly mode
* Create sheet metal parts in Assembly mode
* Mirror parts in Assembly mode (create a new part)
* Replace components automatically by creating interchangeablitiy groups. Create assembly features, existing only in Assembly mode and intersecting several components
* Create families of assemblies, using the family table
* Simplify the assembly representation
* Use the **Move** and **Multiply** commands for assembly components
* Use Pro/PROGRAM to create design programs that allow user entries to program prompts to alter the design model

The process of creating an assembly is accomplished by adding components (parts) to a base component (parent part) using a variety of constraints. A **placement constraint** specifies the relative position of a pair of surfaces on two components. The **Mate**, **Align**, **Insert**, and **Orient** commands and their variations are used to accomplish this task.

Lesson 14

Assembly Constraints

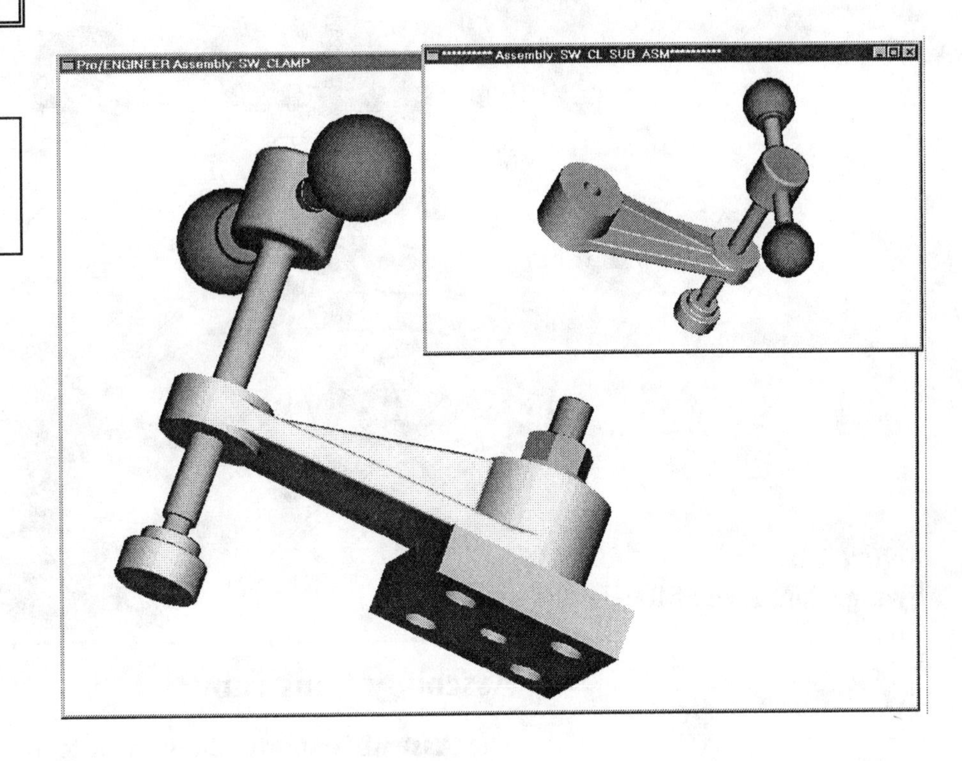

Figure 14.1
Swing Clamp

OBJECTIVES

1. **Assemble components to form an assembly**
2. **Create a subassembly**
3. **Understand and use a variety of assembly constraints**
4. **Redefine a component constraint**
5. **Modify a constraint**
6. **Check for clearance and interference**

EGD REFERENCE
Engineering Graphics and Design
by L. Lamit and K. Kitto
Read Chapters: 23
See Pages: 832-838, 845-846

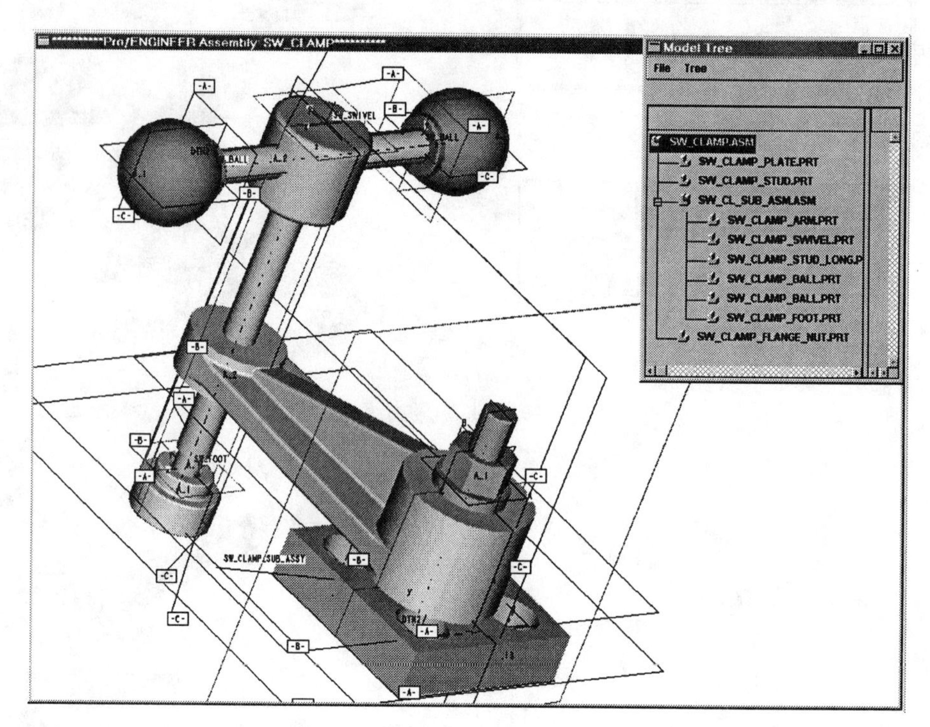

Figure 14.2
Swing Clamp with Model Tree

Assembly Constraints

The **Assembly mode** allows you to place together component parts and subassemblies to form an **assembly** (Figs. 14.1 and 14.2). Assemblies (Fig. 14.3) can be modified, documented, analyzed, or re-oriented. An assembly can be assembled into another assembly, thereby becoming a **subassembly**.

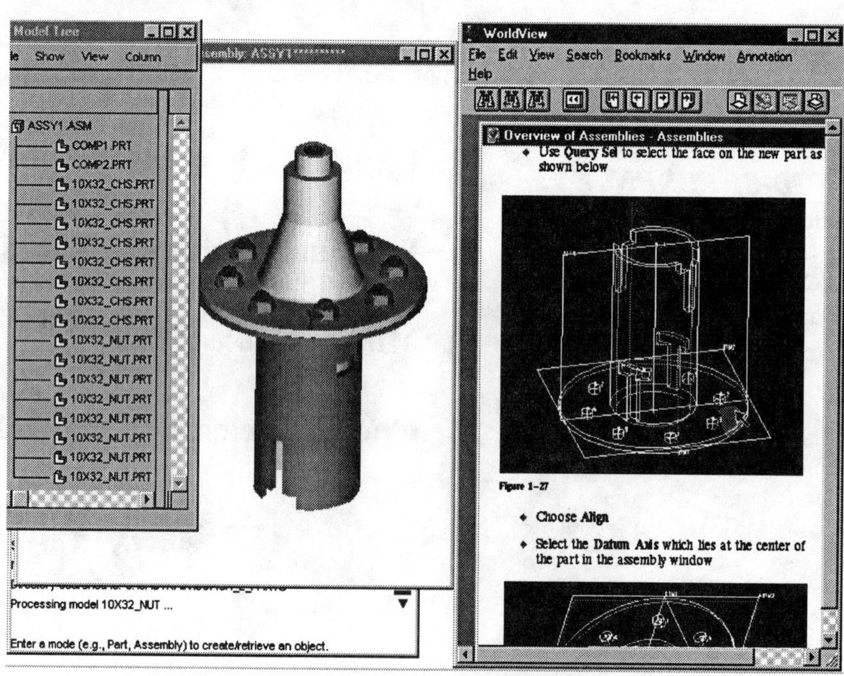

Figure 14.3
COAch for Pro/E, Overview of Assemblies-- Assemblies (Segment 2: Mating Components)

Assembly mode is used for the following functions:

* Placing components into assemblies (Fig. 14.4)
* Exploding views of assemblies
* Part modification, including feature construction
* Analysis

With **Pro/E** you can:

* Place together component parts and subassemblies to form assemblies.
* Remove or replace assembly components.
* Modify assembly placement offsets, and create and modify assembly datum planes, coordinate systems, and crossections.
* Modify parts directly in Assembly mode.
* Get assembly engineering information, perform viewing and layer operations, create reference dimensions, and work with interfaces.

With **Pro/ASSEMBLY** you can:

* Create new parts in Assembly mode.
* Create sheet metal parts in Assembly mode (using Pro/SHEETMETAL).
* Mirror parts in Assembly mode (create a new part).
* Replace components automatically by creating interchangeable groups. Create assembly features, existing only in Assembly mode and intersecting several components.
* Create families of assemblies, using the family table.
* Simplify the assembly representation.
* Use **Move** and **Copy** for assembly components.
* Use Pro/PROGRAM to create design programs that allow users to respond to program prompts to alter the design model.

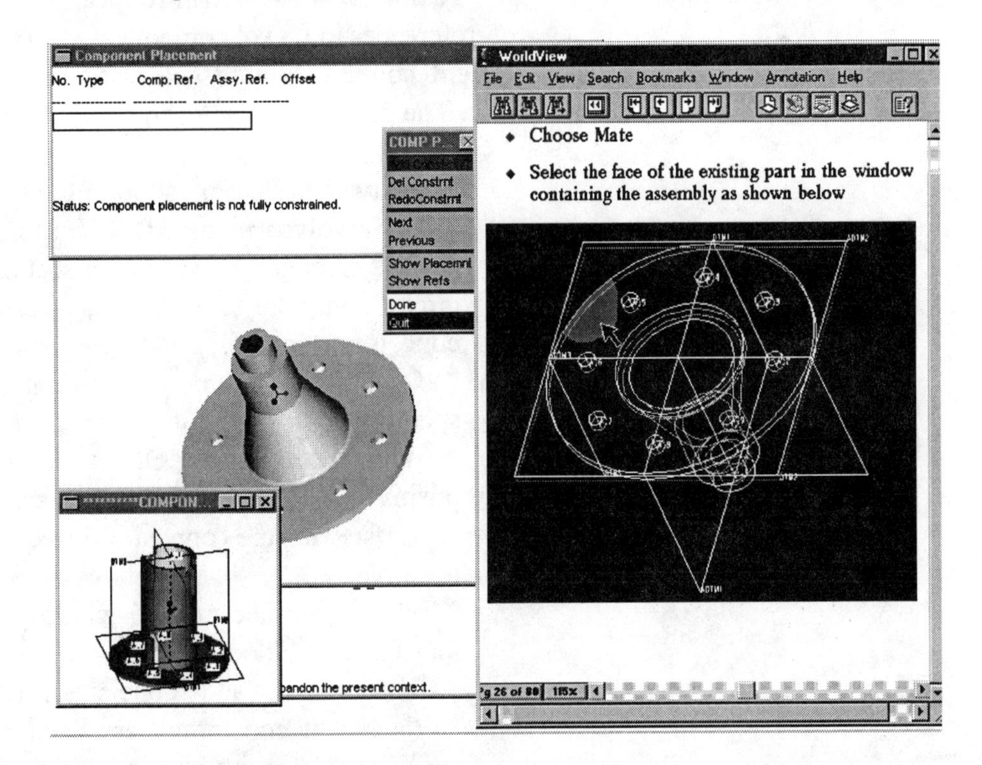

Figure 14.4
COAch for Pro/E, Overview of Assemblies-- Assemblies (Segment 2: Mating Components)

Assembling Components

The process of creating an assembly involves adding components (parts) to a base component (parent part) using a variety of constraints (Fig. 14.5). Components can also be created in the Assembly mode using existing components as references.

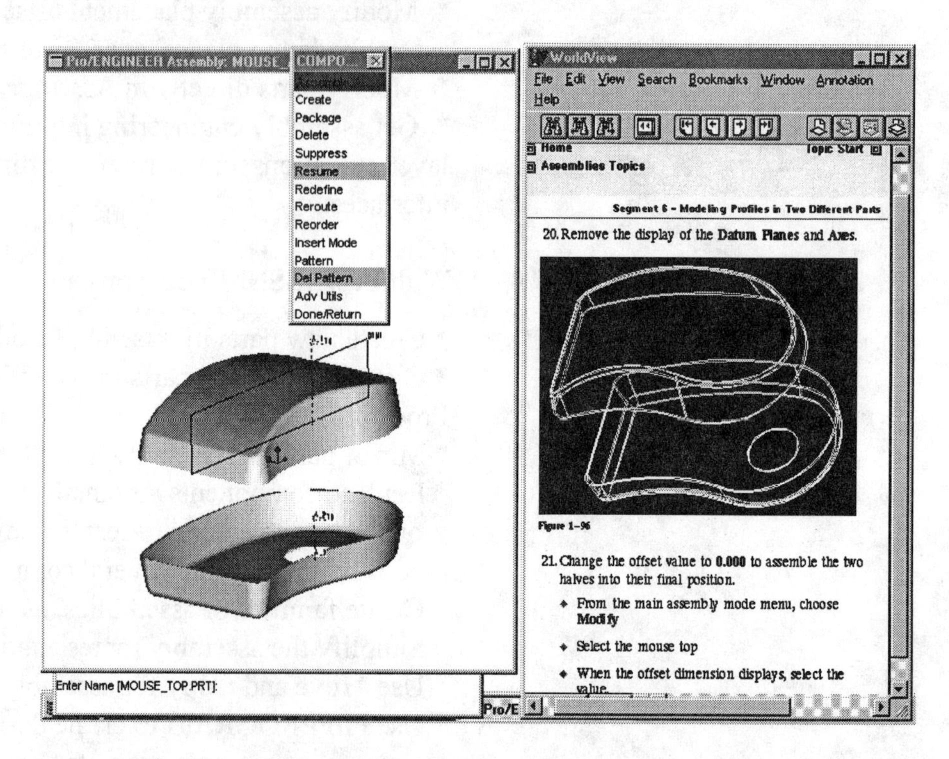

Figure 14.5
COAch for Pro/E, Overview of Assemblies-- Assemblies (Segment 2: Mating Components)

A **placement constraint** specifies the relative position of a pair of references on two components. The **Mate**, **Align**, **Insert**, and **Orient** commands and their variations are used to accomplish this task. The general principles to apply during constraint placement are:

* The two surfaces must be of the same type (for example, plane-plane, revolved-revolved). The term *revolved surface* means a surface created by revolving a section or by extruding an arc or a circle. Only the following surfaces are allowed: plane, cylinder, cone, torus, sphere.
* If you put a placement constraint on a datum plane, you should specify which side of it, yellow or red, you are going to use.
* When using **Mate Offset** or **Align Offset** and entering a positive value, you will be given the offset direction. If you need an offset in the opposite direction, make the offset value negative.
* When a surface on one window is selected, another window may become hidden.
* Constraints are added one at a time.
* Placement constraints are used in combinations in order to specify placement and orientation completely. For example, one pair of surfaces may be constrained to mate, another pair to insert, and a third pair to orient.

Mate (Fig. 14.6) is used to make two surfaces touch one another: coincident and facing each other. When using datums, this means that two yellow sides, or two red sides, will face each other. **Mate Offset** (Fig. 14.6) makes two planar surfaces parallel and facing each other. The offset value determines the distance between the two surfaces.

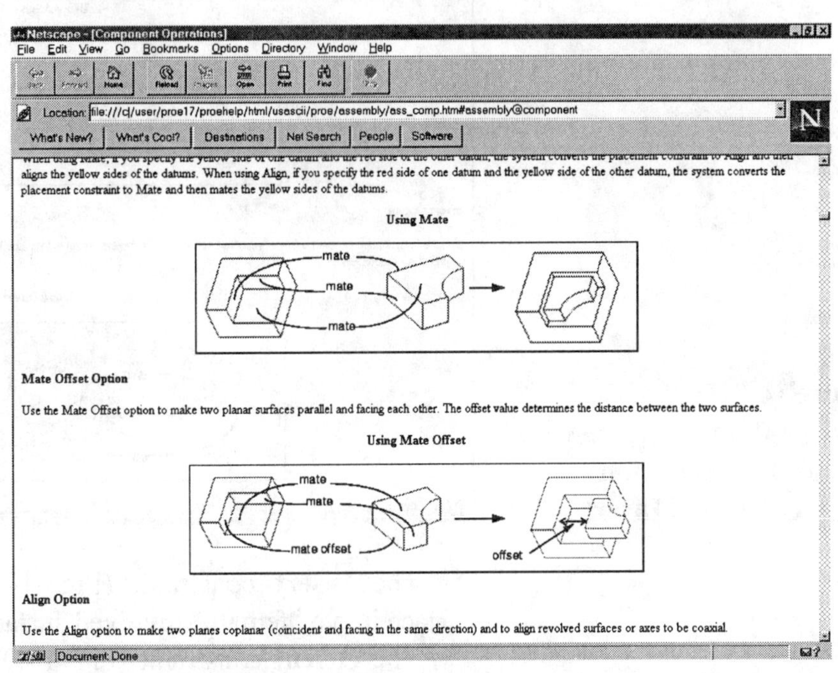

Figure 14.6
Pro/HELP
Online Documentation
Mate and **Mate Offset**

The **Align** command (Fig. 14.7) makes two planes coplanar: coincident and facing in the same direction. The **Align** constraint also aligns revolved surfaces or axes so as to be coaxial. You can also align two datum points, vertices, or curve ends; selections on both parts must be of the same type (that is, if a datum point is selected on one part, only a datum point can be selected on another part). The **Align Offset** constraint (Fig. 14.8) aligns two planar surfaces at an offset: parallel and facing in the same direction.

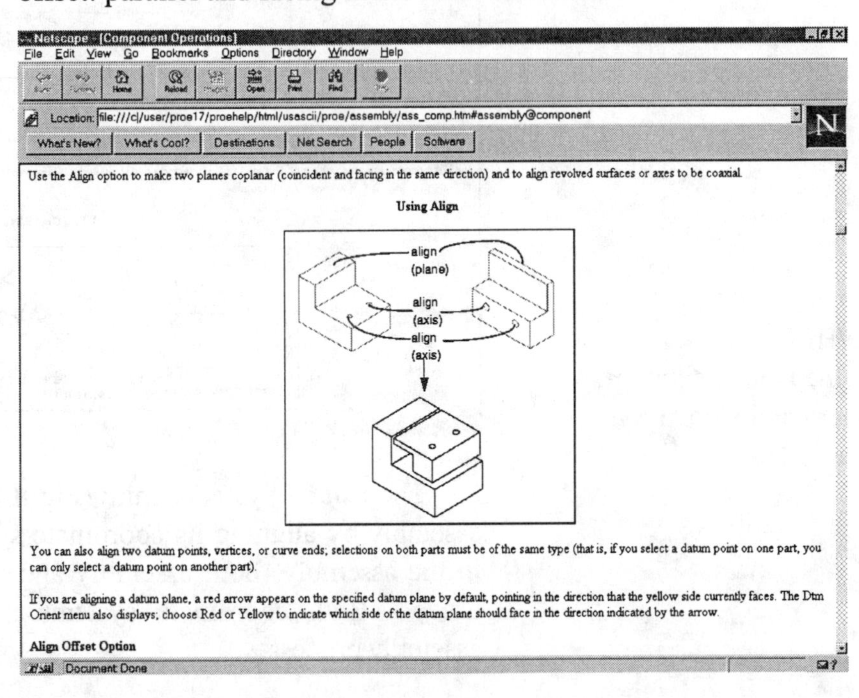

Figure 14.7
Pro/HELP
Online Documentation
Align

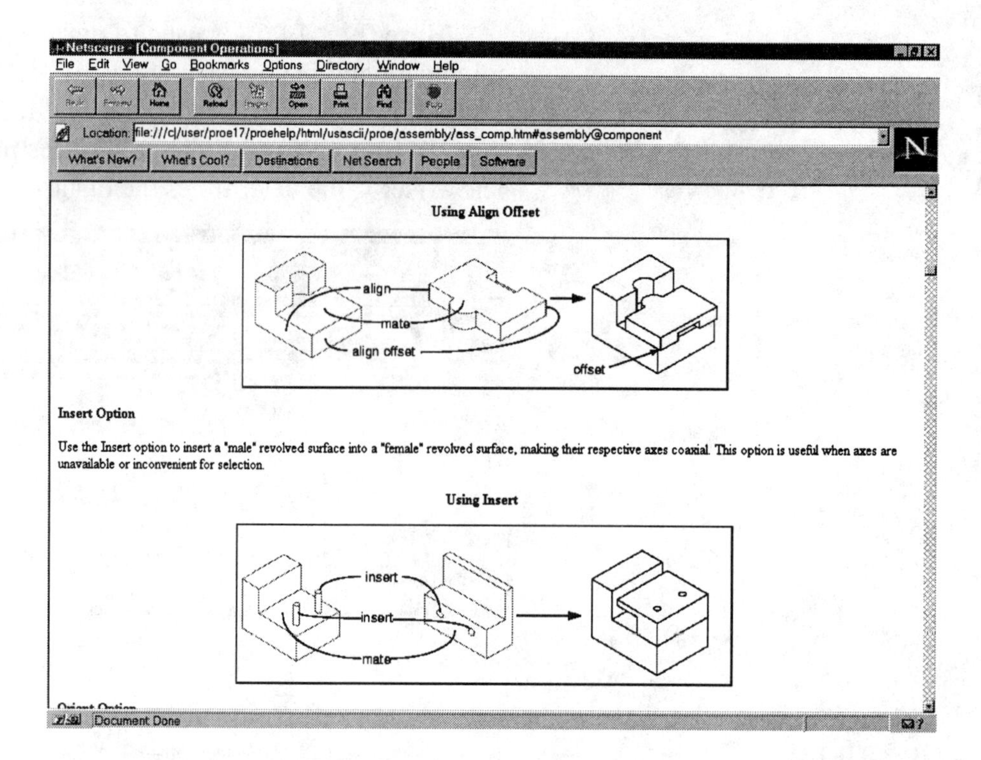

Figure 14.8
Pro/HELP
Online Documentation
Align Offset and **Insert**

The **Insert** constraint (Fig. 14.8) inserts a "male" revolved surface into a "female" revolved surface, aligning axes.

The **Orient** constraint (Fig. 14.9) orients two planar surfaces to be parallel and facing in the same direction; offset is not specified.

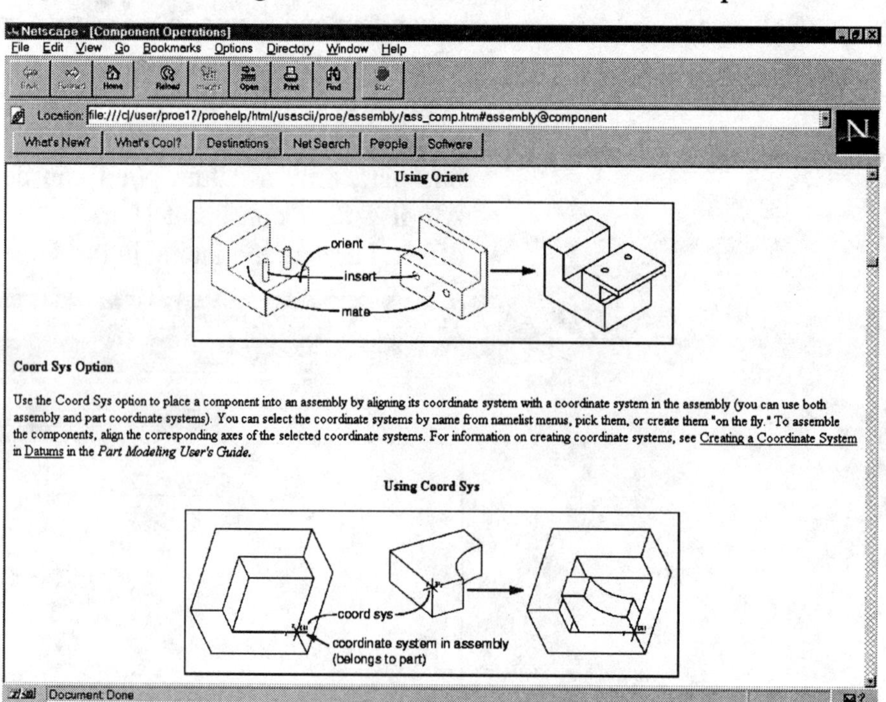

Figure 14.9
Pro/HELP
Online Documentation
Orient and **Coord Sys**

The **Coord Sys** constraint (Fig. 14.9) places a component into an assembly by aligning its coordinate system with a coordinate system in the assembly (both assembly and part coordinate systems can be used). Both coordinate systems must exist before starting the assembly process.

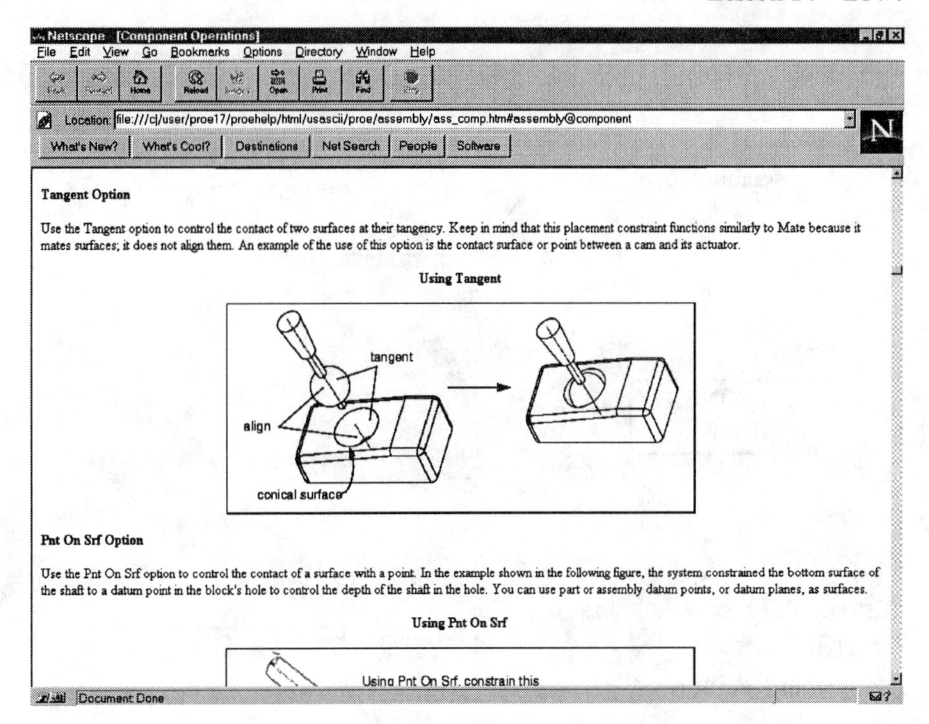

Figure 14.10
Pro/HELP
Online Documentation
Tangent

Coordinate systems can be picked or selected by name from namelist menus. The components will be assembled by aligning **X**, **Y**, and **Z** axes of the selected coordinate systems.

The **Tangent**, **Pnt On Surf**, and **Edge On Surf** constraints are used to control the contact of two surfaces at their tangency, at a point, or at an edge (Fig. 14.10). An example of the use of these placement options is the contact surface or point between a cam and its actuator.

In most cases, a combination of constraints will be required. **Mate**, and **Insert** are required to constrain the two parts shown in Figure 14.11. **Mate**, **Insert**, and **Orient** are another possibility, depending on the parts.

Assemblies can also be displayed *exploded*. The explode distance can be modified to any value.

The completed assembly can be modified by redefining the constraints. An exploded cosmetic view of the assembly can be displayed with or without the reference planes, coordinate system, and hidden lines. The last stage of the project involves putting the assembly into the Drawing mode and displaying the appropriate views.

Component Placement Window

When you choose **Assemble** or **Redefine** from the COMPONENT menu, a *Component Placement window* appears (Fig. 14.11). This window lists the *constraint number*, the *constraint type*, the corresponding *component* and *assembly references*, and the *offset value* (if any) of the component placement constraint.

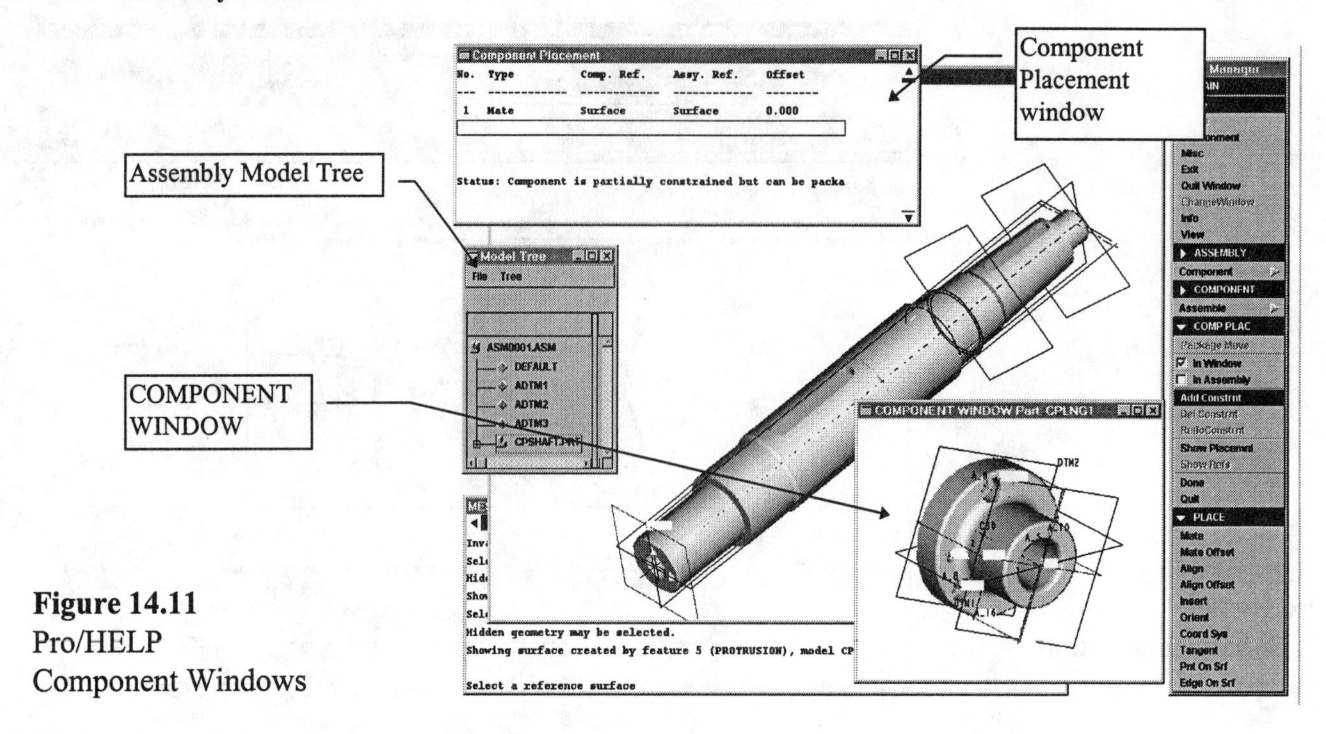

Figure 14.11
Pro/HELP
Component Windows

To assemble the component:

1. Choose **Component** from the ASSEMBLY menu.
2. Choose **Assemble** from the COMPONENT menu, and select the component. The component appears in the COMPONENT WINDOW; the COMP PLAC menu and a corresponding Component Placement window are also displayed. The COMP PLAC menu lists the following options:

   Package Move Dynamically changes the position of the component without parametric constraints

   In Window Shows the component in its own window while you modify its constraints

   In Assembly Shows the component in the assembly window while you modify its constraints

   Add Constrnt Adds a placement constraint for the component.

   Del Constrnt Deletes a placement constraint for the component.

   RedoConstrnt Changes a placement constraint for the component. The CONSTR REDEF menu contains the options:

   Type Redefines the constraint type (mate, align, orient).

   AssemblyRef Redefines the reference in the assembly.

   Comp Ref Redefines the reference on the placed component.

   Show Placemnt Shows the location of the component as it would be with the current placement constraints.

   Show Refs Highlights the current references.
3. Choose **Add Constrnt** from the COMP PLAC menu.
4. Define the placement constraints using the PLACE menu. As you do so, Pro/E automatically updates a line in the Component Placement window corresponding to the constraint.

5. If the component is validly constrained, the Component Placement window displays the message "**Status: The Component is fully constrained and can now be placed**." If you are adding a new component or modifying the constraints for a component, the MESSAGE WINDOW also displays the message "**Component can now be placed**" (Fig. 14.12).

6. Choose **Done** from the COMP PLAC menu to complete the processing. If the component constraints are incomplete or conflicting, the Component Placement window displays the message "**Status: Component has no constraints but can be packaged**" (Fig. 14.13). Pro/E gives you the option to restart, continue placing the component, package the component, or **Quit**. If you choose to **Quit**, the insufficient constraints are erased.

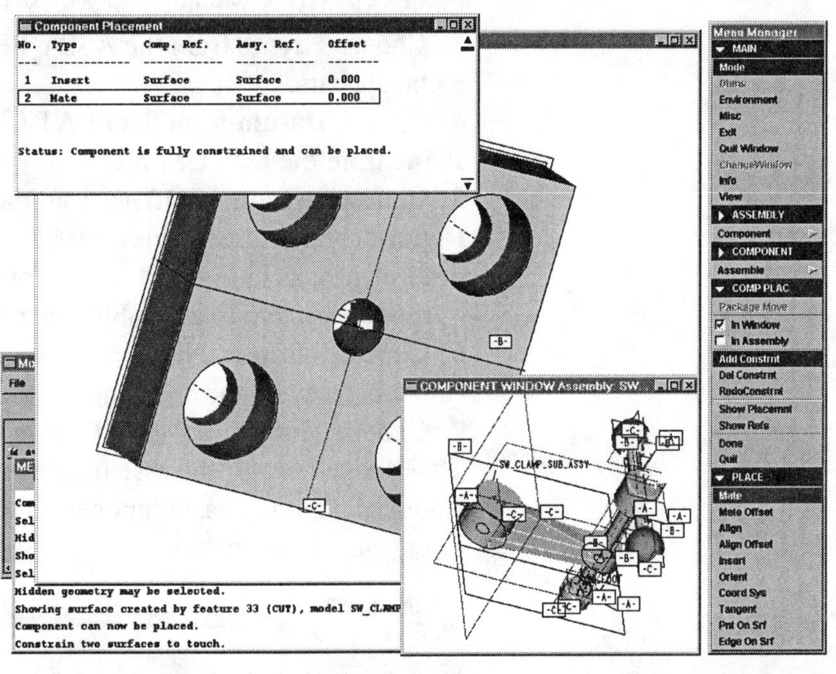

Figure 14.12
Status: **The Component is fully constrained and can now be placed**

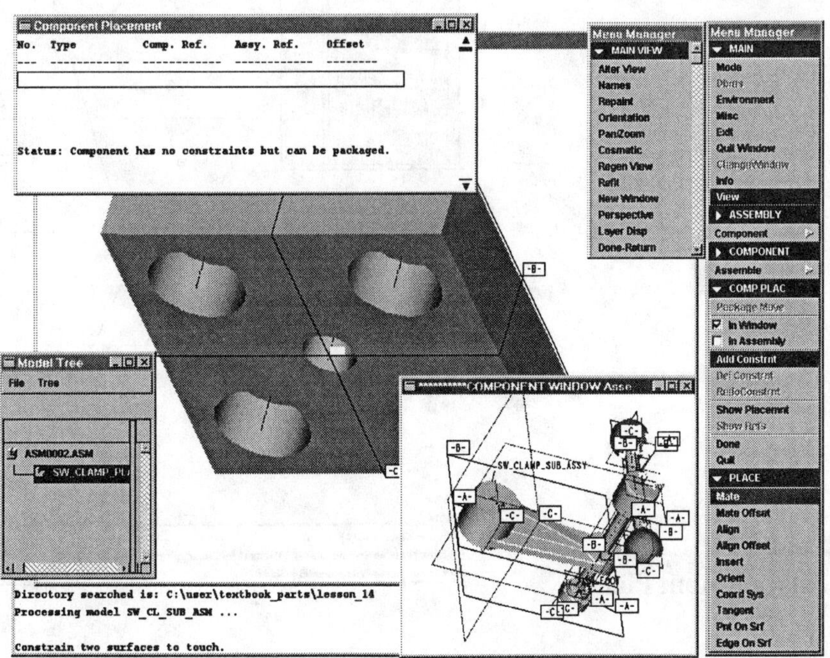

Figure 14.13
Status: **Component has no constraints but can be packaged**

Datum Planes as the First Feature of an Assembly

When you create three default datum planes as the first features in an assembly, you can assemble a component with respect to these planes or create a part in Assembly mode as the first component. When you use datum planes as the first feature, you can create a more flexible design: you can *pattern* the first component you add, and you can *reorder* subsequent components to come before the first component (if the components are not children of the first component). To create a default datum plane assembly feature, follow these steps:

1. Choose **Assembly** from the MODE menu, and then choose **Create** from the ENTER menu.
2. Enter a name for the new assembly, and then choose **Feature** from the ASSEMBLY menu. The ASSY FEAT menu appears.
3. Choose **Create** from the ASSY FEAT menu. The FEAT CLASS menu appears.
4. Choose **Datum** from the FEAT CLASS menu, and then choose **Plane** from the DATUM menu.
5. Choose **Default** or **Offset** from the MENUDTM OPT menu. **Default** creates three planes; **Offset** creates a coordinate system and three planes, as in Figure 14.14. You can use the datums and/or a coordinate system to assemble other components.
6. Choose **Setup** ⇒ **Name** ⇒ **Feature** to change the names of the datum planes and the coordinate system.
7. Choose **Modify** ⇒ **Move Datum** to reposition the new names.

The Model Tree for the assembly can be altered to show the assembly features, components, component features, assembly and component layers, and datum names.

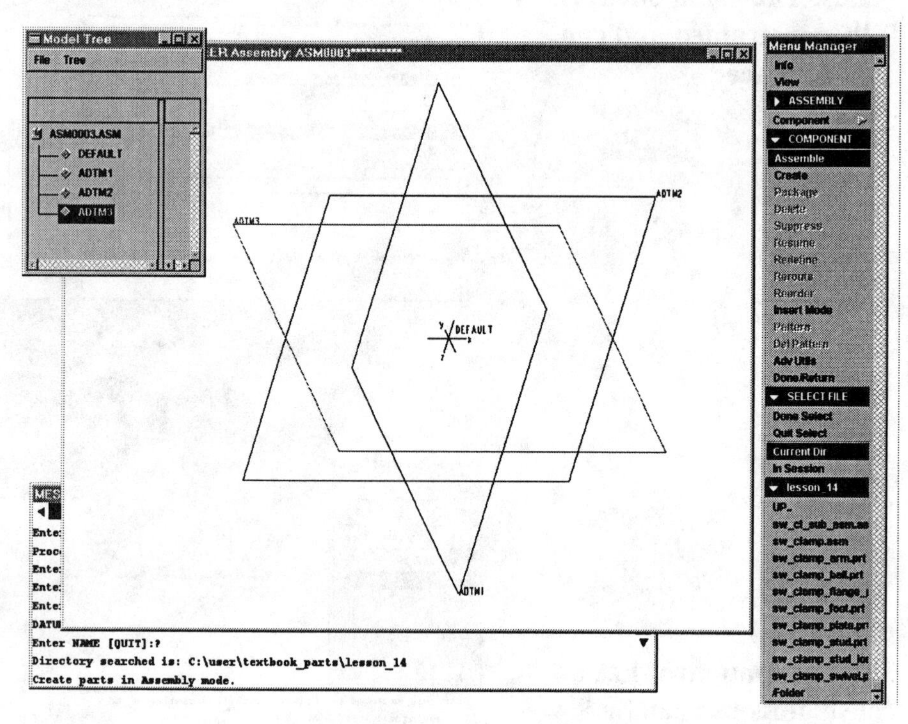

Figure 14.14
Assembly Datum Planes

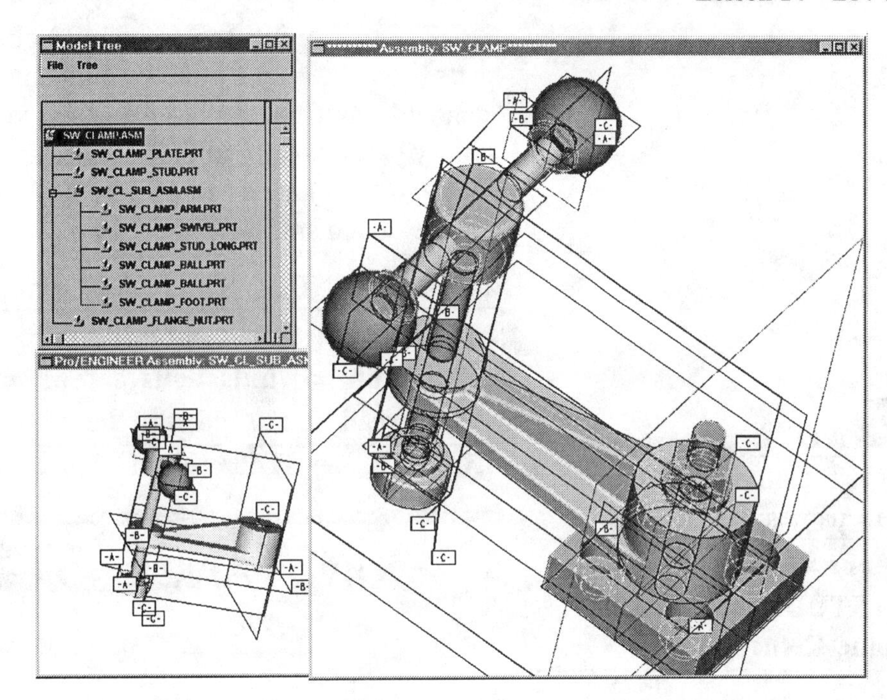

Figure 14.15
Swing Clamp Assembly and Subassembly with Model Tree and Datum Planes

Swing Clamp Assembly

Most of the parts required in this lesson are lesson projects from the text. The **Clamp Foot** and **Clamp Swivel** are from Lesson 5. The **Clamp Ball** is from Lesson 6, and the **Clamp Arm** was created in Lesson 8. If you have not modeled these parts previously, please do so before you start the step-by-step instructions. The other parts required for the assembly (Fig. 14.15) are standard off-the-shelf hardware items that are available on Pro/E by accessing the library. If your system does not have a Pro/LIBRARY license for the Basic and the Manufacturing libraries, model the parts using the detail drawings provided here. The **Flange Nut**, the **3.50" Double-ended Stud**, and the **5.00" Double-ended Stud** are standard parts (Fig. 14.16).

NOTE
Set up the **Swing Clamp** Assembly:

- Units = Inches
- Datum Planes
- Default Coordinate System
- Layers = **ASM_DATUM_PLANE**
- Shading
- No Display Tan
- ☐ Grid Snap

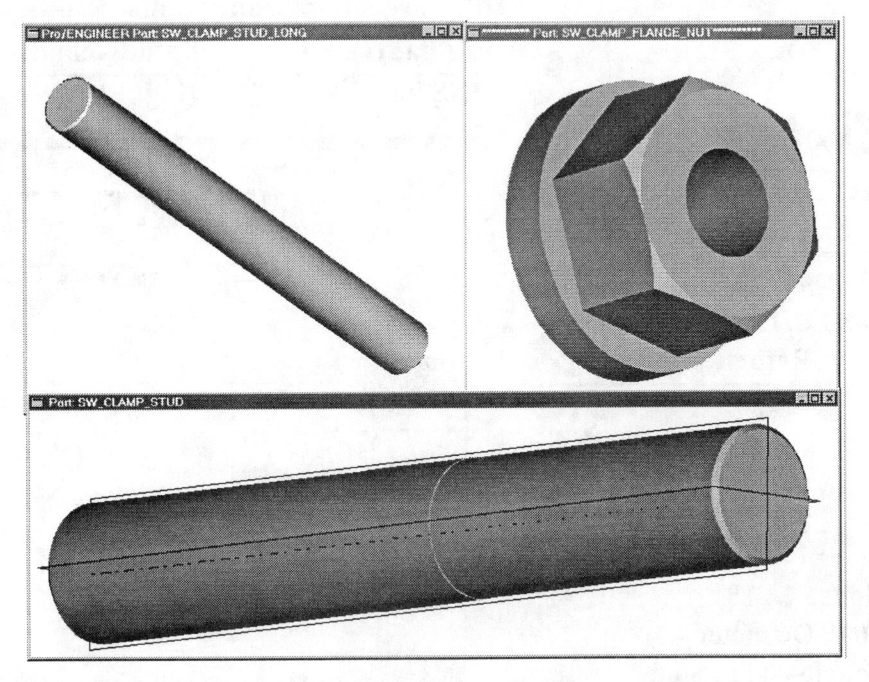

Figure 14.16
Double-ended Studs and a Flange Nut

Before starting the assembly you will be creating the standard parts from the library or by modeling each individually. Save the standard parts under new names in your directory. Do not use the library parts directly in the assembly. Start this process by giving the following commands to access the library and the existing standard double-ended studs:

DOUBLE-ENDED STUD .500" diameter by 3.50" length

Mode ⇒ Part ⇒ Search/Retr ⇒ Pro/Library ⇒ /mfglib ⇒ /fixture_lib ⇒ /nuts_bolts_screws ⇒ st.prt ⇒ SelByParams ⇒ ✓d0,thread_dia ✓ d8,stud_length ⇒ Done Sel ⇒ .5000 ⇒ 3.500

INSTANCE = ST403 (Fig. 14.17)

NOTE

Save the library parts under new names:

Dbms ⇒ Save As ⇒ enter ⇒ STUD35 ⇒ enter ⇒ Done-Return

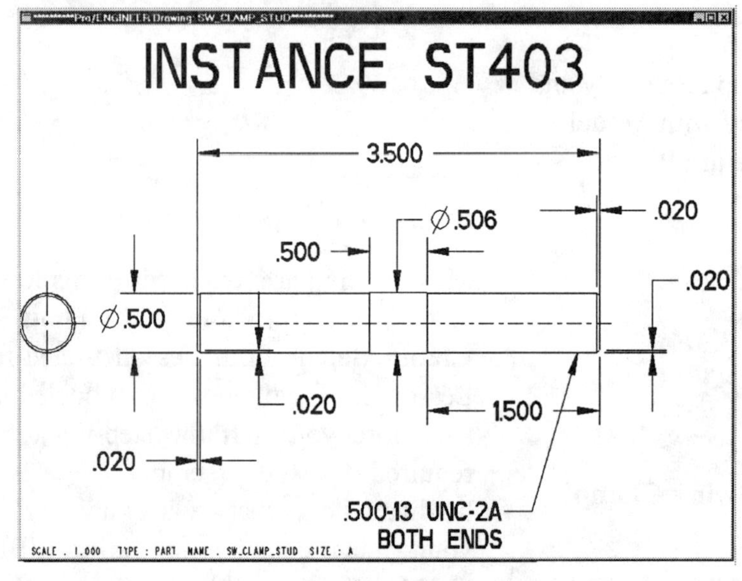

Figure 14.17
3.50" Diameter
Double-ended Stud

DOUBLE_ENDED STUD .500" diameter by 5.00" length

Mode ⇒ Part ⇒ Search/Retr ⇒ Pro/Library ⇒ /mfglib ⇒ /fixture_lib ⇒ /nuts_bolts_screws ⇒ st.prt ⇒ SelByParams ⇒ ✓d0,thread_dia ✓ d8,stud_length ⇒ Done Sel ⇒ .5000 ⇒ 5. 00

INSTANCE = ST406 (Fig. 14.18)

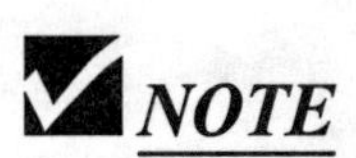

NOTE

Save the library parts under new names:

Dbms ⇒ Save As ⇒ enter ⇒ STUD5 ⇒ enter ⇒ Done-Return

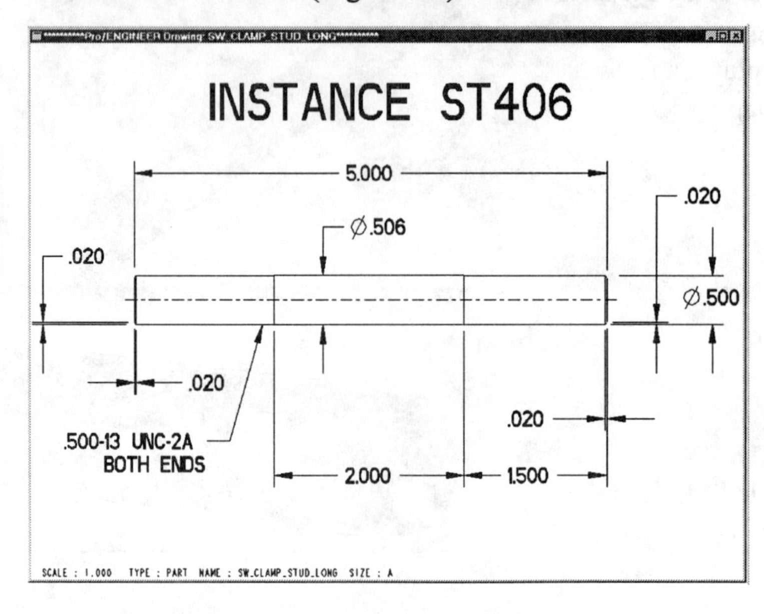

Figure 14.18
5.00" Diameter
Double-ended Stud

The last standard item used for this assembly is a flange nut (Fig. 14.19). Use the following commands to access the part:

FLANGE NUT

Mode ⇒ Part ⇒ Search/Retr ⇒ Pro/Library ⇒ /mfglib ⇒ /fixture_lib ⇒ /nuts_bolts_screws ⇒ fn.prt ⇒ SelByParams ⇒ ✓ d4,thread_dia ⇒ Done Sel ⇒ .5000

INSTANCE = FN7

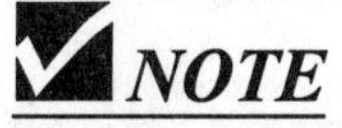

NOTE

Save the library parts under new names:

Dbms ⇒ Save As ⇒ enter ⇒ FLNGNUT ⇒ enter ⇒ Done-Return

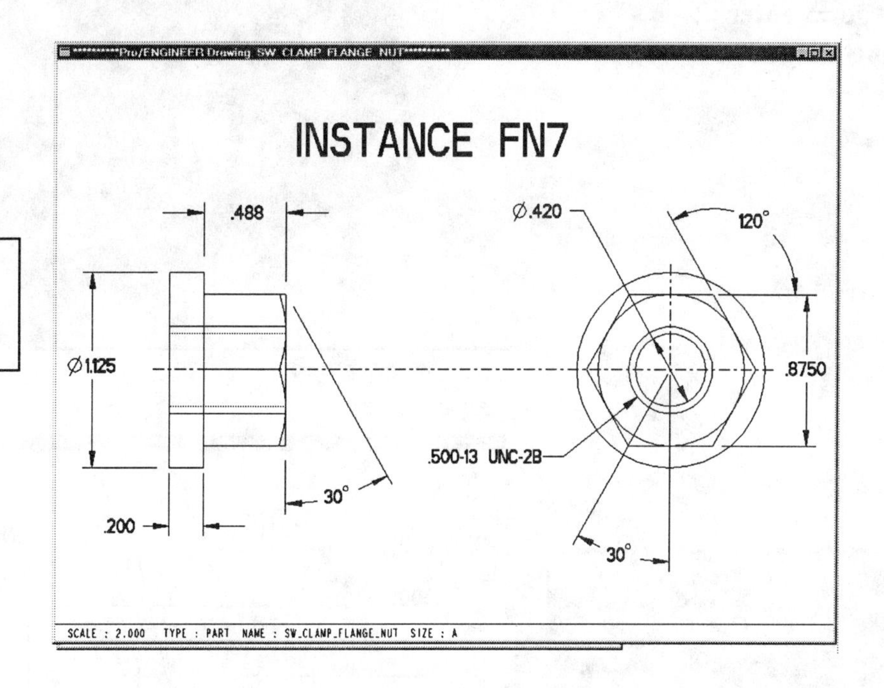

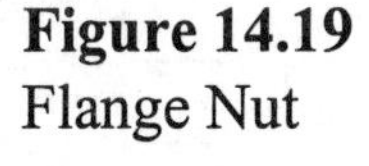

Figure 14.19
Flange Nut

Since you will be creating an assembly using the bottom-up design approach, all of the components must be available before the assembling starts. *Bottom-up design* means that existing parts are assembled one by one until the assembly is complete. The assembly starts with a set of datum planes and a coordinate system. The parts are added to the datum features of the assembly. The sequence of assembly will determine the parent-child relationships between components.

Top-down design is a process describing the design of an assembly where component parts are created in the Assembly mode as the design unfolds. Some existing parts are available, such as standard components and a few modeled parts. The remaining design evolves during the assembly process.

Regardless of the design method, the assembly datums and coordinate system need to be on their own separate layer. Each part should also be placed on a layer as well as the parts datum features.

The **Plate** component is the first component of the main assembly. The Plate is a new part (Fig. 14.20). You must model the plate before creating the assembly. Figure 14.21 provides the dimensions necessary to model the Plate.

Dbms ⇒ Save ⇒ enter
Purge ⇒ enter ⇒ Done-Return

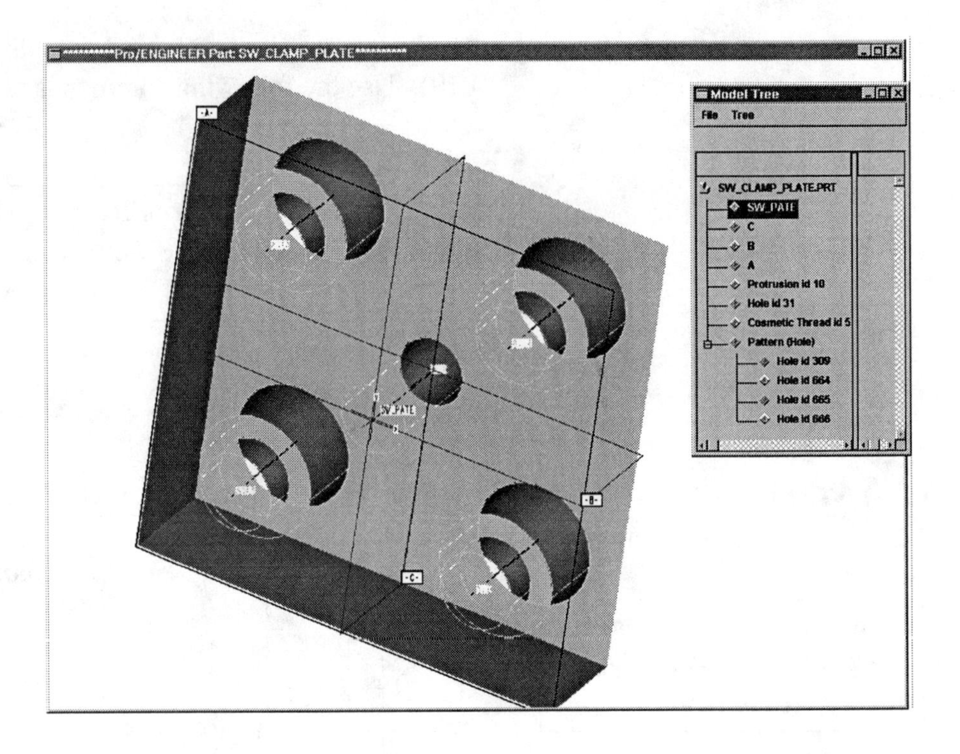

Figure 14.20
Plate

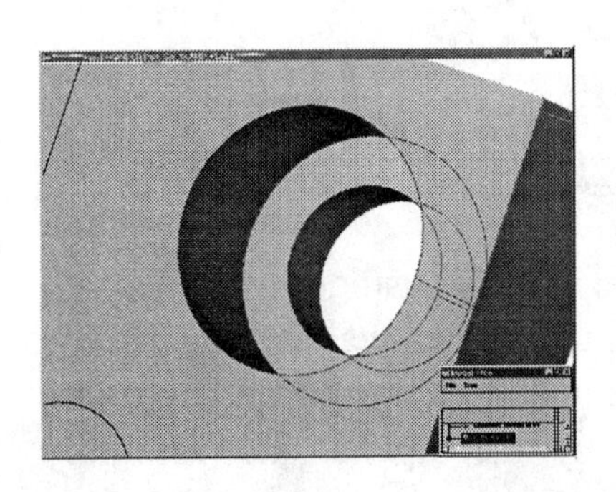

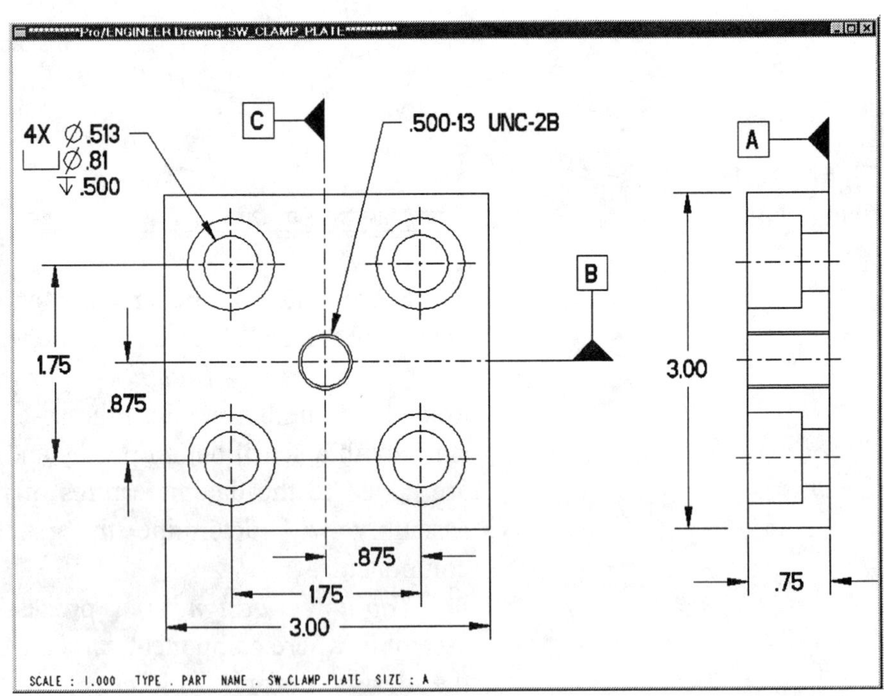

Figure 14.21
Plate Detail Drawing

You now have all nine (eight unique parts, two BALL components are used) parts required for the assembly. A subassembly will be created first. The main assembly is created second. The subassembly is assembled to the main assembly. Start the subassembly (Fig. 14.22) using the following commands:

Assembly ⇒ Create ⇒ CLAMP_SUBASSEMBLY ⇒ enter ⇒ Feature ⇒ Create ⇒ Datum ⇒ Plane ⇒ Offset ⇒ enter ⇒ enter ⇒ enter

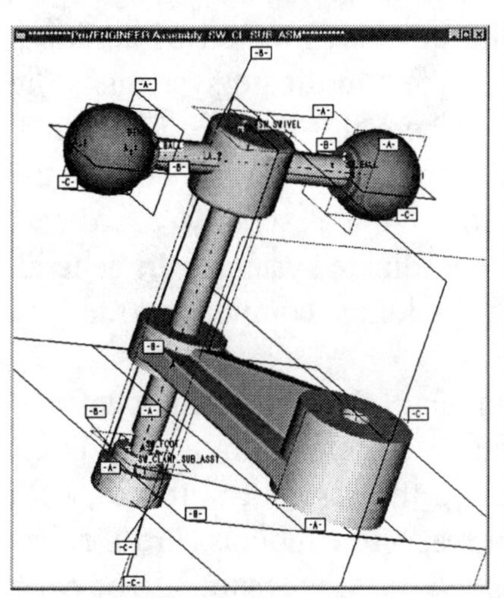

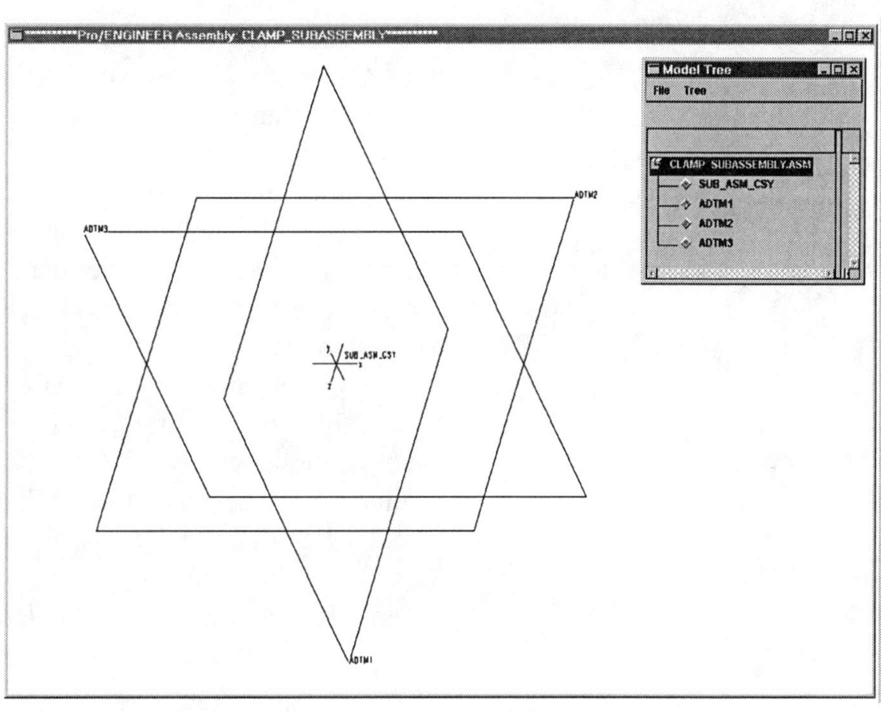

Figure 14.22
Subassembly Datums

Create layers for the datum planes and the coordinate system (Fig. 14.23). The default datums will have default names, i.e., **ADTM1, ADTM2, ADTM3,** provided by Pro/E. You can change the names using **Setup** ⇒ **Name** ⇒ **Other**. The coordinate system's name has been changed in Figure 14.23 to **SUB_ASM_CSY**. The layer for the datums and coordinate system is **SUB_ASM_DATUMS**. You can create your own layering system using unique names, or use the ones provided here.

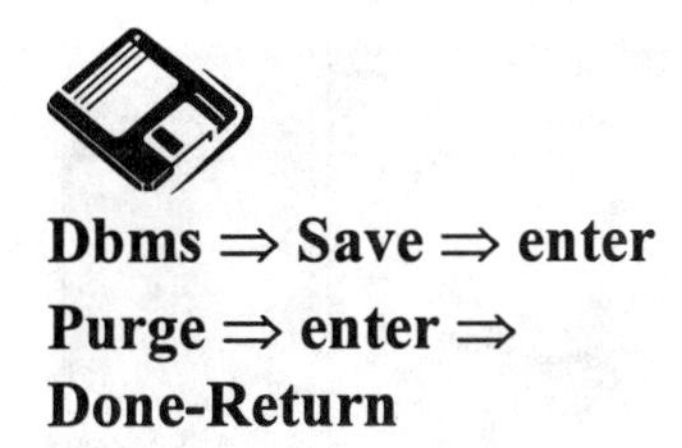

Dbms ⇒ Save ⇒ enter Purge ⇒ enter ⇒ Done-Return

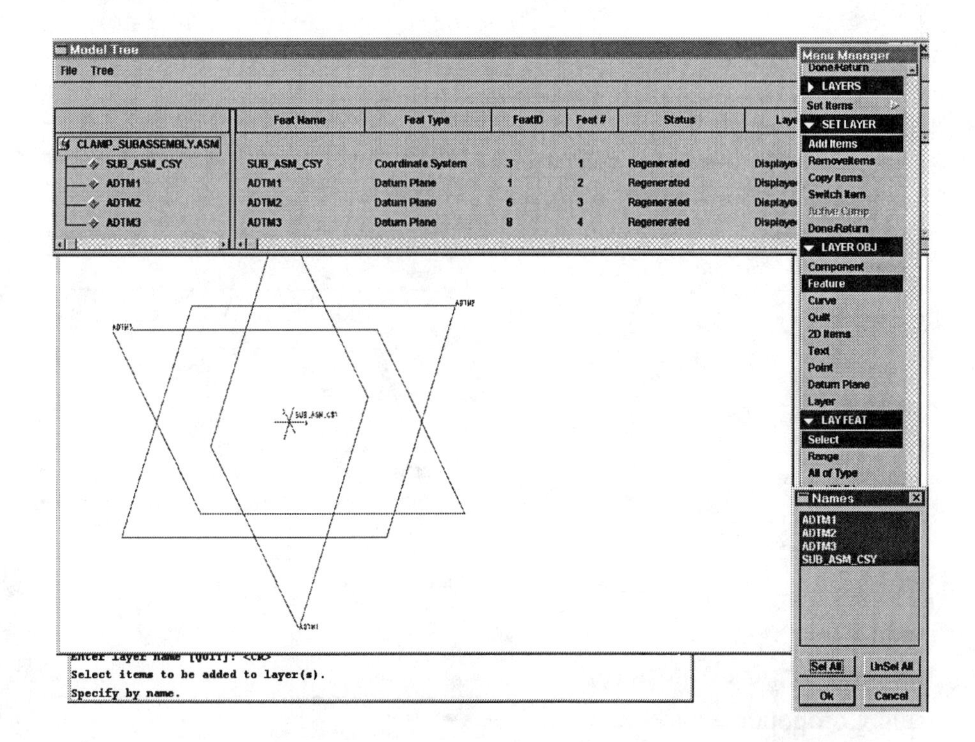

Figure 14.23
Subassembly Layers

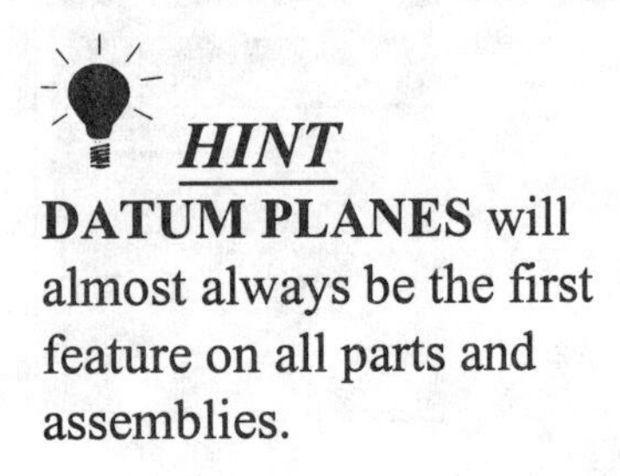

HINT

DATUM PLANES will almost always be the first feature on all parts and assemblies.

The first component to be assembled to the subassembly is the Swing Clamp Arm. The simplest and quickest method of adding a component to an assembly is to match the coordinate systems. The first component assembled is usually where this *constraint* is used, because after the first component is established, few if any of the remaining components are assembled to the assembly coordinate system or for that matter other parts' coordinate systems. In general, different constraints are used when adding components to the assembly model.

The **Coord Sys** (Fig. 14.24) command places a component into an assembly by aligning its coordinate system with a coordinate system in the assembly. Before starting the assembly process, both coordinate systems must exist on their respective models (Pro/E refers to both assemblies and parts as *models*). Make sure all of your models are in the same working directory before starting the assembly process. If you named your models something different than listed here, pick the appropiate part model as requested. After the component window appears, rotate the part model so as to see more clearly the features to be used for constraining. The window with the asterisks is the *active window*: ********MODEL_NAME********. When a feature in one window is selected, another window may become hidden.

Give the following commands (Fig. 14.25 and Fig. 14.26):

Component ⇒ **Assemble** ⇒ (enter a **?** from the keyboard) ⇒ **enter** ⇒ (pick the **SW_CLAMP_ARM** from the **Current Dir**) ⇒ **Coord Sys** ⇒ (pick the coordinate system on the part model and then the coordinate system on the assembly model) ⇒ **Done** ⇒ **Done/Return**

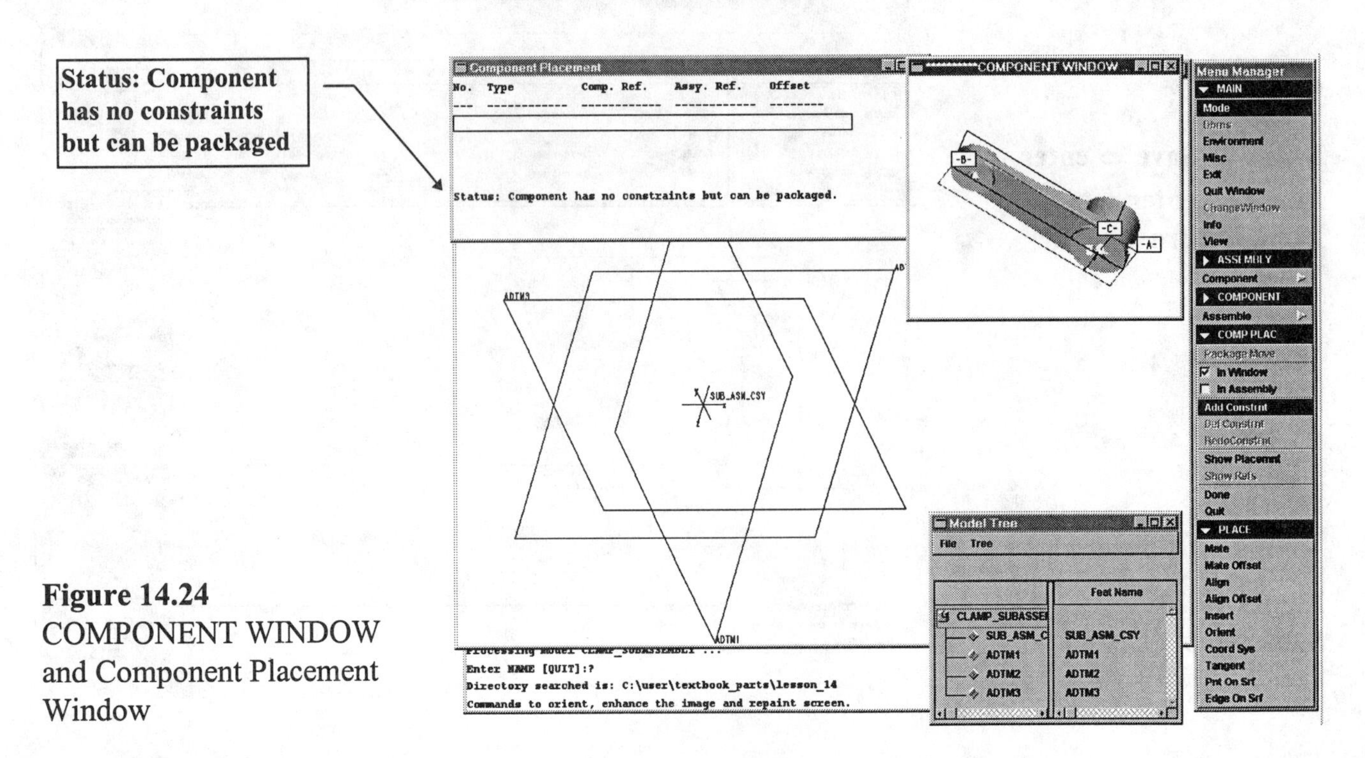

Figure 14.24
COMPONENT WINDOW and Component Placement Window

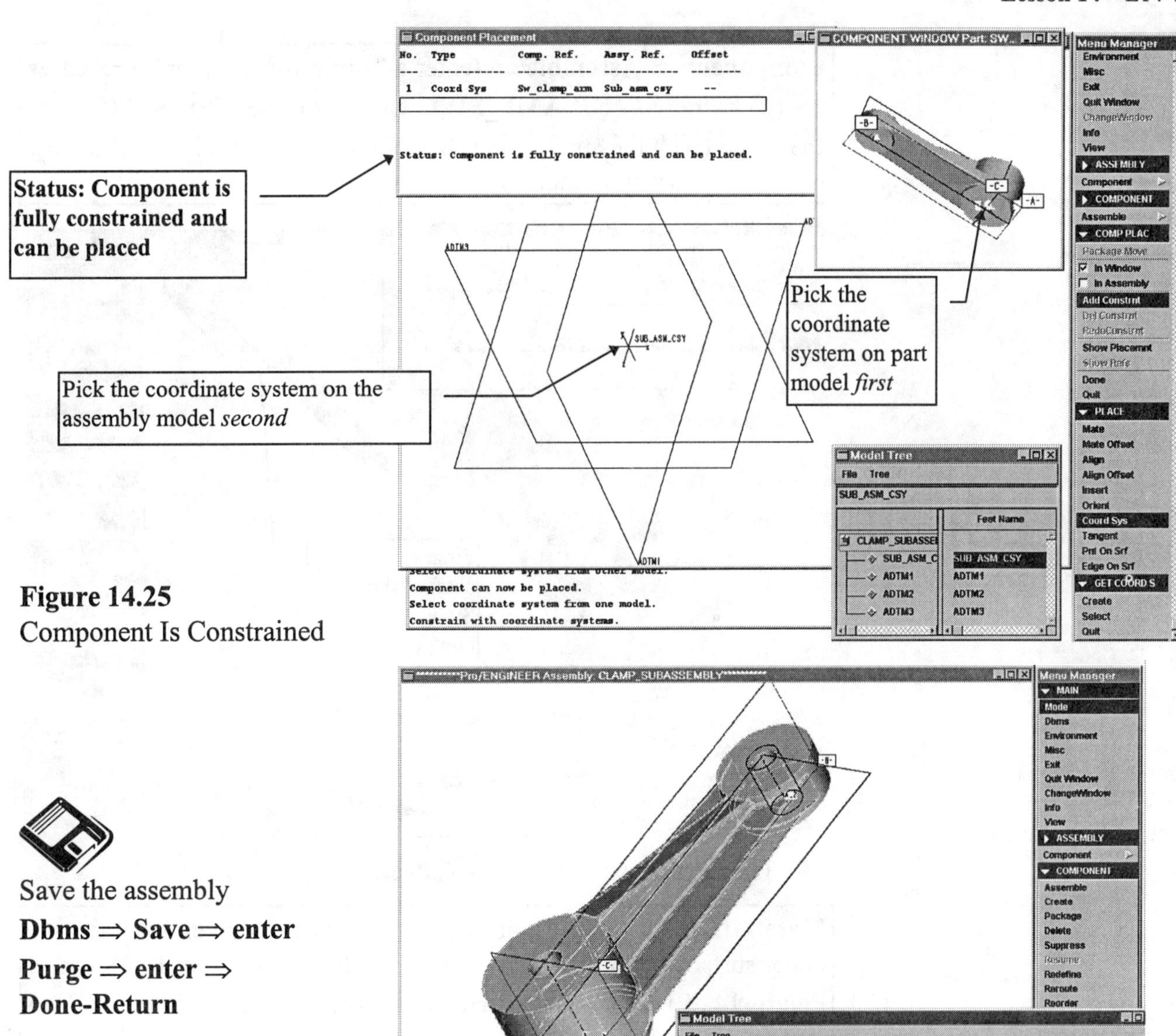

Figure 14.25
Component Is Constrained

Save the assembly
Dbms ⇒ Save ⇒ enter
Purge ⇒ enter ⇒ Done-Return

Figure 14.26
Arm

The next component to be assembled is the **SW_CLAMP_SWIVEL**. Two constraints will be used on this model: **Insert** and **Mate Offset**. *Placement constraints* are used to specify the relative position of a *pair of surfaces/references* between two components. The **Mate**, **Align**, **Insert**, and **Orient** commands are placement constraints. The two surfaces/references must be of the same type.

When using a datum plane as a placement constraint, specify which side to use, yellow or red. When using **Mate Offset** or **Align Offset**, enter the offset distance. The *offset direction* is displayed with a large arrow. If you need an offset in the opposite direction, use a negative value. Give the following commands to assemble the next part to the assembly model as shown in Figure 14.27:

Component ⇒ **Assemble** ⇒ (enter a **?** from the keyboard) ⇒ **enter** ⇒ (pick the **SW_CLAMP_SWIVEL** from the **Current Dir**) ⇒ **Insert** ⇒ (pick the shaft of the Swivel on the part model and then the hole on the assembly model)

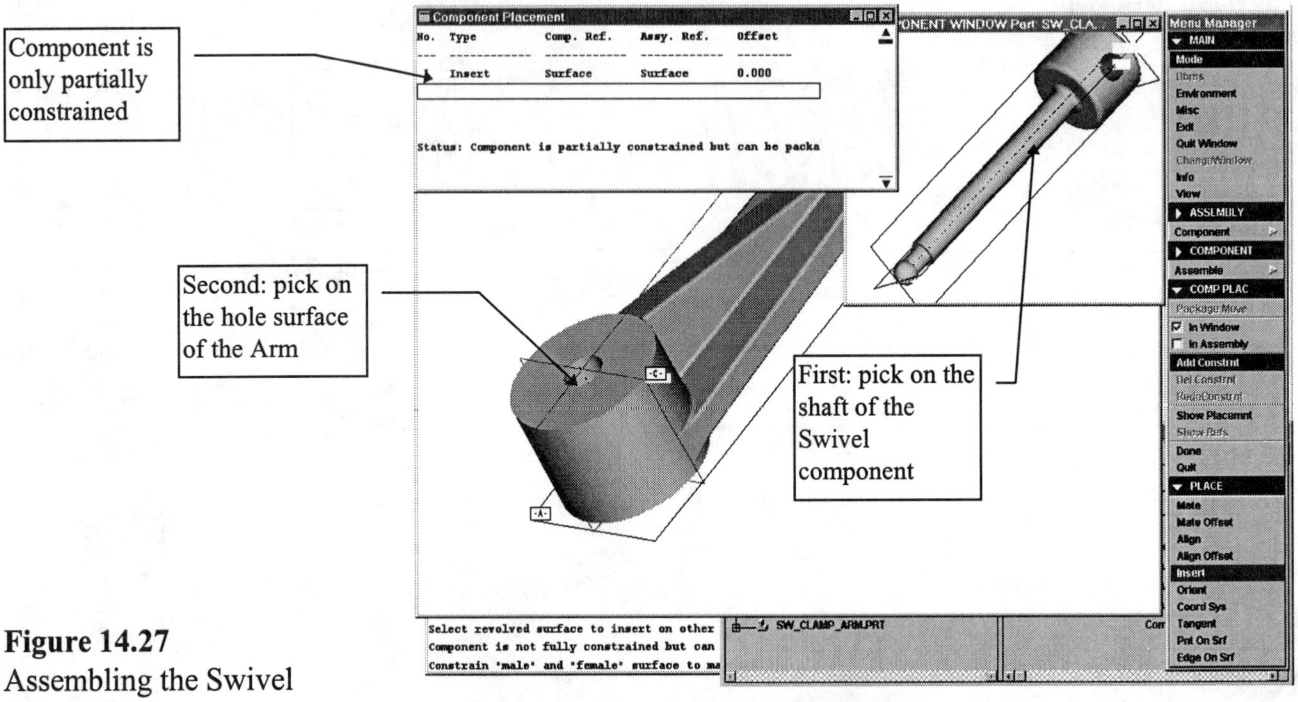

Component is only partially constrained

Second: pick on the hole surface of the Arm

First: pick on the shaft of the Swivel component

Figure 14.27
Assembling the Swivel

The next constraint is **Mate Offset**:

Mate Offset ⇒ (pick the lower surface of the Swivel and then the upper surface on the assembly model, as shown in Figure 14.28 and in Figure 14.29) ⇒ [**Offset (inches) in indicated direction**: **2.00**] ⇒ **enter** ⇒ **Done** ⇒ **Done-Return**

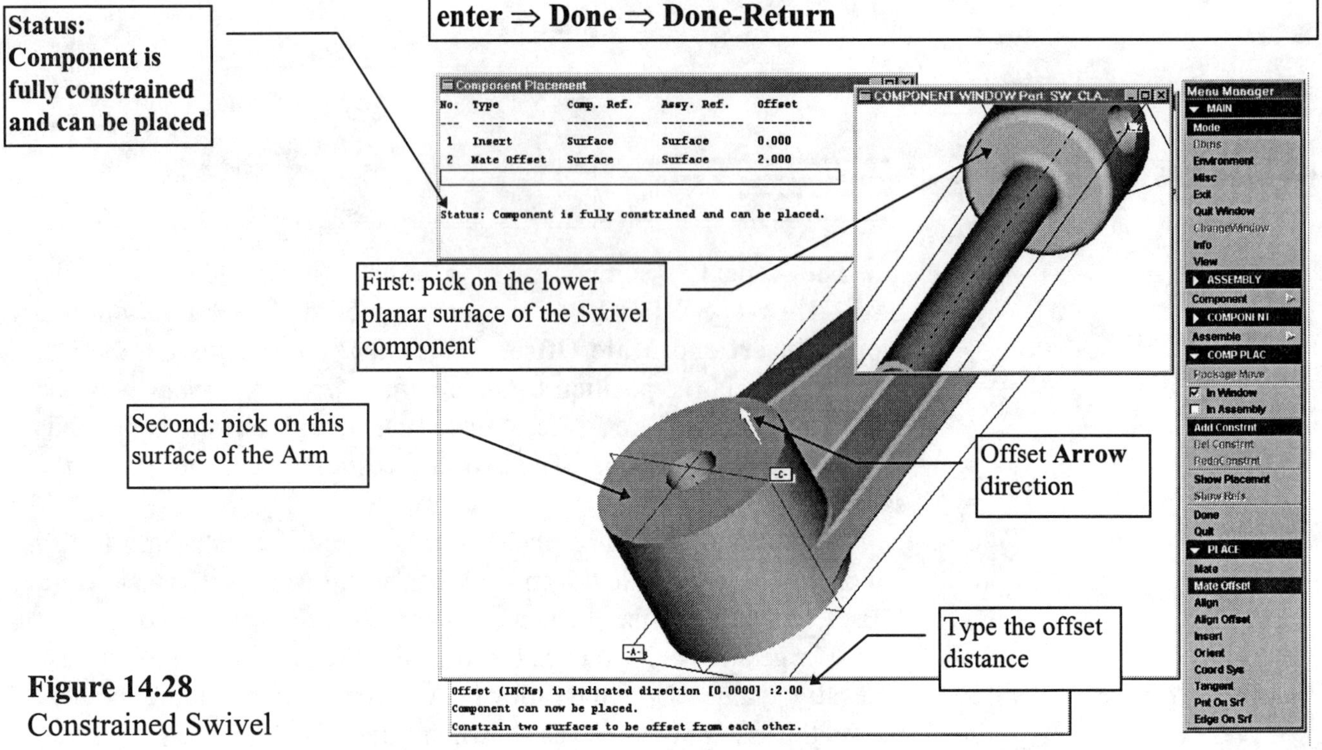

Status: Component is fully constrained and can be placed

First: pick on the lower planar surface of the Swivel component

Second: pick on this surface of the Arm

Offset **Arrow** direction

Type the offset distance

Figure 14.28
Constrained Swivel

Dbms ⇒ Save ⇒ enter
Purge ⇒ enter ⇒
Done-Return

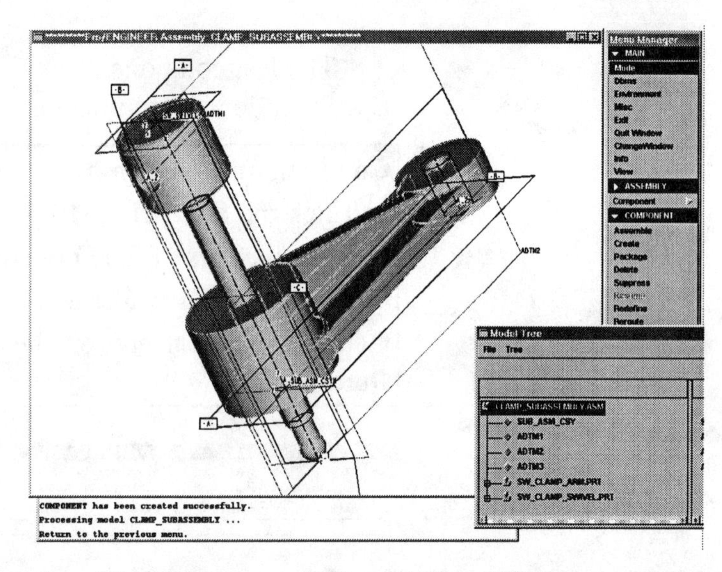

Figure 14.29
Subassembly with Arm and Swivel

Change the offset distance to **1.50** using the following commands:

Modify ⇒ **Mod Assem** ⇒ **Modify Dim** ⇒ **Value** ⇒ (pick the Swivel) ⇒ (pick the **2.00** dimension from the model as in Figure 14.30) ⇒ (type **1.50** at the prompt) ⇒ **enter** ⇒ **Done Sel** ⇒ **Done** ⇒ **Regenerate** ⇒ **Automatic** ⇒ **Done** ⇒ **Done/Return** (Fig. 14.31)

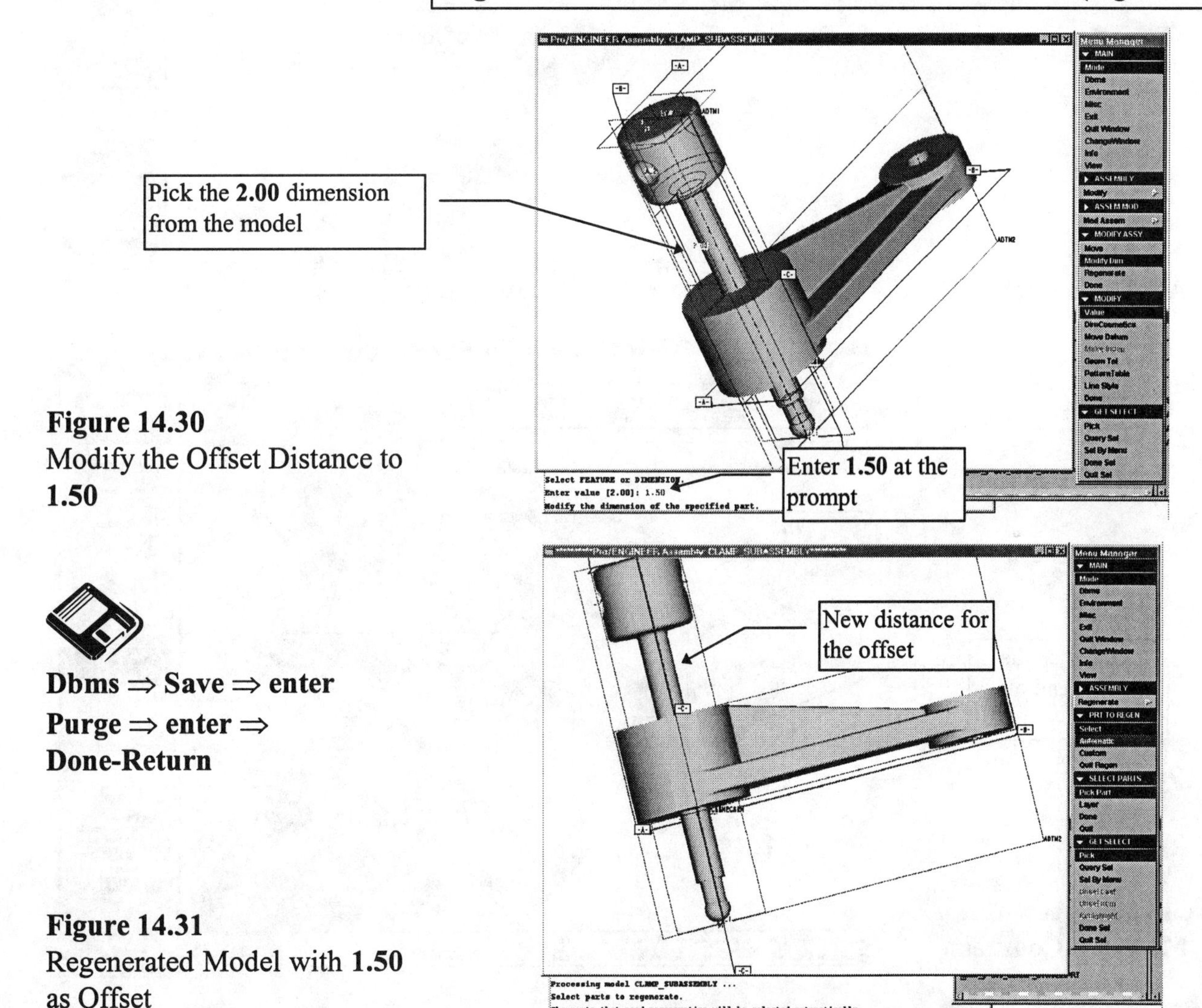

Figure 14.30
Modify the Offset Distance to **1.50**

Dbms ⇒ Save ⇒ enter
Purge ⇒ enter ⇒
Done-Return

Figure 14.31
Regenerated Model with **1.50** as Offset

The last component for the subassembly is the Foot component. Use the following commands:

Component ⇒ **Assemble** ⇒ (enter a **?** from the keyboard) ⇒ **enter** ⇒ (pick the **SW_CLAMP_FOOT** from the **Current Dir**) ⇒ **Align** ⇒ (pick the *axis* of the Foot and then the *axis* on the Swivel as shown in Figure 14.32) ⇒ **Mate** ⇒ (pick the spherical hole of the Foot and then the spherical end on the Swivel as shown in Figure 14.33) ⇒ **Done**

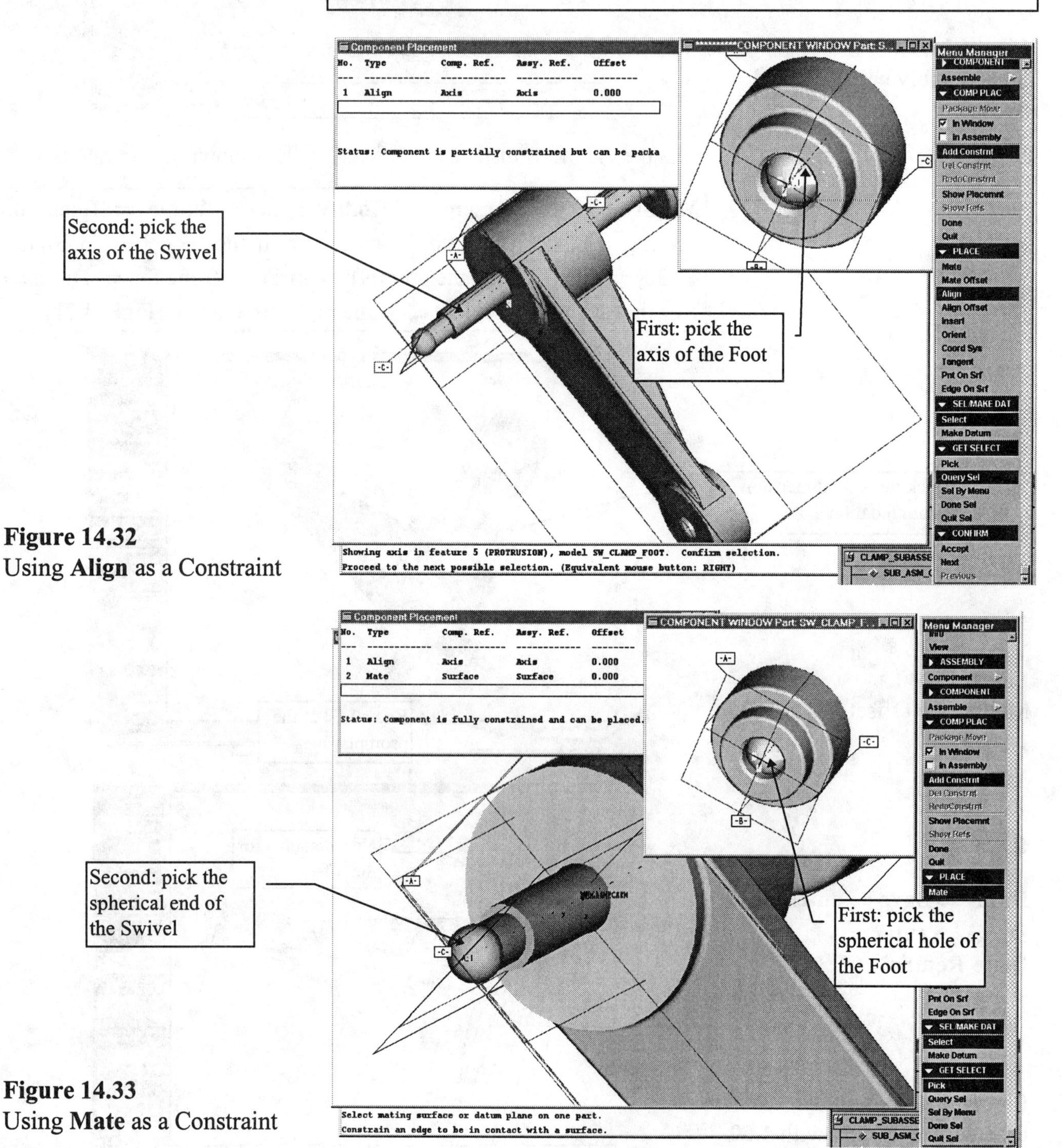

Figure 14.32
Using **Align** as a Constraint

Figure 14.33
Using **Mate** as a Constraint

Well, it doesn't look exactly right yet (Fig. 14.34). A third constraint needs to be added to orient the Foot correctly. Give the following commands (Fig. 14.35):

> **Redefine** ⇒ (pick Foot from the Model Tree) ⇒ **Add Constrnt** ⇒ **Orient** ⇒ (pick lower surface of the Arm as in Figure 14.36) ⇒ **Query Sel** ⇒ (pick top surface of the Foot) ⇒ **Accept** ⇒ **Done**

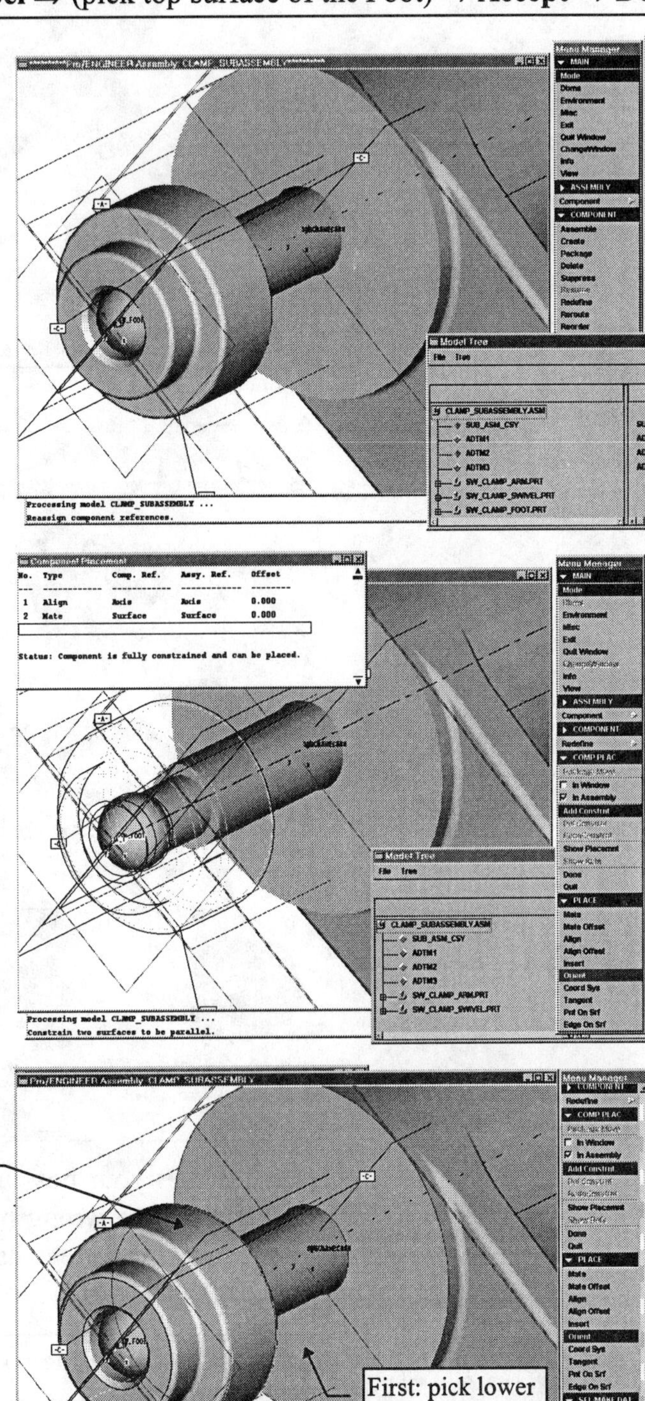

Figure 14.34
Foot Facing the Wrong Direction

Figure 14.35
Using **Orient** as a Constraint

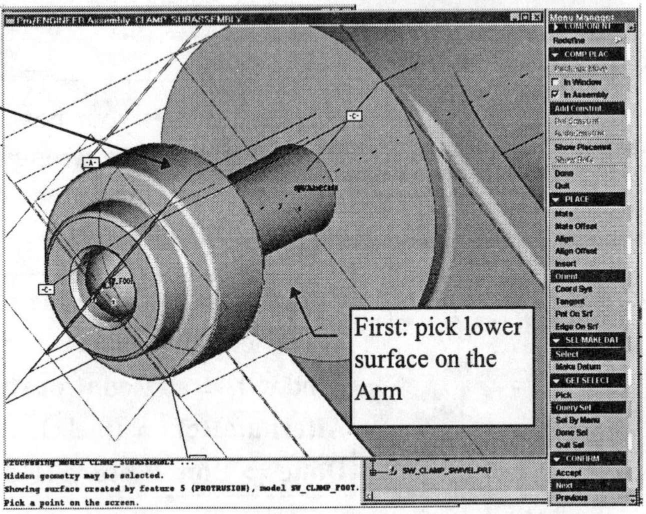

Query Sel to get to the top surface of the Foot

Figure 14.36
Picking Surfaces for **Orient**

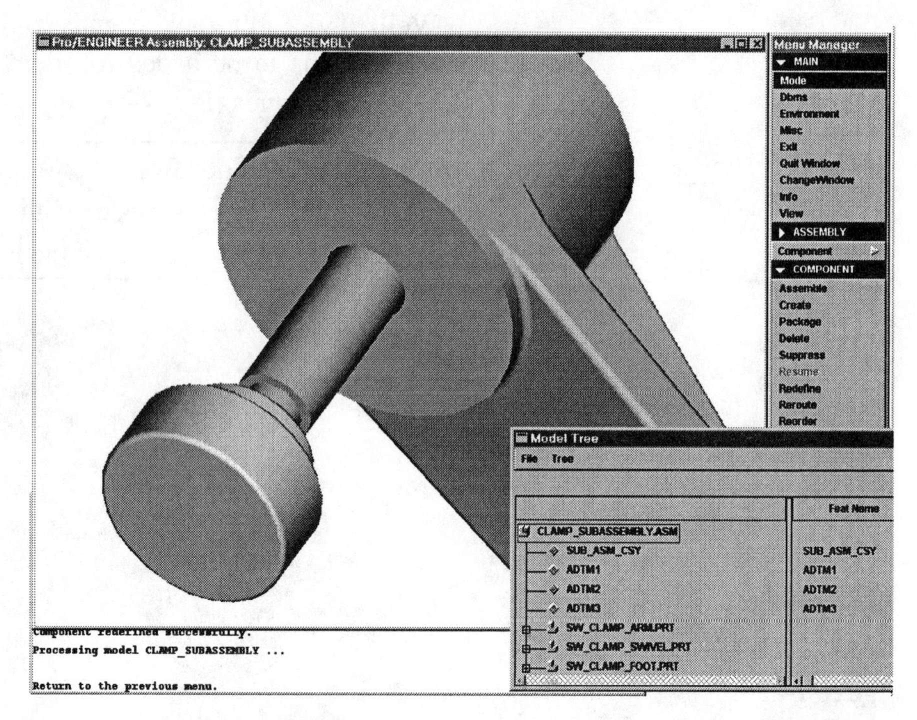

Figure 14.37
Redefined Foot

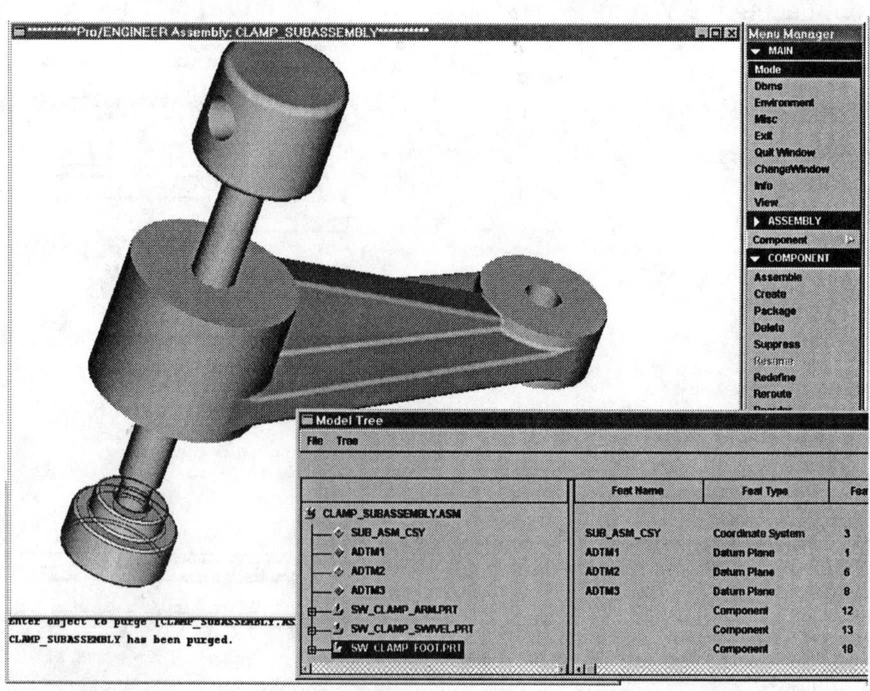

Figure 14.38
Picking Surfaces

Here comes your boss again. Unfortunately *you* put the Swivel and the Foot in the wrong hole! Before the mistake gets any undue attention, let's redefine its placement by giving the following commands (Fig. 14.39):

Component ⇒ **Redefine** ⇒ (select the Swivel from the Model Tree) ⇒ (select the **Insert** constraint from the Component Placement window) ⇒ **RedoConstrnt** ⇒ ✓**AssemblyRef** ⇒ **Done** ⇒ **AlternateRef** ⇒ (pick the other hole as shown in Figure 14.39) ⇒ **Done** ⇒ **Done/Return** (Fig. 14.40)

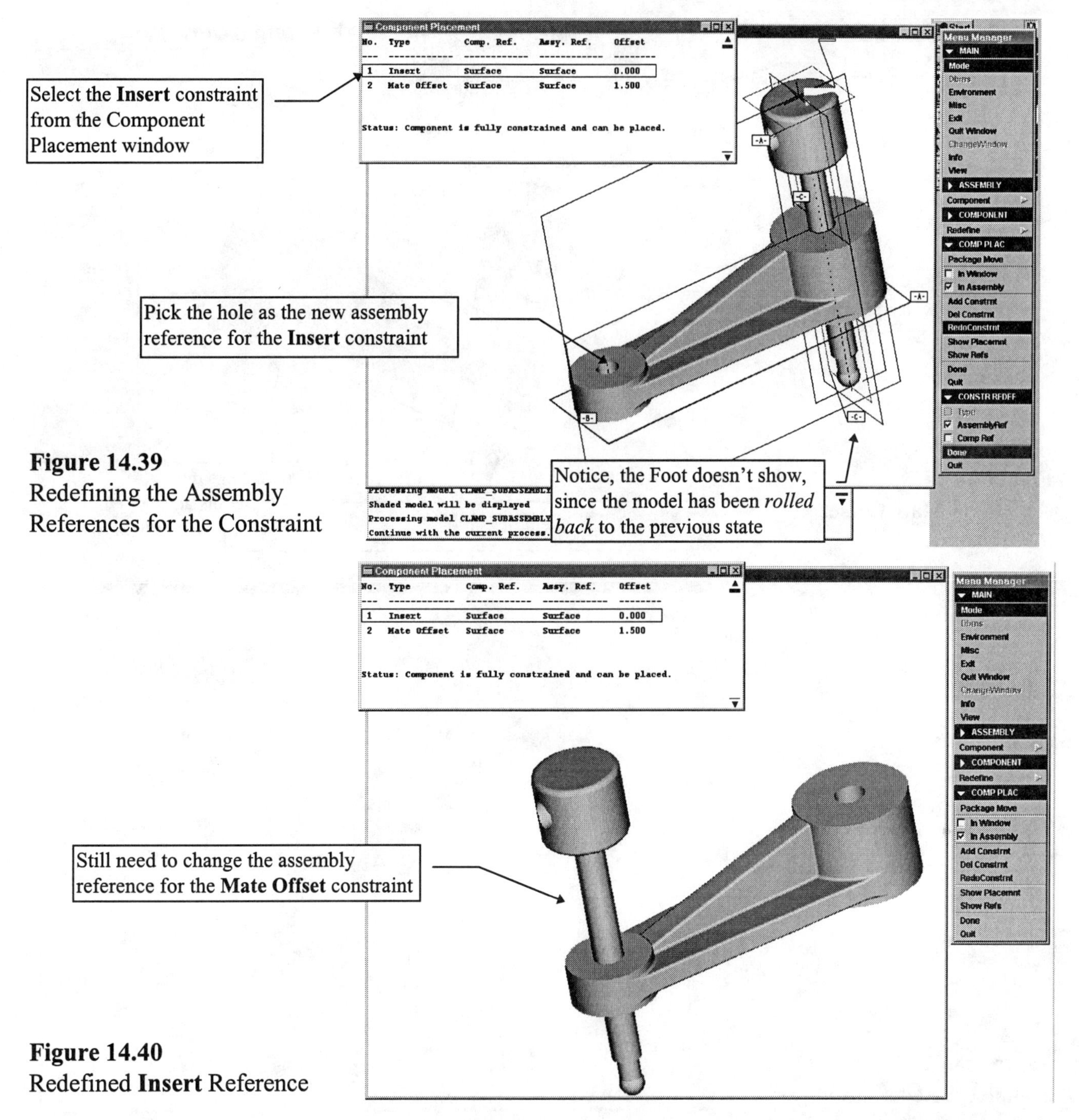

Figure 14.39
Redefining the Assembly References for the Constraint

Figure 14.40
Redefined **Insert** Reference

The Swivel is now in the correct hole, but the **Mate Offset** assembly reference is incorrect, since the bottom of the Swivel head is offset from the top surface of the large-diameter boss on the other side of the part. Complete the redefinition:

Component ⇒ **Redefine** ⇒ (pick the Swivel from the Model Tree) ⇒ (select the **Mate Offset** constraint from the Component Placement window) ⇒ **RedoConstrnt** ⇒ **✓AssemblyRef** ⇒ **Done** ⇒ **AlternateRef** ⇒ (Select offset mating surface as shown in Figure 14.41) ⇒ (Offset [inches] in indicated direction) ⇒ **(**type **2.25** at the prompt) ⇒ **enter** ⇒ **Done** ⇒ **Done/Return** (Fig. 14.42)

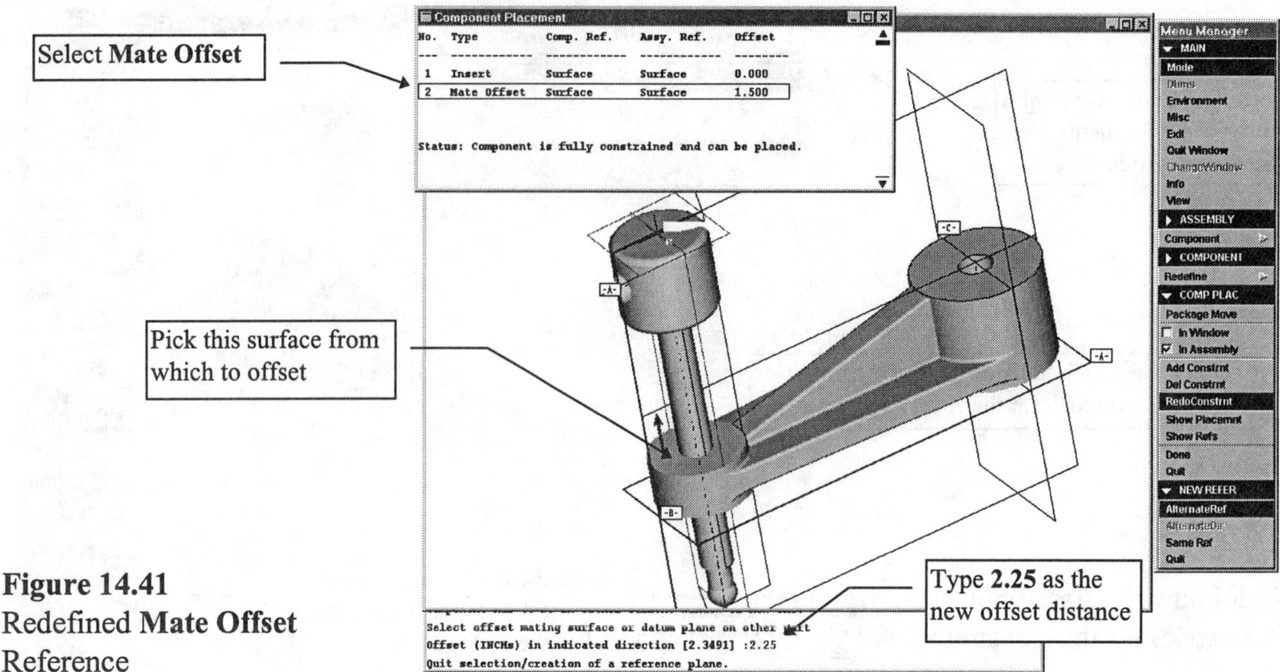

Figure 14.41
Redefined **Mate Offset** Reference

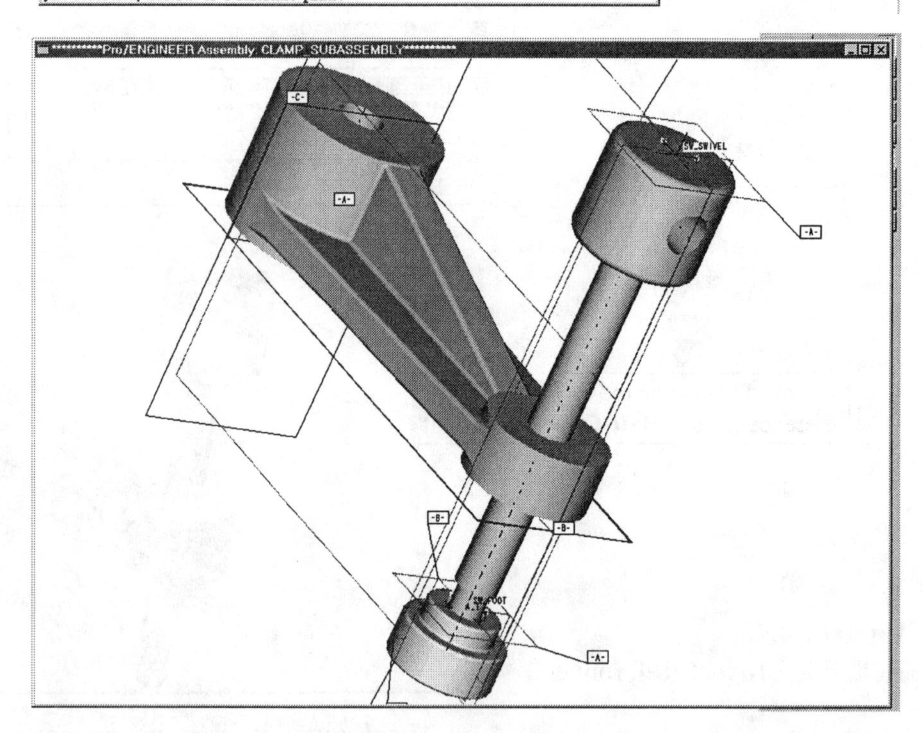

Figure 14.42
Completed Redefinition

As you can see, it is extremely easy to alter your model using **Modify** and **Redefine**. The next step in the assembly of the model is to assemble the **5.00** inch double-ended stud:

HINT

The datum planes were set as geometric tolerance features and renamed when the part was modeled.

Component ⇒ **Assemble** ⇒ (enter a **?** from the keyboard) ⇒ **enter** ⇒ (pick **STUD5** from the **Current Dir**) ⇒ **Insert** ⇒ (select the *revolved surface* of the Stud and then select the *revolved surface- hole* on the Swivel as shown in Figure 14.43) ⇒ **Align Offset** ⇒ (pick the end of the Stud and then pick the datum **C** of the Swivel from the Model Tree) ⇒ **Yellow** (Fig. 14.44) ⇒ (type the offset of **5.00/2** at the prompt) ⇒ **enter** ⇒ **Done** ⇒ **Done/Return** (Fig. 14.45)

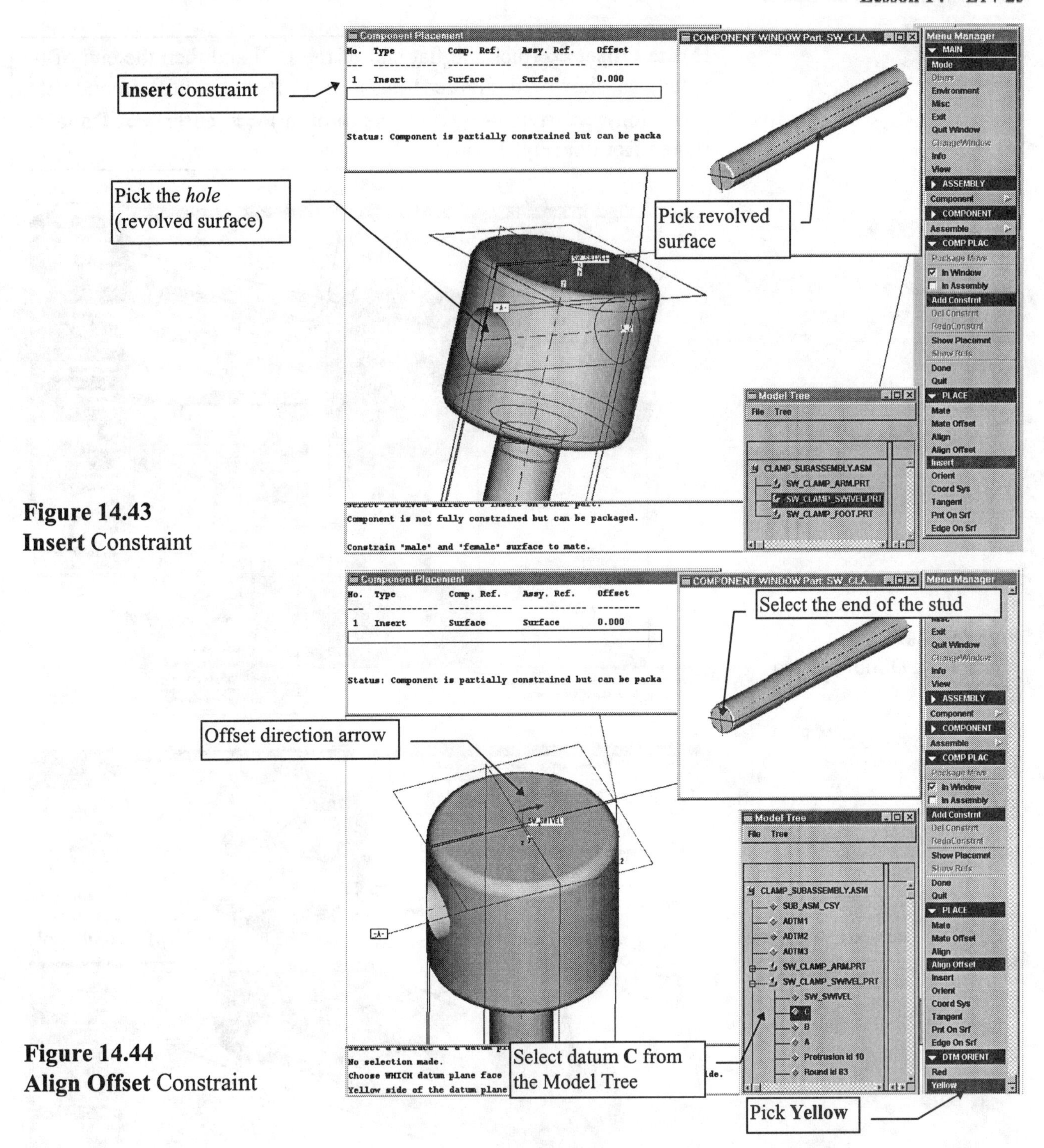

Figure 14.43
Insert Constraint

Figure 14.44
Align Offset Constraint

The two Ball handles are assembled next. Instructions are provided for assembling one Ball, you must assemble the other on your own. Use the following commands:

Component ⇒ **Assemble** ⇒ (enter a **?** from the keyboard) ⇒ **enter** ⇒ (pick the **SW_CLAMP_BALL** from the **Current Dir**) ⇒ **Insert** ⇒ (pick the *revolved hole* of the Ball and then the *revolved surface* on the Stud, as shown in Figure 14.46) ⇒

Mate Offset ⇒ (pick the flat end of the Ball and then the end of the Stud, as shown in Figure 14.47) ⇒ (Offset [inches] in indicated direction) ⇒ (type **-.500** at the prompt) ⇒ **enter** ⇒ **Done** ⇒ **Done/Return** (Fig. 14.48)

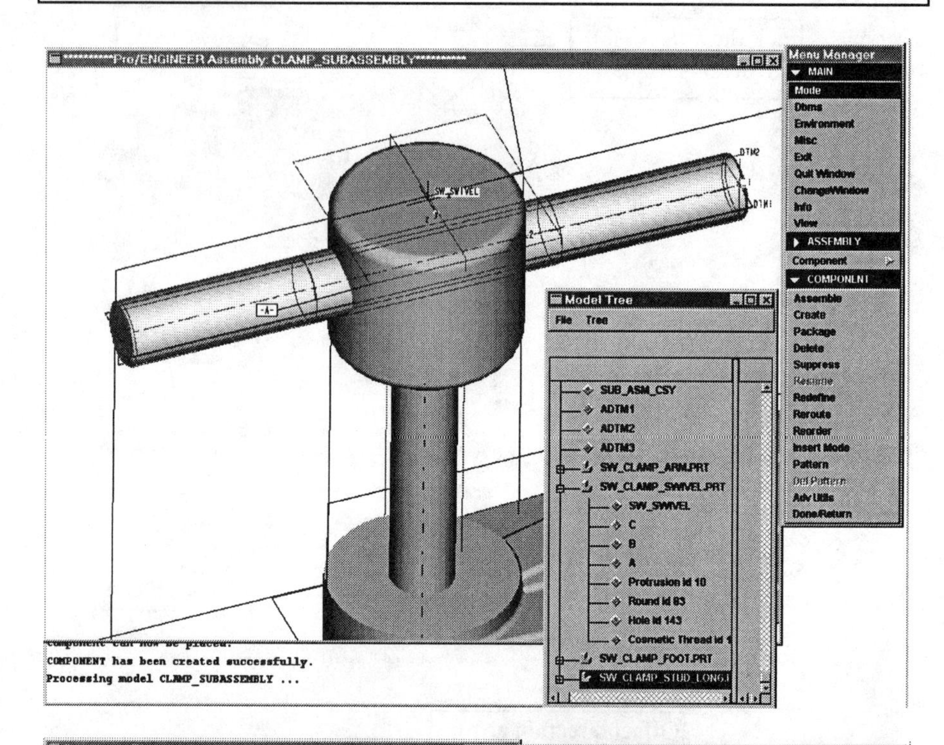

Figure 14.45
Assembled Double-ended Stud

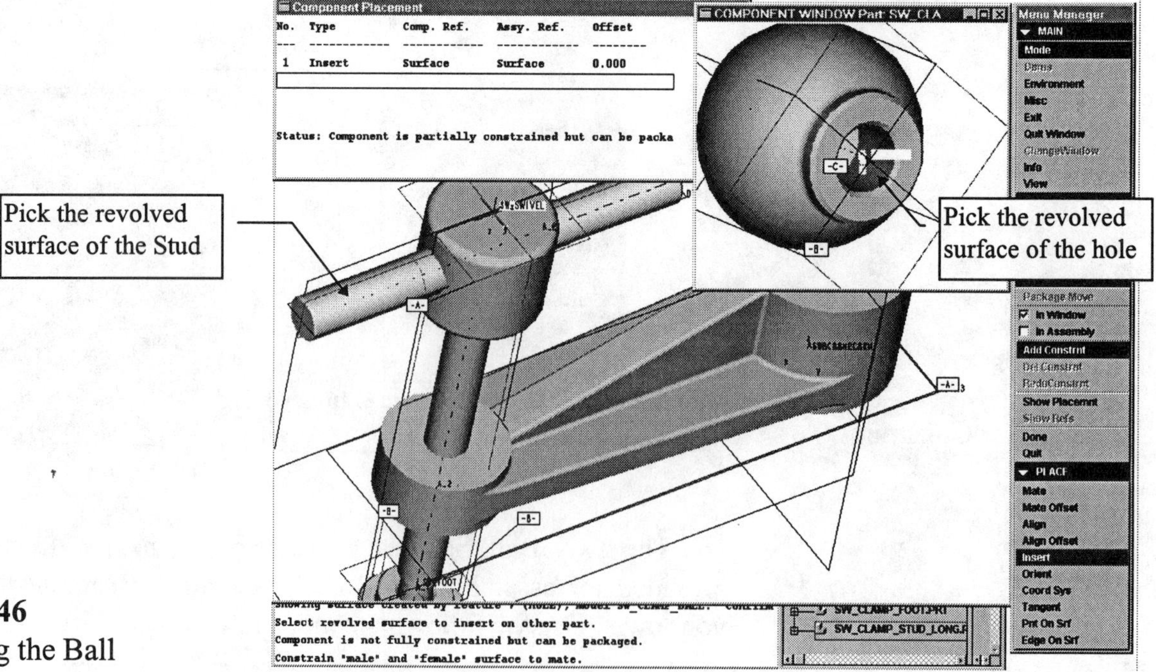

Figure 14.46
Assembling the Ball

Assemble the second Ball as shown in Figure 14.49. The subassembly is now complete. **Save** and **Purge** the model. The subassembly will be added to the assembly in the next set of steps.

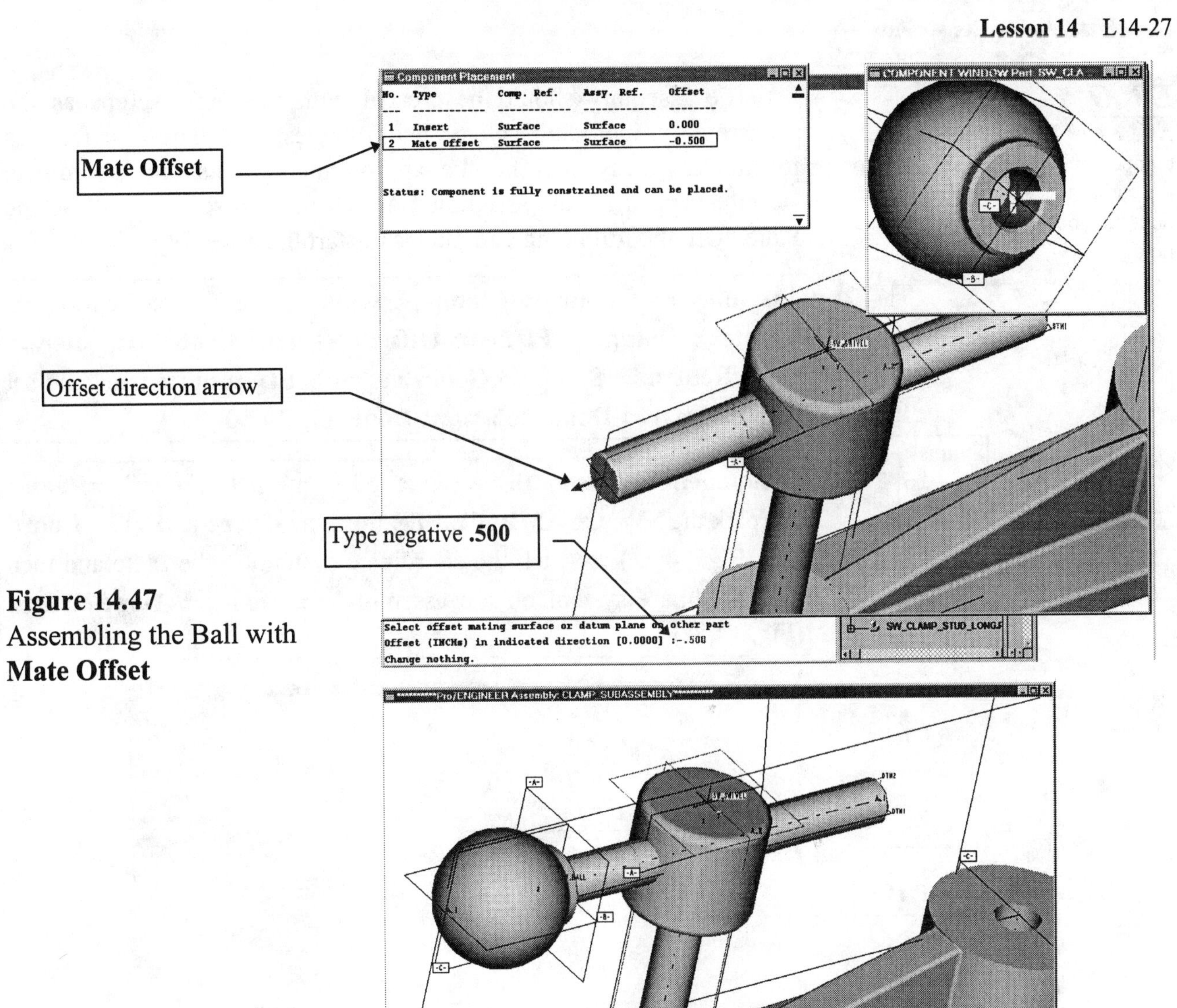

Figure 14.47
Assembling the Ball with **Mate Offset**

Figure 14.48
Ball Assembled on Stud

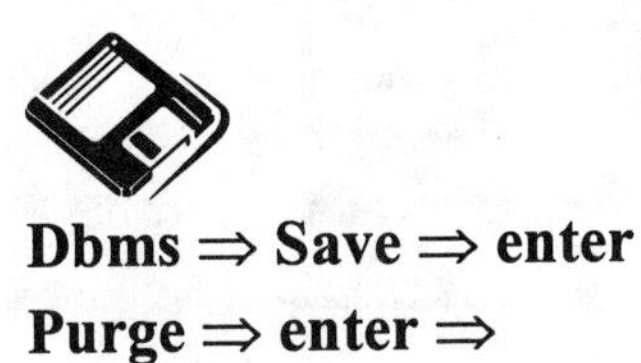

Dbms ⇒ Save ⇒ enter Purge ⇒ enter ⇒ Done-Return

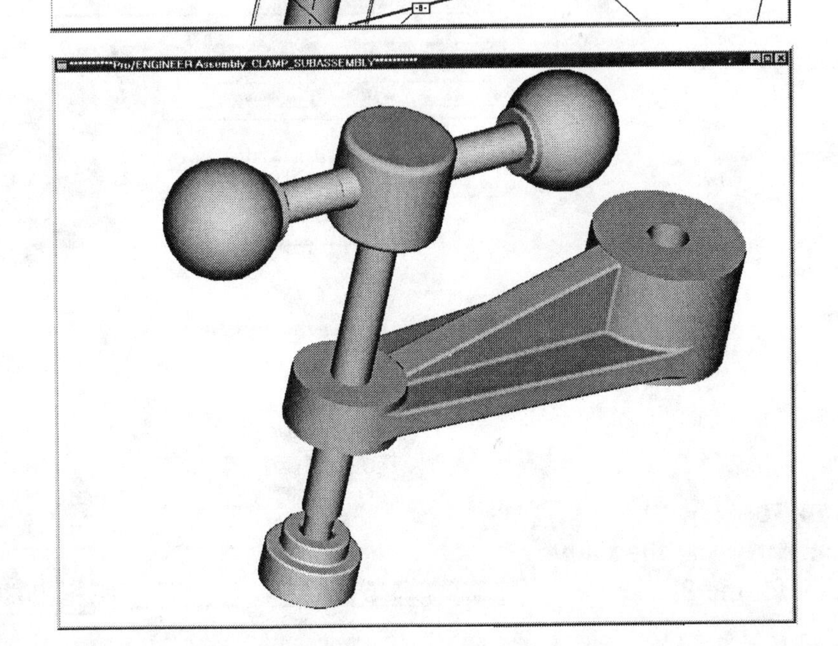

Figure 14.49
Completed Subassembly

ECO

As a minor **ECO**, redefine the Ball component to be offset **.44** from the end of the Shaft so that the Stud does not bottom out in the Ball's hole.

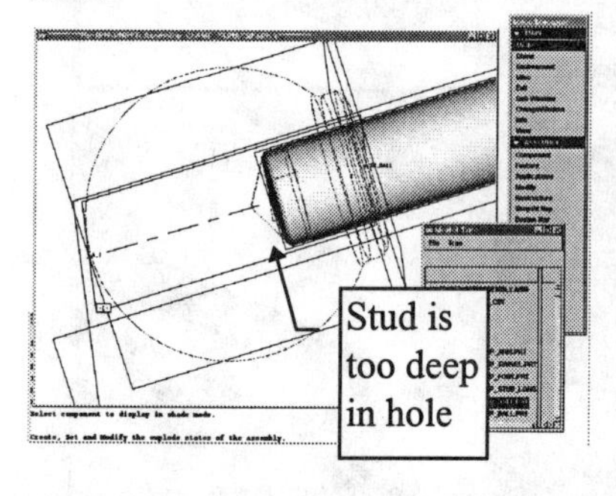

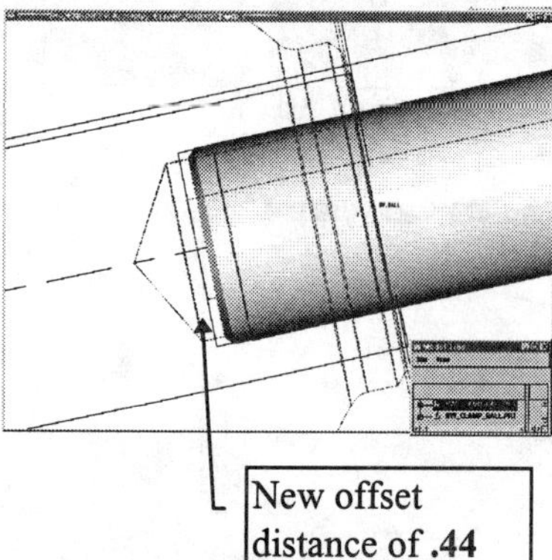

The assembly should be created with the same setups as the subassembly, including layers, datum names, etc. The first features for the assembly will be the appropriate datums and coordinate system. The first part assembled on the main assembly will be the Plate. Use the following commands to start the assembly:

Assembly ⇒ Create ⇒ Clamp_Assembly ⇒ enter ⇒ Feature ⇒ Create ⇒ Datum ⇒ Plane ⇒ Offset ⇒ enter ⇒ enter ⇒ enter ⇒ Done/Return ⇒ Setup ⇒ Geom Tol ⇒ Set Datum (rename and set your datums) **⇒ Done/Return ⇒ Done** (Fig.14.50)

Component ⇒ Assemble ⇒ (enter a **?** from the keyboard) **⇒ enter ⇒** (pick the **SW_CLAMP_PLATE** from the **Current Dir**) **⇒ Coord Sys** (Fig. 14.51) **⇒** (pick the coordinate system on the Plate and then the coordinate system on the assembly) **⇒ Done ⇒ Done/Return** (Fig. 14.52)

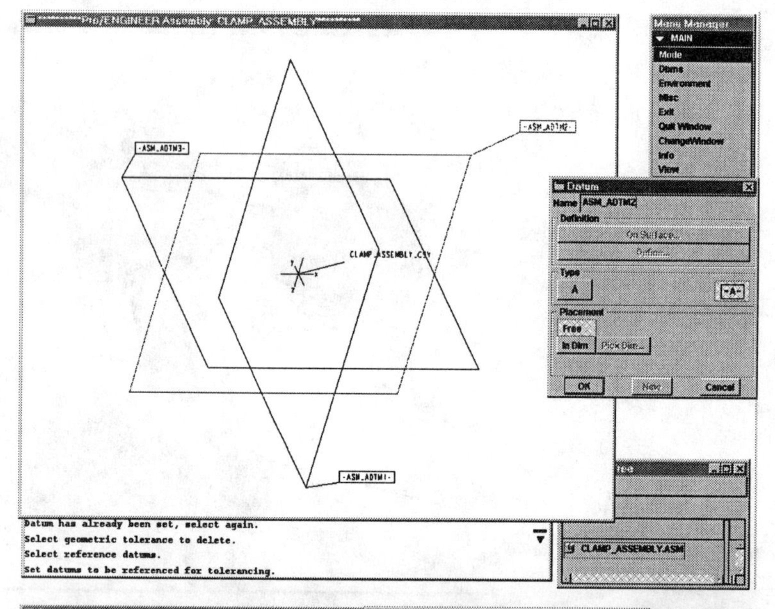

Figure 14.50
Clamp_Assembly Datums and Coordinate System

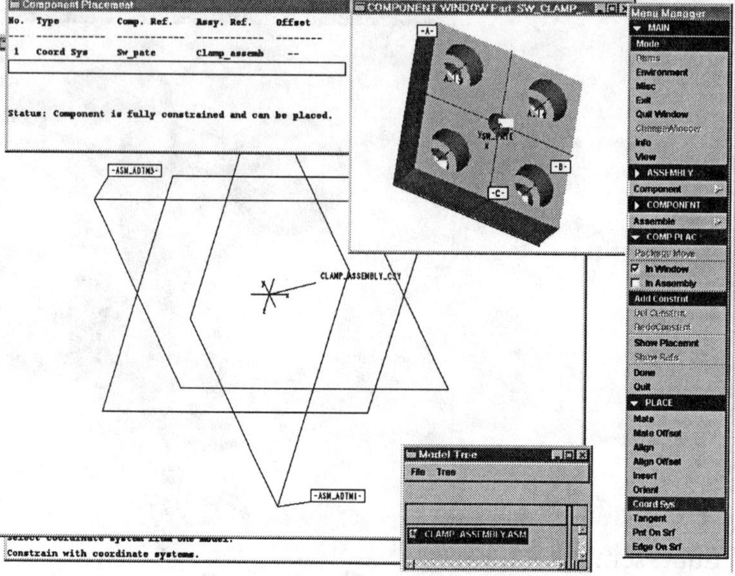

Figure 14.51
Use **Coord Sys** as the Only Constraint for the Plate

HINT

Your datum planes may be called different names. Pick the *yellow side* of both datum planes. To **Orient**, pick a vertical datum on the assembly and subassembly.

Component ⇒ **Assemble** ⇒ (enter a ? from the keyboard) ⇒ **enter** ⇒ (pick the **CLAMP_SUBASSEMBLY** from the **Current Dir)** ⇒ **Mate** (Fig. 14.52) ⇒ (pick the top surface of the Plate and then the bottom circular planar surface on the Arm of the subassembly) ⇒ **Insert** ⇒ (pick the hole in the Plate and then the hole in the Arm) ⇒ **Orient** ⇒ (pick the **ASM_ADTM2** datum plane on the assembly) ⇒ **Yellow** ⇒ (pick then the datum plane **A** of the subassembly) ⇒ **Yellow** ⇒ **Done** ⇒ **Done/Return** (Fig. 14.54)

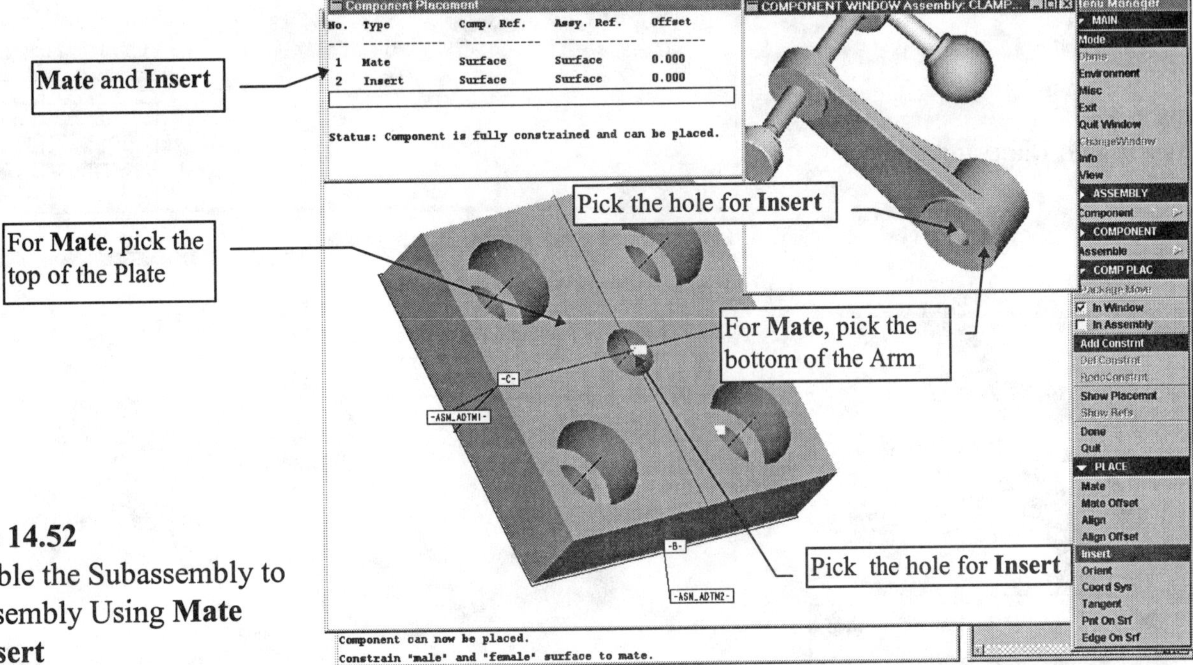

Figure 14.52
Assemble the Subassembly to the Assembly Using **Mate** and **Insert**

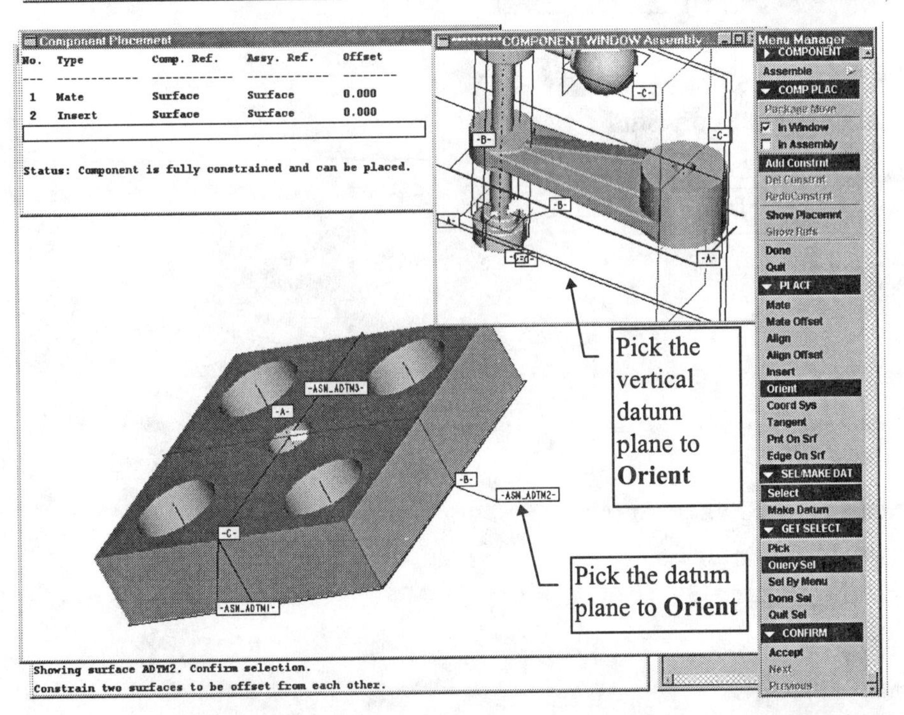

Figure 14.53
Using **Orient**

Assemble the STUD35 and the Flange Nut (FLNGNUT) using constraints learned in this lesson (Fig. 14.55 and Fig. 14.56).

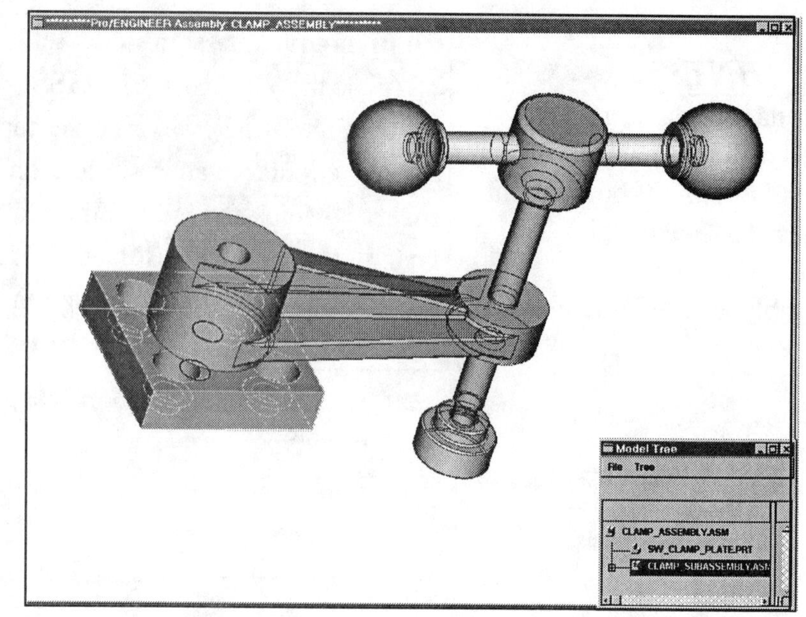

Dbms ⇒ Save ⇒ enter
Purge ⇒ enter ⇒
Done-Return

Figure 14.54
Assembled Plate and Subassembly

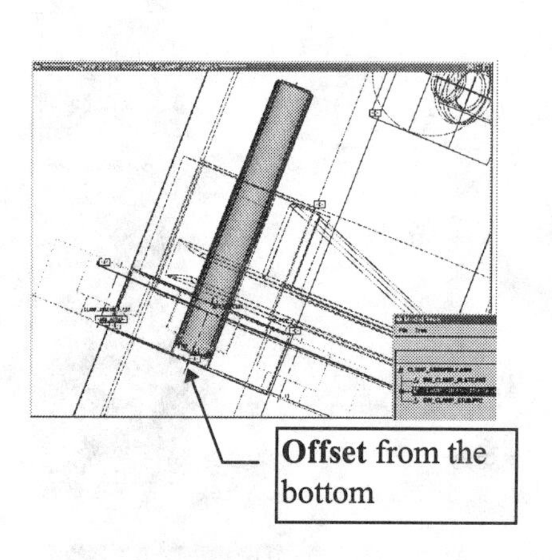

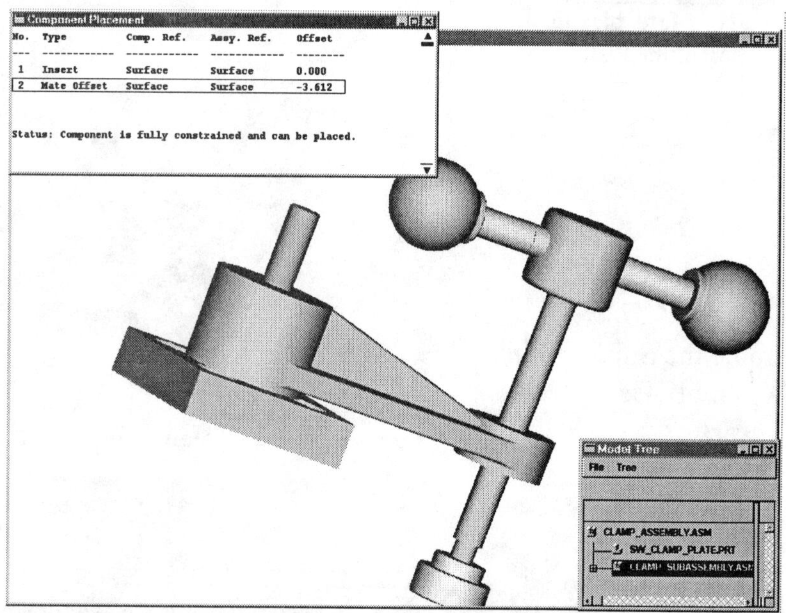

Figure 14.55
Assembled Short Stud

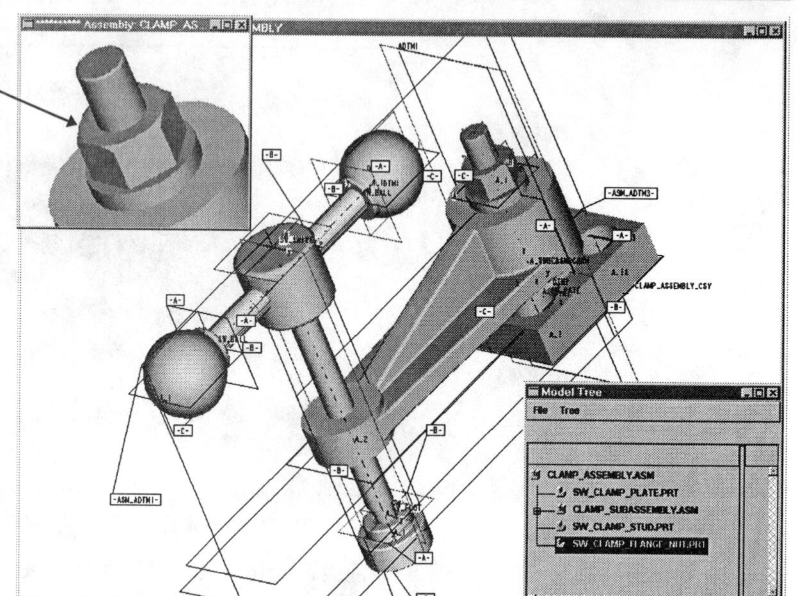

Dbms ⇒ Save ⇒ enter
Purge ⇒ enter ⇒
Done-Return

Figure 14.56
Assembled Swing Clamp

The last thing you want to do before going on to the next lesson is check the assembly using the **Info** command (Fig. 14.57). If you look at the long **5.00** inch stud's detail drawing (see Fig.14.18) you will see that the shaft diameter is greater than **.500** at the center of the stud and each end has a **.500-13 UNC** thread. The hole in the Swivel is **.500**. This means that there should be a slight interference between the two components. Check the clearance and interference between the two component using the following commands (Fig. 14.57):

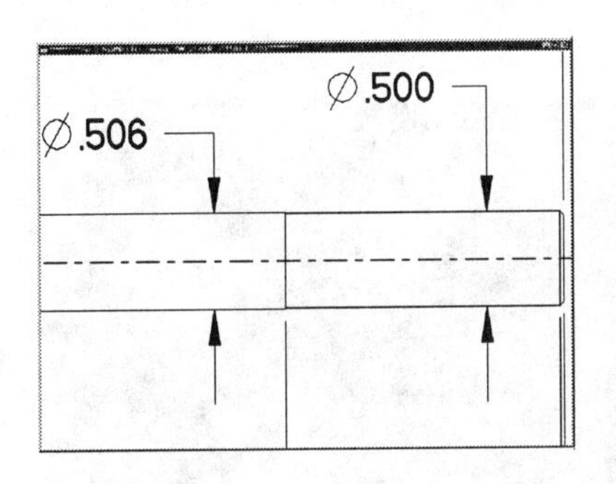

Info ⇒ **Measure** ⇒ **Clear/Intf** ⇒ **Pairs** ⇒ **First** ⇒ **Whole Part** ⇒ **ExcludQuilts** ⇒ (pick the Swivel) ⇒ (pick the **5.00** Stud)

The **Info** Command provides you with a statement as to the status of the two parts:

Interference detected. Volume of Interference is 0.0069

Figure 14.57
Info Command Used to Establish Clearance and Interference

The assembly is now complete. In the next lesson you will learn how to move and rotate components in the assembly, establish views for use in the Drawing mode, create exploded views of the assembly, generate a bill of materials, and change the component visibility.

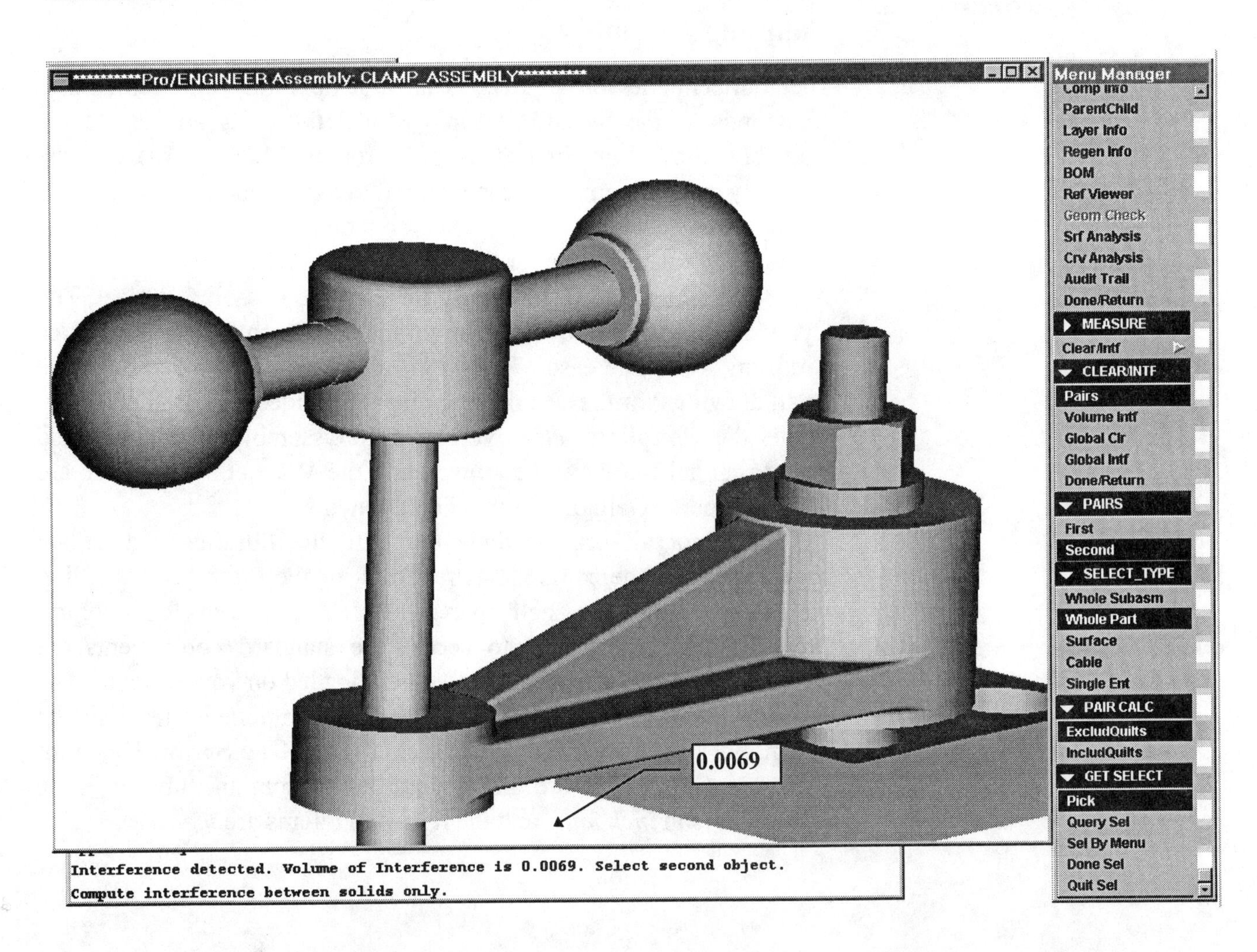

Lesson 14 Project

Coupling Assembly

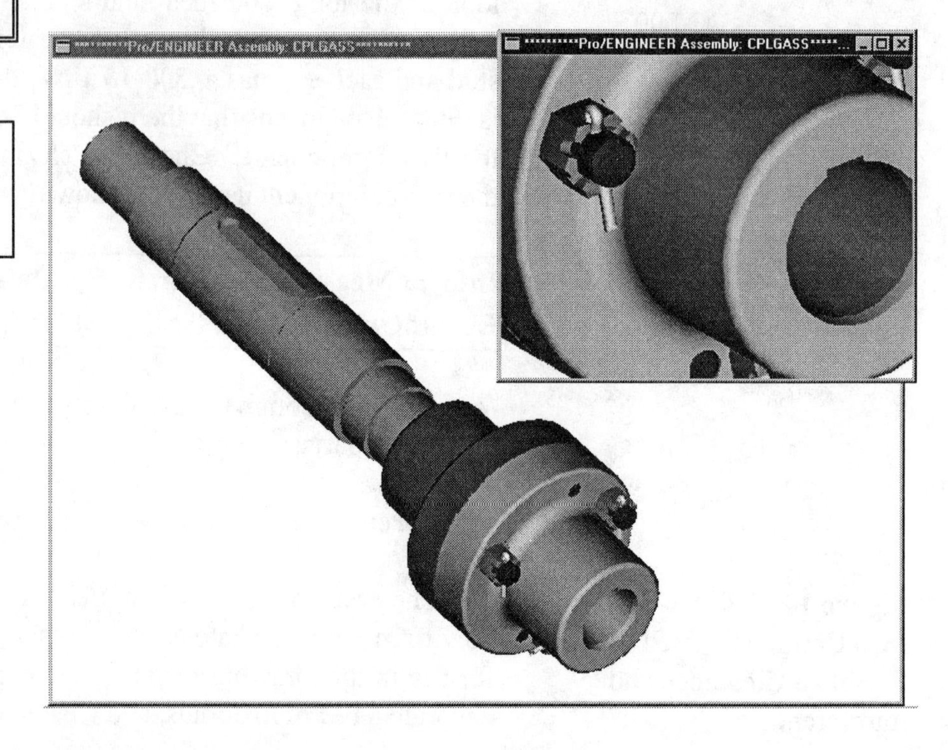

Figure 14.58
Coupling Assembly

Coupling Assembly

The fourteenth **lesson project** is an assembly that requires similar commands as the **Swing Clamp**. Model the parts and create the assembly shown in Figures 14.58 through 14.81. Analyze the assembly and plan out the steps required to assemble it. Use the DIPS in Appendix D to plan out the assembly component sequence and the parent-child relationships for the assembly.

You will use the **Cylinder Shaft** from the Lesson 6 Project. The Cylinder Shaft should be the first component assembled. The **Taper Coupling** from the Lesson 7 Project is also used in the assembly. The detail drawings for the second coupling is provided here in this lesson. Model the **Coupling** *before* you start the assembly. You will also need to model the **Key**, the **Dowel**, and the **Washer**, since there are no library parts available for these components.

Since not all organizations purchase the libraries, details are provided for all the components required for the assembly, including the standard off-the-shelf parts available in Pro/E's library. Pro/LIBRARY commands to access the standard components are provided for those of you who have them loaded on your system. The instance number is given for all standard components used in the assembly. The **Slotted Hex Nut, Socket Head Cap Screw, Hex Jam Nut**, and **Cotter Pin** are all standard parts from the library. The Cotter Pin is in *inch units* and the remaining items are *metric.*

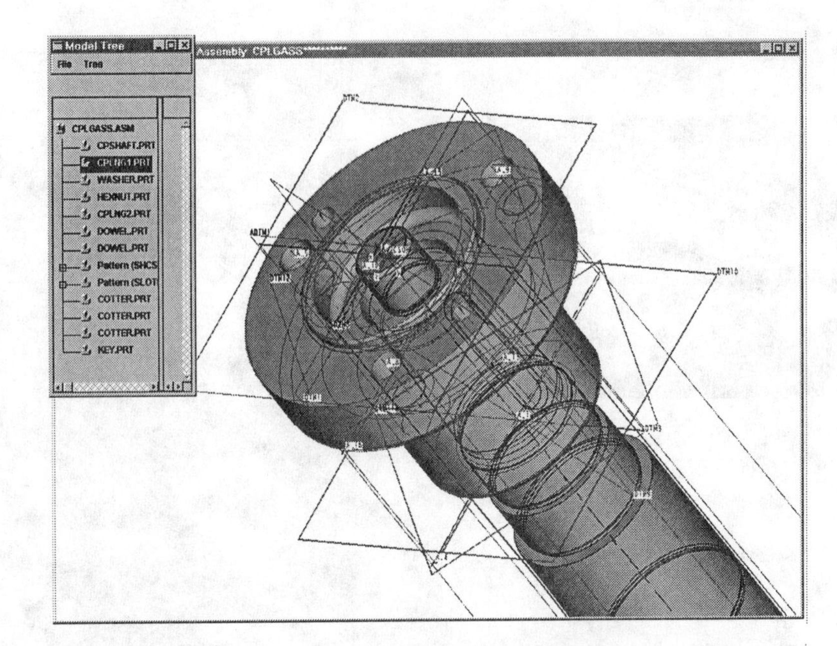

Figure 14.59
Assembling the Cylinder Shaft and Taper Coupling

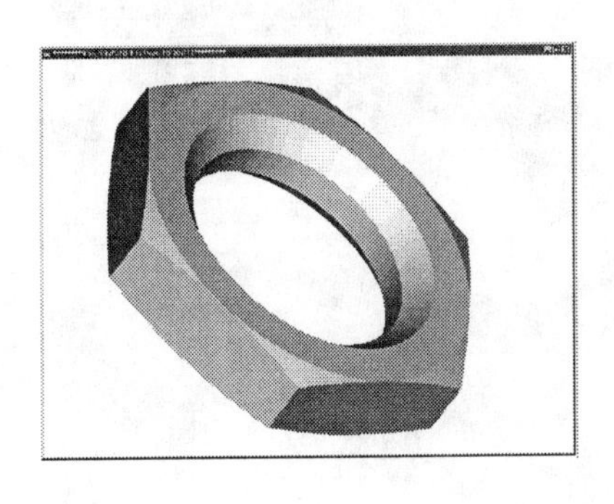

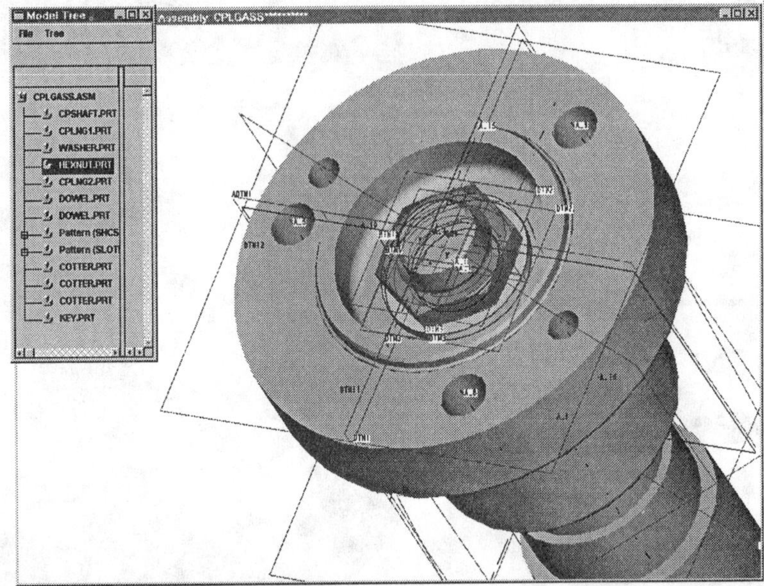

Figure 14.60
Hex Jam Nut and Washer

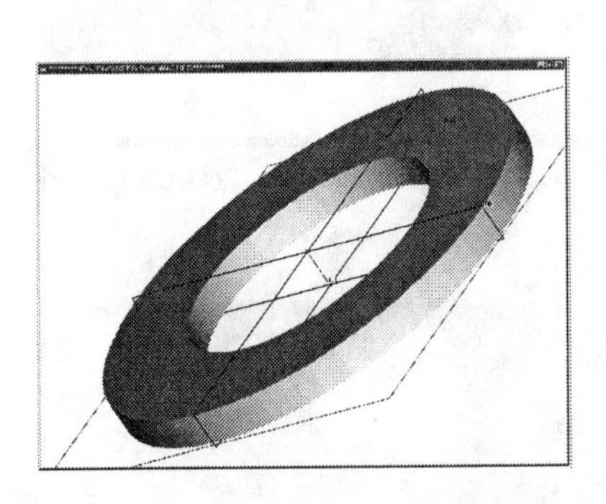

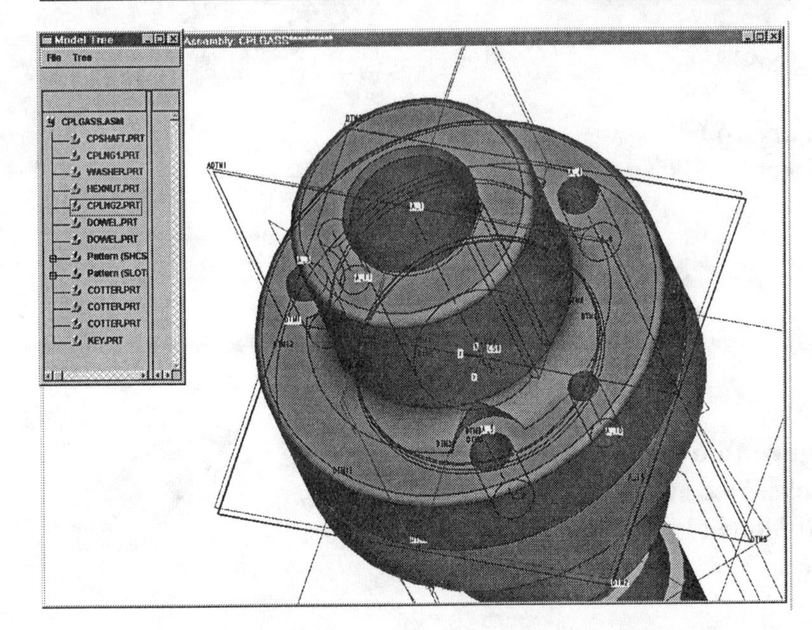

Figure 14.61
Second Coupling

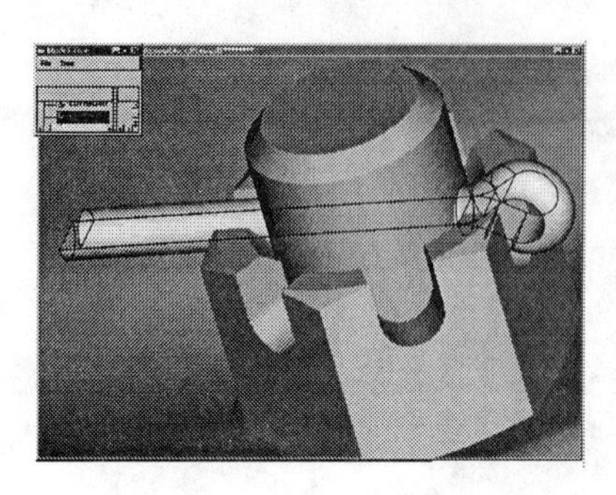

Figure 14.62
Dowel, Slotted Hex Nut, and Cotter Pin

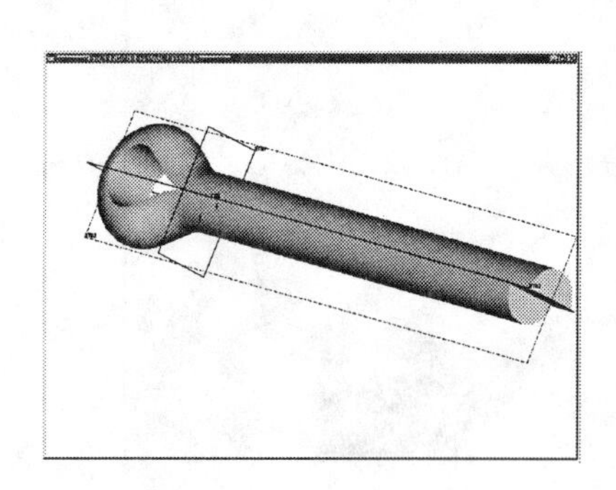

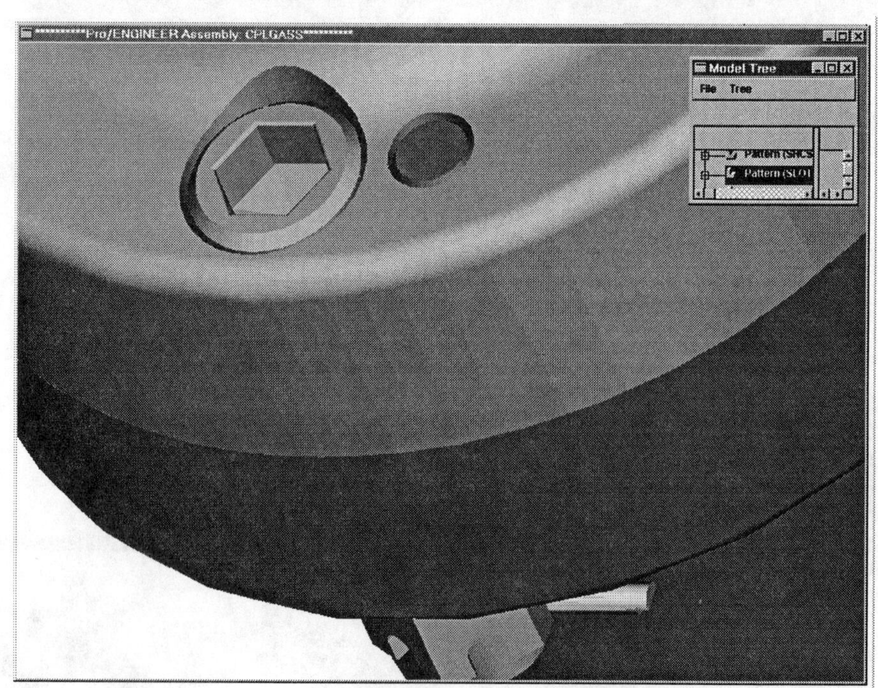

Figure 14.63
Socket Head Cap Screw

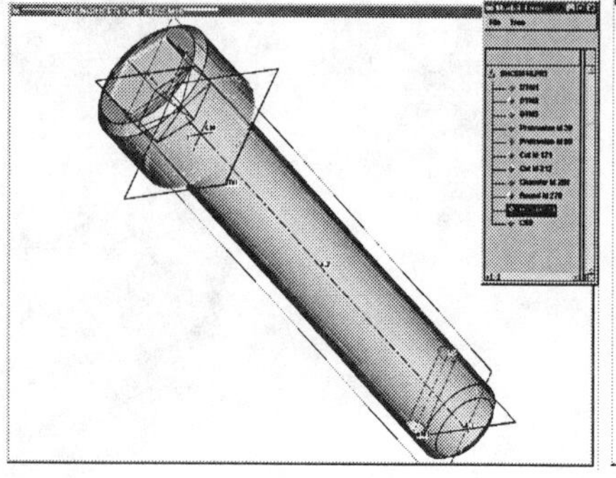

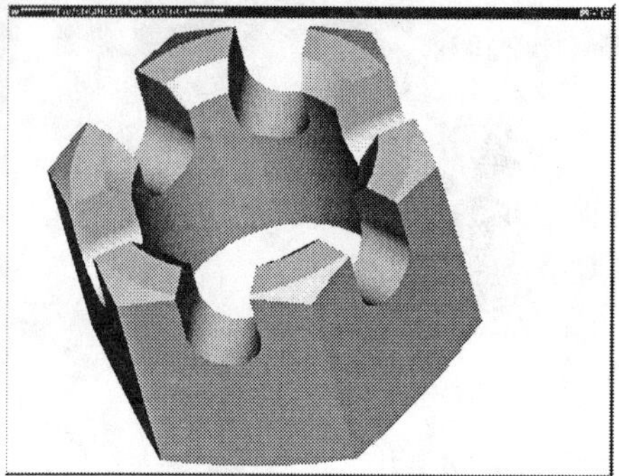

Figure 14.64
Socket Head Cap Screw and Slotted Hex Nut

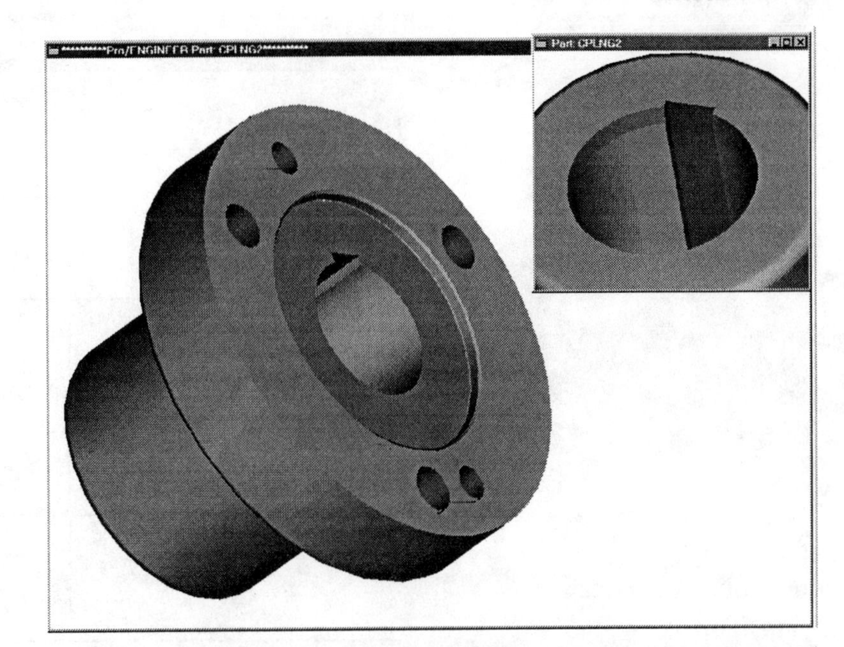

Figure 14.65
Second Coupling, Part Model

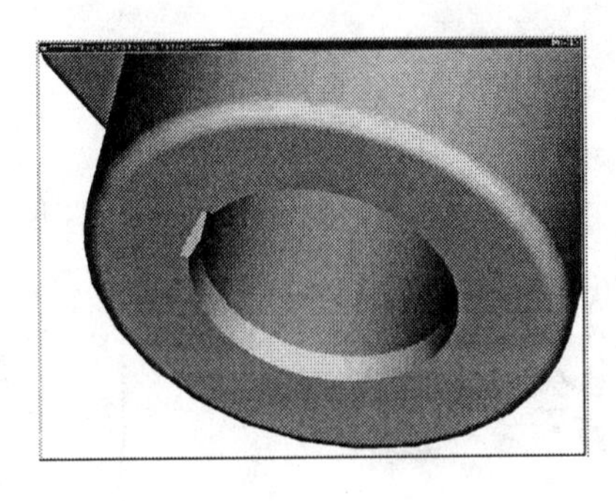

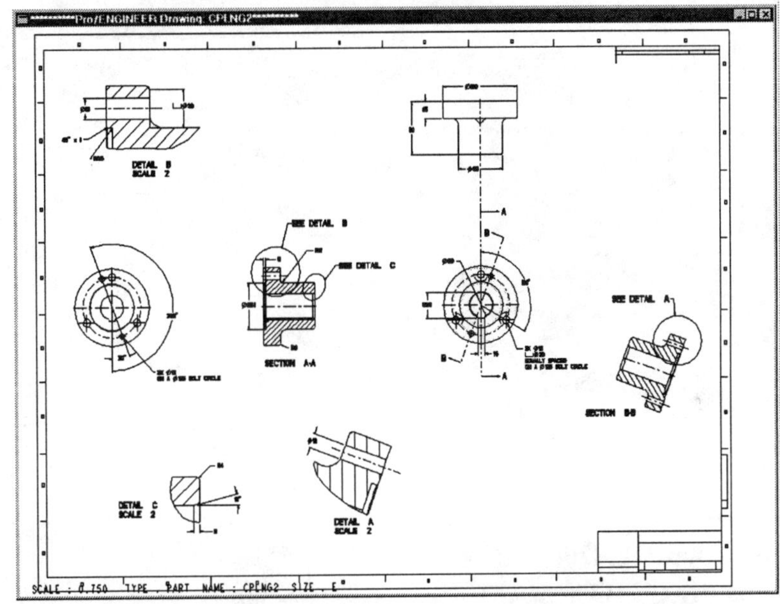

Figure 14.66
Second Coupling, Detail Drawing

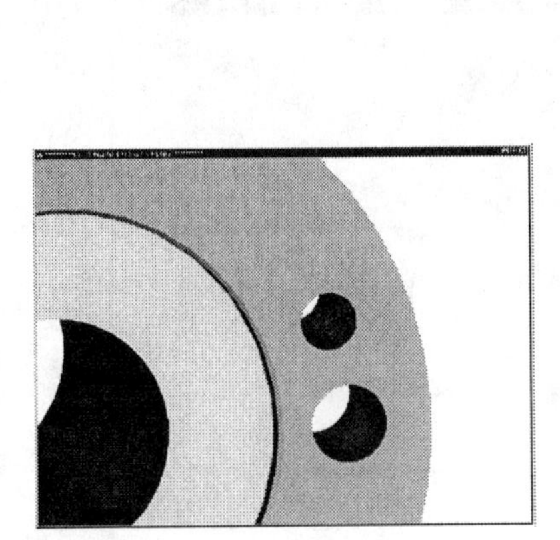

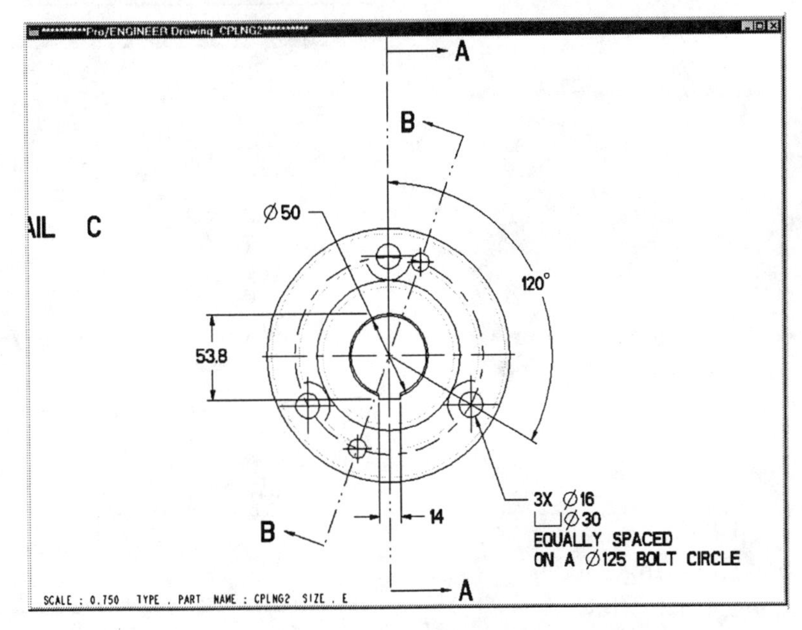

Figure 14.67
Second Coupling, Detail Drawing, Front View

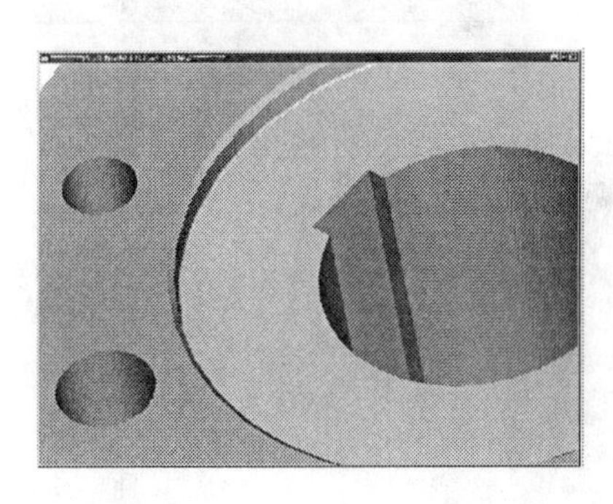

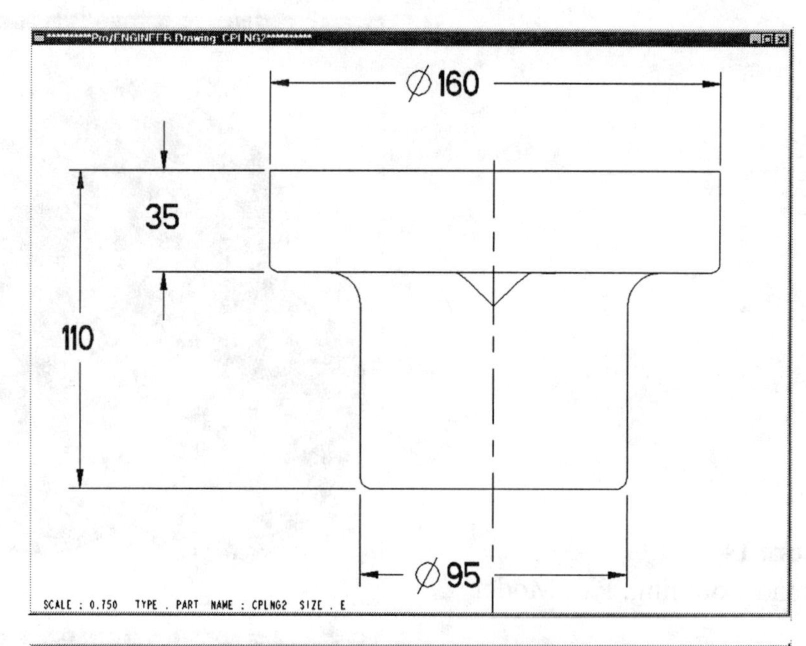

Figure 14.68
Second Coupling, Detail Drawing, Top View

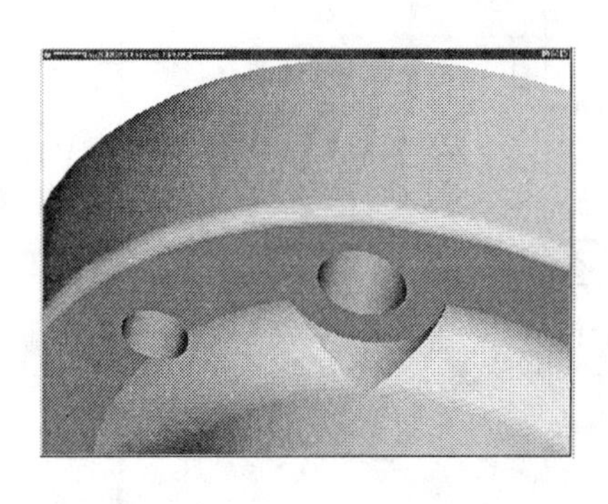

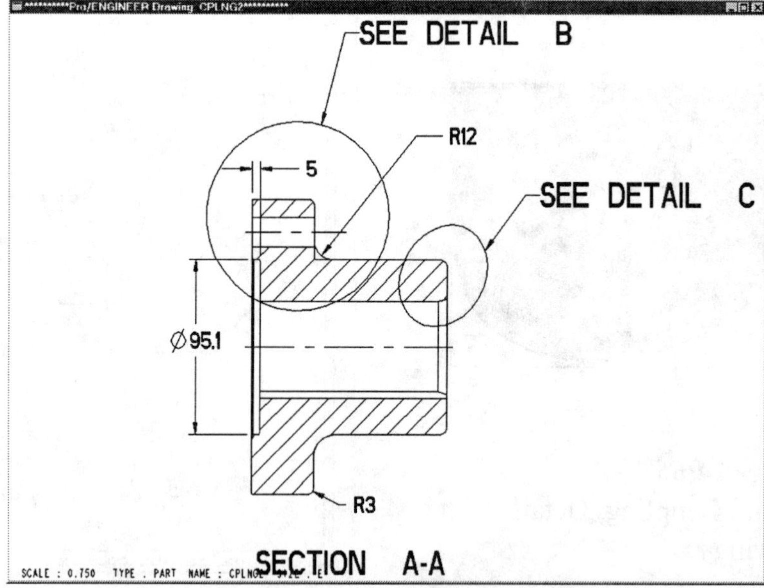

Figure 14.69
Second Coupling, Detail Drawing, **SECTION A-A**

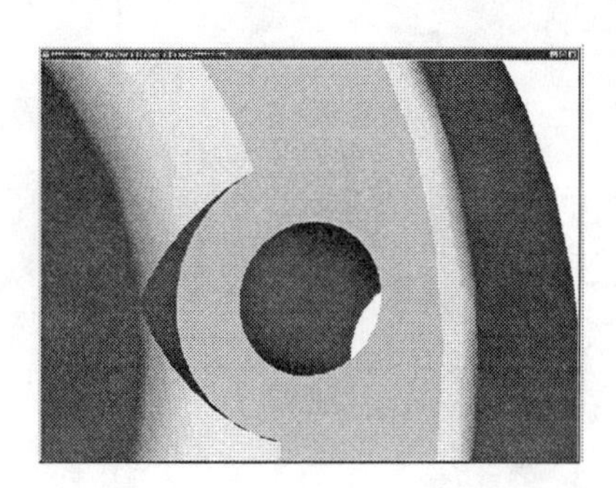

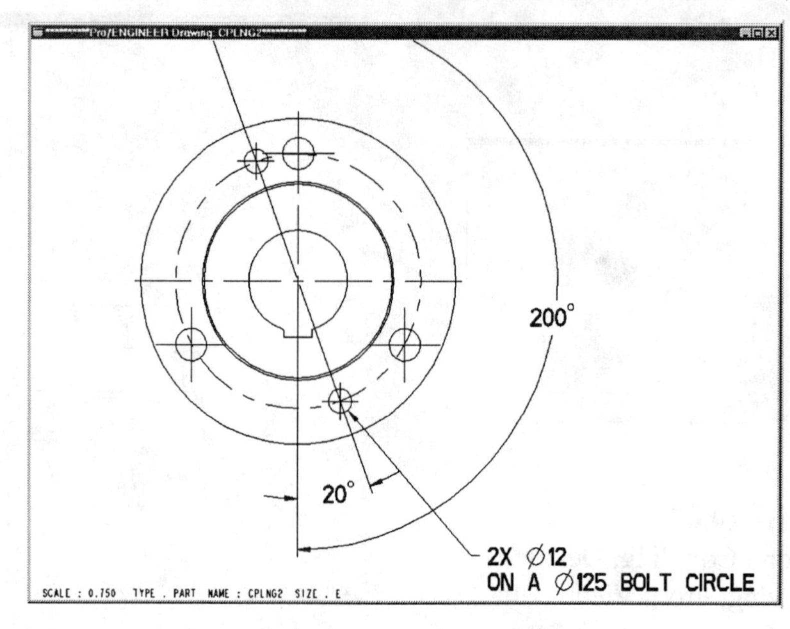

Figure 14.70
Second Coupling, Detail Drawing, Back View

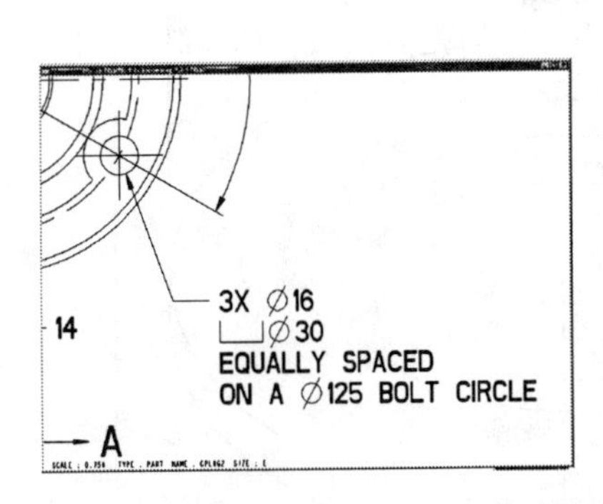

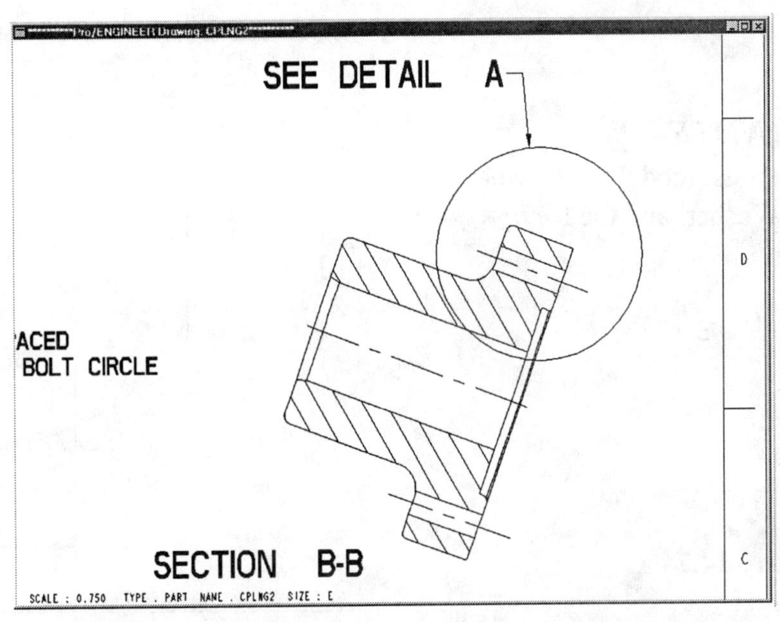

Figure 14.71
Second Coupling, Detail Drawing, **SECTION B-B**

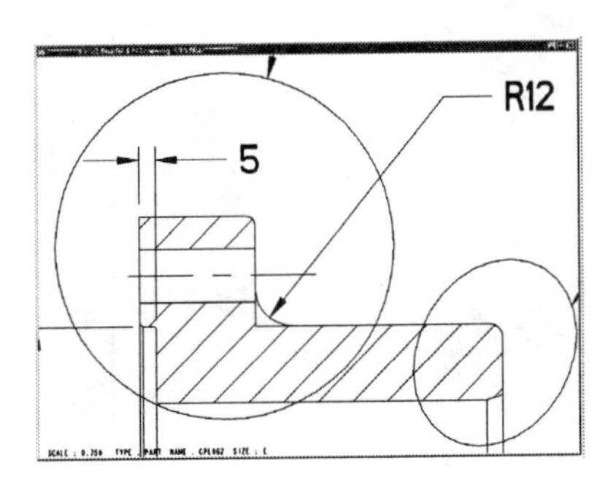

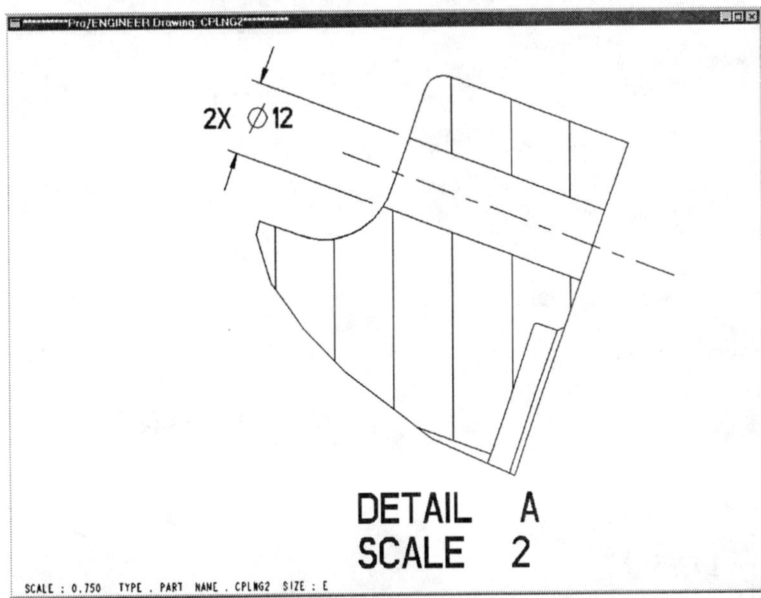

Figure 14.72
Second Coupling, Detail Drawing, **DETAIL A**

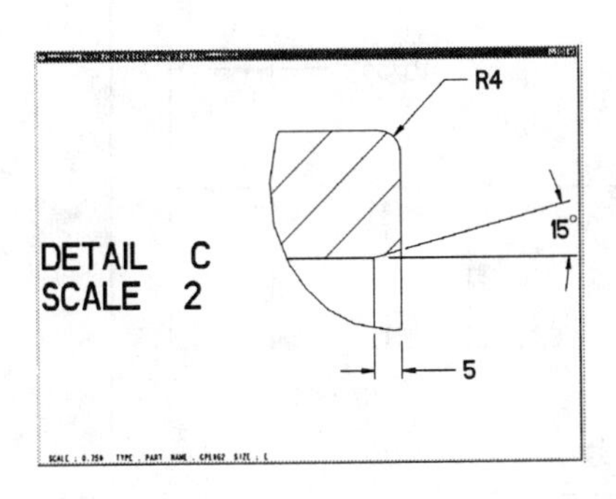

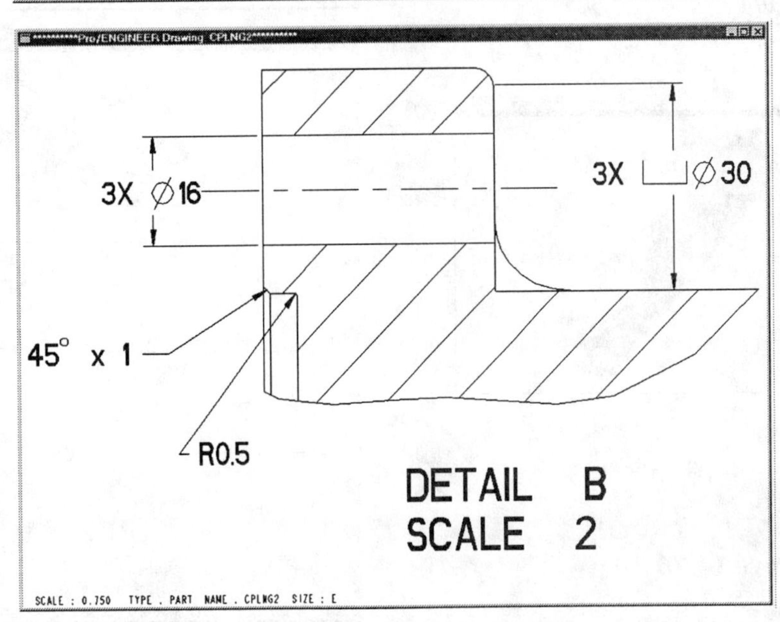

Figure 14.73
Second Coupling, Detail Drawing, **DETAIL B**

NOTE

You must model the Dowel, the Washer, and the Key.

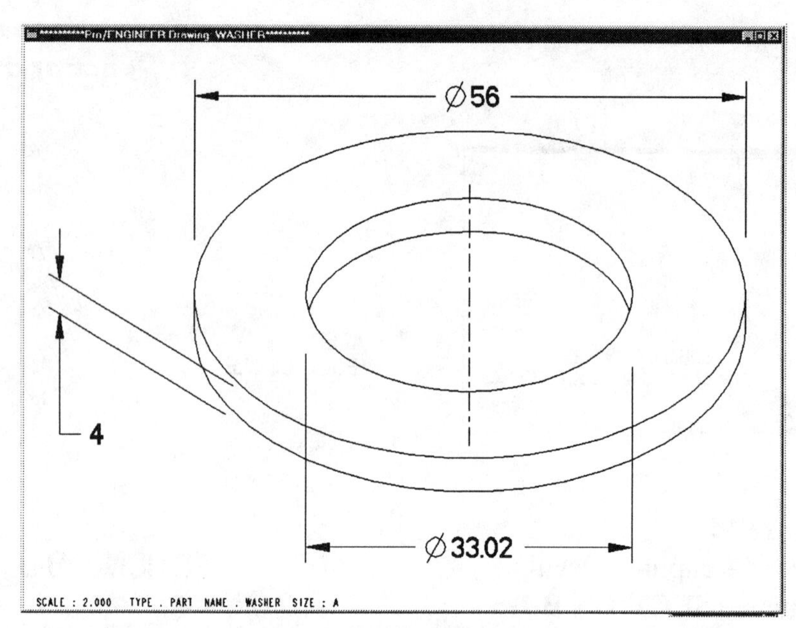

Figure 14.74
Washer

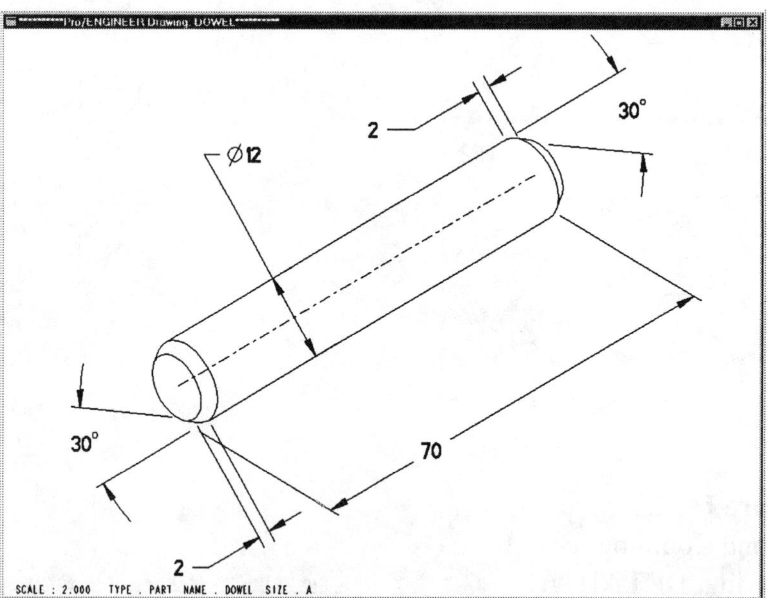

Figure 14.75
Dowel

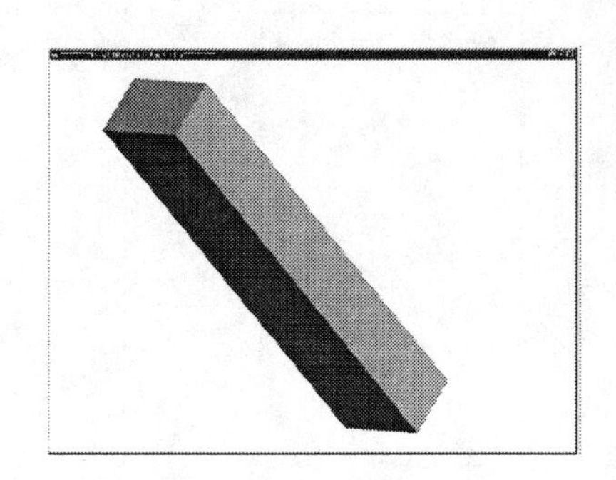

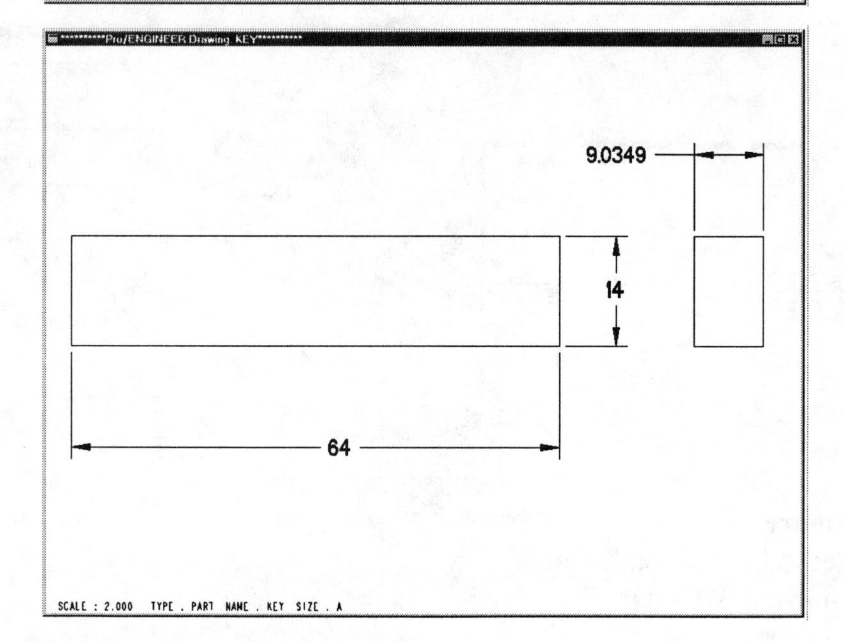

Figure 14.76
Key

SOCKET HEAD CAP SCREW

Part ⇒
Search/Retr ⇒
Pro/Library ⇒
/objlib ⇒ /metriclib ⇒
/sock_hd_scr ⇒ mscs.prt ⇒
SelByParams ⇒
✓NOM_SIZE_THR_PITCH ⇒
Done Sel ⇒ M16X2 ⇒
SelByParams ⇒
✓d5,length ⇒
Done Sel ⇒ 80.000

INSTANCE = MSCS1210

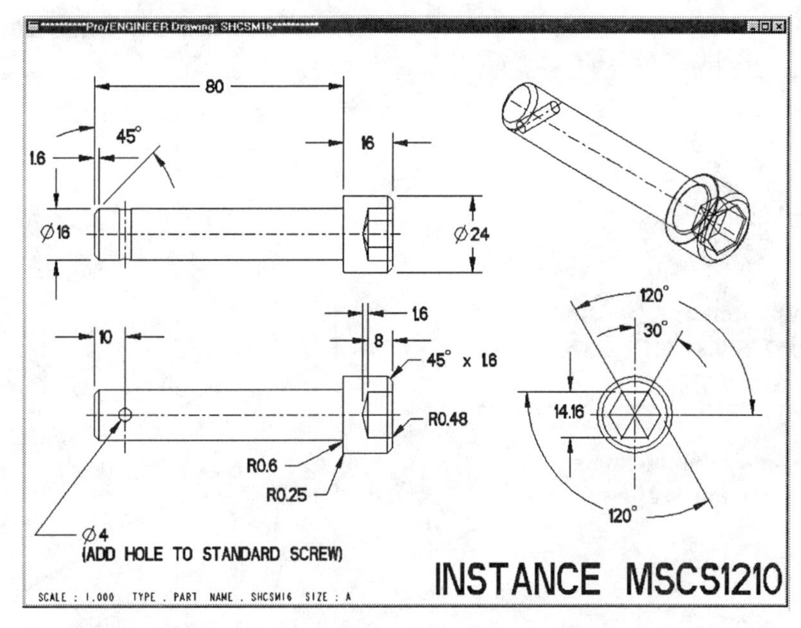

Figure 14.77
Socket Head Cap Screw

SLOTTED HEX NUT

Part ⇒
Search/Retr ⇒
Pro/Library ⇒
/objlib ⇒ /metriclib ⇒
/hex_nuts ⇒
mshn.prt ⇒
SelByParams ⇒
✓NOM_ DIA_THR_PITCH ⇒
Done Sel ⇒ M16X2

INSTANCE = MSHN07

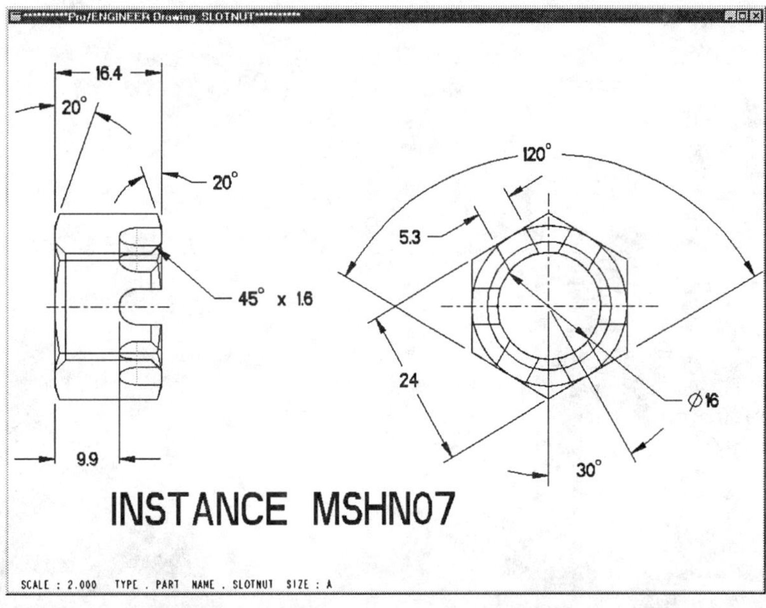

Figure 14.78
Slotted Hex Nut

COTTER PIN

Part ⇒
Search/Retr ⇒
Pro/Library ⇒
/objlib ⇒ /eng_part_lib ⇒
/cot_clvs_pin ⇒ Pina.prt ⇒
SelByParams ⇒
✓NOM_SIZE ⇒
Done Sel ⇒ .1562 ⇒
SelByParams ⇒
✓ d13,1 ⇒
Done Sel ⇒ 1.250

INSTANCE = PNA09L05

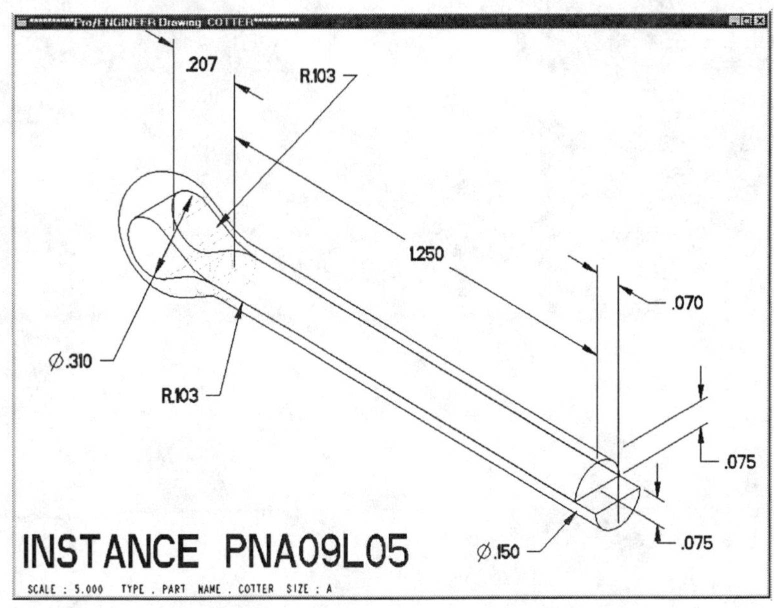

Figure 14.79
Cotter Pin

HEX JAM NUT

Part ⇒
Search/Retr ⇒
Pro/Library ⇒
/objlib ⇒ /metriclib ⇒
/hex_nuts ⇒ pina.prt ⇒
/mhjn.prt ⇒
SelByParams ⇒
✓NOM_SIZE_THR_PITCH ⇒
Done Sel ⇒ M30X3.5

INSTANCE = MHJN10

- **Modify the thickness of the nut to 10mm**

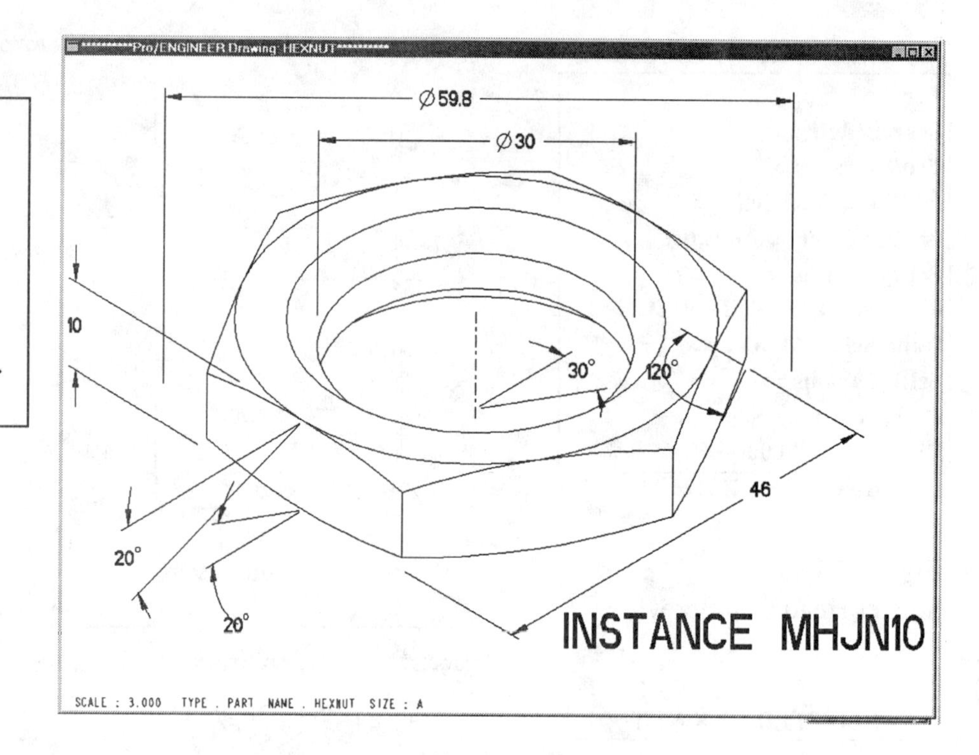

Figure 14.80
Hex Jam Nut

Figure 14.81
Coupling Assembly

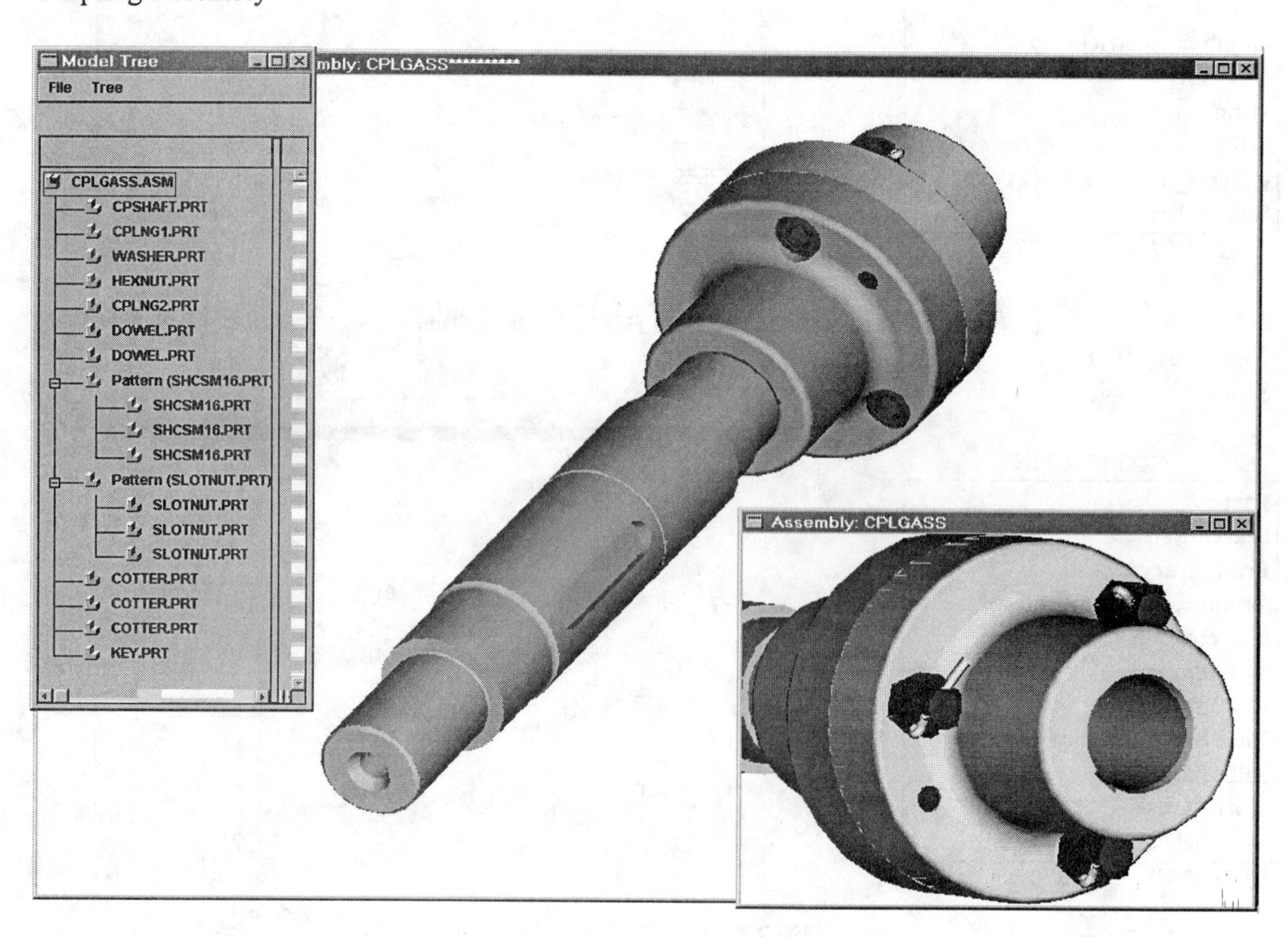

Lesson 15

Exploded Assemblies

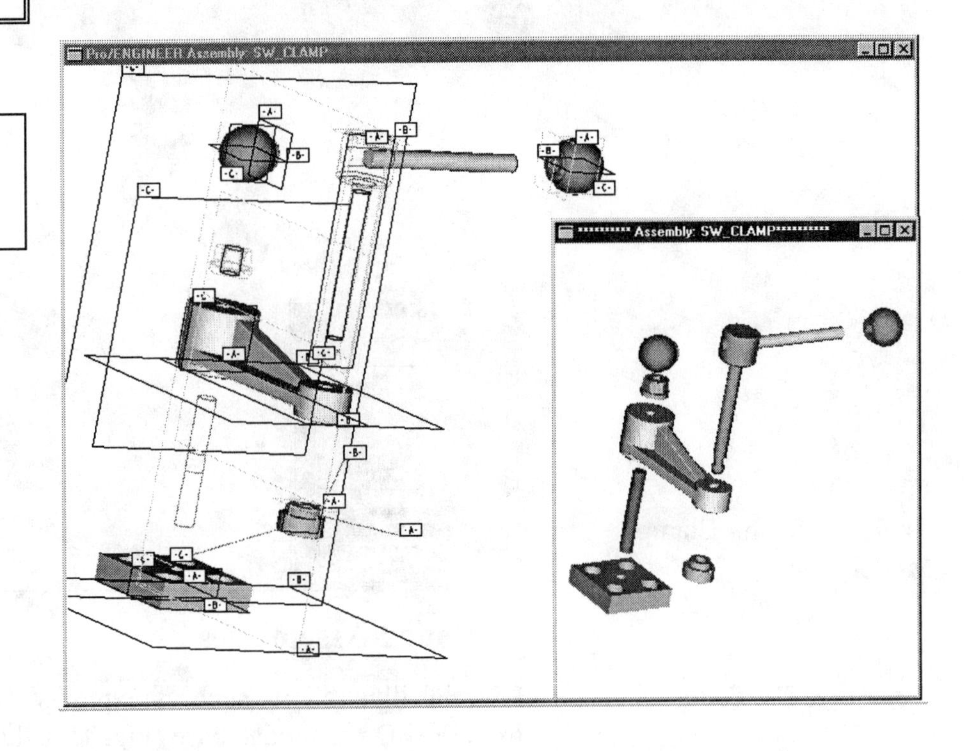

Figure 15.1
Exploded Swing Clamp

OBJECTIVES

1. **Create exploded views**
2. **Edit exploded views**
3. **Create unique component visibility settings**
4. **Move and rotate components in an assembly**
5. **Create shaded and pictorial views of an assembly**
6. **Save named views to use later in Drawing mode**

☑ ***EGD REFERENCE***
Engineering Graphics and Design
by L. Lamit and K. Kitto
Read Chapters: 23
See Pages: 865-866

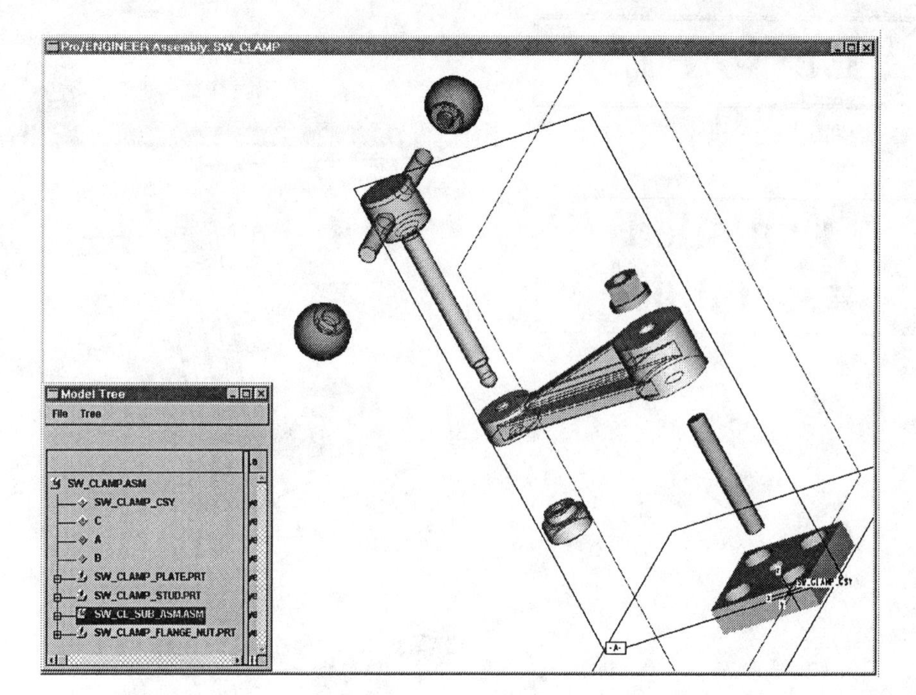

Figure 15.2
Exploded Swing Clamp Model Tree

Exploded Assemblies

Pictorial illustrations such as exploded views are generated directly from the 3D model database (Fig. 15.1 through 15.7). The model can be displayed and oriented in any position automatically. Each component in the assembly can have a different display type: wireframe, hidden line, no hidden, and shading. You may select and orient the part to provide the required view orientation to display the part from underneath or from any side or position. Perspective projections are completely automated with a simple selection from a menu. The assembly can be spun around, reoriented, and even clipped to show the interior features. When assemblies are illustrated, you have the choice of displaying all components and subassemblies, or any combination of the design.

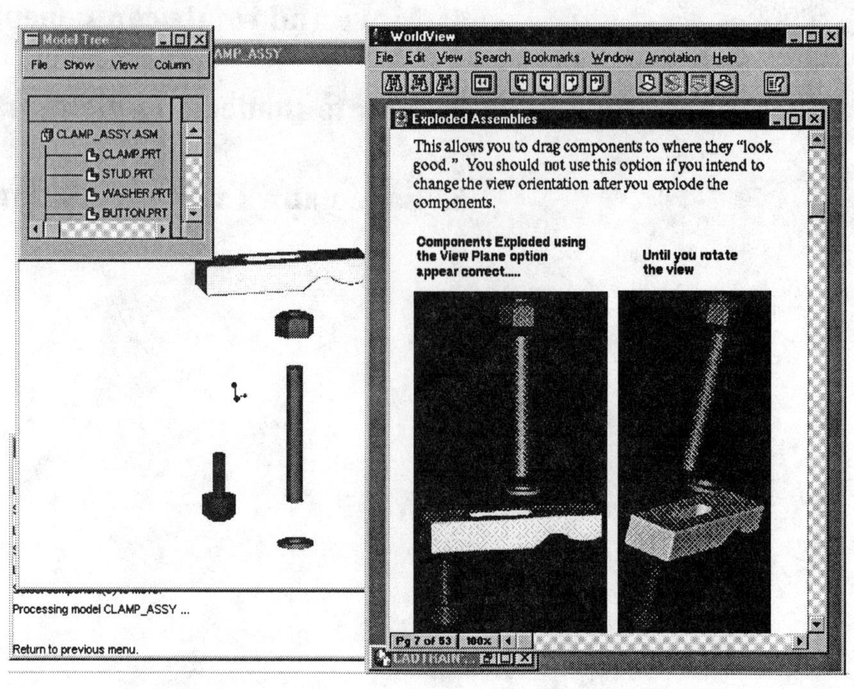

Figure 15.3
COAch for Pro/E, Exploded Assemblies-- Assemblies (Segment 1: Moving Exploded Components)

Creating Exploded Views

Using the **ExplodeState** option in the ASSEMBLY menu, you can automatically create an exploded view of an assembly. Exploding an assembly only affects the display of the assembly; it does not alter actual distances between components. Exploded states are created to define the exploded position of all components. For each explode state, you can toggle the explode status of components, change the explode locations of components, and create explode offset lines. To access this functionality, choose the **ExplodeState** option in the ASSEMBLY menu.

You can define multiple explode states for each assembly, and then explode the assembly using any of these explode states at any time. You can also set an explode state for each drawing view of an assembly. Pro/E gives each component a default explode position determined by the placement constraints. By default, the reference component of the explode is the parent assembly (top-level assembly or subassembly).

To explode components, you use a "drag-and-drop" user interface similar to the **package** functionality. You select one or more components and the motion reference, and then drag the outlines to the desired positions. The component outlines drag along with the mouse cursor. You control the type of explode motion using a Preferences setting.

Two types of explode instructions can be added to a set of components. The children components follow the component being exploded or they do not follow it. Each explode instruction consists of a set of components, explode direction references, and dimensions that define the exploded position from the final (installed) position with respect to the explode direction references.

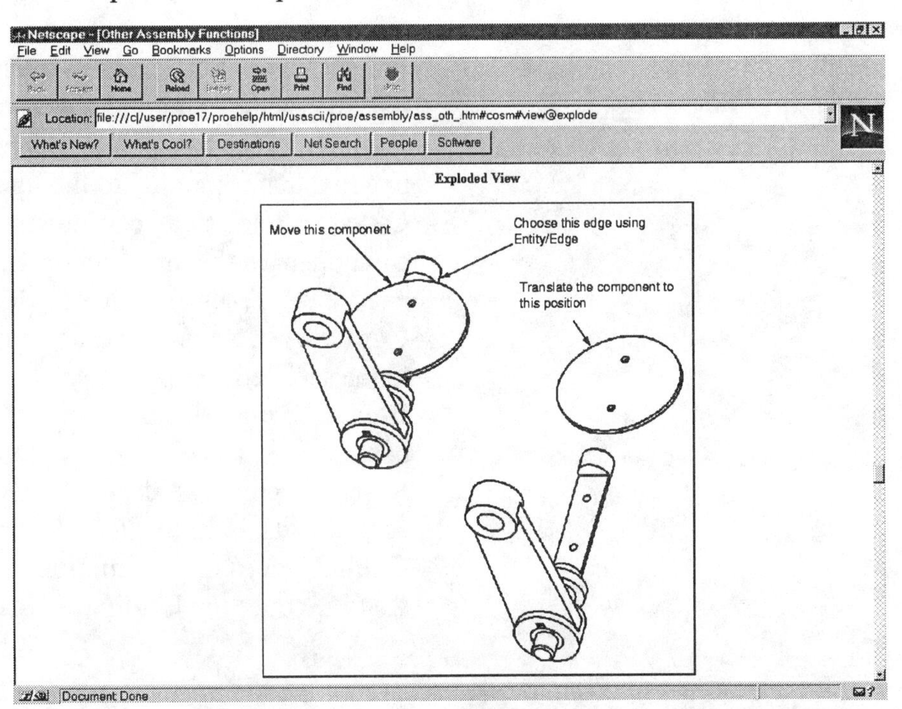

Figure 15.4
Online Documentation Exploded Views

Creating New Exploded Views

To create a new explode state:

1. Retrieve an assembly. Pro/E displays the Model Tree assembly tool.
2. Choose **ExplodeState** from the ASSEMBLY menu. The EXPLD STATE menu lists the following options:

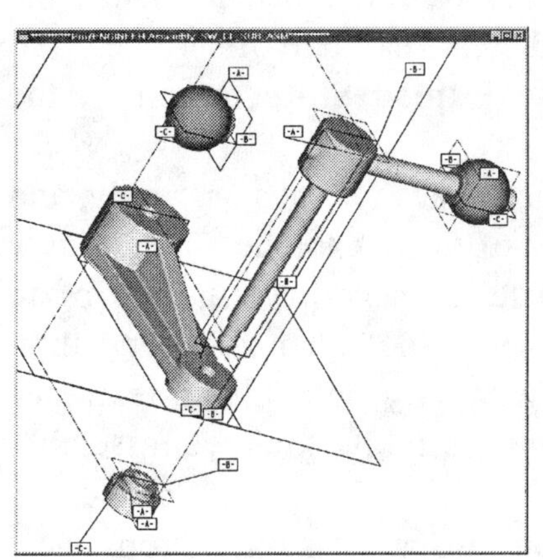

Create Creates a new explode state. Each time Pro/E creates a new state, it creates it with no components exploded.
Set Current Displays the SEL STATE menu, listing available explode states.
Copy Copies an existing explode state using the SEL STATE menu.
Redefine Redefines an explode state using the MOD EXPLODE menu.
Delete Deletes one or more explode states.
List Lists all currently defined explode states in an Information Window.

3. Choose **Create**, and then enter a name for the state. The MTNPREF menu displays the following options:

Preferences Sets up preferences for dragging packaged components.
View Plane Uses the viewing plane as the reference plane (repositions the component in a plane that is parallel to it).
Sel Plane Selects a plane other than the viewing plane as the reference plane (repositions the component in a plane that is parallel to it).
Entity/Edge Selects an axis, straight edge, or datum curve (repositions the component in a line parallel to it).
Plane Normal Selects a plane as the reference plane, and repositions the component in a line that is normal to it.
2 Points Picks two points or vertices (repositions the component in a line that is parallel to the line that connects them).
C-sys Selects a coordinate system axis (repositions the component in the direction of it).
Translate Translates the selected component.
Copy Pos Copies the position instructions from one component to the selected components.
DefaultExpld Places selected components in the default explode position.
Reset Removes all explode instructions from the selected components, even the default explode, and resets their positions.
Undo Undoes the last motion.
Redo Redoes the last motion that was undone.

4. Choose **Preferences.** The MOTION PREFS menu displays the following options:

Trans Incr Specifies the motion increments for translational dragging. Displays the TRANS INCR menu. Choose **Enter** to enter an increment value for the translation, or choose one of the preset values displayed in the menu. To drag the component without apparent incrementing, choose **Smooth.**

Move Type Sets the type of movement (single or multiple components). Displays the MOVE TYPE menu with the following options:

Move One Moves one component at a time.

Move Many Moves multiple components simultaneously.

With Children Highlights the children of the selected component and moves the entire group as one unit.

5. Choose an option from the MOVE TYPE menu, and then specify the motion increments using the **Trans Incr** option. Choose **Done.**

6. Choose one of these options:

Copy Pos to copy the position of selected components.

DefaultExpld to place selected components in the default explode position.

Translate Specify the reference or direction for repositioning the exploded component by choosing **Translate** and one of the following options (the **Entity/Edge** option is highlighted):

View Plane or **Sel Plane** Translation occurs about the component's drag origin in the same plane.

Entity/Edge Translation occurs about the reference line in the plane that is normal to it and contains the drag origin point.

Plane Normal Translation occurs about the component's drag origin in the plane that contains the drag origin point and is parallel to the reference plane.

2 Points Translation occurs about the reference line in the plane that is normal to it and contains the drag origin point.

C-sys Translation occurs about the axis in the plane that is normal to it and contains the drag origin point.

Setting Display Modes for Components

Using the **Comp Display** option in the COSMETIC menu, you can set different visualization (display) modes for components in an assembly. Wireframe, hidden line, no hidden, shaded, or blanked display modes can be assigned to components. The components will be displayed according to their display status in the current display state (that is, blanked, shaded, drawn in hidden line color, and so on). The setting in the ENVIRONTMENT menu controls the display of unassigned components. As a result, they appear according to the current mode of the environment:

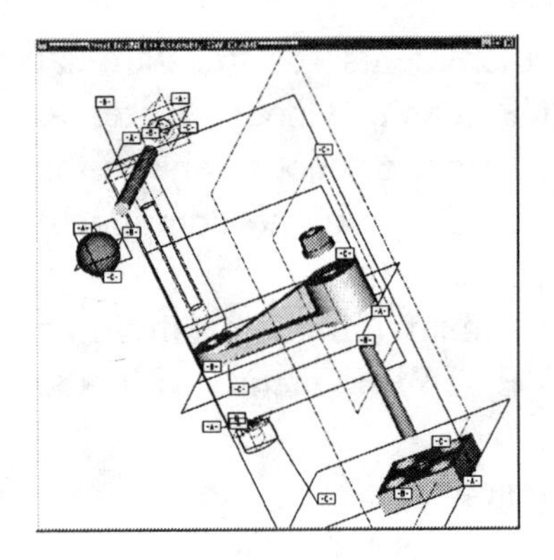

Wireframe Pro/E displays components in front of the assigned component in wireframe and shows all of their edges in white (or the color they have), when you assign one component to display its hidden lines, leave the rest unassigned, or set the ENVIRONMENT menu to **Wireframe**. Pro/E displays the components behind the assigned part in wireframe, but does not display sections of edges masked by the hidden line component.

Hidden Line If you set the ENVIRONMENT menu to **Hidden Line**, Pro/E displays all components and the hidden line component in the same way.

No Hidden If you set the ENVIRONMENT menu to **No Hidden**, Pro/E displays all components in front of the assigned components with no hidden lines, and displays the obscured edges of the hidden line component in gray and its visible edges in white. For components behind the assigned part, visible edges appear, but obscured ones do not. The edges on the other parts that are obscured by the hidden line component do not appear in gray.

Shading If you set the ENVIRONMENT menu to **Shading**, Pro/E displays all unassigned components as shaded, whether they are in front of or behind the hidden line component.

Figure 15.5
Online Documentation Component Display Status

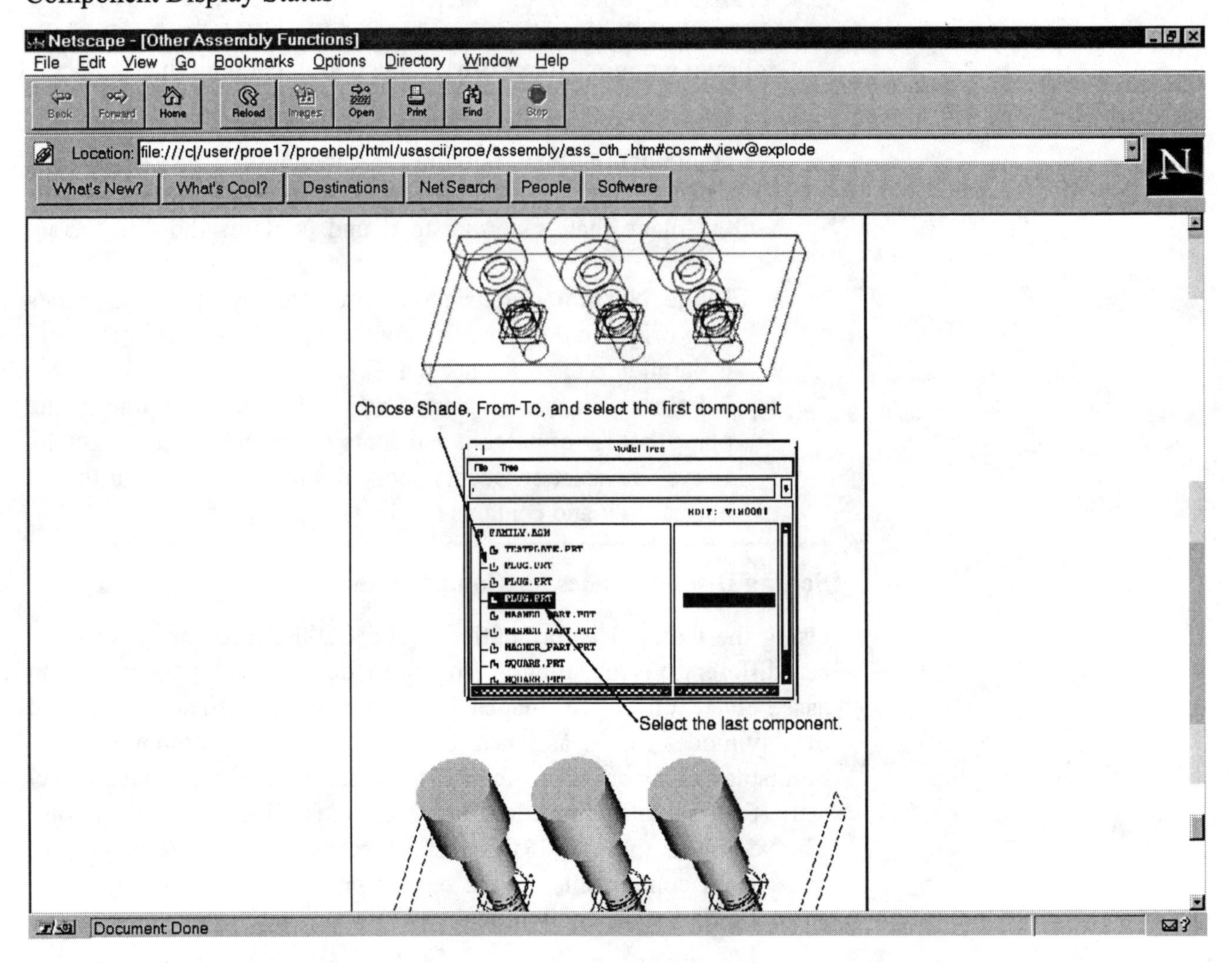

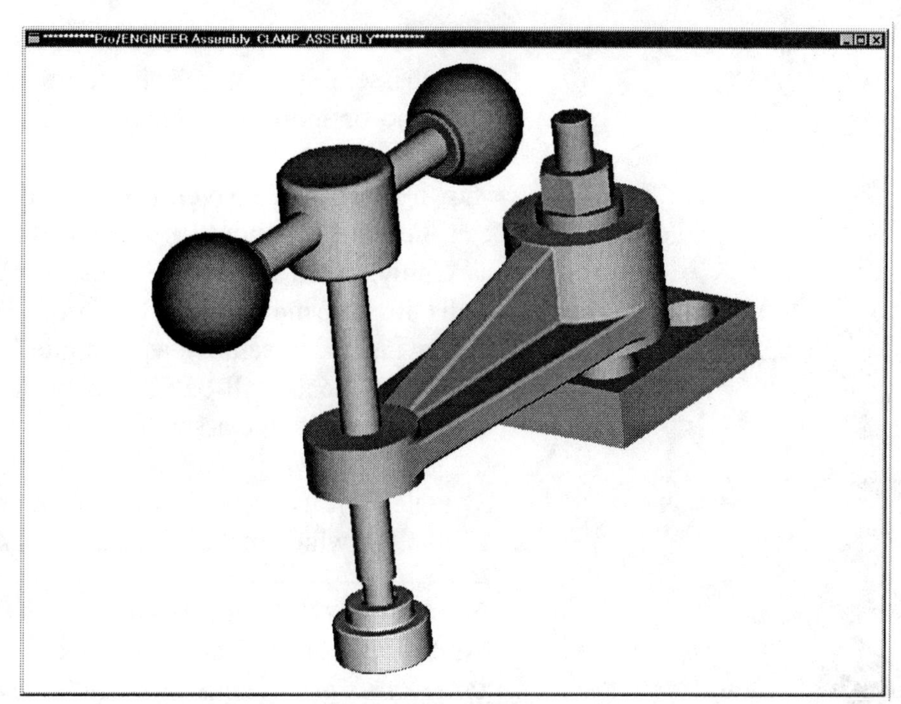

Figure 15.6
Original State of Exploded Swing Clamp

NOTE

Set up the **Swing Clamp** Assembly Exploded View:

- Units = Inches
- Shading
- No Display Tan
- ☐ Grid Snap

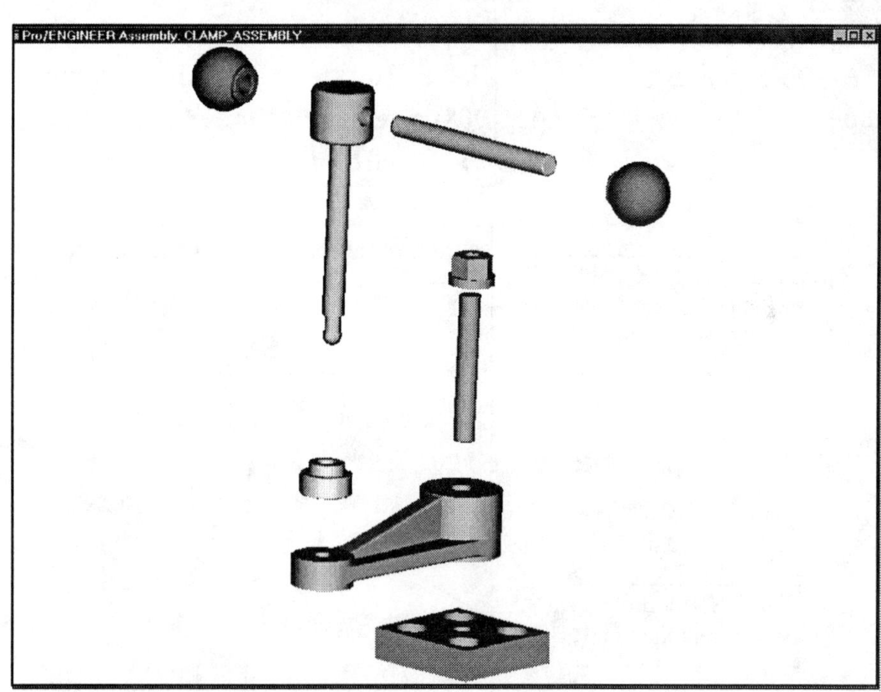

Figure 15.7
Exploded Swing Clamp with Rotated Swivel

Exploded Swing Clamp

In this Lesson, you will use the subassembly and assembly created in Lesson 14 to establish and save new views, exploded views, and views with component display states that differ from one another. You will also be required to move and rotate components of the assembly before cosmetically displaying the assembly in an exploded state. The creation and assembly of new components will not be required. A bill of materials will also be displayed using the **Info** command.

Rotating Components of an Assembly

To rotate an existing component or set of components of an assembly (or subassembly), you select a coordinate system to use as a reference, pick one or more components, and give the rotation angle about a chossen axis of a coordinate system. The Swing Clamp assembly shown here has a **Swivel, Foot, Stud**, and two **Ball** components that can be rotated about the **Arm** during normal operation of the assembly. You will rotate these components so that the Stud and Balls are perpendicular to the Arm. This position looks better when displaying the assembly as exploded (and in its unexploded state). The components will be rotated in the subassembly, and the change will propagate to the assembly.

The following sequence of commands are used to rotate the Swivel, Foot, Stud, and two Ball components about the Arm (Fig. 15.8) all of which make up the *subassembly*:

Dbms ⇒ Save ⇒ enter
Purge ⇒ enter ⇒
Done-Return

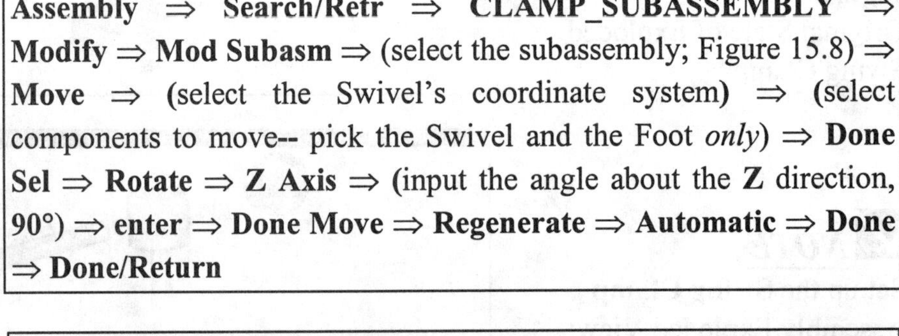

Assembly ⇒ Search/Retr ⇒ CLAMP_SUBASSEMBLY ⇒ Modify ⇒ Mod Subasm ⇒ (select the subassembly; Figure 15.8) **⇒ Move ⇒** (select the Swivel's coordinate system) **⇒** (select components to move-- pick the Swivel and the Foot *only*) **⇒ Done Sel ⇒ Rotate ⇒ Z Axis ⇒** (input the angle about the **Z** direction, 90°) **⇒ enter ⇒ Done Move ⇒ Regenerate ⇒ Automatic ⇒ Done ⇒ Done/Return**

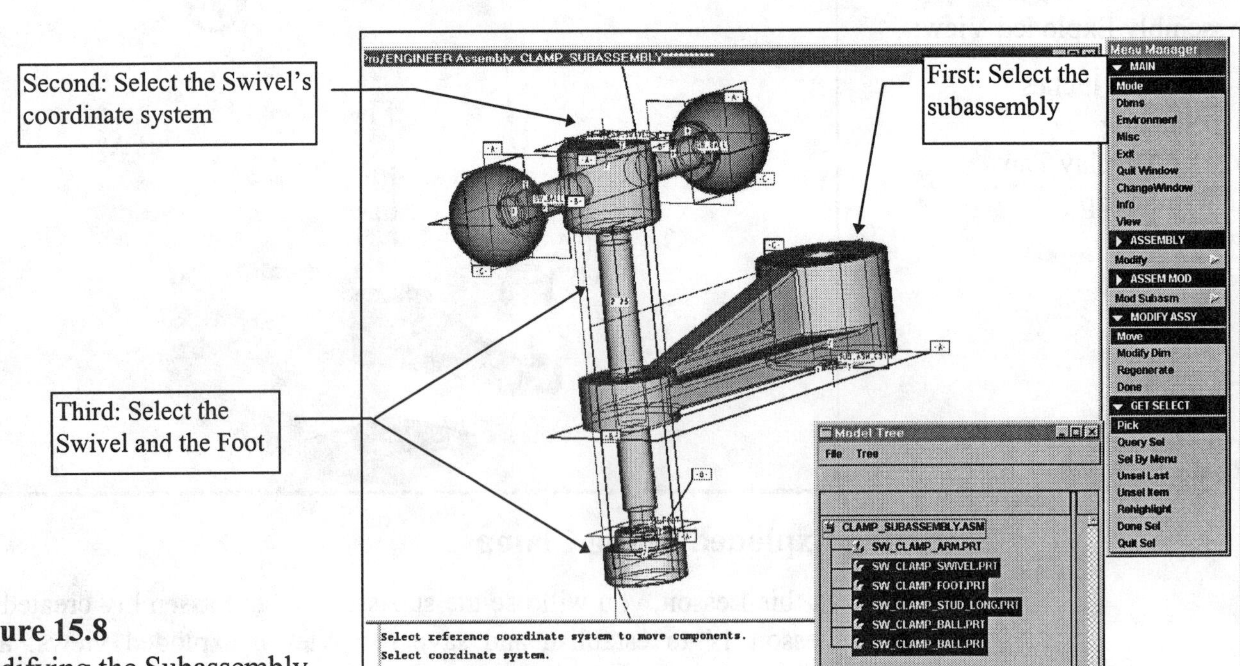

Figure 15.8
Modifying the Subassembly

The Swivel, Foot, Balls, and Stud are now rotated **90°** (Fig. 15.9). Capture this view with the following commands:

View ⇒ Names ⇒ Save ⇒ (enter NAME: **subpict1**) **⇒ enter ⇒ Done-Return**

Dbms ⇒ Save ⇒ enter Purge ⇒ enter ⇒ Done-Return

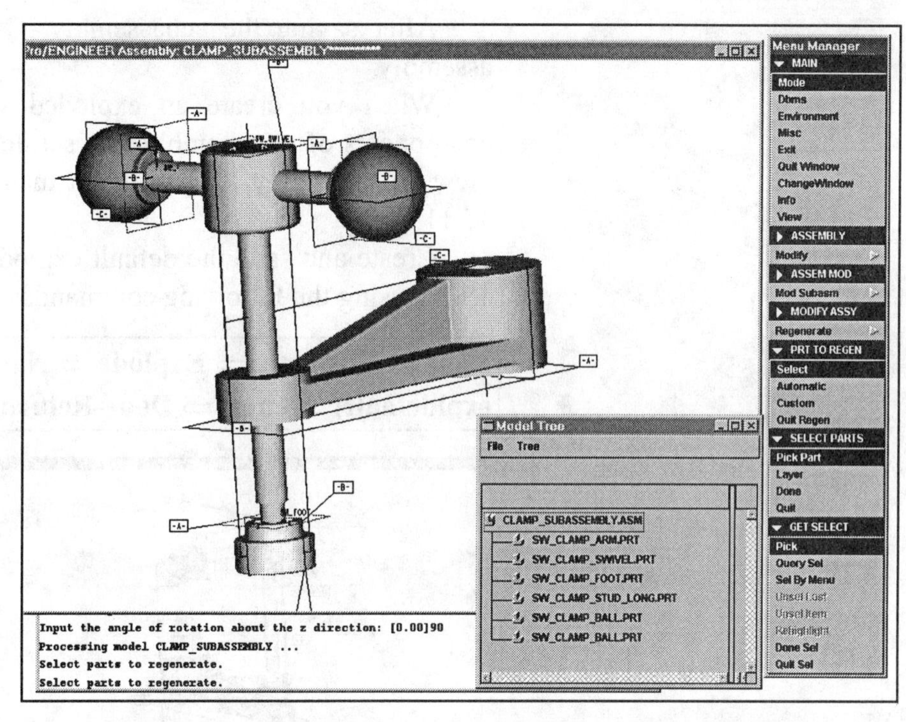

Figure 15.9
Subassembly with Rotated Components

Views: Perspective, Exploded, and Altered Component Displayed State Views

Another type of view that can be created automatically is the perspective view. You will now create a variety of cosmetically altered views. Cosmetic changes to the assembly do not affect the model itself, only how it is displayed on the screen. Use the following commands to create a perspective (Fig. 15.10) and save the view with a unique name:

View ⇒ Perspective ⇒ Default ⇒ Names ⇒ Save ⇒ (enter NAME: **perspec) ⇒ enter ⇒ Done-Return**

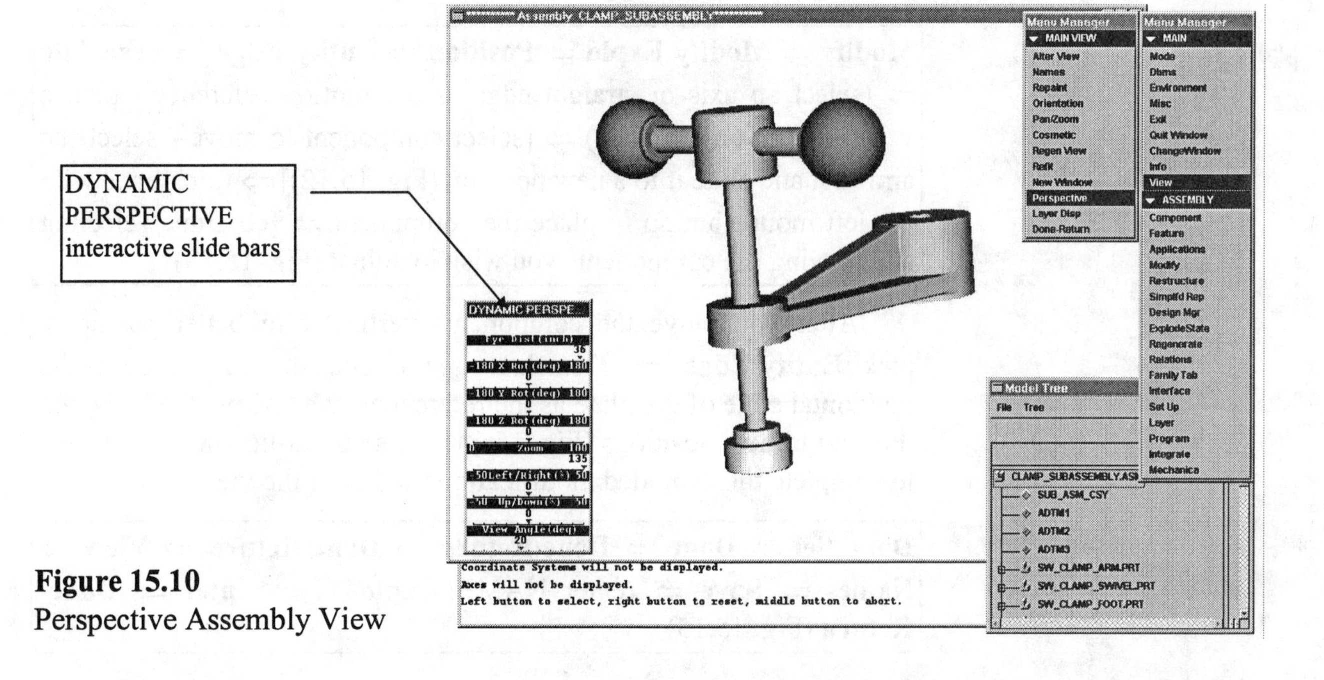

Figure 15.10
Perspective Assembly View

After saving the subassembly, **Quit Window** and retrieve the assembly.

When you create an exploded view, Pro/E moves apart the components of an assembly to a set default distance. Your exploded view will probably look different than the example shown in Figure 15.11.

Create and save the default exploded view of the assembly (Fig. 15.11) using the following commands:

View ⇒ **Cosmetic** ⇒ **Explode** ⇒ **Names** ⇒ **Save** ⇒ (enter NAME: **expldefault)** ⇒ **enter** ⇒ **Done-Return**

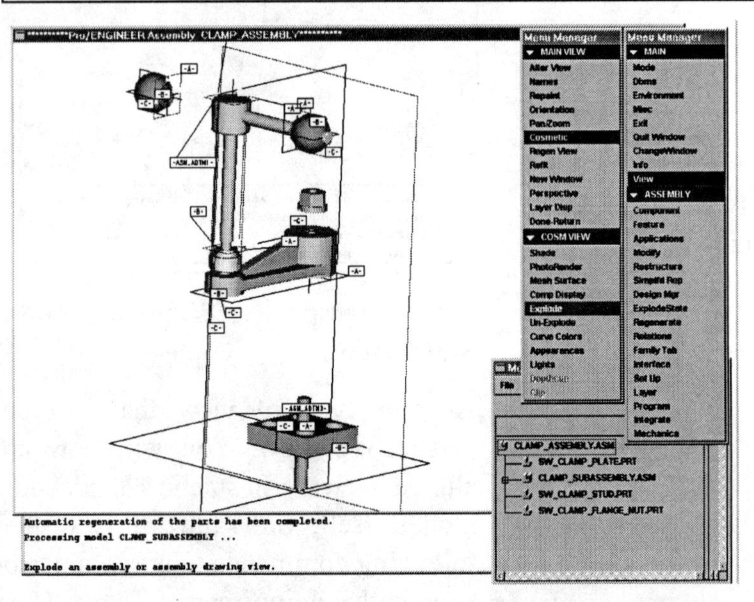

Figure 15.11
Default Exploded Assembly

Since the default exploded view is seldom perfect, Pro/E has a wide variety of commands available to adjust, change, modify, and reorient the exploded view. Using **Modify** from the ASSEMBLY Menu, give the following commands:

Modify ⇒ **Modify Expld** ⇒ **Position** ⇒ **Entity/Edge** ⇒ **Translate** ⇒ (select an axis or straight edge as the motion reference-- pick a vertical edge on the plate) ⇒ [select component to move-- select the arm first and slide it to a new position (Fig. 15.12)] ⇒ (pick once with the left mouse button to place the component ⇒ [continue selecting and moving the components you wish to adjust (Fig. 15.13)]

After you move the components vertically to better positions, pick **Entity/Edge** ⇒ **Translate** again, and this time select the horizontal edge of the plate as the reference edge. Move the balls and the stud to new positions (Fig. 15.14). Use the following commands to complete the exploded modifications and save the view:

Done Sel ⇒ **Done** ⇒ **Done/Return** ⇒ **Done/Return** ⇒ **View** ⇒ **Names** ⇒ **Save** ⇒ (enter NAME: **explod1)** ⇒ **enter** ⇒ **Done-Return** (Fig. 15.15)

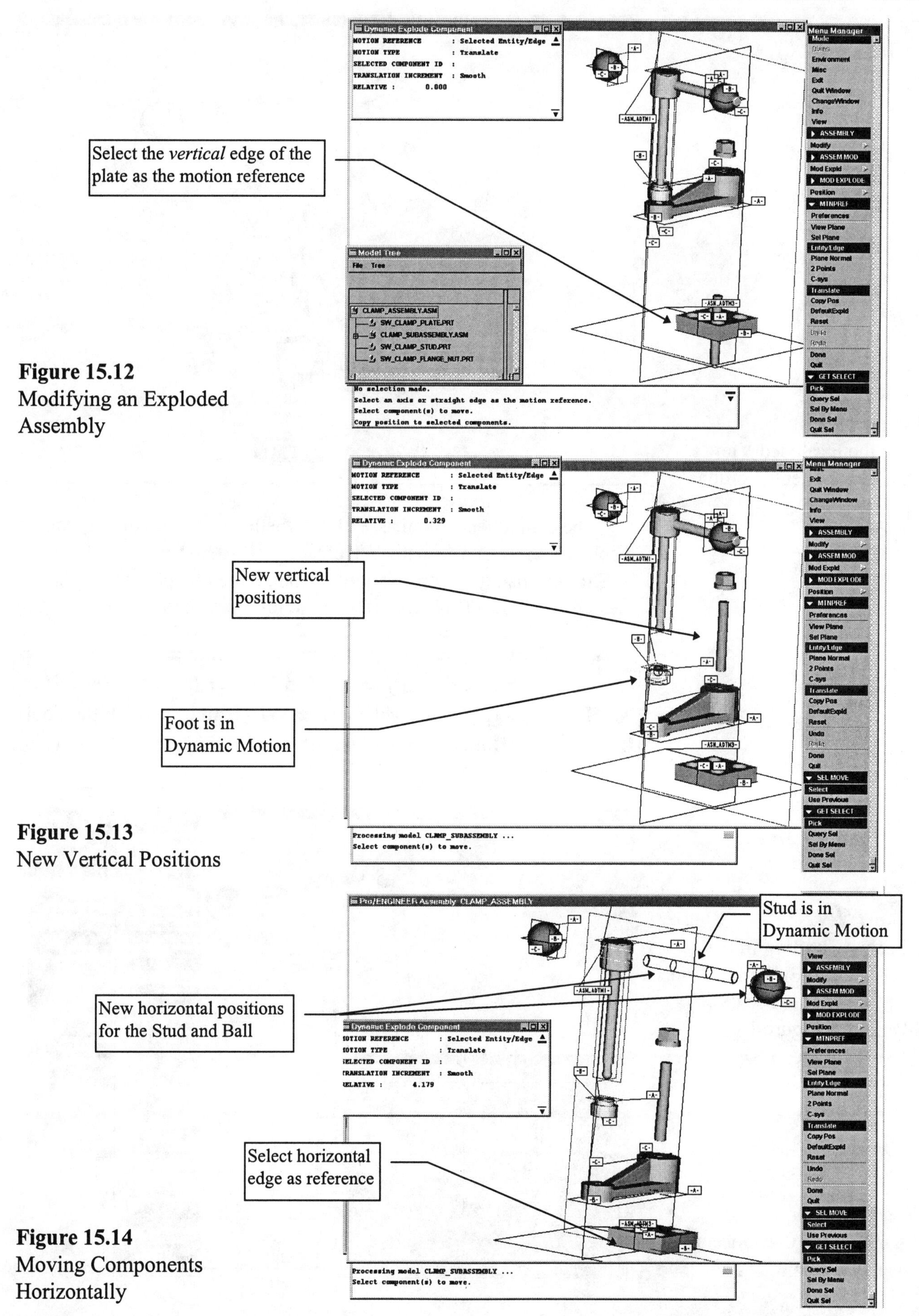

Figure 15.12
Modifying an Exploded Assembly

Figure 15.13
New Vertical Positions

Figure 15.14
Moving Components Horizontally

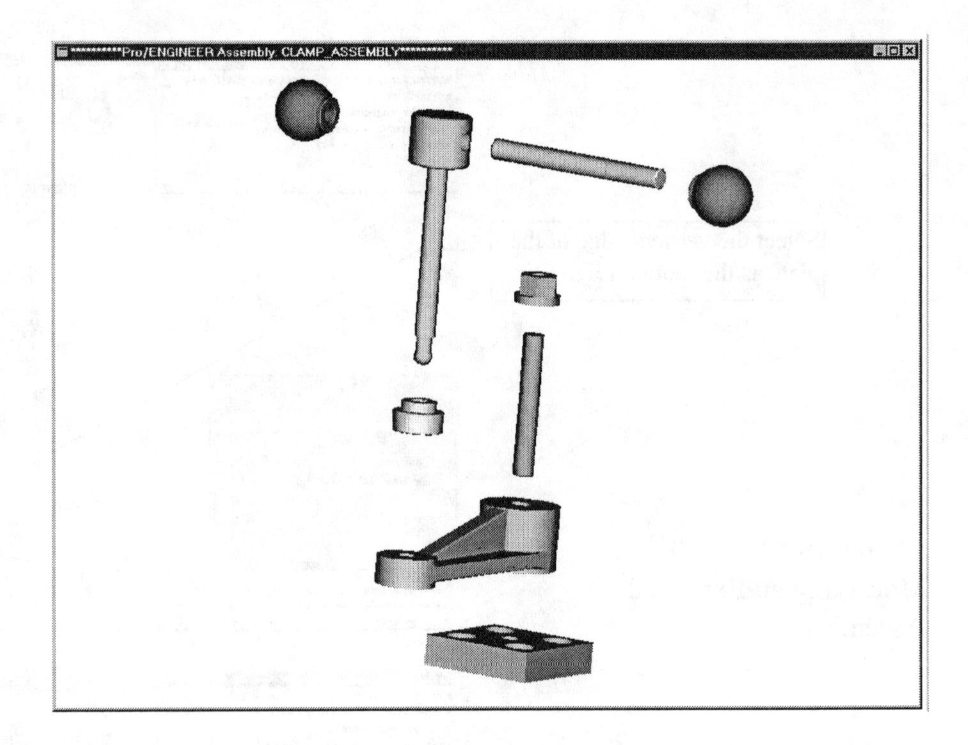

Figure 15.15
Save the Exploded View's New Component Positions

The components of an assembly (whether exploded or not) can be displayed individually with **Wireframe**, **Hidden Line**, **No Hidden**, or **Shading** using **Comp Display**. Change the component display of the exploded view (Fig. 15.16) with the following commands:

View ⇒ **Cosmetic** ⇒ **Comp Display** ⇒ **Create** ⇒ (**VIS0001:**) ⇒ **enter** (to accept the default) ⇒ **Hidden Line** ⇒ (pick the Swivel) ⇒ **No Hidden** ⇒ (pick the right Ball) ⇒ **Wireframe** ⇒ (pick the Foot) **Done Sel** ⇒ **Done** ⇒ **Set Current** ⇒ **✓VIS0001** ⇒ **Done** (Fig. 15.17)

HINT
Check **✓Shading** in the ENVIRONMENT menu and then **Repaint** to see the likeness of Figure 15.17

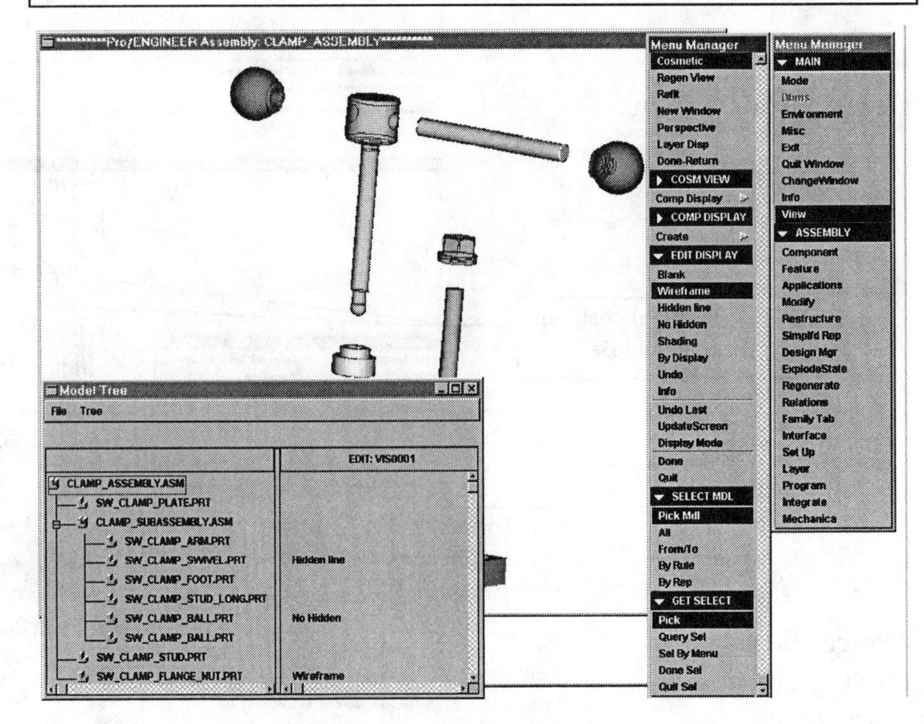

Figure 15.16
Changing the Component Display State

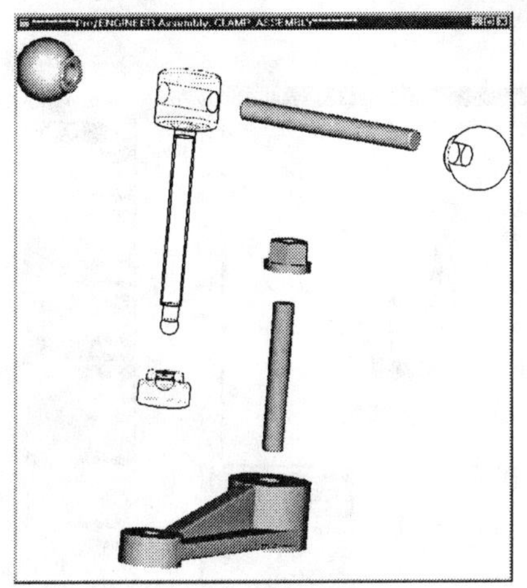

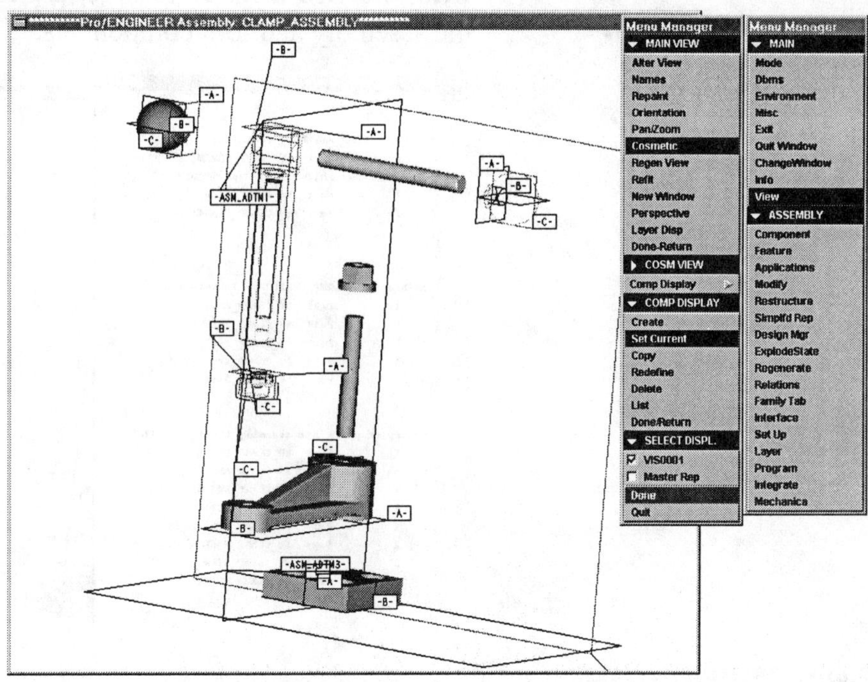

Figure 15.17
Component Display State Is Different for Individual Components

You can see the new component display states (Fig. 15.18) in the Model Tree by doing the following:

(from the Model Tree) ⇒ **Tree** ⇒ **Columns** ⇒ **Add/Remove** ⇒ ▼ ⇒ **Info** ⇒ << << << << << (**add/column** five times) ⇒ ▼ ⇒ **Visual Modes** ⇒ << << (**add/column** two times) ⇒ **Apply** ⇒ **OK** ⇒ (you will need to adjust your column format to see all of the information on the screen)

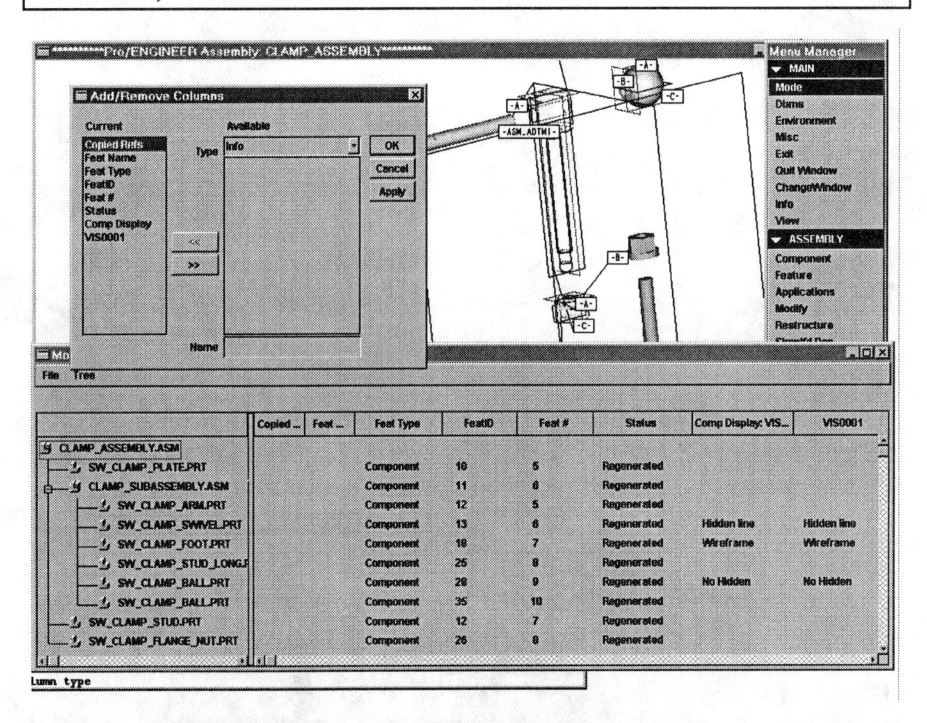

Figure 15.18
Components Display

There are thousands of views, exploded view capabilities, and display states. Experiment with Pro/E commands and enjoy yourself.

A bill of materials can be seen for the assembly by picking **Info ⇒ BOM** (Fig. 15.19). Figure 15.20 shows the subassembly and the assembly and four components.

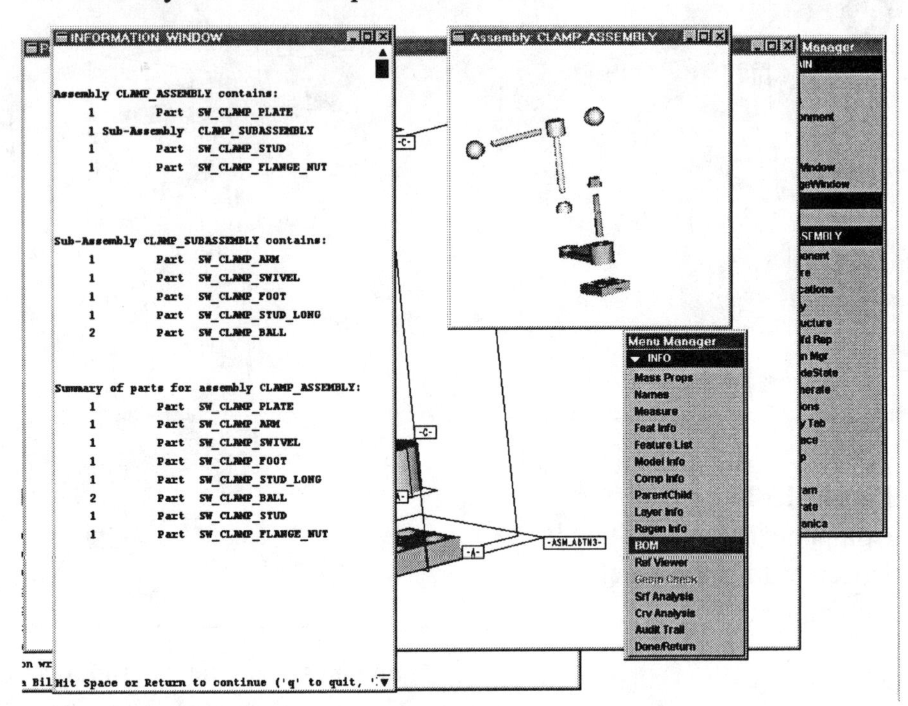

Figure 15.19
BOM

Figure 15.20
Assembly, Subassembly, Foot, Arm, Plate, and Ball

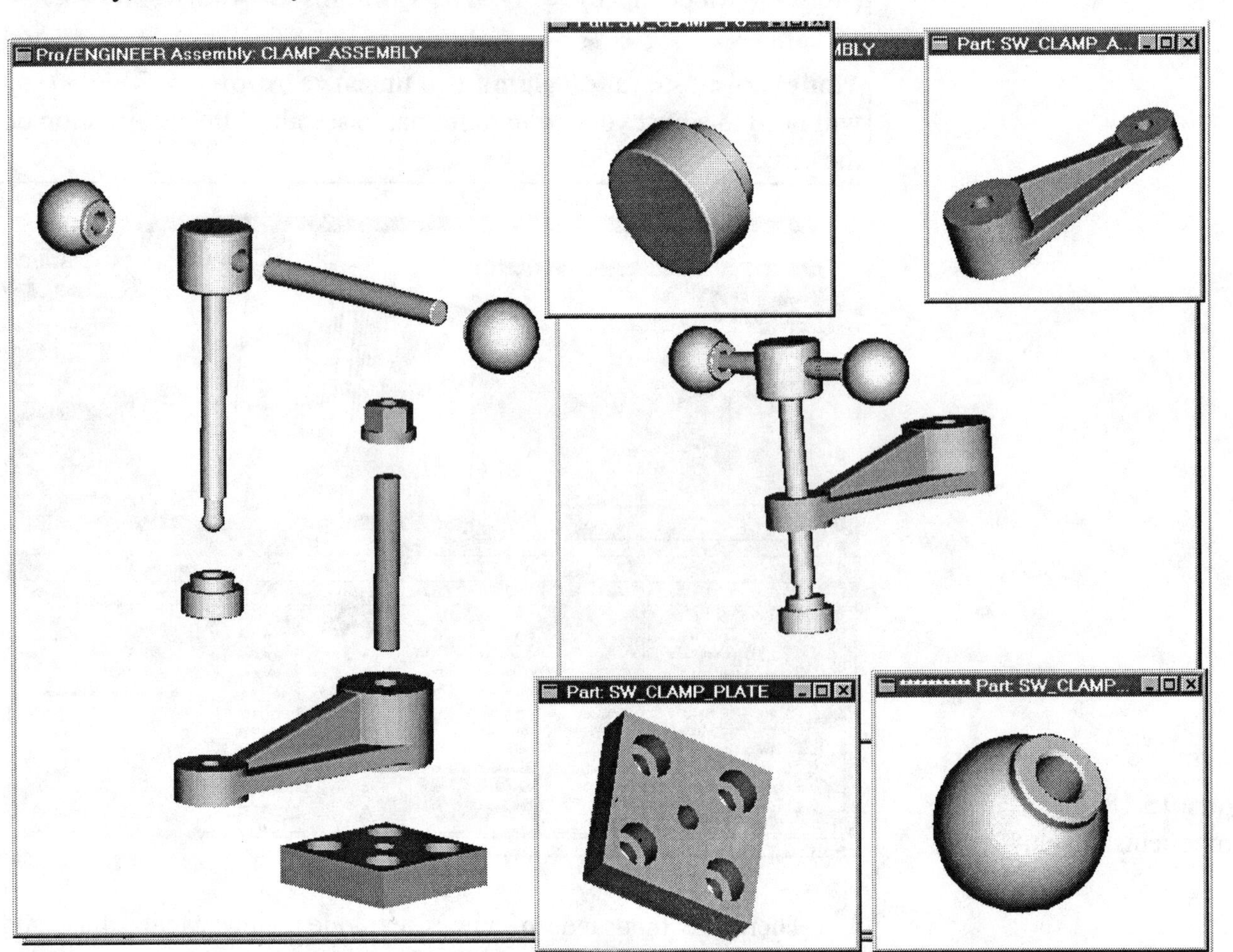

Lesson 15 Project

Exploded Coupling Assembly

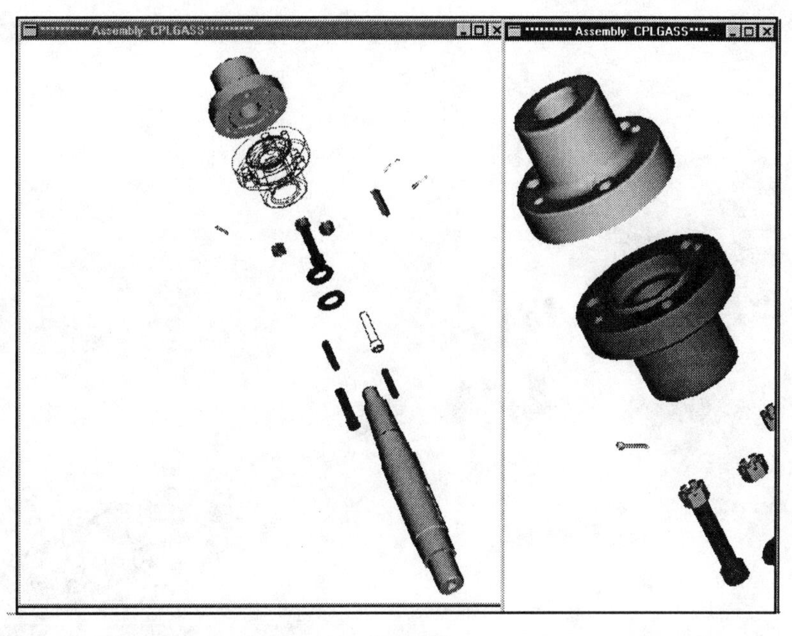

Figure 15.21
Exploded Coupling Assembly

Exploded Coupling Assembly

The fifteenth **lesson project** uses the assembly created in Lesson 14 Project. An exploded view needs to be created and saved for use later in the Draw mode for Lesson 20 Project. A variety of other views are suggested, including a section of the assembly, a perspective view, and an exploded view with a different component display variation. Each component should have its own color. If you did not color the components during the part creation, then bring up each part in the Part mode, define, and set the component with a color. Save three or four unique exploded states (view positions) and component (display visibility) states. You do not need to match the examples (Figure 15.21 through Figure 15.24) provided in this lesson project.

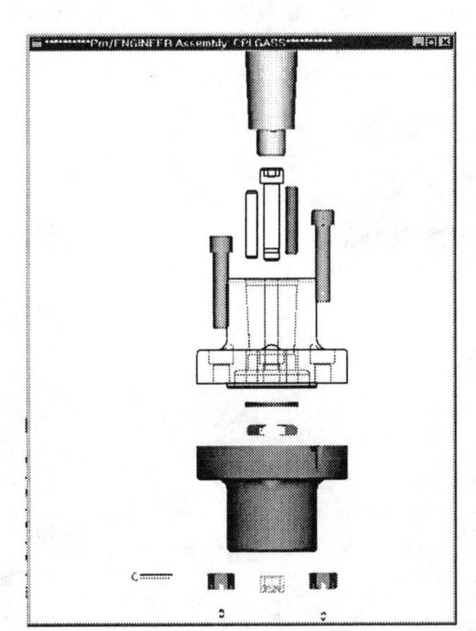

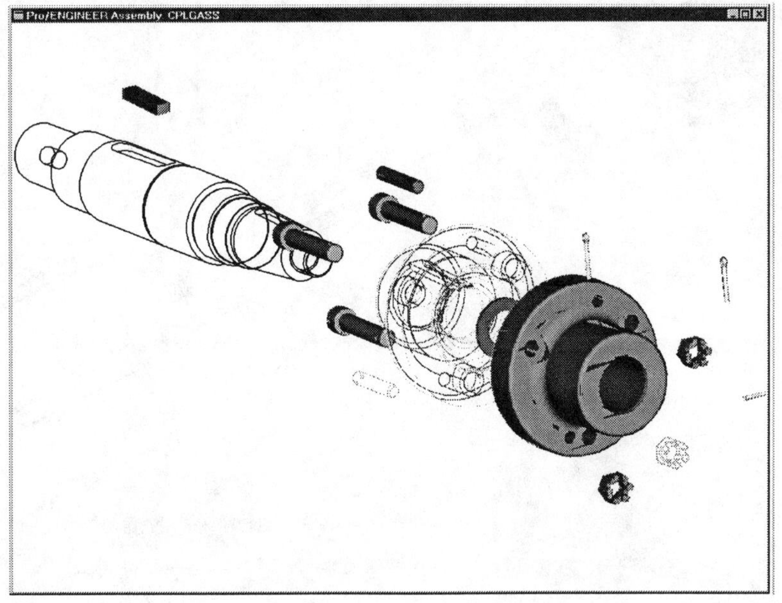

Figure 15.22
Perspective View of Exploded Coupling Assembly with a Variety of Component Display States

Figure 15.23
Shaded Exploded Coupling Assembly

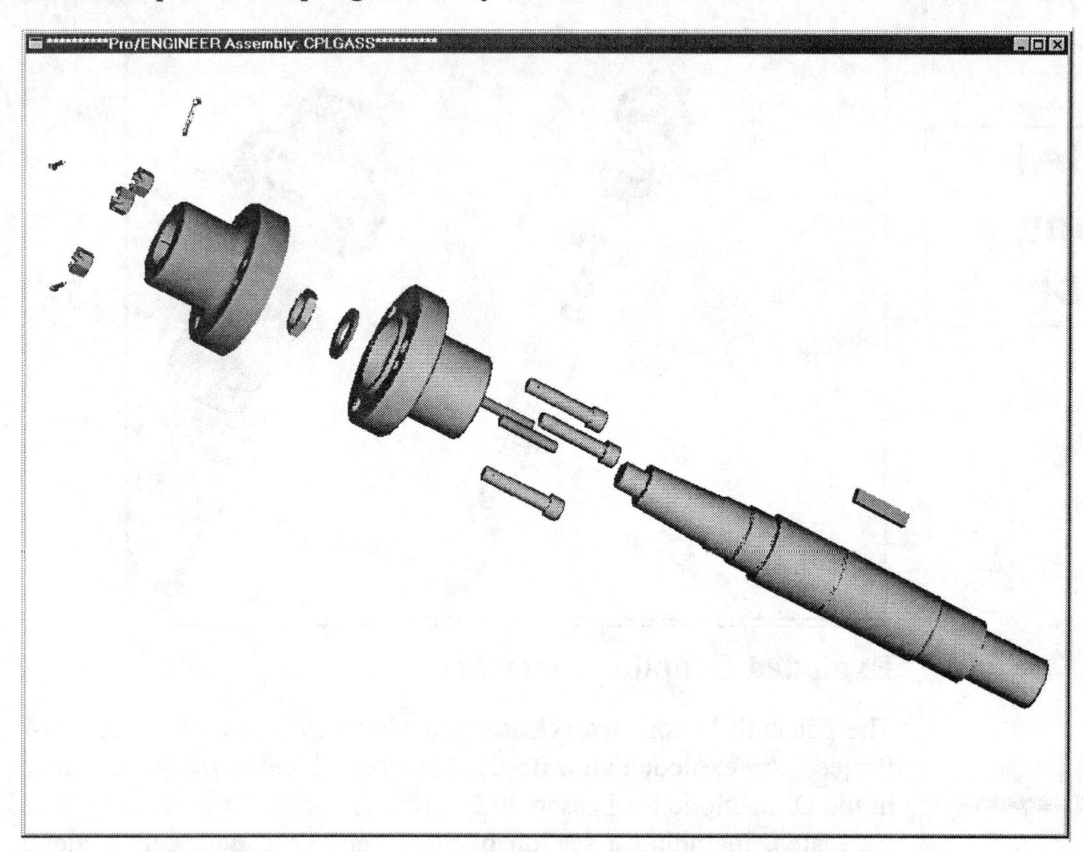

Figure 15.24
Exploded Coupling Assembly with Different Component Display States

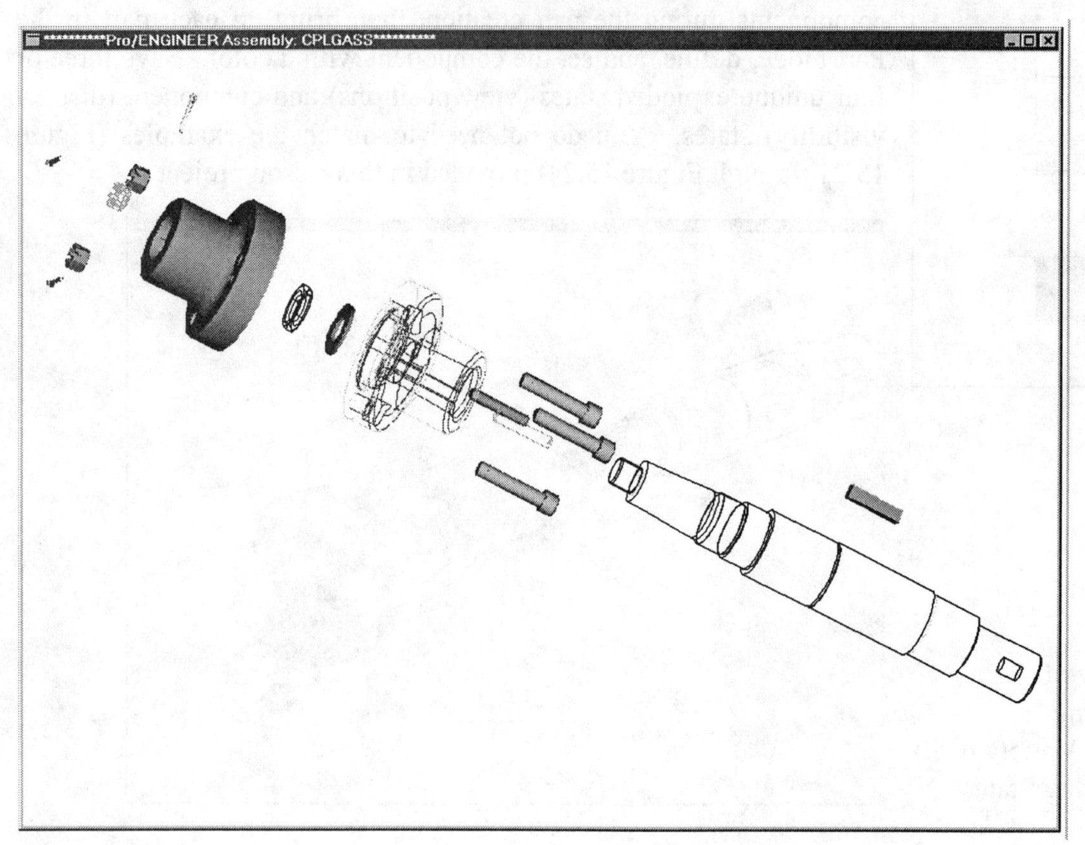

Part Three

Generating Drawings

Lesson 16 Formats, Title Blocks, and Views,
Lesson 17 Detailing
Lesson 18 Sections and Auxiliary Views
Lesson 19 Assembly Drawings and BOM
Lesson 20 Exploded Assembly Drawings

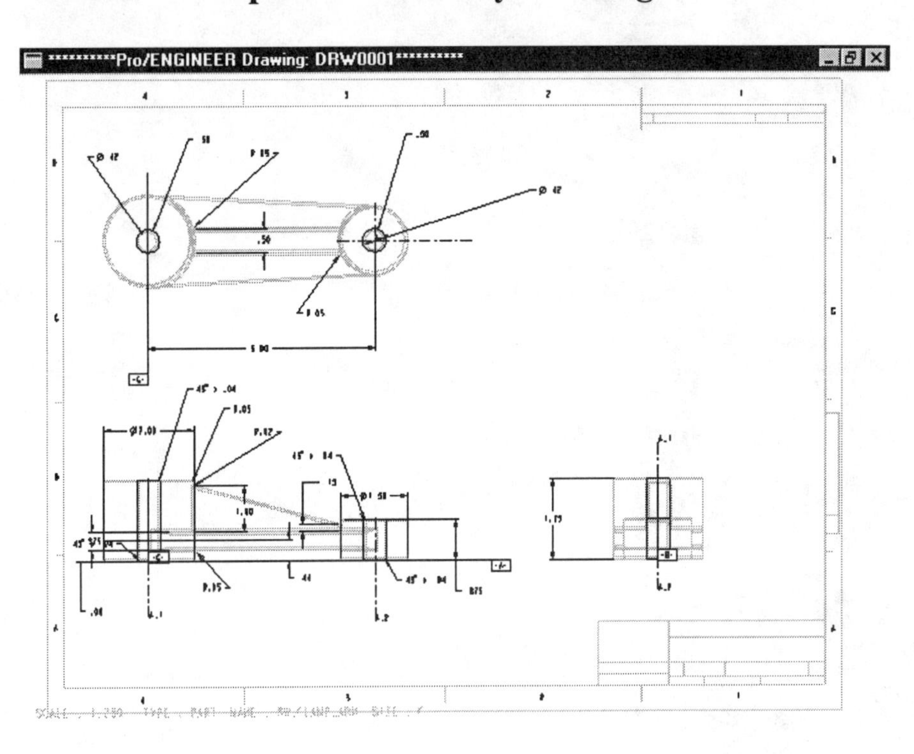

Part Detail Drawing

GENERATING DRAWINGS

The drawing functionality in Pro/E is used to create annotated drawings of parts and assemblies. During drawing creation a variety of options are available, including:

* Add views of the part or assembly to the drawing
* Display existing design dimensions
* Create additional driven or reference dimensions
* Create and insert notes on the drawing
* Add views of additional parts or assemblies
* Add multiple sheets to the drawing
* Create a BOM and balloon the assembly
* Add draft entities to the drawing

Construction of Drawings

Drawings can be created of parts and assemblies. Drawings can be multiview or pictorial, and include sections, auxiliary, detailed, exploded, and broken views. With Pro/DETAIL, ANSI, ISO, DIN, and JIS standard drawings can be created.

Lesson 16

Formats, Title Blocks, and Views

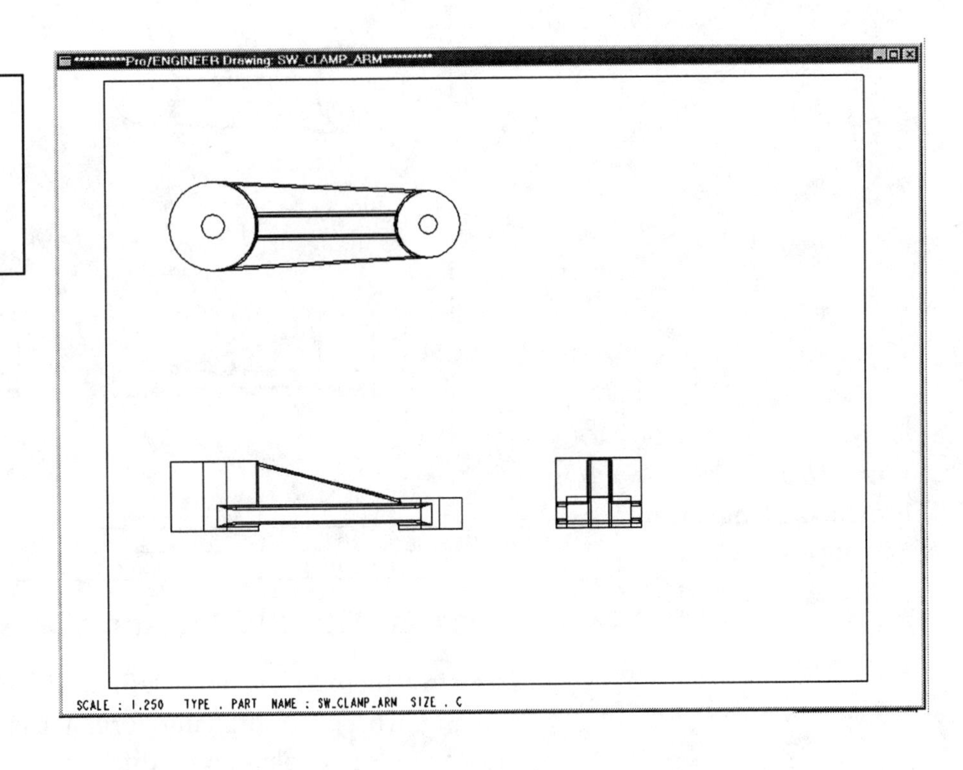

Figure 16.1
Clamp Arm Drawing
Without Drawing Format

EGD REFERENCE
Engineering Graphics and Design
by L. Lamit and K. Kitto
Read Chapter: 10 and 23
See Pages: 314-316

OBJECTIVES

1. **Create drawings with views**

2. **Create and save title blocks and sheet formats**

3. **Change the scale of a view**

4. **Display appropiate views for detailing a project**

5. **Move, erase, and delete views**

6. **Specify a standard-format paper size and units**

7. **Retrieve formats from the format library**

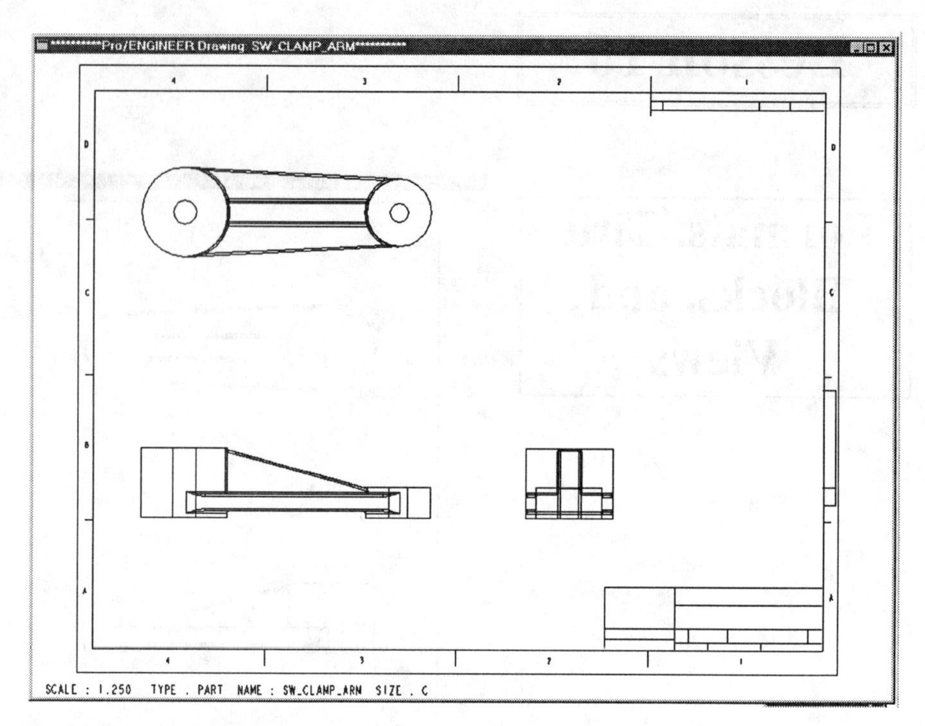

Figure 16.2
Formatted Clamp Arm Drawing

Formats, Title Blocks, And Views

Drawing formats are user-defined drawing sheet layouts (Fig. 16.1 and Fig. 16.2). A drawing format can be used in any number of drawings. It may also be modified or replaced in a Pro/E drawing at any time. **Title Blocks** are standard or sketched line entities that can contain parameters so that the part name, tolerances, scale, etc. will show automatically when the drawing format is retrieved (Fig. 16.3). **Views** created by Pro/E are exactly like views constructed manually by a designer on paper. The same rules of projection are applied; the only difference is that you merely command Pro/E to create the views as needed.

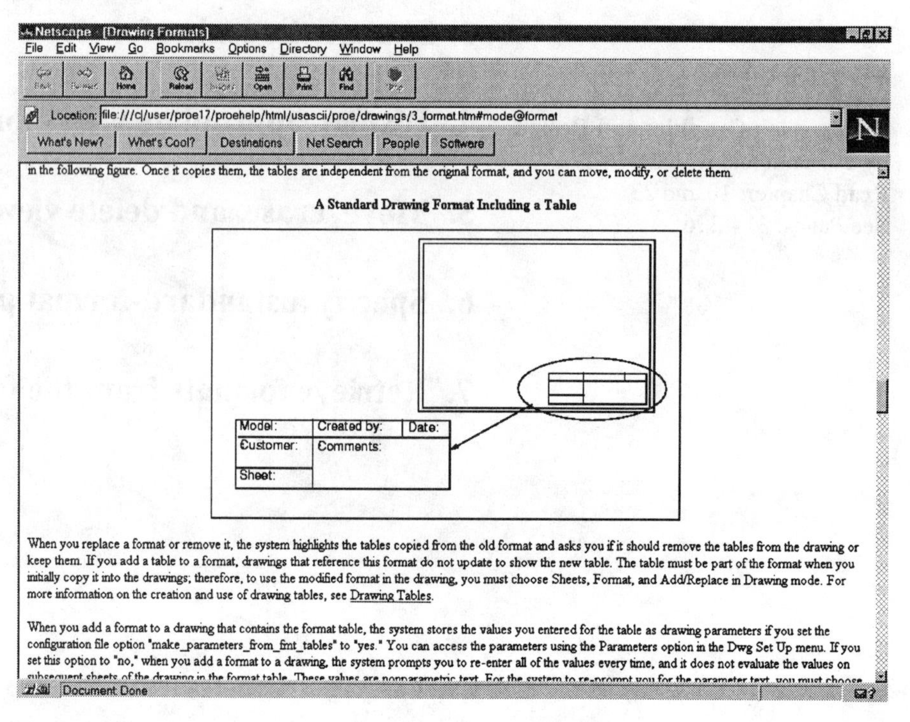

Figure 16.3
Online Documentation Formats

Drawing Formats

There are two types of drawing formats: standard (Fig. 16.2) and sketched. Drawing formats consist of draft entities, not model entities. You can select the desired format size from a list of standard sizes (**A-F** and **A0-A4**), or create a new size by entering values for length and width.

Sketched formats created in Sketcher mode (Fig. 16.4) may be parametrically modified, enabling you to create nonstandard-size formats or families of formats.

Formats can be altered to include: note text, symbols, tables, and drafting geometry, including drafting cross-sections and filled areas.

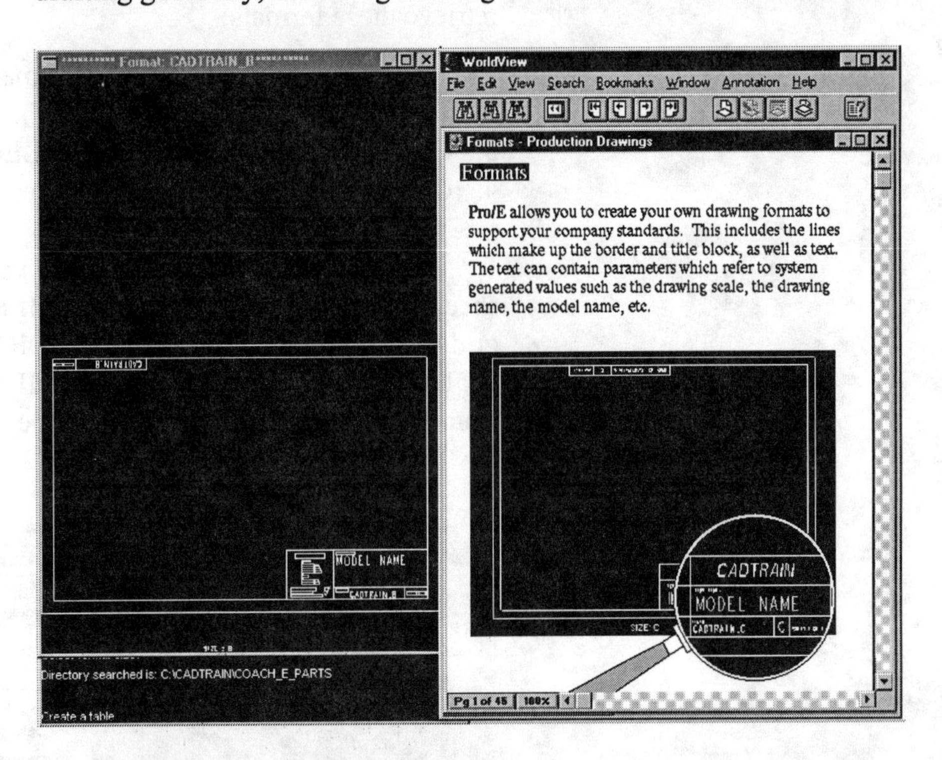

Figure 16.4
COAch for Pro/E, Formats--Production Drawings

With Pro/E (Pro/DETAIL) you are able to do the following in the **Format** mode:

* Create draft geometry, notes.
* Move, mirror, copy, group, translate, and intersect geometry.
* Use and modify the draft grid.
* Enter user attributes (Fig. 16.5).
* Create drawing tables.
* Use interface tools to create plot, DXF, SET, and IGES files.
* Import IGES, DXF, and SET files into the format.
* Create user-defined line styles.
* Create, use, and modify symbols.
* Include drafting cross-sections in a format.

Regardless of whether you use a standard format or a sketched format, the format is added to a drawing created for a set of specified views of a parametric 3D model.

To create a format:

1. Choose **Format** from the MODE menu. The ENTERFORMAT menu is displayed.
2. Choose **Create** from the ENTER menu, and then enter a format name.

Retrieving Pro/E Formats from the Format Library

You can retrieve formats from a format library directory within Pro/E. To enter the format directory in the loadpoint, use the Format Dir option in the SELECT FILE menu. This option is not available unless you have set the "pro_format_dir" configuration file option. To retrieve these formats:

1. Choose **Format** from the MODE menu.
2. Choose either **Search/Retr** from the ENTER menu.
3. Choose **Format Dir** from the SELECT FILE menu.
4. Choose a format from the displayed list.

After specifying a format size, you enter Format mode; and a sheet outline appears on the screen. If necessary, turn on the drawing grid by choosing **Detail** from the FORMAT menu, **Modify** from the DETAIL menu, **Grid** from the MODIFY DRAW menu, and **Grid On** from the GRID MODIFY menu. The grid size depends on the units for the format.

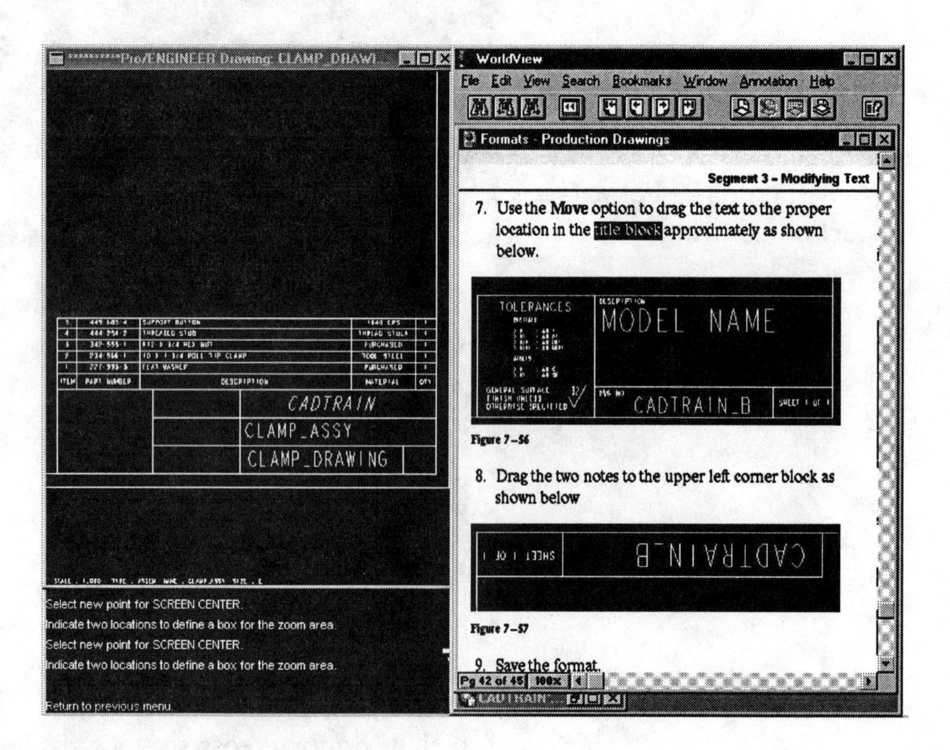

Figure 16.5
COAch for Pro/E, Formats--Production Drawings (Segment 3: Modifying Text)

Specifying the Format Size and Units

You establish the format size by selecting one from a list of standard format sizes or by creating your own size by entering values for the length and width of the format. Once you set the size, you must choose the units of the drawing format. The main grid spacing and format text units depend on the units you select for the format. To select the format units, choose **Inch** or **Millimeter** from the FORMAT UNITS menu. The text height units are in the units of the format: inches or millimeters. The major grid spacing is an inch if the units are inches, and a centimeter if the units are millimeters.

To Specify a Standard-Format Paper Size and Units:

1. Choose an option from the DWG SIZE TYPE menu to determine the orientation of the format sheet:

   Portrait Uses the larger of the dimensions of the sheet size for the format's height; uses the smaller for the format's width.
   Landscape Uses the larger of the dimensions of the sheet size for the format's width; uses the smaller for its height.
   Variable Enters specific values for the height and width of the format.

2. If you select **Landscape** or **Portrait**, choose the desired standard size from the DWG SIZE menu.

A0	841 X 1189	mm	A	8.5 X 11	in
A1	594 X 841	mm	B	11 X 17	in
A2	420 X 594	mm	C	17 X 22	in
A3	297 X 420	mm	D	22 X 34	in
A4	210 X 297	mm	E	34 X 44	in
			F	28 X 40	in

3. Choose the desired units (inches or millimeters) from the FORMAT UNITS menu. The units that you choose supersede those of the standard-format size that you chose.

Sketching the Format

To create format geometry, you use draft geometry. To sketch draft geometry, choose **Detail** from the FORMAT menu, **Sketch** from the DETAIL menu, and any option from the DRAFT GEOM menu.
The sheet outline is the border of the standard drawing format you selected. Because it is the actual border, it may not show up on pen plots unless you use a paper size larger than the drawing size. Everything within the sheet outline border also plots, but you should make an allowance for the plotter hold-down rollers.

When you add a sketched format to a drawing, Pro/E aligns the lower left corner (the origin) of the format to the lower left corner (the origin) of the drawing, and then centers all items in the drawing on the new sheet in the locations that correspond to their positions on the original sheet. If necessary, it adjusts the drawing scale, maintaining relative distance between items.

To create a sketched format (Fig. 16.6):

1. Choose Sketcher from the MODE menu.
2. Choose Create, and then enter a new format name.
3. Sketch the boundary and dimension it.
4. Sketch the title block and dimension it appropriately.
5. Once Pro/E successfully regenerates the drawing format section, save it using Dbms menu options. You cannot save text with a sketched format.

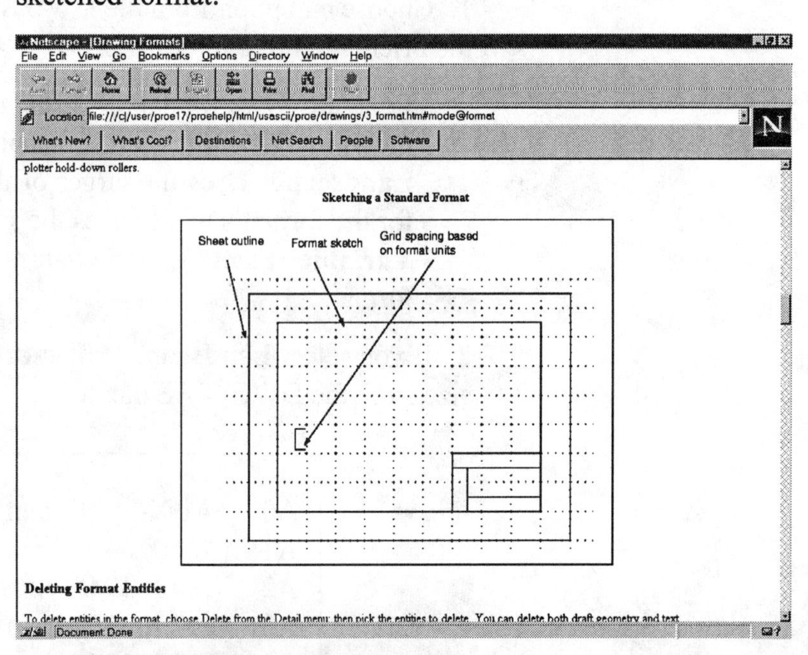

Figure 16.6
Online Documentation
Sketching Formats

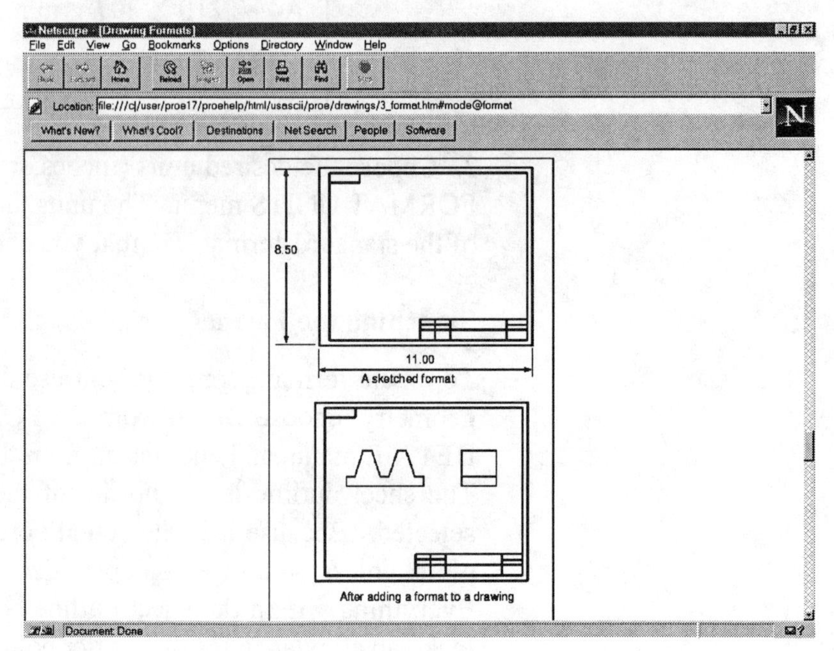

Figure 16.7
Online Documentation
Adding Views to a Sketched Format

Views

A wide variety of views (see Fig. 16.7) can be automatically derived from the parametric model. One of the most common is projection views. Pro/E automatically creates projection views by looking to the left, right, above, and below the picked view location to determine the orientation of a projection view (Fig. 16.8). When conflicting view orientations are found, you are prompted to select the view to be the parent view. A view will then be constructed from the selected view.

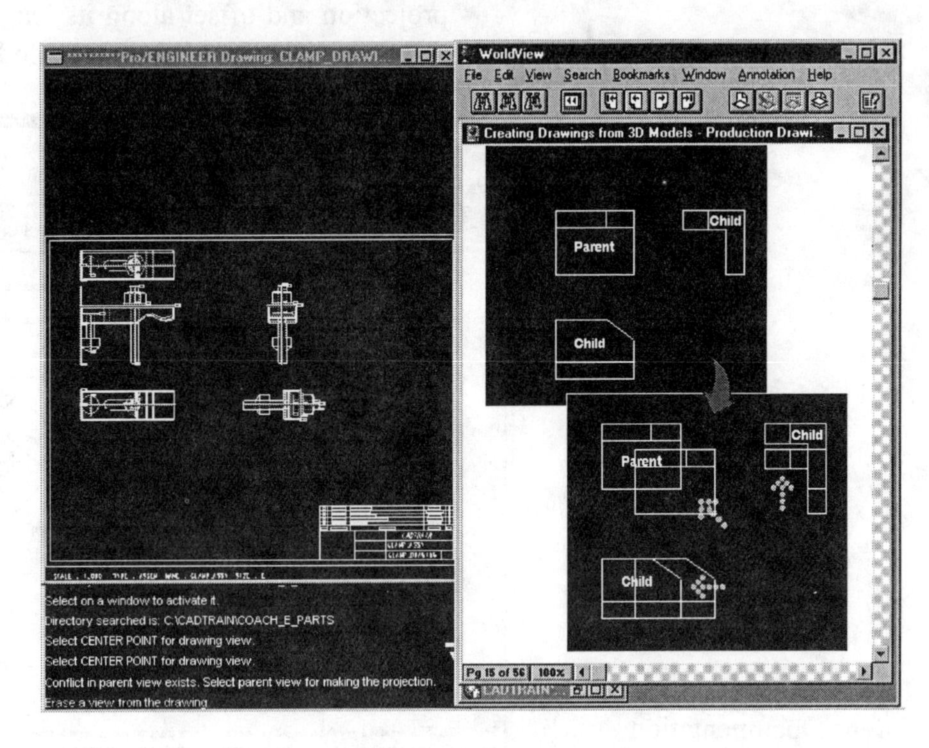

Figure 16.8
COAch for Pro/E, Creating Drawings From 3D Models--Production Drawings (Segment 3: Orthographic Views)

At the time that they are created, projection, auxiliary, detailed, and revolved views have the same representation and explosion offsets, if any, as their parent views. From that time onward, each view can be simplified and restored and have its explosion distance modified without affecting the parent view. The only exception to this are detailed views, which will always be displayed with the same explosion distances and geometry as their parent views. The following view types are available:

Projection Creates a view that is developed from another view by projecting the geometry along a horizontal or vertical direction of viewing (orthographic projection). The projection type is specified by you in the drawing setup file, and may be based on third-angle (default) or first-angle rules.

Auxiliary Creates a view that is developed from another view by projecting the geometry at right angles to a selected surface or along an axis. The surface selected from the parent view must be perpendicular to the plane of the screen (Fig. 16.9).

General Creates a view with no particular orientation or relationship to other views in the drawing. The model must first be oriented to the desired view orientation established by you.
Detailed Details a portion of the model appearing in another view. Its orientation is the same as the view it is created from, but its scale may be different so that the portion of the model being detailed can be better visualized.
Revolved Creates a planar area cross-section from an existing view; the section is revolved **90°** around the cutting plane projection and offset along its length. A revolved view may be full or partial, exploded or unexploded.

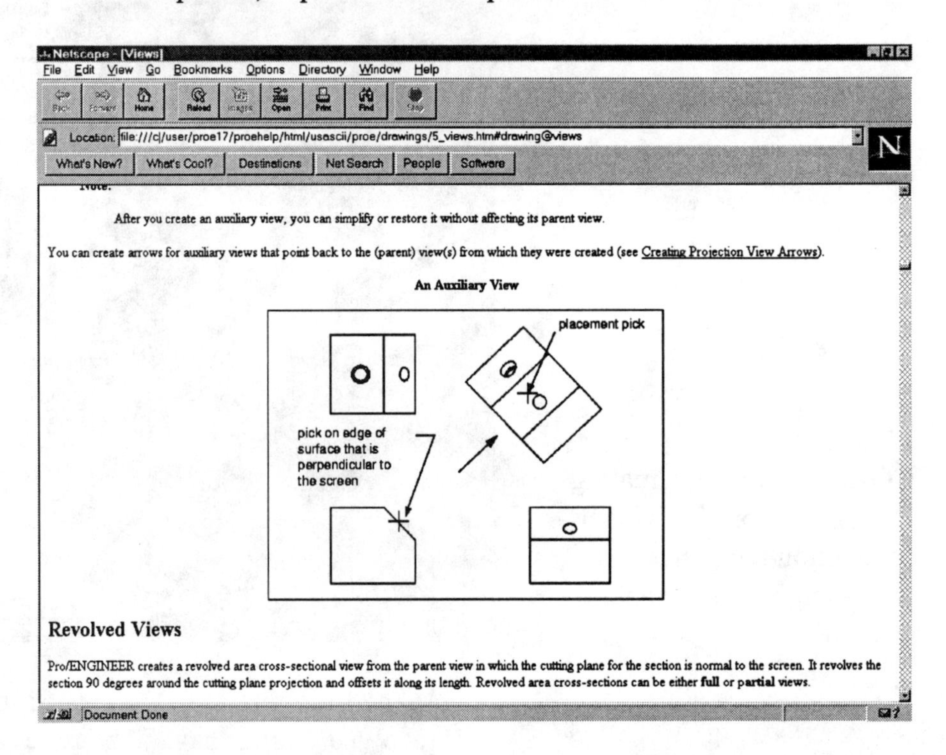

Figure 16.9
Online Documentation Auxiliary Views

The view options that affect how much of the model is visible in the view are:

Full View Shows the model in its entirety.
Half View Removes a portion of the model from the view on one side of a cutting plane.
Broken View Removes a portion of the model from between two selected points, and closes the remaining two portions together within a specified distance.
Partial View Displays a portion of the model in a view within a closed boundary. The geometry appearing within the boundary is displayed; the geometry outside of it is removed.

The options that determine if the view is of a single surface or has a cross-section are:

Section Displays an existing cross-section of the view if the view orientation is such that the cross-sectional plane is parallel to the screen (Fig. 16.10 and Fig. 16.11).
No Xsec Indicates that no cross-section is to be displayed.
Of Surface Displays a selected surface of a model in the view. The single surface view can be of any view type except detailed.

The options that determine if the view is scaled:

Scale Allows you to create view with an individual scale shown under the view. When a view is being created, Pro/E will prompt you for the scale value. This value can be modified later. General and detailed views can be scaled.
No Scale A view will be scaled automatically using a pre-defined scale value that will be in the lower left corner of the screen as "SCALE."
Perspective Creates a perspective general view.

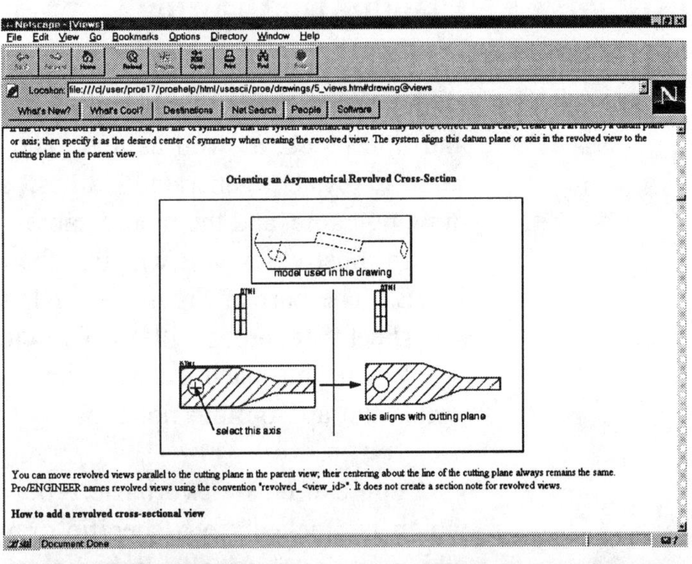

Figure 16.10
Online Documentation Section Views

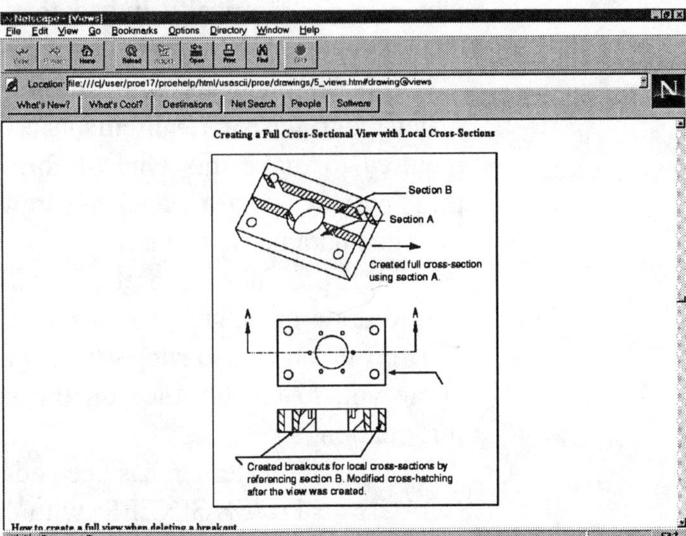

Figure 16.11
Online Documentation Sections

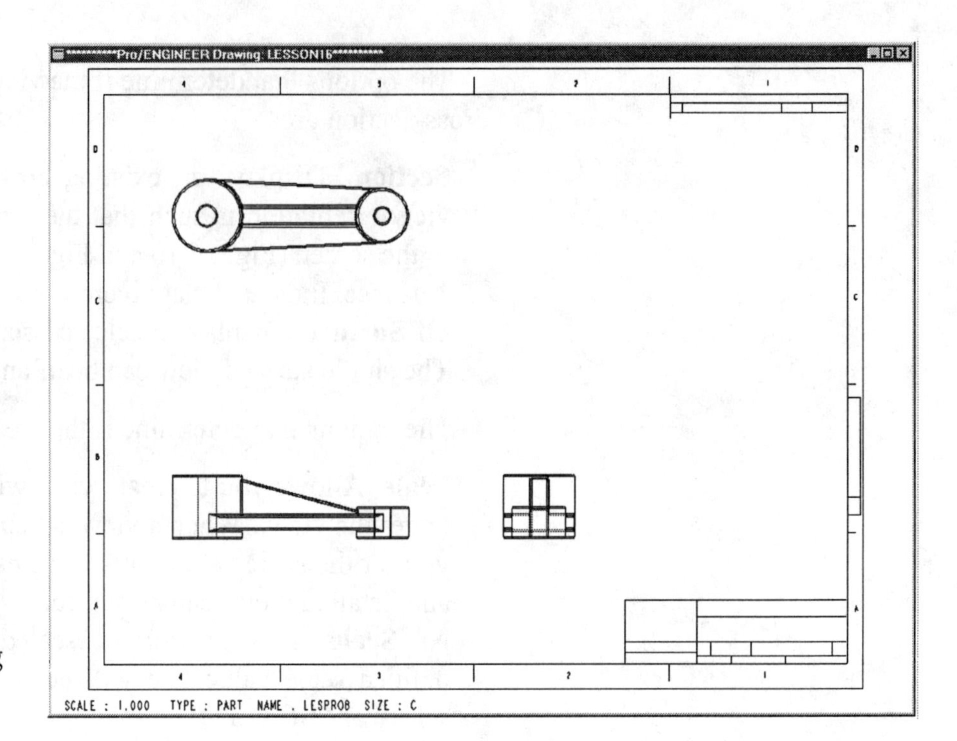

Figure 16.12
Swing Clamp Arm Drawing on a ANSI Standard Drawing Format

Clamp Arm Drawing

Pro/E allows you to create your own drawing formats to support your company standards. This includes the lines that make up the border and title block, as well as text. The text can contain parameters that refer to system-generated values, such as the drawing scale, the drawing name, and the model name.

In most cases you will use the standard formats that come with Pro/E. The part in Figure 16.12 has been placed on a standard "C" size sheet that comes with the Drawing mode.

When you place a format on your drawing, Pro/E automatically writes the appropriate notes based on information in the model you use.

You can create two basic types of formats: a "standard" format, which is "locked" to a specific drawing size, or a sketched format, which is parametrically linked to the size of the drawing and thus changes as the drawing size changes.

This lesson deals with altering an existing standard format to match your format requirements. Once you master the techniques required to make this type of format, you can easily extend these principles to using a sketch to draw the border outlines (to make a sketched parametric format).

Pro/E provides a subset of the 2D Drafting functionality to allow you to draw lines, arcs, splines, etc. to define a drawing border. The **Format** function also supports the complete range of text functions to allow you to create notes on the drawing that serve as title block information.

Much of this lesson has been adapted from CADTRAIN's COACH for Pro/ENGINEER, with their permission.

One of the easiest ways of creating a drawing format is to use an existing format from the Pro/E format library, save it in your directory under a different name, and alter it as required. This method will keep the sheet an ANSI standard size and format with a standard title block. You can add parameters to the sheet so as to display the appropiate information in the title block and at other locations on the sheet where appropiate. Formats have an **.frm** extension and are read-only files.

Retrieve and save for later modification a **"C"** size format using the following commands (Fig. 16.13):

Mode ⇒ **Format** ⇒ **Search/Retr** ⇒ **Format Dir** ⇒ **c.frm** ⇒ **Dbms** ⇒ **Save As** ⇒ **enter** ⇒ (type a unique name for your format: **FORMAT_C**) ⇒ **enter** ⇒ **Quit Window** ⇒ **Format** ⇒ **Search/Retr** ⇒ (pick the new format: **format_c.frm**)

You can make any number of personal formats using this method. Adding parameters to the sheet is the next step in creating your own library of formats that will automatically display the part's name, scale, tolerances, etc.

Figure 16.13
FORMAT_C

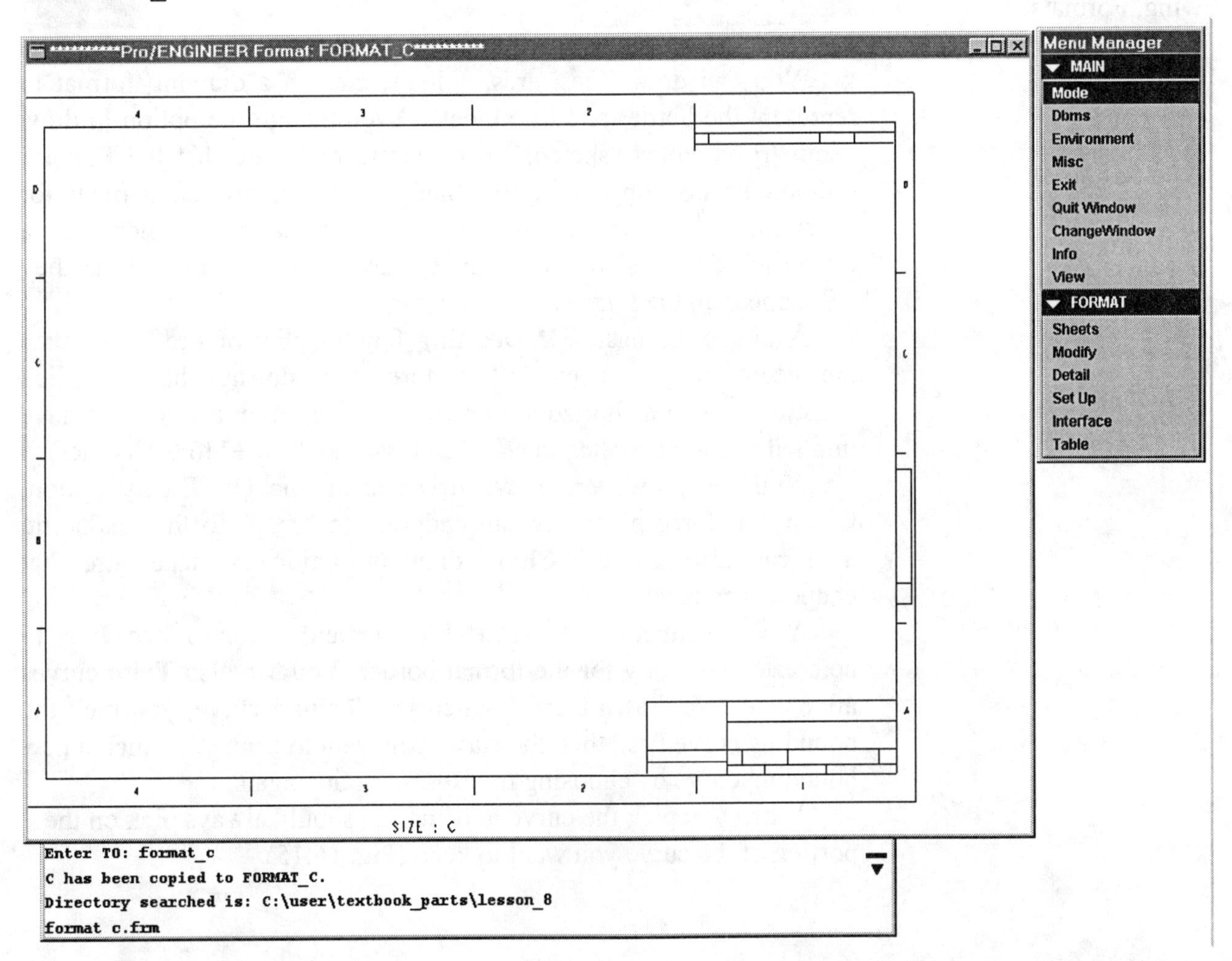

Creating the Format Border

This part of the lesson leads you through creating a "**C**" size, inches format (Fig. 16.14). You can substitute metric values to create the equivalent-size metric format if desired.

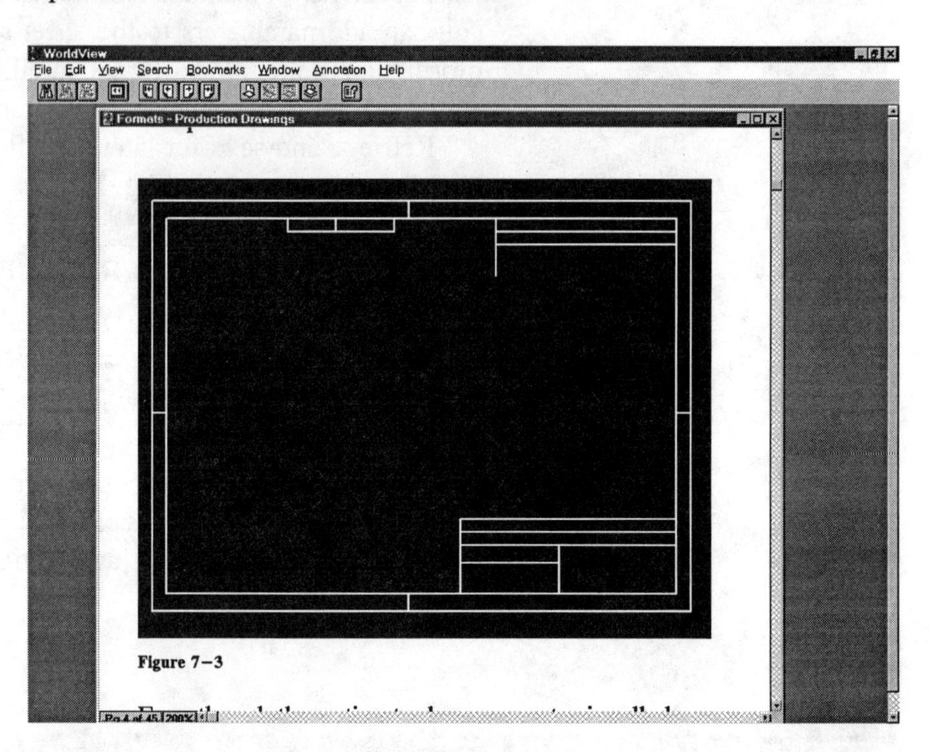

Figure 16.14
CADTRAIN--Production Drawing, Formats

You can draw lines, arcs, splines, etc. on a drawing format to represent the border and title block. Even though the option to draw geometry is called "sketch," you need to be aware that the **Format** option does not support the true Sketcher. No allowance is made for constraining the curves (entities) and regenerating them to a dimension-driven size. You **must** draw the curves exactly as they will appear on the format.

You use the basic **2D Drafting** functionality of Pro/E to create the geometry you need. Therefore, you do not have implied constraints such as horizontal or vertical lines. You also do not have implied endpoint connection (if the curves are "close" to each other).

You can, however, draw curves using the **On Entity** option, which will force picks near an endpoint to "snap" to that endpoint. You can also use the **Chain** drawing option to make sure that endpoints connect.

You use either the **Horizontal** or **Vertical** option to ***force*** lines to be created correctly for the format border. You can also **Trim** curves after you have drawn them. When you **Trim** a curve, you pick the bounding curve first, then the curve you want to trim. You pick a new bounding curve by choosing the **Bound** option again.

When you pick the curve to trim, you should always pick on the portion of the curve you want to keep (Fig. 16.15).

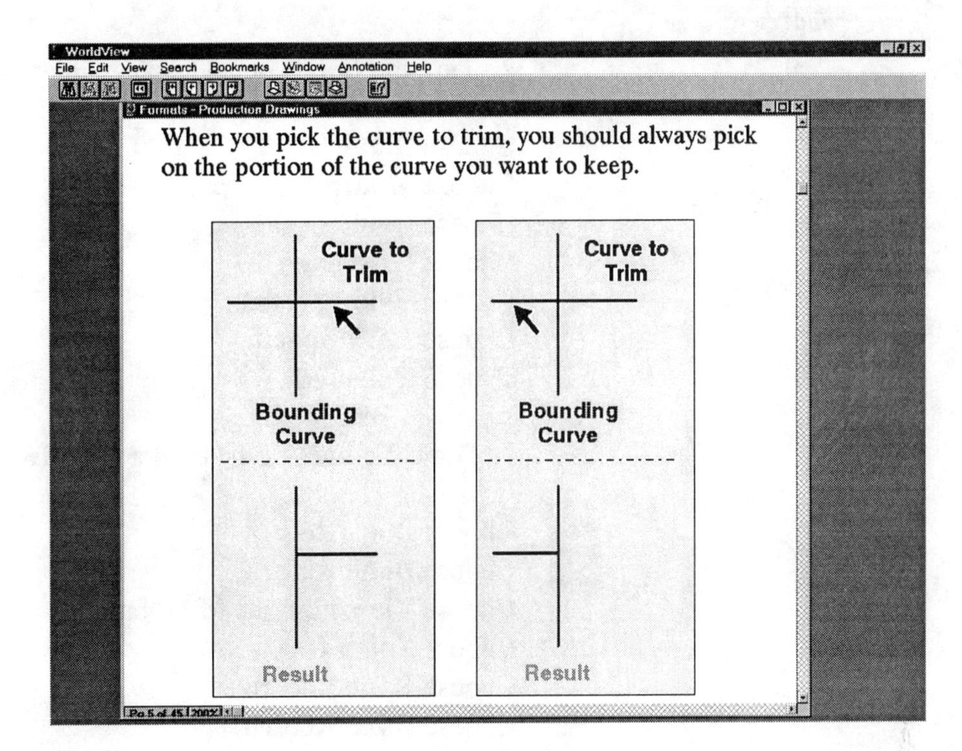

Figure 16.15
CADTRAIN--Production Drawing, Trimming

Draw the lines of a standard **C Size** format using the following commands (Fig. 16.16):

1. Create a new format file.

Choose **Format** from the **Mode** menu
Choose **Create**
Type **my_format_** and ***your initials*** and **enter**.
Choose **C** from the list of sizes
Choose **Inch** as the format unit

HINT

The box that displays represents the edge of the paper. Many companies allow for about **.500** inches inside this box for the border of the format

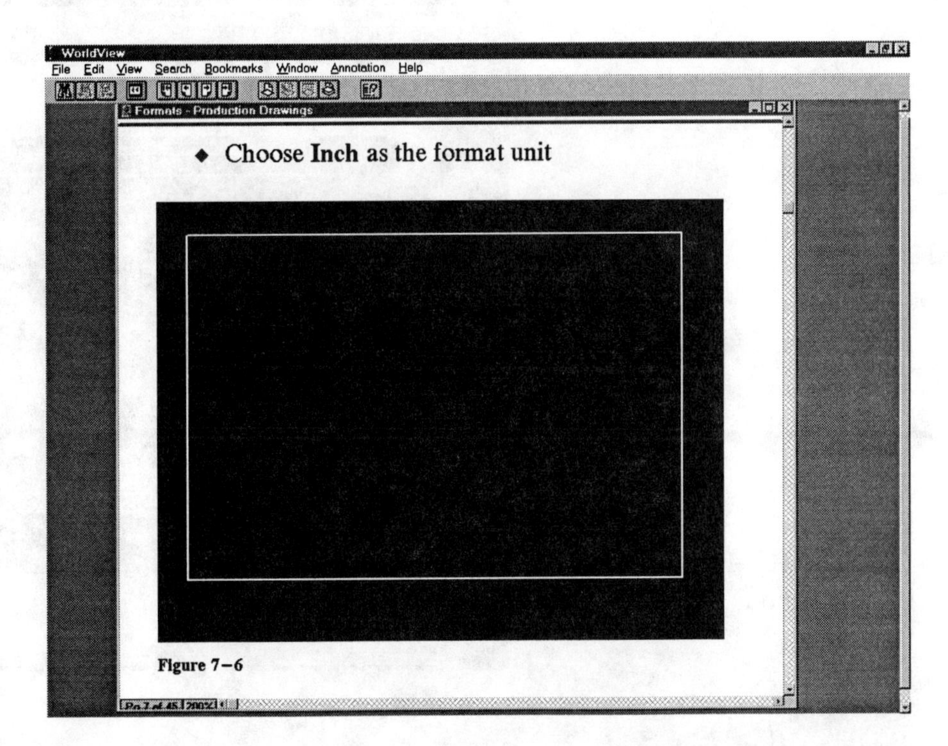

Figure 16.16
CADTRAIN--Production Drawing, Edge of the Paper

2. Turn the **GRID** on and change the spacing to **.50**.

Choose **Detail**
Choose **Modify**
Choose **Grid**
Choose **Grid On**
Choose **Grid Params**
Choose **X&Y Spacing**
Type **.5** and **enter**

3. Draw the lines of the border (Fig. 16.17)

Choose **Done/Return**
Choose **Done/Return**
Choose **View** from the **MAIN** menu
Choose **Zoom Out**
Choose **Done/Return**
Choose **Done/Return**
Choose **Tools** (under the DETAIL menu)
Choose **Offset**
Choose **Ent Chain**
Select the four lines on the screen
Choose **Done Sel**
Type **-.5** and **enter** (the boundary is zero with negative being inward)
Choose **View** from the **MAIN** menu
Choose **Reset**
Choose **Done/Return**

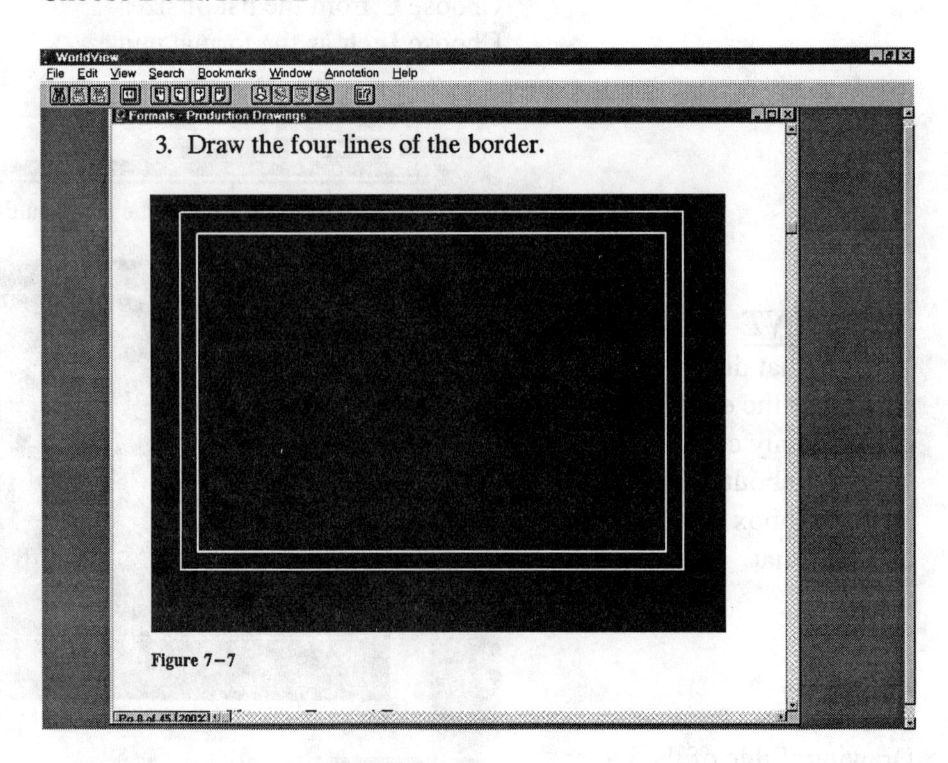

Figure 16.17
CADTRAIN--Production Drawing, Four Lines of the Border

4. Create the lines for the title block.

Select the bottom horizontal line using the **left** mouse button (Fig. 16.18)
Click the **middle** mouse button to finish selection
Type **-.5** and **enter**
Select the line you just created with the **left** mouse button
Click the **middle** mouse button
Type **-1.5** and **enter**
Select the second right vertical line as shown in Figure 16.18
Click the **middle** mouse button
Type **-1** and **enter**
Select the line you just created
Click the **middle** mouse button
Type **-3** and **enter**
Select the line you just created
Click the **middle** mouse button
Type **-2** and **enter**

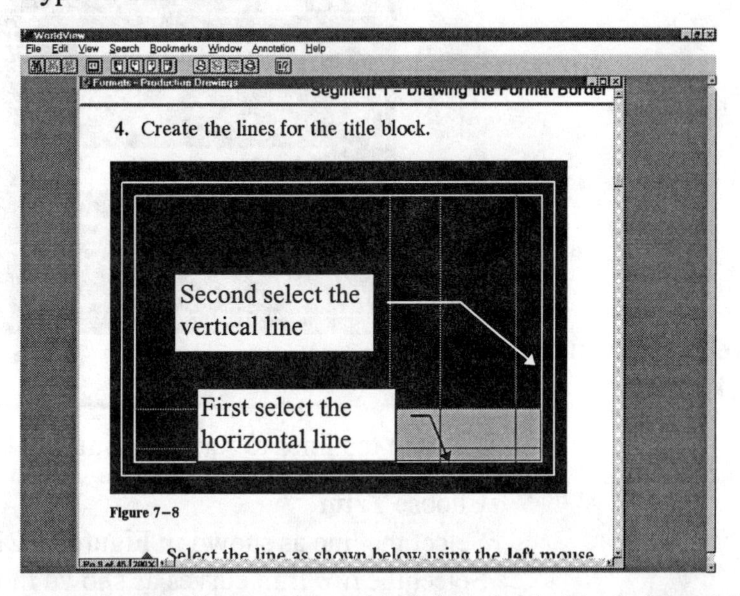

Figure 16.18
CADTRAIN--Production Drawing, Create Lines for the Title Block

5. **Trim** the lines to form the finished title block (Fig. 16.19).

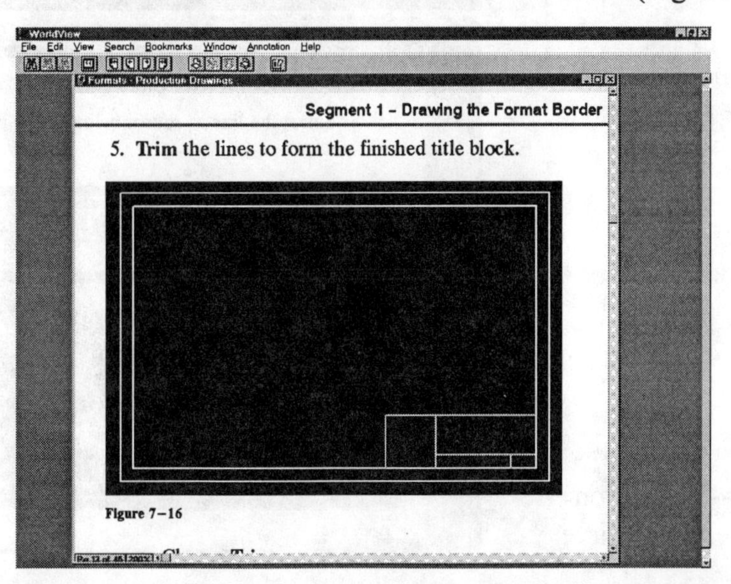

Figure 16.19
CADTRAIN--Production Drawing, Finished Title Block

6. Draw the lines for the drawing number that is inverted at the top left corner of the format.

Choose **Offset ⇒ Ent Chain**
Select the left vertical line (Fig. 16.20)
Click the **middle** mouse button
Type **-1** and **enter**
Select the line you just created
Click the **middle** mouse button
Type **-3** and **enter**
Select the line (Fig. 16.21)
Click the **middle** mouse button
Type **-.5** and **enter**

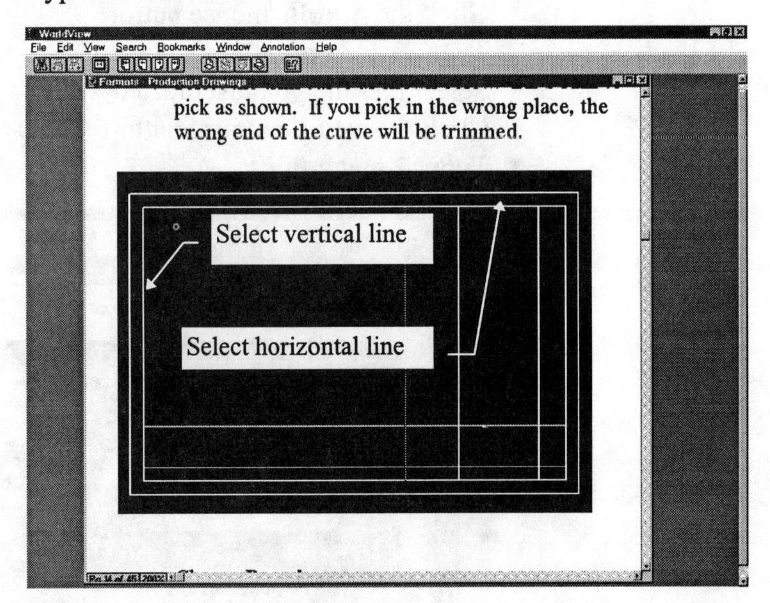

Figure 16.20
CADTRAIN--Production Drawing, Inverted Drawing Number-Block Line

7. Trim the lines to form the finished block.

Choose **Trim**
Select the line as shown in Figure 16.21 as the bounding curve
Select the *two* trim curves as shown in Figure 16.22

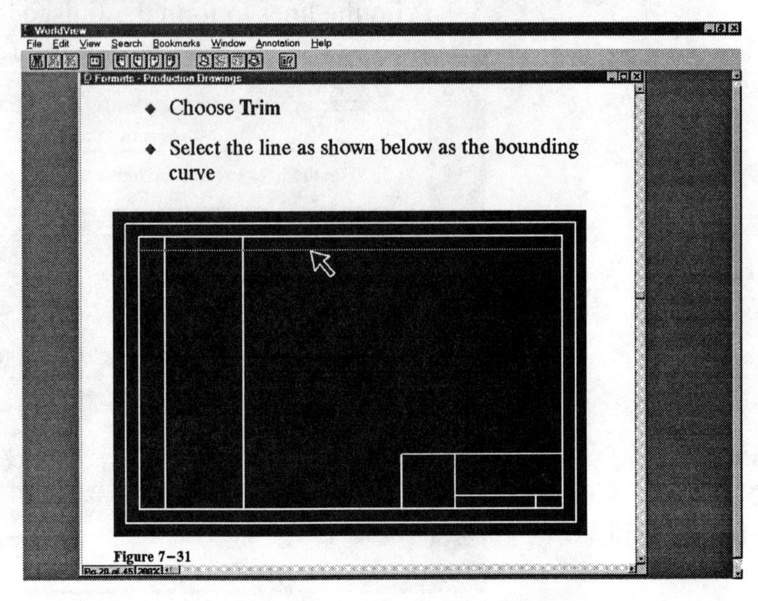

Figure 16.21
CADTRAIN--Production Drawing, Bounding Curve

Choose **Bound**
Select the right most vertical line (Fig. 16.22 and Fig. 16.23) as the bounding curve
Select the trim curve as shown in Figure 16.23
Choose **Done/Return**
Choose **Done/Return**
Choose View ⇒ Repaint ⇒ Done/Return

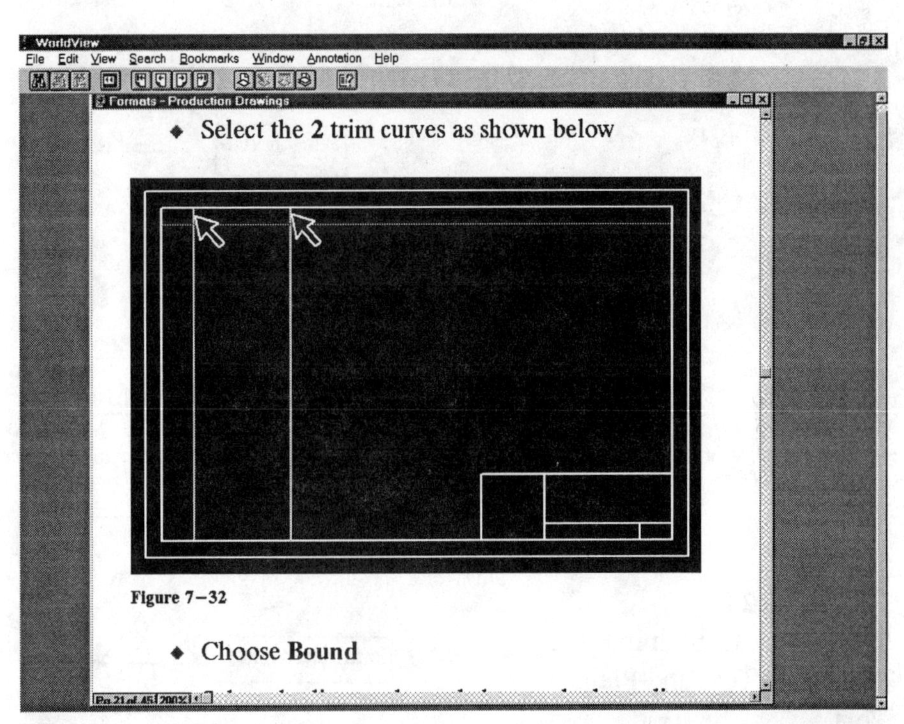

Figure 16.22
CADTRAIN--Production Drawing, Trimming the Block--Trim Curves

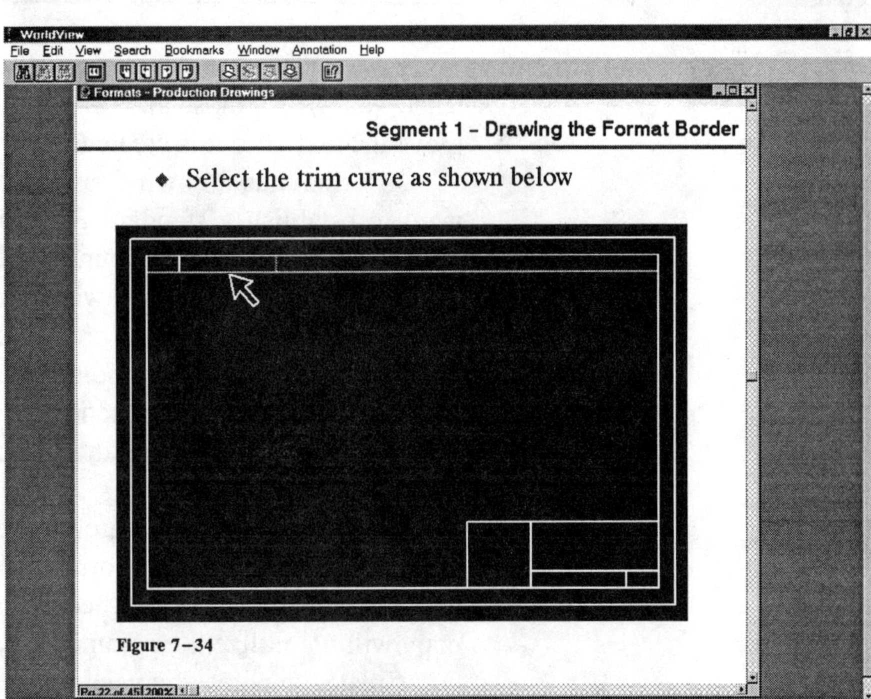

Figure 16.23
CADTRAIN--Production Drawing, Trimming the Block--Final Trim Curve

Next you will add *parameters* to the sheet.

Adding Text to the Title Block and Format

You can add either plain text or text that contains parameters to the format (Fig. 16.24). You should decide which text you require and determine the parameters you will want to call out on the drawing. You can also utilize user-defined parameters; however, these parameters *must* exist in the model you use to create the drawing. If they do not, the parameter title will remain written on the format as plain text.

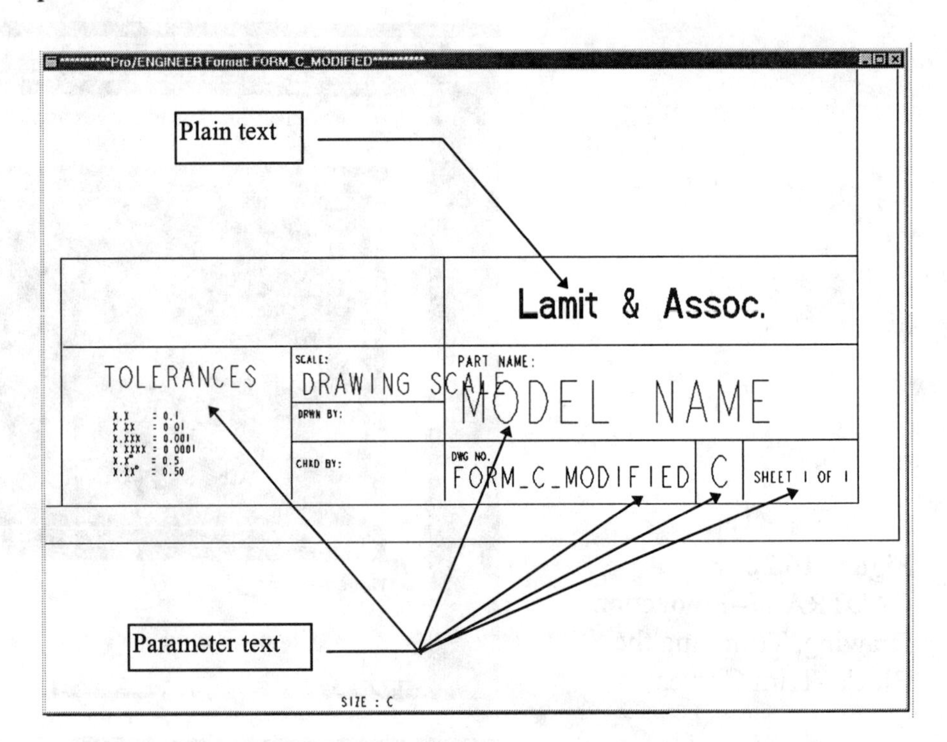

Figure 16.24
Title Block Containing Parameter Text and Plain Text

You can then create the notes almost anywhere on the drawing, since you will almost always want to alter the text parameters and move the text after it is created.

A format has its own Drawing parameter **Set-Up** file. You may want to establish a standard **Set-Up** file for use only with formats. This is especially true if your standards call for text parameters that are different than those used when you detail a drawing.

You may also want to create items such as your company logo and any other special symbols you plan to utilize as symbols (e.g., projection angle and inspection symbols), before you create your formats. You can then add these symbols without the need to redraw them for each format size.

You should also be aware that you can make copies of text on the drawing as you create the format. However, when you make a copy, any parameters in the copy become simple text. This means that the copy will *not* utilize the parametric value.

Create the notes you need on the drawing (Fig. 16.25). The notes initially will all have the same character parameters; you can edit them later. You also do not need to place them anywhere special. You will find it easier to move them into place later in the process.

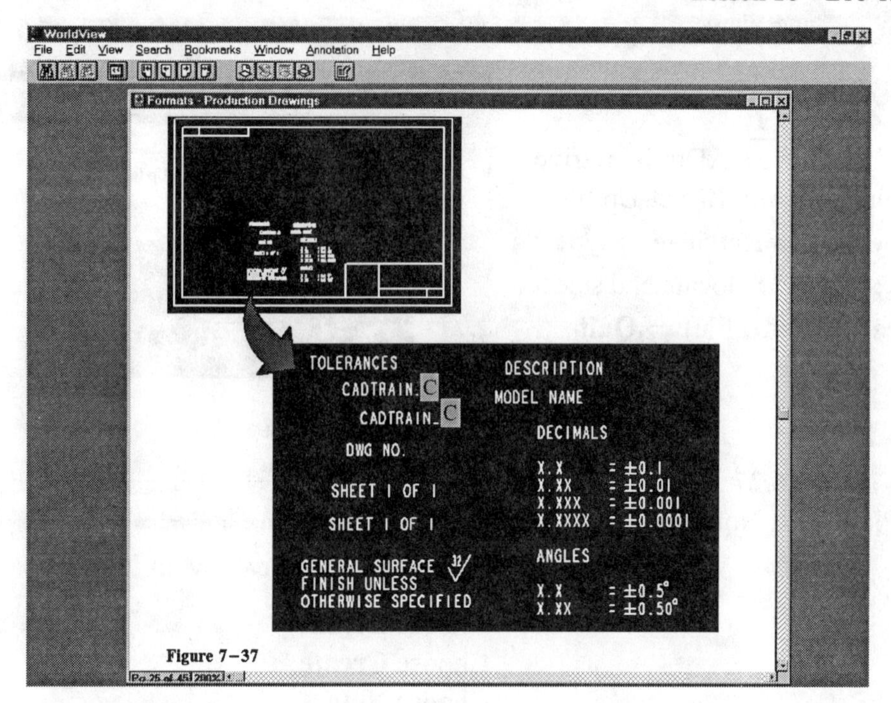

Figure 16.25
Creating Drawing Format Notes

You will be working on one of the formats you previously created or one that you saved under a different name from the Pro/E format directory. When you save this file it will become a drawing format (**.frm**) and can be used to replace a blank drawing format for a project or as the original format retrieved from your directory.

Retrieve the **cadtrain.dtl Set-Up** file if you have CADTRAIN's COAch for Pro/ENGINEER installed on your system, or use a **.dtl** file provided by your instructor. You may also create your own file (Fig. 16.26) by choosing (do not do this at this time):

Set Up ⇒ **Mod Val** ⇒ (edit the file as required) ⇒ **File** ⇒ **Save** ⇒ **Exit** ⇒ **File** ⇒ **Save** ⇒ (enter filename) ⇒ **enter**

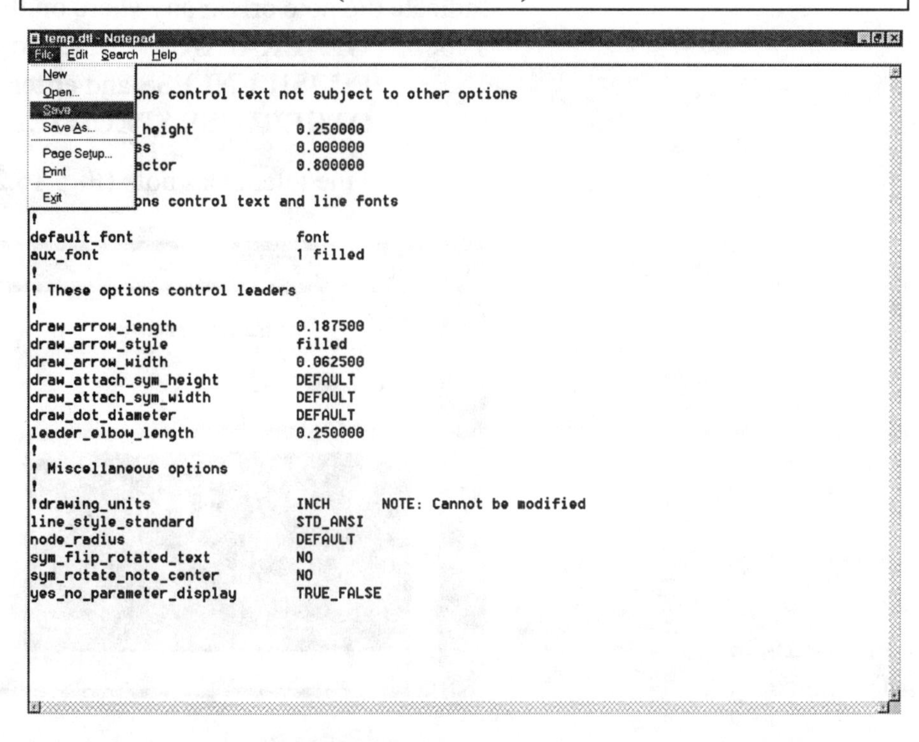

Figure 16.26
.dtl File

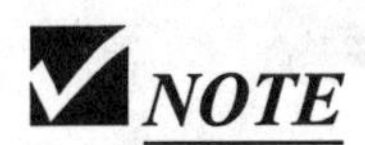

If you have CADtrain, retrive the **cadtrain.dtl** Set-Up file: **Set Up** ⇒ **Retrieve** ⇒ type **?** ⇒ **enter** ⇒ (locate and select **cadtrain.dtl** file) ⇒ **Quit**

1. Create the ***plain text notes*** (Fig. 16.27).

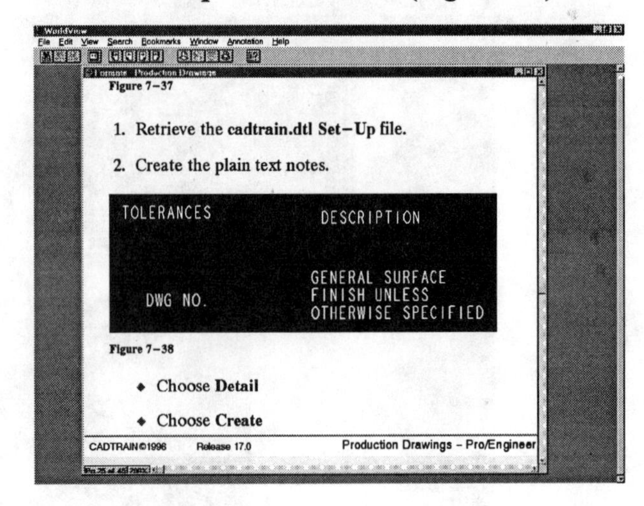
Figure 7–37

1. Retrieve the cadtrain.dtl Set–Up file.

2. Create the plain text notes.

Figure 7–38

- Choose **Detail**
- Choose **Create**

CADTRAIN©1996 Release 17.0 Production Drawings – Pro/Engineer

Figure 16.27
Plain Text Notes

Put your ***Caps Lock*** key on

Choose **Detail**
Choose **Create**
Choose **Note**
Choose **Make Note**
Indicate the note origin anywhere on the screen (pick ounce with the left mouse button
Type: **TOLERANCES** and **enter** twice
Choose **Make Note**
Indicate the note origin anywhere on the screen
Type: **DESCRIPTION** and **enter** twice
Choose **Make Note**
Indicate the note origin anywhere on the screen
Type: **DWG NO.** and **enter** twice
Choose **Make Note**
Indicate the note origin anywhere on the screen
Type: **GENERAL SURFACE** and **enter**
FINISH UNLESS and **enter**
OTHERWISE SPECIFIED and **enter** twice

2. Create the tolerances note (Fig. 16.28).

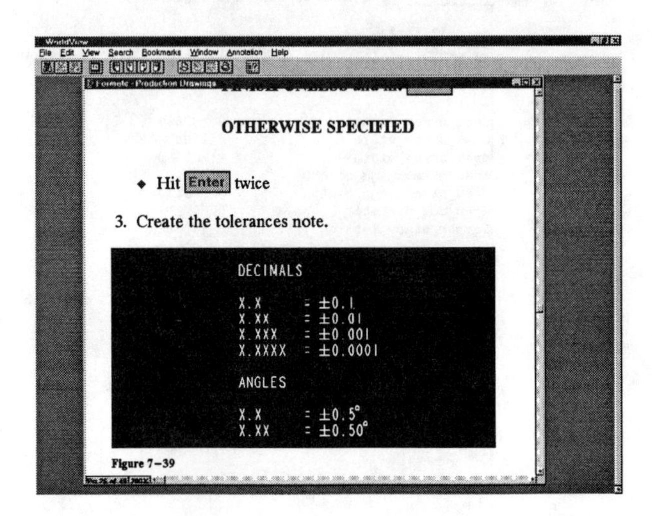
OTHERWISE SPECIFIED

- Hit Enter twice

3. Create the tolerances note.

Figure 7–39

Figure 16.28
Tolerances Note

Choose **Make Note**
Indicate the note origin anywhere on the screen
Type: **DECIMALS** and **enter**

You now need a ***blank*** line. If you just hit **enter** again, Pro/E will assume that you are finished making the note, which you are not.

Press the ***Spacebar*** and then hit **enter**

The ***Spacebar*** counts as a character on the current line and makes Pro/E think that you entered actual text. Therefore, the **enter** is not taken to mean that you are finished making the note.

Hit ***Caps Lock*** to turn it off

This is very important: *You cannot enter the name of a system parameter (such as **&linear_tol_0_0**) in capitals.* Pro/E will not recognize it as a parameter.

Type: **X.X**
Hit the ***Spacebar*** five times
Type: = ***Spacebar***
Choose the ***plus and minus*** symbol ± from the palette
Type: **&linear_tol_0_0** and **enter**
Type: **X.XX**
Hit ***Spacebar*** four times
Type: = ***Spacebar***
Choose the ***plus and minus*** symbol ± from the palette
Type: **&linear_tol_0_00** and **enter**
Type: **X.XXX**
Hit ***Spacebar*** three times
Type: = ***Spacebar***
Choose the ***plus and minus*** symbol [±] from the palette
Type: **&linear_tol_0_000** and **enter**
Type: **X.XXXX**
Hit ***Spacebar*** two times
Type: = ***Spacebar***
Choose the **plus and minus** symbol [±] from the palette
Type: **&linear_tol_0_0000** and **enter**

You now need another *blank* line. If you hit **enter** again, Pro/E will assume that you are finished making the note, and you are not.

Press the ***Spacebar*** and then hit **enter**
Type: **ANGLES** and hit **enter**
Press the ***Spacebar*** and then hit **enter**
Type: **X.X**
Hit ***Spacebar*** five times
Type: = ***Spacebar***
Choose the **plus and minus** symbol ± from the palette
Type: **&angular_tol_0_0** and **enter**

Type: **X.XX**
Hit ***Spacebar*** four times
Type: = ***Spacebar***
Choose the **plus and minus** symbol ± from the palette
Type: **&angular_tol_0_00** and **enter** (**enter** again to finish creating the note)
Choose **Quit**

3. Create the note that will display the model name (Fig. 16.29) in the description area in the title block. Since there is no model at the moment, the text of the name of the parameter displays. ***When this format is placed on a drawing and a view is added (thus adding a model), this text will reflect the name of the model.***

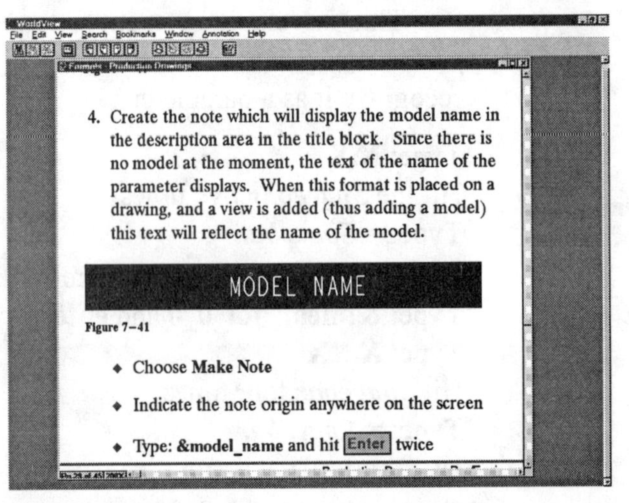

Figure 16.29
Model Name

Choose **Make Note**
Indicate the note origin anywhere on the screen
Type: **&model_name** and **enter** twice

4. Create the note that will display the drawing name in the drawing number area in the title block (Fig. 16.30). Your version of this note will actually show the name of the format that is the current drawing name. When this format is placed on a drawing, the drawing name will replace this text.

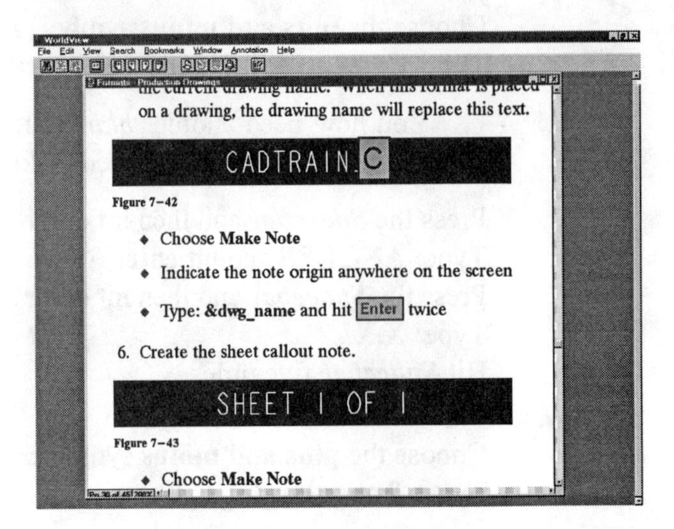

Figure 16.30
Model Name and Sheet Callout

Choose **Make Note**
Indicate the note origin anywhere on the screen
Type: **&dwg_name** and **enter** twice

5. Create the sheet callout note (Fig. 16.30).

Choose **Make Note**
Indicate the note origin anywhere on the screen
Type: **SHEET** (***Spacebar***) **¤t_sheet** (***Spacebar***) **OF** (***Spacebar***) **&total_sheets** and hit **enter** twice.

6. Create another copy of the note that will display the drawing name (Fig. 16.30). This one will be rotated and will be placed in the box in the upper left corner of the drawing (see Fig. 16.25).

Choose **Make Note**
Indicate the note origin anywhere on the screen
Type: **&dwg_name** and **enter** twice

7. Create another copy of the sheet callout note (see Fig. 16.30).

Choose **Make Note**
Indicate the note origin anywhere on the screen
Type: **SHEET** (***Spacebar***) **¤t_sheet** (***Spacebar***) **OF** (***Spacebar***) **&total_sheets** and **enter** twice

8. Add a finish symbol to go with the note you created earlier.

Choose **Done Return**
Choose **Symbol**
Choose **Instance**

Pro/E displays the **Symbol** dialog box (Fig. 16.31).

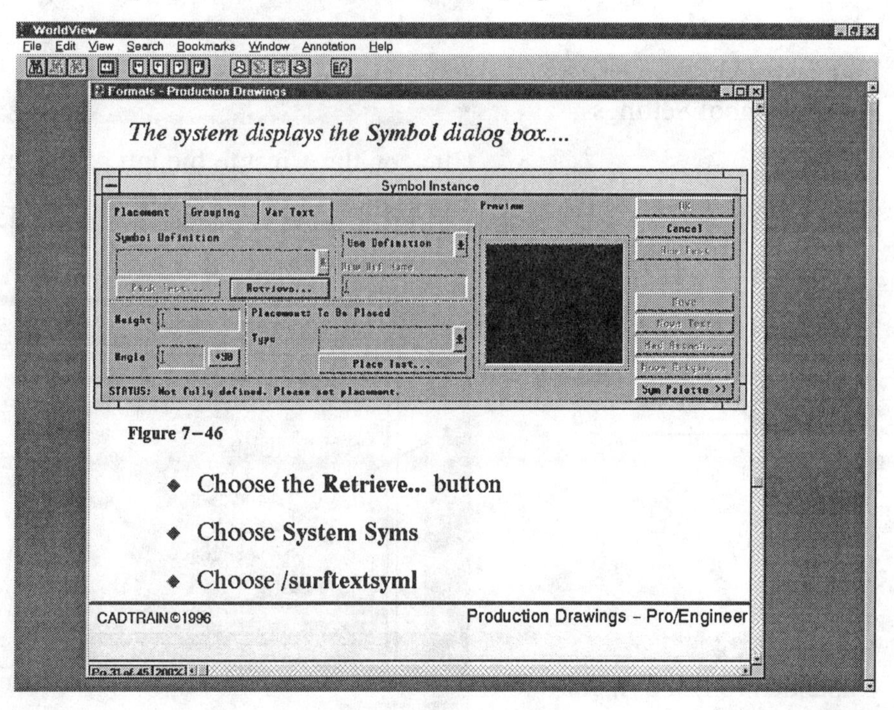

Figure 16.31
Complete Symbol Dialog Box

Choose the **Retrieve** button (Fig. 16.32)
Choose **System Syms** from the menu, not the dialog box
Choose **/surftextsymlib** from the menu
Choose **surftexture.sym** from the menu
Highlight and type **.5** for the symbol **Height**
Choose the **Grouping** Tab from the dialog box (Fig. 16.32)

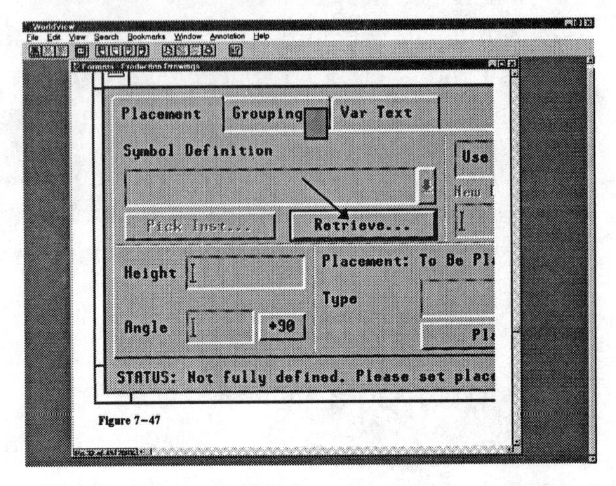

Figure 16.32
Left Side of Symbol Dialog Box

Pro/E now changes the contents of the left side of the dialog box to contain the settings for the finish symbol (Fig. 16.33).

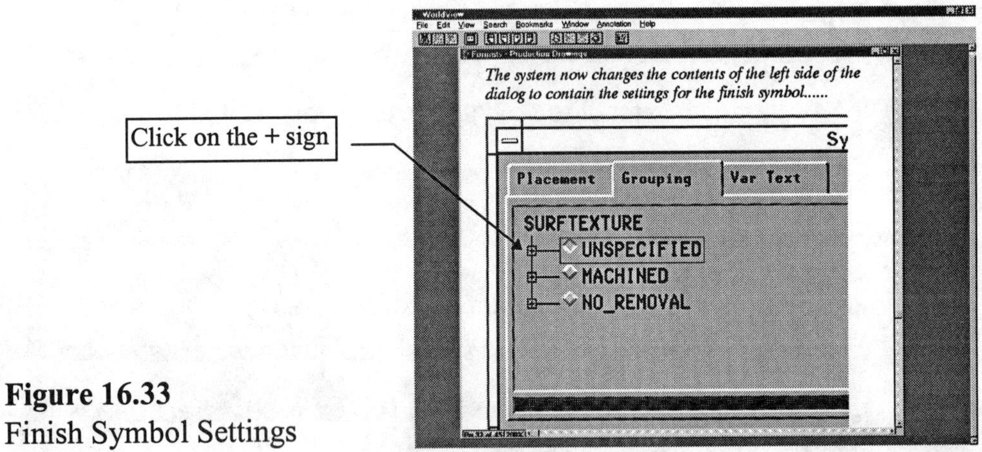

Figure 16.33
Finish Symbol Settings

Click on the + just to the left of the text **UNSPECIFIED**

Pro/E now expands the list under **UNSPECIFIED** to show the options available (Fig. 16.34).

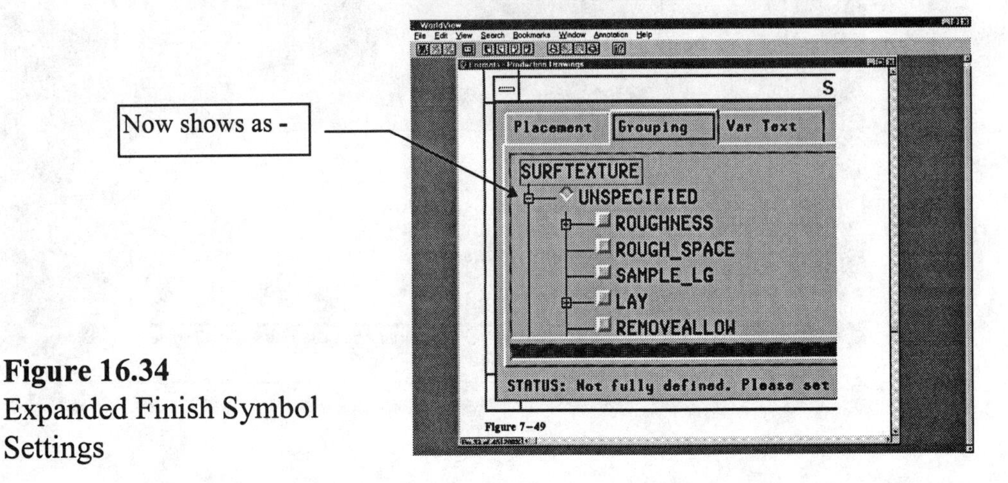

Figure 16.34
Expanded Finish Symbol Settings

Click on the toggle box □ just to the left of the **ROUGHNESS** text (Fig. 16.35).

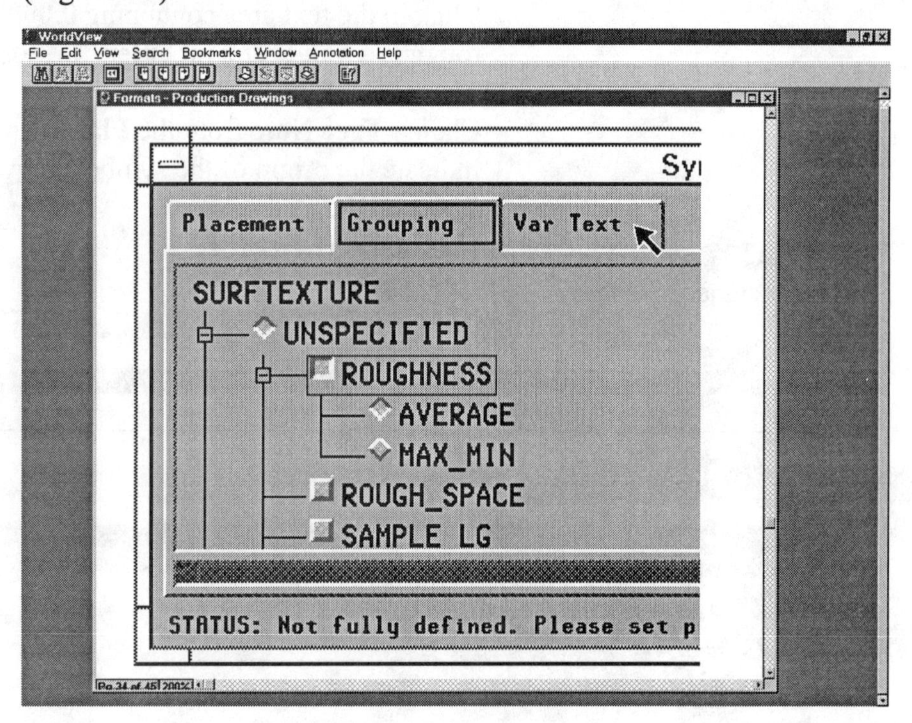

Figure 16.35
Expanded Finish Symbol Roughness

Choose the **Var Text** tab (Fig. 16.36)

Pro/E now changes the contents of the left side of the dialog box to contain the variable text entries for the finish symbol.

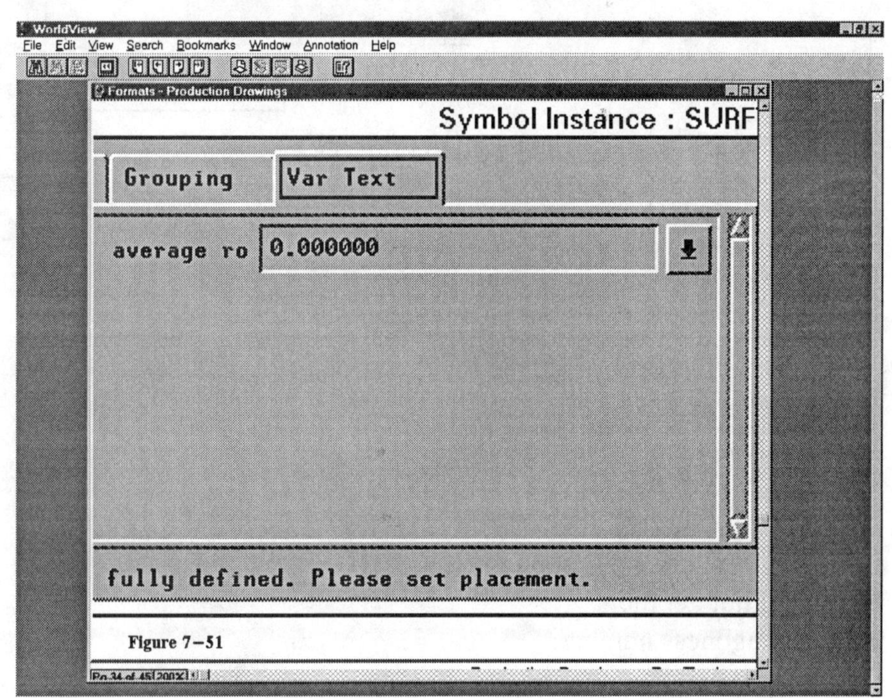

Figure 16.36
Finish Symbol Variable Text

If this symbol had been defined with a series of default parameters, you would be able to select them from an option menu. Since this symbol only has the roughness value as a single number, you need to type it. Figure 16.37 shows the Surface Texture dialog box.

Click in the text area containing **0.0000**, type **32.0000** for the roughness value, and hit **enter**
Choose the **Placement** tab
Choose **Free Note** from the **Placement**: option menu
Indicate the origin of the symbol, anywhere on the screen
Choose **OK**
Choose **Done/Return**

Figure 16.37
Symbol Instance: Surface Texture

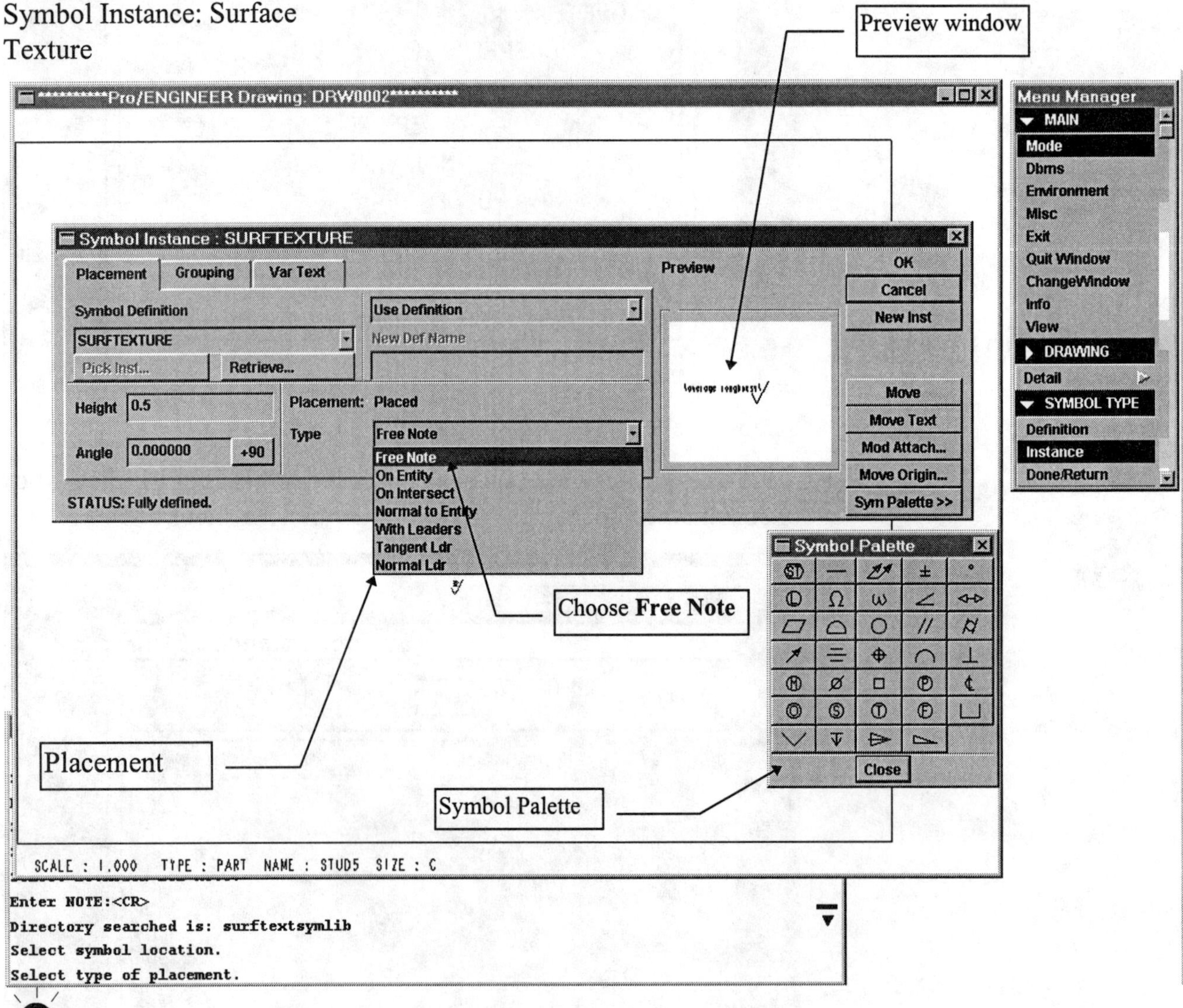

HINT
If you did not get the correct roughness number:
Choose **Modify**
(from the DETAIL menu)
Choose **Value**
Pick the value "**0**"
Type **32** and **enter**
Choose **Done Sel**
Choose **Done/Return**

Modifying Text

Once you create the text, you often need to modify the text parameters. The most common modification is the text height. When you modify the parameters of text, Pro/E treats each text segment ({**1:text1**}) as a separate object. This allows you to set the height, for example, to different values within a text string.

Since this would make selection quite time consuming for complex notes, this option allows you to use the **Pick Many** option to pick the entire note using a rectangle.

You can also use the **Angle** option to rotate text. This allows you to write text "upside-down" along the top of a format, or at **90** or **270** degrees to make it parallel to the right or left border. The **Mirror** option allows you to change text so that it can be read from the back side of the vellum (when you plot it).

Modify the height of the notes and place them at the proper location in the format.

1. Change the height of all of the small "title" text to **.09**.

Choose **Modify**
Choose **Text**
Choose **Text Style**
Select the notes: **DESCRIPTION**, the entire **GENERAL SURFACE** note, and **DWG NO**.
Click the **middle** mouse button to stop selecting notes to change.

(Pro/E now displays the **Text Style** dialog box, which contains all of the parameters that control how text displays, as shown in Figure 16.38)

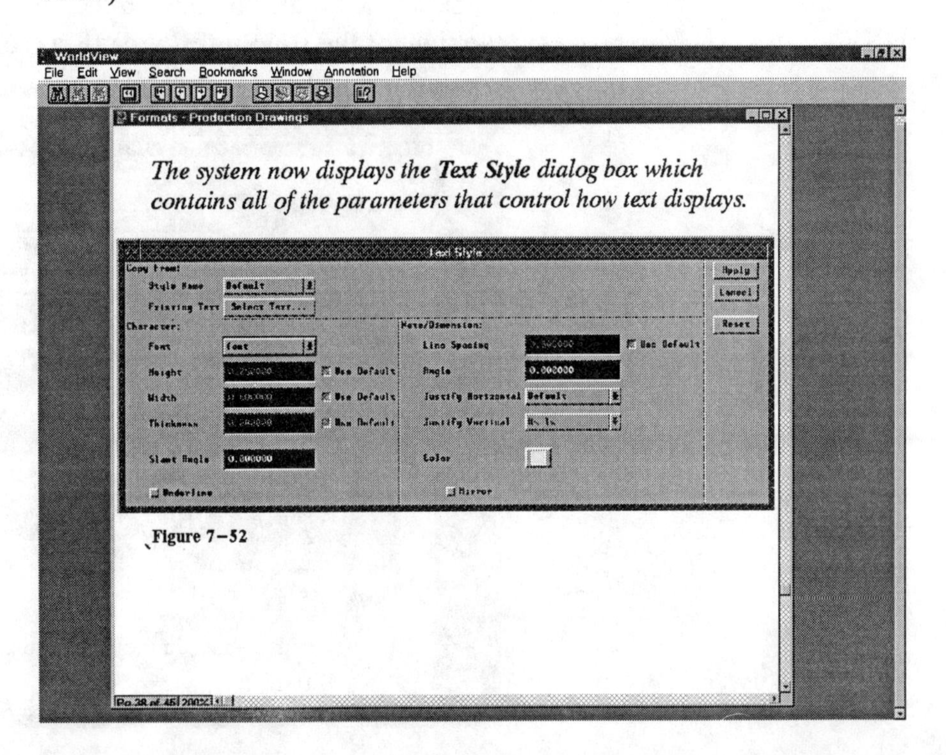

Figure 16.38
Text Style Dialog Box

Click on the toggle box next to the text: □ **Use Default** that is just to the right of the **Height** text area (Fig. 16.39).

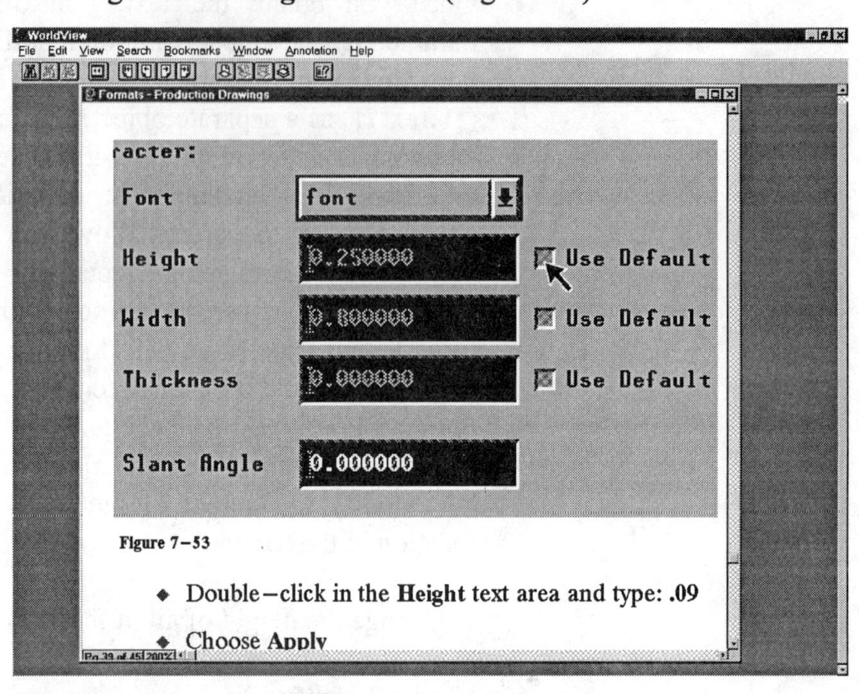

Figure 16.39
Text Height Use Default

Double-click in the **Height** text area and type **.09**
Choose **Apply**
Choose **Close**

2. Change the size of the two **SHEET** callouts to **.09**.

Choose **Pick Many**
Drag a box around both the texts **SHEET 1 OF 1** to select them.
Click the **middle** mouse button to stop selecting notes to change.
Click on the toggle box next to the text: □ **Use Default** that is just to the right of the **Height** text area (Fig. 16.40).

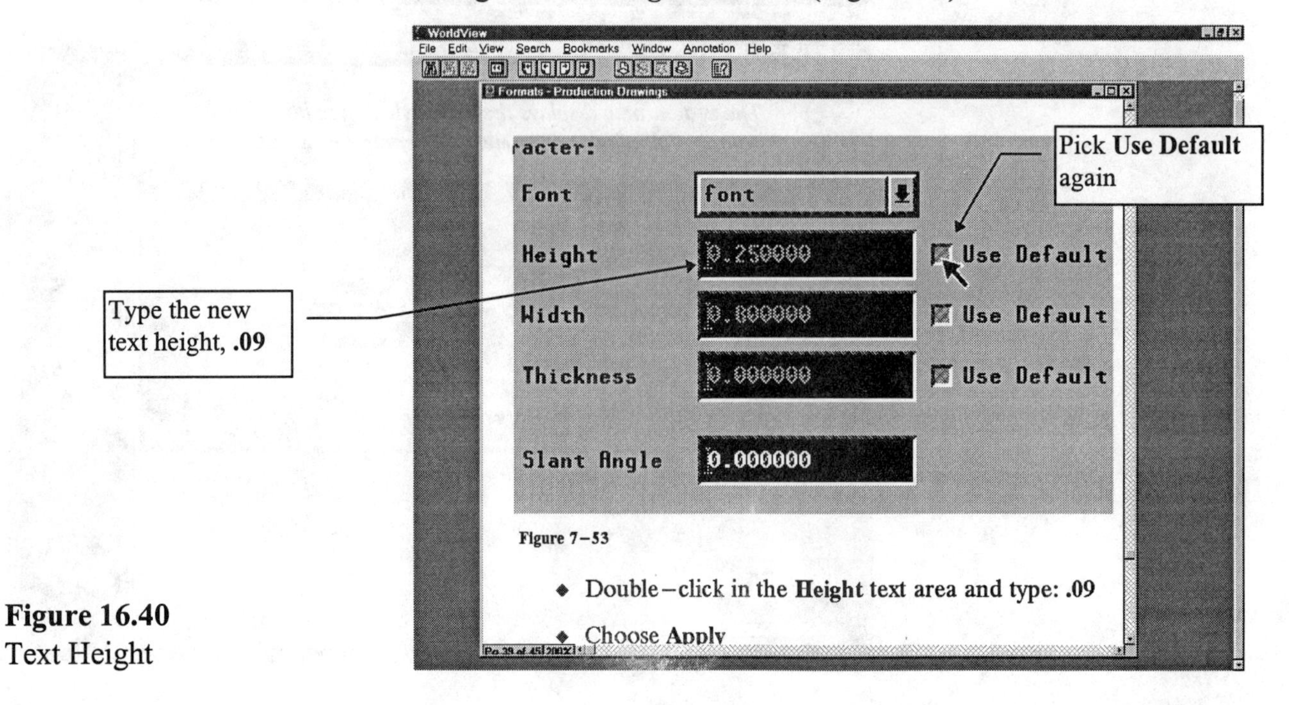

Figure 16.40
Text Height

Double-click in the **Height** text area and type **.09**
Choose **Apply**
Choose **Close**

3. Change the size of the two drawing number notes to **.25**.

Select the two notes that reflect the current name of the format (**MY_FORMAT**...)
Click the **middle** mouse button to stop selecting notes to change.
Click on the toggle box next to the text: □ **Use Default** that is just to the right of the **Height** text area.
Double-click in the **Height** text area and type **.25**
Choose **Apply**
Choose **Close**

4. Change the size of the tolerances note.

Choose **Pick Many**
Drag a box around the complete tolerances note (not the text **TOLERANCES**)
Click the **middle** mouse button to stop selecting notes to change.
Click on the toggle box next to the text: □ **Use Default** that is just to the right of the **Height** text area.
Double-click in the **Height** text area and type **.06**
Choose **Apply**
Choose **Close**

5. Change the size of the **MODEL NAME** text to **.375**.

6. Rotate one of the drawing name notes and one of the sheet callouts.

Select one of the notes that reflects the current name of the format (**MY_FORMAT**) and the sheet callouts note (**SHEET 1 OF 1**)
Double-click in the **Angle** text area (Fig. 16.41).

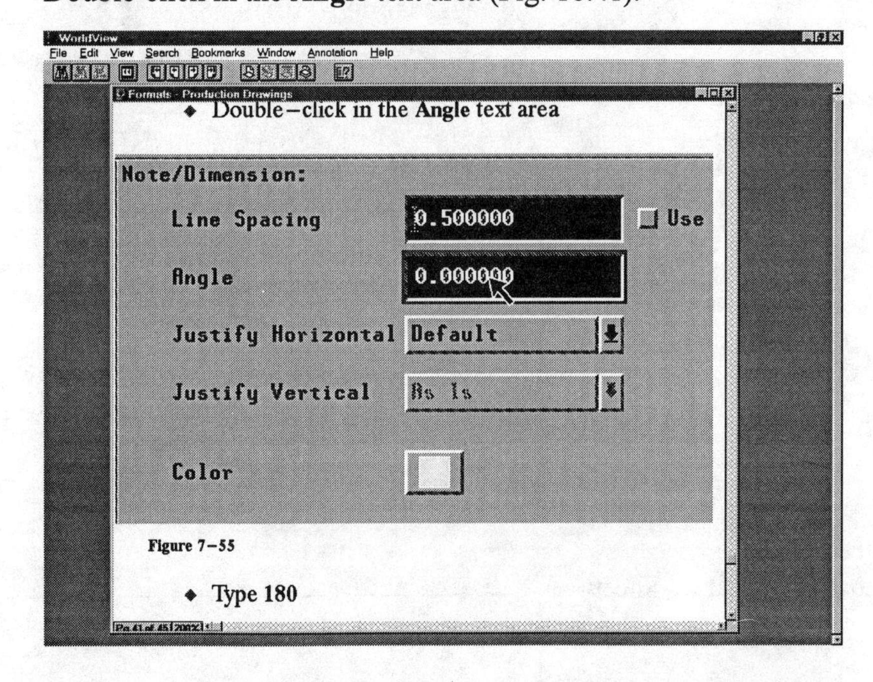

Figure 16.41
Text Angle

Type **180**
Choose **Apply**
Choose **Close**
Choose **Done/Return** ⇒ **Done/Return**

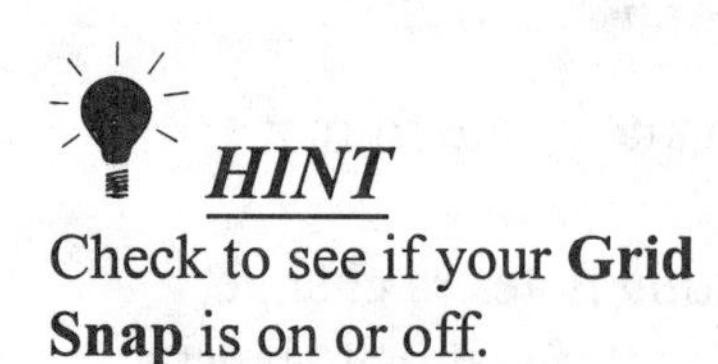

HINT
Check to see if your **Grid Snap** is on or off.

7. Use the **Move** option to drag the text to the proper location in the title block approximately as shown in Figure 16.42.

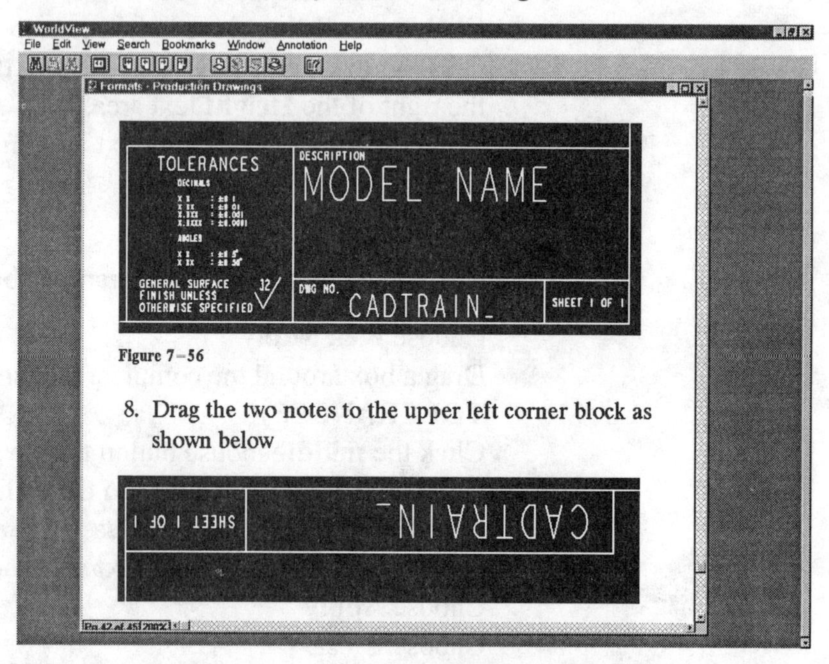

Figure 16.42
Moving the Text to the Title Block

8. Drag the two notes to the upper left corner block, as shown in Figure 16.42). Figure 16.43 shows a format used on a drawing.

9. Save the format.

Choose **Dbms**
Choose **Save**
Hit **enter**
Choose **Done/Return**

Dbms ⇒ Save ⇒ enter
Purge ⇒ enter ⇒
Done-Return

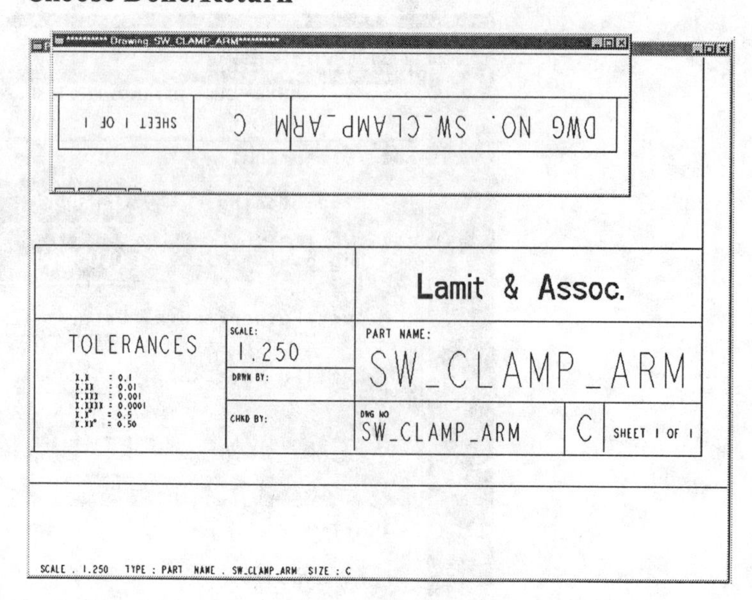

Figure 16.43
Company Formatted Sheet

Views

When you start a drawing, you must follow similar steps in creating a part or an assembly (Fig. 16.44). After the drawing is created, you must add the views required to detail the part.

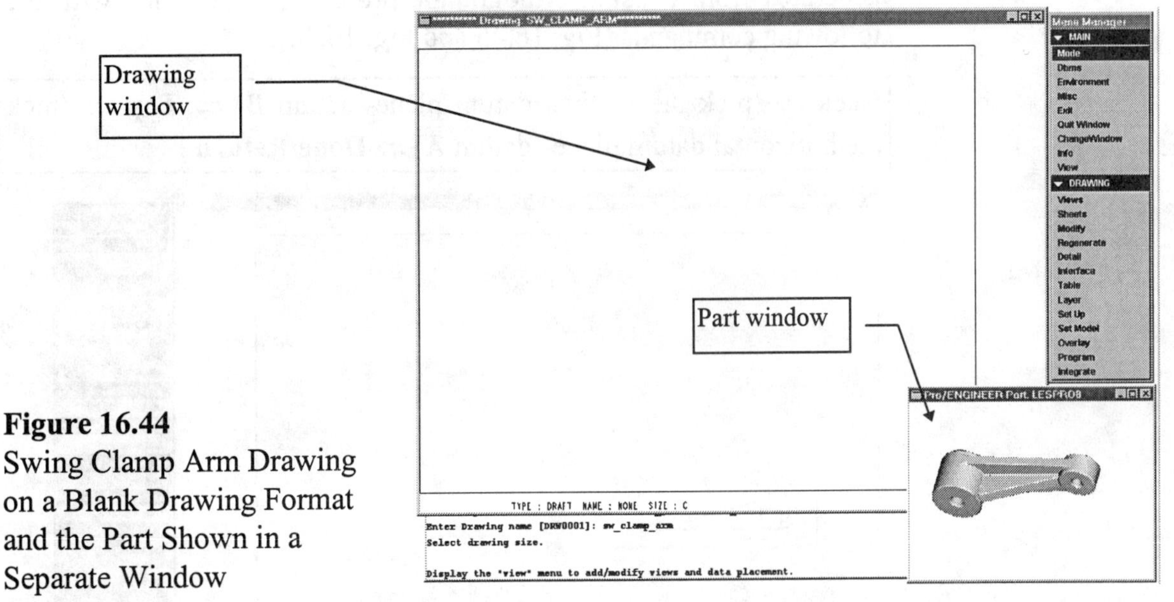

Figure 16.44
Swing Clamp Arm Drawing on a Blank Drawing Format and the Part Shown in a Separate Window

You must be in the same directory as your part. It is actually a good idea to bring up the part in a separate window. Give the following commands (Fig. 16.45):

> **Mode** ⇒ **Part** ⇒ **Search/Retr** ⇒ (pick the name that you gave the Lesson 8 Project) ⇒ (move and reduce the size of the Part window) ⇒ **Mode** ⇒ **Drawing** ⇒ **Create** ⇒ (type a name for your drawing: **SW_CLAMP_ARM**) ⇒ **enter** ⇒ (from the DWG SIZE menu pick **C**) ⇒ **Views** ⇒ **enter** to accept the default (or **?** ⇒ **enter** ⇒ and choose the part model name from the directory list) ⇒ **Done** (to accept the defaults) ⇒ (pick a place on the drawing for the first view, as shown in Figure 16.45)

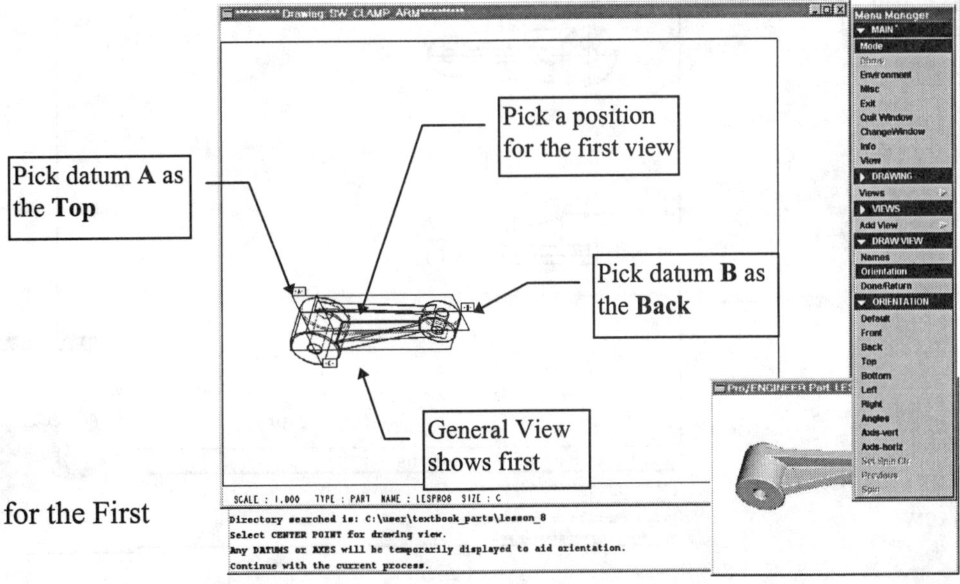

Figure 16.45
Pick a Position for the First View

The next step is to reorient the first view that is now displayed as a ***general view*** in the default orientation. Using the part's planar edges or the datum planes, orient the view to be a *front view* of the part. This first view is important, since all other projected views are generated from it using orthographic projection. Continue with the following commands (Fig. 16.45 and Fig. 16.46):

Back ⇒ (pick the vertical datum plane--datum **B)** ⇒ **Top** ⇒ (pick the horizontal datum plane--datum **A)** ⇒ **Done/Return**

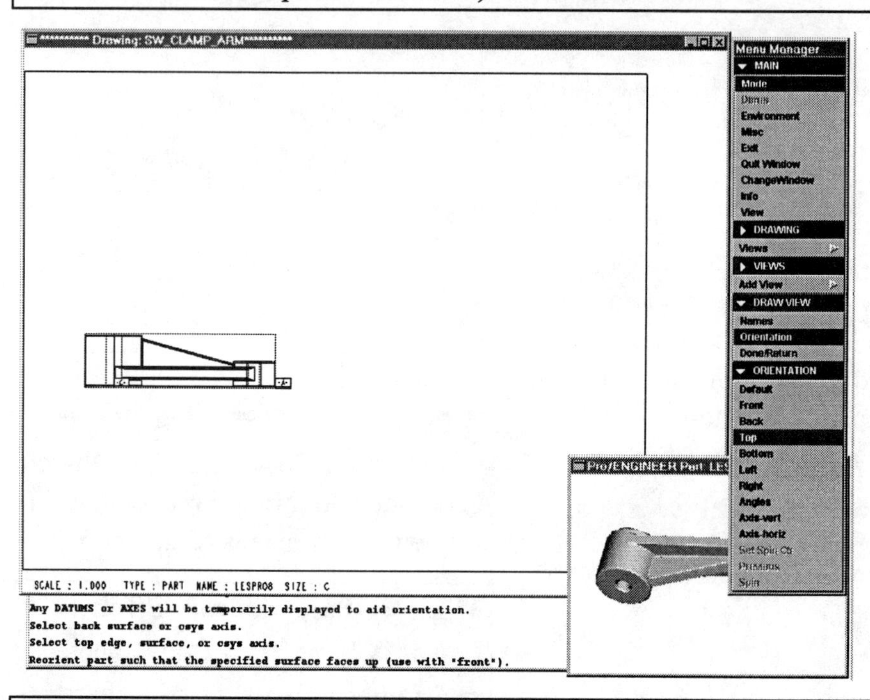

Figure 16.46
First View

Add View ⇒ **Projection** ⇒ **Full View** ⇒ **No Xsec** ⇒ **No Scale** ⇒ **Done** ⇒ (pick a position near where the top view should go, as shown in Figure 16.47)

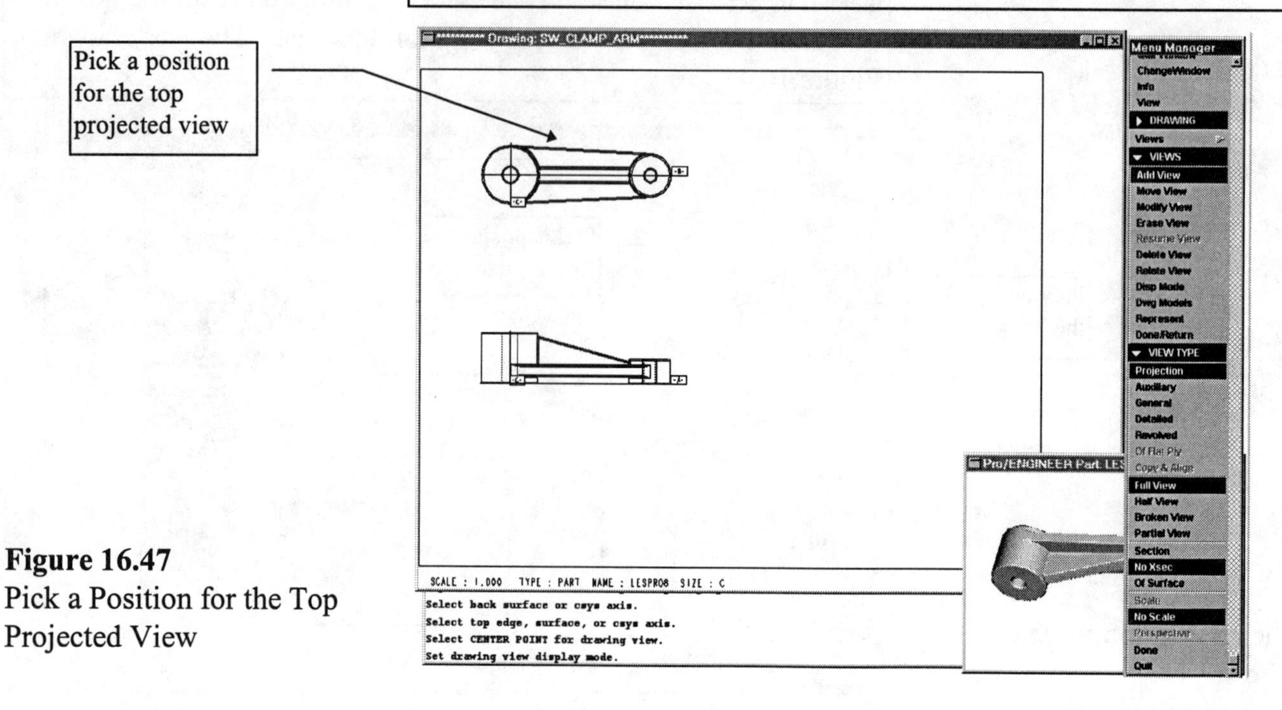

Figure 16.47
Pick a Position for the Top Projected View

Add View ⇒ **Done** (to accept defaults) ⇒ (pick the position for the right side view, as shown in Figure 16.48)

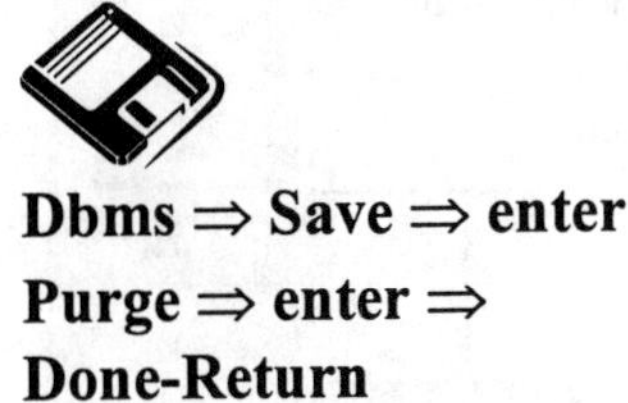

Dbms ⇒ **Save** ⇒ **enter**
Purge ⇒ **enter** ⇒
Done-Return
(this will save your drawing)

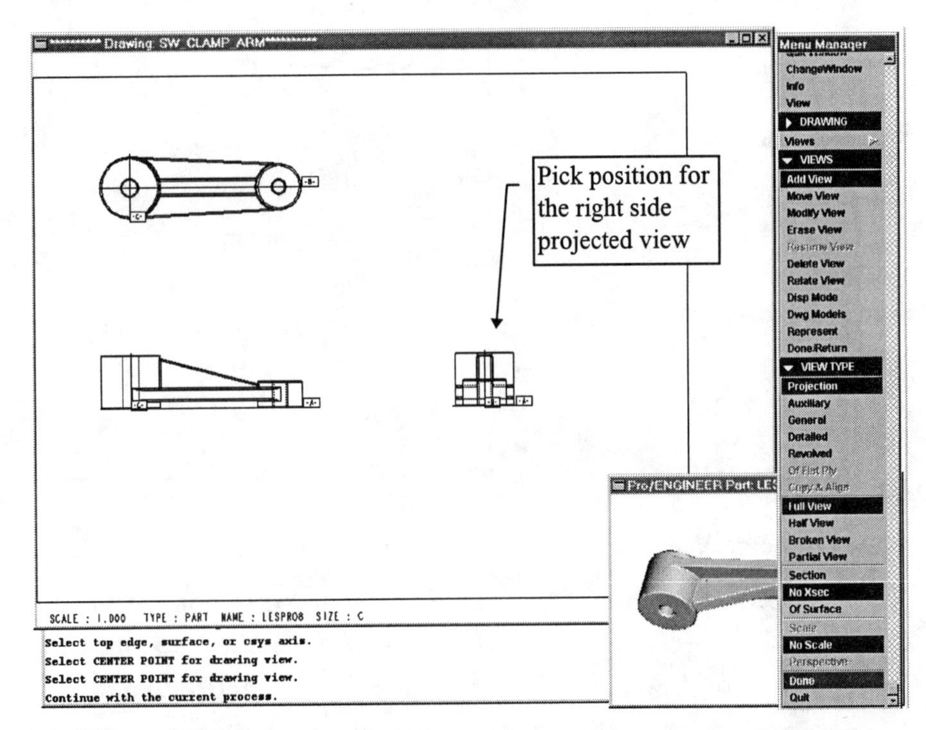

Figure 16.48
Pick a Position for the Top Projected View

Add another projected view, as shown in Figure 16.49 and Figure 16.50: **Add View** ⇒ **Done** ⇒ (pick a position for the new view). Pro/E will prompt you with the following: **Conflict in parent view exists. Select parent view for making the projection**. Select the right side view (Fig. 16.50).

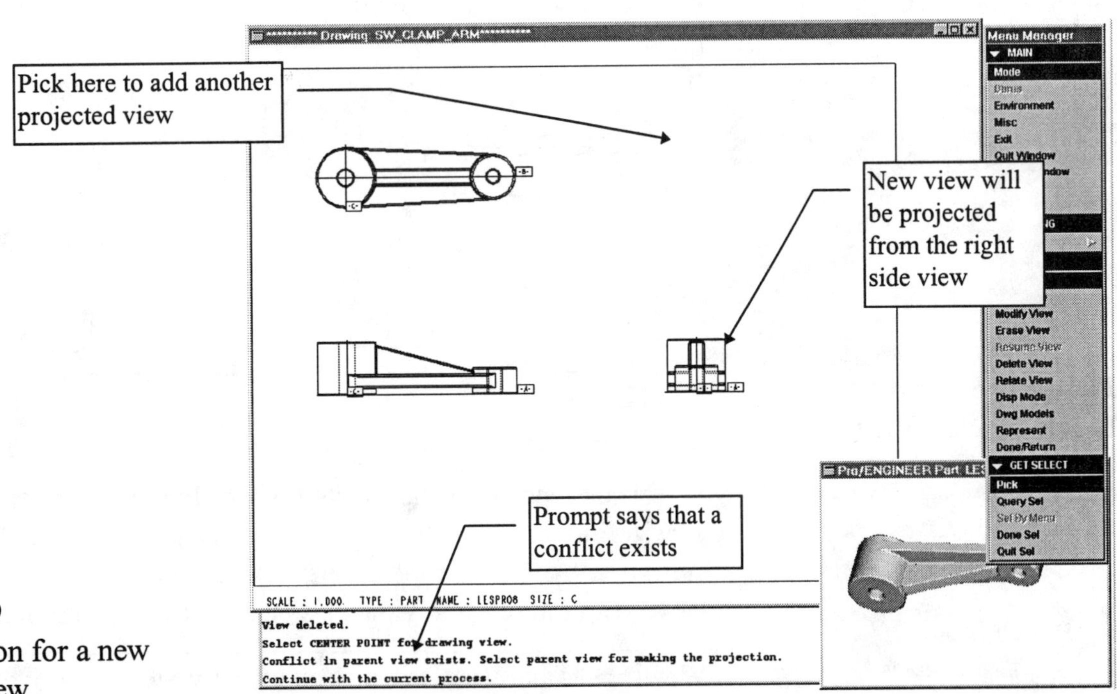

Figure 16.49
Pick a Position for a new Projected View

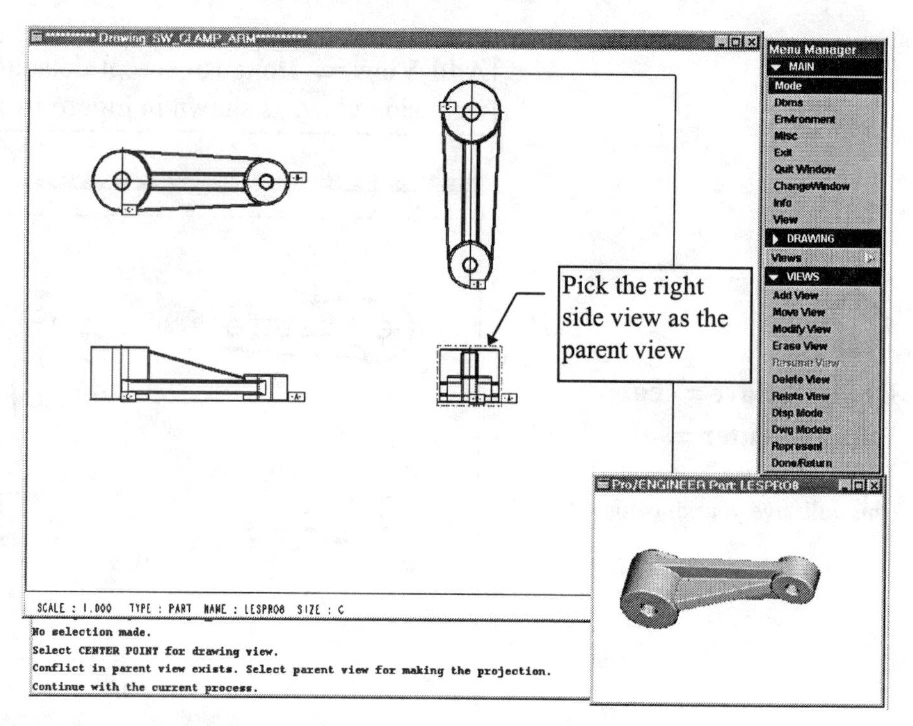

Figure 16.50
New Projected View

Since you really don't want the view just created, remove it by choosing: **Delete View** ⇒ (pick the view) ⇒ **Confirm** (Fig. 16.51) ⇒ **Done Sel** ⇒ **Done/Return**.

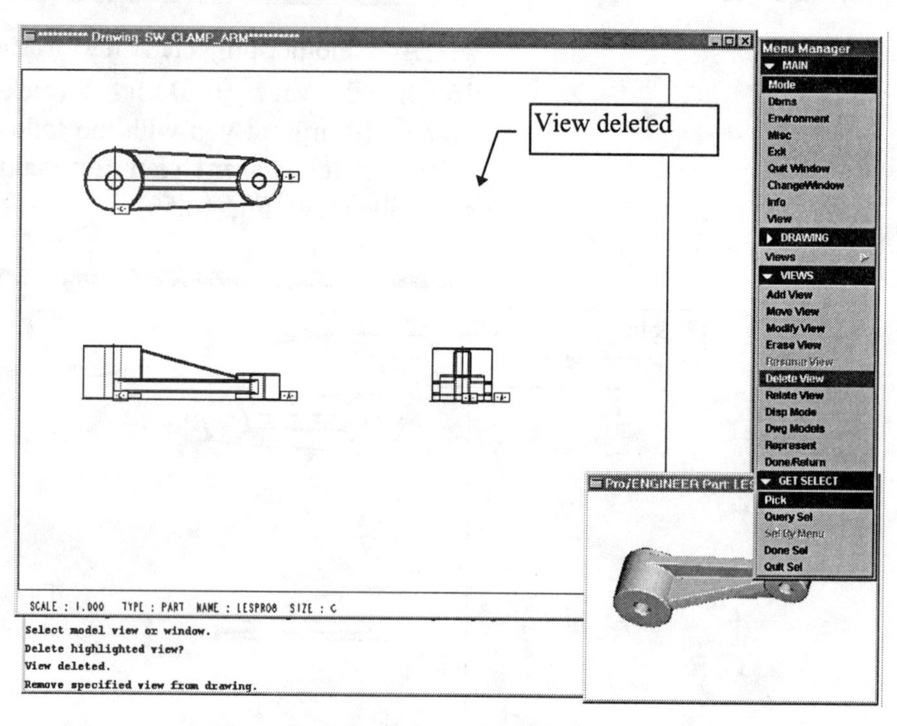

Figure 16.51
Delete the New View

Next, reposition the views on the drawing. After selecting the view to move, it will highlight with dashed lines (Fig. 16.52). If the view is a parent of another view, that view will also highlight and move with the selected view. Use the following commands:

Views ⇒ **Move View** ⇒ (pick the front view) ⇒ (move the view to a new position as in Figure 16.53; try a few positions) ⇒ **Done Sel** ⇒ **Done/Return**

Pick the front view to move

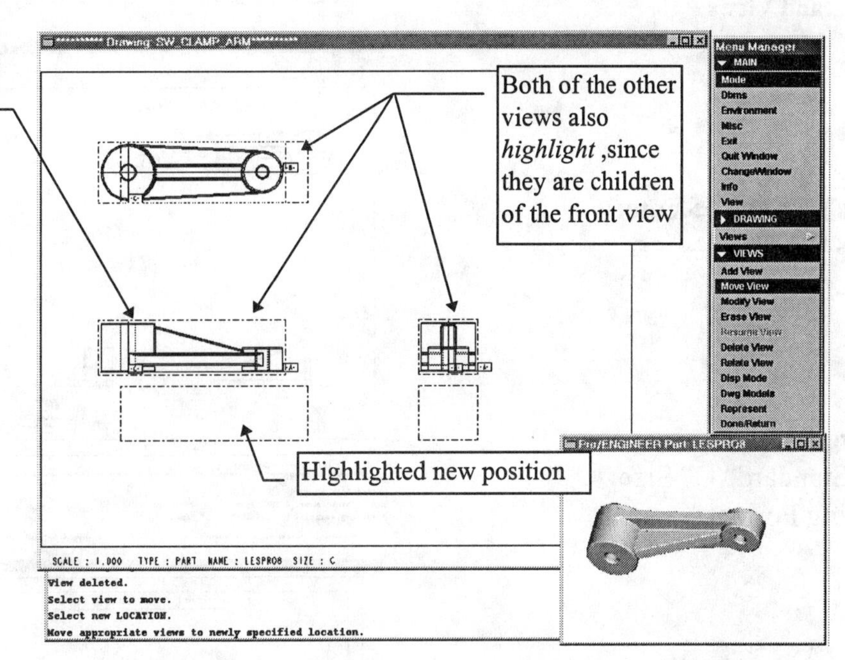

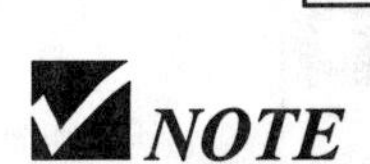

NOTE

The set datum planes are displayed in the old ANSI standard style. You will learn how to change to the new ASME 1994 standard and display the datums correctly in the next lesson.

Figure 16.52
Moving Views

Dbms ⇒ Save ⇒ enter
Purge ⇒ enter ⇒
Done-Return

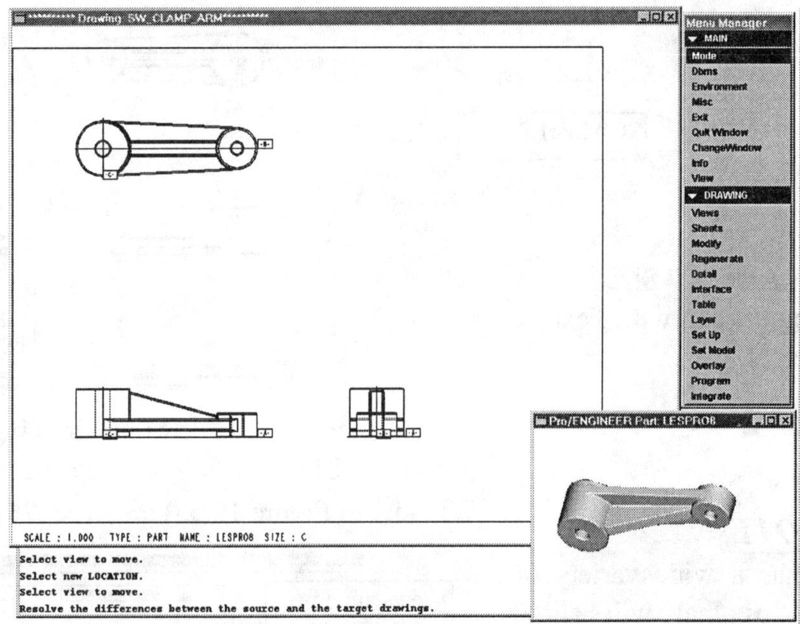

Figure 16.53
New View Position

Change the drawing sheet to a standard ANSI **"C"** size sheet using the following commands (Fig. 16.54):

Sheets ⇒ Format ⇒ Add/Replace ⇒ ? ⇒ enter ⇒ Format Dir ⇒ (pick: **c.frm**)

Now change the drawing format to the one you previously created and saved as a **.frm** for use on your projects. This format should contain parameters you set up and save dwith the format. The example in Figure 16.55 uses the CADTRAIN format and text parameters. Use the following commands:

Sheets ⇒ Format ⇒ Add/Replace ⇒ ? ⇒ enter ⇒ Format Dir ⇒ [pick: ***your_format_name*.frm** (Fig. 16.54)]

Dbms ⇒ Save ⇒ enter

Purge ⇒ enter ⇒ Done-Return

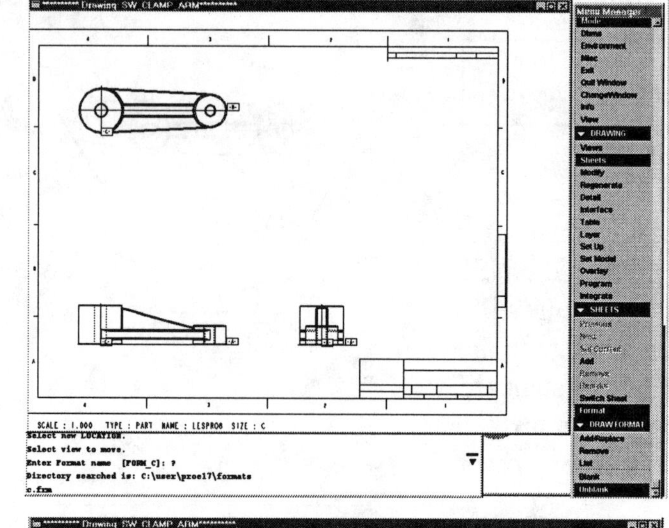

Figure 16.54
New Standard "C" Size Drawing Format

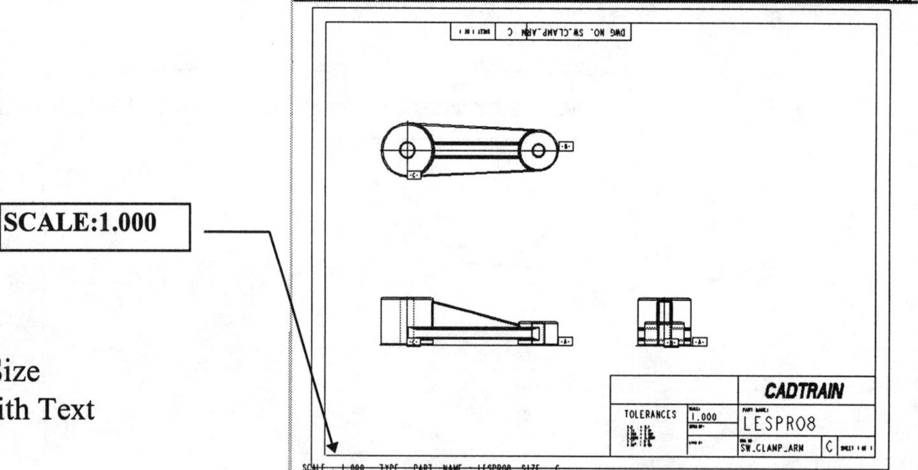

Figure 16.55
CADTRAIN "**C**" Size Drawing Format with Text Parameters

To change the scale of the drawing (*not the part*) pick: **Modify** ⇒ (pick the Scale: **1.000** on the bottom left edge of the Drawing Window in Figure 16.55) ⇒ (type **.75**) ⇒ **enter** (Fig. 16.56).

NOTE
Pro/E has a wide variety of capabilities that will allow you to tailor your drawing to the needs of your company or class.

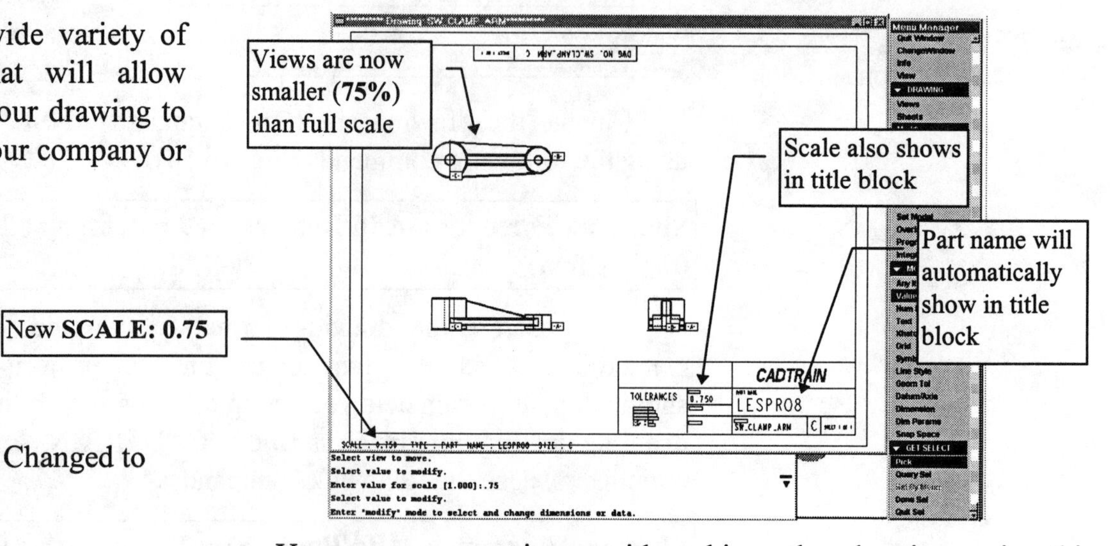

Figure 16.56
Drawing Scale Changed to **SCALE: 0.75**

You can now experiment with making other drawings using this part or other parts, adding, moving, and erasing views, and creating, adding and removing drawing formats.

Lesson 16 Project

Base Angle Drawing

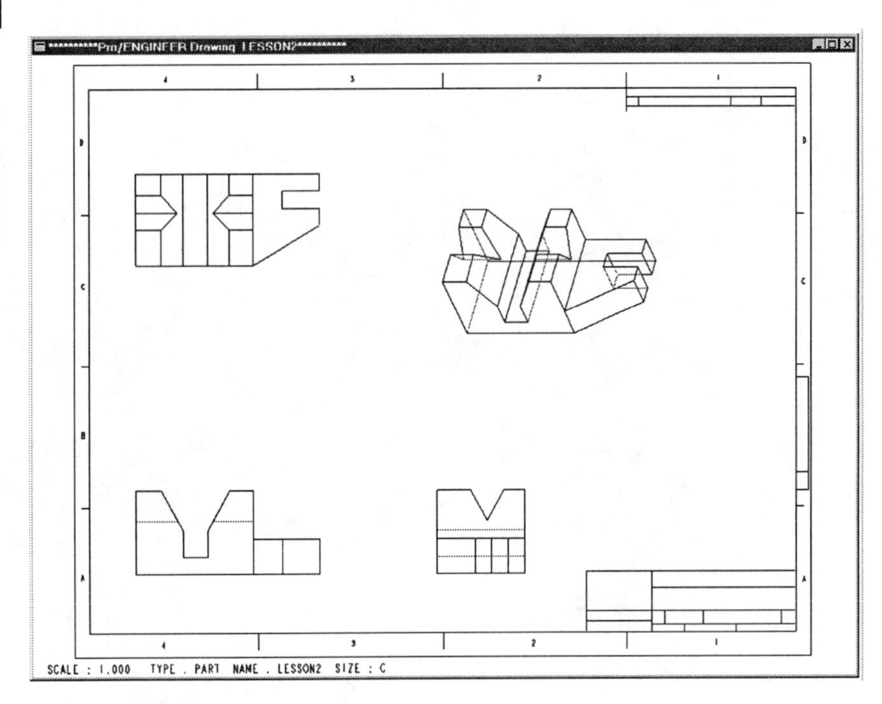

Figure 16.57
Base Angle Drawing Using ANSI (ASME) Standard Format

Base Angle Drawing

The first **lesson project** for the Drawing mode will use the part modeled in Lesson 2. The drawing for the Base Angle (Fig. 16.57) is just one of many lesson parts and lesson projects that can be brought into the Drawing mode and detailed.

Analyze the part, and plan out the sheet size and the drawing views required to display its features for detailing (Fig. 16.58). Use the formats created in the lesson.

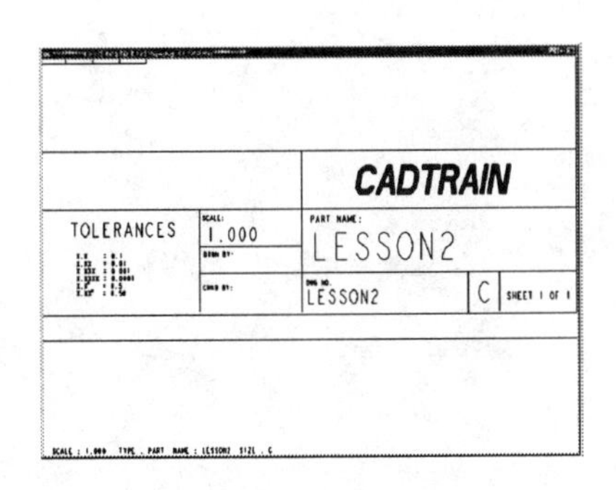

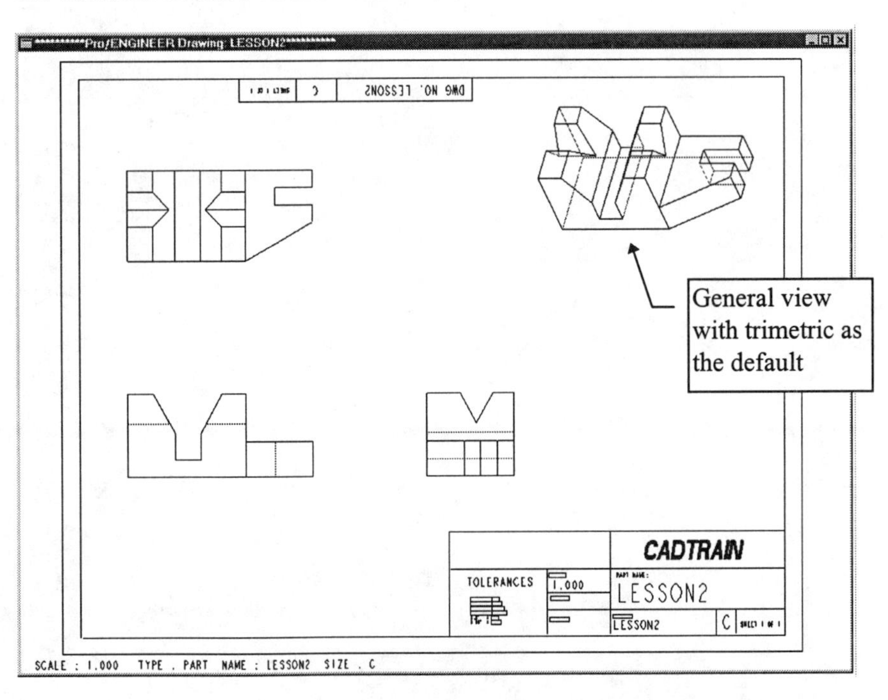

Figure 16.58
Base Angle Drawing Using CADTRAIN Format with Parameters

Lesson 17

Detailing

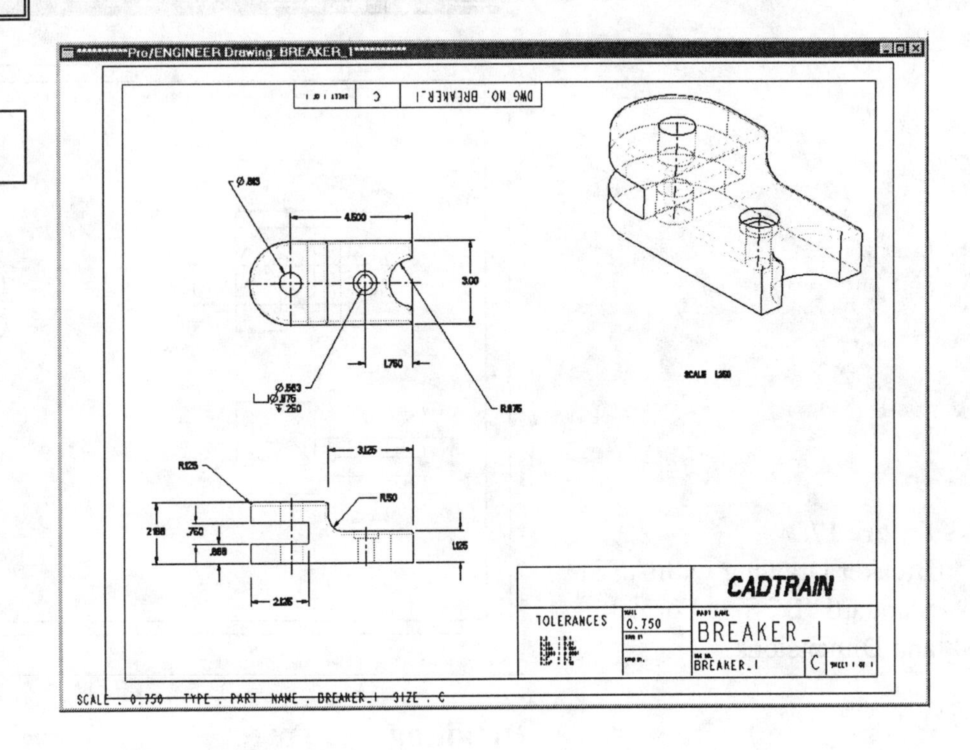

Figure 17.1
Breaker Drawing

OBJECTIVES

1. **Use ASME Y14.5 1994 standards to detail drawings**
2. **Dimension a part**
3. **Create and save .dtl files**
4. **Add geometric tolerancing information to a drawing**
5. **Use Pro/MARKUP to see checker changes**
6. **Move and modify dimensions**

☑ ***EGD REFERENCE***
Engineering Graphics and Design
by L. Lamit and K. Kitto
Read Chapter: 15 and 16
See Pages: 321, 364, 556-559 and 622-626

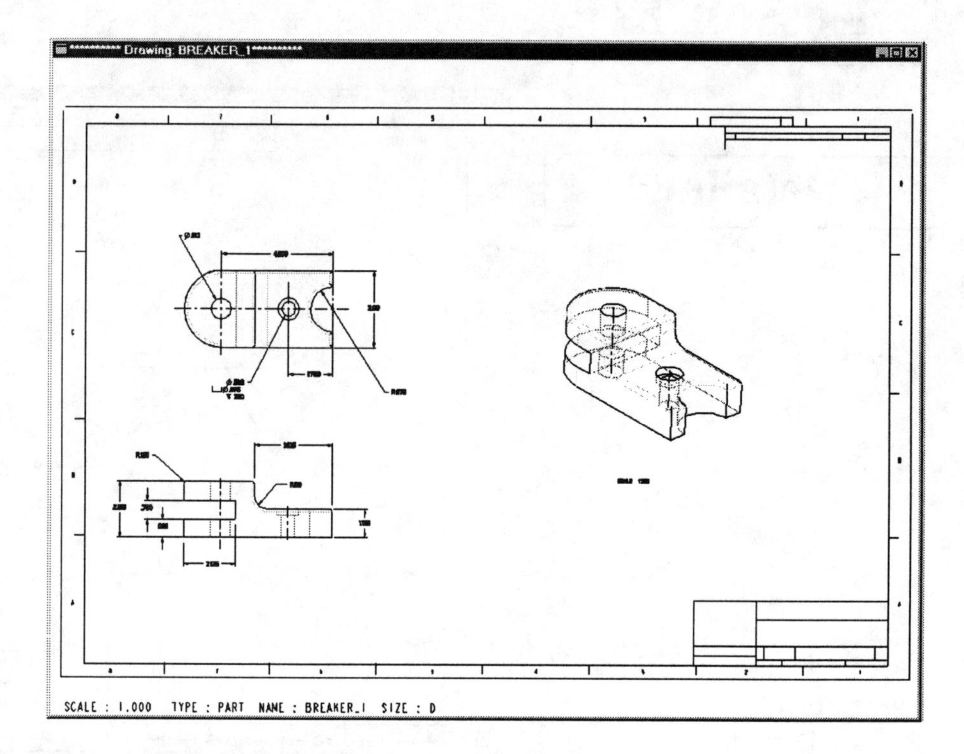

Figure 17.2
Breaker Drawing with ANSI Standard "**D**" Size Format and Dimensions

Detailing

The purpose of an engineering drawing is to convey information so that the part can be manufactured correctly. Engineering drawings use dimensions and notes to convey this information (Fig. 17.1). Knowledge of the methods and practices of dimensioning and tolerancing is essential to the engineer or designer. The multiview projections of a part provide a graphic representation of its shape (*shape description*). However, the drawing must also contain information that specifies size and other requirements.

Drawings are *annotated* with dimensions and notes. Dimensions must be provided between points, lines, or surfaces that are functionally related or to control relationships of other parts. With Pro/E the **design intent** used in the original sequence of feature creation and the selection of dimensions used on the features sketch will determine the dimensions shown on the drawing. You need not create dimensions (unless desired), since the dimensions are already established during the modeling of the part.

Each dimension on a drawing has a **tolerance**, implied or specified. The general tolerance given in the title block is called a **general** or **sheet tolerance**. Specific tolerances are provided with each appropriate dimension. Together, the views, dimensions, and notes give the complete shape and size description of the part. Uniform practices for stating and interpreting dimensioning and tolerancing requirements were established in **ASME Y14.5 1994.**

Dimensioning

Models can be automatically dimensioned in **Drawing** mode (Fig. 17.2). Pro/E displays **dimensions** in a view based on how the part was modeled. The dimension type is selected from options before showing the dimensions on the drawing (Fig. 17.3 and Fig. 17.4). Linear dimensions and ordinate dimensions are two of the options.

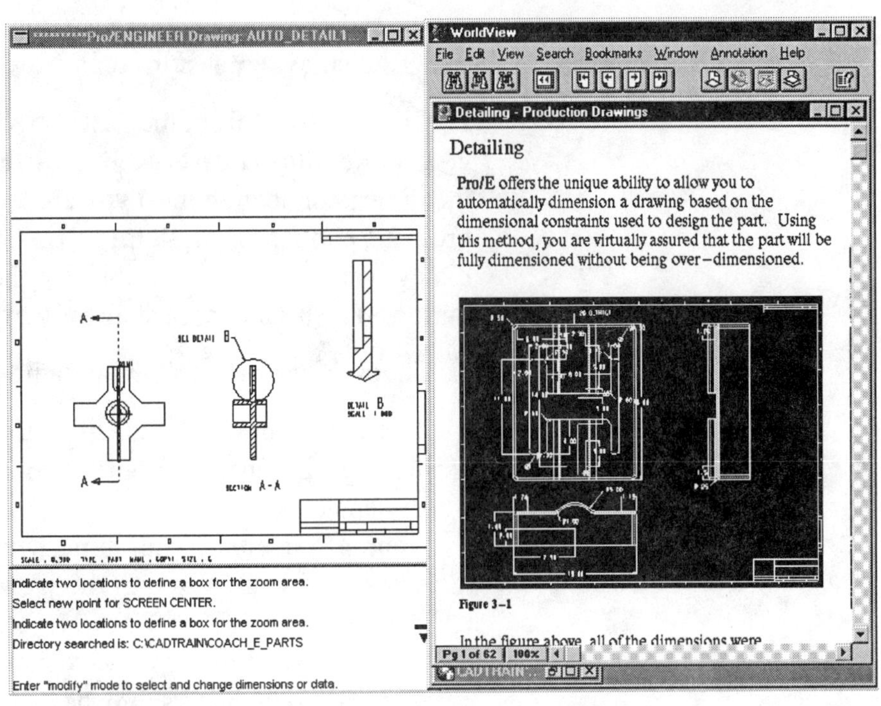

Figure 17.3
COAch for Pro/E, Detailing--Production Drawings

After a part's features are sketched, aligned, and dimensioned, you modify the dimension values to be the exact sizes required for the design. The dimensioning scheme, the controlling features, the parent-child relationships, and the datums used to define and control the part features are determined as you design and model on Pro/E.

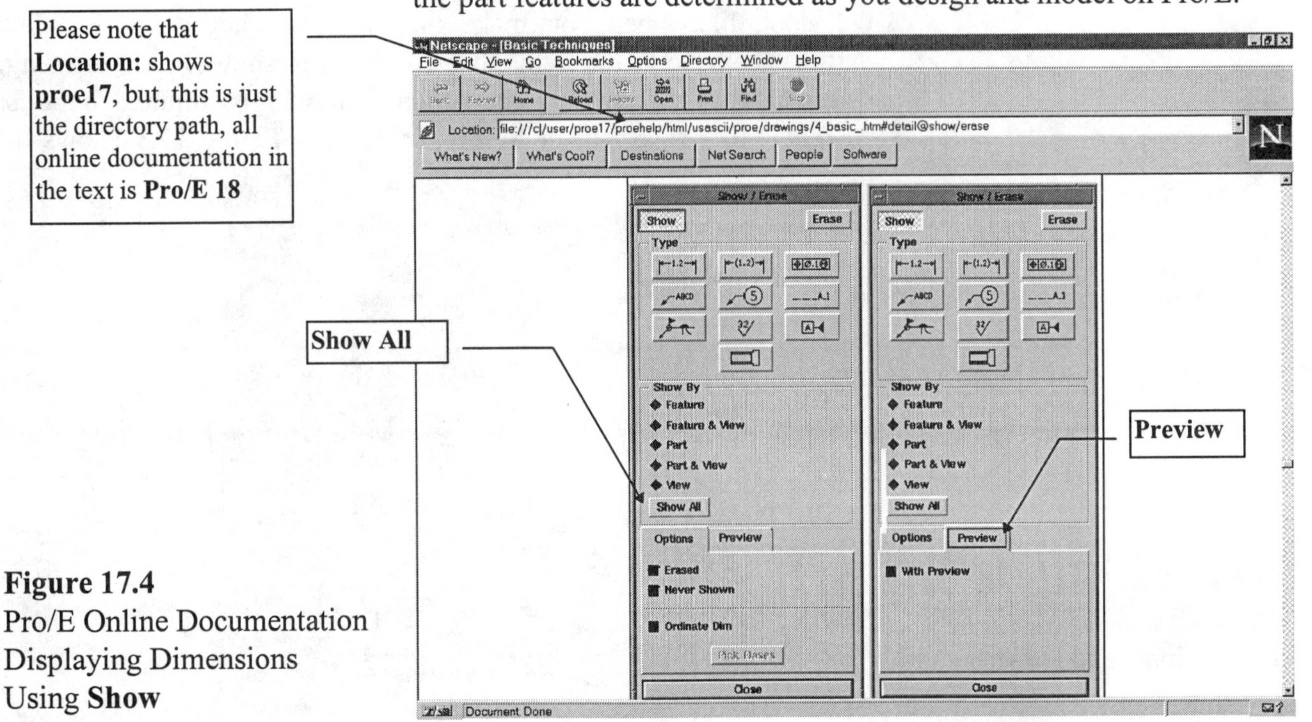

Figure 17.4
Pro/E Online Documentation Displaying Dimensions Using **Show**

When detailing, you simply ask Pro/E to display views needed to describe the part and then display the dimensions used to model the part. These are the same dimensions used in the part design. You cannot underdimension the part or overdimension, since Pro/E displays exactly what is required to model the part. Pro/E will not duplicate dimensions on a drawing. If a dimension is shown in one view, it will not be shown in another view. The dimension, however, can be switched to the other view via detailing options.

To display dimensions on a drawing:

1. Choose **Detail** from the DRAWING menu.
2. Choose **Show/Erase.** The **Show/Erase** dialog box displays. Select the dimension icon in the **Type** area. To specify the dimensions as ordinate, choose the **Ordinate Dim** check box at the bottom of the dialog box, then pick a baseline by choosing the **Pick Bases** command button. The base line must already exist, in this case.
3. Choose one of the following radio buttons:

   Feature Show all the dimensions associated with a particular feature in the appropriate views. Select a feature to be dimensioned.
   Feat & View Show all dimensions for a single feature in a single view. Select a feature in the view where the dimensions are to be displayed.
   Part Shows dimensions associated with part.
   Part & View Shows dimensions associated with a part in a view.
   View Show all the dimensions associated with a selected view. Select the view(s) you would like dimensioned.

To show all the dimensions for the current model, choose the **Show All** command button (Fig. 17.4). To *preview* what the drawing will look like when you make the changes, choose the **Preview** button. To specify the items that you wish to show on the drawing, select **With Preview** and one of the following command buttons: **Accept All, Erase All,** or **Select To Keep**. Figure 17.5 shows all dimensions.

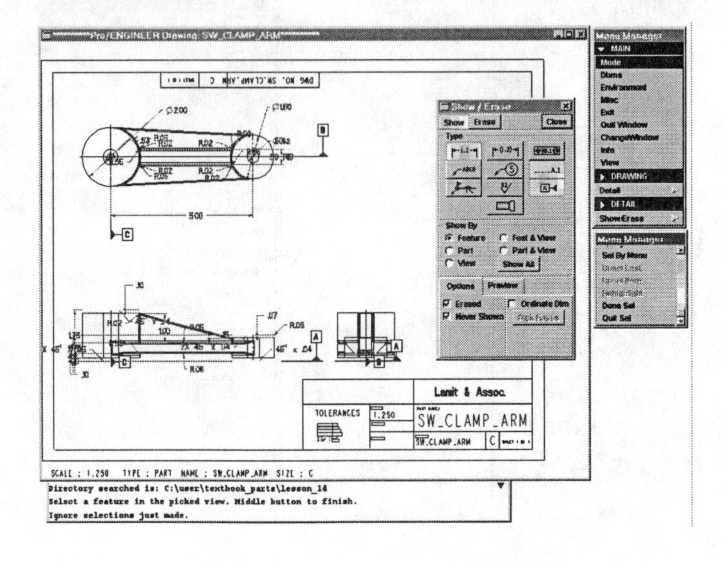

Figure 17.5
Drawing with Axes, Dimensions, and Datums Displayed

The **Clean Dims** option allows you to distribute standard and ordinate dimensions with equidistant spacing along witness lines, displaying them in a more orderly and readable fashion.

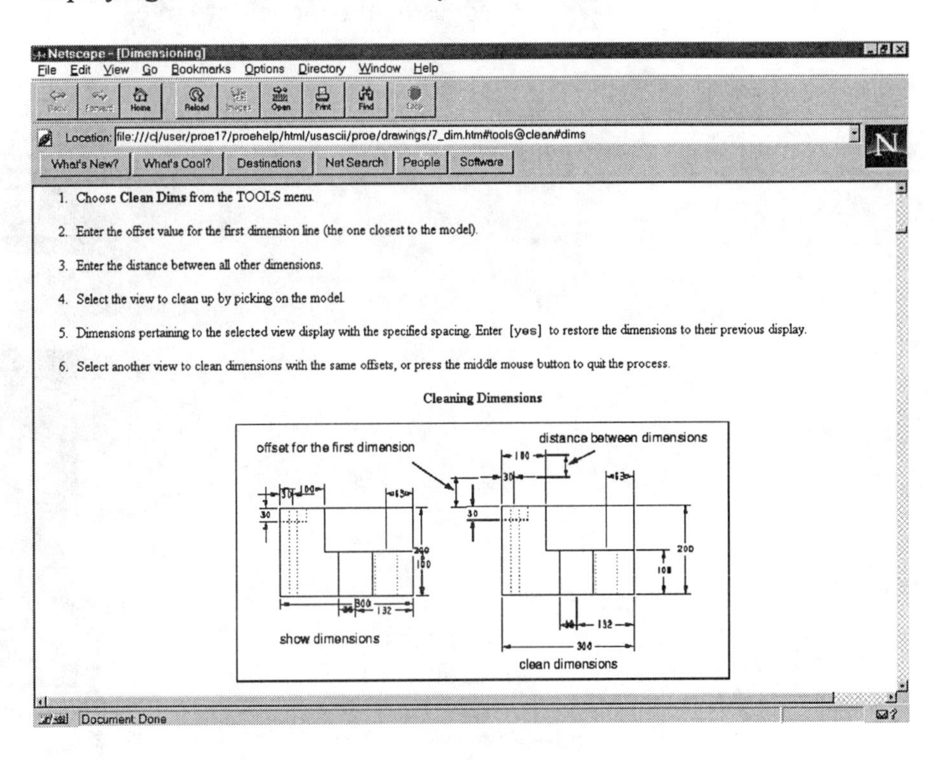

Figure 17.6
Pro/E Online Documentation
Cleaning Dimensions

To clean up the dimension display:

1. Select **Detail ⇒ Tools ⇒ Clean Dims**
2. Enter the offset value for the first dimension line (the one that is closest to the model).
3. Enter the distance between all other dimensions.
4. Select the view to be cleaned up by picking on the model.
5. Dimensions pertaining to the selected view are displayed with the specified spacing. In the event that the new display of dimensions is unsatisfactory, Pro/E will ask if you would like to move the dimensions back to their previous display. Pick **(YES)** to restore.
6. Select another view to clean dimensions with the same offsets, or press the middle mouse button to quit the process.

The cleaned dimensions are usually not in the best positions for each dimension and note. After cleanup, the next step is normally to move and reposition the dimensions to create an ASME standard drawing.

Dimensions can be removed from the display by erasing them. Erasing dimensions does not delete them from the model (driving dimensions, "true part" dimensions, *cannot be deleted*, but reference dimensions and driven dimensions can). Dimensions that have been erased can be redisplayed via the **Show/Erase** dialog box.

Text and Notes on Drawings

Notes (Fig. 17.7) can be part of a dimension, attached to one or many edges on the model, or "free." You can add notes using the keyboard or by reading them from a text file. Notes are created with the default values (height, font, etc.) specified in the drawing setup **(.dtl)** file.

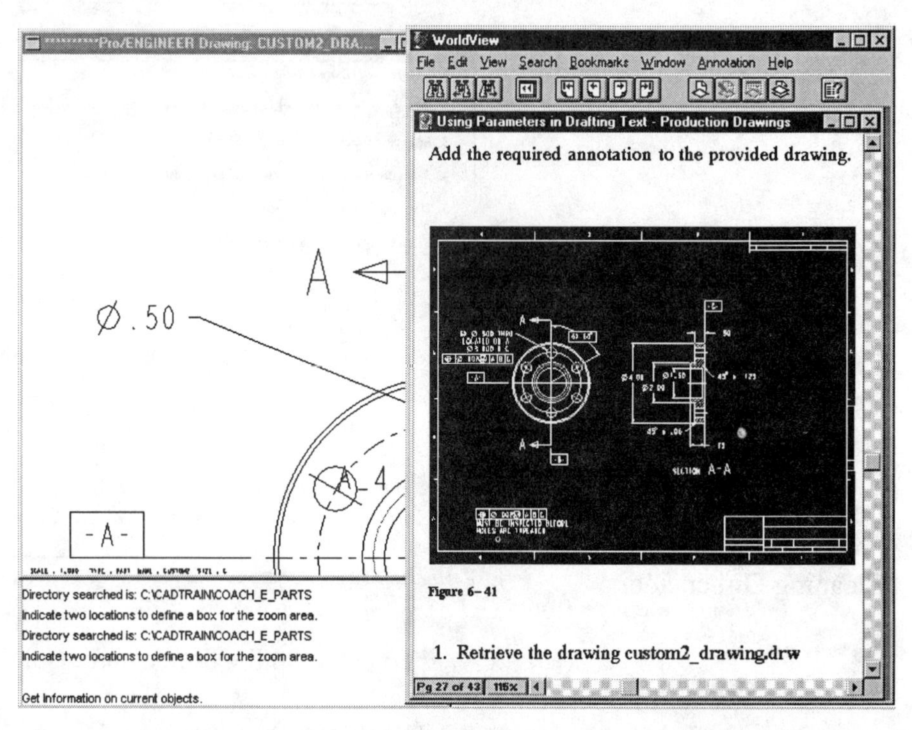

Figure 17.7
COAch for Pro/E, Using Parameters in Drafting Text--Production Drawings (Segment 2: Text Parameters)

To add notes to the drawing, choose **Detail ⇒ Create ⇒ Note ⇒ Make Note**. Notes (Fig. 17.8) can be added with or without leaders, and the style of text can be modified via a dialog box (Fig. 17.9). The following options are available:

No Leader/Leader/On Item Create a note, with or without a leader.
ISO Leader Create a note with ISO standard leader line.
Enter/File Enter the note from the keyboard, or read the note from a text file.
Horizontal/Vertical/Angular Create a horizontal or vertical note, or enter an angular value between **0°** and **359°**.
Standard Create notes with multiple leaders.
Normal Ldr Create a note with a leader that is normal to an entity.
Tangent Ldr Create a note that is tangent to an entity.
Left/Center/Right/Default Create the note text as left-justified, center-justified, or right-justified. **Default** is left-justified.

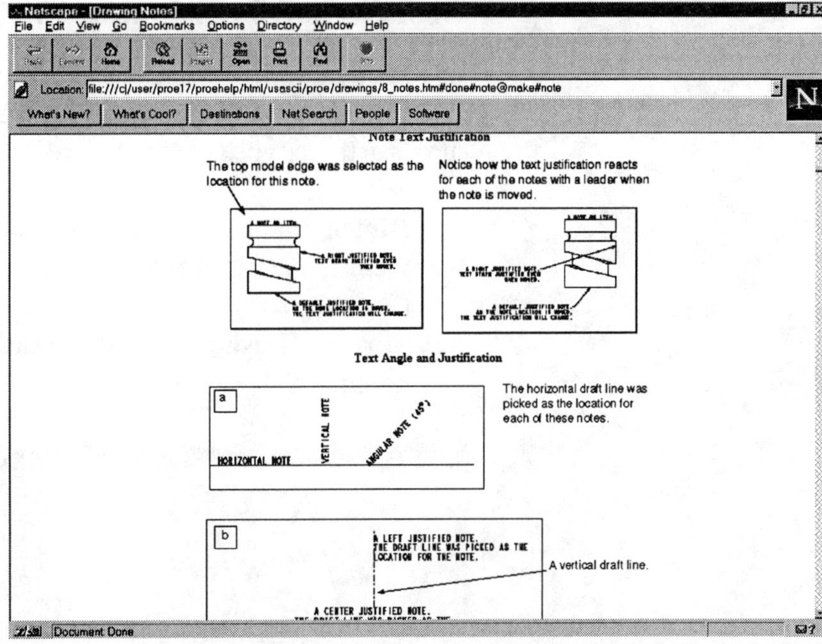

Figure 17.8
Pro/E Online Documentation
Notes

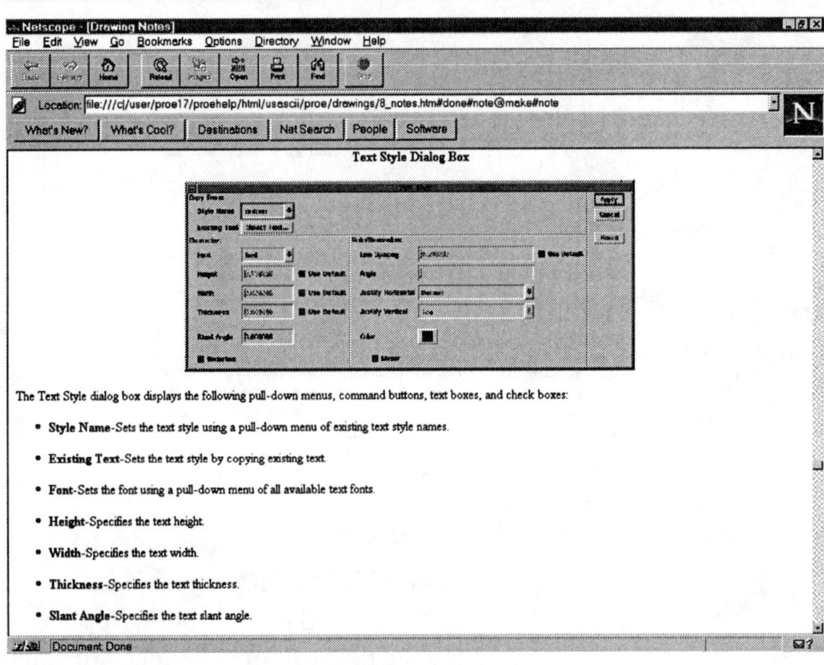

Figure 17.9
Pro/E Online Documentation
Text Style Dialog Box

Adding Text to a Dimension

The **Text** option from the MODIFY DRAW menu allows you to add text to a dimension value (for example, **DIAMETER**, **REF**, and **TYP**), as well as special symbols. You can also define your own special fonts and symbols. To add text:

1. Choose **Text** from the MODIFY DRAW menu.
2. Choose **Text Line** to modify one line of the note or dimension or **Full Note** to modify the whole note or dimension.
3. Pick the dimension to which to add the text.
4. Enter the line or lines of text. Each line must end with **enter**.
5. To complete the text string, follow the line with a carriage return, or exit the editor when finished with the **Full Note**.

Geometric Dimensioning and Tolerancing

The manufacturing of parts and assemblies uses a degree of precision determined by **tolerances**. A typical parametric design system supports three types of tolerances:

Dimensional Specifies allowable variation of size.
Geometric Controls form, profile, orientation, and runout (Fig. 17.10).
Surface finish Controls the deviation of a part surface from its nominal value.

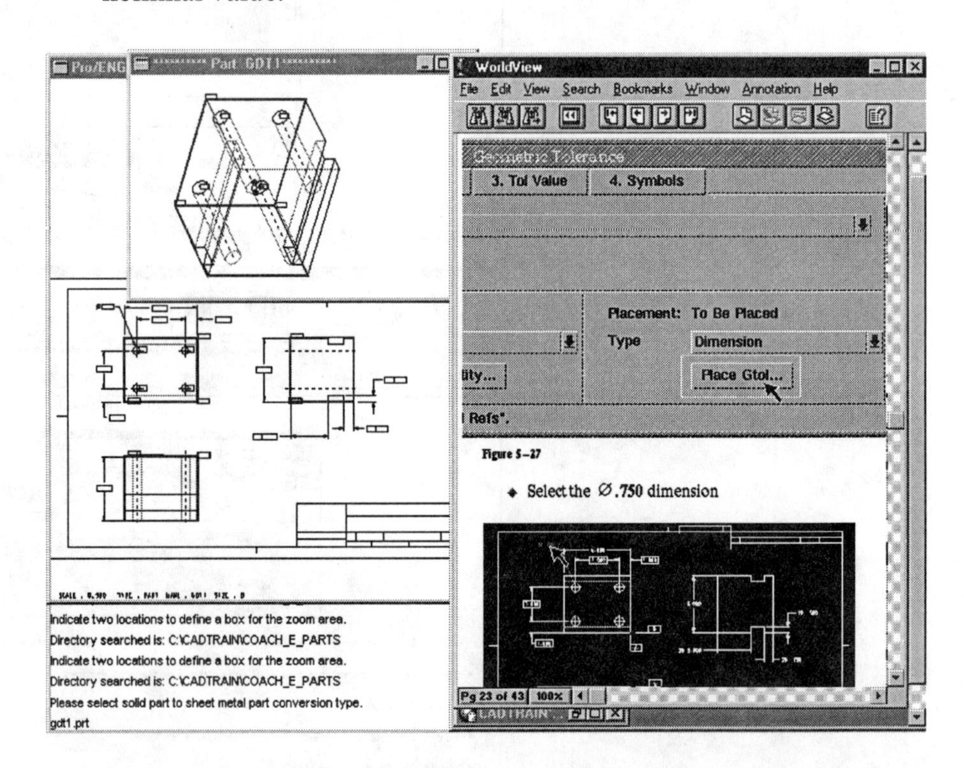

Figure 17.10
COAch for Pro/E, Creating Tolerance Frames--GD&T (Segment 3: Defining a Frame)

When you design a part, you specify dimensional tolerance -- ***allowable variations in size***. All dimensions are controlled by tolerances. The exception applies only to "**basic**" dimensions, which for the purpose of reference are considered to be exact.

Dimensional tolerances on a drawing can be expressed in two forms:

* As **general tolerances** presented in a tolerance table. These apply to those dimensions that are displayed in nominal format, that is, without tolerances.
* As **individual tolerances** specified for individual dimensions.

You can use general tolerances given as defaults in a table or set individual tolerances by modifying default values of selected dimensions. Default tolerance values are used at the moment you start to create a model; therefore, default tolerances must be set prior to creating geometry.

Pro/E recognizes six decimal places for which you can specify the tolerance values. When you start to create a part, the table at the bottom of the window will display the current defaults for tolerances. If you have not specified tolerances, Pro/E defaults are assumed, and the table will look as follows:

*** x.x ±0.1**
*** x.xx ±0.01**
*** x.xxx ±0.001**
*** ANG. ±0.5**

You have a choice of displaying or blanking tolerances. If tolerances are not displayed, Pro/E still stores dimensions with their default tolerances. You can specify geometric tolerances, create "basic" dimensions, and set selected datums as reference datums for geometric tolerancing. ISO tolerances are generated from a table, as shown in Figure 17.11.

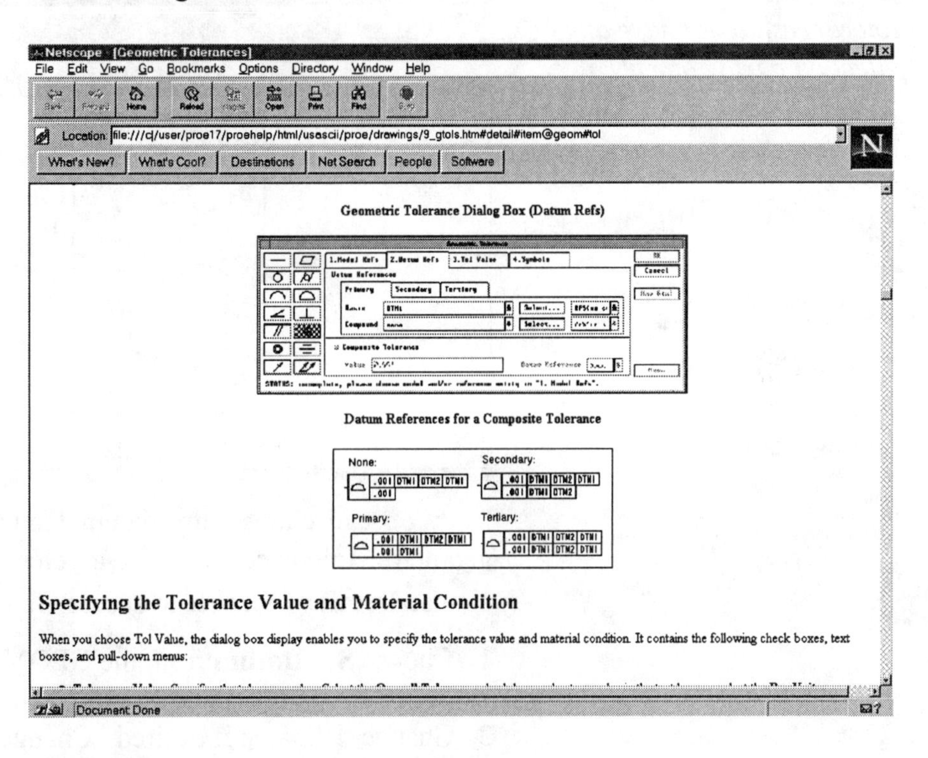

Figure 17.11
Pro/E Online Documentation Geometric Tolerance Dialog Box

The available tolerance formats are:

Nominal Dimensions displayed without tolerances.
Limits Tolerances displayed as upper and lower limits.
Plus-Minus Tolerances displayed as nominal with plus-minus tolerance. The positive and negative values are independent.
±Symmetric Tolerances displayed as nominal with a single value for both the positive and the negative tolerance.
As Is Tolerances as is.

Geometric tolerances provide a method for controlling the location, form, profile, orientation, and runout of features. You add geometric tolerances to the model from the Part mode or the Drawing mode. The geometric tolerances are treated by Pro/E as annotations, and they are always associated with the model. *Unlike dimensional tolerances, geometric tolerances do not have any effect on part geometry.*

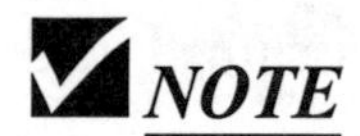

NOTE

Pro/DETAIL enables you to add geometric tolerances to the model from the Draw mode. Note that geometric tolerances can be added in the Part or Draw mode, but are reflected in all other modes. Geometric tolerances are treated by Pro/E as *annotations*, and they are always associated with the model. Unlike dimensional tolerances, *geometric tolerances do not have any effect on part geometry.*

When adding a geometric tolerance to the model, you can attach it to existing dimensions, edges, and existing geometric tolerances, or display them as notes without a leader.

Before you can reference a datum in a geometric tolerance, you must first indicate your intention by "**setting**" the datum (Fig. 17.12). Once a datum is set, the datum name is prefixed and appended by hyphens, and it is enclosed in a rectangle (for the old standards). The ASME Y14.5M 1994 standards display the datum name, as in Figure 17.13 and Figure 17.14. You can change the name of a datum either before or after it has been "set" by using the **Name** item in the SET UP menu from the Part mode

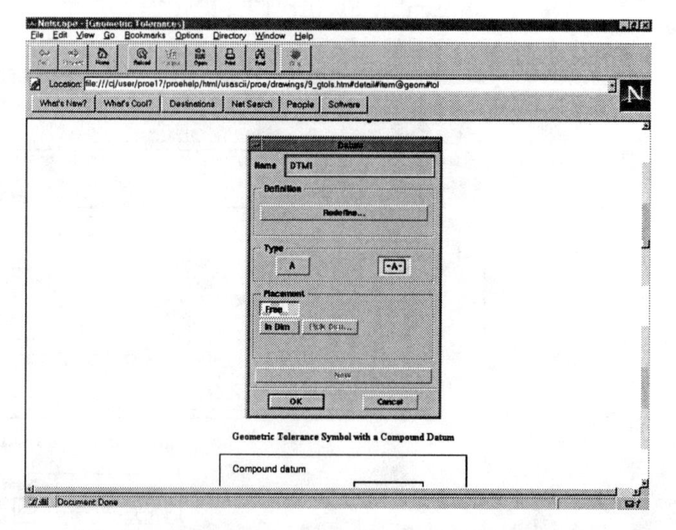

Figure 17.12
Pro/E Online Documentation
Setting Datums

You can choose any datum feature as a reference datum for a geometric tolerance. To set a reference datum in he Draw mode choose **Detail** ⇒ **Create** to access **Geo Tol**:

1. Choose **Set Datum** from the GEOM TOL menu.
2. Select the datum plane or axis to be set.
3. Change the name if desired. Change the type of feature control frame then **Okay**.
4. The datum is enclosed in a feature control frame.

A geometric tolerance for individual features is specified by means of a **feature control frame** (a rectangle) divided into compartments containing the geometric tolerance symbol followed by the tolerance value. Where applicable, the tolerance is followed by a **material condition symbol**. Where a geometric tolerance is related to a datum, the reference datum name is placed in a compartment following the tolerance value. Where applicable, the datum reference letter is followed by a material condition symbol.

For each class of tolerance, the types of tolerances' available, and the appropriate types of entities can be referenced. The available material condition symbols are shown in the dialog box (Fig. 17.11).

HINT

Change your **.dtl** file using **Set Up ⇒ Modify Val ⇒ gtol_datums STD_ANSI** change to **STD_ASME ⇒ File ⇒ Save ⇒ File ⇒ Exit** to see your set datums displayed with the correct ASME Y14.5 1994 symbology (Figure 17.14)

You are guided in the building of a geometric tolerance by Pro/E requests for each piece of required information. You respond by making menu choices, entering a tolerance value, and selecting entities and datums. As the tolerance is built, the choices are limited to those items that make sense, in the context of the information you have already provided. For example, if the geometric characteristic is one that does not require a datum reference, you will not be prompted for one. Other checks are made to help prevent mistakes in the selection of entities and datums.

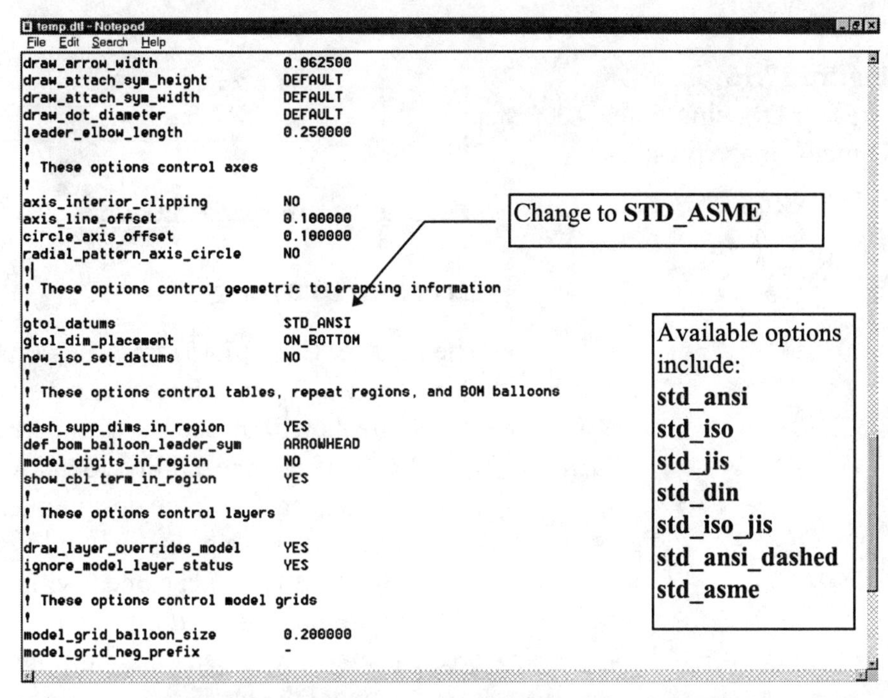

Figure 17.13
Setting Up the Standard

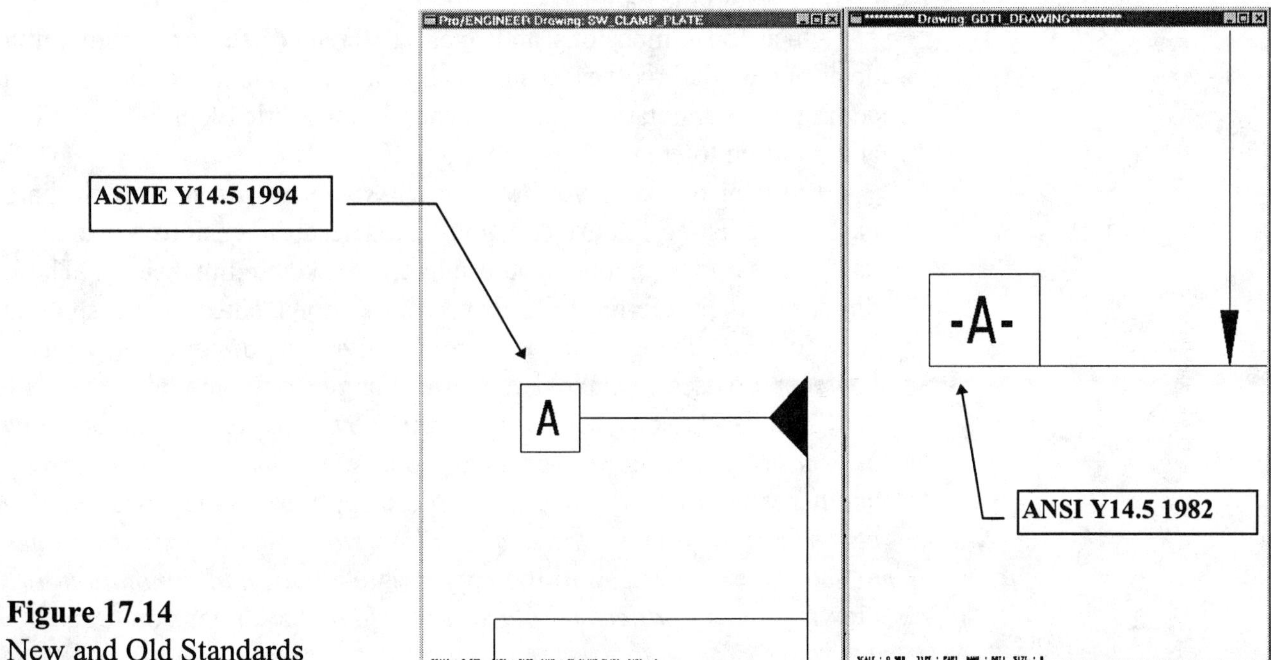

Figure 17.14
New and Old Standards

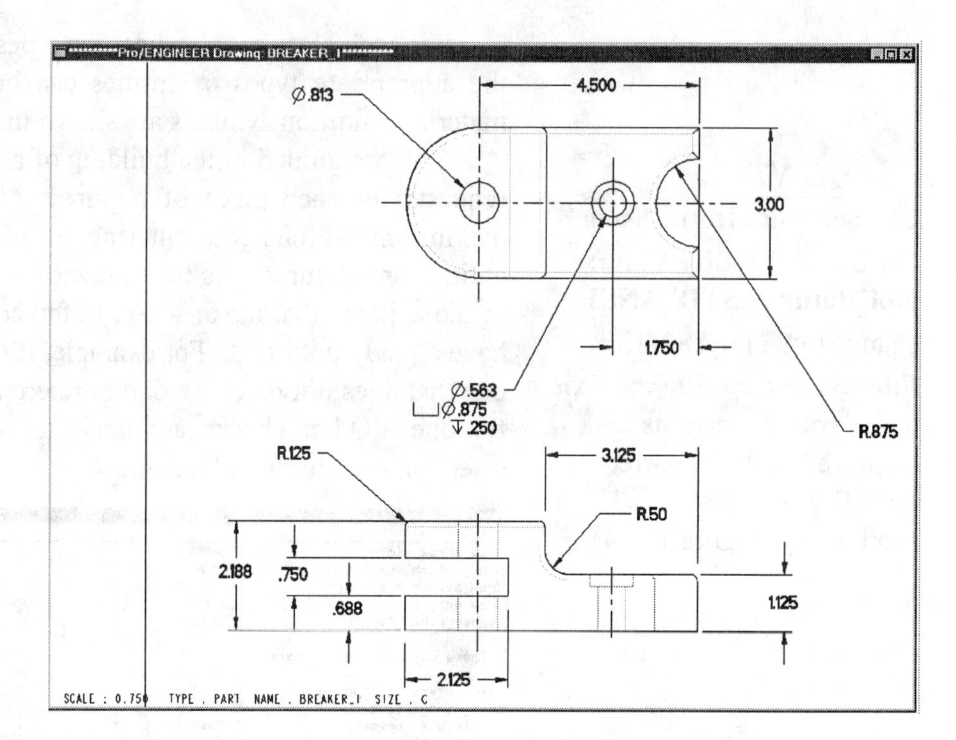

Figure 17.15
Breaker Drawing with Dimensions, Notes, and Centerlines

Breaker Drawing

The Breaker (Fig. 17.15) from Lesson 3 will be detailed in this lesson. Dimensioning, centerlines (axes), annotation, and other drawing requirements will be used to complete the detailing of the model.

Pro/E offers the ability to allow you to dimension a drawing automatically based on the design intent dimensioning scheme used to create the part. Using this method, you are virtually assured that the part will be fully dimensioned without being over-dimensioned. In addition to automatic dimensioning, the system can automatically create centerlines based on Datum Axes and Radial Patterns that were used to model the part.

Once the dimensions and axes are created, the only remaining detailing work involves cosmetically cleaning up their display and adding other annotation, such as general notes, title block information, and required tolerancing information.

Pro/E also allows you to *create associative dimensions* that are based on existing feature constraints. These dimensions are created using the same techniques you use to create sketch dimensions. These dimensions are referred to as *driven* dimensions, since their display is "driven" by changes to the model. *Driven dimensions* cannot, however, be used to make parametric changes to the model.

In this lesson you will learn how to: *create dimensions automatically* using the original constraints, *move dimensions* to another view, move dimensions to more appropriate locations, *display centerlines* on your drawing, *erase dimensions*, *modify extension lines* to show the proper gap to the appropriate edges, *add annotation* to a drawing, and *alter decimal places* and *add text* to dimension text.

The model needs to be brought into the Drawing mode and placed on a format. The format can be a blank format, a Pro/E-provided ANSI standard blank format, or a user-defined format created with parameters that will automatically display title block information and sheet callouts, as was presented in Lesson 16. Use the following commands:

> **Mode** ⇒ **Part** ⇒ **Search/Retr** ⇒ (choose **Breaker** from the directory list) ⇒ (move and reduce the size of the part window) ⇒ **Mode** ⇒ **Drawing** ⇒ **Create** ⇒ (**Breaker_dwg** or some other logical name) ⇒ **enter** ⇒ (from the DWG SIZE menu pick **C**) ⇒ **Views** ⇒ **enter** to accept the default (or **?** ⇒ **enter** ⇒ and choose the part model name from the directory list) ⇒ **Done** (to accept the defaults) ⇒ (pick a place on the drawing for the first view, as shown in Figure 17.16) ⇒ **Front** ⇒ (pick **DTM3**) ⇒ **Top** ⇒ (pick **DTM2**) ⇒ **Done-Return** ⇒ **Add View** ⇒ **Done** (to accept defaults) ⇒ (pick in the area you wish to have a front view, as shown in Fig. 17.17) ⇒ **Done/Return**

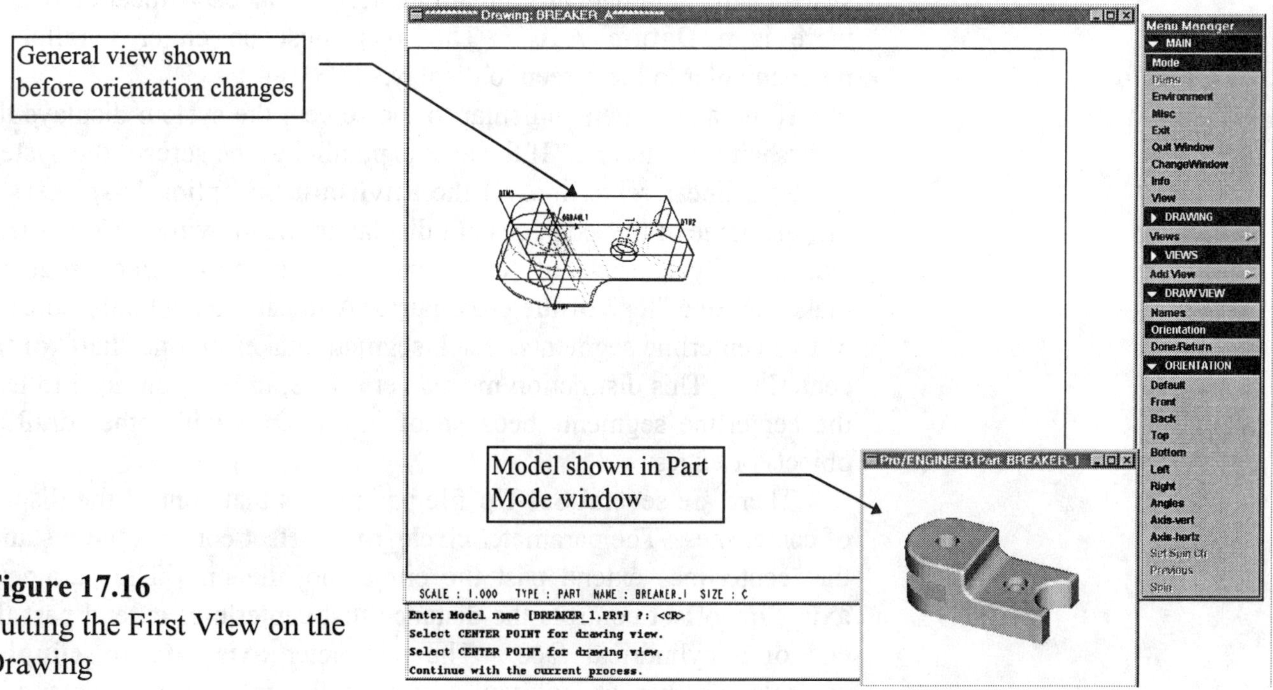

Figure 17.16
Putting the First View on the Drawing

After you have the two views on your drawing, use **Move View** to relocate them to more ideal positions for detailing. Remember, you can reposition the views at any time in the detailing process, even after the dimensions are placed.

Dbms ⇒ Save ⇒ enter Purge ⇒ enter ⇒ Done-Return

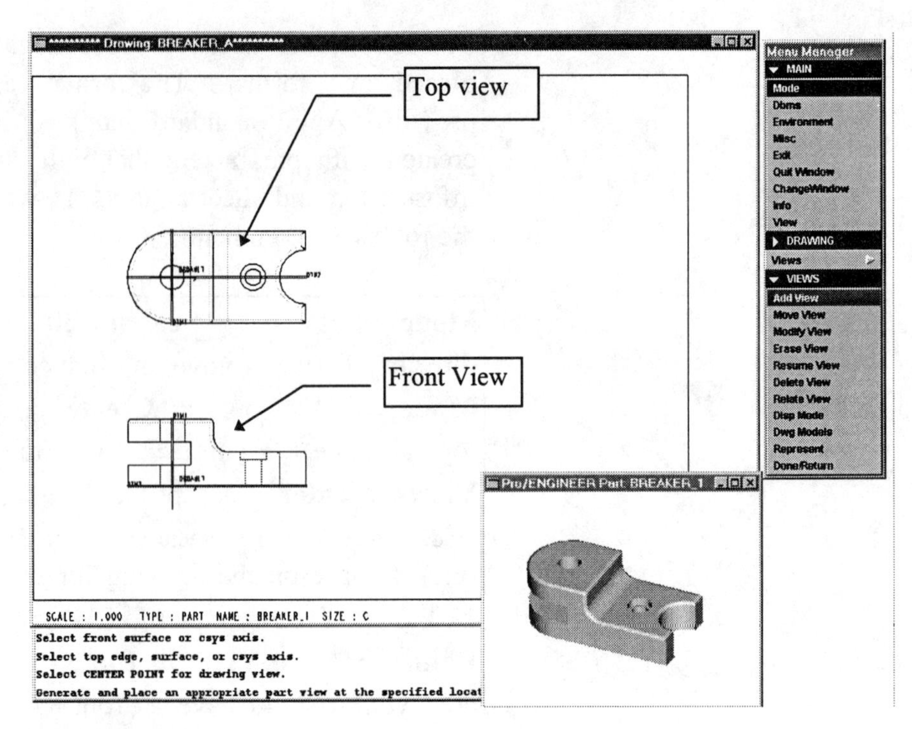

Figure 17.17
Putting the Second View on the Drawing

Showing Axes

Before adding dimensions, you can display the centerlines. The system will automatically create centerlines at each location where there is a **Datum Axis**. The axis must be either parallel or perpendicular to the screen to display.

If the axis is perpendicular to the screen, the system displays the "crosshairs" centerline. If the axis is parallel to the screen, the system creates a linear centerline. If the **Environment** option **Disp Axes** is off, the name of the axis does not display in the drawing. A crosshairs axis actually consists of four centerline segments. Each segment makes up one "leg" of the crosshairs. A linear axis actually consists of two centerline segments. Each segment makes up one "half" of the centerline. This distinction means very little, unless you need to trim the centerline segments because of interference with other drafting objects or edges.

There are several **Set-Up** file parameters that control the display of centerlines. The parameter **circle_axis_offset** controls the distance that centerlines extend past the circle they lie on. The parameter **axis_line_offset** controls the distance that centerlines extend past the end of a cylindrical face. The parameter **axis_interior_clipping** controls whether or not you can trim the interior segments of a centerline. If this parameter is set to **no**, you can only shorten or lengthen the ends of linear and crosshairs centerlines. If the parameter is set to **yes**, you can shorten or lengthen any end of any segment. The parameter **radial_pattern_axis_circle** will display a *bolt circle centerline*.

To show the centerlines/axes for the Breaker, give the following commands:

HINT
Always turn off most of the environment choices so that they do not clutter the drawing views:

- ☐ **Disp DtmPln**
- ☐ **Disp Points**
- ☐ **Disp Axes**
- ☐ **Disp Csys**
- ☐ **Spin Center**
- ☐ **Grid Snap**

Detail ⇒ **Show/Erase** ⇒ **A_1** (radio button on) ⇒ **Show All** ⇒ **Yes** (Fig. 17.18) ⇒ **Close**

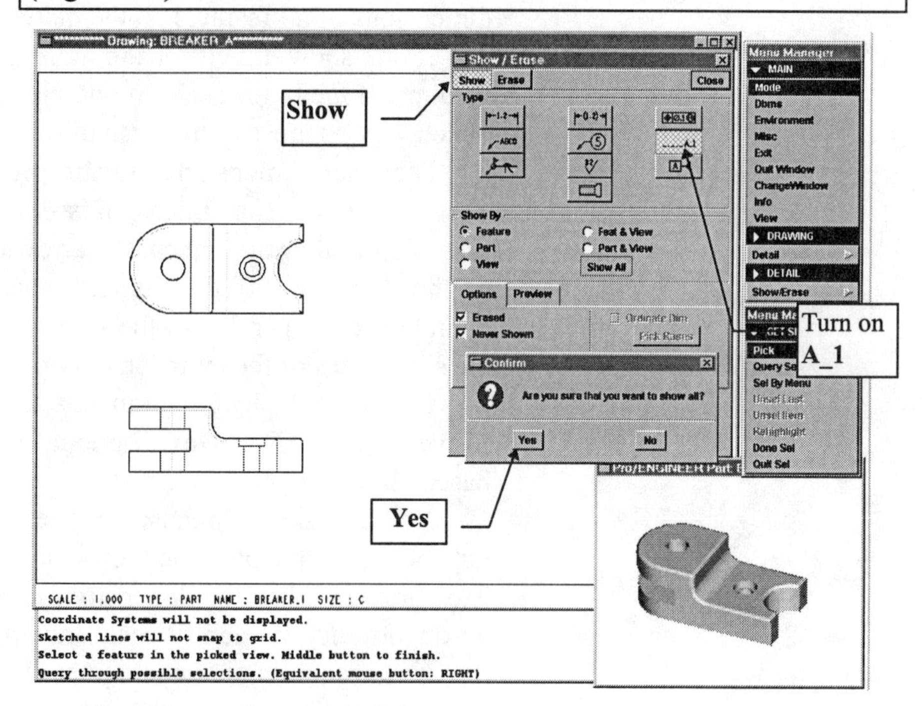

Figure 17.18
Displaying Axes

You can edit an axis with a pop-up menu by *clicking the right mouse button while the cursor is in the main window* (when in the Drawing mode**)**, as shown in Figure 17.19. From the pop-up window, select **Modify Item**, then select the axis to modify. The **Move** option is used to move the end of the axis. Modify the axes on both views to their required positions.

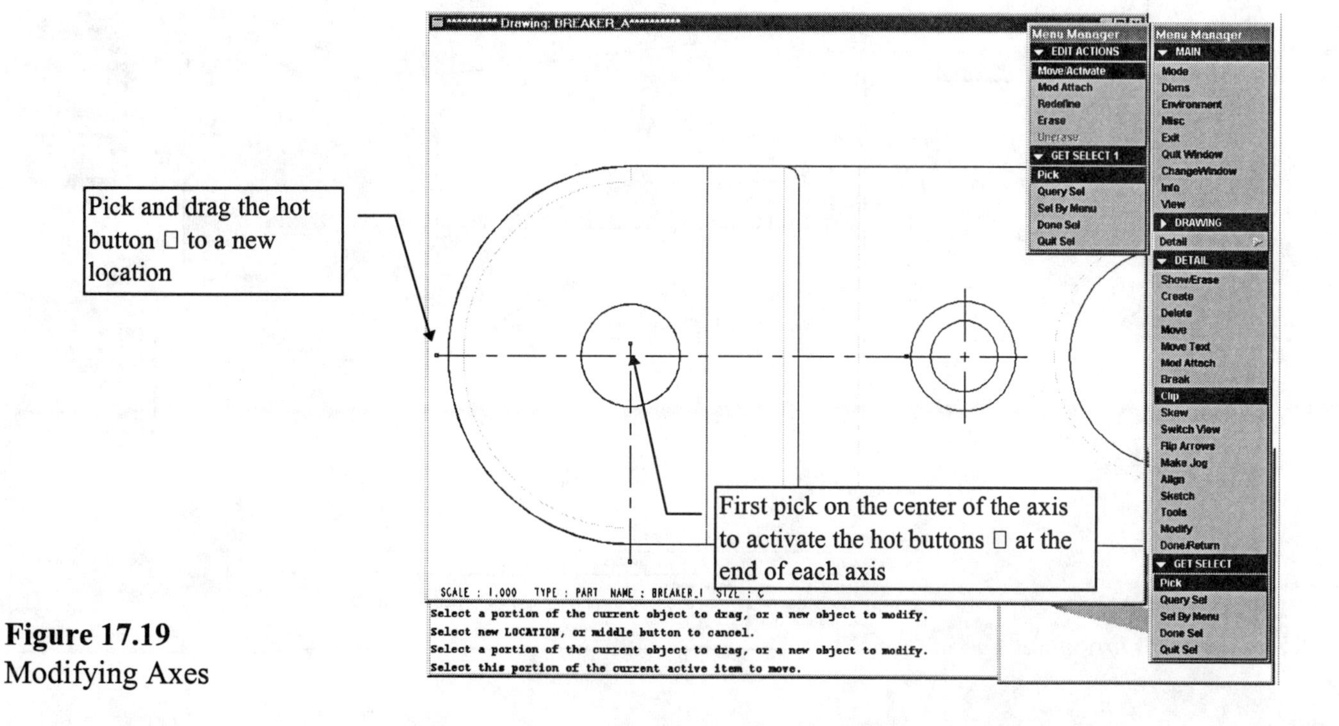

Figure 17.19
Modifying Axes

Show Dimensions

The **Show** option allows you to create dimensions based on the dimensional references and sketch dimensions that were used to create the model. You can show the dimensions of a feature, all of the dimensions of all features in a particular view, or all of the dimensions of all features on the model.

If you show dimensions in more than one view, and a feature can be dimensioned in more than one view, the system attempts to decide which view is most appropriate to show the dimensions in.

Diameter dimensions are displayed differently based upon the view they lie in. If a dimension is shown in a view where the cylinder axis is normal to the screen, the arrows are drawn to the circular edge of the cylinder. If the dimension is shown in a view where the cylinder axis is parallel to the screen, extension lines are added along the silhouettes of the cylinder.

Changing the length of an axis can also be accomplished with the **Move** option. The **Move** option works on either end of the axis, based on which end is selected.

You can show dimensions, axes, datums, etc. at the same time, but for a complex part, the views quickly become cluttered. Showing axes first, and then the dimensions, cosmetic features, etc., gives you an opportunity to modify each drawing entity type separately to see the view requirements more clearly. Figure 17.20 illustrates the dimensions displayed on the drawing using the following commands (your drawing may differ slightly):

Detail ⇒ **Show/Erase** ⇒ (**Dimension** radio button on) ⇒ **Show All** ⇒ **Yes** (Fig. 17.20) ⇒ **Close** ⇒ **Done/Return**

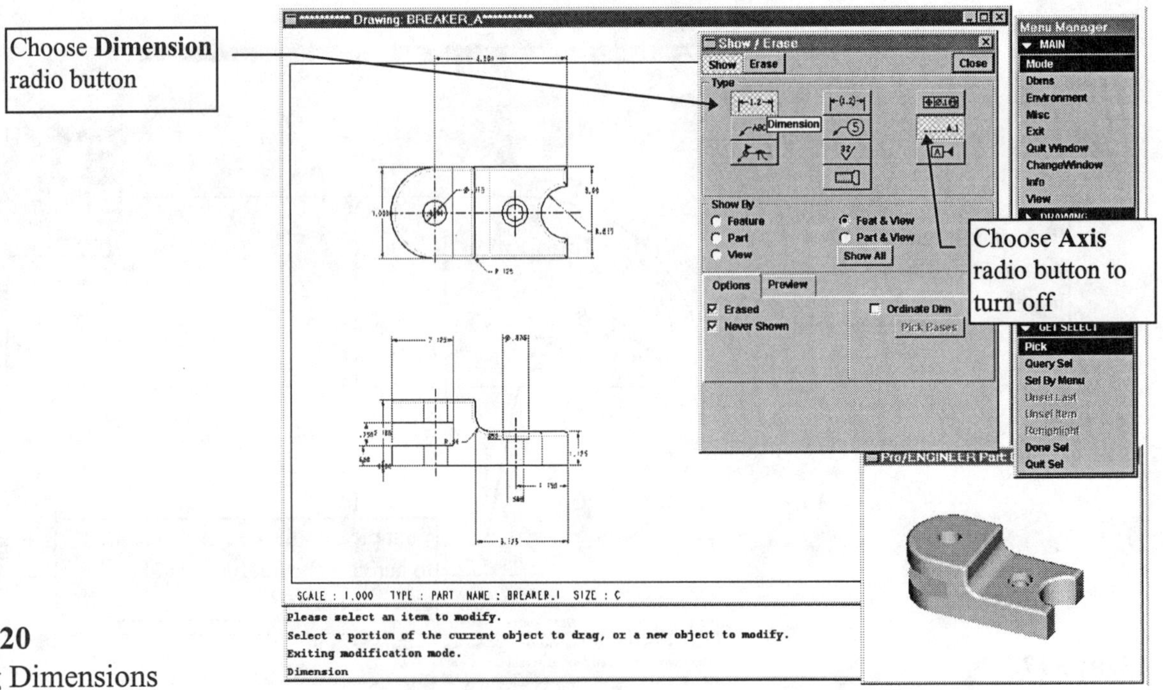

Figure 17.20
Displaying Dimensions

Before modifying the dimension locations and views, change the settings in the **.dtl** file using: **default_font** to **filled** and **draw_arrow_style** to **filled**: **Set Up** ⇒ **Modify Val** ⇒ (make the necessary changes) **File** ⇒ **Save** ⇒ **File** ⇒ **Exit** (Fig. 17.21) ⇒ **Quit**.

HINT

Always set your font and arrow style to filled:

default_font **filled**
draw_arrow_style **filled**

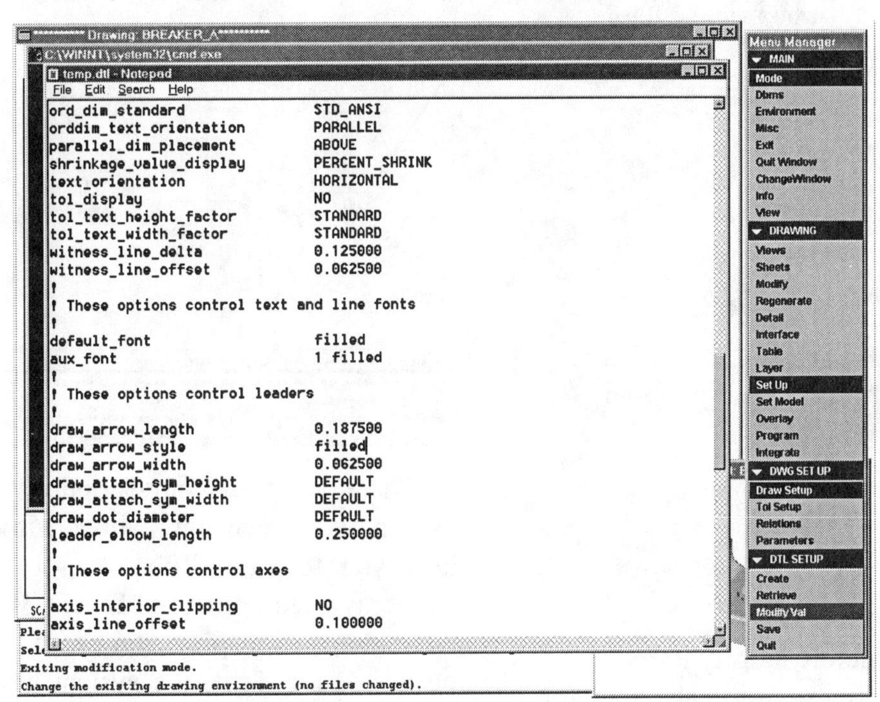

Figure 17.21
Set Up ⇒ Modify Val

Modifying Dimensions and Drawing Entities

When in **Drawing Mode**, you can access a pop-up menu item by clicking the right mouse button while the cursor is in the main window. From the pop-up window, select **Modify Item**, then select the item you wish to modify (Fig. 17.22). Another menu displays the actions that can be performed on this type of item. The **Move** option is the default. To move an item, simply select it, and you will be in dynamic move.

The list of additional actions you can perform on the item is also available in the pop-up menu. To modify another item, simply select it, and the menu will display the actions appropriate to it.

This functionality reduces the number of menus you must navigate. Also, it eliminates the need to traverse the screen to reach the menus, because the pop-up window is available at all times. For example, while the **Show/Erase** dialog box is displayed, you can use the pop-up menu option to move dimensions into the correct location without having to close the dialog box.

Move, **Move Text**, **Clip**, and **Skew** are all available using **Move** (Fig. 17.23). The location of your selection determines how items are moved. For example, if you select the dimension text, you can move the text of the dimension and the leader line. Selection of the leader line allows you to clip that leader line. Picking the witness line allows you to clip that leader line, and picking the other end of the witness line allows you to skew the dimension.

Dimensions and arrows show as **filled**

Click right mouse button

Figure 17.22
Modify Using Pop-up Window

HINT
Before modifying dimensions, make your grid size very small or turn it off: □ **Grid Snap**

There are three move options that allow you to change the position of a dimension. **Move, Move Text,** and **Move Attach** each allow you to alter different aspects of a dimension. The most commonly used option is **Move**. When you **Move** a dimension, the system drags its display along with the cursor in all directions. When you use **Move Text**, the system moves the text only and does not move the extension lines. When you use **Move Attach,** select a note, geometric tolerance, symbol, or surface finish to modify. Select the desired reference options, then select the leader line you want to move, and then the new attach point. Reposition the dimensions and clean up the drawing. Do not concern yourself about dimensions in the wrong view at this time.

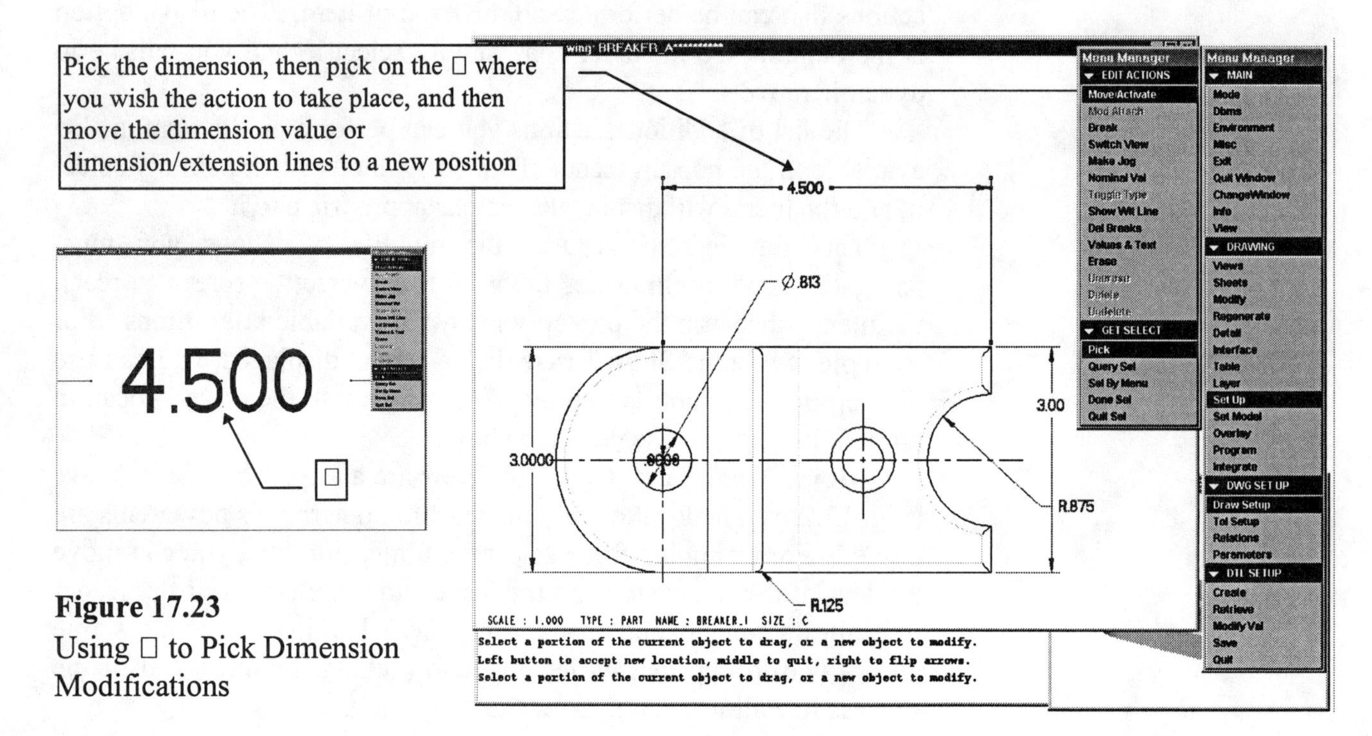

Figure 17.23
Using □ to Pick Dimension Modifications

Clipping Drafting Objects

The **Clip** option allows you to move the endpoint of many drafting object components, such as extension lines and centerlines. This option allows you to "pickup" the end of an extension line and drag it to a new length (Fig. 17.24). Of course, clipping can also be done with the pop-up option **Modify Item** menu.

Dbms ⇒ Save ⇒ enter Purge ⇒ enter ⇒ Done-Return
This will save the ***DRAWING***

One of the most common applications for **Clip** is to "regap" extension lines. When you **Show** feature dimensions, the extension lines are often gapped to a point on the part that is not appropriate for the view in which the dimension is displayed.

Clip allows you to drag the extension line end to a location that is visually more pleasing (and consistent with drafting standards). When you clip the extension line, the system remembers the clip distance and continues to apply it even if the model changes or if you **Erase** and redisplay the dimension.

When you **Clip** extension lines, the system prompts you to pick dimensions to clip. This allows you to clip several dimensions' extension lines together.

To stop picking dimensions, and to indicate which extension line you want to clip, move the cursor near the desired extension line end and click the *middle* mouse button.

If you are clipping multiple dimensions, the system "grabs" the extension line in each dimension that is closest to the cursor when you click the *middle* mouse button.

The system then begins to drag the extension line. To clip the line once you drag it to the desired location, click the *left* mouse button. If you want to abort the clip, click the *middle* mouse button.

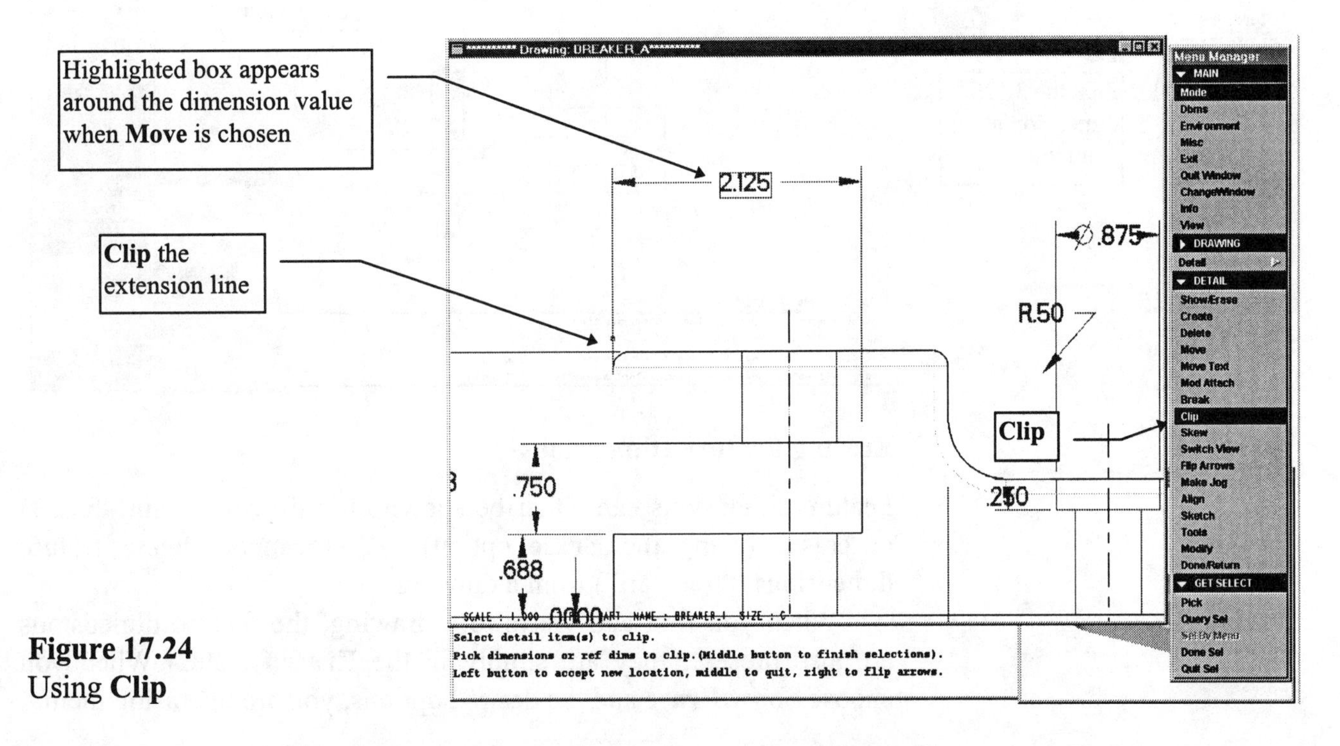

Figure 17.24
Using **Clip**

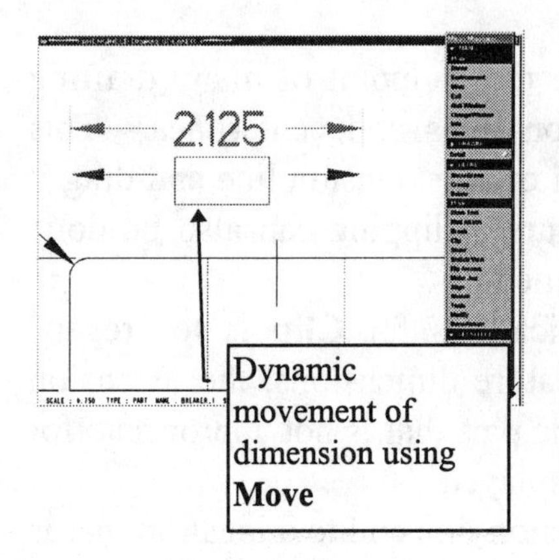

When you use **Move** to move a linear dimension, the system adjusts the leader, arrows, and extension lines in proper relation to the text (which is what you are actually dragging).

When you move the origin of the text from one side of the extension lines to the other, the system automatically flips the leader to the other side of the text.

If you place the text of a dimension between the extension lines and it really does not fit (because the extension lines are too close together), you may find the display of the arrows to be unacceptable.

When you move a radial dimension using **Move**, you can only move the text along the leader and the location where the arrow touches the arc, and thus alter the angle of the leader; however, the short line (stub) between the leader and the text stays at its original length. When you use **Move Text**, you can change only the length of the stub.

Besides moving dimensions, you will also want to switch the view that a dimension appears in after choosing **Show All**. In Figure 17.25, **Switch View** was chosen from the DETAIL Menu (also available after choosing a dimension when the **Modify Item** option is selected from the pop-up window menu).

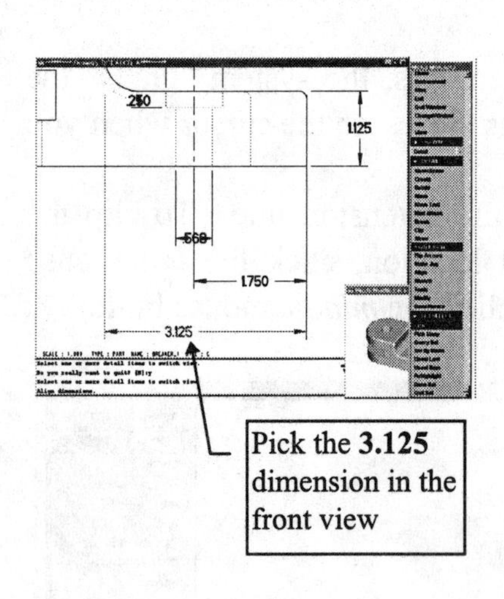

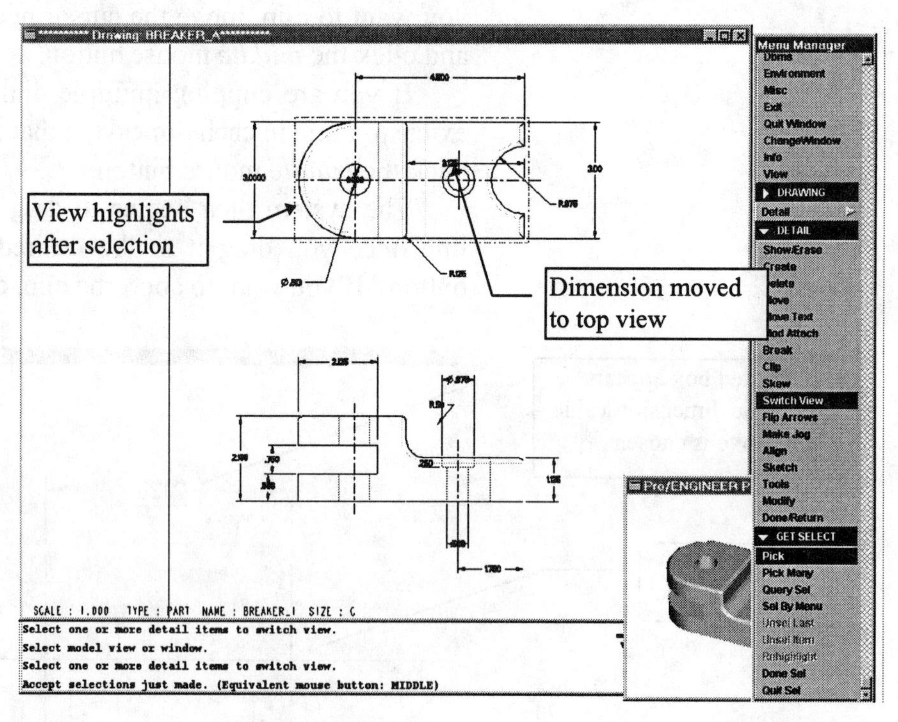

Figure 17.25
Switching Views of Dimensions

Erasing Feature Dimensions

Feature dimensions can either be shown on a drawing (using **Show**) or erased (using the **Erase** option). You cannot **Delete** feature dimensions ("true part") from a drawing.

When you first add views to a drawing, the feature dimensions are also present; they are simply in the **Erased** state. When you choose **Show/Erase** and the desired options, you are unerasing them.

Many times, when you **Show** feature dimensions, there are dimensions that you do not need. This is especially true of dimensions of thickness and pattern number callouts. You may also find that there are other dimensions you do not want.

If you want a dimension, but it appears in the wrong view, you do **NOT** want to erase it. The **Switch View** option allows you to move that dimension to the view where you want it displayed.

If, however, there are actually dimensions that you do not need, then you can **Erase** them. When you choose the **Erase** option, the system prompts you to select dimensions to erase. You use the **middle** mouse button to finish the selecting of the desired dimensions to erase.

If you accidentally erase dimensions, you can redisplay them using the **Show** option. Unfortunately, the **Show** option does not know exactly which dimensions you want to redisplay (from the set of dimensions that you have erased). Therefore, you may have to redisplay several dimensions, and re-erase those you really do not need. Using the **By Feat & By View** option with **Show** (instead of **Show All**) will minimize the number of dimensions to redisplay.

The **Erase** menu gives you a great deal of control over how you select the items to erase. You can choose to select individually a type of drafting object, or erase all objects of a type, in the drawing, in a view, and/or by feature.

If you choose to **Erase All** of an item, the system prompts you to confirm the erasure. If you type **N**, the erasure operation is aborted. If you type **Y**, the items are all erased.

The simplest way to erase drawing items is to use the **Show/Erase** dialog box and pick the **Selected Items** option as the choice. You may now erase items by selectively accept their removal using **Query Sel**.

Using the right button mouse, activating **Modify Item**, and picking the item to modify (as in Figure 17.25, where the **.0000** dimension was chosen) will provide a choice of several actions, including **Erase**. Picking **Erase** will make the dimension *gray out*. You may continue picking items to erase. When the selections are complete, choose **Done Sel** to remove the items from the screen. This method, although faster, provides fewer choices.

When you use the **Create ⇒ Datum ⇒ Plane ⇒ Offset ⇒ enter ⇒ enter ⇒ enter ⇒** method to establish datum planes and a coordinate system in the *Part mode*, you will also be creating three **.0000** dimensions that need erasing in the *Draw mode*. At this time erase the **.0000** dimensions (Fig. 17.26) from the two views and move the dimensions off the face of the views, per ASME standards.

Another aspect to change on the drawing is the number of digits displayed for individual dimensions (Fig. 17.27). In most cases the number of digits is dependent on the tolerance for the dimension.

Change the digits for dimensions displayed with four decimal places to three decimals (Fig. 17.27) by giving the commands:

Detail ⇒ **Modify** ⇒ **Num Digits** ⇒ (type **3** and **enter**) ⇒ (pick the dimension to alter) ⇒ **Done Sel** ⇒ **Done/Return**

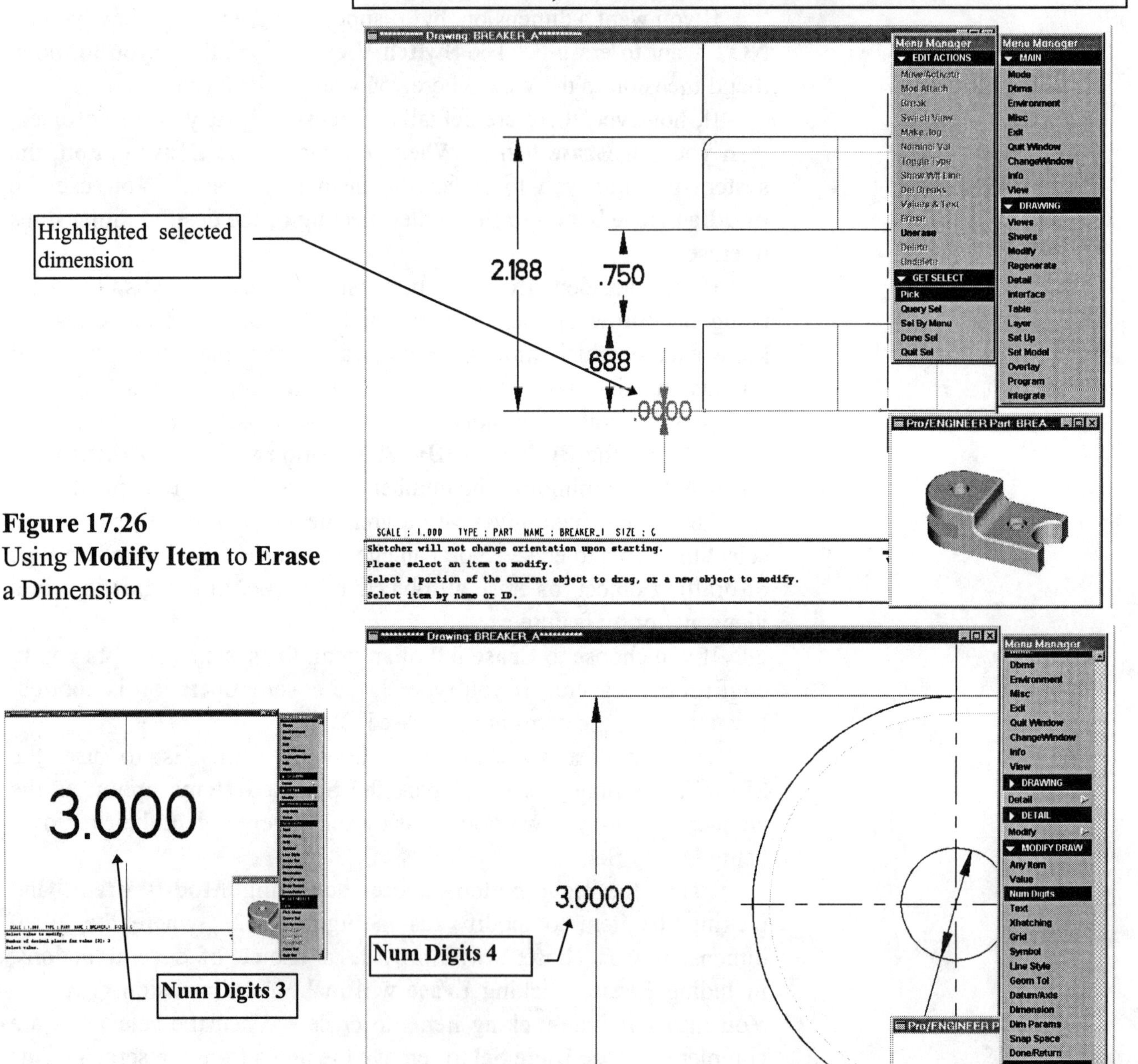

Figure 17.26
Using **Modify Item** to **Erase** a Dimension

Figure 17.27
Changing the Number of Digits

The **Flip Arrows** option is used to change the dimension arrows from inside to outside arrows or outside to inside arrows. Small dimension values will create arrows on top of the dimension value, and radial dimensions sometimes point to the wrong side of the arc, as in Figure 17.28, where the arc radius is dimensioned to the side of the arc that doesn't exist. All three dimensions shown there should be flipped.

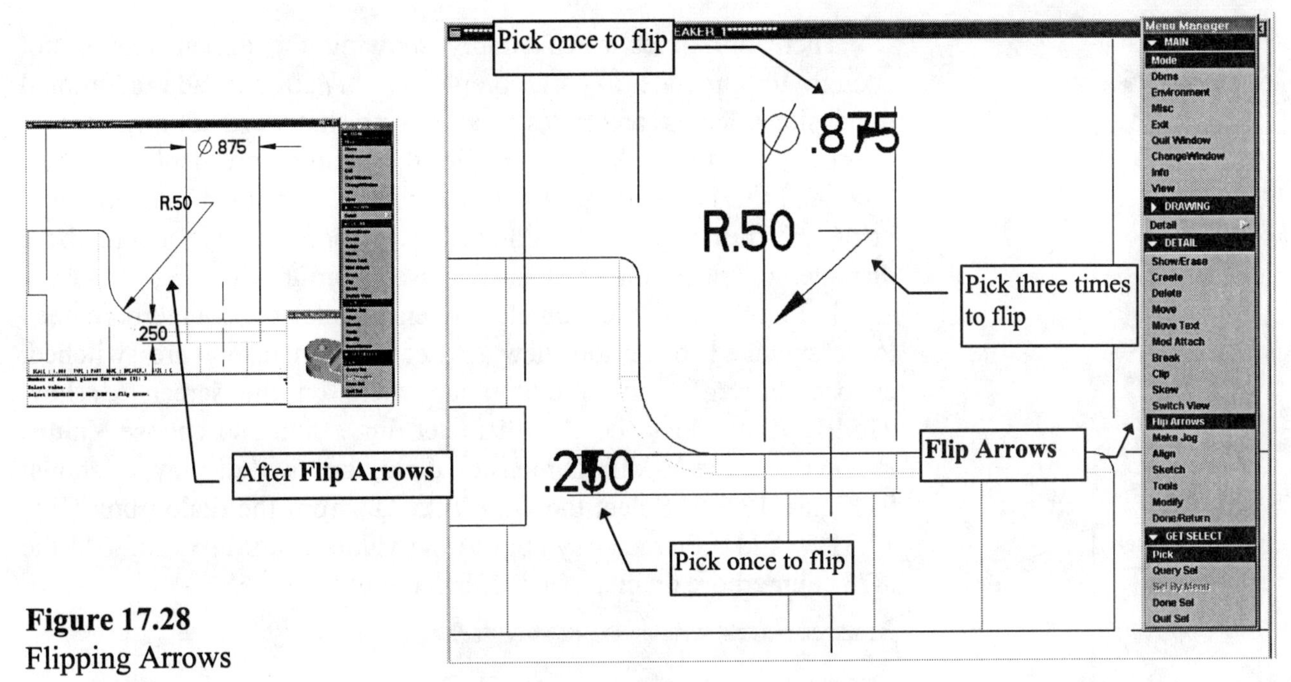

Figure 17.28
Flipping Arrows

To have diameter dimensions point to the outside of the circle with one arrowhead instead of across the diameter using two arrowheads, flip the arrows.

The dimensions are complete (Fig. 17.29), but there are a number of changes you need to make to create a correctly dimensioned detail of the part. In some cases, dimensions need to be combined into notes and reference dimensions added to the drawing.

Figure 17.29
Dimensioned Drawing

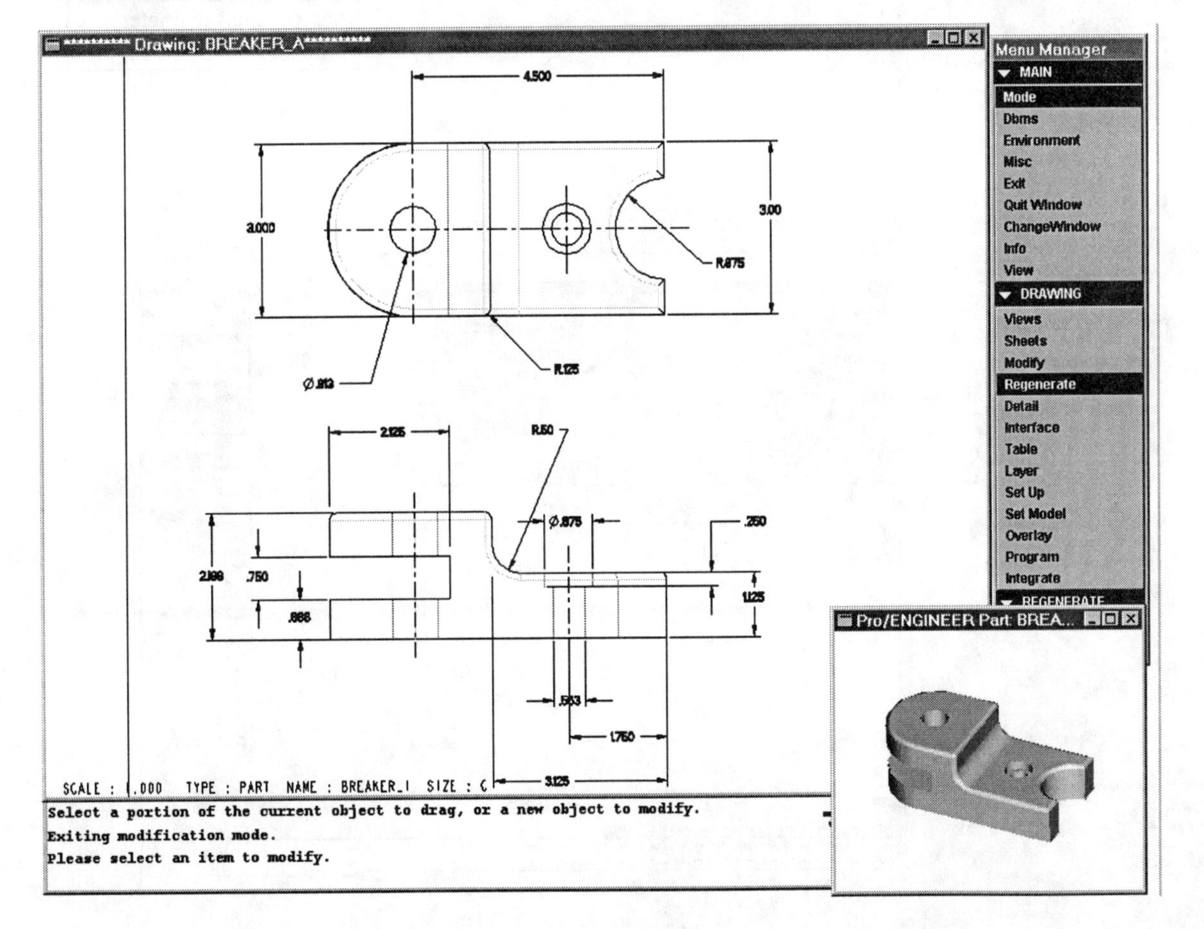

There are times when simply showing the dimensions is not enough to annotate a part completely. You can add additional dimensions, labels, and notes to a drawing that were not a part of its original definition. As an example, the counterbore depth, the thru hole, and the counterbore diameter need be combined into one note using the thru hole as the dimension to modify. The counterbore diameter and depth dimensions are erased from the drawing.

In Figure 17.30 the diameter dimensions for the counterbore need to be switched to the top view. After the dimensions are switched, choose the *right* mouse button anywhere on the screen to show **Modify Item**. Pick the **.563** diameter dimension and choose **Values & Text**, and the Modify Dimension dialog box will display, as shown in Figure 17.30. Select the **Dim Text** tab from the dialog box (Fig. 17.31). Add a diameter symbol to the value as a prefix, and add the **.875** counterbore diameter and **.250** depth to the note.

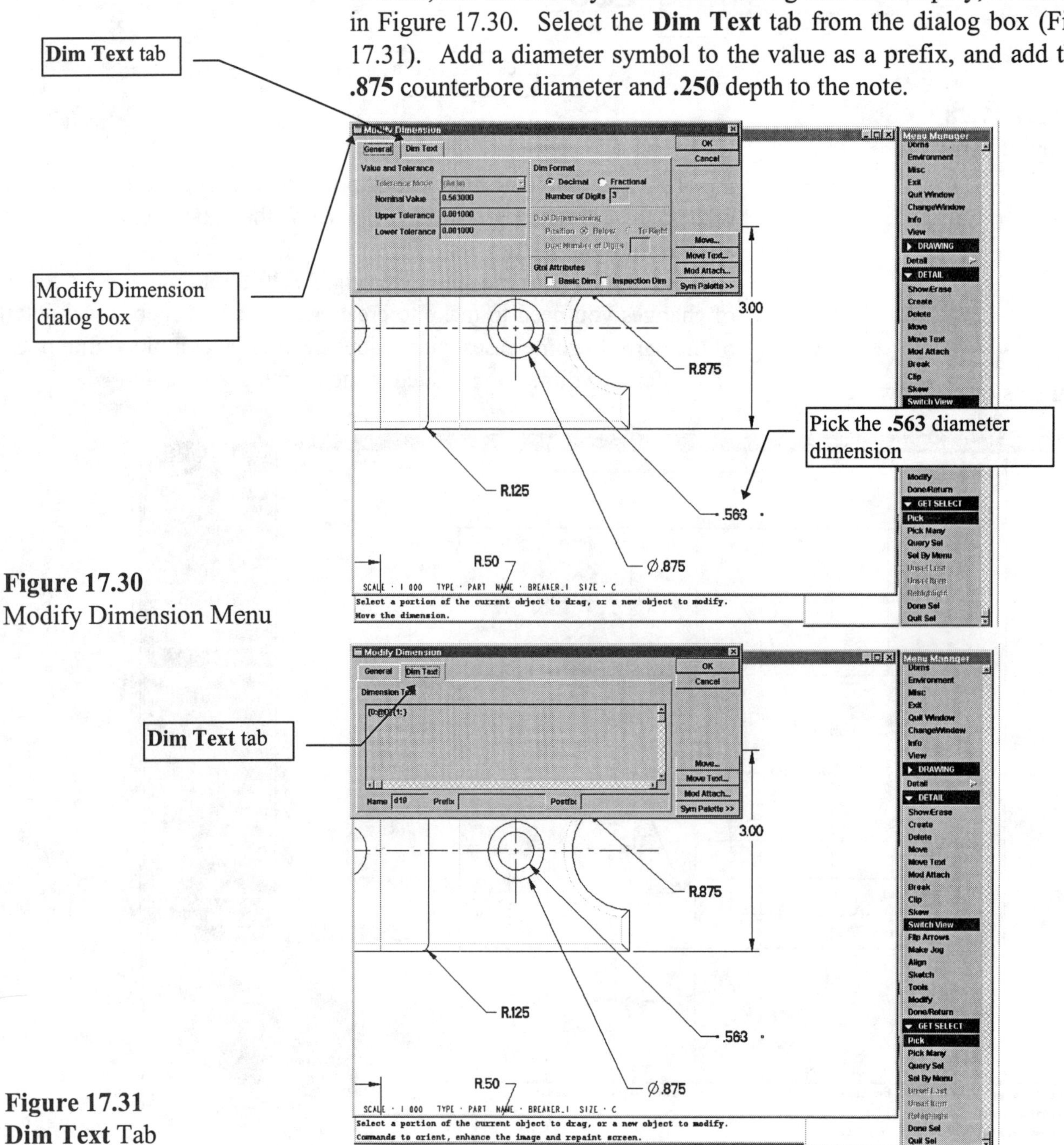

Figure 17.30
Modify Dimension Menu

Figure 17.31
Dim Text Tab

HINT

If instead of entering **.875** and **.250** as text, you enter **&** followed by the proper dim symbols (e.g., **d23**), the dimensions are then parameters which will reflect any modifications and changes.

Choose the **Sym Palette** button, and then place your cursor at the beginning of the text line in the dialog box window or in the **Prefix** subwindow, as shown in Figure 17.32. Pick the diameter symbol ∅ from the palette. This will add the symbol at the beginning of the text line. Add a counterbore symbol on a new text line, along with the **∅.875** dimension. To complete the note, add a third line with a depth symbol and the **.250** dimension. Erase the **∅.875** and **.250** dimensions from the drawing (Fig. 17.33).

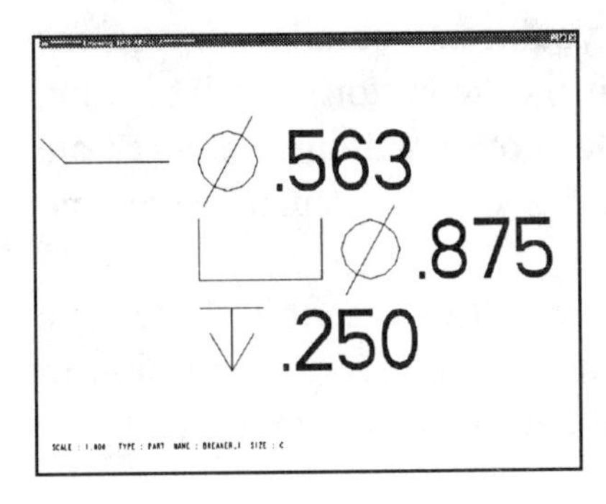

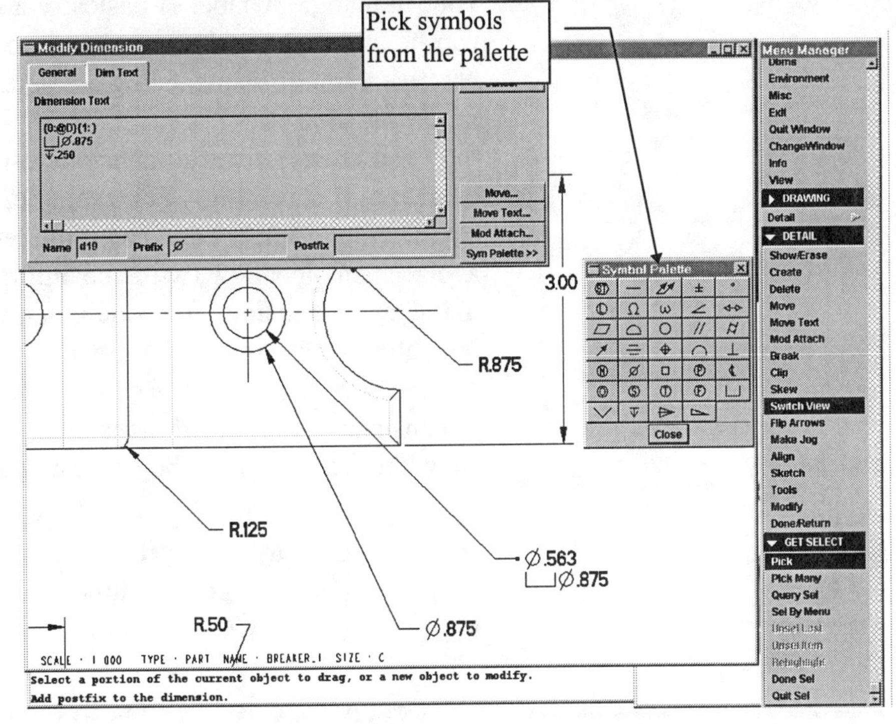

Figure 17.32
Modifying and Adding Text to a Note

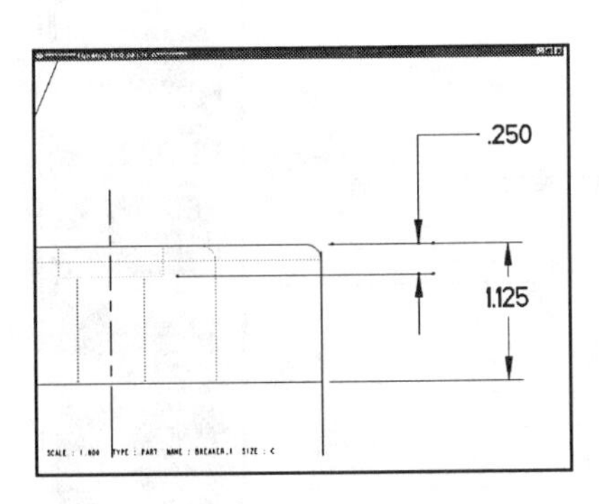

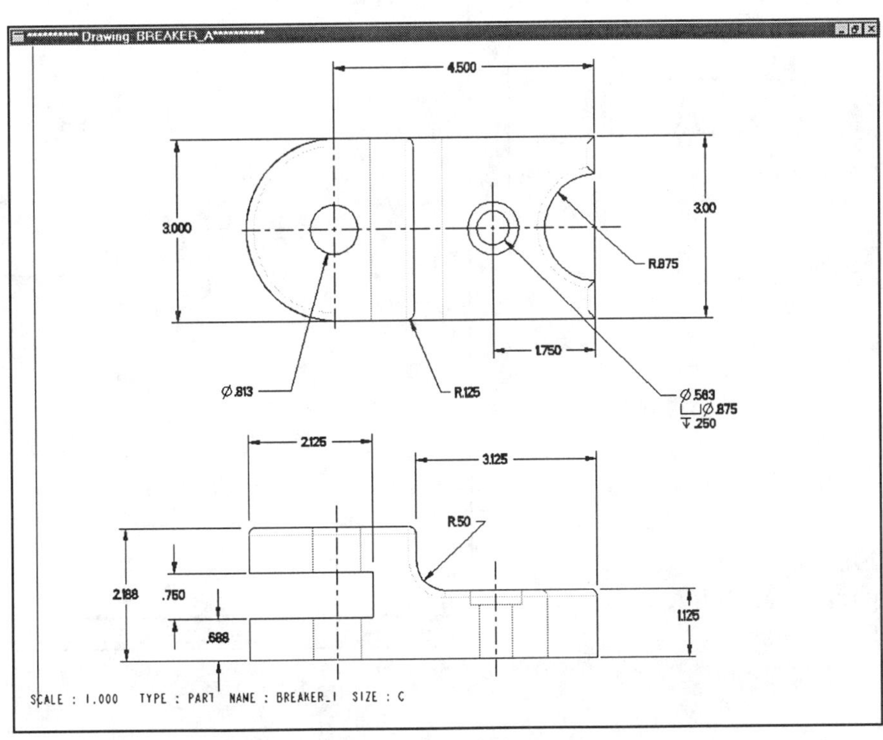

Figure 17.33
Erasing Unneeded Dimensions

Dimensions that you add "manually" are called ***driven dimensions***. Driven dimensions change when the model changes, but they cannot be used to make changes to the model. Only dimensions that you place on the drawing using the **Show** option can ***drive*** changes to the model (and they are referred to as ***driving dimensions***).

Driven dimensions can be erased in the same manner as feature dimensions. In addition, however, driven dimensions can also be deleted, which permanently removes them from the drawing database.

You can also create notes and labels to add to the annotation on your drawing. A label is basically a note that has a leader. Leaders can be attached to edges or drawn to positions in space. It is better practice to modify an existing diameter dimension than it is to create a note with a leader.

You create driven dimensions using the same techniques that are available in sketching. For example, you create parallel dimensions by selecting a linear edge (with the **left** mouse button) and indicating a placement location (with the **middle** mouse button). You create diameter dimensions by double-clicking on a circular edge and indicating a placement location.

Reference dimensions can also be added to a drawing. Reference dimensions are driven dimensions. Create a reference dimension to show the total width of the part using the following commands:

Detail ⇒ **Create** ⇒ **Ref Dim** ⇒ (pick both ends of the part in the front view, and place the dimension below the view, as shown in Figure 17.34) ⇒ **enter** ⇒ **Done/Return** ⇒ **Done/Return**

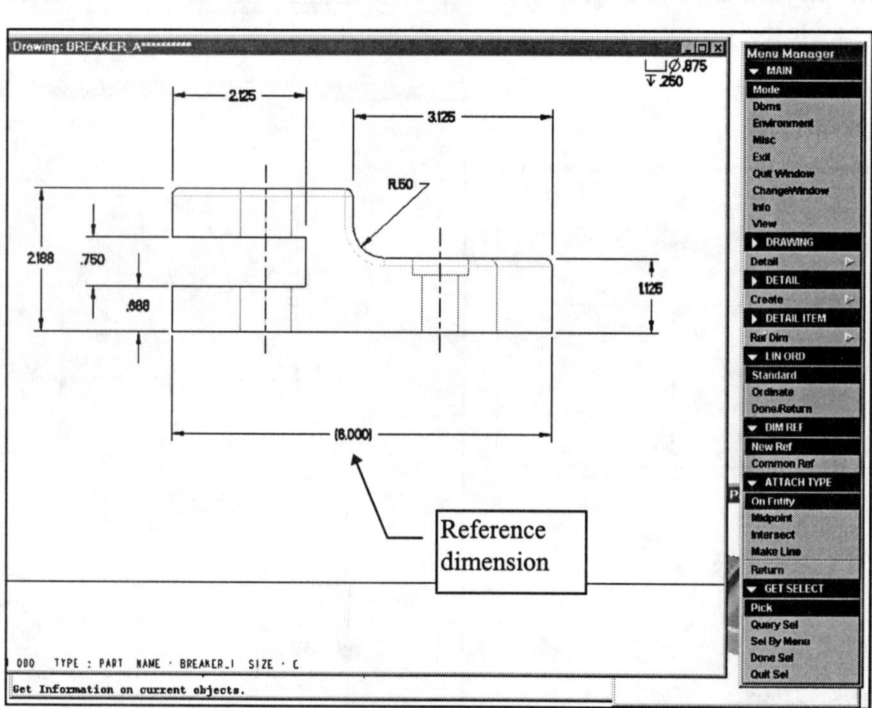

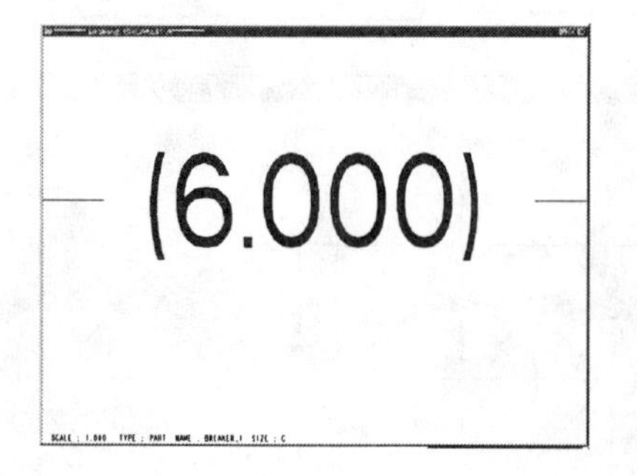

Figure 17.34
Creating a Reference Dimension

The alphabetic characters are created in a font and sized such that they comply with standards for **Geometric Dimensioning and Tolerancing (GD&T)**. These characters can be used for any purpose, but they are designed to be properly displayed in **GD&T** frames.

You may have noticed that up to this point, most of the text you enter is automatically capitalized by Pro/E. For example, you name a section by entering "**a**" but Pro/E displays it as "**A**." When you enter text in labels, notes, or as text appended to a dimension, Pro/E displays exactly the case you enter. So if you enter "material," the system displays "**material**" on the drawing. If you want "**MATERIAL**," you need to type "MATERIAL."

HINT

If you set your **Environment** to **Isometric**, your pictorial will be isometric instead of **Trimetric**.

Since the drawing is almost complete, let us add another general view (pictorial view) in the upper corner of the drawing using the following commands:

Drawing ⇒ **Views** ⇒ **General** ⇒ **Done** ⇒ (pick in the upper right of the drawing, as shown in Figure 17.35) ⇒ **Done/Return** ⇒ **Done-Return**

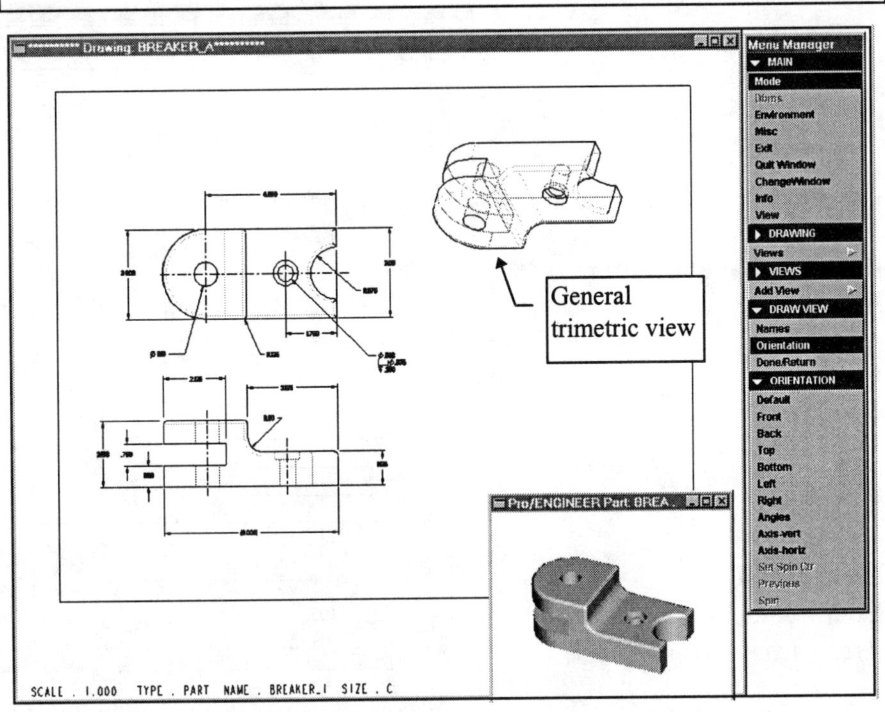

Figure 17.35
Adding a General View

Since you used a blank format to start the drawing, replace the format with a standard ANSI "**C**" size drawing format using the following commands (Fig. 17.36) :

Sheets ⇒ **Format** ⇒ **Add/Replace** ⇒ **?** ⇒ **enter** ⇒ **Format Dir** ⇒ (pick: **c.frm**)

Also, try to use your personal format (Fig: 17.37):

Sheets ⇒ **Format** ⇒ **Add/Replace** ⇒ **?** ⇒ **enter** ⇒ **Format Dir** ⇒ (pick: ***your_filename*.frm**)

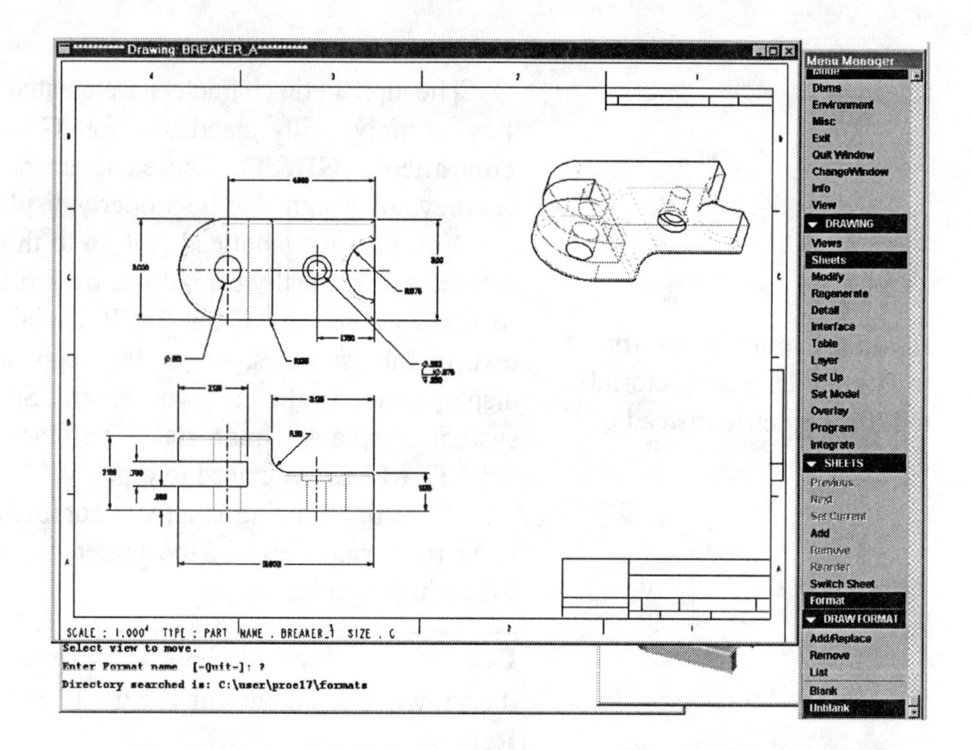

Figure 17.36
ANSI Standard "**C**" Size Format

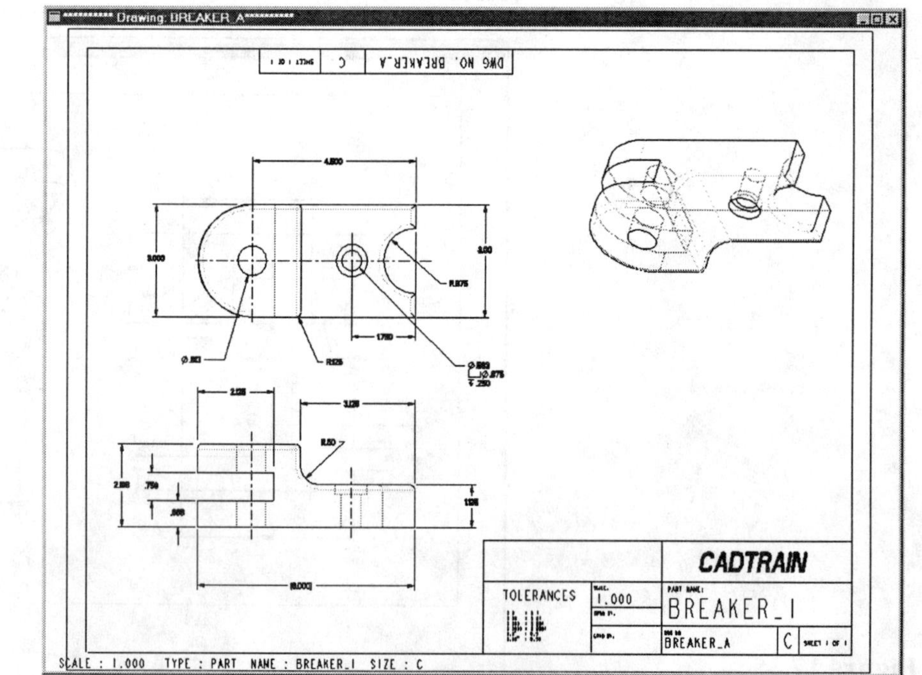

Figure 17.37
Your Personal Format (Here, from CADTRAIN)

Dbms ⇒ Save ⇒ enter
Purge ⇒ enter ⇒ Done-Return

Each view of your model can have a different line display style: **Wireframe, Hidden Line, No Hidden,** or **Default**. The tangent display status can also be established independent of the environment default. **Tan Solid, No Disp Tan, Tan Ctrln, Tan Phantom, Tan Dimmed,** and **Tan Default** are available. Use the following commands to set the display of each view, starting with the general pictorial view:

Views ⇒ Disp Mode ⇒ View Disp ⇒ (pick the general view, as shown in Figure 17.38) **⇒ Done Sel ⇒ No Hidden ⇒ Tan Solid ⇒ Done ⇒ Done Sel ⇒ Done/Return**

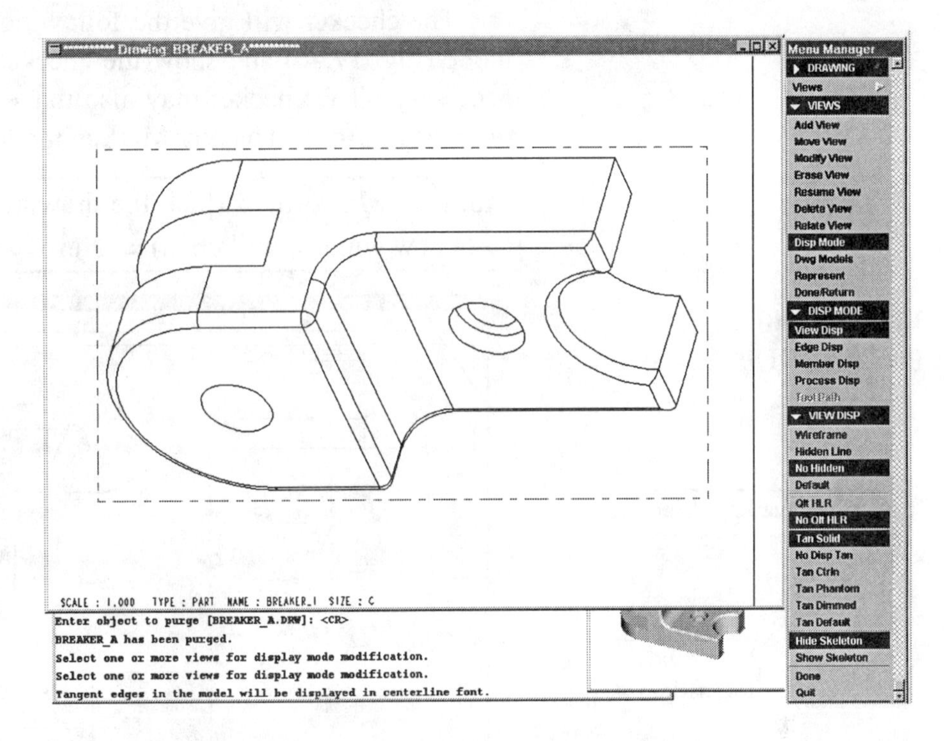

Figure 17.38
Setting the Display Status for a General View: **No Hidden** and **Tan Solid**

Set the front and top views to have **Hidden Line** and **Tan Dimmed** using the following commands (Fig. 17.39):

Views ⇒ Disp Mode ⇒ View Disp ⇒ (pick the top and front views) **⇒ Done Sel ⇒ Hidden Line ⇒ Tan Dimmed ⇒ Done ⇒ Done Sel**

Dbms ⇒ Save ⇒ enter Purge ⇒ enter ⇒ Done-Return

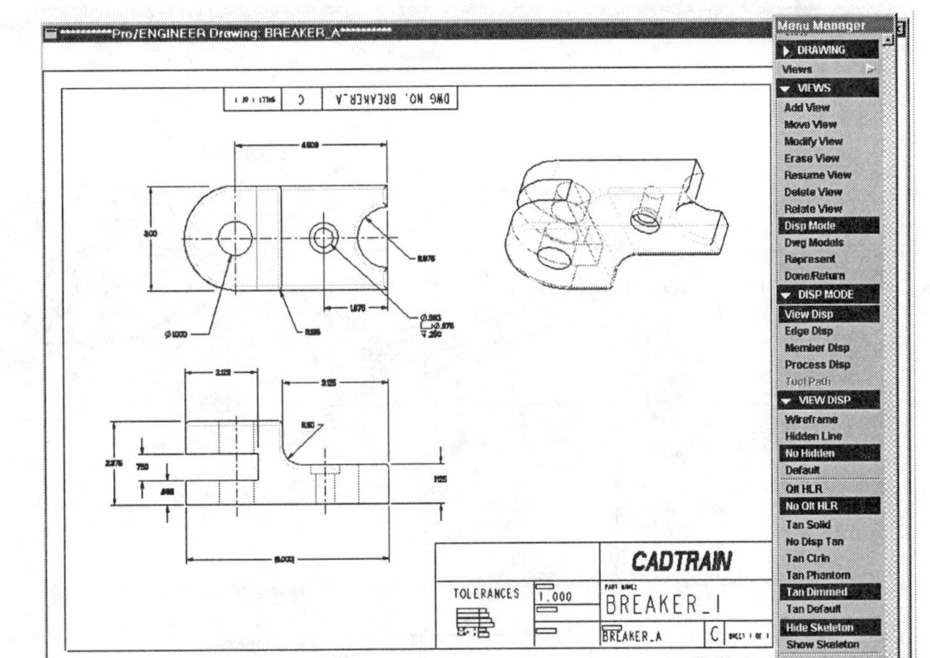

Figure 17.39
Setting the Display Status for the Top and Front Views: **Hidden Line** and **Tan Dimmed**

Since you think the project is complete, print or plot out a copy and submit it to the checker (teacher, boss, design checker). The checker will use **Markup mode** to check and mark up your drawing.

The checker will give the following commands to enter Markup mode (Fig. 17.40) and show the checker changes he or she feels are necessary. The checker may also make some design changes at this time (Fig. 17.41). To enter Markup mode, choose:

Mode ⇒ **Markup** ⇒ (pick the drawing name used for the Breaker) ⇒ (a new window will open) ⇒ **New** ⇒ (type **Brk_Check**) ⇒ **enter**

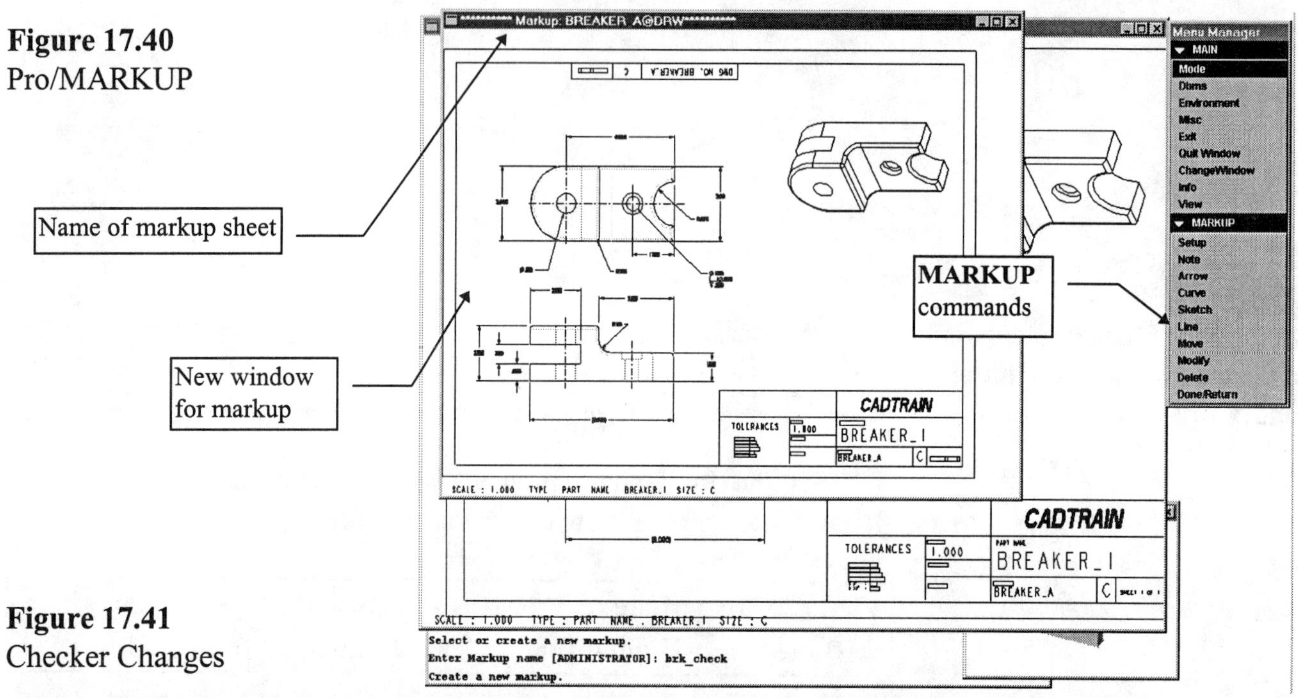

Figure 17.40
Pro/MARKUP

Figure 17.41
Checker Changes

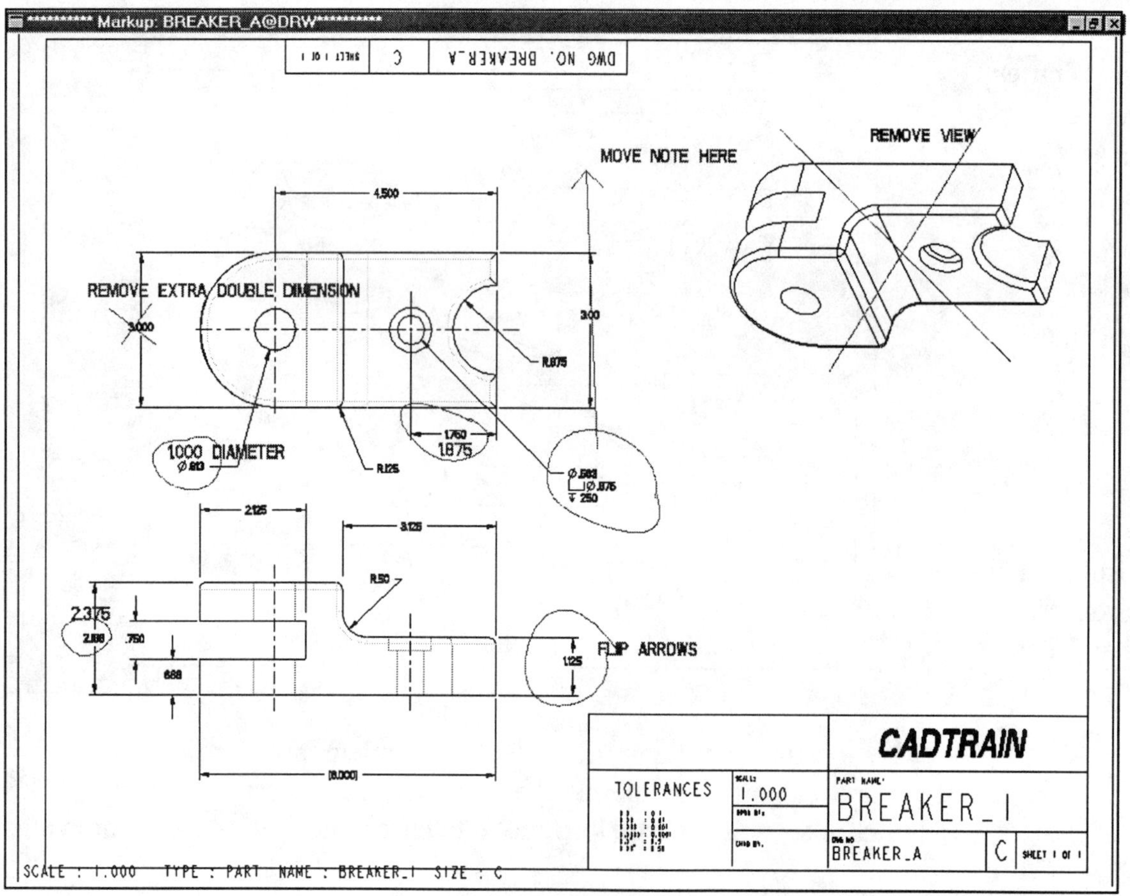

After the checker saves the changes, you can bring up the markup drawing and your Breaker detail drawing. By keeping the two files in session, you can work in your detail window and still see the markup drawing. Since you will also be making some design changes, it is a good idea to keep the part window active to see the changes propagated throughout the model--part, drawing, and assembly (if used in an assembly). All files will be updated to the new design changes after regeneration. If you are starting a new session, use the following commands (Fig. 17.41):

Mode ⇒ Part ⇒ BREAKER ⇒ enter
Mode ⇒ Markup ⇒ BRK_CHECK ⇒ enter
Mode ⇒ Drawing ⇒ BREAKER ⇒ enter

Move the windows for each mode for easy viewing and working, as shown in Figure 17.42. Change the active window to be the drawing, not the part or markup.

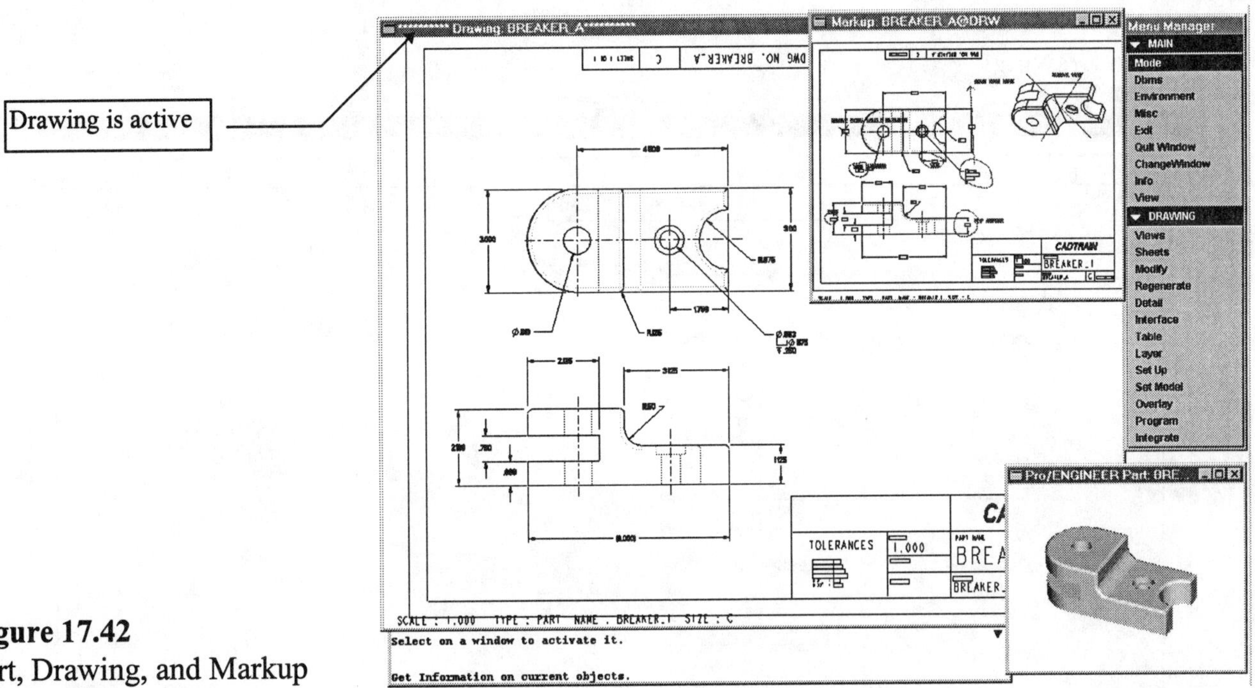

Figure 17.42
Part, Drawing, and Markup

Make the checker changes to the drawing (of course, you could also change the active window to be the Part mode and modify and regenerate the model there). Regardless of the mode or window you are working in, remember to **Regenerate** both the drawing *and* the part (Fig. 17.43).

The completed design can now be saved and plotted, as in Figure 17.44.

Dbms ⇒ Save ⇒ enter
Purge ⇒ enter ⇒ Done-Return

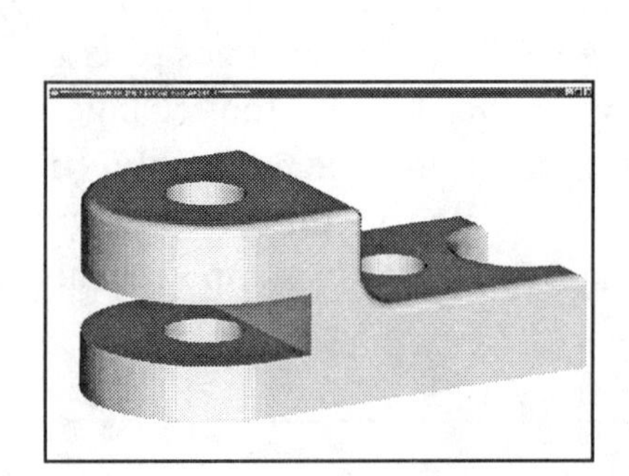

Figure 17.43
Regenerated Checker Changes

Figure 17.44
Completed Drawing

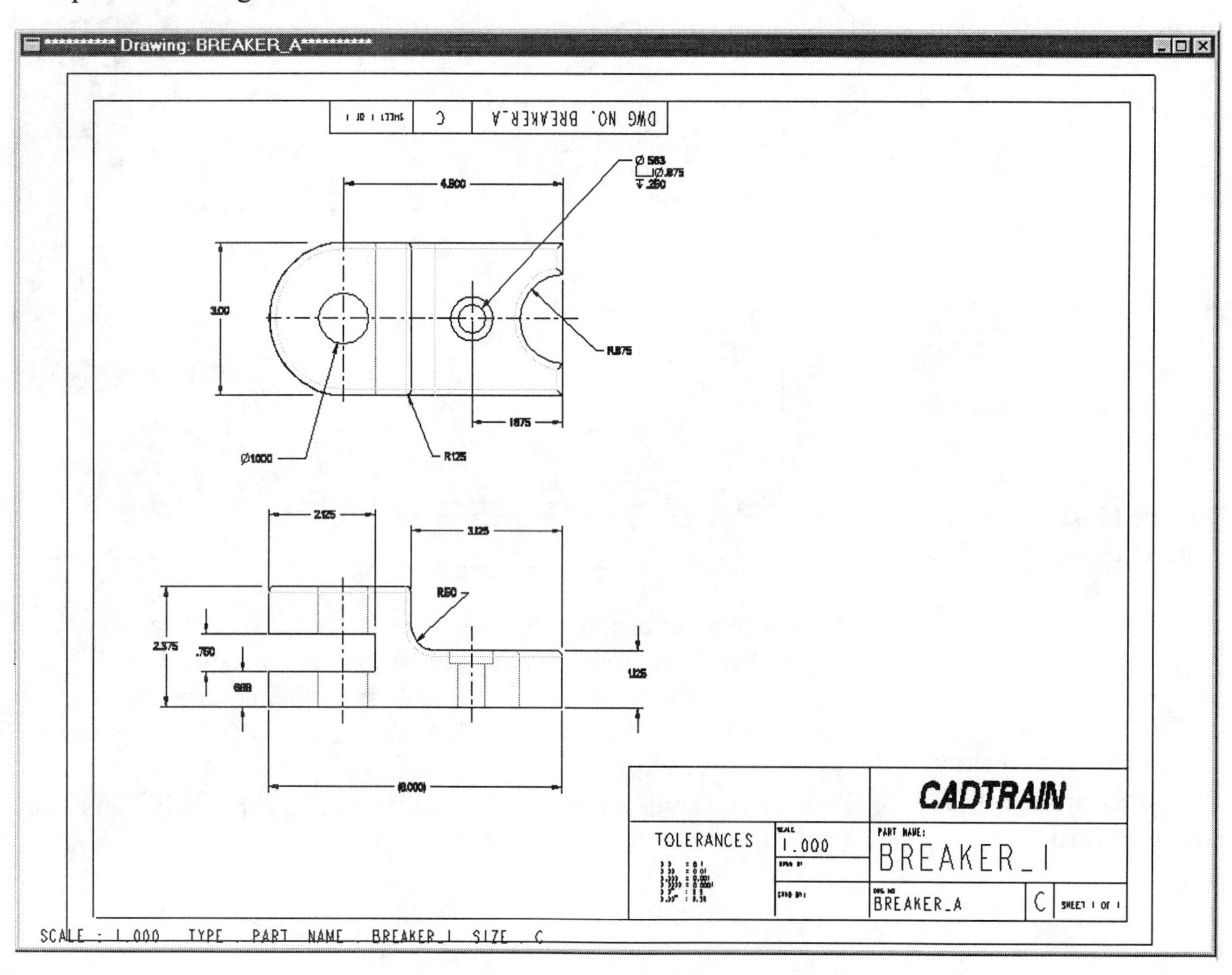

Lesson 17 Project

Cylinder Rod Drawing

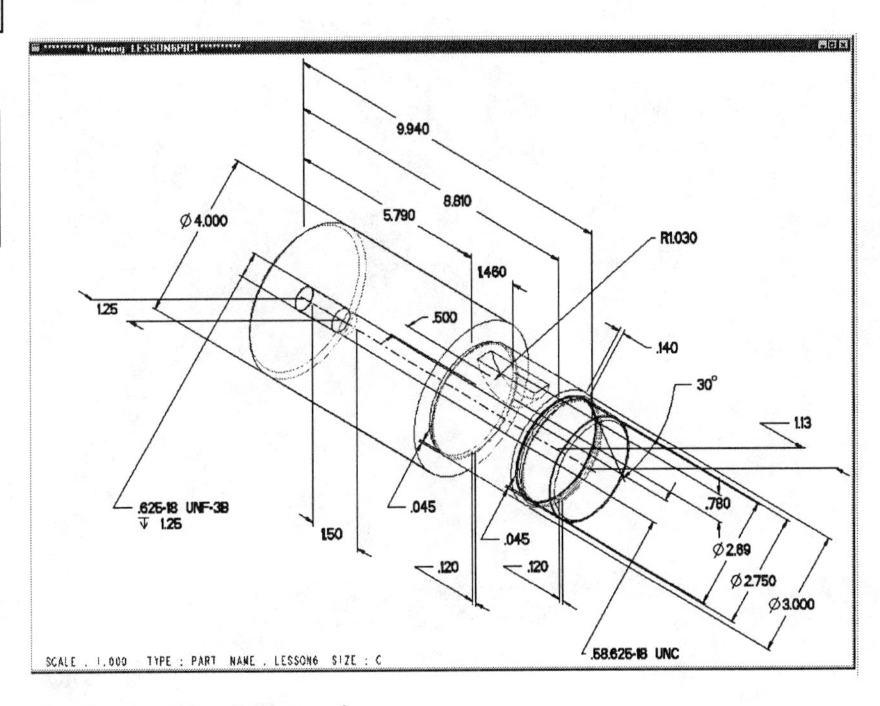

Figure 17.45
Cylinder Rod Pictorial Drawing

Cylinder Rod Drawing

The second **lesson project** for the Drawing mode will use the part modeled in Lesson 6. The drawing for the Cylinder Rod (Fig. 17.45) is just one of many lesson parts and lesson projects that can be brought into the Drawing mode and detailed.

Analyze the part and plan out the sheet size and the drawing views required to display the model's features for detailing (Fig. 17.46). Use the formats created in Lesson 16. Detail the part according to **ASME Y14.5 1994**.

EGD REFERENCE
Engineering Graphics and Design
by L. Lamit and K. Kitto
See Pages: 487, 674-677

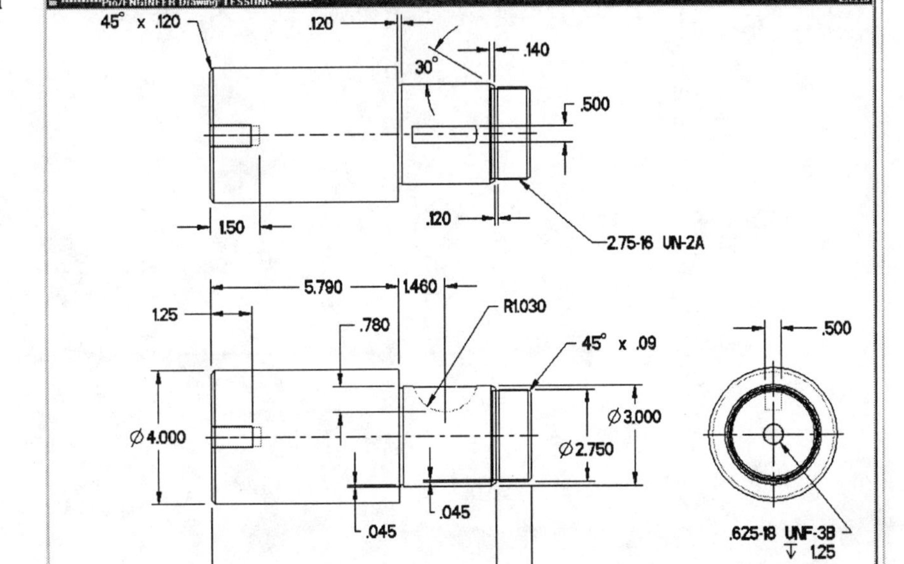

Figure 17.46
Cylinder Rod Drawing Without Format

Lesson 18

Sections and Auxiliary Views

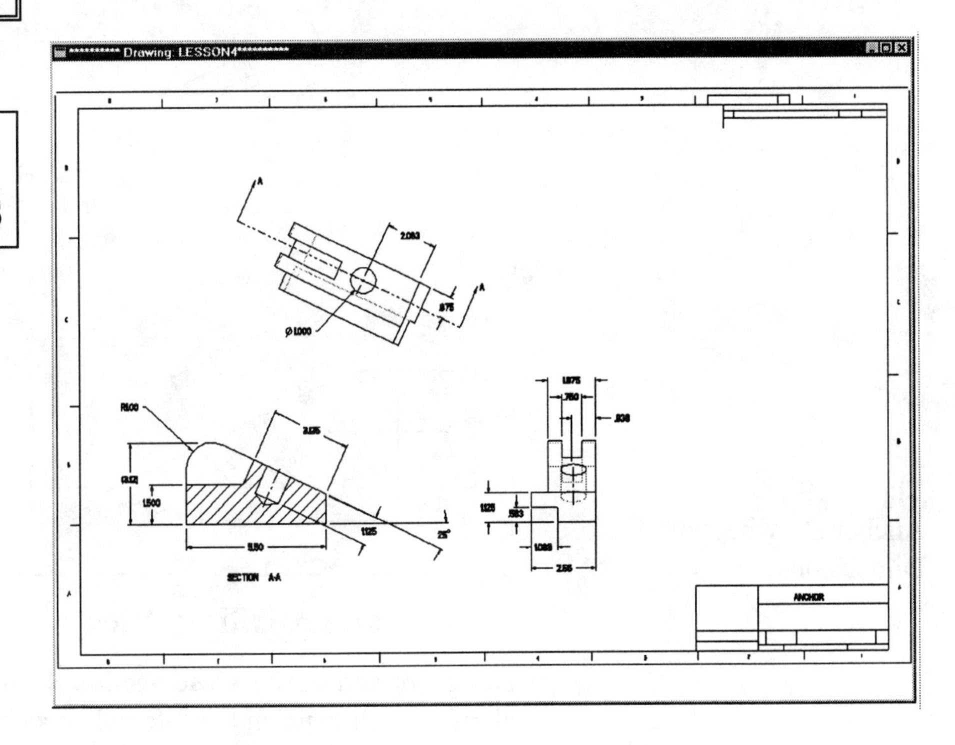

Figure 18.1
Anchor Drawing

EGD REFERENCE
Engineering Graphics and Design
by L. Lamit and K. Kitto
Read Chapters: 12
See Pages: 387-416, 549

OBJECTIVES

1. **Identify the need for sectional views to clarify interior features of a part**

2. **Establish a .dtl file to use when detailing and creating section drawings**

3. **Identify cutting planes and the resulting views**

4. **Create sections along datum planes**

5. **Detail section views**

6. **Develop the ability to produce auxiliary views**

7. **Create detail views**

8. **Create scaled detail views of complicated feature geometry**

9. **Apply standard drafting conventions and linetypes to illustrate interior features**

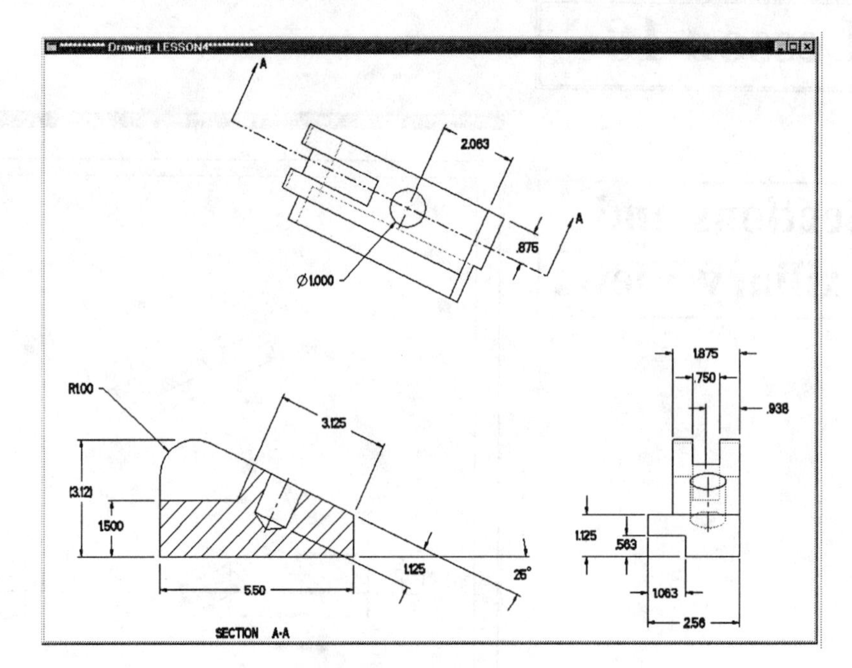

Figure 18.2
Anchor Drawing with Dimensions

Sections and Auxiliary Views

Designers and drafters use **sectional views**, also called **sections,** to clarify and dimension the internal construction of a part. Sections are needed for interior features that cannot be clearly described by hidden lines in conventional views (Fig. 18.1 and Fig. 18.2).

Auxiliary views are used to show the true shape/size of a feature or the relationship of part features that are not parallel to any of the principal planes of projection. Many parts have inclined surfaces and features that cannot be adequately displayed and described by using principal views alone. To provide a clear description of these features, it is necessary to draw a view that will show them *true shape/size.* Besides showing features true size, auxiliary views are used to dimension features that are distorted in principal views and to solve graphically a variety of engineering problems (Fig. 18.3).

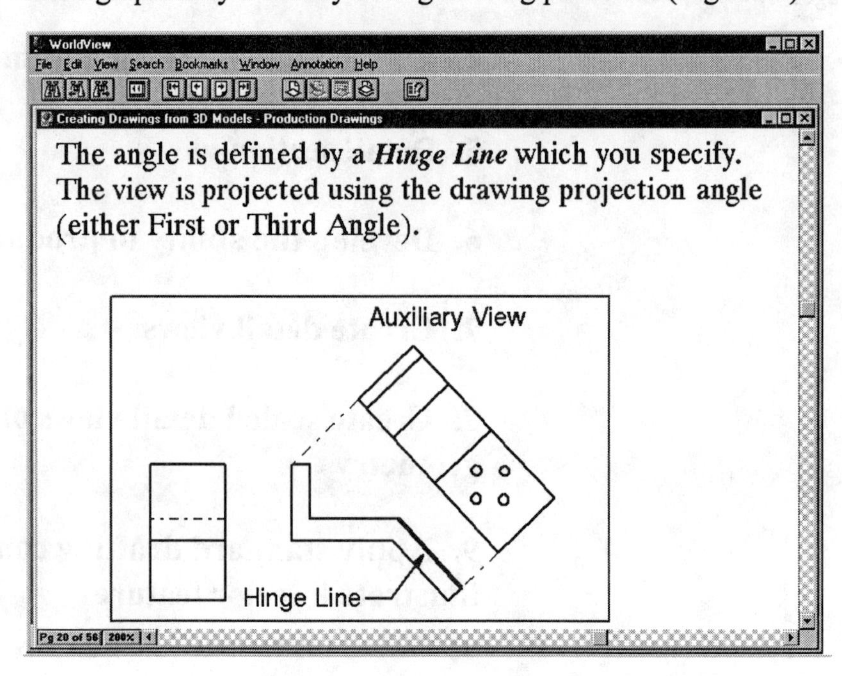

Figure 18.3
COAch for Pro/E, Projecting Auxiliary Views--Creating Drawings (Segment 4: Creating an Auxiliary View)

Sections

A sectional view (Fig. 18.4) is obtained by passing an imaginary **cutting plane** through the part, perpendicular to the **line of sight**. The line of sight is the direction in which the part is viewed. The portion of the part between the cutting plane and the observer is "removed." The part's exposed solid surfaces are indicated by section lines. **Section lines** are uniformly spaced, angular lines drawn in proportion to the size of the drawing. There are many different types of section views.

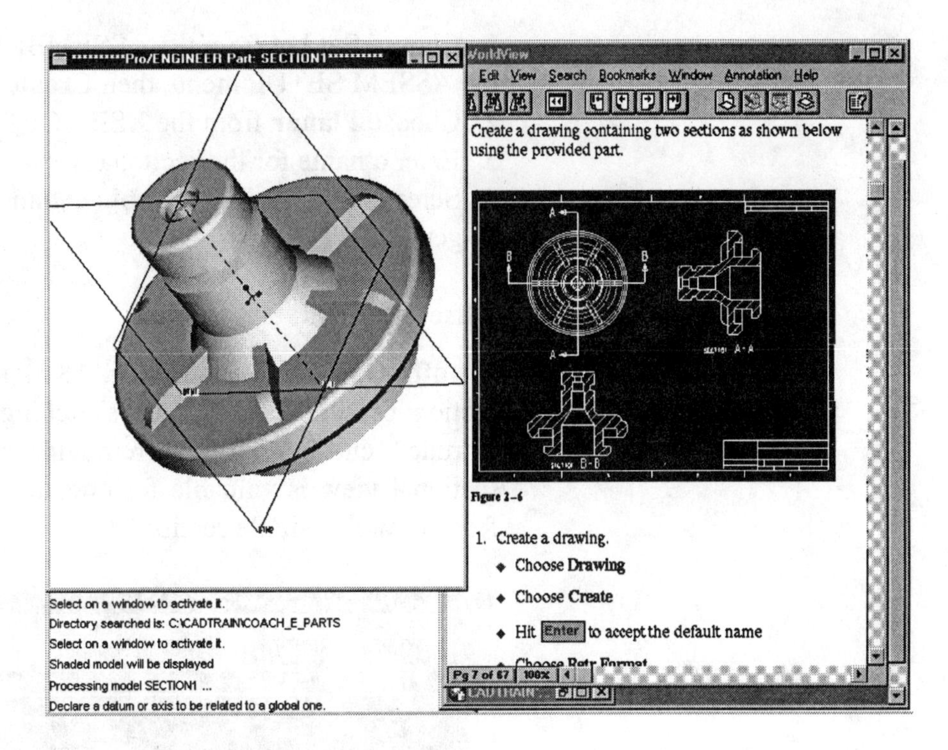

Figure 18.4
COAch for Pro/E, Creating Drawings--Sections (Segment 1: Creating a Planar Section)

Sectional views are slices through a part or assembly and are valuable for opening up the part or assembly for displaying features and detailing in the Draw mode. Part sectional view may also be used to calculate sectional view mass properties. Each sectional view has its own unique name within the part or assembly, allowing any number of sectional views to be created and then retrieved for use in a drawing. A variety of ANSI standard section lining materials can be automatically generated. You have the ability to create a variety of sectional view types:

* Standard planar sections of models (part or assemblies)
* Offset sections of models (part or assemblies)

Planar Sections

Planar sectional views are created along a datum plane. The datum may be established during the creation of the sectional view using the **Make Datum** options, or an existing plane may be selected.

To create a planar sectional view of a part:

1. Choose **X-section** from the PART menu, and **Create** from the CROSS SEC menu.
2. Choose **Planar** from the XSEC OPTS menu, then **Done**.
3. Enter a name for the sectional view, and then select, or make, the datum along which the section is to be generated.

To create a planar sectional view of an assembly:

1. Choose **Set Up** from the ASSEMBLY menu and **X-Section** from the ASSEM SETUP menu, then **Create** from the CROSS SEC menu.
2. Choose **Planar** from the XSEC OPTS menu, then **Done**.
3. Enter a name for the sectional view.
4. Select or create the **assembly** datum along which the section is to be generated.

Offset Sections

An **offset** sectional view (Fig. 18.5) is created by extruding a 2D section perpendicular to the sketching plane, just like creating an extruded cut but without removing any material. This type of sectional view is valuable for opening up the part to display several features with a single section.

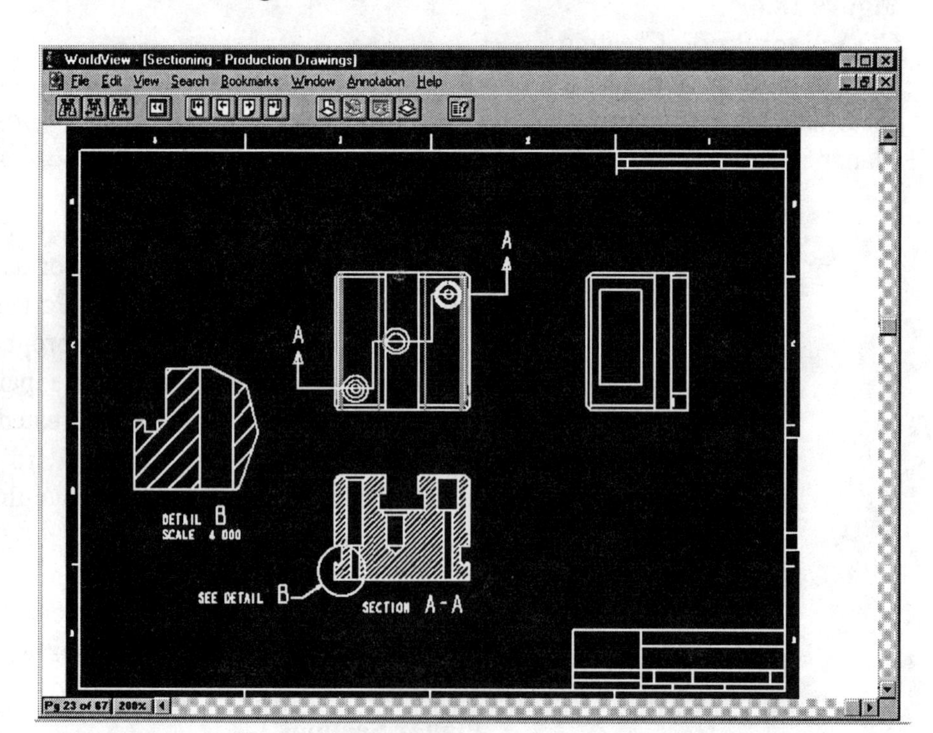

Figure 18.5
COAch for Pro/E, Sections--Section Drawings (Segment 2: Creating an Offset Section)

The sketched section must be an *open section*. The first and last segments of the open section must be straight lines.

To create an offset section:

1. Choose **X-section** from the PART or ASSEM SETUP menu and **Create** from the CROSS SEC menu.
2. Choose **Offset** and **One Side** or **Both Sides** from the XSEC OPTS menu, and then **Done**.
3. Enter a name for the sectional view.
4. Answer the prompts for entering Sketcher. The sketching plane can be created using the **Make Datum** option.
5. **Sketch** the sectional view and **Dimension** it to the model. Choose **Done** when the section has been regenerated successfully.

Auxiliary Views

The proper selection of views, view orientation, and view alignment is determined by a part's features and its natural or assembled position. Normally, the front view is the primary view and the top view is obvious, based on the position of the part in space or when assembled. The choice of additional views is determined by the part's features (Fig. 18.6) and the minimum number of views necessary to describe the part and show its dimensions.

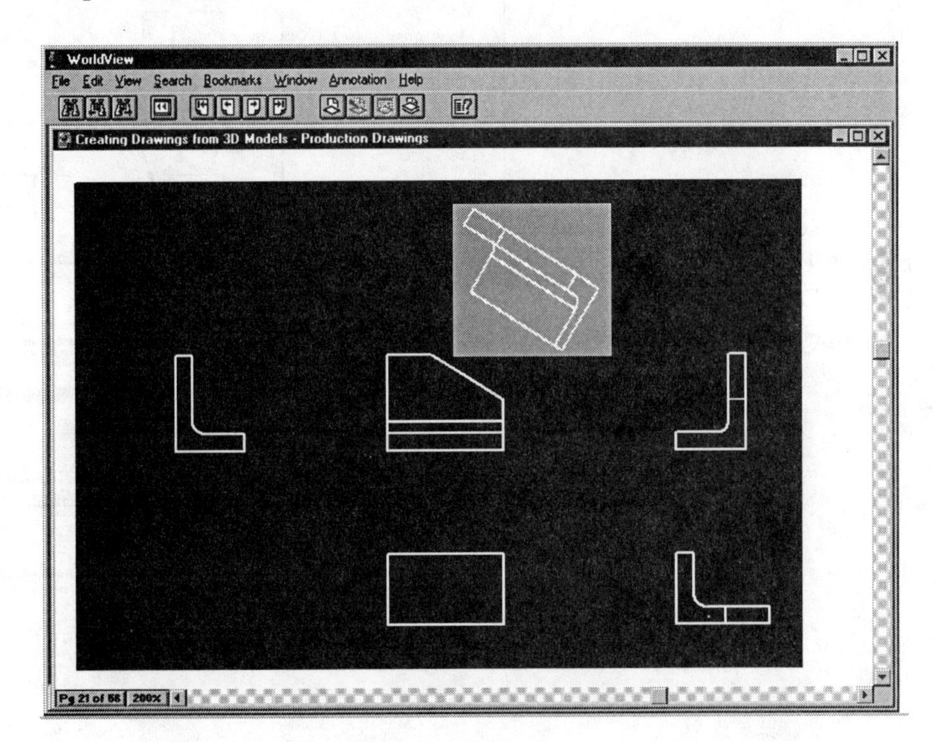

Figure 18.6
COAch for Pro/E, Projecting Auxiliary Views--Creating Drawings (Segment 4: Creating an Auxiliary View)

Auxiliary views are created by making a projection of the model perpendicular to a selected edge. They are normally used to discern the true size and shape of a planar surface on a part. An auxiliary view can be created from any other type of view. Auxiliary views may have arrows created for them that point back at the view(s) from which they were created.

To add an auxiliary view to the drawing, use the following command options:

1. Choose **Auxiliary** and other available options from the VIEW TYPE menu.
2. Choose **Done** to accept the options, or **Quit** to quit the creation of a new view.
3. Pick the location of the new view on the drawing (Fig. 18.7).
4. Pick an edge of, or axis through, or datum plane of, the surface of the model in the view from which the auxiliary view will be developed. If the edge selected is from a view that has a pictorial (isometric-trimetric) orientation, the new view will be oriented as the base feature section was. Otherwise, the view will be oriented with the selected surface parallel to the plane of the drawing.

Detail views will also be discussed in this lesson (Fig. 18.8).

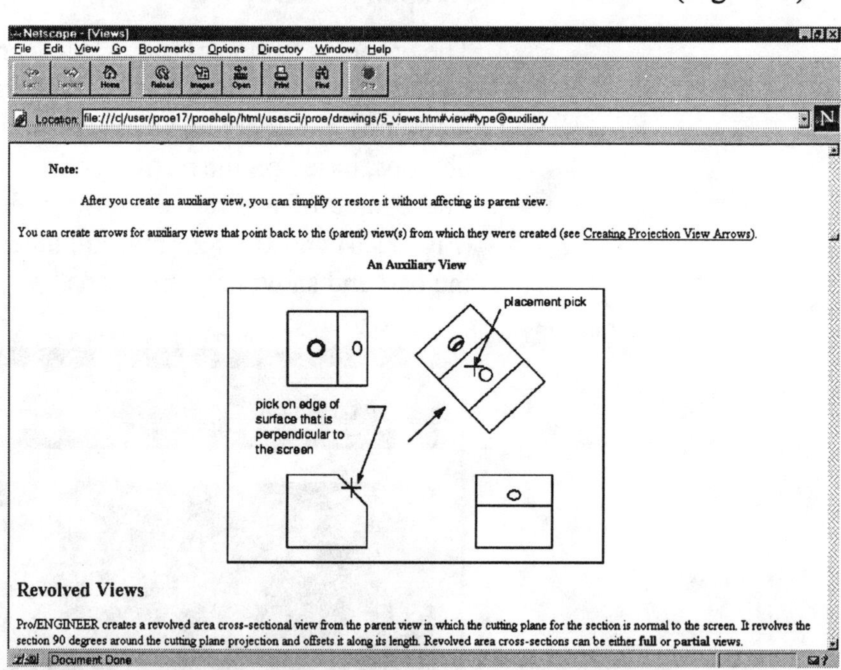

Figure 18.7
Online Documentation
Auxiliary Views

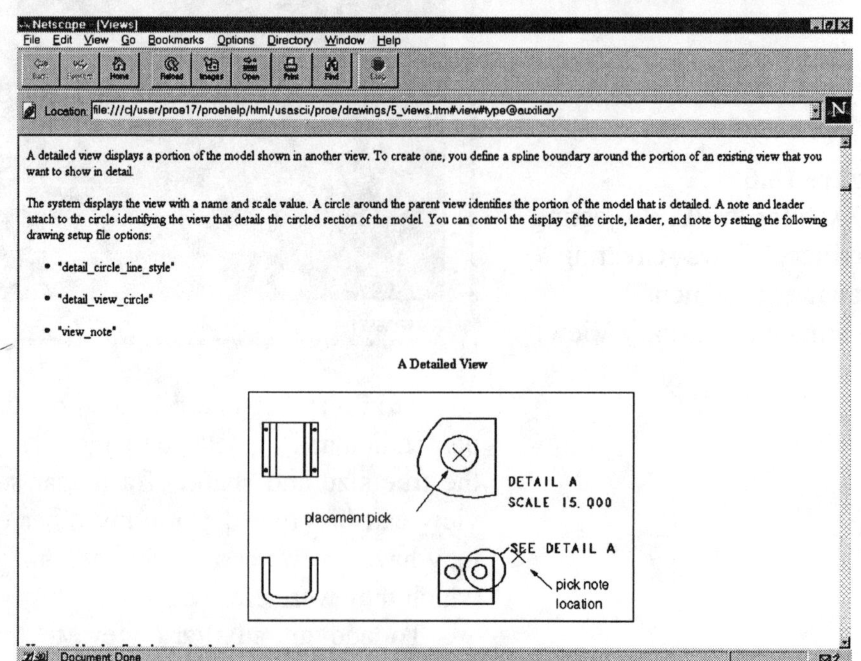

Figure 18.8
Online Documentation
Detailed View

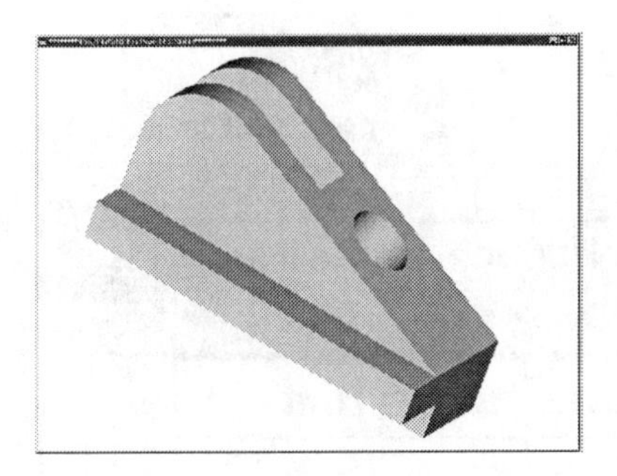

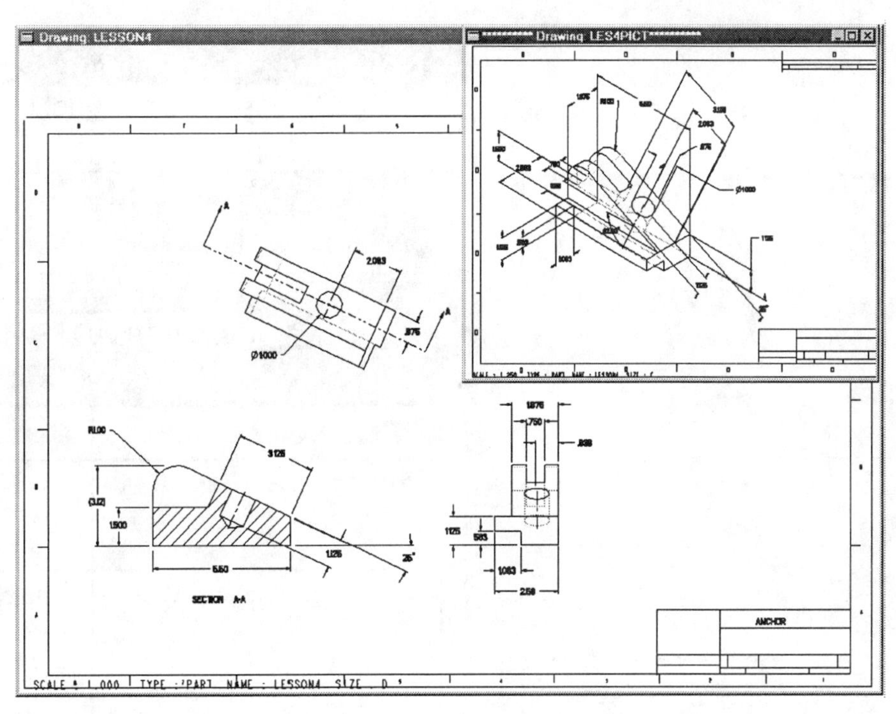

Figure 18.9
Anchor, Orthographic View Detail and Pictorial Detail

Anchor Drawing

Pro/E is able automatically to create fully associative *section views* of solid models (Fig. 18.9). You can create a section view using an existing view on a drawing in Drawing mode or by creating a section in Part mode for retrieval on a drawing. As you were completing some of the lessons in the text, you were instructed to create sections in a number of lesson parts and lesson projects.

Auxiliary views are created, using Pro/E, by making a projection of the model perpendicular to a selected edge. They are normally used to discern the true size and shape of a planar surface on a part, such as the Anchor, which has an angled surface with a hole machined perpendicular to that surface.

You create a section view while looking at the drawing by sketching (with the help of Pro/E) a section line or by selecting **Datum Planes** for the section line to pass through.

The display of the *section line symbol* and the *section view text* is controlled by the **Section Line Display Parameters** in the **Drawing Set-Up** file (**.dtl**).

The sectioning parameters apply to the two basic types of standards: the ANSI (ASME) Standards and the ISO/JIS/DIN Standards. Section line display and the manner in which the view titles (parameters) are created vary based on the standard you chose to use. Refer to the **Pro/ENGINEER Drawing User's Guide**, Appendix A, for a listing of section line and view parameters. The pertinent parameters are shown in Figure 18.10. Figure 18.11 shows the **.dtl** file used by CADTRAIN for its sectioned drawings.

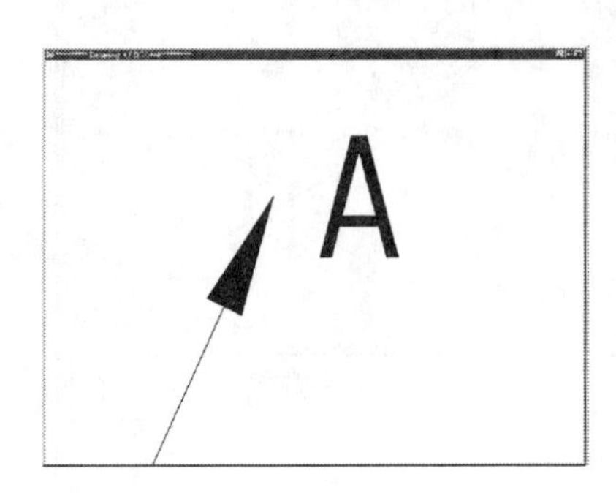

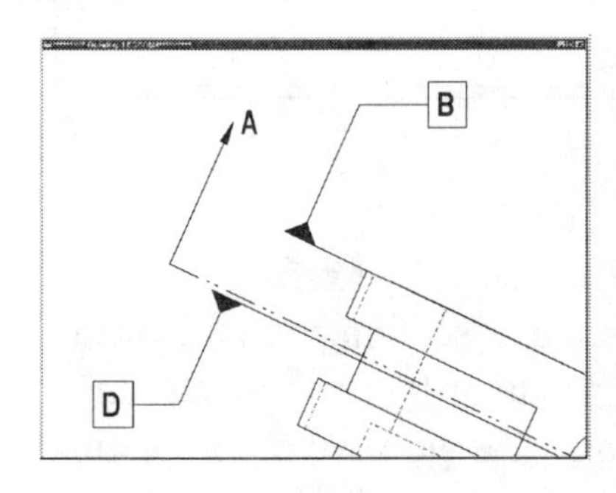

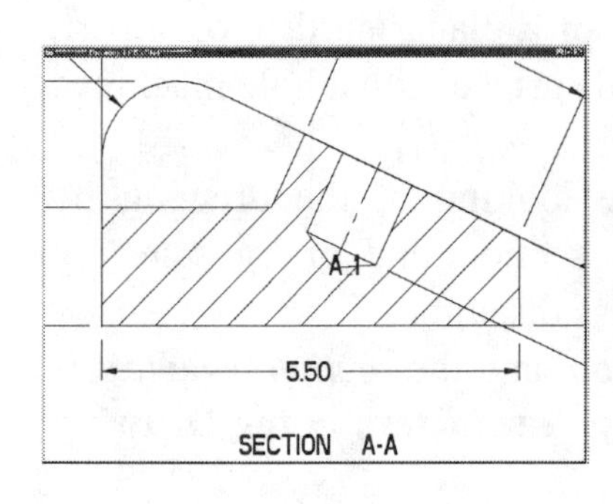

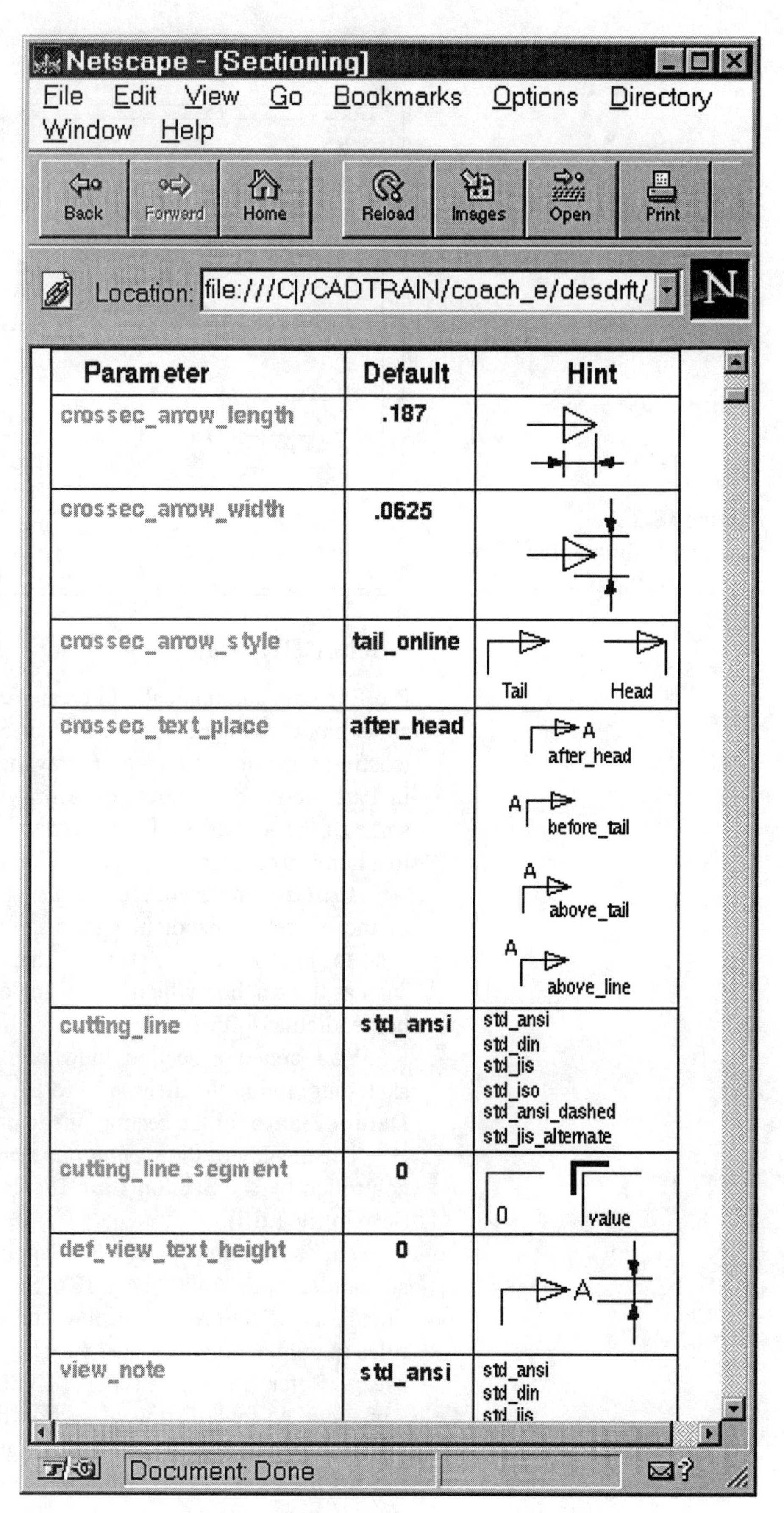

Figure 18.10
COAch for Pro/E, Section Line Symbol and the Section View Text Parameters

Pro/TABLE TM Release 18.0 (c) 1988-95 by Parametric Technology Corporation All Rights Reserved.

File Edit View Format Help

filled

	C1	C2	C3	C4	C5	C6	C7
R21	!						
R22	! These options control cross sections and their arrows						
R23	!						
R24	crossec_arrow_length	0.375000					
R25	crossec_arrow_style	TAIL_ONLINE					
R26	crossec_arrow_width	0.250000					
R27	crossec_text_place	AFTER_HEAD					
R28	cutting_line	STD_ANSI					
R29	cutting_line_adapt	NO					
R30	cutting_line_segment	0.000000					
R31	draw_cosms_in_area_xsec	NO					
R32	remove_cosms_from_xsecs	TOTAL					
R33	!						
R34	! These options control solids shown in views						
R35	!						
R36	datum_point_size	DEFAULT					
R37	datum_point_shape	CROSS					
R38	hlr_for_pipe_solid_cl	NO					
R39	hlr_for_threads	NO					
R40	location_radius	DEFAULT(2.)					
R41	mesh_surface_lines	ON					
R42	thread_standard	STD_ANSI					
R43	hidden_tangent_edges	DEFAULT					
R44	!						

C2R83:

Figure 18.11
.dtl File for drawing Parameters

HINT

If you did not create a section when modeling the Anchor, use the following commands in Part mode:

X-Section ⇒ **Create** ⇒ **Done** ⇒ (type **A** for the section name) ⇒ **enter** ⇒ **Make Datum** ⇒ **Through** ⇒ (pick axis **A_1** of the hole) ⇒ **Parallel** ⇒ (pick **DTM3**) ⇒ **Done** ⇒ **Done/Return**

If you are going to create sections while in the Drawing mode, you may require **Datum Planes**. Since most views on a drawing are orthographic, and since they cannot be changed with **Spin**, you may want to prepare the model with the necessary **Datum Planes** *before* you enter the Draw mode. You were not asked to create a section of the Anchor when you were doing Lesson 4. Therefore, you need to create a section that passes vertically, lengthwise, through the model in Part mode (Fig. 18.12), or you can to create a section in Drawing mode, as described next.

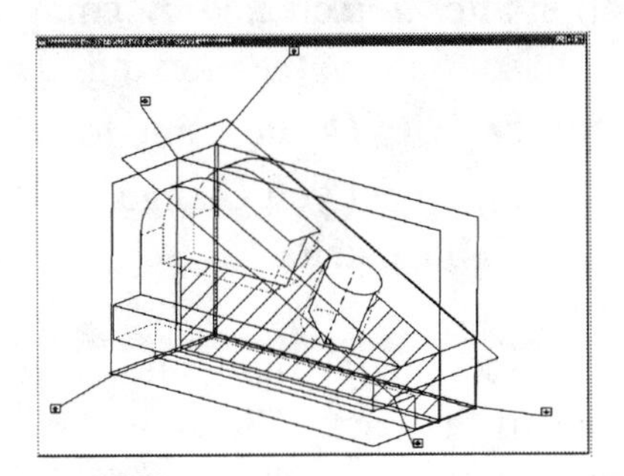

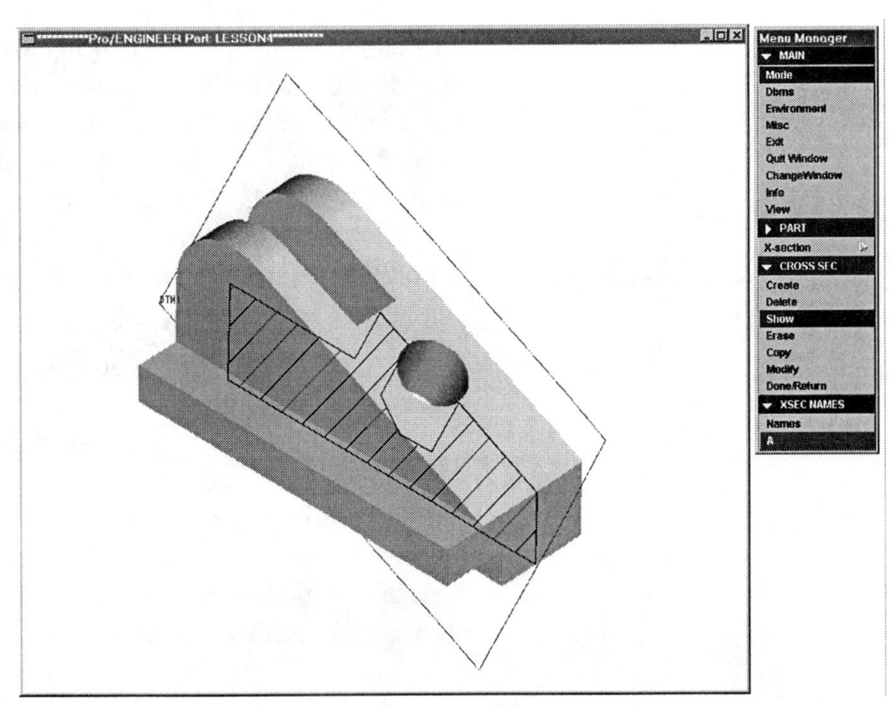

Figure 18.12
Section Created in The Part Mode

In the Drawing mode, the **Planar** section option allows you to create a section that passes straight through a part without any "jogs" (steps) in the section line. **Planar** sections can be defined by retrieving a section, sketching a section line, or selecting a datum plane to define the cut position.

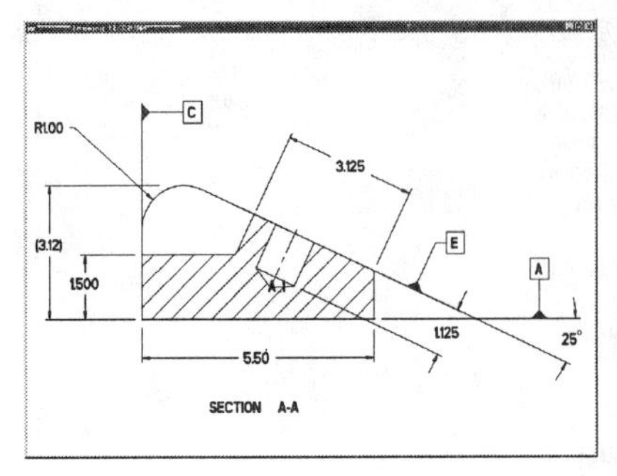

When you want to create the section, while in the Drawing mode, you can select either a planar face or a **Datum Plane** as the "plane" the section cut passes through. Even though you can pick a planar face, this is generally not good practice. The section cut takes place ***at*** the plane you select. If you pick a face, this causes the section to be tangent to it. Not only is this poor drafting practice, it also causes Pro/E to generate (only in some cases) incomplete sections.

When you create a **Planar** section on the drawing, Pro/E first asks you to define the location of the center of the section view. It then displays the orthographic projection to which the section edges (created by the cut) and the crosshatching will be added. You must then select a **Datum Plane** to define the section cutting plane.

Pro/E then prompts you to pick a view in which the cutting is perpendicular to the screen. This is the view in which it will draw the section line and the view where the cut will actually take place. Pro/E displays the section line, with its arrows pointing in the direction it "thinks" you want to view the cut.

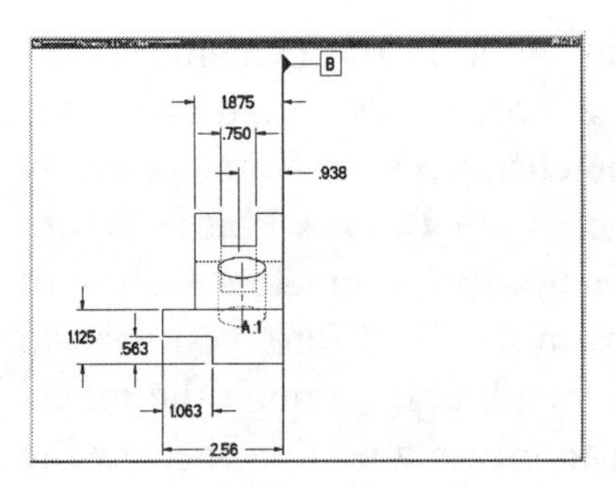

Pro/E also prompts you to enter a **name** for the section. This "name" is actually the section letter that you wish to use. If you are creating your first section and want it to have the letter "**A**" displayed at each arrow and have the title "**SECTION A-A,**" you should enter **A** as the name of the section. It is a common mistake to type **AA**, thinking that you are establishing **SECTION A-A** when in reality you are getting **SECTION AA-AA**. Type one letter only.

After the section has been created, use the right mouse button to get the pop-up menu item **Modify Item** ⇒ (pick the section view) ⇒ **Flip X-sec** (to flip the section identification arrows). The arrow direction does not affect the projection of the section view; which was defined when you indicated the location of the view. It does affect what you see in the "background," behind the cutting plane.

You can change the name of a section after it is created by going to the **Part mode** and retrieving the part used to make the drawing. If you choose the option **X-section** ⇒ **Modify** ⇒ (pick the name to be changed) ⇒ **Name**, enter a new name. The moment you change back to the drawing window, the section name will be updated.

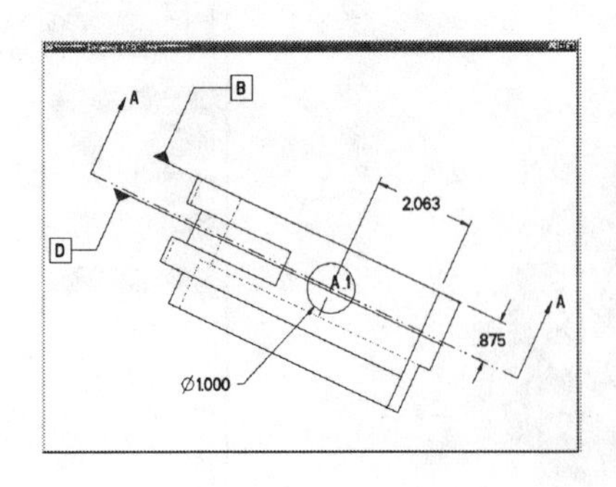

In the following pages, you will be creating a detail drawing of the Anchor. The front view will be a full section. A right side view and an auxiliary view are required to detail the part. Views will be displayed according to visibility requirements per **ANSI** standards, such as no hidden lines in sections. The part is to be dimensioned according to **ASME Y14.5M 1994**. You will use the format created in Lesson 16. Detailed views of other parts will be introduced to show the wide variety of view capabilities of Pro/E.

Using the Anchor model, create a drawing containing a front view section, a right side view, and an auxiliary view, as shown in Figure 18.13. The front view will be changed to a section view after the three views are established.

Dbms ⇒ Save ⇒ enter
Purge ⇒ enter ⇒
Done-Return

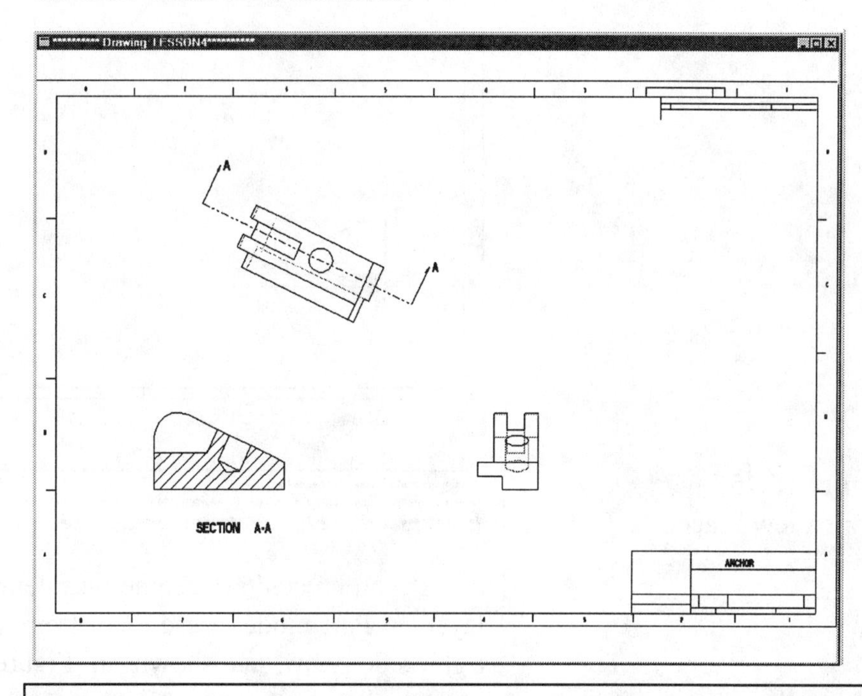

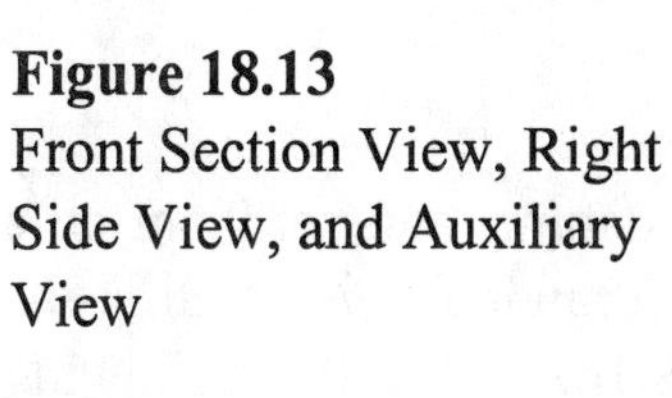

Figure 18.13
Front Section View, Right Side View, and Auxiliary View

Drawing ⇒ Create ⇒ (type **Anchor**) **⇒ enter ⇒ Retr Format ⇒** (?) **⇒ enter ⇒** (pick a previously created **D** size format) **⇒ Set Up ⇒ Retrieve ⇒** (pick your **.dtl** Drawing Set-Up file from the directory list, or choose **Modify Val** and create a new **.dtl** file) **⇒ Views ⇒** (?) **⇒ enter ⇒ Search/Retr ⇒** (locate **Anchor** part) **⇒ Done ⇒** (indicate the approximate center of the view, as shown in Figure 18.14) **⇒ Front ⇒** (pick a front surface) **⇒ Bottom ⇒** (pick the bottom surface of the Anchor) **⇒ Done/Return** (Fig. 18.15)

HINT

Set Up ⇒ Modify Val

draw_text_height	**.25**
default_font	**filled**
draw_arrow_style	**filled**
gtol_datums	**ASME**
allow_3d_dimensions	**YES**

File ⇒ Save ⇒ File ⇒ Exit

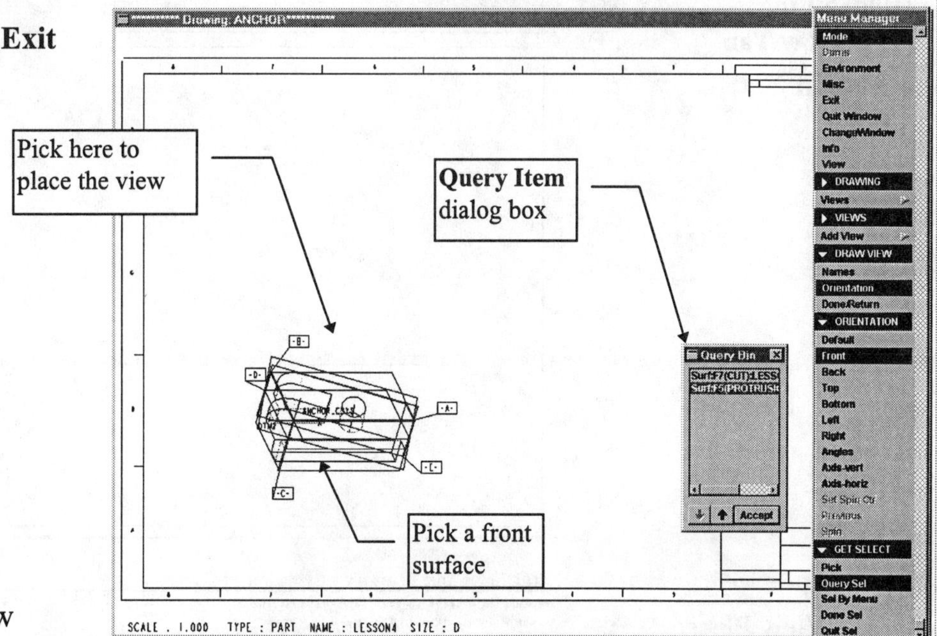

Figure 18.14
Orienting the Front View

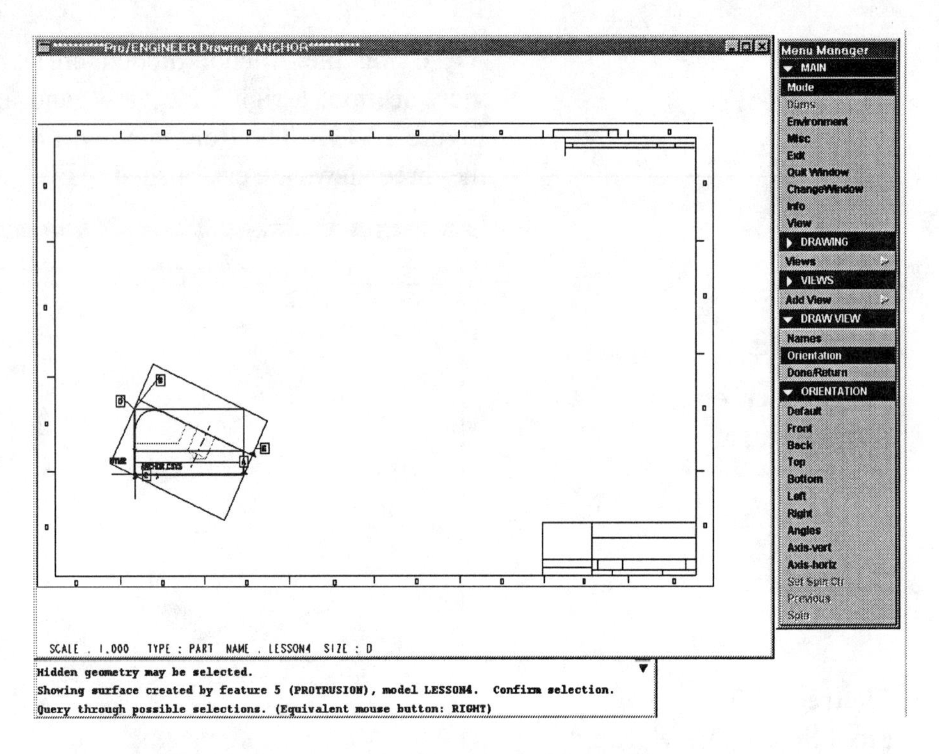

Figure 18.15
Front View Placed

If you created/set geometric tolerance datums and set them on a layer, in Part mode, erase them from your drawing, *after* adding the right side view, as shown in Figure 18.16. Use the following commands:

ENVIRONMENT

Environment ⇒

- □ **Disp DtmPln**
- □ **Disp Points**
- □ **Disp Axes**
- □ **Disp Csys**
- □ **Spin Center**
- □ **Grid Snap**
- **Hidden Line**
- **No Display Tan**

Done-Return

Add View ⇒ **Done** ⇒ (pick a position for the right side view as shown in Figure 18.16) ⇒ **Done/Return** ⇒ **Detail** (from DRAWING menu) ⇒ **Show/Erase** ⇒ **Erase** (radio button) ⇒ (pick the geometric tolerance and datum plane radio buttons) ⇒ **Erase All** ⇒ **Yes** ⇒ **Close**

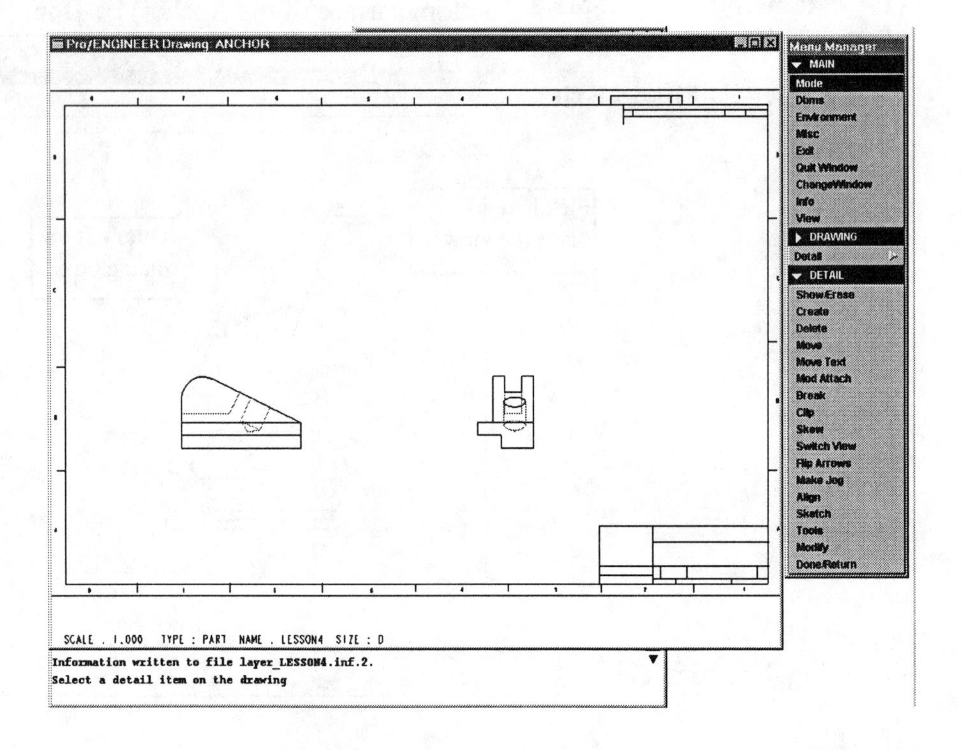

Figure 18.16
Right Side View Placed

Using the following commands, add the auxiliary view as shown in Figure 18.17:

HINT

Every time you add a new view, the set datums will appear in that view. If you don't want to see them, at this stage of the drawing, you must erase them.

Views (from **DRAWING** menu) ⇒ **Add View** ⇒ **Auxiliary** ⇒ **Done** ⇒ (pick a center point for view) ⇒ (select edge in main view) ⇒

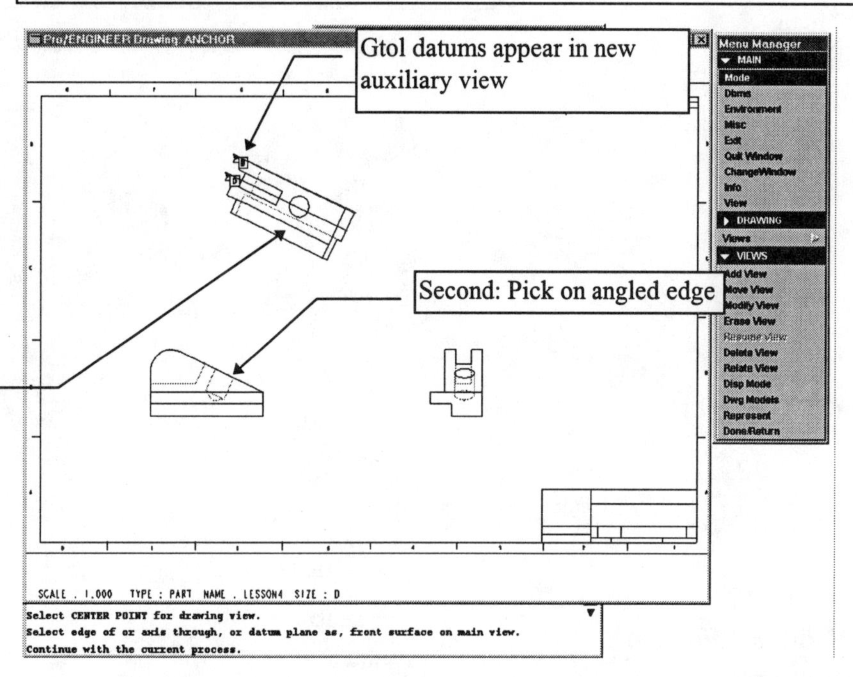

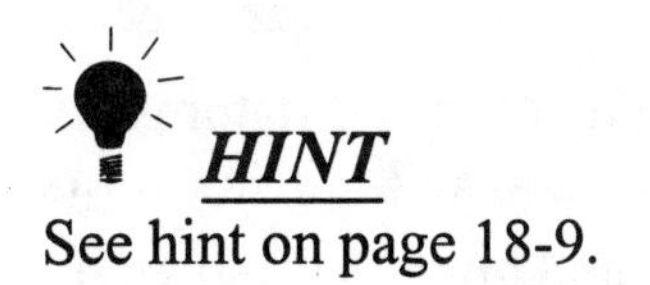

Figure 18.17
Auxiliary View

HINT

See hint on page 18-9.

Now change the front view into a section view. The section "A" was created in the Part mode. Use these commands (Fig. 18.18):

Modify View ⇒ **View Type** ⇒ (pick the front view as the view to be modified) ⇒ **Section** ⇒ **Done** ⇒ **Full** (default) ⇒ **Total Xsec** (default) ⇒ **Done** ⇒ (pick section name from XSEC NAMES menu: **A**) ⇒ (pick the view where for the arrows where the section is perpendicular (Fig. 18.18) ⇒ **Done Sel**

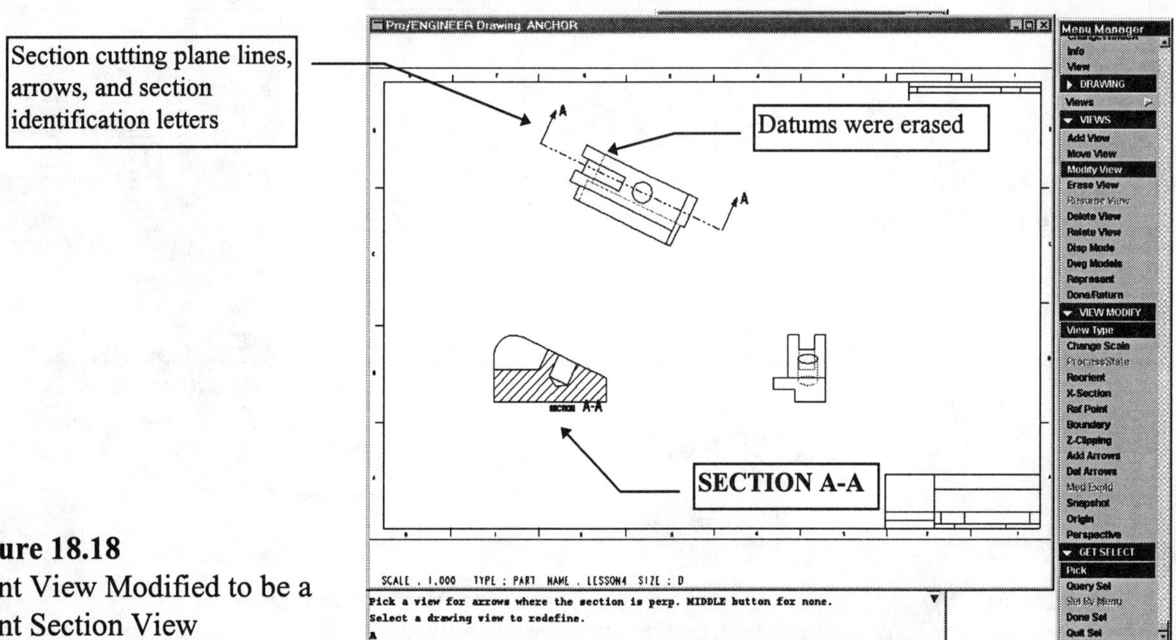

Figure 18.18
Front View Modified to be a Front Section View

Modify the **Disp Mode** of the views to remove all hidden lines:

Disp Mode ⇒ **View Disp** ⇒ (pick the all three views) ⇒ **Done Sel** ⇒ **No Hidden** ⇒ **No Disp Tan** ⇒ **Done** (Fig. 18.19) ⇒ **Done Sel** ⇒ **Done/Return**

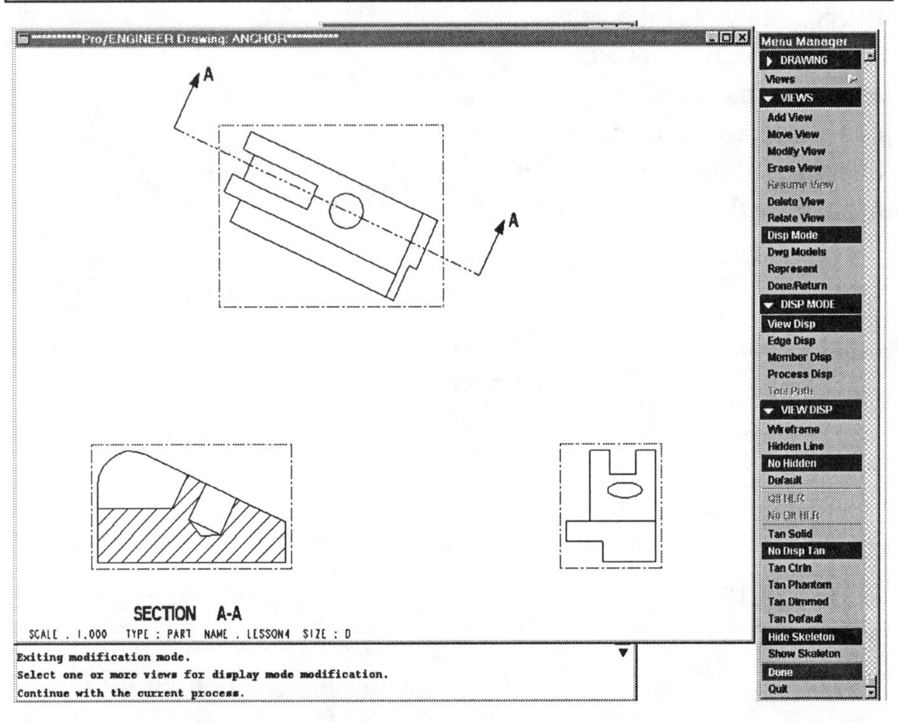

Figure 18.19
No Hidden Line, No Display Tan

Show all dimensions, axis-centerlines, and tolerance datums by choosing **Detail** ⇒ **Show/Erase** ⇒ **Show** ⇒ and picking the three radio buttons for *dimensions*, *datum plane*, and *axis*, as shown in Figure 18.20. Use your right mouse button to edit the dimension, axis, and datum plane name positions until the drawing looks similar to Figures 18.21 and 18.22.

HINT
Change the height of the section lettering to **.375**.

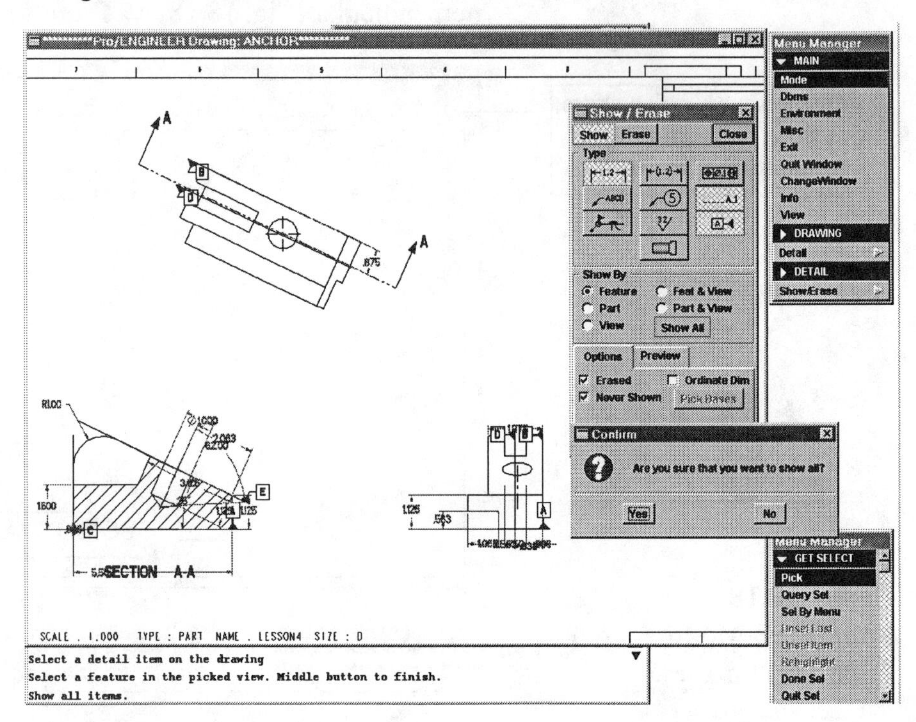

Figure 18.20
Show all Dimensions, Axes, and Datums

Dbms ⇒ Save ⇒ enter Purge ⇒ enter ⇒ Done-Return

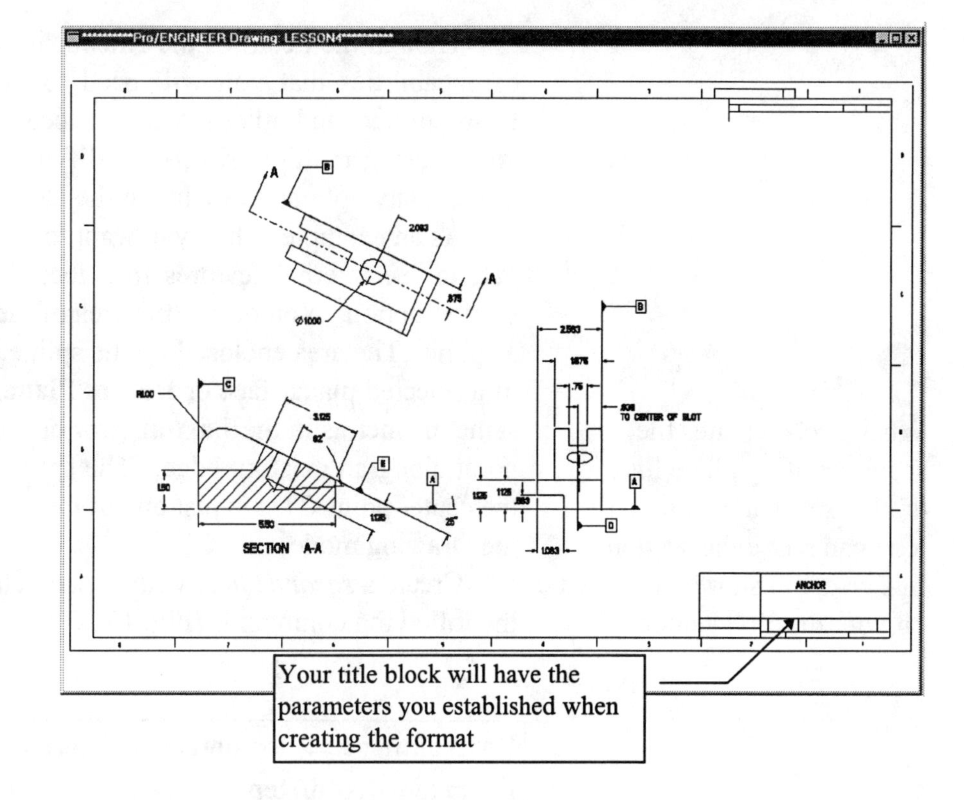

Figure 18.21
Completed Drawing

Figure 18.22
Detail Drawing

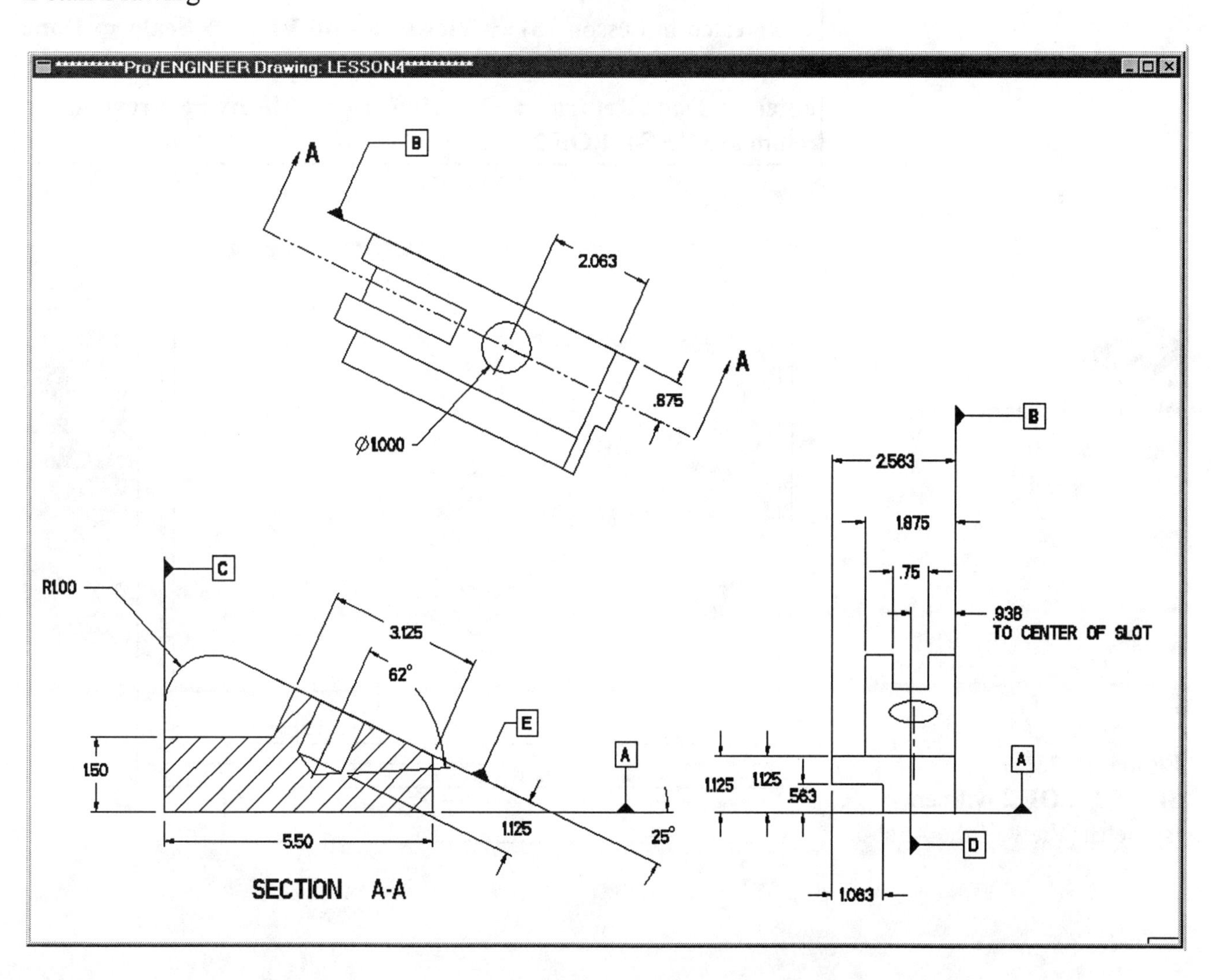

Though the detail of the Anchor is complete, there are a number of capabilities that you will need to master before completing the lesson project and other more advanced projects. Partial views, detail views, using multiple sheets, modifying section lining are just a few of the many options available in the Drawing mode with Pro/E.

As an example, when you want to break away a small portion of a part to see internal features in a local area, you can use the **Local** section option. You define the area of the **Local** breakout by drawing a spline. The area enclosed by the spline is then removed to the depth of a selected planar face or **Datum Plane**. You draw the spline in the same manner as in the **Partial** view option. You must define a center point along an existing edge. The spline must be closed and contain the center point. This is just one of many detailing options available in the Drawing mode.

HINT
When Pro/E creates the section cutting plane line, it draws across the entire part. You can move the section line arrow to shorten it, based on your drafting standard.

Create a *second sheet* with an isometric view of the Anchor using the following commands (Fig. 18.23):

Environment ⇒ **Isometric** ⇒ **Done-Return** ⇒ **Sheets** ⇒ **Add** ⇒ **Format** ⇒ **Add/Replace** ⇒ **?** ⇒ **enter** ⇒ (pick the **C** size format you created in Lesson 16) ⇒ **Views** ⇒ **Add View** ⇒ **Scale** ⇒ **Done** ⇒ (pick a center point for the view) ⇒ (type **1.5** as the scale) ⇒ **enter** ⇒ **Done/Return** ⇒ **Done/Return** ⇒ **Sheets** ⇒ **Previous** (to return to **SHEET 1 OF 2**)

Dbms ⇒ **Save** ⇒ **enter**
Purge ⇒ **enter** ⇒
Done-Return

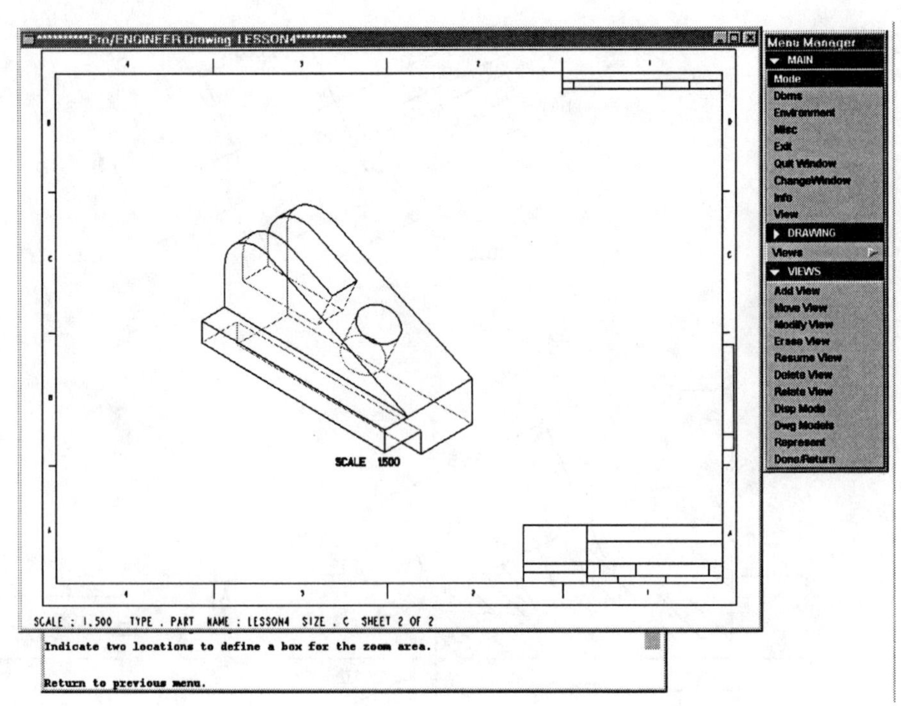

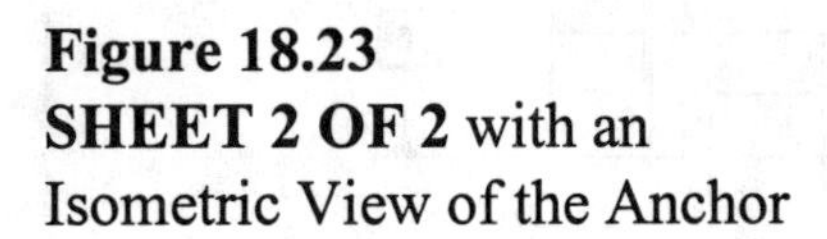
Figure 18.23
SHEET 2 OF 2 with an Isometric View of the Anchor

Create a detailed view with the following commands:

Views ⇒ **Add View** ⇒ **Detailed** ⇒ **Done** ⇒ (pick a open space on the drawing to position the new view, as in Figure 18.24) ⇒ (type **1.5** for the view scale) ⇒ **enter** ⇒ (select the center point for the detail on an existing view, pick the top edge of the hole in the front section view) ⇒ (sketch a spline around the hole--use the left mouse button to sketch the spline and the middle button to end it) ⇒ (enter a name for the detail view: A) ⇒ **enter** ⇒ (pick **Circle** from the BOUNDARY TYPE menu) ⇒ (select the location of the note as shown in Figure 18.25) ⇒ **Done/Return**

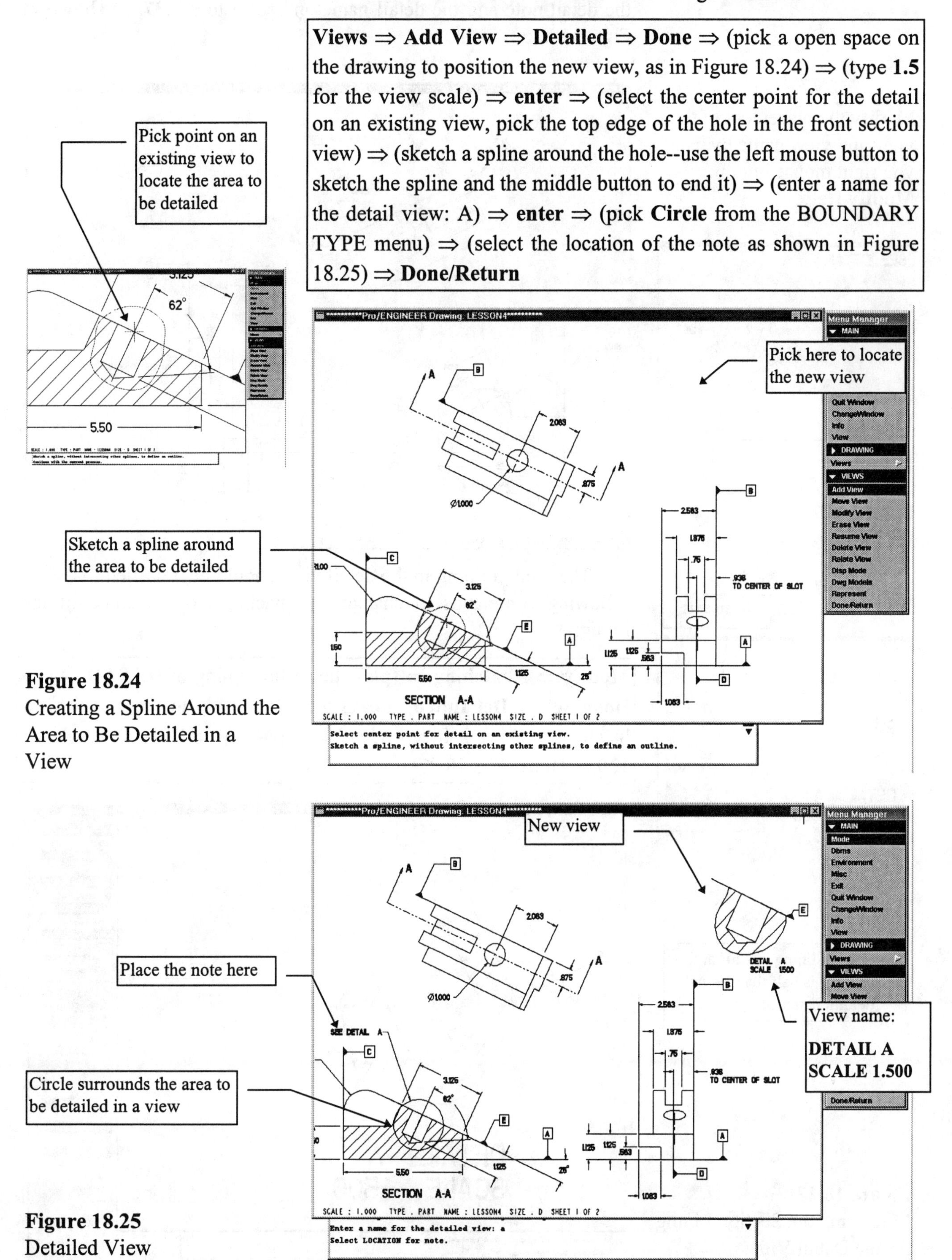

Figure 18.24
Creating a Spline Around the Area to Be Detailed in a View

Figure 18.25
Detailed View

Add an axis to the detail of the hole, and modify the text height of the detail note and the detail name and scale to be **.37**, as shown in Figure 18.26.

HINT
Modify the placement of the view and view name by using you right mouse button: **Modify Item**.

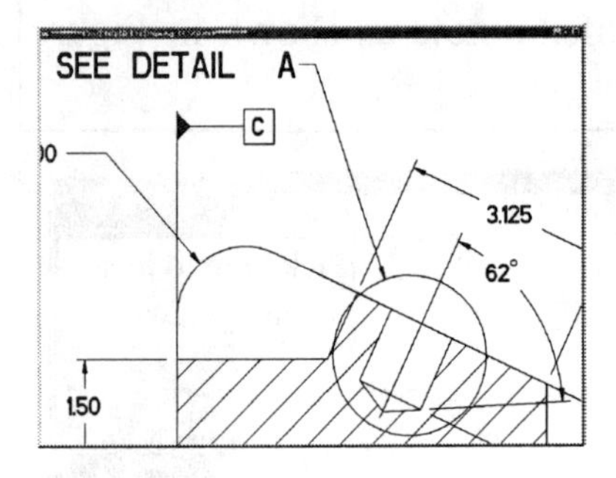

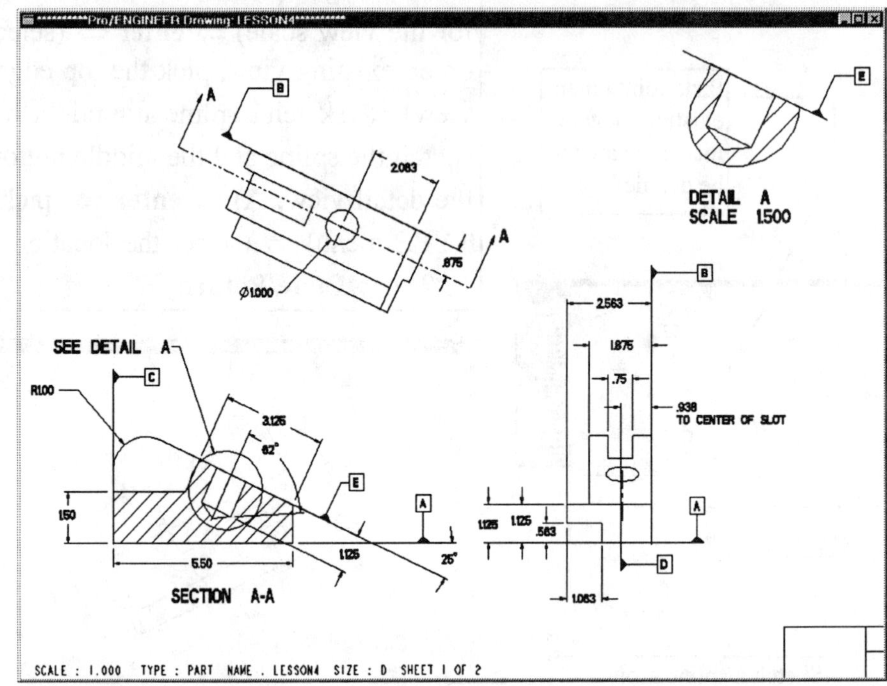

Figure 18.26
DETAIL A

Add an axis

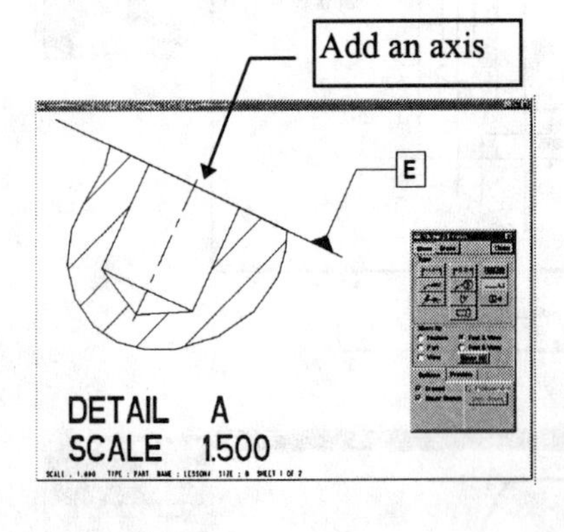

The section lining in the detail view should be modified. Use the following commands to change the spacing and the angle of the lining:

Modify ⇒ **Xhatching** ⇒ (pick the section lining in **DETAIL A)** ⇒ **Done Sel** ⇒ **Det Indep** (breaks the relationship to the parent view hatching, making it independent) ⇒ **Spacing** ⇒ **Half** ⇒ **Angle** ⇒ **120** ⇒ **Done** (Fig. 18.27)

Section lining is **120°** and spacing is half the original default distance

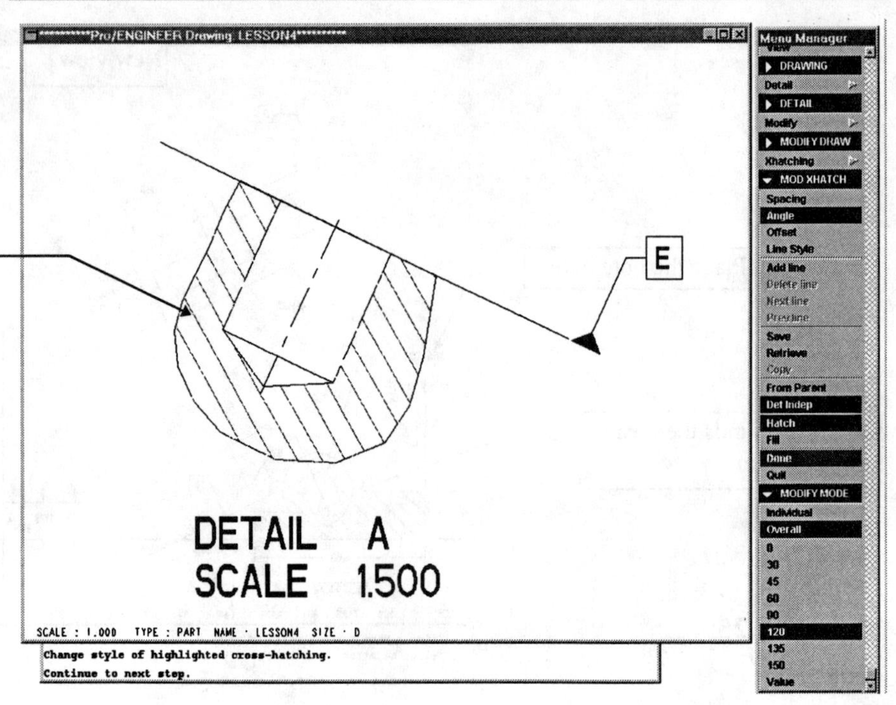

Figure 18.27
Modifying the Section Lining for the Detail View

Switch the hole depth dimension from the front section view to the detail view, as shown in Figure 18.28. The detail is now complete (Fig.18.29).

Dbms ⇒ Save ⇒ enter Purge ⇒ enter ⇒ Done-Return

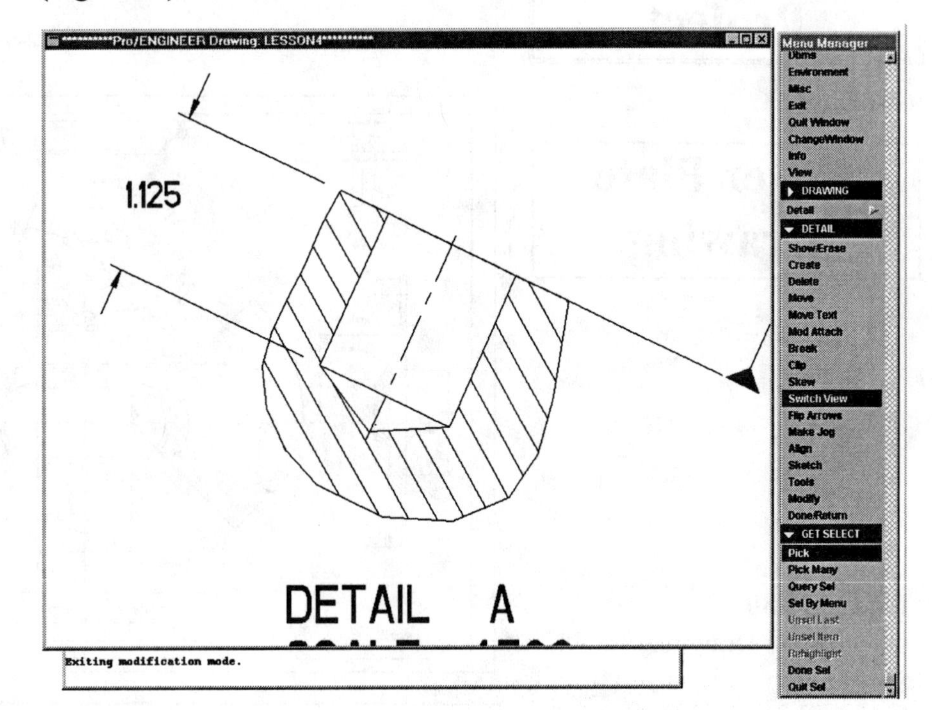

Figure 18.28
Hole Depth Dimension Shown in **DETAIL A**

Figure 18.29
Completed Detail Drawing

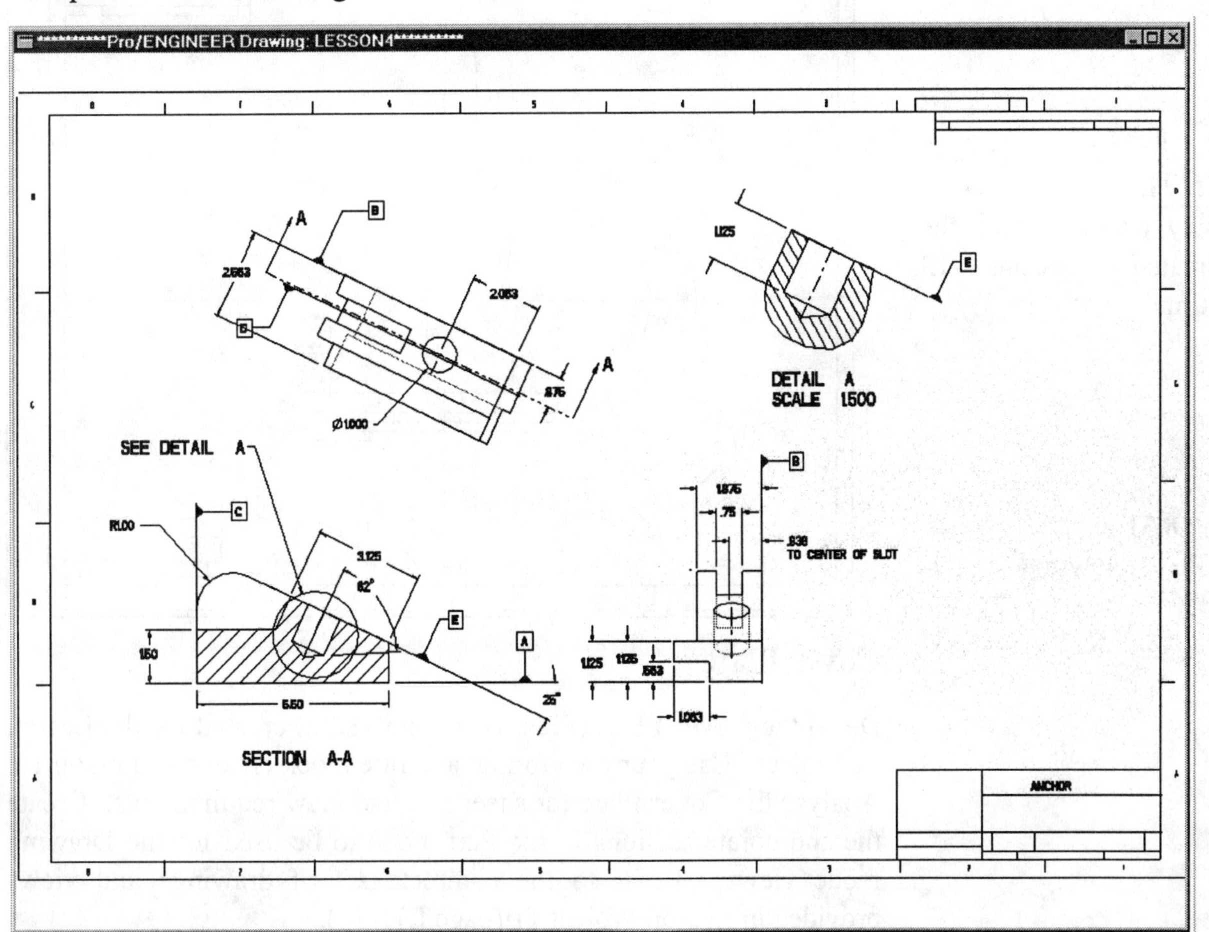

Lesson 18 Project

Cover Plate Drawing

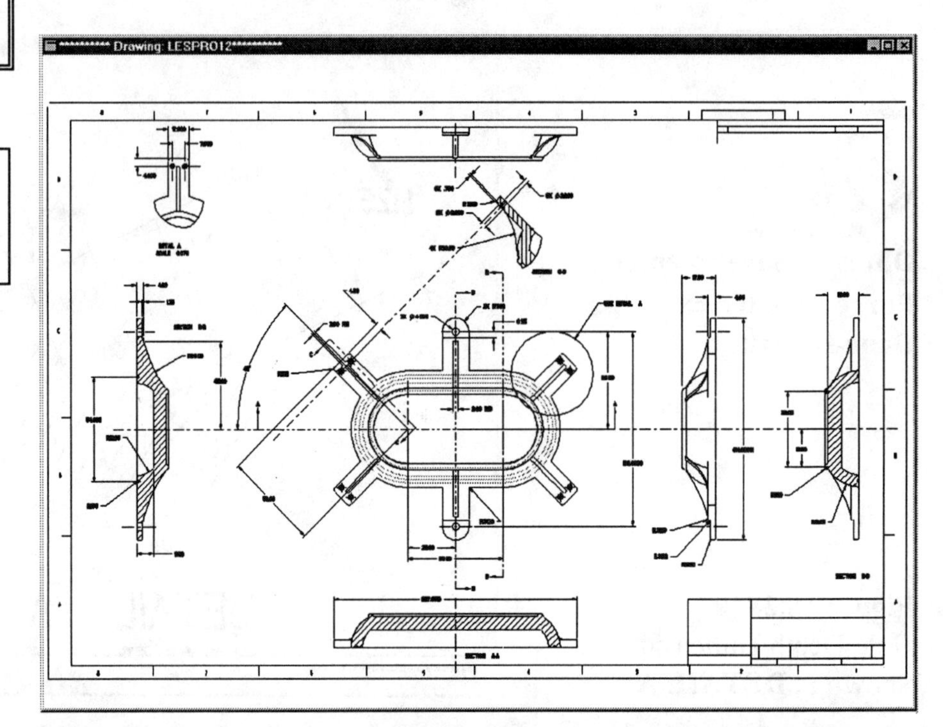

Figure 18.30
Cover Plate Drawing, Sheet One

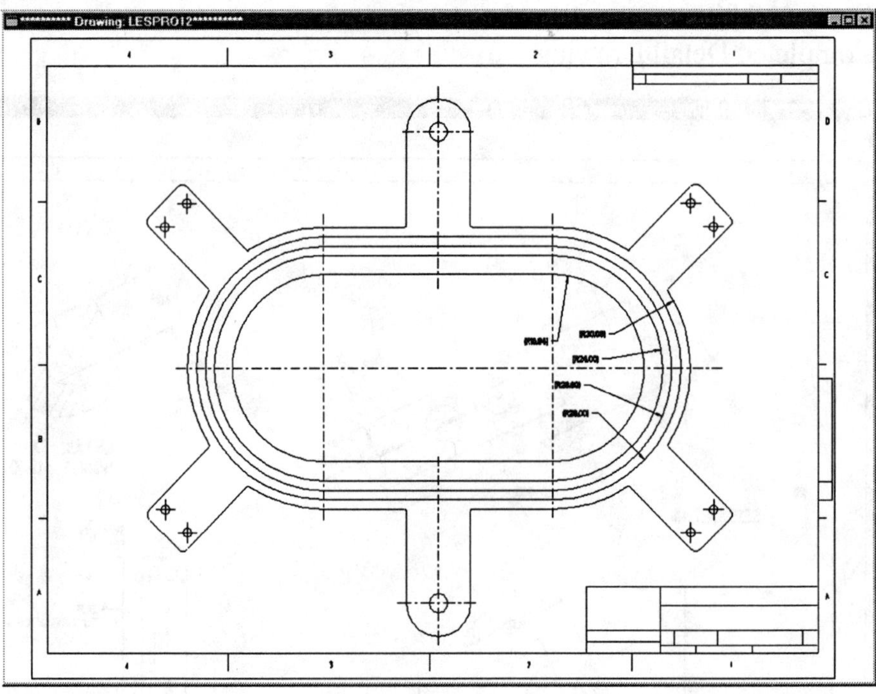

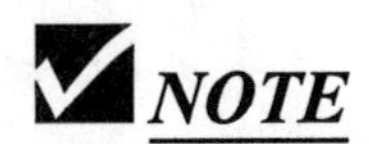

NOTE
You may detail any of the parts created in Lessons 1-13 at this time.

Figure 18.31
Cover Plate Drawing, Sheet Two

Cover Plate

Detail the **Cover Plate** (Figs. 18.31 and 18.32) created for the Lesson 11 Project. Use your own format and title block created in Lesson 16. Analyze the Cover Plate for sheet size and view requirements. Create the appropiate sections in the Part mode to be used for the Drawing mode views. Refer to the complete set of drawings and views provided in Lesson Project 11 (page L11-18).

Lesson 19

Assembly Drawings and BOM

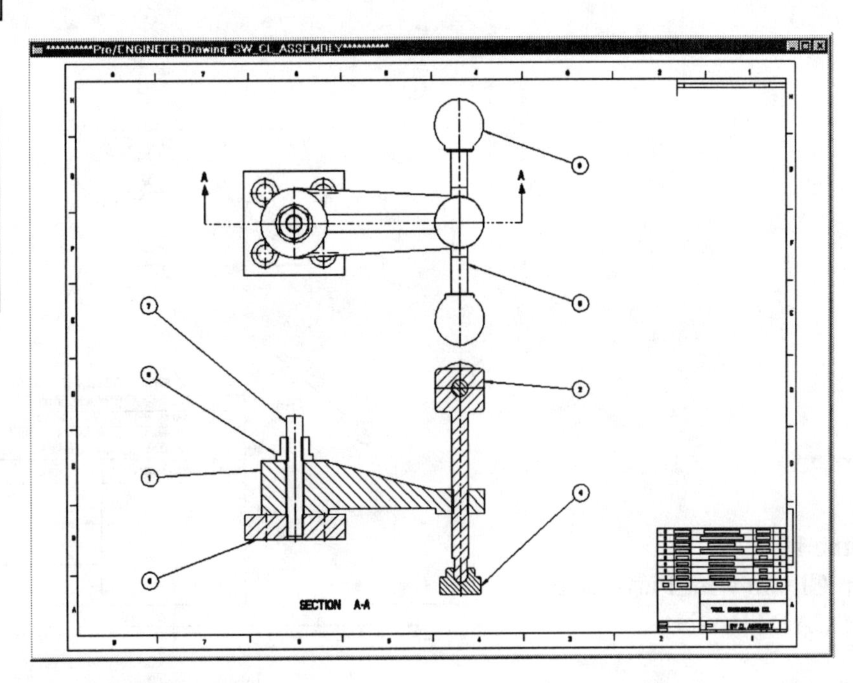

Figure 19.1
Swing Clamp Assembly Drawing

OBJECTIVES

1. **Create an assembly drawing**
2. **Generate a parts list from a bill of materials (BOM)**
3. **Balloon an assembly drawing**
4. **Create a section assembly view and change component visibility**
5. **Add parameters to parts**
6. **Create a table to generate a parts list automatically**

EGD REFERENCE
Engineering Graphics and Design
by L. Lamit and K. Kitto
Read Chapters: 11, 17, 23
See Pages: 358-368, 662-671, 810-846

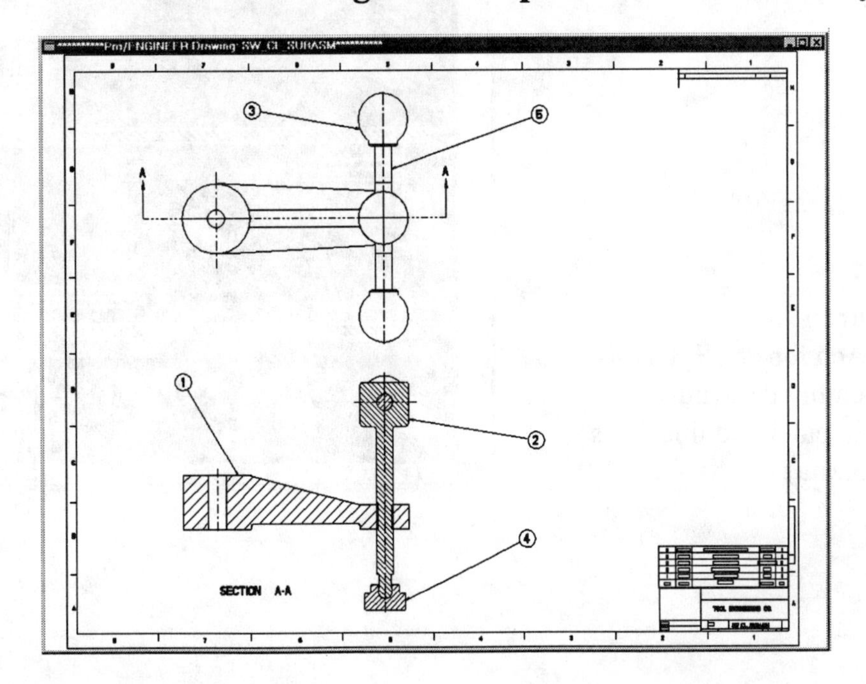

Figure 19.2
Swing Clamp Subassembly Drawing

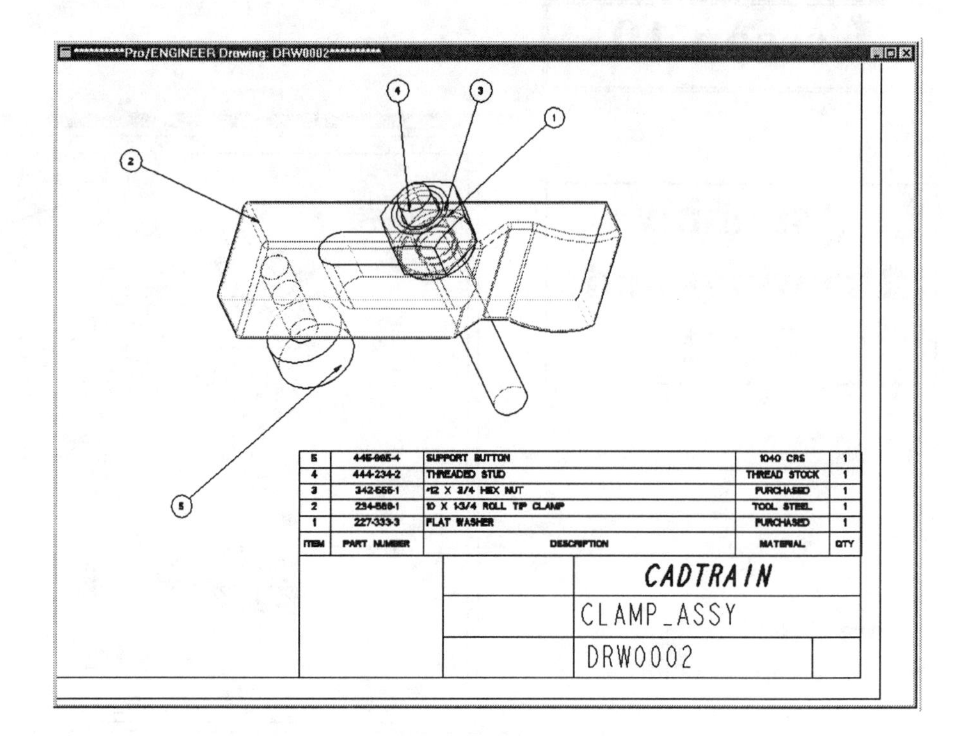

Figure 19.3
Strap Clamp with Parts List from CADTRAIN

Assembly Drawings and BOM

Pro/E incorporates a great deal of functionality into drawings of assemblies (Fig. 19.1 and Fig. 19.2). You can assign parameters to parts in the assembly that can be displayed on a *parts list* in an assembly drawing (Fig. 19.3). Pro/E also automatically generates the item balloons for each component (Fig. 19.4).

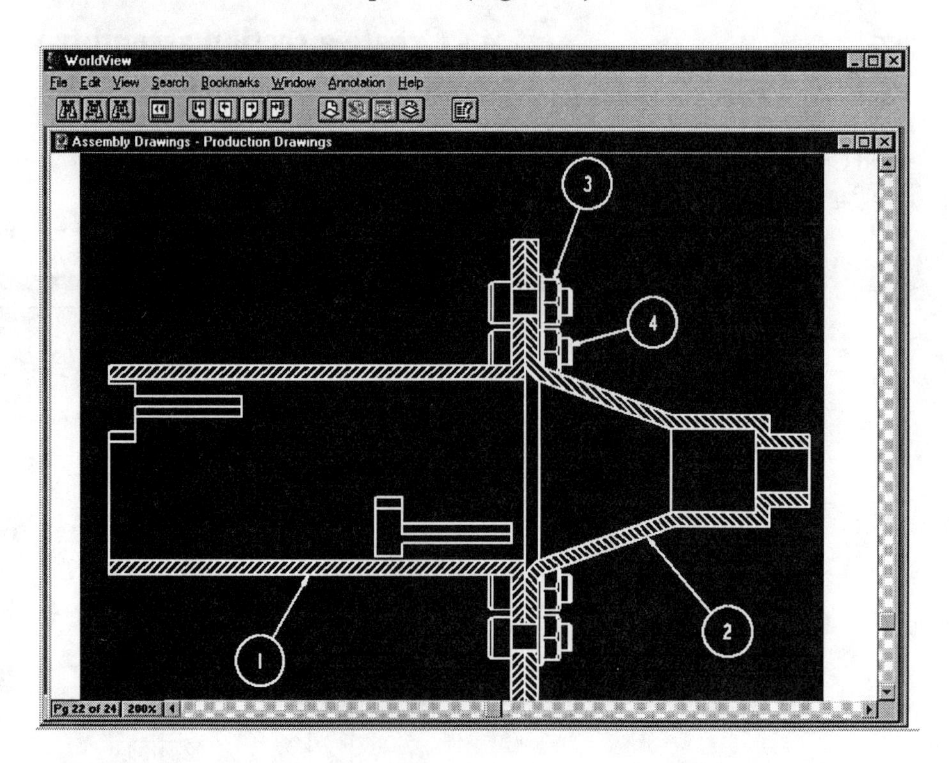

Figure 19.4
COAch for Pro/E, BOM--Assembly Drawings (Segment 3: Adding Parts List Data)

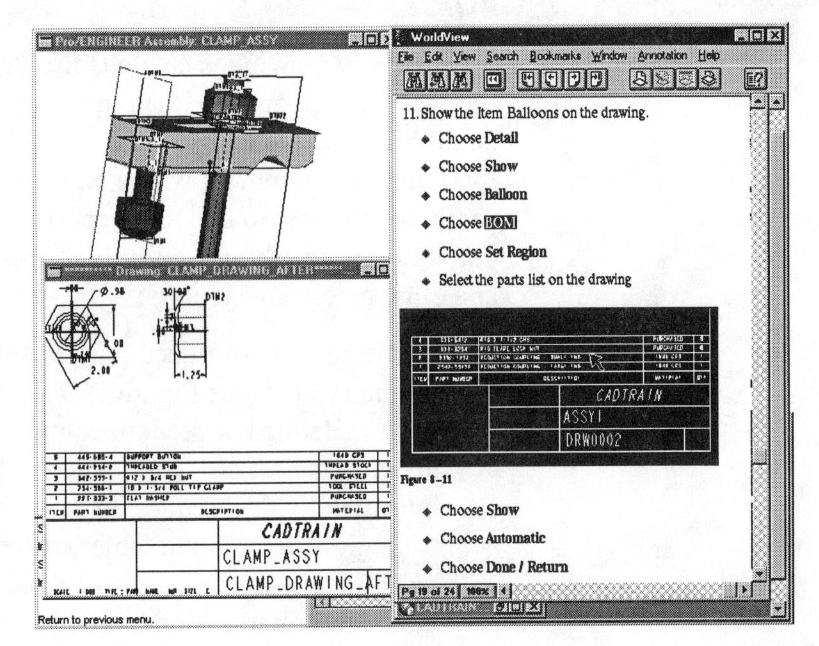

Figure 19.5
COAch for Pro/E, Starting an Assembly Drawing--(Segment 1: Using an Assembly Format)

In addition, there is a variety of specialized capabilities to allow you to alter the manner in which individual components are displayed in views and in sections (Fig. 19.5). The **format** for an assembly is usually (slightly) different than the format used for detail drawings. The most significant difference is the presence of a **parts list**.

An assembly format is provided for you as part of the CADTRAIN tutorial software. As part of this lesson, you will create a set of personal formats and place your standard parts list on them.

A parts list is actually a *Drawing Table object* that is formatted to represent a bill of materials on a drawing. By defining *parameters* in the parts in your assembly that agree with the specific format of the parts list, you make it possible for the system to add pertinent data to the parts list automatically as components are added to the assembly. Pro/E also allows the parts list and parameters to balloon the assembly automatically (Fig. 19.6).

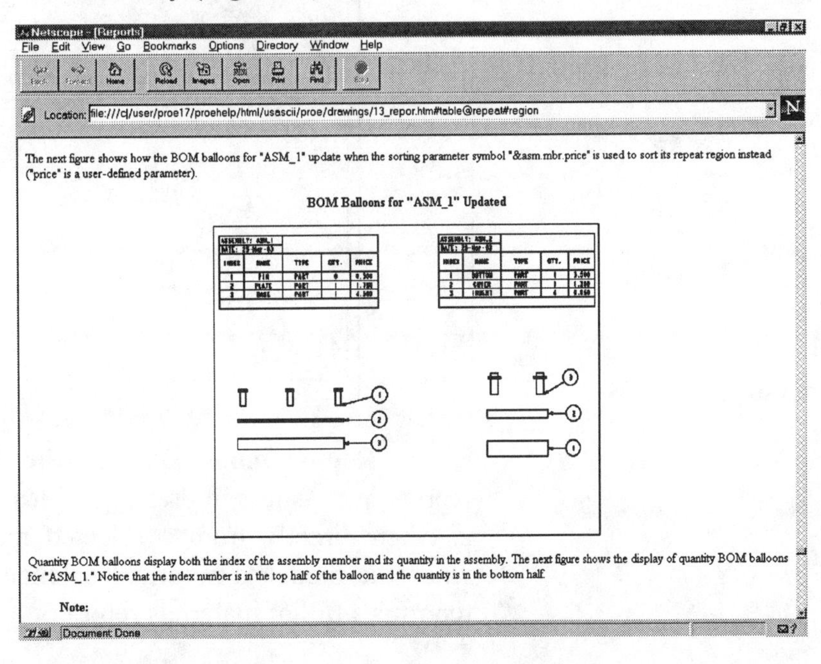

Figure 19.6
Online Documentation BOM Balloons

BOM

An assembly drawing is created, after the assembly is complete. With Pro/REPORT, you can then generate a bill of materials or other tabular data as required for the project. **Pro/REPORT** introduces a formatting environment where text, graphics, tables, and data may be combined to create a dynamic report (Fig. 19.7). Specific tools enable you to generate customized **bills of materials** (BOMs), family tables, and other associative reports:

* Dynamic, customized reports with drawing views and graphics can be created (see Figure 19.7).
* User-defined or predefined model data can be listed on reports, drawing tables, or layout tables. This reported data can be sorted by any individual requested data-type display.
* Regions in drawing tables, report tables, and layout tables can be defined to expand and shrink automatically with the amount of model information that has been requested to be displayed.
* Filters can be added to eliminate specific types of data from displaying in reports, drawing tables, or layout tables.
* Recursive or top-level assembly data can be searched for display.
* Duplicate occurrences of model data can be listed individually or as a group in a report, drawing table, or layout table.
* Assembly component balloons can be linked directly to a customized **BOM** and automatically updated when assembly modifications are made.

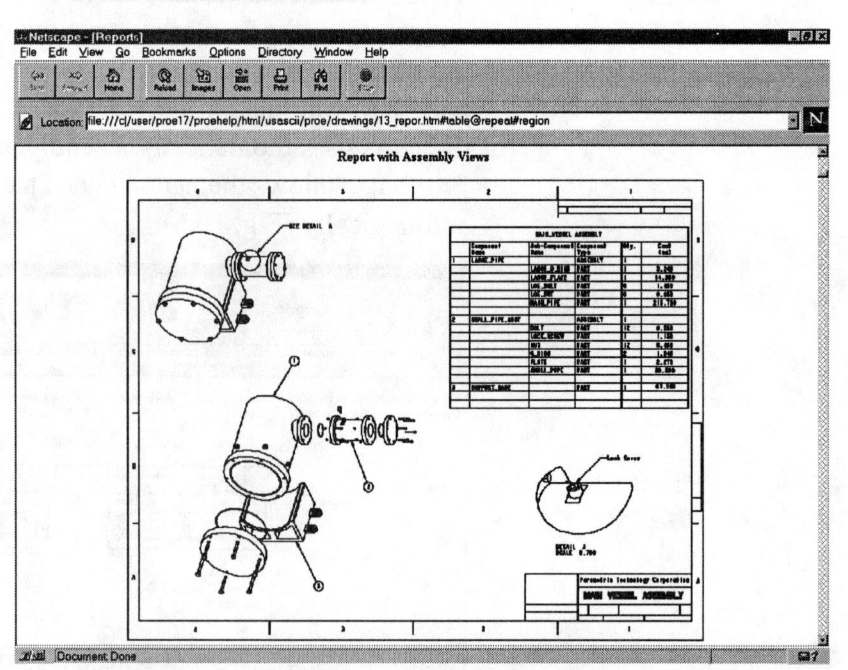

Figure 19.7
Assemblies and BOM

In **Report mode**, data can be displayed in a tabular form on reports, just as it is in drawing tables. The data reported on the tables is taken directly from a selected model and updates automatically when the model is modified or changed. A common example of a report is a bill of materials report or a generic part table.

Including a Bill of Materials in a Drawing

If you do not have Pro/REPORT (Fig. 19.8) and want to add a bill of materials (BOM), create a BOM file in Assembly or Drawing mode using the **Info** option in the MAIN menu. Choose **BOM** from the INFO menu, and add the BOM file to the drawing as a note entered from a file. To format or arrange the information in the BOM, you must use a text editor.

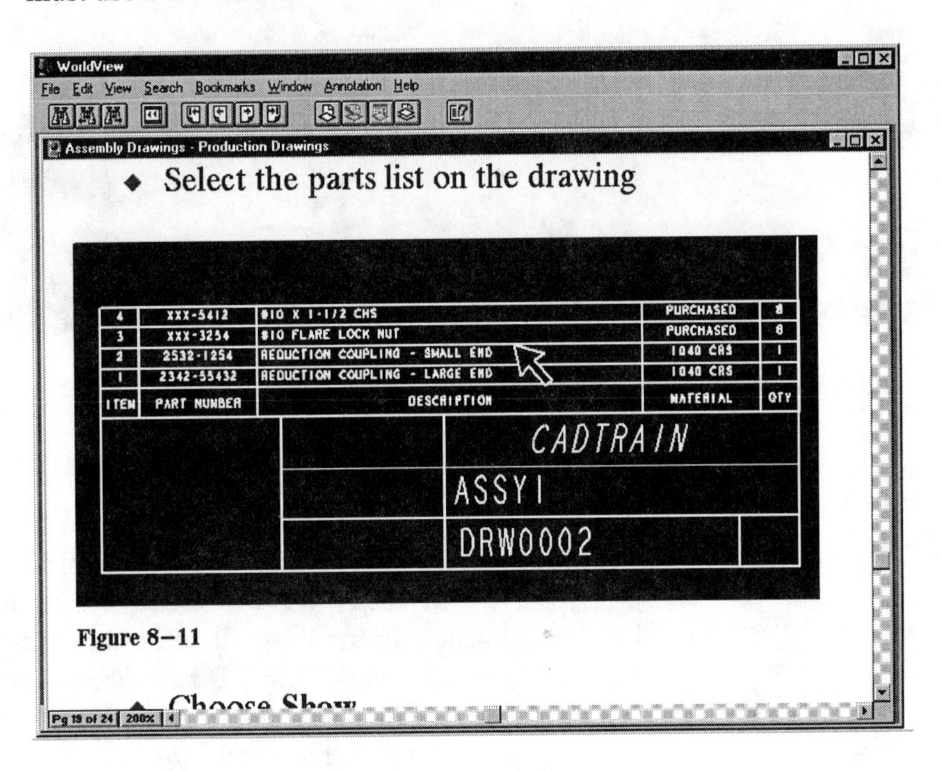

Figure 19.8
COAch for Pro/E, Parts List on Assembly Drawings

A BOM that is added to a drawing as a note is not connected with the BOM file that was used to create the note. If the composition of the assembly changes, you must create a new BOM and add it to the drawing as a new note. You can fully edit the BOM displayed on the drawing as a note without affecting the original BOM file.

To add a BOM to a drawing as a note:

1. Choose **File** from the NOTE TYPES menu.. When adding a BOM to a drawing as a note, justify the note using **Default** or **Left**. If you use **Center** or **Right**, the BOM might format incorrectly on the drawing.
2. Choose **Make Note** when you have finished choosing options
3. Pick a location for the note (BOM) to appear.
4. Enter the name of the BOM file, including the **.bom** full extension.
5. The BOM is displayed on the drawing.

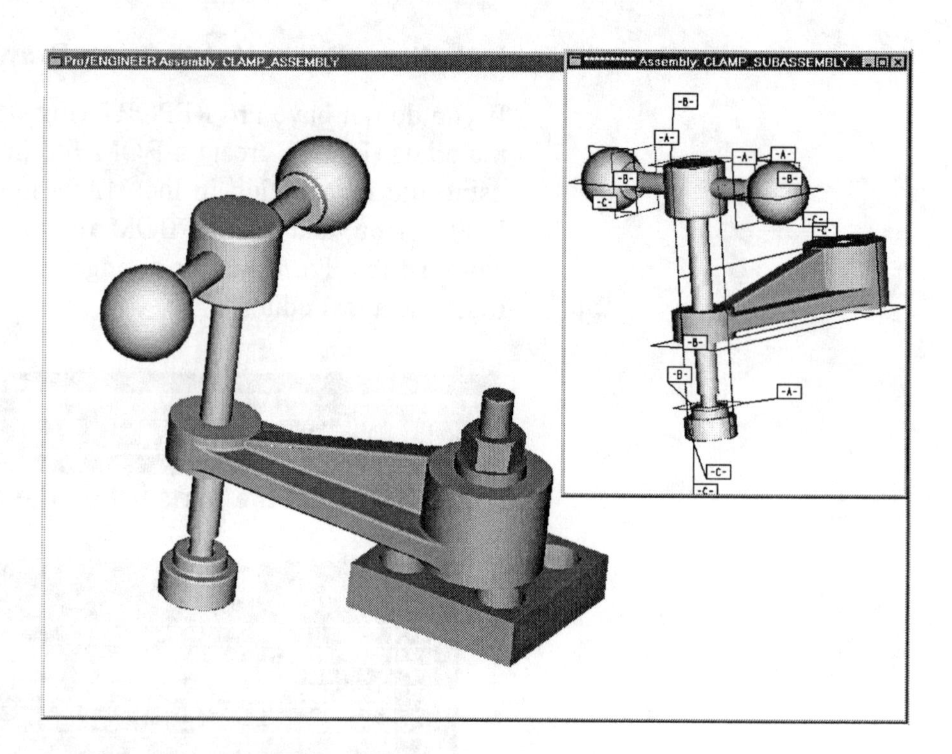

Figure 19.9
Swing Clamp Subassembly and Swing Clamp Assembly

Swing Clamp Assembly Drawing

The format for an assembly is usually different than the format used for detail drawings. The most significant difference is the presence of a parts list. We will create a standard **E** size format and place a standard parts list on it. You should create a set of assembly drawing formats on **B**, **C**, and **D** size sheets at your convenience.

A **parts list** is actually a *Drawing Table object* that is formatted to represent a ***bill of materials*** **(BOM)** on a drawing. By defining parameters in the parts in your assembly that agree with the specific format of the parts list, you make it possible for Pro/E to add pertinent data to the parts list automatically as components are added to the assembly.

After you create an **E** size sheet and parts list table, you will create a new drawing with two views using the assembly format. The Swing Clamp subassembly will be used for the first drawing done in this lesson. The second drawing will use the Swing Clamp assembly. Both drawings use the **E** size format created in the first section of this lesson. The format will have a parameter-driven title block (as in Lesson 16) and an integral parts list.

Using steps similar to those outlined in Lesson 16, where a **C** size format was created and saved, create an "**E**" size format using the following commands:

NOTE
Use a format file name that will identify the format as an assembly format with a parts list, such as **ASM_PTL_E**.

Mode ⇒ Format ⇒ Search/Retr ⇒ Format Dir ⇒ e.frm ⇒ Dbms ⇒ Save As ⇒ enter ⇒ (type a unique name for your format: ***your_format_name***) **⇒ enter ⇒ Quit Window ⇒ Format ⇒ Search/Retr ⇒** (pick the new format: ***your_format_name*.frm**)

Figure 19.10 shows the standard **E** size format available on Pro/E using the format directory.

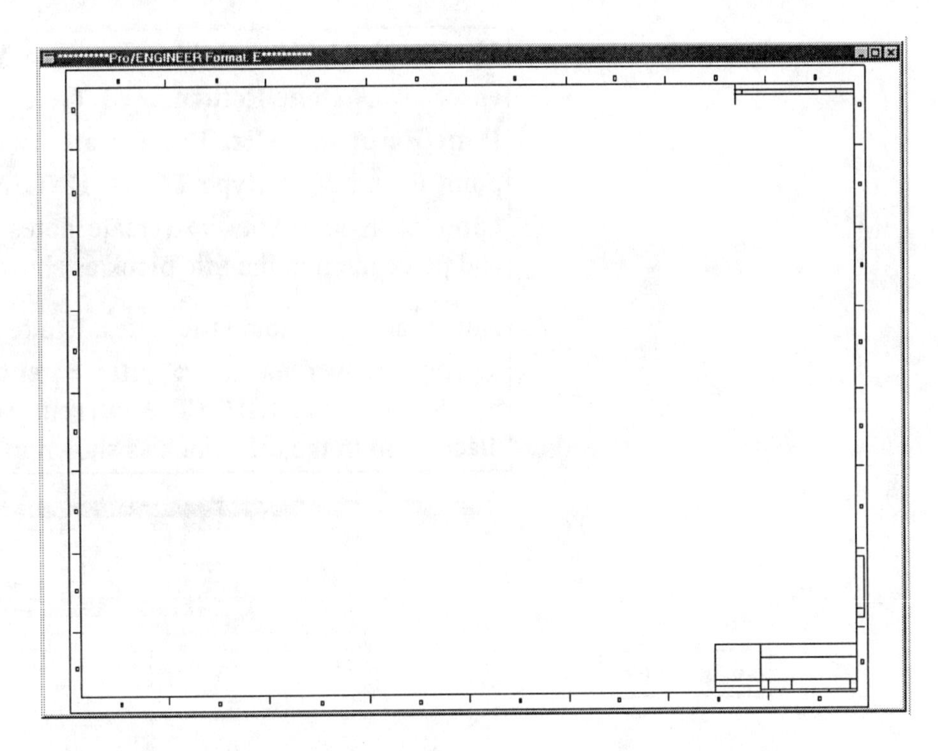

Figure 19.10
Standard **E** Size Format

Modify your **.dtl** file to have filled arrowheads and filled default font, as shown in Figure 19.11.

Set Up ⇒ Modify Val ⇒ (edit the file as required) **⇒ File ⇒ Save ⇒ File ⇒ Exit ⇒ Quit ⇒ View ⇒ Repaint ⇒ Done/Return**

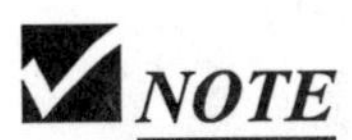

NOTE
Set the following in your **.dtl** file: **Set Up ⇒ Modify Val**

default_font	**filled**
draw_arrow_style	**filled**
drawing_text_height	**.25**
draw_arrow_length	**.25**
draw_arrow_width	**.08**

File ⇒ Save ⇒ File ⇒ Exit

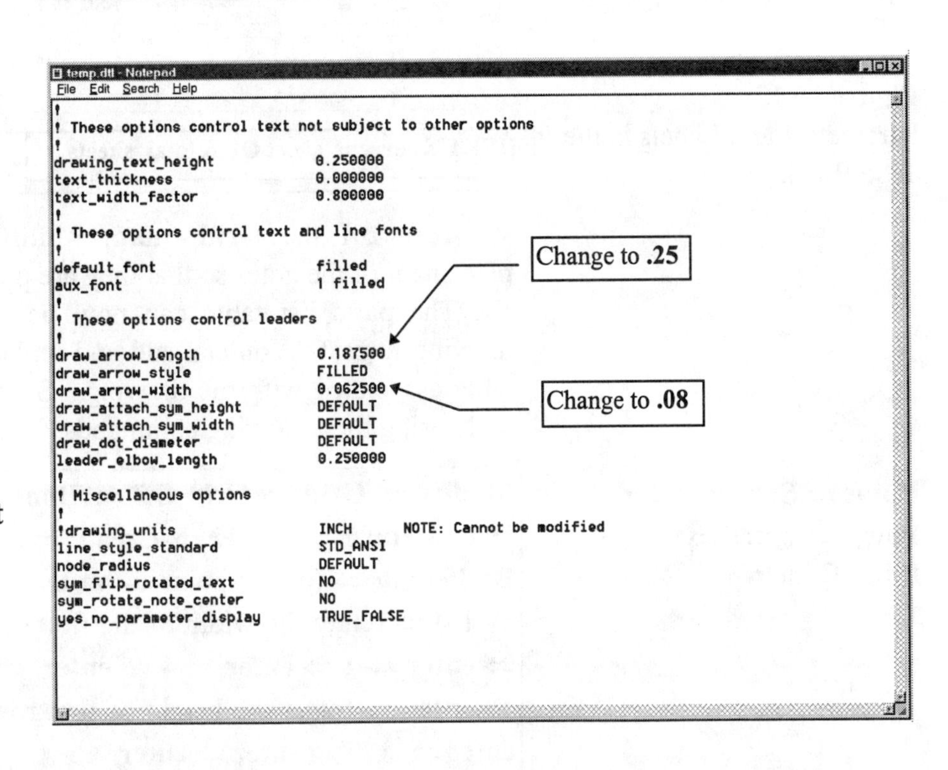
```
temp.dtl - Notepad
File  Edit  Search  Help
!
! These options control text not subject to other options
!
drawing_text_height          0.250000
text_thickness               0.000000
text_width_factor            0.800000
!
! These options control text and line fonts
!
default_font                 filled
aux_font                     1 filled
!
! These options control leaders
!
draw_arrow_length            0.187500
draw_arrow_style             FILLED
draw_arrow_width             0.062500
draw_attach_sym_height       DEFAULT
draw_attach_sym_width        DEFAULT
draw_dot_diameter            DEFAULT
leader_elbow_length          0.250000
!
! Miscellaneous options
!
!drawing_units               INCH        NOTE: Cannot be modified
line_style_standard          STD_ANSI
node_radius                  DEFAULT
sym_flip_rotated_text        NO
sym_rotate_note_center       NO
yes_no_parameter_display     TRUE_FALSE
```


Figure 19.11
.dtl File

Zoom into the title block only and fill in the titles and parameters required to display the proper information. Use the following commands:

Modify ⇒ **Grid** ⇒ **Grid Params** ⇒ **X&Y Spacing** ⇒ (type **.2**) ⇒ **enter** ⇒ **Done/Return** ⇒ **Grid On** ⇒ **Done/Return** ⇒ **Done/Return** ⇒ **Detail** ⇒ **Create** ⇒ **Note** ⇒ **Make Note** ⇒ (pick point for note) ⇒ (type **TOOL ENGINEERING CO.**) ⇒ **enter** ⇒ **enter** ⇒ **Make Note** ⇒ (create notes for **DRAWN,** and **ISSUED,** and place them in the title block as shown in Figure 19.12)

Now make the parametric notes: **Make Note** ⇒ (pick point for note) ⇒ (type **&dwg_name**) ⇒ **enter** ⇒ **enter** ⇒ (create parameter notes for **&scale**, and **SHEET ¤t_sheet OF &total_sheets** and place them in the title block as shown in Figure 19.12)

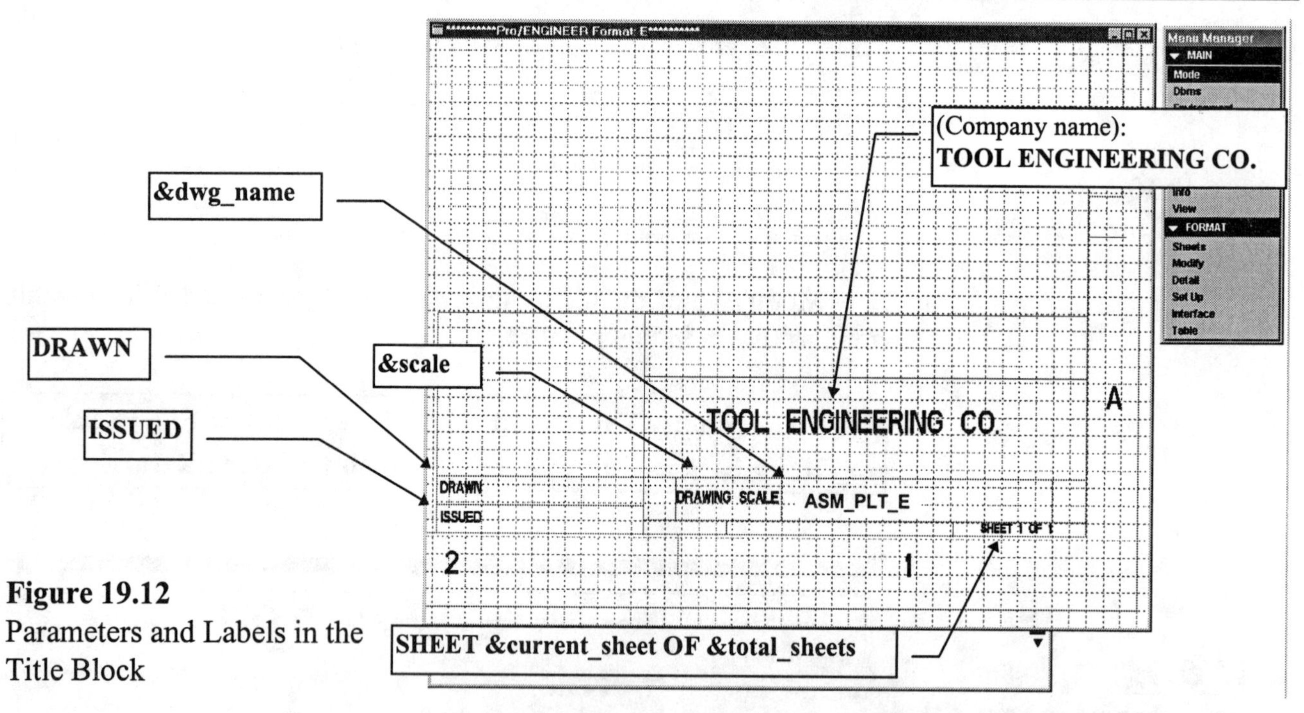

Figure 19.12
Parameters and Labels in the Title Block

Turn off the **Grid Snap**, **Modify** the text height and the placement of the notes so that they are placed similar to Figure 19.12.

The parts list table can now be created and saved with this drawing format. You can add and replace formats and still keep the table associated with the drawing. Start the parts list by creating a table using the following commands:

Dbms ⇒ **Save** ⇒ **enter**
Purge ⇒ **enter** ⇒
Done-Return

Modify ⇒ **Grid** ⇒ **Grid Off** ⇒ **Done/Return** ⇒ **Done/Return** ⇒ **Done/Return** ⇒ **Table** ⇒ **Create** ⇒ **Ascending** ⇒ **Rightward** ⇒ **By Length** ⇒ (pick a point above the title block as shown in Figure 19.13) ⇒ (enter the width of the first column in drawing units) ⇒ **1** ⇒ **enter** ⇒ **1** ⇒ **enter** ⇒ **4** ⇒ **enter** ⇒ **1** ⇒ **enter** ⇒ **.75** ⇒ **enter** ⇒ **enter** ⇒ (enter the height of first row in drawing units) ⇒ **.5** ⇒ **enter** ⇒ **.375** ⇒ **enter** ⇒ **enter**

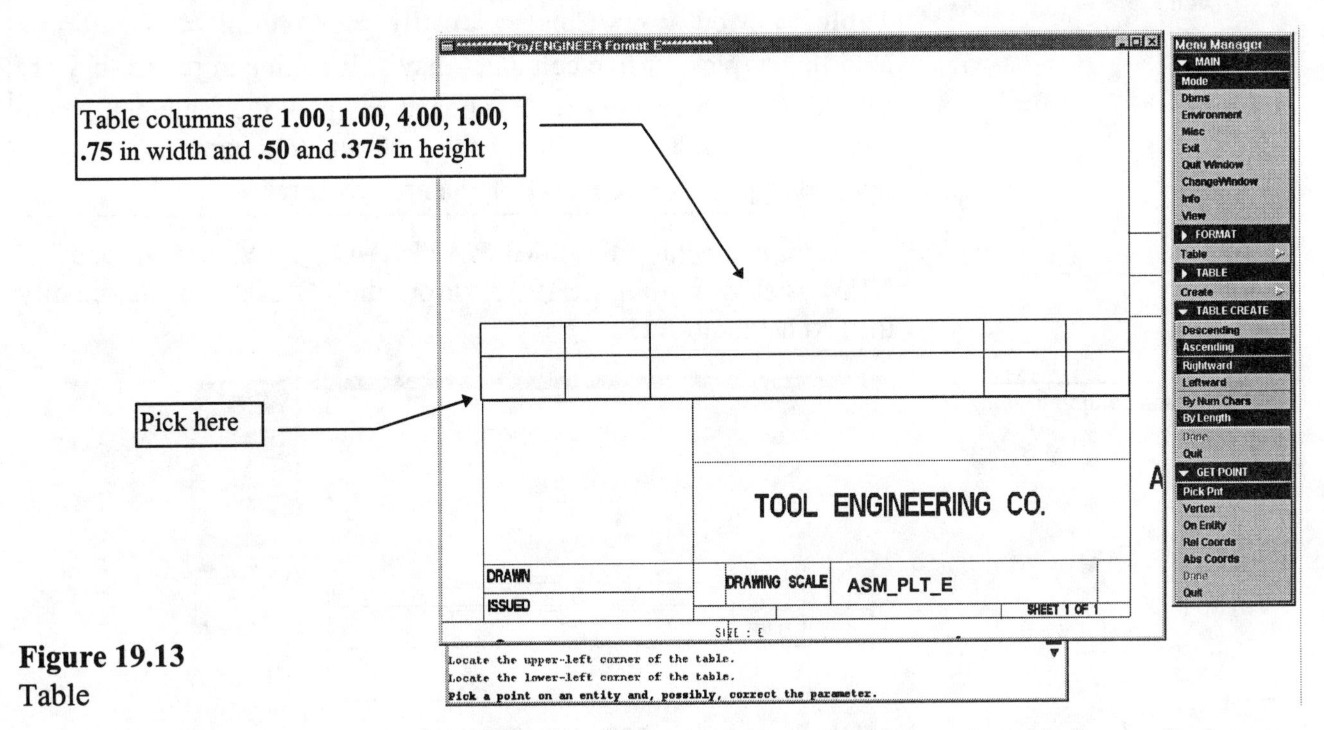

Figure 19.13
Table

Next we will add a **Repeat Region** to the table by continuing with the commands:

Repeat Region ⇒ **Add** ⇒ (locate the corners of the region as shown in Figure 19.14) ⇒ **Attributes** (select the **Repeat Region**) ⇒ **No Duplicates** ⇒ **Recursive** ⇒ **Done/Return**

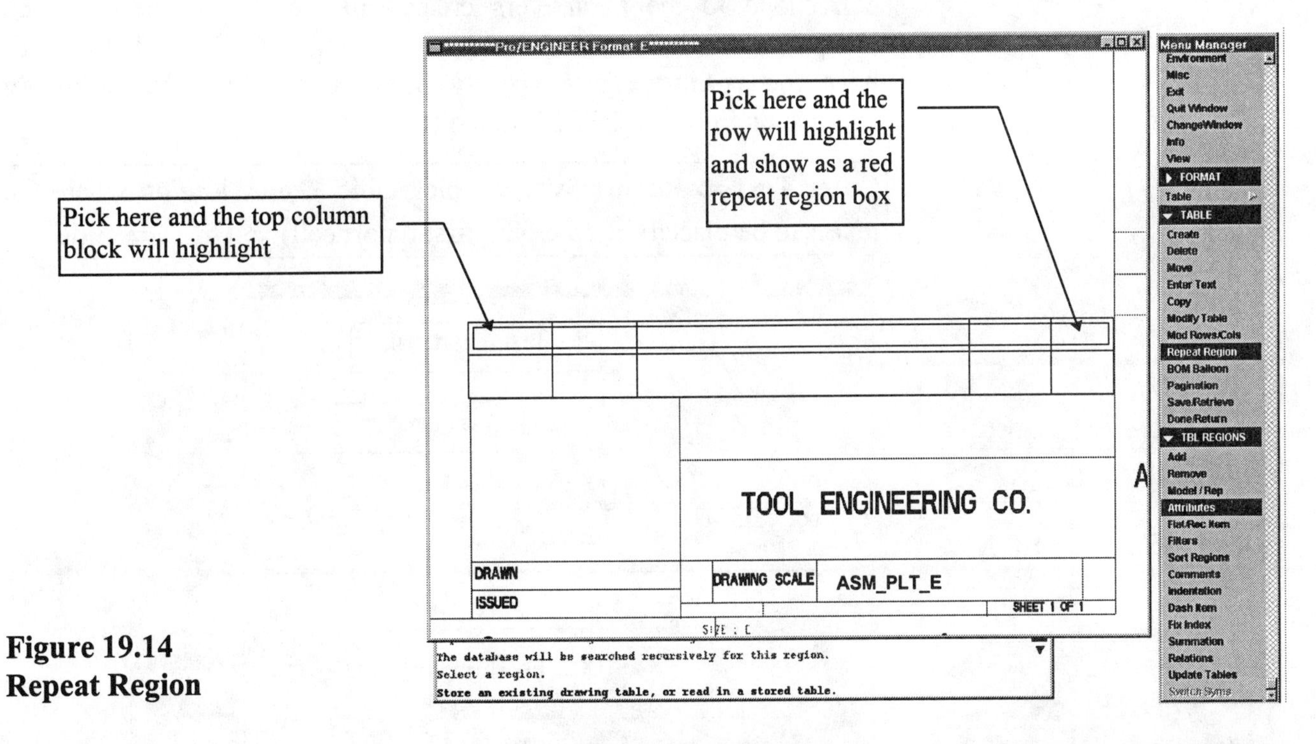

Figure 19.14
Repeat Region

The table must have parameters set in each appropiate block. The column headings should be inserted first, using plain text. Use the following commands:

Table ⇒ **Mod Rows/Cols** ⇒ **Justify** ⇒ **Column** ⇒ **Center** ⇒ **Middle** ⇒ (pick all five columns--they will outline in red as they are selected) ⇒ **Enter Text** ⇒ **Keyboard** ⇒ (pick the table cell where the text is to be placed--choose the **4.00** width column in the lower row) ⇒ (type **DESCRIPTION**) ⇒ **enter** ⇒ **enter**

Continue adding the titles **MATERIAL**, **QTY**, **ITEM**, and **PT NUM** as shown in Figure 19.15. From the FORMAT menu, modify the text height to **.125**.

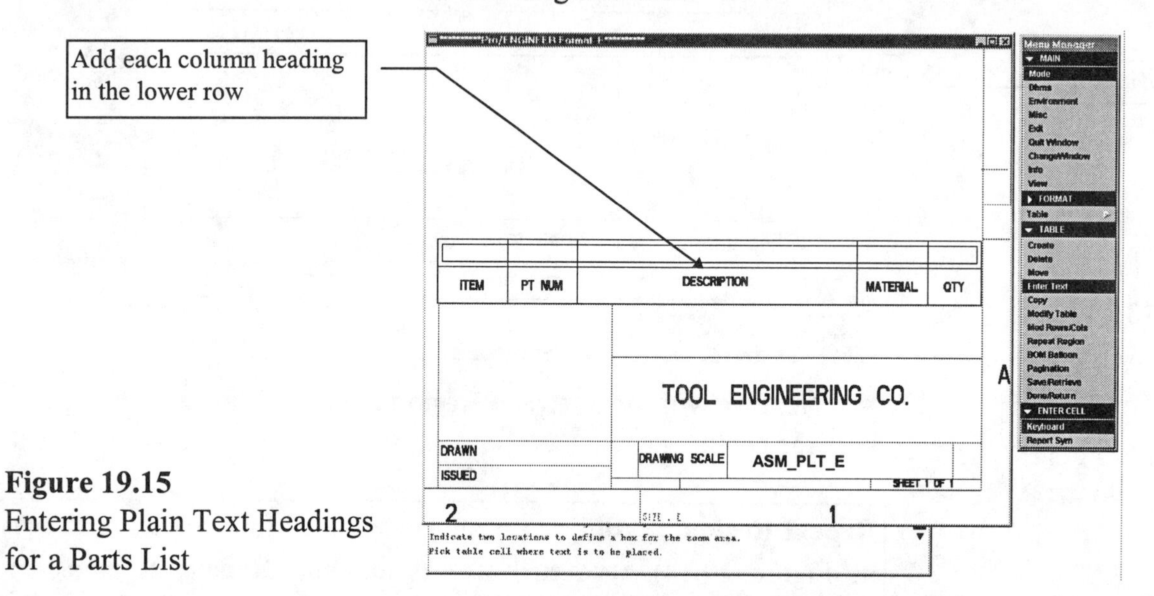

Figure 19.15
Entering Plain Text Headings for a Parts List

The **Repeat Regions** now need to have some of their headings correspond to the parameters created in the Part mode for each component model. The **ITEM** and quantity (**QTY**) columns will have the **rpt.index** and **rpt.qty** parameters. Use the following commands from the TABLE menu (Fig. 19.16):

Enter Text ⇒ **Report Sym** ⇒ (pick table ***Report Region*** where the text is to be placed) ⇒ (pick the first report cell) ⇒ **rpt...** ⇒ **index**

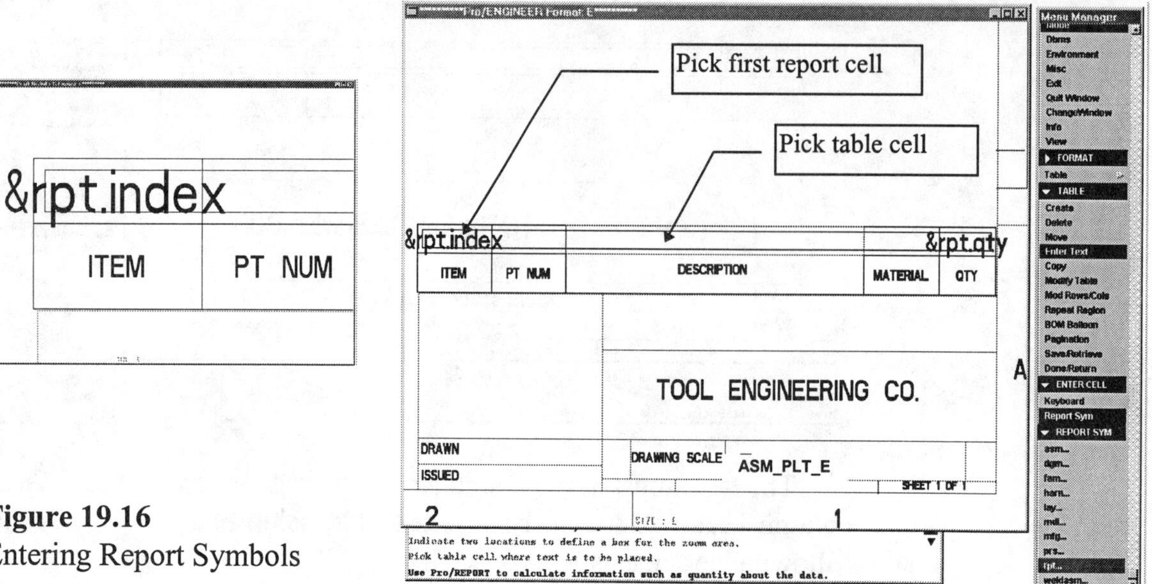

Figure 19.16
Entering Report Symbols

(pick table ***Repeat Region*** where the next text is to be placed) ⇒ (pick the fifth repeat cell) ⇒ **rpt...** ⇒ **qty** ⇒
Now create the parametric **User Defined** text (pick the third report cell where the next text is to be placed--see Figure 19.16) ⇒ **asm...** ⇒ **mbr...** ⇒ **User Defined** ⇒ (enter symbol text--type **DSC**) ⇒ **enter** (pick the fourth report cell--see Figure 19.17) ⇒ **asm...** ⇒ **mbr...** ⇒ **User Defined** ⇒ (enter symbol text--type **MAT**) ⇒ **enter** ⇒ (pick the second report cell) ⇒ **asm...** ⇒ **mbr...** ⇒ **User Defined** ⇒ (enter symbol text--type **PRTNO**) ⇒ **enter** (Fig. 19.18)

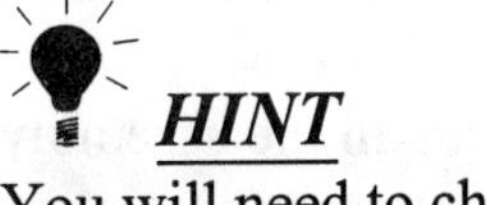

HINT
You will need to change the height of the text in the **Report Regions** to **.125**.

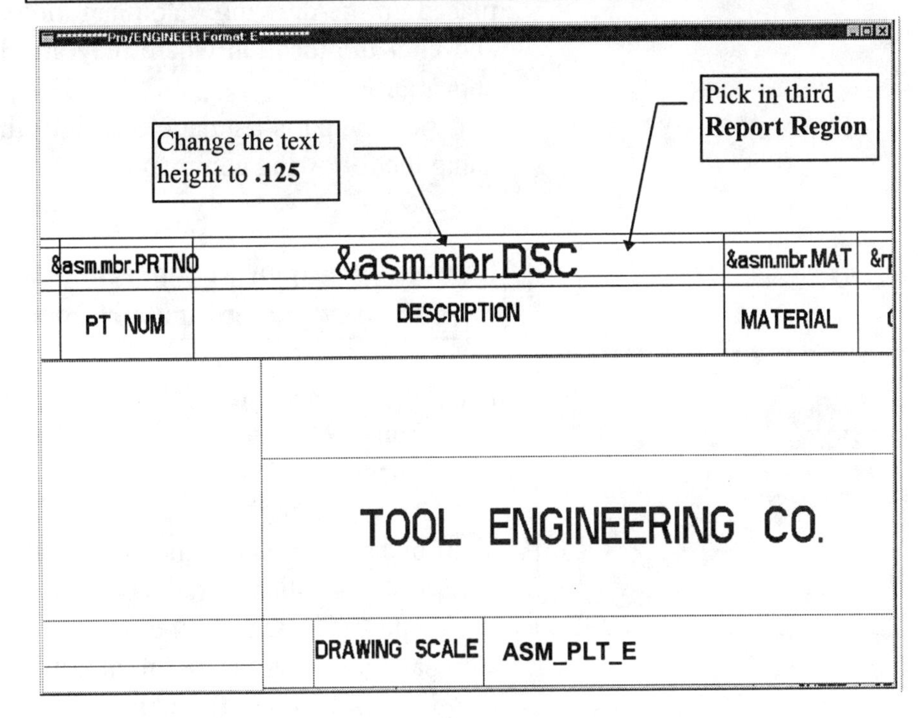

Figure 19.17
Entering Report Region Parameters

Dbms ⇒ **Save** ⇒ **enter**
Purge ⇒ **enter** ⇒
Done-Return

Figure 19.18
Completed Report Region Parameters

Adding Parts List Data

When you save your standard assembly format, the Drawing Table that represents your standard parts list is now included. You must be aware of the titles of the parameters under which this data is stored so that you can add it properly to your parts.

As you add components, Pro/E automatically reads the parameters from them and updates the parts list. You can also see the same effect by adding these parameters after the drawing is created.

Pro/E also creates **Item Balloons** on the first view that was placed on the drawing. You can move these balloons to other views and alter the location where they attach to the part, to improve their appearance.

Add the parts list data to each of the components in the assembly using the following commands:

Retrieve the part **sw_clamp_arm.prt**: **Mode** ⇒ **Part** ⇒ **Search/Retr** ⇒ (Choose **sw_clamp_arm.prt** from the directory list)

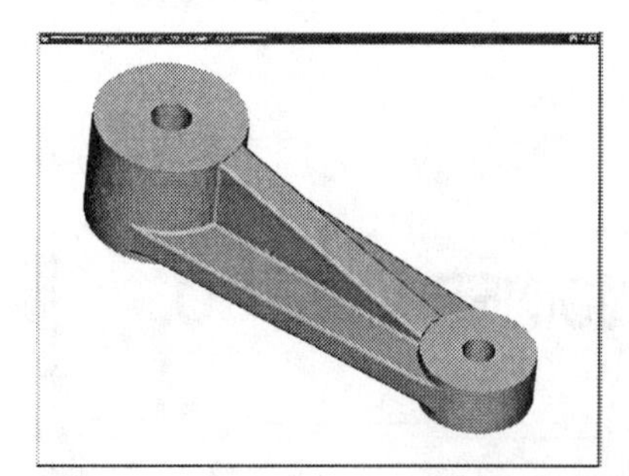

Add the parts list data:

Choose **Relations**
Choose **Add Param**
Choose **String**

You must now enter the *exact* title of the parameter that you previously established in the parts list:

Type **PRTNO** and hit **enter**

You can now enter the part number:

Type **SW101-5AR** and hit **enter**
Choose **String**

You must now enter the *exact* title of the parameter that is part of the parts list:

Type **DSC** and hit **enter**

You can now enter the component description:

Type **SWING CLAMP ARM** and hit **enter**
Choose **String**

You must now enter the *exact* title of the parameter that is part of the parts list:

Type **MAT** and hit **enter**

You can now enter the material:

Type **STEEL** and hit **enter**

Now you must save and quit the window:

Dbms ⇒ **Save** ⇒ **enter** ⇒ **Purge** ⇒ **enter** ⇒
Done-Return ⇒ **Quit Window**

Retrieve the part **sw_clamp_swivel.prt**: **Mode** ⇒ **Part** ⇒ **Search/Retr** ⇒ (Choose **sw_clamp_swivel.prt**)

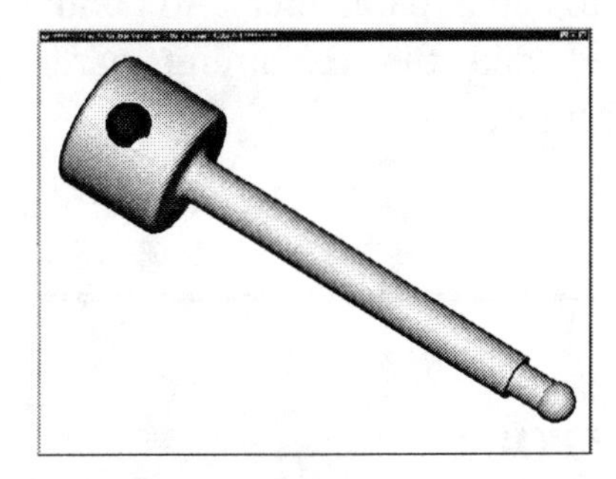

Add the parts list data:
Choose **Relations**
Choose **Add Param**
Choose **String**
Enter the *exact* title of the parameter that is part of the parts list:
Type **PRTNO** and hit **enter**
You can now enter the part number:
Type **SW101-6SW** and hit **enter**
Choose **String**
You must now enter the *exact* title of the parameter that is part of the parts list:
Type **DSC** and hit **enter**
You can now enter the component description:
Type **SWING CLAMP SWIVEL** and hit **enter**
Choose **String**
You must now enter the **exact** title of the parameter that is part of the parts list:
Type **MAT** and hit **enter**
You can now enter the material:
Type **STEEL** and hit **enter**
Now you must save and quit the window:
Dbms ⇒ Save ⇒ enter ⇒ Purge ⇒ enter ⇒ Done-Return ⇒ Quit Window

Retrieve the part **sw_clamp_ball.prt**: **Mode ⇒ Part ⇒ Search/Retr ⇒** (Choose **sw_clamp_ball.prt**)

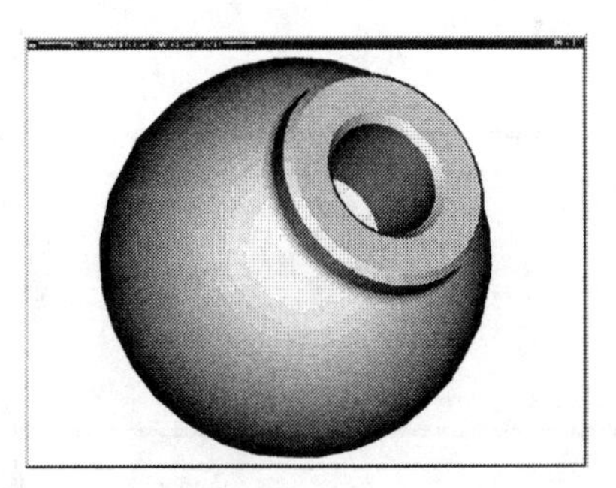

Add the parts list data:
Choose **Relations**
Choose **Add Param**
Choose **String**
Enter the *exact* title of the parameter that is part of the parts list:
Type **PRTNO** and hit **enter**
Enter the part number:
Type **SW101-7BA** and hit **enter**
Choose **String**
Enter the *exact* title of the parameter that is part of the parts list:
Type **DSC** and hit **enter**
Enter the component description:
Type **SWING CLAMP BALL** and hit **enter**
Choose **String**
Enter the *exact* title of the parameter that is part of the parts list:
Type **MAT** and hit **enter**
Enter the material:
Type **BLACK PLASTIC** and hit **enter**
Save and quit the window:
Dbms ⇒ Save ⇒ enter ⇒ Purge ⇒ enter ⇒ Done-Return ⇒ Quit Window

Use the following information to add parameters to both purchased components (standard parts) and the remaining parts required for the subassembly and the assembly:

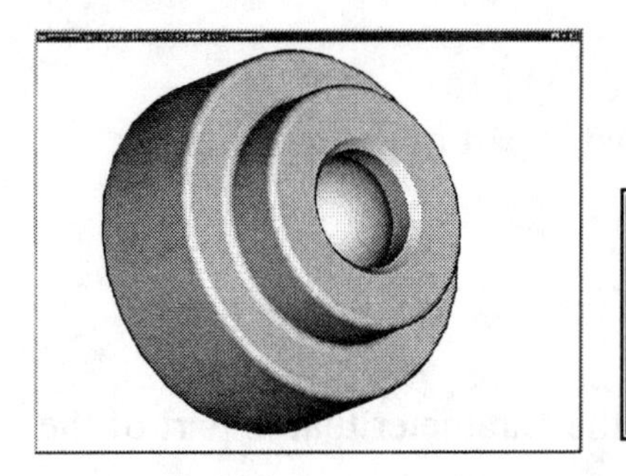

Component	sw_clamp_foot
Part Number	**SW101-8FT**
Description	**SWING CLAMP FOOT**
Material	**NYLON**

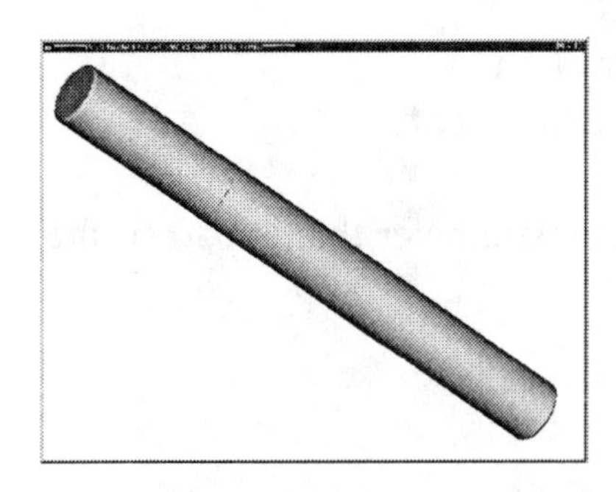

Component	sw_clamp_stud_long
Part Number	**SW101-9STL**
Description	**.500-13 X 5.00 DOUBLE END STUD**
Material	**PURCHASED**

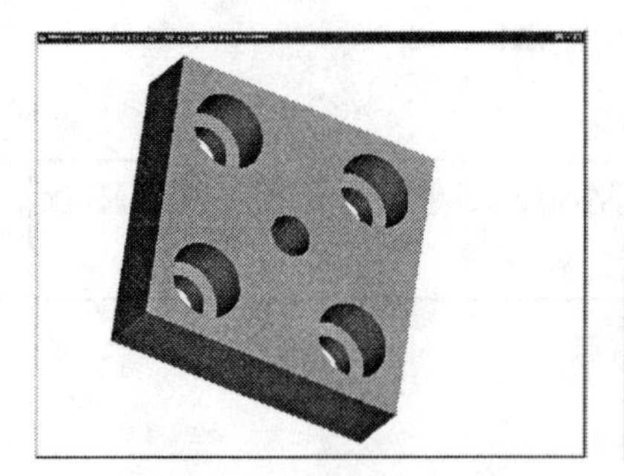

Component	sw_clamp_plate
Part Number	**SW100-20PL**
Description	**SWING CLAMP PLATE**
Material	**1020 CRS**

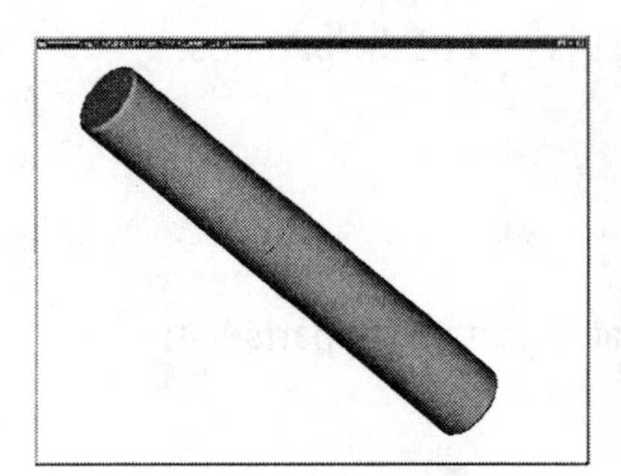

Component	sw_clamp_stud
Part Number	**SW100-21ST**
Description	**.500-13 X 3.50 DOUBLE END STUD**
Material	**PURCHASED**

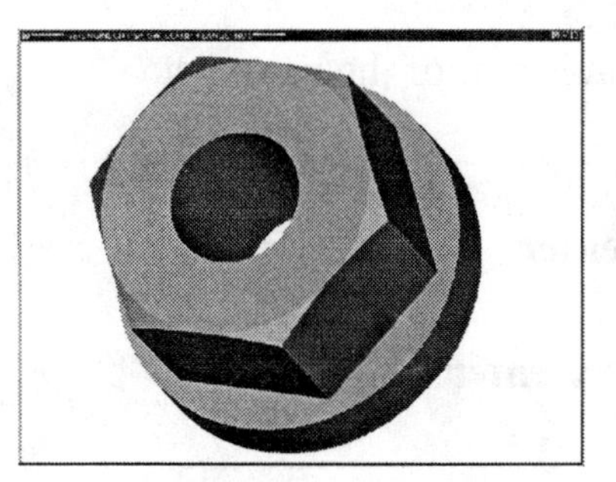

Component	sw_clamp_flange_nut
Part Number	**SW100-22FLN**
Description	**.500-13 HEX FLANGE NUT**
Material	**PURCHASED**

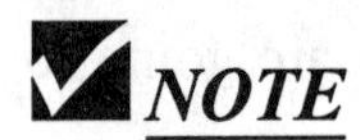

NOTE

Parameters can be added, deleted, and modified in the Part mode, Drawing mode, or Assembly mode. In the Drawing mode and the Assembly mode, choose **Set Up ⇒ Params**. You can also add a *parameters column* to the Model Tree and edit the parameter by highlighting it in the tree and typing a new value.

As you add the required relations to each component, you can see the parameters and check to see if they were input correctly by choosing **Show Rel** from the RELATIONS menu, as shown in Figure 19.19.

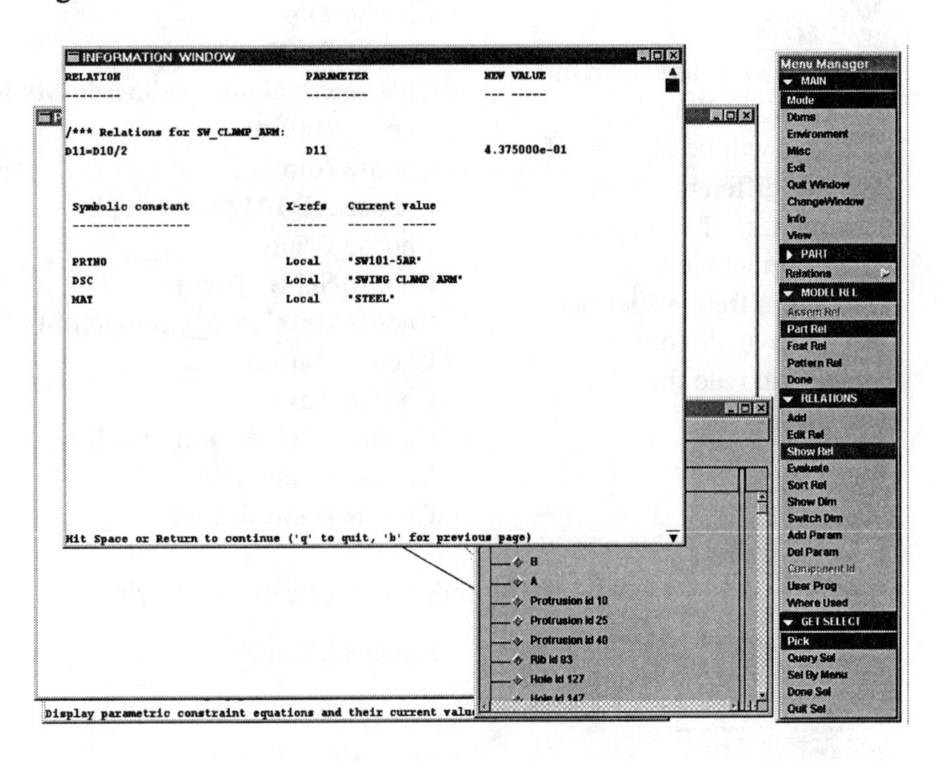

Figure 19.19
Showing Relations for the Arm

Now that the parameters have been established in each part and the assembly format with related parameters in a parts list table has been created and saved on a format, you can create a drawing of the assembly. The parts list will be generated automatically.

Create a new drawing using the following commands (Fig. 19.20):

Mode ⇒ Drawing ⇒ Create ⇒ (type **sw_cl_subasm**) **⇒ enter ⇒ Retr Format ⇒ ? ⇒ enter ⇒** (select the name of the assembly format produced in the previous segment)

NOTE

Set the following in your **.dtl** file:

Set Up ⇒ Modify Val ⇒

default_font	**filled**
draw_arrow_style	**filled**
drawing_text_height	**.50**
draw_arrow_length	**.25**
draw_arrow_width	**.08**
max_balloon_radius	**.50**
min_balloon_radius	**.50**

File ⇒ Save ⇒ File ⇒ Exit

HINT

Use the **Save** option (just under the **Modify Val**) to save the settings to a file name so you can recall them and use on a another drawing.

To get the drawing to appear with the correct style you must modify the values of the **.dtl Drawing Set-Up** file using **Set Up ⇒ Modify Val**.

Create a new drawing using the assembly format, and add two views. Add the top view as the first view. Use the following commands:

Views ⇒ ? ⇒ (pick **sw_clamp_subassembly** as the name of the model you want to use) **⇒ Done ⇒** (indicate the location of the center of the top view)

You are now going to use **datum planes** to establish the view orientation. *You must select assembly datums!* This would be very difficult to accomplish by picking on the screen, so you are going to use **Sel By Menu**:

Choose **Top**
Choose **Sel By Menu**
Choose **sw_clamp_subassembly** from the list
Choose **Datum**
Choose **Name**
Choose **ADTM2** from the list
Choose **Front**
Choose **Sel By Menu**
Choose **sw_clamp_subassembly** from the list
Choose **Datum**
Choose **Name**
Choose **ADTM3** from the list
Choose **Done/Return**
Choose **Done/Return**

HINT

Orient the view as shown in Figure 19.20. Your selections will be different if you have different datum plane names. If you do not get the correct view orientation, **Delete View** and **Add View** again (or pick Default and redo the orientation).

Modify the scale of the drawing to be **2.00**:

Choose **Modify**

Select the drawing scale value (**1.00**) that is just below the drawing:

Type **2.00** and **enter**

Notice that the title block and parts list were filled in as the view was created.

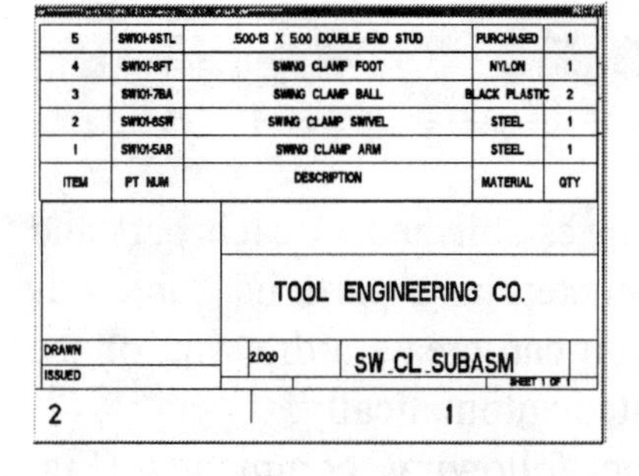

5	SW101-9STL	.500-13 X 5.00 DOUBLE END STUD	PURCHASED	1
4	SW101-8FT	SWING CLAMP FOOT	NYLON	1
3	SW101-7BA	SWING CLAMP BALL	BLACK PLASTIC	2
2	SW101-6SW	SWING CLAMP SWIVEL	STEEL	1
1	SW101-5AR	SWING CLAMP ARM	STEEL	1
ITEM	PT NUM	DESCRIPTION	MATERIAL	QTY

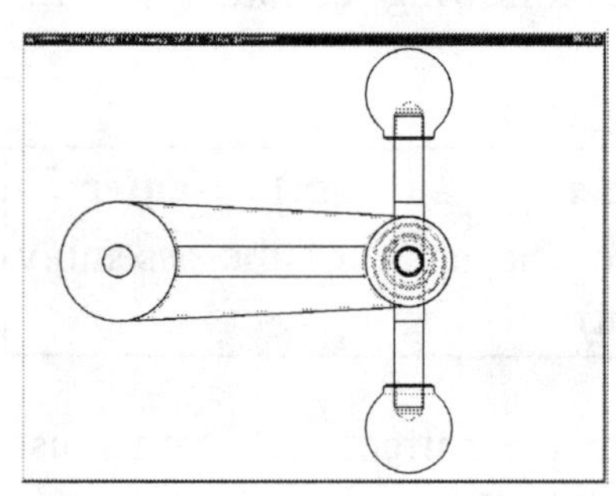

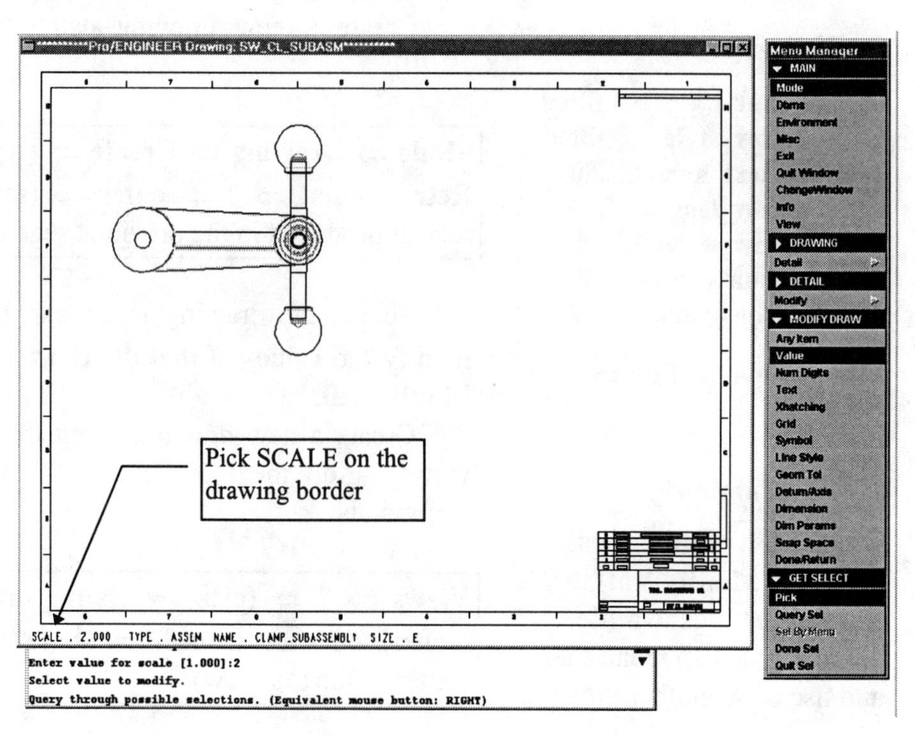

Figure 19.20
Swing Clamp Subassembly Drawing with Top View

HINT

If you haven't already done so, turn off the datums, parts, coordinate system, spin center, axes, and grid snap in the ENVIRONMENT menu.

To erase the set datums use the following commands:

Detail ⇒ **Show/Erase** ⇒ **Erase** ⇒ (pick the datum radio button) ⇒ **Erase All** ⇒ **Yes** ⇒ **Close**

You will need to do this after each view is placed on the drawing or after both views have been established.

The only other view required to show the assembly is the front view. We will be making a front section view using the following commands (Fig. 19.21):

Choose **Views**
Choose **Section**
Choose **Done**
Choose **Done** (since **Full** is the default)

Indicate the center of the view below the top view:

Choose **Create**
Choose **Done** (since **Planar** is the default)
Type **A** and hit **Enter**
Choose **Sel By Menu**
Choose **sw_clamp_subassembly** from the list
Choose **Datum**
Choose **Name**
Choose **ADTM2** from the list

Now select anywhere in the *top* view to define it as the view where the section line, arrows, and section identification lettering will be placed:

Choose **Done/Return**

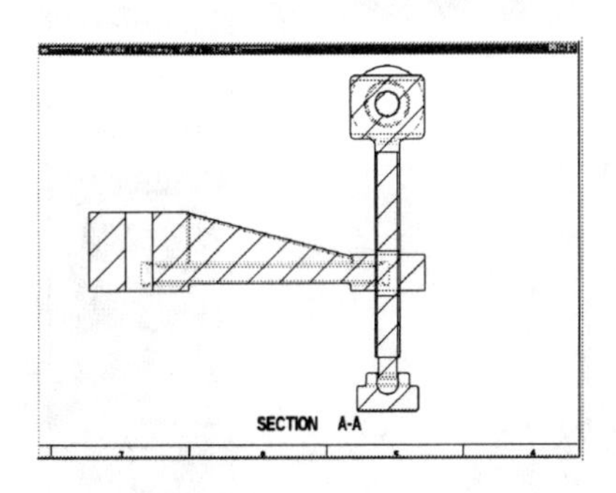

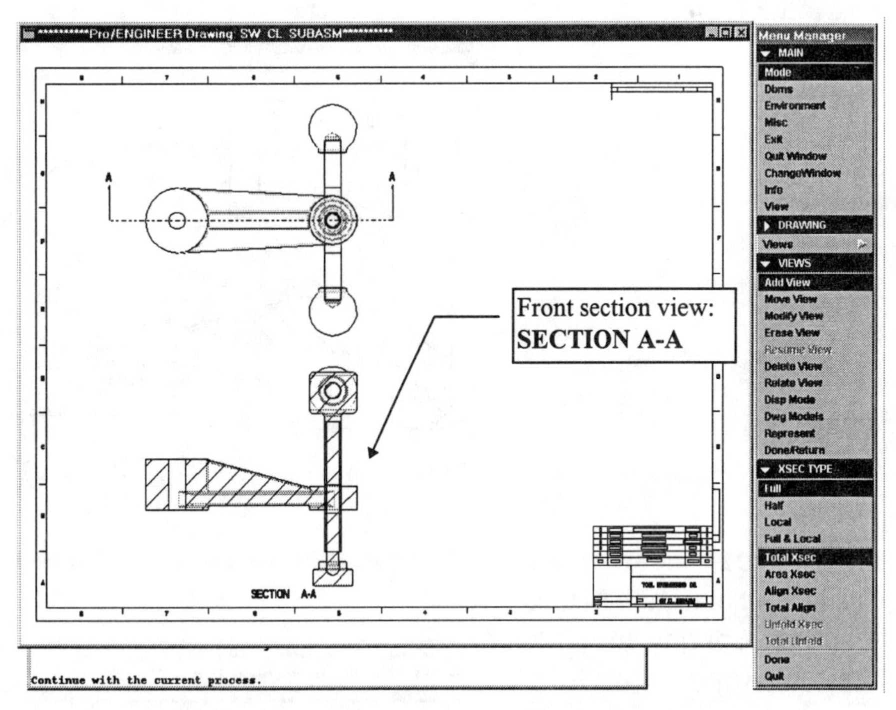

Figure 19.21
Swing Clamp Subassembly Drawing with Top View and Front Section View

Pro/E allows you to alter the display of the section view so an assembly makes more sense and to comply with industry standard practices.

Most companies require that the crosshatching on parts in sections views of assemblies be "clocked" such that parts that meet do not use the same spacing and angle (Fig. 19.22). This makes the separation between parts more distinct. Clean up the section view to comply with industry practices. Change the spacing and the angle on the **Swivel** component (Fig. 19.22):

Choose **Detail**
Choose **Modfy**
Choose **Xhatching**

Select the crosshatching in the section view:

Choose **Done Sel**

Pro/E selects all of the crosshatching in the view as a single object. You can cycle through the portion that lies on each component using the **Next Xsec, Prev Xsec**, and **Pick Xsec** options:

Choose **Next Xsec** until the Swivel **(sw_clamp_swivel)** is selected
Choose **Spacing**
Choose **Half** two times
Choose **Angle**
Choose **135**
Choose **Done**

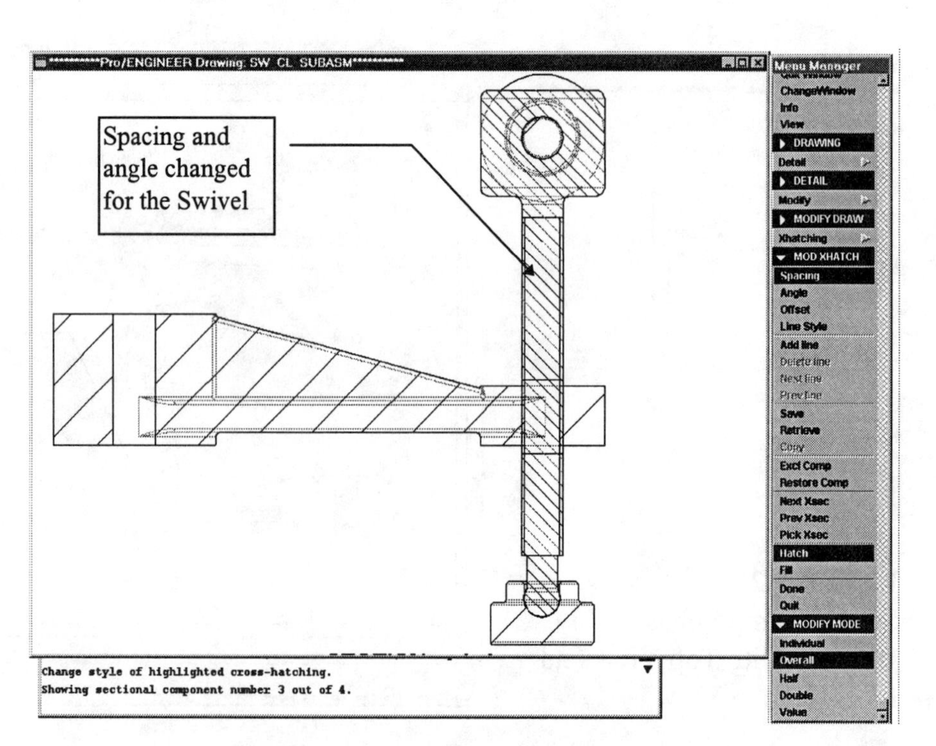

Figure 19.22
Changing the Section Lining Angle and Spacing on Assembly Components

Change the spacing on each component to be similar to that shown in Figure 19.23. While you have the DETAIL menu up, erase the cosmetic threads from both views (Fig. 19.24).

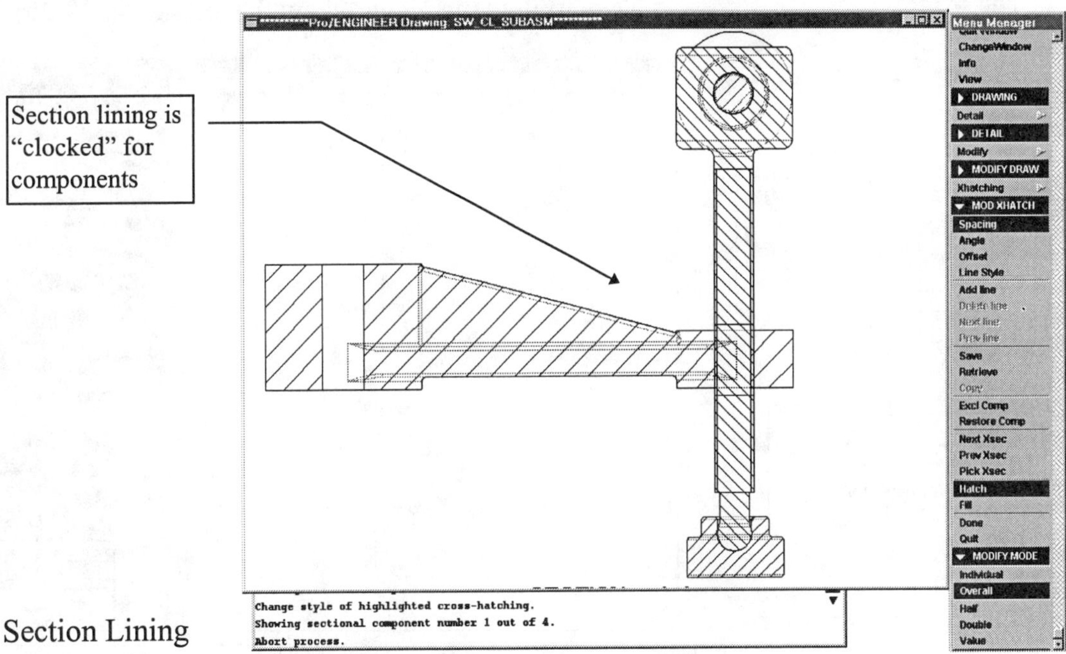

Figure 19.23
Changing the Section Lining

Since hidden lines are usually not shown on a section view, change the display of both views to remove hidden lines and tangent edges:

Choose DRAWING
Choose **Views**
Choose **Disp Mode**
Choose **View Disp**

Select both views:

Choose **Done Sel**
Choose **No Hidden**
Choose **No Disp Tan**
Choose **Done**

With cosmetic threads shown

Cosmetic threads turned off

Cosmetic feature radio button

Figure 19.24
Cosmetic Threads, Hidden Lines, and Tangent Edges Removed from Views

To complete the views, show the axes (Fig. 19.25).

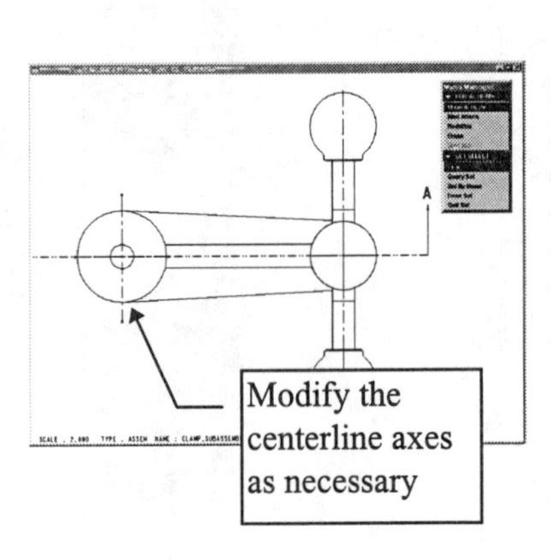

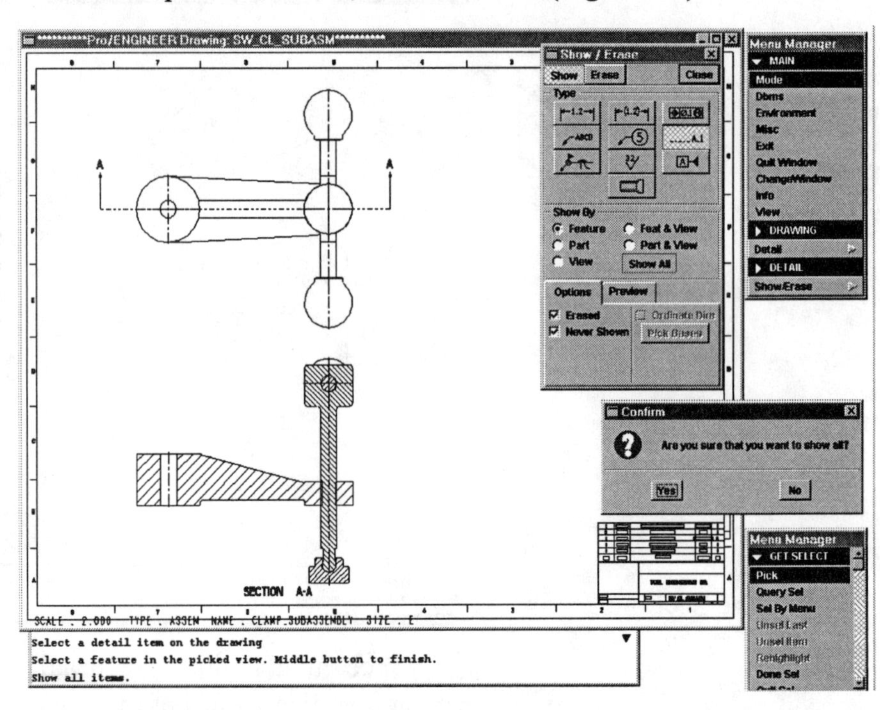

Figure 19.25
Centerline Axes Shown on Assembly Drawing

To finish the assembly drawing, the balloons for each component must be displayed. Show the **Item Balloons** on the drawing using the following commands (Fig. 19.26):

Choose **Table**
Choose **BOM Balloon**
Choose **Set Region**

Select the parts list on the drawing:

Choose **Show**
Choose **Show All**
Choose **Done/Return**

Balloons are displayed in the top view since it was the first view that was created

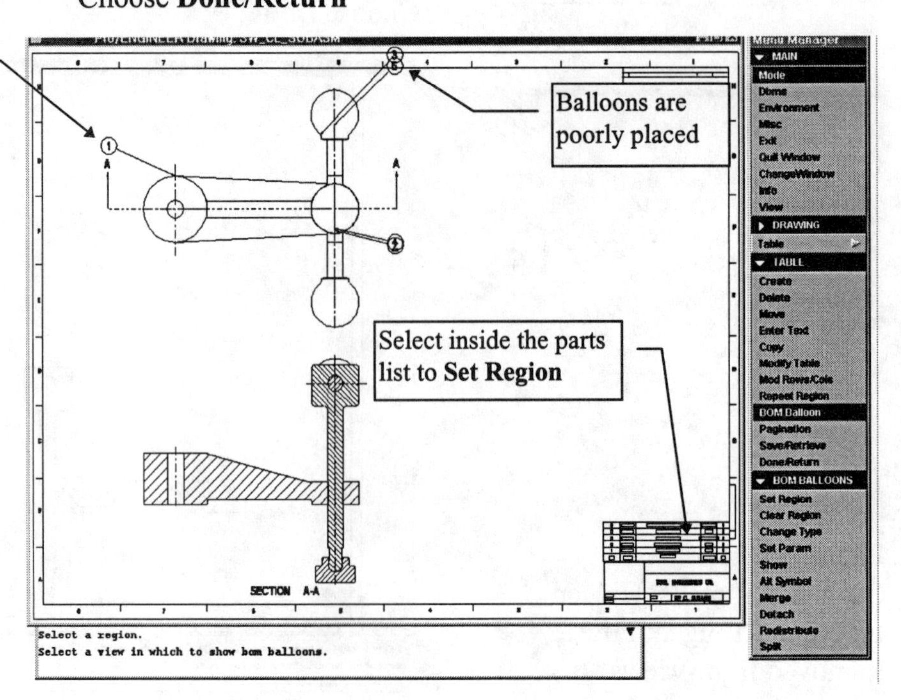

Figure 19.26
Balloons Shown in First View

Switch the display for the Arm, Swivel, and Foot balloons to the front section view (Fig. 19.27). This is similar to switching the view of a dimension:

Choose **Detail** ⇒ **Switch View**
Select a balloon ⇒ **Done Sel**
Select the section view
Select a balloon ⇒ **Done Sel**
Select the section view
Select a balloon ⇒ **Done Sel**
Select the section view
Choose **View** ⇒ **Repaint** ⇒ **Done/Return**

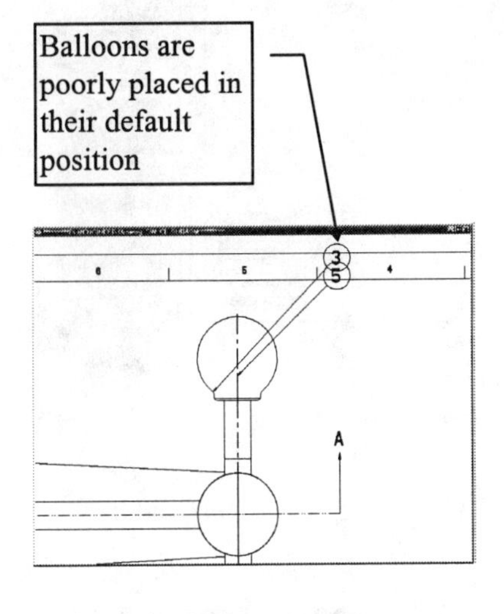

Move the Arm, Swivel, and Foot balloons to the front section view

Figure 19.27
Switch Three Balloons to the Front View

Use the **Detail** ⇒ **Move** option to move the balloons away from each other (a bit). Move the attachment of the balloons (Fig. 19.28 and Fig. 19.29):

Choose **Mod Attach**

Select balloon item **1**:

Choose **Change Ref**

Dbms ⇒ **Save** ⇒ **enter**
Purge ⇒ **enter** ⇒
Done-Return

Change Ref allows you to pick a new edge to place the arrowhead on. Select the edge of the section as the new attachment:

Choose **Done Sel**
Choose **Done/Return**

Repeat the process until all of the balloons are reattached as is shown in Figure 19.29.

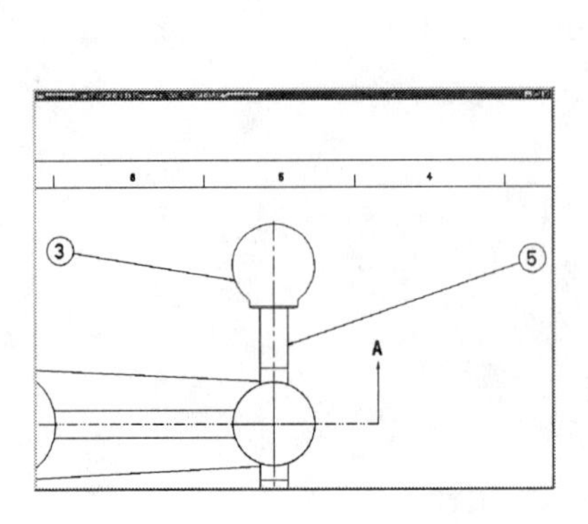

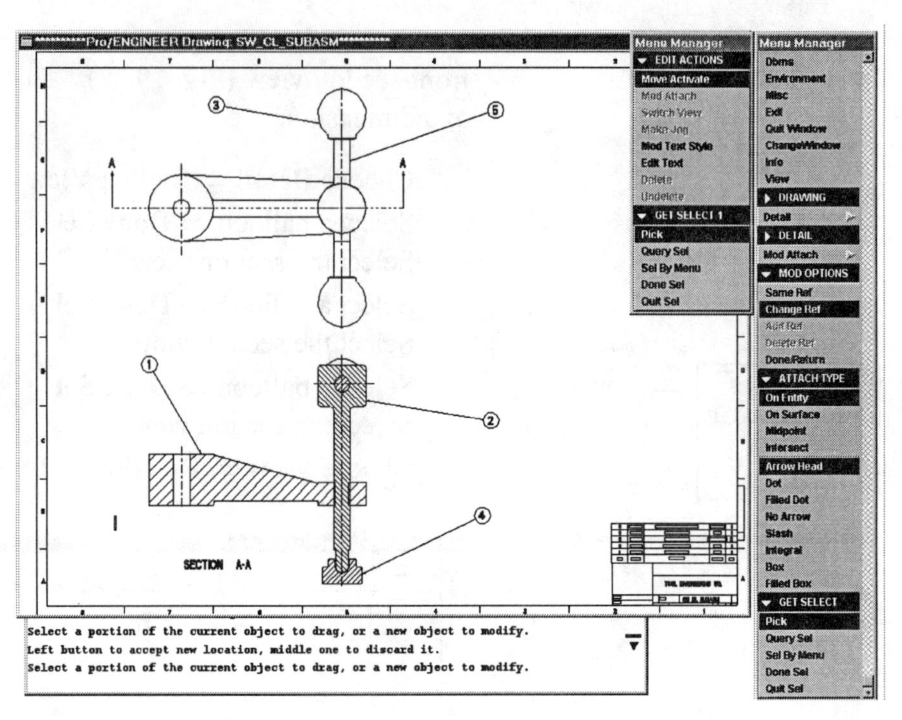

Figure 19.28
Repositions and Modified Attachment Locations for Balloons

Figure 19.29
Completed Drawing

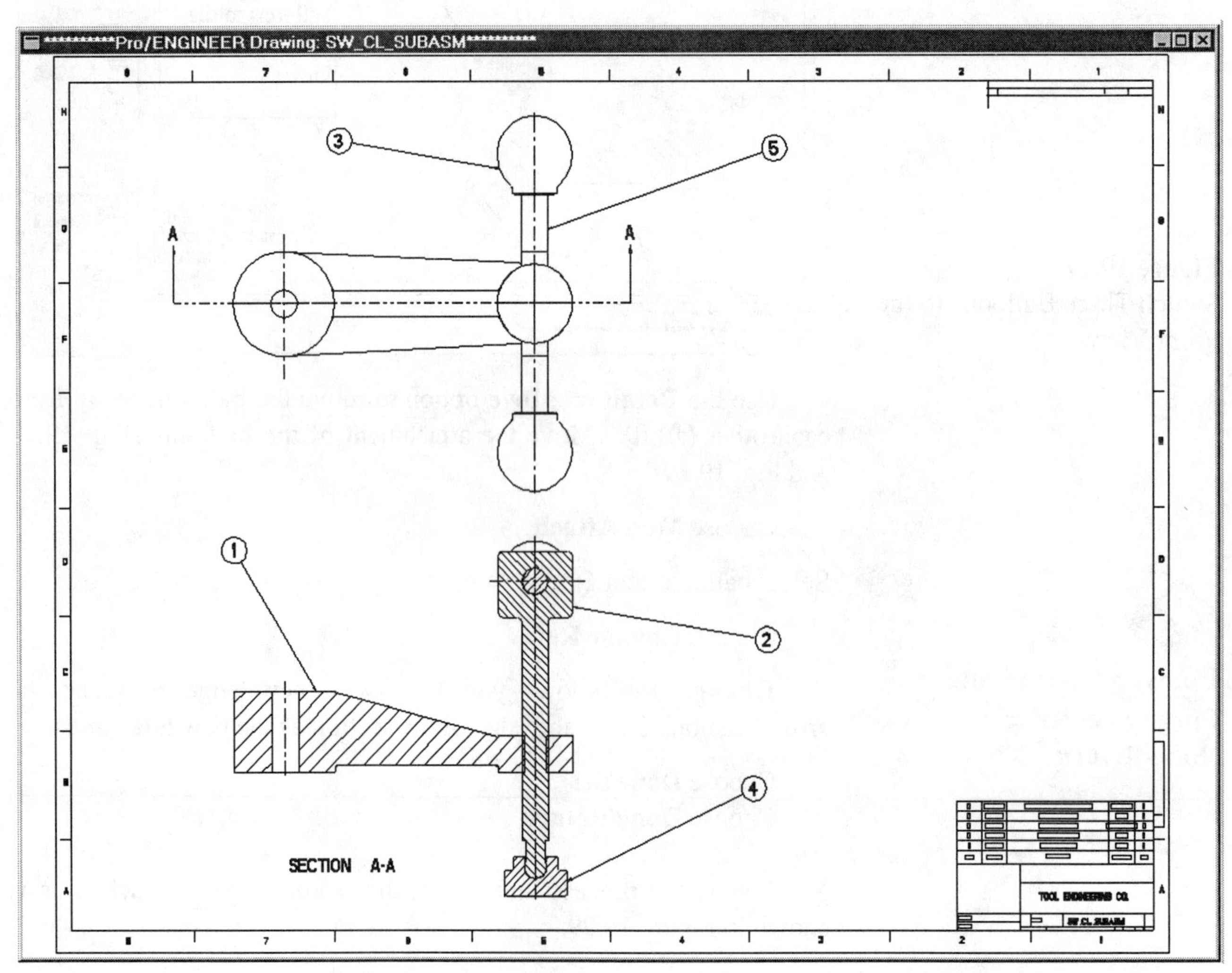

The next drawing we will do is the assembly drawing for the complete Swing Clamp. This assembly is composed of the subassembly in the previous drawing, the **Plate**, short **Stud**, and **Flange Nut**. The drawing will use the same format created for the subassembly. Formats are read-only files that can be used as many times as you want. Create the drawing using the following commands:

Mode ⇒ **Drawing** ⇒ **Create** ⇒ (type **sw_cl_assembly**) ⇒ **enter** ⇒ **Retr Format** ⇒ **?** ⇒ (select the name of the **E** size assembly format you created in this lesson)

To get the drawing to appear with the correct style, you must modify the values of the **.dtl** Drawing Set-Up file using **Setup** ⇒ **Modify Val** and modify the values as before, or you can retrieve the **.dtl** file previously used if you had used the **Save** command just under **Modify Val**.

Create a new drawing using the Assembly format, and add two views. Add the top view as the first view. Use the following commands:

Views ⇒ **?** ⇒ (pick **sw_cl_assembly** as the name of the model you want to use) ⇒ **Done** ⇒ (indicate the location of the center of the top view)

Orient the top view correctly as shown in Figure 19.30, and then add a front section view as was done for the subassembly.

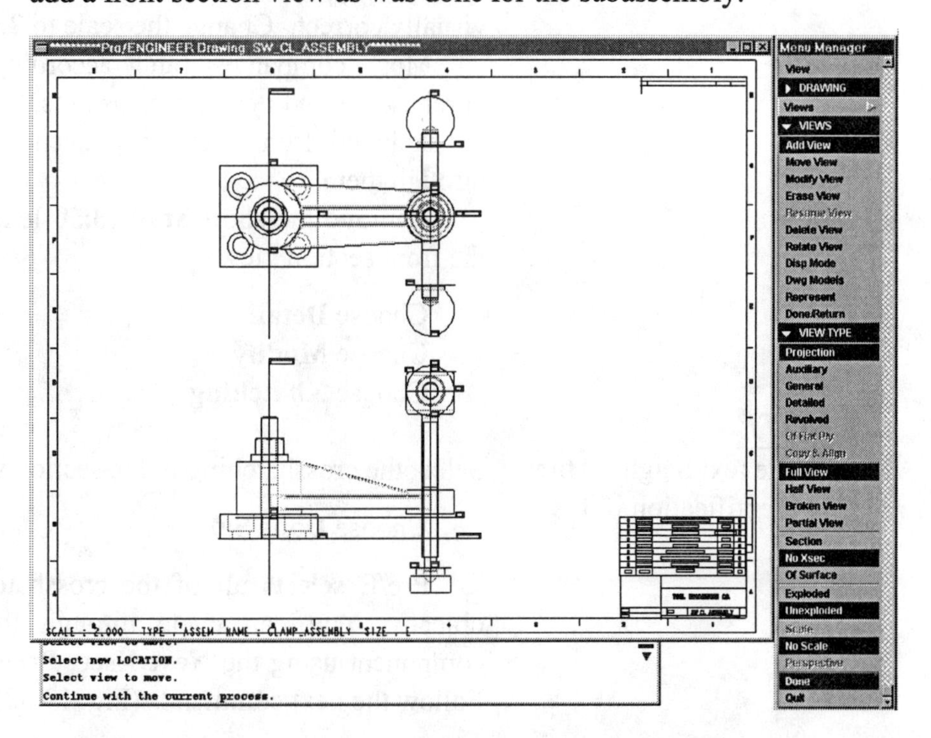

Figure 19.30
Assembly Drawing with Two Views

If you added the front view without choosing **Section** as one of the options as done in Figure 19.30, you must modify the view using the following commands (Fig. 19.31):

Views ⇒ **Modify View** ⇒ **View Type** ⇒ (pick the front view) ⇒ **Section** ⇒ **Done** ⇒ **Done** ⇒ **Create** ⇒ **Done** ⇒ (type **A**) ⇒ **enter** ⇒ (pick a datum plane that passes laterally through the assembly top view, as shown in Figure 19.31) ⇒ (pick the top view for the cutting plane line to be located)

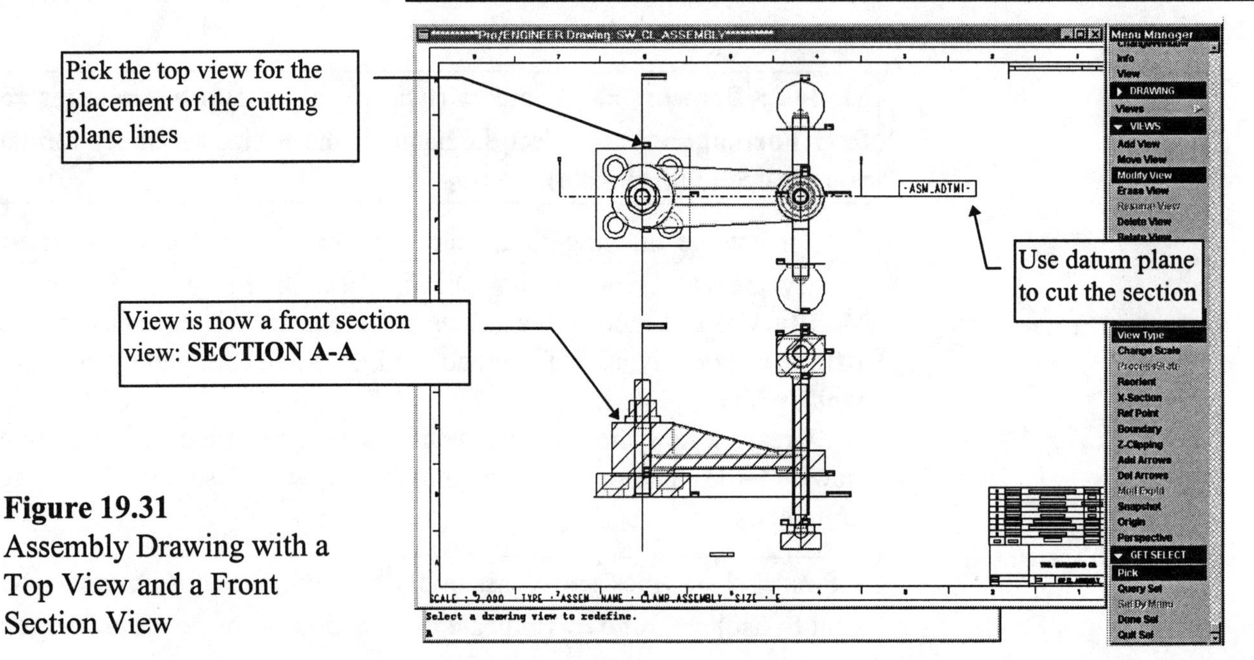

Figure 19.31
Assembly Drawing with a Top View and a Front Section View

Clean up the drawing by removing the cosmetic threads, datum axes, hidden lines, and display of the tangent edges, as you did for the subassembly. Show the centerline axes and modify them to be visually correct. Change the scale to 2.00.

Most companies (and according to standards) require that purchased round items, such as nuts, bolts, studs, springs, and die pins be excluded from sectioning even when the cutting plane passes through them.

Remove the short **Stud** (**3.50** length) and the **Flange Nut** from the front section view:

Choose **Detail**
Choose **Modify**
Choose **Xhatching**

NOTE
Modify the text height of the section identification if it is too small.

Select the crosshatching in the section view:

Choose **Done Sel**

Pro/E selects all of the crosshatching in the view as a single object. You can cycle through the portion that lies on each component using the **Next Xsec**, **Prev Xsec**, and **Pick Xsec** options. Follow the next commands (Fig. 19.32):

Choose **Excl Comp** (this eliminates Xsec of flange nut)
Choose **Next Xsec** (the short stud happens to be next)
Choose **Excl Comp** (this eliminates Xsec of short stud)

Or you can use **Query Sel** *to pick the components:*

Choose **Pick Xsec**
Choose **Query Sel**
Select the short **Stud** ⇒ **Accept**
Choose **Excl Comp**

and then the **Flange Nut** screw:

Choose **Pick Xsec**
Choose **Query Sel**
Select the **Flange Nut** ⇒ **Accept**
Choose **Excl Comp**
Choose **Done** ⇒ **Done Sel** ⇒ **Done/Return** ⇒ **Done/Return**

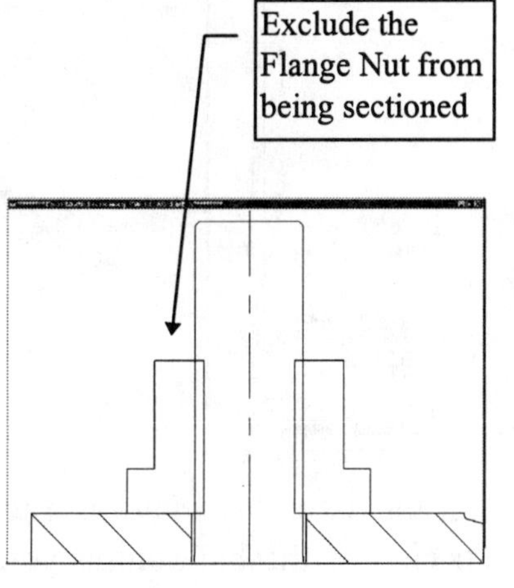

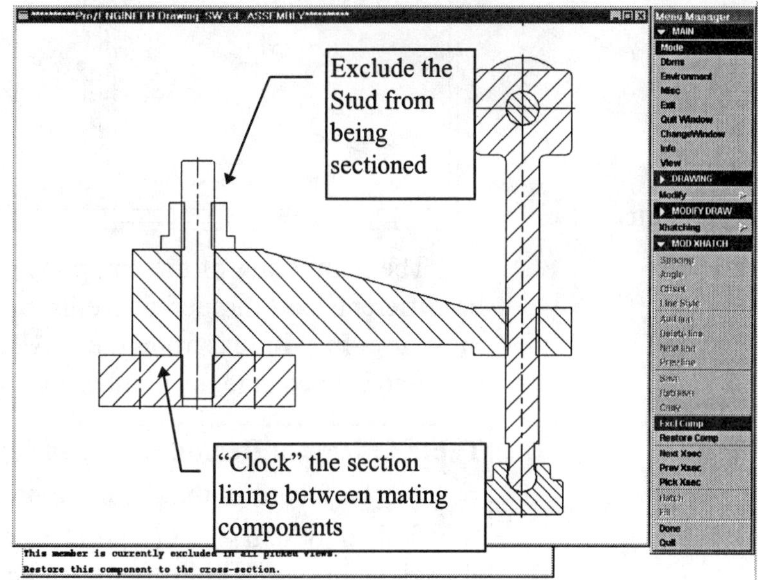

Figure 19.32
Assembly Drawing with Section Excluded from the Stud and Flange Nut

The next step in completing the assembly drawing is to show the balloons. Use the following commands (Fig. 19.33):

Table ⇒ **BOM Balloon** ⇒ **Set Region** ⇒ (select inside the parts list on the drawing) ⇒ **Show** ⇒ **Show All** ⇒ **Done/Return**

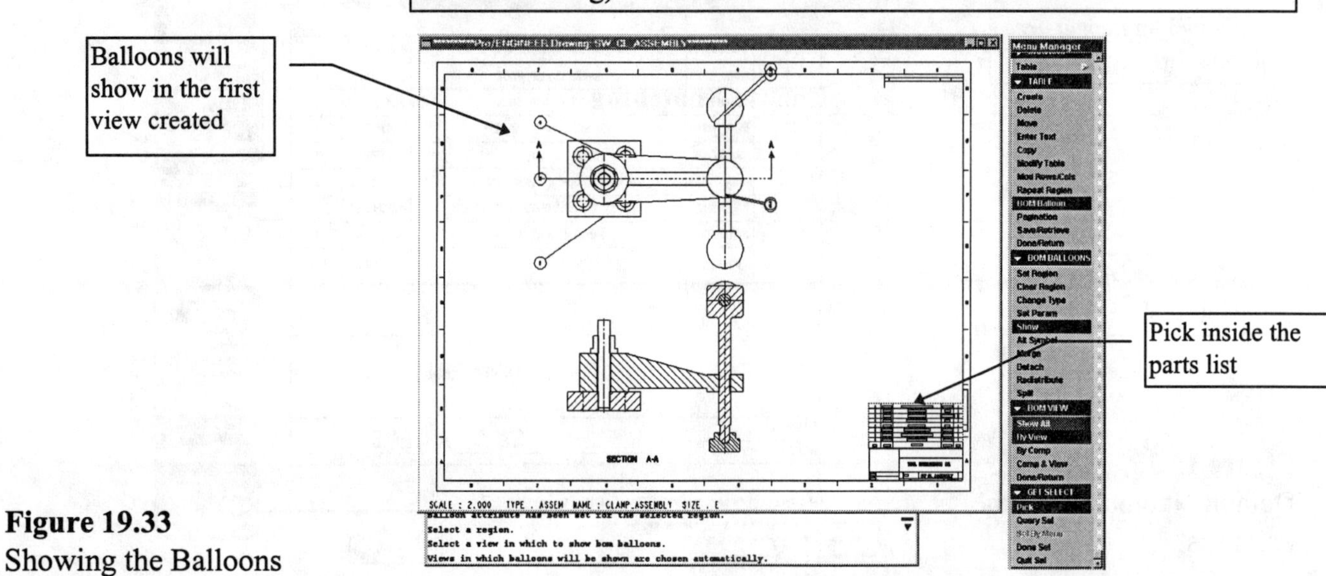

Figure 19.33
Showing the Balloons

Switch some of the balloons to the front view and reposition the others to make a more clear and balanced ballooning scheme (Fig. 19.34). You can line up the balloons by turning on the **Grid Snap** and then moving the balloons to a new position.

Dbms ⇒ Save ⇒ enter
Purge ⇒ enter ⇒ Done-Return

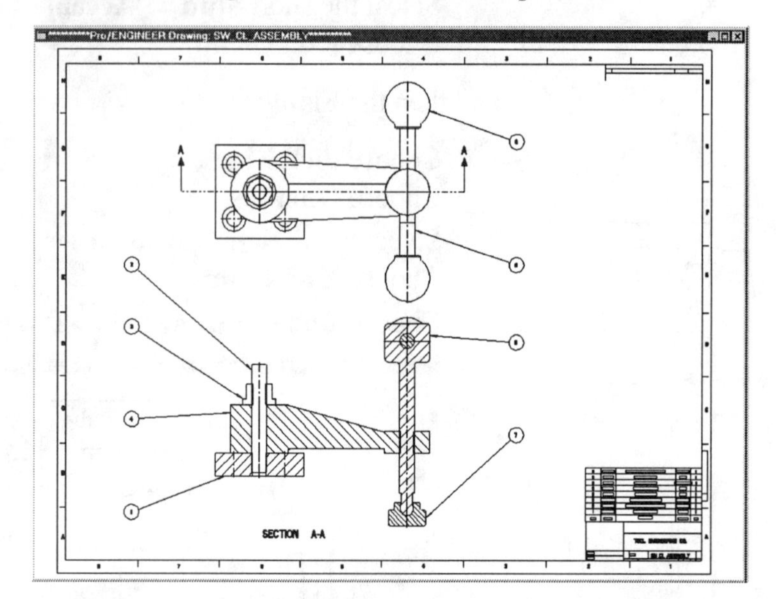

Figure 19.34
Showing the Balloons

The numbering of the components on the assembly may need to be different than the default setting. To change the ballooning you must use **Fix Index** from the TABLE menu. Use the following commands to change the numbering scheme (Fig. 19.35):

Table ⇒ Repeat Region ⇒ Fix Index ⇒ (select the parts list region) ⇒ (select a record in the current repeat region; select the *Arm*, which is defaulted to item **4**) ⇒ (type **1**) ⇒ **enter ⇒** (continue selecting and changing the numbering to the same as the *subassembly* sequence: make the *Plate* **6**, the short *Stud* **7**, and the *Flange Nut* **8**) ⇒ **Done**

When you are done, the parts list will sequence itself (Fig. 19.36) and the ballooning will update automatically (Fig. 19.37).

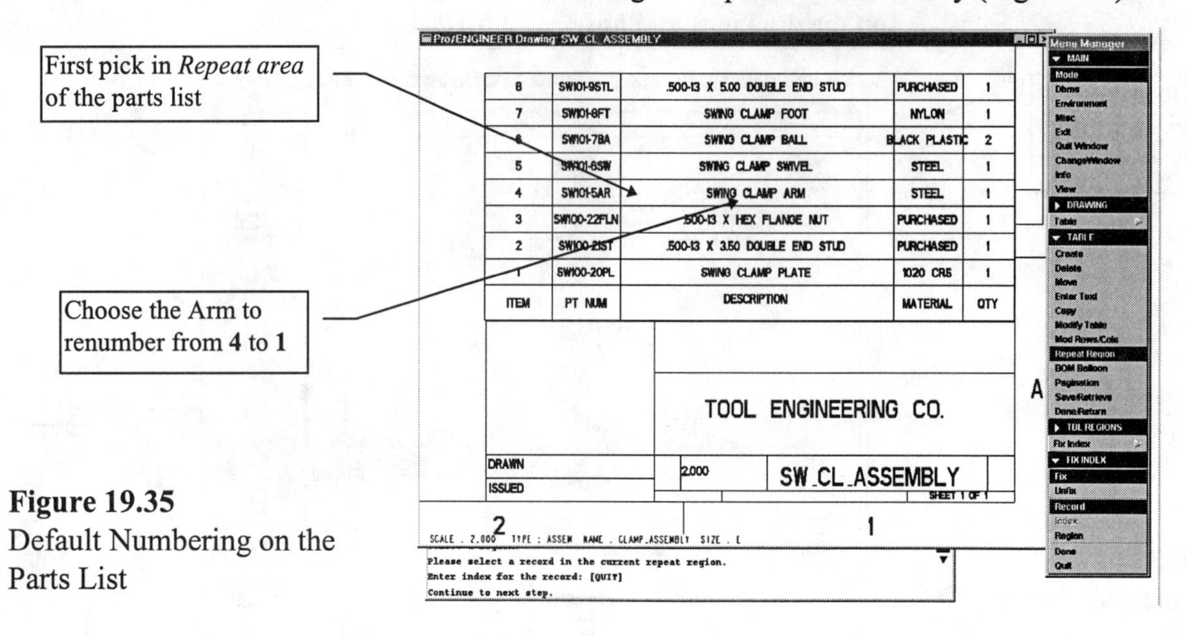

ITEM	PT NUM	DESCRIPTION	MATERIAL	QTY
8	SW101-9STL	.500-13 X 5.00 DOUBLE END STUD	PURCHASED	1
7	SW101-8FT	SWING CLAMP FOOT	NYLON	1
6	SW101-7BA	SWING CLAMP BALL	BLACK PLASTIC	2
5	SW101-6SW	SWING CLAMP SWIVEL	STEEL	1
4	SW101-5AR	SWING CLAMP ARM	STEEL	1
3	SW100-22FLN	.500-13 X HEX FLANGE NUT	PURCHASED	1
2	SW100-21ST	.500-13 X 3.50 DOUBLE END STUD	PURCHASED	1
1	SW100-20PL	SWING CLAMP PLATE	1020 CRS	1

Figure 19.35
Default Numbering on the Parts List

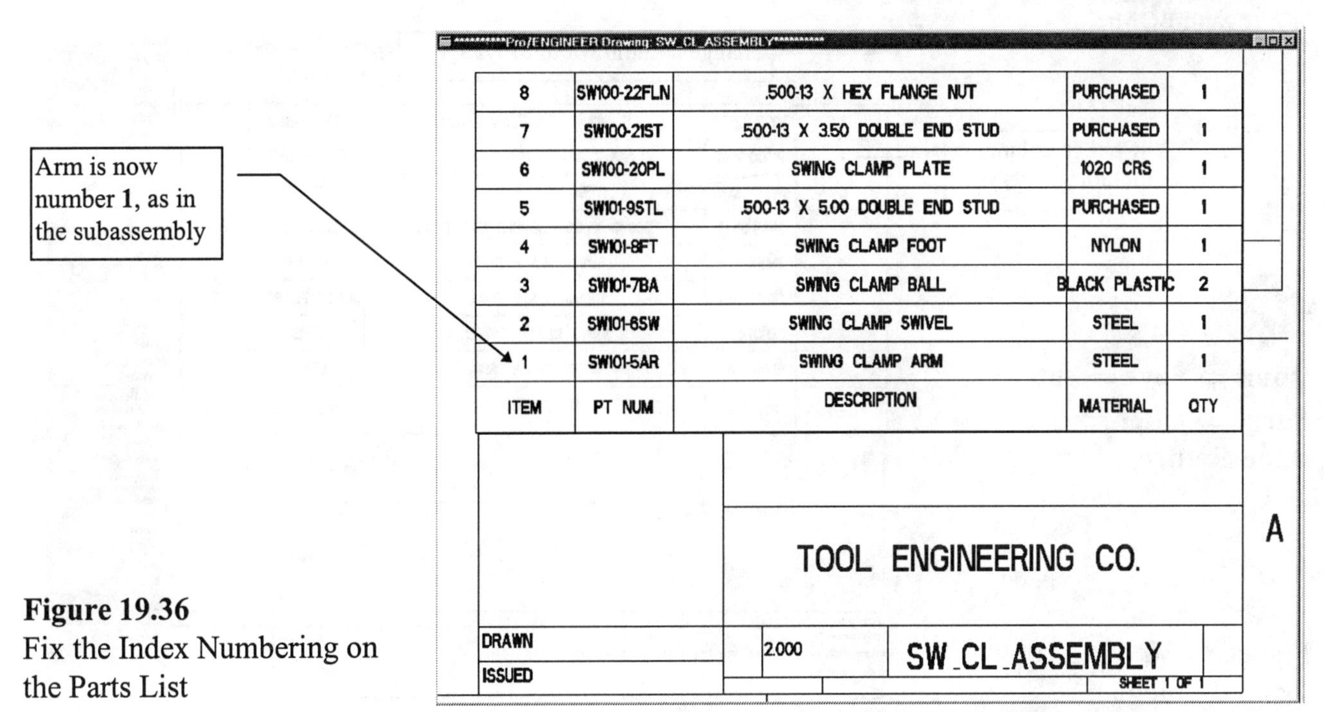

8	SW100-22FLN	.500-13 X HEX FLANGE NUT	PURCHASED	1
7	SW100-21ST	.500-13 X 3.50 DOUBLE END STUD	PURCHASED	1
6	SW100-20PL	SWING CLAMP PLATE	1020 CRS	1
5	SW101-9STL	.500-13 X 5.00 DOUBLE END STUD	PURCHASED	1
4	SW101-8FT	SWING CLAMP FOOT	NYLON	1
3	SW101-7BA	SWING CLAMP BALL	BLACK PLASTIC	2
2	SW101-6SW	SWING CLAMP SWIVEL	STEEL	1
1	SW101-5AR	SWING CLAMP ARM	STEEL	1
ITEM	PT NUM	DESCRIPTION	MATERIAL	QTY

Figure 19.36
Fix the Index Numbering on the Parts List

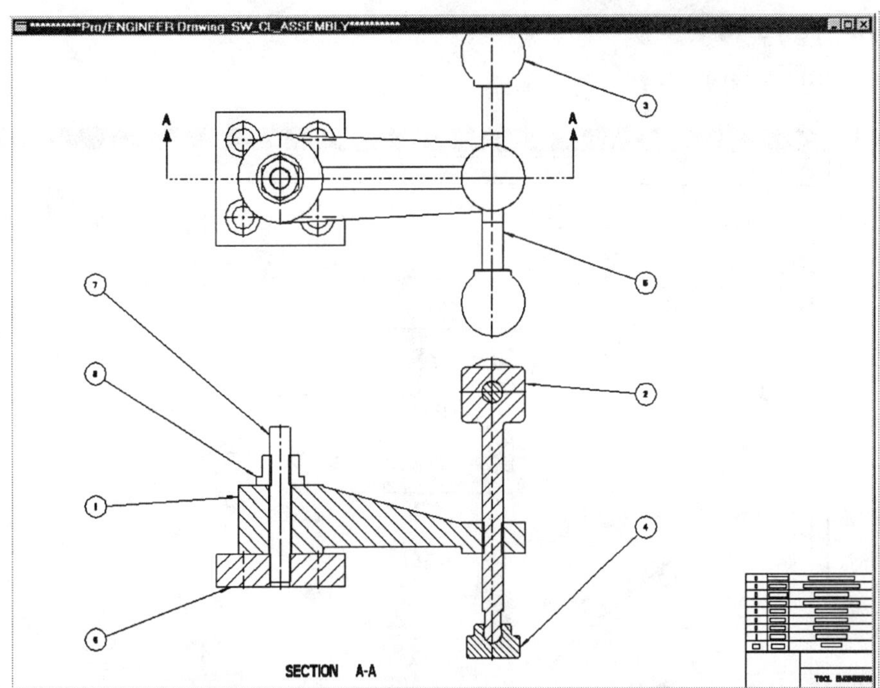

Dbms ⇒ Save ⇒ enter
Purge ⇒ enter ⇒ Done-Return

Figure 19.37
Renumbed Balloons

To change the numbering after it has been fixed, you must **Unfix** first and then **Fix Index** again.

Figure 19.36 shows that the MATERIAL cell column is not wide enough to accommodate the length of the Black Plastic material parameter. You can change the size of the column cell using **Mod Rows/Cols** from the TABLE menu. Use the following commands (Fig. 19.38):

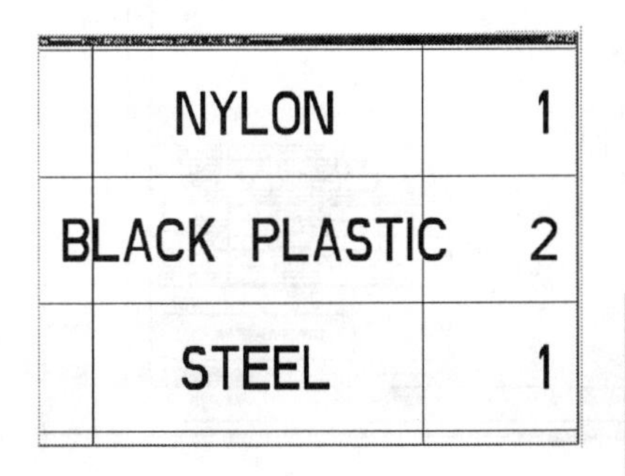

NYLON	1
BLACK PLASTIC	2
STEEL	1

Mod Rows/Cols ⇒ Change Size ⇒ Column ⇒ By Length ⇒ (pick the MATERIAL column) ⇒ (type **1.25**) ⇒ **enter** ⇒ (pick the DESCRIPTION column) ⇒ (type **3.75**) ⇒ **enter** (Fig. 19.39)

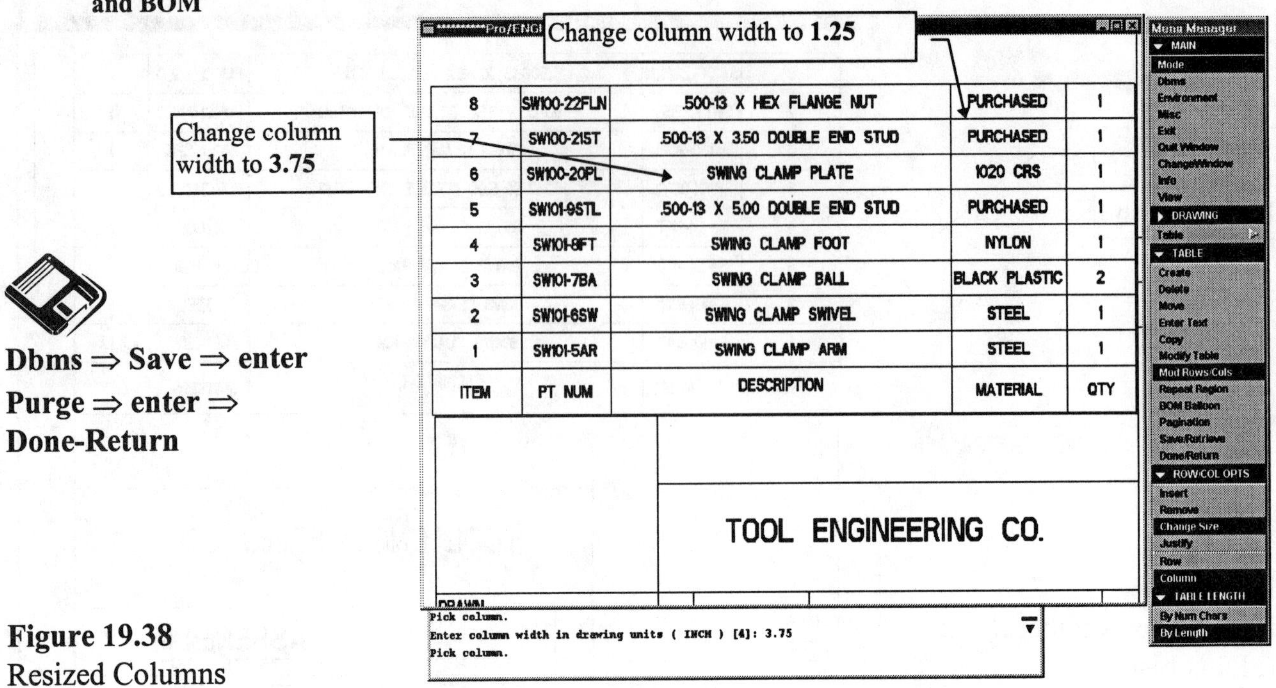

Dbms ⇒ Save ⇒ enter
Purge ⇒ enter ⇒
Done-Return

Figure 19.38
Resized Columns

Figure 19.39
Assembly Drawing

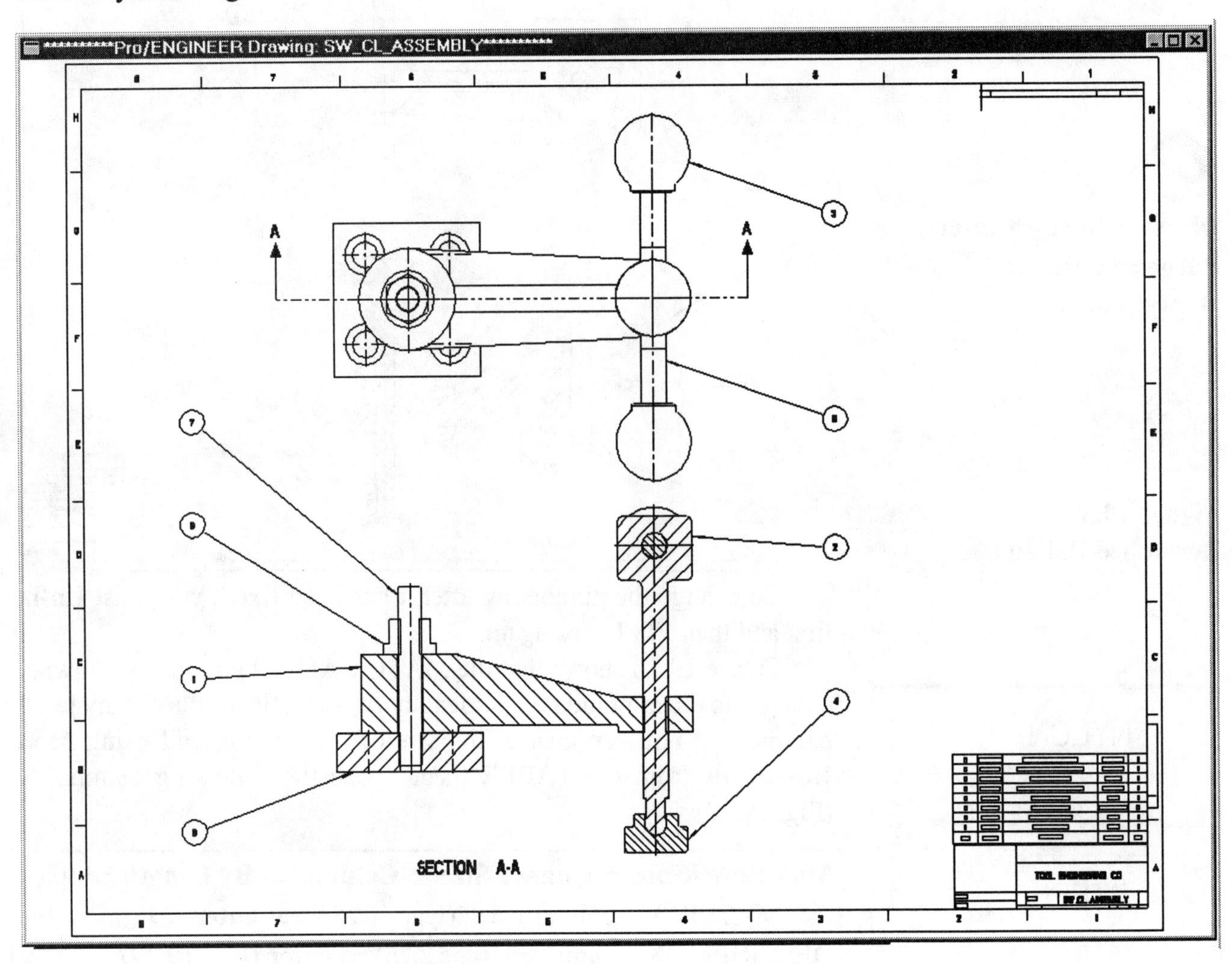

Lesson 19 Project

Coupling Assembly Drawing

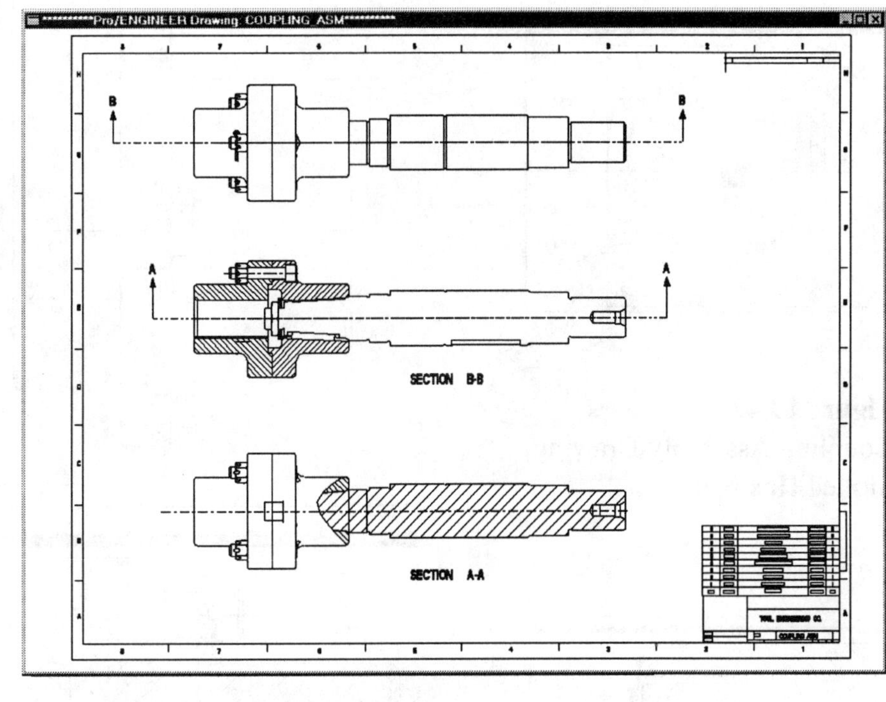

Figure 19.40
Coupling Assembly Drawing

Coupling Assembly Drawing

Create an assembly drawing of the Coupling Assembly using the format made in this lesson. Use Figure 19.40 through 19.44 as examples for this lesson project.

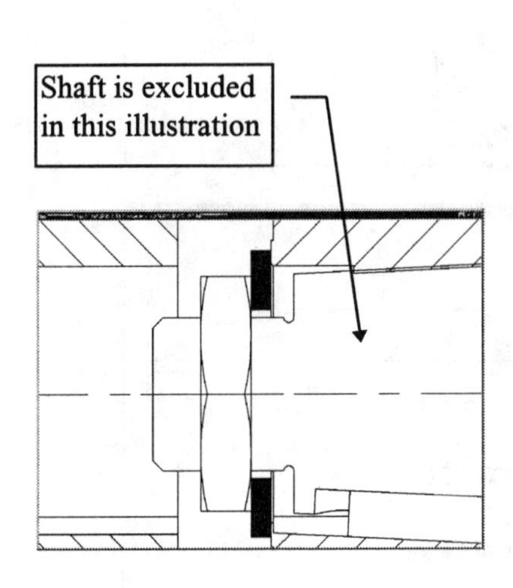

Figure 19.41
Coupling Assembly Drawing Parts List

9	109-2SN	HEX SLOT NUT 16 X 2	PURCHASED	3
8	108-2CP	COTTER PIN .150 X 1.25	PURCHASED	3
7	107-2KY	KEY 14 X 61	PURCHASED	1
6	106-2DW	DOWEL 12OD X 70	PURCHASED	2
5	105-2HN	HEX NUT M30 X 3.5	PURCHASED	1
4	104-2WA	WASHER 33ID X 50OD X 4	PURCHASED	1
3	103-2CP2	COUPLING TWO	1040 CRS	1
2	102-2CP1	COUPLING ONE	1040 CRS	1
1	101-2SH	COUPLING SHAFT	1020 CRS	1
ITEM	PT NUM	DESCRIPTION	MATERIAL	QTY

TOOL ENGINEERING CO.

A

DRAWN

ISSUED

1.000

COUPLING_ASM

SHEET 1 OF 1

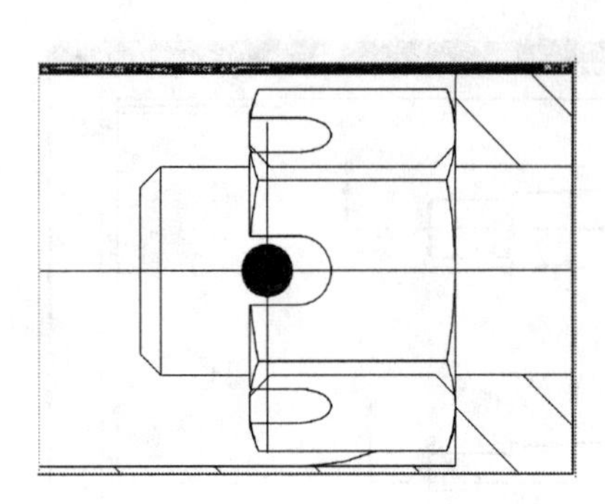

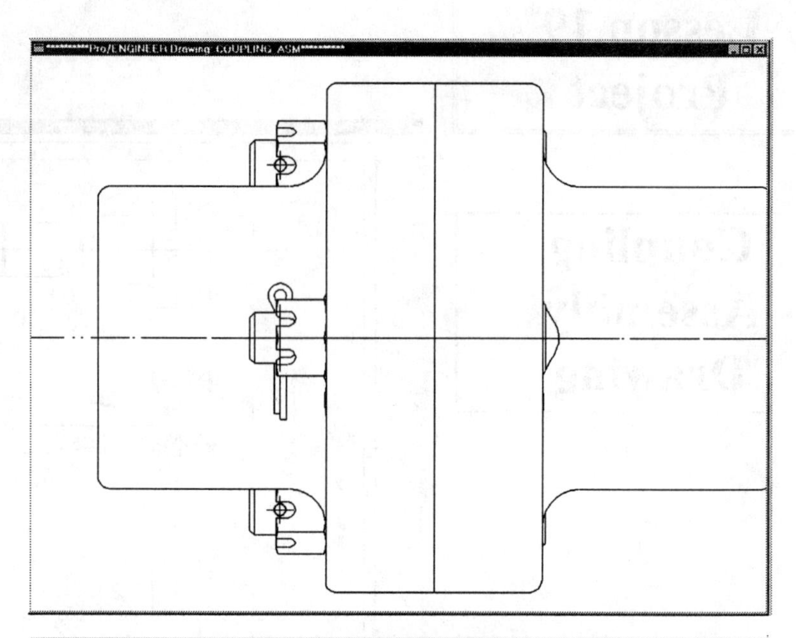

Figure 19.42
Coupling Assembly Drawing, Slotted Hex Nut

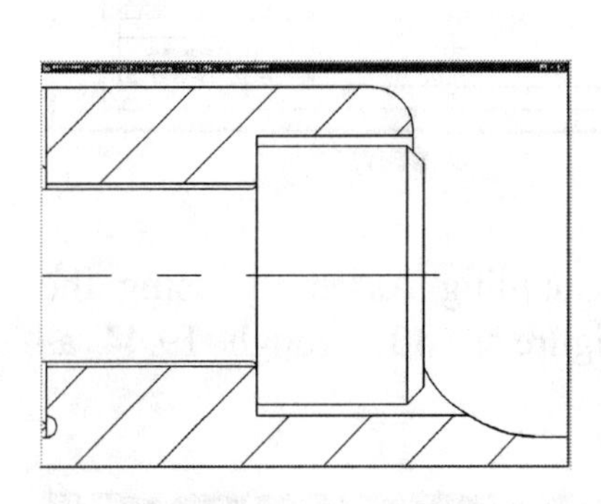

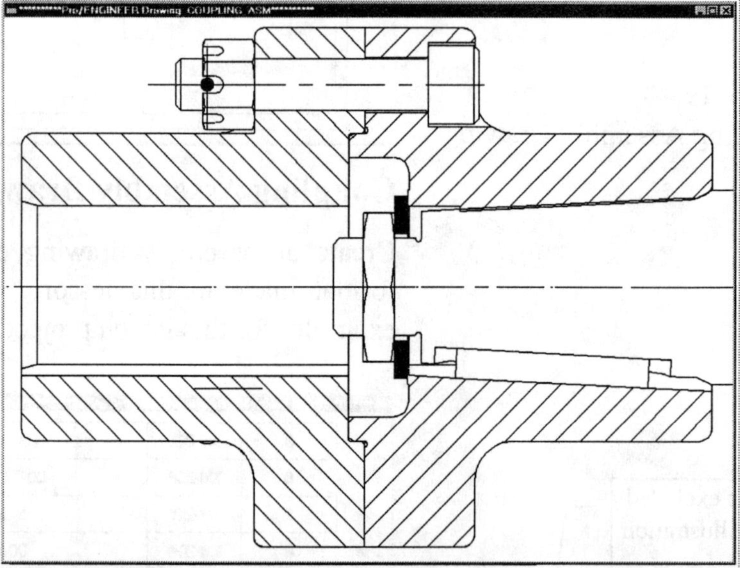

Figure 19.43
Coupling Assembly Drawing, Section Close-up

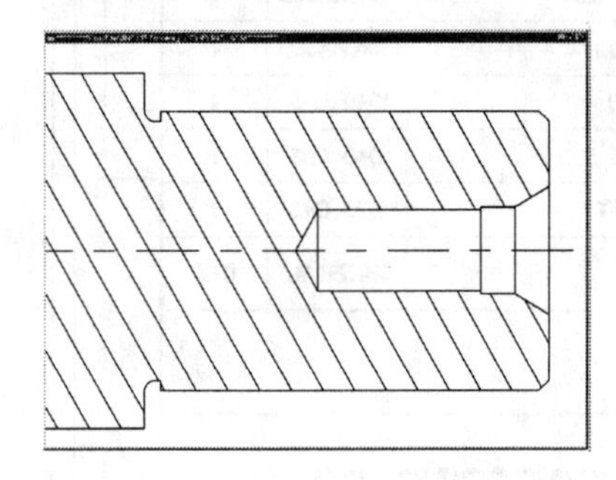

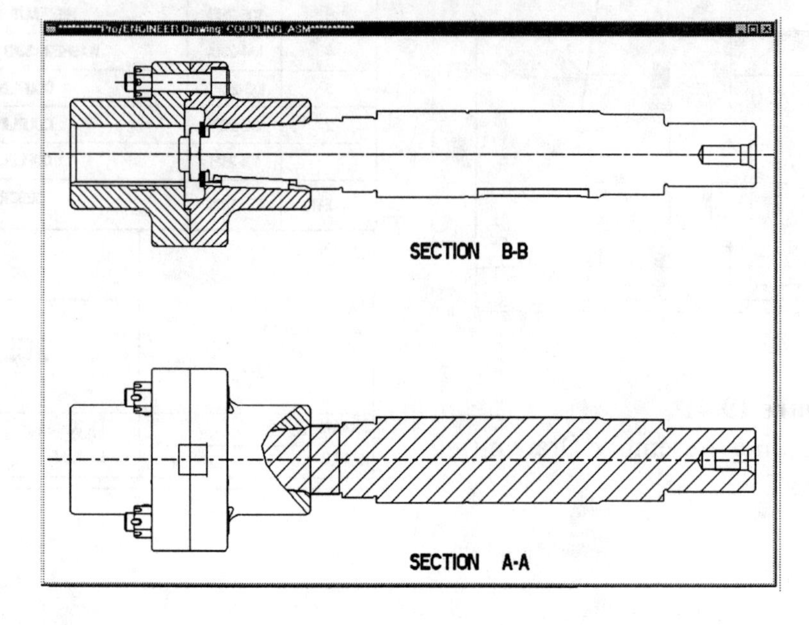

Figure 19.44
Coupling Assembly Drawing, **SECTION A-A** and **SECTION B-B**

Lesson 20

Exploded Assembly Drawings

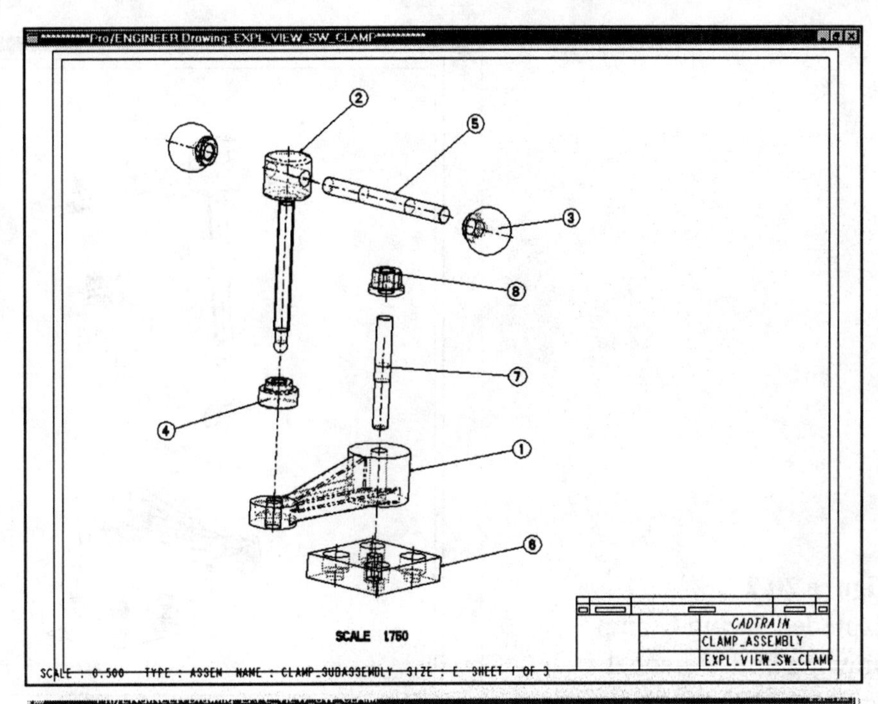

Figure 20.1
Sheet 1 Exploded Swing Clamp Drawing with Standard Format and Balloons and **Sheet 2** Exploded Trimetric View

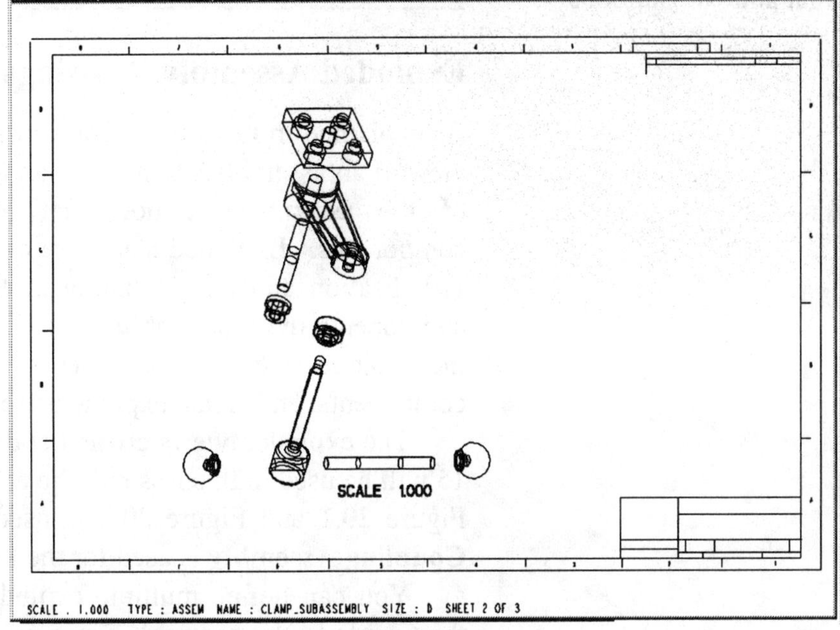

Figure 20.1 (continued)
Sheet 2 Exploded Trimetric View

EGD REFERENCE
Engineering Graphics and Design
by L. Lamit and K. Kitto
Read Chapters: 13, 23
See Pages: 447, 823, 836-837, 841

OBJECTIVES

1. **Create drawings with exploded views**
2. **Use multiple sheets**
3. **Make assembly drawing sheets with multiple models**
4. **Create balloons on exploded assemblies**

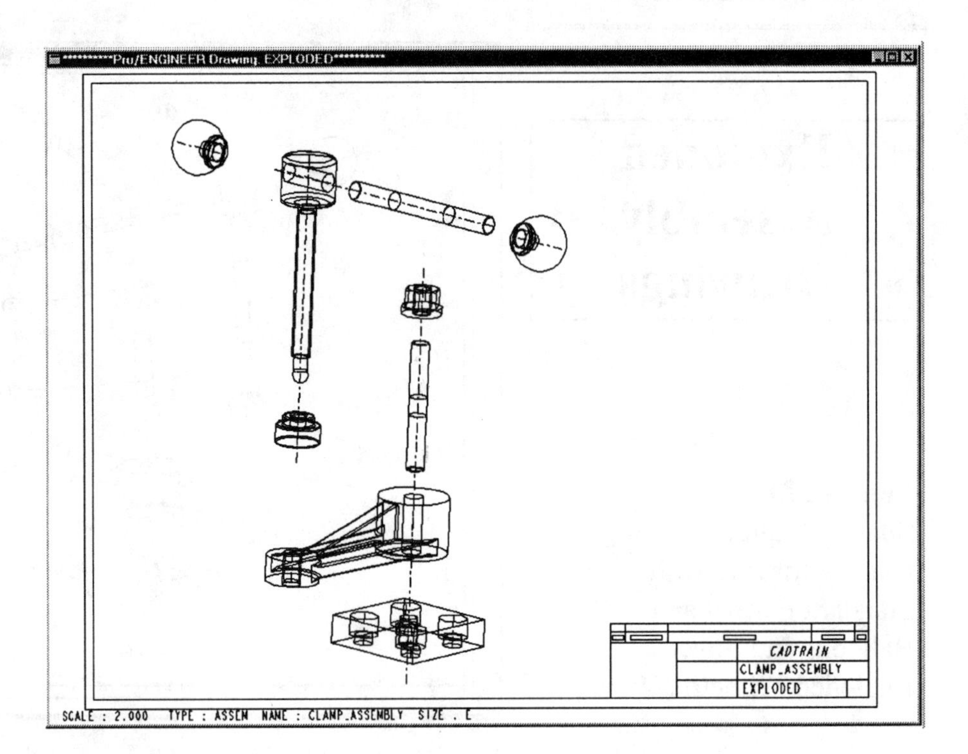

Figure 20.2
Exploded Swing Clamp Drawing with Personal Format and No Balloons

Exploded Assembly Drawings

As explained in Lesson 15, you can automatically create an exploded view of an assembly. Exploding an assembly affects only the display of the assembly; it does not alter actual distances between components. Exploded states are created and saved to allow a clear visualization and understanding of the positional relationship of all components in an assembly. For each explode state, you can toggle the explode status of components, change the explode locations of components, and create explode offset lines.

The exploded views created and saved with the model in Lesson 15 will be used in this lesson. The **Swing Clamp Assembly** shown in Figure 20.1 and Figure 20.2 is used for the lesson model, and the **Coupling Assembly** is used for the Lesson 20 Project.

You can define multiple explode states for each assembly, and then explode the assembly using any of these explode states at any time (Fig. 20.3). You can also set an exploded state for each drawing view of an assembly. If the exploded views previously created are not exactly the exploded state you wish to use in this lesson, bring up the model (assembly) and create new exploded views and save them for use in this lesson model and lesson project.

Since this is the last lesson model and lesson project for the current text, you will be required to create a complete documentation package for the two assemblies. A documentation package contains all models and drawings required to manufacture the parts and assemble the components. Your instructor may change the requirements, but, in general, create and plot the following:

Part Models for all swing clamp components
Detail Drawings for each nonstandard component, for example, the Clamp Arm, Clamp Swivel, Clamp Foot, and Clamp Ball
Detail Drawings for standard components that have been altered--show only the dimensions required to alter the component, for example, the socket head cap screw in the coupling assembly
Assembly Drawing using standard orthographic ballooned views
Exploded Assembly Drawing of the ballooned assembly
Exploded Sub Assembly Drawing of the ballooned subassembly

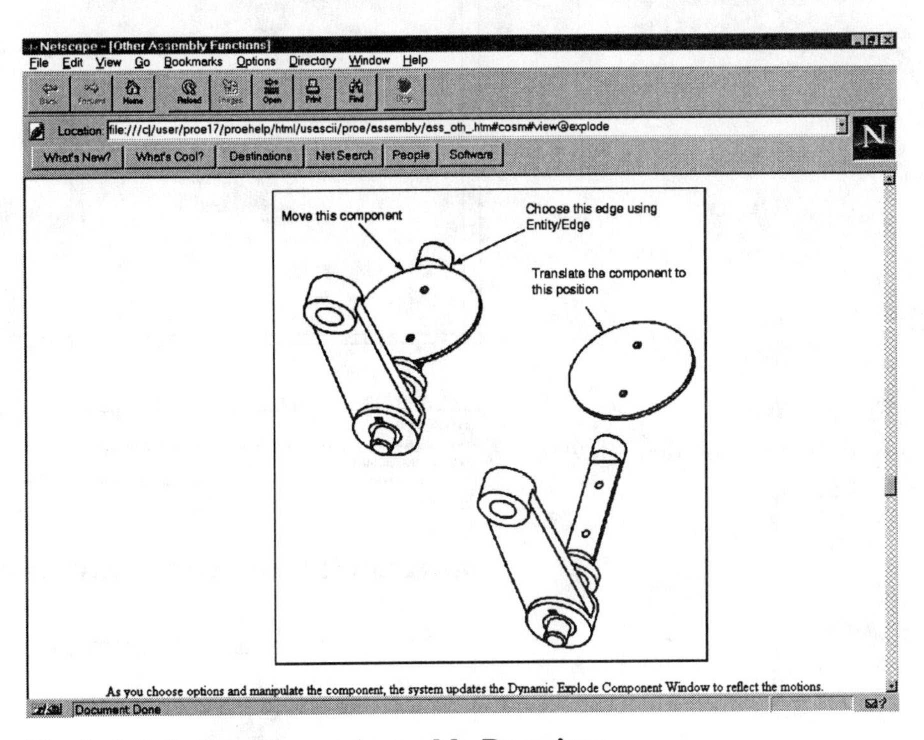

Figure 20.3
Online Documentation
Exploded Assemblies

Exploded Swing Clamp Assembly Drawings

The process required to place an exploded view on a drawing is similar to that provided in Lesson 19 for adding assembly orthographic views to a drawing. Use the following commands:

> **Mode** ⇒ **Drawing** ⇒ **Create** ⇒ (type **EXPL_VIEW_SW_CLAMP)** ⇒ **enter** ⇒ **Retr Format** ⇒ **?** ⇒ **enter** ⇒ **Format Dir** ⇒ **d.frm** ⇒ **Views** ⇒ **?** ⇒ (pick the Swing Clamp assembly from the directory) ⇒ **Exploded** ⇒ **Scale** ⇒ **Done** ⇒ (pick the center of the drawing, as in Figure 20.4) ⇒ **Done** ⇒ (type **.75** as the scale) ⇒ **enter** ⇒ **Names** ⇒ (pick one of the saved exploded views) ⇒ **Done/Return** ⇒ **Done/Return** ⇒ **View** ⇒ **Modify View** ⇒ **Change Scale** ⇒ (pick the view) ⇒ (type **1** as the new scale) ⇒ **enter** ⇒ **Done/Return**

The drawing will have only one view. If you wanted to add a second sheet with a different exploded view orientation (Fig. 20.5), use the following commands:

Drawing ⇒ **Sheets** ⇒ **Add** (sheet number **2** has been added to the drawing) ⇒ **Views** ⇒ **Exploded** ⇒ **Scale** ⇒ **Done** ⇒ (pick the center of the new drawing sheet) ⇒ **Done** ⇒ (type **1** for scale) ⇒ **Done/Return** (to keep the trimetric default orientation)

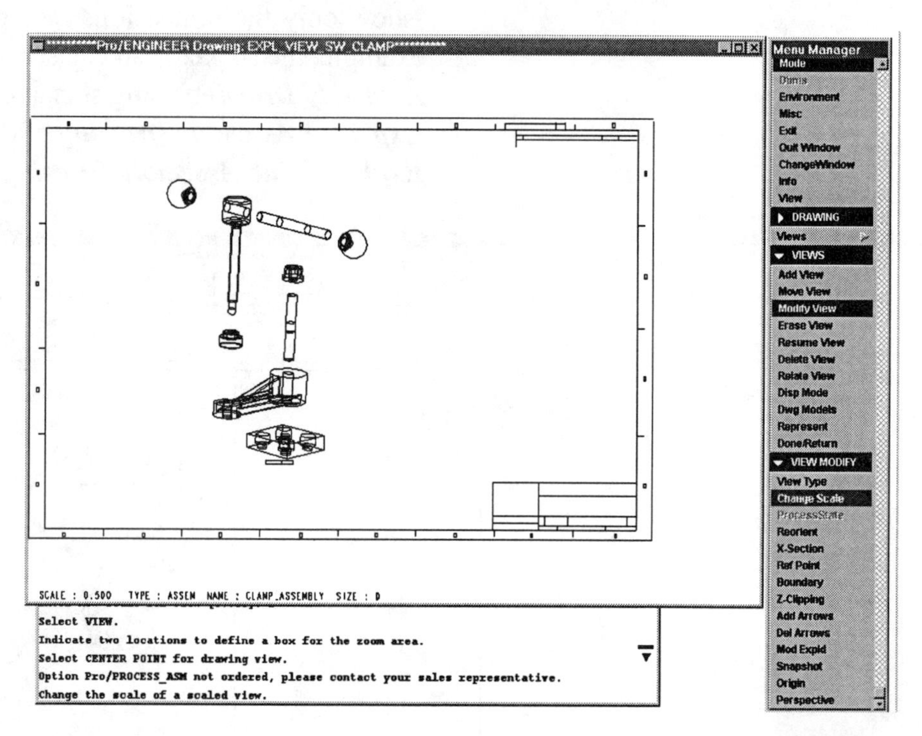

Figure 20.4
Adding an Exploded View to a Drawing

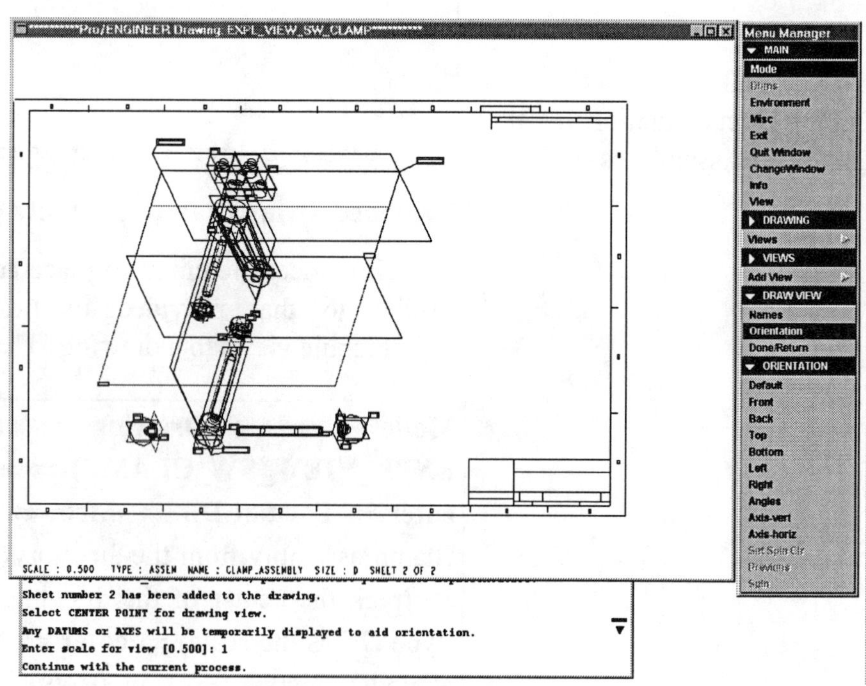

Figure 20.5
Adding a Second Sheet and a New View

Next, add another sheet and a different model. Use the **Swing Clamp subassembly** (Fig. 20.6). Modify the scale of the drawing after the view is placed. Use the following commands:

Sheets ⇒ **Add** (sheet number **3** has been added to the drawing) ⇒ **Format** ⇒ **Add/Replace** ⇒ (type **C**) ⇒ **enter** ⇒ **Views** ⇒ **Dwg Models** ⇒ **Add Model** ⇒ **?** ⇒ **enter** ⇒ (pick the Swing Clamp subassembly from the directory list) ⇒ **Done/Return** ⇒ **Views** ⇒ **Exploded** ⇒ **Scale** ⇒ **Done** ⇒ (pick the center of the new drawing sheet) ⇒ **Done** ⇒ **Names** ⇒ (pick a saved view name from the list) ⇒ **Done/Return** ⇒ **Done/Return** ⇒ **Views** ⇒ **Modify** ⇒ **View** ⇒ **Change Scale** ⇒ (pick the scale on the window) ⇒ (type **1.25** for scale) ⇒ **enter**

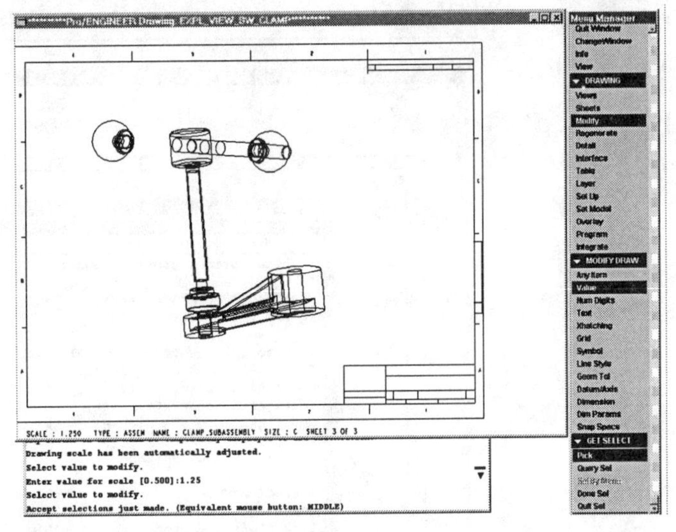

Figure 20.6
Sheet 3 with Exploded View of Subassembly

Show all of the axes for the components on the first sheet. To switch between sheets, choose: **Sheets** ⇒ **Previous**. Use the following commands:

Detail ⇒ **Show/Erase** ⇒**Show** (radio button) ⇒ **A_1** (radio button) ⇒ **Show All** (radio button) ⇒ **Yes** (radio button) ⇒ **Close** (radio button) ⇒ (press right mouse button anywhere on drawing--**Modify Item**) ⇒ (pick an axis to be modified) ⇒ (modify each centerline-axis to stretch between components that are in line, as in Figure 20.7)

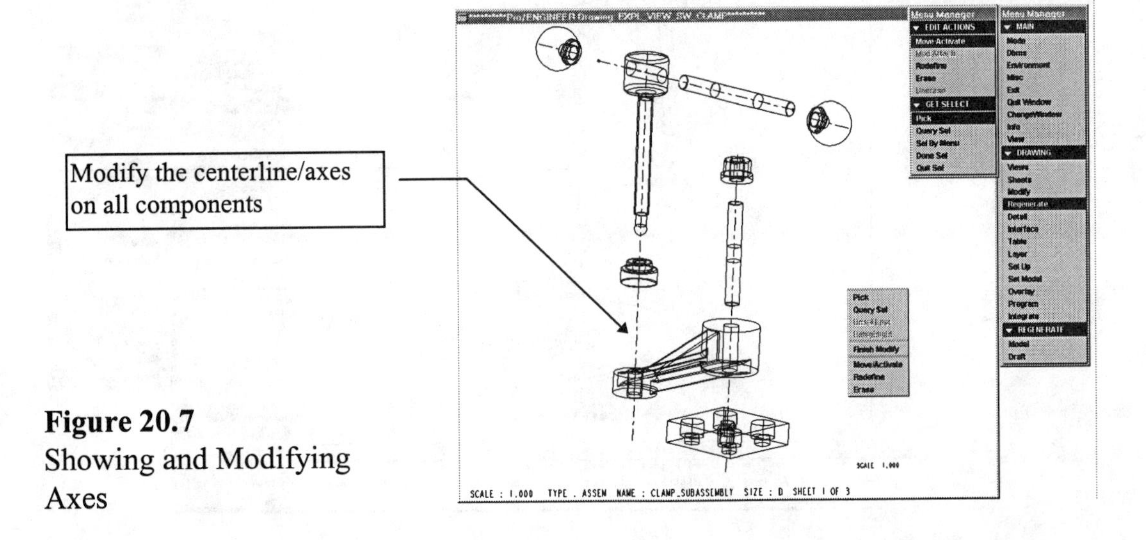

Figure 20.7
Showing and Modifying Axes

To add balloons to one of the sheets not using a parametric title block, you need to create each balloon separately. Create balloons for the components on the first sheet. The balloons added must correspond to the component balloons on the assembly drawing completed in Lesson 19. Before creating balloons, change the **.dtl** file to have **1.00** diameter balloons and **.500** height for lettering (Fig. 20.8). Use the following commands:

Detail ⇒ Create ⇒ Balloon ⇒ Leader ⇒ Make Note ⇒ (pick the Swing Clamp Arm) **⇒ Done Sel ⇒ Done ⇒** (pick the location for the note) ⇒ (type **1**) ⇒ **enter ⇒ enter ⇒ Make Note ⇒** (continue until all eight components are ballooned, as in Figure 20.9)

NOTE

To change the defaults on balloons:

Drawing ⇒ Setup ⇒ Modify Val ⇒

drawing_text_height .50
max_balloon_radius .50
min_balloon_radius .50

File ⇒ Save ⇒ File ⇒ Exit

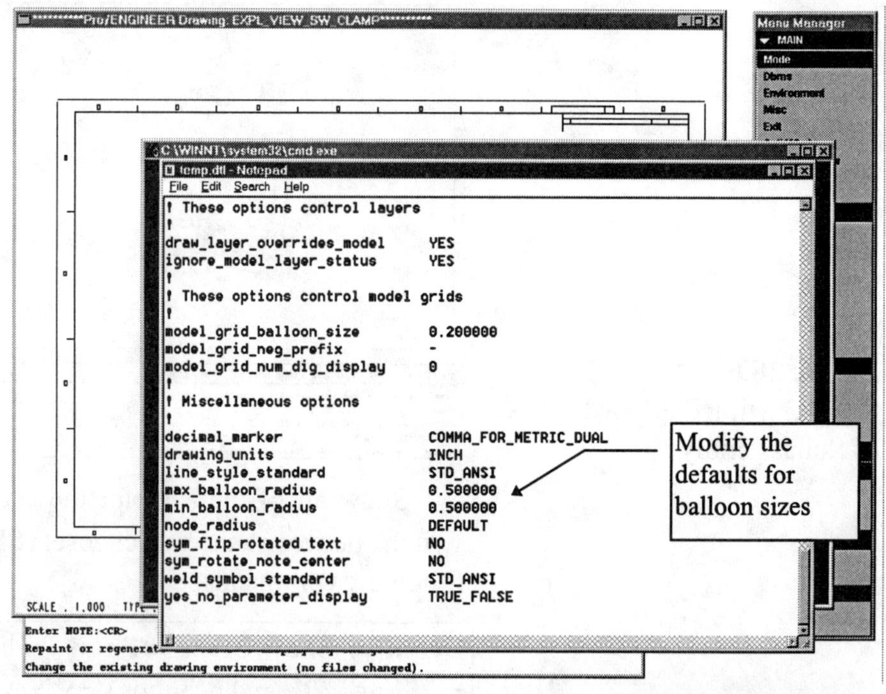

Figure 20.8
Modify Val

Dbms ⇒ Save ⇒ enter
Purge ⇒ enter ⇒
Done-Return

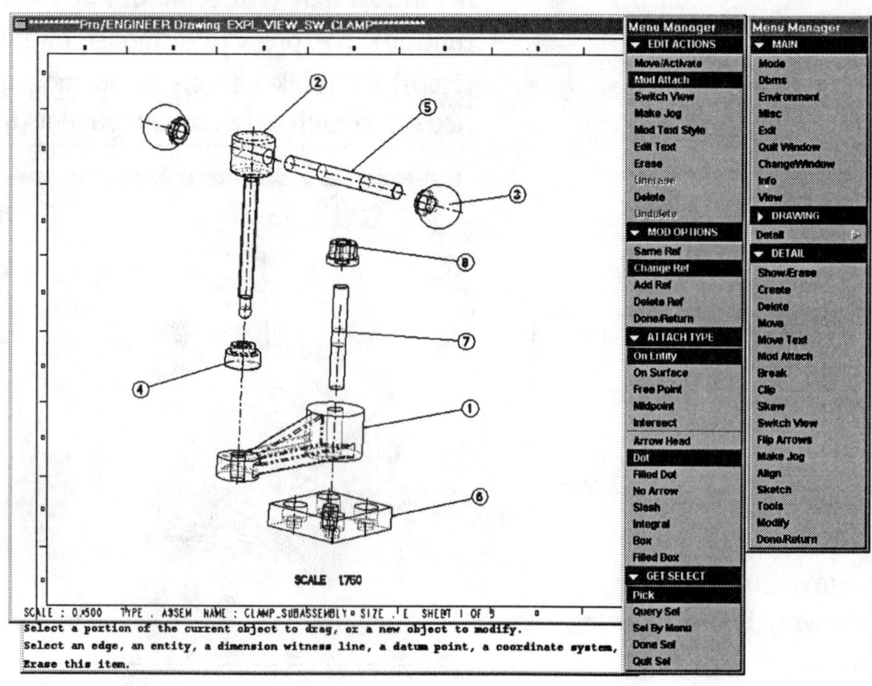

Figure 20.9
Ballooned Drawing

After the ballooning is complete, move the balloons and their attachment points to clean up the drawing. Use the following commands to modify the attachment point from edge to surface, and change the arrow to a dot (Fig. 20.10):

Modify Item (right mouse button) ⇒ (pick balloon **3**) ⇒ **Mod Attach** ⇒ **On Surface** ⇒ **Dot** ⇒ (pick a place on the ball's surface) ⇒ **Done/Return** ⇒ **Done Sel**

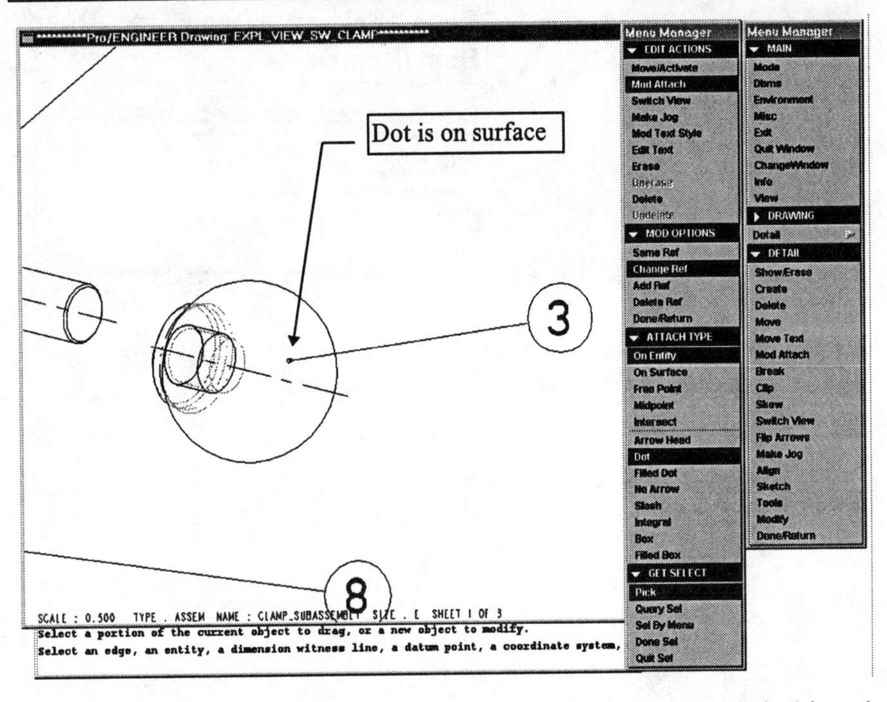

Figure 20.10
Modifying Balloons

Complete the drawing by filling in the title block or switching the sheet format with your parametric format created in Lesson 19 (Fig. 20.11).

NOTE
Change the view display to show as **Hidden Line** and **Tan Phantom**.

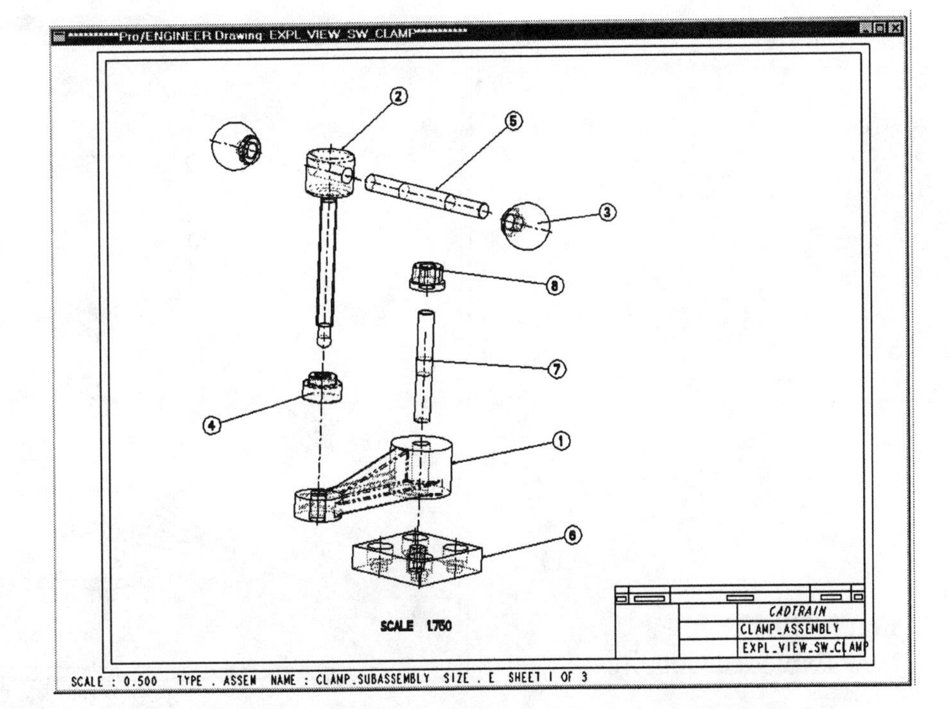

Figure 20.11
Ballooned Drawing

You can modify the exploded drawing without going back to the Assembly mode and editing the model (Fig. 20.12 and Fig. 20.13).

Use the following commands to modify the drawing:

Views ⇒ **Modify View** ⇒ **Mod Expld** ⇒ (pick the view) ⇒ **Redefine** ⇒ **Position** ⇒ **Entity/Edge** ⇒ (select a vertical axis) ⇒ (pick the Arm) ⇒ (slide the Arm to a new position) ⇒ (continue moving the components until they appear as in Figure 20.13; you will need to choose a horizontal axis to move the Balls and long Stud) ⇒ **Done Sel** ⇒ **Done** ⇒ **Done/Return** ⇒ **Done/Return** ⇒ **Done/Return**

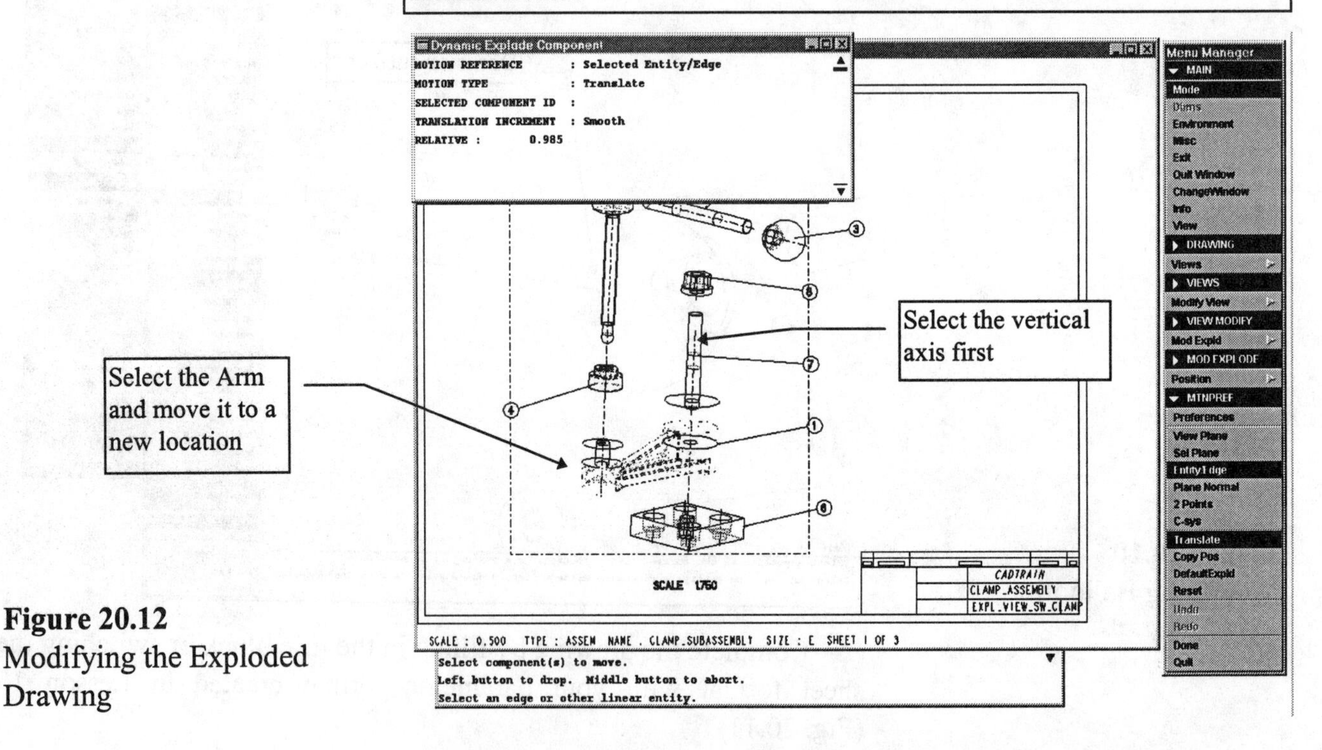

Figure 20.12
Modifying the Exploded Drawing

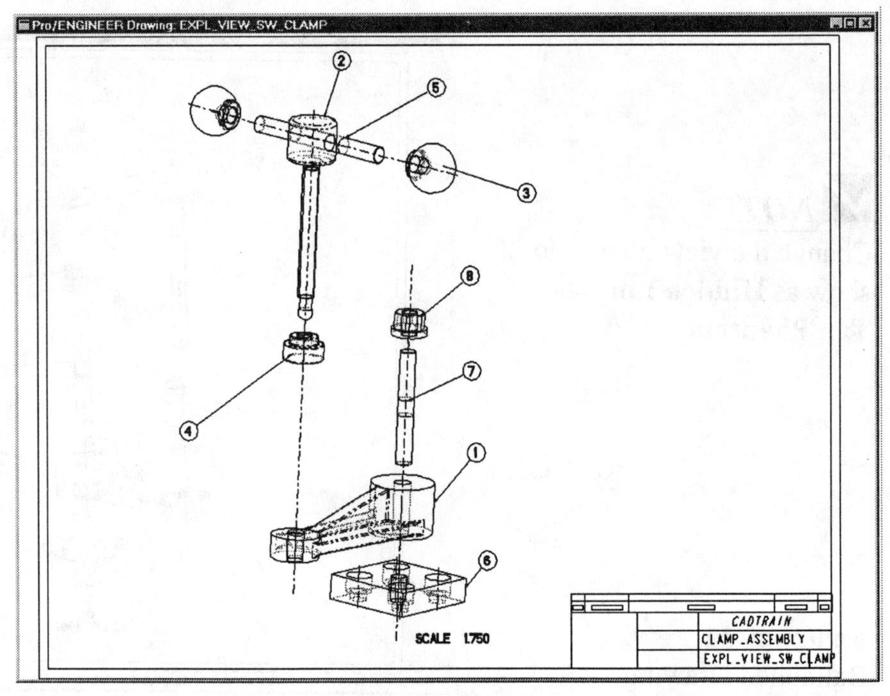

Figure 20.13
New Exploded Condition

Lesson 20 Project

Exploded Coupling Assembly Drawing

Exploded Coupling Assembly Drawing

Create a complete documentation package for the Coupling Assembly. A documentation package contains all models and drawings required to manufacture the parts and assemble the components. Some of the items listed here have been created in other lessons. Create or extract existing models and drawings and plot the following:

Part Models for all coupling assembly components
Detail Drawings for each nonstandard component, for example, the Coupling Shaft
Detail Drawings for standard components that have been altered; show only the dimensions required to alter the component, for example, the socket head cap screw in the coupling assembly
Assembly Drawing and ***Parts List*** (***BOM***) using standard orthographic ballooned views
Exploded Assembly Drawing of the ballooned assembly
Exploded Subassembly Drawing of the ballooned subassembly

Figure 20.14
Exploded Coupling Assembly Drawing

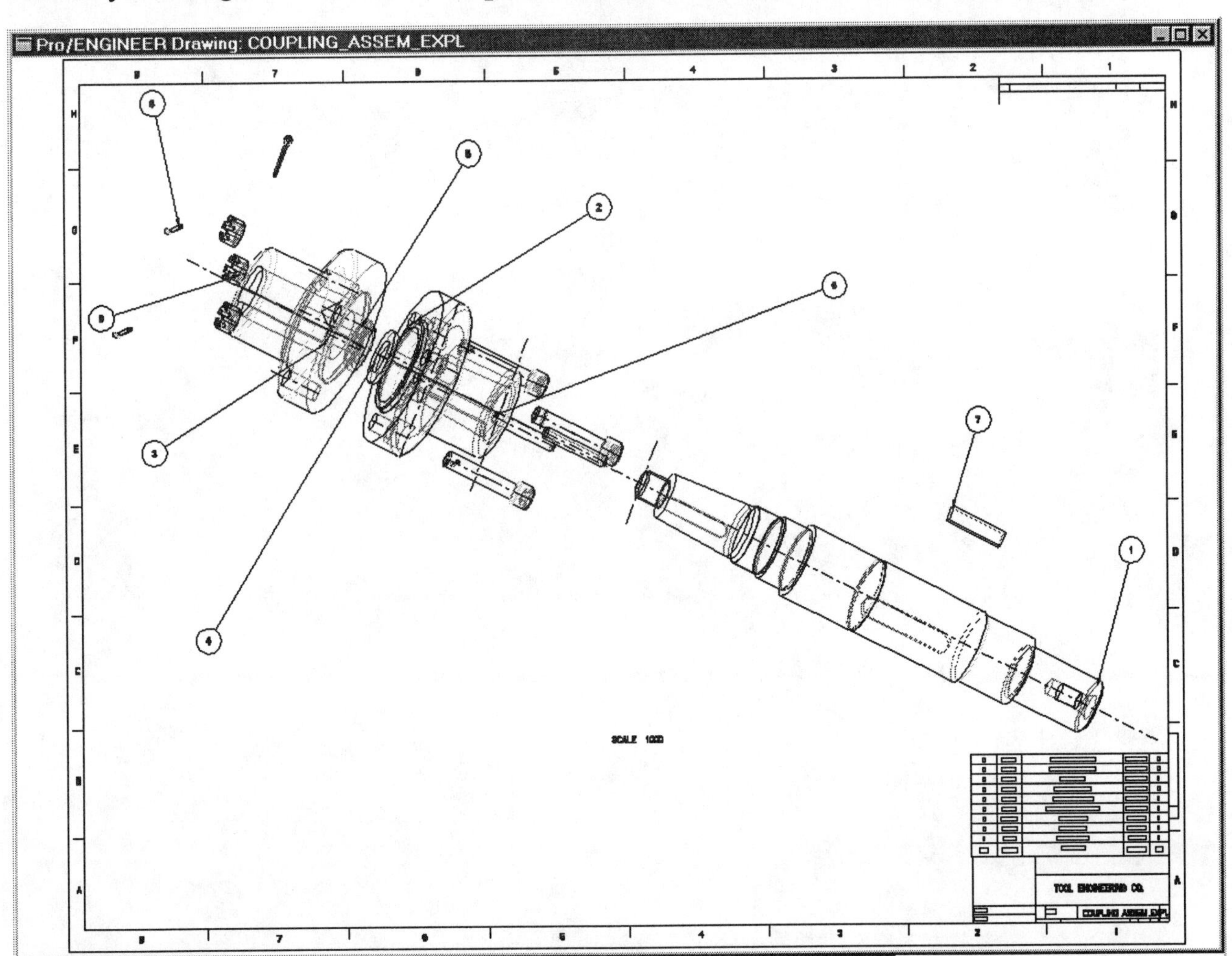

Appendixes

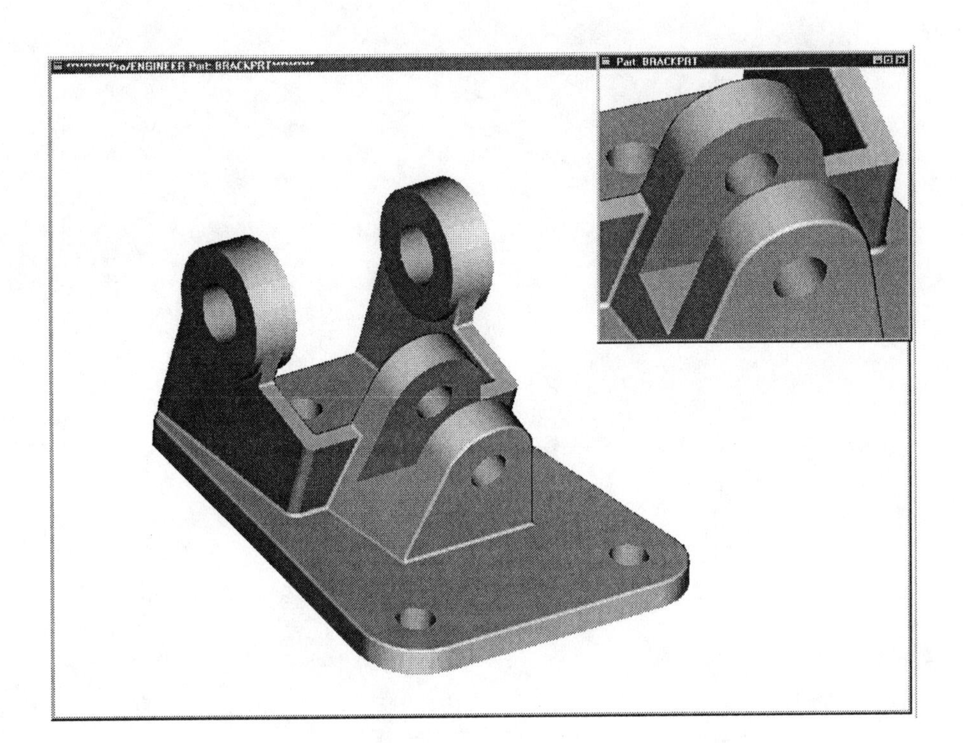

Advanced Project

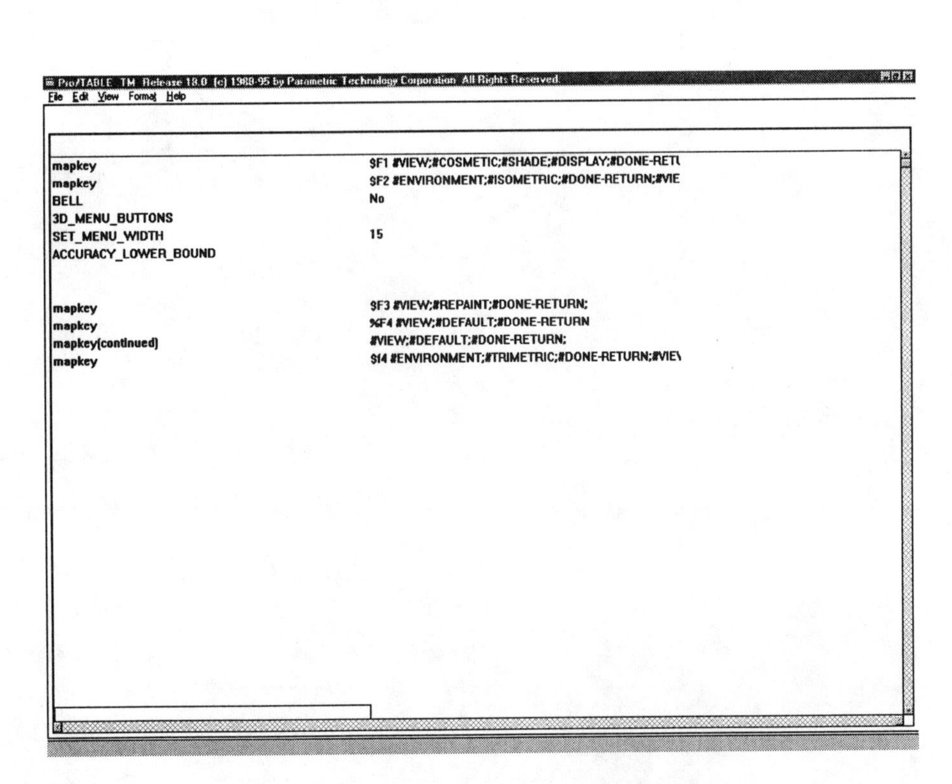

Config.pro File

Appendix A

Advanced Projects

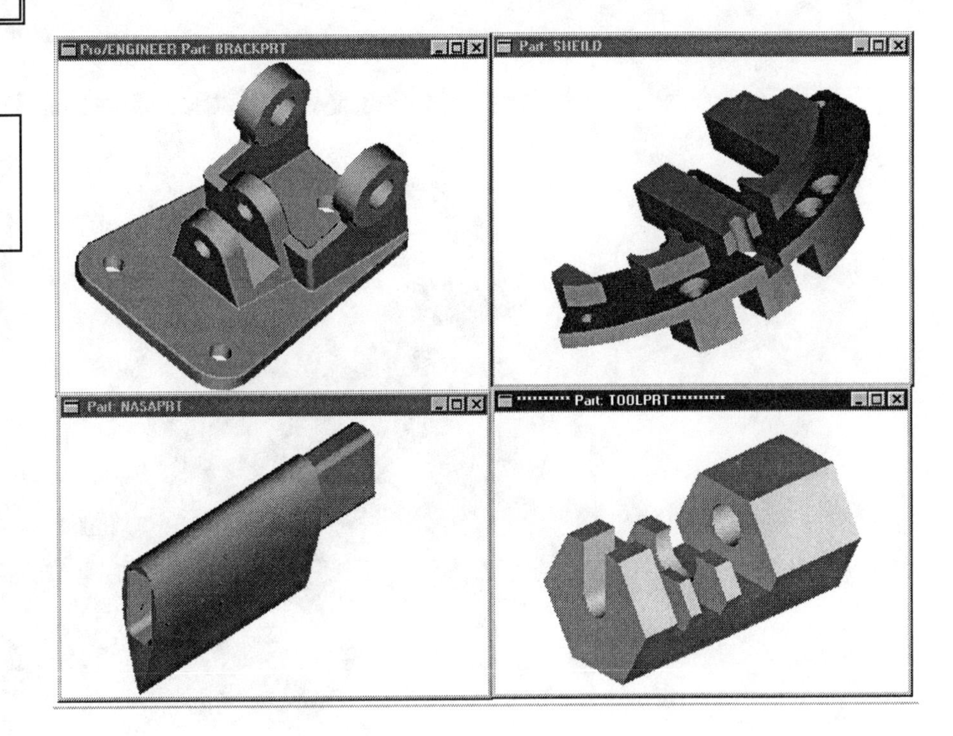

Figure A.1
Advanced Parts

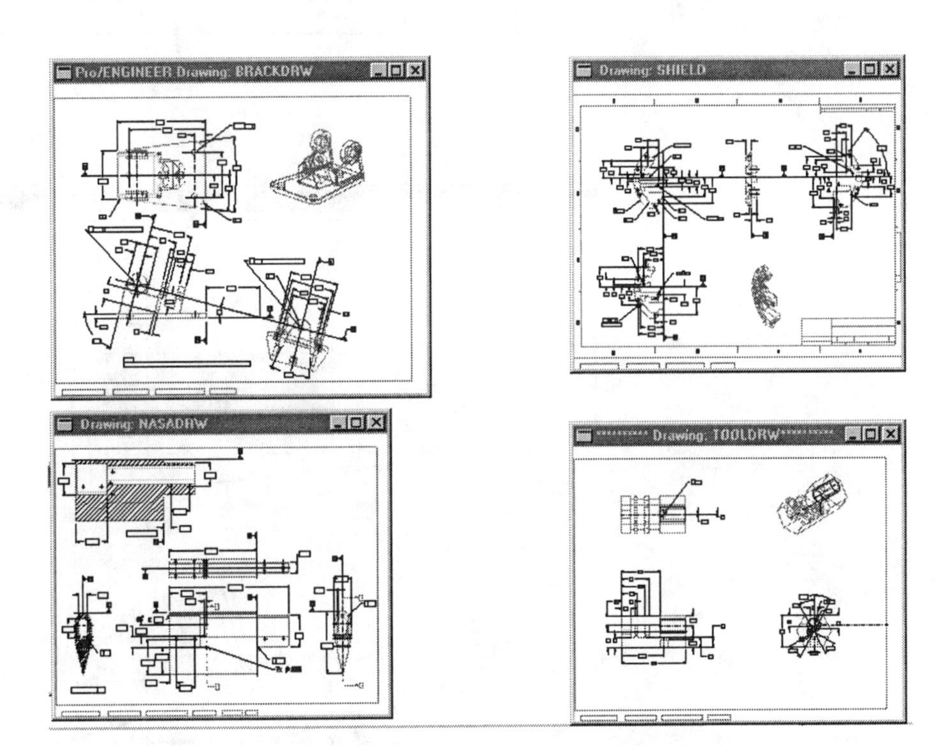

Figure A.2
Advanced Part Drawings

Bracket

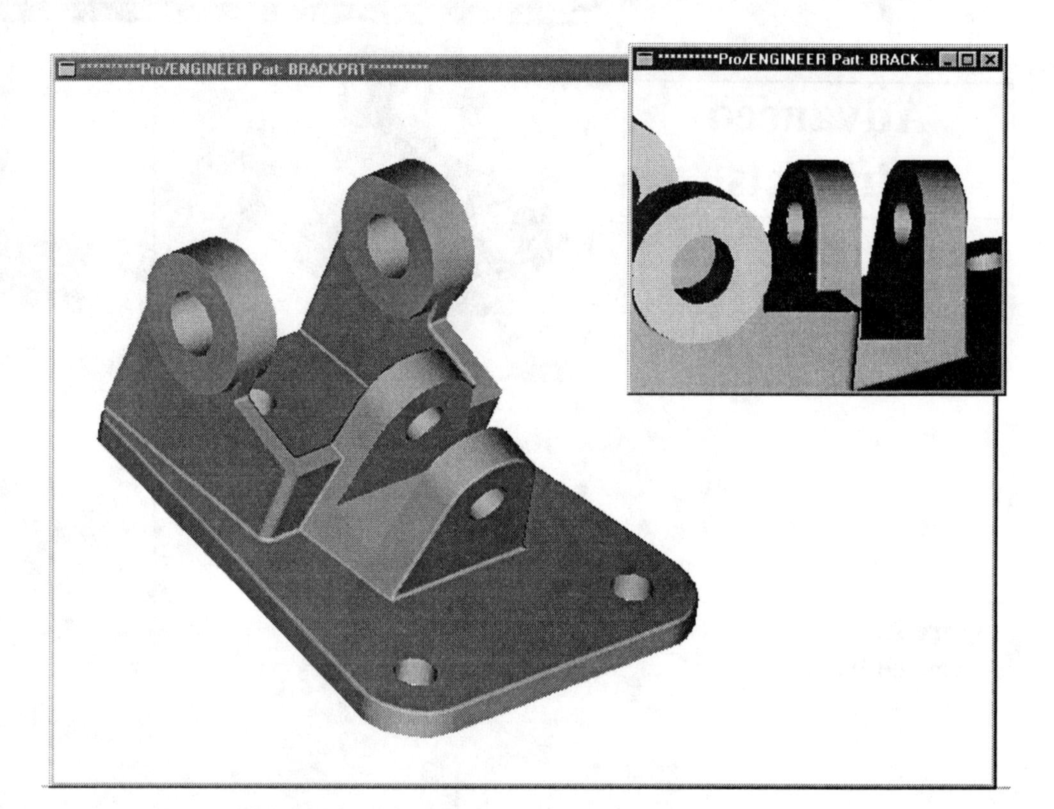

Figure A.3
Bracket Part Model

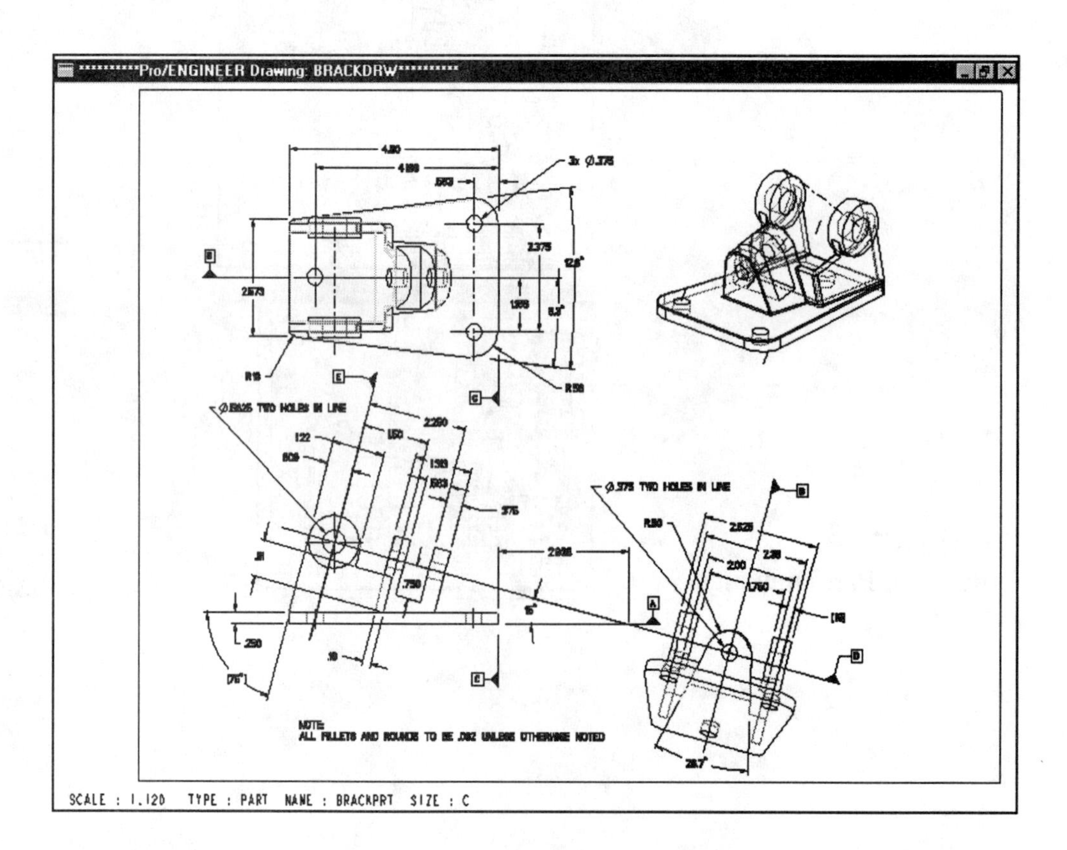

Figure A.4
Bracket Drawing

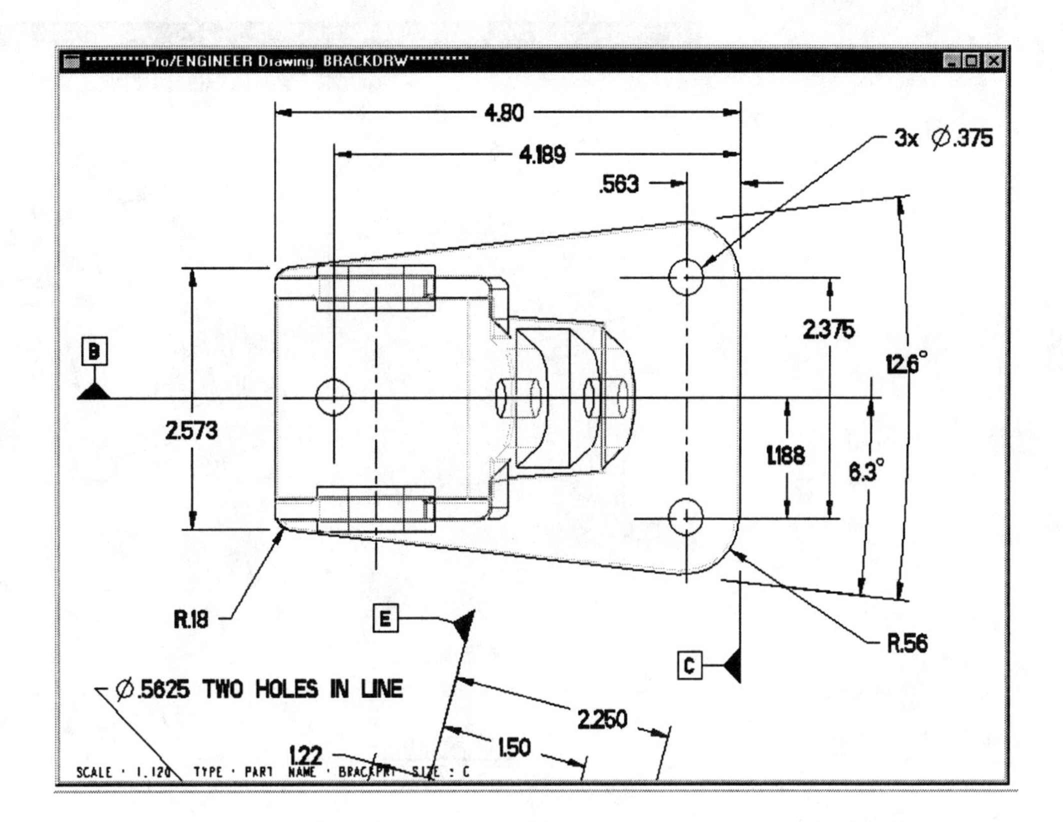

Figure A.5
Bracket Drawing, Top View

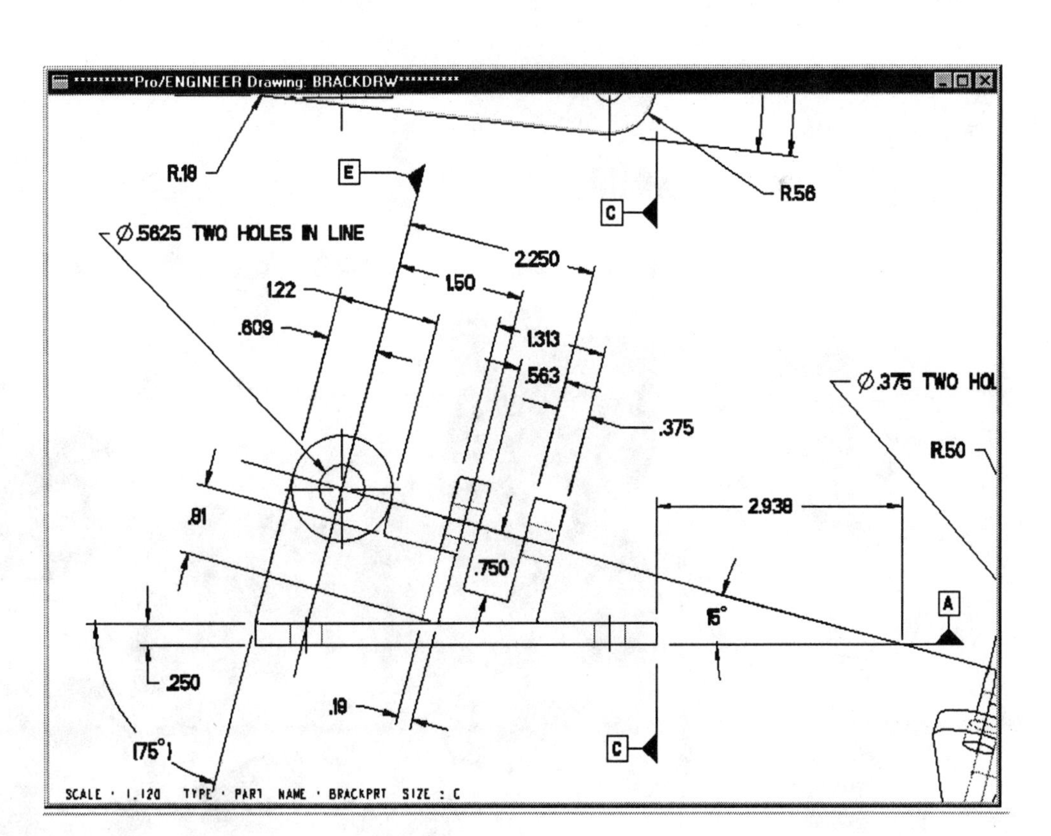

Figure A.6
Bracket Drawing, Front View

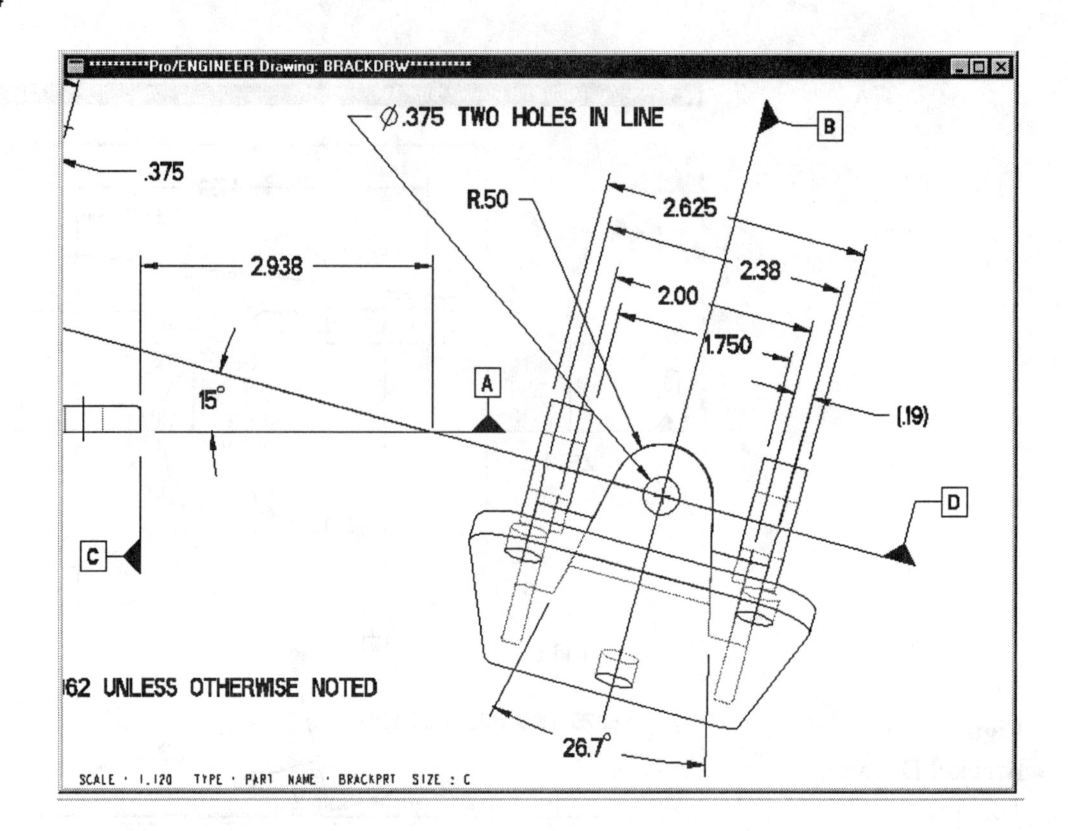

Figure A.7
Bracket Drawing, Auxiliary View

Shield

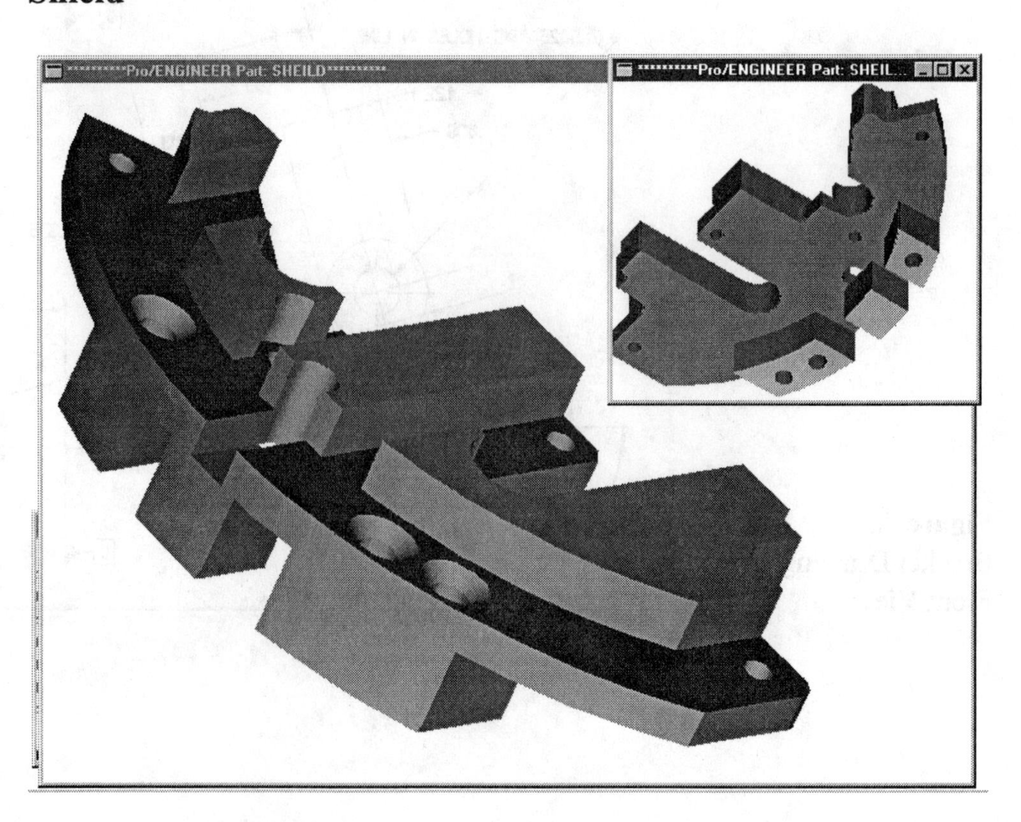

Figure A.8
Shield Part Model

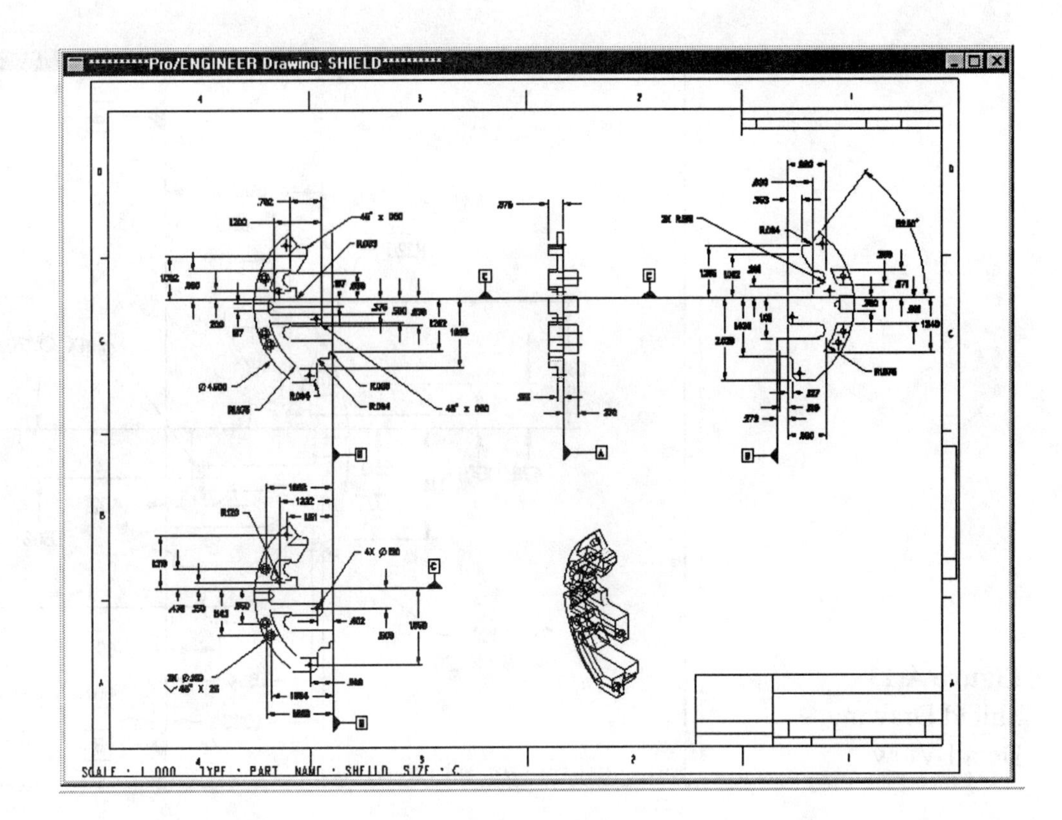

Figure A.9
Shield Drawing

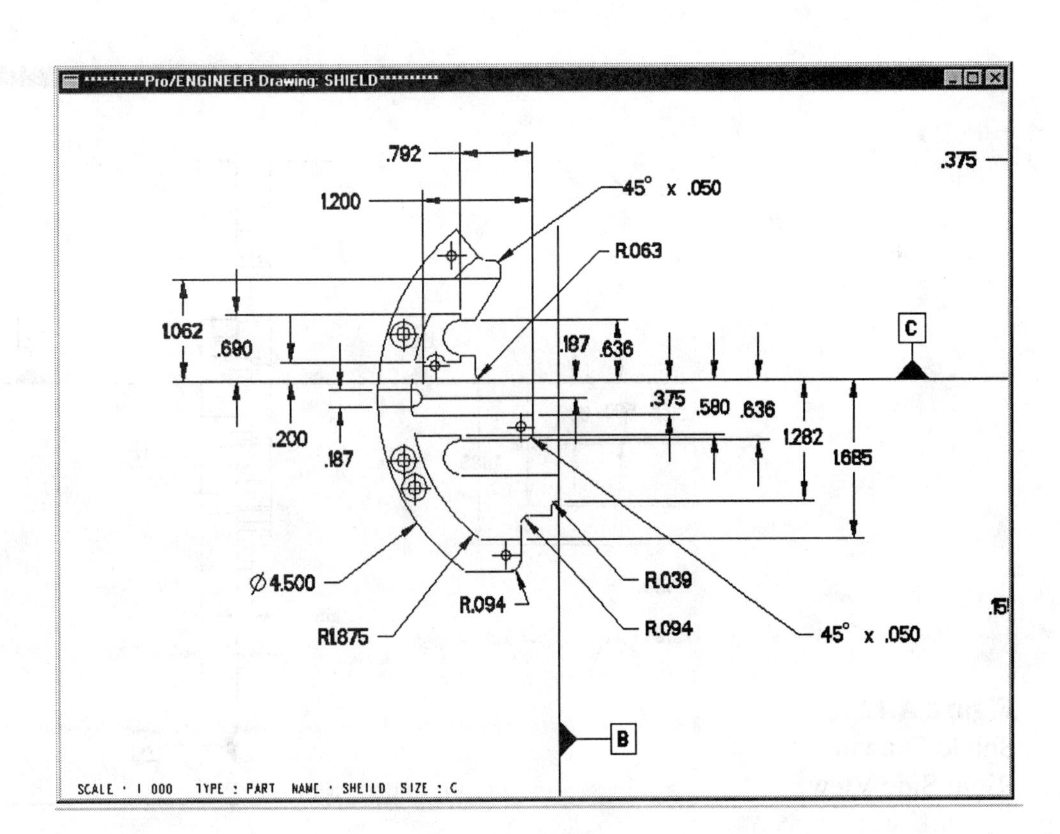

Figure A.10
Shield Drawing ,
Front View

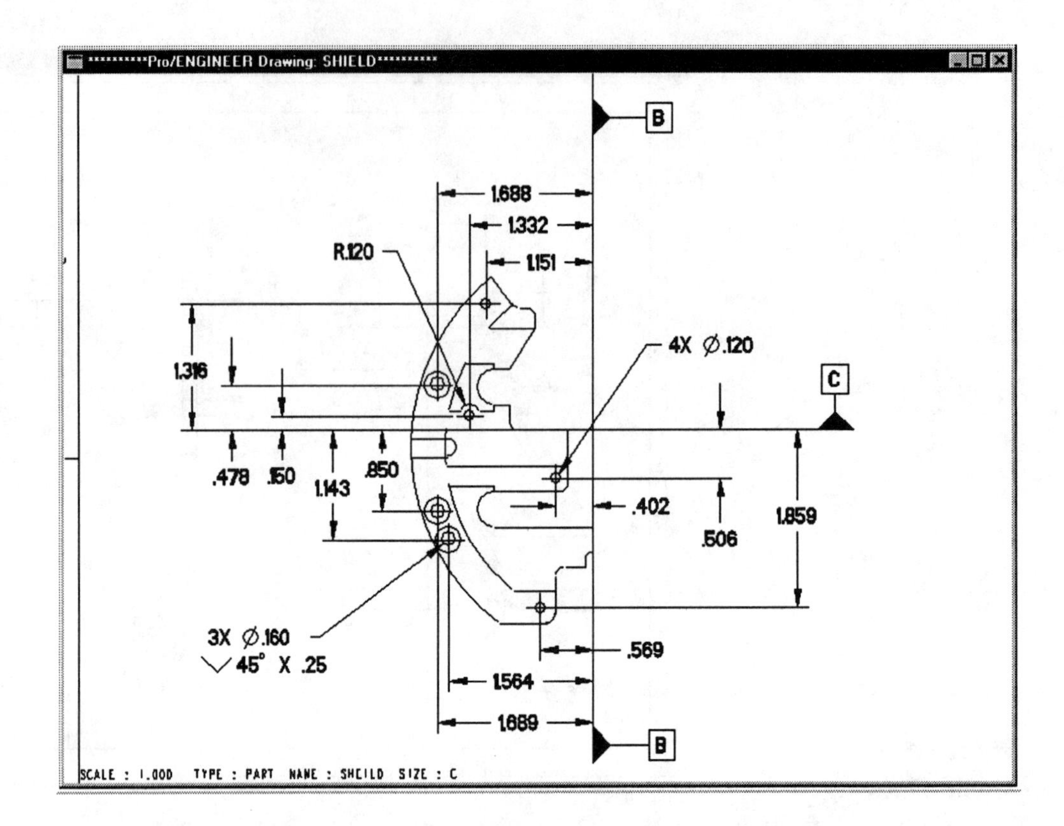

Figure A.11
Shield Drawing, Detail View

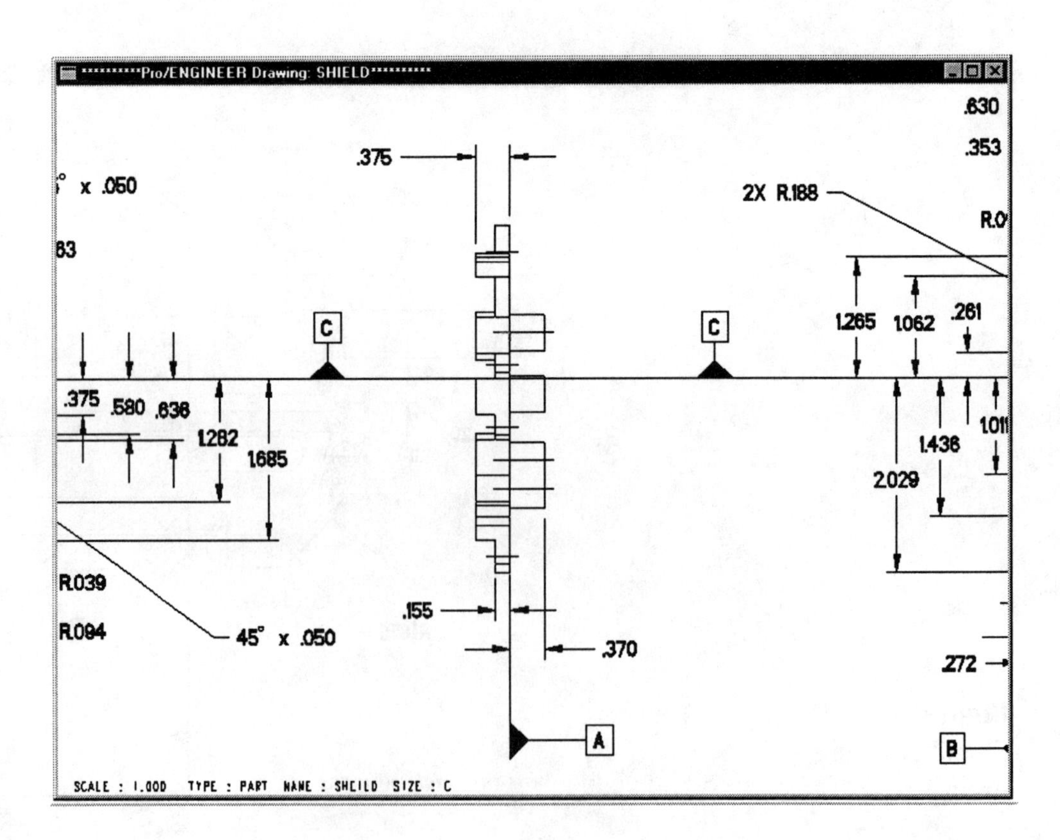

Figure A.12
Shield Drawing, Right Side View

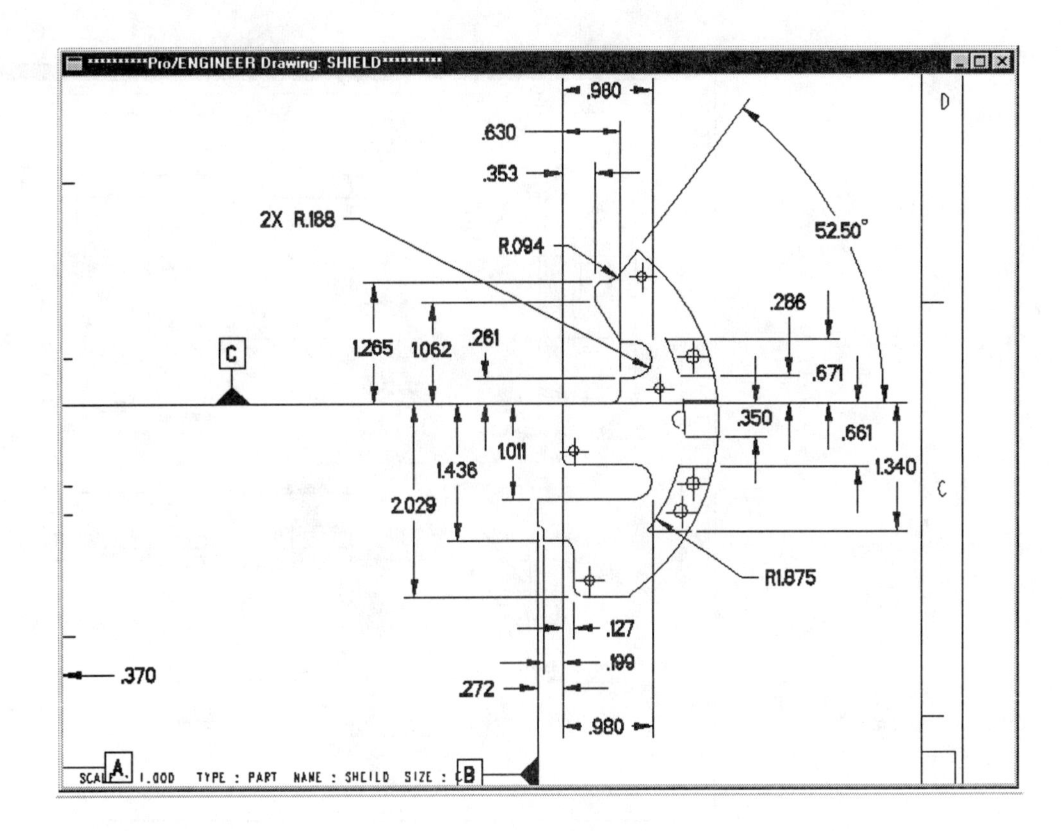

Figure A.13
Shield Drawing, Back View

Air Foil

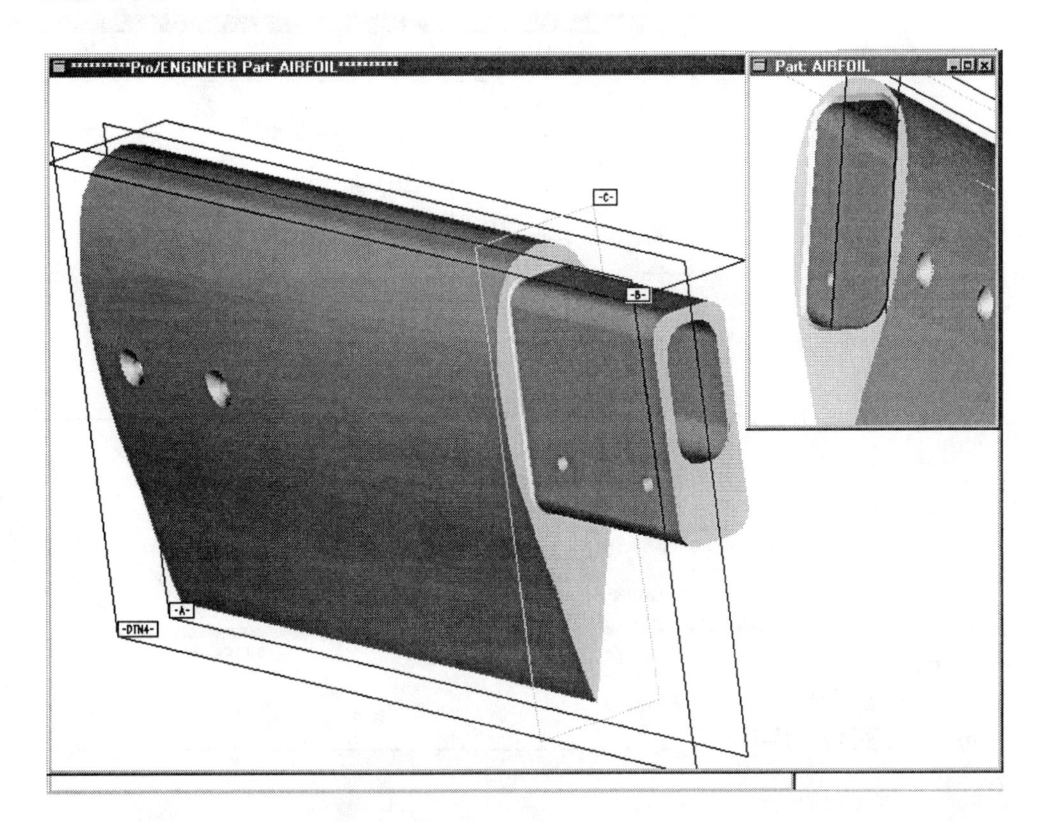

Figure A.14
Air Foil Part Model

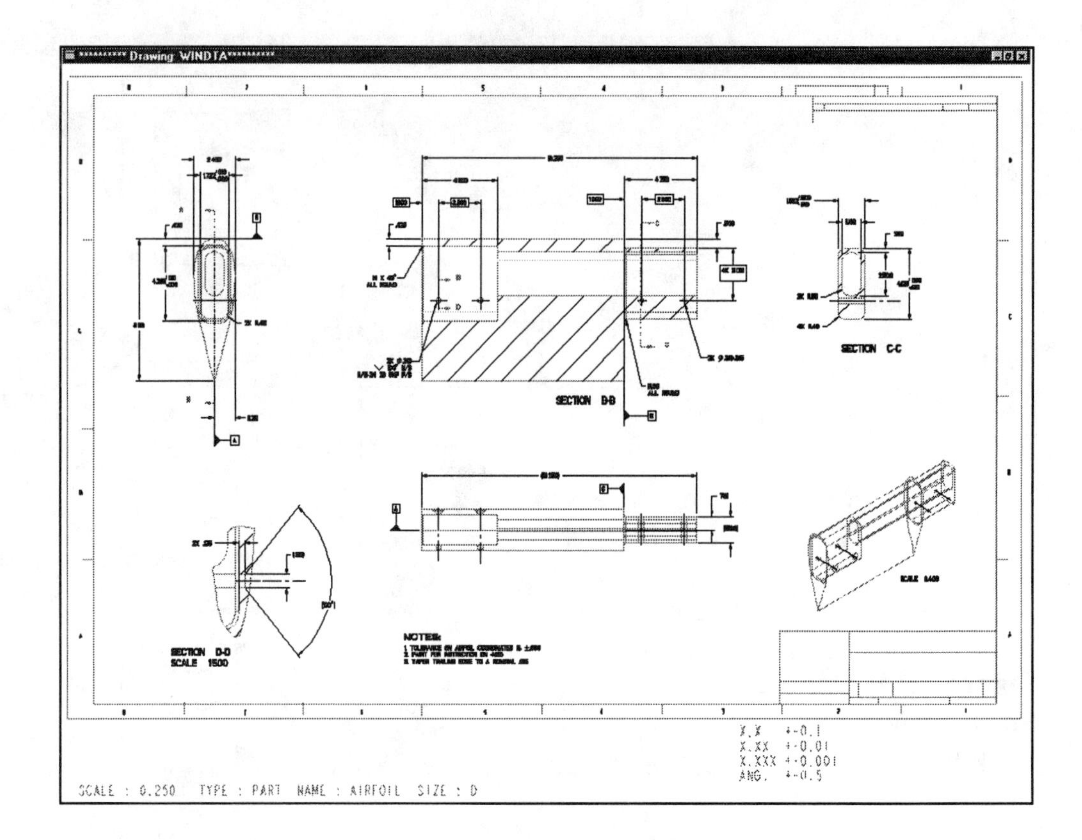

Figure A.15
Air Foil Drawing,
Sheet One

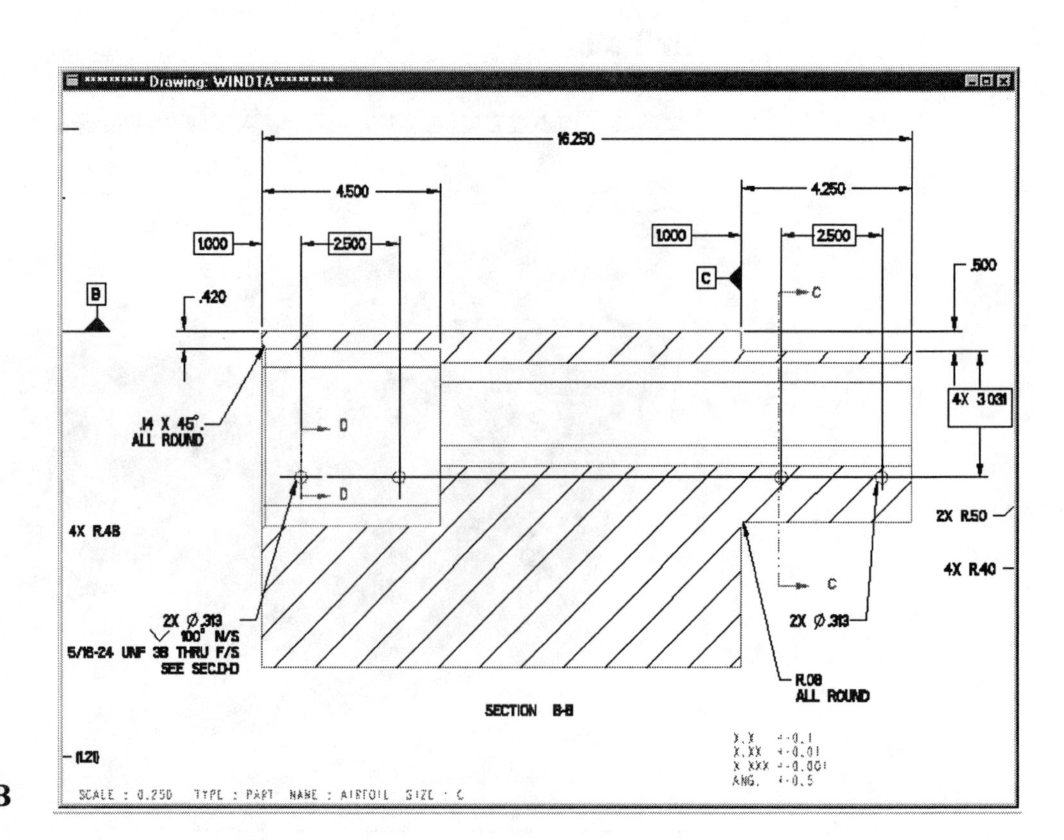

Figure A.16
Air Foil Drawing,
Front **SECTION B-B**

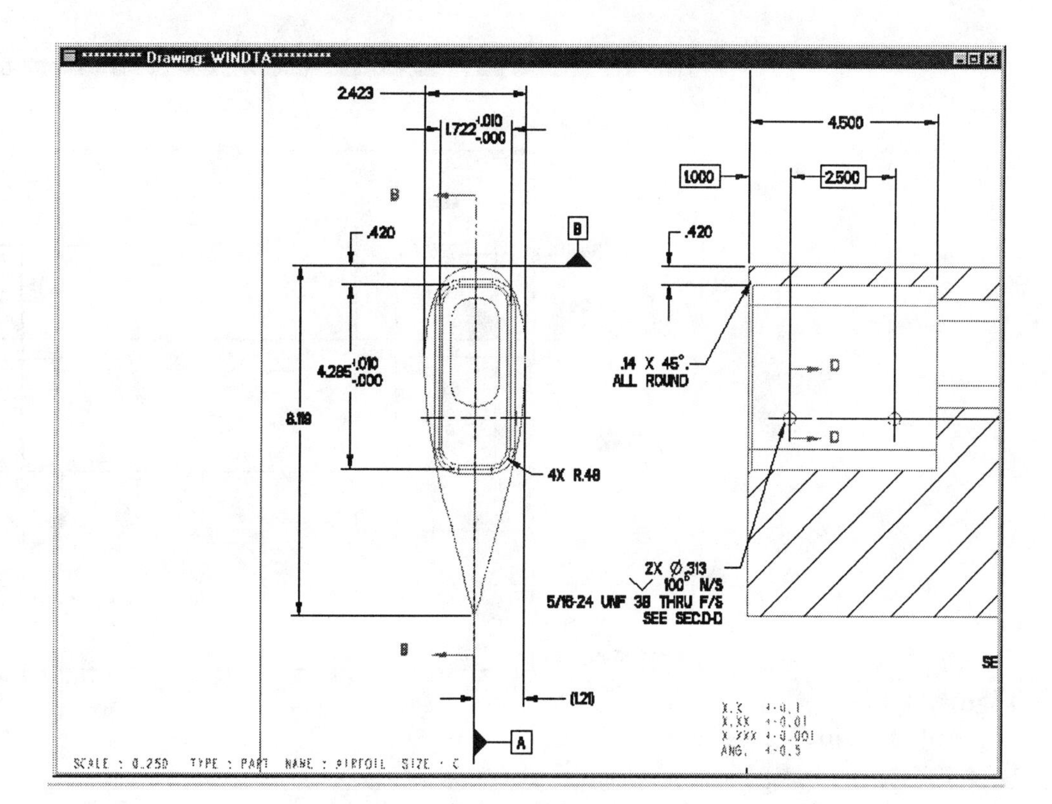

Figure A.17
Air Foil Drawing,
Left Side View

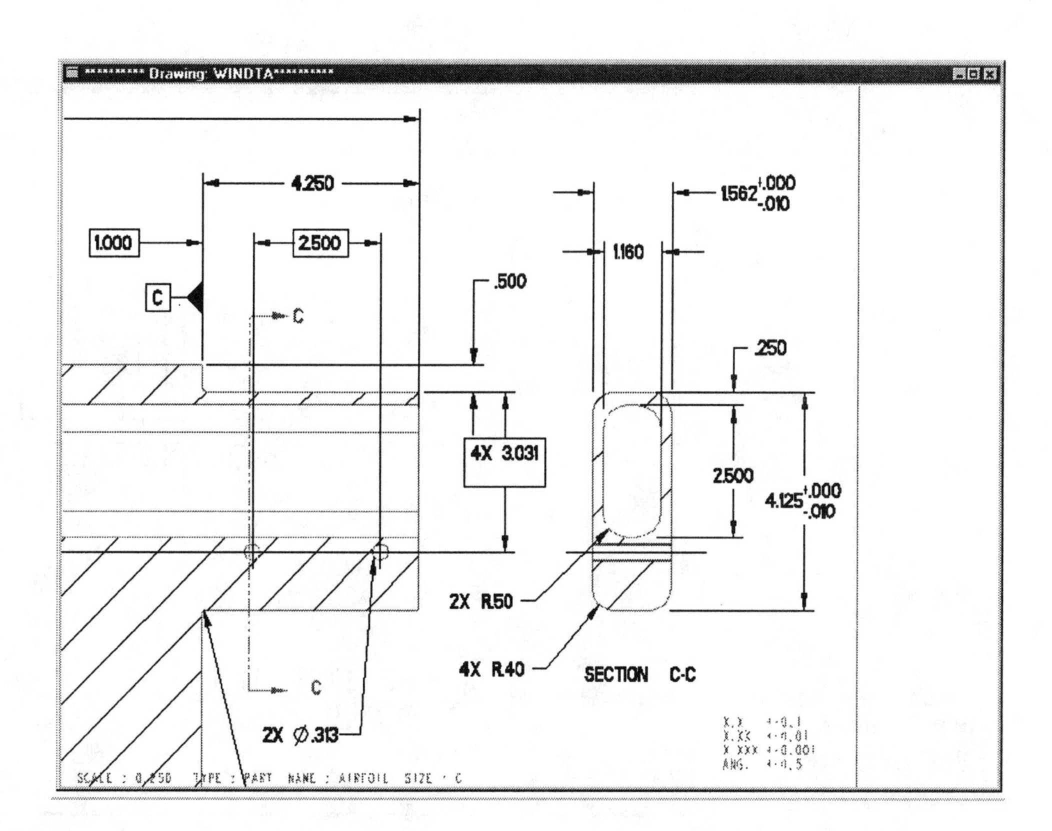

Figure A.18
Air Foil Drawing,
SECTION C-C

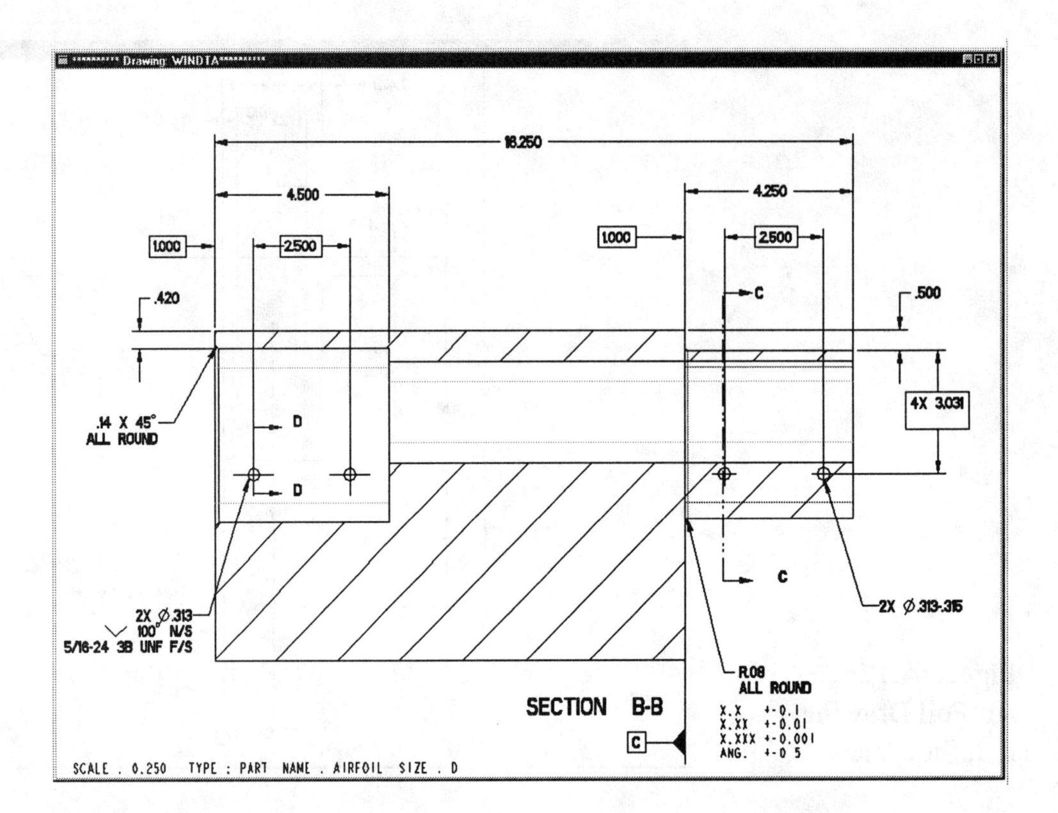

Figure A.19
Air Foil Drawing,
SECTION B-B

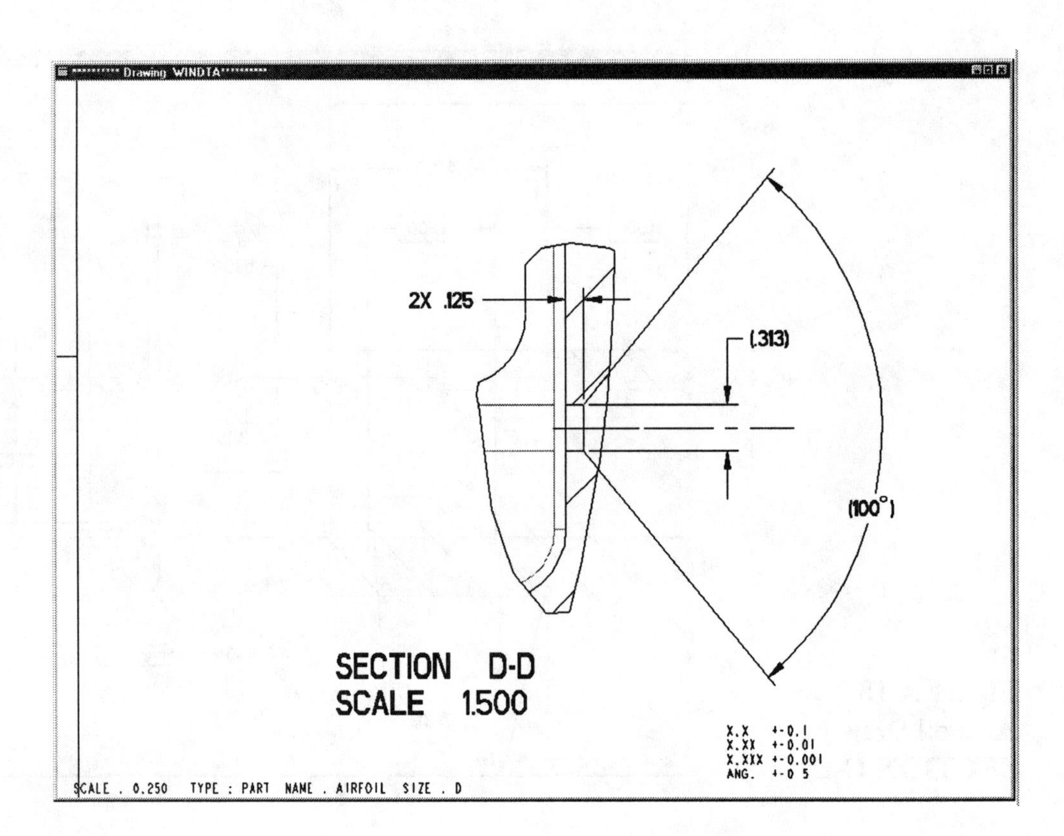

Figure A.20
Air Foil Drawing,
SECTION D-D

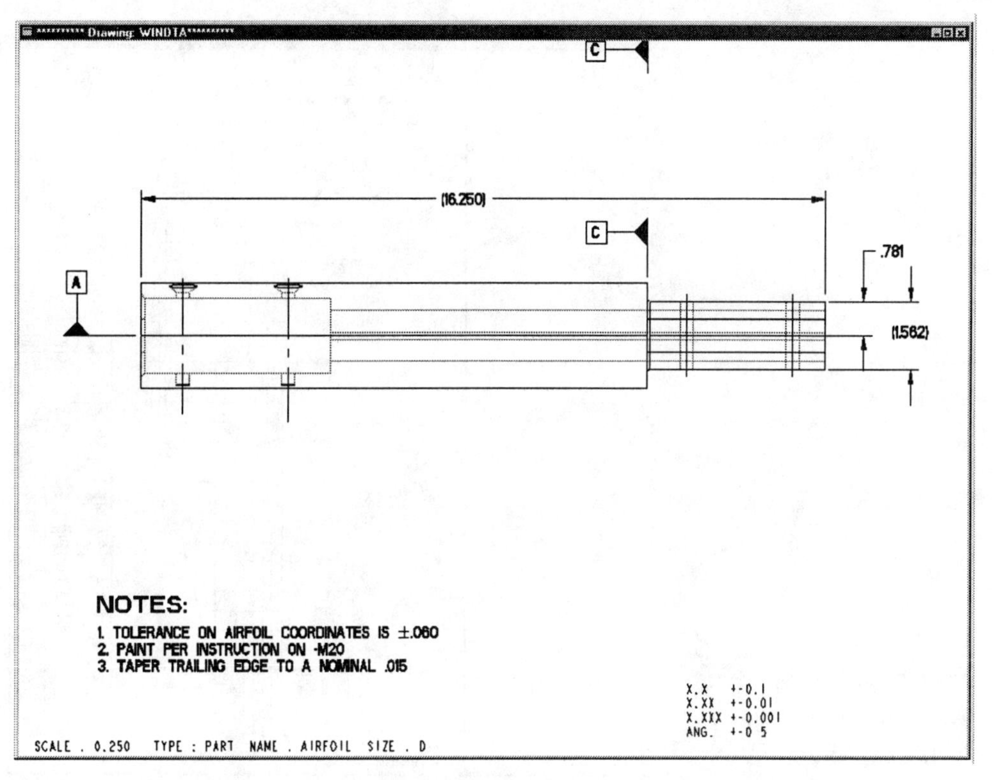

Figure A.21
Air Foil Drawing, Bottom View

Drawing: AIRFOIL_TABLE

X down	+Y to right	-Y to left
.000	.000	.000
.008	.098	.098
.032	.239	.239
.099	.393	.393
.191	.542	.542
.354	.711	.711
.510	.832	.832
.832	.969	.969
1.159	1.062	1.062
1.480	1.122	1.122
1.803	1.155	1.155
2.450	1.168	1.168
3.097	1.133	1.133
3.743	1.063	1.063
4.067	1.012	1.012
4.390	.949	.949
5.035	.801	.801
5.684	.640	.640
6.333	.477	.477
6.815	.357	.357
7.138	.278	.278
7.460	.201	.201
7.785	.144	.144
8.109	.017	.017
8.119	.000	.000

TYPE : DRAFT NAME : NONE SIZE : A

Figure A.22
Air Foil Drawing, Sheet Three Table

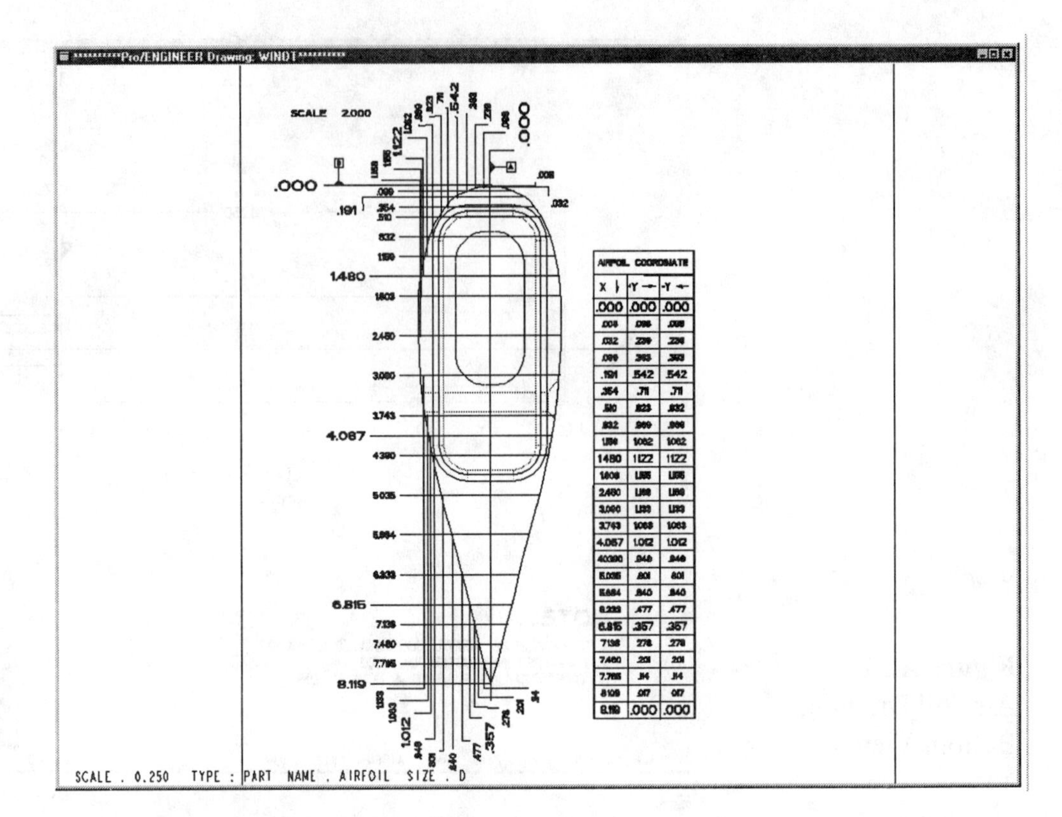

Figure A.23
Air Foil Drawing,
Table and Section

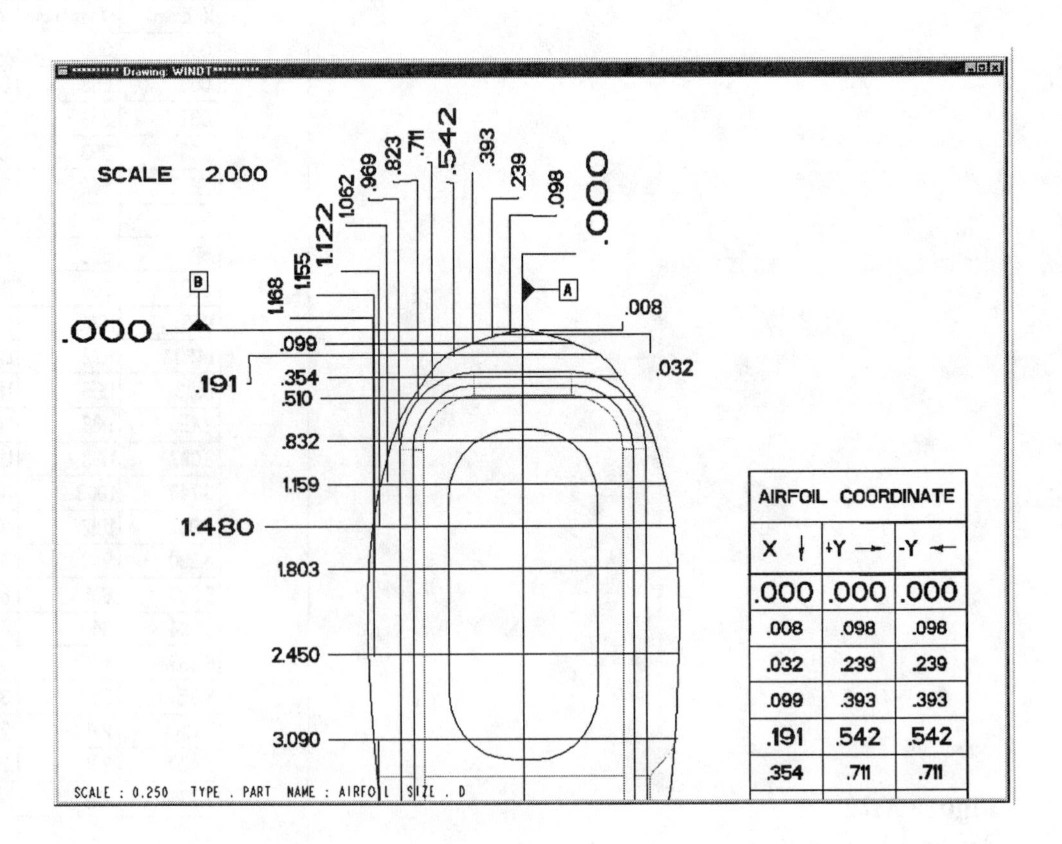

AIRFOIL COORDINATE		
X	+Y	-Y
.000	.000	.000
.008	.098	.098
.032	.239	.239
.099	.393	.393
.191	.542	.542
.354	.711	.711

Figure A.24
Air Foil Drawing
Upper Table

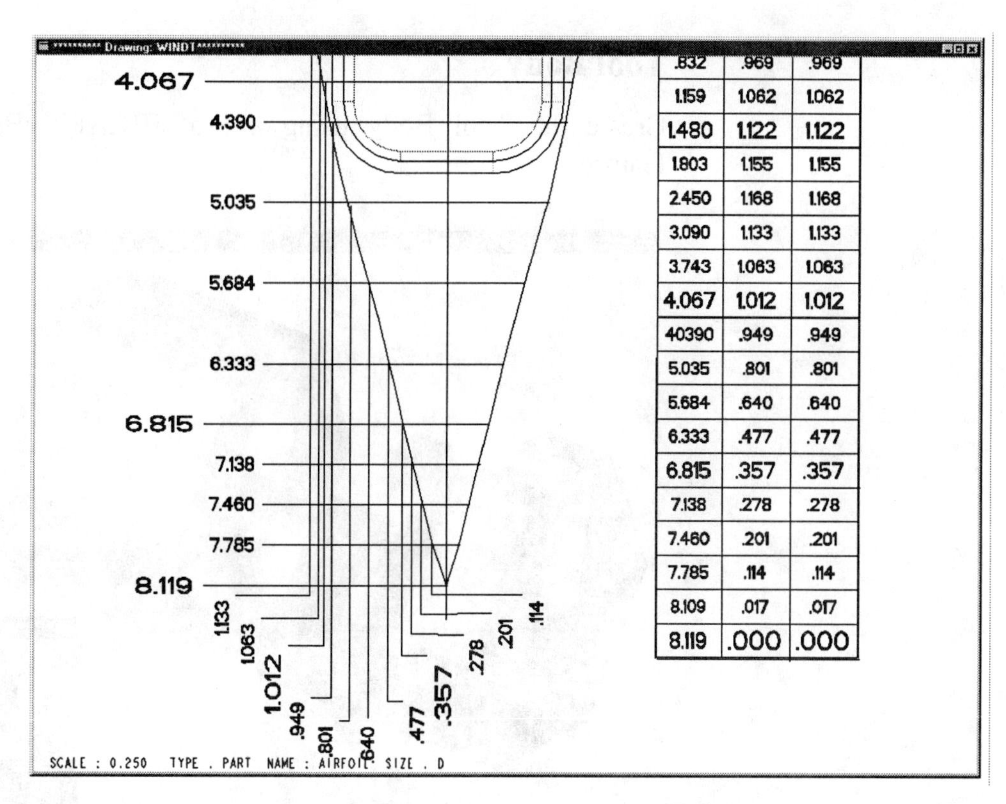

.832	.969	.969
1.159	1.062	1.062
1.480	1.122	1.122
1.803	1.155	1.155
2.450	1.168	1.168
3.090	1.133	1.133
3.743	1.063	1.063
4.067	1.012	1.012
40390	.949	.949
5.035	.801	.801
5.684	.640	.640
6.333	.477	.477
6.815	.357	.357
7.138	.278	.278
7.460	.201	.201
7.785	.114	.114
8.109	.017	.017
8.119	.000	.000

Figure A.25
Air Foil Drawing, Lower Table

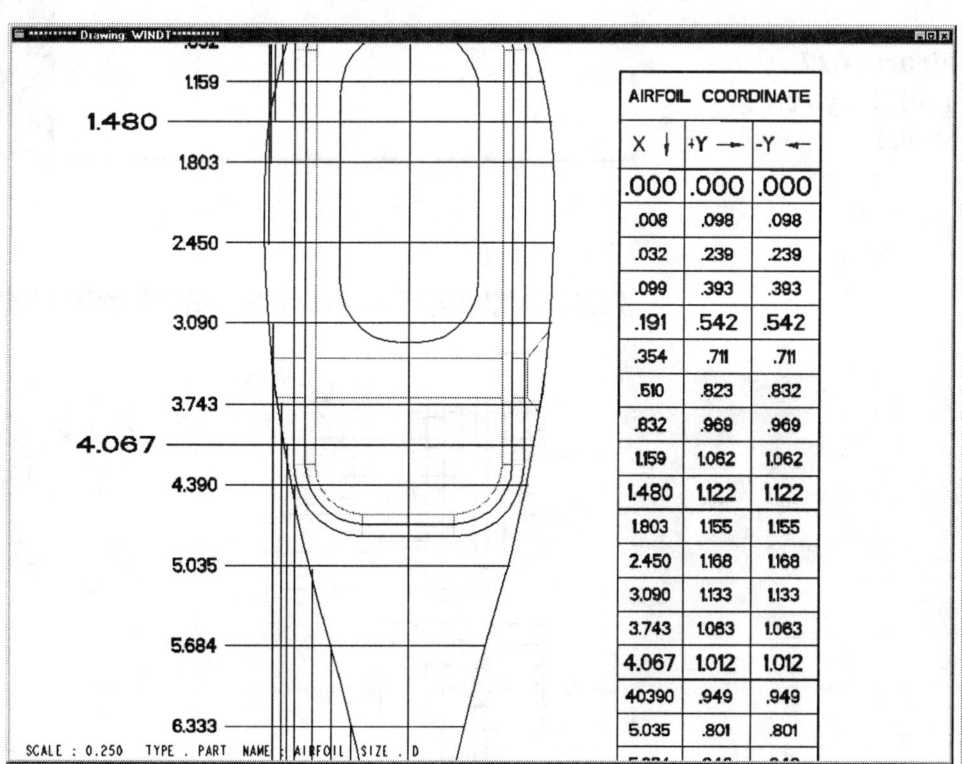

AIRFOIL COORDINATE		
X ↓	+Y →	-Y ←
.000	.000	.000
.008	.098	.098
.032	.239	.239
.099	.393	.393
.191	.542	.542
.354	.711	.711
.510	.823	.832
.832	.969	.969
1.159	1.062	1.062
1.480	1.122	1.122
1.803	1.155	1.155
2.450	1.168	1.168
3.090	1.133	1.133
3.743	1.063	1.063
4.067	1.012	1.012
40390	.949	.949
5.035	.801	.801

Figure A.26
Air Foil Drawing, Middle Table

Tool Body

Create the Tool Body using a Pro/LIBRARY Basic Library protrusion feature.

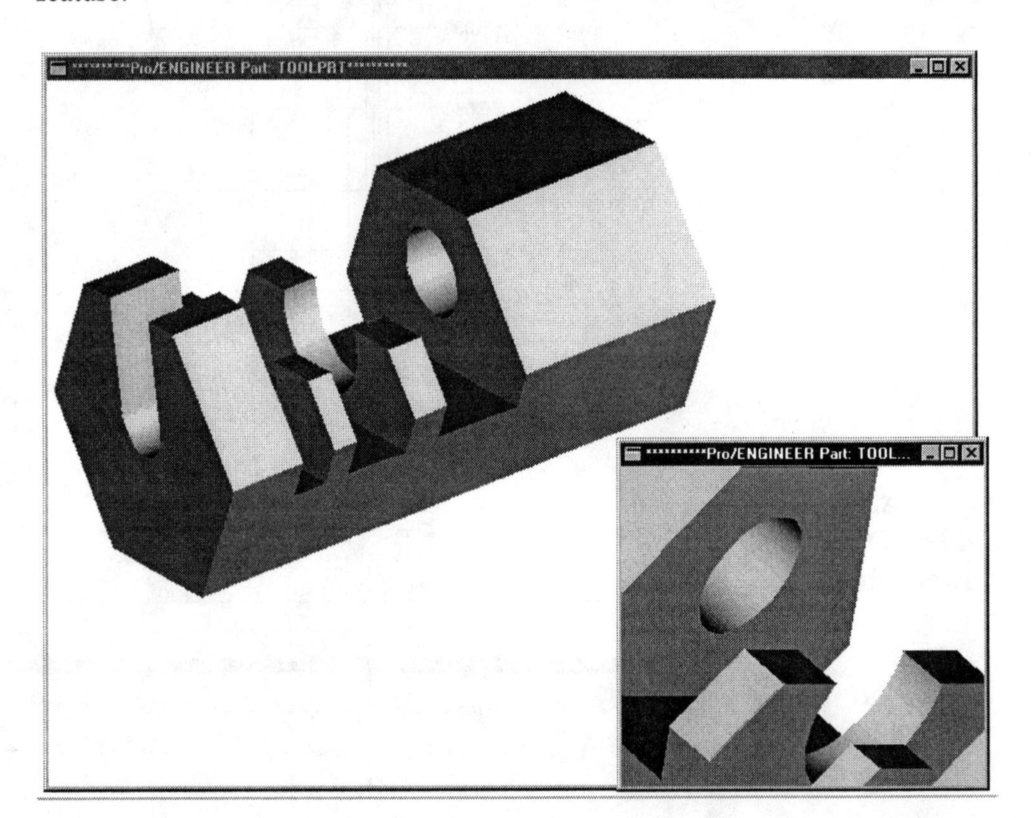

Figure A.27
Tool Body Part Model

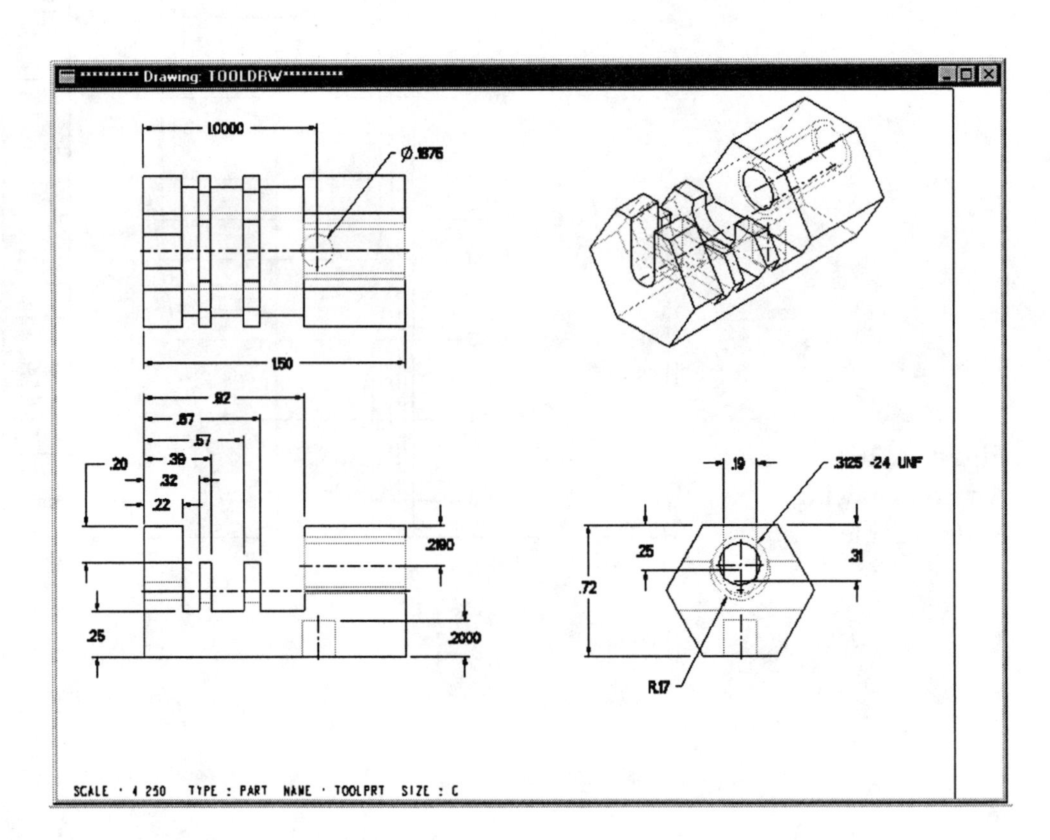

Figure A.28
Tool Body Drawing

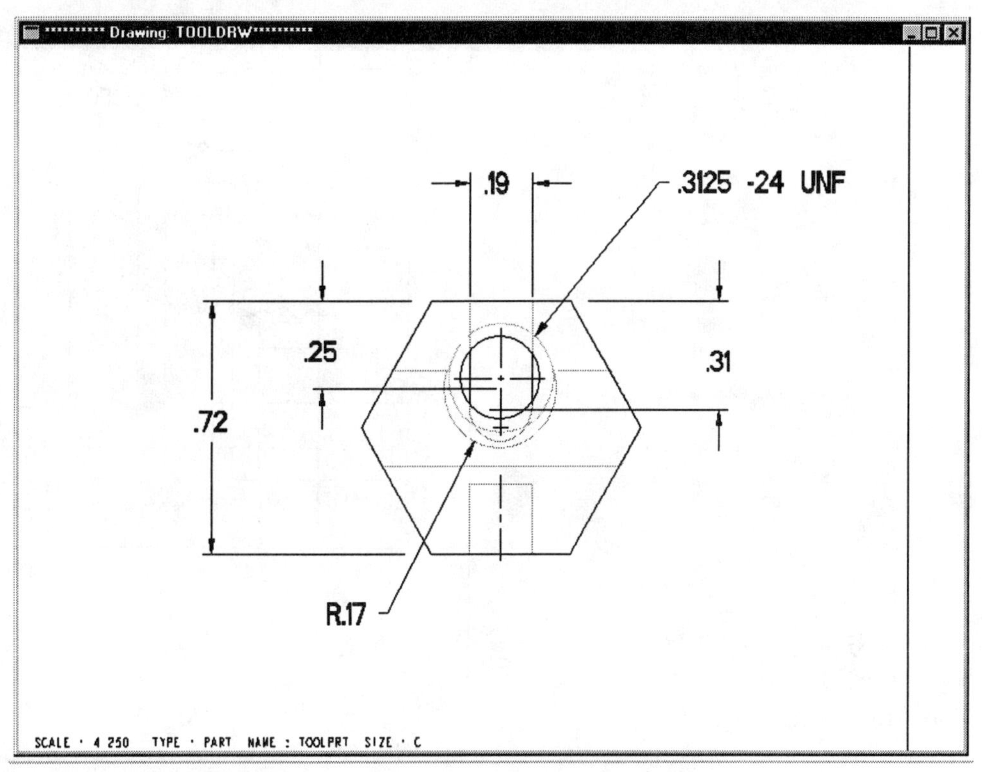

Figure A.29
Tool Body Drawing, Right Side View

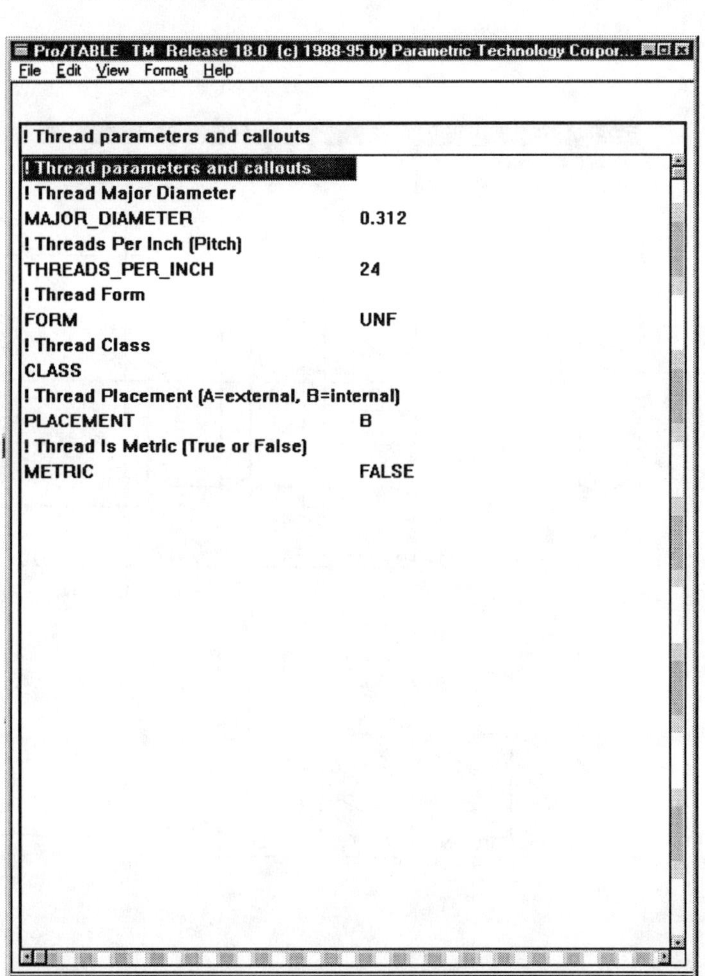

Figure A.30
Tool Body Thread Table

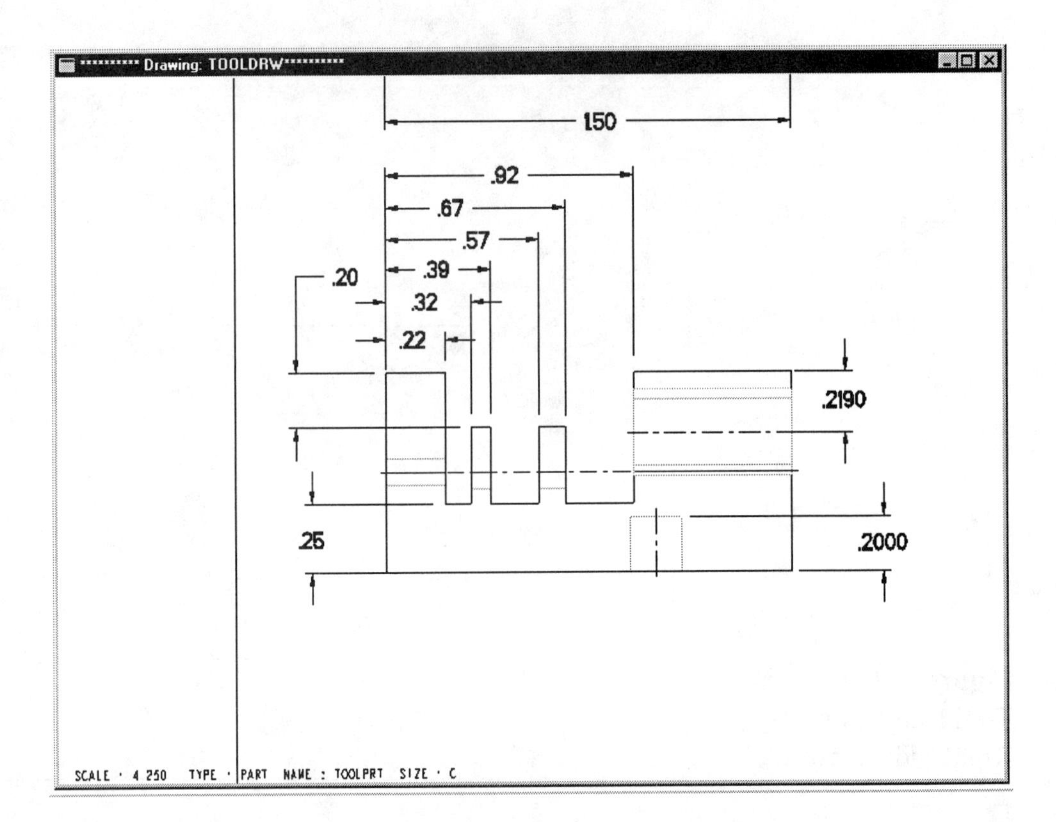

Figure A.31
Tool Body Drawing,
Front View

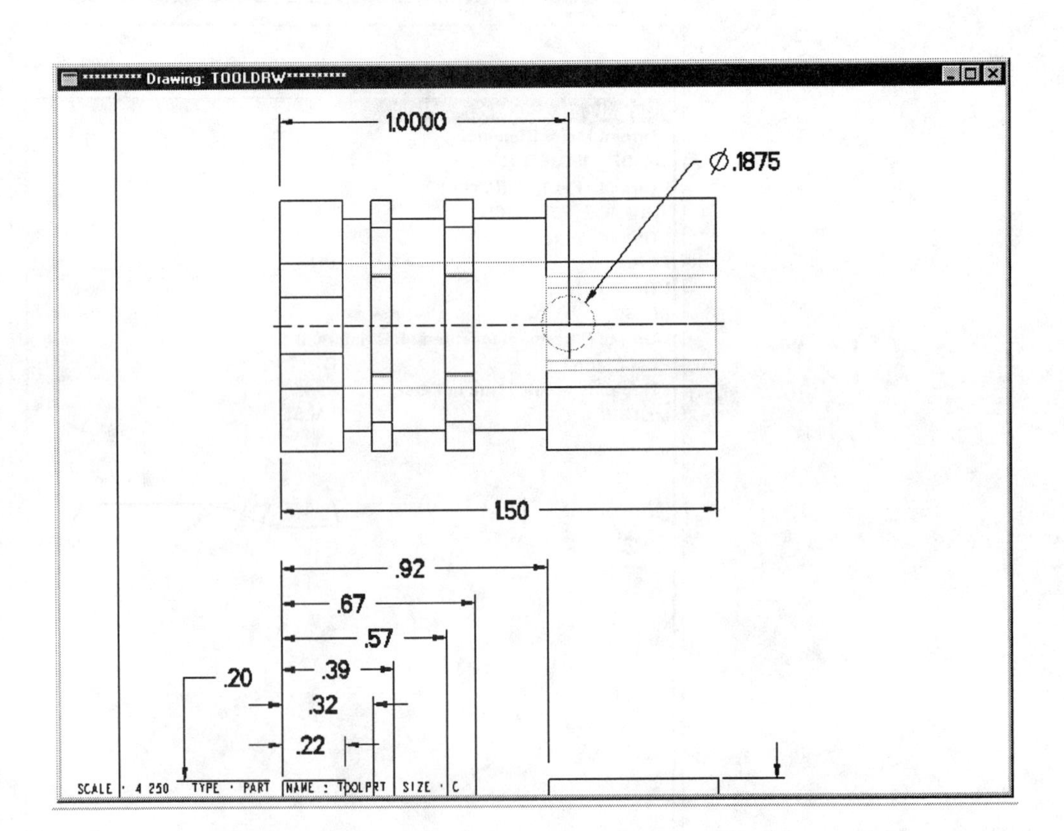

Figure A.32
Tool Body Drawing,
Top View

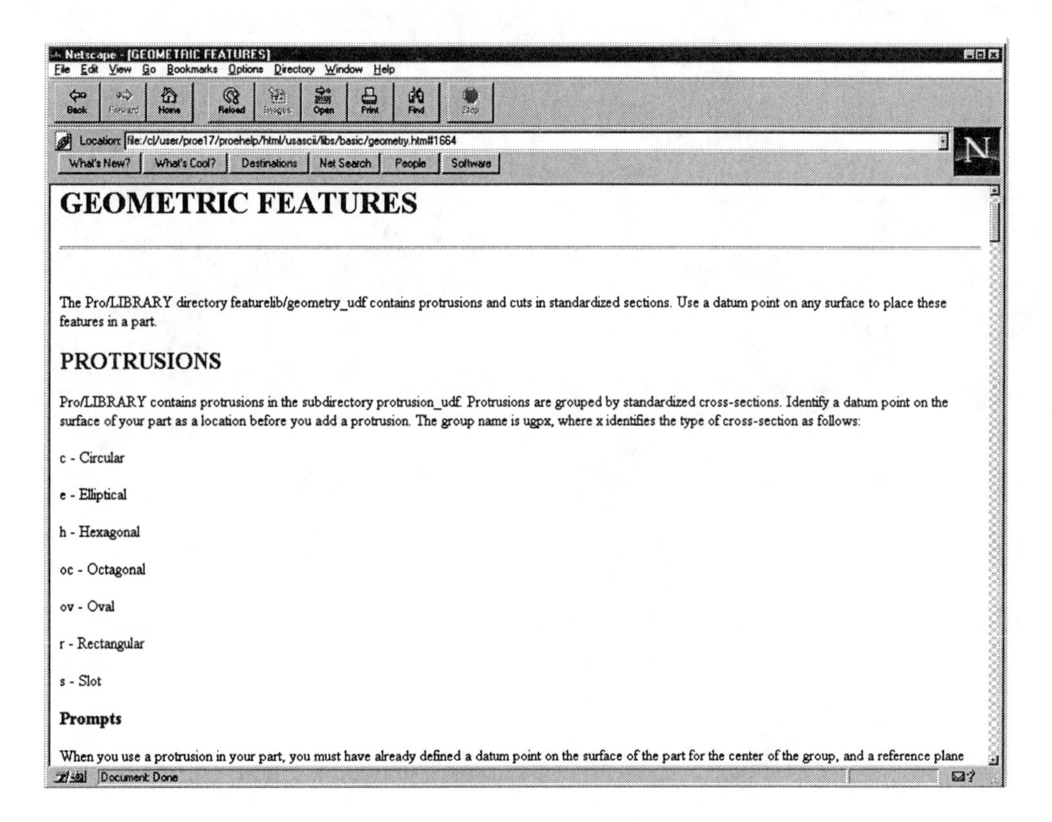

GEOMETRIC FEATURES

The Pro/LIBRARY directory featurelib/geometry_udf contains protrusions and cuts in standardized sections. Use a datum point on any surface to place these features in a part.

PROTRUSIONS

Pro/LIBRARY contains protrusions in the subdirectory protrusion_udf. Protrusions are grouped by standardized cross-sections. Identify a datum point on the surface of your part as a location before you add a protrusion. The group name is ugpx, where x identifies the type of cross-section as follows:

c - Circular

e - Elliptical

h - Hexagonal

oc - Octagonal

ov - Oval

r - Rectangular

s - Slot

Prompts

When you use a protrusion in your part, you must have already defined a datum point on the surface of the part for the center of the group, and a reference plane

Figure A.33
Pro/LIBRARY Basic Library Protrusions

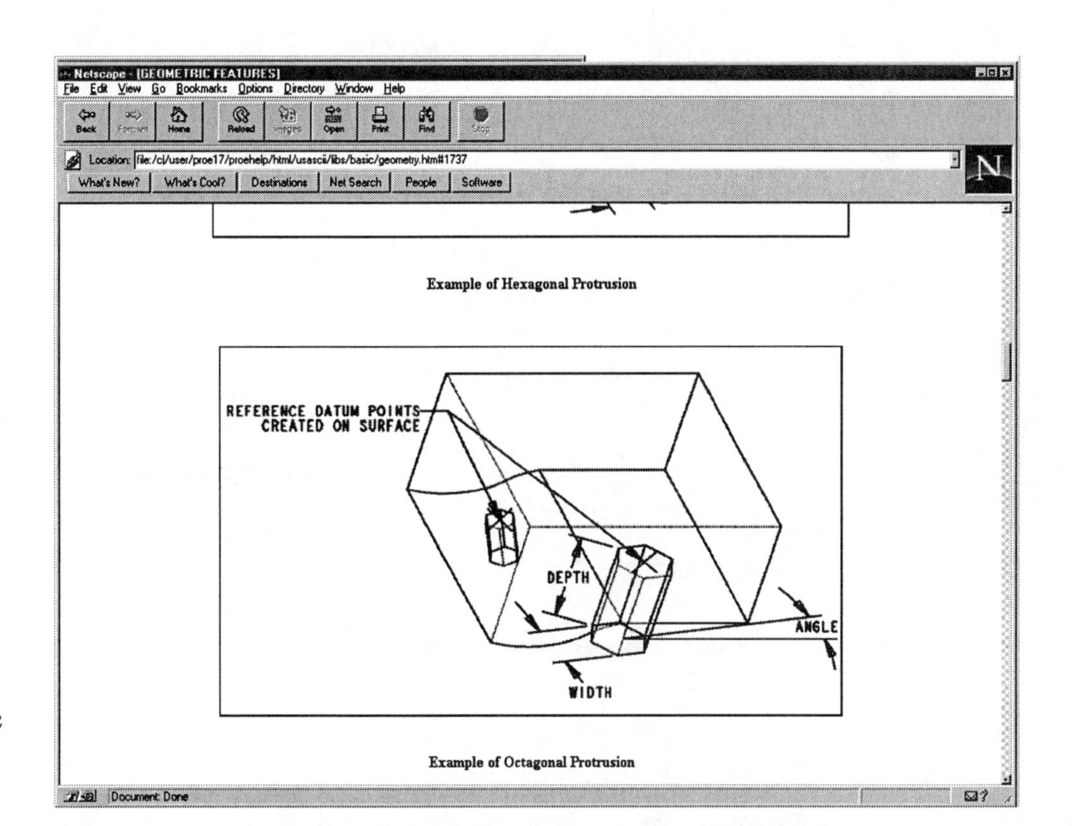

Figure A.34
Pro/LIBRARY Basic Library Protrusions, Hexagons

Appendix B

Config.pro Files and Mapkeys

CONFIG.PRO FILES

The following is an example of various entries in a config.pro file.

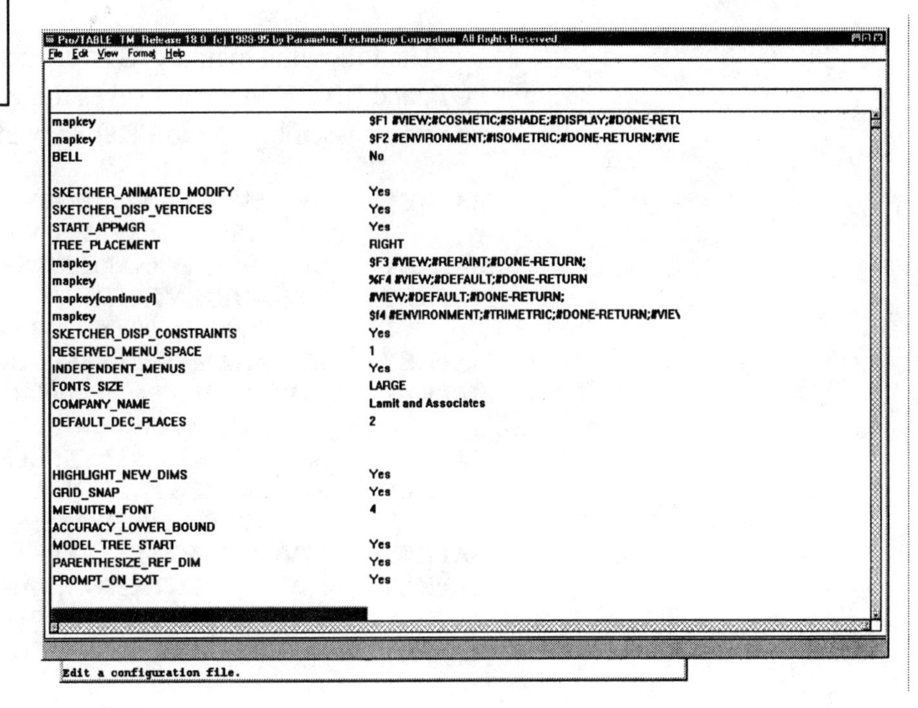

MAPKEYS

Mapkey entries can be added to a *config.pro* file at any time in the design process. These keyboard macros will help streamline the design process. Feel free to experiment with your own macro creation. If you find yourself using a set of commands for a project over and over, it may be to your advantage to create a mapkey macro that will automate the process. The following provides a few suggestions for mapkey creation from Parametric Technology Corporation:

```
MAPKEY   DR  #DONE-RETURN
MAPKEY   DT  #ENVIRONMENT; #DISP  DTMPLN; %DR
MAPKEY   RP  #VIEW; #REPAINT; %DR
MAPKEY   SC  #SKETCH; #LINE; #CENTERLINE
MAPKEY   SA  #SKETCH; #ALIGNMENT
MAPKEY   SD  #VIEW; #COSMETIC; #SHADE;\
             #DISPLAY; %DR
MAPKEY   SN  #VIEW; #ORIENTATION; #SPIN
MAPKEY   SW  #RELATIONS; #SWITCH DIM
MAPKEY   TI  #GEOM TOOLS; #INTERSECT
MAPKEY   TT  #GEOM TOOLS; #TRIM
MAPKEY   TB  #GEOM TOOLS; #TRIM; #BOUND
MAPKEY   VT  #VIEW; #NAMES; #RETRIEVE; #TOP; %DR
MAPKEY   VR  #VIEW; #NAMES; #RETRIEVE; #RIGHT; %DR
MAPKEY   VF  #VIEW; #NAMES; #RETRIEVE; #FRONT; %DR
MAPKEY   VB  #VIEW; #NAMES; #RETRIEVE; #BACK; %DR
MAPKEY   VD  #VIEW; #ORIENTATION; #DEFAULT; %DR
```

If the **MAPKEY** input is too long for one line, add a backslash

```
MAPKEY    VL  #VIEW; #NAMES; #RETRIEVE; #LEFT; %DR
MAPKEY    ZO  #VIEW; #PAN/ZOOM; #ZOOM OUT
MAPKEY    ZP  #VIEW; #PAN/ZOOM; #PAN
MAPKEY    ZR  #VIEW; #PAN/ZOOM; #RESET
```

The following mapkey **'SV'** creates a **FRONT**, **RIGHT**, **LEFT**, **TOP**, and **BACK** view and saves the view for later retrieval. Orient the part or assembly to the **FRONT** view before typing **'SV'**.

```
MAPKEY    SV   %S1V; %S2V; %S3V; %S4V:\
               %S5V; %S6V; %S7V; %S8V; %S9V
MAPKEY    S1V  #VIEW; #NAMES; #SAVE; FRONT
MAPKEY    S2V  #ORIENTATION; #ANGLES;\
               #VERT; -90 ; #DONE/ACCEPT
MAPKEY    S3V  #NAMES; #SAVE; RIGHT
MAPKEY    S4V  #ORIENTATION; #ANGLES;\
               #VERT; 180; #DONE/ACCEPT
MAPKEY    S5V  #NAMES; #SAVE; LEFT
MAPKEY    S6V  #ORIENTATION; #ANGLES;\
               #VERT; 90; #DONE/ACCEPT
MAPKEY    S7V  #NAMES; #SAVE; BACK
MAPKEY    S8V  #ORIENTATION; #ANGLES; #VERT;\
               180; #HORIZ; 90; #DONE/ACCEPT
MAPKEY    S9V  #NAMES; #SAVE; TOP
```

The following mapkey **'DD'** will create default datum planes and name them **TOP**, **FRONT**, and **SIDE**. **'DD'** should be typed immediately after typing in the name of a new part or new assembly. Use **'DP'** to create default datum planes (without renaming them).

```
MAPKEY    DP   #FEATURE; #CREATE; #DATUM; #PLANE;\
               #DEFAULT; #DONE
MAPKEY    DD   %DP; %D2D; %D3D; %D4D; #DONE
MAPKEY    D1D  #FEATURE; #CREATE; #DATUM; #PLANE
MAPKEY    D2D  #SET UP; #NAME;\
               #SEL BY MENU; #NAME; #DTM3; FRONT
MAPKEY    D3D  #SEL BY MENU; #NAME; #DTM2; TOP
MAPKEY    D4D  #SEL BY MENU; #NAME; #DTM1; SIDE
```

Appendix C

Glossary

Active Window The window to which the displayed menus apply. To change the active window, choose **Change Window** from the MAIN menu.

Align Aligns a sketched entity with a part or assembly edge and is used as an assumption when solving the section.

ASCII American Standard Code for Information Interchange. It is a set of 8-bit binary numbers representing the alphabet, punctuation, numerals, and other special symbols used in text representation and communications protocols.

Attributes (Feature) The various characteristics of a feature. For example, a blind hole will have the following attributes: position - how the hole is placed (linearly, radially, coaxially, etc.); *section* - the cross section that defines the shape of the blind hole; *intersection* - simple or complex.

Attributes (User) The attributes that are added by a user to supply a description of the object beyond the geometric definition. For example, stock number, price, and cost per unit are all user attributes.

Base Feature The very first feature created for a part.

Base Member The very first component of an assembly.

Child An item, such as a view, part, or feature, that is dependent on another item for its existence See also **PARENT**

Collinear Two or more objects that occur along the same axis.

Configuration File A special text file that contains default settings for many Pro/ENGINEER functions. Default environment, units, files, directories, etc. are set when Pro/ENGINEER reads this file when it is started. A configuration file can reside in the startup directory to set the values for your working session only, or it can reside in the *load point directory* to set values for all users running that version of Pro/ENGINEER. Also known as the *config.pro* file.

Dependent Parameter A parameter (dimension or user defined) in a relation that is defined as a function of other parameters and values. A relation has one and only one dependent parameter, and it ***always*** appears on the left side of the equal sign.

Dimmed Option An option that appears highlighted in gray when a menu is displayed. You cannot choose menu items when they are dimmed.

Entry Menu A menu that appears when you first enter a mode of Pro/ENGINEER. This menu allows you to create, retrieve, list, or import (in Part mode) an object.

Flip Arrow An arrow that appears on surfaces and edges of features to specify in which direction an operation should occur. If the arrow is pointing in the proper direction, choose **Okay**. If not, choose **Flip** and the arrow will be flipped **180°**. Then choose **Okay**.

Independent Parameter A parameter (dimension, value, relation) in a relation that is used to specify the value of the dependent parameter. Independent parameters always appear on the right side of the equal sign.

Information Window A Pro/ENGINEER window that displays information: object lists, mass properties, BOM, etc.

Load Directory The directory where Pro/ENGINEER is loaded.

Macro Keys The keyboard function keys or key sequences for which you predefine a menu option or sequence of menu options. This allows you to pick these keys to perform frequently used menu sequences.

Main View The first view added to a drawing.

Main Window The large, primary window created by Pro/ENGINEER.

Menu A list of options presented by Pro/ENGINEER that you select using the mouse or predefined macro keys.

Mode An environment in which Pro/ENGINEER allows you to perform closely related functions (e.g., Drawing, Sketcher).

Model A part or an assembly.

Object A Pro/ENGINEER object can be a drawing, a part, an assembly, a layout, etc.

Parent An item that has other items dependent upon it for its existence. For example, the base feature will have all other features dependent upon it. If a parent is deleted, all dependent items (**children**) will be deleted.

Part Type A part can be either standard or nonstandard. The part type affects the number of times a part can be assembled.

Pattern A method of feature creation in which one construction feature is used to create several related features.

Pick To select an item on screen by clicking the left mouse button.

Placement Plane The plane on which a construction feature is located or is referenced for placement.

Regenerate A menu option that recreates a model, incorporating any changes that have been made since the last time the model was stored.

Startup Directory The directory from which you started Pro/ENGINEER.

Selection Box A special rubberbanding box used to select items. Used most frequently for zooming in and selecting draft items.

System Editor The text editing functions available within your operating system.

Toggle A menu option that lets you switch between two settings, for example: **Flip Arrow.**

Trail File A record of all the menu picks, screen picks, keyboard entries, etc. that occur during a Pro/ENGINEER session. The trail file can be run to recreate a work session.

Working Directory The working directory is where you are using **Dbms** commands to store and retrieve your Pro/ENGINEER files. It may be in your startup directory or the directory you changed to using the **MISC** menu command **Change Dir**.

Work Session The period between when you start and stop Pro/ENGINEER.

View A particular display of the model. View parameters include view orientation matrix, center, and scale.

Zoom Magnify or reduce your view of the current object on the screen.

Appendix D

Design Intent Planning Sheets

This appendix provides a variety of sketching formats for planning your design. The *design intent* of a feature, a part, or an assembly (and even a drawing) should be established before any work is done with Pro/ENGINEER. Here we have provided a number of different formats to sketch and plan your feature, part, assembly, or drawing. Copy the sheets so that you have a number available for each lesson in the text.

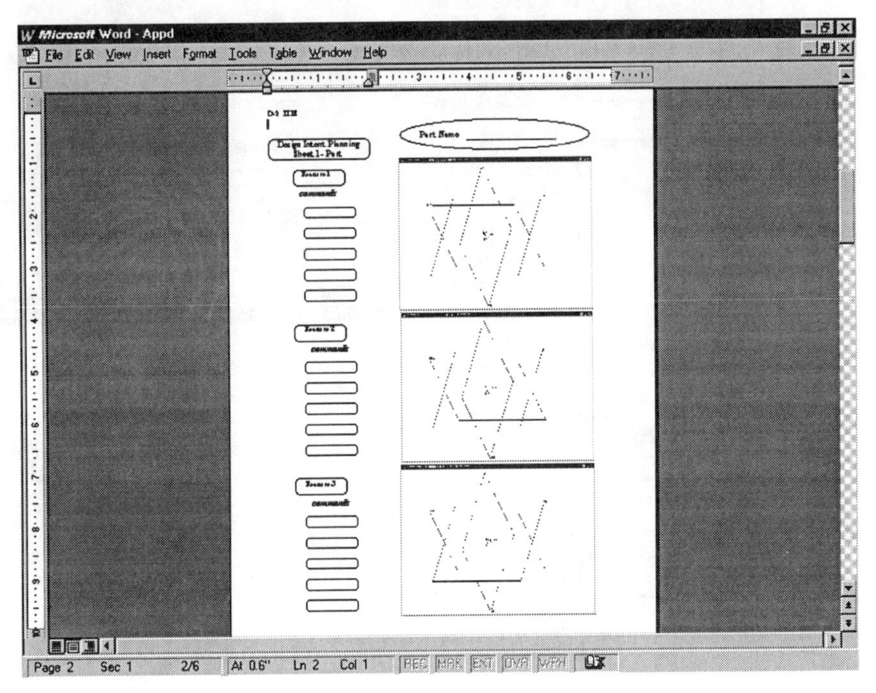

This appendix contains the following sheets:

1. Part Planning Sheet--Trimetric **(DIPS1)**
2. Part Planning Sheet--Pictorial with Axes **(DIPS2)**
3. Assembly Planning Sheet--Trimetric **(DIPS3)**
4. Assembly Planning Sheet--Pictorial with Axes **(DIPS4)**
5. Drawing Planning Sheet--No Format **(DIPS5)**
6. Drawing Planning Sheet--Format **(DIPS6)**
7. Feature/Sketch Planning Sheet--3 Sketches **(DIPS7)**
8. Feature/Sketch Planning Sheet--2 Sketches **(DIPS8)**

Part Name ______________________

Design Intent Planning Sheet 1--Part

Feature 1

commands

Feature 2

commands

Feature 3

commands

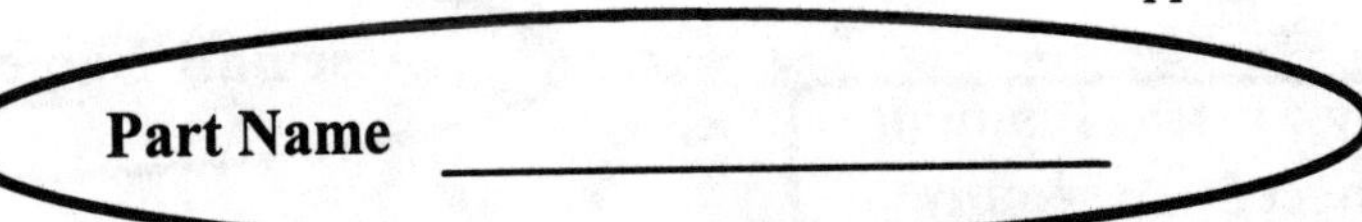

Design Intent Planning Sheet 2--Part

Feature 1

commands

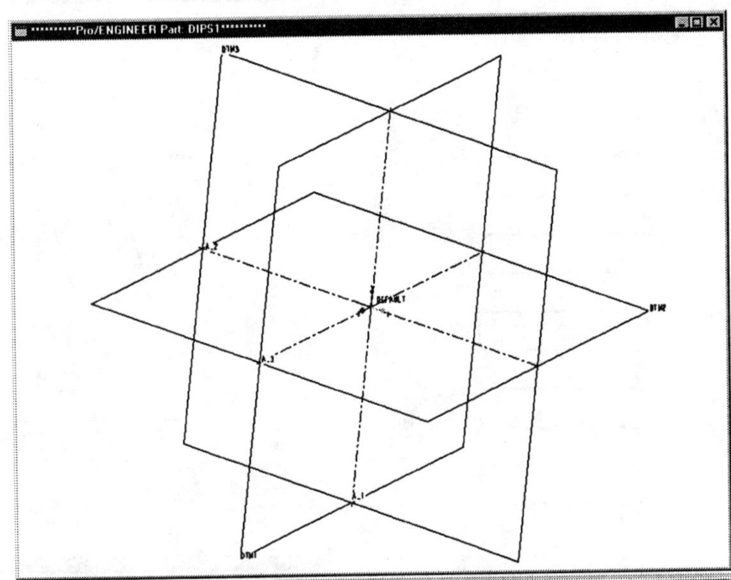

Feature 2

commands

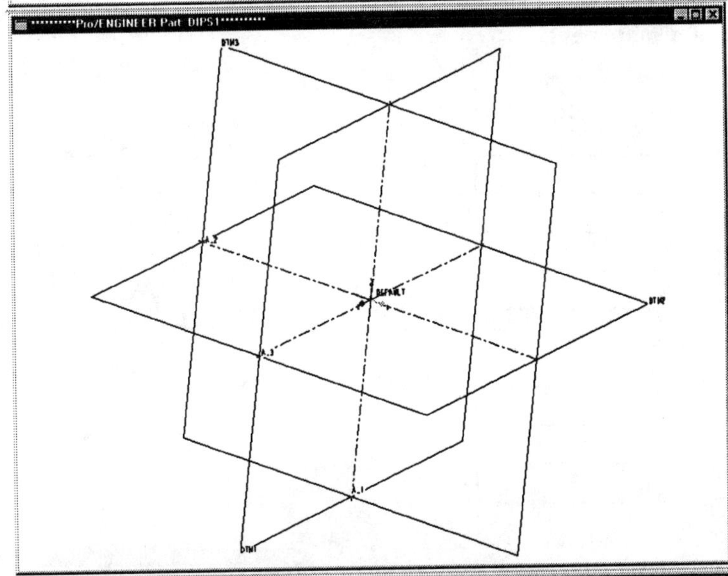

Feature 3

commands

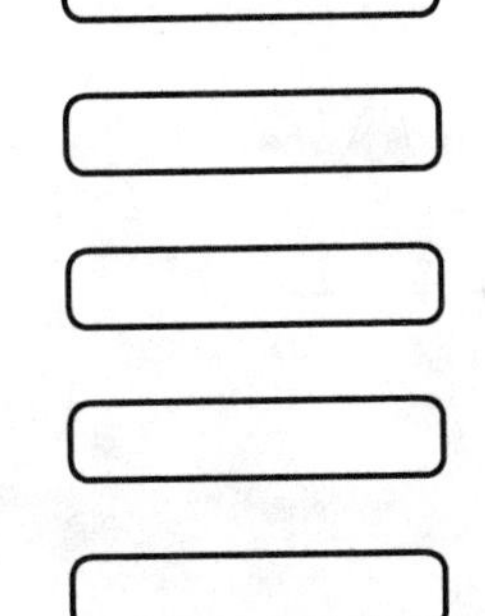

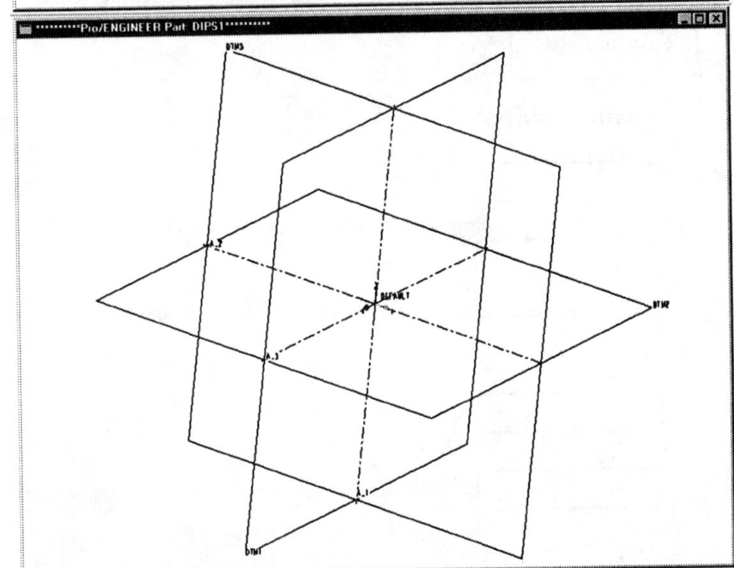

Design Intent Planning Sheet 3--Assembly

Assembly Name ____________________

Component

constraints

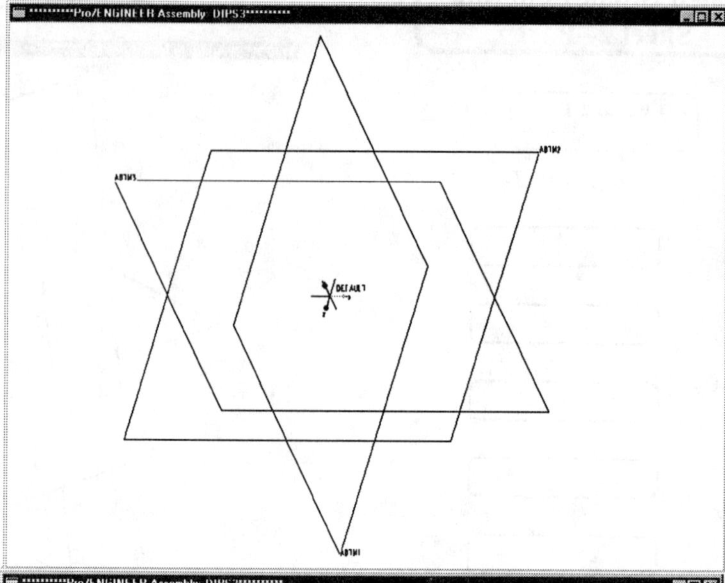

Component

constraints

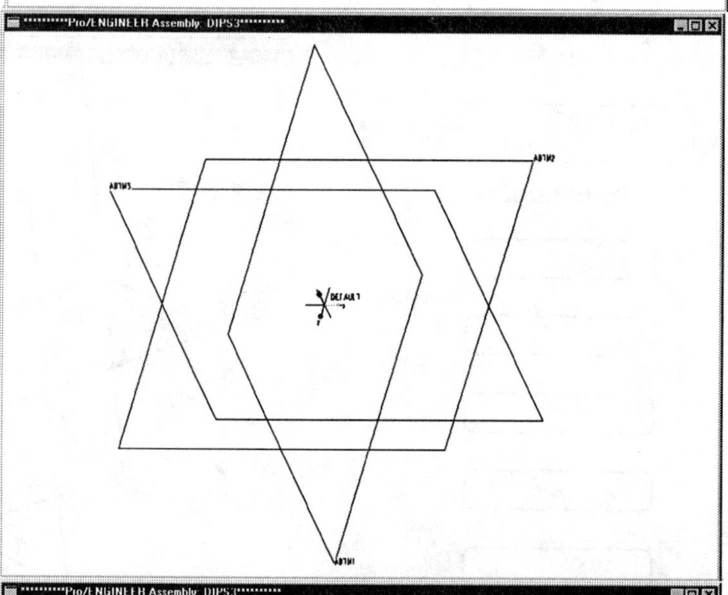

Component

constraints

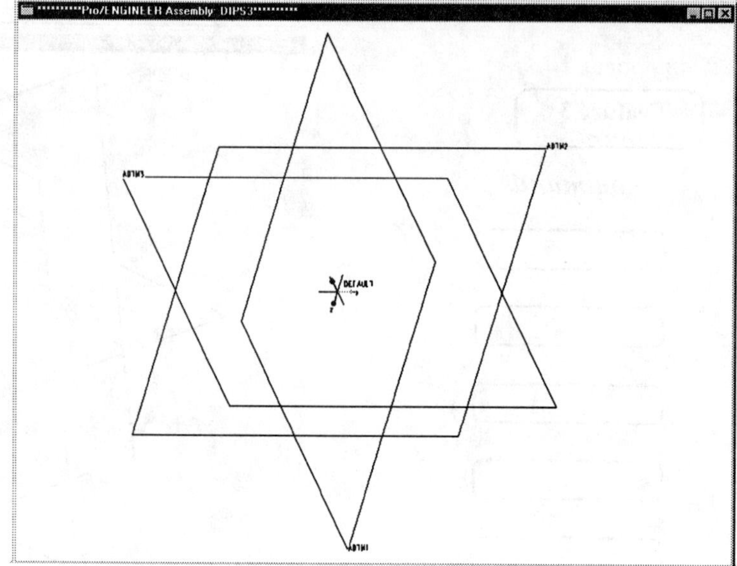

Design Intent Planning
Sheet 4--Assembly

Assembly Name ____________________

Component

constraints

Component

constraints

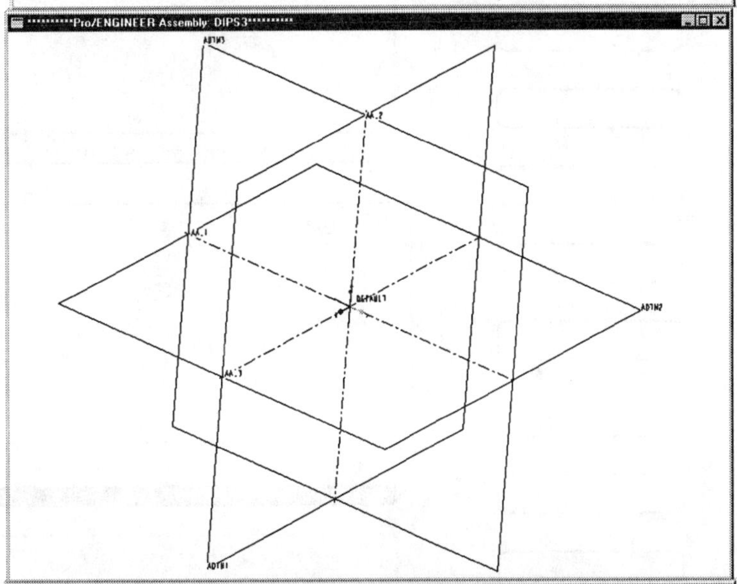

Component

constraints

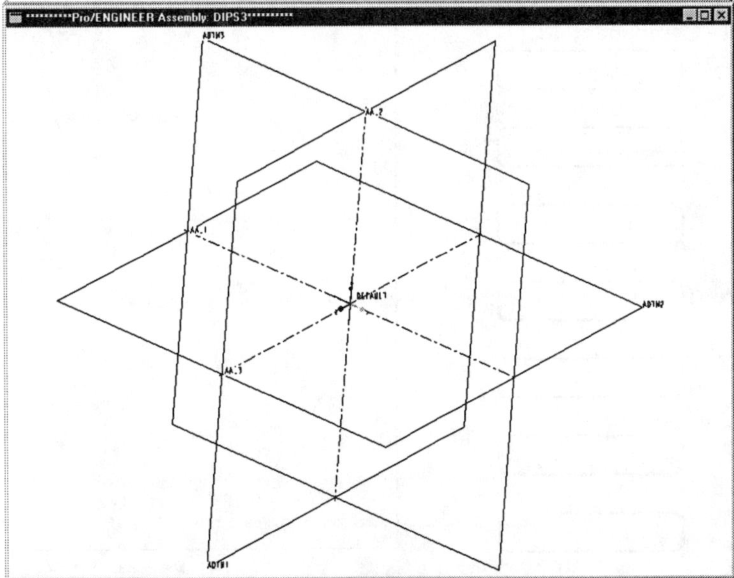

Design Intent Planning Sheet 5--Drawing

Drawing Name ______________________

Part/Assembly

Specifications

Pro/ENGINEER Drawing: DIPS5

TYPE · DRAFT NAME · NONE SIZE · C

NOTES:

Pro/ENGINEER Drawing: DIPS5

TYPE · DRAFT NAME · NONE SIZE · C

Design Intent Planning Sheet 6--Drawing

Drawing Name ____________

Part/Assembly

Specifications

NOTES:

Design Intent Planning Sheet 7--Feature/Sketch

Part Name ____________________

Sketch/section 1

Sketch/section 2

Sketch/section 3

Part Name ______________________

Design Intent Planning
Sheet 8--Feature/Sketch

Sketch/section 1

**********Pro/ENGINEER Part: DIPS7**********

DTM3

DTM2

DTM1

Sketch/section 2

**********Pro/ENGINEER Part: DIPS7**********

DTM3

DTM2

DTM1

Index

N

O

P

T

U

V

W

X